FIELD CROPS
Production and Management

Volume I

Srinivasan Jeyaraman, M.Sc.(Ag), Ph.D.
Former Dean,
Anbil Dharmalingam Agricultural College and Research Institute (TNAU),
Tiruchirapalli, Tamil Nadu
Former Director
Centre for Soil and Crop Management Studies,
Tamil Nadu Agricultural University, Coimbatore, Tamil Nadu
Former Principal
Institute of Agriculture (TNAU), Kudumiyanmalai,
Pudukkotai district, Tamil Nadu

OXFORD & IBH PUBLISHING CO. PVT. LTD.
New Delhi

FIELD CROPS : PRODUCTION AND MANAGEMENT

Oxford & IBH Publishing Company Pvt. Ltd.
113-B Shahpur Jat,
Asian Games Village Side
New Delhi 110 049, India

Fax: (011) 4151 7559
Email: oxford@oxford-ibh.in

ISBN 978-81-204-1794-6

Printed at Chaman Enterprises, New Delhi.

1-R16-07

Preface

Crop cultivation is a dynamic process of taking different decisions at different time as the cultivation practices changes with change in agro climatic conditions, soil types and locations. It is necessary to achieve 'more crop per drop' through adoption of improved crop/water management systems. The science of crop production embraces the knowledge to perform the various operations at the farm in a skillful manner to enhance input and output efficiency.

The open global trade of agricultural produces leads to competitive agriculture throughout the world. The crop products produced in one part of the world are made available to the other part of the world. Fair market price for agricultural produce can be achieved only through yield maximization per unit area with high quality. This can be done through adopting scientific principles in crop production and management for maximum profit with minimum production costs. Documentation of basic principles and practices for sustainable agriculture is necessary to enrich the knowledge of the students of agriculture. This book will be a ready reckoner for under graduate students of agriculture, extension officers and personnel involved in agro based industries. This book is mainly a compilation of scientific information available from different journals, papers, bulletins, books and reports. The new chapters on hybrid rice, System of Rice Intensification, basmati rice, rice cultivation in salt affected soils, rice ratooning management practices, coconut, oil palm, jatropha, potato, chinese potato, sweet potato, aroids, tapioca, tea, turmeric, betel vine, aromatic and medicinal plants are included in this book to meet the competitive examinations at national level in agronomy subject since these crops are included in syllabus of different State Agricultural Universities in India. Each chapter covers with standard headings such as importance, origin, distribution, climate, soils, varieties, growth stages, seeds and sowing, intercultivation, weed management, manures and fertilizers application, water management, harvest, yield and utilization. I hope that this book will be helpful to the students and extension officers in fulfilling the objective of holistic knowledge of crops production *i.e.*, soil, climatic requirement, crop production and crop protection technologies for increasing the income and sustainability to the production system.

I am thankful to Prof Dr. R. Rajagopal, former Professor and Head, Department of Plant Breeding and Genetics; Prof Dr. K. Annadurai, Agronomist, Prof Dr. S. Avudaithai, Professor and Head, Department of Agronomy; Dr. R. Arulmozhian, Professor (Horticulture); Dr. N.

Thavaprakaash, Dr. P.M. Shanmugam, Dr. T. Ramesh, Dr. S. Somasundaram, Dr. S. Anandha Krishnaveni, Dr. S. Rathika, Assistant Professors (Agronomy); Dr. R. Neelavathi, Assistant Professor (Horticulture); Dr. M. Shanmuganathan, Assistant Professor (Plant Breeding and Genetics) and Dr. M. Surulirajan, Assistant Professor (Plant Pathology) for their kind help in writing the book. The author is thankful to fellow agronomists and other scientists of Tamil Nadu Agricultural University who have helped me during the course of preparation of the manuscript. The author is thankful to the Publisher, Oxford & IBH Publishing Co. Pvt. Ltd., New Delhi for bring out this book in time.

I am highly grateful to the beloved Prof. Dr. K. Ramasamy, Vice Chancellor, of Tamil Nadu Agricultural University, Coimbatore who has kindly given the foreword to this book.

S. Jeyaraman

TAMIL NADU AGRICULTURAL UNIVERSITY

Dr. K. Ramasamy, Ph.D.,
Vice - Chancellor

Coimbatore - 641 003
Tamil Nadu, India

FOREWORD

Food, feed and fibre production have to be increased to meet the requirement of ever increasing population. The land area under cultivation is dwindling due to urbanization in cultivable lands. In view of the burgeoning human population, horizontal expansion of crop cultivation is impossible. Therefore, there is a very limited scope for extension of cultivated area but it is necessary to produce more food, feed, fodder, fuel and fiber to fulfill the future requirements from the existing land area with the available resources. This challenging scenario, demands to increase the productivity, increase in yield per hectare per crop, increase in the number of crops per hectare per year, reduction of post-harvest losses besides focusing on ecological approaches for sustainable agriculture. The water availability for crop production is shrinking due to global warming coupled with low rainfall and increase in temperature. Under such conditions, the crop productivity can be achieved with the use of short duration high yielding varieties which are resistant to pest and diseases and other abiotic stresses such as drought, flood, salinity and alkalinity; diversity of crops and varieties to suit a wide array of the soil and climatic conditions, timely operations and appropriate techniques on efficient use of inputs such as seed, bio-fertilizers, green manures, green leaf manures, balanced application of organic manure and fertilizers, irrigation water, herbicides and pesticides based on requirement at various crop growth stages without any detrimental effects on soil quality. Use of low cost technology, eco-friendly inputs and non-monetary inputs such as cultivar, time of sowing, maintenance of plant population, timely inter-cultivation, crop rotation play an important role in cost reduction in crop production which really benefit the farmers. Good agricultural practices include choice of crops and varieties, time of sowing/planting, input management based on crop growth stages, cropping systems involving legumes, bio-fertilizers green manures, green leaf manures, shade management, balanced application of organic and mineral fertilizers and other agro-chemicals, water-saving measures, method and scheduling of irrigation, micro irrigation and fertigation, integrated pest and disease management resulting in safe and healthy food and non-food agricultural products, while taking into account economic, social and environmental sustainability.

• **Tel: Off.: +91 422 2431788** • **Res.: +91 422 2430887** • **Fax: +91 422 2431672** • **Email: vctnau@tnau.ac.in**

Vice - Chancellor Coimbatore - 641 003
Tamil Nadu Agricultural University Tamil Nadu, India.

This book on 'Field Crops: production and management' is planned in two volumes to cover the above scientific principles and practices on different field crops. The Volume-I deals with special reference to food crops under cereals, millets, pulses, forage crops, green manure and green leaf manure crops. The second volume deals with special reference to commercial crops under oilseeds, fibre crops, sugar and starch crops, spices and condiments, narcotics and beverages, tuber crops, aromatic and medicinal crops. Each crop is dealt in sub-titles *viz*., vernacular name, importance, history, origin, distribution, botany, climate, soil, varieties, growth stages of crops, season, land preparation, seeds and sowing, after cultivation, weed management, manures and fertilizer application, water management, cropping system, plant protection, harvest, post-harvest technology, yield and quality.

Prof.S.Jeyaraman, a reputed agronomist, has worked in different field crops and has vast practical knowledge on crop production with farming background. He has taught B.Sc.(Ag.) and B.V.Sc. Courses in Agronomy. Besides, the author has wide experience in different capacities as Dean, Anbil Dharmalingam Agricultural College and Research Institute (TNAU), Tiruchirapalli, Tamil Nadu; Director, Centre for Soil and Crop Management Studies, Tamil Nadu Agricultural University, Coimbatore, Tamil Nadu and Principal, Institute of Agriculture (TNAU), Kudumiyanmalai, Pudukkotai District, Tamil Nadu. The author has included special chapters/topics covering the emerging issues in crop production such as basmati rice, hybrid rice cultivation, rice cultivation in salt affected soils, rice ratoon management, system of rice intensification, sustainable sugarcane initiative, aroids, aromatic and medicinal crops, tea husbandry and biofuel crops such as jatropha. This book has been prepared based on syllabus of different State Agricultural Universities, ICAR Junior Research Fellowship and ARS–NET Examinations which will be very useful to the students to prepare for Competitive Examinations. I am confident that this text book in two volumes would be a valuable resource for the students of under graduate and post graduate degree programmes in the field of Agriculture and Horticulture, Diploma in Agriculture courses and Extension officers. I congratulate Prof.S.Jeyaraman, a dedicated Agronomist, for bringing out an excellent educational treatise for the benefit of students and farming community.

(K.RAMASAMY)

Place : **Coimbatore – 641 003.**
Date : **13.01.2016**

Contents

CHAPTER

1

Classification of Crops and their Economic Importance

Crop is a plant cultivated for economic purposes. An alternative crop is either a plant new to a region such as chickpea, canola, or a minor crop such as millet or buckwheat, which shows increased production or economic promise. Crops are classified based on various factors:

A. AGRICULTURAL CLASSIFICATION OR AGRONOMIC CLASSIFICATION

Cereals, grain/seed legumes, oil seeds crops, root and tuber crops, sugar crops, latex and rubber crops, pasture and forage crops, and fiber crops are classified under agronomic crops or field crops. According to the use of plants and plant products to man, the grouping is made as follows:

1. Cereals: The word 'cereal' is derived from 'Ceres', the ancient Roman goddess of harvest who is the 'giver of grain'. Cereals are cultivated grasses grown for their edible starchy dry grains. The grain of cereals is technically a type of single, dry, indehiscent fruit called caryopsis, which has a hard outer pericarp fused to the seed coat. The endosperm predominantly consists of starch. The cereals belong to the family *Poaceae / Gramineae* are grown for their fruits or one seeded fruit known as caryopses. Cereals are grown under a wide range of climate, ranging from very cold temperate to very hot tropical conditions. In the tropics, rice and maize crops dominate. Oats and rye are the predominant cereals of cold temperate regions, whereas wheat and barley predominate in warm temperate climates. Cereal crops harvested green for forage, silage or grazing are classified as fodder crops. Examples: rice, wheat, maize.

A few millet crops are sometimes grouped under industrial crops. Examples: broom sorghum and sweet sorghum when grown for syrup.

2. Millets: Millets are small seeded grains which are used for food, feed and forage purposes. The millets are classified into two groups viz., major millets and minor millets or small millets. The major millets are sorghum, pearl millet and finger millet in India. The small millets also refers to a group of small-seeded cereal crops. Small millets may also be called minor millets. The important small millets grown in India are foxtail millet, kodo millet, common or proso millet, little millet and barnyard millet.

3. Pulses: Seeds of legumes which are rich in protein and used as food. *e.g*., pigeonpea, chickpea. Pulses are annual leguminous crops yielding grains or seeds used for food and feed purposes. The term 'pulses' is limited to crops harvested for dry grain only, excluding crops harvested green for forage, used for grazing, as green manure, and also crops harvested green for food (green beans, green peas, etc.), which are considered vegetables. They exclude those used mainly for extraction of oil, *e.g.,* soybeans. They also excluded from this group should be those leguminous crops whose seeds are used exclusively for forage purposes, such as alfalfa and clover.

4. Oilseeds: The crops produce seeds which are rich source of edible and industrial oil. Examples are sesame, groundnut. Temporary oil-bearing crops are annual crops and are usually called oilseeds. These are annual plants whose seeds are used mainly for extraction of culinary and industrial oils, excluding essential oils. Permanent oil-bearing crops are perennial plants whose seeds (kapok), fruits or mesocarp (olives) and nuts (coconuts) are used mainly for extraction of culinary or industrial oils and fats. Consequently, dessert or table nuts, such as walnuts, are excluded because although they are rich in oil content, they are not used mainly for extraction of oil. Both cotton seed and cotton lint (but not seed cotton) are considered by FAO to be primary crops and are classified in the oil crops and fibre crops groups. This is because seed cotton is a mixture of both food (seed) and non-food (fibre).

5. Green-manure crops are plants grown which are incorporated *in situ* during their latter growing stages to supply nutrients and organic matter to improve the soil quality. Examples are clover, vetch, daincha, sunnhemp, etc.

6. Forage crops: Plants that are used for feeding the domestic animals reared in a farm. e.g., para grass, rhodes grass, lucerne and berseem. Fodder crops are cultivated forage crops which are cut and stall fed to livestocks. *e.g.,* fodder maize, fodder cowpea, Bajra-Napier hybrid grass. Pasture crops are plants grown or managed as vegetable feed for grazing animals. They are classified as either native or improved species, grasses or legumes, and may be fed fresh or dry or in processed form. Examples: Cenchrus grass, para grass, napier grass, stylo, siratro and pea blue. Fodder crops may be classified as temporary or as permanent fodder crops. The temporary fodder crops are cultivated and harvested like any other crop. Temporary fodder crops are grown intensively with various cuttings per year. They contain three major groups of fodder: grasses, including cereals harvested green; legumes, including pulses harvested green; and root crops that are cultivated for fodder. All can be fed to animals as green feed; as hay, i.e. crops harvested dry or left to dry if harvested green; or as silage products. The permanent fodder crops relate to land used permanently (five years or more) for herbaceous forage crops, either cultivated or growing wild (wild prairie or grazing land). They may include some areas of forest lands that are used for grazing. Soiling crops are plants harvested as fresh green biomass before attaining the maturity to feed the livestock in stalls. Examples are maize, teosinte, oats, cowpea, berseem. Silage crops are grown for preservation in a succulent condition by partial fermentation in air tight container. Silage crops are grasses grown, cut, fermented and preserved before being fed to animals. Silage or ensilage is a method of preservation of green fodder through fermentation to retard spoiling.

7. Fibre crops: Fibre crops are annual crops yielding vegetable fibres, mostly soft fibres, which are utilized by the textile industry to produce first thread and yarn, and, from these, innumerable fabrics are manufactured. The fibre is extracted from the bark, leaves, or other organs including the husk of coconut. There are different kinds of fibres based on the part of plant used as fibre. *e.g.*, seed fibre (cotton), bark fibre (jute, mesta, roselle and flax), leaf fibre (pineapple and agave). Examples are abaca, jute, kenaf, maguey and ramie. The primary fibre crops are cotton, jute and flax.

8. Sugar and Sweetener crops: Sugar crops are those crops cultivated primarily for the manufacture of sugar, secondarily for the production of alcohol (food and non-food) and ethanol. Sugar and syrups are also produced from the sap of certain species of maple trees, maize and sweet sorghum.

9. Narcotics: The word 'Narcotics' seems to originate from the Greek word *'Narkotikos'* which includes plants that produce a state of lethargy, torpor (numbness) or sleep. Crops used for stimulating numbing, drowsing or relishing effect. *e.g.*, tobacco, ganja.

10. Beverage crops: Plants used for preparation of mild, agreeable and simulating drinks including fruit juices, tea, coffee, cocoa, toddy, beer and wine. They supply water which is essential to human nutrition. Some of these drinks also provide vitamins and minerals. Others have stimulating or relaxing effects. Examples: Seed (cacao, coffea) and Leaf (tea, yerba mate - *Ilex paraguariensis*)

11. Spices and condiments: Spices are natural plant or aromatic vegetable products of tropical origin that are used in a pulverized state, primarily for seasoning or garnishing foods and beverages. These are characterized by pungency, strong odour and sweet or bitter taste. The *spices* cannot be grouped as foods since these contain less nutrititive value. They stimulate the appetite and increase the secretion and flow of gastric juices. So these are commonly known as '*food adjuncts*'. The aromatic value of the spices is due to the presence of the essential oils. All aromatic vegetable products that are used for flavouring foods and drinks are known as spices. Examples of spices are pepper, cinnamon, cardamom, cloves, ginger, turmeric, nutmeg, mace, vanilla, chillies, garlic, onion and coriander. Condiments are spices or other flavouring substances which possess sharp taste and are commonly added to food after it has been cooked. Turmeric is used as an important condiment in India.

B. CLASSIFICATIONS BASED ON ECOLOGICAL FACTORS

1. Classification based on climatic types

Climate type	Crops
Tropical crops	Rice, sugarcane, banana, cocao, cashew, mango, papaya, and pineapple.
Tropical high levels	Coffee, potato, wheat.
Tropical rainforest	Cacao, rubber
Sub tropical crops	Rice, cotton, citrus, dates, figs, olives, orange, and litchi.
Humid subtropics	cotton, oil palm
Intermediate humid continental climate	Maize, wheat.
Intermediate continental climate	Apple, grapes, orange
Temperate crops	Wheat, barley, rye, oats, sugarbeet, sunflower, apples, plums, almond cherries, peaches, maples, cottonwoods
Polar crop	All pines, pasture grasses

2. Classification of crops based on sunlight intensity

Plant group	Adaptation	Crops
Sciophytes	Shade loving plants	Turmeric, tobacco, coffee, betel vine, black pepper, cacao, gingers, and many orchids
Heliophytes	Sun loving plants	Rice, maize, sugarcane, sunflower, wheat, cotton, banana, etc. This plant classification also applies to the xerophytic plants.

3. Classification of crops based on photoperiod

Plant group	Bright sunshine hours	Crops
Long day plants	$\geq$12 hours	Wheat, barley, mustard, oats, rye, sugar beet, orchard grass, red clove, timothy grass, sweet cloves, chicory, lettuce, radish.
Day neutral plants	Unaffected due to sunshine hours	Rice, maize, cotton, tobacco, carrot, pepper, cucumber, buckwheat, pea, sunflower, strawberry.
Short day plants	$\leq$ 12 hours	Greengram, Blackgram, Sugarcane, sudan grass, soybean, sweet potato, spinach, orchid, cactus, *Bryophyllum*.

4. Seasonal classification: Crops are grouped under the season in which their major field duration falls.

(i) ***Kharif*** crops (June-September): Crops with short days for flowering. *e.g.* rice, maize, castor, groundnut, cotton, sweet potato, black gram, green gram, cowpea, sorghum, sesame, guar, jute, sunnhemp, pigeon pea, soybean, okra

(ii) ***Rabi*** crops (October–February): Crops require long days for flowering. *e.g.* wheat, barley, oats, mustard, potato, bengalgram, peas, linseed, lentil, berseem

(iii) **Zaid** or **Summer** crops (February-June): pumpkin, muskmelon, watermelon, bottle gourd, sponge gourd, blackgram, green gram, cowpea, cucumber, chilly, tomato, sunflower.

This classification is not a universal one. It only indicates the period when a particular crop is raised. Examples are *kharif* rice, summer rice, *kharif* maize, *rabi* maize, summer pulse, etc.

5. Classification of crops based on edaphic / soil factors

Edaphic plant group	Adaptation	Crops
Halophytes	On saline soils	Sugar beet, lucerne, karnal grass, almond, sunflower, castor, chillies, rice, finger millet, cotton, cowpea,
Psammophytes	On sandy soils	Succulents, cali, *Casurina*, *Prosopis juliflora,*
Lithophytes	On rock surface	Ferns,
Chasmophytes	On rock crevices	Algae, ber
Oxylophytes	On acid soils	Potato, tea, rice,
Calciphytes	On basic soils -containing high calcium	*Asparagus, Cenchrus* grass
Crypsophytes	on gypsum soils	

6. Classifications according to ecological / natural adaptation or habitat

(i) **Hydrophytes:** Hydrophytes are known as aquatic plants or water-loving plants or bog plants. Aquatic plants grow in water and swampy areas while bog plants grow in soils too wet or waterlogged soil for ordinary plants. Hydrophytes grow entirely or partly submerged, or floating on the water surface, or with their roots anchored to the ground in swamps as their natural habitat due to presence of modified roots called 'knee pneumatophore'. Hydrophytes are characterized by the sponginess of tissues. Stomata may be numerous and are located on the upper surface of the leaves. Hydrophytes produce roots even in its water bodies or into saturated soil condition. Hydrophytic plants grow habitually in water or in very wet soils where oxygen is deficient as a result of excessive water content. It may grow entirely submerged, partly submerged or floating, or anchored to the ground in bogs, swamps, waterlogged soil or beside the edges of ponds, lakes or streams. Examples are Azolla, kangkong (*Ipomea aquatica*), bulrush (*Cyperus* spp.), lotus, water lily, mangrove species, taro (*Colocasia esculenta*), lowland rice, water hyacinth family (*e.g. Monochoria vaginalis*), water lily (*Nymphaea* spp.), papyrus and umbrella plant (*Cyperus* spp.), lotus (*Nelumbo nucifera*), and bakawan (*Rhizophora mucronata*) and other mangrove species.

(ii) **Mesophytes:** Mesophytic plants are called moist-loving plants, belong to the terrestrial plants. A terrestrial plant which is adapted to moderate conditions for growth, *i.e.* not too dry and not too wet (e.g. corn and most commercially-grown crops). Mesophytic plants intermediary between hydrophytes and xerophytes. Mesophytes have stomata mostly on under leaf surfaces. Root hairs are abundant. Root's length and volume often equal or exceed the top growth. There are two groups in mesophytes *viz.*, (i) True Mesophyte which wilt permenantly after loosing 25% of their total water content and (ii) Xerophyte mesophyte which wilt permanently after loosing from 25 to 50% of their water content. Examples are corn, sugarcane, fruit trees and vegetables.

(iii) **Xerophytes:** Plants are adapted to conditions with little or no water. Xerophytes are capable of growing in a dry habitat or prolonged drought condition without injury. Xerophyte plants wilt permanently only after losing 50 to 75% of their total water content. Examples: adelfa, bromyliads, euphorbias, cacti and many succulents. The pygmy cedar (*Peucepyllum*) can live without soil water. It obtains its water need from the water vapor in the air alone, replenishing its supply during the night. The caper plant (*Capparis spinosa*) of the Sahara seems to have the same ability. The caper has one of the deepest root systems among plants. Uncultivated caper plants are more often seen hanging, draped over soil and rocks but their vegetative canopy covers soil surfaces which help to conserve soil water reserves.

(iv) **Halophytes:** Halophytic plants are called salt loving plants which can tolerate growing under saline conditions or in natural habitats which are excessively rich in salts. Examples are mangrove species, coconut, cashew, jackfruit and tamarind have varying levels of tolerance to saline conditions. The common table salt is in fact used as a fertilizer for coconut.

(v) **Lithophytes:** Lithophytic plants are adapted to growing on rocks or in rocky terrain with little/ scant humus, absorbing nutrients from the atmosphere, rain, and decaying matter which accumulate on the rocks. Vanda, Ascocenda, Ascocentrum, and Trudelia orchids can be grown as lithophytes. Dendrobium grows in pots filled with gravel or stone.

Epiphytes: Epiphytic plants, also called air plants and tree dwellers which are adapted to grow aboveground on another plant but is not parasitic, usually deriving only physical support from the host and obtaining nourishment from the air and other sources. Some have roots that take moisture and minerals leached from the canopy of trees and others catch rain and debris in special hollow leaves. The most common epiphytes belong to the pineapple (bromyliad), orchid, and fern families. These are called 'air plant' or 'tree dweller'. Sometimes the plant grows as an epiphyte which later becomes rooted to the soil, it is especially called a *hemiepiphyte*. Example is strangler fig (balete tree, ficus group).

Parasites: Parasitic plant grows on another plant (host) from which it takes part or all nourishment either partially or entirely (*e.g. Cassytha*, *Loranthaceae* (mistletoe family), *Rafflesia*; *Neottia* and *Corallorhiza* orchids).

Saprophytes: Saprophytic crops refer to the mushrooms. Plants grow on decaying organic matter and has no green tissue from which they obtain their food. This classification applies to the mushrooms, which are fungi.

Acidifuge or calcicole plants: These plants are called chalk-loving, lime-loving and acid-escaping plants, are plants that prefer calcareous or alkaline soils or soils with pH above 7.0. Examples are alfalfa, blazing star (*Chamaelirium luteum*) and southern redcedar (*Juniperus silisicola*).

Calcifuge or acidicole plants: These plants are called chalk-escaping, lime-hating, acidophilous, acid-loving, and acid soil plants, are those that prefer acidic soils or soils having pH levels below 7.0 but do not tolerate alkaline or calcareous soils. Examples are the rhododendrons and azaleas which have a low lime requirement and can live in soils with pH levels of 4.0 or less.

Phreatophytes: Phreatophytic plants are adapted to arid conditions by growing long roots which obtain water from underground reserves. The mere presence of these plants indicate a stable supply of underground water and such knowledge has been applied by digging wells close to them. Examples are the mesquite (*Prosopis*), cottonwood tree (*Populus*) and California fan palm (*Washingtonia filifera*).

Psychroxerophytes: Drought resistant plants that grow in cold territory.

Neutrophils: Neutrophilus plants can tolerate either acidic or alkaline soils.

Metallophytes (metal-tolerant plants) are plants adapted to natural habitats with toxic levels of metals such as Ni, Co, Cr and Mn. Examples are *Myristica laurifolia, Shorea tenuiramulosa, Rinorea bengalensis, Phyllanthus balgooyi, and Walsura monophylla.*

7. Classification based on crop effect on soil

(i) Soil depleting or exhaustive crops: *e.g.*, maize, sunflower, sesame
(ii) Soil conserving or cover crops: *e.g.*, calopogonium, vettiver grass
(iii) Soil building or restorative crops: *e.g.*, green manure and leguminous crops

C. CLASSIFICATION BASED ON CULTURAL PRACTICES

1. Classification based on place of origin:

(i) **Native crops:** Crops are grown within the geographical limits of their origin *e.g.*, rice, barley, blackgram, greengram, castor, sugarcane and cotton are of Indian origin.

(ii) **Exotic or introduced crops:** Crops grown even beyond their site of origin. *e.g.*, tobacco, potato, jute, maize are introduced to India from other countries.

2. Classification of crops based on carbon dioxide fixation in photosynthesis

(i) **C_3 plants:** A plant in which the first product of CO_2 fixation is the 3 carbon compound (phosphoglyceric acid). *e.g.* rice, wheat, barley, etc,.

(ii) **C_4 plants:** A plant in which the first product of CO_2 fixation is the 4 carbon compound (oxalo acetic acid). Thses plants are photosynthetically more efficient than C_3 plants. *e.g.* sugarcane, maize, etc,.

(iii) **CAM Plants:** The Crassulacean Acid Metabolism (CAM) system is prevalent in desert plants where CO_2 is fixed at night. In the CAM plants, there is diurnal fluctuation of acidity in thickened leaves. The CAM plants are adapted to more or less constant acidity. *e.g.* Cactus plants.

3. Classification based on tillage requirements

(i) **Arable crops:** Crops requiring preparatory tillage. *e.g.*, Rice, sugarcane, maize, potato, tobacco.

(ii) **Non arable crops:** Crops that do not require preparatory tillage. *e.g.*, Rice fallow pulses.

4. Classification of crops based on water requirement and permanent wilting point

Plant group	Permanent wilting point (% of field capacity)	Crops
Hydrophytes	< 20	Rice, water lily, water hyacinth, water chestnut
True mesophytes	20–75	Most of field crops, fruit and vegetable crops.
Xerophyte mesophytes	25–50	Sorghum, millets
Xerophytes	50–75	Succulents and cactus

5. Classification of crops based on usable products

(i) Food crops: Cereals, millets, pulses, legumes, fruits, vegetables and nuts

(ii) Feed crops: Forage crops such as grasses

(iii) Cash crops or industrial or commercial crops: Cotton, sugarcane, tobacco, jute, groundnut, castor, gingelly, tapioca

(iv) Food adjuncts: Turmeric, cumin, garlic, spices and condiments, beverages and narcotics - food and industrial usage.

Note: It is possible that one crop which has been included as a food crop may also figure as an industrial crop. *e.g.* maize or tapioca.

6. Classification of plants based on rangeland

(i) **Grasses** are plants with long narrow leaves and hollow stems.

(ii) **Grass-like plants** look like grasses, but have solid, often triangular stems (not hollow) without joints. Veins in the leaves are parallel. Examples are sedges and rushes. The sedges, grasses, and rushes, which make up almost all of the grass-like plants in temperate ecosystems, comprise three separate families: the Cyperaceae (the Sedge Family; Poaceae (the Grass Family, Gramineae), and Juncaceae (the Rush Family)

Poaceae (Grasses family)	Cyperaceae (Sedges family)	Juncaceae (Rushes family)
Stems round (terete)	Stems usually 3-angled (or round, 4-angled, lenticular)	Stems round (terete)
Stems with solid nodes and hollow internodes	Stems usually with solid pith	Stems with solid pith
Leaf sheaths open	Leaf sheaths closed	Leaf sheaths open
Leaves 2-ranked	Leaves 3-ranked or spiral	Leaves 2-ranked
Fruit a grain with papery palea, lemma, and glumes	Fruit an achenewith bristles, bracts, may have tubercle	Fruit a capsule with tiny dust-like seeds; capsule surrounded by 6 scale-like structures (tepals)
Examples: *Panicum repens, Echinochloa* sp, *Brachiaria mutica, Saccharum giganteum*	Examples: *Cyperus rotundus, C. papyrus, C. giganteus, C. textilis, C. pangorei, C. esculentus*	Examples: *Juncus repens Juncus megacephalus Juncus effusus Juncus roemerianus*

(iii) **Forbs** are herbaceous (non-woody) plants that have broad leaves and showy flowers. Most of the range weeds are forbs.

(iv) **Shrubs** are woody plants and do not have a main trunk, instead they have several main stems. Some plants can take both a tree and a shrub form depending on soil and topographic conditions.

(v) **Weed** is a designation that can be given to any plant that grows where it is not wanted or interferes with the growth of desirable plants.

7. Classification based on range of cultivation and market location

(i) **Garden crops:** Crops are grown on a small scale. *e.g.*, kitchen garden, flower garden, backyard garden, etc.,

(ii) **Field crops:** Seasonal crops grown in a large scale. *e.g.*, rice, wheat, and cotton.

(iii) **Plantation crops:** Crops are perennial in nature and grown on a large scale in estates under intensive culture where products are sold in distant markets rather than for local consumption. *e.g.*, coconut, oil palm, tea and coffee.

D. GENERAL CLASSIFICATION OF CROPS

1. Crop Classifications according to growth habit

(i) **Herbs:** Succulent plants with soft / non-woody stems, less fibrous compared to woody plants. Examples: banana, tomato.

(ii) **Vines:** Herbaceous climbing or twining plants without self-supporting stems. Examples: charantia (ampalaya), cucumber, luffa (patola), pole sitao, yam.

(iii) **Lianas:** Woody climbing or twining plants which depend on other plants for vertical support to climb up to the top of the canopy. These climbers often form bridges between the forest canopy. Examples: Climbing bamboo, grape, rattans, passion fruit.

(iv) **Shrubs:** Small trees or tree-like plants, generally less than 5 meters in height but by other authorities it is restricted to small, erect, woody plants which produce several trunks from the base. Examples: Barbados cherry, pink jasmine, *Lantana*.

(v) **Trees:** Plants having erect and continuous growth with a large development of woody tissue, with a single distinct stem or trunk, reaching a height of 5 meters or more. Examples: durian, mango, tamarind.

2. Classification based on ontogeny or life cycle

(i) **Annual crops:** Crop plants that complete their life cycle (from seed to seed) within a season or year. They produce seed and die within the season. Examples are rice, wheat, tobacco.

(ii) **Biennial crops:** Crop plants having a life span of two consecutive growing seasons or two years to complete life cycle. During the first season they produce vegetative parts and store food. In the second season they produce flowers and seeds. Vegetative development usually in the form of a rosette in first year while the reproductive development, flower stalks form in a process known as bolting in second year. Examples are banana, carrot, sugar beet, onion, cabbage.

(iii) **Perennial crops:** Crop plants that live for three or more years. These may be seed bearing or non-seed bearing. Examples are perennial grasses, alfalfa, oil palm, coconut, rubber, mango.

(iv) **Ephemeral plants:** Ephemeral plants can take the advantage of water resources and temperature conditions to rapidly complete their life-cycle once in three or four months which dies off completely leaving the dormant seeds / roots / underground organs remain alive and produce plants from underground organs or seeds in the next/following season. **Ephemeral** plant has two or more generations in a year. Examples are tomato, groundsel (*Senecio vulgaris*), *Drosera indica, Habenaria viridiflora, Dopatrium junceum, Lindernia crustacea.*

3. Crop classification based on plant habit

(i) **Determinate crops:** Flowering is confined to a specific period of time in these crops, at the end of vegetative growth. Example: rice

(ii) **Indeterminate crops:** Flowering is continuous and not confined to a specific period in these crops, as flowering overlaps with vegetative growth. Example: Redgram, groundnut

4. Classifications according to leaf retention

(i) **Evergreen** are plants that retain living leaves throughout the year. Examples: pines, banana, papaya, palms and most tropical plants.

(ii) **Deciduous** are plants which naturally shed off or lose leaves at approximately the same time annually for extended periods during winter in temperate climate or during the dry season in the tropical climate. Natural leaf shedding is pronounced in deciduous trees of temperate regions. Examples: defoliation occurs during summer months in *Delonix regia*. Some plants exhibit partial defoliation during drought periods.

5. Classifications according to mode of reproduction

(i) **Sexual** plants that develop from a seed or spore after undergoing union of male and female gametes. Examples: palms and ferns.

(ii) **Asexual** plants which reproduce by any vegetative means without the union of the sexual gametes or by apomixis. Examples: bread-fruit, mangosteen

6. Classifications according to mode of pollination

(i) **Naturally self-pollinated crops:** Both pollen and embryo sac are produced in the same floral structure or in different flowers but within the same plant. Examples: rice, pulses, okra, tobacco, tomato.

(ii) **Naturally cross-pollinated crops:** pollen transfer in these plants is from the anther of one flower to the stigma of another flower in a separate plant, although self-pollination may reach 5 percent or more. Examples: corn and many grasses, avocado, grape, mango, many plants with unisexual or imperfect flowers.

(iii) **Both self- and cross-pollinated crops:** These plants are largely self-pollinated but varying amounts of cross-pollination occur. Examples: cotton and sorghum.

7. Plants classification based on level of lignification: Lignin is an indigestible portion of cell walls that impregnates cellulose to form wood

(i) **Herbaceous plants** = Non-woody plants. Plant dies back to ground each year.
All annuals are herbaceous.
All grasses and forbs are herbaceous

(ii) **Woody plants** are plants with lignified stems which includes trees and shrubs. A shrubby plant is woody, considerably branched and usually less than ten meters tall. An arborescent plant is one that is tree-like.

(iii) **Suffrutescent plants** are plants with a woody base but herbaceous stems which die back to ground every year.

8. Classification of plants based on modified plant structures

(a) Underground stems and shoots

(i) **Rhizomes:** Horizontal stems that grow at or below the soil surface

(ii) **Tubers:** The tips of rhizomes, which become enlarged with the storage of food

(iii) **Bulbs:** Large buds, each consisting of a small stem and numerous fleshy, storage leaves.

(iv) **Corms:** Stems that superficially resemble bulbs but consist mostly of stem tissue. The leaves are usually smaller and thinner (bractlike and papery) than those of bulbs.

(b) Other modified stems

(i) **Tendrils** are modified aerial organs for climbing. (Leaves can also form tendrils)
(ii) **Runners/ (Stolons)** are creeping stems that grow horizontally on the soil surface and often give rise to new plants at the nodes.
(iii) **Thorns** are modified twigs that grow in the axils of leaves; sometimes they are branched. The epidermis of stems and leaves can also be very 'thorny'. Examples are the thorns on roses and blackberry bushes are modified epidermis cells, called prickles.

(c) Specialized Leaves

(i) Bud scales are protect buds of woody plants.
(ii) Spines
(iii) Plantlets are propagules which can form new plants.
(iv) Showy bracts are leaves used to attract pollinating agents.

(d) Carnivorous plants

Carnivorous plants are plants that derive some or most of their nutrients (but not energy) from trapping and consuming animals or protozoans, typically insects and other arthropods. Carnivorous plants have adapted to grow in places where the soil is thin or poor in nutrients, especially nitrogen, such as acidic bogs and rock outcroppings.

9. Classification of plants based on root structures

There are two main types of roots according to origin of development and branching pattern *viz.*, taproot system and fibrous system. The primary root which develops from a radicle and becomes dominant is called a taproot, as in carrot. Roots that develop from other roots are generally called lateral roots; those that arise from other plant organs rather than the root, such as from stems or leaves, are called adventitious roots. The specialized taproot and fibrous roots are as follows:

(i) **Storage roots** (fleshy and tuberous roots): These consist of thickened roots due to the accumulation of high-energy storage compounds, usually starch. These are further subclassified into fleshy and tuberous roots. Examples of crops with fleshy roots are the carrot, ginseng (Panax) and sugar beet while those with tuberous roots are the sweet potato, cassava and yam bean
(ii) **Aerial roots**: These are adventitious roots that are common in many epiphytes such as in the monocot plants belonging to the arum or gabi family (Araceae) and orchid family (Orchidaceae). In general, these fibrous roots remain aerial, *i.e.,* they do not enter the soil. Depending on the plant species, aerial roots perform special functions such as water retention, photosynthesis, and support.
(iii) **Contractile roots:** These fibrous roots contract vertically to pull the plant downward into the soil. Examples are the plants that form bulbs (*e.g.* lily) or corms (*e.g.* Gladiolus). These roots are also found in ginseng (Panax). Contractile roots can shrink more than 50% in only a few weeks.

(iv) **Haustoria:** These are specialized roots in parasitic plants that penetrate the tissues of a host plant, as in witchweed (Striga) and broomrape (Orabanche).

(v) **Prop roots:** These are aerial roots that arise from a stem and subsequently sink into the soil to provide additional support to the plant such as in corn and banyan tree (*Ficus benghalensis*).

(vi) **Pneumatophores:** These are specialized roots in some plants that grow in watery places and which function mainly for oxygen absorption. The raised pneumatophores of many mangrove species and the 'knees' or knee pneumatophores of bald cypress serve as entry of atmospheric oxygen which diffuses to roots growing in anaerobic soils. The 'knees' are cone-shaped extension of the root that protrude from the ground. The oxygen from the air diffuses to the pneumatophores through aerenchyma cells which compose as much as 80% of these roots.

(vii) **Buttress roots:** These are enlarged, often thickened roots that spread horizontally from the base of trees to provide additional support. In tropical trees like fig (Ficus) that are shallow rooted, large buttress roots are formed at the base of their trunks.

10. Classification of crops based on root depth

Shallow rooted	Moderately deep rooted	Deep rooted	Very deep rooted
Rice	Wheat	Maize	Sugarcane
Potato	Castor	Cotton	Citrus
Cauliflower	Ground Nut	Sorghum	Coffee
Cabbage	Pea	Bajara	Apple
Lettuce	Bean	Soybean	Grape Vine
onion	Chilli	Sugar Beet	Safflower
	Tobacco	Tomato	Lucerne/alfalfa

E. BOTANICAL CLASSIFICATION

Crop plants are grouped into different families. Some of the family names have new names such as *Compositae* (*Asteraceae*), *Cruciferae* (*Brassicaceae*), *Gramineae* (*Poaceae*), *Guttiferae* (*Clusiaceae*), Labiatae (*Lamiaceae*), *Leguminosae* (*Fabaceae*), *Palmae* (*Arecaceae*) and *Umbelliferae* (*Apiaceae*).

(i) **Euphorbiaceae**

Castor – *Ricinus communis*

Tapioca – *Manihot esculuntum*

(ii) **Liliaceae**

Onion – *Allium cepa*

Garlic – *Allium sativum*

(iii) **Poaceae** (*Gramineae*)

Festucoideae (sub family)

Bread wheat – *Triticum aestivum*

Barley – *Hordeum vulgare*

Oats – *Avena sativa*

Rye – *Secale cereale*

Panicoideae (sub family)

Maize – *Zea mays*

	Sorghum	– *Sorghum bicolor*
	Sweet sorghum	– *Sorghum saccharum*
	Sudan grass	– *Sorghum sudanense*
	Sugarcane	– *Saccharum officinarum*
	Kodo millet	– *Paspalum scrobiculatum*
	Bulrush	– *Pennisetum americanum*
	Pearl millet	– *Pennisetum glaucum*
	Cat-Tail millet	– *Pennisetum typhoides*
	Fox Tail millet	– *Setaria italica*
	Little millet	– *Panicum miliare*
	Common millet	– *Panicum miliaceum*
	Barnyard millet	– *Echinochloa colona*
	Finger millet	– *Eleusine coracana*
	Golden Timothy	– *Setaria anceps*
	Setaria grass	– *Setaria sphacedata*
	Oryzoideae (sub family)	
	Rice	– *Oryza sativa*
(iv)	**Asteraceae** (Compositae)	
	Sunflower	– *Helianthus annuus*
	Safflower	– *Carthamus tinctorius*
	Niger	– *Guizotia abyssinica*
(v)	**Brassicaceae** (Cruciferae)	
	Rape	– *Brassica napus*
	Mustard	– *Brassica juncea*
(vi)	**Cannabinaseae**	
	Hemp	– *Cannabis sativa*
(vii)	**Chenopodiaceae**	
	Sugarbeet	– *Beta vulgaris*
(viii)	**Fabaceae-Grain legumes**	
	Groundnut	– *Arachis hypogea*
	Field bean	– *Lablab purpureus* var *lignosus*
	Garden bean	– *Lablab purpureus* var *typicus*
	Redgram, Pigeon pea	– *Cajanus cajan* L. Mill.
	Chickpea, Bengal gram	– *Cicer arietinum*.L
	Sunhemp	– *Crotolaria juncea*
	Soyabean	– *Glycine max* Merr.
	Clusterbean	– *Cyamopsis tetragonoloba*
	Lentil	– *Lens culinaris* (*L.esculenta Moench.*)
	Grass pea	– *Lathyrus sativus* L.
	Common pea	– *Pisum sativum* L. var *purpureus*
	Horsegram	– *Macrotyloma uniflorus*
	Cowpea	– *Vigna unguiculata* L. Walp.
	Blackgram	– *Vigna mungo* L.Hepper.
	Greengram	– *Vigna radiata* L.Wilazek

(ix) **Linaceae**
Flax / Linseed – *Linum usitatissimum*

(x) **Malvaceae**
Cambodia cotton – *Gossypium hirsutum*
Egyptian cotton – *Gossypium barbadense*
Uppam cotton – *Gossypium herbaceum*
Karunkanni cotton – *Gossypium arboreum*
Kenaf – *Hibiscus cannabinus*
Roselle – *Hibiscus sabdariffa*
Bhendi/ Okra – *Hibiscus esculentus*

(xi) **Solanaceae**
Chillies – *Capsicum annuum*
Tobacco – *Nicotina tabacum*
Potato – *Solanum tuberosum*

(xii) **Agavaceae**
Sisal – *Agave sisalana*

(xiii) **Musaceae**
Banana – *Musa sapientum*
Plantain – *Musa paradisiaca*

(xiv) **Palmae**
Betel palm (Arecanut) – *Areca catechu*
Coconut – *Cocos nucifera*
Oil palm – *Elaeis guineensis*

(xv) **Pedaliaceae**
Sesame – *Sesamum indicum*

(xvi) **Tiliaceae**
Tossa jute – *Corchorus olitorius*
White jute – *Corchorus capsularis*

(xvii) **Utricaceae**
Ramie – *Buehmeria nivea*

(xviii) **Zingiberaceae**
Turmeric – *Curcuma domestica*
Ginger – *Zingiber officinalis*

(xix) **Apiaceae**
Carrot – *Daucus carota*
Celery – *Apium graveolens*

(xx) **Cucurbitaceae**
Squash – *Cucurbita pepo*
Pumpkin – *Cucurbita moschata*
Cucumber – *Cucumis sativus*
Watermelon – *Cucurbita lanatus*

(xxi) **Lamiaceae**
Mint – *Mentha arvensis*

F. SPECIAL PURPOSE CROPS

Crop cultivated for specific purpose in addition to its economic yield.

(i) **Augmenting crops:** Crops are sown to supplement the yield of main crops. The supplement crops are known as augmenting crops. e.g., mustard with berseem,

(ii) **Alley crops:** Arable crops are grown in alleys formed by trees or shrubs, established to hasten soil fertility restoration, enhance soil productivity and reduce soil erosion. These are known as alley crops. Such crops should have slight shade tolerance and should be non-trailing. Examples are sweet potato, black gram, turmeric and ginger in between the rows of Eucalyptus and *Subabul* while grasses in between Hedge lucerne.

(iii) **Border / Barrier crops / Guard crops:** The crops grown around the field boundaries of the plot or field of another crop in narrow strips to protect the main crop from the trespassing cattle menace or restrict the speed of wind. *e.g., Sesbania grandiflora* around betelvine crop; castor and kenaf in the outer ridges of sugarcane; safflower around pea and sorghum around maize.

(iv) **Bonus crop:** An additional crop taken in intercropping or mixed cropping systems without any extra inputs. Redgram and castor are broadcasted in groundnut.

(v) **Brake crops:** Crops are grown to break the continuity of the ecological situation of the field in crop sequence or crop rotation. Sugarcane-rice rotation breaks the continuity of weeds, pests due to variations in host ranges and changing of ecological situations.

(vi) **Cash crops:** Cash crop is a high value marketable crop which is grown for sale to earn hard cash. These crops are processed for their economic products at industries. Examples are sugarcane, cotton, jute, sugarbeet.

(vii) **Industrial crops:** Plants grown to provide materials for industrial processing and production of non-food products such as biofuel, sugar, rubber, starch, industrial oil, aromatic compounds, steroids, medicinal drugs, organic pesticides, tannins and dye. This is a special classification based on the method of processing and the nature of the product (non-food) and not on the part of the plant which is harvested and used as raw material. Based on these criteria, both agronomic and horticultural crops can be classified as industrial crops. Corn (grain crop) and legume seed crops (e.g. soybean) can be classified as industrial if they are grown primarily for industrial processing to produce biofuel or industrial oil. Examples are cotton, sugarcane, tobacco, groundnut, castor, gingelly and tapioca.

(viii) **Paira crops / Relay crops:** Crop plants sown a few days or weeks before the harvesting of the standing mature crop. These crops are grown on residual moisture and nutrients without preparatory tillage. The standing crop and latter sown (paira) crop will be in the same field for a brief period and remain as single crops for the rest of the duration. Examples are blackgram, greengram, *khesari* bean in rice.

(ix) **Catch crop/ Emergency crop:** A quick growing crop incidentally planted and harvested between two major crops in consecutive seasons to utilize residual fertilizer and soil moisture. It may be a contingency crop grown to replace a major crop which has failed. It is often used as a green manure or to provide supplemental livestock feed; also called emergency crop.

(x) **Companion crop:** Any crop which is planted close to the main crop to complement the latter's growth and production, or to maximize utilization of space because they do not compete. Examples are Black pepper planted with live coconut, cacao, *Gliricidia sepium* to serve as trellis. Castor and maize grown in turmeric field.

(xi) **Cover crops:** Crop plants are able to protect the soil surface from the erosion though their ground covering foliage and / or root mats or loss of moisture due to leaching and erosion by wind and water. Examples are groundnut, marvel grass, cenchrus grass, sweet potato and para grass. Leguminous vines such as improved pasture and forage crops are excellent cover crops. As legumes they can enrich the soil fertility by fixing atmospheric nitrogen.

(xii) **Contingent crops:** Crops raised in the event of failure of a main crop that has failed due to biotic or climatic or management hazards and utilize the remaining period of the season. They are generally of very short duration, quick growing, fast bulking, harvestable or usable at any time of their field duration. Examples are greengram, blackgram, cowpea, sunflower, spinach, coriander and onion.

(xiii) **Exhaustive crops:** The soil fertility gets exhausted when aggressive nature of crop is grown. Examples are sunflower, sesame, brinjal, maize and linseed. These are plants which on growing leave the field exhausted.

(xiv) **Ley crops:** Forage crops grown for grazing in rotation with field crops. Example is *Cenchrus* grasss rotated with field crops in Kangayam tracts, Tamil Nadu.

(xv) **Mulch crops:** These crop plants are grown to conserve soil moisture from bare ground by their foliage, trailing habits and self-seeding nature. Examples are cowpea, coriander.

(xvi) **Nurse crops:** A companion crop which nourishes the main crop by way of nitrogen fixation and/or adding the organic matter into the soil. Nurse crop refers to the small grain crop in some way protected the new seedlings or nursed it along in some way until it was strong enough to establish for itself. Crops help in the nourishment of the other crops by providing shade and act as climbing sticks such as mustard in peas, sorghum in cowpea, Tephrosia, Glyricidia in tea. Leguminous or deciduous plants shed their leaves and enrich the soil fertility. Tall crops such as coconut, pigeon pea, castor, etc., nourish shade-loving plants such as turmeric, ginger and corm. Cowpea intercropped with cereals or new plantations of fruit trees.

(xvii) **Opportunity cropping:** The practice of placing an option on use of stored soil moisture while deciding whether or not to raise or establish a crop.

(xviii) **Plant crop or first harvest crop or stub crop or first cycle crop:** Crop plants refer to the first harvest after sowing or planting the crop in perennial or multicut crop. Examples are sugarcane, cotton, pigeonpea, napier grass, para grass, guinea grass and marvel grass.

(xix) **Restorative crops:** Crops assist in enrichment, restoration or amelioration of the soil in addition to its economic yield. Examples are legumes. These fix the atmospheric nitrogen in root nodules and shed their leaves during ripening stage to restore soil conditions.

(xx) **Smother crops:** Crop plants are able to smother or suppress the population and growth of weeds with their dense foliage developed due to quick growing ability, or branching or procumbent or trailing habits. Smother crop is grown for the purpose of eliminating any undesirable plant through physical or its alleopathic effects. Examples are cowpea and sweet potato, calapogonium in coconut.

(xxi) **Rubber crops:** Plants grown for the production of latex which is processed into the industrial product called rubber. Examples are para rubber tree, Castilla rubber, guayule.

(xxii) **Biofuel crops or Energy Crops**: Plants grown for the production of fuel or biodiesel from vegetable oils. Examples: sugarcane, cassava, corn, coconut, castor bean, Jatropha. Energy crops are harvested for biodiesel, bioethanol and biomass crops.

(xxiii) **Root crops:** Crop plants whose surplus or reserve foods are stored primarily in enlarged roots. A tuberous root is a thickened secondary root as in arrowroot, cassava, sweet potato and yam bean. A fleshy root is usually an enlarged primary root, as in carrot, ginseng (*Panax* spp.) and sugar beet. In radish, the fleshy root consist mainly of the hypocotyl. Examples are tapioca, sugar beet, sweet potatoes.

(xxiv) **Tuber crops**: Crop plants possessing enlarged underground reproductive portion of stems which has high carbohydrate. A tuber is an enlarged tip of an underground stem with leaves reduced to scales or scars subtending the auxillary buds, as in white potato and yam. The "eyes" represent buds in nodes, arranged in spiral pattern from base to the apical end of the tuber. Example is potato. Aerial tubers are called tubercle.

(xxv) **Track crops:** Crops grown for distant markets requiring heavy transport.

(xxvi) **Trap crop or Decoy crop:** Plants grown to attract certain insect pests or parasites because they are favourite hosts. These act as decoys to lure pests away from the main crop. These make pest control easier because the insects are concentrated on a few plants. Examples of trap crops are as follows :

(a) Basil (*Ocimum basilicum*)- green, loopy tomato caterpillar.
(b) Garlic (*Allium sativum*)- greenfly.
(c) Chive (*Allium schoenoprasum*)- greenfly and cutworms.
(d) Corn (*Zea mays*)- cotton bollworm.
(e) Marigold (*Calendula officinalis*)- caterpillars and cutworms.
(f) Nasturtium (*Tropaeolum majus*)- aphids.
(g) Tarragon (*Artemisia dracunculus*)- loopy caterpillar.
(h) Tobacco (*Nicotiana tabacum*)- cotton bollworm.
(i) *Orobanche* and *Striga* are trapped through solanaceous and sorghum crops respectively.

CHAPTER

2

Cereals and Millets

Cereals are annual cultivated grasses grown for edible starchy grains / seeds. In general, they provide the main concentrated carbohydrate food for man and for livestock. All cereals are members of the grass family Poaceae (or grass family) and further classified into tribes within the family. Some common cereal groups are:

(i) **Triticeae:** Wheat (*Triticum* spp), Barley (*Hordeum* spp), Rye (*Secale cereale*), Triticale (*Triticosecale*)

(ii) **Oryzeae:** Rice (*Oryza sativa*), African rice (*O. glaberrima*), Wild rice (*Zizania palustris*)

(iii) **Aveneae:** Oat (*Avena sativa*)

(iv) **Andropogoneae:** Maize (*Zea mays*), Sorghum (*Sorghum bicolor*), Adlay (*Coix lacryma-jobi*)

Six great cereals of the world are rice, wheat, maize, barely, rye and oats. These form the principal sources of food for man and animals. Wheat and barely are the most important cereal grains of the warm temperate regions and high rainfall. Rye and oats are the dominant cereals of the colder regions with high rainfall and low fertility areas. Rice and maize dominate in the tropical zones with assured irrigations. Millets are grown in areas with limited water supply for irrigation.

The term cereals is derived from '*Ceres*', the Roman goddess of harvest. In ancient Rome at every harvest, great festival in honour of *Ceres* was celebrated and she was worshipped as the giver of grain. Wheat and barley were generally the offerings to the goddess and these were called *Cerealia munera*. Subsequently, the grains used for food, especially for making bread were called *Cerealia* or cereals. Most of the cereals are herbaceous annuals, although rye has a tendency towards a perennial habit. The stem or culm is often erect, usually cylindrical and always hollow, except at the nodes. The cereals are characterized by the shallow fibrous root system, tillering habit, sheathing leaf bases and inflorescence which are panicles. The primary root developed from the radicle is generally short lived; often the adventitious and fibrous root system which persists and is functional during the whole life time of the crop plant. The plant assumes a tufted appearance through this tillering (stooling) habit. The tillering habitat though a common habitat in the cereals, may be suppressed as in the case of maize or Indian sorghum. The culms are generally enclosed within the sheathing leaf bases and the jointed nodes may be distinctly exposed or completely enveloped. Ligules are almost invariably present except in the case of barnyard millet. Spiklets are the ultimate units of inflorescence in *Poaceae* which may be borne on panicles. Panicles are free and lax or compact or in some compressed to spiciform one. Flowers in the spiklet may vary from one as in the case of paddy to many as in the case of finger millet and wheat. The fruits generally termed as grains are the caryopsis, which are either fully enveloped by the glumes as in the case of paddy or partly exerted as in the case of sorghum or fully exposed as in the case of maize. All the cereals supply food to man and straw to animals. The flour or the meal of the grains is generally made use of and this is made up mostly of starch. The whole grains are used as food as in case of rice. Grains are also popped as in maize, and rice, or used as pressed or processed food as in case of rice and oats. In a light form, they go as cereal foods or breakfast cereals.

Rice is a staple food for nearly half of the population of the world. It contains larger proportion of starch than all the other cereals. Wheat, a very valuable cereal contains a good proportion of proteins besides the starch and comes second in the popularity. Maize has gained importance in all tropical regions of the world. In the colder regions of Europe and Russia, rye takes the place of wheat. Barley is important mainly as malt food and oats is beaten and processed in the form of light food. A large percentage of the world's population subsists mainly on wheat, rice or maize each accounting for nearly a quarter of the world's total cereal supply. Barely, oats and rye together make up the rest.

Characters of cereals

- Most of the cereals are herbaceous annuals
- Stem or culm often erect, cylindrical, hollow except at nodes
- Tillering habit with shallow fibrous root system
- Leaves alternate, distichous with parallel venation and sheathing leaf base
- Presence of ligules and lodicules
- Infloresence is panicle or spike
- Stamens usually three (but six in rice)
- Fruit is a caryopsis

The cereals are cultivated in major parts of the world due to the following reasons

- Greater adaptability
- Easy for cultivation
- Giving more yield per unit area due to tillering habit
- Grains are compact, dry and can be easily handled
- Grains can be easily separated from the plants
- Grains have high nutritive value with higher percentage of carbohydrates, sufficient protein (7 to 10%), fats, vitamins and minerals.

Pseudocereals: There are few species of plants other than those of Poaceae, which produce small grains which have a similar grain chemical composition and used for food as in the case of cereals. These are termed as *Pseudocereals*. Buckwheat (*Fagopyrum esculentum*), grain amaranth (*Amaranthus sp* – Amaranthaceae), and Quina (*Chenopodium* *quinoa* – Chenopodiaceae) are considered under pseudocereals. There are certain botanically unrelated plants whose seeds are also used in a similar manner as that of cereals.

Millets: Millets are small grained cereals. The term *millet* is applied to a variety of small seeds, which are of minor importance as food. Millets are coarse seeded, annual cereal grasses used for food, feed and forage. The term 'coarse grains' is also used to describe the millets. The millets are the staple food of the poor, the working classes and the farming community. The whole grain is used in soups, stews or as a cooked cereal, popped; roasted or sprouted grains. Millets contain no gluten and hence not suitable for making bread, but they are good for people who are gluten-intolerant. However, the flour from the millets is blended in 15 to 40% with the refined wheat flour to prepare biscuits. The millets also provide substantial quantities of palatable fodder for cattle. Small millet grains are nutritionally rich. These grains are particularly low in phytic acid and rich in iron and calcium. The millets are classified into two groups viz., major millets and minor millets or small millets. The major millets are sorghum, pearl millet and finger millet in India. Small millets may be defined as millets cultivated for their small grains which are borne on short, slender grassy plants. The small millets also refers to a group of small-seeded cereal crops. Small millets may also be called minor millets. The important small millets grown in India are foxtail millet (*Setaria italica*), kodo millet (*Paspalum scrobiculatum*), common or proso millet (*Panicum miliaceum*), little millet (*Panicum sumatrense*) and barnyard millet (*Echinochloa colona*). Small millets grains are nutritionally rich in iron and calcium and low in phytic acid. They provide palatable fodder for cattle.

Factors limiting small millets productivity: Production of small millets is subject to wide fluctuations and the area is declining, except in the case of finger millet. The major constraints limiting small millets production are:

- These crops are often grown in uneven marginal lands, poor in fetility, shallow and gravelly, with low moisture retention capacity.
- These crops are grown under rainfed conditions in low rainfall arid regions.
- Improved crop management practices are not adopted by the farmers due to socio-economic constraints.
- There is no organized programme for production and supply of seeds of improved varieties.
- There is no ready market for the disposal of surplus produce at a remunerative price.
- There is lack of extension and development support.

CHAPTER

3

Rice–*Oryza sativa* L. (2n=24)

Family: Poaceae

Vernacular names: Rice, paddy (English); Riz (French); *Chaval, Dhan* (Hindi); *Dhanya, Vrihi, Nivara* (Sanskrit); *Sali, Dhan* (Bengali); *Nel, Arishi* (Tamil); *Nellu* (Malayalam); *Bhatto, Nellu, Bhatta, Akki* (Kannada); *Dhanyamu, Odalu, Biyyamu* (Telugu); *Shalichokha, Bhata, Corava, Damgara* (Gujrati); *Bhat, Tandulamul, Dhanarmul* (Marathi). The word rice is derived from the Tamil word '*arisi*' while the word paddy is derived from the Malaysian word '*Padi*', meaning rice. Sometimes, both the words 'rice' and 'paddy' are used interchangeably. Generally, rice plant is also known as paddy. However, the word 'paddy' refers the rice grain with husk. Therefore, rice is a part of paddy.

Importance: Rice is the most important staple food crop of the world in terms of area, production and the number of people depending on the crop. Globally rice ranks second to wheat in terms of area. Rice has shaped the culture, diets and economics of thousands of millions of people. Rice provides more calories than any other cereal crop. Rice provides 20 % of global human per capita energy and 25 % of per capita protein. Rice also provides minerals, vitamins and fibre. Rice is used to prepare snacks, desserts and special foods. Some of the alcoholic beverages are also made from rice.

Origin: There are two cultivated species in rice *viz.*, *Oryza sativa* is originated in South East Asia (India, Bangladesh, Myanmar, China) and African rice, *Oryza glaberrima* originated in West Africa.

Differentiation between *O. sativa* and *O. glaberrima*: There are discrete differences between *O. sativa* and *O. glaberrima* (Table 1). *O. sativa* has more secondary branches on the panicles, and longer and smoother ligules than *O. glaberrima*. A typical *O. glaberrima* has glabrous (hairless) spikelets and leaf blades while *O. sativa* cultivars are mostly pubescent, although most cultivars in the United States are glabrous. The seed of *O. glaberrima* has longer dormancy than that of *O. sativa* . *O. sativa* is cultivated as an annual agricultural crop, but botanically it is a perennial plant, while *O. glaberrima* is annual both botanically and agronomically.

Table 1 Comparison of domesticated cultivars of O. sativa and O. glaberrima

Character	*O. sativa*	*O. glaberrima*
Ecotypes	Many	Few
Distribution	Worldwide	Endemic to West Africa
Ligule	Long and soft	Short and tough
Panicle branches	Many	Few
Frequency of glabrous varieties	Low	High
Varietal differentiation	Highly variable	Limited variation

Source: Oka, 1991

Races of rice and their eco-geographic distribution: There are three races of *Oryza sativa* in the world *viz.*, (i) *indica* type (long grain) is distributed in the tropics and sub-tropics, (ii) *japonica / sinica* type (round grain) is distributed in the cooler subtropics and temperate regions and (iii) *javanica* type (medium grain) is distributed in Indonesia and neighbouring countries. The characteristic features of the rice races are furnished in Table 2. The Indica rice is grown in warm climate zone of Indo-China, India, Pakistan, Thailand, Brazil and Southern U.S.A. The Japonica is mostly grown in the temperate regions and sub-temperate regions, in Mediterranean climate zones like: Egypt, Morocco, Turkey, *etc.*, in cold climate zone of Northern China, Korea, Japan and California, USA; in high altitude areas: Nepal, Bhutan (Asia), Rwanda, Burundi (Africa), and in the Southern Cone of Latin America: Chile, Argentina, etc. The Javanica is grown in Indonesia only.

Table 2 Differences between three rice races

Character	*Indica*	*Japonica*	*Javanica*
Climatic zone	tropical	temperate	equatorial
Photoperiod	very sensitive	not sensitive	not sensitive
Vegetative period	long	short	Very long
Response to manuring	low	high	low
Lodging	susceptible	resistant	resistant
Seed dormancy	present	absent	---
Grain shape	long, narrow	short, thick	coarse, broad, thick
Awns	usually absent	sometimes present	present
Plant colour	light green	dark green	light green
Plant height	tall	short	tall
Leaves	narrow to broad, light green leaves	narrow dark green	Broad, stiff, light green
Tiller number	many	moderate	few
Flag leaf	long, drooping	short, erect	Long, wide, semidrooping
Shedding of grain	susceptible	resistant	resistant
Ear size	Large	Large	small
Ear weight	Light	Heavy	heavy
Endosperm	translucent	chalky	---
Breakage during milling	high	low	low
Yield potential	medium	high	low
Market price	high	low	medium

Distribution of cultivated species of rice: There are two most important cultivated species of paddy namely i) *Oryza sativa* and ii) *Oryza glaberrima*. Of these, *Oryza sativa* is cultivated in Asia, Africa, Europe, North America, Central America, South America, Australia, and Oceania. There are 111 rice growing countries in the world. Important countries are India, China, Indonesia, Japan, Bangladesh, Pakistan, Vietnam, Philippines, Thailand, Korea, Australia, USA, Spain, Portugal, Italy, France, Romania, Russia, Myanmar and Brazil. Rice occupies one tenth of arable land of the world and more than one-third area of Asian rice countries. *Oryza glaberrima* is grown in West Africa. In India, rice is cultivated in Andhra Pradesh, Karnataka, Kerala, Tamil Nadu, Gujarat, Maharashtra, Rajasthan, Haryana, Punjab, Western Uttar Pradesh, Uttrakhand, Himachal Pradesh and Jammu & Kashmir, Bihar, Chhattisgarh, Jharkhand, Madhya Pradesh, Odisha, Eastern Uttar Pradesh, West Bengal, Assam and North eastern states. It is grown from below sea-level (Kuttanad area of Kerala) up to an elevation of 2000 metres in Jammu & Kashmir, hills of Uttarakhand, Himachal Pradesh and North-Eastern Hills (NEH) areas. Andhra Pradesh is known as the 'Rice bowl of India'. Andhra Pradesh is the major exporter of rice to the world.

Botany: The rice is an annual, freely tillering, semi-aquatic grass with a cylindrical jointed stem (culm), about 50 to 150 cm tall, but may go up to 5 m in floating rice. As a general rule, taller varieties have a greater penetration and lateral spread in the root system than the shorter varieties. Similarly, varieties with good tillering have a well-developed root system. The stem of rice, popularly known as haulm or culm, is generally erect, cylindrical, hollow at the internodes and solid at the nodes. Rice plants begin to develop tillers around two weeks after transplanting and continue developing more tillers for three to five weeks thereafter. The maximum production of tillers is generally attained 30 to 40 days after transplanting. The internodes are the shortest at the base, becoming progressively longer. The leaves are borne alternatively on the stem. Leaf consists of leaf sheath, leaf blade, ligule and auricles, the former encircling the whole or part of the internode. The leaf blade is long, narrow, 30 to 50 cm or more in length and 1-2 cm broad. The lamina of the uppermost leaf below the panicle ('flag' or 'boot') is wider and shorter than the others. The rice inflorescence is a loose terminal panicle, 7.5 to 38.0 cm long. Panicle is erect or drooping, open or compact. The spikelets open between 10 a.m. and 12 noon in the cultivated varieties. Self-pollination is the rule, natural cross-pollination also occurs, varying from 0.1 to 4.0%. In some varieties cross pollination goes even up to 10 %. It takes around 7 days to complete the anthesis of all spikelets in a panicle, starting from the top and progressing downwards. The period from flowering to full ripeness of all the grains in a panicle is usually about 30 days. The mature rice grain is a caryopsis. Grain elongated or round enclosed by the lemma and palea. The lemma and palea together are known as the 'hull'. Rice harvested with the hull is called 'rough rice' or 'paddy', while that with the hull removed is known as 'brown', 'husked' or 'cleaned' rice. Hull constitutes about 20% of rough rice.

Rice growth terminologies: *Emergence* refers to the radicle or primary root and coleoptile, the protective cover for the first leaf, both emerge from the imbibed seed in five to seven days after planting. *Seedling stage* is from emergence to appearance of first tiller, seminal roots form during this period. Leaves develop at the rate of one for every 3-4 days during early stages. *Tillering stage* refers to development of tillers from buds found inside leaf sheaths and located just above the point of attachment of the sheath to the main culm. The first primary tiller is established when the first complete leaf of the tiller is visible *i.e.*, sheath, collar and blade. The *crown structure* occurs at the juncture of the shoot and secondary root system that develops at the base of the coleoptile. The crown may be almost at the seed if placement is shallow and no mesocotyl elongation occurred or it will be located where the mesocotyl joins the coleoptile inside the crown, nodes form simultaneously with each early leaf, with successively younger nodes forming above older nodes. All shoot growth (*i.e*, primary leaf, complete leaves and primary tillers) originates from these nodes within the crown until internode formation begins. The process of internode elongation is sometimes referred to as '*Jointing*' although it is possible for roots to be produced at stem nodes above the crown, most will originate at crown nodes. Primary tillers also originate from crown nodes, although they can emerge from main stem nodes above the crown. *First internode* or *first green ring* refers to the first stem node and the upper most crown node becomes clearly separated by an internode visible in properly dissected stems. This green ring may be observed in early stages of each additional internode. Early maturing cultivars reach first internode about 40 days after emergence and early season cultivars require approximately 50 days, while 55 days may be required for mid-season cultivars. *Panicle initiation* refers to the panicle, or inflorescence which can be detached microscopically 35 to 40 days before it is visible above the flag leaf sheath. Visual panicle initiation occurs 11 days after the differentiation of panicle primordial initiation. *Panicle differentiation* refers to the panicle reaching one-eighth of an inch long and appears as a small tuft of hairs inside the stem. Just above the crown, at panicle differentiation, the rachis, rachila, and florets are microscopically visible. *Boot stage* begins when the flag leaf sheath first appears above the collar of the penultimate leaf (*i.e.*, the immediately preceding leaf) and ends when it extends two inches above. Panicle development is completed during the boot stage of growth. Booting occurs 16 days after visual panicle initiation (The sheath of the flag leaf swells). This swelling of flag leaf sheath is called booting. Rice is at *first head* when the tip of the panicle appears above the flag leaf sheath on the main culm. *Full head* refers to the entire panicle being pushed through the flag leaf sheath by the full elongation of the topmost internode. *Heading* is emergence of panicle out of the flag leaf sheath. *Flowering* or anthesis or blooming, begins with opening of spikelets. *Flowering* means 50% panicle emergence. Flowering occurs for 25 days after the visual panicle initiation regardless of variety. Rice is a self-pollinated crop. The florets open from 0900 to 1500 hours depending on variety and weather. They open early on light days and late on humid days. When the starch or endosperm kernel is in a highly liquefied state, grains are considered to be in the *milk stage* of development. At milk grain stage, caryopsis (starch portion) is first watery and later turns milky. As the kernel reach maximum, grain fill and begin to dry, the consistency becomes similar to that of bread dough and is referred to as the *dough stage*. The dough stage can be divided into *soft dough* and *hard dough*. At dough grain stage, the milky portion turns dough soft at first and hard dough later. *Ripening* refers to flowering to maturity, which takes 25 to 35 days regardless of variety. At physiological maturity, grain fill is complete and the average moisture content is in the range of 30%. Average moisture of main stem grain should be between 15 and 18%, with the entire crop averaging about 18 to 21% moisture when harvested at maturity. At mature grain stage, grain colour changes from green to yellow.

Growth stages: The growth of the rice plant is divided into three phases *viz.*, i) Vegetative (germination to panicle initiation); ii) Reproductive (panicle initiation to flowering) and iii) Ripening (50%flowering to mature grain). Each phase of rice growth can be further divided into different stages. The vegetative phase has four growth stages namely i) germination to emergence, ii) seedling, iii) tillering and iv) Stem elongation. The reproductive phase has four stages namely, i) panicle initiation, ii) booting, iii) heading and iv) flowering stage. The ripening phase has three stages namely i) milking, ii) grain filling and iii) maturity stage.

(i) Germination to emergence stage: If the rice seeds are soaked in water, root is initiated after 50 hours of sowing, root and first leaf at 75 hours after sowing, root and three leaves formed at 100 hours and roots with three well developed leaves formed within 125 hours. The optimum temperature for germination is 30 to 32°C. Ten days after germination the plant becomes independent as the seed reserve is exhausted.

(ii) Seedling stage: The seedling stage starts right after emergence and lasts until just before the first tiller appears. During this stage, seminal roots and up to five leaves develop. Seminal roots – any of the adventitious roots that grow from the base of the stem during early seedling growth and take over the functions of the radicle.

(iii) Tillering stage: Tiller is a shoot that develops from axillary or adventitious buds at the base of a stem. Rice seedling with 1-2 leaves has ability to recover from transplanting shock. However, transplanting of rice seedlings is done with 4-5 leaves. Seedling with 4-5 leaves influences tiller numbers. The 6th to 9th leaf emerges during tillering influence shoot elongation and spikelets formation.

(iv) Stem elongation stage: Leaves of 10 to 12 emerge during stem elongation period influence the degree of ripening.

(v) Panicle initiation stage: The initiation of the panicle primordium at the tip of the growing shoot marks the start of the reproductive phase. The panicle primordium starts to differentiate about 40 days after seeding and becomes visible to the naked eye about 10 days after initiation. At this stage, 3 leaves will still emerge before the panicle finally emerges. The panicle becomes visible as a white feathery cone 1.0-1.5 mm long. It occurs first in the main culm and then in tillers where it emerges in uneven pattern. It can be seen by dissecting the stem.

(vi) Booting stage: Booting refers the bulging of the leaf sheath due to increase in size of the young panicle and its upward extension inside the upper leaf sheath.

(vii) Heading stage: It is also known as the panicle exsertion stage. Heading is marked by the emergence of the panicle tip from the flag leaf sheath. The panicle continues to emerge until it partially or completely protrudes from the sheath.

(viii) Flowering stage: Flowering occurs a day after heading and about 25 days after visual panicle initiation regardless of variety.

(ix) Milking stage: In this stage, the grain starts to fill with a white, milky liquid, which can be squeezed out by pressing the grain between the fingers.

(x) Grain filling stage: During this stage, the milky portion of the grain first turns into a soft dough and later into a hard dough. The grains in the panicle begin to change from green to yellow. Senescence of tillers and leaves is noticeable.

(xi) Maturity stage: The individual grain is mature, fully developed, hard, and has turned yellow. 90-100% of the filled grains have turned yellow and hard. The upper leaves are now drying rapidly although the leaves of some varieties remain green. A considerable amount of dead leaves accumulate at the base of the plant.

The growth and development phases of short, medium and long duration are presented in Table 3.

Table 3 Growth and development phases and stages for different duration of rice

Growth phases and stages	Short duration (105 days)	Medium duration (135 days)	Long duration (160 days)
I Vegatative phase			
i) Germination stage	3-5	3-5	3-5
ii) Seedling stage	20-25	30-35	35-40
iii) Active tillering stage	30-35	40-50	55-60
iv) Maximum tillering stage	34-45	50-65	70-85
II Lag phase	---	65-75	85-100
III Reproduction phase			
i) Panicle initiation stage	45-50	75-80	100-105
ii) Mid heading stage	50-55	80-85	105-110
iii) Booting stage	55-65	85-95	110-120
iv) Heading stage	65-75	95-105	120-130
IV Ripening phase			
i) Milky stage	75-77	105-110	130-135
ii) Soft dough stage	77-87	110-120	135-145
iii) Hard dough stage	87-100	120-130	145-155
iv) Maturity stage	100-105	130-135	155-160

Climate: Rice is indigenous to the humid area of tropical and subtropical regions. Rice is grown in the latitudes of 53°N to 35°S up to an altitude of 2000 m from MSL. In both the tropics and the temperate regions, the level of incidence of solar radiation primarily determines rice yield per hectare. In the tropics, when adequately managed, the dry season crop usually produces higher yields than the wet season crop because it receives more sunlight. Maximum rice yield could be achieved with accumulated sunshine hours of 1000 hours during 130-140 days of crop period with 220 to 240 hours in the last 30 days, *i.e.*, flowering to ripening. Solar radiation at reproductive stage has the greatest effect on rice yield which is followed by ripening stage. Solar radiation of 300 cal cm^{-2} day^{-1} is required during reproductive stage. The optimum light intensity for rice is 32.3 to 86.1 klux. Rice is generally a short day plant. Thus, long day delays flowering in rice. The optimum photoperiod is 10 hours for short day sensitive varieties. However, long day and day-neutral rice varieties are available for cultivation. Rainfed rice cultivation is limited to areas where annual rainfall of 750 to 1000 mm over a period of 3 to 4 months is received as it does not tolerate desiccation. Low sunlight associated with high rainfall during ripening causes low yields. Too little or too much rainfall at any stage of growth can cause partial or total failure. The lowland rice tends to be concentrated in flat lowlands, river basins and deltas. The cardinal minimum, optimum and maximum temperatures for rice at different growth stage are furnished in Table 4. The most sensitive stages against the temperature stress are panicle initiation stage (24 days before heading), reduction division stage (12-14 days before heading) and anthesis (0 day before heading) stage.

Table 4 Cardinal temperatures at different growth stages of rice

Growth stage	Critical temperature °C		
	Minimum	Optimum	Maximum
Germination	10	20-30	45
Seedling emergence and establishment	12-13	25-30	35
Rooting	16	25-28	35
Leaf elongation	7-12	31	45
Tillering	9-16	25-31	35
Panicle initiation	15		-
Panicle differentiation	15-20	-	38
Anthesis	20	25 to 30	35
Grain formation	17-20	27-31	-
Ripening	12-18	20-25	30

Transplanting under relatively low air temperature, windless, cloudy and rainy conditions favours root growth of rice. The optimum water temperature for rooting is 32°C. The water temperature influences rice growth till the panicle primordia initiation only and thereafter the effect of water temperature decreases. Once the seedlings form root, a wide range of temperatures between day and night is advantageous for tillering *i.e.*, 35°C in the day time and 15°C in the night seem to be optimum for increasing the number of tillers. But, when the rice plants attain neck node differentiation stage, the effect of the range of temperatures between day and night disappears and the optimum temperature is likely to be 30°C and the same lasts up to the heading time. The optimum temperature for rice growth is 20 to 30°C. *Indica* rice varieties are better adapted to high optimum temperature while *japonica* rice varieties require low optimum temperature for ripening. For practical purposes, 14-7 days before heading, commonly referred as the booting stage, is considered the most sensitive to low temperatures. The second most sensitive stage is heading or flowering. High temperatures exceeding 35°C at anthesis and lasting for more than one hour or two hours cause spikelet sterility. Rice is most sensitive to high temperature at heading and next sensitive stage at about 9 days before heading. Low temperatures of 15-20°C reduce growth and cause spikelet sterility. Flowering in rice is affected below 40% relative humidity (RH) and is favoured at 70 to 80%. Lower RH of 60 % induces a decrease in moisture content, loss of chlorophyll and faster senescence of leaves. Cold weather (< 15.5°C) present during the reproductive stages causes panicle blanking or blighting. Individual florets or the whole panicle may be white when emerging. Blast occurs when minimum temperature is below 23°C with relative humidity at ≥ 90%.

Soil: The term rice soils or paddy soils are commonly used for soils where lowland rice is grown. Paddy soil is not a pedological term, but, signifies the land use. Paddy soils occur in low-lying areas that are indicated naturally or water is let into the lands by gravitational force usually from tank or river. These are common in wetland soils. Wetlands are defined as having free water at or near the surface for at least major part of the growing season of arable crops. The floodwater is sufficiently shallow to allow the growth of a crop or natural vegetation rooted in the soil. Free surface water may occur naturally or rainfall, run off, or irrigation water contained by field bunds, puddled plain layers or traffic pans.

Wetlands have at least one wet growing season, but may be dry, moist or without surface water in other seasons. Wetland soils may therefore alternately support wetland and upland crops depending on crops and water supply. There are three types of water saturation in rice soils *viz.*, (i) *Endosaturation* in which entire soil is saturated with water, ii) *Episaturation* in which upper soil layers are saturated but underlined by unsaturated subsoil layers and (iii) *Anthric-saturation* in which a variant of episaturation with controlled flooding or puddled surface soil. Most Histosols are wetland soils. Practically all arid soils are upland soils.

Rice is grown on a variety of soils and texture ranging from sandy loam to heavy clays. Heavy alluvial soils of river valleys and deltas are better suited to rice than light soils as they permit puddling, less percolation and ideal condition of a creamy mud. The optimum pH for rice is 5.5 to 6.5 under dry condition which becomes 7.0 to 7.2 under flooding. However, rice is grown up to a pH of 10. Soil salinity or alkalinity and water salinity injures rice and are characterized by areas of stunted, chlorotic plants in the field. Under severe conditions, leaves turn from yellow to white and plants die. Salt affected areas usually have dead or dying plants in the centre or on high spots, with stunted yellow or white plants surrounding that area and green, less affected plants in lower areas. 'Straight head' is a physiological disorder associated with sandy soils, fields with arsenic residues or fields having large amounts of plant residue incorporated into the soil before flooding.

Rice seasons in India based on hydrologic, edaphic and cultural conditions:

(i) Aus rice: early rice (April-May to August-September): Short duration rice varieties are grown.
(ii) Boro rice: winter (Dec-Jan to Apr-May)
(iii) Transplanted *Aman* / Autumn (Jun- Jul to Nov-Dec) rice: Long duration *indica* rice varieties adapted to tropics of 5 to 6 months are grown
(iv) Deep water rice (broadcasted *Aman*)
(v) Upland rice (dryland)

Rice hybrids and varieties:

Table 5 Four maturity groups in rice

Maturity group	Duration (days)
Early	105-110
Medium	130-140
Long	150-160
Very long	170-180

Table 6 State wise rice hybrids released in India

S.No.	State	Hybrids
1	Andhra Pradesh	APRH-1, APRH-2, PHB-71, PA-6201, PA-6444, RH-204, Suruchi, DRRH-1, GK-5003, PAC-837, US-312, DRRH-3, NK-5251
2	Bihar	KRH-2, PA-6201, Ganga, JKRH-401
3	Chhattisgarh	Indra Sona, Suruchi, HRI-157, DRH-775, PAC-837
4	Delhi	PRH-10
5	Gujarat	Suruchi, KRI-157, PAC-835, PAC-837, DRRH-3, NK-5251

6	Goa	KRH-2
7	Haryana	PRH-10, Ganga, HKRH-1, PHB-71, RH-204, Suruchi, DRRH-2, Sahyadri-4
8	Karnataka	KRH-1, KRH-2, PHB-71, PA-6201, PA-6444, RH-204, Suruchi, GK-5003, PAC-837, HRI-157, US-312, NK-5251
9	Maharashtra	KRH-2, PA-6444, Suruchi, Sahyadri 1, Sahyadri-2, Sahyadri-3, Sahyadri-4, NK-5251
10	Madhya Pradesh	PA-6201, JRH-4, JRH-5, JRH-8, HRI-157, DRRH-3
11	Odisha	KRH-2, PA-6201, PA-6444, Ganga, Suruchi, Rajlaxmi, Ajya, JKRH-401, PAC-835, DRRH-3
12	Punjab	PRH-10, Ganga, PHB-71, PA-6129, Sahyadri 1
13	Pondicherry	KRH-2, PA-6129, HRI-157
14	Rajasthan	KRH-2, RH-204
15	Tamil Nadu	MGR-1, KRH-2, CORH-2, ADTRH-1, PHB-71, PA-6201, RH-204, DRRH-2, CORH-3, PA-6129, US-312, NK-5251
16	Tripura	KRH-2, PA-6201, PA-6444
17	Uttar Pradesh	KRH-2, Pant Sankar Dhan-1, Pant Shankar Dhan 3, Narendra Sankar Dhan-2, PHB-71, PA-6201, PA-6444, PRH-10, Ganga, Narendra Usar Sankar Dhan-3, Sahyadri-4, HRI-157, US-312, DRRH-3
18	Uttarakhand	PA-6444, Ganga, RH-204,Pant Shankar Dhan 1, Pant Sankar Dhan-3, DRRH-2
19	West Bengal	KRH-2, CNRH-3, PA-6201, DRRH-2, JKRH-401, Sahyadri-4, DRH-775, US-312
20	Jharkhand	DRH-775
21	Jammu & Kashmir	PA-837

Source: Directorate of Rice Research, Rajendranagar.

Table 7 Rice varieties in Tamil Nadu .

Variety	Duration (days)	1000 grain weight (g)	Grain yield (t/ha)	Grain colour	Grain type
ADT 42	115	24.8	5.5	White	Long slender
TKM 9	100-105	25.1	5.0	Red	Short bold
PKM 2	110-115	22.1	3.2	Dull white	Medium bold
Co 37	115	22.5	5.0	White	Medium slender
ADT 36	110	20.6	4.0	White	Medium
IET 1444	115	19.7	4.5	White	Medium
PY 2	115	18.4	4.5	White	Medium
ADT 43	110	15.5	5.9	White	Medium
ADT 41 (JJ 92)	105-110	24.2	4.7	White	Extra long slender
ASD 20	110	-	6.7	White	Long slender
TPS 1	110-115	24.3	4.8	Dull white	Short bold

PMK 1	110-115	24.0	3.2	Dull white	Short bold
ASD 16	110-115	24.2	5.6	White	Short bold
ASD 17	105	23.8	5.4	White	Short bold
ASD 18	105-110	21.8	5.9	White	Medium
ADT 37	105	23.4	6.2	White	Short bold
ADT 45	110	---	6.1	White	Medium slender
IR 50	105	20.4	6.0	White	Long slender
IR 36	120	21.0	5.0	White	Medium
TRY 2	115-120	---	5.4	White	Long slender
MDU 3	120-125	23.1	5.0	White	Long slender
ADT 39	120-125	18.0	5.6	White	Medium
IR 64	115-120	23.1	6.1	White	Long slender
MDU 4	120-125	22.9	5.9	White	Long slender
Co 46	125	---	6.0	White	Long slender
TKM 11	110-120	---	3.0	White	Long slender
TKM 12	115-120	---	3.0	White	Medium slender
IR 20	130-135	19.0	5.0	White	Medium slender
Bhawani	130-135	21.5	5.0	White	Long slender
ASD 19	127-132	18.4	5.8	White	Short slender
TKM 10	135	23.2	2.5	White	Long slender
ADT 38	130-135	21.0	6.2	White	Long slender
ADT 46	135	---	6.2	White	Long slender
TPS 2	125-130	23.5	4.6	White	Short bold
AU 2	140-145	27.0	5.2	White	Long bold
Co 45	135-140	-	5.5	White	Long slender
TPS 3	135-140	23.2	5.2	White	Short bold
TRY 1	135-140	24.0	5.2	White	Medium
Co43	135-140	20.0	5.2	White	Medium
MDU 2	130-140	16.4	4.5	White	Medium
ADT 40	145-150	25.2	5.0	White	Short bold
PY4 (Jawahar)	145-150	24.8	5.3	White	Medium
Paiyur 1	150	15.7	5.0	White	Fine
Ponmani	155-160	-	5.3	White	Short bold
Co 25	160-170	-	3.0	White	Medium slender
Co 40	160-175	-	7.3	White	Short bold
Anna 4 or PMK (R) 4	100-105		3.7	White	Long slender

Table 8 Indigenous high quality non-aromatic rice (landraces or pure line selection from land races) types in India

State	Non-aromatic landraces types
Andhra Pradesh	Aswani Karthi Vadlu, Atragada, Kichidi Sannalu, Krishna Katakalu, Maharajabhogam, Punasa Akkullu
Assam	Aijani (Mahsuri), Guniribora (Glutinous), Kakua Bao, Lalahu, Manoharsali, Rangaduria
Bihar	Black Gora, BR13, BR14, BR34, Brown Gora, Kessore, Kolaba, T141, NP130
Gujarat	Kolamba 42, Sathe 34-36, Sukhvel 20, Zeerasal 280, Zinya 31
Haryana	Jhona 221, Jhona 349
Himachal Pradesh	Dundar 43, Lalnakanda, Phulpattas 72, Ramjawain 100
Jammu and Kashmir	CH988, CH1039, Niver, Siga, T137, T138
Karnataka	Allur Sanna, Antersal, Bangarutheega, Dodda byra, Kare kagga, S1092, SR26B
Kerala	Cheruvirippu, Pokkali, Ptb2, Ptb12, Ptb15, Ptb18, Ptb21, Ptb33, Red Triveni, Vytilla
Madhya Pradesh	Nungi, Pandhri Luchai 16, Safari 17, Sariya, Laloo 14
Maharashtra	Bhura rata, Chimansel, Kala rata, Kolam, Kolamba 184, Kolipi, Zinya 149
Odisha	Bahyayahuynda, FR13A, FR43B, Nonabokra, Rangolata, Ratnachudi, T90 (Machha Kanta), T141 (Soruch Inamali)
Punjab	Jhona 349, Lalnakanda, Phulpattas
Rajasthan	Batika, Dhaniasal, Pathria, Segra, Sutar, Tukri
Tamil Nadu	ADT 8, ADT 10, Anaikomban, Co25, GEB24, Kar Samba, Katti Samba, Kichdi Samba, TKM6
Uttar Pradesh/ Uttaranchal	Adamchini, Anjee, Bambasa, Bansi, Dehula, Jaisuria, Jalamagna, N22, NP130, Tapachini, T100, T136
West Bengal	Badshah pasand, Badshahbhog, Basmati, Gopalbhog, Govindabhog, Kaminibhog, Kataribhog, Randhunipagal, Sitabhog

Source: Siddiq *et al.*, 2005

Golden rice: The traditional rice varieties do not produce provitamin-A (beta-carotene). Golden rice produces provitamin-A (beta-carotene) which is useful to alleviate vitamin A (retinol) deficiencies in the diets of poor people in developing countries. Vitamin A deficiency (VAD) is the leading cause of preventable blindness in children. VAD in pregnant women causes night-blindness. Transgenic technologies are used to develop golden rice. Dr. Ingo Potrykus of the Swiss Federal Institute of Technology in Zurich and Dr. Peter Beyer of the University of Freiburg in Germany have developed golden rice plants by inserting two daffodil plant genes and one gene from the bacterium *Erwinia uredovora* in the rice genome which are required for the production of beta-carotene in rice. These three genes produce the enzymes necessary to convert geranylgeranyl diphosphate (GGDP) to produce provitamin-A. The inserted genes are controlled by specific promoters such that the enzymes and the provitamin-A are only produced in the rice endosperm. The resulting plants appear normal except that after milling (to remove the brown bran), their grain is a golden yellow colour due to the presence of provitamin-A. Golden rice is considered as a 'Trojan horse' for the poor people to gain acceptance for genetically modified crops. Further, a normal daily intake of 300 grams (0.7 pounds) of golden rice provides only 8 % of the vitamin A needed daily. Even if 'golden rice' help to prevent vitamin A deficiency, it may not be socially accepted and replace white coloured rice.

Super rice: Scientists at the International Rice Research Institute, Philippines have developed 'Super Rice' a high-yielding rice of the future which increases harvests by 25 percent. It is far less bushy as each plant consists of only 10 stems or so in comparison with 20 to 25 of the traditional rice plant. Besides that, a single super rice plant can produce 2,500 grains of rice compared to 1,500 grains from conventional plant.

Herbicide Tolerance Rice: As herbicide tolerance was often due to a single gene, the idea has been to create rice plants with the mammalian P450 enzyme that could detoxify several of these herbicides and make these rice plants tolerant to herbicides. Now, a transgenic rice plant with human gene CYP2B6 not only give good yields but also shows high herbicide tolerance capacity. They could detoxify several herbicides such as thiocarbamates, oxyacetamides and 2, 6-dinitroanilines.

Rice growing environments: Rice can be grown in different environments, depending upon water availability. Generally, rice does not thrive in a waterlogged area, yet it can survive and grow herein and it can also survive flooding. The details of classification rice based on water depth are furnished as follows:

1. Irrigated rice ecosystem: Rice is grown under irrigated conditions in the states of Punjab, Haryana, Uttar Pradesh, Jammu & Kashmir, Andhra Pradesh, Tamil Nadu, Sikkim, Karnataka, Himachal Pradesh and Gujarat. Irrigated rice is grown in bunded (embanked) paddy fields in upland conditions as well as in lowland conditions.

2. Rainfed rice: The rainfed ecosystem may be broadly categorized into upland and lowland ecologies.

2.1. Rainfed upland rice: The upland rice is also known as Ghaiya rice, well known for its drought tolerance. The upland rice fields are generally dry, unbunded, and directly seeded. Land utilized in upland rice production can be low lying, drought-prone, rolling, or steep sloping. The upland rice areas lie in the eastern zone comprising Assam, Bihar, Chhattisgarh, Odisha, eastern Uttar Pradesh, West Bengal and North-eastern Hill region. In the rainfed upland rice, there is no standing water in the field after few hours of cessation of rain. Because of uneven topography, the upland rice accounts for about 6.0 million ha (13.5 per cent of the total rice area). Productivity of upland rice is very low and stagnant for long. As against the present national average productivity of about 1.9 tonnes per ha., the average yield of upland rice is only 0.90 tonnes per ha.

2.2. Rainfed lowland rice: Rainfed lowland rice farmers are typically challenged by poor soil quality, drought/flood conditions, and erratic yields. Flooding occurs during the wet season from June to November. The rice varieties are chosen for their level of tolerance to submergence. Lowland rice area is mostly located in the eastern region comprising Assam, Bihar, Chhattisgarh, Odisha, eastern Uttar Pradesh and West Bengal. Lowland rice area is about 14.4 million ha., which accounts for 32.4 per cent of the total area under rice. The average productivity of lowland rice ranges from 1.0 to 1.2 tonnes per ha. as against the national average of 1.9 tonnes per ha. The lowland rice ecology depending on the water regimes may be further categorized into three sub-ecologies as detailed below:

2.2.1. Shallow lowland rice: water depth is below 50 cm.

2.2.2. Semi-deep-water rice: water depth is between 50 and 100 cm.

2.2.3. Deep water rice: water depth is more than 100 cm in the field. The soil is poor in nitrogen, phosphorus and organic matter but is rich in potassium. Deep water rice areas are prone to seasonal floods and duration of which varies from year to year. Deep water rice is planted where standing water of 100 cm for more than half of the growth duration and sometimes with complete submergence of the crop. The low lying areas so called stagnant water areas and the tidal swamp areas are of deep water rice growing regions. Un-controlled water regimes and deep water rice are found in deltas and river valleys of India, Bangladesh, Myanmar, Thailand, Vietnam, Kampuchea and Indonesia. India has large deep water rice areas in West Bengal, Assam, Odisha, Bihar, Andhra Pradesh and Tamilnadu. Problems in deep water rice culture are drought damage at germination and the seedling stage because the crop is direct seeded into dry soil, high weed competition at the seedling stage, poor stand establishment and high seedling mortality due to sudden flood. Submergence of rice at various growth stages leads to lodging susceptibility. Rice varieties should tolerate submergence and should have ability to elongate if the water level increases. Traditional varieties of 130 cm plant height are grown. Maturity of deep water rice takes 240 days in West Bengal. Dry seeding of deep water rice is fairly common, although transplanting or double transplanting is occasionally practiced. In West Bengal, seeds are broadcasted in April-May on dry soil. The seedlings are pulled after 30 DAS and transplanted in clumps. Thirty days afterwards the already planted seedlings are pulled out and replanted again in the field. Limited quantity of fertilizer or pesticide is applied to deep water rice. Weeds are primarily controlled by harrowing following germination of rice and weeds. Flooding at 50 to 60 days after seeding prevents reinfestation of aquatic weeds.

3. Floating rice culture: Floating rice is grown where maximum water depth ranges from one to six metres for more than half of the growth duration. These areas cover the flood plains of the Ganges and Brahmaputra in India. In West Bengal, there are about 20,000 ha of low lying area where water accumulates slowly but steadily from adjoining flood areas during monsoon season (June-September) and increases water depth 3 m. The length of time that flood water remains in the field varies from place to place and year to year in the same area. In floating rice areas, the flood duration is 3 to 4 months in India. Problems of floating rice culture are similar to that of deep water rice culture. Duration of submergence, turbidity of water and temperature of water all affect the floating rice crop. Floating rice varieties increase plant height of 20 to 25 cm in a day. Floating rice plants elongate upto 6 m, forming a dense mat on the water surface. Varieties should tolerate submergence for at least 7 to 21 days. Duration of varieties ranges from 150 to 270 days. These are photoperiod sensitive varieties. The un-germinated rice seeds are normally broadcasted (60-30 kg/ha) on dry soil before the receipt of sowing rains for floating rice. In certain areas pre-germinated seeds are broadcasted in puddled soil. The seedlings are grown as a dry land crop for 4 to 20 weeks before flooding occurs. Drought damage and weed competition are often limiting factors in getting an optimum stand of rice. Harvesting is done after the flood water recedes. Occasionally, harvesting is done with boats if the crop matures before the floodwater recedes. Floating rice yields 0.4 to 1.0 t/ha.

4. Coastal saline rice: Rice areas close to east and west coasts suffer from salinity. Andhra Pradesh, Kerala, Odisha, Tamil Nadu and West Bengal on the east coast and Gujarat and Maharashtra on the west coast have sizeable saline rice areas. Total rice area under coastal salinity rice is estimated to be about 1 million ha, which accounts for 2.3% of area under rice cultivation. The coastal saline soils often show deficiency of iron and zinc, which cause chlorosis and reduced tillering. Average yield in coastal saline area is about 1 tonne as against the average national yield of 1.9 tonnes per ha.

5. Cold/ hill rice: Such rice areas lie in the hill regions comprising Jammu and Kashmir, Uttaranchal and Northeastern hill states. Area under rice in cold/hill region is about one million ha which accounts for 2.3% of total area under rice. Average yield is about 1.1 t/ha as against the average national yield of 1.9 t/ha. Major productivity constraints of these areas are low temperature, blast, drought spell and very short span of cropping season. Because of the rolling topography in these areas, bench terracing is being followed, which limits the use of fertilizers and improved agronomical practices. In these areas the crop is invariably affected by low temperature in the early stage and sometimes at the flowering stage resulting in sterility and hence, reduced yields.

Systems of rice cultivation: The systems of rice culture have been developed to suit specific environments and socio economic conditions of the farmers. The classification of rice culture is as follows:

1. Classification based on source of water supply

(i) Rainfed

(ii) Irrigated

2. Classification based on land and water management practices

(i) Lowland (wetland preparation of fields)

(ii) Upland (dryland preparation of fields)

3. Classification based on water regime

(i) Upland rice with no standing water

(ii) Lowland (*Bulu*) rice with 5-50 cm of standing water

(iii) Deep water rice with 51 cm to 5-6 metres of standing water

4. Classification based on varietal types

(i) Upland rice: plants of medium to tall height (130-150 cm)

(ii) Lowland rice: plants semidwarf, medium to tall (100 cm to 2 m)

(iii) Deep water rice: plants of medium to tall (120 –150 cm to 2-3 m height with rising water level)

(iv) Floating rice: Tall plants (>150 cm to 5-6 m tall with rising flood water level)

5. Classification based on water depth is furnished in Table 9.

Table 9 Types of rice culture based on water depth

S.No.	Types of rice culture	Method of planting	Maximum water depth (cm)
1	Upland rice (10 %) Dry rice Semidry rice	Broadcast or drilled into dry soil	No standing water
2	Rainfed low land rice (30%)	Broadcast into dry soil or puddled soil	0-50
2.1	Shallow rainfed	------	5-15
2.2	Medium deep rainfed	------	16-50
3	Irrigated lowland rice (45%)	Transplanted	5-15
4	Deep water rice (11%)	Broadcast onto dry soil	51-100
5	Floating rice (4%)	Broadcast onto dry soil	101-600

Note: Figure in brackets indicates % of land area under each system.

Rice seasons in India based on rainfall distribution: There are three seasons for growing rice in India *viz.*, autumn, winter and summer seasons.

(i) Autumn rice is known as pre-*kharif* rice. The sowing of pre-*kharif* rice is taken up during May to August. However, the time of sowing slightly differs from state to state according to weather condition and rainfall pattern. It is harvested in September-October. Autumn rice crop is known as '*Aus*' in West Bengal, 'Ahu' in Assam, '*Beali*' in Odisha, '*Bhadai*' in Bihar, '*Virippu*' in Kerala and '*Kuruvai/ kar/ Sornavari*' in Tamil Nadu.

(ii) Winter rice is known as *Kharif* rice. It is the main rice growing season in the country. The sowing time of winter (*kharif*) rice is June-July and it is harvested in November-December. Winter rice is known as '*Aman*' in West Bengal, '*Sali*' in Assam, '*Sarrad*' in Odisha, '*Agahani*' in Bihar and Uttar Pradesh, '*Sarava*' in Andhra Pradesh, '*Mundakan*' in Kerala and '*Samba/Thaladi*' in Tamil Nadu. About 84% of the country's rice crop is grown in this season and generally, medium to long duration varieties are grown in this season.

(iii) Summer rice is called as *Rabi* rice. It is known as '*Boro*' in Assam and West Bengal, '*Dalua*' in Odisha, '*Dalwa*' in Andhra Pradesh, '*Punja*' in Kerala and '*Navarai*' in Tamil Nadu and '*Garma*' in Bihar. The sowing time of summer rice is November to February and harvesting time is March to June. The area under summer rice is only 9% and early maturing varieties are mostly grown in this season.

The sowing and harvesting periods of rice in different states of India are furnished in Table 10.

Table 10 Sowing and harvesting periods of rice in different states of India

Region / State	Autumn		Winter		Summer	
	Sowing	Harvesting	Sowing	Harvesting	Sowing	Harvesting
Northern Region						
Punjab	May-Aug	Sep-Nov	-	-	-	-
West U.P.						
Himachal Pradesh	June-July	Sep-Nov	-	-	-	-
Jammu & Kashmir	-	-	Apr-July	Sep-Dec	-	-
Haryana	-	-	-	-	-	-
Western region						
Gujarat	-	-	Jun-Aug	Oct-Dec	-	-
Maharashtra	-	-	Jun-July	Oct-Dec	-	-
Rajasthan	-	-	July-Aug	Oct-Dec	-	-
North-East Region						
Assam	Feb-Apr	June-July	June-Aug	Nov-Dec	Dec-Feb	May-June
Eastern Region						
Bihar	May-July	Sep-Oct	July-Sep	Nov-Dec	Jan-Feb	May-June
East M.P.	June-Aug	Mid Sep - Mid Dec	-	-	-	-
Odisha	May-June	Sep-Oct	June-Aug	Dec-Jan	Dec-Jan	May-June
East U.P.	May-July	Sep-Nov	July-Aug	Nov-Dec	Jan-Feb	Apr-June
West Bengal	Mar-June	July-Nov	Apr-June	Nov-Dec	Oct-Feb	Apr-May
	May-June		July-Aug			
Southern Region						
Andhra Pradesh	Mar-April	July-Aug	May-June	Nov-Dec	Dec-Jan	Apr-May
Karnataka	May-Aug	Sep-Dec	June-Oct	Nov-Mar	Dec-Feb	Apr-July
Kerala	April-June	Aug-Oct	Sep-Oct	Jan-Feb	Dec-Jan	Mar-Apr
Tamil Nadu	Sonavari		Early Samba		Late Samba	
	Apr-May	July-Aug	July-Aug	Nov-Dec	Oct-Nov	Mar-Apr
Early Kar	Apr-May	July-Aug	Samba			
Kar	May-June	Aug-Sep	July-Aug	Dec-Jan		
Navarai	Jan-Feb	Apr-May				
			Thaladi / Pishanam		Late thaladi	
Kuruvai	June-July	Sep-Oct	Sep-Oct	Dec-Jan	Oct-Nov	Jan-Feb

SEED RATE

Direct seeded rice

(i) Rainfed upland rice - 75-100 kg per ha

(ii) Puddled lowland rice - 80-100 kg per ha

(iii) Semidry upland rice - 100 kg per ha

Transplanted rice

(i) Short duration rice varieties - 60 kg per ha
(ii) Medium duration rice varieties - 40 kg per ha
(iii) Long duration rice varieties - 30 kg per ha

Seed treatment

(i) Treat the seeds in Carbendazim or Pyroquilon or Tricyclozole solution at 2 g/l of water for 1 kg of seeds. Soak the seeds in water for 10 hrs and drain excess water.

(ii) This wet seed treatment gives protection to the seedlings up to 40 days from seedling disease such as blast and this method is better than dry seed treatment.

(iii) If the seeds are required for sowing immediately, keep the soaked seed in gunny in dark and cover with extra gunnies and leave for 24hrs for sprouting.

(iv) Seed treatment with *Pseudomonas fluorescens*: Treat the seeds with talc based formulation of *Pseudomonas fluorescens* 10g/kg of seed and soak in 1lit of water overnight. Decant the excess water and allow the seeds to sprout for 24hrs and then sow.

(v) Seed treatment with *Azospirillum*: Three packets (600 g/ha) of *Azospirillum* and 3 packets (600g/ha) of Phosphobacteria or 6 packets (1200g/ha) of *Azophos*. The bioinoculants are mixed with sufficient water wherein the seeds are soaked overnight before sowing in the nursery bed (The bacterial suspension after decanting may be poured over the nursery area itself).

(vi) Seed treatment (for diseases) with carbendazim 50% WP @ 2 g/kg seed or *Trichoderma/Pseudomonas* @ 5-10 g/ha of seed for seed or soil borne diseases and carbosulfan 2 g/kg of seed for root nematodes or as per local recommendations. In termites endemic areas, seed treatment with chlorpyriphos 20% EC @ 10000 ml/ha along with 10% solution of gum arabica or imidacloprid 200 SL (20%) @ 0.25 litre/100 kg seed along with 10% solution of gum Arabica in 3.75 litre of water just before sowing.

Note:
Biocontrol agents are compatible with biofertilizers
Biofertilizers and biocontrol agents can be mixed together for seed soaking
Fungicides and biocontrol agents are incompatible.

Direct seeding techniques

(i) Wet seeding: Pre-germinated seeds are broadcasted on to the puddled fields without much standing water. Field is prepared by puddling and properly levelling. Stand establishment is affected often due to poor land preparation, weed competition and poor water control.

(ii) Dry seeding: It requires dry land preparation. Land is tilled with plough before or after receipt of sowing rains. Seed is directly sown either by broadcasting, line sowing or drilling. Dry seeding commence after rains results in poor seedling emergence and low yields.

(iii) Water seeded rice: For water seeding, precise water control is a must and more seeds are required than for transplant method. Good seed viability is essential. Pre-soaked seeds for 8-24 hours are sown with air craft or air plane into fields flooded to a depth of 7.5 to 15 cm water.

Direct sowing under dry conditions: On receipt of rains during the months of May-July, repeated ploughings should be carried out so as to conserve the moisture, destroy the weeds and break the clods. A seed rate of 80 to 100 kg/ha should be adopted. Seeds are treated with biofertilizer (*azospirillum*). Seeds are soaked for 10 hours and drained. Seeds are further hardened with 1% KCl to induce drought tolerance. Then seeds are treated with fungicide 24 hours before sowing. Seeds can be sown either by broadcasting or drill sowing with bullock drawn seed drill with 20 cm interrow spacing under dry condition. Seeds are sown behind the country plough or drilled with *gorru* or seed drill. The depth of sowing should be 3 cm and the top soil can be made compact with *guntaka*. If the seeds are sown in moist soils, the covering will be difficult in clay soils. Pre-monsoon sowing is advocated so that the germination will be uniform on receipt of soaking rains. The pre-germinated seeds should not be used for seeding in dry soil because the seeds will dry up and subsequently die if it does not rain immediately after sowing.

Direct sowing under wet condition: Direct seeding greatly reduces labour cost for establishing a stand of rice. The land is prepared with puddling and leveling practices. On levelled well puddled soils, pre-germinated seeds can be sown by broadcasting under wet field condition. Low tillering varieties require higher seeding rates than high tillering varieties. It is not advisable to broadcast seeds while the water is too muddy because the seeds will be covered by mud upon setting down, thus causing high mortality of rice emergence. Direct seeding requires much better control of irrigation and drainage than transplanting. Newly germinated seedlings will die within a few days if the soil surface dries out or if they are completely covered with water. Make channels around the field for drainage if rain falls after the seeding. Drain all excess water from the field before broadcasting seed.

In general, the direct seeding requires less labour. There is no need to prepare a nursery, care for it, and pull the seedlings. The direct-seeded plants mature 7 to 10 days earlier than transplanted rice. In direct seeding, the seeds are exposed to birds, rats, and snails. There is greater crop-weed competition because rice plants and weeds are of similar age. Plants tend to lodge more because there is less root anchorage.

Rice nursery practices and management: Healthy rice seedlings are to be produced to have large panicles and finally high grain yield. Healthy seedlings should be short and thick stature. Healthy seedlings should be with uniform growth. Uniform seedlings can be obtained with seeds having a specific gravity of 1.13 for non-glutinous variety and 1.10 for glutinous varieties. Secondly, uniform seedlings can be produced through uniform sowing (spacing). An optimum spacing can safely be obtained in most cases by sowing 64 gm of seeds (with a 1000 grain weight 29 gm) per m^2 in the seed bed. Thirdly, proper fertilizer application is also necessary for raising short and thick seedlings. Fourthly, pre-germinated seeds can be sown to produce uniform seedlings growth. Soak the seeds in water for one day at room temperature of 30 to 32°C in tropical countries. Make the plant to develop more roots than leaves with controlled shallow irrigation. Healthy seedlings should be free from insects and diseases. There are three methods of raising rice seedlings *viz*., the wet bed, dapog bed and the dry bed. The wet bed is widely practiced.

(a) Wet nursery bed method: Select 20 cents (800 m^2) of land near the water source for raising rice seedlings per hectare. Apply 1000 kg of FYM or compost. Spread the manure uniformly on dry soil. Flood the nursery area one or two days before ploughing and allow the water to soak in. Keep the soil under shallow submergence. Allow the water to a depth of 2.5 cm and puddle well. Before last puddling, apply 40 kg of DAP for 20 cents of nursery area. Basal application of DAP is recommended when seedlings are to be pulled out in 20 to 25 days after sowing. If seedlings are to be pulled out after 25 days, application of DAP is to be done 10 days prior to pulling out the seedlings. In clayey soils, where root snapping is a problem, DAP has to be applied @ 1 kg/cent on 10 days after sowing. Form seedbeds by marking plots, 2.5 m broad with channels 30 cm wide in between. Collect the mud from channel and spread on the seedbed or drag a heavy stone along the channel to lower it, so that the seedbed is at a higher level. Level the surface of the seedbed so that water drains into the channel. Sow the pre-germinated seeds in raised beds. Seedlings absorb and store phosphorus and utilizes even in the later stages of crop growth. Application of phosphorus to nursery is economical. If nursery is applied with phosphorus, the application of 30% of phosphorous required for the main fields is adequate to realize high rice yield. *Vesicular Arbuscular Mycorrhiza* (VAM-fungi) application @ 50g/sq meter penetrates the roots of a vascular plant in order to help them to capture nutrients especially phosphorous from the soil. These fungi are scientifically well known for their ability to uptake and transport mineral nutrients from the soil directly into host plant roots.

Germination will be affected in place where there is water stagnation. Drain the water 18 to 24 hours after sowing. If there are pockets of water stagnation, drain the water into the channel. Allow enough water to saturate the soil from the 3rd to 5th day. From 5th day onwards, increase the quantity of water to a depth of 1.5 cm depending on the height of the seedling. Afterwards, maintain 2.5 cm depth of water. Apply any one of pre-emergence herbicidies *viz.*, Butachlor @ 2 lit/ha., Thiobencarb @ 2 lit/ha., Pendimethalin @ 2.5 lit/ha, Anilophos @ 1.25 kg/ha on 8 DAS to control weeds in the wet nursery bed. Keep a thin film of water and allow it to disappear. Avoid drainage of water to control the germinating weeds.

Forty litres of spray fluid of Endosulphan (35 EC) @ 80 ml or Monocrotophs (36 WSC) @ 40 ml is nursery for spraying 800 m^2 or 20 cents of nursery area. Apply carbofuran (3G) 3.5 kg for 20 cents necessary to control army worm, thrips, green leaf hopper, case worm, nematodes and tungro virus disease. Edifenphos @ 40 ml or Copper oxychloride @ 100 g or Carbendazim @ 40 g is sprayed to control the diseases in nursery. If seedling shows symptom of nitrogen deficiency and if growth is not satisfactory, apply urea @ 500 g per cent of nursery in 7 to 10 days prior to pulling. If DAP is applied 10 days prior to pulling, urea application is not necessary. One kg of gypsum may be applied one day prior to pulling of seedlings if nursery soil is too clayey.

The normal age of seedling is 18-20 days, 25-30 days and 35-40 days for short, medium and long duration rice varieties. Pulling of seedlings is done at appropriate time. Wash the roots with water if too much mud is found sticking to roots. Tie the seedlings of 5 to 8 cm diameter with soft materials such as banana twine or paddy straw and keep the root portion submerged in water. Do not allow the seedlings to dry as it hinders separation of seedlings during transplanting.

(b) Dapog method: The land is prepared as in the wet bed method. The banana leaves or plastic sheets are used to cover the soil or a concreted area is used as a bed. The dapog bed should be about 1.5 m wide and its length will depend on the area to be planted. 80 m^2 raised bed area is required for one ha. Banana bracts are placed on seed with bamboo pegs to keep the leaves in position. Seed rate is 80-100 kg/ha. Sow the pre-germinated seeds @ one kg per m^2. Dapog seedlings are ready for transplanting on 9 to 14 days after sowing. For dapog seedlings, the seedbed is cut into convenient sizes and the seedlings mats are rolled with roots outward.

Advantages of Dapog method

- Labour savings because of the bed is easily made.
- Short period for raising seedlings.
- Relative ease in transport of seedlings because a mat of seedlings can be rolled.
- Require small seed bed area.
- Seedlings recover faster after transplanting.
- Less expensive.

Disadvantages of Dapog method

- Require more seeds.
- Needs better water management.
- Seedlings are less competitive against weeds.

(c) Dry nursery bed method: It is practiced in rainfed lowland areas where there is insufficient water to irrigate the seed beds. Irrigation is provided to keep the soil moist but not inundated. Water the seedlings 2 to 3 times daily. The seed rate is similar to that of the wet bed. Apply FYM or compost @ 2.0 to 2.4 tonnes for 20 cents. Plough the land to get an optimum tilth. 20 cents or 800 m^2 nursery area is required to produce seedlings for one hectare. Sow the seeds @ 100 g/m^2. Cover the seeds with well decomposed FYM. Sprinkle the water both morning and evening hours. Hand weeding is done on 15 to 20 DAS.

The features of the three methods of raising rice seedlings for transplanting are furnished in Table 11.

Table 11 Comparison of three methods of raising rice seedlings for transplanting

Particulars	Method of sowing		
	Wet bed	**Dry bed**	**Dapog**
Seed rate (kg/ha)	60 SD	60 SD	100
	40 MD	40 MD	
	30 LD	30 LD	
Seed bed area (m^2/ha)	800	800	80
Seed soaking time (hrs)	24	24*	24
Incubation time (hrs)	24-48	24*	36-48
Sowing to transplanting (days)	20-35	20-35	9-14
Transplanting depth (cm)	1.5-3.0	1.5-3.0	1.5
Seedlings per hill	2-3	2-3	5-10

* Not recommended if the soil is dry. SD means short duration; MD means medium duration and LD means long duration.

Optimum age of seedlings for quick establishment: Optimum age of the seedlings is 18-22 days for short, 25-30 days for medium and 35-40 days for long duration varieties.

Pulling of rice seedlings: Keep the seed bed moist or flooded when pulling the seedlings. Pull out the seedlings at the appropriate time (3rd or 4th leaf stage). These seedlings can produce more tillers, provided enough care is taken during the establishment phase through thin film of water management and perfect leveling of main field. Transplanting after 5th and higher order leaf numbers will affect the performance of the crop growth and grain yield. Then they are called as 'aged seedlings'. Special package is needed to minimize the grain yield loss while planting those aged seedlings. Grasp two or three seedlings at a time and as close to the base as possible and pull them gently at an angle about 30 degrees from the horizontal soil surface. If too much soil sticks to the roots, wash by soaking the roots in water, striking the seedlings against a hard material will injure roots and the seedlings. Make the seedlings into bundles of 5 to 8 cm in diameter. Use any soft material like banana or Abaca or paddy straw for tying the rice seedlings. Keep the seedlings from drying out by keeping the seedling roots immersed in water and do not expose to sun.

Root dipping: Prepare the slurry with 5 packets (1000 g) per ha of Azospirillum and 5 packets (1000g/ha) of Phosphobacteria or 10 packets of (2000g/ha) of Azophos inoculant in 40 lit. of water and dip the root portion of the seedlings for 15-30 minutes in bacterial suspension and then transplant the seedlings.

Management of aged seedlings

Plant 5 to 6 seedlings per hill
Adopt closer spacing
Apply 25 % extra nitrogen at basal
Spray 20 ppm NAA at tillering stage
Terminal clipping of leaves

MAIN FIELD PREPARATION

(a) Main field preparation for dryland rice cultivation: The field is prepared to have a fine tilth, taking advantages of summer rains and early monsoon showers during the months of May-June. Repeated ploughings should be carried out so as to conserve the moisture. Destroy the weeds and break the clods. Apply gypsum @ 1 t/ha basally whenever the soil crusting and soil hardening problem exists. Apply FYM or compost @ 12.5 t/ha. Application of 750 kg of enriched FYM with 50 kg P_2O_5 per ha can be applied as basal dose in clay soil.

(b) Main field preparation for wet rice cultivation: Small leveled fields are bordered by dikes or bunds to retain water on the field surface. Plough the land during summer to economise the water requirement for initial preparation of land. Early land preparation incorporates the crop residue to ensure good decomposition of plant material which prevents early-season nitrogen deficiency. If early land preparation is not possible, decomposition will not be at an advanced stage at planting time. Since the soil's microorganisms (bacteria, fungi, *etc.*) that decompose crop residue are competing with rice plants for nutrients, particularly nitrogen, the rice plant will show nitrogen deficiency. If this situation arises, an additional 10 to 20 units of nitrogen may be required when the base fertilizer is applied at or near planting.

Wet rice requires a well puddled soil. Apply 12.5 t/ha of FYM or compost. Spread the manure uniformly on the dry soil before letting in water. Flood the field one or two days before wet ploughing (puddling) and allow water to soak in. Keep the surface of the field covered with water to a depth of 2.5 cm. When the fields are flooded, the soil is ploughed mainly in order to work in the stubble and weeds. Subsequent secondary tillage operation consists of puddling. Puddling refers to the breaking down of the soil aggregates at near saturation into ultimate fine soil particles. For a farmer, puddling is mixing soil with water to make it soft for transplanting and impervious to the percolation of water. Only the soils with more than 20% clay particles are prone to puddling. High clay content favours puddling. Kaloinite clays are more difficult to puddle than montmorillonite clays. Similarly sodium (Na) saturated clays puddle more than calcium (Ca) saturated clays. Maximum puddling occurs at moisture contents between field capacity and saturation. Puddling has the beneficial effect in well aggregated soils only. Puddling is not effective on soils with impervious layers or high ground water table. Puddling and subsequent flooding differentiate lowland rice soils chemically and pedologically from other arable soils. An important difference between a dry land and puddle lowland soils is the presence of the reduced soil layer in the puddle soil system. The degree of puddling depends on tillage implement and intensity of puddling. Rotary implements generally are better for puddling than ploughs because the rotary motion continuously changes the direction of the shear stress. Nevertheless country ploughs, mould board ploughs, disc harrows, angular puddlers and rototillers are also effective.

The advantages of puddling are i) improved weed control, ii) case of transplanting, iii) plough pan or compacted layer reduces percolation of water and leaching of nutrients, iv) reduced soil condition improves the soil fertility by means of increased solubility of nutrients and v) reduced draft requirement for tillage operations.

The disadvantages of puddling are i) high water requirements (150 to 200 mm) for puddling a wetland field, ii) impeded root developments due to hard pan formation below the puddled layer, iii) hinderness to regeneration of soil structure and iv) incorporation of crop residues helps in regeneration of soil structure. Land leveling for a uniform grade of 0.2% slope or less provides the following: (1) necessary early drainage in the spring for early soil preparation which permits early seeding, (2) uniform flood depth which reduces the amount of water needed for irrigation, and (3) the need for fewer levees. Two different systems are used to produce rice with early flood culture. One system is 'continuous flood'; another is 'pinpoint flood'. In the continuous flood system, sprouted seed are dropped into a flooded field that is maintained until near harvest. In the pinpoint system, dry or preferable sprouted seed are dropped into floodwater. The field is drained after 24 hours and left dry for 3 - 5 days to provide oxygen and allow the roots to anchor or 'peg' to the soil. Then the flood is re-established and maintained until near harvest. For the rice plant to continue growth, a portion of the plant must be above water by at least the fourth leaf stage. When a 'delayed flood' is used, fields are drained after water seeding for an extended period (usually 3 or 4 weeks) before the permanent flood is applied. With this system, fertilization timing and water management after the initial drainage are similar to dry-seeded systems.

Incorporation of green leaf manure: If FYM or compost is not applied, green leaf manure @ 6.25 t/ha can be applied. Maintain 5 cm depth of water. The green leaf manure incorporated into the soil by cage wheel ploughing with tractor or a power tiller mounted cage wheels or green manure trampler. Trample all the green matter which is exposed above the soil. In case of in situ ploughing of green manure crop, apply phosphotic and potassic fertilizer at the time of incorporation. In case of green leaf manuring, spread manure first on the ploughed field and then apply the fertilizer over to green leaf manure and then incorporate. Maintain 2.5 cm of water over the puddled soil and allow the green manure to be inside the puddled layer for a minimum of 7 days in the case of soft plants like sunhemp or *dhaincha* or 15 days for green manure plants which are more fibreous.

Digging corners, trimming and plastering of bunds: Dig the corners of field which are not covered by ploughing. Cut off 2.5 cm of soil from the top and sides of the bunds to remove the weeds along with their seeds and to destroy the eggs of insect pests by using spade. If bunds are very broad, trim them to a width of 15 cm and height of 15 cm so that rats do not harbour in the bunds. If rat burrows are noticed, insert pellets of 0.5 g or 0.6 g Aluminum phosphide and plug the exit holes. Apply mud paste to the sides and top of the bund to a thickness of 2.5 cm and plaster it, using the flat surface of the spade. Plastering the bunds helps in checking weed growth and prevents harbouring of insect pests. Leveling board is used for leveling of the main field. Field leveling is useful for better water management and uniform growth of rice crop in the whole field.

Transplanting specification

Seedlings per hill: Two or three seedlings per hill are adequate.

Transplanting of rice: Farmers prefer transplanting despite its cost because the field interaction such as weeding, spraying and applying fertilizer are easier. Maintain a water level of 2.5 cm for easy transplanting. When the field is completely drained during transplanting, soil mud may cling to the fingers making transplanting inefficient. Plant the rice seedlings by holding the seedlings close to the base, protecting the roots with thumb, index and middle fingers while inserting into the soil.

Depth of transplanting: Plant seedlings at 1.5 to 3.0 cm deep. The tiller buds formed at the basal node are not suppressed in case of shallow plantings. If seedlings are planted deeper, the old roots suffer and new roots develop in about 2 to 4 days after planting from a node close to the soil surface. This is called node rooting. This may delay the normal absorption of nutrients from the soil, thus delaying the recovery of seedlings. Seedlings planted too shallow will not be firmly held erect by the soil especially when the wind is strong. Shallow planting enhances greater number of tillers. Further, the growing point comes to the ground surface under shallow planting. Diurnal variation in temperature favours tiller production. The shallower the depth of transplanted seedlings, the easier to absorb nutrients and consequently, the more number of tillers will be produced. Tillers normally develop 5 to 10 days after transplanting. Deep planting delays tillering.

Techniques of transplanting: Rice seedlings are transplanted by two techniques *viz.*, (i) Random transplanting with no definite spacing between hills. Less labour cost in this method is an advantage and (ii) Straight row transplanting with definite spacing between hills. There are three methods in straight row planting using (i) planting wire and board, (ii) markers and (iii) mechanical transplanter. Tie each of wire or rope to the bamboo poles stretch the rope along the first base line and drive bamboo poles into the ground at each end of the first base line. Distribute the seedling bundles throughout the field. Set the planting board at the both ends of the planting rope for proper row spacing. Then plant the seedlings. When the first row has been planted, move the planting rope to the next mark of each of the planting boards. Move backwards to plant subsequent rows. Extra seedlings will be planted in alley ways for replanting in the missing hills 7 to 10 days later. Maintain two cm water depth up to seven days after transplanting until the plant recovers. **Advantages of straight row planting** are (i) weeding is done in two direction with a rotary weeder or by hand is possible and the weeding operations are easy, (ii) proper spacing makes for uniform plant growth and (iii) other field operations are made easier. **Disadvantages** are (i) labour cost is high. About 150 to 180 man hours are required to plant one hectare, (ii) harvesting is delayed for 7 to10 days compared to direct seeding and (iii) seedlings are exposed to possible injury during handling.

Plant density and geometry: Plant the seedlings at the correct age and at optimum spacing. Rice seedlings are transplanted at 20 days for short duration and 25-30 days for medium duration rice cultivars in the 4th or 5th leaf stage. The thumb rule is allowing one week in nursery to one month length of crop duration. Gap filling is done within 7 to 10 days after planting. The planting management of rice seedling of different crop duration with varying soil fertility is presented in the Table 12.

Table 12 Plant density and geometry for short, medium and long duration rice varieties

Soil/ Duration	Medium and low fertility			High fertility		
	Short	**Medium**	**Long**	**Short**	**Medium**	**Long**
Number of seedlings per hill	2-3	2	2	2-3	2	2
Depth of planting	3	3	3	3	3	3
Spacing (cm)	15×10	20×10	20×15	20×10	20×15	20×20
Hills/m^2	66	50	33	50	33	25

Distinguished features of transplanting and direct seeding techniques are furnished in Table 13.

Table 13 Comparison between transplanting and direct seeding

Criterion	Transplanting	Direct seeding
Land preparation	lowland	lowland or upland
Seed rate (kg/ha)	30-60	80-100
Nursery area	required	not required
Seed exposure to predators	less	high
Plant population	maintained	can not be maintained
Crop weed competition	less	greater
Lodging	less	more
Water management	not very critical	critical
Crop maturity	longer	7-10 days earlier
Potential yield	high	low

TRANSPLANTED RICE

Advantages

- Plant population is maintained
- Inputs applied to nursery area is lesser compared to broadcasting
- Nursery can be raised in advance in anticipation of water release from dams / tanks or receipt of rains.

Disadvantages

- Labour cost is high
- Seedling damage and transplanting shock

Weed management: Weed problem in rice is more acute in direct seeded rainfed rice than in transplanted lowland rice. In direct seeded rice culture, rice plants compete with weeds from the time they emerge. The maximum weed menace occurring in transplanted rice as compared to direct seeded rice may be attributed to the age gap between transplanted rice seedlings and emerging weeds. Transplanted seedlings have a greater competitive advantage over weeds, that emerge after transplanting. Inadequate land preparation, poor water management and increased fertilizer use leads to severe weed problems. The common weed species include grasses, sedges and broadleaved plants in rice field are:

(a) Direct sown upland rice

Sedges: *Cyperus rotundus,*
Bullostylos sp.

Broad leaved: *Eclipta alba*
Marsilia quadrifoliata
Trianthema portulacastrum
Amaranthus viridis
Ammania bacifera
Ludwigia parviflora

Grasses: *Echinochloa colona*
Paspalaum sp.
Panicum sp.
Dactyloctenium aegypticum

(b) Direct seeded lowland rice

Echinochloa crusgalli
Ipomea sp
Monochoria vaginalis

(c) Direct seeded in puddled soils

Grasses: *Echinochloa crusgalli*
E.colona

Sedges: *Cyperus difformis*
Fimbristylis miliacea
Scirpus sp.

Broad leaved: Eclipta alba
Marselia quadrifolia
Monochoria vaginalis
Ludwigia parviflora

Crop yield loss due to weeds in various systems of rice culture

Direct seeded upland rice = 50-60 %
Direct seeded puddled rice = 30-35 %
Transplanted rice = 15-20 %

Critical periods of weed competition

Timeliness of the weeding operation is more important than the number of weeding. Weed free period for rice irrespective of weeding techniques for different methods of growing rice are as follows:

Direct seeded upland rice = 60 DAS (days after sowing)
Direct seeded lowland rice = 50 DAS
Direct seeding on puddled rice = 35 DAS
Transplanted rice = 35 DAT (days after transplanting)

Time of application of herbicides in rice

Pre-sowing = 3 days before sowing
Pre-emergence = 0-4 DAT (days after transplanting)
Post-emergence = 21-30 DAT (days after transplanting)

The nutrient removal pattern of rice under different systems of cultivation under weed free condition and weed infested condition is presented in Table 14.

Table 14 Nutrient removal of rice under different systems of cultivation

System of rice culture	Nutrient removal of rice (kg/ha) on 35-50 DAT					
	Weed free condition			Weed infested condition		
	N	P	K	N	P	K
Direct seeded upland rice	12.9	2.0	16.8	5.4	0.9	11.3
Lowland transplanted rice	38.0	10.5	70.0	42.0	7.5	42.0

The herbicide recommendation for different rice systems is furnished in Table 15.

Table 15 Herbicide recommendations for different rice systems

System of rice culture	Time of application	Chemical name	Dosage kg ai per ha	Trade name
Direct seeded upland rice	Pre-emergence (8 DAE)	Thiobencarb or Pendimethalin	1.5 1.25	Saturn 50 EC Stomp 30 EC
	Post-emergence	Propanil	1.5	Stamp-F34
Direct seeded lowland rice without puddling	Pre-emergence (8 DAE)	Thiobencarb Pendimethalin	1.5 1.25	Saturn 50 EC Stomp 30 EC
	Post emergence	Propanil	1.5	Stamp-F34
Direct seeded puddled rice	Pre-emergence (8 DAE)	Thiobencarb Pendimethalin	1.5 1.25	Saturn 50 EC Stomp 30EC
	Post emergence	Propanil	1.5	Stamp-F34
Lowland transplanted rice Nursery	3 days before sowing or 8 days after emergence (DAE)	Butachlor Thiobencarb Oxyflurofen	1.0 1.0 0.1	Machete 50EC Saturn 50EC Goal 24 EC
Mainfield	Pre-emergence 3-4 DAT	Butachlor Thiobencarb Anilophos Oxadiazon	1.25 1.25 0.40 0.60	Machete 50EC Saturn 30 EC Arozin 30 EC Ronstar 24 EC

Management of weeds in transplanted rice: Weed management in transplanted rice is easy since the normal cropping practices *viz.*, land preparation, cultivars selection, plant spacing, flooding and fertilizer application reduce the number of weeds.

(i) Land preparation: Field is ploughed to incorporate the weeds left from the previous crop or fallow period into the soil. Harrow 2-4 times and puddle the soil to destroy weeds which are not controlled by ploughing. Level the field to have uniform flooding and water depth, which in turn will help to suppress the weed growth.

(ii) Cultivar selection: A taller cultivar producing a large number of tillers will compete better with weeds than shorter (semi-dwarf) cultivar with fewer tillers.

(iii) Plant spacing: Closer spaced plants compete effectively against weeds. Spacing of transplanted rice with 15 × 15 cm minimizes weed competition.

(iv) Flooding: Flood the fields 2-3 days after transplanting. Maintain 5 to 10 cm standing water continuously throughout the season. In areas, where weed problems are severe or land preparation and the water supply are inadequate, additional weed control may be necessary.

(v) Fertilizer application: In transplanted rice, control the weeds before top dressing of fertilizer so that the fertilizer will benefit the crop and does not stimulate weed growth.

Principles of applying herbicides to transplanted and wet seeded rice

(i) Granular pre-emergence treatments should have minimum 5 cm standing water.

(ii) Stagnating thin film of water one week time after application of herbicide

(iii) Thorough coverage of herbicide application is required

(iv) Wet seeded rice field water level should not be above 25% of rice plant height.

Considerations in applying herbicides to dry seeded rainfed upland rice

(i) Pre-emergence herbicides may cause stand reduction if followed by heavy rains.

(ii) If pre-emergence-herbicides alone are not adequate, a post emergence herbicide treatments must be combined with hand weeding or inter row cultivation.

(iii) Pre-emergence herbicides should be applied only to moist soil.

Manures and fertilizer application: Apply FYM or compost @ 12.5 t/ha. Spread the manure uniformly over the dry soil before letting in water for puddling. Only 2 to 3 kg N per ton of straw or rice stubbles are incorporated. It provides organic matter and other nutrients, especially potassium. If FYM or compost is not applied, apply green manure @ 6.25 t/ha. *Sesbania rostrata* forms nitrogen fixing nodules both on its roots and stems and has 5 to 10 times more nodules than most of the legumes. The stem nodules allow *S. rostrata* to fix nitrogen in flooded paddies. Spread green manure evenly on the puddled field and incorporate properly to a depth of 15 cm. When green manure is used, apply rock phosphate to supply 50 kg P_2O_5 along with green manure to hasten decomposition. Maintain 2.5 cm of water and allow 7 to 10 days for decomposition before transplanting. Incorporation of 6 to 8 tonnes green matter per ha to the rice soil is equivalent to 40 kg N per ha through inorganic fertilizers. Generally lime is not recommended for rice production unless the pH of the soil is 4.9 or lower. Over liming can induce zinc deficiency in rice.

Biofertilizers

Azospirillum: *Azospirillum* increases grain and straw yield of rice, when it is applied by adopting three methods. *viz*., seed treatment, root dipping and soil application. Any single method of treatment alone is ineffective.

(i) **Seed treatment:** The required quantity of seeds (60 kg) to plant one ha are to be soaked in 50 to 60 litres of water containing one kg of innoculum for 24 hours. Then allow the seeds to sprout by incubation for another 24 hours and sow the sprouted seeds in the nursery.

(ii) **Root dipping:** Dip the root portion of the rice seedlings required to plant one ha in 400 litres of water containing 2 kg of inoculum for 20 minutes and transplant. Alternatively, apply 2 kg of inoculum in the nursery just before pulling out the seedlings and then pull out the seedlings and leave the seedlings in the nursery for 30 minutes and transplant.

(iii) **Soil application:** Mix 2 kg of peat based Azospirillum innoculam (10 pockets per ha) with 25 kg soil or powdered FYM. Broadcast this mixture in the main field just before planting. *Pseudomonas fluorescens* (Pf 1) at 2.5 kg/ha mixed with 50 kg FYM and 25 kg of soil and broadcast the mixture uniformly before transplanting.

Azolla: *Azolla* is a union of an aquatic and nitrogen fixing blue green algae (BGA). Incorporating one crop of azolla is equal to application of 30 kg N per ha. It is raised as a dual crop by inoculating azolla @ 250 kg/ha on 7 DAT. Allow azolla to grow in field for 20 to 25 days to give biomass yield of 10 t/ha. It is incorporated in field at weeding of rice crop. Azolla can also be applied as green manure @ 5 t/ha before transplanting in the main field.

Blue Green Algae (BGA): 10 kg of powdered soil based blue green algae flakes has to be applied to one ha. Broadcast the powdered algae flakes 10 days after transplanting. Maintain a thin film of water in the field. BGA has good performance in salt affected soils with pH 7.5 to 10.

Phosphobacteria: It is applied as that of *azospirillum*. *Phosphobacterium* is compatible to apply with Azospirillum.

Silica Solubilizing Bacteria (SSB): Soil application of SSB @ 2 kg/ha + Rice husk ash @ 2 t/ha + 50% of recommended K to save K fertilizer in rice.

The recommended fertilizer dose in major rice growing states is furnished in Table 16.

Table 16 Recommended fertilizer dose in major rice growing states

<table>
<tr><th rowspan="2">Name of State / Union Territories</th><th rowspan="2">Season</th><th rowspan="2">Variety</th><th colspan="3">Recommended dose (kg/ha)</th></tr>
<tr><th>N</th><th>P</th><th>K</th></tr>
<tr><td rowspan="2">Andhra Pradesh</td><td>Kharif</td><td>HYV</td><td>100</td><td>60</td><td>40</td></tr>
<tr><td>Rabi</td><td>HYV</td><td>120</td><td>60</td><td>40</td></tr>
<tr><td rowspan="4">Assam</td><td rowspan="2">Kharif</td><td>HYV</td><td>40</td><td>20</td><td>20</td></tr>
<tr><td>Local</td><td>20</td><td>10</td><td>10</td></tr>
<tr><td rowspan="2">Rabi</td><td>HYV</td><td>60</td><td>30</td><td>30</td></tr>
<tr><td>Local</td><td>20</td><td>10</td><td>10</td></tr>
<tr><td rowspan="2">Bihar</td><td rowspan="2">Kharif</td><td>Irrigated</td><td>80</td><td>40</td><td>20</td></tr>
<tr><td>Rainfed</td><td>60</td><td>30</td><td>20</td></tr>
<tr><td rowspan="2">Jharkhand</td><td>Kharif</td><td></td><td>80</td><td>40</td><td>20</td></tr>
<tr><td>Rabi</td><td></td><td>70</td><td>40</td><td>20</td></tr>
</table>

Gujarat	*Kharif*	HYV	100	27	-
		Local	75	25	-
	Rabi	HYV	120	25	-
Haryana	*Kharif*	HYV	150	60	60
		Basmati	60	30	30
Himachal Pradesh	*Kharif*	HYV	90	40	40
		Local	50	25	25
Jammu & Kashmir	*Kharif*	HYV	80	45	20
		Local	50	25	25
Karnataka	*Kharif*		100	50	50
	Rabi		120	62	62
Kerala	*Kharif*		90	45	45
	Rabi		70	35	35
Madhya Pradesh	*Kharif*		120	60	60
Chhattisgarh	*Kharif*	HYV	80	50	30
		Local	40	30	20
Maharashtra	*Kharif*		100	50	50
Manipur	*Kharif*	HYV	61	51	38
		Local	32	23	15
	Rabi	HYV	66	51	38
		Local	32	23	15
Mizoram	*Kharif*	HYV	60	40	40
		Local	40	20	20
	Rabi	HYV	60	40	40
		Local	20	10	10
Odisha	*Kharif*	HYV	60	30	30
		Local	40	20	20
	Rabi	HYV	80	40	40
Punjab	*Kharif*	HYV	125	30	30
Rajasthan	*Kharif*	HYV	80-120	40-60	30-45
Tamil Nadu	*Kharif*		150	60	60
	Rabi		120	38	38
Tripura	*Kharif*	HYV	70	35	35
		Local	45	22	22
	Rabi	HYV	100	50	50
Uttar Pradesh	*Kharif*	HYV			
		Early	130	60	50
		Medium	150	75	60
		Dwarf	120	60	50
		Local	75	40	40
	Rabi	HYV	150	75	75
Uttaranchal	*Kharif*	HYV	120	60	40
West Bengal	*Kharif*		60	30	30
	Rabi		100	50	50
Pondicherry	*Kharif*	HYV	120	38	38
	Rabi		150	50	50

Time of fertilizer application

Short duration varieties

A. Light soils

Time	N (kg/ha)	P_2O_5 (kg/ha)	K_2O (kg/ha)
Basal	25	50	25
35 DAS	25	0	0
45 DAS	25	0	0
55 DAS	25	0	25
70 DAS	25	0	0
Total	125	50	50

Note: DAS stands for days after sowing

B. Heavy soils

Time	N (kg/ha)	P_2O_5 (kg/ha)	K_2O (kg/ha)
Basal	50	50	25
35 DAS	25	0	0
50 DAS	25	0	25
70 DAS	25	0	0
Total	125	50	50

Medium duration varieties:

Time	N (kg/ha)	P_2O_5 (kg/ha)	K_2O (kg/ha)
Basal	30	60	30
40 DAS	30	0	0
55 DAS	30	0	0
70 DAS	30	0	30
100 DAS	30	0	0
Total	150	60	60

Long duration varieties:

Time	N (kg/ha)	P_2O_5 (kg/ha)	K_2O (kg/ha)
Basal	25	60	30
40 DAS	25	0	0
55 DAS	25	0	0
70 DAS	25	0	0
90 DAS	25	0	30
115 DAS	25	0	0
Total	150	60	60

Method of fertilizer application: Fertilizer applied before transplanting should be mixed thoroughly with the soil to prevent N losses to atmosphere by action of air and to put the fertilizers nearer to the roots. Do not top dress in water immediately after transplanting while the leaves are wet condition, since the fertilizer stick to leaves and may cause leaf burn. The dissolved fertilizers will be lost to the air when the droplets dry up. Similarly do not top dress if a heavy rain is impeding as the fertilizer might to washed out from the field. Under special circumstances, apply nitrogen in 4 or more splits or use slow release nitrogen fertilizers such as neem cake coated urea and coal tar coated urea in soils with high percolation as in red soils. N recovery ranges 30 to 50% in tropics. So an average of 15 to 25 kg rough rice is produced per kg N applied. Rice derives 69% of its total N need from soils and the rest from applied fertilizers or manure.

Neem treated urea and coal tar treated urea: Blend the urea with crushed neem seed or neem cake 20% by weight. Powdered neem cake to pass through 2 mm sieve before mixing with urea. Keep it overnight before use or urea can be mixed with gypsum in 1:3 ratio or urea can be mixed with gypsum and neem cake at 5:4:1 ratio to increase the nitrogen use efficiency. For treating 100 kg urea, take one kg coal tar and 1.5 lit of kerosene. Melt coal tar over a low flame and dissolve it in kerosene or use one litre liquid coal tar. Mix urea with the solution thoroughly in a plastic container using a stick. Allow it to dry in shade on a polythene sheet. This can be stored for a month and applied basally.

Nitrogen management: For increasing the number of tillers, a nitrogenous top-dressing needs to be given only during active and maximum tillering stages to have productive tillers. The final top dressing should be 5 to 7 days before panicle initiation with flood water not more than 5 cm deep. Nitrogen absorbed at early growth stages is used to produce more straw than grain. Nitrogen absorbed at later growth stages is used to produce more grain than straw. The total nitrogen uptake reaches to a maximum of 170 mg N per plant. There are two peaks related to the amount of nitrogen absorbed by the plant. The first peak appears on 23 DAS while the second peak appears at 19 to 9 days prior to heading. In practice, nitrogen top dressing is applied at panicle initiation when the young panicle grow about 1-2 mm long which is about 23-25 days before heading. The absorbed nitrogen at this time is efficiently used to increase spikelet number, panicle size, panicle weight, lodging resistance by affecting the length and diameter of internodes, dry matter accumulation in the basal portions and the strength of shoots. Nitrogen utilization is high when it is applied into the deep reduced soil layer and not in the oxidized zone. On degraded rice soils, where H_2S gas develop, urea or NH_4Cl is preferred over NH_4SO_4. Soils with montmorillomite clays tend to have higher nitrogen recovery rates than those with kaolinite clays or allophane. There is greater response to nitrogen fertilizer during the dry season than during wet or rainy season. This is because sunlight is more abundant during the dry season. Grain yield increase as a result of nitrogen application is more in high yielding varieties than in tall *indica* varieties, regardless of the season of planting or amount of nitrogen used. Application of nitrogen fertilizer to tall varieties will increase their height and tendency to lodge. Grain yield may actually decrease with increase in nitrogen fertilizer application because of lodging and shading of leaves.

Use Leaf Colour Chart (LCC) for need-based nitrogen fertilizer application to rice: LCC measures the green colour intensity of rice leaves. A standaridized leaf colour chart is used for assing leaf N status.

The leaf colour chart contains six gradients of green colour from yellowish green (No.1) to dark green (No.6). It is calibrated with the chlorophyll meter and used effectively for guiding nitrogen application in rice fields. LCC observations are taken from the beginning of tillering *i.e.*, 14 days after transplanting in transplanted rice or 21 DAS in direct seeded rice at 7 to 10 days interval up to booting stage. Observe the leaf colour in the fully opened third leaf from the top as index leaf which best reflects the N status of rice. Compare the colour of the middle portion of the leaf with the LCC. Match the leaf colour with the colours in the LCC chart during morning hours from 8 to 10 am. The last reading is taken when the crop starts to heading (initiation of flowering). Observation is taken at least 10 disease-free rice plants or hills randomly in a field with uniform plant population. Select the top most fully expanded leaf from each hill or plant. The colour of a single leaf is measured by placing the middle part of the leaf in front of a colour strip for comparison. Measure the leaf colour under the shade of your body, because direct sunlight affects leaf colour readings. If possible, the same person should take LCC readings at the same time of the day every time. Place the middle part of the leaf on a chart and compare the leaf colour with LCC shades. Determine the average LCC reading for the selected leaves. When the leaf colour falls between two shades, the mean value is taken as the reading, *e.g.* 2.5 for colour between 2 & 3. Do not detach or destroy the leaf. Repeat the process at 7 to 10 days intervals or at critical growth stages (early tillering, active tillering, panicle initiation, and first flowering) and apply N as needed. Time of fertilizer application is decided by LCC score. LCC critical value is 3.0 in low N response cultures like White Ponni (*e.g.* varieties with light green foliage/ direct seeded rice) and 4.0 in other cultivars and hybrids. When 6/10 observations of LCC show less than the critical colour value, the amount of N to be applied at different growth stages for semi-dwarf indica varieties as furnished in Table 17. No urea should be applied if colour of leaves is equal to or darker than LCC shade 4.

Table 17 Nitrogen to be applied based on LCC at different growth stages of rice

For transplanted (TPR)		**Dry season**	**Wet season (cloudy)**
Early growth stage	14-21 DAT	35 kg/ha of N	30 kg/ha of N
Rapid growth stage	28-42 DAT	45 kg/ha of N	30 kg/ha of N
Late growth stage	49-flowering	30 kg/ha of N	20 kg/ha of N
For direct wet-seeded rice (WSR)			
Early growth stage	21-48 DAS	35 kg/ha of N	30 kg/ha of N
Rapid growth stage	35-49 DAS	45 kg/ha of N	30 kg/ha of N
Late growth stage	56 DAS-flowering	30 kg/ha of N	20 kg/ha of N

Time and method of phosphorus application: Phosphorus fertilizer should be applied as basal dressing. Top dressing and placement of P is effective in P fixing soils. As phosphorus becomes usually more easily available under reduced conditions, rice crop can make very good use of soft rock phosphates. Diammonium phosphate (DAP) is superior as basal dose than single super phosphate (SSP). Ammonium polyphosphate is superior than DAP and SSP. Mixing of SSP with fresh cow dung reduce P fixation @ 50%. Incorporation of P fertilizers is conducive for greater P fixation. Dipping roots of rice seedlings in SSP / DAP - soil - water slurry saves 50% P use for rice. Foliar spray of 1% urea + 2% DAP + 1% MOP at PI and 10 days later for all varieties can be followed.

Secondary and micronutrients: Application of gypsum @ 500 kg/ha basally with NPK in the main field in non-calcareous heavy soils. A combination of gypsum (3 t/ha) and green manure (6.25 t/ha) has been found equally in reclaiming saline and alkali soils with pH 8.5 to 9.7 rapidly brings ESP from 30 to 50% in the very first season. Application of lime @ 2.5 t/ha before last ploughing to acid soils and continue to each crop up to the fifth crop.

Application of $ZnSO_4$ @ 40 kg/ha mixed with 50 kg of sand uniformly on the levelled field for every 3 crop seasons is optimal for normal or neutral soils. Do not incorporate in the soil as there will be zinc phosphate interaction. If $ZnSO_4$ is not applied basally, spray 0.5% $ZnSO_4$ thrice on 20, 30 and 40 DAT for short duration varieties and 30, 40 and 50 DAT for medium and long duration rice varieties. Zinc response is influenced by pH and season. Zinc availability is a constraint in intensively cultivated lands, saline and alkaline soils. In saline and alkaline soils, $ZnSO_4$ @ 100 kg/ha is applied to make allowance for high Zn fixation, followed by normal application of Zn after three successive crop seasons.

Sulfur may be applied @ 30 to 40 kg/ha through gypsum, phospho-gypsum, ammonium sulphate, elemental shulphur, etc. Sulfide toxicity can be overcome with midseason drainage at mid tillering stage (25–30 DAT/DAS), avoiding flooding and maintain moist conditions for 7–10 days and dry plough field after harvest to oxidize S and Fe.

Ferrous sulphate (19-20.5% Fe), Fe-EDTA (9-12% Fe), Fe-EDDHA (10% Fe), besides FYM (0.15% Fe), poultry and piggery manure (0.16% Fe), sewage sludge are used as sources for correcting Fe chlorosis. Seed treatment is done with 2% $FeSO_4.7H_2O$ solution / slurry to overcome Fe deficiency. Foliar sprays (2-3) of 1-2% $FeSO_4.7H_2O/FeNH_4SO_4$ (pH 5.2) solution or of chelates at weekly interval at early stage of deficiency are successful. Combination of green manure (GM) or organic manures with foliar spray of un-neutralized 1% $FeSO_4.7H_20$ / $FeNH_4SO_4$ (pH 5.2) solution can be done to correct Fe deficiency. Seed treatment is done with Ca peroxide @ 50–100% seed weight to avoid Fe toxicity.

Borax, granubor and boric acid are efficient sources to correct boron deficiency. Basal soil application of borax @ 1-2 kg/ha is superior to foliar sprays. Soil application has residual effect for 1-2 seasons. For hidden deficiency spray 0. 2% boric acid or borax at pre flowering or flower head formation stages

Soil application of $MnSO_4.4H_2O$ @ 40-50 kg/ha or foliar spray 3 to 4 times @ 0.5-1.0% $MnSO_4$ solution (5-15 kg Mn /ha) at tillering stage is carried out to correct manganese deficiency in rice

Al toxicity may be managed with planting tolerant cultivars such as IR43, CO37 and Basmati 370 which accumulate less Al and absorb Ca and P efficiently. Liming of soil with $CaCO_3$ preferably dolomite lime to supply Mg @ 2-4 t/ha to neutralize soil acidity and replace exchangeable Al. The sub-soil acidity is reclaimed by leaching with soluble source of Ca like gypsum / phosphogypsum / SSP / lime. Incorporate 1 t/ha of reactive rock phosphate to supply phosphorous. Soil mulching and / or green manuring / organic manuring prevents water loss and phytotoxicity of Al.

Rice absorbs 100 kg silicon (Si) per ton of grain. Si-deficient plants are susceptible to lodging with soft, droopy leaves and culms, lower leaves with yellow / brown necrotic. The critical concentration for Si is 40 mg Si per kg soil (1 M Na acetate 4.0 pH). Si deficiency occurs in old and strongly weathered, leached acid soils, and due to removal of rice straw and excessive use of N. Si deficiency is not yet common in intensive irrigated rice systems of tropical Asia. Silicon deficiency can be overcome with recycling rice straw (5–6% Si), and rice husks (10%), applying rice hull ash and balanced nutrient use of NPK, application of granular silicate fertilizers such as Ca silicate @ 120 to 200 kg/ha; K silicate @ 40 to 60 kg/ha, application of basic slag @ 2-3 t/ha once in two years, or fly ash (23% Si), foliar spray of Si @0.1-0.2% with sodium silicate is recommended to improve silicon nutrition.

Nutritional disorders of macro, secondary and micronutrients in rice

Nitrogen deficiency symptoms first appear at the leaf tip and progress along the mid rib until the entire leaf is dead. Narrow and short leaves which are erect and become yellowish green as they age. Young leaves remain green. Nitrogen deficiency induces intense chlorosis in old leaves and dies at later stage when older leaves become straw coloured.

Phosphorous deficiency symptom shows narrow and short leaves that are erect and dirty dark green. Young leaves remain healthier than older leaves which turn brown and die. Reddish or purplish colour may develop on leaves of varieties that tend to produce anthocyanin pigment. Phosphorous deficiency occurs in soils of low or high pH acid soils, acid sulphate soils, calcareous soils and alkali soils.

Potassium deficiency symptom shows bluish green leaves when young, yellowish at the interveins on lower leaves, starting from the tip and eventually drying to a light brown colour. Leaves are short, droopy and dark green. Brown spots develop on leaves. Irregular nectotic spots may develop on panicle. Potassium deficiency causes weakening of the straw which results in lodging. Potassium deficiency is often associated with iron toxicity which is common in acid and acid sulphate soils.

Calcium deficiency symptom shows the tip of upper leaves becomming white, rolled and ovaled.

Magnesium deficiency shows intervenal chlorosis characterized by an orangish yellow colour on lower leaves with white tips. Leaves are wavy droopy.

Sulphur deficiency symptom shows general chlorosis in the whole plant. Older leaves do not dry quickly. Yellowness initially on leaf sheaths, which proceeds to leaf blades. Whole plant becomes chlorotic at the tillering stage. H_2S causes sulphide toxicity on sandy, well drained , degraded paddy soils and on poorly drained organic soils. Sulphur deficiency occurs in lowland rice. Sulphide toxicity is known as *Akiochi* diseases in Japan.

Zinc deficiency shows the chlorotic symptom especially at the base mid ribs of the younger leaves. Brown blotches and streaks appear in lower leaf. Two to three weeks after transplanting, high rates of fertilizer aggravates the disorder and drainage alleviates it. Zinc deficiency occurs in neutral to alkaline soils particularly in calcareous soils. Zinc deficiency is referred as *khaira* disease in India.

Iron deficiency symptom shows chlorotic and then whitish over the entire leaf while iron toxicity symptom shows brown spots on the lower leaf from tips to the leaf base. Then leaves turn into orangish purplish brown. Iron toxicity is referred as '*bronzing*' in Sri Lanka, '*Akagare type* 1' in Japan and '*Akikochi*' in Korea. There is an interaction between iron and potassium in the plant. Rice exhibiting bronzing symptoms has low in potassium. Interveinal tiny brown spots appear in lower leaves starting from the tips and spreading to the leaf base. It is a serious problem for rice grown in acid sulphate soils. Application of urea, lime, phosphorus and potassium helps in amelioration of iron toxicity. Iron toxicity symptoms occur in one to two weeks after transplanting in sandy lateritic soils and one to two months after transplanting in peaty or boggy soils. Iron deficiency occurs in neutral to alkaline soils especially in calcareous soils. Iron toxicity is referred as *bronzing* disease in Sri Lanka.

Manganese deficiency symptom shows intervenal chlorosis on the leaves, chlorotic streaks spreading downward from tip to the base of the leaves, which become dark brown and necrotic. Manganese toxicity deficiency symptom shows brown spots in the veins of the leaf blade and the leaf sheath, especially on lower halves. Manganese deficiency occurs in degraded soils while its toxicity occur in manganese mined areas.

Boron deficiency symptom shows the tips of emerging leaves becoming white and rolled as in the case of calcium deficiency. The first symptom of boron toxicity is a yellowish white colour discolouration of the tips of lower leaves, which appears about 6 weeks after transplanting. As the disorder progresses, chlorosis occur at the tips of the margins of older leaves. 2 to 4 weeks later elliptical dark brown blotches / spots appear in the affected areas. Boron toxicity symptom shows chlorosis at the tip of older leaves especially along the margins. Irrigation water with boron content more than 2 ppm causes boron toxicity in soils. High potassium content aggravates boron toxicity. High boron content is diluted with good quality surface water with less or no boron content. Internal drainage minimizes boron accumulation.

Copper deficiency symptom shows bluish green leaves which become chlorotic near the tips. New leaves fail to unroll and maintain a needle like appearance of the entire leaf.

Aluminum toxicity retards root growth than that of the shoots. This reduces nutrient and water uptake and finally decrease drought tolerance. Interveinal chlorosis is orangish yellow.

Silicon deficiency symptom shows soft and droopy leaves. Application of silicate slag to degraded soils and peaty soils is beneficial.

H_2S toxicity, K, Mg and Si deficiency are referred as *Akiochi* disease in Japan and Korea. K deficiency and Fe toxicity is referred as *Akagare* I disease while Zinc deficiency is known as *Akagare* II disease. Salt injury causes chlorosis, stunted growth, reduced tillering and whitish leaf tips.

Water management: Water management practices in farmers field play a pivotal roll for maximum yield by effective utilization of the related inputs and environment. The managerial aspects involve control over excess or limited water supply considering the physiological and climatological aspects. Water requirement for rice includes the water requirement on soil, climate, plant and management, needed to raise seedlings, land preparation and to grow the crop from transplanting to harvest. The water requirement for transplanted short duration rice (105 days) is 900 mm, medium duration rice (135 days) is 1100 mm and long duration (160 days) is 1300 mm. Additional 40% water is required when rice crop is raised under direct sown condition. Of the total water required for the crop, 3% is used for the nursery; 16% for land preparation of the main field and 81% for irrigating the main field crop. The daily consumptive use of rice varies from 6 to 10 mm and the total water requirement is 1250 mm. Of this, 180 to 380 mm is lost by evaporation, 200 to 500 mm by transpiration and 200 to 700 mm by percolation. Out of the total water required for the crop, 40 mm is used for nursery, 200 mm for land preparation of the main field and 1000 mm for the field irrigation of the crop.

Water management in nursery is important for good germination and healthy seedlings. A shallow depth of 2 to 3 cm is adequate at the time of sowing. Water is to be drained 18 to 24 hours after sowing. Water stagnation in low patches is to be drained in to the channel. Water is let into saturate the field from third to fifth day. From fifth day onwards, the quantity of water applied is increased to a depth of 2.5 cm. A water depth of 10 cm can be maintained during time of land preparation. At the time of transplanting shallow depth of 2.5 cm is adequate since high depth of water will lead to deep planting resulting in reduction in tillering. Seven days after planting, 5 cm of water is necessary to facilitate development of new roots and crop growth. The field bunds are strengthened to minimise the seepage losses and to prevent the crab holes. Basal application of fertilizer can be done on the previous day of planting with a thin film of water (0.5 cm). At this time, there is no need to impound more water. Water can be saved by this practice instead of flooding, besides increasing fertilizer use efficiency. Rain water can be conserved by impounding rain water upto 10 cm depth in the field. Water management practices in main field of rice crop include continuous flooding and intermittent irrigation. Pre-planting submergence atleast 15 days prior to planting has been found to improve the yield. This allows adequate time for decomposition of residues and removal of toxic organic substances by drainage. The rice yield is high mainly because of high tiller number. The percolation loss is low under shallow flooding. Shallow flooding of 5 cm is the best practice since the weeds of grasses and sedges are controlled and increase in fertilizer use efficiency. The distribution and stability of granular pesticides are improved at moderate water depth.

Deep continuous flooding (10 to15 cm) has the potential to produce optimum yield. Plant height will increase whereas the tiller number, lodging resistance and yield will decrease. Net seepage and percoalation requirements are increased. The grasses and sedges are controlled effectively. However, the broadleaf weeds are relatively unaffected by flooding depth. Mid-season drainage may improve the respiratory function of roots, stimulate vigourous growth of roots, checks the development of unproductive tillers and prevents root rot and sheath rot diseases considerably. The field has to be drained completely for the control of brown plant hopper whenever the incidence is noticed. Intermittent irrigation can be practiced in areas with limited water supply and in areas served by pump irrigation. Here the water losses are minimum and available rainfall is effectively utilized.

Giving moisture stress for 2 or 3 days during the tillering phase is quite essential for the *Navarai* (Jan-May) season crop in Tami Nadu, because the vegetative phase of the crops pass through the winter months of January-February. By giving stress, the mineralization of nitrogen is increased in the surface layers. Avalilability of other nutrients also increased because of improvement in the soil temperature. Giving moisture stress during reproductive stage would prove harmful as this stage will pass through summer months where the atmospheric stress (higher air temperature and evaporation) is also more. Under intermittent irrigation, good crop yield can be obtained. This is achieved with more productive tillers, short plant height and high lodging resistance. However, scheduling irrigation after disappearance of water in field favours the germination and growth of weeds, which can be controlled by application of herbicides. Poor nitrogen uptake and some losses of nitrogen occur in soils due to alternate aerobic and anaerobic conditions. It is also less profitable to apply high levels of nitrogenous fertilizer with lower level of water supply. Split application of nitrogen and use of slow release nitrogenous fertilizers such as neem-coated urea can be recommended under such conditions to increase fertilizer use efficiency. In red loamy soil, application of urea super granules and irrigating to a depth of 5 cm water once in 3 days after disappearance of ponded water registered higher yield compared to the application of prilled urea. Intermittent irrigation minimises formation of toxic substances in the soil and reduces the drainage problem.

Rice can also be grown when soil moisture is maintained at saturation level without serious effect on yield. The irrigation efficiency is generally high when this practice is used on heavy soils. When soil moisture is 70 to 80% saturation rice yield start to decline. When it drops to 50% saturation, the yield decreases from 33 to 50% as that of continuous submerged conditions. If water supply is inadequate, irrigation can be given on priority basis depending on the physiological conditions of the crop. The crop demand for water varies during different crop growth stages. The critical stages in rice for water are tiller initiation stage (0-20 days) and panicle initiation to flowering (40 to 65 days from planting).

A shallow depth of water (2-3 cm) is adequate at planting since high depth of water lead to deep transplanting resulting is reduction in tillering. When the soil moisture remains close to saturation, new roots will develop within the first four days after planting. On third or fourth day after planting pre-emergence herbicides can be applied maintaining 2 cm depth of water. Efficiency of herbicides will be affected by increasing the water level. Seven days after planting 5 cm of water is necessary to facilitate tiller production and promote firm root anchorage in soil. During vegetative growth phase, water depth should be gradually increased from 3 to 5 cm during the succeeding 8 week period after planting. Draining water during 20-30 days after planting *i.e.,* active tillering stage is harmful irrespective of the depth of submergence. Water depth is to be maintained at 5 cm level from panicle initiation to flowering stage. Excess water more than 5 cm at booting stage, leads to reduction in growth of panilces, delay in heading, weakening of stem and lodging.

Rice is most sensitive to moisture stress from 20 days before and 10 days after heading. In contrast to vegetative phase stress during the reproductive phase causes very drastic effects on yield because potential recovery of the plant is very low. Water deficiency during reproductive phase causes serious reduction in yield due to impairing panicle formation, heading, flowering, fertilization and increased spikelet sterility. Water should be gradually drained in the field from 21-25 days after flowering till it is finally drained at about two weeks before harvest. Undue delay of terminal drainage prolongs ripening and promotes unfavourable harvesting conditions such as lodging. The drainage management in rice fields is generally neglected despite its negative impact on the yield.

Water saving methods: There are a few water saving methods in rice cultivation under water scarcity condition. These are summer ploughing, quick land preparation, formation of inner bunds near the main bunds and strengthening of field bunds. Drainage for a day or two during the beginning of maximum tillering stage check the development of ineffective tillers and stimulate vigorous root growth and increase respiratory function of roots by aerating the soil. At primordial initiation and booting stages, 5 cm water depth will reduce rat menace. Excess water of more than 5 cm at booting stage leads to reduction in the growth of panicles and delay in heading, stem become weak and lodge. The field has to be gradually drained 21 to 25 days after flowering.

Summer ploughing helps to reduce the water requirement for land preparation conserves sub-soil moisture and reduce weed infestation. Use of aged seedlings and irrigating the field after formation of hairline cracks reduces water requirement. Perfect leveling is necessary to increase irrigation efficiency and to reduce the water requirement. Percolation rates vary from 50 to 60 % of the total water expense. Three to four puddling with use of cage wheel mounted on a tractor will reduce percolation loss of water. Highest saving of water occurs when the loam sandy soils has been compacted to bulk density of 1.84 g cm^{-1}. To ensure uniform application of irrigation water, fields can be divided into small plots of 40 to 60 cents by providing small bunds. Lining of earthern channels with sands, cement slabs (6:1) of 5 cm thickness is found to be better in improving the conveyance efficiency in field irrigation channels, both in terms of economics and durability. The duration and seepage loss is only 3.25 % as against 15% in unlined earthern channels. There is 12% water saving and 14% time saving when rice field is irrigated through irrigation channels directly as that of field to field irrigation. At the time of transplanting, a shallow depth of 2 to 3 cm water is adequate since high depth of water will lead to deep planting resulting in reduction of tillers. Irrigation once in 3 to 4 days in *kharif* season and once in 5 to 6 days in *rabi* season is adequate. Irrigating rice crop once in 3, 5 or 7 days has no influence on grain yield.

Conservation of rain water by impounding rain water up to 10 cm depth in the field can be done. Maintenance of shallow depth (5 cm) of submergence is sufficient to maintain rice productivity as achieved in greater depths of submergence (7.5 cm and 10 cm).

Scheduling irrigation after disappearance of water under scarce condition will have more weed growth but can be controlled by application of herbicides. Intermittent irrigation can be practiced in areas with limited water supply and in areas served by pump irrigation. Hence, the available rainfall is effectively utilized. If water supply is inadequate irrigation can be given on priority basis depending on the physiological condition of crop. The critical growth stages of rice for moisture stress are tiller initiation, panicle initiation and flowering stages. Mid-season drying or drainage is done for two days in the late tillering stage and prior to the early panicle formation stage. At this stage, the number of panicles is fixed and the requirement of water by the rice crop is minimal. Mid-season drying or drainage is advocated in soils where rapid soil reduction takes place. It changes the root zone temporarily to an oxidised stage and stimulate root growth besides preventing the incidence of root-rot diseases. Field has to be drained gradually 21 to 25 days after flowering till it is finally drained at about 14 to 21 days before harvest. Termination of irrigation three weeks before harvest saves 160 mm of water. Too early draining of water leads to increase in immature grains and broken kernals. Too late draining is conducive for crop lodging. Rotational irrigation through distribution systems and excellent farmers cooperation will be very helpful in saving irrigation water.

Cropping systems: The cropping pattern in different agro-climatic zones has been adopted by the farmers after long experience based on suitability of soil, profitability, availability of market and industrial infrastructure and quantum of water available. Rice-Rice-Rice; Rice-Rice-Cereals (other than rice such as Ragi, Maize and Jowar); Rice-Rice-Pulses; Rice-Groundnut; Rice-Wheat; Rice-Wheat-Pulses; Rice-Toria-Wheat and Rice-Fish farming system are the popular rice based cropping patterns being followed in the country.

Pests and diseases management: The major rice insect pests are yellow stem borer, brown plant and white backed plant hopper, leaf folder, gundhi bug, gall midge, mealy bug, root weevil, black bug, blue beetle, rice hispa, caseworm and thrips. The major rice diseases are rice blast, bacterial leaf blight, sheath blight, false smut, brown spot, sheath rot, stem rot and rice tungro virus. The major rice nematodes are root knot nematode and white tip nematode. Majority of insects population can be monitored by fixing and positioning of pheromones or light traps (one for at least 5 ha) and bird perches (40 to 50 /ha) at appropriate stage of crop.

Table 18 Economic Threshold Level (ETL) of major pests of rice crop stage wise

Crop stage	Pest/Disease	Economic Threshold Level (ETLs)
Nursery	Yellow stem borer	1 egg-mass/m^2
	Root-knot nematode	1 nematode/g soil
	BLB: Kresek Phase	2-3 plants/m^2
	Green leafhopper	1-2 insects/hill in Rice Tungro Disease (RTD) endemic areas
	Gall midge	1 silver shoot (gall)/sqm.

Early to late tillering	Leaf-folder	2 Fully damaged leaves (FDL) with larva/hill
	Stem borer	2 egg-mass/m^2 or 10% dead heart or 1 moth/m^2 or 25 moths/ trap/week
	Gall midge	1 gall/m^2 or 10% Silver shoot
	Brown planthopper / WBPH	Brown planthopper/WBPH
	Green leafhopper	10-15 insects/hill (in RTV endemic areas 2 insects/hill)
	Rice hispa	2 adults or 2 dead leaf /hill
	2 adults or 2 dead leaf / hill	2 FDL/hill
	Swarming caterpillar	1 damaged tiller/hill or 2 larvae/ m^2
	Blast	3-5 lesions/leaf
	Brown spot	2-3 spots/leaf & 2-3 infected plants/ m^2
	Sheath blight	Lesions of 5-6 mm in length & 2-3 infected plants/m^2
	Sheath-rot	Lesion length 2-3 mm on sheath & 3-5 infected plants/ m^2
	Bacterial leaf blight	2-3 infected leaves/m^2
	Tungro	1 Tungro infected plants/m^2 & 2 GLH/hill (in fungus endemic areas)
Panicle initiation to booting	Stem borers	2 egg-mass/m^2 or 1 moth/m^2 or 25 moths / trap / week
	Leaf-folders	2 FDL/hill
	Brown Plant hopper / White backed plant hopper - BPH/WBPH	15-20 hoppers/hill
	Swarming caterpillar/cut worm	1 damaged tiller/hill or 2 larvae/ m^2
	Neck blast	2-5 neck infected plants/m2
	Sheath-rot	5 infected plants/m2
	Sheath blight	5% or more tillers affected
Flowering to milky grain	Gundhi bug	2 bugs/hill
	Climbing cutworm / Rice ear cutting caterpillar/ Armyworm	4-5 larvae/sqm.
	Rice panicle mite	No ETL. If mite appeared in previous season, it requires prophylactic control measures in the current season.

Table 19 Varieties resistant / tolerant to various insect pests and diseases

Insect pests	Resistant/tolerant varieties
Stem borer	Ratna, Sasyasree, Vikas, HKR 46, NDGR 21, Pantdhan 6, VLK 39, Prahlad, Birsadhan 201, Bhudeb Ainesh, Matangini, Radha, Sudha, Amulya, Bhagirathi, Jogan, Mandira, Nalini, Sabita, VL16 and VL 206
Gall midge	Bhadrakali, Pavitra, Panchami, Triguna, lndursamba, Shiva, Vasundhara, Mahamaya, Ratnagiri 3, Erra Mallelu, Kavya, Oragallu, Sneha, Bhuban, Shaktiman, Abhaya, Divya, Ruchi, Vibhava, Kshira, Lalat, MDU 3, Pothana, Suraksha, Tara, Rashmi, Karna Mahavir, Neela, Rajendradhan 202, Sarsa, Udaya, Pratap, Daya, Dhanya Lakshmi, Kunti, IR 36, Asha, Samalei, Samariddhi, Pusa, Surekha, Phalguna, Vikram, Shakti, Jyoti, Kakatiya, Kanchan and Birsa Dhan 202.
Brown plant hopper	Vijetha, Chaitanay, Krishnaveni, Pratibha, Vajram, Makom, Pavizham, Mansarovar, CO 42, Jyoti, Chandana, Nagarjuna, Sonasali, Rasmi, Neela, Annanga, Daya, Bhadra, Karthika, Aruna, Remya, Kanakam, Bharathidasan, Remya, Triguna, IET 8116, Rajendra Mahsuri-l, Pant dhan 11, Rajshree, Bhudeb and Hanseshwari
White backed plant hopper	HKR 120, HKR 126, HKR 228, PR 108, Menher, Pant dhan 10, Pant dhan 11, Mahananda and Hanseshwari
Green leaf hopper	Vikramarya, Nidhi, IR 24, Radha, Mahananda and Kunti
Blast	Rasi, Vikas, Krishna Hamsa, Tulasi, IR 64, Aditya, Swarnadhan, Himalaya 1, Himalaya 2, Himalaya 2216, Pant dhan 10, HKR 228 and PNR 519
Bacterial leaf blight	Ajaya, IR 36, IR 64, Swarna, Bhumbleshwari, PR 111, PR 113, PR 114, PR 115, PR 116, PR 118, Rajendra Basmati, Pant dhan 11, Govind, Radha, Kamini, Pant dhan 10, Jayshree, Kanchan and improved sambha masouri.
Rice tungro disease	Vikramarya, Nidhi, Amulya, Dinesh, Lakshmi and Nalini
Sheath blight	PR 108, Bhudeb Dinesh, Jogan, Mandira, Nalini, Neeraj and Sabita

Table 20 Varieties with resistance to more than one pest or disease

Variety	Source/States	Resistant to
Udaya	Odisha	BPH, GM, GLH, RTD and RKN
Suraksha	Andhra Pradesh, Odisha, and West Bengal	GM, BPK, WBPH and BL
Vikramarya	Andhra Pradesh	GM, GLH and RTD
Shaktiman	Odisha, and West Bengal	GM, BPH, WBPH and BL
Rasmi	Kerala	GM, BPH and BL
Daya	Odisha	GM, BPH, GLH and BLB
Samalei	Odisha, and Madhya Pradesh	GM, BPH, GLH and BL
Bhuban	Odisha	GM and BLB
Kunti	West Bengal	GM and BL
Lalat	Odisha	GM, BPH, GLH and BL
Sneha	Odisha	GB and RTD

Note: GM- Gall Midge; BPH- Brown Plant Hopper; WBPH- White Backed Plant Hopper; GLH- Green Leafhopper; BL- Blast; RTD- Rice Tungro Disease; RKN- Root-Knot Nematode; GB-Gundhi Bug; BLB-Bacterial Leaf Blight.

Table 21 Crop stage / pest vis-à-vis IPM practices

Pests	Pesticides
I. Insects	
Nursery	
Gall midge	Carbofuran 3% CG @ 25000-66600 g/ha or carbosulfan 6% G @ 16700 g/ha or carbosulfan 25% EC @ 800-1000 ml/ha or Spray of NSKE 5% or Neem oil 3 % or chlorpyriphos @ 20EC 2ml/litre of water
Stem borer	Cartap hydrochloride 4% granules @ 18.75 kg/ha or cartap hydrochloride 50% SP @ 1000 g/ha
Main field	
Stem borer	Carbofuran 3% CG @ 25000-66600 g/ha or cartap hydrochloride 4% granules @ 18750 g/ha or cartap hydrochloride 50% SP @ 1000 g/ha or monocrotophos 36 % SL @ 625-1250 ml/ha
Leaf folder	Spray cartap hydrochloride 4% granules @ 18750-25000 g/ha or cartap hydrochloride 50% SP @ 1000 g/ha or monocrotophos 36 % SL @ 625-1250 ml/ ha or chlorpyrifos 1.5% DP @ 25000 g/ha or Neem oil 3 % or chlorpyriphos @ 20EC 2ml/litre of water or Spraying of 5% NSKE
Brown plant hopper and WBPH	Spray of imidacloprid 70% WG @ 30-35 g/ha or imidacloprid 30.5% m/m SC @ 60-75 ml/ha or ethofenoprox 10% EC @ 500-750 ml/ha or acephate 75% SP @ 666-1000 g/ha or buprofezin 25% SC @800 ml/ha or Spraying of 5% NSKE
Gall midge	Application of carbofuran 3% CG @ 25000-66600 g/ha or fipronil 0.3% GR @16670-25000 g/ha at 20 days after transplanting
Hispa	Spray quinalphos 25% gel @ 1000 ml/ha or chlorpyrifos 20% EC @ 1250 ml/ha
Caseworm	Spray carbaryl 10% DP @ 25000 g/ha or Neem oil 3 % or chlorpyriphos @ 20EC 2ml/litre of water
Black bug	Spray of Acephate 75 SP @ 1g/l or Chlorpyriphos 20 EC @ 2.5 ml/l targeting the bottom of the hills
Ear head bug	Spray of fenthion 100EC @ 1ml/l or malathion 50EC 1ml/l or dust malathion 5 D or carbaryl 10D @ 25kg/ha
II. Diseases	
Nursery	
Blast	Spray carbendazim 50% WP @ 250-500 g/ha or isoprothiolan 40% EC @ 750 ml/ ha or tricyclozole 75% WP @ 300-400 g/ha or tricyclazole 70% WG @ 300 g/ha or Pseudomonas @ 2%
BLB	Spray of streptomycin sulphate 9% + tetracycline hydrochloride 1% SP @ 100-150 ppm.
Main field	
Blast	Spray carbendazim 50% WP @ 250-500 g/ha or isoprothiolan 40% EC @ 750 ml/ ha or tricyclazole 75% WP @ 300-400 g/ha or tricyclazole 70% WG @ 300 gm/ ha or Pseudomonas *fluorescens* -Pf1 @ 2%. Soil application of *Pseudomonas* @ 2.5 kg/ha.
Bacterial leaf blight	Reduce nitrogen application and apply if needed only small dose of N in more split doses, chemicals as recommended earlier. Soil application of *Pseudomonas fluorescens* -Pf1 @ 2.5 kg/ha.
Sheath blight	Apply validamycin 3% L @ 2000 ml/ha or hexaconazole 5% EC @ 1000 ml/ha or propiconazole 25% EC @ 750 ml/ha or propiconazole 10.7% + tricyclazole 34.2% SE @ 500 ml/ha or or Pseudomonas @ 2% or Spraying of 5% NSKE. Soil application of *Pseudomonas fluorescens* –Pf1 @ 2.5 kg/ha.
Sheath rot	Spraying of 5% NSKE

Nematode Management Practices: Seed treatment and foliar spray of *Pseudomonas fluorescens* -Pf1 suggested for disease management also helps in controlling rice root and white tip nematodes. White tip nematode (*Aphalenchoides besseyi*) is controlled with sun drying of seeds for 6 hours for 4 days and pre-sowing of nursery bed treatment with Carbofuran 3% G @ 50000 g/ha, if nematode population crosses the ETL. Root knot nematode (*Meloidogyne graminicola*) is controlled with rotation with the crops like sweet potato, sunflower, cowpea, sesamum, and onion. Soil application of carbofuran 3% CG @ 50 kg/ha is also recommended.

Rat Management Practices: Application of bromodiolone (0.005% a.i) in baits six weeks after transplantation. The residual live burrows may be treated with second application of bromodiolone (0.005%). The above control operations with rodenticides except Zinc phosphide (as rodents develop bait shyness) may be repeated if the rodent population exceeds working index. The optimum period for undertaking control operation is six weeks after transplantation. Zinc phosphide (2.5%) in baits may be applied. For getting effective control, application of Zinc Phosphide with ISI mark in 10 g pouches (Popped rice: poison- 49:1) or Bromodialone cake or trapping with bow trap (Thanjavur bow) @ 100 no./ha is recommended.

Harvest: Harvest the crop as soon as the grain matures when panicles turn into golden yellow in colour usually about 30 days after flowering. Harvesting rice crop at a grain moisture content of 21 to 25% or when 90% of the grains are firm and do not have a greenish tint or the presence of one to two green grains in panicles, 80% of panicles with 80% of ripened spikelets is considered to be optimum stage for harvesting the rice crop. When 80% of the panicles turn straw colour, the crop is ready for harvest. Even at this stage, the leaves of some of the varieties may remain green. This ensures maximum grain yield and better milling characteristics. Early maturing rice cultivars are usually harvested one month after full flowering and late maturing achieves at about 6 weeks after full flowering. At the time of harvest, the spikelets should be straw coloured. Pre harvest spraying rice crop at the optimum stage of maturity with 20% NaCl solution or 5% KCl solution or 1% Gramaxone or Reglone 20% EC at 500 lit per ha to hasten maturity to escape monsoon rains and harvesting can be done after 48 hours for uniform maturity. Drain the water from the field 7 to 10 days before the expected harvest date as draining hastens maturity and improves harvesting conditions. The crop should be cut with serrate edged sickles by hand quite close to the ground and left in the field for a day to dry. Timely harvesting ensures good grain quality and consumer acceptance, since grain is less likely to break when milled. If harvest is delayed, grain may be lost due to damage by rats, birds, insects, shattering and lodging. Too early harvest results in more of immature grains. Immature grains are prone to milling breakage. Leaving rice crop in the field for prolonged duration causes sun checks in the grains. Similarly, leaving the harvested leaves in the field for drying also causes cracks in the grains. Alternate wetting and drying cycles due to sun and dew induce cracks in the mature grains. To avoid crack formation due to the above factors, rice crop should be harvested at the optimum stage of maturity. After the harvest, the plants should be collected in bundles and stacked for threshing. Threshing is done soon after harvesting when the weather is dry. Threshing is separation of the grain with its enclosing husk from the stalk. It may be done by hand beating the sheaves against a hard surface or allowing the sheaves to the trodden by animals or by tractor on a beaten earth surface or rubbing with bare human feet in hills, etc. Foot operated pedal thresher and power driven stationary threshers are also in use. Winnowing is usually done by shaking and tossing the paddy backwards and forwards on a basket work tray with a narrow rim. The grain falls on a mat and the husk, chaff and dust are carried away by the wind. Winnowing can be done by hand or by power operated winnowers. After winnowing the grains are dried in the sun.

Yield: Maximum recorded rice grain yield is 13.2 t/ ha in Japan and 8.5 t/ha in India. The biological yield of *indica* and *japonica* varieties are 30.6 and 22.7 t/ha. Harvest index of semi dwarf of *indica* and *japonica* varieties are around 0.45 to 0.55 and 0.40 to 0.49 respectively. Thousand grain weight of rice ranges from 19 to 26 g. Small panicle has only 30 grains while large panicle has 400 grains. Maximum grain number with 30,000/m^2 with 400 panicles/m^2 can be achieved to get high yield.

The rice productivity status has been classified in to following groups namely:

(i) High productivity - yield more than 2,500 kg/ha

(ii) Medium productivity - yield more than 2,000-2,500 kg/ha

(iii) Medium-Low productivity - yield more than 1,500-2,000 kg/ha

(iv) Low productivity - yield in between 1,000-1,500 kg/ha

(v) Very Low productivity - yield less than 1,000 kg/ha

Storage: The dried rice grains should be stored in a ventilated godown in an air tight container. In this case, periodical disinfection measure is either by spraying with malathion / nuvan or fumigations with aluminium phosphide tablets depending on the level of infestation should be undertaken. Maintaining godown hygiene an important factor for avoiding the multiplication of insects. The old gunny bags / receptacles should be thoroughly disinfected. As the milling quality of infested grain is poor, periodical inspection and disinfection measure should be carried out. Infected grains will be soft and powdered while milling. Grain must be dried to at least 14% moisture (wet basis) and seed grain should be dried to12% moisture before storage. When sun drying, the grain should be spread in thin layers, 2-5 cm, and turned every 1-2 hours. When sun-drying seed, the grain should be turned more often and not exposed to temperatures above 42°C. If high temperatures occur the seed should be dried in the shade. Where grain is to be stored in bags, the bags should bestacked on pallets at least 50 cm away from the walls. The safe moisture content required for different storage periods of rice is presented in Table 21. A rule of thumb for seed is that the life of the seed will be halved for every 1% increase in moisture content or a 5°C increase in storage temperature.

Table 22 Safe moisture content required for different storage periods.

Storage period	moisture % for safe storage	Potential problems
2 to 3 weeks	14–18%	Molds, discoloration, respiration loss
8 to 12 months	12-13%	Insect damage
> One year	9 % or less	Loss of viability

Processing

(i) Parboiling: Soft grain rice varieties with pre-disposed cracks in it, invariability break during milling. The total rice recovery will be poor in such cases. The pre-existing cracks or defects in the grains can be sealed or healed by a process known as parboiling. The term parboiling means partial cooking of grain with husk intact. In parboiling, rice grain is soaked to saturation moisture level, steamed to gelatinize the starch and then dried to a moisture level of 14%. Improper parboiling and drying lead to breakage during milling.

Various parboiling process exist. In most of the rice mills, the rice grain is soaked in water with or without pre-steaming for 12 to 36 hours, water is drained and then steamed for 10 to 20 minutes at normal atmospheric pressure. This rice grain is mostly sun dried. Because of prolonged soaking, certain fermentative changes occur in soak water resulting in off-smell. The grain and polished rice also pick up the off-smell. Besides the smell grain constituents leach out leading to 1-2% of soaking loss. The smell as well as leaching loss can be avoided by treating the soak water with some chemicals or reducing the soaking time by hot soaking the rice grain.

In CFTRI, Mysore method, rice grain is soaked in hot water at 65°C for 4 to 5 hours in soaking cum steam vessels, water drained and steamed as above. These grains are dried with the use of mechanical dryers. Recently, the Paddy Processing Research Centre, Thanjavur, developed a short soaking cum tempering (SST) method. Here, rice grain is soaked at 70°C for one hour only and then water is drained. This rice grain is left in the same vessel for about four hours to attain a moisture content sufficient enough for complete parboiling. After this paddy grain is steamed as usual.

(ii) Drying: Since the moisture content of this parboiled grain is about 30% as against 35 to 40% in other methods, it is dried faster by 1 to 1½ hours. Soon after parboiling, the hot grains should be discharged and dried immediately. Two stages drying with a tempering phase, when the grain moisture crosses 16% reduces milling breakage considerably. On no account, hot dried grain should be cooled as it induces cracks in the grain. The dried parboiled grain should be kept in heap atleast for 4 hours prior to milling. This again reduces milling breakages.

(iii) Milling: The husk and bran should be removed to make the rice consumable. The removal of husk termed as dehusking / shelling / hulling should be done using rubber rollers. After removal of husk, the left over portion of the paddy is called *brown rice* or *cargo rice* or *dehusked rice* which needs to be polished or whitened to remove the pericarp and the bran (alourone) layers. For removing the bran layers *i.e.*, polishing energy coated polishers should be used. For polishing, parboiled rice, huller can also be used. During polishing, husk should not be mixed with brown rice. On milling, the grain gives approximately: husk 20%, whole kernels 50%, broken kernels 16%, bran and meal 14%. The husked or hulled rice is usually called brown rice, and this is then milled to remove the outer layers, after which it is polished to produce white rice.

Quality characters: Rice quality depends on many factors: moisture content, chalkiness, and proportions of brokens, shape, length, width and cooking characteristics. The rice grain is classified on the basis of grain length (short, medium or long): head rice content (percentage of brokens) and milling process methods (regular milled or parboiled). A secondary classification is based on cooking quality which is primarily measured in terms of amylose content. The amylose content determines whether the grain in hard or soft; sticky (glutinous) or non-sticky (non-glutinous). Of the different types or rice, long and medium grain rice with intermediate amylose content (non-glutinous) dominate production, consumption and trade. Low-income consumers are willing to pay a premium for long grain rice with higher amylose content, in contrast to high-income consumers who prefer lower amylase rice. The long and medium grain rice is commonly called *indica* rice while the short grain and bold variety commonly referred to as *japonica* rice. This short grain and bold rice type with low amylose content becomes sticky (waxy) after cooking making it suitable for eating with chopsticks, a common practice among Japan and Korean consumers. Glutinous rice consumed as a stable food in Myanmar, Thailand, Laos and Vietnam. It is consumed in small quantities as dessert during festive occasions or is fermented into rice wine. In south Asia, parboiled long grain rice with medium to high amylose content and mild aroma is a common stable food for middle-and high income consumers (the poor cannot afford the choice). The cooked parboiled rice is firm, fluffy and non–sticky and mixes well with curry.

Grain endosperm structure

Glutinous type: Soft endosperm that contains dextrin instead of starch. It will become sticky, when cooked. It is useful for special preparations like '*puttu*'. It is used to prepare chop sticks for eating in Japan, China.

Non-glutinous type: It contains starchy endosperm when cooked they remain flaky (non- sticky). Non-glutinous types are cultivated in worlds 90% of area.

Grain size is classified based on the kernel length

(i) Long grain rice has a kernel length of 6.6 mm or more.
(ii) Medium grain rice has a kernel length of 6.2 mm or more but less than 6.6 mm.
(iii) Short grain rice has a kernel length of less that 6.2 mm.

Grain shape classification based on length / breadth ratio (L/B) ratio

Long grain rice

(i) Husked rice or parboiled husked rice with a length/width ratio of 3.1 or more.
(ii) Milled rice or parboiled milled rice with a length/width ratio of 3.0 or more.

Medium grain rice

(i) Husked rice or parboiled husked rice with a length/width ratio of 2.1-3.0.
(ii) Milled rice or parboiled milled rice with a length/width ratio of 2.0-2.9.

Short grain rice

(i) Husked rice or parboiled rice with a length/width ratio of 2.0 or less.
(ii) Milled rice or parboiled milled rice with a length/width ratio of 1.9 or less.

Grain classification size and shape based on a combination of the kernel length and the length/width ratio

(i) Long grain rice has either a kernel length of more than 6.0 mm and with a length/width ratio of more than 2 but less than 3, or a kernel length of more than 6.0 mm and with a length/width ratio of 3 or more.

(ii) Medium grain rice has a kernel length of more than 5.2 mm but not more than 6.0 mm and a length/width ratio of less than 3.

(iii) Short grain rice has a kernel length of 5.2 mm or less and a length/width ratio of less than 2.

Classification of milled rice: Milled rice (white rice) may be further classified into the following degrees of milling:

(i) Undermilled rice is obtained by milling husked rice but not to the degree necessary to meet the requirements of well-milled rice.

(ii) Well-milled rice is obtained by milling husked rice in such a way that some of the germ and all the external layers and most of the internal layers of the bran have been removed.

(iii) Extra-well-milled rice is obtained by milling husked rice in such a way that almost all of the germ, all of the external layers and the largest part of the internal layers of the bran, and some of the endosperm, has been removed.

The details of rice quality aspects are furnished in Table 23 to Table 28.

Table 23 Rice grain quality and moisture management

Activity	Grain moisture content (%)	Primary loss
Harvest	20-24	Spoilage and discoloration if left too long in the field Shattering if too dry
Threshing	20-24	Spoilage and discoloration if threshing is delayed. Grain damage and/or fissuring if too dry. If too wet, grain retained in unthreshed panicle.
Drying	14	Spoilage if drying is delayed for several days Drying rate too low: reabsorption of moisture Drying rate too high: fissuring of grain Temperature too high: fissuring of grain
Storage	18% MC for temporary storage for 3 weeks 14% MC for long-term storage of rough rice 10% MC for long-term storage of seed	Spoilage and discoloration if grain is too wet
Milling	14	Declining milling recovery and head rice if too dry

Source: http://aciar.gov.au/files/node/2140/pr100chapter4.pdf

Table 24 Rice grain (brown rice or milled rice) dimension characteristics based on length, width and thickness and shape based on the length and width ratio

Grain characteristics	FAO scale for milled rice	USDA worker's scale for brown rice	TRTP-IRRI scale for brown rice
Grain length class (mm)			
Extra long	7.0 and over	-----	Over 7.50
Long	6.0-6.99	6.6-7.5	6.61-7.50
Medium	5.0-5.99	5.5-6.6	5.51-6.60
Short	less than 5.0	less than 5.5	less than 5.51
Grain shape class (length width ratio)			
Slender (long)	over 3.0	over 3.0	over 3.0
Medium	-----	2.1-3.0	2.1-3.0
Bold	2.0-3.0	less than 2.1	1.1-2.0
Round (short)	less than 2.0	-----	less than 1.1

Source: IRRI and IBPGR, 1980

Table 25 Grain characteristics based on Gel consistency

Gel Consistency	Gel length (mm)
Soft	61-100
Medium	41-60
Hard	26-40

Table 26 Grain characteristics based on amylose content

Category	Amylose content (%)
Waxy	1-2
non-waxy	> 2
Very low amylose	2-9
Low	10-20
Intermediate	20-25
High	25-33

Table 27 Grain characteristics based on Translucency

Scale	Descriptive
1	Clear
5	Intermediate
9	Opaque

Table 28 Grain characteristics based on Chalkiness (white belly, white center or white back).

Scale	Descriptive
0	None
1	Small (less than 10% of kernel areal)
5	Medium (10.20% of kernel areal)
9	Large (more than 20% of kernel areal)

Utilization of by-products of rice industry: The inedible outer husk is only removed in 'brown rice' while the husk, bran and germ are removed in 'white rice'. The brown rice is rich in vitamin B1, B2, B3, B6 and iron as compared to polished white rice. The lipid content in brown rice ranges 1.9 to 3.9% which is present in the outer layers. Broken grains account about 10% of total polished rice. The broken grains are used for preparations of breakfast dishes, feed and industrial starch. Rice husk constitutes the largest by-product of rice milling and one fifth of the paddy by weight consists of rice husk. Husk to grain ratio accounts for 15 to 29% depending on variety. Rice husk has a considerable fuel value for a variety of possible industrial uses. Rice husk is tough because of its silica-cellulose content. The silica content in husk is the highest among plant offal. It contains 15 to 18 percent silica, therefore, it is a potent source of silica for the manufacture of silicates or in glass manufacture. Husk is used as fuel, feed, additives, soil mulch, absorbent, abrasive, degreaser, bedding or litter, carriers, pressing aids, water purifier, paper boards, activated char, for paper making, etc,.

Rice bran is a by-product of rice milling industry. It contains 18-20 % of fat, 14-15 % protein and to some extent of minerals and vitamins. Rice bran is used in cattle and poultry feed, defatted bran, which is rich in protein, can be used in the preparation of biscuits and as cattle feed. The de-oiled bran contains about 1 to 3% oil only. Rice bran oil is a potential source of vegetable oil in Japan, Myanmar, Thailand and India. True raw rice bran contains 12 to 18% oil and parboiled rice bran contains 20 to 25% oil. Rice bran oil has high capacity to control serum cholesterol level. Rice bran oil contain substances like *oryzand* and *scalene* which are of importance to human health. Refined oil can be used as a cooling medium. The crude rice bran oil can be used for manufacturing of soap, enamel paints, varnishes, detergent, metal soap and squalene (for skin disease) can be extracted from crude bran oil. Rice bran wax, a byproduct of rice bran oil is used in industries.

Rice straw is used as cattle feed, used for thatching roof and in cottage industry for preparation of hats, mats, ropes, sound absorbents, straw board and used as litter material.

The ready to eat products such as popped and puffed rice, instant or rice flakes, canned rice and fermented products are produced.

Rice export prospects: India now exports rice to a large number of countries in the World. Basmati rice is exported to more than 80 countries mainly to Gulf and European Countries such as Sweden, Netherlands, Belgium and Italy. The Indian Basmati rice varieties have special pleasant aroma, long slender grain with soft texture on cooking. Indian parboiled rice has good demand in countries like Bangladesh, Saudi Arabia, Russia, Singapore, Israel, USA, etc; whereas in some African countries, consumers prefer yellow tinted parboiled rice. India also exports paddy to some countries like Indonesia, Srilanka, Russia etc. India also exports non-Basmati rice, brown rice, and broken rice to many countries. The major markets of rice are furnished in Table 29.

Table 29 The major export markets of Indian rice in the world

Type of rice	Countries where exported
Basmati rice	Saudi Arabia, Kuwait, UK, USA, Belgium, Canad Germany, Netherlands, Italy, Oman, Qatar etc.
Parboiled Rice	Saudi Arabia, Russia, Bangladesh, Egypt ARP, Singapore, Srilanka, United Arab Emits, Yemen Republic, Malaysia, Maldives, Oman etc.
Non Basmati (excluding Parboiled)	Malaysia, Singapore, South Africa, Bangladesh, Indonesia, Philippines, USA, *etc.*
Paddy (Rice in the husk)	Australia, Germany, Srilanka, Myanmar, Malaysia, South Africa, Saudi Arabia.
Brown Rice (Husked)	Germany, Sri Lanka, Japan, South Africa, Saudi Arabia, Australia, USA.
Broken Rice	Ethiopia, France, Kuwait, Malaysia, Oman, South Africa, Saudi Arabia, Singapore, UAE, USA.

HYBRID RICE

An eminent Chinese agronomist, Prof. Yuan Longping has developed the first hybrid rice varieties in the 1970s and he is known as the Father of hybrid rice. Dr. E.A. Siddiq, Hyderabad, India is widely referred to as the 'Father of Hybrid Rice in India'.

Hybrid Rice Seed Production

Hybrid rice is the first generation progeny (F1) obtained by crossing two genetically different varieties (parents) of rice is called 'Hybrid'. Since rice is self-pollinated, cytoplasmic male sterile (CMS) parent is used as female parent, which is normally called 'A' line. The fertility restoring line which is called 'pollinator' to the female parent is known as male parent. It is generally referred to as 'R' line, and is used for hybrid seed production. The hybrid combines the desirable characters from CMS line and R line. They exhibit vigour for several quantitative characters including yield. The hybrid seed is purchased or procured afresh every year/ season for raising the commercial crop. The harvested grains from hybrid crop should not be used for planting the next crop.

Selection of field: Previous crop should not be of rice. If previous crop is rice, irrigate the field and there by the dropped seeds will germinate which can be puddled in. If the previous crop is having dormancy means, we must be careful to see that all the dropped seeds are all germinated and puddled in.

Isolation distance: If time isolation is to be followed, there should not be any rice crop nearby within 100 meters, in the process of flowering while the crop in seed production plot is in flowering. There must be a difference of 30 days in flowering for the nearby crop. To ensure purity of hybrid seed and avoid pollination by unwanted pollen isolation is a must.

(i) Space isolation: An isolation distance of over 100 m is required. Within this range, no other rice varieties should be grown except pollen parent.

(ii) Time isolation: The heading stage of varieties grown within 100 m around the seed production field should be over 25 days earlier or later than that of the CMS line.

(iii) Barrier isolation: Topographic features like hills, wood lot, tall crops as vegetative barrier like maize, sesbania, sugarcane, etc. to a distance over 30 m and artificial obstacles (plastic sheets above 2 m in height) will provide better isolation.

Season: April - May and Dec - January month of sowing.

Seed rate

A line or female parent: 20 kg/ha

B line or R line or male parent: 10 kg/ha

Nursery: Female and male (A and R) seedlings have to be raised in separate seed beds. Apply 2 kg DAP/cent to the nursery. Adopt 1 kg/cent or (25 g/m^2) of nursery for both A line and R line while raising the R line 5 kg seeds can be raised on the same date when A line is raised. The rest 5 kg can be sown five days after first sowing. Sparse sowing is done to obtain healthy and vigorous seedlings with 2-3 tillers at the time of planting. Seedlings of 20 to 25 days old have to be planted. Another way to economize on seed rate is to plant only one or at the most two multi-tillered seedlings per hill, instead of 4-5 seedlings/hill in case of high yielding varieties. The cultivation of hybrid rice by SRI method requires only 5 kg seeds/ha.

Row ratio: Row ratio refers to the ratio of number of rows of male parent to that of the female parent in the hybrid seed production field. Practically a row ratio of 2:8 (R:A) is currently widely used for hybrid rice seed production. Under good seed production management, the row ratio can be increased to 2:10 or even 2:12. Making the row direction nearly perpendicular to the direction of wind prevailing at heading stage will facilitate cross pollination.

2B: 8A for CMS multiplication

2R: 10A for hybrid seed production

The principles include are i) R line should have enough pollen to provide and ii) the row direction should be nearly perpendicular to the direction of winds prevailing at heading stage to facilitate cross pollination. Practically, a row ratio of 2:8 is currently widely used in indica hybrid seed production.

Generally, the R line is transplanted with two to three seedlings per hill and separated by a spacing of 15cm from plant to plant, 30cm from one row of restorer to another and

20cm from CMS line. The MS line is transplanted with one to two seedlings per hill with a spacing of 15×15 cm.

Number of seedlings per hill

2 seedlings/ hill for female parent
3 seedlings/ hill for male parent

Spacing
Male: Male = 30 cm
Male: Female = 20 cm
Female: Female = 15 cm
Plant : Plant = 15 cm

Weed management: Butachlor @ 2.5-3.0 kg should be mixed in 50-70 kg of sand and apply in one ha area after 5-6 days of transplanting. Then hand weeding is recommended depending on field condition.

Manures and fertilizer application: Application of 12.5 tonne FYM/ha and a fertilizer dose of 150-60-60 kg/ha is recommended. Entire phosphorus is applied as basal dressing. One third dose of nitrogen and potassium is equally applied as basal dressing, tillering stage and boot leaf stage.

Water management: Maintain a thin film of water for initial 30 days and later increase the water level to 4 to 5 cm when the crop reaches maximum tillering stage. Drain out water for 4-5 days after maximum tillering stage so that emergence of late tillers can be suppressed. Drain out water completely 10 days before harvest.

Synchronization: Synchronization in flowering can be attained by the following measures.

(i) Seeding interval: The parental lines differing in their growth duration can be sown on staggered dates in the nursery beds, so that they come to flowering at the same time in the main field where hybrid seed is to be produced. This is called 'staggered' or 'differential' sowing. The R line is sown in three splits *i.e.*, 3, 5 and 7 days after sowing of A line. However, the nursery of both the parents is transplanted on the same date. The nursery of R line sown on two dates is transplanted in alternate hills in the same rows. The seeding date for synchronization between male and female parents is determined based on growth duration difference. In general, the days of staggering slightly vary from location to location and season to season.

ii) Through fertilization: Depending upon the environmental conditions, synchronization of two parents can be adjusted by foliar spray of nitrogenous/phosphatic fertilizers. The spray of 2% urea to early parent delays flowering by 2-3 days and use of phosphatic fertilizer to late parent enhances flowering by 2-3 days. However, the dose of the fertilizers will depend upon the difference in growth duration and responsiveness of the parental lines.

GA3 application: GA3 application enhances panicle and stigma exertion, adjusts plant height of seed and pollen parents, speed up the growth of late tillers and increased the effective tillers, sets uniform panicle layer, increase the flag leaf angle, increase 1000 grain weight, reduces unfilled grains, enhances seed setting and seed yield. The dose of 60 g/ha in 500 litre of water at 5 to 10% of heading in two split doses on consecutive days. This hormone does not dissolve in water and hence it should be first dissolved in 70% alcohol (1 g of GA3 in 25-40 CC of alcohol). The ideal time for spraying is from 8 to 10 am and from 4 to 6 pm.

Flag leaf clipping: The flag leaves should be clipped off when the main culms are in booting or pre-emergence of panicle stage for free movement and wide dispersal of pollen grains to produce high seed yield. About half to two-third portion of flag leaf from the top should be removed.

Supplementary pollination: Rice is self-pollinated crop and hence there is need for supplementary pollination for enhancing out-crossing. Generally supplementary pollination is carried out at 30 minutes intervals 5 times daily both morning and evening during peak anthesis (10-12 am and 2 to 4 pm) until no pollen remains on the R line. It is not needed when the wind is greater than moderate breeze. In this operation, the pollen parent plants are shaken which helps in shedding and dispersal of pollen grains over the A line. This can be done either by rope pulling or by shaking the pollen parent with the help of two bamboo sticks. The first supplementary pollination should be done at peak anthesis time when 30 to 40% of the spikelets are open and anthers are fully exerted. This process is repeated three to four times during the day at flowering phase with an interval of 30 minutes. This process should be done for 7-10 days during flowering period.

Rouging: Rouging is done in both A line and R line at vegetative phase based on morphological characters of leaf shape and pigmentation and plant type, at flowering phase based on absence/presence of awns, the pollen shedders and other off-types and at maturity based on grain characteristics. The off types in A and R lines should be removed 2 to 3 times before heading and daily from initial heading to dough stage. Pollen shedders are to be removed along with tillers. In A line seed set may not exceed 40%. If plants having a setting of 70 to 80% means they are rogues and they have to be removed before harvest.

Plant protection: Normal IPM practices recommended to rice crop may be followed for hybrid rice seed production. Spraying or dusting during anthesis (10-12 a.m. and 2-4 p.m.) will affect pollination, resulting poor seed set.

Harvesting and processing: Drain out water from the field when grains in the lowest portion of the panicle are on the dough stage (about 20 days from 50% flowering). Allow the grains to harden. Harvest 30-35 days after flowering when stalks still remain green to avoid grain shedding. The male parent should be harvested first at physiological maturity and after thorough inspection harvesting is done for female lines. Moisture content of paddy should be 20 to 24% at harvest. Thresh as early as possible preferably a day after harvest. Care should be taken to avoid the admixture of male line with female line while harvesting. The female parent should be threshed and winnowed separately in well cleaned threshing floor. Dry the seeds gradually under shade until the moisture content is brought down to 12%, for safe storage. The seed should not be dried under direct sunlight between 12 to 3 pm during hot sunny days. After drying, the seed should be cleaned and then graded by using appropriate sieves. After grading, the seeds may be given seed treatment and packed in gunny or cloth bags and stored under ideal storage condition. The seed should be tagged with labels both inside and outside the bags.

Seed yield: The hybrid seed yield is 1.5 to 2.0 t/ha. The seed yields are higher in dry season as compared to wet season. Hence large scale seed production is generally taken in dry season only.

The seed production from hybrid rice is only about 20 % of conventional rice.

Hybrid Rice Grain Production Technology

The Hybrid rice released/notified in India is presented in Table 1. Hybrid rice have a 15 per cent yield advantage (1 to 1.5 tons / ha) over the best conventional varieties. Growing hybrid rice is a complex process and especially agronomic management of hybrid rice differs considerably from that of conventional varieties, primarily because of heterosis.

Table 30 Hybrid rice released/notified in India

S.No.	Rice Hybrids	Year of release	Duration (Days)	Yield (t/ha)	Developed by	Recommended for
1	APRH 1	1994	130-135	7.14	APRRI, Maruteru (ANGRAU), Hyderabad	Andhra Pradesh
2	APRH 2	1994	120-125	7.52	APRRI, Maruteru (ANGRAU), Hyderabad	Andhra Pradesh
3	MGR 1	1994	110-115	6.08	TNAU, Coimbatore	Tamil Nadu
4	KRH 1	1994	120-125	6.02	VC Farm , Mandya, UAS, Bangalore	Karnataka
5	CNRH 3	1995	125-130	7.49	RRS, Chinsurah (W.B.)	West Bengal
6	DRRH 1	1996	125-130	7.30	DRR, Hyderabad	Andhra Pradesh
7	KRH 2	1996	130-135	7.40	VC Farm , Mandya, UAS, Bangalore	Bihar, Karnataka, T. N., Tripura, Maharashtra, Haryana, Uttarakhand, Odisha, W. B., Pondicherry, Rajasthan
8	Pant Sankar Dhan 1	1997	115-120	6.80	GBPUAT & T, Pantnagar	Uttar Pradesh
9	PHB 71	1997	130-135	7.86	Pioneer Overseas Corporation, Hyderabad	Haryana, U.P., T. N., A.P., Karnataka
10	CORH 2	1999	120-125	6.25	TNAU, Coimbatore	Tamil Nadu
11	ADTRH 1	1999	115-120	7.10	TNRRI, Aduthurai (TNAU)	Tamil Nadu
12	Sahyadri	1998	125-130	6.64	RARS, Karjat (BSKKV)	Maharashtra
13	Narendra Sankar Dhan 2	1998	125-130	6.15	NDUAT & T, Faizabad	Uttar Pradesh
14	PA 6201	2000	125-130	6.20	Bayer Bio-Science, Hyderabad	A.P., Karnataka, Bihar, Odisha, M.P., U. P., W. B., T. N., Tripura
15	PA 6444	2001	135-140	6.11	Bayer Bio-Science, Hyderabad	U. P., Tripura, Odisha, A. P., Karnataka, Maharashtra, Uttarakhand

16	Pusa RH 10	2001	120-125	4.35	IARI, New Delhi	Haryana, Delhi, Western U.P. and Uttarakhand
17	PRH-122R (Ganga)	2001	130	5.64	Paras Extra Growth Seeds Ltd., Hyderabad	Bihar, Odisha, Punjab, U.P., Uttarakhand, Nagaland, Haryana
18	RH 204	2003	120-126	6.89	Parry Monsanto Seeds Ltd., Bangalore	A. P., Karnataka, T. N., Haryana, Uttarakhand , Rajasthan
19	Suruchi 5401	2004	130-135	5.94	Mahyco Ltd., Aurangabad	Haryana, A. P., Karnataka, Gujarat, Odisha, Chattisgarh , Maharashtra
20	Pant Sankar Dhan 3	2004	125-130	6.12	GBPUAT & T, Pantnagar	Uttarakhand
21	Narendra Usar Sankar Dhan 3	2005	130-135	5.15	NDUAT &T, Faizabad	Saline & alkaline areas of U.P.
22	DRRH 2	2005	112-116	5.35	DRR, Hyderabad	Haryana, Uttarakhand, W.B. T. N.
23	Rajlakshmi (CRHR 5)	2005	130-135	5.84	CRRI, Cuttack	Boro areas of Aasam, Odisha
24	Ajay (CRHR 7)	2005	130-135	6.07	CRRI, Cuttack	Irrigated areas of Odisha
25	Sahyadri 2	2005	115-120	6.50	RARS, Karjat (BSKKV)	Maharashtra
26	Sahyadri 3	2005	125-130	7.5	RARS, Karjat (BSKKV)	Maharashtra
27	HKRH-1	2006	139	9.41	RARS,Karnal (CCSHAU)	Haryana
28	CORH-3	2006	115	6.6	TNAU, Coimbatore	Tamil Nadu
29	JKRH 401	2006	125	6.22	JK Agri. Genetics Ltd. Hyderabad	Bihar, Odisha, W.B., U. P.
30	KJTRH 2	2006	N.A.	N.A.	RARS, Karjat (BSKKV)	Maharashtra
31	Haryana Shankar Dhan-1 (HKRH-1)	2006	139	9.40	HAU, Haryana RARS,Kaul (CCS,HAU.)	Haryana
32	HRI-152 (IET 18815)	2007	120 (Mid early)	NA	Bayer Bio-Science, Hyderabad	Punjab & T. N.
33	JRH-4	2007	110-115	7.50	JNKVV, Jabalpur	Madhya Pradesh
34	JRH-5	2007	105-108	7.50	JNKVV, Jabalpur	Madhya Pradesh
35	Indira Sona	2007	120-125	7.0	IGKKV, Raipur	Chhattisgarh
36	PA 6129	2007	115-120	6.58	Bayer Bio-Science, Hyderabad	Punjab, T. N., Pondichery

37	GK -5003	2008	128	6.04	Ganga Kaveri Seeds Pvt. Ltd. , Hyderabad	A. P., Karnataka
38	Sahyadri - 4	2008	115-120	6.80	RARS, Karjat (BSKKV)	Haryana, W. B., U.P. , Maharashtra, Punjab
39	JRH- 8	2008	105-110	7.50	JNKVV, Jabalpur	Madhya Pradesh
40	DRH - 775	2009	97	7.70	Methelix Life Sciences, Pvt. Ltd. Hyderabad.	Bihar, Chhattisgarh, Jharkhand, M.P., U.P., W. B., Uttarakhand,
41	HRI -157 (IET 19511, 91H97226) (Arize Prima)	2009	130-135	6.50	Bayer Bio-Science, Hyderabad	Chhattisgarh , Gujarat, Bihar, Jharkhand, Odisha, A.P.,T.N.. Maharashtra, Karnataka, M. P.,U. P., Tripura
42	PAC 835 (PAC 80035) (IET 18178) Hybrid	2009	130	5.60	Advanta India Ltd., Hyderabad	Odisha, Gujarat
43	PAC 837 (PAC 80037) (IET 19746) Hybrid	2009	130	6.30	Advanta India Ltd., Hyderabad	Gujarat, Chhattisgarh, J&K, A. P., Karnataka
44	NK - 5251	2009	128	6.65	Syngenta India Ltd., Secundrabad	A. P., Gujarat, Karnataka, T. N., Maharashtra,
45	DRRH- 3	2009	131	6.07	DRR, Hyderabad	A. P., Gujarat, M. P., Odisha, U.P.
46	US - 312	2010	125-130	5.76	Seed Works International, Hyderabad.	A. P., Bihar , Karnataka, T. N., U. P., W. B.
47	CRHR-32	2010	125	5.43	CRRI, Cuttack, Odisha	Bihar, Gujarat
48	INDAM 200-017	2010	120-125	6.60	Indo-American seeds, Hyderabad	Odisha, A.P.,Chattisgarh, Gujarat Maharashtra,
49	27P11	2010	115-120	5.67	PHI Seeds(P) Ltd.	Karnataka, Maharashtra
50	VNR 2245 (IET 20716) (VNR-204)	2011	90-95	6.83	VNR Seeds Pvt. Ltd., Raipur	Chhattisgarh, T. N.
51	VNR 2245 (IET 20735) (VNR-202)	2011	100-105	5.75	VNR Seeds Pvt. Ltd., Raipur	U. P., T. N., Uttarakhand, W. B., Maharashtra,
52	Shyadri-5 (Hybrid)	2011	110-115	NA	RARS, Karjat (BSKKV)	Konkan Region of Maharashtra
53	CO (R) H-4	2011	130-135	7.34	TNAU, Coimbatore	Tamil Nadu

54	Hybrid CO 4	2012	130-145	7.34	NA	Tamil Nadu, Gujarat, Maharashtra, Uttarakhand, U.P., Chhattisgarh, W.B.., Bihar.
55	US 382 (IET 20727)	2012	125-130	6.70	Seed Works International Pvt. Ltd. , Hyderabad-34.	Tripura, M. P., Karnataka
56	27P31 (IET 21415)	2012	125-130	8 to 9	PHI Seeds Pvt. Ltd. Hyderabad-82.	Jharkhand, Maharashtra, Karnataka, Tamil Nadu, Uttar Pradesh, Bihar, Chhattisgarh, Madhya Pradesh, Odisha
57	27P61 (IET 21447)	2012	132	6.70	PHI Seeds Pvt. Ltd. Hyderabad2.	Chhattisgarh, Gujarat, A. P., Karnataka, Tamil Nadu
58	25P25 (IET 21401)	2012	110	6.70	PHI Seeds Pvt. Ltd. Hyderabad.	Uttarakhand, Jharkhand, Karnataka
59	Arize Tej (HRI 169) (IET 21411)	2012	125	7. 0	Bayer Bio Science Pvt. Ltd, Hyderabad – 81. ,	Bihar, Chhattisgarh, Gujarat, Andhra Pradesh, Tamil Nadu and Jharkhand
60	PNPH 24 (IET 21406)	2012	120-130	5.8-6.9	Nuziveedu Seeds Limited, Medchal Mandal, Ranga Reddy	Bihar, W. B., Odisha
61	PNPH 924-1 (IET 21255)	2012	125-135	NA	Nuziveedu Seeds Limited, Medchal Mandal, Ranga Reddy	W. B., Assam
62	NK 5251 (IET 19738)	2012	NA	NA	NA	T. N., A. P., Karnataka, Maharashtra, Gujarat
63	CR Dhan 701	2012	140-145	5.0	NA	Bihar, Gujarat
64	JKRH 3333 (IET 20759)	2013	135-140	5.98	JK Agri Genetics Ltd, Hyderabad.	W. B., Bihar, Chhattisgarh, Gujarat, A. P.
65	RH-1531(Frontline Gold) (IET 21404)	2013	118-125	NA	Devgen Seeds & Crop Technology, Hyderabad	M. P., U. P., A. P., Karnataka, Maharashtra
66	CO 4 (IET 21449) (TNRH 174)	2013	130-135	7.3-7.5	TNAU, Coimbatore	T. N., Gujarat, Maharashtra
67	Arize Dhani	2013	NA	NA	Bayer Bio-Science, Hyderabad	Odisha

68	27P52 (IET 21433)	2013	NA	NA	PHI Seeds Pvt. Ltd. Hyderabad-82.	A.P, Chhattisgarh, Gujarat, Odisha, Uttarakhand
69	27P63 (IET 21832)	2013	132-135	6.0-6.5	PHI Seeds Pvt. Ltd. Hyderabad.	A.P, U.P., Chhattisgarh, Karnataka,
70	KPH - 199	2013	95-100	NA	Kaveri Seed Company Limited, Secunderabad	A.P, Chhattisgarh, M.P.
71	KPH - 371	2013	95-100	6.8	Kaveri Seed Company Limited, Secunderabad	Chhattisgarh, Jharkhand, Karnataka, Kerala
72	VNR 2375 PLUS (IET 21423) (VNR – 203)	2013	130	NA	VNR Seeds Pvt. Ltd., Raipur-492099	Bihar, Karnataka, Punjab Maharashtra, Uttarakhand,
73	US 305 (IET 21827)	2013	NA	NA	Seed Works International Pvt. Ltd. Hyderabad.	A.P., T. N., Maharashtra
74	US 314 (IET 21777)	2013	89	NA	Seed Works International Pvt. Ltd. Hyderabad.	A.P., Bihar, W.B., Uttarakhand
75	Ankur 7434	2014	NA	NA	Ankur Seeds, Pvt. Ltd.	Chhattisgarh
76	PAC 807	2014	NA	NA	Advanta India Ltd., Hyderabad	Chhattisgarh
77	CSR 43	2014	NA	NA	Uttar Pradesh	Uttar Pradesh
78	Arize 6444 Gold (HRI-174) (IET-22379)	2015	130-135	NA	Bayer Crop Science, Hyderabad	Assam, Chhattisgarh, Odisha, Uttar Pradesh, Bihar, Meghalaya, Karnataka and Tamil Nadu

Source: Directorate of Rice Development, Patna

Seeds and sowing: Seed rate is 12-15 kg/ha. Seed density in nursery is 25-30 g/m^2. Spacing of 15 × 15 cm or 25 cm × 10 cm is recommended. The optimum plant density is about 30-37 hills/m^2 for hybrid rice and 45-60 hills/m^2 for conventional rice. The number of seedlings per hill is one or two.

Fertilizer management: Hybrid rices are more tolerant of fertilisers than conventional rice. The dry matter production at different growth stages showed different patterns for hybrid rice and conventional rice. Hybrid rice has more dry matter accumulation in the early (transplanting to panicle initiation) and middle growth stages (panicle initiation to heading); conventional rice has more in late growth stages (heading to maturity). Thus, more fertilizer should be applied at early stages for hybrid rice. Further, the combination of organic and inorganic fertilizers in a ratio of 7:3 or 6:4 is favourable for hybrid rice. Hybrid rice absorbs more potassium and may need additional potassium fertilizer. The NPK ratio of 1-0.52-1.47 is required for hybrid rice. The highest yield is obtained with the application of 100 kg N/ha applied in two equal proportions at 10 and 40 DAT. Application of 150 kg N/ha as prilled urea produces high grain yield. The substitution of green manure nitrogen to the extent of 25 to 50 kg N/ha is also found to produce similar yield as that of 150 kg N/ha prilled urea.

Water management: Maintain a thin film of water for initial 30 days and later increase the water level to 4-5cm when the crop reaches maximum tillering stage. Drain out water for 4-5 days after maximum tillering stage so that the emergence of late tillers can be suppressed. Drain out water completely 10 days before harvest.

Harvesting and Threshing: Drain out water from the field when grains in the lowest portion of the panicle are on the dough stage (about 20 days from 50% flowering). Allow the grains to harden. Harvest 30 to 35 days after flowering when stalks still remain green to avoid grain shedding. The grain moisture content of paddy should be 20 to 24% at harvest. Thresh as early as possible preferably a day after harvest. Dry gradually under shade until the moisture content is brought down to 12%, which ensures better milling quality and storage. Under good management conditions the hybrid can out yield the best high yielding varieties by 1.0 to 1.5 t/ha.

SYSTEM OF RICE INTENSIFICATION

The system of rice intensification (SRI) is originated in Madagascar and has been first synthesized in 1983 by Father Henri de Laulanı´e, a French Jesuit priest. He has been trained as an agronomist at the National Agronomic Institute in Paris, France. He is popularly known as 'Father of SRI'. Father Henri de Laulanıe has transplanted very young rice seedlings of just 14 days old seedlings; using a fairly wide spacing (25 × 25 cm) of single seedlings in a square pattern to facilitate mechanised weeding. The rice is grown in moist soil with intermittent irrigation. Laulanı`e observed tremendous increases in tillering, rooting and subsequent number of grains. Seedlings should be transplanted only 1–2 cm deep in the mud, ensuring that the roots are laid in a horizontal position so that the root tips can easily resume their downward growth. Weeds need to be controlled regularly, starting about 10 days after transplanting. Mechanical rotary weeding, which ensures a churning action and thereby soil aeration, appears to be an important factor. Locally available sources of organic nutrients such as compost in particular are used instead of external inputs such as mineral fertilizers and other agricultural chemicals.

Key points in modified System of Rice Intensification:

(i) Perfect leveling with laser land levelers for good inputs management

(ii) Use of profuse tiller bearing rice varieties

(iii) Seed rate is 5 kg/ha. Seed treatment with *Pseudomonas fluorescens* @ 10 g/kg of seed

(iv) Nursery area requirement is 2.5 cents for mat nursery

(v) Young seedlings 10 to 14 days and certainly less than 15 days with a soil clump (along with seed sac) attached to their roots

(vi) Seedlings per hill @ 1 seedling/hill or 1-2 seedlings per hill

(vii) Wider spacing and square planting at 25 × 25 cm or even wider

(viii) Weed suppression using a mechanical cono / rotary weeder or motorized rotary weeders 3-4 times after transplanting with 10-15 days interval in between the hills in perpendicular directions to aerate the soil and control the weeds

(ix) Use of organic manure, vermicompost or FYM or mixtures of organic and inorganic fertilizers. Use Leaf Colour Chart (LCC) for need-based nitrogen fertilizer application and crop protection with biopesticides and biocontrol agents

(x) Alternate wetting and drying method of water management practice with a depth of 2.5 cm up to flowering with intermittent irrigation but mostly aerated condition rather than flooded condition and maintain 2-3 cm of standing water after panicle initiation.

Importance of system of rice intensification (SRI): The system of rice intensification is a set of principles and ideas that translate into a combination of agroeconomic practices, which might differ depending on agro-ecological and cropping system conditions. SRI can reduce water requirements, increase land productivity, and promote less reliance on chemical fertilizers, pesticides, herbicides, and other agrochemicals.

The SRI is considered as Climate-Smart Agriculture. SRI method is highly useful for sustaining rice productivity against climate change is attributable to two things *viz.*, i) improvements in crop root systems, with larger, better roots and less degeneration than under flooded, hypoxic conditions; and ii) soil biodiversity and biological activity. The beneficial effects that soil microorganisms can have when residing within rice plants as symbiotic endophytes. The soil health is maintained with high use of organic source of manures in SRI method. The seed requirement is less. There is water saving up to 40% and improved input use efficiency. The grain yield is 10 to 20% high in SRI over conventional method of planting.

The water requirement is reduced and there is high crop water use efficiency. This benefits the ecosystems and people in competition with agriculture for scarce water supplies. There is less use of inorganic fertilizer and less reliance on agrochemicals for crop protection which enhances the quality of both soil and water. SRI is tolerant to abiotic stresses such as drought, storm damage and extreme temperatures. Methane emission from rice fields are determined mainly by water regime and organic inputs. Flooded rice paddies are a major source of CH_4. The methane gas emission is reduced in SRI method. The flooding causes methane emission since organic inputs stimulate methane emissions as long as fields remain flooded. The mid-season drainage and intermittent irrigation can reduce methane emission by 40%. However, keeping soil nearly saturated conditions may promote N_2O release. About 15 to 20% of the benefit gained by decreasing methane emission is offset by the increase of N_2O emission. Soil organic carbon declines after a shift from flooded system to non-flooded system. SRI enhances the growth and health of roots and of soil biota by keeping soil moist but not flooded. The soil is mostly aerobic, not continuously saturated and thereby aerating the soil frequently and enhancing the soil organic matter content due to addition of high organic manures.

Package of System of Rice Intensification Practices

Soils: SRI methods worked best on the most fertile soils and not in saline or acidic soils.

Season: Dry season with assured irrigation is more suitable for SRI method than conventional practice of rice cultivation. It is difficult in areas with heavy downpour during crop establishment period.

Varieties: Hybrids and varieties with heavy tillering are suitable for SRI method of cultivation.

Profuse tillering rice varieties may be used in SRI instead of shy tillering varieties.

Mat nursery management: The seedlings can be produced using mat nursery technique. A mat nursery area of 2.5 cents or 100 m^2 is required for planting one hectare. The raised beds of 1 × 5 m^2 are made and a polythene sheet is spread over the shallow raised bed to prevent roots growing deep into soil. The wooden frames of 0.5 m long, 1 m wide and 4 cm deep divided into 4 equal segments are placed over the polythene sheet and filled to a height of 4 cm with a mixture of local soil (70–80%), decomposed farmyard manure (15-20%), rice husk (5-10%) and powdered di-ammonium phosphate 1.5 kg or 2 kg 17-17-17 NPK fertilizer per 100 m^2. Four m^3 of soil mix is needed for each 100 m^2 of nursery. *Azospirillum* @ 2 kg, *Pseudomonas fluorescens* @ 750 g and mycorrhizal fungi @ 5 kg for 100 m^2 can be applied in the nursery area. Seed rate is 5 kg/ha is needed for planting one hectare of land under SRI. Seeds are treated with Carbendazim @ 2 g/kg of seed. Soak the seeds for 24 h, drain and incubate the soaked seeds for 24 hours. Sprouted seeds and radical (seed root) grows to 2-3 mm long are spread uniformly on the soil. Sow the pre-germinated seeds weighing 90 to 100 g/m^2 (100 g dry seed may weigh 130g after sprouting) uniformly. Cover them with dry soil to a thickness of 5 mm. Water is sprinkled immediately using rose to soak the bed and then later as and when needed (twice or thrice a day) to keep it moist all the time. Protect the nursery from heavy rains for the first 5 days after sowing (DAS) and continue watering until 14 DAS. If seedling growth is slow or leaves are yellow, spray 0.5% urea + 0.5% zinc sulfate solution at 8-10 DAS. Seedlings reach 18-20 cm height at 12-14 DAS depending on the local weather conditions, pest and diseases. Protect the nursery from heavy rains for the first 5 DAS. At 6 DAS, maintain thin film of water all around the seedling mats. Water should be drained 2 days before lifting the seedling-mats. Then remove the wooden frame. Seedling-mats that are approximately 12 to 14 days old should be lifted from the mat nursery and taken to main field for transplanting. Care should be taken to prevent any harm to seedlings while pulling them from nursery or at the time of transplanting. A metal sheet is inserted 4-5 inches below the seedbed and seedlings scooped along with soil without any disturbance to their roots. Lift the seedling mats and transport them to main field.

Advantages of mat nursery

(i) Reduced nursery area and easy to manage: 100 m^2 to plant 1 ha

(ii) Robust young seedlings (18-20 cm tall with 4 leaves) produced within 12-14 days after sowing.

(iii) Easy transportation of seedling-mats to main field and convenient for machine transplantation

(iv) Easy separation of seedlings for transplanting with minimum root damage.

Main field preparation: Puddled lowland prepared as described in transplanted section. Perfect leveling is the pre-requisite for proper water management and good crop stand.

Transplanting techniques: Transplant 12 to 14 days old single rice seedling (3-4 leaf stage to preserve potential for tillering and rooting ability) within 30 minutes of pulling out of seedlings in square planting with spacing of 25 x 25 cm at a shallow depth of 2 to 3 cm. The tiller buds formed at the basal node are not suppressed in case of shallow plantings. Therefore, the seedlings should be transplanted at 2 to 3 cm depth. Shallow planting ensures quick establishment and more tillers. The deeper planting delays and inhibits tillering. The seedlings with 3-4 leaves stage have great potential for profuse tillering and root development. Fill up the gaps between 7th and 10th DAT. Young seedlings are unable to withstand heavy downpour and local flooding at transplanting of rice seedlings. When mature, however, rice plants under SRI management are considered to be more resilient to storms and cyclones. Gap filling may be required to compensate for up to ten per cent of seedlings lost in the early stages of growth.

Weed management: The manual operated cono weeder / rotary weeder or power operated two row rotary weeder is used with forward and backward motion on either direction of the rows and column three to four times during the growing period at 7 to 10 days interval from 10-12 days after transplanting rice to bury the weeds, to churns up the soil with small toothed wheels and as well to aerate the soil. It saves labour for weeding, aerates the soil and root zone, prolongs the root activity, and improves the grain filling through efficient translocation and ultimately the grain yield. There is a substantial beneficial effect from using the rotary weeder and that this relates not to soil aeration but to pruning the lateral roots of the rice plants. Pruning the lateral roots encourages the rice plants to develop a deeper root system that can access water and some nutrients from lower soil layers. The manual weeding is also essential to remove the weeds closer to rice root zone. Normally, herbicide application is not recommended for SRI. Under special circumstances, the herbicides such as Butachlor @ 1.25 kg/ha or Anilophos @ 0.4 kg/ha as pre-emergence application or pre-emergence application of herbicide mixture viz., Butachlor 0.6 kg + 2,4 DEE 0.75 kg/ha, or Anilophos + 2, 4 DEE 'ready-mix' at 0.4 kg/ha followed by one hand weeding on 30 to 35 DAT will have a broad spectrum of weed control. Any herbicide has to be mixed with 50 kg of dry sand on the day of application (3-4 DAT) and applied uniformly to the field with thin film water on the 3rd DAT. Water should not be drained for next 2 days from the field (or) fresh irrigation should not be given. If pre-emergence herbicide application is not done, hand weeding has to be done on 15th DAT. 2,4-D sodium salt (Fernoxone 80% WP) @ 1.25 kg/ha dissolved in 625 litres with a high volume sprayer, three weeks after transplanting or when the weeds are in 3 to 4 leaf stage can be sprayed to control emerged weeds.

Manures and fertilizer application: The manures and fertilizer application for rice under SRI is done as per transplanted rice. The application of organic compost or animal manure is a highly desirable, conditioned by the availability of local resources. Green manure and farm yard manure application will promote positive microbial activity in the rhizosphere and enhance the growth and yield of rice in this system approach. Under sodic soils, during rotary weeding, apply Azophosmet @ 2.2 kg/ha and Pink Pigmented Facultative Methylotrophs (PPFM) as foliar spray @ 500 ml/ha. Top dressing of nitrogenous fertilizers is done based on the colour of the leaf using leaf colour chart (LCC) in SRI method.

Water management: Irrigation is given only to moist the soil in the early period of 10 days. The irrigation is given to a depth of 2.5 cm after development of hairline cracks in the soil until panicle initiation and to a depth to 5.0 cm after panicle initiation one day after disappearance of ponded water. The field is to be drained 15 days before harvest. The water requirement is 1100 mm in conventional method while 700 mm in SRI method through intermittent irrigation. There is 30% saving in irrigation water of 400 mm.

Cropping systems: Dual cropping of rice-azolla reduces the weed infestation. Summer ploughing and cultivation of irrigated dry crops during post-rainy periods reduces the weed infestation.

Plant protection measures: Application of *Pseudomonas fluorescens* @ 5 kg/ha is effective in controlling rice blast disease. The incidence of sheath blight, leaf folder, brown plant hopper and the rat damage is less in SRI method.

Yield: The rice grain yield is 7 t/ha in the wet season and 5.6 t/ha in the dry season. The comparison of conventional and SRI practices of rice cultivation is presented in Table 31.

Table 31 Comparison of conventional and SRI practices of rice cultivation

Particulars	Conventional method	SRI method
Nursery area per ha	20 cents	2.5 cents
Bed size	1.5 m width and length as per field	1 x 5 m^2
Number of Beds	According to length	8 Beds
Seed rate	24-30 kg/ha	5 kg
Seed rate/m^2	75 to 90 g	75 g (375 g / bed)
DAP	2 kg/cent	760 g (95 g /bed)
Urea	-----	400 g
Seedling age (days)	20-30 days old, or even 40 days old seedling	12 to 14 days old seedling and certainly less than 15 days
Number of seedlings per hill	Transplant (2-4) seedlings in clumps per hill	Transplant seedlings singly, one per hill
Plant spacing, cm	15 x 10 and 20 x 10 cm (varieties); 20 x 10 cm (hybrid) or and random planting	25 cm x 25 cm
Planting geometry	Rectangular	Square
Mechanical weeding / soil stirring	Herbicide + one HW or Two HW	Two cono / rotary weeder + one HW (no herbicide)
Nutrient application	Use chemical fertilizers to enhance soil nutrients. Blanket N, P & K or Soil test based recommendation.	Apply as much organic matter to the soil as possible; Recommended dose of P & K as per soil test, use of LCC for N application
Water management	Maintain paddy soil continuously flooded, with standing water throughout the growth cycle	Keep paddy soil moist, but not continuously saturated, so that mostly aerobic soil conditions prevail

BASMATI RICE

Importance: Basmati rice is known for its fragrance and unique flavour. It is popularly known as 'Queen of Rice' and 'Pearl of Rice'. All aromatic rice is not basmati rice. India is the largest cultivator, consumer and exporter of basmati rice. India enjoys monopoly over basmati in the world markets. Basmati rice is exported to more than 80 countries mainly to Saudi Arabia, United Arab Emirates, Iran, Kuwait and others Gulf and European Countries United Kingdom, Sweden, Netherlands, Belgium, Italy. Half the quantity of basmati exported from the country is *sela* (parboiled) basmati mainly to the gulf countries.

Features of Basmati rice: Basmati has originated from Vasumati which means earth recognized by its fragrance; while the full exposition of the word is from Hindi *'Bas'* originating from *Prakrit Vas* which has a Sanskrit root-*Vasay* connoting aroma; and *mati* from may up meaning in grained from the origin. Common usage has changed *Vas* to *Bas* while joining *Bas* and *mayup* the latter changed to *mati*. An alkaloid 'pandamarilactione' is the cause of fragrance. This alkaloid is present in the leaves of 'Pandanus' also. Basmati rice is characterized by extra-long, superfine slender grains having a length to breadth ratio of more than 3.5, sweet taste, soft texture, delicate curvature and an extra elongation with least breadth-wise swelling on cooking. Basmati rice gives pleasant flavour after cooking. Basmati rice are poor yielders, prone to lodging and to the onslaughts of pests. In India, Basmati rice is characterized by extra-long, superfine slender grains having a length to breadth ratio of more than 3.5, sweet taste, soft texture, delicate curvature and an extra elongation with least breadth-wise swelling on cooking. These superfine best quality of Basmati rice are most preferred specially for Biryani and Pulao preparation on special occasion and also meant for high premium value in the national and international market.

Origin: Basmati rice originated in India.

Distribution: Basmati rice is grown in Punjab, Haryana, Rajasthan, Jammu and Kashmir, Himachal Pradesh, Uttaranchal and Uttar Pradesh.

Climate: The rainfall requirement is 700 mm to 1100 mm. The mean temperature ranges from 16.4 to 32.1°C. Relative humidity ranges from 65 to 92%.

Soil: Basmati type rice s is mostly cultivated in alluvial and *tarai* soils. The soils are neutral to slightly alkaline and calcareous. Soils are loam to clay loam.

Basmati rice varieties: Basmati 370 (Punjab Basmati), Type-3 (Dehraduni Basmati rice), Taraori Basmati (HBC-19 or Karnal Local), Basmati 217, Ranbir Basmati, Basmati 370, Basmati 385, Basmati 386, Punjab Basmati –1, Punjab Basmati 2, Punjab Basmati 3, Kasturi (IET 8580), Haryana Basmati– 1 (IET – 10367), Mugad Sugandha (IET 13549), Mahi Sugandha (IET –12601), Pusa Basmati– 1121 (IET – 18004), Improved Pusa Basmati – 1 (IET – 18990), Vallabh Basmati-22 (IET 19492), Pusa Basmati –1 (IET 10364), CSR 30, Vasumati 1 (IET 15391), IET 15392, IET 13846, IET 13548, IET 13549, IET 14131, IET 14132, Yamini (IET 14720), IET 15833, Pusa Punjab Basmati 1509, Pusa Basmati – 6 (IET –18005), Pusa Basmati 1121. Agricultural and Processed Food Products Export Development Authority (APEDA) has identified varieties such as kalanamak, tilakchandan and jeerabati (Uttar Pradesh), kala jeera (Odisha), katrani (Bihar), ambemohar (Maharashtra), govindbhog and badshahbhog (West Bengal), dubraj, badshahbhog and jawaphool (Chhattisgarh) and kala joha (Assam), which could be harnessed and developed for their export potential. The growth, yield and quality characteristics of Basmati rice varieties are furnished in Table 32 and Table 33.

Table 32 Characteristics of export quality traditional basmati varieties

Character	Karnal Local	Pakistani Basmati	Basmati 370	Type 3
Grain Yield (t/ha)	2.13	2.12	2.23	2.19
Plant height (cm)	178	180	165	164
Duration (days)	155	155	145	145
Tillers /m^2	285	286	277	278
Grains /Panicle	138	139	140	142
Lodging score	9	9	9	9
Length of rice (mm)	7.25	7.27	6.84	6.83
Grain length of Cooked rice (mm)	16.24	16.23	13.91	13.06
Amylose content (%)	23.46	23.52	23.48	23.39
Aroma	Strong	Strong	Strong	Strong

Table 33 Growth, yield and quality characteristics of dwarf scented varieties

Character	Pusa Basmati 1	Kasturi	Haryana Basmati	Ranbir Basmati
Plant height (cm)	87.0	98.0	116.0	115.0
Total duration (days)	135	120	140	120
Yield (kg/ha)	4660	4240	4075	3000
Milling (%)	70.0	70.0	72.0	70.0
Head Rice (%)	55.0	61.0	60.0	54.0
Kernel length (raw) (mm)	7.20	6.94	6.65	7.13
L / B ratio	3.99	3.85	3.91	377
Kernel length (cooked) (mm)	13.9	12.4	12.4	11.9
Elongation ratio	1.93	1.79	1.86	1.67
Aroma	Present	Present	Present	Present

Methods of crop establishment: The crop can be established either by direct sowing or by transplanting.

Direct seeded rice

Season: Direct sowing of can be done during second fortnight of June.

Seeds and sowing: Direct seeding is done only in medium to heavy textured soils in unpuddled condition. The seed rate is 20 to 25 kg seed per ha. Seeds are sown with tractor drill at 20 cm row spacing.

Weed management: For controlling weeds, apply Stomp 30 EC @ 1.0 litre/acre within three days after sowing. If needed, apply Bbispyribac (Nominee Gold 10 SC) @ 100 ml per acre or Segment 50 DF (Azimsulfuron) @ 16 g per acre as post emergence herbicide at 30 to 35 days after sowing. Spray these herbicides uniformly by mixing them in 150 to 200 litres of water per acre using flat fan or flood jet nozzle for spray. Use Bbispyribac (Nominee Gold 10 SC) when the crop is infested with *swank* and paddy *mothas* are present in the field.

Manures and fertilizer application: Apply 60 kg N per ha in 3 equal splits at 3, 6 and 9 weeks after sowing. Apply P and K only when the soil test shows deficiency of these nutrients.

Water management: To fulfill the water need of the crop, apply irrigation at 5 to 7 days interval depending upon the soil type. The interval may be adjusted with rainfall. Stop irrigation 10 days before harvesting.

Transplanted rice

Season: Basmati crop is generally a Kharif (May to November) crop. The optimum time for nursery sowing is in June. The time of transplanting is a crucial factor in determining the yield and quality of the *Basmati*/aromatic varieties. Some varieties are photoperiod sensitive. These varieties flower when a specific day length is reached. Punjab Mehak 1, Punjab Basmati 2, Punjab Basmati 3 and Pusa Basmati 1121 may be transplanted in first fortnight of July. Optimum time of transplanting for Pusa Punjab Basmati 1509, Basmati 386 and Basmati 370 is second fortnight of July.

Nursery management: A nursery area of 1000 sq m is sufficient to transplant one hectare. Plough nursery field twice under dry condtion. Apply wherever possible decomposed and powdered farm yard manure @ 500 kg/1000 sq.m. Puddle two to three times subsequently by ploughing in standing water of 2-3 cm, preferably at an intervals of five days. Level the field perfectly after final puddling and prepare seedbeds of 1.0 to 1.5 m width and of convenient length leaving 30 cm wide channels in between the beds. Fertilize the seedbeds with 5 to 10 kg N, 5 kg of P_2O_5, and 5 kg of K_2O for every 1000 sq m. area of the nursery before the final levelling. In zinc deficient areas, apply zinc sulphate @ 3-4 kg/100 sq.m. In calcareous soils, the dry nursery iron chlorosis (seedling yellowing) can be prevented either by spraying 2% ferrous sulphate solution 2 to 3 times at weekly intervals or by frequent inundations. Application of $ZnSO_4$ @ 20-25 kg /ha is essential for CSR 30 if grown under moderate sodic soil. Use 30 kg of dry seeds per 1000 sq m. Dip the seed in Carbendazim (Bavistin 50 WP) @ 0.05% (5 g) + Streptocycline @ 0.01% (one gram dissolved in 10 litre of water) for 12 hours and smear the seeds with talc formulation of *T. harzianum* @ 15 g/ kg of seed immediately before sowing. Soak the seeds in gunny bags for 24 hours in water. Subsequently, incubate for 48 hours under sheaves of straw with occasional sprinkling of water. Broadcast the germinated seeds uniformly over the seed bed and keep the beds moist for 4 to 5 days. Gradually raise and maintain water to a depth of 2-5 cm.

Seed and nursery treatment: (i) Seed treatment with streptocycline at 0.3 g + Carbendazim at 1.0 g/kg seed (ii) *Trichoderma harzianum* + *Pseudomonas fluorescens* mixture- seed treatment (at 10 g/ Kg) + seedling dip (at 10 g/L) + soil application (at 2.5 kg/ha each) (iii) Spray of 3% neem oil in nursery after 15 DAS.

Age of Seedlings: Seedlings of Basmati/aromatic rice varieties are ready for transplanting when they attain 5 to 6 leaf stage or are 20 to 25 days old after sowing for good tillering. The seedling root dip in Carbendazim (Bavistin 50 WP) @ 0.2% or *Trichoderma harzianum* @ 15 g/litre of water for 6 hours before transplanting.

Planting methods: Transplant two to three seedlings per hill in lines at 20 x 10 cm for high yielding dwarf vareities and 20 x 20 cm for tall statured traditional basmati rice at a shallow depth of 2-3 cm in a well puddled field. In the late transplanted crop the spacing may be reduced to 15 cm x 15 cm (44 hills/sq. metre) to overcome the reduction in grain yield. Increase the number of seedlings from 2-3 to 5-6 per hill in case of delayed planting.

Weed management: Hand weed twice depending on the level of weed infestation at intervals of three weeks starting from 20 days after transplanting. Any one of the following herbicides *viz.*, Anilophos (0.6 kg a.i. ha^{-1}) in combination with 2,4 D EE (0.53 kg a.i. ha^{-1}), Butachlor (1.5 kg a.i. ha^{-1}) in combination with 2,4 D EE (0.4 kg a.i. ha^{-1}), Pertilachlor (0.4 kg a.i. ha^{-1}), Pertilachlor (0.4 kg a.i. ha^{-1}) in combination with 2,4 D EE (0.4 kg a.i. ha^{-1}) is recommended to control weeds.

Manures and fertilizer application: For traditional tall basmati cultivars, apply basally 30 kg P_2O_5, 30 kg of K_2O and 25 kg /ha of $ZnSO_4$. Apply nitrogen @ 60 kg N/ha in three splits 50% as basal; 25% at tillering and 25% at panicle initiation stages. For high yielding dwarf basmati varieties, apply basally 50 kg P_2O_5, 40 kg of K_2O and 25 kg /ha of $ZnSO_4$. Apply nitrogen @ 90 kg N/ha in three splits 50% as basal; 25% at maximum tillering and 25% at panicle initiation stages. The field should be drained prior to topdressing with nitrogen and irrigated after 24 to 36 hours. If green manure crop has been grown and incorporated in the main field before rice planting, the quantity of nitrogen may be reduced by 25%.

Water management: Maintain 2-5 cm water throughout the growing season with the mid- season drainage at tillering stage. The crop should not suffer any water stress particularly during flowering. Stop impounding water about a fortnight before harvesting to facilitate easy harvesting.

Cropping systems: The common Basmati rice based cropping systems are *viz.*, Basmati Rice-Wheat/Sunflower, Basmati Rice-Wheat-*Sathi,* maize/Summer *Moong,* Basmati Rice-Mentha, Basmati Rice-Berseem (seed), Basmati Rice-Celery-Bajra (Fodder).

Pest management: The major insect pests occur in basmati rice are stem borers, leaf folder, rice hispa and diseases are blast and foot rot. The insect pests can be controlled by spraying cartap hydrochloride (Mortar 75 SG) @ 170 g or Chlorantraniliprole (Coragen 20 SC) @ 60 ml or monocrotophos (Monocil 36 SL) @ 560 ml/acre or Quinolphos or Triazophos at 0.2%; spray carbendazim+ mancozeb at 0.2% for sheath blight; Spray of streptocycline 15g + copper oxychloride 500g/ha for bacterial leaf blight. Release of egg parasitoid *Trichogramma japonicum* at 20 trichocards/ha twice (First after the appearance of adult stem borer/dead hearts in the field followed by second after 10 days interval). Spray of *Pseudomonas fluorescens* thrice at 0.2% concentration commencing from 45 days after transplanting at 10 days interval to manage sheath blight, blast and sheath rot diseases. Three foliar sprays of neem oil at 3% to manage diseases and insects. Fix bird perches @ 40 to 50 nos. /ha.

Harvest: Basmati rice matures in about 140 days after seeding. Basmati/Aromatic rice varieties should be harvested as soon as they mature *i.e.* when the ears are nearly ripe and the straw has turned yellow. Delayed harvesting may cause over-ripening and shattering of grains. The harvested crop should preferably be threshed on the same or next day of harvest. The delayed threshing causes high shattering losses, reduced head rice recovery and ultimately reduces the market price of paddy.

Yield: The average paddy yield is 3.0 to 4.0 t/ha.

Quality: The minimum acceptable quality limits in Basmati rice is furnished in Table 34

Table 34 Minimum acceptable quality limits in Basmati rice

Component feathers	Minimum acceptable limit
Aroma	Strong aroma at the time of cooking and in the cooked rice
Appearance and texture of cooked rice	Flaky, absolutely no hard core after cooking
Kernel dimensions:	
Length (L)	6.65 mm
Breadth (B)	1.7-2.0 mm
L / B ratio	3.5 mm
Milling recovery	40 %
Amylose content	20-23 %
Gel consistency	Soft of medium
Kernel length after cooking	12 mm
Elongation ratio	1.8

Source: Proceedings of brain–storming on "India Aromatic Rice: Present status and future needs " held at IRRI Liaison office, New Delhi, 29-30 March, 1997.

RICE CULTIVATION IN SALT AFFECTED SOILS

Rice is categorized as salt-sensitive crop but has enough variability for tolerance. Rice is the only crop which can be grown on extremely diverse conditions.

Rice is a preferred crop in salt-affected environments because of the following reasons *viz.*,

(i) Coastal saline areas are affected with sea water inundation, excessive rains and river outflow during wet or rainy season. Under such situation, a crop tolerant to water logged condition is to be chosen which must tolerate salt-stress and short spell of submergence

(ii) Sodic areas (inland) are facing water logging due to poor hydraulic conductivity, high ESP and usually high pH. The crop selected for cultivation has to tolerate water logged condition, high pH and salt-stress.

(iii) Saline areas (inland or coastal) usually have high soil salinity, less water availability for irrigation and sometimes available with poor quality underground water. The crops to be grown tolerate high soil salinity and preferably less water requiring crop if ample amount of fresh water is not available. If water is available through canal or other irrigation source, other crops including rice can be grown.

Rice salt stress tolerance can be managed with use of low rates of inorganic fertilizers and high organic manures to have stronger seedlings and higher grain yields; older seedlings at transplanting are more tolerant to salt stress, seedling handling with reduced root injury since it decreases passive salt uptake after transplanting; high transplanting density/number of seedlings per hill to compensate for seedling mortality or older seedlings.The reclamation of salt affected soils is a prerequisite in taking up profitable rice cultivation.

All saline soils can be reclaimed easily, if good quality water is available. Since the salts in these soils are soluble in nature, using quality water they can be solubilized and leached off from the field. In the absence of good quality water, it becomes necessary to manage saline soils for better growth of plants. Providing drainage in water logged areas also helps to reduce salt accumulation. In non-calcareous sodic soils, using gypsum as an amendment is by far the cheapest way. The calcium in gypsum ($CaSO_45H_2O$) replaces the sodium in the soil exchange complex and providing adequate surface drainage takes care of the removal of the harmful sodium from the site. It is sufficient to add 50 % of the calculated quantity of gypsum to derive the full benefit provided green leaves such as that of vedanarayana leaves (*Delonix regia*), neem leaves, casuariana needles, are used. The fields are to be provided with strong bunds and adequate surface drainage facilities. Field is to be puddled shallowly with good quality irrigation water preferably from tank or river. The suggested quantity of gypsum is to be evenly spread and a superficial ploughing is to be given with pounding of water of 10 to 12 cm. The water is to be drained after 4 to 5 days. Ponding of water and drainage can be repeated without allowing the soil to dry. Green leaf manure at the rate of 5 t/ha is to be incorporated in the soil. After 5 to 6 days, the field is to be leveled as for the ordinary rice cultivation. Since salt affected soils particularly sodic soil is deficient in nitrogen, 25 % additional nitrogen is applied in addition to the recommended dose. Just before planting 15 to 20 kg per ha of zinc sulphate must be applied as rice in sodic soils is prone to zinc deficiency. The age of seedlings for short and medium duration rice is 25-30 and 35-40 days respectively. Number of seedlings to be planted at the rate of 4 to 5 seedlings per hill. On the third day of planting, herbicide application effectively controls the weeds. Other cultural practices are similar to the rice grown in normal soil.

Salt tolerant rice varieties: The salt tolerant rice varieties released for commercial cultivation in India are CSR10, CSR13, CSR22, CSR23, CSR27, CSR30, CSR36; Lunishree, Vytilla 1, Vytilla 2, Vytilla 3, Vytilla 4, Vyttila 5; Vyttila 6; TRY 1; TRY (R) 2; TRY (R) 3; Panvel 1, Panvel 2, Panvel 3; Sumati, Jarava; Bhutnath; Usardhan 1, Usardhan 2, Usardhan 3; CSRC(s)7-1-4. The particulars of salt tolerant rice varieties are furnished in

Table 35 to Table 37.

Table 35 Particulars of salt tolerant rice varieties

Parameter / Variety	CSR 10	CSR 13	CSR 23
Parentage	M40-431- 24-114/ Jaya	CSR1 / Bas.370 / / CSR5	IR64 // IR4630-22-2-5-1-3/ IR 964- 45-2-2
Plant height (cm)	85	105	115
Maturity duration (days)	120	145	130
Grain type	Short Bold	Long Slender	Long Slender
Salinity tolerance (dS m-1)	11	9	10
Sodicity tolerance (pH)	10	10	10
Grain yield in Normal soils (Qtl/ha)	60	60	65
Grain Yield in Salt affected soils (Qtl/ha)	30	35	40
Recommended States	Haryana	Uttar Pradesh, Haryana, Gujarat, Maharastra	Uttar Pradesh, Haryana, Gujarat, Maharastra, Kerala, Tamil Nadu, West Bengal

Table 36 Particulars of salt tolerant rice varieties

Parameter /Variety	CSR 27	CSR 30	CSR 36	CSR 43
Parentage	NONA BOKRA / IR565-33-2	BR4-10 / Basmati 370	CSR13/ Panvel 2// IR36	KDML105/IR4630-22-2-5-1-3/ IR 20925-33-3-1-28 KDML105/IR4630-22-2-5-1-3/ IR 20925-33-3-1-28
Plant height (cm)	115	155	110	95
Maturity duration (days)	120	155	135	110
Grain type	Long Slender	Basmati Type	Long Slender	Short Bold
Salinity tolerance (dS m-1)	10	7	11	7.0
Sodicity tolerance (pH)	9.9	9.5	9.8	10.0
Grain yield in Normal soils (Qtl/ha)	65	30	65	60
Grain Yield in Salt affected soils (Qtl/ha)	40	20	40	35
Recommended States / Areas	Haryana and Uttar Pradesh,	Haryana, Punjab, Uttar Pradesh	Haryana, Uttar Pradesh Pondicherry	Uttar Pradesh

Table 37 Particulars of salt tolerant rice varieties in Tamil Nadu

Parameter / Variety	TRY 1	TRY (R) 2	TRY(R) 3
Parentage	IR578-172-2-2/BR-1-2-B-1	IET 6238 / IR 36	ADT 43 / Jeeraga Samba
Maturity duration (days)	135-140	115	135
Grain type	Medium	long slender grain	Medium bold grain
1000 grain weight (g)	24.0	22.8	23.0
Grain Yield in Salt affected soils (t/ha)	5.2	4-5	5.8
Special features	white rice, suitable for making flaked rice (*Aval*) and puffed rice (*pori*)	white rice, tolerates drought condition	white rice, resistant to leaf folder, stem borer, brown plant hopper, blast, brown spot, sheath rot and sheath blight

Rice Ratooning Management Practices

Ratoon cropping is defined as cultivation of the regrowth of rice crop from the stubbles after the harvest of crop or managed for next crop cultivation from the regrowth of the stubbles. The word ratoon probably originated from the Latin word 'retofisus', which means to cut down or mow. Ratoon as a basal sucker for propagation, such as in banana, sugarcane, and pineapple. Rice ratooning depends on the ability of dormant buds on the stubble of the first crop to remain viable. Root vigour and distribution also affect ratooning. *Rice ratooning* is one practical way to increase rice production per unit area and per unit time. However, ratoon rice matures one month earlier than plant crop. Rice ratooning is practiced in India, Japan, USA, Philippines, Brazil, Colombia, Thailand and Taiwan. In India, research on rice ratooning is practiced in Karnataka, Andhra Pradesh, Assam, Bihar, Kerala, Odisha, Tamil Nadu, Uttar Pradesh and West Bengal.

Advantages of ratooning in rice: Crop duration in ratoon rice is reduced compared to the main plant crop. Ratoon crop takes about 30 days less for maturity compared to the plant crop. There is reduction in cost of cultivation. Input requirement is less in ratoon rice compared to main planted rice crop. There is no seed cost since the ratoon crop is produced from the crop stubbles. Panicle production is 70% of the ratoon rice tillers.

Disadvantages of ratoon rice: Grain yield per unit area is very poor in ratoon rice as compared to the main plant crop. Rapid leaf senescence of main plant crop causes low ratoon rice yields. Pest and disease occurrence and their spread are common both in main and ratoon rice crop. Ratooning rice does provide juicy green leaves on which pests can continue to feed and multiply. Ratoon crop promotes the outbreak of pests in the following season. The beneficial insects can not survive under rice ratooning practices.

Management of rice ratoon crop: Total non-structural carbohydrates (TNC) in the stems of the main crop form a storehouse of energy for the second crop. There is a strong correlation between TNC in stems at main crop harvest and ratoon crop yields. Good sunlight and cooler night temperatures contribute significantly to high TNC levels at harvest. Ratoon tillers develop from basal auxiliary buds that exist on the stubble of the main crop plant.

The growth and vigour of ratoon tillers depend on carbohydrate reserves of the stubbles and the root system after harvest of the main crop. Plant with thick stems / culms store more carbohydrate than those with thin culms. Harvest time, cutting height, fertilizer application, irrigation management, plant protection and weed control for the main crop naturally have a bearing on growth and yield of the ratoon crop.

Rice varieties: Rice varieties suitable for ratooning in rice are Bhavani, Co 37, ACM 8, ACM 10, ADT 36, ADT 16, PKM 1, IR8, IR42, IET 9668, IR 20, CR1009 and Ponni. Bhavani is more suitable for ratooning which has recorded grain yield of 4 t/ha. It yields around 58% of main crop yield.

Hybrids offer outstanding ratoon crop performance: One of the benefits of growing hybrid rice is the ability to achieve outstanding ratoon crop grain yields without sacrificing quality. Grain yields have been recorded as high as 6400 kg/ha in one commercial field, with an average range of 3200 to 5600 kg/ha. Some tips for getting the high yield in ratoon rice are i) harvest main crop on dry ground to avoid combine rutting, ii) apply ratoon crop nitrogen to dry ground and flood up immediately after main crop harvest and Fail mowing of the main crop stubble has increased ratoon crop yields substantially in some fields.

Stubble management: The optimum height of stubbles should be 20 cm height at the time of crop harvest for ratooning rice. Row planting of seedling is must for ratooning rice because the stubbles are uniform in distance after the harvest. There is an advantage to maintain the stubble height at 20 cm during harvest of the crop when the ratoon potential of rice variety is high. When the grain yield potential of rice variety is low, it doesn't make any difference whether stubble height is high or low. Cutting low takes longer to harvest because you are putting all that straw through the combine. Mowing after harvesting is an alternative to cutting low with a combine.

Spacing: Optimum spacing for ratoon rice is 20 cm x 20 cm or 25 cm x 25 cm.

Nutrient management: Application of fertilizer immediately after the harvest is must. Complete basal application of fertilizers leads to higher grain yield compared with split application of fertilizers. Ratoon crops yield is high at nitrogen level of 75% of the main crop. P and K do not influence ratoon rice crop grain yield. 120 kg N per ha in split doses produce maximum grain yield.

Water management: Field should not be flooded until new ratoon tillers are 10-15 cm height. Irrigation to 5 cm depth and one day after disappearance can be practiced in ratoon rice crop.

Growth regulator: Application of growth regulators such as IAA, NAA, GA3 and 2,4-D enhance the growth and yield of the ratoon rice crop.

Pest management: Ratoon rice favours multiplication of insects such as borer, gall midge, leafhopper and diseases such as grassy stunt virus and yellow dival virus during off season. It is advisable to leave some decomposing organic material in the field or around the fields to provide food and refuge for natural defenders.

Harvest: The best time of harvest of main crop for raising a good ratoon crop is when its colour are still green. Stalks should be cut before the main crop is fully matured. Harvest is done at 18-20% grain moisture content. Cutting height at 15 to 20 cm is considered on optimum cutting height above ground for ratoon rice. Main crop cutting height should leave 1.5 cm of the stem above the water, submerged stubbles may rot, longer stubble may produce weak tillers.

Yield: In general, ratoon rice produce lower grain yield.

Economics of ratoon rice: There is no expenditure for preparatory cultivation, seedling cost and less irrigation charges. There is a saving of 30 to 45 % in cost of cultivation. Ratoon rice requires less water, input and cost.

REFERENCES

Ahuja, S.C., Panwar, D.V.S., Uma Ahuja and K.R. Gupta. 1995. Basmati rice - the scented pearl, CCS Haryana Agricultural University, Hisar, Haryana, P 63.

Anand Prakash, Bentur, J. S., Srinivas Prasad, M. Tanwar, R. K. Sharma, O. P. Someshwar Bhagat, Mukesh Sehgal, Singh, S. P. Monika Singh, Chattopadhyay, C. Sushil, S. N. Sinha, A. K., Ram Asre, Kapoor, K. S. Satyagopal, K. and P. Jeyakumar. 2014. Integrated Pest Management for Rice. National Centre for Integrated Pest Management, LBS Building, IARI Campus, New Delhi. p. 43

Anon. 2002a. Marketable Surplus and Post Harvest Losses of Paddy in India-2002, Directorate of Marketing & Inspection, Nagpur.

Anon. 2002b. Agmark Grading Statistics, 2001-02, Directorate Marketing and Inspection, Faridabad.

Barah, B., 2009. Economic and ecological benefits of System of Rice Intensification (SRI) in Tamil Nadu. Agricultural Economics Research Review 22, 209–214.

Berkhout, E. and D. Glover. 2011. The Evolution of the System of Rice Intensification as a Socio-technical Phenomenon: A report to the Bill & Melinda Gates Foundation, Wageningen, NL: Wageningen University and Research Centre.

Catling, D. 1992. Rice in deep water. The MacMillan Press Ltd., London, United Kingdom. 542 pp.

Chang T. T. 1976. The origin, evolution, cultivation, dissemination and diversification of Asian and African rices. Euphytica, 25: 435-444.

CPG. 2012. Crop Production Guide. Department of Agriculture, Government of Tamil Nadu, Chennai and Tamil Nadu Agricultural University, Coimbatore.

Diwakar. M.C. 2014. Status Paper on Rice. Directorate of Rice Development. Govt. of India, 250A, Patliputra Colony, Patna, Bihar.

Dobermann, A., 2004. A critical assessment of the system of rice intensification (SRI). Agricultural Systems 79, 261–281.

Grist, D.H. 1986. Rice. Longman, London, United Kingdom. 599 p.

IARI. 1980. High yielding basmati rice – problems, progress and prospects. Indian Agricultural Research Institute. Research Bull 30: 47

Johnson, G.I., Le Van To, Nguyen Duy Duc and M.C. Webb. 2000. Quality assurance in agricultural produce. ACIAR Proceedings 100. Australia.

Jones J. W. 1926. Hybrid vigour in rice. J. Am. Soc. Agron., 18: 423-428.

Khush, G.S., 1997. Origin, dispersal, cultivation and variation of rice. Plant Molecular Biology 35: 25–34.

Krishnaiah K. and N. Shobharani. 1997. Rice research for attaining self-sufficiency. Rice India, 7: 29-32.

McDonald, A., Hobbs, P. and S. Riha. 2006. Does the system of rice intensification out-perform conventional best management? A synopsis of the empirical record. Field Crops Research. 96, 31–36.

Muhammad Ashfaq, Muhammad Saleem Haider, Iqra Saleem, Muhammad Ali, Amna Ali and Sumaira Aslam Chohan. 2015. Basmati–Rice a Class Apart (A review). J. Rice Res. 3(4): 1-8

Muniyappa V., Nagaraju, M. Mahadevappa and K.T. Rangaswamy. 1988. Rice-ratooning. Proceedings of a workshop held at Bangalore, India, 21-25 April 1986, pp. 219-225.

Nene, Y. L, 1998. Basmati rice: a distinct variety (cultivar) of the Indian subcontinent, Asian Agri-history 2 (3): 175-188.

Oka H. I. 1988. Origin of cultivated rice. Japan. Sci. Soc. Press, Tokyo. 245 p.

Oka, H.I. 1991. Genetic diversity of wild and cultivated rice. In Khush, G. S. and G. H. Toenniessen (ed), Rice Biotechnology: 55-81, IRRI. 320 p.

Satyanarayan, A., Thiyagarajan, T.M., and Norman Uphoff. 2007. Opportunities for water saving with higher yield from the system of rice intensification. Irrigation Science. 25: 99-115

Sharma S. D. and U. P. Rao. 2004. Genetic Improvement of Rice Varieties of India. Vol I and Vol II.

Sheehy, J. E., Peng, S. B., Dobermann, A., Mitchell, P. L., Ferrer, A., Yang, J. C., Zou Y. B., Zhong, X. H. and J.L. Huang. 2004. Fantastic yields in the system of rice intensification: Fact or fallacy? Field Crops Research 88, 1-8.

Sheehy, J. E., T. R. Sinclair, and K. G. Cassman. 2005. Curiosities, nonsense, non-science, and SRI. Field Crops Research 91:355–356

Shobha Rani, N., B. Mishra, G.S.V. Prasad, U. Prasada Rao, S.V. Subbaiah, K.Muralidharan and I.C. Pasalu, 2001. Basmati Rice – Heritage of India, DRR Technical Bulletin 2001-02, DRR, Hyderabad, pp: 28

Siddiq, E. A. 1990. Export prospects of Indian Basmati Rice. Indian Farming 40(9): 45-47.

Siddiq E. A., S. Saxena and S. S. Malik. 2005. In: B. S. Dhillon, S. Saxena, A. Agrawal, R. K Tyagi (eds.), Plant Genetic Resources: Food Grain Crops. Indian Society of Plant Genetic Resources, New Delhi, Narosa Publishing House, New Delhi, India, pp. 27-57.

Smith, C.W. and R.H. Dilday. 2003. Rice: origin, history, technology, and production. John Wiley & Sons, Hoboken, New Jersey, United States. 642 p.

Stoop, W. A., and A. H. Kassam. 2005. The SRI controversy: A response. Field Crops Research 91:357–360.

Thanwalee Sooksa-nguan, Janice E. Thies, Phrek Gypmantasiri, Nantakorn Boonkerd and Neung Teaumroong. 2009. Effect of rice cultivation systems on nitrogen cycling and nitrifying bacterial community structure. Applied Soil Ecology 43. 139–149

Vergara, B.S. and T.T. Chang. 1983. The flowering response of the rice plant to photoperiod - a review of the literature. International Rice Research Institute (IRRI), Los Baños, Laguna, Philippines. 61 p.

Virmani S. S. and J. B. Edwards. 1983. Current status and future prospects for breeding hybrid rice and wheat. Adv. Agron., 36: 145-214.

Willem A. Stoop, Abdoulaye Adam and Amir Kassam. 2009. Comparing rice production systems: A challenge for agronomic research and for the dissemination of knowledge-intensive farming practices. Agricultural Water Management 96 (2009) 1491–1501

Zhao, Limei, Wu, Lianghuan, Li, Yongshan, Animesh, Sarkar, Zhu, Defeng and Uphoff, Norman. 2010. Comparisons of Yield, Water Use Efficiency, and Soil Microbial Biomass as Affected by the System of Rice Intensification. Soil Science and Plant Analysis. 41: 1, 1-12

http://www.rkmp.co.in
http://www.agritech.tnau.ac.in/agriculture/agri_cropproduction_rice_sri.html
www.knowledgebank.irri.org/
http://edis.ifas.ufl.edu/AG123
http://www.lsuagcenter.com/NR/rdonlyres/7312C0C3-BE82-409C-A24D-C93838A334ED/97775/pub2331RiceProductionHandbook2014completebook.pdf

CHAPTER

4

Wheat–*Triticum aestivum* L. (2n=42)

Family: Poaceae

Vernacular names: *Gehun* (Hindi), *Kothumai* (Tamil), *Gothambu* (Malayalam), *Godumalu* (Telugu), *Gahu* (Marathi), *Kanak* (Punjabi).

Importance: Wheat is the world's most important cereal crop in terms of both area cultivated (232 million ha) and amount of grain produced (595 million tons). It is the staple food in more than 40 countries. India stands first in area and second in production next to China in the world. It has extensive industrial use also. Wheat grain is a good livestock and poultry feed. Most of the wheat milling by-products especially bran is utilized in preparation of commercial livestock feeds. In some areas in the USA, extensive acreages of wheat are grazed each year while the plants are in the juvenile stages. The same fields are later harvested for grain. Bread wheat contributes approximately 95% to total production, 4% comes from Durum wheat while Emmer wheat share remains only 1%. Bread wheat is mostly consumed in the form of *chapatti* whereas *T. durum* wheat is most suitable for making macaroni, noodles, semolina and pasta products.

History: It is believed that Aryans brought wheat grains to India, since then it has been cultivated in India. In 1944, Norman E. Borlaug had brought high yielding wheat strains or varieties into Mexico from all over the world. He attempted to produce new wheat possessing rust resistance. In 1953, Borlaug had developed semi-dwarf spring wheat. The discovery of Japanese Norin 10 dwarf wheat paved the way for the development of Mexican wheat. The introduction of Norin 10 genes into the Mexican programme has resulted in the development of four Mexican semi-dwarf and dwarf varieties *viz.*, Sonara 63, Sonara 64, Mayo 64 and Lerma Roja 64A, all are bread wheat. By the end of the 1960s, Mexican wheat produces grain yield of more than 2620 kg per ha and Mexico became self-sufficient in wheat. The wheat variety 'Gains' which is a cross between supreme cultivar 'Norin 10' and American cultivar 'Brevor' in 1961, opened an era of modern wheat production. In 1963, the Rockefeller funded research programme in Mexico has been reorganized and renamed the International Centre for the Improvement of Maize and Wheat (the Spanish acronym is CIMMYT). Four principal commercial spring wheat varieties *viz.*, Sonara 63, Sonara 64, Mayo 64 and Lerma Roja 64A, have been brought to India for large scale cultivation during 1963-64. These varieties are high yielding, non-lodging and responsive to nitrogen fertilizer. India had placed an import order for 18000 tonnes of Mexican wheat in 1965. Indian scientists under the dynamic leadership of M.S. Swaninathan have achieved developing wheat varieties in five years. This dramatic change in productivity is known as 'Green Revolution' for which Borlaug won the Nobel Peace Prize in 1970 - not so much for the technology that produced the HYVs, but more for this humanitarianism as he helped to feed the hungry world at a time when widespread famine has been predicted. In 1967, six additional varieties Kalyansona, Safed lerma, Chotti lerma, PV 18, Sonalika and Sharbati Sonara have been released to farmers for cultivation. The yield potentials of these semi dwarf varieties are as high as 6 to 7 t/ha. Hybrid wheat cultivation is popular since 1974 in India.

Origin: The center of origin of wheat is South West Asia (Syria and Asia Minor). The origin and growth habits of different wheat types are furnished in Table 1.

Table 1 Origin and growth habits of different wheat types

Common name	Botanical name	Origin and distribution	Growth habit
Common bread wheat Hexaploid (2n = 42)	*Triticum aestivum L.*	Southwest Asia Europe, Asia (Iran, Israel)	Winter and Spring
Club wheat (2n= 42)	*Triticum compactum*	Southwest Asia, USA	Winter and Spring
Short wheat, Indian dwarf wheat (2n= 42)	*Triticum sphaerococcum* Perceival.	India, Turkey, Iran, Israel, Morocco	Spring
Durum wheat Macaroni wheat Tetraploid (2n = 28)	*Triticum durum* Desf.	Abyssinia, Ethiopia Mediterranean, Iran, Israel, North and South America, Australia	Spring
Emmer or Khapli wheat (2n = 28)	*Triticum dicoccum* Schuble	Abyssinia, Iran, Israel, Russia, Syria, Jordan	Winter and Spring
Einkorn wheat Diploid (2n = 14)	*Triticum monococcum* L.	Turkey	Winter and Spring

Distribution: Wheat is the world's most widely cultivated crop. Globally, India is the second largest producer of wheat, next to China with maximum area under wheat. However, in terms of productivity, it is ranked thirteenth and marginally less relative to world average. The other major wheat producing countries are the USA, Russia, France, Pakistan, Germany, Australia, Canada, United Kingdom, Ukraine, Turkey and Argentina. Wheat yield is 4 to 5 tonnes per ha in the developed countries while the average yield is less than one tonne per hectare in the developing countries. High average wheat yield of 5-7.5 t/ha is common in Denmark, UK, Netherlands, France, Germany and China. The U.S.A, Canada and Australia are leading export countries for wheat. Wheat accounts for one-third of India's food grain production. Wheat is the second most important crop after rice in India. Wheat is mainly grown in Uttar Pradesh, Madhya Pradesh, Punjab, Rajasthan, Haryana, Bihar, Gujarat and

Maharashtra. The crop is raised under irrigated condition in Punjab, Haryana, parts of Uttar Pradesh, Maharashtra, Rajasthan and Madhya Pradesh. It is generally raised as a rainfed or *Barani* crop in Bihar, eastern Uttar Pradesh and West Bengal. It is grown on terraced fields cut across the steep slopes in hilly areas. About 91.5% of the wheat produced in six states *viz*., Uttar Pradesh, Punjab, Haryana, Madhya Pradesh, Rajasthan and Bihar. Bread wheat, durum wheat and dicoccum wheat contributes 95, 4 and 1% to total wheat grain production. The wheat growing zones in India are furnished in Table 2.

Table 2 Wheat growing zones in India

Zone	Area covered	Area m ha (%)
Northern Hills Zone (NHZ)	Western Himalayan regions of J&K (except Jammu and Kathua distt.); H.P. (except Una and Paonta Valley); Uttarakhand (except Tarai area); Sikkim and hills of West Bengal and N.E. States	0.8 (2.9%)
North Western Plains Zone (NWPZ)	Punjab, Haryana, Delhi, Rajasthan (except Kota and Udaipur divisions), Western UP (except Jhansi division), parts of J&K (Jammu and Kathua districts), HP (Una dist. and Paonta valley) and Uttarakhand (Tarai region)	11.3 (40.1%)
North Eastern Plains Zone (NEPZ)	Eastern UP, Bihar, Jharkhand, Odisha, West Bengal, Assam and plains of NE States	9.5 (33.2%)
Central Zone (CZ)	MP, Chattisgarh, Gujarat, Rajasthan (Kota and Udaipur divisions) and UP (Jhansi division)	5.2 (18.1%)
Peninsular Zone (PZ)	Maharashtra, Karnataka, Andhra Pradesh, Goa, plains of Tamil Nadu	1.6 (5.4%)
Southern Hills Zone (SHZ)	Hilly areas of Tamil Nadu and Kerala comprising the Nilgiri and Palani hills of southern plateau	0.1 (0.4%)
	Total	28.5 (100%)

Botany: Wheat is an erect annual, which grows 30 to 120 cm. The wheat stem (culm) is erect and cylindrical, the nodes being solid, whereas the internodes are hollow, but in a few these are solid, filled with pith. There are two sets of roots, three to six seminal or seedling roots developing from the embryo and the adventitious roots form from the coleoptile's base and lower nodes. The roots are mostly adventitious and fibrous representing the permanent root system. Both the types of roots have been found to function till the plant matures. Wheat is a monoecious plant with perfect flowers. Inflorescence, called the 'head', 'spike' or 'ear' is a distichous spike of spikelets. Flowering begins at the middle third of the spike and continues towards the basal and apical parts in 3 to 5 days. All spike-bearing tillers eventually flower almost simultaneously. The wheat spikelets have anthesis during the warm part of the day between 9 a.m. and 2 p.m. while the peak is being from 10 a.m. to 1 p.m. Wheat is normally self-pollinated; cross-pollination is 1 to 4%. Pollen is largely shed within the floret. Stigmas remain receptive for 4 to 13 days. Pollen is viable for up to 30 minutes only. Grains in the centre of the spike and in the proximal florets tend to be larger than the other ones. Physiological maturity is reached when the flag leaf (uppermost leaf) and spikes turn yellow and the moisture content of the fully formed grain has dropped to 25-35%. The wheat grain is a dry, one-seeded, indehiscent fruit, known as a caryopsis. The fruit or the grain is oval in shape and the apex (or the stigmatic end) has a tuft of hairs usually called a brush. Varieties with the awn are called 'bearded varieties' as differentiated from the awnless or 'non-bearded varieties'. The presence of an awn is considered a primitive character. The general range of kernel weight is between 30 to 45 g for bread wheat, whereas in case of durum wheat it is between 35 to 55 g. The grain may be creamy white, amber, red or purple in colour depending upon the variety.

Climate: Wheat is a temperate cereal. It is grown in the temperate zones between 30 to 60°N and 25 to 40°S latitudes with elevations up to 3000 m from the mean sea level. It is a cool season crop. Cool weather during vegetative development and warm weather for maturity is deemed ideal for wheat. Agronomically, the favourable wheat climate is described as having mild winters, warm summers with high radiation without heavy cooling summer rains, water supply more provided by sufficient soil water capacity. It is grown where annual average rainfall ranges from 250 to 1750 mm. However, three quarters of that land area in wheat are grown in areas receiving rainfall between 375 and 875 mm of annual precipitation. Wheat is a drought resistant crop. It is a long-day plant. Winter varieties are chill demanding, long-day sensitive and frost resistant. At their early stages of development they are highly resistant to frost (-20°C). At heading and flowering, frost hardiness is gradually lost. The minimum, optimum and maximum cardinal temperature for wheat crop is 3 to 4°C, 20 to 25°C and 32°C respectively. Germination of wheat occurs at temperatures of 4–37°C, the optimum being 12–25°C. The optimum temperature for leaf development and tiller production is 20 to 25°C and 15 to 25°C respectively. For floral induction, spring types usually require temperatures between 7°C and 18°C for 5–15 days, while winter types require temperatures between 0°C and 7°C for 30–60 days. The low temperature of 0-5°C next-to-ear emergence and pollination reduce grain setting. The optimum temperature for flowering is 20 to 22°C. The minimum, optimum and maximum temperature for fertilization is 10°C, 18 to 24°C and 32°C. Wheat requires an optimum average temperature of 14-15°C at the time of ripening. Temperatures above 25°C during anthesis may cause sterility and at ripening tend to depress grain weight. In all varieties, grain weight and grain yield are higher at 15 to 20°C although protein and gluten contents seem to be highest at 20 to 25°C. Grain weight decreases proportionally when the temperature rises above 17°C. The grain weight and grain composition decreases at high temperature of 25 to 30°C. The favourable soil temperature for growth and yield is 18 to 20°C. In order to avoid frost damage at heading, the 20% probability of frost (0 to 2°C) in the spring is selected as the criteria for the planting time. When winters are severe and prolonged, only spring varieties can be sown. The occurrence of rust diseases is observed at 9-13°C with relative humidity of 70%. The crop requires 100 frost free days during cropping period.

Vernalization refers to a cold requirement period of cool temperature that enhances or even enables the apex of a tiller to initiate an inflorescence under appropriate day-length condition. For floral induction, spring types usually require temperatures between 7 to 18°C for 5 to 15 days, while winter types require temperatures between 0 to 7°C for 21 to 40 days to switch from vegetative growth to flowering (anthesis) to produce grain. Vernalization requirements for winter varieties may be entirely substituted by growing under short-day conditions (at 21°C day temperature and 16°C night temperature) for a similar period of time. In field, vernalization begins with daily minimum temperature of about 10 to 13°C for fast floral initiation. Vernalization occurs during grain ripening can be reversed by exposure to temperature of 27 to 33°C for a week. Mostly, spring varieties are day neutral, more frost sensitive than winter ones and require low chilling. Spring wheat does not require low temperature chilling requirements for entering reproductive phase and are less tolerant to low temperature. Similarly it must pass vegetative growth in short days to reach earing with long days.

Day length sensitivity varies tremendously between wheat genotypes. It can be grown both under short day and long day photoperiod. In general, wheat is a long day plant. Long days enhance ear initiation and development while short days delaying both. The day length at heading is between 12 and 14 hours. Light saturation of a single wheat leaf under good growing conditions may be assumed with 4 to 5 klux whereas this value for a closed wheat community will increase to about 70 to 90 klux.

Leaf wetness can be due to rain, dew or fog. It is usually measured in terms of duration in hours. Wheat plant requires 50-60% humidity for their growth, but at the time of maturity, crop requires less humidity. The relative humidity of > 90% induces spore germination and sustains infection. Cloudy weather with high humidity and low temperature is conducive for rust attack.

Soils: Wheat grows on a variety of soils. Well drained loamy soil is desirable. Highly porous and excessively drained soils or heavy soils with bad soil structure are not suitable. The largest area under wheat in India is in the alluvial soils of the Indo-Gangetic areas where the soil has generally a loamy texture, good structure and moderate water holding capacity with pH ranging from 6.0 to 8.5. The threshold level for salinity ECe is 6.0 dS/m. Wheat is also grown in black alluvium to heavy clay soil and in the hill soils, which are generally acidic with pH ranging from 5.0 to 7.0.

Season: In temperate climates, the cultivated varieties of wheat are classified based on growing season into two main types, namely 'spring wheat' and 'winter wheat'. The spring wheat is characterized by their short growing season of at least 100 days and do not require very low temperatures in the early phases of their growth. They are sown in March-May and harvested in the late summer from August to September. The winter wheat is of long duration and need relatively low temperatures during the initial stages of development. The crop is sown in October-November and reaches maturity in the early summer of the next year, *i.e.*, from May-July. In India, wheat is essentially a winter or *rabi* crop. It is only in the hills of south India (Nilgiris) while wheat is grown in winter as well as in summer season. In the hills, the growing season is the longest, *i.e.*, nine months, extending from October–November to June–July, but the area under the crop is very small. The period of growth is 150 to 160 days in the northwestern plains and becomes shorter towards the northeast and the south, lasting for only 120 to 140 days. In Tamil Nadu, wheat is grown during winter season from November to February. The optimum date of sowing is second or third week of November for dwarf wheat while second fortnight of October for tall traditional wheat. Wheat is sown up to 25th December in northwestern plain zone, up to 10th December in northeastern zone and central zone and up to 30th November in Peninsular zone. The problems encounters in late sowing are (i) the germination and early growth of the seedlings is slow because of initial low temperature and (ii) the shortened growing season.

Varieties: The popular wheat varieties in India are presented in Table 3 and Table 3.1.

Table 3 Wheat varieties in different regions of India

Zone	Production condition	Varieties (potential yield in q/ha)
Northern Hills Zone	TS-IR-high fertility	VL 804 (46.2), VL 738 (47.1), HS 240 (45.0)
	TS-RF-low fertility	VL 804 (43.0), VL 738 (47.1), HS 240 (26.0), TL 2942(trit) (32.1)
	ES-RF-low fertility	VL 829 (48.9), HPW 251 (49.5), VL 616 (39.0)
	LS-RI-medium fertility	HS 490 (49.7), VL 892 (48.9), HS 420 (28.2), HS 295 (27.5)
	High altitude areas	HS 365 (26.8), VL 832 (28.6), SKW 196 (24.0)
North Western Plains Zone	TS-IR-high fertility	HD-3086 (Pusa Gautami), HD-3043 (Pusa Chaitanya), HD 2967 (Pusa Sindhu Ganga), HD 2894 (Pusa Wheat 109), HD 2851 (Pusa Vishesh), DBW 17 (63.0), PBW 550 (62.0), PBW 502 (60.0), PBW 343 (61.0), WH 542 (61.5), UP 2338 (55.1), HD 2687 (51.2), PDW 291(d) (58.0), PDW 233 (d) (48.5), WH 896(d) (58.0)
	LS-IR-medium fertility	HD 3059 (Pusa Pachheti), WR 544 (Pusa Gold), PBW 590 (57.2), WH 1021 (52.8), DBW 16 (45.0), PBW 373 (43.9), RAJ 3765 (43.8), UP 2425 (44.7)
	TS-RF-low fertility	PBW 396 (39.4), PBW 175 (48.1), Kundan (31.2)
North Eastern Plains Zone	TS-IR-high fertility	HD 2824 (Poorva), HD 2733 (VSM), DBW 39 (55.2), CBW 38 (57.9), Raj 4120 (50.0), K 307 (51.9), HD 2824 (65.0), PBW 443 (52.7), PBW 343 (61.0), HD 2733 (56.2), HUW 468 (52.8), K 9107 (55.0), HP 1761 (47.0)
	LS-IR-medium fertility	HD 2985 (Pusa Basant), HW 2045 (Kaushambi), DBW 14 (42.1), NW 2036 (41.9), HD 2643 (47.7), NW 1014 (40.0), HP 1744 (39.4), Halna (35.0)
	TS-RF-low fertility	HD 2888 (Pusa Wheat 107), HD 2888 (38.3), K 8027 (37.0)
	LS-RF-low fertility	K 9465 (30.0), K 8962 (23.5)
Central Zone	TS-IR-high fertility	HI 1544 (68.2), GW 366 (77.9), GW 322 (48.9), GW 273 (50.2), DL 803-3 (50.3)
	LS-IR-medium fertility	MP 1203 (52.7), HD 2864 (51.5), HD 2932 (57.8), MP 4010 (43.5), DL788-2 (43.0)
	TS-RF-low fertility	*HI 1531 (30.1), HI 1500 (30.3), HW 2004 (20.2), Sujata (17.8), *HI 8627(d) (39.8), HD 4672(d) (30.0)
Peninsular Zone	TS-IR-high fertility	GW 322 (48.9), HUW 510 (47.3), DWR 162 (54.9), HD 2189 (45.7), MACS 2971(dic) (56.2), HI 8663(d) (62.0),
	LS-IR-medium fertility	PBW 533 (53.1), HD 2932 (53.6), HD 2833 (45.0), Raj 4083 (56.6), NIAW 34 (37.6)
	TS-RF-low fertility	*PBW 596 (46.0), K 9644 (17.4), HD 2781 (21.0), NI 5439 (30.6), AKDW 2997-16(d) (10.0)
Southern Hills Zones	TS-RI-medium fertility	HW 1085 (48.3)
Marginal areas	Salinity-alkalinity condition	KRL 210 (49.3), KRL 213 (43.9), KRL 19 (45.0), KRL 1-4 (49.0)
Very late sowing	IR-Medium fertility	Raj 3765 (43.8), PBW 373 (43.9), Halna (35.0)
Summer sowing		HS 375 (40.4)

Source: http://www.dwr.in/

*also suitable for restricted irrigation, (d)=durum wheat, (Dic)=dicoccum wheat, TS=Timely Sown, LS=Late Sown, ES=Early Sown, IR=Irrigated, RF=Rainfed, RI=Restricted irrigation

Table 3.1 Wheat varieties recommended at different sowing conditions in India

Wheat type	**Production conditions**			
	Normal sown	**Late sown**	**Rainfed**	**Sodic soils / Others**
Northern Hills Zone (NHZ) Western Himalayan regions of J&K (except Jammu and Kathua distt.); H.P. (except Una and Paonta Valley); Uttarakhand (except Tarai area); Sikkim and hills of West Bengal and N.E. States				
Bread wheat	VL, 907, VL 738, VL 804, HS 240, HW 5207	HS 295, HS 420	VL 907, VL 738, HPW 42, HS 365, VL 829, VL832 SKW 196	HS 375 (For summer sowing)
Triticale	-	-	DT 46	-
Southern Hills Zone (SHZ) Hilly areas of Tamil Nadu and Kerala comprising the Nilgiri and Palni hills of southern plateau				
Bread wheat	HUW 318, HW 1085, HW 2044	-	-	
North Western Plains Zone (NWPZ) Punjab, Haryana, Delhi, Rajasthan (except Kota and Udaipur divisions), Western UP (except Jhansi division), parts of J&K (Jammu and Kathua districts), HP (Una dist. and Paonta valley) and Uttarakhand (Tarai region)				
Bread wheat	DBW 17, PBW 550, PBW 502, PBW 343, WH 542, UP 2338, HD 2687, HD2967	WH1021, PBW 373, UP 2425, RAJ3077, DBW16, RAJ 3765, PBW 590	PBW299, PBW 175, WH 533, PBW 396	RAJ3077, KRL-19, KRL 210, KRL 213
Durum	PBW 34, PDW 215, PDW 233, WH 896, PDW 291, PDW 314	-	-	
North Eastern Plains Zone (NEPZ) Eastern UP, Bihar, Jharkhand, Odisha, West Bengal, Assam and plains of NE States				
Bread wheat	CBW 38, Raj 4120, K 0307, NW 1012, HUW 468, PBW443, HD2733, HD2824, K 9107, HD 2967, DBW 39	HD2643, HP 633, HP1744, NW1014, HW 2045, DBW 14, NW2036, HD 2985	HDR77, K8962, K 9465, K8027, HD 2888, MACS 6145	RAJ3077, KRL-19, KRL 210, KRL 213
Central Zone (CZ) MP, Chhattisgarh, Gujarat, Rajasthan (Kota and Udaipur divisions) and UP (Jhansi division)				
Bread wheat	GW 190, GW 273, DL803-3, GW 322, GW 366, HI 1544	GW 173, DL 788-2, MP 4010, HD 2932, MP 1203, HD 2864	HW 2004, JWS 17, HI 1500, HI 1531, Sujata	RAJ 3077, KRL-19, KRL 210, KRL 213
Durum	HI 8381, HI 8498, MPO 1215	-	HD 4672, HI 8627	-
Peninsular Zone (PZ) Maharashtra, Karnataka, Andhra Pradesh, Goa, plains of Tamil Nadu				

Bread wheat	DWR162, MACS 2496, GW 322, Raj 4037, NIAW 917, UAS 304, MACS 6222, MACS 6273	DWR 195, HD 2501, NIAW 34, HUW 510, HD 2932, HI 977, HD 2833, PBW 533, Raj 4083, AKAW 4627	K9644,HD2781, PBW596,HD 2987	–
Durum	MACS 2846, HI 8663, UAS 415	-	AKDW 2997-16	-
Dicoccum	DDK 1025, DDK 1029, DDK 1066	-	-	-

Growth stages: Wheat emerges about a week after sowing. The radicle emerges first and the coleoptile emerges 4–6 days after germination. The growing point or tip of the stem, remains underground through the dormant period. There are two distinct root systems in wheat, the seminal roots arising directly from or below the seed and the adventitious roots that arise from the nodes of the stem above the seed. Growth becomes upright, the point at which nodes of the stem start to become visible as the stem length increases. As the stem elongates, the developing head at the tip of the stem eventually emerges and flowering takes place. Flowering begins at the middle third of the spike and continues towards the basal and apical parts in 3–5 days. All spike-bearing tillers eventually flower almost simultaneously. Wheat is normally self-pollinated; cross-pollination is 1–4%. Pollen is largely shed within the floret. Stigmas remain receptive for 4–13 days. Pollen is viable for up to 30 minutes only. The grain begins growing immediately after flowering and reaches its maximum size (not weight) within about 2 weeks. The maximum weight occurs about 4 weeks after flowering. This period is determined largely by temperature and can extend up to 12 weeks in areas where the weather is cool. Physiological maturity is reached when the flag leaf (uppermost leaf) and spikes turn yellow and the moisture content of the fully formed grain has dropped to 25–35%. The length of the total growing period of spring wheat ranges from 100 to 130 days while winter wheat needs about 180 to 250 days to mature. The growth stage of wheat is presented in Table 4 and Table 5.

Table 4 Growth stages in spring wheat

Stage	Description	Days from sowing		
0	Emergence	5-7	5-7	5-7
1	Crown root initiation	18-22	20-25	20-25
1.5	Tiller begins	15-25	15-30	15-30
2	Collars visible	30-32	35-38	40
3.0	Peak tillering	35	40-45	45
3.5	Jointing stage	35-40	45-55	45-60
4	Booting / flag leaf stage	45-50	65-70	75
5	Heading	51-55	75	85
6	Flowering/ Anthesis	58-60	78-80	95
7	Milky stage	61-65	85-90	105
8	Soft dough stage	70	100	120
9	Hard dough stage	90	115	135
10	Maturity	95-100	125	145

Table 5 Growth stages of winter wheat

Growth Stage	Description	Days from sowing
0	Emergence of coleoptile	5-7
1	Crown root initiation stage, tillers development	20-25
2	Leaf sheaths elongate and form a false stem; collars visible.	26-45
3	Clump elongation; first internode visible (jointing stage)	46-60
4	Tip of flag leaf visible (boot stage)	70-75
5	Pendulence elongates; inflorescence emerges (heading)	85-90
6	Flowering (anthesis)	85-95
7	Anthesis complete; grain filling begins, milking stage; lower leaves turn yellowish colour	100-105
8	Grain is soft dough; flag leaf has lost all green colour (dough stage)	105-110
9	Ripening; grain hard, but will not crack; inflorescence has lost all green colour; upper most node still green	110-115
10	Maturity; grain cracks and easily separated from chaff	115-120

Growth Stage Scales: There are several scales that describe the growth and development of cereal crops, including wheat. The Zadoks and BBCH scales are commonly used in Europe, while the Feekes scale is more widely used in the United States. Each scale differs in the level of detail it uses to describe the stages of crop growth. Management decisions in wheat production are growth-stage dependent. Applying fertilizers, herbicides and fungicides are most effective and profitable when applied at specific times during wheat development. The Feekes scale is a numerical scale. It begins at Feekes 1.0 (which describes emergence) and ends in 11.4 (which describes a mature plant that is ready for harvest). The description Feekes scale for wheat crop is furnished in Table 6. The key growth stages in wheat for yield determination are furnished in Table 7. The approximate freezing injurious temperature for wheat is furnished in Table 8.

Table 6 Description of Feekes scale for wheat growth stages

Stage	Scale	Description
Tillering	1	One shoot, first leaf through coleoptile.
	2	Beginning tillering: one shoot, one tiller.
	3	Tillers formed, leaves often twisted spirally. Main shoot and six tillers. In some varieties, plant may be creeping.
	4	Start of erection of pseudo-stem; leaf sheaths beginning to lengthen.
	5	Pseudo-stem (formed by sheaths of leaves) strongly erected.
Stem extension	6	1st node of main stem visible at base of shoot
	7	2nd node of main stem; next to last leaf visible
	8	Flag leaf (last leaf) visible but still rolled up; ear beginning to swell.
	9	Ligule of flag leaf just visible.

Boot (late jointing)	10	Sheath of flag leaf completely grown out; ear swollen but not yet visible.
Heading	10.1	awns visible, heads emerging, first spikelet of head just visible.
	10.2	¼ of heading process completed.
	10.3	½ of heading process completed.
	10.4	¾ of heading process completed.
	10.5	All heads out of sheath.
Flowering	10.51	Beginning of flowering.
	10.52	Flowering complete to top of head.
	10.53	Flowering complete at base of head.
	10.54	Flowering complete; kernel watery ripe.
Milk	11.1	Milky ripe.
Dough	11.2	Milky rip; contents of kernel soft but dry.
Physiological maturity	11.3	Kernel hard (difficult to divide with thumbnail)
Ripening	11.4	Ripe for cutting. Straw dead.

Source: James Herbek and Chad Lee, 2006.

Table 7 Key growth stages in wheat for yield determination

Critical yield component	Determined by
Tiller and head number	Jointing (Feekes 3; Zadoks 23 to 29)
Head Size	Mid to late tillering (Feekes 3; Zadoks 23 to 29)
Kernel number per head	Jointing (Feekes 6; Zadoks 31)
Kernel Size	Beginning at flag leaf (Feekes 8; Zadoks 37) and continuing through grain fill

Source: James Herbek and Chad Lee. 2006.

Table 8 Freeze injury in wheat

Growth Stage	Feekes	Zadoks	Injurious temp (2 hrs)	Symptoms	Effect on yield
Tillering	1-5	20-29	-11°C	Leaf chlorosis; burning of leaf tips; silage odor; blue cast to fields	Slight to moderate
Jointing (stem extension)	6-7	31-32	-4°C	Death of growing point; leaf yellowing or burning; lesions, splitting, or bending of lower stem; odoor	Moderate to severe
Boot (late jointing)	10	41-49	-2°C	Floret sterility; spike trapped in boot; damage to lower stem; leaf discoloration; odour	Moderate to severe
Heading	10.1-10.5	50-58	-1°C	Floret sterility; white awns or white spikes; damage to lower stem; leaf discoloration	Severe

Flowering	10.51-10.54	60-71	-1°C	Floret sterility; white awns or white spikes; damage to lower stem; leaf discoloration	Severe
Milk	11.1	75	-2°C	White awns or white spikes; damage to lower stems; leaf discoloration; shrunken, roughened, or discolored kernels	Moderate to severe
Dough	11.2	85	-2°C	Shriveled, discolored kernels; poor germination	Slight to moderate

Source: Chad Lee et al., 2006.

For maximum wheat yields, proper management and favorable weather are necessary during these key growth stages that affect yield determination. The details of management considerations in wheat according to the Feekes growth stages are furnished in Table 9.

Table 9 Wheat management considerations by Feekes growth stages

Feekes Growth Stage	Zadoks scale	Management Considerations
1.0	10	First leaf through coleoptile - Check stands for emergence and uniformity. Check for weeds and apply herbicides if necessary. Begin monitoring for various aphid species (continue through season). Check seedlings for Hessian fly feeding damage.
2.0	21	Main shoot and 1 tiller - Make early nitrogen applications to enhance tillering in thin stands. Avoid excess nitrogen
3.0	26	Main shoot and 6 tillers -
4.0	30	Scout for insect and disease problems. Check stands for heaving caused by freezing/thawing cycles. Decide whether post-emergence weed control is warranted.
5.0	30	Make spring top-dress nitrogen applications. Apply herbicides as needed for weed control.
6.0	31	Jointing stage - 1st detectable node - Cutoff for nitrogen applications to avoid leaf injury. Cutoff for some growth regulator herbicides, like 2, 4-D and dicamba.
7.0	32	2nd detectable node - Scout for insect and disease problems.
8.0	37	Flag leaf visible - Apply fungicides to protect flag leaf from foliar diseases if necessary. Cutoff for any further herbicide applications unless harvest aid treatments are needed
9.0	39	Flag leaf ligule and collar visible
10.0	45	Boot leaf swollen - Determine if fungicide applications for glume blotch management are needed. Check risk for Fusarium head blight (scab), Check for armyworm feeding. Consider control measures if armyworm feeding is clipping heads.
10.1	50	First spikelet of head visible - Flowering starts

10.2	52	¼ of head visible
10.3	54	½ of head visible
10.4	56	¾ of head visible
10.5	58	Head completely emerged
10.51	60	Flowering usually begins in middle of head. Apply fungicides to suppress Fusarium head blight if necessary
10.52		Flowering completed at top of head.
10.53		Flowering completed at bottom of head.
	64	½ of flowering complete
	68	Flowering completed
10.54	71	Kernel (caryopsis) watery ripe
11.1	75	Medium milk - Milky ripe
11.2	85	Soft dough - Mealy ripe: kernels soft but dry.- Soft dough
11.3	91	Kernel hard (hard to split by thumbnail) - Physiological maturity. No more dry matter accumulation
11.4	92	Kernel hard (cannot split by thumbnail) - Ripe for harvest. Straw dead.

Source: Managing wheat by growth stage, Purdue Extension publication ID 422.
http://weedsoft.unl.edu/documents/GrowthStagesModule/wheat/wheat.htm#
http://sanangelo.tamu.edu/agronomy/wheat/whtmang.htm

Land preparation: Wheat crop requires a well-pulverized, but compact seed-bed for proper germination. Plough twice with an iron plough or two to three times with country plough to a fine tilth. Form beds of size of 10 m^2 or 20 m^2 depending on the availability of water. Use a bund former to economise cost of production. The land can also be ploughed with mould board plough deeply followed by 2-3 harrowing should be done. In recent times, zero-till sowing and furrow irrigated raised bed methods are practiced. For timely wheat sowing after rice harvest and also to reduce *Phalaris minor* and *Avena sativa* menace with normal land preparation, seeds are drilled into soil without land preparation. Furrow irrigated raised bed (FIRB) system of planting developed in Yucatan Valley of Mexico is being adopted by farmers only recently. In this method, raised beds (75-90 cm) alternating with furrows (30 cm) are prepared and 2-3 rows of wheat are planted on the bed. The crop is irrigated in furrows.

Seeds and sowing: Seed rate of 100 kg per ha is recommended for the medium sized grains (38 to 44 g for 1000 seeds); 125 kg per ha for bold seeded varieties (45 g for 1000 seeds) and 25% higher seed rate (125-150 kg/ha) is recommended for late-sown and salinity condition to overcome poor tillering due to low temperature. Seed treatment is done with *T. viride* @ 4 g / kg seed in combination with carboxin 75 WP @ 1.25 g/kg seed or tebuconazole 2 DS @ 1.0 g/kg of seed 24 hours before sowing. Seed treatment with 0.5 ppm IAA helps to withstand moisture stress condition. A row spacing of 22.5 cm is recommended for dwarf wheat and sow the seeds continuously (solid sowing) after application of fertilizers to a depth of 4-5 cm. When sowing is delayed a closer spacing of 15-18 cm should be adopted. Wheat is sown by seed drill or ferti-seed drill, behind local plough, dibbling and broadcast method. Avoid deep sowing and thinning of the seedlings. Seeds can be sown at the depth of 3-4 cm since the coleoptile length of high yielding dwarf wheat varieties is about 5 cm. Therefore, seeds of these varieties should be covered not by more than 4 cm soil to ensure uniform and good germination. Deep sowing delays the emergence of seedlings by 2-3 days and heading by 5-6 days.

Weed management: Weeds like wild oats *(Avena fatua)*, canary grass (*Phalaris minor*), Cynodon grass, *Convolulus* sp are serious and difficult to manage. Dicot annual weeds like *Chenopodium album*, *Carthamus oxyacanthus*, *Melilotus sp*. and *Lathyrus aphaca* commonly occur in wheat. *Phalaris minor* is the major weed of wheat which is controlled with spray of Isoproturon (Tolkan/ Graminon/Arelon) @ 1 kg/ha as pre-emergence spraying 3 days after sowing followed by one hand weeding on 35th day after sowing. Both *Phalaris minor* and wild oat can be achieved with the application of Methabenthiazuron @ 1.5 to 2.0 kg/ha as pre-emergence while Metoxuron (Dosanex/ Hexamar/ Hilnex) @ 1.75 kg/ha as post emergence on 30 to 35 DAS. If herbicide is not applied give two hand weeding on 20th and 35th day after sowing.

Both grassy and broadleaf weeds

- Sulfosulfuran @ 25.0g a.i./ha in 250-300 liters of water /ha.
- Metribuzin @ 175 g a.i./ha in at least 500 liters of water /ha.
- A mixture of Sulfosulfuran at 25g/ha and metsulfuron methyl @ 4 g/ha in 250-300 liters water /ha.
- Combination of 2, 4-D and isoproturon at the recommended doses of each or Isoguard-plus @1.2 kg a.i./ ha 30-35 days after sowing.
- Sulfosulfuron (Leader) (33.3 gm/ha) and Metribuzin (Sencor) (250 gm /ha) to control both grassy and non-grassy weeds
- Pendimethalin (Stomp) @ 1.0 kg a.i/ha on 0-3 days after sowing as pre-emergence controls both grasses and broadleaved weeds.

Only grassy weeds

- Clodinafop @ 400 g/ha (60g a.i./ha) in 250-300 liters of water / ha.
- Fenoxaprop-ethyl @ 80-120g a.i./ha in 250-300 liters of water / ha.

Only broadleaf weeds

- 2, 4-D @ 500 g a.i./ha in 250-300 liters of water /ha.
- Metsulfuron methyl @ 4 g a.i. / ha in 250-300 liters of water /ha.

Manures and fertilizers application: Spread 12.5 t/ha of FYM or compost during last ploughing. The quantity and time of application recommended in different wheat growing zones of India and Tamil Nadu is furnished in Table 11 and Table 12.

Table 11 Fertilizer recommendation in wheat growing zones and sowing conditions

Zones / Sowing conditions	Quantity and time of application
North Western Plains Zone (NWPZ) and North Eastern Plains Zone (NEPZ)	
Irrigated timely sown	150:60:40 of N, P_2O_5 and K_2O kg/ha 1/3 N +P + K at sowing and 2/3 at first node stage i.e. 35-40 DAS

Irrigated late sown	120:60:40 of N, P_2O_5 and K_2O kg/ha 1/3 N +P + K at sowing and 2/3 at first node stage i.e. 35-40 DAS
Rainfed Dwarf varieties	60:30:20 of N, P_2O_5 and K_2O kg/ha at the time of sowing
Rainfed *Deshi* varieties	25:20:15 of N, P_2O_5 and K_2O kg/ha at the time of sowing
Northern Hills Zone (NHZ), Central Zone (CZ), Peninsular Zone (PZ), Southern Hills Zone (SHZ)	
Irrigated timely sown	120:60:40 of N, P_2O_5 and K_2O kg/ha 1/3 N +P + K at sowing and 2/3 at first node stage i.e. 35-40 DAS
Irrigated late sown	90:60:40 of N, P_2O_5 and K_2O kg/ha 1/3 N + P_2O_5 + K_2O at sowing and 2/3 at first node stage i.e. 35-40 DAS
Rainfed	60:30:20 of N, P_2O_5 and K_2O kg/ha at the time of sowing

Table 12 Fertilizer requirement in wheat in Tamil Nadu (kg/ha)

Planting condition	**Time of sowing**	**N**	**P_2O_5**	**K_2O**
Irrigated	Timely sown	120	60	40
Irrigated	Late sown	80	40	20
Rainfed	Timely sown	40	20	-

The most suitable timing for three splits is ⅓ at sowing, ⅓ at tillering and ⅓ ear initiation. Nitrogen is also applied in four splits *viz.*, sowing, crown root initiation stage, tillering and heading. Nitrogen must be applied early to ensure good early tillering and high yields. Nitrogen application at the time of tillering promotes the formation of new shoots. The nitrogen top dressing at the early heading stage mainly influences the number of grains per earhead. Wheat plants can absorb substantial amounts of the nitrogen applied after ear emergence. Peak nitrogen uptake occurs at crown root initiation, tillering, jointing and heading and soft dough stages. Wheat absorbs only 15% of total P uptake in first two weeks of growth. The critical period for P requirement is first 5 to 6 weeks. P has major impacts on tillering and rooting. Phosphorus is to be placed near the seed row since the root development in wheat is mostly confined to 2 to 5 cm from seed row and to a depth of 8 cm. Further, P deficiencies earlier in the season are more detrimental than those occurring later. The entire phosphorus and half the nitrogen should be placed 5 cm below the seed at the time of sowing. To realize full effect, phosphorous is placed 5 cm below and 5 cm to the side of the seed with the help of fertilizer cum seed drill. Phosphorus deficiency shows reddish violet discoloration of leaves in young wheat plants. Potassium fertilizers may be broadcasted before sowing. Potassium deficiency appears first on the lower leaves, progressing towards the top as the severity of the deficiency increases. Typical symptom of potassium deficiency is yellowish to brown discoloration at the tips and margins of the leaves. The deficiency level in micro nutrients in wheat is Zn: 46%, B: 17%, Mo: 12%, Fe: 11%, Cu: 5% and S: 38% at the national level. Soil application of $ZnSO_4$ @ 25 kg/ha once in three years corrects or 2-3 foliar spray of 0.5% $ZnSO_4$ at 15 days after sowing to rectify Zinc deficiency. Manganese deficiency symptom shows necrotic streaks on the leaf lamina. In Mn deficient soil, spray 0.5% $MnSO_4$ solution 2-4 days before first irrigation and two to three sprays afterwards at weekly intervals on clear sunny day. In sulphur deficient soils, SSP, Cosavet-90 WDG (10 kg/ha) or gypsum (250 kg/ha) can be applied to increase yield and protein content.

Water management: Water requirement for wheat is 350 to 420 mm. The soil moisture should be about 50 to 70% of the field capacity to enable proper germination and establishment of the seedlings. Water stagnation should be avoided at the time of germination. The crop requires 4 to 6 irrigations depending on the soil type and rainfall. The critical stages for irrigation are crown root initiation, late tillering, late jointing, flowering and milk stage (Table 13). Crown root initiation and flowering are sensitive to moisture stress. The crown-root-initiation (CRI) stage occurs about 3 weeks after sowing. Crown roots of wheat generally develop within 2 cm below the soil surface. The crown node starts swelling about 17 days after sowing and crown root develops from this node approximately 21days after sowing (DAS). Irrigation on 21 DAS is essential for high productivity. Tillers develop from crown root has direct relevance to the number of tillers and final heads is supposed to be the most critical period for irrigation. This means, the crown root initiation (CRI) stage and heading stage are the critical growth stages to moisture stress. A dry period of a few weeks is necessary for ripening, but a water deficit during floral initiation reduces grain number and spikelets per ear. Irrigation can be given at 50% depletion of available moisture from top 30 cm soil layer. The irrigation depth of 7-8 cm is followed for wheat crop. The IW/CPE ratio of 0.75 can be practiced in alluvial soil for optimum wheat yield.

Table 13 Critical growth stages of wheat for irrigation

Number of irrigations	Crop growth stages	Schedule of irrigation DAS
One irrigation	Crown root initiation	20-25
Two irrigations	Crown root initiation, booting (late jointing)	20-25, 70-75
Three irrigations	Crown root initiation, booting, milking	20-25, 70-75, 100-105
Four irrigations	Crown root initiation, tillering, flowering, milking.	20-25, 40-45, 90-95, 110-115
Five irrigations	Crown root initiation, tillering, flowering, milking and dough	20-25, 40-45, 70-75, 110-115, 120-125
Six irrigations	Crown-root initiation, tillering, jointing, flowering, milking and dough stage	20-25, 40-45, 70-75, 90-95, 110-115, 120-125,

Cropping Systems: Intercropping of wheat with linseed / pea / gram / lentil in 4:2 ratio and with mustard in 8:2 ratio are also practiced. Some of the crop rotations are namely, sorghum or maize-*toria*-wheat-green gram, pigeonpea-wheat-black gram, pearl millet-wheat-green gram, paddy-wheat-cowpea or black gram, soybean-wheat-green gram, maize-*toria*-wheat-green gram, paddy-wheat-green gram/ blackgram, maize-potato-wheat-cowpea/mungbean, sugarcane-wheat, cotton-wheat rotations, pearl millet--wheat, sorghum-wheat and jute-wheat. Wheat is also grown as a companion crop between rows of sugarcane and potato. The growing of rainfed wheat mixed with chickpea, lentil, mustard, linseed, barley and safflower is quite common. The wheat based cropping systems prevalent in different states of India is presented in Table 14.

Table 14 State-wise wheat based cropping system in India

S.No.	State	Cropping System
1	Assam	Maize-Wheat, Sugarcane-Wheat, Pigeon Pea-Wheat
2	Bihar	Rice-Wheat, Maize-Wheat, Sesame –Wheat
3	Jharkhand	Rice-Wheat
4	Gujarat	Groundnut-Wheat, Maize-Wheat, Rice-Wheat, Cotton-Wheat, Pigeon Pea-Wheat
5	Haryana	Rice-Wheat, Sorghum-Wheat, Cotton-Wheat, Bajra-Wheat, Maize-Wheat
6	Himachal Pradesh	Maize-Wheat
7	Jammu & Kashmir	Rice-Wheat, Maize-Wheat
8	Karnataka	Groundnut-Wheat
9	Madhya Pradesh	Rice-Wheat, Sorghum-Wheat, Soybean-Wheat, Cotton-Wheat,
10	Chattishgarh	Soybean-Wheat, Rice-Wheat, Sorghum-Wheat, Cotton-Wheat
11.	Maharashtra	Soybean-Wheat, Bajra-Wheat, Rice-Wheat, Cotton-Wheat
12.	Odisha	Sesame-Wheat,
13.	Punjab	Rice-Wheat, Cotton-Wheat, Maize-Wheat,
14.	Rajasthan	Sorghum-Wheat, Maize-Wheat, Bajra-Wheat
15.	Uttar Pradesh	Rice-Wheat, Bajra-Wheat, Sorghum-Wheat, Sugarcane-Wheat
16.	West Bengal	Rice-Wheat

Pest and diseases management: Insects, brown mite, nematodes and diseases affect wheat crop. The major insect pests are aphid, army worm/cut worm, American pod borer, Pink stem borer, shoot fly and termites while the diseases are brown rust, yellow/stripe rust, black rust, loose smut, flag smut, karnal bunt, hill bunt, powdery mildew, *Helminthosporium* leaf spot, *Alternaria* leaf blight, head scab/*fusarium* head blight and foot rot. The nematodes infesting wheat are seed gall nematode, cereal cyst nematode and root knot nematode. The rust resistant/tolerant wheat varieties grown in different zones/ states of India are presented in Table 15.

Table 15 Rust resistant/tolerant wheat varieties

Zone	Varieties
Northern Hill Zone (Hills of J & K, H. P. and U. P.), North Hill Zone (High Altitude)	HS 420; HS 277; HS 295; HPW251; VL 892
North Western plain Zone (Punjab, Haryana, Western UP, Northern Rajasthan and foot hills of HP & J & K)	CPAN 3004;; WH 542, HD 2687; PBW 550; WH 896; WH 1105; HD 2964;; 8804
North Eastern Plain zone (Eastern UP, Bihar, West Bengal)	HP 1102; UP 262; HUW 206; HP 1102; K 8804
Central Zone (MP, Gujarat and southern Rajasthan)	WH 147; GW 190; H 1977GW273; GW322;; DL 803-3; Lok 1
Peninsular Zone (Maharashtra and Karnataka)	HD 2501, MACS 2496; DWR 162; HW 971; HW 2022

Use yellow sticky traps for aphids and blue sticky traps to monitor thrips and other pest incidence @ 4-5 traps/acre or use light trap @ 1/acre and operate between 6 pm and 10 pm. Install pheromone traps @ 4-5/acre for monitoring adult moths activity (replace the lures with fresh lures after every 2-3 weeks). Erecting of bird perches @ 20/acre for encouraging predatory birds such as King crow, common mynah, etc. Spray of Quinalphos 25% EC @ 400 ml in 200-400 l of water per acre or Thiamethoxam 25% WG @ 20 g in 200 l of water per acre is recommended to control aphids. Spray of Dichlorvos 76% EC@ 112.8-150.4 ml in 200-400 l of water per acre is recommended for control of Pink borer (leaf eating caterpillar). Spray of Cypermethrin 10% EC@ 220 ml in 200-320 l of water/ acre or Phorate 10% CG @ 7500 g per acre is recommended to control shoot fly. Spray of Quinalphos 25% EC @ 640 ml in 200-400 l of water per acre is recommended for control of Brown mite. Spray of Carbaryl 50% WP @ 800 g in 200 l of water/acre or Trichlorfon 5% GR @ 300 g per acre or Trichlorfon 5% DUST @ 300 g per acre or Trichlorfon 50% EC@ 300 ml per acre or Dichlorvos 76% EC@ 112.8-150.4 ml in 200-400 l of water per acre is recommended for control of Army worm/Cutworm. Spray of Quinalphos 25% EC @ 640 ml in 200-400 l of water per acre is recommended to control American pod borer), armyworm. Termite is controlled with spray of Thiamethoxam 30% FS @ 1.32 kg per 40 kg seeds or Chlorpyrifos 20% EC @ 3 – 4 ml/kg seed and 0.8-1.2 l/acre as soil application. Seed treatment with *Pseudomonas fluorescens* 1.75% WP @ 5 g/kg seed and foliar spray of *Pseudomonas fluorescens* 1.75% WP @ 5 g/l or Benomyl 50 % WP @ 2g/kg seeds or Carbendazim 50% WP @ 2 g/kg seeds or Carboxin 75% WP @ 2 -2.5 g/kg seeds or Tebuconazole 2% DS @ 0.2 kg/10 kg seed or Carboxin 37.5% + Thiram 37.5% DS @ 3.0 g/kg of seeds and application of neem cake@ 80 kg/acre is recommended for control of loose smut and flag smut. Seed treatment is done with Carboxin 75% WP@2 -2.5 g/kg seed or Tebuconazole 2% DS@0.2 kg/10 kg seed for control of flag smut and bunt diseases. Soil application of Carbofuran 3% CG @ 25 kg/acre is recommended for control of nematodes.

Harvest: Harvesting is done when leaves and stems turn yellow and become fairly dry but before it is dead ripe. It is done manually using serrated sickle. The crop is harvested at a seed moisture content of 18 to 20%. The cut stem with ears are bundled up and hanged to dry for 3-4 days. Threshing is done either beaten with stick or trampling the straw with bullocks. Use mechanical threshers to reduce the cost of threshing and winnowing. Wheat is harvested at 18 to 24% grain moisture content. The moisture content for safe storage is 10-12 %. The time of harvest in different wheat growing zones of India is presented in Table 16.

Table 16 Time of harvest of wheat in different growing zones of India

S.No.	Wheat growing zones	Time of harvest
1	North East Plain Zone	Second fortnight of March to mid-April
2	North West Plain Zone	Second fortnight of April
3	Central Zone	End of February to March
4	Peninsular Zone	Second fortnight of February to first fortnight of March
5	Hilly zone	May-June

Yield: The crop produces grain yield of 4-6 t/ha and straw of 7-8 t/ha from dwarf wheat varieties under irrigated conditions. Under rainfed conditions 2-2.5 t grain and 3-3.5 t straw/ ha may be obtained. The harvest index for high yielding varieties is 40 to 45% while for the local tall varieties is 25 to 30%. The grain moisture content is 10 to 12% for safe storage.

Processing: The traditional village stone namely the '*chakkis*' is used for grinding of wheat to make wheat flour. Presently the roller mill is used to separate wheat into its constituent parts. The thoroughly cleaned wheat grains are moistened slightly with water to toughen the bran and prevent it from being pulverized. The tempered or conditioned wheat is run between the first pair of corrugated roller that crakes the grain and partially flattens it. The crushed material goes to a sifter or bolter equipped with a series of inclined sieve, the topmost is relatively coarse and progressively finer sieves towards the bottom. The medium sized granular particles, mainly endosperm, retained on the other sieves, are known as 'middlings'; while the finer particles that pass through the fine silk bottom sieve are known as 'break flour'. The purified middlings (free from bran particles) are then passed through smooth reduction rolls for final granulation. Flour is sifted out after each reduction, the last reduction unit yielding pure white flour, free of branny matter and is ready for packing. The loss of thiamine and nicotinic acid is the least in the hand-driven *chakki atta* and roller mill *atta*. There is a popular belief among *chapati* eaters that *chapaties* prepared from hand driven *chakki atta* are sweeter and more palatable than those made from mill flour. The diastatic enzymes like the vitamins, may be destroyed in milling because of the high temperature. Flour yield from milling is approximately 70 to 74%.

Classification of wheat flour

(i) White flour is made from the endosperm only.
(ii) Brown flour includes grain's germ and bran,
(iii) Whole grain or whole meal flour is made from the entire grain, including the bran, endosperm, and germ.
(iv) Germ flour is made from the endosperm and germ, excluding the bran.

Wheat quality: The wheat grain or the wheat kernel is one seeded fruit, called caryopsis. A wheat kernel is about 5-8 mm in length and 2.5-4.5 mm in width. The average wheat kernel is comprised of more than 83% endosperm (energy for plant growth, carbohydrate and protein for people), 8% bran (protects seed) and seed coat material, 6.5% aleurone (fibre, B-vitamins, minerals) and 2.5 to 3.0% Germ (embryonic tissue and scutellum which nourishes seed, contains Vitamins-E and B, Antioxidants). Bran is removed during milling. Chemically, whole wheat grain consists of starch (60-70%), protein (10-12%) and minerals (1.4-2.3%), pentosans (6.0-9.5%), cellulose (2.5-3.3%), sugars and fats. Starch is the major component of wheat, being 53 to 62% of the grain and 62 to 71% of flour. Starch is composed of amylose and amylopectin. The amylose content of most wheat is between 25 and 30%, but in wheat preferred for Japanese udon noodles, the amylose content is closer to 20%. Starch granules can be mechanically damaged during the milling process and, this in turn, increases the water absorption of the flour and influences fermentation time during baking. Starch degradation caused by the enzyme α-amylase. When starch is heated in excess water, the granules swell and a paste is formed as the starch gelatinizes. Starch gelatinisation is important in baking as it assists in forming and stabilising the loaf crumb structure. It is also implicated in staling of baked goods. Starch swelling properties influence the appearance and textural properties of noodles, particularly white salted noodles, but are also important in other products such as batters. The dough properties are largely determined by variations in protein content and protein quality. Pentosans modifies the bread making property of flour, water absorption and gluten formation. It is capable of absorbing 12 to 13 times their own weight of water. Alpha amylase enzyme is present in embryo or germ of wheat kernels which is a critical component for various properties of dough and flour. It affects bread making property, gas retention, dough handling and bread texture.

Wheat is classified based on growing season, bran colour, kernel hardness and trade/ market. Wheat is classified a winter or spring based on growing seasons; red or white wheat based on bran colour; hard wheat and soft wheat based on kernel hardness. Kernel hardness is an indication of how much protein will be in wheat grain or flour. Hard wheat has medium to higher protein flour stronger gluten strength while Soft wheat contains lower protein flour weaker gluten strength. Red wheat produces dark and quite bitter bran which limits its use for whole wheat products and for products using flour that requires a bright and white appearance. The bran from white wheat does not impart a bitter flavour to grain products. The terms 'red' and 'white' wheat are used to identify the colour of the kernel and not of the colour of the flour that is eventually milled from those kernels. Red varieties generally exhibit more dormancy than white varieties. White wheat is more suited to areas that are dry during ripening and harvest and are favoured for the manufacture of certain types of flat bread and noodles.

There are five classes of wheat in India based on trade/market namely,

(i) Indian Medium Hard Wheat (IMHW) for *chapati* and related products

(ii) Indian Hard Wheat (IHW) for bread

(iii) Indian Soft Wheat (ISW) for biscuit

(iv) Indian *durum* Wheat (IDUW) for pasta and traditional products and

(v) Indian *dicoccum* Wheat (IDIW) for traditional and pasta products

There are six commercial wheat grades in the United States based on trade/market as

(i) Hard Red Winter Wheat has wide range of protein content, good milling and baking characteristics. It is used to produce bread, rolls and, to a lesser extent, sweet goods and all-purpose flour.

(ii) Hard Red Spring Wheat contains the highest percentage of protein, making it excellent bread wheat with superior milling and baking characteristics.

(iii) Soft Red Winter Wheat is high yielding, but relatively low protein. It is used for flat breads, cakes, pastries, and crackers.

(iv) Hard White Wheat is closely related to red wheat's. It has a milder, sweeter flavor, equal fiber and similar milling and baking properties. It is used mainly in yeast breads, hard rolls, bulgur, tortillas and oriental noodles.

(v) Soft White Wheat is used in much the same way as Soft Red Winter (for bakery products other than bread). It produces flour for baking cakes, crackers, cookies, pastries, quick breads, muffins and snack foods.

(vi) Durum Wheat is used to make semolina flour for pasta production.

There are six commercial Australian wheat grades *viz*.,

(i) Prime Hard Wheat has minimum protein content of 13%. It is hard-grained type. It has excellent milling quality and high dough strength

(ii) Hard Wheat has minimum protein content of 11.5% with superior milling quality.

(iii) Premium White Wheat has minimum protein content of 10% with high milling quality.

(iv) Standard White Wheat is comparable with Soft Red Winter and Soft White Wheat.

(v) Soft Wheat has weak doughs with low water absorption.

(vi) Durum wheat has good semolina yield with high yellow pigment levels.

Wheat kernel is classified as soft and hard based on hardiness (endosperm texture) and white and red based on colour. Hard and soft wheat has both white and red colour ecotypes. Soft grained wheat will produce more fine meal than harder grained wheat. Soft wheat is soft, opaque and less compact endosperm compared to hard wheat. Soft wheat starch contains water soluble protein (friabilin) located on the surface of the starch granule. Soft wheat contains 8 to 10 % protein which is used to prepare biscuits, cookies and cakes. The hard wheat contains 10 to 13% protein which is used to prepare bread, chapaties, pasta products, noodles, hard rolls, crackers, etc. Hard wheat has more protein, including more gluten, which makes it purposeful to bake bread, while soft wheat has a much lower protein content, which when milled produces 'cake flour' for sweet biscuits and cakes. Hard starch granules are larger and jagged shaped so that they fit tightly together making the kernel strong and hard to crack or break. Soft starch granules are small and round and break apart easily.

Hardness in wheat is related to the degree of adhesion between starch and protein. Wheat hardness is defined as a mechanical property of the individual wheat kernel or its resistance to crushing/grinding. Wheat hardness is influenced especially by genetic factors, but can be influenced also by environment and factors like moisture, lipids, pentosans and protein content. The presence or absence of a friablin protein is responsible for hardness or softness of wheat grains.

Wheat Hardness Index (WHI) = [peak height on hardness tester (Brabender units) + load value (g)]/[0.140 mm sieve threw weight (g) × 2]

Particle Size Index (PSI) value is calculated as follows:

PSI % = [0.075 mm sieve threw weight (g)/sample weight (g)] × 100

The following equation models can predict the flour yield of wheat varieties:

Flour yield (Hard wheat) = 36.2 – 0.30 TKW + 17.4 KW – 0.26 TW

Flour yield (Soft wheat) = 113.0 + 0.15 KL – 0.68 TW

where, TKW = 1000-kernel weight; KW = kernel width; TW = test weight and KL = kernel length.

Grain and flour protein content indicate potential water absorption, dough strength and extensibility. Wheat contains a protein called gluten. On dry weight basis, gluten consists of 75 to 80% protein and 5 to 10% lipids. The baking quality/ bread making quality is assessed based on gluten content of flour. Gluten is responsible for spongy texture of bread and other baked products. The gluten contains proteins insoluble in water and alcohol, *i.e.,* the prolamin storage proteins. These prolamins are divided into the monomeric gliadins and the polymeric glutenins. Gluten is a combination of two proteins, gliadin and glutenin. Glutenin is an alcohol insoluble protein fraction of wheat gluten while Gliadin is an alcohol soluble protein fraction of wheat gluten. Glutenin is said impart solidity to the gluten and gliadin is responsible for softness and stickyness. Wheat varieties with a high gliadin-glutenin ratio tend to have viscous, extensible doughs that are often suitable for cookie (biscuit) making. Those having a low gliadin-glutenin ratio have more elasticity and strength, which are desired for bread making.

Indian wheat flours are divided into 3 groups based on gluten content *viz.*, strong, medium strong and weak.

(i) **Strong gluten varieties**: These are most suitable for bread and for expression of high loaf volume, attractive brownish, crust, silky and small granular structure of crumbs. *e.g*: HI 977, NI 5439, HW 657, DWR 39.

(ii) **Medium strong gluten varieties**: These makes excellent chapattis. It allows good dough consistency, soft with high puffing and keeping quality. *e.g*: Sujata, Mukta, LOK 1,

WH 291, UP 2338, WL 711, WH 542, HP1209, C 306, HD 2403, HD 2189, HW 147, HI 1077, Kundan, Raj 3077.

(iii) **Weak gluten varieties**: These are suited for biscuits and cookies. *e.g*: Sonalika, HD 2285, HD 2278, HD 2501, K 8027, DC 787-3, PBW 175.

Utilization: Wheat grain is a staple food used to make flour for breads; cookies, cakes, breakfast cereal, pasta, juice, noodles and for fermentation to make beer, alcohol, vodka or biofuel. Bread is one of the most common staple foods prepared from wheat flour. Wheat-based noodles are a traditional food widely consumed throughout Asia. White-salted noodles are made by mixing flour, water and salt into a dry, crumbly dough. Udon are a type of white salted noodle popular in Japan. Yellow alkaline noodles are made from flour, water, salt and an alkaline salt solution (*kan sui*) mixed to form a dry, crumbly dough which is then sheeted with rolls to form a dough sheet. Instant noodles are an important class of alkaline noodles. Starches are often added to instant noodle dough to produce a chewier noodle texture. Biscuits are produced from flour and/or starch, fats (shortening) and other food ingredients. Biscuit flour may be milled from soft or hard wheat or a mixture of soft and hard wheat. Cakes are typically produced from flour, sugar, eggs and shortening, including emulsifiers, as well as other food ingredients. Cake flours may be milled from both hard and soft wheat. Pasta is a traditional food consumed in southern Europe that has now gained universal popularity. It is produced from a mixture of durum semolina and water, sometimes with the addition of eggs or flavour/colour components, which is extruded through a dye. Couscous is a major staple food in North Africa, particularly in Morocco, Algeria and Tunisia. Couscous may be produced commercially by carefully mixing durum semolina and water into crumbly dough mass.

Bread wheat has a high gluten content and hard grain which is good for baking qualities (elasticity). Bread wheat is used to prepare chapattis, tandoori, roti, naan, puri, samosa, noodles, laddu, bread, biscuit, cake, buns and pastry. Durum wheat is a soft grain used to prepare *suji* or *rawa*, *halwa,* chapatis, pasta products like macaroni and vermicelli, noodles, smack food, snack pellets, etc,. Dicoccum wheat is used to prepare laddu, sweet pan cake, pasta products like macaroni, vermicelli, etc. Compactum wheat is used to prepare biscuits, cakes, noodles, ice cream cones, puffed and flaked breakfast foods, soup thickeners, etc. In some cases the green crop is utilized as hay or silage. Wheat starch is used in the fabric and textile industry.

Wheat grain is a more widely traded product than wheat flour due to the following main reasons: (i) lower import tariffs on wheat grain as compared with wheat flour and ii) logistical and quality issues with transportation and shorter shelf life of flour as compared with wheat grain. The refined, enriched flours are made from the endosperm only. Kazakhstan, the Europe and Turkey have been the three main wheat flour exporters while the Brazil, Libya, Uzbekistan, Tajikistan, Iraq and Afghanistan are the wheat flour importers. Wheat flour is mostly used for bread production. Wheat straw often used as part of the roughage fed to ruminants and is used extensively for livestock bedding. Wheat straw is also used in making paper, wall board or packing material.

Constraints in wheat production

1. Biotic stresses

(i) Diseases: Stripe rust is more prevalent on north western parts and hills. Leaf rust occurs in all parts of the country. The stem rust is prevalent in Central and Peninsular parts.

Leaf Blight is more prevalent in north eastern plains followed by peninsular, central and low in north western parts of the coutry. Karnal bunt is a serious problem in northwestern plains. The powdery mildew is an emerging problem in northwestern plains. Aphids and termites are observed in a few pockets.

(ii) Weeds (Broad and narrow leaved): weeds such as *Phalaris minor* and wild oats *Chenopodium*, Rumex sp., *Medicago* sp are common while *Malva parviflora* occur more in zero tillage fields.

2. Abiotic stresses

(i) **Drought:** Central, peninsular and northeastern parts - 3.5 to 4 mha

(ii) **Heat:** Central, peninsular and northeastern parts - 3.5 mha in north eastern plains and 3-4 mha in central and peninsular parts.

(iii) **Suppressive soils / Soil health:** Salinity, alkalinity spread in about 2.5-3.0 mha in Northwestern plains, northeastern plains and Central parts of the country. Nutrient deficient soils *viz.*, 46% Zn, 17% B, 12% Mo, 11% Fe and 5% Cu. 38% S occur in wheat growing areas. Water logging occurs in 3.2 m ha in northwestern and northeastern plains.

REFERENCES

Ali, A., Atkins, I., Rooney, L. and K. Porter. 1969. Kernel dimensions, weight, protein content and milling yield of grain from portions of the wheat spike. Crop Science. **9:** 329-330.

Anderson, W.K. and J.R. Garlinge. (eds). 2000. The wheat book: principles and practice, Western Australia.

Bettge, A.D. and C.F. Morris. 2000. Relationships among grain hardness, pentosan fractions, and end-use quality of wheat. Cereal Chemistry. **77:** 241-247.

Bowden, W.M., 1959. The taxonomy and nomenclature of the wheats, barleys, and ryes and their wild relatives. Canadian Journal of Botany 37: 657–684.

Bushuk, W. 2000. Wheat Production, Properties and Quality. American Association of Cereal Chemists Inc. St. Paul MN. USA.

Chad Lee, James Herbek, David Van Sanford, and William Bruening. 2006. Cultural Practices. In: A Comprehensive Guide to Wheat Management in Kentucky. University of Kentucky, USA. p.13-19.

CIMMYT, 1985. Wheats for more tropical environments. A proceedings of the international symposium. CIMMYT, Mexico D. F. , Mexico. 354 pp.

Dixon, J., Braun, H.J., Kosina, P. and J. Crouch (eds). 2009. Wheat Facts and Futures. CIMMYT, Mexico.

Faměra, O. Hrušková, M. and D. Novotná. 2004. Evaluation of methods for wheat grain hardness Determination. Plant Soil Environ, 50(11): 489–493

Faridi, H. and Faubion, J.M. 1995. Wheat End Uses Around the World. American Association of Cereal Chemists, MN, USA.

Faruqi, N.Y.Z. and H.P. Singh. 1981. Chapati Making Properties of Indian Wheats – A review. Agricultural Marketing Vol.XXIV (1), pp 105.

Heyne E.G. 1987. Wheat and wheat improvement. American Society of Agronomy. Haworth Press Inc, Canada.

Hoseney, R.C. 1987. Wheat Hardness. Cereal Foods World**.** 32: 320-322.

Hruškova M. and Švec I. 2009. Wheat hardness in relation to other quality factors. Czech J. Food Sci., 27: 240–248.

James Herbek and Chad Lee. 2006. Growth and development. In: A Comprehensive Guide to Wheat Management in Kentucky. University of Kentucky, USA. p.6-12.

Khan, K. and P.R. Shewry. 2009. Wheat: Chemistry and Technology. Fourth edition. American Association of Cereal Chemists, St. Paul, MN, USA.

Kirby, E.J.M. and M. Appleyard. 1984. Cereal development guide, Coventry, UK: Arable Unit, National Agriculture Centre.

Klatt, A.R. (Editor), 1988. Wheat production constraints in tropical environments. CIMMYT, Mexico D. F. , Mexico. 410 pp.

Quisenberry, K.S. and L.P. Reitz. (Editors), 1967. Wheat and wheat improvement. American Society of Agronomy, Madison, Wisconsin, United States. 560 pp.

Saunders, D.A. and G.P. Hettel. (Editors), 1994. Wheat in heat-stressed environments: irrigated, dry areas and rice-wheat farming systems. CIMMYT, Mexico D. F. , Mexico. 402 pp.

Turnbull, K.M. and S. Rahman. 2002. Endosperm texture in wheat. J. Cereal Sci., 36: 327–337.

http://www.dwr.in/

http://farmer.gov.in/imagedefault/pestanddiseasescrops/wheat.pdf

http://www.cimmyt.org/english/docs/facts/whtfacts09.pdf

http://www.agri-outlook.org/

http://www2.ca.uky.edu/agc/pubs/id/id125/id125.pdf

http://www.dpi.nsw.gov.au/__data/assets/pdf_file/0008/516185/Procrop-wheat-growth-and-development.pdf

www.wheatfoods.org

http://www.namamillers.org/education/

http://www2.ca.uky.edu/agc/pubs/id/id125/02.pdf

http://www.slideshare.net/CIMMYT/presentations

http://www.iasri.res.in/

http://knowledgecenter.cimmyt.org/

CHAPTER

5

Barley–*Hordeum vulgare* L. (2n=14)

Vernacular names: *Jau* (H), *Barli arisi* (T), *Jaba* (Bengali).

Importance: Barley is considered to be the most ancient cultivated grain. It is a major staple food in several regions of the world. It is used for malt, beverage or food. It is considered fourth largest cereal crop in the world with a share of 7% of the global cereal production. The United Arab Emirates, Bhutan, Oman, Bahrain and Nepal import huge quantities of barley from India.

Origin: The hulless with short awn barley is originated in Nepal-Tibet-China region whereas the hulled awned type barley is originated in Abyssinia (Ethiopia) and Fertile Crescent area of the Near East (Mediterranean region).

Distribution: Barley is grown in Russia, USA, Canada, China, India, Turkey, France, Spain, United Kingdom, West Germany, Poland, Denmark, Iran, Iraq, Algeria, Ethiopia, Morocco and Australia. In India, it is grown in Uttar Pradesh, Bihar, Rajasthan, Haryana, Punjab, Madhya Pradesh, Karnataka and Odisha in plains and Himachal Pradesh, *Uttarakhand*, Jammu and Kashmir in the hills.

Botany: Barley is an herbaceous annual with a few tillers. It grows 0.5 to 1 m in height. The roots are adventitious and fibrous. The stem is with five to eight nodes. Leaves are linear. The auricles occur at the junction of the blade and sheath and consist of two claw-like appendages that appear to clasp the stem. Inflorescence is a terminal cylindrical spike or head. It is a normally self-pollinated crop, but cross-pollination occurs up to 10%. Fruit is a caryopsis. The thousand grain weight ranges from 45 to 58 g. The husk content is 10 to 11%. The protein content ranges from 9 to 12%. Barley is classified in different groups based on morphological features such as arrangement of spikelets, presence or absence of awns and adherence of chaff to grains. The seed weight is 25.64 g per 1000 seeds.

(a) **Classification based on arrangement of spikelets on rachis** *viz.,*

(i) Six-rowed barley *(Hordeum vulgare* L. emend.Lam): All florests of the spikelets are fertile; the central kernel is bigger than the laterals, ii) Irregular row barley *(Hordeum distichon* L. emend.Lam): Only the central florets are fertile, while the laterals are sterile and iii) Two-rowed barley (*Hordeum irregulare* E. Aberg and Wiebe): Central florets are fertile, whereas the lateral florets are distributed irregularly on the spike and are sterile.

(b) **Classification based on the presence or absence of awns in grains** *viz.,* i) Awned type and ii) Awnless type. Awnless types are also called hooded because of the hood shaped structure that develops in place of awn. Awned type are further sub-grouped into i) Smooth awned and ii) Rough awned.

(c) **Classification based on adherence of chaff to grains** *viz.,* i) Hulled (*syn.* husked) type: (ii) hull-less (*syn.* naked) type. In case of hulled barley the husk, *i.e.* floral glumes also called chaff remains attached to grains resulting in poor flour making. In the hull-less type the husk readily falls after threshing and naked grains free of chaff can be collected. Hulless barley is preferred over hulled types because it allows easier removal of the hull and has high digestible energy of the grain, especially for swine and poultry.

Growth stages: Germination (Braiding stage) starts with appearance of coleoptile at soil surface. The seedling emerges from the soil 5–6 days after germination. The seedling stage is up to 12 to 20 DAS. Tillering stage begin with appearance of stem from the coleoptile at the ground surface. Secondary tillers of 2 to 5 are produced in spring barley and five to ten tillers in winter barley. Tillering stage occur during 35 to 40 DAS. The jointing, booting or shooting stage refers to rapid elongation of internode in the main stem and a leaf develops from each node. The period of jointing stage lasts up to 55-65 DAS. Heading or ear emergence stage refers to the appearance of the whole earhead or emergence of collar of earhead from the sheath of the flag leaf. This stage lasts up to 70-75 DAS. The time of earhead emergence may differ from three to four weeks between early and late varieties. The anthesis begins in the central florets at 80 DAS. The milk stage lasts up to 85 DAS. Ripening stage lasts up to 100 DAS. At this stage grain filling and grain development starts resulted into hard dough stage. Physiological maturity refers to 95 to 100 % of grain germination in three or four days. At maturity, plant is completely dry. The grain is hard and horny and the stems are yellow. Maturity stage is attained on 100 and 135 DAS depending on varieties. The plant turns yellowish, loose stiffness and become droopy at maturity and becomes ready for harvest. The growth stages of barley are furnished in Table 1.

Table 1 Growth stages in barley

Stage	**Crop duration (days)**	
Sowing and germination	5	7
Seedling emergence and early growth	20	20-25
Tillering, initiation of ear primordia	30	30-35
Stem elongation and formation of ear primordia	45	50-60
Flag leaf, floret reduction, booting	60	65
Ear emergence	70	80
Flowering and grain initiation	75-85	95
Grain formation	90	107
Maturing of the grain	100	122
Total	105	136

Climate: Barley is a temperate cereal crop. It grows reasonably well in temperate as in sub-tropical regions of the world. It is best suited to dry conditions than hot humid areas. The crop can be grown up to an altitude of 4000 m in Himalayas from mean sea level. It is grown in areas receiving rainfall of 300 to 1000 mm of rainfall where drought is encountered. It is more drought-escaping, due to its early maturity, than drought-tolerant. Barley is a long day plant. A certain level of vernalization at 2°C, hastens awn emergence. The cardinal minimum, optimum and maximum temperature of 2°C, 15 to 25°C and 30°C respectively are required for growth and yield. It cannot tolerate frost at any stage of growth and incidence of frost at flowering at highly detrimental for yield. The optimum temperature at the sowing should be around 22- 24°C. The crop requires around 12-15°C during growing period and around 30°C at maturity. It generally requires cool weather during early growth and warm and dry weather at maturity. Floret numbers are reduced at higher temperature of 24°C in comparison to 18°C. The favourable soil temperature for growth and yield of barley is 18 to 20°C. Barley is susceptible to winter injury than wheat. Dry weather is best for growth during reproductive phase of barley. High temperature and humidity favour diseases such as rust, mildew and scab.

Table 2 Cardinal temperatures at different stages of barley

Development stage	Temperature (°C)		
	Minimum	**Optimum**	**Maximum**
Sowing and germination	2-4	20-25	27
Tillering, initiation of ear primordia	-----	< 8	-----
Beginning of stem elongation and formation of ear primordia	-----	< 9	-----
Flag leaf, floret reduction, booting	-----	< 14	-----
Flowering and grain initiation	-----	< 17	-----
Grain formation	-----	< 19	-----
Maturing of the grain	-----	19	-----

Soils: The well-drained sandy loam, loamy or clayey loam soils are suitable for barley production. Barley is also the most salt-tolerant among cereal crops, but it does not tolerate acid soils. The desirable soil pH is 6.0 to 8.5. It is very sensitive to waterlogging.

Season: Sowing time is 15th October to 15th November. The sowing dates of barley for different zones is presented in Table 3.

Table 3 Barley sowing dates for different zones

Zone	Sowing Time	Recommended dates
Northern hills zone	Normal	25th Oct. to 11th Nov
	Late	25 th Nov. to 0Ist Dec
North western plains zone	Normal	05 th Nov. to 15 th Nov
	Late	10 th Dec. to 16 th Dec
North eastern plains zone	Normal	15 th Nov. to 25 th Nov
	Late	10 th Dec. to 16 th Dec.
Central zone	Normal	12 th Nov. to 18 th Nov
	Late	02 th Dec. to 10 th Dec

Varieties: Cultivar choice depends on season, soil type and utility. Not all barley cultivars produce grain acceptable for malting. Maltsters prefer grains that are plump, well-matured and protein around 10 to 12%. The particulars of barley varieties grown in different ecological zones of India are presented in Table 4.

Table 4 Barley varieties grown in different ecological zones of India

Barley types	Variety	Year	Production Condition	Zone
Malt Barley	Alfa 93	1994	Irrigated (TS)	NWPZ
	BCU 73	1997	Irrigated (TS)	NWP, NEPZ, PZ
	RD 2503	1997	Irrigated (TS)	
	K 551	1997	Irrigated (TS)	
	DL 88	1997	Irrigated (TS)	
	DWR 28	2002	Irrigated (TS)	
	DWRUB52	2006	Irrigated (TS)	NWPZ
	RD2668	2006	Irrigated (TS)	NWPZ
	DWRB 73	2011	Irrigated (LS)	NWPZ
	DWRUB64	2011	Irrigated (LS)	NWPZ
	DWRB 73	2011	Irrigated (LS)	NWPZ
	DWRUB 64	2012	Irrigated (LS)	NWPZ
	DWRB 92	2013	Irrigated (TS)	NWPZ
	DWRB 91	2013	Irrigated (LS)	NWPZ
	BH 946	2014	Irrigated	NWPZ
	HUB 113	2014	Irrigated	NEPZ
Feed Barley (Irrigated)	RD 2035	1994	Irrigated (TS)	NWPZ
	RD 2503	1997	Irrigated (TS)	NWPZ
	RD 2552	1999	Irrigated (TS)	NWPZ, NEPZ
	BH 902	2010	Irrigated (TS)	NWPZ
	PL426	1995	Irrigated (TS)	Punjab
	K329	1996	Irrigated (LS)	Uttar Pradesh
	K508	1996	Irrigated (TS)	Uttar Pradesh
	K 409	1997	Irrigated (TS)	Uttar Pradesh
	Narendra Barley-2 (NDB940)	1999	Irrigated (TS)	Uttar Pradesh
	BH 393	2001	Irrigated (TS)	Haryana
	RD2592	2003	Irrigated (TS)	Rajasthan
	PL751	2006	Irrigated (TS)	Central Zone
Feed Barley (Rainfed)	RD 2508	1997	Rainfed & Irrigated (LS)	NWPZ, NEPZ
	K 560	1997	Rainfed (TS)	NEPZ
	K 603	2000	Rainfed (TS)	NEPZ
	RD2624	2003	Rainfed (TS)	NWPZ
	RD2660	2006	Rainfed (TS)	NWPZ
	PL419	1995	Rainfed (TS)	Punjab
	Getanjali (K1149)	1997	Rainfed (TS)	Uttar Pradesh
	JB58	2004	Rainfed (T.S.)	Madhya Pradesh
	Jawahar Barley 1	2009	Irrigated (TS)	Madhya Pradesh

Dual Purpose Barley	RD 2035	1994	Irrigated (TS)	NWPZ
	RD 2552	1999	Irrigated (TS)	NWPZ, NEPZ
	RD2715	2008	Irrigated (TS)	Central zone
	BHS380	2010	Rainfed (TS)	NHZ
	UPB 1008	2011	Rainfed	Northern Hills
	VLB 118	2013	Rainfed	Northern Hills
	BHS 400	2014	Rainfed	Northern Hills
Saline/ sodic soils tolerance	DL 88	1997	Irrigated (TS & L S)	NWPZ
	RD 2552,	1999	Irrigated (TS)	NWPZ, NEPZ
	Narendra Barley-1 (NDB209)	1999	Irrigated (TS)	Uttar Pradesh
	Narendra Barley-3 (NDB1020)	2001	Irrigated (TS & LS)	Uttar Pradesh
	NDB1173	2004	Irrigated (TS)	NWPZ, NEPZ
Nematode resistance	RD 2035	1994	Irrigated (TS)	NWPZ
	RD 2052	1991	Irrigated (TS)	Rajasthan
Cold Tolerance/ rust resistance	HBL 113	1994	Rainfed (TS)	N Hills Zone
	HBL 276	1999	Rainfed (TS)	N Hills Zone
	BHS 352	2003	Rainfed (TS)	N Hills Zone
	HBL316	1993	Rainfed (TS)	H.P.
	VLB56	2004	Rainfed (TS)	Uttarakhand
	VLB85	2008	Rainfed (TS)	Uttarakhand
	PRB502	2009	Rainfed (TS)	Uttarakhand
	Sindhu (NBL11)	1999 (2005)	Rainfed (TS)	J & K
	Nurboo	1999 (2005)	Rainfed (TS)	J & K

Source: http://www.dwr.in/

Land preparation: Two to three ploughings are given to have a fine tilth. Cultural practices for barley are similar to those for wheat.

Seeds and sowing: Seed rate and spacing for barley in different growing environments is furnished in Table 5. Continuous line sowing with a row spacing 25 cm is required. The depth of sowing is 2 to 3 cm.

Table 5 Seed rate and spacing for barley in different growing environments

Production condition	Seed rate (kg/ha)	Spacing (cm)
Irrigated Timely sown	100	23
Irrigated Late sown	125	18
Rainfed Plains	100	23
Rainfed Hilly Region	100	23

Weed management: Two hand weeding are given on 20-25 DAS and 35-45 DAS. 2,4-D Na salt or 2,4 D amine salt @ 0.75 kg/ha can be applied as post emergence to control weeds. Broad leaved weeds can also be controlled with the application of Algrip 20 WP

(Metsulfuron Methyl) @ 20 g + 500 ml surfactant or Affinity 40DF (Carfentrazone ethyl) @ 50g per ha in 500 litre of water at 40-45 DAS. Isoproturon or methabenzthiazuron or metaxuron @ 1.5 kg/ha as post-emergence is effective to control grasses.

Manures and fertilizer application: Farm yard manure or compost @ 12.5 t/ha is to be applied basally during last ploughing. Blanket application of 60-30-0 and 30-20-0 kg per ha of N, P_2O_5 and K_2O are recommended for irrigated and rainfed condition respectively. Half of N and entire P_2O_5 are to be applied basally. The remaining half of N is applied after first weeding on 20-25 DAS for irrigated crop. The full quantity of N and P is applied basally for the rainfed crop. The fertilizer recommendation in different growing environments of barley is presented in Table 6.

Table 6 Fertilizer recommendations in different growing environments of barley

Zone/State	ecosystem	N : P : K (kg/ha)
Northern Hill Zone	Rainfed	40:20:20
NWP Zone and NEP Zone	Irrigated timely sown	60:30:20 (feed barley) NEP Zone 90:40:20 (malt barley)
	Irrigated Late sown	60:30:20
	Rainfed	40:20:20
Dual Purpose in Plains and Hills	Irrigated/ rainfed	75:30:20 (plains) Plains and Hills 60:30:20 (hills)

Water management: Barley is generally grown as rainfed crop. Generally two or three irrigation are recommended when there is irrigation facility. Depending upon the water availability, irrigation may be given at suitable growth stages. Crown Root Initiation (25-30 DAS), Active tillering (30-40 days) and flowering (65-70 days) are critical crop growth stages for moisture stress. The crop is irrigated with IW/CPE ratio of 0.9 to produce maximum grain yield.

Cropping systems: Barley + chickpea and barley + lentil intercropping are widely practiced. It is generally grown in rotation with pearl millet (*bajra*), maize, rice, cotton, groundnut, green gram and moth bean. The crop rotations followed in respect of barley is furnished in Table 7.

Table 7 Barley based crop rotations prevailed in different states of India

State	Crop rotation
Bihar	Sugarcane-Barley, Sesame-Barley, Pigeonpea-Barley
Jharkhand	Sugarcane-Barley, Sesame-Barley, Pigeonpea-Barley
Haryana	Sugarcane-Barley
Himachal	Pradesh Maize-Barley, Rice-Barley
Jammu & Kashmir	Rice-Barley
Punjab	Rice-Barley, Maize-Barley, Cotton-Barley
Rajasthan	Cluster bean-Barley
Uttar Pradesh	Maize-Barley, Rice-Barley, Sorghum-Barley, Pigeonpea-Barley, Sugarcane-Barley, Bajra-Barley

Pests and diseases management: Seed treatment with Chlorpyriphos 20 EC or Formathion 25 EC @ 6 ml /kg of seed to control termite damage. Seed treatment with Carbendazim (Bavistin) @ 2g or carboxin (Vitavax) @ 0.2g/kg or Tebuconazol (Raxil-2 DS) @ 1g per kg seed is quite effective to control rusts, covered smut and stripe disease. Foliar spray of Dithane M 45 @ 0.2% at the initiation controls the stripe disease disease.

Harvest: Harvesting is to be done as the crop matures. Barley is considered to be physiologically mature at approximately 35% moisture in the seeds. After physiological maturity, 10 or 15 days are required to harvest barley with combine in temperate dry lands. If this duration is exceeded, crop will get too dry and then cause shattering at harvest. Threshing is done either by trampling bullocks or with threshers. The crop is harvested by hand using a sickle or by combine. Threshing of malting and 'naked' barley requires special care to avoid too much broken seed.

Yield: Grain yield of 2.0 to 3.5 t/ha and straw yield of 4.0 to 5.0 t/ha can be obtained. The grain has to be dried to 14% moisture or less for safe storage. On an average, 18 to 24 and 2.4 to 3.5 t/ha of green fodder and grains, respectively can be produced from dual purpose barley crop.

Utilization: Barely is the principal grain used in malting for brewing beer and other liquors like whisky, brandy etc. A good malting barley contains a high percentage of starch and relatively low amounts of proteins. American malters prefer the six row barley over two row barley. Two-rowed barleys are widely used for brewing. A decoction of barley in water, 'barley water', is valued medicinally for the treatment of the inflammation of the membranes of the chest and feverish disorders. Barley gruel is a commonly fed to sick persons. Barley flour mixed with wheat flour is used for preparation of chapattis. Young plants provide nutritious green fodder for cattle. It is grown for forage purposes. Good quality hay is prepared from barley. Barley straw is also used for animal bedding and as cover material for hut roofs.

REFERENCES

Akar, T, Avci, M. and F. Dusunceli. 2004. Barley: Post-harvest operations. http://www.fao.org/fileadmin/user_upload/inpho/docs/Post_Harvest_Compendium_-_BARLEY.pdf

Briggs, D.E. 1978. Barley. Chapman & Hall, London, United Kingdom. 612 p.

Bhatty RS. 1999. β-glucan and flour yield of hulless barley. *Cereal Chemistry* 76:314-315.

Cavallero A, Empilli S, Brighenti F and A.M. Stanco. 2002. High (10 → 3,1 → 4)-β-Glucan barley fractions in bread making and their effects on human glycemic response. *Journal of Cereal Science* 36: 59-66.

Goyal, A, Pradhan, S. and V.K. Bajaj. 2012. Screening of salt tolerance in different varieties of *Hordeum vulgare*. *World Applied Sciences Journal* 16 (7): 926-932.

Jaiswal, S.K., Pandey, S.P., Sharma, S., Prasad, R., Prasad, L.C. Verma, R.P.S. and AK Joshi. 2010. Diversity in Indian barley (*Hordeum vulgare*) cultivars and identification of genotype-specific fingerprints using microsatellite markers. *Journal of Genetics* 89: 46-54.

Kumar, D., Narwal, S., Verma, R.P.S., Kharab, A.S., Kumar, V. and I. Sharma. 2012. Genotypic variability in β-glucan and crude protein contents in barley genotypes. *Journal of Wheat Research* 4(2):61-68.

Kumar, D., Kumar, R., Verma, R.P.S., Verma, A. and I. Sharma. 2013. Recent trends in breeder seed production of barley (*Hordeum vulgare*) in India. *Indian Journal of Agricultural Sciences* 83(5): 576-578.

Nevo, E., 1992. Origin, evolution, population genetics and resources for breeding of wild barley, Hordeum spontaneum, in the fertile crescent. In: Shewry, P.R. (Editor). Barley: genetics, biochemistry, molecular biology and biotechnology. CAB International, Wallingford, United

Kingdom. pp. 19–44.

Rasmusson, D.C. 1985. Barley. American Society of Agronomy, Madison, Wisconsin, United States. 522 p.

Ram, S. and R.P.S. Verma. 2002. β-Glucan content and wort filtration rate of Indian barleys. *Cereal Research Communications* 30: 181-186.

Sarkar, B., Verma, R.P.S. and B. Mishra. 2008. Genetic diversity for malting quality in barley (*Hordeum vulgare* L.). *Indian Journal of Genetics* 68 (2): 163-170.

Singh, D.P., Babu, K.S., Mann, S.K., Karwasra, S.S., Kalappanavar, I.K., Singh, R.N., Singh, A.K. and S.P. Singh. 2010. Integrated pest management in Barley (*Hordeum Vulgare*). *Indian Journal of Agricultural Sciences,* 80: 437-442.

Slafer, G.A., Molina-Cano, J.L., Savin, R., Araus, J.L. and I. Romagosa. 2002. Barley science: recent advances from molecular biology to agronomy of yield and quality. Food Products Press, New York, United States. 565 pp.

Verma, R.P.S., Sharma, R.K. and B. Mishra. 2005. Future of barley for malt, feed and fodder in India. Directorate of Wheat Research, Karnal -132001. *Technical Bulletin No.* 9, p 28.

Verma, R.P.S., Kharub, A.S., Kumar, D., Sarkar, B., Selvakumar, R., Singh, R., Malik, R., Kumar, R. and I. Sharma. 2011. Fifty years of coordinated barley research in India. Directorate of Wheat Research, Karnal-132001. *Research Bulletin No.* 27: 46.

Verma, R.P.S., Kumar, V., Sarkar, B., Kharub, A.S., Kumar, D., Selvakumar, R., Malik, R. and I. Sharma. 2012. Barley cultivars released in India: Names, parentages, origins and adaptations. Directorate of Wheat Research, Karnal, India. *Research Bulletin No.* 29: 26.

Vishnu Kumar, Anil Khippal, Jogendra Singh, R. Selvakumar, Rekha Malik, Dinesh Kumar, Ajit Singh Kharub, Ramesh Pal Singh Verma and Indu Sharma. 2014. Barley research in India: Retrospect & prospects. Journal of Wheat Research 6(1):1-20.

von Bothmer, R., Jacobsen, N. and C. Baden. 1995. An ecogeographical study of the genus Hordeum. 2nd Edition. Systematic and ecogeographic studies on crop genepools 7. IBPGR, Rome, Italy. 129 pp.

CHAPTER

6

Oats–*Avena sativa* L. (2n=21)

Vernacular name: *Jaie, Javi, Jau* (Hindi); *Oats arisi* (Tamil); *Yavalu* (Telugu); Avoine (French); Avena (Spanish); Avena (Italian); Hafer (German).

Importance: The grain is used almost exclusively for feeding horses and dairy cattle. The crop is of minor importance in India, mainly used as a fodder crop. The health effects of oat rely mainly on the total dietary fibre and β-glucan content. The β-glucans exhibits an antioxidant property is included in the soluble dietary fibre fractions of oats that participates in glucoregulation and causes a decrease in serum cholesterol levels in humans. The consumption of oats is recommended in diet for hypercholesterolemic patients

Origin: Oats is originated in Western Europe.

Distribution: Oats is grown in Russia, USA, Canada, Poland, West Germany, China, Turkey, Canada, India, Australia, South Africa, Argentina, France, Finland, Spain, Italy, Sweden, United Kingdom. In India, oats is grown in Himachal Pradesh, Jammu and Kashmir, Madhya Pradesh, Punjab, Rajasthan, Haryana, Uttar Pradesh, Maharashtra and West Bengal.

Botany: A sub–erect annual; roots are shallow and adventitious, three to five culms with hollow nodes; leaves are long and narrow, apex drooping, Oats can be distinguished from other cereals by the bluish tint of the foliage, which is also devoid of auricles at the base of the leaf blades. The leaf sheath firmly envelops the internode (closed type) – a feature unusual in grasses. The inflorescence is a branched panicle, the branches either equally distributed on all sides (spreading type) or falling to one side ('side' or 'horse mane' oat). It may be either erect or drooping. Self-pollination occurs while cross-pollination seldom exceeds one per cent. The grain, a caryopsis, is more slender than that of wheat, usually greyish yellow and firmly invested by a hull (lemma and palea). There may be two or one kernel (seed) per spikelet. The oat kernel is a caryopsis which is also known as 'groat'. As the grain matures, the lemma and palea remain attached to the kernel forming the 'hull' or 'husks'. The hull generally comprises 25 to 30% of the total weight of the grain.

Climate: Oats is a temperate cereal. It is best suited to cool and humid climate. It grows up to an altitude of 600 m from MSL. The crop requires 200 mm of rainfall. Oats is a long day plant. The cardinal minimum, optimum and maximum temperature for germination is 2°C, 25°C and 30°C respectively. The optimum temperature for growth and yield of oats is 24 to 26°C. The optimum temperature during grain filling is 22°C. It requires 60% of relative humidity to produce good grain yield. Awn emergence is hastened at vernalization level of 2°C. Oat is the most frost-sensitive of cereals. Its heat requirements are less than those of wheat. The crops with increasing order of tolerance to frost are oats, barley, wheat and rye.

Soils: Well drained loamy soils are best suited for oats but grown in all types of soils with a pH of 5.8 to 6.5.

Field preparation: Two to three ploughings are given to have a fine tilth.

Season: Oats is a *rabi* season crop. The crop growing period is November-December to February-March. The best sowing period for Spring oats is 15th January to 15th March while for Winter oats is 15th September to 15th October.

Varieties: NP 1, NP 2, NP 3, NP 27, NP 101, Western 2, Western 11, Algerian 19, Kent, JARI 114, Fresh Hullen are grain types while Brumker 10 is a fodder type variety.

Seeds and sowing: Seed rate is 80 to 100 kg/ha. Continuous line with row spacing of 22.5 cm at depths of 2 to 4 cm is recommended for grain purpose. Broadcasting is done for forage purpose.

Weed management: Two hand weeding are given on 20 and 35 days after sowing.

Manure and fertilizer application: Basal application of farm yard manure or compost 12.5 t/ha of along with 50-20-0 of N, P_2O_5 and K_2O kg/ha is adequate for oats crop.

Water management: Oats is grown as both irrigated and rainfed crop. Three irrigations are recommended for the irrigated crop.

Cropping systems: Oats based crop rotation and inter-cropping systems are jowar-oat-maize; maize-oat-maize; cowpea-oat+ mustard–maize+cowpea and jower+cowpea-oat+lucerne.

Harvest: Oats requires 3 to 4 months to mature. Harvesting is started when seeds are 90% mature and moisture is below 33%. At this stage, the stem shows green colour, 1/3 panicle green, 1/3 panicle green mature and 1/3 panicle mature. Harvesting should be done soon after maturity as delay causes shattering of grains. Two to three cuttings are taken for fodder purpose. Threshing is done either by trampling with bullocks or with threshers. Hulls constitute 25 to 30% of the total grain weight. The moisture content of the grain should be 14% or less for safe storage.

Yield: Grain yield is 500 to 600 kg per ha. Straw yield is 1.5 to 2.0 tonnes per ha.

Utilization: Oats is grown for both grain and forage purposes. The palatability of oats is the best of the small grains. Oats is used for medicinal purposes and is a good source of protein, fibre and minerals.

REFERENCES

CPG. 2012. Crop Production Guide. Department of Agriculture, Government of Tamil Nadu, Chennai and Tamil Nadu Agricultural University, Coimbatore.

Mushtaq Ahmad, Gul-Zaffar, Z. A. Dar and Mehfuza Habib. 2014. A review on Oat (Avena sativa L.) as a dual-purpose crop. Scientific Research and Essays. 9(4): 52-59

Suttie, J.M. and S.G. Reynolds. 2004. Fodder Oats: A World Overview. FAO, http://www.fao.org/docrep/008/y5765e/y5765e00.htm.

Welch, R.W. 1995. The Oat Crop: Production and Utilization. Chapman and Hall, UK. 584 p.

CHAPTER 7

Rye–*Secale cereale* L. (2n=14)

Vernacular names: Rye, Grain rye, (English); *Seigle* (French); Centeno (Spanish); Segale (Italian); Roggen (German).

Importance: Rye is important as a food grain, and pasture crop. Rye grains are used for bread making, feeding livestock and producing alcoholic beverages. Rye is often grown either for its grain or as a cover crop to reduce erosion and compaction, or as a green manure.

Origin: Rye is originated in southern Europe to Syria, Armenia, Iran, Turkestan and the Kirghis Steppe.

Distribution: The leading rye grain producers are the Russia, Poland, Germany, Czechoslovakia, Hungary, Ukraine, Denmark, Austria, Canada, USA and Turkey.

Botany: Rye is tufted annual grass, 1 to 2 m high with a tendency towards a perennial habit, putting forth new plants from the stubble. In external appearance, it resembles wheat. The stems of the rye plant are larger and longer than wheat. The inflorescence is awned having two flowered spikelets. It is cross-pollinated crop and relies on wind-borne pollen. Fruit a caryopsis (grain), oblongoid, 4.5–10 mm × 1.5–3.5 mm. The mature grain is more slender than that of wheat and usually greyish yellow. There are two cultivated species in rye. *Secale cereale* L. is grown in USA while *Secale fragile* Bieberst. is grown in Southwestern Asia.

Growth stages: Rye germinates within 4 days at a soil temperature of 4–5°C, and more rapidly at higher temperatures. At the appearance of the fourth leaf, tillers and roots are formed to anchor the plant. In each spike 40–45 spikelets are initiated, 30–35 of which bear 1–2 grains, resulting in 45–55 grains per spike. Flowering lasts 3–5 days for a spike and 8–12 days for a rye crop. Rye is cross-pollinated by wind. The post-floral period for grain-filling is 4–5 weeks. The period from sowing to harvesting varies from 4 to 6 months. The duration of growth is largely dependent on temperature during reproductive development. The particulars of different growth stage of rye are furnished in Table 1.

Table 1 Growth stages in Rye

Stage	Crop duration (days)
Sowing and germination	7
Seedling emergence and early growth	11
Tillering, initiation of ear primordia	12-48
Stem elongation and formation of ear primordia	49-67
Flag leaf, floret reduction, booting	68-79
Ear emergence	80-86
Flowering and grain initiation	87-106
Grain formation	107-131
Maturing of the grain	132-152
Total	160

Climate: It is a temperate cereal crop. It is the most winter hardy of all cereals. Its chief characteristic is its ability to survive temperatures too low for wheat production. Winter rye is mainly grown between 40 and 65°N. Spring rye are occasionally grown at high elevations up to 3000 m from MSL in subtropical and tropical areas. Winter rye is a long-day plant. They are less sensitive to daylength and do not need vernalization. Their flowering and seed set are satisfactory at a daylength of 12–13 hours. It is grown in regions where the average rainfall ranges from 500 to 750 mm. It is tolerant to drought. The cardinal minimum, optimum and maximum temperatures for growth are 2°C, 15 to 25°C and 30°C respectively. Tillering, shoot growth and flower initiation require low temperatures of 10 to 15°C. It requires exposure of cool temperature of 13°C or lower to induce flowering. Flowering is favoured by dry and sunny weather. The mean daily temperature must not exceed 20°C during reproductive development. The optimum temperature for ripening is 16 to 20°C. Rye requires more chilling than wheat; for this reason the equatorial limit of rye is at higher latitude than that of wheat.

Soil: Rye grows well on well-drained light loams and sandy soils than on heavy clay soils. It is able to tolerate soil acidity. It does best at a pH 5.0 to 7.5. Rye is a salt tolerant crop. It tolerates up to a soil salinity of 11.4 dS m^{-1}. Each unit increase in salinity above 11.4 dS m^{-1} reduces the grain yield by 10.8%. Bread quality decreased slightly with increasing levels of salinity. Straw yield was more sensitive to salinity than the grain yield. Some varieties can tolerate waterlogging while others do well in dry soils.

Land preparation: Two to three ploughings are given to have a fine tilth.

Varieties: Dakold, Rosen, Petkus, Pierre, Caribou, Imperial, Emerald.

Seeds and sowing: Seed rate is 75 to 100 kg/ha. Continuous line sowing with a row spacing of 22.5 cm is recommended for grain purpose. Broadcasting is done for forage purpose.

Weed management: Two hand weedings are given on 20 and 35 days after sowing.

Manure and fertilizer application: Basal application of FYM or compost @12.5 t/ha along with 50-20-0 of N, P_2O_5 and K_2O kg/ha is recommended.

Water management: It is a drought tolerant crop. It has an extensive root system and adjusts scarcity of moisture. Rye uses about 20 to 30% less water per unit of dry matter formed than wheat. Rye is grown as both irrigated and rainfed crop. Three irrigations are recommended for the irrigated crop.

Pest and diseases management: Ergot, a fungal disease is a prevalent in cereal rye crops. Ergot produces black growths called sclerotia, which are visible in the heads of the rye

Harvest: The moisture level in rye should be from 15 to 22% at harvest. It can be harvested and threshed in one operation with a combine, or swathed and later threshed. To reduce shatter loss when direct combining, begin harvest at about 22% moisture and follow by drying. Threshing is done either by trampling with bullocks or threshers. Grain moisture should be 12% for long term storage. For best forage quality, cut rye between early heading and the milk stage of growth.

Yield: Average rye grain yield varies from 3 to 4 t/ha and straw yield from 4 to 5 t/ ha. When green chopped in the boot stage, rye can produce dry matter yield of 3 to 5 t/ha.

Utilization: The rye is harvested for grain, with the remainder used as pasture, hay, or as a cover crop. Rye grain is milled to prepared flour, alcoholic beverages, paper, etc. Rye is considered inferior to wheat in production of high-volume pan breads, because its dough lacks essential elasticity and gas-retention properties. Rye is usually mixed with 25 to 50% wheat flour for bread making. The grain is also used as feed for cattle. Rye is used as good pasture crop and livestock bedding material. It can also be grown as a cover or green manure crop.

REFERENCES

Bushuk, W. 2001. Rye production and uses worldwide. Cereal Foods World 46:70-73.

Casey, P.A. 2012. Plant guide for cereal rye (Secale cereale). USDA-Natural Resources Conservation Service, Plant Materials Center, Elsberry, MO.

Deodikar, G. B. 1963. Secale cereale Linn. Indian Council of Agricultural Research, New Delhi.

http://plants.usda.gov/plantguide/pdf/pg_sece.pdf

http://plants.usda.gov/java/

http://www.hort.purdue.edu/newcrop/afcm/rye.html

CHAPTER

8

Triticale–*Triticale hexaploide* L. (2n=42, 56)

Importance: Triticale is used as feed grain and in the milling industry. It produces higher forage and silage yields than oats, barley, wheat or rye. It is grown on a cover crops to prevent runoff and erosion.

History: The first hybrid made between wheat and rye in Scotland in 1875. The name triticale (trit-ah-kay-lee) has been created by combining *Triticum* and *Secale,* the genus names for wheat and rye.

Distribution: Triticale is grown in Russia, France, Australia, Poland, USA, Germany, Bulgaria, Brazil, Portugal, Spain and South Africa.

Botany: Triticale resembles wheat both in plant type and kernel characteristics. The main difference lies in its greater vigour relative to wheat and its larger spike and kernel size. The culms are thick, hence resistant to lodging. The inflorescence (spike) of triticale resembles that of wheat, rather than rye. Triticale is a self-pollinated crop.

Climate: The solar radiation level of 76 W m^{-2} produces excellent growth of triticale. It exhibits vigorous growth and normal floral initiation at a day temperature of 22 to 24°C and a night temperature of 10°C. The 17°C day/night temperature is optimum for seed development. They are also more frost hardy than wheat.

Soil: Soil requirements for triticale are similar to wheat, but it is found to be considerably more tolerant to low pH of 6.0 to 6.5. Grain yields of triticale under field conditions are unaffected by soil salinity upto 7.3 dS m^{-1} electrical conductivity of the saturated soil extracts in the root zone. Each unit increased in salinity above 7.3 dS m^{-1} reduces grain yield by 2.8%. Triticale is a salt tolerant crop.

Season: The rainfed crop is sown in October, while irrigated crop is sown in mid November.

Land preparation: Two to three ploughings are given to have a fine tilth.

Varieties: The promising triticale varieties are Triticale 70-2, Triticale DTS-703, Triticale DTS-551, TL 419, TL 2942, DT 46, ST-69-1, Armadillo PPV-13, Amphidiploid-1, Triticale arm, PM-312,T4, and PC-202.

Seeds and sowing: Seed rate is 75 to 100 kg/ha. Continuous line spacing with a row spacing of 22.5 cm at 5-8 cm depth is recommended for grain purpose. Broadcasting is done for forage purpose.

Weed management: Two hand weedings are given on 20 and 35 days after sowing.

Manure and fertilizer application: Basal application of farmyard manure or compost @ 12.5 t/ha along with 50-20-0 of N, P_2O_5 and K_2O kg/ha is recommended.

Water management: Triticale is grown in regions receiving the average annual rainfall from 500 to 750 mm. Five to seven irrigations are recommended for the irrigated crop. It is a drought tolerant crop.

Harvest: Harvesting should be done soon after maturity. Threshing is done through trampling with bullocks or threshers. The crop matures in 120-150 days when grown for grain purpose depending on variety and time of sowing. It is harvested at late boot stage and early-heading stages (25% heads emerged) for fodder purpose.

Yield: Triticale has a grain yield potential of 4 to 5 t/ha. The green fodder yield is 7.5 t/ha. The harvest index is 16%.

Utilization: Triticale flour is used alone or blended with wheat for bread making by small scale farmers. Flakes, bread, cakes are made from triticale flour. Triticale flour is suitable for making *chapaties*. Gluten protein in flour protein is 50% as against 78% in wheat. It produces high levels of alpha-amylase activity which allows performing well in malting and brewing. The grain is used as a feed for livestock and poultry.

CHAPTER

9

Maize–*Zea mays* L. (2n=20)

Vernacular name: Corn, Indian corn, American corn (English), *Makkacholam* (Tamil), Makka (Hindi), *Makka jonna* (Telegu), *Bhutta* (Bengali).

Importance: Maize is the third most important cereal grain after wheat and rice. Globally, maize is known as 'queen of cereals' because it has the highest genetic yield potential among the cereals. The term "maize" is derived from the ancient word mahiz from the Taino language, a now extinct Arawakan language, of the indigenous people of pre-Columbian America. The name 'maize' is derived from the South American Indian Arawak name '*mahiz*'. Maize is used as a staple human food particularly in developing countries and as feed for livestock and raw material for many industrial products in developed countries. In India, maize is used as human food (25%), poultry feed (49%), animal feed (12%), industrial (starch) products (12%), beverages and seed (1% each). USA, Brazil, Argentina and India are the major exporters of maize while Japan, South Korea, Mexico are the major importers. Maize is largely exported from India to Bangladesh, Malaysia, Vietnam, Indonesia and the United Arab Emirates.

History: In 1492 AD, when Columbus reached the northern Antilles near San Salvador, it has been inhabited by the indigenous Tahino people who called their staple crop as mahiz. The Spaniards took enough of these giant grains for distribution on their voyage and propagated the name mahiz throughout. The word 'maize' is commonly used in English while 'maiz' in Spanish. For the English settlers in the New World, maize was a new crop and they lacked a proper word to call maize. So they called it Indian corn, which later became corn. Corn in England means wheat; in Scotland and Ireland, it refers to oats. Corn mentioned in the Bible probably refers to wheat or barley. In Southern Africa, maize is commonly called mielie (Afrikaans) or mealie (English), words derived from the Portuguese word for maize, milho. In United States, Canada and a few other countries, corn and maize are one and the same, and is meant for the plant that produces kernels used for cooking. However, the term corn is preferred over maize for food products that are made from it, such as corn flour, corn starch, cornmeal etc. However, with regard to commodity trading, corn is only to refer to maize, and does not include any other grains. In culinary, the corn is generally mentioned as sweet corn, popcorn, cornflakes, baby corn, dent corn, flint corn, flour corn, and waxy corn. Maize is preferred in formal, scientific, and international usage. 'Maize' is used by agricultural bodies and research institutes such as the FAO and CSIRO.

Origin: Maize is originated in Mexico, Andean highlands of Peru, Bolivia and Ecuador.

Distribution: Maize is grown in USA, Canada, Brazil, Mexico, Argentina, China, India, Indonesia, France, Ukraine, and South Africa. The maize is cultivated throughout the year in all states of the country for various purposes including grain, fodder, green cobs, sweet corn, baby corn, popcorn in peri-urban areas. The predominant maize growing states that contributes more than 80 % of the total maize production are Andhra Pradesh (20.9%), Karnataka (16.5%), Rajasthan (9.9%), Maharashtra (9.1%), Bihar (8.9%), Uttar Pradesh (6.1%), Madhya Pradesh (5.7%), Himachal Pradesh (4.4%).

Botany: Maize is a tall varying in height from 1 to 4 m herbaceous annual crop. The roots are adventitious and fibrous. Brace roots or prop roots or stilt roots form from the bottom nodes. Leaves are large, narrow, opposing leaves (about a tenth as wide as they are long), borne alternately. Stem is cylindrical and solid often single stalked, rarely tillering. The number of leaves per plant varies from 12 to 20. It is normally a monoecious plant having two types of inflorescence. The male inflorescence called the 'tassel' is terminal on the main axis and is a much-branched panicle. The female inflorescence is known as the 'cob' or 'ear'. The overlapping sheaths of modified leaves which cover the inflorescence, forming the 'husk' of the ear. The ovary is surmounted by a long slender style ranging from 36 to 40 cm in length. These slender styles of the many spikelets grow rapidly and emerge from the top of the husk and these are together generally called as the 'silk'. Tassels appear 1 to 3 days prior to silking. A single anther may shed 2500 pollen grains. Maize is predominately cross-pollinated. The fruit is a caryopsis. The embryo in maize popularly called the 'germ' which contains oil; the oil percentage may be even up to 50% of the germ. Besides oil, the embryo contains proteins, minerals and sugars. Corn has naked seeds, but can't disperse them due to husk covering

The principal maize groups with their peculiarities, distribution and importance are furnished as follows:

1. Dent corn (*Zea mays* var *indentata* Sturt)**:** Dent is the most widely grown type of maize in the United States and northern Mexico. It is characterised by a depression or dent in the crown of the seed. The starch at the sides of the seeds is corneous while the soft starch extends to the narrow base or tip. Rapid drying and shrinkage of the soft starch results in the characteristic denting. Dent corn can be stored for long period.

2. Flint corn (*Zea mays* var *indurata* Sturt.)**:** Flint maize is predominant in Europe, Asia, Central America and South America. Seeds of flint maize are hard and contain little soft starch. The seeds are smaller than those of dent corn and therefore it is well adapted to poultry feedings.

3. Sweet corn (*Zea mays saccharata* Sturt.): This type contains a sweetish starchy endosperm is characterized by a translucent, corny appearance when immature and a wrinkled condition when dry. The ears are picked green for table use and canning. Sweet corn is grown in Mexico and USA.

4. Flour corn (*Zea mays* var *amylacea* Sturt.): Flour maize has kernels which are composed entirely of soft starch. They usually develop no dents. It is grown in USA, South America and South Africa.

5. Pop corn (*Zea mays* var *everta* Sturt.): The grains are usually small, pointed and contain a higher percentage of hard starch than flint maize. The grain expands explosively on heating forming good pops. It is grown in Mexico and USA.

6. Pod corn (*Zea mays* var *tunicata* Sturt.): Pod corn is an unusual type of maize, each kernel of which is enclosed within the lemma and palea which are well developed. This is probably one of the earliest domesticated types and is considerable interest in studies of the origin of maize. It is not grown commercially but cultivated in South America.

7. Waxy corn (*Zea mays* var *ceritina* Kulesh.): Waxy corn has waxy appearance on the kernels. The starch is gummy and has some of the characteristic of tapioca. Waxy corn is grown in northern Myanmar, Philippines, eastern China and Manchuria.

Growth stages: Germination is the sprouting stage which comes about one week after sowing. The adventitious root system develops from the first stem node below the soil surface and takes over the main root function approximately 10 days after emergence. Growing point of the plant should be 1 to 1½ inches below the soil surface. Permanent (nodal) root system will begin developing at this point. If seed is planted too shallow the root system will have a difficult time becoming established. As the growing point is below the ground, young maize plants are susceptible to damage from waterlogging, especially when combined with high temperatures. Three weeks after emergence the growing point is at the soil surface. The first leaf of maize emerges from the soil usually 4–6 days after planting. All the leaves the plant will ever produce are formed by a single growing point below the ground during the first 2 to 3 weeks. The first leaf of maize emerges from the soil usually 4–6 days after planting. The collar of third leaf occurs approximately 10 to 14 days after emergence. Growing point is just below the ground surface. A hail or light freeze will cause little long term damage to the plant. However, flooding while the growing point is below ground can kill the plant, especially if temperatures are high. Plant develops seminal roots up to three leaf stage. The coleoptile emerges from the soil 6 to 10 days after sowing while the shoot meristem remains below the soil surface. The particulars on growth stages of maize is presented in Table 1.

(i) Seedling stage is from 7-14 DAS and the plants have 2-4 leaves at this stage. Collar of 6th leaf occurs approximately 3 weeks after emergence. Collar of 8th leaf occurs approximately 4 weeks after emergence. New leaf is emerging about every 3 days in vegetative stage. Maize plant typically develops 20-21 leaves but the lowermost leaves are damaged by expansion of the stalk and often disintegrate. There may be only 14 to 16 intact leaves at the time of pollination. It produces silks about 65 days after emergence, and matures about 120 days after emergence.

(ii) Grand growth stage is knee height stage of the plants which arrives about 35 to 45 days after sowing. The shoot meristem and the tassel primordium emerge above the soil surface at 6th leaf stage and the shoot meristem fully emerge 15 cm above soil layer at 8 leaf stage. The first 5 to 6 lower leaves may senesce and cease to be functional.

(iii) Tasseling stage is called as flower initiation stage at which the tassels or male flowers come out. The tassel emerges about 16th leaf stage. Tassels fully emerge and pollen sheds 40 to 50 days after emergence, with the length of time depending on variety and environmental conditions. The tassel begins to differentiate when about 5th leaf stage.

(iv) Silking stage is also known as cob initiation stage at which the female flowers or cobs are formed. Silking stage involving the formation of the female flowers or cobs is the first reproductive stage and occurs 2-3 days after tasseling stage. This stage begins when any silks are visible outside the husk. These are auxillary flowers unlike tassels that are terminal ones. Pollination occurs when these new moist silks catch the falling pollen grains. Cobs, husks and shanks are fully developed by day

7 after silking. These are auxiliary flowers unlike tassels that are terminal ones. Usually they are formed in the axis of 11th to 13th leaf whereas male flowers or tassels are formed at the apex after 14 or 15 leaves have come out. Tasseling/ Flower initiation stage is the stage at which the tassels or male flowers appear. Each spike of the tassel sheds pollen for about eight days. Generally the maize plant would have attained its full height by this stage.

(v) Soft dough stage is called as milky stage which commences after pollination and fertilization are over. Grains start developing but they do not become hard. The silks on the top of the cob remain partially green at this stage and the covering of the cobs also remain green. This is the best stage for using green cobs for table purposes. Soft-dough/milky stage commences after pollination and fertilization is over. Grains start developing but they do not become hard. This soft dough stage is noticed by the silks on the top of the cob which remain partially green at this stage. The covering of the cobs also remains green. Blister stage occurs 10 to 14 days after silking. Kernels are watery at blister stage. Milk stage occurs 18 to 22 days after silking.

(vi) Hard dough stage occurs 30 to 40 days after silking. Most kernels have dented at 55% moisture. Generally the crop takes two months from anthesis to physiological maturity. Approximately 30 days after silking the plant has reached the maximum dry weight, a stage called physiological maturity. Moisture content is approximately 30 to 35% at black layer formation. Kernels are at 30-35% moisture and have attained 100% of dry weight. Grain needs to be at 13-15% moisture for safe storage.

(vii) Maturity stage shows that the leaves get dried; silks get dried completely and become very brittle. Harvesting is done at this stage.

Table 1 Growth stages of maize

Phenological stage	**Stage**	**Duration (days)**		
Planting/sowing	0.0	0	0	0
Emergence		5	5	5
First leaf		7-9	7-10	9-12
Two leaves (seedling establishment)	0.5	10-12	10-13	7-14
Four leaves	1.0	18	20	20
Six leaves	1.5	24	25-30	30
Eight leaves	2.0	28-30	33-36	36
Ten leaves	2.5	36	45	44
Twelve leaves (ear forming)	3.0	42	47	50
fourteen leaves	3.5	48	50	56
Sixteen leaves (tassel visible)	4.0	54	56	60
Silks emerging- anthesis	5.0	58	60	70-80
kernals in blister	6.0	65	71	90
Kernals in dough	7.0	75	80	100
Kernals begining to dent	8.0	90	95	115
Kernals fully dented	9.0	100	105	120
Physiological maturity	10.0	110	112	125
Harvest	---	115	120	130

Climate: Maize is essentially a warm season crop. It is grown in the warmer parts of temperate, humid subtropical and in highlands. It is grown in the latitude 58ºN to 40ºS up to an altitude of 3000 m from MSL. It grows well with 750 to 1250 mm rainfall during its life period. It is a short day plant. The optimum light for maize is 32.3 to 86.1 klux. It requires a cardinal minimum, optimum and maximum temperature of 10ºC, 21 to 27ºC and 35ºC respectively under adequate moisture supply. It grows well with an optimum day temperature of 24ºC and a night temperature of above 15ºC. The minimum and optimum temperature for germination is 7°C and 21°C respectively. Both the pollen and silks are very sensitive to high temperatures of >30°C. Temperature of 38ºC plus water stress at tasselling and silking prevent seed set while temperature of 15.6ºC and lower greatly retards flowering and maturity. If temperature remains above 35ºC during noon hours for several days, pollen is destroyed and the yield is drastically reduced. Extremely high temperatures coupled with low humidity are not conducive for pollination. The optimum soil temperatures for germination, early seedling growth are 15 to 27°C and at tasselling 21 to 30°C. The maize plant cannot withstand temperatures below about –2°C. It is sensitive to frost and moderately sensitive to chilling. It cannot withstand frost at any stage of its life cycle. Frost injury results in chlorotic bands and drying of leaf margins as well as tips. In advanced stage, the crop may give a burnt appearance. Hail storms are most harmful during the stages of tasseling and silking.

Soil: Well drained sandy loamy to silt loamy soils are desirable. The pH value ranges from 5.5 to 7.5. Saline and alkaline soils may be avoided.

Season: Maize can be grown in *kharif, rabi* and *zaid* (spring and summer) seasons. In irrigated areas, it is most desirable to complete the sowings 10 to 15 days before the onset of the rain. This practice will enhance 15% higher grain yield than the crop sown with or after the onset of rains. The crop can be raised under rainfed condition if there is well distribution of 500 mm rainfall during the crop season. The last week of June to first fortnight of July during *kharif*, last week of October to first week of November during *rabi* and first week of February during spring season are the optimum time of sowing to avoid flowering from heavy rains during *kharif* and low temperature should not coincide with flowering. It is desirable to complete the sowing 12-15 days prior to onset of monsoon.

Varieties and hybrids: The maturity groups of maize is presented in Table 2. The particulars of varieties and hybrids of maize are provided in Table 3 to Table 8.

Table 2 Maturity groups of maize

Maturity group	Duration (days)	Yield potential (t/ha)
Kharif season : Full season	100–110	4.5–6.5
Medium	90–100	3.5–4.5
Early	85–90	2.5–3.5
Extra early	75–85	2.0–2.5
Rabi (winter) season : Full season	100–110 160–180	10.0–12.0 12.0

Table 3 Special type of maize varieties in India

Corn type	Cultivars
Quality Protein Maize (QPM)	H:HQPM 1 & HQPM 5 (all states of India), HQPM 7, Vivek QPM 9 (Peninsular India), Shaktiman1,2,3 & 4 (Bihar)
Baby corn	H:HM-4, Prakash C: VL Baby Corn 1, Co 1, Him 123, Early Composite, VL 64, PEHM-1 & PEHM-2
Sweet corn	H:HSC1 for J&K and HP C:Madhuri, Win orange, Priya, Almora
Pop corn	C: Jawahar, Amber popcorn, Pearl popcorn, VL Almora popcorn
Green-eared Corn	Harsha, Ashwini, Varun, Rohini, Megha
High Starch Corn	Ganga 111, Histarch, Deccan 103, Deccan 105,Trishulata, Sheetal, paras.
High Oil	HOP-1, HOP-2.
Fodder maize	C: African tall, PFM-66, J 1006 & Pratap chari-6

Table 4 Maize hybrids (H) and composites(C) varieties of different maturity groups for different states for *kharif* season

States	Extra early maturity	Early maturity
Delhi	H:Vivek 17 &21, PMH 2	H:PAU 352, PEH 3, Parkash, X 3342
Punjab	H:Vivek 17& 21, PEEH 5	H:PAU 352, PEH 3, JH 3459, Parkash, PMH 2, X 3342
Haryana	H:Vivek 17 &21, PMH 2, PEEH 5	H:HHM 1, PAU 352, Pusa Early Hybrid 3, JH 3459 Parkash, X 3342
Uttar Pradesh	H:Vivek 5, 15, 17, 21 & 27 PMH 2,	H: JH 3459, Parkash, PEH 2, X 3342 C: Pusa Composite 4,
Rajasthan	H:Pratap hybrid 1, Vivek4 & 17,	H: PEHM 2 ,Parkash ,Pro 368, X 3342 C: Pratap Makka 3, Aravali Makka 1, Jawahar Makka 8, Amar, Azad Kamal, Pant Sankul Makk 3,
Madhya Pradesh	H:Vivek4 & 17	H: PEHM 2, Parkash, Pro 368, X 3342 C: Jawahar Makka 8, Jawaharcomposite 12, Amar, Azad Kamal, Pant Sankul Makk 3, Chandramani, Pratap Makka 3
Gujarat	H:Vivek 4 &17	H: PEHM 2, Parkash, Pro 368, X 3342 C: Jawahar Makka 8, Pant Sankul Makka 3, Pratap Makka 3, G M 2,4 & 6 Aravali Makka 1, Narmada Moti
Andhra Pradesh	H:Vivek 9, 15, 17& 27, PEEH 5	H:PEHM 1, PEHM 2, DHM 1, BH- 2187, Parkash, JKMH 1701, X 3342
Tamil Nadu	H:Vivek 9, 15, 17, 21& 27, PEEH 5	H:PEHM 2 , Parkash, X 3342, JKMH 1701
Maharashtra	H:Vivek 9, 15,17, 21& 27, PEEH 5	H: PEHM 1& 2, Parkash, X 3342, JKMH 1701
Karnataka	H:Vivek 9, 15, 21& 27,PEEH 5	H: PEHM 2 , Parkash, X 3342, JKMH 1701 C: NAC 6002

Jammu & Kashmir	H: Vivek 15, 21, 25 &33, PEEH 5 C: Pratap Kanchan 2, Shalimar KG 1 & 2, Vivek 35, and 37	H: Vivek33, Parkash, JKMH 1701, X 3342 C: C 8,14 & 15
Uttarakhand	H: Vivek 5, 9, 21& 25PEEH 5 C: Pratap Kanchan 2,Vivek 35 and 37	H:Vivek hybrid 33, Vivek hybrid 23, Parkash
Bihar	H: Vivek 27 C: D 994	H: Parkash, X 3342 C: Dewaki, Birsa Vikas Makka 2
Jharkhand	H: Vivek 27 C: D 994,	H: Parkash, X 3342 C: Dewaki, B V M 2, B M 1
Odisha	H: Vivek 27 C: D 994,	H: Parkash, HIM 129, X 3342
West Bengal	H:Vivek 27	H: Parkash, X 3342
Himachal Pradesh	H:Vivek 15, 21& 25, PEEH 5	H: Parkash, X 3342
NEH Region	H:Vivek 21& 25, PEEH 5	H: Parkash , JKMH 1701, X 3342
Chhattisgarh	H:Vivek 27	H: Parkash, X 3342
Assam	-	H:Parkash, X 3342

Table 5 Maize hybrids (H) and composites (C) varieties of different maturity groups for different states for *kharif* season

States	Medium maturity	Late maturity
Delhi	H:HM4, HM 8 10, DK 701	H:PMH 3, Buland, NK 61, Pro 311, Bio 9681, Seed Tech 2324
Punjab	H:HM4, HM 8& 10, DK 701	H:PMH 3, PMH-1 , Buland, Pro 311, Bio 9681 , NK 61, Pro 311, Bio 9681, Seed Tech 2324
Haryana	H:HM 2, HM 4,8 &10 DK 701	H:PMH 3, Buland, ,HM 5, NK 61, Pro 311, Bio 9681, Seed Tech 2324
Uttar Pradesh	H:HM 8 & 10, Malviyahybrid makka 2, Bio 9637 , DK 701	H:PMH 3, Buland, Pro Agro 4212, Pro 311, Bio 9681, NK 61,Seed Tech 2324
Rajasthan	H: HM 10, NK 21 C: Pratap Makka 5	H: Trishulata, Pro 311, Bio 9681, Seed Tech 2324
Madhya Pradesh	H: HM 10, NK 21 C: Pratap Makka 5	H: Trishulata , Pro 311, Bio 9681, Seed Tech 2324
Gujarat	H: HM 10, NK 21 C: Pratap Makka 5	H:Trishulata , Pro 311, Bio 9681, Seed Tech 2324 C: G M 3,
Andhra Pradesh	H:HM 8& 10, DHM111,DHM117	H:DHM113,Kargil 900 M, Seed Tech 2324, Pro 311, Bio 9681, Pioneer 30 v 92, Prabal, 30 V 92,

Tamil Nadu	H:HM 8& 10, COHM 4	H:COHM 5, Prabal , Pro 311, Bio 9681, Seed Tech 2324, 30 V 92,
Maharashtra	H:HM 8& 10	H: Prabal, Pro 311, Bio 9681, Seed Tech 2324, 30 V 92,
Karnataka	H:HM 8& 10	H:Nithya Shree,EH434042, DMH 1, DMH 2,Bio 9681, Prabal, Pro 311, Bio 9681, Seed Tech 2324 C: NAC 6004, 30 V 92
Jammu & Kashmir	H: HM 10 C: C 6	-
Uttarakhand	H: HM 10 C: Bajaura Makka	-
Bihar	H:HM 9, Malviya hybridmakka 2	H: Pro 311, Bio 9681, Seed Tech 2324, 30 V 92, 900 M C: Hemant, Suwan & Lakshmi
Jharkhand	H:HM 9, Malviya hybridmakka 2, DK 701	H: Pro 311, Bio 9681, Seed Tech 2324 C: Suwan
Odisha	H:HM 9, DK 701, DMH 115, Pro 345 Malviya hybridmakka 2,	H: , Pro 311, Bio 9681, Seed Tech 2324 , PAC 705
West Bengal	H:Malviya hybrid Makka2	H: Pro 311, Bio 9681, Seed Tech 2324
Himachal Pradesh	C: Bajaura Makka, Pratap Makka 4	H: Pro 311, Bio 9681, Seed Tech 2324
NEH Region	C: Pratap Makka 4	H: Pro 311, Bio 9681, Seed Tech 2324 C: NLD white
Chhattisgarh	C: Pratap Makka 5	H:PEHM 1, Pioneer 30 V 92 & 30 R 26, Bio 9681, Pro 4640 & 4642,
Assam	H: DK 701 C: Pratap Makka 4	C: Vijay ,NLD white,

Table 6 Maize hybrids (H) and composites (C) varieties of late maturity groups for different states for *rabi* season

States	**Late maturity**
Delhi	H: PMH 3, Buland, NK 61, Pro 311, Bio 9681, Seed Tech 2324, HM11, HM8
Punjab	H: PMH 3, PMH-1 , Buland, Sheetal , Pro 311, Bio 9681 , NK 61, Pro 311, Bio 9681, Seed Tech 2324, HM11, HM8
Haryana	H: PMH 3, Buland, ,HM 5, NK 61, Pro 311, Bio 9681, Seed Tech 2324, HM11, HM2, HM1, HM8
Uttar Pradesh	H: PMH 3, Buland, Pro Agro 4212,Pro 311, Bio 9681, NK 61,Seed Tech 2324, HM8
Rajasthan	H: Pro 311, Bio 9681, Seed Tech 2324, HM8
M.P.	H: Pro 311, Bio 9681, Seed Tech 2324
Gujarat	H: Pro 311, Bio 9681, Seed Tech 2324 C: G M 3, Ganga safed 2

Andhra Pradesh	H: Kargil 900 M, Seed Tech 2324, Pro 311, Bio 9681, Pioneer 30 v 92, Prabal, 30 V 92, 900 M
Tamil Nadu	H: COHM 5, Prabal , Pro 311, Bio 9681, Seed Tech 2324, 30 V 92, 900 M
Maharashtra	H: Prabal, Pro 311, Bio 9681, Seed Tech 2324, 30 V 92, 900 M
Karnataka	H: Nithya Shree, DMH 1, DMH 2, 900 M, Bio 9681, Prabal, Pro 311, Bio 9681, Seed Tech 2324 C: NAC 6004, 30 V 92
Bihar	H: Rajendra Hybrid 2, Rajendra Hybrid 1, Pro 311, Bio 9681, Seed Tech 2324, 30 V 92, 900 M C: Hemant, Suwan and Lakshmi
Jharkhand	H: Pro 311, Bio 9681, Seed Tech 2324 C: Suwan
Odisha	H: Pro 311, Bio 9681, Seed Tech 2324 , PAC 705
West Bengal	H: Pro 311, Bio 9681, Seed Tech 2324
Himachal Pradesh	H: Pro 311, Bio 9681, Seed Tech 2324
NEH Region	H: Pro 311, Bio 9681, Seed Tech 2324 C: NLD white
Chhattisgarh	H: PEHM 1, Pioneer 30 V 92 & 30 R 26, Bio 9681, Pro 4640 & 4643, 900 M
Assam	C: NLD white,

Table 7 List of hybrids (H) and composites (C) varieties of different maturity groups for different states for *spring* season

States	Extra early maturity	Early maturity
Delhi	H:Vivek 17 &21, PMH 2	H: PAU 352, PEH 3, Parkash, X 3342
Punjab	H:Vivek 17& 21, PEEH 5	H:PAU 352, PEH 3, JH 3459, Parkash, PMH 2, X 3342
Haryana	H:Vivek 17 &21, PMH 2, PEEH 5	H: HHM 1,PAU 352, Pusa Early Hybrid 3, JH 3459 Parkash, X 3342
Uttar Pradesh	H: Vivek 5, 15, 17, 21 & 27 PMH 2,	H: JH 3459, Parkash, PEH 2, X 3342, C: Pusa Composite 4, Gaurav, Azad Uttam, Surya, Kiran, Tarun
Rajasthan	H:Pratap hybrid 1, Vivek 4 & 17,	H: PEHM 2 ,Parkash ,Pro 368, X 3342 C: Pratap Makka 3, Aravali Makka 1, Jawahar Makka 8, Amar, Azad Kamal, Pant Sankul Makk 3, Mahi Kanchan, Mahi Dhawal
M.P.	H: Vivek4 & 17	H: PEHM 2, Parkash, Pro 368, X 3342 C: Jawahar Makka 8, Jawahar composite 12, Amar, Azad Kamal, PantSankul Makk 3, Chandramani, Pratap Makka 3
Gujarat	H: Vivek 4 &17	H: PEHM 2, Parkash, Pro 368, X 3342 C: Jawahar Makka 8,Pant Sankul Makka 3, Pratap Makka 3, G M 2,4 & 6Aravali Makka 1, Narmada Moti

Andhra Pradesh	H: Vivek 9, 15, 17& 27, PEEH 5	H:PEHM 1, PEHM 2, DHM 1, BH- 2187, Parkash, JKMH 1701, X 3342
Tamil Nadu	H: Vivek 9, 15, 17, 21& 27, PEEH 5	H: PEHM 2 , Parkash, X 3342
Maharashtra	H:Vivek 9, 15,17, 21& 27, PEEH 5	H: PEHM 1& 2, Parkash, X 3342 , C:, JKMH 1701
Karnataka	H:Vivek 9, 15, 21& 27,PEEH 5	H: PEHM 2 , Parkash, X 3342, C: NAC 6002
Jammu & Kashmir	H: Vivek 15, 21, 25 &33, PEEH 5 C: Pratap Kanchan 2, Shalimar KG 1 & 2	H: Vivek33, Parkash, JKMH 1701, X 3342, C: C 8,14 & 15
Uttarakhand	H: Vivek 5, 9, 21 & 25 PEEH 5 C: Pratap Kanchan 2	H:Vivek hybrid 33, Vivek hybrid 23, Parkash
Bihar	H: Vivek 27 C: D 994, Gujarat Makai 6	H: Parkash, X 3342, C: Dewaki, Birsa Vikas Makka 2
Jharkhand	H: Vivek 27,. C: D 994,	H: Parkash, X 3342, C: Dewaki, B V M 2, B M 1
Odisha	H: Vivek 27, C: D 994,	H: Parkash, HIM 129, X 3342
West Bengal	H: Vivek 27	H: Parkash, X 3342
Himachal Pradesh	H: Vivek 15, 21& 25, PEEH 5	H: Parkash, X 3342
NEH Region	H: Vivek 21& 25, PEEH 5	H: Parkash ,JKMH 1701,X 3342
Chhattisgarh	H:Vivek 27	H:Parkash, X 3342
Assam	-	H:Parkash, X 3342

Field preparation: Plough the field twice with an iron plough and three or four times with a country plough to get a fine tilth. Form ridges and furrows of 6 m long and 60 cm apart. Form irrigation channels across the furrows. If ridges and furrows are not made, form beds (20 m^2) depending on the availability of water. Use a bund former or ridge plough to economise cost of production.

Seeds and sowing: Seed rate of 15 kg per ha for hybrids and 20 kg per ha for composites is recommended. Spacing is 60 cm x 20 cm with a plant population of 83,333 per ha. Treat the seeds with Bavistin + Captan in 1:1 ratio @ 2 g/kg seed for Turcicum leaf blight (TLB), Banded leaf and sheath blight (BLSB), Maydis leaf blight (MLB), etc., Apran 35 SD @ 4g/kg seed for Brown stripe downy mildew (BSDM), Captan 2.5 g/kg for Pythium Stalk Rot, Imidachlorpit @ 4g/kg or Fipronil @ 4 ml/kg seed for termite and shoot fly. Seeds treated with fungicides may be treated with three packets (600 g/ha) of Azospirillum. Place one seed per hole in case of hybrids and two seeds per hole in composites.

Method of sowing

(i) **Raised bed sowing:** The raised bed sowing is best for maize during monsoon both under excess moisture as well as limited irrigation availability conditions. There is saving of 20-30% irrigation water in raised bed planting method. The furrows will act as drainage channels and crop can be saved from excess soil moisture stress under temporary excess soil moisture/water logging due to heavy rains.

ii. **Flat bed sowing:** Flat bed sowing can be done using seed-cum-fertilizer planters under both irrigated and rainfed areas. If bed system of planting is followed, open furrows 6 cm deep at a distance of 60 cm apart. Dibble the seeds at a depth of 3-4 cm along the furrow in which fertilizers are placed and covered with soil.

iii. **Ridges and Furrow sowing:** Maize can be grown in ridges and furrows under irrigated conditions. In the case of ridge planted crop, open a furrow 6 cm on the side of the ridge at two thirds the distance from the top of the ridge.

After cultivation: Leave only one healthy and vigorous seedling per hole and remove the others on the 7th or 8th day of sowing. Gap filling is done by dibbling the seeds at the rate of two seeds per hill which is followed by irrigation. Demolish the original ridges. Earth up and form new ridges so that plants come directly on the centre of the ridges. This will provide additional anchorage to the plants.

Weed management: Hoe and hand weed on the 17th or 18th day of sowing, if herbicide is not applied. Pre-emergence application of Atrazine (Atratraf 50 wp, Gesaprim 500 fw) @ 1.0-1.5 kg a.i/ha in 600 litre water, Alachlor (Lasso) @ 2-2.5 kg a.i/ha, Metolachlor (Dual) @ 1.5-2.0 kg a.i/ha, Pendamethalin (Stomp) @ 1-1.5 kg a.i./ha on 3 days after sowing for control of many annual and broad leaved weeds. This is followed by one hand weeding on 30-35 days after sowing. If pulse crop is to be raised as intercrop, do not use Atrazine. For maize + soybean intercropping system, spray pre-emergence herbicide Alachlor @ 4.0 lit per ha or Pendimethalin @ 3.3 1it/ha on 3rd day after sowing as spray. Apply herbicide when there is sufficient moisture in the soil. Do not disturb the soil after herbicide application. For areas where zero tillage is practiced, pre-plant application (10-15 days prior to seeding) of non-selective herbicides viz., Glyphosate @ 1.0 kg a.i./ha in 400-600 litre water or Paraquat @ 0.5 kg a.i./ha in 600 litre water is recommended to control the weeds. Under heavy weed infestation, post-emergence application of Paraquat can also be done as protected spray using hoods.

Manures and fertilizers application: Apply farmyard manure or compost or composted coir pith @ 12.5 t/ha evenly on the unploughed field along with 10 packets of Azospirillum (2 kg/ha) and incorporate in the soil. A fertilizer dose of 135-62.5-50 N, P_2O_5 and K_2O kg/ha is recommended for maize crop. Full dose of phosphorus, potash and zinc and 10% N should be applied as basal. Apply the fertilizer mixture along the furrows evenly and cover to a depth of 4 cm with soil. Place the fertilizer mixture along the furrows evenly and cover to a depth of 4 cm with soil. When Azospirillum is used as seed and soil application, application of 100 kg of N per ha i.e., 25 % reduction of the total nitrogen is recommended.

Table 8 Fertilizers recommended for different maturity groups in India

Nutrients	Early season group (85-90 days)	Medium group (90-100 days) and Full season group (100-110 days)	Winter maize group (100-110 days)
N	80 to 100 kg/ha	120 kg/ha	150 to 180 kg/ha
P_2O_5	60 kg/ha	60 kg/ha	60 kg/ha
K_2O	40 kg/ha	40 kg/ha	40 kg/ha

Application of FYM @ 10 t/ha, 10-15 days prior to sowing supplemented with 150-180 kg N, 70-80 kg P_2O_5, 70-80 kg K_2O and 25 kg $ZnSO_4$ per ha is recommended. Full doses of P, K and Zn should be applied as basal preferably drilling of fertilizers in bands along the seed using seed-cum-fertilizer drills. Nitrogen should be applied in 5 splits @ 20% as basal (at sowing), 25% at four leaf stage, 30% eight leaf stage, 20% at tasseling stage and 5% at grain filling stage for high productivity and nitrogen use efficiency. Nitrogen can also be applied in 3 splits *viz*., sowing, knee-high and flowering stage. Top dressing of nitrogen can be done by opening a furrow 5 cm depth along the base of the ridge (10 cm away from the base of the plants) with a hand hoe or stick. Place half of the dose of nitrogen on the 25th day of sowing along the furrows evenly and cover it with soil. Second top dressing is to be done with the remaining quarter of nitrogen on the 45th day of planting as furrow application. The nitrogen can also be applied in four splits @ 20% N at 4th leaf stage, 30% N at 8th leaf stage, 30% N at flowering stage and 10% N at grain filling stage or in five splits @ 10 % N should be applied as basal, 20% N at 4th leaf stage, 30% N at 8th leaf stage, 30% N at flowering stage, 10% N at grain filling stages for higher nitrogen use efficiency.

Deficiency symptoms

Nitrogen deficiency: Older leaves turn yellow, older leaves show drying at tips which progress along mid veins and stalks become slender.

Phosphorus deficiency: Leaves are purplish green during early growth. Growth is spindly, slow maturity and irregular ear formation.

Potassium deficiency: Leaves show yellow or yellowish green streaks and become corrugated, tips and marginal scorch. Tips end in ears are poorly filled. Stalks have short internodes. Plants are weak and may fall down.

Magnesium deficiency: Older or lower leaves are the first to become chlorotic at margins and between veins. Necrotic or chlorotic spots are seen in leaves.

Zinc deficiency: Older leaves have yellow streaks or chlorotic stripping between veins. In several cases, unfolding of young leaves may be white or yellow.

Iron deficiency: Interveinal chlorosis is observed in leaves. The entire crop may exhibit bleached appearance.

Application of micronutrient: Micronutrient mixture @ 12.5 kg/ha of with sand to make a total quantity of 50 kg/ha is to be applied. Apply the mixture over two thirds in the top of ridges, if ridge planting is followed. If bed system of sowing is followed, broadcast the micronutrient mixture. Do not incorporate the micronutrient mixture in the soil. Application of $ZnSO_4$ @ 25 kg/ha for zinc deficient soils is recommended.

Water management: Maize requires 450 to 600 mm of water for its growth and development in temperate climate and up to 900 mm or more depending on evaporative demand under irrigated condition in dry climates. Crop requires 6 to 8 mm per day during silking and soft dough stage. Flowering (tasselling, silking) and grain filling are the critical stages for irrigation. Irrigation is to be a given at 50% depletion of available moisture in most soil. The optimum IW/CPE ratio is 0.8 for maize crop. Maize is susceptible to moisture stress throughout its life cycle. Water logging for 36 hours will injure plants. Regulate irrigation according to the growth phase of the crop. The water use efficiency will be 2.5 kg grain per m^3 of water used. The water use rate is 6 to 8 mm per day at full canopy stage.

Table 9 Irrigation scheduling for maize

Growth stage	Duration (days)	No. of irrigation		Irrigation frequency	
		HS	LS	HS	LS
Germination stage	1 to 14	1 1	1 1	at sowing 4 DAS	at sowing 4 DAS
Vegetative stage	15 to 39	3	4	12 days interval	10 days interval
Flowering stage	40 to·65	2	4	12 days interval	10 days interval
Maturity stage	66 to 95	1	2	10 days before harvest	10 days before harvest

Note: HS= Heavy soil, LS= Light soil

Cropping systems: Intercropping system of maize + soybean, maize + cowpea and maize + blackgram is recommended for higher net return in the red lateritic soils whereas maize + redgram intercropping systems is ideal for vertisols. The important cropping rotations are viz., maize-potato /wheat/berseem/barley/oats for one year, maize-toria -wheat for one year, maize-potato-wheat for one year, Maize-toria-sugarcane for 2 years, maize-wheat-sugarcane 2 years, maize-wheat-cotton-berseem 2 years, maize-senji-sugarcane-cotton 2 years, maize-wheat-sorghum-sugarcane 3 years and maize-potato-sugarcane-wheat 3 years.

Pests and diseases management: The insect-pests that affect maize crop are stem borers, shoofly, armyworm, jassids, thrips, white ants, pyrilla, grasshoppers, grey weevil, hairy caterpillars, root worms, earworms and leaf miner are more serious, though the spectrum varies in different agro-ecological regions. Several diseases such as seed and seedling blights, foliar disease, downy mildews, stalk rots and leaf sheath blight occur in various parts of the country.

Table 10 Seed treatment practices

Disease/insect-pest	Fungicide/Pesticide	Rate of application (g/kg seed)
Turcicum leaf blight, Banded leaf blight and sheath blight	Bavistin + Captan in 1:1 ratio	2.0
Brown stripe downy mildew (BSMD)	Apran 35 SD	4.0
Pythium stalk rot	Captan	2.5
Termite and shoot fly	Imidachlorpit	4.0

The biological control practices are:

(i) Add *Trichoderma harzianum* formulation 2.0% WP in furrows at the time of sowing prior mixing with FYM @ 10 g/kg FYM & incubated for 10 days in moist condition for Charcoal rot (Post flowering stalk rot)

(ii) Seed treatment with *Trichoderma harzianum* 2.0% WP @ 20 g/kg of seeds for control of *Rhizoctonia solani* f. sp. *sasakii* (Banded leaf & sheath blight)

(iii) Combined application of mustard & tobacco dust @ 2.5q/ha (ETL – 2 cyst/g of soil) for cyst nematode

(iv) Application of *Trichogramma chilonis* @ 1,60,000/ha. on 7 and 15 days old crop onwards at weekly interval for various pathogens.

The need based and judicious application of fungicides is most important component of integrated pest management practices are:

(i) Seed treatment with Thiram 75% WS @ 25 to 30 g/kg seed.

(ii) Seed treatment with fungicide Metalaxyl-M 31.8% ES @ 2.4 ml/kg of seed in the endemic areas of downy mildews.

(iii) Foliar spray at first appearance of leaf blight with Mancozeb 75% WP @ 1.5 to 2 kg/l of water followed by 2 to 4 applications at 10 days interval if needed.

(iv) Spray of Mancozeb 75% WP @ 1.5 to 2 kg/l of water or Zineb 75% WP @ 1.5-2 kg/ha at first appearance of pustule of Polysora rust or Common rust and three sprays of fungicide at 15 days interval are recommended if needed.

(v) Foliar spray of Mancozeb 75% WP @ 1.5 to 2 kg/l at very first appearance of symptoms of downy mildews

vi. Soil drenching of bleaching powder containing 33% chlorine @ 10 kg/ha at pre-flowering stage if symptoms of bacterial rot appears (Pre-flowering)

Rodent management practices are:

(i) Rodent problem appears generally at milky stage and seed maturity stage

(ii) Practice burrow smoking using natural smoking materials in burrow fumigator for 2-3 min. for each burrow

(iii) Application of 0.005% bromadiolone in ready to use form (wax blocks) or loose bait in packets near rodent burrows

(iv) Apply 2% Zinc phosphide poison baits when the rodent infestation is very high.

Harvest: Harvesting can be done for grain/kernel taking into consideration of the average duration of the crop and with maturity symptoms, viz., (i) the sheath covering the cob will turn yellow and dry at maturity and (ii) the seeds become fairly hard having 25 to 30% moisture. Harvesting is done with break-off or cut and remove the cobs alone along with the sheath. Remove the sheath to separate the cob. The husk is removed from the cobs, and cobs are dried in sun for 7-8 days. Carry out harvest operations at a single stage. Dry the cobs in the sun till the grains are dry. Use mechanical threshers for threshing or beat with a stick and separate the grains from the cob. Clean the seeds by winnowing. The harvested green stover is cut into bits with a chaff cutter or chopping knife and feed the cattle. Do not dry maize fodders as it is good cattle feed when green.

Yield: The grain yield potential of maize in high land / transitional land, mid-altitude / subtropical land and tropical / lowland is approximately 5.0 to 6.0, 7.0 to 10.0 and 5.0 to 5.5 t/ha respectively. Maize exhibits 50% harvest index in winter season in contrast to 35 to 40% in summer season.

Table 11 Yield of maize in different climatic regions

Region	Crop duration (days)	Mean yield (t/ha)	Maximum yield (t/ha)	Harvest index (%)	Kernel number per m^2
Temperate	160	3.5	22	54	3360
Subtropics	135	1.8	12	47	3000
Tropics	112	1.0	8	38	2750

Utilization: Maize is mainly utilized for direct human consumption in developing countries and for livestock feed in developed countries. The germ is the source of maize 'vegetable oil' (total oil content of maize grain is 4% by weight). The endosperm occupies about two thirds of a maize kernel's volume and accounts for approximately 86% of its dry weight. The endosperm of maize kernels can be yellow or white. Maize is also utilized as a source in industrial by-products like glucose, dextrose, starch, alcoholic beverages, bio-fuel, processed food, corn oil.

Baby corn: Cobs removed within 3 to 5 days after their emergence/after flowering is called baby corns. The baby corn cobs are not allowed to fertilize and set seed. Fresh baby corn has a crisp texture and a subtle, slightly sweet corn flavour. The tiny ears of baby corn are simply immature ears from regular-sized corn plants. The plants of baby corn varieties tend to produce more ears per plant than other corn varieties. Sweet corn and field corn varieties may also be suitable for baby corn production. Most of the baby corn varieties will produce 2–3 ears per plant; however, quality of the third ear may not be adequate. The baby corn ears should be 2–4 inches long and 1/3–2/3 inch in diameter at the base, or butt end. Kernels should be uniform in shape and petite in size, with rows neatly aligned and ends evenly tapered. The immature ears are harvested before pollination and before any sugars have accumulated in the kernels. It may be advantageous, however, to use sweet corn types because they tend to be easier to hand-harvest. Sweet corn varieties have ears that are easier to break off from the stalk. The benefit of using field corn types is lower seed cost. Its green ears are used for roasting and boiling and also consumed as food at dough stage.

Differences between maize and teosinte

- (i) Teosinte plants are branched and produce many ears, while corn plants produce a single upright stem with ears.
- (ii) In teosinte the primary lateral branches are long ones, but in maize they are short ones.
- (iii) In teosinte the primary lateral inflorescence is usually male, whereas in maize the primary lateral inflorescence is usually female.
- (iv) Ear structure: maize ears have 8 to 12 rows of seeds, while teosinte only has two rows.
- (v) The teosinte ear possesses only about 5 to 12 kernels, each sealed tightly in a stony casing. Collectively, the kernel and its stony casing are known as a 'fruitcase'. Teosinte individual fruitcases become the dispersal units. Protected within its casing, the teosinte kernel can survive the digestive tracks of birds and grazing mammals, enabling the seed to be easily dispersed. The maize kernels are naked without adequate protection from predation and are easily digested by any animal that consumes them. Each segment of the teosinte fruit can produce only one seed because one of the paired potential seeds (spikelet primordia) is aborted. Both of these spikelets are fertile in corn.
- (vi) In teosinte the outer glumes are very hard, whereas in maize they are soft and external.
- (vii) In teosinte the glumes cover the seeds, whereas in maize the kernels are usually exposed.
- (viii) In teosinte the kernel are embedded into the deep cupules in the rachis, whereas in maize the kernels are held in place by cupules that are not too deep.
- (ix) Teosinte kernel are fragile, but they are not so in maize.
- (x) Teosinte seeds are small ones, those of maize are small, but are usually twice the size of wild races.
- (xi) What looks like a seed of teosinte, is actually the seed surrounded by a hard 'fruitcase'. In modern corn, the fruitcase is reduced and develops into part of the corn cob.

REFERENCES

Beadle, G. W. 1939. Teosinte and the origin of maize. Journal of Heredity 30: 245–247.

Beadle, G. W. 1977. The origin of *Zea mays*. In C. A. Reed (ed.), Origins of Agriculture, 615–635. Mouton Press, The Hague, The Netherlands.

Beadle, G. W. 1980. The ancestry of corn. Scientific American. 242: 112–119.

Doebley, John. 2004. The genetics of maize evolution. Annu. Rev. Genet. 38: 37-59.

Eubanks, M. W. 2001. The mysterious origin of maize. Economic Botany 55: 492–514.

Kato Y., T. A. and A. R. Lopez. 1990. Chromosome knobs of the perennial teosintes. Maydica 35: 125–141.

Manglesdorf, P.C. 1974. Corn: Its Origin, Evolution and Improvement. Harvard University Press; Cambridge, Massachusetts. 262 p.

Nirupma Singh, R.Ambika Rajendran, Meena Shekhar, S.L. Jat, Ramesh Kumar and R.Sai Kumar. 2010. Rabi Maize Opportunities Challenges, Directorate of Maize Research, Pusa Campus, New Delhi -110 012, Technical Bulletin, No. 9: 32 p.

Piperno, D. R. and D. M. Pearsall. 1998. The Origins of Agriculture in the Lowland Neotropics. Academic Press, San Diego, CA, USA.

Sharma, R. C. and Lal, S. 1998. Maize diseases and their management. Indian Farming. 48:92-96

Smith, Bruce D. 1995. The Emergence of Agriculture. Scientific American Library. New York. 230 pp.

Smith, C.W., Betrán, J. and E.C.A. Runge. (eds). 2004. Corn: origin, history, technology, and production. John Wiley & Sons, Hoboken, New Jersey, USA. 949 pp.

White, P.J. and L.A. Johnson. (eds). 2003. Corn: chemistry and technology. American Association of Cereal Chemists, St. Paul, Minnesota, USA . 892 pp.

Wilkes, H. Garrison. 1967. Teosinte: The Closest Relative of Maize. Bussey Inst., Harvard Univ.: Cambridge. 159 pp.

Wilkis, H. G. 1977. Hybridization of maize and teosinte, in Mexico and Guatemala and the improvement of maize. Economic Botany. 31: 254-293.

http://graincrops.ca.uky.edu/files/corngrowthstages_2011.pdf

http://agron-www.agron.iastate.edu/Courses/agron212/readings/corn_history.htm

https://en.wikipedia.org/wiki/Maize

http://dictionary.cambridge.org/dictionary/english/corn

http://maize.teacherfriendlyguide.org

CHAPTER

10

Sorghum–*Sorghum bicolor* L. (2n=20)

Vernacular name: Great millet, sorgo, kafir corn, guinea corn, milo (English); *Jowar* (Hindi); *Cholam* (Tamil); *Jonna* (Telegu); *Jola* (Kannada); *Jowari, Jondhla* (Marathi); *Juara* (Oriya).

Importance: The word sorghum is derived from the Italian word 'Sorgo' which means rising above i.e., above its growth other crops in the field. It is also known as 'Great millet'. It is called 'Camel of crops' because of its exceptional ability to tolerate drought. Sorghum is the one of the four major food grains in the world after rice, wheat and maize. It's a staple food crop of poor in arid and semiarid areas of the world.

Origin: Sorghum is originated in the region of the northeast Africa comprising Ethiopia, the Egyptian-Sudan border and East Africa. Indian sub-continent is its secondary center of origin.

Distribution: It is grown extensively in India, Sudan, Nigeria, China and Manchuria, Asia Minor, Iran and south European countries, Japan, Korea, Argentina, United States of America and Australia. Among the sorghum growing countries, India ranks the first in acreage but second in production, USA being the largest producer. In India, the major sorghum producing states are Maharashtra which occupies 49% share in total production followed by Karnataka (21%), Madhya Pradesh (9%), Rajasthan (7%), Andhra Pradesh (4%), Uttar Pradesh (3%) and Gujarat (2%).

Botany: The height of the plant varies from 0.5 to over 4 m. The stems are erect and solid. The culms are made up of 7-18 nodes and internodes. The leaf number on main stem varies from 7 to 24. Most of the roots are confined to the upper 15 cm of soil and the penetration is from 0.5 to 0.75 m. The inflorescence of sorghum is a panicle. The panicles are commonly known as 'heads'. They may be compact or loose. The peduncle (the uppermost internode which bears the inflorescence) commonly known as the 'neck' may be straight or curved downwards (goose-necked). The blooming takes place chiefly in early morning, the maximum being between 12 midnight and 2 a.m. and sometimes anthesis may continue up to 8 a.m. to 10 a.m. It is self-pollinated crop although cross pollination up to 6% is reported depending on the extent of openness of the panicle. The plant is propagated through sexual seed. The fruit is a free caryopsis between the glumes. It is commonly called the grain or 'seed'.

Classification:

Classification of sorghum based on use as

(i) grain sorghum which is grown for grain
(ii) forage sorghum grown mainly for fodder, hay and silage
(iii) grass sorghum is wild sorghum e.g. Sudan grass
(iv) broom corn sorghum used for making brooms
(v) waxy sorghum has waxy endosperm and is used for starch manufacture
(vi) sweet sorghum for ethanol production, making sorghum syrup and silage

Snowden classification of cultivated sorghum based on spiklet types: There are seven basic spiklet types namely wild types, shatter cane, bicolor, guinea, caudatum, kafir and dura.

(i) Wild type: Grain is small, linear, oblong and symmetrical dorso-ventrally, completely covered by the glumes racemes fragile and spiklets deciduous.
(ii) Shatter cane: It is similar to above, but grains are large and round occasionally slightly exposed at the tip, spiklets deciduous.
(iii) Bicolor: Bicolor race consists of several distinct sub-races. Sudan grass, sorgo, broomcorn and bicolor are the main ones. The long glumes clasping the grain, elongate seed and open panicles are considered to be primitive characters; glumes clasping the grain, which may be completely covered or exposed as much as ¼ of its length at the tip; spiklets persistent.
(iv) Guinea: Guinea race is basically a West African race. A secondary centre is found in East Africa, primarily in Malawi. It is most convenient to divide the race into three subraces based on the seed size *viz.*, *conspicium* (large seeds), *guinea* (medium seeds) and *maragaritiferum* (small seeds). The grain is flattened dorso-ventrally, sublenticular in outline, twisting at maturity nearly 90° between gaping involutes glumes that are from nearly as long as to longer than the grain.
(v) Caudatum: The caudatum race with its characteristics turtle backed grains is dominant in Sudan, Chad, Nigeria and Uganda. The grain is asymmetrical, the side next to the lower glume flat or in extreme cases even somewhat concave, the opposite side rounded and bulging; the persistent style often at the tip of a beak pointing toward the lower glume; glumes ½ length of the grain or less.
(vi) Kafir: The kafir race is a major race in East Africa from Tanzania Southward. The name is derived from the Arabic for 'unbeliever' or pagan refers to the blacks who grow it. The grain is approximately symmetrical, more or less spherical; glumes clasping and variable in length.
(vii) Durra: The name durra is derived from the Arabic for sorghum (or millets). The durras are dominant in the Ethiopia. Durra is drought resistant or at the least drought evading. Durras are known as milo in the United States. The grain is rounded, obvate, wedge shaped at the base and broadest slightly above the middle; glumes very wide.

Growth stages: The growth stages for sorghum of different crop duration are furnished in Table 1.

(i) Seedling stage: Seedling emergence takes 3–10 days. Leaves\nodes develop at the rate of one in 3 to 6 days. Early maturity hybrids typically produce 15 leaves per plant, while medium and late maturity hybrids produce 17 and 19 leaves each.

(ii) Tillering stage: Tillering occurs when the plants are in the 4 to 6 leaf stage.

(iii) Booting stage: About 6 to 10 days before flowering, the boot forms a bulge in the sheath of the flag leaf (uppermost leaf). This stage is called boot leaf stage.

(iv) Flowering stage: Sorghum usually flowers 55 to 70 days. The maximum flowering takes place on 3rd or 4th day after panicle exertion. It takes 6 days for the whole inflorescence to complete flowering. Individual panicles start flowering from the tip downwards and flowering may extend over 4 to 9 days.

(v) Milky stage: The soft dough stage occurs approximately 15 to 25 days after flowering and the seed attains 50% of its weight. Kernels reach their maximum volume approximately 10 days after flowering. The seed is soft and a white milky fluid appears when the seed is squeezed.

(vi) Hard dough stage: When the seed is in the hard dough stage, the grain cannot be squeezed with the fingers and approximately 75% of the seed weight has been reached.

(vii) Physiological maturity: The physiological maturity attains 25 to 45 day period after flowering and 30 to 35 days after fertilization. Grain moisture content at physiological maturity is between 25 and 35% moisture. Grain harvest can begin at approximately 20 to 24% moisture.

Table 1 Growth stages of sorghum

Stage	**Growth stage**	**Duration (days)**				
0	Emergence- coleoptile visible at soil surface after sowing	3-5	3-5	3-5	3-5	3-5
0	Seedling -collar 3rd leaf visible	7-10	7-10	7-10	10-12	10-12
2	Collar 5th leaf visible	12-15	12-15	15-17	15-20	15-20
3	Head forming growing point differentiation	25-30	25-30	30-35	35-40	45
4	Flag leaf visible	35-40	45	45-48	50	55
5	Booting – Head extended into flag leaf sheath	50	45-50	50-55	58	65
6	Half bloom	60	55-60	60-65	70	75-78
7	Soft kernel (milky stage)	65-70	60-80	82	84	87
8	Hard kernels	85	80-90	95	95-99	100-105
9	Harvest	90-95	100	105	114	120

Climate: Sorghum is essentially a crop of the tropics and dry temperate areas of the world. It is grown in 40°N to 40°S up to an altitude of 1500 m from MSL. It is suited to regions receiving annual rainfall of 600 to 1000 mm. A rainfall of 500–800 mm evenly distributed over the cropping season is normally adequate for cultivars maturing in 3–4 months. Sorghum tolerates waterlogging and can also be grown in areas of high rainfall. It is a short day plant. Sorghum requires solar radiation of 400 to 450 cal cm^2 day^1. It is an extremely drought resistant crop next to pearl millet. It tolerates heat and dry conditions better than corn. It is susceptible to frost. The cardinal temperature at different growth phases of sorghum is presented in Table 2.

Table 2 Cardinal temperature for growth and development of sorghum

Critical level	Germination	Vegetative phase	Reproductive phase
Minimum	7-10°C	15°C	13°C
Optimum	18-21°C	26-30°C	22-35°C
Maximum	38°C	40°C	40°C

Soils: Sorghum is grown in red sandy, red loamy, alluvial and coastal alluvial soils as well as on mixed black and red and medium black soils. It grows well in pH range of 6.0 to 8.5 as it tolerates considerable salinity and alkalinity.

Hybrids, varieties/composites of sorghum: The particulars of sorghum hybrids / varieties in India are furnished in Tables 3.

Tables 3 State wise hybrids/varieties of sorghum

State	Season/uses	Hybrids/varieties
Andhra Pradesh	Kharif	**Hybrids:** CSH-13, MLSH-296, CSH-16, JKSH- 22, PSH-1, CSH-18, ASH-1, CSH-21, & CSH-23, CSH-5, CSH-6, CSH-9, CSH- 16; ASH-1, ICMH-451 and ICTP-8203. **Varieties:** SPV-462, CSV-15, Nandyal Tella Jona-3, Palem-2, CSV-23, CSV-17, CSV-18, CSV-20 & Kinnerea, PSV-1, CSV-14 R, PJ-890, SJ-092, 122, 2169, Palem-2, N-13, N-14, APS-1, ICMV-221 and SPV-462.
	Rabi	**Hybrids:** CSH-12 R, CSH-13 R, CSH-15 R and CSH-19 R. **Varieties:** CSV-216 and SPV-1626.
	Fodder	**Hybrids:** Hara Sona and Pantchari-5.
	Sweet sorghum	CSH-22 SS.
Bihar	Kharif	**Hybrids:** CSH-16, CSH-1, CSH-2, CSH-3, CSH-5, Swarna hybrid-, BR-1 & BR-2. **Variety**- CSV-15
	Fodder	CSH-20-MF, MP Chari & Pusa Chari-1
Gujarat	Kharif	**Hybrids:** CSH-13, CSH-16, CSH-17, JKSH-22, CSH-18, CSH-21, CSH-23, GJ-37, GJ-39, GJ-40 and GJ-41. **Varieties:** GJ-38, GJ-39, CSV-15, GJ-40, GJ-41, RSV-9 (CSV-19 SS), CSV-23, CSV-17 and CSV-20.
	Rabi	**Hybrids:** CSH-13 R and CSH-19 R. **Variety:** SPV-1626.
	Fodder	**Hybrids:** Hara Sona, Pusa Chari Hybrid-106, Pant Chari-5, CSH-20-MF, Gujarat Fodder Sorghum-5, CSH-24. JF-4, CSH-5 and CSH-6.
	Sweet sorghum	CSH-22 SS.
Haryana	Kharif	**Hybrid:** CSH-16 **Varieties:** CSV-15, SSV-84 and CSV-23.
	Fodder	**Hybrids:** Hara Sona, Haryana Chari-308, Pusa Chari Hybrid-106, Pantchari-5, CSH-20-MF, Haryana Jowar-513 and CSH-24.

Karnataka	Kharif	**Hybrids:** CSH-13, MLSH-296, CSH-216, CSH-21 and CSH-23. **Varieties:** – CSV-15, ICSV-745, SPV-462, RSSV-9, CSV-17, CSV-18, CSV-20 and DSV-6.
	Rabi	**Hybrids:** CSH-15 R, CSH-19-R and DSH-4 R. **Varieties:** DSV-5, DSV-4, CSV-216 and SPV-1626.
	Fodder	Hara Sona
	Sweet sorghum	CSH-22 SS
Madhya Pradesh	Kharif	**Hybrids:** CSH-5, CSH-16, CSH-9, CSH-14, CSH-13, CSH-16, ICI-501, CSH-17, CSH-18, CSH-21, CSH-22 SS and CSH-23. **Varieties:** JK-22, JK-741, JK-938, JK-1041, SPV-15, CSV-15, SPV-1022, CSV-15, Jawahar Jowar-938, Jawahar, Jowar-1041, CSV-17 and CSV-20.
	Fodder	Hara Sona and Pantchari-5
Maharashtra	Kharif	**Hybrids:** CSH-9, CSH-13, CSH-14, CSH-16, MLSH-296, ICI-501, CSH-18, Mahabeej-7, SPH-840, CSH-21, CSH-23 and SPH-1567. **Varieties:** CSH-9, CSH-14, CSV-15, Parbhani Sweta, PVK-809, CSV-17, CSV-18 and CSV-20.
	Rabi	**Hybrids:** CSH-15 R and CSH-19 R. **Varieties:** CSV-216 (Phule Yashoda), RSLG-262 (Maulee), Parbhani Moti, Uttara, SPV-1626, Vasudha, AKSV-13 R, Phule Chitra, M-35, Prabhani Moti, Prabhani Sweta, Phule Yashoda and Phule Maulee.
	Fodder	**Hybrids:** Hara sona, Pusa chari hybrid-106 and Pusa chari-5.
	Sweet sorghum	**Hybrids:** CSH-22 SS, **Varieties:** Phule Amrita and AKSSV-22.
Rajasthan	Kharif	**Hybrids:** CSH-13, CSH-16, CSH-17, CSH-18, MLSH-296, ICI-501, SPH-837, CSH-23, CSH-16, CSH-17, CSH-6, CSV-15, CSH-9, Partap Jowar-1430, SPH-837 and SPV-245. **Varieties:** CSV-15, Partap jowar-1430, CSV-23 and CSV-17.
	Fodder	Hara sona, Haryana chari-308, Pant chari-5 and CSH-20-MF.
	Sweet sorghum	CSH-22 SS
Tamil Nadu	Kharif	**Hybrids:** CSH-13, MLSH-296, CSH-16, ICI-501, CSH-17, CSH-1, CSH-5, CSH-6, COH-3 and COH-4. **Varieties:** CSV-15, BSR-1, SPV-462, Paiyur-2, CO (S) 28, K-11, CSV-2, CSV-17, CO(S) -28, CO(S)-26, APK-1, Paiyur-1, Paiyur-2, CO-10, CO-18, CO-19, CO-21, CO-25, CO-26, K-4, K-5, K-6, K-7, K-8, K-9, K-10, K-11, K Tall, BSR-1 and IS 3541.

	Rabi	**Hybrid:** CSH-19 R. **Varieties:** BSR – 1, APK – 1, Paiyur-2, CO(S)28, K-11 and SPV-1626
	Summer	CO(S) 28 and K-11
	Fodder	Hara sona, Pant chari-5 and CO (FS) 29.
	Sweet sorghum	CSH-22 SS
Uttar Pradesh	Kharif	**Hybrids:** CSH-13, CSH- 16, CSH-18, CSH-23. **Varieties:** CSV-15, Bundela, CSV-23, CSV-17 and CSV-20
	Fodder	Hara sona, Haryana chari-308, Pant chari-4, Pant chari-5 and Pusa chari-106
	Sweet sorghum	CSH-22 SS

Season: Sorghum is sown in 3rd week of June to 1st week of July with onset of monsoon during kharif season; 15 September-15 October during *rabi* season and 3rd week of January to 1st week of February during summer season.

Seeds and sowing: Good quality seeds are collected from disease and pests free fields. Seed rate is 7.5 and 10.0 kg/ha for transplanted and direct sown under irrigated condition whereas 15 kg/ha for direct sowing under rainfed condition. Treat the seeds 24 hours prior to sowing with Carbendazim or Captan or Thiram 75 @ 3 gm per kg of seed. Treat the seeds with 2 % KH_2PO_4 for 6 hours and shade dry. Dissolve 20 g of KH_2PO_4 in one litre of water for soaking 5 kg of seeds. Dissolve 0.5 g of gum in 20 ml of water. Add 4 ml of chlorpyriphos 20EC or monocrotophos 35 WSC or phosalone 35EC. To this, add one kg of seed, pellet and shade dry. Seed hardening ensures high germination. The seeds are pre-soaked in 2% KH_2PO_4 solution for 6 hours in equal volume and then dried back to its original moisture content in shade and are used for sowing. Seed treatment with phosphate solubilizing bacteria (PSB) @ 50 g/kg seed and azotobactor @ 25 g/kg seed can also done. In the case of pure crop of sorghum, maintain the seed rate of 10 kg/ha. In the case of intercrop of sorghum with pulse crop, a seed rate of sorghum at 10 kg and pulse crop at 10 kg/ha are required. In the case of pure crop of sorghum sow the seeds with a spacing of 45 x 15 cm. Maintain one plant per hill. If shoot fly attack is there, remove the side shots and retain one healthy shoot. Sow the seeds over the lines where fertilizers are placed. Sow the seeds at a depth of 2 cm and cover with soil. In the case of sorghum intercropped with pulses, sow one paired row of sorghum alternated with a single row of pulses. The spacing between the row of sorghum and pulse crop is 30 cm.

Nursery practices: For raising seedlings to plant one hectare, select 7.5 cents (300 m^2) near water source where water will not stagnate. Apply 750 kg of FYM or compost for 7.5 cents nursery and apply another 500 kg of FYM for covering the seeds after sowing. Spread the manure evenly on the unploughed soil and incorporate by ploughing or apply just before last ploughing. Provide three separate units of size 2 m x 1.5 m with 30 cm space in between the plots and all around the unit for irrigation. Excavate the soil from the interspace and all around to depth of 15 cm to form channels and spread the soil removed on the bed and level. Make shallow rills, not deeper than 1 cm on the bed by passing the fingers vertically over it. Broadcast 7.5 kg of treated seeds evenly on the beds. Cover by leveling the rills by passing the hand lightly over the soil. Apply 2.5 kg of Lindane 10 % dust over the seed beds and all around to prevent ants carrying seeds away and ensure that the seeds are not sown deep as germination will be affected. For water management in nursery, provide one inlet to each nursery unit. Allow water to enter through the inlet and cover all the channels till the raised bed are wet and then cut of water. First irrigation is given immediately after sowing and life irrigation on the fourth day. Subsequently once in five days. Four to five irrigations are required depending on soil types in the nursery. Do not keep the seedlings in the nursery for more than 18 days. If older seedlings are used, establishment and yield are adversely affected. Do not allow cracks to develop in the nursery by properly adjusting the quantity of irrigation water. Pull out the seedlings when they are 15 to 18 days old. Prepare slurry with 5 packets (1 kg per ha) of azospirillum in 40 litres of water and dip the root portion of the seedlings in the solution for 15 to 30 minutes and then transplant the seedlings. Let in water through the furrows. Plant one seedling per hill. Plant the seedlings at a depth of 3 to 5 cm. Plant the seedlings on the side of the ridge, half the distance from the top of the ridge and the bottom. Maintain a spacing of 45 x 15 cm. Transplanted crop has a few advantages. Main field duration is reduced by 10 days. Shoot fly, which attacks direct sown crops during the first 3 weeks and which is difficult to control can be effectively and economically controlled in the nursery itself. Seedlings which show chlorotic and downy mildew symptoms can be eliminated, thereby incidence of downy mildew in the main field can be minimised. Optimum population can be maintained as only healthy seedlings are used for transplanting. Seed rate can also be reduced by 2.5 kg/ha.

Main Field Preparation

Land preparation: Plough the field with an iron plough once and twice with a country plough, Sorghum does not require fine tilth, since it adversely affects germination and yield in the case of direct sown crop. Form ridges and furrows using a ridger 6 m long and 45 cm apart. For irrigation channels across the furrows. Alternatively, form beds of size 10 m^2 or 20 m^2 depending on the availability of water.

Field preparation depends on the system of sorghum sowing viz.,

(i) sowing on a flat surface, or
(ii) using ridge-and-furrow system, or
(iii) on a broad bed-and-furrow system.

If sowing is done on a flat surface, the land should be leveled after final ploughing using bullock-drawn or tractor-drawn levelers. In ridge and furrow system, ridges are made using either tractor drawn or animal drawn ridge ploughs. Broad beds and furrows are prepared by an animal-drawn ridger, mounted on a tool carrier, or by tractor-drawn implements with ridgers. The top of the bed is 1.2 m wide and the distance from the center of one furrow to the center of the next furrow is 1.5 m. The depth of furrows should be 15 cm or more. The broadbed-and-furrow system has many advantages over flat sowing. It helps in draining off excess water in the field and soil; provides more soil aeration for plant growth; greater *in-situ* moisture conservation; easier for weeding and mechanical harvesting.

After cultivation: Thin the seedlings and gap fill with the seedlings thinned out on 10-15 days after emergence and the second at 20-25 days after sowing. Maintain a spacing of 15 cm between plants after the first hand weeding on the 23rd day of sowing. Thin the pulse crop to a spacing of 10 cm between plants for all pulse crop except cowpea for which spacing is maintained at 20 cm between plants.

Weed management: Sorghum is slow growing in early stages and is adversely affected by weed competition. Therefore, keep the field free of weeds upto 45 days. Apply the pre-emergence herbicide Atrazine 50%WP @ 500 g ai/ha on 3 days after sowing as spray on the soil surface, with 900 lit of water per ha. Then one hand weeding may be given on

30 to 35 days after sowing. If pulse crop is to be raised as an intercrop in sorghum, do not use atrazine. Hand hoeing and weeding is taken on the 10th day after transplanting, if herbicides are not used. Hand hoeing and weeding are taken up between 30-35 days after transplanting and between 35-40 days for a direct sown crop. The root parasite witch weed (*Striga asiatica* and *Striga hermonthica*) reduces sorghum grain yield, between 15 to 30%. Per-emergence application of atrazine at 0.5 kg ai./ha and post emergence application of 2,4-D at 2 kg ai/ha around 30 days after planting provides excellent control of *Striga* weed. Striga flowering can begin within 2 weeks and seeds begin to mature 2-4 weeks later. The seed capsules may contain 400-500 seeds and a single plant may produce 20,000 seeds.

Manures and fertilizers application: Apply FYM or composted coir pith @ 12.5 t/ha along with azospirillum (10 packets) @ 2 kg/ha on the unploughed field and incorporate the manure in the soil by working a country plough.

(i) Fertilizers application for direct sown crop: Fertilizer dose of 90-45-45 kg of N, P_2O_5 and K_2O per ha is recommended. Apply half of the dose of N and full dose of P_2O_5 and K_2O as basal dressing. In the case of bed planted crop, mark lines to a depth of 5 cm and 45 cm apart. Place the fertilizer mixture at a depth of 5 cm along the lines. Cover the lines up to 2 cm from the top before sowing. In the case of sorghum raised as a mixed crop with a pulse crop (blackgram, greengram or cowpea), open furrows 30 cm apart to a depth of 5 cm. Apply fertilizer mixture in the two lines in which sorghum is to be raised and cover up to 2 cm. Skip the third row in which the pulse crop is to be raised and place fertilizer mixture in the next two rows and cover up to 2 cm with soil. When Azospirillum is used apply only 75% of recommended N for irrigated sorghum. Mix 12.5 kg of micronutrient mixture with enough sand to make up the total quantity of 50 kgs. Spread the mixture evenly on the beds. Basal application of 25 kg $ZnSO_4$ per ha for the zinc deficient soils is recommended. Basal application of $FeSO_4$ @ 50 kg/ha along with FYM @ 12.5 t/ha for iron deficient soils is recommended.

(ii) Fertilizers application for transplanted crop: Fertilizer dose of 90-45-45 kg of N, P_2O_5 and K_2O per ha is recommended. Apply half the dose of N and full dose of P_2O_5 and K_2O basally before planting. In the case of ridge planted crop, open a furrow of 5 cm depth on the side of the ridge at two thirds the distance from the top of the ridge and the bottom and place the fertilizer mixture along the furrow and cover with soil up to 2 cm. Apply the remaining 50% of N along the furrows on the 15th day of planting and irrigate. Soil application of azospirillum @ 10 packets (2 kg/ha) after mixing with 25 kg of FYM + 25 kg of soil may be carried out before sowing/planting. Mix micronutrient mixture @ 12.5 kg/ha with enough sand to make a total quantity of 50 kg. Apply the mixture over the furrows and on top one third of the ridges. If micronutrient mixture is not available, mix 25 kg of zinc sulphate with sand to make a total quantity of 50 kg and apply on the furrows and on the top one third of the ridges. Soil application of 25 kg/ha of ferrous sulfate at the final ploughing is recommended in soils low in available iron. Foliar application of 0.5-1.0% ferrous sulfate solution corrects iron deficiency if noticed in a standing crop.

Deficiency symptoms

Zinc: Deficiency symptoms first appear in the newly formed leaves at 20 to 30 days age. Older leaves have yellow streaks or chlorotic striping between veins.

Iron: Interveinal chlorosis will be observed. If the deficiency continues the entire leaf including the veins may exhibit chlorotic symptom. Newly formed leaves exhibit chlortic symptoms. The entire crop may exhibit bleached appearance, dry and may die.

Water management: Water requirement is 400 mm for grain production and 500 mm for green fodder production. Irrigation is given at 50% available soil moisture. The IW/CPE ratio of 0.4 to 0.6 is recommended for grain production while 0.9 for green fodder production. Adjust irrigation schedule according to the weather conditions and depending upon the receipt of rains. The critical growth stages of sorghum in relation to water requirement are viz., initiation of grand growth stage (20-25 DAS), flag-leaf stage or boot stage (50-55 DAS), flowering stage (70-75 DAS) and grain-filling stage 90-100 DAS. Spraying 3% kaolin (30 g in one litre of water) during periods of moisture stress will mitigate the ill effects.

Cropping systems: Sorghum can be intercropped with pigeon pea, green gram, cowpea, soybean and sunflower. Sorghum-based intercropping systems such as Sorghum + pigeonpea (2:1 or 3:3); Sorghum+ soybean (3:6 or 2:4) and Forage sorghum + cowpea/clusterbean (2:2) are promising in different states of India. After *kharif* sorghum, a sequence crop like chick pea, safflower and mustard are found most suitable in in *rabi* season. Intercropping of soybean with sorghum in the ratio 4:2 is recommended for *kharif* and summer seasons. Intercropping of sunflower with sorghum in 4:2 ratio is recommended under rainfed conditions during northeast monsoon for black soils. Paired row planting of sorghum and sow one row blackgram/cowpea as intercrop or fodder sorghum + fodder cowpea at 3:2 ratio can be practiced in rainfed black soils. A sequence crop in rabi following sorghum in kharif is found to be profitable in those areas which receive rainfall above 700 mm and having moisture retentive medium to deep black soils. The promising sorghum based crop rotations are sorghum-wheat, sorghum-wheat-moong, sorghum-gram, sorghum-potato/wheat, sorghum-raya, sorghum-potato/rape-wheat/tobacco, sorghum-cotton-groundnut, sorghum-wheat-cowpea/pearlmillet, sorghum-wheat-greengram, sorghum-pea/oat/berseem, sorghum-gram or barley, sorghum-lentil, sorghum-cotton, sorghum-sorghum(*rabi*), sorghum-tobacco, ground nut-sorghum (*rabi*) and sorghum-ragi-groundnut. The following techniques are suggested to make sequence cropping system economical and feasible.

(i) *Kharif* sorghum crop should be harvested at its physiological maturity to gain about one week's time in planting the winter crop.

(ii) Practice of minimum tillage needs to be adopted. It helps to gain time, minimizes land preparatory costs and prevents soil moisture loss.

(iii) Sowing of winter crop should be drilled without much opening of the soil.

(iv) Inter cultivation should be done at appropriate time to minimize weeds and soil water loss.

Pest and diseases management: The important sorghum insect pests are shoot fly, stem borer, midge, white grub, armyworm, cutworms, grasshopper, pyrilla, shoot bug, earhead caterpillars, earhead bug, spider mite, aphids and the diseases are grain mold, charcoal rot, downy mildew, anthracnose, loose smut, ergot, grain smut, red leaf spot, rust, etc. Set up light traps @ 1 trap/acre 15 cm above the crop canopy for monitoring and mass trapping insects. Light traps with exit option for natural enemies of smaller size should be installed and operate around the dusk time (6 pm to 10 pm). Apply Carbofuran 35 @ 20 kg/ha in seed rows to safe guard against shoot fly incidence. Stem borer is controlled with application of Endosulfan 4G or 3D in the whorls @ 8 and 12 kg/ha at 20 and 35 days after emergence. Dusting of Malathion 10D @ 20 kg/ha to control Head bug at pre-bloom and 50% flowering stage. Seed treatment is done with Thiram or Captan @ 3 g per kg or metalayl (Apronxl) @ 3 ml/kg of seed to control grain mold and charcoal rot. Sugary disease is managed by 2 sprays of propaconazile (TILT 25 EC) @ 5.0 ml/lit at flowering and 15 days later on. Grain mold can be controlled with Aureofungin @ 30 g/10 l of water) + captan (@ 3%. Leaf spot such as rust can be controlled by spraying Dithane M 45 @ 3%.

Harvest: Physiological maturity can be determined by the black (dark) spot at the bottom of the grain. The *Kharif* sorghum should be harvested at its physiological maturity to avoid grain mold damage. Sorghum grains are usually harvested at a grain moisture content of 20 to 24%. The leaves turn yellow and a dried up appearance when the crop matures. The grains are hard and firm. At this stage, harvest the crop by cutting the earheads separately. Cut the straw after a week, allow it to dry and then stack. Sometimes, the stem is cut at 10 to 15 cm above ground level and afterwards separate the earheads and stack the straw. Dry the earheads. Thresh the grains using a mechanical thresher or by drawing a stone roller over the earheads or by using cattle and dry the produce and store. The threshed grain should be cleaned and dried in sun for 6-7 days to reduce the moisture content down to 12-15% for safe storage. Hundred seed weight is 3.0 to 3.5 g.

Yield: The average grain yield ranges between 2.5 to 3.5 t/ha, and that of hay between 15.0 to 17.0 t/ha under assured water supply. With improved cultural practices, it is possible to harvest nearly 5.0 t/ha of grain and about 10.0 to 12.5 t/ha of dry straw under irrigated conditions.

Utilization: Sorghum is a major source of staple food for the humans and for livestock feed. The grain is processed into starch paste, flour, dextrose, dextrose syrup, gluten meal, gluten feed, edible oil and alcoholic beverages. It is used in the brewing industry to make beer. The stems and foliage are used for green chop, hay, silage, and pasture.

RATOON SORGHUM

Varieties for ratooning: CO 25, CO 26, CSH 5, K tall.

Ratooning technique: Harvest the main crop leaving 15 cm stubbles. Remove the first formed two sprouts from the main crop and allow only the later formed two sprouts to grow. Allow two tiller per hill. The duration of the ratoon crop is about 15 days less than the main crop.

Weed management: Remove the weeds immediately after harvest of the main crop. Hoeing and weeding are carried out twice on 15th and 30th day after cutting.

Manures and fertilizers application: Apply N @ 100 kg/ha in two split doses. Apply the first dose on 15th day after cutting and the second on 45th day after cutting. Apply P_2O_5 @ 50 kg/per ha along with first application of nitrogen.

Water mangement: Irrigate immediately after cutting the main crop. Irrigation should not be delayed for more than 24 hours after cutting. Irrigate on 3rd or 4th day after cutting. Subsequently irrigate once in 7 to 10 days. Stop irrigation on 70 to 80 days after ratooning.

Harvest: Harvest the crop when the grain turns yellow.

RAINFED SORGHUM

Soil: Rainfed sorghum is grown under wide range of agroclimatic conditions in Vertisols, Entisols, Inceptisols and Alfisols. Post-rainy season sorghum is largely confined to Vertisols. Yield can be increased up to one tonne per ha with an increase in soil depth and moisture storage in the post rainy season. The yield difference between 'shallow' and a 'deeper' soils is almost one tonne per ha.

Varieties: CO 21, CO 25, CO 26, K tall, COH 3.

Land preparation: It is done 5 days before to 10 days after daily precipitation is equal to 0.5 PET.

Seeds and sowing: Seed rate is 15 kg/ha. Soak the seeds in 2% KH_2PO_4 (20 g in one litre of water) or 500 ppm of CCC (one ml in one litre of water) for six hours and shade dry the seeds for 5 hours. Use 350 ml of solution for soaking one kg of seed. The seed is pelletised with 15 g of chloropyriphos in 150 ml of gum and shade dried. Treat the seeds with three packets of azospirillum (600 g) and 3 packets of phosphobacterium. In the main field, apply azospirillum 2 kg (10 pockets/ha) with phosphobacteria 2 kg with 25 kg FYM + 25 kg soil. Sow the seeds well before onset of monsoon in Vertisols at 5 cm depth by seed drill or by country plough. Sow the hardened seeds at 5 cm depth with seed cum fertilizer drill to ensure uniform depth of sowing. Sow the sorghum seeds over the line where the fertilizers are placed. Sow the seeds at a depth of 5 cm and cover the soil. Sow the seeds with the spacings of 15 cm in the paired rows spaced 60 cm apart. Sow the pulse seeds to fall 10 cm apart in the furrows between the paried rows sorghum. Spacing for pure crop is 45 x 15 cm.

Manures and fertilizer application: Apply FYM or composted coir pith @ 12.5 t/ha or enriched FYM @ 750 kg/ha along with 40-20-0 kg of N, P_2O_5 and K_2O per ha as basal dressing.

Weed management: Keep sorghum field free of weeds from second week after germinations till 5th week. If sufficient moisture is available spray atrazine @ 500 g per ha as pre-emergence application within 3 days after the receipt of the soaking rainfall for sole sorghum whereas Pendeimethalin @ 3.0 1it/ha for sorghum intercropped with pulses.

Yield: Grain yield potentials of up to 5.5 t/ha in Entisols and Vertisols and of 3 to 4 t/ha in Alfisols can be obtained. The average yields of sorghum in areas with favorable rainfall and soils are around 2 to 2.5 t/ha, while the post rainy season sorghum grown under unfavorable conditions yield around 500 kg/ha.

Sorghum injury/Sorghum effect: The growth and yield of crops following sorghum are depressed to certain extent and this is referred as *'sorghum injury'* or *'sorghum effect'*. Sorghum stubbles ploughed into the soil contain sugars, which encourages microbial activity. The bacteria in the soil use all the available nitrogen for their growth and crops that follow are starved of nitrogen. Structure of the soil gets unfavourable when the sugars decompose actively and this also affects the following crop. This is only a temporary problem. This can be avoided with intercropping of legumes with sorghum and application of more nitrogenous fertilizers to the succeeding crop.

Sorghum poisoning: Young sorghum plants and also ratoon sprouts from sorghum stubbles are poisonous and bring about death of the animals which graze them. These contain cynogenic glucoside called 'dhurrin' particularly in the plants which are stunted or affected by the drought. They can be used for feeding after drying or ensiling. The glucoside hydralyses in the stomach of the cattle and produces hydrocyanic acid, or prussic acid which brings death to the animals. Young sorghum seedlings below fifty days old contain a high content of hydrocyanic acid (HCN) and it decreases from the early stages progressively till the plants attain the boot stage. Side tillers and axillary branches contain a high percentage of the acid as compared to main shoots. The glucoside content decreases as the sorghum plant makes growth. The toxic level of HCN is above 200 ppm. Plants of more than 75 days old are nearly free of the glucoside.

REFERENCES

CPG. 2012. Crop Production Guide. Department of Agriculture, Government of Tamil Nadu, Chennai and Tamil Nadu Agricultural University, Coimbatore.

Damania, A.B. 2002. The Hindustan centre of origin of important plants. Asian Agri-History 6(4): 333–341

de Wet, J.M.J. 1977. Domestication of African cereals. Afr. Econ. History 3:15.

Dial, H.L. 2012. Plant guide for sorghum (*Sorghum bicolor* L.). USDA-Natural Resources Conservation Service, Tucson Plant Materials Center, Tucson, AZ.

Doggett, H. 1988. Sorghum. Longmans Scientific and Technical Publishers, UK, IDRC, Canada.

Gerik, T., Bean, B. and R. Vanderlip. 2003. Sorghum Growth and Development, Texas Cooperative Extension Service. USA.

GOI. 2014. Status Paper on Coarse Cereals. Directorate of Millets Development, Jaipur and Ministry of Agriculture, Government of India, New Delhi. p 216

Harlan, J.R. and J.M.J. de Wet. 1972. A simplified classification of cultivated sorghum. Crop Science 12:172–176.

House, L.R. 1985. A guide to sorghum breeding. International Crops Research Institute for the Semi-Arid Tropics. Patancheru, Andhra Pradesh 502 324, India: 216 pp.

House, L.R. 1987. Sorghum-present status and future potential. Outlook on Agriculture 16(1):21–27.

ICRISAT. 2004. Sorghum, a crop of substance. Patancheru 502 324, Andhra Pradesh, India: International Crops Research Institute for the Semi-Arid Tropics. 97 pp.

Kimber, C.T. 2000. Origins of domesticated sorghum and its early diffusion to Indian and China. Pages 3–98 in Sorghum origin, history, technology and production (Smith CW and Frederiksen RA, eds.). New York, USA: John Wiley & Sons Inc.

Peacock, J.M. and G.L. Wilson. 1984. Sorghum. The physiology of tropical field crops (Goldsworthy PR and Fisher NM, eds.). Chichester, UK: John Wiley and Sons Ltd. pp 249–279.

Reddy, B.V.S., Ramesh, S. and P. Sanjana. 2004. Sorghum and its improvement, CRC press, LLC, USA

Vavilov, N.I. 1992. Origin and geography of cultivated plants (Dorofeev VF, ed.). Cambridge University Press, Cambridge, UK. 332 pp.

Vanderlip, R. L. 1993. How a Grain Sorghum Plant Develops, Kansas State University, USA.

http://agmarknet.nic.in/profile-jowar.pdf

CHAPTER

11

Pearl Millet–*Pennisetum typhoides* *Stapf and Hubbard* (2n=14)

Synonyms: *Pennisetum americanum* Schum. *Pennisetum glaucum* (L). R. Br.

Family: Poaceae

Vernacular names: Pearl millet, Bulrush millet, Cat-tail millet, Spiked millet, Candle millet (English); *Cumbu* (Tamil); *Bajra* (Hindi); *Sajja,* (Telugu); *Sajje* (Kannada); *Bajri* (Marathi).

Importance: The word *Pennisetum* is derived from two Latin words '*Penna*' meaning feather and '*seta*' the bristle and hence called 'the feather like bristles'. It is considered to be poor man's food. It is a C4 plant.

Origin: The primary centre of origin is a long belt in the Sahel from Senegal to Sudan as well as in the central highlands of the Sahara.

Distribution: It is mostly cultivated under rainfed conditions in the arid and semi-arid regions of country. It is widely grown in India, China, Australia, Nigeria, Niger, Mali, Tanzania, Sudan, Senegal, Pakistan, Sudan and Egypt. In India, it is grown in Rajasthan, Maharashtra, Gujarat, Haryana, Madhya Pradesh, Punjab, Uttar Pradesh, Andhra Pradesh, Karnataka and Tamil Nadu.

Botany: Pearl millet is an erect annual, tillering and growing up to 3 m in height. Roots arise from second and third nodes above the soil surface are known as 'brace' or 'prop' roots. Tillers arise from the basal nodes. Secondary tillers are produced from the upper nodes of stems. The stem is solid (woody), rounded to oval. The stem is thinner with 1 to 3 cm in diameter. The node is slightly swollen and bears the ring of adventitious root. The leaves are 90 to 100 cm in length, and 5 to 8 cm in width. The inflorescence is a terminal spike, almost cylindrical, ranging from 15 cm to 1 m in length and 2 to 5 cm in width. The number of spikelets may be from 800 to 3000 per spike. Anthesis occurs throughout the day and night with the peak between 8.00 p.m. to 2.00 a.m. It is a highly cross-pollinated crop. The pollinating agent is wind. The flowers are protogynous. The grain is a caryopsis which is mostly grey and rarely yellow. The seed is 3-4 mm long and 2-2.5 mm wide. The colour of grain varies from whitish-yellow to grey or dull light blue, while the embryo has a reddish tinge. Seed weight is 10-15 g/1000 kernels.

Growth stages: Basal tillering occurs 2 to 6 weeks after sowing. Secondary aerial tillers produce 2–3 leaves and a small inflorescence which may contribute 15 to 50% grain yield. It takes 15 to 20 days from inflorescence differentiation to flowering. The spike emerges about 10 weeks after sowing. The grain-filling period normally takes 22 to 25 days. The inflorescence is a cylindrical spike densely packed with the spikelets. The number of spikelets may be from 800 to 3000 per spike. The length of the spike may vary from 15 to over 60 cm. Two to three days after the emergence of the inflorescence, the bluish-white styles begin to protrude out of the glumes. A panicle continues shedding pollen for about 3 days. The anther emergence continues throughout the day and night. The anthesis occurs between 8 am and 2 pm with a peak at about 10 am. The time required for complete stigma emergence varies from 2 to 3 days and they remain receptive for next two to three days. The stigmas remain receptive for one to two days to receive pollen from other plants for fertilization. It is a highly cross-pollinated species. Wind is the cross pollinating agent. The seed is a caryopsis and its shape varies from globular to conical shape. The seed colour varies from ivory to purplish black, with light to deep gray being the most common seed colour. Sometimes, pearl millet grains germinate in the inflorescence when harvesting is delayed in prolonged wet weather. Seed weight ranges from 4-12 g per 1000 grains.

The different growth stages of pearl millet are presented in Table 1 and Table 2.

Table 1 Growth stages in Pearl millet

Growth stage	**Crop duration**			
Coleoptile emergence	0	0	0	0
3rd leaf stage	6	6	6	6
5th leaf stage	14	14	15	15
Panicle initiation	21	22	28	28
Flag leaf visible	28	33	43	43
Boot stage	35	36	47	47
50% stigma emergence	40	42	53	50
Milk stage	49	49	61	59
Dough stage	56	58	69	68
Physiological maturity	65	70	75	80

Table 2 Types of root, initiation and effectiveness

Types	**Initiation time**	**Effective up to**
Primary	within 4 days	45-60 days
Adventious	8-12 days	Maturity
Crown or collar	30-40 days	Maturity

Climate: Pearl millet is well adapted to an altitude range of 800 to 1,800 m from MSL. It is adapted to hot climates, and is even more resistant to drought than sorghum.

The length of the growing period of pearl millet is 90 days under rainfed conditions. It is suitable for areas with annual mean rainfall of 400 to 1500 mm. It grows well under warm climatic conditions with 500-600 mm of rainfall during the growth period. It does not tolerate waterlogging condition. It is the most drought tolerant crop among the cereals and millets. It is a short day plant. The cardinal minimum, optimum and maximum temperature requirement is 15ºC, 22 to 30ºC and 45ºC respectively. No germination occurs below 12°C. The optimum temperature for tiller production and development is 21–24°C, and for spikelet initiation and development about 25°C. High temperatures at early stages induce flowering, while low temperatures promote ergot incidence. Soil temperatures must reach 12°C for germination to begin. Pearl millet development begins at a base temperature around 12°C. The optimum temperature for root elongation is 32°C. It requires temperatures of 31 to 35°C for a good photosynthetic activity. Tillers production is high at temperature of 25°C. The rate of leaf production is high at temperature of 18°C to 30°C. Grain setting is optimum from 22 to 25°C and declines at temperatures below and above this range, while grain mass steadily declines with increasing temperatures from 19 to 31°C. The grain set is low at temperature <13°C during the booting stage.

Soil: It prefers well drained sandy loam. However, it is also grown in black cotton soils, alluvial soils and red soils. It is sensitive to acidic soil and waterlogging condition. It is tolerant to salinity (4 to 8 dSm^{-1}) and alkalinity.

Seasons: It is mainly grown in rainy (*kharif*) season (June-September). It is also cultivated post-rainy (*rabi*) season (November-February) and during summer (February-May).

Varieties: The hybrids and varieties of pearl millet recommended in different states of India are presented in Table 3.

Table 3. Hybrids and varieties of pearl millet recommended in different states of India

State	Hybrids	Varieties
Andhra Pradesh & Tamil Nadu	HB 4, HB 5, PHB 10, PHB 14	WCC 75, CO 7, X 6, K 3, K 4
Karnataka	BJ 104, MBH 110	RCB 2
Gujarat & Maharashtra	PHB 10, PHB 14, BJ 104, MBH 110, GHB 526, PB 180	Pusa Composite-383, PPC 6, Jawahar *Bajra* variety-2
Rajasthan	BK 104, BK 560, BD 111, BD 163, CM 46, ICMH 356, GHB 526*, PB 180	Pusa Composite-383, Pusa Bajri-266, CZ-IC-923,
Haryana and Punjab	HB 5, PHB 10, PHB 14, PBH 47, MH 67, MH 169, MH 190, MH 208, HHB 67, BJ 104, BK 560, BD 111, CM 46, BD 183, HHB 68, HC 10, HHB 117	HS-1. PCB-164
Madhya Pradesh and Chattisgarh	BJ 104, BK 560, BD 111, BD 163, CM 46, CM 5, ICMH 356	RCB 2, Vijay Composite
Bihar & Jharkhand	PHB 10, MH 143, MH 169, BJ 104. BK 560, BD 111, BD 163, ICMH 356	------
Uttar Pradesh & Uttarakhand	HB 5, PHB 10, PHB 14, PHB 47, MBH 110, MH 36, BJ 104, BK 560, BD 111, BD 163, CM 46, ICMH 356	Pusa Composite-383, PSB 8, Jawahar *Bajra* variety-2

Nursery management: Select 7.5 cents near a water source for raising seedlings to plant one ha. Water should not stagnate. Plough the land and bring it to fine tilth. Apply 750 kg of FYM or compost over the unploughed soil and incorporate by ploughing. In each cent, mark 6 plots of size 3 × 1.5 m with 30 cm channel in between the plots and all around. Form the channel to a depth of 15 cm. Spread the earth excavated from the channel on the beds and level. Dissolve one kg of common salt in 10 litres of water. Drop the seeds into the salt solution. Remove the ergot and the Scletoria affected seeds which will float. Wash the seeds in fresh water 2 to 3 times to remove the salt on the seeds. Dry the seeds in shade. Treat the seeds with 3 packets (600 g) of Azospirillium. Apply 600 g of carbaryl 5% against ants and phorate 10G or carbofuran 3 G 600 g mixed with 2 kg of moist sand spread on the beds and work into the top 2 cm of soil to protect the seedlings from shoot fly infestation. If the seed bed is not treated before sowing protect the nursery by applying any one of the insecticides namely Endosulfan 35EC @ 12ml; Methyl Demeton 25EC @ 12ml; Demethoate 30EC @ 12ml on the 7th and 14th day of sowing by mixing in 6 litres of water. Open small rills not deeper than one cm on the bed by passing the fingers over it. Sow 3.75 kg of seeds in rills in 7.5 cents (0.5 kg per cent). Cover the seeds by smoothing out the rills with the hand. Sprinkle 500 kg of FYM or compost evenly and cover the seeds completely with hands. Provide an inlet to each unit of 5 plots to allow water in the channels. Allow water to enter the channel and rise up in it. Turn off the water when the raised bed is wet. Irrigate immediately after sowing and then on third day after sowing. Subsequently irrigation is given once in 5 to 7 days depending on soil type. Ensure that cracks do not develop in the nursery. This can be avoided by properly adjusting the quantity of irrigation water. The seedlings should not be kept in nursery for more than 18 days. As they mature establishment and yield are affected adversely.

Main Field Management

Field preparation: Plough with a an iron plough twice and with country plough twice. Bring the soil into fine tilth. Chisel the soils having hardpan formation at shallow depths with chisel plough at 0.5 m interval first in one direction and then in the direction perpendicular to the previous one, once in three years. Apply 12.5 tonnes per ha FYM or composted coir pith besides chiseling to get an additional yield of 30%. Form ridges and furrows (using 3 ridger 6 m long and 45 cm apart). If ridge planting is not followed, form beds of the size 10 m^2 or 30 m^2 depending upon the water availability. Form irrigation channels to conserve soil moisture under rainfed conditions. Sow the seeds in flat and form furrows between crop rows during intercultivation on third week after sowing.

Direct sowing: Seed rate is 3.75 to 5 kg/ha. Soaking of seeds either in 2 % potassium chloride (KCl) or 3% sodium chloride (NaCl) for 16 hours or 2% KH_2PO_4 or 0.5% brassinolide for 16h with a seed to solution ratio of 1:0.06 followed by 5 hours shade drying improves germination and stand. This management can be used both for dryland agriculture as well as garden land. Spacing recommended is 45 x 15 cm for all varieties. Seeds are sown at 2.5 to 3 cm depth. If pulse is intercropped, adopt a spacing of 45 cm x 15 cm for pearl millet and 30 × 15 cm for cowpea and 30 × 10 cm for other pulses. The spacing for paired row planting is (30 cm + 60 cm) × 15 cm. One paired row of pearl millet is alternated with a single row of pulse crop. In the furrows in which fertilizers have been applied, place 5 kg of seed allowing them to fall 4 to 5 cm apart (use higher seed rate of 5 kg to offset mortality).

Transplanted crop: Pull out the seedlings when they are 15 to 18 days old. Spacing recommended is 45 x 15 cm for all varieties. Plant seedlings on the side of the ridge half way from the bottom. Depth of planting should be 3 to 5 cm. Root dipping with biofertilisers can also be done. Prepare the slurry with 5 packets per ha of azospirillium in 40 litres of water and dip the roots of the seedlings 15 to 30 minutes before planting.

Thinning and gap filling: In direct sown crop, first weeding is done at the time of irrigation. Gap filling and thinning of the crop is done to a spacing of 15 cm between plants, cowpea crop to 20 cm between plants and other pulse crops to 10 cm between plants.

Weed management: For transplanted and direct sown crop, Atrazine is sprayed 50 WP @ 500 g/ha on third day after transplanting of seedlings. Then one hand weeding on 30 to 35 days after transplanting may be given. If herbicide is not used, hand weed on 15th day and again between 30 and 35 days after transplanting.

Manure and fertilizer application: Spread 12.5 t/ha of FYM or compost or composted coir pith uniformly on unploughed soil. Incorporate the manure by working the country plough and apply azospirillium to the soil @ 10 packets per ha (2 kg). Apply fertilizers of 70-30-35 kg N, P_2O_5 and K_2O per ha. Apply 50% of the recommended nitrogen and full dose of phosphorous and potassium basally. For transplanted crop, open a furrow more than 5 cm deep on the side of the ridge (1/3 distance from the bottom) place the fertilizer and cover. For the direct sown crop, mark the lines more than 5 cm deep for the recommended spacing of each variety. Place the fertilizer below 5 cm depth and cover up to 2 cm from the top before sowing. In the case of intercropping with pulses, mark lines more than 5 cm deep, 30 cm apart in the beds. Apply fertilizer only in the rows in which pearl millet seeds is to be sown and cover up to 2 cm. When azospirillium is used for seeds, seedlings and soil, apply only 50 kg N per ha for variety and 60 kg N per ha for hybrid, i.e., reduce 25% N of soil test recommendations. Apply 12.5 kg per ha of micronutrient mixture with enough sand to make 50 kg and apply on the surface just before planting, after sowing and cover the seeds. Broadcast the mixture on the surface to seed line. If micronutrient is not available, apply 25 kg of zinc sulphate per ha. Mix the chemical with enough sand to make 50 kg and apply as above. Top dress the remaining quantity of nitrogen (50%) on 15 days after transplanting for transplanted crop and 30 days after sowing for direct sown crop. In transplanted crop, open a furrow 5 cm deep with a stick or hoe at the bottom of the furrow, place the fertilizer and cover. In the case of direct sown crop, apply the fertilizer in band. If intercropped with pulses, apply fertilizer to pearl millet crop only. Irrigate the crop after application of fertilizer.

Water management: Pearl millet produce optimum grain yield with irrigation based on 75 % depletion of available moisture. Water requirement is 300 to 350 mm. It is irrigated with an IW/CPE ratio of 0.6. Heading and flowering are the critical growth stages of water demand. Water requirement of pearl millet (300 to 350 mm) is much lower than maize, sorghum and finger millet (500-600 mm). It requires, on an average 140-150 mm of water/ tonne of grain produced.

Cropping Systems: The most important Pearl millet based cropping system are *viz.*, pearl millet–barley, pearl millet–wheat, pearl millet–chick pea, pearl millet–field pea, pearl millet–potato, pearl millet–potato–wheat, pearl millet–toria–wheat, pearl millet–wheat–moong (green gram), pearl millet–wheat–jowar (fodder) and pearl millet–wheat–pearl millet (fodder). In rainfed areas, pearl millet is raised mixed with blackgram, greengram and sesame. Intercropping of groundnut, greengram or castor with pearl millet is common. Pearl millet based intercropping practiced in different states are furnished in Table 4.

Table 4 Pearl millet based intercropping practiced in different states of India

State	Intercropping systems
Rajasthan	Pearl millet + cluster bean/ mothbean/ sesame
Haryana	Pearl millet + Green gram/ sesame
Gujarat	Pearl millet + Green gram/ sesame
U.P.	Pearl millet + Green gram/ sesame
M.P.	Pearl millet + Black gram/ soybean
Delhi	Pearl millet + Pigeonpea/ groundnut / castor
Maharashtra	Pearl millet + Moth bean / Pigeonpea
Karnataka	Pearl millet + Pigeonpea
Tamil Nadu	Pearl millet + cowpea / sunflower

Pests and diseases management: The important insect pests in pearl millet are shoot fly, stem borer, cutworms, white grubs, grass hopper, grey weevil and chafer beetle and diseases are downy mildew, smut, ergot and rust. Fishmeal trap @ 10/ha, seed treatment with imidacloprid 600 FS @ 8.75 ml/kg seed or Carbaryl @ 200 mg/kg followed by dusting of fenvelerate 0.4% @ 20 kg/ha or spraying of NSKE 5% at ear head stage are recommended for the management of pest complex of pearl millet such as shoot fly and stem borer. Application of carbofuran 3% or quinalphos 5% granules @ 12 kg/ha at the time of sowing is recommended to control white grub. Seed treatment can also done with clothianidin 50 WDG @ 7.5 g/kg seed or imidacloprid 600 FS @ 8.75 ml/kg seed against white grub. Grass hopper, grey weevil and chafer beetle are controlled with dusting of quinalphos 1.5% or methyl parathion 2% @ 25 kg/ha at the time of pest appearance. Seeds are treated with metalaxyl @6g/kg of seed to prevent the infestation by downy mildew. Seeds are also treated with 5% carbofuran 3G or chlorpyriphos @4 ml /kg to protect the seed from shoofly infection/attack. Seeds are dry dressed with bavistin @2g/kg of seed to protect against seed borne pathogens and soil borne pathogen. Bifenthrin @ 5 mg /kg of seed is also recommended for seeds storage.

Harvest: The crop attains physiological maturity 30-35 days after 50% flowering and the seed moisture at this stage will be around 25 to 30%. The symptoms of grain maturity show the leaves turning yellow and present a dry appearance. The ear heads are harvested when 80 % of the ear heads are physiologically matured, where the grain moisture content will be around 20%. It is harvested by hand, either by picking the ear heads/panicles or by harvesting whole plants. The grains are separated either by beating the ear heads by sticks or trampling by bullocks. The moisture content of seed at the time of threshing will be 15-18%. The optimum grain moisture content for safe storage is 12 to 14%.

Yield: An irrigated crop gives 4.0-5.0 t/ha of grain and 10.0 t/ha of straw while rainfed crop yields about 1.2-1.5 t/ha of grain and 7.0-7.5 tons/ha of straw. The harvest index of landraces is low (0.15–0.20), attaining 0.35 in improved cultivars, and up to 0.45 in dwarf hybrids.

Utilization: Pearl millet is the staple food in parts of tropical Africa and India. The stems are widely applied for fencing, thatching and building, as fuel and as a poor-quality fodder. Split stems are used for basketry.

REFERENCES

Andrews, D.J. and K.A. Kumar. 1992. Pearl millet for food, feed and forage. Advances in Agronomy 48: 89–139.

Begg, J.E. 1965. The growth and development of a crop of bulrush millet (*Pennisetum typohoides* S&H). Indian Journal of Agricultural Sciences 65: 341-349.

Brunken, J., de Wet, J.M.J. and J.R. Harlan. 1977. The morphology and domestication of pearl millet. Economic Botany 31: 163–174.

Burton, G.W. 1969. Breaking dormancy seeds of pearl millet, *Pennisetum typhoides*. Crop Science 9: 659-664.

CPG. 2012. Crop Production Guide. Department of Agriculture, Government of Tamil Nadu, Chennai and Tamil Nadu Agricultural University, Coimbatore.

D'Andrea, A.C., Klee, M. and J. Casey. 2001. Archaeological evidence for pearl millet (*Pennisetum glaucum*) in sub-saharan West Africa. Antiquity 75:341–348

Dendy, D.A.V. (editor), 1995. Sorghum and millets: chemistry and technology. American Association of Cereal Chemists, St. Paul, Minnesota, United States. 406 pp.

GOI. 2014. Status Paper on Coarse Cereals. Directorate of Millets Development, Jaipur and Ministry of Agriculture, Government of India, New Delhi. p 216

Gregory, P.J. and G.R. Squire.1979. Irrigation effect on roots and shoots of pearl millet (*Penisetum typhoides*). Experimental Agriculture. 15: 161-168.

Gupta, S.C. 1999. Seed production procedures in sorghum and pearl millet. Information Bulletin no. 58. ICRISAT, Patancheru 502 324, Andhra Pradesh, India. 16 pp.

Khairwal, I.S., Rai, K.N., Andrews, D.J. and G. Harinarayana. 1999. Pearl millet breeding. Science Publishers, Enfield, New Hampshire, United States. 511 pp.

Khairwal, I.S. 1980. Seed dormancy in pearl millet. Food Farming and Agriculture 12: 128-129.

Maiti, R.K. and F. R. Bidinger. 1981. Growth and development of pearl millet plant. Research Bulletin No. 6. ICRISAT, Hyderabad.

Pearson, C.J. (ed). 1985. Pearl millet, special issue. Field Crops Research 11(2–3): 111–290

Rachie, K.O. and J.V. Majmudar. 1980. Pearl millet. Pennsylvania State University Press, University Park, United States. 305 pp.

Zohary, D. and M. Hopf. 2000. Domestication of plants in the old world: the origin and spread of cultivated plants in West Asia, Europe and the Nile Valley. Oxford University Press, New York

CHAPTER

12

Finger Millet–*Eleusine corocana* (L) Gaertn (2n=36)

Synonyms: *Eleusine indica* (L) Gaertn (2n= 18)

Vernacular name: Finger millet, Birdsfoot millet, African millet (English); *Keppai, Kelvaragu, Ragi* (Tamil); *Mathuri, Ragi* (Malayalam); *Chodi* (Telugu); *Mandua, Mandika, Marawah* (Hindi); *Marwa* (Bengali); *Nagli, Nachni* (Marathi); *Nagli, Bavto* (Gujarati); *Mandia* (Oriya); *Mandhuka, Mandhal* (Punjabi).

Importance: The generic name 'Eleusine' is derived from the Greek goddess of cereals 'Eleusine'. The common name 'finger millet' is derived from the finger-like branching of the panicle. Finger millet is a staple food crop in the majority of drought prone areas in the world especially in Africa and South Asia. Finger millet, one of the oldest crops in India is referred as '*nrttakondaka*' in the ancient Indian Sanskrit literature, which means 'Dancing grain'. It ranks sixth in production after wheat, rice, maize, sorghum and pearl millet in India.

Origin: *Eleusine indica* is originated in India whereas *Eleusine coracana* domesticated in Ethiopia.

Distribution: Finger millet is grown in India, China, Nepal, Uganda, Zambia, Tanzania, Kenya, Ethiopia, Sudan, Nigeria, Somaliland, Japan, Malaysia, Sri Lanka. In India it is grown in Karnataka, Tamil Nadu, Andhra Pradesh, Orissa, Gujarat, Maharashtra, Bihar, Uttar Pradesh and Himachal Pradesh.

Botany: Finger millet is an erect, tufted annual growing to 60-120 cm height with profuse tillers. The stem is compressed, elliptic and leaves are linear with a distinct mid-rib; ligule has a fringe of hairs. Inflorescence borne on a long peduncle, form the end of which four to five spikes radiate in a whorl called finger, with an odd one a little lower down the whorl and called the thumb. There are five types of panicle shapes in finger millet, namely the fisty (in curving so deep that gives a fisty appearance), the top-curved (The top of the each finger is curved towards the inner cavity, 5 to 10 cm length), the incurved (practically closes the central cavity, 4 to 7 cm length), the open (fingers are erect, 8 to 15 cm length), and Cock's comb (just like the comb of a cock). The average number of spikelets in a finger is 67 to 73. Each spikelet has 4-5 flowers and may take 6-8 days to complete flowering. Flowering takes place simultaneously in all fingers. Anthesis is between 1 a.m. and 5 a.m. It is a self-pollinated crop in general but cross pollination also occur through wind and insects. The spikelets possess 3-8 seeds which are tiny in size and generally reddish brown in colour. There are two cultivated types of finger millet viz., i) Indian ecotype (E. coracana). It has long fingers, bold grain, stiff straw, photosensitive and uneven grain maturity phase and ii) African ecotype (E. africana). It has short fingers, small grains and photosensitive.

Growth stages: The growth stages of Finger millet are furnished in Table 1.

Table 1 The growth stages of Finger millet

Crop stage	Duration in days	
Germination stage	1-5	1-5
Vegetative phase	15-20	20-30
Flowering phase	21-55	37-63
Maturity stage	56-100	78-120
Total	100	120

Climate: It is a tropical crop, grown from sea level to 2000 m elevation. It is grown in areas receiving mean annual rainfall of 500 to 1500 mm. The crop possesses good drought tolerance. It's a short day plant and requires an optimum photoperiod of at least 12 hrs. The minimum temperature required for germination is 8-10°C. A mean temperature of 26-29°C is optimum for growth. The night temperature below 18°C and day temperature above 32°C inhibit flowering. The optimum temperature for growth and yield is from 23 to 32°C. The optimum day/night temperature is 27 to 32°C and 22°C respectively. The grain yield decrease drastically below 20°C and above 38°C. It is highly sensitive to frost.

Soil: Finger millet can be grown on a variety of soils including laterite soils, red soils, acid soils, etc., but reasonably fertile, well drained sandy loamy soils with pH of 5.5 to 9.5. It is tolerant to salinity and alkalinity. It cannot tolerate water logging.

Season: Finger millet is grown during December-January and April-May under irrigated condition while June-July and September-October under rainfed condition.

Varieties: The crop flowers in 60-80 days, and matures in about 120-130 days depending on the tract and the variety.

Table 2 State-wise finger millet varieties in India

State	Season	Varieties
Andhra Pradesh	Kharif	Padmawathi, Maruti, Bhairabi, Champavathi, Chilika, Bharathi, Shri Chaitanya, Kalyani, Godawari, AKP-2, SURAJ, Simhadri, Ratnagiri, Gouthami, Sapthagiri.
	Rabi	Bhairabi, Bharathi & Maruti
Bihar	Kharif	Birsa Marua-2, BR-407
Chhattisgarh	Kharif	Bhairabi, Chilika, GPU-67, Saura, VL-149, PR-202, Ratnagiri, GPU-28, HR-374 & VL-147
Gujarat	Kharif	Chilika, GPU-45, Saura, Gujarat Nagali-2, Gujarat Nagali-3.
Jharkhand	Kharif	Birsa Marua-2, GPU-45 & GPU-67.

Karnataka	Kharif	VL Ragi-146, Akshaya, Champavathi, GPU-28, MR-1, Bhairabi, GPU-26, GPU-45, GPU-48, L-5, Divya (MR-6), Rathana, ML-365, KMR-301 & GPU-67, Indaf-8, Indaf-9, HR-911, PR-202, MR-1, MR-6, L-5, GPU-26, GPU-28, GPU-66, GPU-45, VR-708 & OEB-10.
	Rabi / Summer	Bhairabi, GPU-28, GPU-26, KMR-301 & ML-365, Indaf-5, Indaf-7, Indaf-15 & Indaf-9, HR-911, GPU-26 & GPU-48.
Madhya Pradesh	Kharif	Bhairbi, Chilika, GPU-45
Maharashtra	Kharif	Bhairabi, GPU-45 & GPU-67.
	Rabi	Bhairabi.
Odisha	Kharif	Subra, Chilika & Saura.
	Summer	Subra.
Tamil Nadu	Kharif	Champavathi, Chilika, Co(Ra)-14, Paiyur (Ra)-2, GPU-67, GPU-28, CO-7, CO-10, CO-11, CO-12, CO-13, CO-14, Paiyur (Ra)-2, K-567, Indaf-5, Indaf-7, Indaf-9, Paiyur-1, PR-202 and TRY-1.
Uttar Pradesh	Kharif	KM-65 & Champavathi
Uttrakhand	Kharif	KM-65, Champavathi, VL Mandua-315, VL Mandua-324, PRM-1, GPU-67, VL-146, VL-149, VL-315, VL-324, PRM-1 & PRM-2.

Nursery management: For raising seedlings to plant one ha of main field select 12.5 cents (500 m^2) of nursery area near a water source, where water does not stagnate. Plot size of 3.0 m x 1.5 m size is preferred. Provide 30 cm space between plots for irrigation. Excavate the soil from the inter space and all round to a depth of 15 cm to form channels and spread the soil removed from the channel on the bed and level it properly. The ratio of nursery to main field area ranges between 1:30 and 1:10 i.e., between 300 to 1000 m^2 of bed for each hectare of field to be planted. Bed size is 4 m × 1.0 m. Mix 37.5 kg of superphosphate with one tonne of FYM or compost and spread the mixture evenly on the nursery area. Plough two or three times with a mould board plough or five times with a country plough. Seed rate is only 3-4 kg per ha. Mix the seeds in a polythene bag to ensure a uniform coating of seeds with Thiram or captan 4 g/kg, carbendazim 2 g/kg of seeds. Treat the seeds at least 24 hours prior to sowing. Seed treatment with Azospirillum may be done @ 3 packets per/ha i.e., 600 g/ha. Mix Lindane 10% D @ 1.5 kg with 2 kg of sand and apply all around and on the beds to prevent ants carrying away the seeds. Finger millet seeds treated with 1 % urine solution of pregnant cow increases grain yield. Make shallow rills not deeper than one cm on the beds by passing the fingers vertically over them. Broadcast 4kg of treated seeds evenly on the beds. Cover the seeds by levelling out the land lightly over the soil. Spread 500 kg of powdered FYM over the beds evenly to cover the seeds which are exposed and compact the surface lightly. Do not sow the seeds deep as germination will be adversely affected. Allow water to enter through the inlet and cover all the channels around the beds. Allow the water in the channels to raise till the raised beds are wet and then cut-off water. First irrigation is given immediately after sowing and life irrigation on the fourth day. Seeds germinate in 3 days. Adjust the frequency of irrigation according to the soil type once in five days. Do not allow cracks to develop in the nursery bed by properly adjusting quantity of irrigation water. Apply 300 g of carbryl 10% dust on the foliage on the 10th day of sowing over the nursery area. Spray any one of the following insecticides viz., Methyl demeton 25EC @ 20 ml, Dimethoate 30EC @ 20 ml mixed in 10 litres of water using a high volume sprayer if dusting is not done. Spray any one of the following fungicides viz., Carbendazim 5 g or Edifenphos 10 ml or IBP 10 ml in 10 litres of water using a high volume sprayer to cover the nursery area on noticing initial infection of blast. The sprayable insecticides and fungicides can be applied simultaneously. Pull out seedlings on the 15th to 20th day after sowing for planting or 3 to 5 weeks with plant height of 12-15 cm. Transplanting of finger millet has a few advantages over direct seedlings. Shortening of growing season, better weed control, better plant stand and high grain yields

Main field management

Land preparation: Field is prepared with three to four ploughing to get fine tilth. Form beds and channel of convenient size depending on irrigation source.

Transplanting the seedlings: Plant 2 seedlings per hill. Plant the seedlings at a depth of 3 cm. Plant 18 to 20 days old seedlings. Adopt a spacing of 15 x 15 cm for planting. Prepare slurry with 5 pockets (1000 g) per ha of Azospirillum in 40 litres of water and dip the root portion of the seedlings in the solution for 15 to 30 minutes and transplant.

Intercultivation: Finger millet cannot tolerate weed competition in the early stages of its growth. Hence, early and thorough weeding is essential. Thinning is done when finger millet is 5 cm height.

Weed management: Hoe and hand weed on the 15th day after planting in light soils and 17th day after planting in heavy soils and subsequently on 30-32 days respectively. Allow the weed to dry for 2 or 3 days after hand weeding before giving irrigation. Do not adopt hoeing and hand weeding if herbicide is applied. Hand weeding twice or thrice in direct broadcast or row seeded stands of finger millet at 10 day interval from the three weeks of sowing. Apply Butachlor @ 2.5 lit/ha or Fluchloralin 2 lit/ha or Pendimethalin @ 2.5 litre per ha or Isoproturon @ 0.5 kg ai/ha or Oxyflorofen @ 0.1 lit a.i./ha with 900 lit of water per ha. Apply the herbicides when there is sufficient moisture in the soil or irrigate immediately after application of herbicide. If pre-emergence herbicide is not applied hand weed twice on 10th and 20th day after transplanting. For rainfed direct seeded crop, apply post emergence herbice, 2, 4-DEE or 2, 4-D Na salt at 0.5 kg/ha on 10th day after sowing depending on the moisture availability.

Manures and fertilizer application: Sheep penning is practiced with 2000 to 4000 sheeps per acre per night. Apply FYM/compost @ 12.5 t/ha as basal dressing. Fertilizer dose of 60-30-30 kg N, P_2O_5 and K_2O per ha is recommended. Half dose of nitrogen and entire P_2O_5 and K_2O are applied as basal dressing. Top dressing of 20 kg N per ha is done on 30 day after transplanting or 40 day after sowing. $ZnSO_4$ @ 12.5 kg/ha, Borax @ 5 kg/ha, $CuS0_4$ @ 50 ppm on 15 days after planting can be done.

Water management: Water requirement is 350 to 400 mm for finger millet. Irrigation can be given with an IW/CPE ratio of 0.6. The best timing for irrigation is at weekly intervals and at the rate of three acre inches per irrigation.

Cropping systems: Finger millet + pigeon pea in 8-10 : 2 or finger millet + field bean in 8: 1 for Karnataka and Tamil Nadu; finger millet + field bean in 6 : 2 row proportion for Bihar; Finger millet+soybean (90:10 crop mixtures) for Gadhwal region of Uttarakhand, Finger millet + moth bean / black gram (4:1) for Kolhapur are recommended.

Harvest: Finger millet crop does not mature uniformly and hence the harvest is to be taken up to in two stages. When the earhead on the main shoot and 50% of the earheads on the crop turn brown, the crop is ready for the first harvest. Finger millet normally matures in 3 to 5 months after sowing. The ripe heads may be removed directly from the standing plants in the field or the plants can be cut at their base, tied into sheaves and stacked up to dry. After the earheads is thoroughly dried, the grain is threshed by beating by the sticks, treading under the feet of animals or by working a stone roller over them. First harvest is done by cutting all earheads which have turned brown. Dry, thresh and clean the grains by winnowing. Second harvest is done on seven days after the first harvest by cutting all the ear heads including the green ones. Cure the grains to obtain maturity by heaping the harvested earheads in shade for one day without drying so that the humidity and temperature increase and grains get cured. Dry, thresh and clean the grains by winnowing and store the grains in gunnies.

Yield: Grain yield is 650 and 2000kg/ha in rainfed and irrigated condition respectively. The proportion of husk in grain is 5 to 13% by weight. The grains can be successfully stored for up to 50 years. One gram of finger millet grains contains 300 to 450 numbers of seeds.

Utilization: Finger millet flour is mixed with sugar, salt, spices and milk and taken as finger millet malt. The green grain may be roasted and eaten as a kind of vegetable. The grain is generally used in the form of the whole meal for preparation of traditional foods, such as roti (unleavened breads or pancake), mudde (dumpling) and ambali (thin porridge). The presence of higher levels of dietary fiber and other protective healthy nutrients in the whole meal-based finger millet preparations helps in the nutritional management of diabetes.

REFERENCES

de Wet, J.M.J., Prasada Rao, K.E., Brink, D.E. and M.H. Mengesha. 1984. Systematics and evolution of Eleusine coracana (Gramineae). American Journal of Botany 71(4): 550–557.

Hilu, K.W. and J.M.J. de Wet. 1976. Domestication of Eleusine coracana. Economic Botany 30: 199–208.

Hilu, K.W. and J.M.J. de Wet. 1976. Racial evolution in Eleusine coracana ssp. coracana (finger millet). American Journal of Botany 63(10): 1311–1318.

Hulse, J. H., Laing, E. M., and O. E. Pearson. 1980. Sorghum and the millets: Their composition and nutritive value. London: Academic Press.

Riley, K.W., Gupta, S.C., Seetharam, A. and J.N. Mushonga. 1993. Advances in small millets. Oxford & IBH Publishing, New Delhi, India. 557 pp.

Rachie, O.K. and V. L. Peters. 1977. The Eleusines: A review of the world literature. pp 179. Patancheru, A.P. 502 324 India. ICRISAT.

ICMR. 2010. Nutrient requirements and recommended dietary allowances for Indians. A report of the Expert Group of the Indian Council of Medical Research.

Patel, J. C., Dhirawani, M. K., and R.D. Dharne. 1968. Ragi in the management of diabetes mellitus. Indian Journal of Medical Sciences, 22, 28–29.

Pradhan, A., Nag, S. K., and S. K. Patil. 2010. Dietary management of finger millet controls diabetes. Current Science, 98(6), 763–765.

Umapathy, P. K., and A. Kulsum. 1976. Ragi: A poor man's millet. Journal of the Mysore University Section A, XXXVII, 45–48.

Bhag Mal, S. Padulosi and S. Bala Ravi, (editors). 2010. Minor Millets in South Asia: Learnings from IFAD-NUS Project in India and Nepal. Bioversity International, Maccarese, Rome, Italy and the M.S. Swaminathan Research Foundation, Chennai, India. 185 p.

CHAPTER

13

Foxtail Millet–*Setaria italica* (L) Beauv. (2n=18)

Vernacular name: Italian millet, Siberian millet, German millet (English); *Kakum, Kangni* (Hindi); *Tenai* (Tamil); *Navane* (Kanada); *Korra* (Telugu); *Thina* (Malayalam); *Kaon* (Bengali); *Kang* (Gujarati); *Kang, Rala* (Marathi); *Kanghu, Kangam, Kora* (Oriya); *Kangni* (Punjabi).

Importance: The name foxtail millet refers the ear head resemblance with the tail of fox. Foxtail millet is fairly tolerant of drought; it can escape some droughts because of early maturity. Due to its quick growth, it can be grown as a short-term catch crop.

Origin: Foxtail millet is native of China.

Distribution: It is cultivated in China, Japan, India, Nepal, Russia and USA. In India, it is grown in Andhra Pradesh, Maharashtra, Gujarat, Rajasthan, Karnataka and Tamil Nadu.

Botany: Foxtail millet is an annual, growing erect to about 0.75 to 1 metre, tillering; stems are slender; internodes are hollow and shorter at base becoming long above, leaves narrow, linear-lanceolate in shape, usually leafy up to the panicle; leaves are linear, leaf sheath often glabrous, ligule in a narrow fringe of hairs; Inflorescence is a spike, terminal, 8 to 32 cm long, dropping, usually compact, sometimes loose, and cylindrical. Flowering occurs top-down on each of the stemmed branches. There are two maxima of flowering during a day, one between 10 p.m. and 12 mid-night and the other between 6 a.m. and 8 a.m. It is a self-pollinated crop and natural crossing is up to 0.05%. It bends quite a bit at maturity due to heavy weight of earhead. The common colour of grain is buff.

Climate: It is grown both in the tropics and sub tropics up to an altitude of 1500 m from MSL. It is grown in areas receiving mean annual rainfall of 500 to 750 mm.

Soils: It is grown on a variety of soils, sandy loamy to clay loamy, red soils and black soils. It has a high level of tolerance to salinity.

Season: It is cultivated during June- July and September-October seasons.

Varieties: The crop produces seed in 75 to 90 days. The foxtail millet varieties in different states of India are presented in Table 1.

Table 1 Foxtail millet varieties in different states of India

State	Varieties
Andhra Pradesh	Pant Setaria-4, TNAU-186 and Krishnadevaraya, Srilaxmi.
Chhattisgarh	Pant setaria-4.
Karnataka	Pant setaria-4, HMT-100-1 and SIA-326, PS-4 and TNAU-186
Madhya Pradesh	Pant setaria-4.
Maharashtra	Pant setaria-4.
Orissa	Pant setaria-4.
Rajasthan	SR-16, SR-1 and SR-51.
Tamil Nadu	TNAU-43, CO (Te)-7, PS-4, K-2, K-3, CO-4, CO-5, CO-6, TNAU-196, TNAU-186. CO (Te)-7.
Uttar Pradesh	Pant setaria-4, PRK-1
West Bengal	Pant setaria-4

Land preparation: Plough field thoroughly using a small iron plough or country plough to get fine tilth.

Seeds and sowing: Seed rate is 10 to 12.5 kg per hectare. Spacing is 25 cm x 10 cm. Seeds are sown in lines, broadcasting, or sown in line with seed drills. Seed treatment is done with 2 g of thiram or carbendazim per kg of seeds.

After cultivation: Thinning is done on 20 day after sowing.

Weed management: Hand weeding on 20 days and 35 days after sowing. Pre-emergence application of Isoproturon 0.5 kg a.i/ha which is followed by one hand weeding on 25-30 DAS or post–emergence application of 2, 4-D Sodium salt @ 0.75 kg a.i./ha on 20th DAS followed by one hand weeding.

Manures and fertilizer application: Apply FYM or compost @ 12.5 t/ha and 44-22-0 kg of N, P_2O_5 and K_2O per ha as basal dressing.

Harvest: The matured crop is with golden yellow earheads. Entire plants are cut or the earheads alone, dried and threshed under the feet of cattle.

Yield: Grain yield of 1.5 to 2.0 t/ha can be obtained.

Utilization: It is grown for hay due to its fast growth, short and fine stems. Hay harvesting can begin at boot stage, when ear heads are beginning to emerge, or from late boot to early bloom stage for optimum quality. It may act as a laxative to horses unless the hay is mixed with other roughage, and can damage a horse's liver, kidneys, and bones due to the glucoside setarian. It is used as a weed-suppressing smother crop.

REFERENCES

Krishnamurthy, L., H. D. Upadhyaya, C. L. L. Gowda, J. Kashiwagi, R., Purushothaman, S. Singh, and V. Vadez. 2014. Large variation for salinity tolerance in the core collection of foxtail millet (*Setaria italica* (L.) P. Beauv.) germplasm. Crop and Pasture Science 65(4): 353–361.

Sheahan, C.M. 2014. Plant guide for foxtail millet (*Setaria italica*). USDA-Natural Resources Conservation Service, Cape May Plant Materials Center, Cape May, NJ.

CHAPTER

14

Little Millet–*Panicum miliare Lank* (2n=36)

Synonyms: *Panicum sumatrense* Roth

Vernacular name: Little millet (English); *Kutki, Savan*, *Sharan* (Hindi); *Samai* (Tamil); *Samalu* (Telegu); *Suan* (Oriya); *Sama* (Bengali); *Gajrao; Kuri* (Gujarati); *Same, Save* (Kannada); *Sava* (Marathi); *Swank* (Punjabi); *Samalu* (Telugu); *Chama (Malayalam).*

Importance: The specific name '*miliare*' is derived from the old Latin '*milium*' which means 'millet'. The term 'little millet' is because of the size of the grains. Little millet is grown to a limited extent in India, It occurs wild in northern India and southeastern Asia. It produces some grain and useful fodder under very poor conditions. It is described as a 'quick growing, short duration millet which withstands both drought and water logging'.

Origin: Little millet is originated in India, Sri Lanka and Southeast Asia.

Distribution: Little millet is grown in India, Myanmar, Malaysia, China, Sri Lanka and Pakistan. In India, it is grown in Tamil Nadu, Karnataka, Andhra Pradesh, Maharashtra, Gujarat, Madhya Pradesh, Odisha, Bihar and Uttar Pradesh.

Botany: It is an annual, growing erect and slender stem 45-100 cm height with solid or hollow internodes and distinctly swollen nodes. It tillers profusely. Leaves are linear and slender with leaf sheath hairy at junctions and short-ligule. The inflorescence is an open panicle, contracted or thyrsiform bearing numerous spikelets. Each spikelet has 2 glumes and 2 lemmas of which the second lemma bears hermaphrodite flowers containing 2 lodicules and 2 stamens. The ovary has bifid style and plumose stigmas. The anthesis of the spikelets is from 9 am to 12 am noon and close within about 6 minutes. Self-pollination is the rule. Natural cross pollination is up to 0.05%. The fruit, a caryopsis, is globular and enclosed firmly by a lemma and palea. The grains may be creamy white, yellow, red or black. The major features which distinguish between little millet (*Panicum sumatrenese*) and common millet (*Panicum milliaceum*) is furnished in Table 1

Table 1 Distinguished features between little millet and common millet

Characters	Little millet (Samai)	Commom millet (Panivaragu)
Plant	Smooth (Glabrous)	Hairy (Pubescent)
Grain	Small, very shiny glumes	Longer, less hairy glumes
Panicle	Lax	Very lax
Spiklet	Dense	Sparse

Climate: It is grown both in the tropics and sub tropics and even at an altitude of 2100 m from MSL. It is grown in areas receiving annual rainfall of 250 to 500 mm. It is resistant to drought.

Soils: A wide variety of soils mostly grown on marginal, sub-marginal fertilie soils and hilly slopes.

Season: It is mostly raised during June-July and September as rainfed crop.

Varieties: Little millet varieties grown in different states of India are presented in Table 2.

Table 2 Little millet varieties grown in different states of India

State	Varieties
Andhra Pradesh	Tarini (OLM 203)
Bihar	Tarini & Kolab (OLM-36).
Chhattisgarh	Kolab, Sabara, OLM-20, JK-1, VG-1 and TNAU-63 OLM-208 & OLM-217
Gujarat	OLM-208, OLM-217, Kolab and TNAU-63.
Karnataka	Tarini, Kolab and TNAU-63.
Madhya Pradesh	Kolab, Sabara and Jawahar Kutki-36, JK-8, JK-36, CO.2, PRC-3.
Odisha	OLM-217, OLM-208, OLM-20, Tarini, Kolab and Sabara.
Tamil Nadu	TNAU-63, CO-3, Paiyur-2 and K-1, CO-2, CO-3, CO-4, OLM-203. Paiyur-1 and Paiyur-2.

Land preparation: Plough the field thoroughly 2 or 3 times using a small iron plough or country plough to fine tilth.

Seeds and sowing: Seed rate is 10 to 12.5 kg/ha. Seed treatment is done with Thiram or carbendazim @ 2 g/kg of seeds. Seed are sown by broadcasting, line sowing or with seed drills. Spacing is 25 x 10 cm.

After cultivation: Thinning is done on 15th day after sowing.

Weed management: Hand weeding is to be carried out on 20 and 40 days after sowing. Pre-emergence application of Isoproturon 0.5 kg a.i / ha which is followed by hand weeding on 25 to 30 DAS in assured rainfall areas. The post–emergence application of 2, 4-D Sodium salt 0.75 kg a.i./ha on 20th DAS followed by one hand weeding is also recommended.

Manures and fertilizers application: Basal application of 12.5 t/ha of FYM or compost, 44-22-0 kg N, P_2O_5 and K_2O per ha is recommended.

Harvest: Entire crop is cut, dried and threshed under the feet of cattle.

Yield: Grain yield is 700 to 1300 kg/ha under rainfed condition.

Utilization: Little millet is used as grain and fodder.

REFERENCES

CPG. 2012. Crop Production Guide. Department of Agriculture, Government of Tamil Nadu, Chennai and Tamil Nadu Agricultural University, Coimbatore.

CHAPTER

15

Kodo Millet–*Paspalum scrobiculatum* L (2n=40)

Vernacular name: Ditch millet (English); *Varagu* (Tamil); *Kodon, Kodra* (Hindi); *Kodra* (Marathi), *Arikelu, Arika* (Telegu); *Kodo* (Bengali); *Kodra* (Gujarati); *Harka* (Kannada); *Kodra* (Marathi); *Kodua* (Oriya).

Importance: Kodo millet is said to be poisonous after rain. This could be due to a fungal infection. Winnowed clean healthy grain seems to pose no health problem.

Origin:vKodo millet is originated in India.

Distribution:vKodo millet is cultivated in India, China and Uganda. Kodo millet is said to occupy the largest area of any small millet in India. It is grown for grain in Uttar Pradesh, Chhattisgarh, Gujarat, Madhya Pradesh, Maharashtra, Tamil Nadu, Karnataka and Andhra Pradesh.

Botany: Kodo millet is an annual, erect and growing 45-90 cm in height. The tillers are produce profusely up to eighteen. Leaves are thick and stiff and linear to linear-lanceolate. Ligules are membranous and hairy. Both leaves and stems are purple in colour. The inflorescence a panicle with two to eight spikes. Spikelets have a single flower. Spikelets begin to open after midnight and very early in the morning between 2.30 a.m. and 3.00 a.m. and continue till sunrise. The plant is a self-pollinated crop and there is absence of cross-pollination. The grain matures thirty to thirty-five days after flowering.

Climate: It is adapted to very hot and dry climate. It grows well in areas receiving mean annual rainfall of 400 to 500 mm. It is a highly drought resistant crop.

Soils: It thrives well in sandy loamy to loamy soil. It comes even in gravelly soils. It is resistant to alkalinity

Season: It is grown during June-July and September-October months.

Varieties: Kodo millet varieties grown in different states of India are presented in Table 1.

Table 1 Kodo millet varieties grown in different states of India

State	Varieties
Andhra Pradesh	JK-48
Chhattisgarh	JK-13 and JK-48, JK-41, JK-76 and GPUK-3
Gujarat	RK 65-18, GK-2, JK-48 and JK-65, GK-1 and GK-2.
Karnataka	JK-13, JK-48 and RBK-155, GPUK-3, RBK-155 and DPS-48
Madhya Pradesh.	JK-13, JK-48, JK-65, JK-106, JK-21, JK-48, RBK-155, JK-439, RK 65-18 and RBK-155, K-106 & JK-76.
Tamilnadu	Vamban and JK-13, K-1, CO-2, CO-3, APK-1
Uttar Pradesh	KK-2, JK-13, JK-65 & RK 65-18

Land preparation: Plough the field thoroughly using a small iron plough or country plough to fine tilth.

Seeds and sowing: Seed rate is 10 to 12.5 kg/ha. It is sown as broadcasting, line sowing or seed drills. Seed treatment is to be carried out with thiram or carbendazim @ 2 g/kg of seeds. Spacing is 25 × 10 cm.

After cultivation: Thinning is done on 20th day after sowing.

Weed management: Hand weeding is to be carried out on 20 and 40 days after sowing.

Manures and fertilizers application: Apply FYM or compost @ 12.5 t/ha and 44-22-0 kg of N, P_2O_5 and K_2O per ha as basal dressing.

Water management: It is mostly raised as rainfed crop.

Harvest: Entire crop is cut and threshed under the feet of cattle.

Yield: The grain yield of 1500 to 1800 kg/ha can be obtained. The husk accounts for 40% of grain weight i.e., It gives 60% milling percentage.

REFERENCES

CPG. 2012. Crop Production Guide. Department of Agriculture, Government of Tamil Nadu, Chennai and Tamil Nadu Agricultural University, Coimbatore.

CHAPTER

16

Proso Millet–*Panicum miliaceum* L. (2n=36)

Vernacular name: Hog millet, Common millet, broomcorn millet, Russian millet, French millet (English), *Panivaragu* (Tamil); *Chena, Barri* (Hindi); *Variga* (Telugu); *Cheena* (Bengali); *Cheno* (Gujarati); *Baragu* (Kannada); *Vari* (Marathi); *Bachari* (Oriya); *Cheena* (Punjabi).

Importance: The term 'millet' has initially been used for this 'common millet', which had its ancestral origin from the Mediterranean basin. The botanical name of the species *miliaceum* is derived from the Latin word *milium* which means millet. This millet has a very short duration and hence is suitable as a catch crop. It has a very low water requirement and is able to evade drought by its quick maturity.

Origin: Proso millet is originated in Central and eastern Asia. Proso millet has domesticated in Manchuria.

Distribution: Proso millet is grown in Europe, Italy, China, Russia, Kazhakhistan, Mongolia, Manchuria, Africa, America, Japan, India, Bangladesh, Sri Lanka, Nepal, Kenya, Zimbabwe and Ethiopia. In India, it is grown in Andhra Pradesh, Maharashtra, Gujarat, Rajasthan, Tamil Nadu, Karnataka, Uttar Pradesh and Bihar.

Botany: It is an erect annual growing up to 75-100 cm in height; medium to dark purple pigmentation may be noted in plant parts. The stems are slender and leafy up to panicle. The inflorescence is 14-40 cm long oblong panicle having erect hairy branches. The spikelets are solitary or sometimes in pairs, generally 3-4.5 mm long and flattened. Flowers open between 10 a.m. and 12 noon as the day's temperature rises. It is a self-pollinated crop, but natural cross-pollination may exceed 10%. Grains are olive brown in colour. The fruit is a caryopsis. The 1000-seed weight is 4.7 to 7.2 g.

Growth stages: It matures in 90 days. Emergence of the seedling is usually in 4–8 days after sowing. During the vegetative phase, which is usually completed 16–20 days after sowing, tillering occurs and the inflorescence primordia are initiated. Flowering takes place on 40 to 45 days after sowing. Flowering proceeds from top to bottom. The flowers are normally self-fertilized, but cross-fertilization frequently exceeds 10%. The period from flowering to grain maturity has a duration of about 20–30 days. At grain maturity the lower part of the inflorescence as well as the stem and leaves are still green. Proso millet follows the C_4-cycle photosynthetic pathway.

Climate: This is essentially a crop of the temperate regions, but it also grown in the tropics, sub tropics and on high ground in tropical winters. It is a cold weather crop. It is grown up to an altitude of 2700 m from MSL and areas receving annaual rainfall of 250 to 400 mm. It is a drought resistant crop. It can withstand water stagnation. Proso millet seeds germinate well at temperatures of 10–45°C, with the highest rate at temperatures between 35°C and 40°C.

Soils: Proso millet is grown on a wide variety of soils, marginal and sub marginal soils.

Season: It is grown during September-October under rainfed condition. Proso millet generally matures between 60-90 days after planting.

Varieties: Proso millet varieties grown in different states of India are presented in Table 1.

Table 1 Proso millet varieties grown in different states of India

State	Varieties
Andhra Pradesh	TNAU-151 and TNAU-164
Bihar	TNAU-151 and TNAU-164
Karnataka	GPUP-8, GPUP-21, TNAU-145, GPUP-8 and GPUP-21. TNAU-151 and TNAU-164.
Maharashtra	TNAU-164
Rajasthan	Pratap chena-1.
Tamil Nadu	GPUP-21, CO (Pv)-5, K-1, K-2, CO-2, CO-3, CO-4 TNAU-145, TNAU-151, TNAU-164 and CO (Pv)-5.
Uttrakhand	PRC-1, TNAU-164 and TNAU-151

Land preparation: Land is ploughed with iron plough or country plough for 2-3 times.

Seeds and sowing: Seed rate is 10-12 kg/ha. Seed treatment is done with Thiram or carbendazim @ 2 g/kg of seeds. Spacing is 25 × 10 cm. Seeds are sown by broadcasting method. Thinning is done 15 days after sowing.

Weed management: Hand weeding is to be carried out on 20 and 40 days after sowing. Pre-emergence application of Isoproturon 0.5 kg a.i / ha on 3 DAS which is followed by hand weeding on 25 to 30 DAS in assured rainfall areas. The post–emergence application of 2, 4-D Sodium salt 0.75 kg a.i./ha on 20th DAS followed by one hand weeding is also recommended.

Manures and fertilizers application: Apply FYM or Compost @ 12.5 t/ha and 44-22-0 kg of N, P_2O_5 and K_2O per ha as basal dressing.

Water management: It is mostly raised as rainfed crop. If water is available irrigation can be given.

Harvest: The seeds with top of panicle ripe and shatter before lower seeds get mature. Hence, the harvest is carried out before its full maturity and dried to prevent shedding of ripped grains. Threshing is done by treading with cattle. Husk accounts 30% of grain weight i.e., the milling percentage is 70%.

Yield: Grain yield is 700-1500 kg/ha. It should be stored at 13% moisture content or less.

Utilization: It is a grain crop in India and fodder crop in USA. It is also grown as catch crop.

REFERENCES

Cardenas, A., Nelson, L. and Neild, R., 1984. Phenological stages of proso millet. University of Nebraska, Lincoln, United States. 8 pp.

CPG. 2012. Crop Production Guide. Department of Agriculture, Government of Tamil Nadu, Chennai and Tamil Nadu Agricultural University, Coimbatore.

CHAPTER

17

Barnyard Millet

Barnyard grass - *Echniochloa crusgalli* L.Beauv. (2n = 42)
Jungle rice - *Echinochloa colona* (L.) Link. (2n = 36)
Vernacular names: Barnyard millet (English); *Sanwa, Jhangora* (Hindi); *Kuthiraivalli* (Tamil); *Banti* (Marathi); *Oodalu* (Kannada); *Udalu, Oodalu, Kodisama* (Telugu); *Shyama* (Bengali); *Khira* (Oriya); *Swank* (Punjabi). The *Echinochloa colona* (L.) Link is commonly known as Jungle rice, jungle grass, jungle rice grass, water grass, Shama millet, Deccan grass, millet-rice, awnless barnyard grass, small barnyard grass and corn panic grass. The *Echinochloa crusgalli* (L.) Beauv., is commonly known barnyard grass, Japanese millet, cockspur grass, cockspur panicum, barnyard millet, summer grass, water grass, billion dollar grass and chicken panic grass.

Importance: It is the fastest growing of all the millets. In India, Japan and China, it is often used as a substitute for rice when the paddy crop fails. It is a fair source of protein, which is highly digestible and is an excellent source of high dietary fibre with good amounts of soluble and insoluble fractions. It is an appropriate food for patients intolerant to gluten causing celiac disease.

Origin: *Echinochloa crusgalli* is native to Europe and India while *Echinochloa colona* is native to Africa and tropical Asia.

Distribution: *Echniochloa crusgalli* is distributed in China, Japan whereas in *E. colona* is distributed in India, Central Africa, Tanzania, Malawi, Tanzania and Malawi. In India, it is cultivated in Uttar Pradesh, Uttarakhand, Jharkhand, Madhya Pradesh, Maharashtra, Gujarat, Madhya Pradesh, Andhra Pradesh, Karnataka and Tamil Nadu

Botany: Barnyard millet is an annual, growing up to 0.50 to 0.75 m in height. Its leaves are flat, glabrous or slightly hairy without ligule. Inflorescence is a panicle. The inflorescence has unawned spikelets in 3-5 rows. The opening of the individual flowers is between 5 a.m. and 10 a.m. The maximum number of flowers opens between 6 a.m. and 7 a.m. Self-pollination is the general rule. The grain is caryopsis and white or yellow in colour. The plant produces 5000-7000 seeds/plant.

Table 1 Difference between **E. colona** and **E.colona** var. frumentacea

Character	E. colona (2n=36,48,54,72)	E.colona var. frumentacea (2n=36,54)
Rachis	Flat and triquetrous	Tetraquetrous
Arrangement of spikes	Bilateral, alternating	Whorled to spiral
Disposition of the spikes	Almost verical and adpressed to rachis	Horizontal to ascending divergent
Spikelets	Solitary or in twos, pedicelled	Always in groups of three to five sub-sessile

Climate: It is a cosmopolitan crop of both temperate and tropical regions. It is cultivated in areas spread in 45°N to 40°S. It is grown in areas receiving annual rainfall of 400 to 1200 mm. It is adapted to full sunlight or partial shady conditions. It can grow at elevations from sea level to 2000 m. In arid regions, it occurs in ponds, freshwater swamps, moist, loamy soils, in marshes, seepage sites, and in the mud and water of lakes, ditches and floodplains. It is adapted to wet sites and waterlogged conditions. It will tolerate poor drainage and flooding, although does not tolerate dry periods. It can grow in cooler regions, but is better adapted to regions having an annual average temperature range of 14-28°C. It favours warm regions, requiring a frost free period of 160-200 days/year.

Soils: A wide range of soils with marginal and sub marginal fertility. It is most common in loams, silts, and clays. It tolerates a pH range of 4.8-8.2.

Land preparation: Land is ploughed with iron plough or country plough for 2-3 times to get a fine tilth.

Season: It is one of the fastest growing crops of all the millets, attainting maturity in 90 to 100 days. It is a cold season crop. It is grown during September-October.

Varieties: Barnyard millet varieties grown in different states of India are presented in Table 2.

Table 2 Barnyard millet varieties grown in different states of India

State	Varieties
Bihar	VL-Madira-181 and VL-207
Gujarat	VL-Madira-172
Jharkhand	VL-Madira-181 and VL-207
Karnataka	VL-Madira-172, VL-Madira-181, VLM-181, VLM-172 and VL-207 and RAU-11 VLM- 29.
Madhya Pradesh	VL-Madira-181 and VL-207
Rajasthan	VL-207
Tamilnadu	VL-Madira-181, CO-1, K-1 and K-2.
Uttar Pradesh	VL-Madira-181 and VL-207
Uttrakhand	VL-Madira-172, VL-207, VLM-172, VLM-29, VL-207, PRJ-1 PRJ-1.

Seeds and sowing: Seed rate is 10 kg/ha. Spacing is 25 × 10 cm. It can be vegetative propagate since it possess a prostrate habit by rooting at its nodes and producing new shoots.

After cultivation: Thinning is done on 15th day after sowing.

Weed management: Hand weeding is to be carried out on 20 and 40 days after sowing. Pre-emergence application of Isoproturon 0.5 kg a.i/ha which is followed by hand weeding on 25 to 30 DAS in assured rainfall areas. The post–emergence application of 2, 4-D Sodium salt 0.75 kg a.i./ha on 20th DAS followed by one hand weeding is also recommended.

Manures and fertilizers application: Application of FYM or Compost @ 12.5 t/ha and 44-22-0 kg of N, $P_2O_5$and K_2O per ha as basal dressing is recommended.

Water management: It is cultivated under rainfed conditions.

Harvest: The earheads are harvested, dried and threshed under the feet of cattle. The grain and straw are of poor quality. Husk accounts for 55% of grain weight.

Yield: The grain yield is 1000 to 1500 kg/ha.

Utilization: It has edible stalks which are most favoured fodder for cattle. In the U.S.A. it is grown primarily for forage, and can produce up to eight harvests a year.

REFERENCES

Barrett, S.C., and B.F. Wilson. 1983. Colonizing ability in the *Echinochloa crus-galli* complex (barnyard grass). II. Seed biology. Canadian Journal of Botany 61:556-562.

British Columbia Government, Ministry of Agriculture and Food. 1998. Barnyardgrass (*Echinochloa crus-galli*). Website:
http://www.agf.gov.bc.ca/croplive/weedguid/barnyard.htm

FAO. 2002a. *Echinochloa colona* (L.) Link. Grassland Index. Website: http://www.fao.org/WAICENT/FAOINFO/AGRICULT/AGP/AGPC/doc/GBASE/Data/Pf000226.HTM

FAO. 2002b. *Echinochloa crus-galli* (L.) Beauv. Grassland Index. Website: http://www.fao.org/WAICENT/FAOINFO/AGRICULT/AGP/AGPC/doc/GBASE/Data/Pf000227.HTM

University of California. 1998. The Grower's Weed Identification Handbook. Cooperative Extension University of California, Division of Agriculture and Natural Resources, Publication 4030. 311 pp

Vengris, J., M. Drake, W.G. Colby, and J. Bart. 1953. Chemical composition of weeds and accompanying crop plants. Agronomy Journal 45:213-218.

Vengris, J. 1966. Competition between barnyardgrass and alfalfa. Agronomy Journal 58:478-479.

Vengris, J., E.A. Kacperska-Palacz, and R.B. Livingston. 1966. Growth and development of barnyardgrass in Massachusetts. Weeds 14:299-301.

Vengris, J., F.R. Hill, and D.L. Field. 1966. Clipping and regrowth of barnyardgrass. Crop Science 6:342-344.

http://www.hear.org/gcw/

http://www.huntana.com/feis/plants/graminoid/echcru/all_frames.html

http://www.csdl.tamu.edu/FLORA/image/k4509000.htm

http://www.oardc.ohio-state.edu/seedid/ then select species

http://kaweahoaks.com/html/barnyardgrass.html

http://www.csdl.tamu.edu/FLORA/image/k4509100.htm

http://www.ppws.vt.edu/~sforza/weeds/echcg.html

CHAPTER

18

Pulse Crops

The word '*Legume*' comes from Latin word '*Legere*' meaning 'to gather' and indicate that the seeds are collected by hand instead of being threshed from the plant as in cereal grains. The term 'legume' refers to the plants whose fruit is enclosed in a pod which contains the seeds. The legume pod is a one-celled seed container formed by two sealed parts called valves. Legume pods always split along the seam which connects the two valves. This characteristic is called dehiscent, from the Latin word meaning to gape or burst open. Some pods are winged or indehiscent (meaning the pods do not split open at maturity). The legume fruit (pod) develops from a simple carpel and usually dehisces (opens along a seam) on two sides. Examples of crops that bear legume fruits include alfalfa, clover, peas, beans, lupins and peanuts. A peanut is not a nut in the botanical sense but it is an indehiscent legume. The green pods of legume crops that are used for culinary purposes are known as 'legume vegetables' namely; cowpea, cluster bean, garden pea, french bean, pigeon pea, bengal gram, faba bean and lablab bean. The whole plant of legume crops used as green fodder namely; cowpea (*Lobia*), cluster bean (*Guar*), clovers (*Senji*), alfalfa (lucerne) are called 'Fodder legumes'. Legume crops used for oil extraction namely; soybean and groundnut are called 'oil seeds legumes' and cluster bean (*Guar*) for extraction of gum. The grain legumes are leguminous species with edible seeds which includes the pulse grains and high protein leguminous oilseeds such as soybean and peanut.

The term Pulse is derived from the Latin 'puls' meaning thick soup or potage. The word '*Pulse*' is defined as the split cotyledons of dry legume seed. The word 'pulse' has been used from biblical times to describe legumes that bear edible dry seeds that are directly consumed by man. Pulses are dry seeds of leguminous plants which are distinguished from leguminous oil seeds by their low fat content. Pulses are used only after de-husking and splitting. Strictly, the definition applies to pigeonpea, lathyrus, chickpea, greengram, blackgram and lentil. Pulses are known for their high protein and essential amino acids but with low fat content. Pulses can be consumed as whole or split, ground in to flours or separated into fractions such as protein, fibre and starch. Pulse is boiled in water, softened, macerated and used as soup. Presently all legumes are used as dried, fried or boiled, powdered or macerated into soup. Grain legumes belong to the family Fabaceae (alternatively Leguminosae). Leguminous plants that produce edible parts are generally referred to as 'food legumes'. Most of the pulse crops assigned to *kharif* season (rainy season) thrive in uplands as rainfed crops-source of moisture from rain water. The *rabi* season (post-rainy season) pulses largely grow on conserved moisture with supplemental irrigation competing with highly remunerative oil seeds as well as cereals crops. Summer pulses are grown with supplemental irrigation. All *rabi* pulses have a chromosome number of 2n=14 except chickpea which has 16 whereas *kharif* pulses possess 2n=22 except horsegram and field bean which have 24.

FAO recognizes 11 primary pulses globally which are as follows:

1. Dry beans (*Phaseolus spp*. now in *Vigna spp*)
   - Kidney bean, haricot bean, pinto bean, navy bean (*Phaseolus vulgaris*)
   - Lima bean, butter bean (*Phaseolus lunatus*)
   - Azuki bean, adzuki bean (*Vigna angularis*)
   - Mung bean, golden gram, green gram (*Vigna radiata*)
   - Black gram, *urd* (*Vigna mungo*)
   - Scarlet runner bean (*Phaseolus coccineus*)
   - Ricebean (*Vigna umbellata*)
   - Moth bean (*Vigna acontifolia*)
   - Tepary bean (*Phaseolus acutifolius*)
2. Dry broad beans (*Vicia faba*)
   - Horse bean (*Vicia faba var. equina*)
   - Broad bean (*Vicia faba var. major*)
   - Tic bean (*Vicia faba var. minor*)
3. Dry peas (*Pisum spp.*)
   - Garden pea (*Pisum sativum var. hortense*)
   - Field pea (*Pisum sativum var. arvense*)
4. Bengal gram, garbanzo (*Cicer arietinum*)
5. Dry cowpea, black-eyed pea, blackeye bean (*Vigna unguiculata*)
6. Pigeon pea, *Arhar/Tur*, cajan pea, congo bean, gandules (*Cajanus cajan*)
7. Lentil (*Lens culinaris*)
8. Bambara groundnut, earth pea (*Vigna subterranea* (L.) Verdc)
9. Vetch, common vetch (*Vicia sativa*)
10. Lupins (*Lupinus spp.*)
11. Minor pulses, including:
    - Lablab, hyacinth bean (*Lablab purpureus*)
    - Jack bean (*Canavalia ensiformis*), sword bean (*Canavalia gladiata*)
    - Winged bean (*Psophocarpus teragonolobus*)
    - Velvet bean, cowitch (*Mucuna pruriens var. utilis*)
    - Yam bean (*Pachyrrizus erosus*)

Classification of pulses based on area of production in India:

(i) Major pulses: Bengal gram (*Chana*); Pigeon pea (*Arhar*); Green gram (Moong bean); Black gram (*Urd* bean); and Lentil (*Masur*).

(ii) Minor pulses: Kidney bean (Moth bean); Field pea (*Matar*); Horse gram (*Kulthi*); Cow pea (*Lobia*); French bean (*Rajmash*); Grass pea (*Lathyrus/Khesari*); Lablab bean (*Sem*); and Faba bean (*Baqla*).

Classification of pulses based on growth features

(i) Dwarf or bush varieties (which do not require climbing support and mature early)
(ii) Climbing varieties (which take long duration to mature and require support)

Importance: Pulses form an important source of human food next to their cereals. In India, people are mostly vegetarian, depending largely on cereals and pulses as their staple food. Pulses provide the main source of dietary proteins and calories. India is the largest pulses producing country and consumer in the World. The significant producer and exporter of pulses to India are Canada, Myanmar, Australia, Tanzania, France and USA. Canada alone accounts for 35% of global pulse trade. India is the major pulse growing country of the world accounting roughly for one third of the total world area under pulses and one fourth of the world production pulse. In the tropics, cereals on an average account for about 68% of total plant protein consumption; legume seed accounts 18.5%, tubers, nuts, fruit and vegetables accounts 13.5%.

Grain legumes are synergistic with cereals, roots and tubers in the farming systems. They intensify cropping systems by utilizing under-exploited system niches as rotation, double- and inter-crops. Grain legumes also diversify farming systems, making them more nutrient-efficient, resilient and sustainable. Legumes also break pest, disease and weed cycles of other crops, and extend soil-protective land cover.

Legumes have ability to fix atmospheric nitrogen through a symbiotic relationship with certain bacteria known as rhizobia which are found in root nodules of these plants. Legumes convert atmospheric nitrogen into nitrogenous compounds useful to plants. Root nodules containing the bacteria *Rhizobium* fix free nitrogen for the plants and in return, the legumes then supply the bacteria with valuable carbon produced by photosynthesis. Legume nitrogen fixation starts with the formation of a nodule. The rhizobia bacteria in the soil invade the root and multiply within its cortex cells. The plant supplies all the necessary nutrients and energy nodules are visible with the naked eye. In the field, small nodules can be seen 2 to 3 weeks after planting, depending on legume species and germination conditions. When nodules are young, they are usually white or gray inside. As nodules grow in size, they gradually turn pink or reddish in colour, indicating nitrogen fixation has started. The pink or red color is caused by leghemoglobin (similar to hemoglobin in blood) that controls oxygen flow to the bacteria. If white, grey, or green nodules predominate, little nitrogen fixation is occurring as a result of an inefficient rhizobia strain, poor plant nutrition, pod filling, or other plant stress. Pink or red nodules should predominate on a legume in the middle of the growing season. The legume fixes the atmospheric nitrogen for their own needs and for soil enrichment, thereby reducing the requirement of fertilizer nitrogen in crop production. The quantity of nitrogen fixed for different legume crops are furnished in Table 1.

Table 1 Quantity of nitrogen fixed by legume crops

Crops	Quantity of nitrogen fixed (kg/ha)
Pigeon pea	41-91
Greengram	61
Cowpea	65-80

Cluster bean	130
Soybean	65
Chickpea	103
Groundnut	24
Pea	50-80
Lentil	35-75

Nitrogen harvest index values (seed nitrogen / total plant nitrogen) for cowpea, soybean, groundnut and chickpea is 0.61, 0.75, 0.80 and 0.73 respectively indicate the nitrogen economy of pulse or legume crops. Pulse crops have deep penetrating tap root system that helps to utilize the limited available soil moisture more efficiently than many other crops including cereals that contribute substantially to the loosening up of the soil. Pulses such as redgram, horsegram, mothbean, lathyrus and lentil are drought tolerant.

Pigeon pea, chickpea, black gram, green gram and horse gram have a seed coat accounting for 12-15% of the total weight of the grain where as it is in the range of 8-11% for lentils, French bean, kidney bean, pea, soybean and cowpea. On an average, pulses (including soybean) contain 11% seed coat, 2% embryo and 87% cotyledons. The embryo has two parts known as hypocotyl and plumule.

Legumes are the cheapest option for improving human nutrition. The seeds of legumes are higher in protein than any other food crop and are close to animal meat in quality. Pulses are often called 'poor man's meat' because of inexpensive source of high-quality protein. In general, pulses contain 20 to 30% protein, 60% carbohydrates, 1.0 to 2.5% fat and are fairly good sources of thiamine, nicotinic acid, calcium and iron. The cotyledon contains about 90% of protein and minerals of the whole seed. The pulse proteins are rich in lysine and show deficiency of tryptophan and sulphur containing amino acids like methionine and cystine, a reverse situation exists in the cereals proteins. Seed coat accounted for 32 to 50% of calcium of the whole seed.

The protein provided by cereals and pulse mixture is equivalent to skimmed milk in terms of its biological value. Pulses help to eradicate protein malnutrition, especially among children and nursing mothers. Pulses mixed with cereals in 3:7 proportions enhance the overall nutritive value of proteins. Although grain legumes are considered primarily as sources of protein, they also provide energy particularly those with high oil content. The average energy supply of cowpea, pigeon pea, soybean and groundnut is 13.9, 14.8, 18.0 and 22.9 MJ kg ha^{-1}, compared with an average of 15 MJ kg ha^{-1} for cereals in tropics. The major form of phosphorus fraction in pulses is phytin phosphorus, which accounts for 40 to 50% of total phosphorus. Pulses are a good source of vitamins. There will be a loss of 20 to 35% during cooking and a further loss of 10 to 15% during storage. Germination and cooking improve protein digestibility. '*Kabuli*' type of bengalgram has higher protein content than '*Desi*' type. Black gram is superior in its nutritive value among the pulse crops. The digestibility coefficient, biological value and protein efficiency ratio varied from 60 to 90%, 45 to 78% and 0.7 to 1.1 % respectively. Blackgram and redgram are deficient in methionine, trytophan, threomine and lysine. The presence of saponins, glycosides, tanins, alkaloids with phytin in hemicellulose substance inhibit the action of digestive enzyme Trypsin in different pulses adversely affect the pulses digestibility. Many legumes contain gums for thickeners such as gum arabic, guar gum and tragacanth gum. Germinated seeds of pulses contained increased amounts of carotene, ascorbic acid, pantothenic, biotin, nicotinic acid, thiamine, riboflavin and vitamin B12.

Reasons for poor yield of pulses: Grain legumes have either origniated or being cultivated on marginal lands. Pulses have their deep root systems, nodulation and nitrogen fixation capacity. These hardly enjoyed the inputs of irrigation and fertilization. Growth habit of grain legumes like *Phaseolus*, *Vigna* and *Pisum* are prostrate/ spreading / twining which near maturity pods becomes very close to or on to the ground. This type of growth creates a tight leaf canopy that does not allow air exchange and is inefficient in utilizing sunlight. Growth and development are suppressed or delayed due to self shading. *Vicia*, *Cicer* and *Cajanus* have erect growth habit with different degrees of branching and herbaceous woody stems. Most of them have indeterminate growth habit except a few improved varieties. The plants continue to grow vegetatively even after the start of reproductive development. Here the competition for available metabolites happens between pod/seed development and vegetative growth. Consequently many flowers and pods abscise / abort for lack of nutrition and the potential yield is diminished. There may be 30 to 50% flower shedding. The already formed pods do not develop to their full potential. Uneven maturity occurs in indeterminate cultivars that makes difficult in harvest, lowers the quality of produce and necessitates several pickings of pods. Raising of plant population in indeterminate crop varieties restricts branching and helps in shattering the flowering and fruiting periods. All the grain legumes have a high rate of photorespiration, typical of C3 plants. Selection for lower rates of photorespiration should be carried out. The C4 and CAM (Crassulacian Acid Metabolism plants) are not found in the grain legumes.

Factors limiting pulse crop production

i) Ecological factors: 92% of area under pulse cultivation is rainfed. More than 50% of area under pulses is sown in post rainy (*rabi*) season, largely on limited moisture. Pulses are sensitive to excess moisture, water logging, acidity, salinity and alkalinity. Frost and low temperature during the night cause heavy damages to *rabi* pulses particularly to chickpea whereas continuous rain invites more pest and diseases both in *kharif* and *rabi* pulses.

ii) Lack of proper agronomic management: Pulse crops can utilize the limited soil moisture and nutrients. Pulses are grown with poor management practices such as inadequate seed rate, non availability of rhizobium culture, improper sowing time, defective method of sowing and uneven distribution of seeds due to broadcasting.

iii) Varietal features: Non availability of seeds of improved varieties, indeterminate growth, no response to good management, flower and pod drop due to limitation of nutrients, hormonal imbalance, low or high temperatures, etc.

iv) Socio-economic factors: Pulses have subsidiary status as catch crops or intercrops. Hence, limited resource is allocated for pulses production. There is a low economic return from pulse crop. Pulses are not remunerative as the high yielding cereals, potato, sugarcane, cotton and tobacco. Pulses lack price policy in parity with cereals.

Anti-nutritional and toxic factors: Legumes/pulses contain anti-nutritional and toxic substances such as phytohaemagglutinins, protease inhibitors, lectins, goitrogens, phytates, saponins, favism factors, lathyrism factors, amylase inhibitors, tannins, polyphenolic compounds, aflatoxins, amines, cyanogenic glycosides, pyrimidine glycosides, protease inhibitors, oestrogens, antivitamins, flatulence factors, non-protein amino acids, quinolizidine alkaloids, allergens and lysinoalanine. The effect of different bioactive phytochemicals of pulses on physiological functions are presented in Table 2. Lathyrism is a paralytic disease affecting the lower limbs. The disease has been associated with consumption of kesari dhal (*Lathyrus sativus*) is commonly noticed in poor families who regularly eat considerable quañtity of the dhal. However, lathyrism develops only when the consumption of dhal is high (300 g daily) and the diet does not contain adequate quantities of cereals and is used for long time (six months or more). The toxic substance of *Lathyrus sativus* responsible for lathyrism has been identified as *selenium*. Favism is a disease characterized by haemolytic anaemia which affects certain individuals following the ingestion of fresh or cooked broad beans (*Vicia faba*). The anti- nutritional factors elimination can be achieved either by selection of plant genotype with low levels of such factors or through post-harvest processing (germination, soaking, heating, boiling, leaching, fermentation, extraction etc.). Panching and roasting of pulse seeds improved nutritive value of proteins and biological value.

Table 2 Bioactive phytochemicals in pulses

Bioactive components	**Food sources**	**Biological effects / health effects**
Lectins or Hemagglutinnins or Phytohemagglutinnins	Red kidney bean (*Phaseolus vulgaris*), lentil,	Substance agglutinate red blood cells and destroy them, induces red blood cell clumping, can impair the integrity of the intestinal epithelium and thus alter the absorption and utilization of nutrients; affect protein utilization
Lathyrogens	Chickpea	Disrupts collagen structure
Goitrogens	Beans	Causes goitre by limiting iodine absorption
Tannin	Lentil, Redgram, chickpea, *Phaseolus aureus*	binds or precipitates proteins and amino acids, inhibit the activities of trypsin, decrease the protein quality of foods, decrease feed intake and protein digestibility
Cynogenic glycosidases	Lima bean (*Phaseolus lunatus*) and *Vicia sativa*	Cyanogenetic glycosides releases hydrocyanic acid (HCN) on hydrolysis
a-Amylase inhibitors	Peas, Beans	Slows starch digestion
Trypsin inhibitors	Red kidney bean Hyacinth beans, lima bean, Soyabean, Redgram, Cowpea.	trypsin inhibition affect protein utilization
Protease inhibitors	Soyabean, kidney bean, cowpea, common beans, lima bean, cowpea, and lentil	Reduces digestibility and bioavailability of nutrients
Amylase inhibitors	Common beans (*Phaseolus vulgaris*)	Reduces utilization of dietary starch and protein, Potentially therapeutic in diabetes. Can reduce the digestibility and biological value of dietary proteins
Phytates/ Phytic acid	Soyabean, lentil, Pea, chickpea, cowpea, common bean	cause mineral ions deficiency in animals and human, acts as anticarcinogen, it has antioxidant activity and protects DNA damage
Phenolic compounds	Red kidney beans Lentil, Blackgram, soybean	Antioxidant activity, Has been inversely associated with the risk of colon cancer

Phytosterols	Chickpea, Pea, Kidney bean	Lowers serum cholesterol level
Saponins	Soybean, chickpea, black gram, moth bean, broad beans, peas, pigeon pea, sword bean, jack bean, Lucerne (alfalfa)	Hypocholesterolaemic effect, Anti-cancer activity, Saponins cause nausea and vomiting; cause abdominal pain, vomiting and diarrhea; reduce cholestrol level and reduce the risk of heart disease;
Flavonoids, isoflavones	Soyabean, Beans, Lentil, Chickpea,	reduction in osteoporosis, car- diovascular disease and prevention of cancer

REFERENCES

Khokhar, S. and Chauhan, B.M. 1986. Anti-nutritional factors in moth beans (*Vigna aconitifolia*): Varietal difference and effects of methods of domestic processing and cooking. J. Food Sci. 51(3): 591-594.

Singh, J. and Basu, P.S. 2012. Non-Nutritive bioactive compounds in pulses and their impact on human health: An Overview. Food and Nutrition Sciences, 2012, 3, 1664-1672.

CHAPTER

19

Redgram–*Cajanus cajan* L. Millsp. (2n=22)

Synonyms: *Cajanus indicus* Spreng, *Cajanus bicolor* DC., *Cajanus flavus* DC.

Vernacular name: Redgram, Pigeonpea, Congo bean, Angola pea, No eye pea, Gungo pea, (English), *Arhar*, *Tur* (Hindi), *Thovarai* (Tamil) and *Kandulu* (Telugu).

Importance: India alone accounted for about 80% of total world's production and 90% of total world's consumption of Red gram. It is the second most important legume of India. Redgram is consumed as fresh green peas in Caribbean, Latin America, Puerto Rico, Trinidad, India, Kanya, Tanzaniya and Zambia. In India, late maturity types of vegetable redgram is grown in kitchen backyards or as bund crop. The top of the plants with fruit provide excellent fodder and are also made into hay and silage. Redgram is used as shade crop in cacao and turmeric as wind breaks and for anti-erosion purpose. The dried stalks are used for firewood, thatching and baskets.

Origin: India is the primary center of origin while Africa is the secondary center of origin of pigeonpea. *Cajanus* is derived from a Malay word '*katschang*' or '*katjang*' meaning pod or bean.

History: The Sanskrit word for redgram/pigeonpea is *adhaki* as per Charaka (c. 700 BC) and Susruta (c.400 BC). Amarsimha (c. 200 BC), in his *Amarkosa*, has mentioned '*adhaki*' and '*tuvarika*' as names of pigeonpea. The word *adhaki* originated most likely from the word *ardha,* meaning 'one-half' or 'split into two parts'. Dry whole pigeonpea seed is rarely consumed; only the dhal is commonly consumed. One of the two common names used for pigeonpea in the Indian subcontinent is '*arhar*' which has come from the word '*adhaki*'. The second common name for pigeonpea is the Sanskrit, '*tuvara*' or '*tubara*' which means astringent. The green seed, which has been consumed in Gujarat for centuries, has an astringent taste. The word *arhar* is common in northern India, and *tuvara* (with variants *tuvarika*, *turri*, *tur*, etc) in southern India.

Distribution: Redgram is widely cultivated in USA, Hawai, West Indies, India, Australia, Kenya, Uganda, Tanzania, Namibia and Malawi. In India it is mainly cultivated in Uttar Pradesh, Madhya Pradesh, Bihar, Maharashtra, Andhra Pradesh and Tamil Nadu.

Botany: Cultivated redgram is an annual, 0.6 to 1.5 m in height, deep tap root system. Pigeonpea with its deep root system (>150 cm) can break hard pans in plough layer, and hence called '*biological plough*'. Leaves are trifoliate and spirally arranged. Leaflets are lanceolate to narrow elliptic, hairy on both sides. Inflorescence is raceme 4 to 12 cm long, and has 3 to 4 seeds. Hundred seeds weight 11 to 13 gm. The flowers are self-compatible and usually self-pollinated and 20% cross pollination can occur due to the visit of bees and other insects. The majority of flowers open between 11 a.m. to 3 p.m. an often remain open for 6 hours. Rain at flowering reduces fertilization, it is essential to emasculate before 9 a.m. on the day before the flowers open and they may be hand pollinated at the same time for hybridization. Pod formation begins at 12 to 14 weeks in early varieties and they require 5 to 6 months to reach maturity while late varieties require around 9 to 12 months to reach maturity. The conversion ratio from fresh green pods to dried pods is about 3.3. Redgram has 3% out crossing at 14 m and 9 m guard rows can be provided as an adequate barrier. Redgram attains a height of 25 cm from ground in 30 days. Two botanical ecotypes have been recognized in redgram viz., (i) *Cajanus cajan* var. *flavus* is early maturing, shorter plants with yellow standards and green glabrous pods, which are light coloured when ripe and are usually three seeded. These are *Tur* cultivars and are extensively cultivated in Peninsular India and (ii) *Cajanus cajan* var *bicolor* is perennial, late maturing, large, bushy, plants, with dorsal side of red or purple streaks and hairy pods blocked with maroon or dark coloured with 4-5 seeds. These are *Arhar* cultivars and are grown in north India.

In a fully developed bud, anthers surround the stigma and dehisce a day before the flower opens. Anthesis in pigeonpea starts from 0600 and continues till 1600. The peak anthesis period is between 0900 and 1000 hours. Fertilization occurs on the day of pollination. The extent of crosspollination ranges from 3 to 40%, with an average of 20%. A plant produces many flowers of which only 10% set pods. In the cotyledons, synthesis of starch and protein starts about 17 days after pollination and continues for 14 days. In each raceme 1 to 5 pods may mature, and rarely up to 10. Pods colour may be green, purple, dark purple, or mixed green and purple. Pods with deep constrictions in shape are beaded, while others are somewhat flat. The number of seeds per pod range from three to five in cultivated varieties. The 100-seed weight ranges from 2.8 to 22.4 g with the cultivated varieties ranging from 7.0 to 9.5 g.

Climate: Redgram is cultivated between 30ºN and 30ºS up to an altitude of 1800 m from MSL although about 92 percent of total world production is in India. It can be grown between 14°N and 28°N latitude, with a temperature ranging from 26° to 30°C in the rainy season (June to October) and 17° to 22°C in the post-rainy (November to March) season. The amount of daily global solar radiation varies from 400 to 430 cal cm^{-2} day^{-1} in the rainy season and 380-430 cal cm^{-2} day^{-1} in the post rainy season. Pigeonpea is very sensitive to low radiation at pod development, therefore flowering during the monsoon and cloudy weather, leads to poor pod formation. Mean annual rainfall ranges from 600 to 1400 mm, of which 80 to 90% is received in the rainy season. It is cultivated all over the country with exception of areas which are excessively wet or severe frost. It is intolerant of shade and tolerates only moderate competition. It does best in full sun on bare ground but can grow with side shade or broken shade from trees and a low cover of grass and forbs. Redgram is short day plant. The cardinal minimum, optimum and maximum temperature for redgram is 10ºC, 24 to 29ºC and 35ºC respectively. Redgram is susceptible to frost damage at all stages of growth. Sowing of redgram should not be made at soil temperature less than 19°C. The germination and emergence is enhanced by soil temperature of 26 to 43°C and particularly 29 to 36°C. It prefers warm weather (max 30-40°C) during germination and pre-flowering growth, and a lower temperature (max 25-30°C) during the flowering and pod filling stages. The optimum temperature for flowering and pod setting is 24°C. Temperature is the dominant factor influencing the number of days from emergence to floral initiation. The number of days from emergence to floral initiation is 24 days at a day/night temperature of 24/16°C and 36 days at 32/24°C. The time from floral initiation to flower opening varried from 40 days under an 8 hour photoperiod at 24 /16°C to 22 days under 16 hour photoperiod at 32/24°C. Increased the day/ night temperature regimes from 24/16°C to 32/24°C reduced the duration of floral bud development by 12 days under an 8 hours photoperiod and 9 days under a 16 hours photoperiod. Average annual rainfall of 600 mm to 1000 mm is most suitable for redgram growth and yield. High yields are obtained when there is good rainfall during the first 2 months of growth followed by a dry period during flowering and harvesting. It is a drought resistant crop with deep root system. Rainfall of 625 mm per year is adequate for good growth under semiarid conditions. Redgram is less suitable for wet tropics and sensitive to frost. It is fairly resistant to wind and is sometimes planted in double rows as a wind break. It is however sensitive to sea spray and does not thrive if planted near the seashore. Late maturity varieties are short day plants while early varieties are photosensitive plants.

Soil: Redgram can be grown on almost all soil types provided the soil is not markedly deficient in lime. It will not tolerate waterlogging. It thrives best on deep loamy soils free from excessive soluble salts. It tolerates pH from 4.5 to 8.4 and some varieties tolerate 6 to 12 mmhos/cm of salinity. It is extremely drought resistant crop.

Season: Ten maturity groups have been identified under Indian conditions which have been combined into four categories: extra early (120 days), early (145 days), medium (185 days) and late-maturing (200 days) cultivars.

North West Plain Zone (Punjab, Haryana, Western UP, Delhi, estern/Rajasthan):
2nd fortnight of May to 1st fortnight of June.

North East Plain Zone (Assam, West Bengal, Bihar, Jharkhand, eastern and central UP):
Early: 1st fortnight of June
Late: 1st fortnight of July
Pre-*rabi*: 1st fortnight of September

Central Zone (Gujrat , Maharashtra, Rajasthan, MP) :
Rainfed: 1st fortnight of July (onset of monsoon)
Irrigated: 2nd fortnight of June

South Zone(Orissa, AP., Tamil Nadu, Karnataka):
Onset of monsoon: 2nd fortnight of June

Land preparation: Land preparation for redgram requires at least one plowing during the dry season followed by 2 or 3 harrowings. The summer plowing helps in minimizing the weed flora and to conserve moisture. Subsoiling on lands having hardpan permits both deep proliferation of roots and great infiltration of water to the deeper layers of the soil. The seedbed should have a moisture content of about 40 to 50% of the available water to ensure quick and adequate germination. Adequate provision for surface drainage is an important consideration in the seed bed preparation of redgram. Organic manure may be applied 2-4 weeks before sowing. In acidic soils 2-4 t ha-1 of lime is incorporated 3-4 weeks before sowing to neutralize the acidity.

Varieties: Pigeon pea varieties with salient features recommended for different states of India are furnished in Table 1 to Table 5.

Table 1 Pigeon pea varieties recommended for different states of India

State	Type		
	Early (120-150 days)	**Medium (150-180 days)**	**Late (> 180 days)**
UP (Central & Western)	PA 3, T 21, Prabhat, UPAS 120, Pusa 84, Pusa 74, Manak, Pusa 33, Pusa 993, Pusa 855, TT 5, ICPL 151	MA 6, Mukta, Paras, Sharda, Pant A 3	T 7, T 17, NP (WR) 15, Pusa 33, Pusa 55, Bahar (1258), MAL 13, KA 32-1 (Amar), Narendra Arhar-1, Pusa 9, Gwalior 3, Azad,
Punjab, Haryana, Delhi	PA 1, T 21, Prabhat, UPAS 120, Pusa 84, Pusa 74, Manak, ICPH 8, Sagar (H 77-208), Pusa 33, Pusa 992, Pusa 855	Mukta, Paras, Sharda	NP (WR), Pusa 55
Bihar, Jharkhand, Eastern UP	Prabhat, UPAS 120, Pusa Ageti, Pusas 74, Pusa 84	Mukta, ICPL 85063, BR 65, BR 183, MA 6, Birsa Arhar 1	T 7, T 17, NP (WR) 15, AS 71-77, MAL 13, Azad (K91-25), Pusa 9, DA 11 (Sharad), Bahar, Basant
West Bengal, Orissa & Assam	T 21, Prabhat, Pusa Ageti, Pusa 74, Pusa 84, TT 5, BS 1	BR 65, BR 183, Mukta, ICPL 85063, C 11, WB 20 (105)	Sweta (B7), Chuni (B 517), T 7, T 17, NP (WR) 15, MAL 13, Pusa 9, Bahar
Rajasthan	T 21, Prabhat, UPAS 120, Manak, Pusa Ageti, J 9-19, Pant A 1, Pant A 2, Sagar, Pusa 74, Pusa 33,	Sharda, Mukta, Paras	NR (WR) 15, Gwalior 3
Madhya Pradesh and Chattisgarh	T 21, Prabhat, UPAS 120, Pusa Ageti, Vishakha 1, J 9-19, Pusa 33	Sharda (S 8), No. 148, Mukta, MA 3, KM 7, BDN 1, BDN 2, C 11, Paras, ICPL 87119, JA 3, JA 4	T 7, T 17, NP (WR) 15, Kanke-3
Gujarat	T 21, Pusa Ageti, Prabhat, Vishakha 1, TAT 10, J 9-19, Pusa 84, Pusa 74	Sharda, Mukta, ICPL 87119, BDN 1, BDN 2, C 11, ICPL 871, GTH 1	NP (WR) 15, Gwalior 3
Maharashtra	T 21, Pusa Ageti, Prabhat, Vishakha 1 (TT 6), TAT 10, J 9-19, AKT 8811	Sharda, Mukta, No. 148, BDN 1, BDN 2, C 11, ICPL 87119, BSMR 175, BSMR 736, MA 3, Malviya Vikalp, KM 7	NP (WR) 15, Gwalior 3,

Andhra Pradesh	Pusa Ageti, Prabhat, CORG 9701, T 21, ICPL 87, ICPL 151, Hy 5, ICPL 84031 (Durga)	Sharda, PDM 1, GS 1, Hy 3A, Hy 3C, Hy 4, ICPL 87119, ICPL 8863, ICPL 85063 (Laxmi), LRG 30 (Palanadu), LRG 36, LRG 38, ICPL 332, C 11, PT 221	SA 1
Tamil Nadu	Pusa Ageti, Prabhat, CORG 9701, ICPL 87, Co 1, Co 2, Co 4, Hy 5	Sharda, PDM 1, GS 1, Co 5, Co 6, Hy 3A, Hy 3C, Hy 4, ICPL 87119, BDN 2, PT 221	SA 1
Karnataka	Pusa Ageti, Hy 5, ICPL 151 (Jagriti), CORG 9701, ICPL 87 (Pragati)	T 21, Sharda (S 8), Hy 3C, GS 1, KPL 87, ICPL 87119, C 11, TS 3, ICPL 8863 (Maruthi), PT 221	SA 1

Table 2 Pigeonpea varieties features in different states of India

Variety	Yield (q/ha)	Suitability	Major characteristics
UPAS 120	11-15	Punjab, Haryana, North Rajasthan, West U.P.	Indeterminate. Semi-spreading, synchronous in maturity, tolerant to pod borer
Bahar	25-30	U.P., Bihar, W.B.	Compact, resistant to sterility mosaic
Amar (KA 32-1)	16-20	Uttar Pradesh	Compact, resistant to SMD
Narendra Arhar 1	20-22		Resistant to SMD and tolerant to wilt and phytophthora blight
Pusa 9	22-26	U.P., Bihar, W.B.	Indeterminate, resistant to SMD and Alternaria, suitable for Pre-rabi also
MAL 13	2465	U.P., Bihar, W.B.	Tolerant to wilt, PB, SME
NDA 98-1 (NDA2)	2443	U.P., Bihar, W.B.	Resistant to wilt
MA 6	2281	Bihar, U.P.	Resistant to wilt and SMD
Paras (H 82-1)	15-20	Haryana	Indeterminate
AI 201	15-16	Punjab	Indeterminate
PUSA 992	14-18	Delhi, Punjab, Haryana	-
TJT501	18.60	MS, Guj. & MP.	Tolerant to pod borer, pod fly
Malviya Vikalp (MA 3)	20-22	M.P., Gujarat, Maharashtra	Spreading, constricted pod resistant to pod fly
Jawahar (JKM 7)	18-20)	M.P., Gujrat, Maharashtra	Tolerant to wilt and Phytophthora blight
BSMR 175	11-12	Maharashtra	White seeded, wilt and sterility mosaic resistant
JA 4	16-18	Madhya Pradesh	Pod borer tolerant
Asha (ICPL 87119)	16-18	M.P., Gujarat, Maharashtra	Indeterminate, spreading, bold seeded, bold seeded, wilt and SMD resistant

BSMR 736	12-18	Maharashtra	Red seeded, wilt and SMD resistant
Co 6	9-10	Tamil Nadu	Indeterminate, tolerant to pod borer
Laxmi (ICPL 85063)	18-20	Andhra Pradesh	Semi-spreading, Suitable for Rabi planting
GT 100	16-18	Gujrat	Determinate, white bold seeded, tolerant to pod borer and pod fly
AKT 8811	13-14	Maharashtra	Moderately resistant to wilt & pod borer
CORG 9701	11-12	A.P., Tamil Nadu, Karnataka, Orissa	Short duration
Durga (ICPL 84031)	8-10	Andhra Pradesh	Determinate
Vamban 2	10-12	Tamilnadu	Medium duration, indeterminate
Vamban 3	10-12	Tamilnadu	Short duration
GTH 1	14-15	M.P., Gujarat, Maharashtra	Medium duration

Table 3 Hybrids released for different states of India

Hybrid	Parentage	Year of release (by)	Suitable for
GMS based hybrids			
ICPH 8	Ms Prabhat (DT) x ICPL 161	1991 (ICRISAT)	Central Zone
PPH 4	Ms Prabhat x AL 688	1994 (PAU)	Punjab
CoPH 1	Ms21 x ICPL 87109	1994 (TNAU)	Tamil Nadu
CoPH 2	Ms Co 5 x ICPL 83027	1997 (TNAU)	Tamil Nadu
AKPH 4101	Ms Prabhat NDT x AK 101	1997 (PKV)	Central Zone
APKH 2022	AKMS 2 x AK 2	1998 (PKV)	Maharashtra
GMS based hybrids			
GTH 1	2003 (GAU)		Gujarat

Table 4 Pest resistant / tolerant varieties of pigeonpea

Variety of Pigeonpea	Pest Resistance	Area for which recommended
ICPL-1	Pod borer	Gujarat, Maharashtra and A.P.
ICPL-269	Pod borer	Maharashtra and Northern Telangana of A.P.
ICPL-329	Pod fly	U.P., West Bengal, A.P., Tamil Nadu and Gujarat
ICPL-372	Pod fly	U.P., West Bengal, A.P., Tamil Nadu and Gujarat
ICPL-85066	Pod fly	Gujarat, Maharashtra and A.P.
Hy-2C	Pod borer	Madhya Pradesh, Karnataka and A.P.
BDN-1 - 2	Pod borer, Wilt	Madhya Pradesh, Karnataka and A.P.
Bahar	Pod borer, Sterlity Mosaic	Central India
Birsa Arhar-1	Wilt	Central and Eastern zone
Vishakha, TT-6	Wilt	Central India

Maruthi, ICPL-8863, TS-3R	Wilt	Central and South India
BSMR-736, Asha (ICPL-87117)	Wilt + SMD + PSB	Central and South India
C-11	Wilt	All zones
PDA 89 - 2E, PDA 92 - 3E	Pod fly	U.P., Maharashtra and M.P.
PDA-93 - 1E	Pod borer	U.P., Maharashtra and M.P.
PDA-88 - 2E	Pod fly	Orissa, Maharashtra and M.P.
MA-91 – 2	Pod fly	Orissa, Maharashtra and M.P.
Mukta	Pod fly	U.P., West Bengal, A.P. , T.N, and Gujarat
UPAS 120	Pod borer	U.P., Maharashtra and M.P.
JA-3 & JA-4	Pod borer	Central India

Table 5 Pigeon pea varieties tolerant to diseases

Specific trait	Varieties
Wilt	BDN 1, BDN 2, C 11, TT 6 Maruthi, BSM 6736, Sharda, Amar, Narendra Arhar; Asha BSMR, 736, BSMR 853, JKM 7, Maruti , C11
Sterility mosaic disease (SMD)	Bahar, HY 3C, Pusa 9, Azad, ICPL 366, ICPL 87051, Amar, BSMR 175, BSMR 763, BSMR 736, BSMR 853, MA 3, Pusa 9, Co 5, MA 6, MAL 13, NDA 1, Asha,
Wilt + Sterility mosaic disease (both)	Narendra Arhar 1, Asha (ICPL 87119), DA 11, BMSR 853, MA 3, Amar, Asha BSMR 736,
Alternaria blight	WB 20 (105), Pusa 9, DA 11, DA9
Phytopthora blight	KM 7, DA 11, Pusa 9, Narendra Arhar 1
Short duration	UPAS 120, PUSA 992, VLA1, Co 7, PAU 881

Table 6 Seed rate and variety particulars of Redgram in Tamil Nadu.

Varieties	Seed rate (kg/ha)		Duration (days)		Hundred seed weight (g)	Seed yield kg/ha	
	Pure crop	Mixed crop	50 % flowering	Maturity		Rainfed	Irrigated
SA 1	10	5.0	120-130	180	8.5	1250	
CO 3	25	12.5	90-95	130	7.2	1180	1400
CO 4	25	12.5	90-95	130	8.5	980	1750
CO 5	25	12.5	70-75	110	8.0	760	-
CO 6	8	3.0	120-130	180	8.8	900	-
Vamban 1	25	12.5	70	100	8.8	840	1200
BSR 1	-	-	100-110	180	12.0	1.0 kg / plant	-

Note: SA 1, CO 3, CO 4 and BSR 1 are used for bund planting @ 50 gm per 100 m.

Growth stages: When sown under optimal moisture and temperature (29°C-36°C), the seed testa splits open near the micropyle on the 2nd day. The tip of the radical elongates and emerges from the seed coat. On the 3rd day the hypocotyl appears as an arch and continues to grow upward. The hypocotyl turns light purple. The seedling epicotyl elongates 3-7 cm before the first trifoliate leaf emerges. Growth is moderately slow during the first 2 to 3 months of life during which time seedlings are not competitive with grass and weeds; afterwards redgram competes well with vegetation equal or lower in height. Seed development is visible 7 days after pollination. A pod is formed 15-20 days after fertilization. Seeds reach physiological maturity in 30 days and are ready for harvest at a lower moisture content in 40 days. There is little or no shattering of mature pods in the field. Redgram requires 65-80 days to flower and 50-75 additional days to create mature seeds.

Table 7 Growth stages of long and short duration redgram varieties

Type	Days to 50 % flowering	Days to 75 % flowering	Days to maturity
Long duration	100	107	150
Short duration	83	93	125

Land preparation: A deep ploughing with a mold-board plough followed by two to three cross-harrowings, and proper leveling should be done to ensure uniform irrigation and good drainage. The entire basal dose of fertilizer should be mixed in top 15 cm of soil during land preparation.

Seeds and Sowing: Seeds are sown either by broadcasting, line sowing or dibbling. Dibble the seeds with recommended spacing. The seed rate and spacing varies with varieties, hybrids and cropping systems.

Table 8 Quantity of seed required for different redgram varieties (kg/ha)

Varieties	Vamban 2	LRG 41	Co(Rg) 7	VBN (Rg) 3	APK 1
Sole Crop	8	8	15	15	15
Mixed Crop	3	3	5	5	5

[BSR 1 (Bund planting) 50 g/100 metre]

Table 9 Spacing for sole, inter crop and bund crop for different redgram varieties

Variety	Spacing for pure crop	spacing for intercrop / bund crop
Co 2	45 cm x 30 cm	----
Co 3	45 cm x 30 cm	----
Co 4	45 cm x 30 cm	----
Co 5	45 cm x 30 cm	----
Co 6	90 cm x 30 cm	240 cm x 30 cm
SA 1	90 cm x 30 cm	240 cm x 30 cm
BSR 1, SA 1	----	60 cm

Treat the seeds with Carbendazim or Thiram @ 2 g/kg of seed 24 hours before sowing (or) with talc formulation of *Trichoderma viride* @ 4g/kg of seed (or) *Pseudomonas fluorescens* @ 10 g/kg seed. Fungicide treated seeds should be again treated with a rhizobium culture. There should be an interval of at least 24 hours after fungicidal treatment for giving biocontrol agents / rhizobium culture treatment. Bio-control agents are compatible with biofertilizers. First treat the seeds with biocontrol agents and then with rhizobium. Fungicides and biocontrol agents are incompatible. Three packets of rhizobial culture are sufficient for treating seed required for one ha. The culture slurry may be prepared with 500 ml of rice gruel. 10% of sugar/gur/jaggery solution is made by adding 50 gm of sugar or gur to 500 ml of water. Boil the solution for 15 minutes to dissolve the sugar in water. Cool the sticky solution to room temperature. Mix three packets of rhizobial culture into the cooled sticky solution thoroughly. The slurry solution helps rhizobium stick to the seed surface and also enhances the survival of rhizobium. Dry the bacterial culture treated seeds in shade for 15 minutes before sowing. The inoculated seeds should not be exposed to sunlight. Seeds after inoculation with rhizobium should be sown as soon as possible preferably on the same day. Inoculated sown seed should be covered with soil. The inoculated seeds should not be allowed to come in direct contact with chemical fertilizers. If the seed treatment is not carried out apply 10 packets (2 kg) of Phosphobacteria (*Bacillus egaterium*) and 10 packets (2 kg) of *Pseudomonas fluorescens* with 25 kg of FYM and 25 kg of soil before sowing. In very acid soils, it is advisable to sow the inoculated seeds along with lime, dolomite or neutralized super phosphate. Sowing of redgram can be done on ridges when soil internal drainage is poor which leads to waterlogging. Seeding depths of 2.5 to 5 cm are recommended. The seeds take nearly five days for germination and the seedlings begin to appear above the ground by the fifth day under favourable conditions.

Seed hardening will help redgram to tolerate drought stress. Calcium chloride 20 g per litre (c. 0.2%) is to be prepared. One kg of seeds per liter should be immersed for one hour and dry it under shade for 7 hours to harden the seeds. Dry the seeds under shade for seven hours. Treat the seeds with 100 g rhizobium, 100 g PSB and 4 g trichoderma per kg seed by sprinkling jaggery solution and then microbial powder on the seeds.

Nipping: Nipping off 5-6 cm top growth in red gram at 20-30 days after transplanting (DAT) will allow side branches which in turn increase pods yield.

Weed management: Redgram is very sensitive to weed competitions in the first 60 days of growth. When protected in this period, the crop makes rapid growth and weeds do not affects its growth thereafter. Spray Fluchloralin 1.5 lit per ha or Pendimethalin 2 lit per ha three days after sowing mixed with 900 litres of water which is followed by irrigation. Post emergence application of Imazethapyr @ 60 g ai/ha on 15 DAE of weeds (2-3 leaves stage of weeds) and quizalofop ethyl @ 50 g ai/ha on 20 DAE of weeds (2-3 leaves of weeds) are recommended for controlling broad leaved and grassy weeds respectively, At the time of herbicide application, there should be sufficient soil moisture. Then one hand weeding may be given on 30 to 35 days after sowing. If herbicide is not applied, two hand weeding are given on 20 and 35 days after sowing.

Manures and fertilizer application: Farm yard manure / Compost @ 12.5 t/ha or vermicompost @ 5.0 t/ha is applied as basal dressing. Application of 12.5-25-0 and 25-50-0 kg of N, P_2O_5 and K_2O per ha is recommended for rainfed and irrigated conditions respectively. Seedlings in early stages depend on soil-nitrogen and do respond to a starter dose of 15 to 20 kg N/ha. In the later stages of plant growth, most of the nitrogen requirement is derived from fixation in nodules. Soil application of 25 kg of sulphur as gypsum (110 kg/ha) or 2% urea in two sprays at flower commencement and 15 days after may be given. At the time of first appearance of flowering, spray DAP @ 2% and a second spray is given 15 days after the first spray.

Growth regulators: Apply 40 ppm NAA (40 mg/litre) at flower initiation. This may be advantageous when mixed with urea and sprayed. NAA can be mixed with fungicide and insecticide. Use of TIBA @ 150 ppm (150 mg/litre) as a foliar spray at flowering initiation is effective in increasing grain yield.

Water management: Water requirement for redgram crop is 400 mm which yields about 910 kg of grain per ha. Though it is a drought tolerant crop, it responds well to irrigation during summer seasons. Irrigate the crop immediately after sowing, third day after sowing, bud initiation, 50% flowering and pod development stages. 50% increase in grain yield with irrigation at 60% depletion of available soil moisture compared to irrigation at 100% depletion of available moisture. Water stagnation should be avoided. Ridge planting has proved advantageous as it ensured quick disposal of rain water allowing no stagnation. In general, redgram is grown as rainfed crop and the rains received during the growing season itself are sufficient to allow the crop to complete its life cycle without water stress. The moisture stored in the soil profile during monsoon (rainy) period is used by the crop in the post-monsoon period.

Cropping systems: The promising sole and intercropping systems of redgram is presented in Table 9. Early-maturing (100-120 days) genotypes are grown as a sole crop. It is generally grown during the *kharif* season (June-July) either as sole crop or as an intercrop with cereals, like sorghum, pearl millet or oilseeds like groundnut. The practice of mixed cropping has been common with long duration varieties of Redgram under rainfed conditions. It can be one of the alternate crops grown in *rabi* (post rainy) season to utilize the residual soil moisture and nutrients. The optimum plant populations for the *rabi* crop is three to four times higher than that normally used in *kharif* season because of the influence of low temperature and short photoperiod. Intercropping of redgram + groundnut in 1:6 ratio is recommended during *kharif* season. Multitier cropping of *agathi* (*Sesbania grandiflora*) with 1 x 1 m spacing forms first tier, redgram with 45 cm x 20 cm forms second tier, cotton with 45 x 20 cm forms third tier and blackgram with 30 cm x 10 cm forms fourth tier is recommended for rainfed black soil areas recording more than 300 mm rainfall during the cropping period.

Table 10 Particulars of redgram based cropping systems

Inter crop	Sowing pattern	Ratio of component crops
Cereal combinations		
Sorghum + Pigeonpea	Paired rows at 30:30:60 cm.	2:1
Pearl millet + Pigeonpea	Paired rows at 30:30:60 cm.	2:1
Maize + Pigeonpea	Paired rows at 40:40:80 cm. or Uniform rows at 60 cm.	2:1
Pigeonpea + upland Rice	Uniform rows at 60-75 cm.	2:2
Legume combinations		
Pigeonpea + Groundnut	Uniform rows at 75-90 cm.	2:2
Pigeonpea + Soybean, Mung Bean or Black Gram or Chickpea	Uniform rows at 75 cm. or Uniform rows at 50 cm.	2:1
Long duration redgram varieties (one row) + groundnut (six row)	Rainfed condition	1:6
Short duration redgram varieties + groundnut	Rainfed condition	1:4

Crop protection: Among insect pests gram pod borer (*Helicoverpa armigera*) and pod fly (*Melanagromyza obtusa*) cause severe damage and associated yield losses across pigeon pea growing regions of the country. Pod fly causing 2.5 to 86.8% of grain losses in different parts of the country was notable. The estimates of avoidable losses due to pod borer complex, mainly pod fly and *H.armigera* are 43.5 and 30.2%, respectively. Among diseases, *Fusarium* wilt (*Fusarium udum*) in Central and Southern states followed by sterility mosaic (virus transmitted by eriophyid mite *Aceria cajani*), and SMD and *Phytophthora* blight *(Phytophthora drechsleri f.sp.cajani)* diseases in the North east plain (Uttar Pradesh) cause substantial yield losses to the crop. Unchecked weeds also cause 21-97% yield loss in pigeon pea.

Integrated Pest Management practices

- Grow trap crops such as marigold (*Chrysanthemum* spp.) on the border and in between rows as an inter-crop.
- Use of neem seed kernel extract 5% against pod borer in SDP and ESDP is quite effective.
- Spray HNPV (nuclear polyhedrosis virus that infects *H. armigera*) 250 LE (larval equivalent) ha-1 on noticing egg and first instar larva of *H. armigera* (2-3 eggs or 1 larva per 5 twigs of plants, which is the economic threshold level).
- Installing bird perches @ 50/ha for attracting insectivorous birds that feed on larvae of *Helicoverpa* pod-borer.
- Sprays of monocrotophos 36 EC (0.04%) (1ml L-1 water) followed by Nimbicidine (0.3%) was best for the control of pigeonpea pests. Cypermethrin (0.004%), and endosulfan 35EC (0.07%) (2 ml/l water) are also effective for pod borers. Spraying should be done @ 600-1000 litres of water ha^{-1} with knapsack sprayer or 200-300 litres of water ha^{-1} with power sprayer.
- Seed dressing with Ridomil MZ @ 3g kg-1 seed. Two foliar sprays of Ridomil MZ at 15 days intervals starting from 15th day after germination, if there is continuous rain (high humidity), and risk of infection to control Phytophthora blight.
- Spray acaricides Kelthane or Morestan or Metasystox at 0.1% to control the mite vectors in the early stages of plant growth which favours the spread of Sterility mosaic disease.
- Pheromone traps for *Helicoverpa armigera* 12/ha.
- Apply any one of the following insectcides viz., Azadirachtin 0.03%WSP 2500-5000 g/ha, *Bacillus thuringiensis* serovar kurstaki (3a,3b,3c) 5%WP 1000-1250 g/ha, Dimethoate 30%EC 1237 ml/ha, Emamectin benzoate 5%SG 220 g/ha, Indoxacarb 15.8%SC 333 ml/ha, Chlorantraniliprole 20 E 150ml/ha, NPV of *H. armigera* 2% as 250-500, Spinosad 45%SC 125-162 ml/ha, NSKE 5% twice followed by triazophos 0.05%, Neem oil 2%, Phosalone 0.07%, (Spray fluid 625 ml/ha) etc., to control pod borers, plume moth, pod fly, pod bug.
- Gram pod borer management practices includes (i) Setting up of pheromone traps @ 5 traps/ha before the initiation of flowering and collection and destruction of moths caught in the traps, (ii) Spray application of phosalone 0.07 % or endosulfan 0.07% or profenofos 0.05 % or cypermethrin 0.005% or the combination with *Bacillus thuringiensis* var. kurstaki three times at fortnightly interval commencing from flowering affords protection, (iii) Dusting of endosulfan 4% or carbaryl 10% dust @ 25 kg/ha once at initiation of flowering controls the pest on Bengal gram and (iv) Spray application of *Ha*NPV at dusk @ 250 larval equivalent /ha.
- Plume moth may be controlled with spray application of phosalone 0.07% or endosulfan 0.07% or profenofos 0.05% to control

- Spray Methyl demeton 25 EC 500 ml/ha or Dimethoate 30 EC 500 ml/ha to control aphids.
- Apply *P. fluorescens* (or) *T. viride* – 2.5 kg / ha + 50 kg of well decomposed FYM or sand at 30 days after sowing to control wilt.
- Drenching with Carbendazim @ 1 gm/ lit to control root rot.
- Rogue out the infected plants in the early stages of growth.
- Spray Fenazaquin @ 1 ml/lit on 45 and 60 DAS as prophylactic spray to manage sterility mosaic.
- Seed treatment with *Pseudomonas fluorescens* and *Trichoderma viride* (5g + 5g) and Soil application of *Pseudomonas fluorescens* or *Trichoderma viride* @ 2.5 kg/ ha at the time of sowing to control Cyst nematode.

Harvest: Redgram should be harvested when 75 to 80% of the pods are at physiological maturity. It can be harvested about 40 days after anthesis when pods turn brown and are dry and when the dry mass of the seed is low and moisture content is between 20 and 24 %. Sometimes, the whole crop is cut down when about two-thirds or three quarters of the pods are matured. Traditionally the crop is harvested by cutting the stem at the base, with an axe or sickle. The harvested plants are tied in bundles and transported to a threshing floor. The plants are stacked in upright bundles to dry. The pods and grain are separated by beating the dry plants with sticks or by using a thresher. Hand picking of the mature pods can also be done. This allows the crop to flower and pod for a second or sometimes a third harvest. Hand picking may not be economical beyond a second flush. When hand picking of pods is not feasible, the upper branches with mature pods are cut (good for determinate types). The seeds and chaffs separated by winnowing.

Yield: Red gram may yield 1.0-1.5 tonnes (*kharif*), 2.0-2.5 tonnes (*rabi*) of grain/ ha, 5.0-6.0 tonnes sticks, 0.8-1.0 tonnes of dry leaves and 0.2-0.3 tonnes of pod husk/ha. Crop residue yield is ranging 20-25 quintal per ha. In general, early-maturing (100-120 days) genotypes have a higher harvest index with an average of 34% compared to medium-maturing genotypes at 24%.

Post-harvest technology: After threshing and winnowing, the seeds are cleaned and thoroughly dried to a moisture content of 10% before being stored. The ripe dry seeds are boiled and eaten as pulse. In India, the ripe dry seeds are split and made into dhal, which may be prepared either by a dry or wet method. In the dry method, the dry seeds are placed in the sun for 3 to 4 days and are then split in a mill and this process is repeated for 3 to 4 times. In the wet method, the seeds are soaked in water for 6 to 10 times mixed with red earth overnight, then dried in the sun, after which the red earth is removed by sieving and the seeds are finally split into '*dhal*' in a hand mill. The split '*dhal*' is then cleaned by repeated winnowing and sieving to remove the hulls and broken pieces and is then treated with castor or sesame oil to preserve its quality, prevent insect attack and give it an attractive appearance. The yield of '*dhal*' is about 66% by the dry method and 80% by the wet method.

Redgram is susceptible to insect infestation during storage and periodic fumigation of the storage chamber with carbon bisulphide, phosphine, ethylene dibromide or methyl bromide is recommended. Slurry treatment of the seeds with carbendazim @ 2g using 5 ml of water kg^{-1} of seed (or) dress the seeds with halogen mixture (Pure $CaOCl_2$ + $CaCO_3$ + leaf powder (*arappu*) of *Albizzia amara* at (5:4:1) ratio) @ 3g/kg of seed (or) Treat the seeds with neem oil/ groundnut oil or leaf powder of tobacco / notchi / neem / *Albizzia amara* (arappu) or fruit rind powder of *Sapindus laurifolius* (Poochi kottai) or *Acacia concinna* (Soapnut powder) @ 1:100 ratio. Treat the seeds with turmeric rhizome powder (or) neem leaf powder @ 1:50 powder to seed ratio against bruchid infestation as eco-friendly seed treatment.

Seed Storage: The seeds can be stored in gunny or cloth bags for short term storage (8-9 months) with seed moisture content of 8 to 9% in polylined gunny bag for medium term storage (12-15 months) with seed moisture content of 8 to 9 % and in 700 gauge polythene bag for long term storage (more than15 months) with seed moisture content of less than 8%. Many factors such as seed moisture, relative humidity, temperature and infestation by stored grain pests influence the viability of seed during storage and reduce the quality of seed. The seeds damaged by bruchids do not germinate well resulting in poor plant stand and consequently yield, and economic loss. The seed can also be solarized for couple of days before storing them in shade. High temperatures (~65°C) in polythene bags due to sunrays will kill any living insect pest. Use new gunny bags lined with polythene to store seed in them. In case of old bags, disinfect them with 0.1% malathion 50 EC or with fenvelrate 20 EC. Dip the old gunny bags in this solution for 10-15 minutes and dry properly in shade before storing seed. Grain earmarked for sowing should be mixed with 5% malathion dust at 250 g 100-kg-^{1} seed. In case of insect attack, fumigate the seed in store with aluminium phosphide (30g celphos t-^{1} seed or 7-10 tablets of celphos 28 m-^{3}) for a period of one week. Seed can also be treated with 7.5 ml rapeseed oil or groundnut oil per kg of seed. By this way, the seed can be kept safe for 8-9 months.

Utilization: Pigeon pea is cultivated either as a food crop (dried pea or vegetable pea), cover crop, forage crop, green manure crop, nurse crop, a windbreak hedge, as a host for lac insects, and as food for silk worms. When it is used as pulse, the dried seeds are used as human food. When used as a vegetable, redgram should be harvested when the seeds are fully grown but just before they lose their green colour. The stalks are used for fuel, thatch, and basketry. Redgram is mainly eaten in the form of split pulse as 'dal' or'dhal'. The outer covering of its seed together with part of the kernel provides a valuable feed for milch cattle. The husk of pods and leaves obtained during threshing constitute a valuable cattle feed.

Transplanting in redgram

Seedling preparation: Seed rate is 3.0 kg/ha to prepare red gram plants for one hectare land. To raise the seedlings, polythene bags of size 15 cm × 10 cm (l × b) with 200 μ guage should be used. At lower side of bags three-four holes are to be opened for draining of excess water. Polythene bags can be filled with FYM / Vermicompost, Soil and Sand in 1:1:1 ratio. Sow the seeds @ 2/poly bag at 1 cm depth and keep the bags under partial shade and watering should be done at regular intervals. The age of seedlings for transplanting is 30 to 40 days. Plant hardening is essential before transplanting. Seedlings can also be raised in the field to avoid transportation cost.

Transplanting of seedling in main field: Under water logging condition, form furrows before digging pits. Apply inorganic fertilizers @ 25:50:25 kg NPK /ha at 20-30 days after planting as urea, DAP and potash around the seedlings

After deep ploughing and tilling, pits of 15 sq cm size should be opened at 5 × 3 ft in deep-medium soils (2904 plants per acre) for sole crop under irrigated condition or 6 × 3 ft in deep soils (2420 plants per acre) for inter crop under irrigated condition. And 5 × 3 ft in deep soil (2904 plants per acre) for sole crop under *rainfed* condition. Fill the pits with soil and compost in equal proportion before 15 days of transplanting.

Transplant one seedling per pit. In intercrops, prior to transplanting sow the crops (black gram/green gram/soybean) then red gram plants have to be transplanted into pits (as said above). By opening ridges and furrows with 5 and 6 ft interval in sole and intercrop, respectively. Separate the bag before transplanting. Transplanting is done with 25-30 days old polybag seedlings in pits prepared using power tiller operated post hole auger. Irrigation is given immediately after transplanting. The advantages of transplanting the seedlings are viz., i) reduced main field duration, ii) adequate plant population and iii) increased yield per unit area.

Lay out for drip fertigation: The recommended lateral spacing is 150 cm and plant spacing is 90 x 60 cm for long duration for drip fertigation while the lateral spacing is 100 cm and plant spacing is 60 × 30 cm for short duration. Power tiller operated heavy duty auger digger can be used to dig holes for planting seedlings.

Nipping (removal of top 5 cm) the plants may be done on 20 DAT which is equivalent to the seedling age of 50 DAS to arrest the terminal growth and to enhance branching in the plant.

Manures and fertilizer application: Farm yard manure / Compost @ 12.5 ton/ha or vermicompost @ 5.0 ton/ha should be added three weeks before transplanting. By this, water holding capacity and microbial population increases due to which nitrogen fixation will also occur at faster rate. Application of zinc sulphate@ 5 kg/ha mixed with farm yard manure @ 100 kg/ha / vermicompost @ 50kg/ha is recommended to control zinc deficiency while gypsum @ 200 kg/ha to overcome sulphur deficiency. DAP and MOP application should be made at 20-30 DAT in ring method. Soil application of sulphur @ 20kg/ha through SSP, gypsum or pyrite enhances seed yield.

Fertigation schedule (Drip irrigation once in 3 days at 100 % CPE) is given on 14, 21, 28, 35, 42, 49, 56, 63, 77, 84, 91, 98, 105, 112, 119 DAS. Bio-fertigation using liquid biofertilizer, rhizobium and azophosmet each@ 750 ml /ha, TNAU pulses tonic @ 1 % once at peak flowering and humic acid @ 2 litres per irrigation and sea weeds extract @ 2.5 lit can be given for first five irrigations. The fertigation schedule for redgram 100 % dose with normal fertilizers and with water soluble fertilizers are furnished in Table 11 and Table 12.

Table 11 Fertigation schedule for redgram 100 % dose with normal fertilizers

Stage	Duration	Fertilizer Form	Fertilizer grade			Dose / ha /day	Total Qty kg/ ha	Nutrients (kg/ha)		
			N	P	K			N	P	K
15-45 DAT	30	MAP	12	61	0	1.57	47	5.64	28.67	0
		UREA	46	0	0	0.33	10	4.6	0	0
46-75	30	MAP	12	61	0	1.17	35	4.2	21.35	0
		UREA	46	0	0	0.33	10	4.6	0	0
76-105	30	SOP	0	0	50	0.83	25	0	0	12.5
		UREA	46	0	0	0.33	10	4.6	0	0
106-135	30	SOP	0	0	50	0.83	25	0	0	12.5
	30	UREA	46	0	0	0.17	5	2.3	0	0

Table 12 Fertigation schedule for redgram 100 % dose with water soluble fertilizers

Stage	Duration	Fertilizer Form	Fertilizer grade			Dose / ha / day	Total Qty kg/ha	Nutrients (kg/ha)		
			N	P	K			N	P	K
15-45 DAT	30	MAP	12	61	0	1.33	40	4.8	24.4	0
		UREA	46	0	0	0.33	10	4.6	0	0
46-75	30	MAP	12	61	0	0.90	27	3.24	16.47	0
		PF	19	19	19	0.83	25	4.75	4.75	4.75
76-105	30	PF	19	19	19	0.83	25	4.75	4.75	4.75
		SOP	0	0	50	0.60	18	0	0	9
	30	UREA	46	0	0	0.27	8	3.68	0	0
106-135	30	SOP	0	0	50	0.47	14	0	0	7
		UREA	46	0	0	0.17	5	2.3	0	0

Yield: The average redgram yield is 700 kg/ha. The achievable redgram yield under drip fertigation is 1500 to 1750 kg/ha while the per plant yield ranges from 125 to 250 grams depending on the duration of varieties.

REFERENCES

Center for New Crops and Plants Products. 2002. *Cajanus cajan* (L.) Millsp. Purdue University. https://hort.purdue.edu/newcrop/duke_energy/Cajanus_cajun.html

Faujdar Singh and D.L. Oswalt. 1992. Pigeonpea Botany and Production Practices. International Crops Research Institute for the Semi-Arid Tropics. Patancheru, Andhra Pradesh 502 324.

Morton, J.F. 1976. The pigeon pea (Cajanus cajan Millsp.), a high protein tropical bush legume. HortScience 11(1):11–19

Mullen, C.L., Holland, J.F. and L. Heuke. 2003. Cowpea, lablab, and pigeon pea. Agfact P4.2.21. NSW Agriculture, Orange, New South Wales.
http://www.dpi.nsw.gov.au/__data/assets/pdf_file/0006/157488/cowpea-lablab-pigeon-pea.pdf

Nene, Y.L. and V.K. Sheila. 1990. Pigeonpea: geography and importance. In: Y.L. Nene, S.H. Hall, and V.K. Sheila. The Pigeonpea. CAB International, Wellingford, UK. p. 1-14.

Shanower, T.G., and J. Romeis. 1999. Insect pests of pigeonpea and their management. Annual Review of Entomology 44: 77-96

Sheahan, C.M. 2012. Plant guide for pigeonpea (Cajanus cajan). USDA-Natural Resources Conservation Service, Cape May Plant Materials Center. Cape May, NJ. 08210.

Skerman, P.J., Cameron, D.G. and F. Riveros. 1988. Tropical forage legumes. FAO Plant Production and Protection Series 2. Food and Agriculture Organization of the United Nations. 692 p.

Smartt, J. 1976. Tropical pulses. Longman Group Limited, London. 348 p.

Stevens, J.M. 2012. Pea, pigeon- *Cajanus cajan*. Institute of Food and Agricultural Sciences (IFAS), Univ. of FL. Extension. Publication #HS641.
http://edis.ifas.ufl.edu/mv108#

Sundaraj, D.D., and G. Thulasidas. 1980. Botany of field crops. The Macmillan Company of India Ltd. New Delhi. 496 pp .

Thorp, S., Davison, M., Frost, O. and M. Pickstock. 2012. New agriculturist: the potential of pigeonpea. Wren Media Ltd., Suffolk, UK. http://www.new-ag.info/99-5/focuson/focuson3.html

van der Maesen, L.J.G. 1990. Pigeonpea: origin, history, evolution, and taxonomy. In: Y.L. Nene, S.H. Hill, and V.K. Sheila. The Pigeonpea. CAB International. Wellingford, UK. p. 15-46.

CHAPTER

20

Chickpea–*Cicer arietinum* L. 2n=14 for *Desi and* 2n=16 for *Kabuli*

Family: Fabaceae (Leguminoceae)

Vernacular name: Chickpea, Bengal gram, Garbanzo bean, Garbanzo Garbarizos, Bengal gram or Gram (English), *Chana* (Hindi), *Kondai kadalai* (Tamil), *Chhloe* (Punjabi), *Chola* (Bengali), *Kadala* (Malayalam), *Kadale* (Kannada), Garbanzo (Spanish), Pois chiche (French), Kichar, Chicher (German) and Nakhut, Nnohut (Turkey, Afghanistan).

Importance: The scientific name *Cicer arietinum* has been derived from the Roman word, '*cicer*' owing to the resemblance of the seed to the head of a '*ram*' and the word '*arietinum*' derives its name from '*aries*' meaning 'ram'. In India about 75% of chickpea is consumed as dhal. Chickpea seeds are sometimes sprouted similarly to mung beans and used as a Vitamin C rich vegetable. The young fresh plant shoots are sometimes used as a vegetable similar to spinach in Asia. Chickpea is exported to USA, U.K, Canada, Saudi Arab, UAE, Srilanka, Malaysia, etc.

Origin: Chickpea is known to have originated in Western Asia (Asia Minor/Turkey) and then dispersed in two directions. Large seeded types (weighing more than 26 g/100 seeds) are Owl-head shaped or rounded chickpeas of white salmon colour have developed in the west along Mediterranean region whereas small seeded types (15 g/100 seeds) are of irregular shape and of various colours like yellow, brown, black, green, etc., probably have dispersed in eastern and southern parts of Asia including India. In the Indian subcontinent, large seeded chickpeas of white salmon colour are popularly referred to as *kabuli* types while the small seeded cultivars of different colours are known as *desi* type. Both *kabuli* and *desi* type chickpea are indeterminate growth habit.

Distribution: India is the premier chickpea growing country and it accounts for 76% of the total area and 75% production in the world. Other important chickpea producing countries are Pakistan, Turkey, Mexico, Burma and Ethopia, Iran, Greece and Portugal. In India the major chickpea growing states are Rajasthan, Madhya Pradesh, Uttar Pradesh and

Haryana, accounting 80% of acreage and 84% of total production in India. Other chickpea growing states are Madhya Pradesh, Uttar Pradesh, Rajasthan, Punjab, Maharashtra, Bihar, Karnataka, West Bengal, Gujarat and Andhra Pradesh.

Botany: Chickpea is an erect or spreading, much branched, annual herb, 25 to 50 cm tall, all parts covered with glandular hairs. It has a tap root system. Bengal gram has imparipinnate leaves, serrated leaf margin, presence of oil glands throughout the plant body, flowers axillary, solitary cyme, inflated pods having one to two seeds only and branching from the base of the stem (inverted umbrella shape). Chickpea flowers are complete and bisexual, and have papilionaceous corolla. They are white, pink, purple or blue in colour. Chickpea is naturally self-pollinated species with less than 2% natural out crossing. Self pollination takes place one or two days before opening of the flower. The flowers open in the forenoon and close in the afternoon. Every flower opens on two to three successive days. Chickpea stigma is receptive for 36 hours commencing 24 hours before opening and pollen grains remained viable for 8.5 hours at 27.8°C at 95% relative humidity and for nine days at 7.7°C at zero percent relative humidity. Half open flower is the best source of mature pollen. Lowest flowers open first with usually one open on each branch. Flowering continues on each plant for more than a month. On cloudy and wet days, little pollination takes place and empty pods results. Seeds are angular, 0.5 to 1.0 cm in diameter with pointed beak and small hilum. Germination is hypogeal. The number of pods per plant varies between 30 and 150, depending on the environmental conditions and the genotype. The number of seeds per pod ranges from one to two, with the maximum being three. The seeds are ram head or owl's-head shaped, and the surface may be smooth or wrinkled. The seed mass varies from 0.10 to 0.75 g/seed. Based on seed size and colour, cultivated chickpeas are of two types viz.,

(i) Macrosperma (*kabuli* type or white gram): The seeds of this type are large (100-seed mass >25-40 g), round or ramhead, and cream-coloured. The plant is medium to tall in height, with large leaflets and white flowers, and contain no anthocyanin.

(ii) Microsperma (*desi* type or brown gram): The seeds of this type are small (100-seed mass >17-25 g),and angular in shape. The seed colour varies from cream, black, brown, yellow to green. There are 2-3 ovules per pod but on an average 1-2 seeds per pod are produced. The plants are short with small leaflets and purplish flowers, and contain anthocyanin. The desi types account for 80-85% of chickpea area. The splits (*dal*) and flour (*besan*) are invariably made from desi type.

The characteristic feature of desi and kabuli types are presented in Table 1:

Table 1 Chickpea ecotypes

Characters	Desi type	Kabuli type
Plant character	Plant short	Plant tall
Leaves	leaflet small in size, pale green / dark green leaves, flowers purple or white in colour	leaflet large in size, light green leaves, flowers white in colour
Seeds	Small, irregular shape seeds of varying colours	Bold seeds, smooth, cream to white in colour
Seed coat	Seed coat wrinkled	Seed coat somewhat smooth

Nodulation pattern: Plants nodulated within 14 days from sowing. Nodules will form within 15 to 21 days after sowing under rainfed cool season. In short days nodule growth continues until 45 days. Nitrogenase activity starts 4 days after the appearance of first nodules. Nitrogenase activity per plant increases until 45 days for *desi* varieties and 65 days for *kabuli* varieties. The decline in nitrogenase activity is assoicated with flowering which begins about 30 days after sowing. Nodule efficiencies decline when nodlues are less than 14 days old. N_2 fixation is maximum between 24ºC and 33ºC. Pod filling is well advanced and leaves are senescing on 70 days after sowing. Rapid senescence of chickpea nodules soon after flowering will result in low nitrogen fixation during pod filling. At that time the plant requires rapid nitrogen uptake during which the remobilization of already fixed nitrogen within the plant may not satisfy. Hence, foliar application of DAP @ 2% can be given to meet nitrogen and phosphorus requirement during pod filling stage. When root rot sets in 45 days remove all the affected plants and spray the remaining plants with aqueous solution of urea (2%), jaggery (2%) and bavistin (0.1%) immediately and again 10 days afterwards.

Climate: Chickpea is a temperate crop, but is grown more in the subtropical areas of the world. It is also cultivated tropical areas where it is grown under cool temperatures. Chickpea is grown within areas having 600 to 1500 mm average annual rainfall. Among the cool season grain legumes, chickpea is believed to be the most drought resistant. Both *kabuli* and *desi* type chickpea are long day plants. Germination and emergence are tolerant of a broad range of temperatures from 10 to 35ºC. Chickpea seeds germinate at an optimum temperature (28-33°C) and moisture level in about 5-6 days. Emergence falls from 35 to 55% at 10ºC and 40ºC respectively. Faster germination occurs at a temperature range of 31.8°C to 33°C. The optimum temperature for seedling growth is 25 to 27ºC which confirms with a broad plateau of photosynthetic rates from 19 to 28ºC. An increase in temperature of 1ºC will reduce the time to 50% emergence by 0.26 to 0.33 days. Root development is slow below about 20ºC. Nitrogen fixation changes little from 15 to 25ºC but is low at 30ºC. Nodules do not form at 33ºC and die at 35ºC. In north India, maximum and minimum temperature at flower formation is 25 to 31ºC and 10 to 14ºC respectively and day light is usually 11.5 to 12.5 hours. Temperatures below 15ºC cause low fruit set whereas high temperature of 35ºC increase flower shedding and pod shedding which account for 20% and 20 to 40% of the flowers and pods formed respectively. The duration of grain filling decreases by 4 to 6 days for 1ºC increase in temperature from 15 to 25ºC.

Soil: Chickpea is grown in a wide range of soil textural classes. Medium to heavy textured soils are preferred. It is grown in *Vertisols* of Maharashtra, and Decan plateau and *Inceptisol* of upper Ganges basin. Heavy soils are favoured in the wet and dry climate regions of India because of their high residual moisture content in the post monsoon growing period. It is grown sandy loams in western parts of the Punjab, Haryana and Rajasthan. A soil pH of 6 to 9 is favourable for growth of chickpea. Growth and yield is affected under sodic soils. The crop is susceptible due to poor aeration which is caused by soil compaction, heavy soil texture and excess water table. Under these conditions the incidence of wilt diseases may increase. Prepare the field to fine tilth in alluvial soil and coarse tilth in *Vertisol* is preferred since a fine tilth may reduce germination. The crop is sometimes also sown directly into moist uncultivated soil following wet rice. Extremely high fertility leads to luxuriant vegetative growth and may depress grain yield through poor pod set.

Season: Chickpea is grown as a cold weather (*rabi*/winter season) crop both in north and south India. Seeds are sown during October. The sowing time in different states vary as furnished in Table 2.

Table 2 Sowing time of chickpea in different states of India

State	Rainfed	Irrigated
Punjab and Haryana	Second to third week of October	Fourth week of October to 15 November
Rajasthan	First to 15 October	15-30 October in south and south-eastern Rajasthan; 25 October-15 November in Ganganagar district
Bihar and Gujarat	15 October to first week of November	First fortnight of November
Madhya Pradesh Maharashtra, Uttar Pradesh, Karnataka, Jammu-Kashmir and Tamil Nadu	First fortnight of October	End of October to first week of November

Varieties: The Chickpea varieties grown in India are furnished in Table 3 to Table 5.

Table 3 Chickpea varieties released for different states of India

State	Varieties
Andhra Pradesh	Bharati (ICCV 10), JG 11, Kranti (ICCV 37), Phule G 95311 (K), Sweta (ICCV 2), MNK 1
Assam	JG 74, KWR 108, Radhey, Udai (KPG 59), Pusa 372
Bihar	Gujarat Gram 4, KWR 108, Pusa 256 (BG 256), Pusa 372 (BG 372), Pusa 1003(BG 1003)(K), Udai (KPG 59), RAU 52
Chhattisgarth	JG 16, JG 315, Jawahar Gram Kabuli 1 (JGK 1), KAK 2, Pusa 372(BG 372), Pusa Chamatkar (K) (BG 1053), RSG 888, Udai (KPG 59)
Delhi	Alok (KGD 1168),DCP 92-3, Karnal Chana 1 (CSG 8962), Pusa 372 (BG 372), Pusa Chamatkar (K) (BG 1053), RSG 888, Udai(KPG 59)
Gujarat	Dharwad Pragati (BGD 72), Gujarat Gram 1, Gujarat Gram 2, JG 16 (SAKI 9516), Jawahar Gram Kabuli 1 (JGK 1), Pusa 372 (BG 372), Pusa 391 (BG 391), Vikas, Vijay, Vishal, Vishwas
Haryana	Alok (KGD 1168), DCP 92-3, GPF 2, GNG 1581, Haryana chana 1, Haryana Kabuli Chana 1, Haryana Kabuli Chana 2, Karnal Chana 1, PBG 1, Pusa 329 (BG 329), Pusa 362 (BG 362), Pusa 372 (BG 372), Pusa 547 (BG 547), Pusa Chamatkar (K) (BG 1053), Rajas, RSG 888, RSG 963, Samrat (GNG 469), Udai (KPG 59), Vardan
Jharkhand	Birsa Chana 1, Gujarat Grant 4, KWR 108, Pusa 256 (BG 256), Pusa 372 (BG 372), Pusa kabuli 1003 (BG 1003), Udai (KPG 59)
Karnataka	Annegeri 1, Bharati (ICCV 10), Chaffa, JG 11, Phule G 95311 (K), Sweta (ICCV 2)
Madhya Pradesh	Dharwad Pragati, Gujarat Gram 1, JG 16 (SAKI 9516), JG 130, JG 218, JG 315, JG 322, JG 74, JGG 1, Jawahar Gram Kabuli 1 (JGK 1), Pusa 391(BG 391), Vishwas (Phule G 5), Vijay, Vishal
Maharashtra	Annegeri 1, Dharwad Pragati, Gujarat Gram 1, Jawahar Gram Kabuli 1 (JGK 1), JG 16 (SAKI 9516), KAK 2(K) (PKV 1), Pusa 372 (BG 372), Pusa 391 (BG 391), Sweta (ICCV 2), Phule G 12, Vihar (K), Vijay, Vikas, Virat (K), Vishal, Vishwas (Phule G 5)

Punjab	Alok (KGD 1168), DCP 92-3, GPF 2, GNG 1581, Hare Chhole 1, Haryana Chana 1, Karnal Chana 1, L 551, PBG 3, PBG 5, Pusa 256 (BG 256), Pusa 329 (BG 329), Pusa 362 (BG 362), Pusa 372 BG(372), Pusa 547 (BG 547), Pusa Chamatkar (K) (BG 1053), Rajas, RSG 963, Samrat, Udai (KPG 59), Vardan
Rajasthan	Abha, Abhar, Abhilasha, Akash, Ankur, Anvita, aparna, Arpan, Arpita, Aruna, Asar, Asha, Alok (KGD 1168), GNG 1581, GPF 2, Harayana Chana 1, Karnal Chana 1, PBG 1, Pusa 329 (BG 329), Pusa 362 (BG 362), Pusa 372 (BG 372), Pusa 547 (BG 547), Pusa Chamatkar (K) (BG 1053), RSG 44, RSG 888, RSG 963, Rajas, Samrat, Udai (KPG 59), Vardan
Uttar Pradesh	Alok (KGD 1168), DCP 92-3, GPF 2, GNG 1581, Gujarat Gram 4, Haryana Chana 1, JG 16 (SAKI 9516), JG 315, Karnal Chana 1, KWR 108, Pant G 186, Pragati, Pusa 256 (BG 256), Pusa 372 (BG 372), Pusa 547 (BG 547), Pusa kabuli 1003 *(BG 1003), Rajas, Sadabahar, Sadbhawana, Samrat (GNG 469), Surya, Udai (KPG 59), Vardan, Vishwas,
West Bengal	Anuradha, JG 74, KWR 108, Mahamaya 1 (B 105), Mahamaya 2 (B 108), Pusa 256 (BG 256)
Tamil Nadu	Bharti (ICCV 10), JG 11, Phule G 95311, Co 2, Co 3, Co 4

Table 4 Chickpea varietal features grown in India

Variety name	Yield (q/ha)	Suitability	Major characteristics
Pusa 256 (BG 256)	18-20	NWPZ, NEPZ, CZ	Medium tall, semi-spreading plants, attractive and bold seeds, tolerant to wilt and *Ascochyta* blight
Pusa 372 (BG 372)	18-22	NWPZ, NEPZ, CZ	semi-spreading plants, moderately resistant to wilt, *Ascochyta* blight and root rot, small seeds
RSG 888	21.0	NWPZ	Semi – Spreading, small seeded, drought tolerant, twin podded genotype, tolerant to dry root rot
Gujarat Gram 4 (GCP 105)	18-20	NEPZ	Medium tall, semi-erect, tolerant to wilt, pinkish flower, smooth, round and brown seeds
HK 2 (HK 94-134)	22	NEPZ	Tolerant to wilt and root diseases, suitable for irrigated areas.
Pusa Kabuli 1003 (BG 1003:K)	18	NEPZ	White large seeds, tolerant to wilt, medium tall
Uday (KPG 59)	20	NWPZ, NEPZ	Semi-erect, medium tall plants, small seeded, tolerant to wilt, root rot, stunt and pod borer
Ganguar (GNG 1581)	23-24	NWPZ	Medium height (= 50cm)with semi erect plant type, tolerant against diseases like wilt, *Ascochyta* Blight, Stunt and root rot and possesses a good level of protein (21.88 percent) also.
Rajas(Phule G 9425 -9)	19	NWPZ	------
Pusa 547 (BGM 547)	18	NWPZ	------
Anvita (RSG – 931)	20	NWPZ	Double podded high yielding chickpea variety, resistance to dry root rot, wilt and nematodes in addition to less damage by pod borer

Karnal Chana 1 (CSG 8962)	20-22	NWPZ	Tolerant to mild salinity, resistant to wilt, small brown seeds
DCP 92-3	19	NWPZ	Genotype is semi erect, me dium tall, lodging resistant, wilt resistant Seeds yellowish brown and medium large
Samrat (GNG 469)	21	NWPZ	Plant erect and tall, seeds brown and large, tolerant to wilt and resistant to *Ascochyta* blight
KER 108	20-23	NWPZ, NEPZ	Medium tall and semi-erect plants, seeds are brown and large
Pusa 362	24	NWPZ	Medium tall, tolerant to wilt and large seeds
Pusa C hamatkar (BG 1053)	17	NWPZ	Large seeded kabuli, tolerant to wilt
Aadhar (RSG-963)	20	NEPZ	Although it was released for late sown condition but performs well under normal sown as well as rainfed condition. It has good level of resistance to wilt, dry root rot and pod borer.
Pusa Shubhra (BGD 128)	18	CZ	......
SAKI 9516 (JG 16)	20.0	CZ	Medium seed size, resistant to wilt, tolerant to collar rot, BGM and stunt
Gujarat Gram 1 (GCP 101)	18	CZ	Medium tall, semi – erect plants, resistant to wilt, medium large and dark brown seeds
BGD 72	18	CZ	Semi-erect plants, medium large seeds, tolerant to wilt
Pusa 391 (BG 391)	17-18	CZ	Seeds are large and dark brown, moderately resistant to wilt and root rot
Vijay (Phule G 81-1-1)	20	CZ	Resistant to wilt, tolerant to low moisture stress, small brown seeds
JAKI 92-18	18	CZ	Semi-spreading, bold seeded and resistant to wilt
Digvijay	16-18	Maharashtra	Yellowish brown large seeds, wilt resistant
KAK 2 (PKV kabuli 2)	18	CZ	Semi-spreading plants, large and white seeded
IPCK 02-29 (Shubra kabuli)	19-22	Central India	Moderately resistant to F.udum, medium tall and typically erect, medium maturity duration, seeds are large (34-35 g/100 seed weight).
JG 11	15	SZ	Semi- spreading, large seeds, resistant to wilt and moderately resistant to root rot
Vihar (Phule G 95311)	18.0	SZ	Large seeded kabuli, wilt resistant
Co 2	9.8 (R)	Tamil Nadu	Crop duration is 90 days. Hundred seed weight is 14-15 g.
Co 3	10.0 (R)	Tamil Nadu	Crop duration is 85 days. Hundred seed weight is 30-32 g.

Note: CZ-Central Zone, SZ –South Zone, NEPZ – North Eastern Plane Zone, NWPZ- North Western Plain Zone

Table 5 List of moderatery resistant chickpea genotypes against pests and disease

Pest/Disease	Tolerant/resistance genotype
Vascular wilt	Avrodhi, BG 212, KPG 59, BGD 72, Pusa 391, Vijay, Vishal, Rajas, BG 256, Gujrat Gram, GNG1581, GNG 1292,Virat, GJ 3, RSG88, JG315, JG74, JG11, JG322, JG63, JG12, JG14, JG16, JGK1, JGK2, JGK3, KAK2, Subhra BG 1053, HK2, JAK1 9218, Phule G9531, RSG 931, RSG 963, CSG 8962, DCP 92-3, JSC 55, JSC56, HK 05-169
Dry root rot	JSC 37, JG 25174, CSJ 556, MPJG 89- 1155, MPJG 89- 9023, CSJ 592, Rajas, JS 2000-07, ICCC 32, GL769
Ascochyta blight	C235, GS43,CG558, Himachal channa1, Himachal channa 2, IPC 08-11, 23094, GNG 1581, GPF2, PBG5, Gaurav (H75-35), BG 267, Vardhan, Samrat
Botrytis grey mould	BG 276, GL90159, GL 9-1071, GL92162, HK94-134, IPCK 2004- 29, ICC38, ICC 202, ICC1069, IPC 2004-52, IPC 2000-06, NDG 10-11, Phule G 07112

Land preparation: To reduce the incidence of soil borne diseases, one deep ploughing during summer is recommended. To prepare field for sowing, one harrowing after rainy season followed by 2 ploughing with cultivator is sufficient. At the time of field preparation, Methyle parathion dust @ 20-25 kg/ha should be applied to control the infestation of termites. Chick pea is highly sensitive to soil aeration. This imposes a restriction for its cultivation on heavy soils and calls for special care in seedbed preparation. Very fine and compact seedbed is not good for chickpea. It requires a loose and well aerated seedbed.

Seeds and sowing: The seed rate is 45 kg per hectare for small seeded desi type and 90 kg per hectare for large seeded *kabuli* type. Soak the seeds in 1% aqueous solution of KH_2PO_4 for 3 to 4 hours at one third volume of seeds and quickly air dry in shade. Treat the seeds with carbendazim or Thiram 2 g per kg of seeds or *Trichoderma virde* or *Pseudomonos fluoresecens* at 4 g per kg of seed. Treat the seeds 24 hours before sowing. Then treat the seeds with rhizobial culture. Fungicide treated seeds should be again treated with bacterial culture. There should be an interval of atleast 24 hours after fungicidal treatment for giving the bacterial culture treatment. Three packets of rhizobial culture is sufficient for treating the seeds required for one ha. Use rice gruel as a binder. Dry the bacterial culture treated seeds in shade for 15 minutes before sowing. Dibble the seeds by adopting the spacing of 30 cm x 10 cm. Sowing depth is 5 cm.

Cropping systems: Chickpea based cropping systems include sorghum / pearl millet-chickpea, Pearl millet-gram (irrigated uplands), Sorghum-gram (irrigated uplands), Maize-gram (irrigated upland and lowland), Paddy/ cotton-chickpea, Paddy-gram (irrigated lowland), Clusterbean-chickpea, Maize-chickpea, Onion-chickpea, Groundnut – chickpea, Soyabean-chickpea and Kharif fallow-chickpea (in barani areas). The promising chickpea based intercropping systems are presented in Table 6.

Table 6 Chickpea based intercropping systems

Intercropping system	Row ratio
Chickpea + mustard	6:1
Chickpea+wheat	2:1
Chickpea+safflower/linseed	3:1
Chickpea+autumn sugarcane	2:1

Weed management: The critical weed free period for chickpea is 45 days from sowing. Spray fluchloralin 1.5 lit/ha or pendeimethalin 2.5 lit/ha as pre-emergence spray on third day after sowing followed by one hand weeding on 25 to 30 days after sowing. If herbicide is not applied, give two hand weeding on 15 and 30 days after sowing.

Manures and fertilizer application: FYM @ 12.5 t/ha with 12.5-25.0-0 kg N, P_2O_5 and K_2O per ha are applied as basal dressing under rainfed condition while 25-50-0 kg of N, P_2O_5 and K_2O per ha are applied as basal dressing under irrigated condition. Chickpea is very efficient in taking up phosphorus from low phosphorus soils but it can benefit from phosphorus applications. Placement of phosphatic fertilizers below the seed increases fertilizer use efficiency better than broadcast method. In case of zinc deficiency, spray 0.5% $ZnSO_4$ and 0.25% lime mixture.

Water management: The consumptive water use ranges from 250 to 400 mm for a seed yield of between 900 and 3000 kg/ha. It needs 15% soil moisture by volume in the root zone, extending as deep as 50 cm in sandy loam soil. The supplementary irrigation is recommended during the vegetative growth period of chickpea in light soils with a low water-holding capacity or during the latter period of vegetative growth and early pod-filling stages in heavy deep soils. Chickpea production can be increased by applying two irrigations at pre-flowering and pod-formation stages. Chickpea meets its water requirement from conserved moisture in deep soils (more than 150 cm depth). It responds well to irrigation in absence of winter rains or enough stored moisture. Irrigation at pod filling increased the proportion of effective pods. There is adverse effect of irrigation on growth in heavy soils due to of poor aeration. The crop does recover when the soil aeration is improved by tillage. Irrigation water with a conductivity of 10 dSm-1 can reduce chickpea yield by 50%.

Growth regulators: Seed soaking with 0.2% succinic acid, 0.5% KH_2PO_4, 0.05 % $CuSO_4$ for 5 to 6 hours increase seed yield. 0.2% Cycocel or 25 ppm Ascorbic acid or 5 ppm TIBA, Trichodobenzoic acid increase the pod and seed yield.

Nipping: Nipping terminal shoot on 20 and 30 DAS reduces the excessive growth, encourages branching and pod set and consequently yields. Nipping can also be attained through spraying of 75 ppm TIBA (Tri-iodo-benzoic acid). Nipping encourage early onset of flowering. If the growing buds are nipped during the pre-flowering stage, arrests excess vegetative growth, favours branching, increases leaf area index, pods and grain yield.

Crop protection: The major pests are gram pod borer, cut worm, termite, black bean aphid, white grub, semi looper, cutworm and tobacco caterpillar while the major disease are vascular wilt, dry root rot, collar rot, *Ascochyta* blight, Botrytis Grey mould, rust, and *Stemphylium* blight. The nematode affecting chickpea are root knot, reniform, root lesion and chickpea cyst nematode. The estimated loss (%) due to specific pests is 10-90% with gram pod borer, 5-30% with gram cutworm, 5-15% with termites, 5-10% with semilooper, 20-25% with wilt/root rot, 5-10% with *Ascochyta* blight and 5-10% with Botrytis grey mould.

Table 7 Economic Threshold Levels (ETLs)

Pest insects	**Stage of the crops**	**Economic threshold Levels (ETLS)**
Cut worm (*Agrotis ipsilon* Huflnagel and *Spodoptera exigua* Hubner)	Seedling stage	One larvae/ square meter under the soil near cut plant.
Termite (*Odontotermes obesus* or *Microtermes obesi*)	Seedling stage	5 damaged plants/sq.meter

White grub (*Phyllophaga implicita*)	Seedling stage	5 grubs/sq. meter
Gram pod borer (*Helicoverpa armigera*)	Vegetative/ reproductive	5 to 8 eggs or 2 early instar /10 plants or one mature larvae (more than 1cm in length)/10 plants or I meter row
Semilooper (*Autographa nigrisigna*)	Vegetative phase	2 larvae/10 plants
Wilt & root rot (*Fusarium oxysporum*)	Seedling/vegetative	5-10% plants infested
Rodents	Before podding	25 live burrows/ha
Nematodes	Vegetative phase	1-2 larvae/g of soil

Integrated pest management practices

- Seed treatment is with *Trichoderma viride* @ 4 -10 g/kg of seed.
- Incorporation of *Trichoderma viride* @ 5 kg/ha multiplied on decomposed FYM @100kg/ha under moist soil condition in wilt/root rot affected areas.
- Apply well decomposed FYM or Neem cake/ Mahua cake @ 500 kg/ha or nematode prone areas.
- Fix bird percher in field @ 20/ha. Bird perches should be removed just after maturity/ harvesting of the crop.
- Sow marigold as trap/ disease indicator crop on border or in between the crop rows.
- Spray crude NSKE 5% or Azadirachtin 0.03% (300 ppm) Neem oil based WSP @ 2500 to 5000 ml /ha at pre-flowering stage at 15 days interval.
- Spray *Bacillus thuringiensis* var. *kurstaki* (HD-1, serotype 3a, 3b or H-3a, 3b, Strain -52 or 0.5% WP serotype 3a, 3b, 3c, Strain DOR Bt-1) against Pod borer @1 (Kg/ha).
- Spray Beauveria bassiana 1% WP (Strain No: NBRI – 9947) @ 3 kg/ha
- Spray *Ha*NPV 2.0% AS (Strain No. IBH-17268 or Strain No. BIL/HV-9) @ 250 LE (POB 5x1011/ml)/ha + 0.5% Jaggery + 0.1 % fabric whiteners (tinopal, blue etc)/ ha on noticing 1st instar larvae or eggs of pod borer (3 sprays at weekly intervals in evening hours).
- Conserve *Campoletis*, lady bird beetles, *Chrysopa*, Stinkbugs, Reduviid bug, Predatory wasps and spiders by intercropping with coriander.
- For reducing wilt, treatment of seed with 1g Carbendazim + 2g Thiram or 4g of *Trichoderma viride* per kg of seed.
- Gram pod borer (*H. armigera*) can be managed with the use of sex pheromone trap to monitor moth population. The ETL is 1-2 larvae per m row length. The IPM strategies include, timely sowing to exploit host avoidance, intercropping with mustard, barley and linseed, use of trap crops like *Vicia sativa* and African giant marigold, application of Nuclear Polyhedrosis Virus (NPV) @ 250 LE/ha or *Bacillus thuringensis* (Bt) @ 1-1.5 kg/ha, erection of perches @ 20-30 perches/ha to attract insectivorous birds, spraying with 5% Neem Seed Kernel Extract (NSKE) or Achook @ 1.25 l/ha or 0.07% Endosulfan or 0.04% Monocrotophos or 0.004% Fenvelarate at 15-20 days interval.

Table 8 Chemical control of insects and diseases in Chickpea

Insecticides	**Insect**	**Dose per hectare**
Azadirachtin 0.03% (3000 ppm)	Pod borer	2500-5000 ml
Chlorpyriphos 20EC	Pod borer, Cut worm	2,500 ml
Chlorpyriphos 20 EC	Termite (seed treatment)	15- 30 ml/kg of seed
Quinalphos 25EC	Pod borer	1000 ml
Deltamethrin 2.8% EC	Pod borer	400 – 500 ml
Monocrotophos 36% SL	Pod borer	625ml
Ethion 50 EC	Pod borer	1000 – 1500 ml
Emamectin benzoate 5% SG	Pod borer	220 ml
Novaluron10% EC	Pod borer	700 ml
Chlorantraniliprole18.5%SC	Pod borer	125 ml
Fungicides	Diseases	Dose per hectare
Trichoderma viride	Wilt/Root rot	5 g/kg seed.

Seed treatment with *Trichoderma viride* @ 10g/kg of seeds helps to reduce root knot, lesion nematode and wilt problems also.

Rodents control in Chickpea: Lesser bandicoot: *Bandicota bengalensis* (Gray) (throughout India). Nibbles the germinating seeds and at maturity damages the pods/seeds

Apply bromadiolone @ 0.005% in ready to use form (wax blocks) or loose bait in packets near rodent burrows. Apply 2% Zinc phosphide poison baits when the rodent infestation is very high. Practice pre-baiting in case of ZNP poison baiting. Don't apply ZNP poisons more than one time in a crop season as rodents develop bait shyness to this poison.

Harvest: The plant is harvested at physiological maturity which occurs 35 to 40 days after anthesis and when 70 to 80% pods present creamy colour appearance or when all the pods are mature. The crop can be harvested when leaves turn reddish-brown, leaves start senescence and start shedding. Plants are either plucked out by hand or cut with sickle at the base of the plant by manual labour, using a sickle or by using a combine harvester. The crop is allowed to dry in sun on threshing floor for about 5-6 days. Thereafter threshing is done either by beating the plant with stick or by trampling by bullocks or with a thresher.

Yield: Desi type chickpea crop yields about 2.0-2.5 t grain/ha while kabuli types yield 2.5-3.0 t/ha under irrigated conditions. Under rainfed conditions, the crop yields are generally 30-50% of the irrigated crop. The seed must be properly dried before storage. The ideal seed moisture level is 10-12% for short-term storage (up to 8 months).

Utilization: The popped grain is known as '*pottu kadalai*' in Tamil while the broken seed as *dhal* (Hindi) and '*kadalai paruppu*' (Tamil). The flour is used in various food preparations. Sometimes whole mature seeds are soaked in water, boiled and then eaten. The whole seeds may be roasted in hot pans and eaten similarly to groundnuts. Broken seeds and the residue from dhal production may be used for livestock and poultry feeding. The straw left after the seeds have been harvested and is a valuable forage crop. The yield of chickpea dhal is about 75 to 84% from whole seed. Green pods and tender shoots used as vegetable. An acrid liquid from glandular hairs collected contains 94% malic acid and 6% oxalic acid. It is used as vinegar and medicine.

REFERENCES

Bishaw, Z. and van Gastel, A.J.G. 2007. Seed production of cool-season food legumes: faba bean, chickpea, and lentil. International Center for Agricultural Research in the Dry Areas, Aleppo, Syria

Chandrashekar, K. Om Gupta, Suhas Yelshetty, O. P. Sharma, Someshwar Bhagat, C. Chattopadhyay, Mukesh Sehgal, Arpana Kumari, N. Amaresan, S. N. Sushil, A. K. Sinha, Ram Asre, K. S. Kapoor, K. Satyagopal and P Jeyakumar. 2014. Integrated Pest Management for Chickpea. pp. 43. Directorate of Plant Protection, Quarantine & Storage CGO Complex, NH IV, Faridabad- 121001

Gaur, P.M., Tripathi, S., Gowda, C.L.L., Ranga Rao, G.V., Sharma, H.C., Pande, S. and Sharma, M. 2010. Chickpea Seed Production Manual. International Crops Research Institute for the Semi-Arid Tropics. Patancheru 502 324. 28 pp.

GOI. 2002. Marketable surplus and post harvest losses of gram in India. Directorate of marketing and inspection branch head office, Nagpur – 440001

IBPGR, ICRISAT, and ICARDA. 1993. Descriptors of chickpea (Cicer arietinum). IBPGR, Rome, Italy; ICRISAT, Patancheru, India; and ICARDA, Aleppo, Syria. 31 pp.

Joshi, PK., Parthasarathy Rao, P., Gowda, C.L.L., Jones, R.B, Silim, S.N., Saxena, K.B., and Jagdish Kumar. 200I. The world chickpea and pigeonpea economies: facts, trends, and outlook. International Crops Research Institute for the Semi-Arid Tropics. Patancheru 502 324, Andhra Pradesh, India: 68 pp.

http://legume.ipmpipe.org

http://wiki.bugwood.org/PIPE:Legume

http://www.apsnet.org/ - Compendium of Chickpea & Lentil

CHAPTER

21

Blackgram–*Vigna mungo* (L.) Hepper (2n=22)

Family: Fabaceae (Leguminoceae)

Vernacular name: Blackgram (English), *Urd bean, Urd* (Hindi), *Uzhundu* (Tamil), *Minumulu* (Telegu), *Uddu,* (Kannada) *maash* (Nepali), *uzhunnu* (Malayalam), *urdu bele* (Tulu), *biri dali* (Oriya), *adal* (Gujarati) and *mashkali* (Bengali).

Importance: In India, approximately 50% of the blackgram seeds is used for the production of dhal. It is sometimes grown as a forage or green manure crop. It differs from other pulses in its peculiarity of attaining a mucilaginous pasty character when soaked in water. In south India, it is consumed in variety of popular dishes like *vada, idli, dosa, halwa, imarti* in combination with other food grains.

Origin: Blackgram is native to the northeastern India-Myanmar region of Asia.

Distribution: Blackgram is cultivated in India, Thailand and Australia. In India it is grown in Madhya Pradesh, Uttar Pradesh, Punjab, Maharashtra, West Bengal, Andhra Pradesh, Tamil Nadu and Karnataka.

Botany: *Vigna mungo* is divided in to two sub species viz., *V. mungo* var. *niger* includes varieties which mature early and have bold and black seeds. *V. mungo* var. *viridis* includes varieties having small seeded late maturing types with brown, grey or olive-green seeds. Blackgram is a twining herb, annual, plant densely hairy and stem slightly ridged, covered with brown hairs, hairs pointing downwards; much branches from the base. The leaves are large, trifoliate and are also hairy, generally with a purplish tinge. The leaflets are 5-10 cm long and ovate. Inflorescence consists of a cluster of 5-6 flowers at the top. Inflorescence is an axillary raceme; fruit a legume densely hairy, seeds ovoid, generally black. The flowers are axillary and are yellow in colour. The pods are cylindrical, erect, separate between the seeds, about 4 to 6 centimeters in length and covered with long ferruginous hairs. There are four to ten seeds in a pod. The seeds are black or very dark brown, oval, and the cotyledons are white. The flowers of black gram start opening early in the morning and are completely open between 7 a.m. and 8 a.m. Self-pollination is the rule.

Climate: Blackgram is essentially a tropical crop. It is grown from 30ºS to 30ºN latitude with an altitude up to 1500 m from sea level. It is resistant to high temperature of

25 to 35°C. Prolonged cloudy weather is detrimental to growth. It cannot withstand frost. It is grown in areas with an average annual rainfall of less than 900 mm. It is relatively drought tolerant crop. Rain at flowering time has a very adverse effect on seed yield. It is a short day plant. It is grown in areas receiving annual rainfall of 900 mm.

Soils: It is grown well on black cotton soil, red and alluvial soils. Heavy clayey loam soils that are water retentive are preferred for raising rice fallow crops.

Seasons: Blackgram can be grown during June-August and September-October under rainfed condition while it can be raised as irrigated crop during February-March. Rice fallows crop can be taken during January.

Varieties: The particulars of blackgram varieties grown in India are furnished in Table 1 to Table 3.

Table 1 Blackgram varieties grown in different states of India

Variety Name	Yield (q/ha)	Suitability	Major characteristics
NDU 99-2	10-12	NHZ	Resistant to MYMV, suitable for kharif season
Pant Urd 31	10-12	Uttarakhand and Himachal Pradesh	Main stem bearing, resistant to MYMV, suitable for intercropping in kharif and spring season
Pant Urd 40	10-12	For Plain and lower hill of Uttarakhand	Main stem bearing, resistant to MYMV, suitable for kharif season
Uttara	12-13	NWPZ and NEPZ	Resistant to MYMV, suitable for kharif season
Ujala	8-10	Orissa	Resistant to MYMV and CLS, suitable for kharif and rabi season
KU 309	10-12	U.P.	Tolerant to MYMV
Birsa Urd 1	10-12	Bihar and Jharkhand	Tolerant to MYMV
WBU 108	10-12	NWPZ, SZ and NEPZ	Resistant to MYMV
Narendra Urd 1	10-12	U.P.	Black, large seed, resistant to MYMV
WBU 109	10-12	NEPZ	Resistant to MYMV, suitable for spring season
KU 91-2	10-11	NEPZ	Resistant to MYMV, suitable for spring season
Mash 338	8-10	Punjab	Tolerant to MYMV
Mash 414	9-10	Punjab	Resistant to MYMV, suitable for spring season
Mash 1008	10-12	Punjab	Resistant to MYMV, suitable for spring season
KU 300	10-12	NWPZ	Resistant to MYMV, suitable for spring season
KU 96-3	8-10	CZ	Resistant to YMV, suitable for kharif season
AKU 15	10-12	Maharashtra	Tolerant to PM
GU 1	10-12	Gujarat	Tolerant to PM and CLS
Barkha	9-10	CZ	Large seed, tolerant to CLS
AKU 4	10-11	Maharashtra	Tolerant to MYMV and PM, suitable for rabi season

TAU 2	10-11	Maharashtra	Medium large seed, tolerant to PM
TPU 4	7-8	CZ	Erect, medium large seed
Jawahar Urd 3	10-12	M.P.	Medium seed size, tolerant to MYMV and CLS
TAU 1	12-14	Vidarbha region of Maharashtra	Purplish black, large seed, tolerant to powdery mildew
IPU 2-43	10-12	SZ	Resistant to MYMV and PM, suitable for kharif season
Vamban 5	8-10	Tamil Nadu	Resistant to PM and MYMV
LBG 625 (Rashmi)	10-12	A.P.	Suitable for rice fallows
Vamban 4	10-12	Tamil Nadu	Tolerant to MYMV and PM
Vamban 3	8-10	Tamil Nadu	Dull black seed, resistant to MYMV and tolerant to Powdery mildew
WBG 26	9-10	SZ	Resistant to PM
LBG 685	10-12	A.P.	Resistant to wilt
KU 301	10-12	SZ	Resistant to MYMV and tolerant to PM, suitable for rabi season
LBG 623	10-12	A.P.	Photo insensitive, large shining black seed, tolerant to MYMV
KBG 512	7-8	Tamil Nadu	Tolerant to stem fly and leaf spot
Prasad	8-9	Orissa	Tolerant to MYMV, suitable for kharif and rabi season
TU 94-2	12-13	SZ	Resistant to MYMV, tolerant to PM, suitable for rabi season
LBG 611	12-14	A.P.	Tolerant to wilt
LBG 20	12-14	A.P.	Tolerant to MYMV and PM, suitable for rice fallows
LBG 402	10-12	SZ(Rabi)	Tolerant to wilt and PM, suitable for rice fallows
LBG 17	10-12	SZ	Erect, greenish black large seed, tolerant to PM
LBG 709	10-12	A.P.	Photosensitive resistant to wilt and tolerant to YMV, suitable for rabi uplands and rice fallows

Table 2 Blackgram varieties for specific traits

Trait	**Varieties**
Short duration for spring / summer cultivation (60-65 days)	WBU 109, PDU 1, Pant U 31, KU 300, KU 92-1
Resistant to powdery mildew for rabi season.	LBG 625, LBG 709, LBG 645, CO 5, Vamban 1, Vamban 2, Vamban 3, Vamban 4, LBG 623, WBG 26
MYMV resistant	WBU 109, Pant U 19, Pant U 30, Pant U 31, Pant U 40, UG 18, PDU 1, Azad U 1, Sekhar 3, Sekhar 2, Azad U 2, Uttara, Narendra U 1, Mash 1008, IPU 02-43

Table 3 Blackgram varieties grown in Tamil Nadu

Varietal features	Duration (days)	50 % flowering	100 seed weight (g)	Grain yield (kg/ha)	
				Irrigated	Rainfed
TMV 1	65-70	35-40	5.7	1320	----
CO 4	70	35-40	5.7	1040	640
CO 5	70-75	30-35	5.1	1270	740
KM 2	60-65	30-35	4.0	----	690
T 9	65-70	30-35	4.0	1000	----
VBN 1	60-65	30-35	4.6	850	700
VBN 3	65-70	30-35	4.5	900	775
VBN (Bg) 4	75-80	35-40	4.8	900	780
VBN (Bg) 5	65-70	30-35	4.0	820	836
TNAU (Blackgram) VBN 6	65-70	30-35	3.8-4.0	890	850
TNAU (Blackgram) VBN 7	65-70	30-35	3.8-4.0	980	880
ADT 2	70-75	30-35	4.7	970	----
ADT 3	70-75	35-40	3.6	720	----
ADT 4	60-65	30-35	5.0	600	----
ADT 5	60-65	30-35	3.6	1320	----

Land preparation: Prepare the land to fine tilth and form beds and channels.

Seeds and sowing: The seed rate for pure crop, rice fallow and intercrop is 20, 25 and 10 kg/ha respectively. Seed treatment with fungicide/ bio- control agents and insecticides will protect the crop from diseases and pests.

- Seeds are treated with imidacloprid @ 5 ml/ kg of seed.
- The treated seeds are again treated with a slurry of carbendazim or thiram 75% WP @ 2 g dissolved in 5 ml of water per kg of seeds (or)
- The seeds may be treated with talc formulation of *Trichoderma viride* @ 4g /kg of seed or *Pseudomonas fluorescens* @ 10g/kg of seed.
- Bio control agents are compatible with bio fertilizers
- Fungicides and bio control agents are incompatible.

Seed treatment will protect seedlings from seed borne pathogens. After 24 hours fungicide treatment with seeds, three packets of rhizobial strains (200 g /packet) are to be mixed with seeds required for one ha. Use rice gruel as binder. Dry the bacterial culture treated seeds in shade for 15 minutes before sowing. Spacing for pure crop is 30 x 10 cm. Dibble the seeds adopting a spacing of 30 x 10 cm in wetlands/garden lands.

Cropping systems: It can be intercropped with crops such as cotton, sorghum, pearl millet, green gram, maize, soybean, groundnut, etc. The blackgram based cropping systems prevalent in irrigated areas are viz., maize–wheat-urdbean, rice–wheat–urdbean, rice–potato–jute+urdbean, pigeonpea+urdbean–wheat–urdbean, maize–toria–urdbean, maize–potato–urdbean, maize–potato, sugarcane+urdbean, maize (cobs)–peas (pods)–urdbean, urdbean-wheat–mungbean, urdbean–sarson–urdbean, maize-rapseed-mung/urdbean, urdbean-mustard-mungbean/urdbean and potato-wheat- urdbean.

Weed management: Alachlor / Fluchloralin may be applied to the irrigated crop @ 1.5 lit/ha or pendimethalin @ 2.0 lit/ha as pre-emergence. Spray the herbicide 3 days after sowing followed by one hand weeding on 25-30 days after sowing. If herbicide is not applied, two hand weeding are given on 15 and 30 DAS.

Manures and fertilizers application: Apply 12.5-25-0-10 and 25-50-0-20 kg N,P_2O_5, K_2O and S as basal dressing in rainfed and irrigated conditions respectively. Spray DAP @ 2% at flowering stage in rice fallow crop. DAP @ 2% spray fluid can be prepared by dissolving 2 kg DAP in 10 litres of water. Keep it overnight period or minimum of 12 hours. Transfer supernatant liquid and filter it. Then make up the volume to 100 litres of spray fluid. Soil application of 40 kg sulphur through gypsum + 25 kg $ZnSO_4$ + 15 kg borax ha^{-1} is recommended for irrigated crop. Soil application of micronutrient mixture @ 5 kg/ha as Enriched FYM (Prepare enriched FYM at 1:10 ratio of MN mixture & FYM ; mix at friable moisture and incubate for one month in shade).

Water management: Irrigation at 50% available moisture is optimum for nodulation, growth and yield. It is highly responsive to irrigation. Critical stages for irrigation are pre-flowering, flowering and grain filling stage (45 DAS). Flowering begins at 30 DAS and flowering is completed in all plants by 38 days. Avoid water stagnation at all stages. It can withstand water stress conditions. Draining in wetlands will improve blackgram yields as water logging restricts the oxygen supply to the plant root system and inhibits nodulation and nitrogen fixation. Irrigate immediately after sowing followed by life irrigation on the third day. Subsequently, irrigate at interval of 10 to 15 days depending upon soil and climatic conditions. For wetland bunds, pot water daily for a week after sowing. Short duration blackgram varieties can be grown utilizing residual moisture in soils with a high water holding capacity. e.g., rice fallow pulse. Foliar spraying of 2% KCl + 100 ppm boron/ boric acid during dry spell as mid-season management practice in black gram during *Rabi* season is recommended to increase the yield over KCl spray alone in rainfed condition.

Multi bloom technology: The crop is sown during early summer (Jan.-Feb.) as normal crop and fertilizer is applied as per the recommendation for irrigated crop. In addition to that, top dressing of nitrogen is done with an extra dose of 25 to 30 kg N/ha through urea. Since pulses have indeterminate growth habit and continue to produce new flushes, top dressing will be done on 40-45 days after sowing. The crop completes its first flesh of matured pods during 60-65th day; further their second new flush within 20-25 days. Therefore two flushes of pods can be harvested at a time within the duration of 100 days.

Crop protection: Mungbean yellow mosaic and leaf crinkle virus during *kharif*, mungbean yellow mosaic virus during spring in northern India and powdery mildew in southern coastal parts of the country during winter season cause considerable seed yield losses. Defoliators e.g., hairy caterpillars and semilooper are common pests during the vegetative stage. Activity of thrips starts at the bud stage and pose serious problems when the crop attains peak flowering leading to heavy flower drop. Seed treatment and foliar application of *Beauveria bassiana* + *Pseudomonas fluorescens* on 30 DAS @ 5 g each is recommended for the management of stem fly-root rot complex in blackgram. *Neem* cake @ 5 q/ha as soil application in mungbean also reduces the nematode population and increases the yield. Seed soaking in carbosulfan 25 EC @ 1% solution for 4 hours reduced the nematode population. Seed treatment with neem oil @ 10ml/kg seed reduces the nematode population. Application of carbofuran 3G @ 1.5 kg/ha reduces the nematode population. Control the rats by using 2% zinc phosphide poison bait (96 parts bait material, 2 parts zinc phosphide, 2 parts edible oil) or use single dose anticoagulant like Bromadiolone 0.005% (93 parts crushed wheat/maize/jowar/bajra or flour, 3 parts sugar/jaggery, 2 parts edible oil and

2 parts Bromadiolone 0.25%). Treat (coat) the pulses with coconut or groundnut or mustard oil depending upon the availability and cost, @ 250 to 500 ml per quintal to protect pulses from the infestation of pulse beetles.

Harvest: The plants are after pulled up along with roots, stacked to dry for 3 to 7 days. Threshing can be done by beating with sticks or trampling under the feet of oxen (bullocks, buffaloes). When grown on rice fields the pods are normally hand picked. The plants grazed for a day or two and then ploughed in. The seed is thoroughly cleaned, dried in the sun to 11 to 12 % moisture or less and stored.

Yield: The seed yield is 800 to 1200 kg/ha. It provides crop residues about 1.25 to 1.75 t/ha.

Processing: Blackgram is processed into dhal by wet or dry methods. In the dry method, the seeds is given a coating of vegetable oil (1-2 g/100 g of seed) and left over night. It is then spread in the sun to dry for 3 to 6 hours after which it is sprayed with water and then left to dry for 3 to 4 days. Then the seeds are passed through a roller or other suitable machine. In the wet method, the seeds are soaked in water. This softens the seed coat which is removed easily by passing it through a roller. The yield of dhal is approximately 70%.

Utilization: Blackgram seed flour is used in the preparation of *idli*, *dosa*, etc. The vines or haulms left after harvesting are also used for feeding to cattle. The different features of blackgram and greengram is presented in Table 4.

Table 4 Distinction between blackgram and greengram

Characters	Blackgram (Vigna mungo)	Greengram (Vigna radiata)
Stem	Mostly spreading or trailing	Mostly erect or sub-erect
Leaves	Mostly yellowish green	Mostly green or dark green
Leaf colour	Dark green	Pale green
Hairiness	Densely hairy	Sparsely hairy
Hair colour	Ferruginous (Reddish brown)	Slightly brownish
Pods	Short in length, erect with long hairs, less shattering	Long, radiating small hairs present, highly shattering
Seeds	Large oblong with square ends	Small, round
Seed coat	Normally black in colour, testa with no ridges	Normally green in colour, fine wavy ridges
Cotyledons	White pasty when chewed	Cream, not pasty, broken into bits
Hilum	Concave	Flat
Water logging	Not tolerable	Tolerable

REFERENCES

Duke, J.A. 1981. Handbook of legumes of world economic importance. Plenum Press, New York. 293 pp.

GOI. 2006. Post-harvest profile of black gram. Department of Agriculture and Cooperation, Directorate of Marketing and Inspection. Directorate of Marketing and Inspection. Nagpur – 440001

Singh, B.B., Dixit, G.P. and Katiyar, P.K. 2010. *Vigna* Research in India (25 Years of Research Achievements). All India Coordinated Research Project on MULLaRP, Indian Institute of Pulse Research, Kanpur-208 024

Smartt, J. 1976. Tropical pulses. Longman Group Limited, London. 348 pp.

CHAPTER

22

Greengram–*Vigna radiata* (L.) Wilczek (2n=24)

Family: Fabaceae (Leguminoceae)

Vernacular name: Greengram, Mungbean, Golden gram (English), *Moong* (Hindi), *Mung* (Oriya), *Pachaipayaru* (Tamil) and *Pachapesalu* (Telegu).

Importance: Mungbean is a pulse crop grown principally for its protein rich edible seeds.

Origin: Greengram is native to the Indo-Myanmar region.

Distribution: Greengram is cultivated in India, Myanmar, China, Philippines, Australia, Thailand, Iran, Pakistan, Vietnam, Korea, Indonesia, Sri Lanka and USA. Green gram production in India is 55% of world area and 45% of the world production. In India, mungbean ranks third among the pulse crops after chickpea and redgram. Greengram is grown in Andhra Pradesh, Orissa, Maharashtra, Madhya Pradesh, Rajasthan, Bihar, Gujarat, and Tamil Nadu.

Botany: Greengram is an erect or semi-erect herbaceous annual, with slight tendency for twinning in the upper branches; leaves trifoliate with long petioles; leaflets entire, ovate; flowers, ten to twenty crowded in axillary racemes; pod longer than in black gram, with short hairs; seeds globular. The leaves are 5-10 cm long trifoliate with long petioles. The flowers open between 6 a.m. to 8 p.m. Self-pollination is the rule. It produces 6-10 cm long hairy pods which are round, slender and used to bear about 7-11 seeds in them. The seeds are small and nearly globular. The colour of seed is usually green, but yellow brown or purple brown seeds also occur. The hilum is white, more or less flat. Germination of greengram is epigeal.

Growth stages in greengram:

Stages	Crop duration (days)
Emergence	5-7
Seedlings	20
Vegetative stage	30

Flowering stage	30-40
Pod initiation stage	40-45
Soft dough stage	45-60
Hard dough stage	60-75
Physiological maturity	75-90

Climate: Greengram can be grown in latitude from 30ºS to 30ºN up to an altitude 1800 to 2000 m from MSL. It is grown in areas receiving an annual rainfall 600 to 1000 mm. It is adapted to the drier and warmer climates of lowland tropics and subtropics. It is a short day plant and is thermo sensitive. Flowering is being delayed by long photoperiods while hastened by high temperature. Temperature beyond 40ºC is harmful to the crop, while 27-35ºC is the optimum temperature for seed germination and plant growth. It is sensitive to low temperature and is killed by frost. The crop is favoured by dry weather during pod ripening to facilitate seed harvest and prevent seed damage.

Soils: It is cultivated in well drained loamy or sandy loamy soil types. Deep loam soils are moisture retentive and light soils facilitate aeration, infiltration and internal drainage. It can be grown on acidic laterite soils with high fixation capacity of phosphorus.

Season: It is grown during June-July and September-October under rainfed conditions whereas it is raised during February-March under irrigated condition. It is also raised as rice fallow crop during January.

Varieties: The particulars of blackgram varieties grown in India are furnished in Table 1 and Table 3.

Table 1 Green gram varieties suited to different state of India

Variety Name	Yield (q/ha)	Suitability	Major characteristics
Pant Mung 6	12-15	NEHZ	Resistant to MYMZ, PM, CLS and LCV
Shalimar Mung 1	9-10	Jammu & Kashmir	Shining seed, medium large
Sweta	10-11	NHZ	Resistant to MYMV
Pusa 672	9-10	NHZ	Medium large shining green seed, tolerant to MYMV
Pant Mung 5	12-15	U.P & Uttarakhand	Resistant to MYMV, suitable for all season
HUM 16	10-12	NEPZ	Resistant to MYMV, suitable for Summer season
Sukumar (WBM 29)	9-10	West Bengal	Suitable for rabi season, small seed
OBGG 52	8-9	Orissa	Tolerant to MYMV, suitable for all season
Pratap (SG-1)	10-12	Assam	Suitable for early and late situation
Pant Moong 4 (UPM 92-1)	12-15	NEPZ	Resistant to MYMV, dull green seed
Narendra Mung 1	10-12	U.P.	Tolerant to MYMV,
TMB 37	11	NEPZ	Tolerant to MYMV, suitable for spring season
IPM 99-125 (Meha)	10-12	NEPZ	Resistant to MYMV, suitable for spring season
HUM 12	11-13	NEPZ	suitable for Summer season

HUM 6	10-11	U.P.	Tolerant to MYMV, suitable for spring season
HUM 2	10-12	U.P.	Resistant to MYMV, suitable for spring season
Malviya Jankalyani	10-12	NEPZ	Erect, shining green large seed, tolerant to MYMV, suitable for summer season
Samrat	10-12	U.P.	Erect, synchronous shining green seed, highly resistant to MYMV, suitable for spring season
MUM 2	11-12	NWPZ	Resistant to MYMV, shining green seed,
IPM 02-3	10	NWPZ	Erect, resistant to MYMV, suitable for kharif season
Satya	12-14	NWPZ	Erect, resistant to MYMV, suitable for kharif season
MH 96 -1	10-11	Haryana	Resistant to MYMV,
ML 818	10-12	Punjab	Tolerant to MYMV, BLS and CLS
RMG 268 (Durga)	8-9	Rajasthan	Photothermo-insensitive, shining green seed
ML 613	10-12	Punjab	Tolerant to MYMV and CLS
Ganga 1	10-12	Rajasthan	Moderately resistant to MYMV
PAU 911	12-14	Punjab	Resistant to MYMV, suitable for kharif season
Basanti	10-12	Haryana	Resistant to MYMV, suitable for kharif and spring season
SML 668	11-13	Punjab	Tolerant to MYMV, suitable for spring / summer season
Ganga-8 (Gangotri)	9-10	NWPZ	Suitable for kharif season
Pusa Bold 1	11-12	NWPZ	Erect, synchronous, large seed, tolerant to MYMV, suitable for spring season
SML 134	11-12	Punjab	Suitable for spring/summer
AKM 9911	10-12	Maharashtra	Tolerant to PM
GM 4	15	Gujarat	Tolerant to MYMV
PKV Mung 8802	12-14	Maharashtra	suitable for kharif season, tolerant to MYMV and MB
TARM 18	10-12	Maharashtra	Resistant to PM
TARM 1	10-12	Maharashtra	Resistant to PM, suitable for rabi season
JM 721	12-14	Madhya Pradesh	Resistant to PM and MYMV
Gujarat M 3	10-12	Gujarat	Erect, shining green small seed, tolerant to MYMV
TARM 2	10-12	Maharashtra	Small seed, tolerant to PM
BM 4	10-12	CZ	Large green seed, semi- erect bushy type
AKM 9904	10-12	Maharashtra	High yielding, multiple disease resistance
HUM 1	10-12	CZ & SZ	Suitable for spring and kharif season
Pusa 9531	9-10	CZ	Suitable for summer season

KKM 3	8-10	Karnataka	
TM 96-2		Andhra Pradesh	Resistant to MYMV, suitable for rice-fallow
COGG 912	8-9	SZ	Suitable for kharif season
OUM 11-5	7-8	SZ	Suitable for kharif season
WGG 37	6-8	Andhra Pradesh	Erect tall, resistant to MYMV
LGG 460	10-12	Andhra Pradesh	Resistant to MYMV
Madhira 295	12-14	Andhra Pradesh	Tolerant to PM
Madhira Pesara 347	11-12	Andhra Pradesh	Tolerant to MYMV, suitable for kharif and rabi season
Warangal 2	10-12	Andhra Pradesh	Suitable for kharif, rabi and summer
Pusa 9072	9-10	SZ	Tolerant to PM, suitable for rabi season
LGG 410	10-12	Andhra Pradesh	Resistant to leaf spot, suitable for rice – fallow
LGG 450	11-13	Andhra Pradesh	Tolerant to MYMV and pre- harvest sprouting
LGG 407	12-14	Andhra Pradesh	Resistant to MYMV

Note: CZ-Central Zone, SZ –South Zone, NEPZ – North Eastern Plane Zone, NWPZ- North Western Plain Zone

Table 2 Greengram varieties in Tamil Nadu

Particulars	**ADT 3**	**CO 6**	**Co (Gg) 7**	**VBN (Gg) 2**	**VBN (Gg) 3**	**VRM (Gg) 2**	**Paiyur 1**
Duration (days)	66	65-70	60-65	65-70	65-75	60-70	85-90
Days to 50 % flowering	36	40-45	40 -43	40-45	40-45	40-45	45-50
100 seed weight (g)	2.6	2.8-3.5	3.5-4.0	3.6-3.9	2.8-3.5	----	3.5
Grain yield (kg/ha) (rainfed)	500	900	980	750	775	1100	742
Irrigated	----	1050	----	980	880	----	----
Rice fallow	500	----	----	----	----	----	----

Table 3 Greengram varieties of for specific traits

Specific traits	**Varieties**
Short duration for spring / summer cultivation (60-65 days)	Samrat, Pusa Vishal, OUM 11-5, Meha, Part Moong 5, SML 668, HUM 1, HUM 16, PM 2-3, IPM 02-14
Large seed size (>4g/100 seeds)	Pusa Vishal, Pant Moong 5, SML 668, HUM 16
Resistant to powdery mildew for *rabi* season.	TM 96-2,Vamban 2, Vamban 4, TARM 18, BPMR 145
MYMV resistant	Samrat, Meha, Pant mung 1, Pant mung 2, Pant mung 3Pant Moong 4, MH 2-15, Narendra Moong 1, Pant Moong 6, HUM 1, HUM 12, CO6, ML 267, ML 337, IPM 2-3, IPM 02-14
Tip blight & root rot	CO4

Land preparation: Prepare the land to get fine tilth and form beds and channels.

Apply lime @ 2 t/ha along with FYM or composted coir pith @ 12.5 t/ha to tide over the soil surface crusting.

Seeds and sowing: The seed rate for pure crop, rice fallow and intercrop is 20, 25 and 10 kg/ha respectively. Treat the seeds with carbendazim or Thiram @ 2 g/kg of seed for the prevention of soil borne diseases. Fungicide treated seed should again be treated with three packets of bacterial culture. Take 50 g of molasses and mix it with half litre of water and 250 g/acre of rhizobium and mix 10 kg of seed thoroughly. The treated seed should be dried in shade for 2-3 hours before sowing. There should be an interval of at least 24 hours after fungicidial treatment for giving the bacterial culture treatment. Then treat seeds with *Trichoderma* @ 4 g/kg of seeds for the prevention of soil borne diseases. This can be done just before sowing. *Trichoderma* treated seeds should again be treated with bacterial culture immediately whereas *Trichoderma* treated seeds should not be treated with fungicides. Dry the bacterial culture treated seeds in shade for 15 minutes before sowing. Spacing for pure crop is 30 x 10 cm. Dibble the seeds adopting a spacing of 30 cm x 10 cm in wetlands / garden lands. It germinates within 3 to 4 days after sowing with 50% available soil moisture. For rice fallows, broadcast uniformly the seeds in the standing crop 5 to 10 days before the harvesting of rice crop when soil moisture is optimum so that the seeds should get embedded in the waxy mire. On wetlands bunds dibble the seeds with 30 cm spacing. The optimum depth of sowing ranges from 3 to 5 cm.

Cropping system: The greengram based cropping systems prevalent in irrigated areas are viz., Rice-wheat-mungbean, Rice-wheat-jute + mungbean, Rice-potato-jute + mungbean, Maize-wheat-mungbean, Maize-rapeseed- mungbean, Maize–Toria–mungbean, Maize–Potato–mungbean, Maize– Potato – sugarcane + mungbean, Maize (cobs) –Peas (pods) – mungbean, Maize-Potato- wheat - mungbean, Maize–Potato–Potato-mungbean, Pigeonpea + mungbean– wheat - mungbean, Pigeonpea-wheat-mungbean, Urdbean-wheat-mungbean, Urdbean-mustard-mungbean, Urdbean–Potat –wheat, Potato-wheat- mungbean and Cotton–mungbean.

After cultivation: Thinning of excess seedlings 10 to 15 days after sowing is an essential operation to get the balanced growth and fruiting of individual plants.

Weed management: Greengram needs weed free condition for 35 days during the wet season and 30 days during the dry season. If herbicides are not applied, give two hand weedings on 15th and 30th day after sowing. Spray Fluchloralin @ 1.5 lit/ha pre-emergence spraying 3 days after sowing followed by one hand weeding on 30 days after sowing. Pre-emergence application of pendimethalin 1.0 kg/ha on 3 DAS followed by post - emergence application of either quizalofop ethyl 50 g/ha on 30 DAS for late emerging grassy weeds or imazethapyr @ 60 g/ha on 30 DAS for late emerging broad leaved weeds are recommended for irrigated green gram.

Manures and fertilizer application: Application of nitrogen fertilizer to greengram will reduce the amount of nitrogen fixed by rhizobial organisms. Early development of the greengram plant is dependent upon nitrogen supplied from the soil, after nitrogen stored in the germinating seed is exhausted and before nodules are fully functioning. Supplying nitrogen during this period will enhance early plant growth and yield. Additional nitrogen may be desirable during the pod filling stage to supplement contributions from declining rhizobial activity. In soils defficient in both N and P, application of DAP may be beneficial in obtaining maximum benefit. Small quantity of nitrogen is applied as starter fertilizer. Nitrogen fertilization at the pod initiation or pod filling stage with 2% urea or 2% DAP spray increases seed yield by 15%. Application of phosphorus fertilizer induces earlier flowering, hastens maturity and promotes growth of fibrous and lateral roots. There will be 15% increase in grain yield of greengram through both basal and foliar application of phosphorus over basal application alone. Phosphorus deficiency leads to formation of dark green leaves and later become blotchy in appearance. Potassium is generally inadequate in sandy soils and soils of humid tropics. Potassium deficiency causes malformation of leaves and browning of leaf veins on the undersurface of the leaflets. Yellowing of leaf margin occurs at the advanced stage of severe potassium deficiency. Calcium is a key element in nodule formation and symbiotic nitrogen fixation. $CaCO_3$ (lime) neutralizes soil acidity and improves the availability of other nutrients. FYM @ 12.5 tonnes per ha with 12.5–25-0 and 25-50– 0 kg of kg of N, P_2O_5 and K_2O per ha is recommended under rainfed and irrigated conditions respectively. Spray 2% DAP at the time of first appearance of flowers and then second spray 15 days later. DAP spray solution can be prepared by dissolving 2 kg DAP in 10 lit of water. Keep it for overnight. Filter the superannuated solution, which is to be mixed with 90 lit of water and then sprayed. Soil application of 40 kg sulphur through gypsum + 15 kg borax ha^{-1} is recommended for irrigated crop. Application of $ZnSO_4$ @ 25 kg/ha along with 50 kg well powdered FYM is recommended to correct Zinc deficiency. If the soil is deficient in iron, application of Fe_2SO_4 @ 5 kg/ha along with 50 kg well powdered FYM is recommended to correct this disorder. If symptoms appear at later stages, foliar application of 0.5% ferrous sulphate is recommended.

Water management: Greengram is a drought tolerant crop. Irrigation at 50% available moisture is optimum for nodulation, growth and yield. Water requirement for greengram is 3.2 mm/day. Greengram root penetrates 12 to 17.5 cm. It is highly responsive to irrigation. Critical stages for irrigation are pre-flowering, flowering and grain filling stage (45 DAS). Flowering begins at 30 DAS and flowering is complete in all plants by 38 days. Avoid water stagnation at all stages. It can withstand water stress conditions. Spraying of 6% Kaoline or 2% KCL + 100 ppm borax during vegetative stage is desirable to overcome drought if there is moisture stress. Draining wetlands will improve greengram yields as waterlogging restricts the oxygen supply to the plant root system and inhibit nodulation and nitrogen fixation. On dry soils, pre-soaking of seeds prior to sowing will reduce soil crusting and assist in obtaining rapid seed germination. Irrigation at 0.75 bar atmospheric pressure until flowering and 0.5 bar during post flowering improves nodulation and seed yield. Irrigate immediately after sowing followed by life irrigation on the third day. Subsequently, irrigate at interval of 10 to 15 days depending upon soil and climatic conditions. For wetland bunds, pot water daily for a week after sowing. Short duration greengram varieties can be grown utilizing residual moisture in soils with a high water holding capacity. eg., rice fallow pulse.

Growth regulators: Flower shedding is greater with high temperatures and desiccating winds at the time of flowering. Flower shedding is 40 to 70% in greengram. Foliar application of NAA (1-Napthalene Acetic Acid) @ 40 ppm (40 mg/one lit of water) reduce flower shedding / increase pod setting and seed yield. Spray NAA solution one week after first flowering since greengram flowers over a long period. Use TIBA @ 150 ppm as foliar spray at flower initiation to check vegetative growth and to increase seed yield of pulse crops.

Multi-bloom technology: The crop is sown during early summer (Jan.-Feb.) as normal crop and fertilizer is applied as per the recommendation for irrigated crop. In addition to that, top dressing of nitrogen is done with an extra dose of 25 to 30 kg N/ha through urea. Since pulses have indeterminate growth habit and continue to produce new flushes, top dressing will be done on 40-45 days after sowing. The crop completes its first flesh of matured pods during 60-65th day; further their second new flush within 20-25 days. Therefore two flushes of pods can be harvested at a time within the duration of 100 days.

Crop protection: Stem fly, white fly and mite can be controlled with seed treatment with dimethoate 30 EC 5 ml/kg of seed; spray with any one of the following (Spray fluid 250 l/ha) Methyl demeton 25 EC @ 500 ml/ha, Dimethoate 30 EC @ 500 ml/ha (Imidachloprid) or Wettable sulphur @ 1.5 kg/ha. Apids, Spotted pod borer and Blue butterflies may be controlled with any one of the dust chemicals such as Endosulfan 4D @ 25 kg/ha, Quinalphos 1.5 D @ 25 kg/ha, Phosalone 4 D @ 25 kg/ha, Carbaryl 5D @ 25 kg/ha or spray any one of the following (spray fluid 500 l/ha) Endosulfan 35 EC @ 1.0 l/ha or Monocrotophos 36 SL @ 500 ml/ha. Insecticide application should be avoided when the activity of coccinellid predator (both grubs and adults) is observed.

Powdery mildew disease can be controlled with spray NSKE 5% or Neem oil 3% twice at 10 days interval from initial disease appearance. This is also controlled with spray of Carbendazim 250 g or Wettable sulphur 2500g/ha or botanical spray Eucalyptus leaf extract 10% at initiation of the disease and 10 days later. Rust disease can be controlled with spray of Mancozeb @ 1000g/ha or Wettable sulphur @ 2500g/ha. Leaf spot disease may be controlled with spray of Carbendazim @ 250 g/ha. Yellow mosaic and Leaf Crinkle vector, Leaf Curl vector can be controlled with spray of Monocrotophos 500 ml or Methyl demeton 500 ml/ ha and repeat after 15 days, if necessary. For seed crop, the plants affected by leaf crinkle should be periodically removed up to 45 days after sowing since the leaf crinkle virus is seed borne. The infected plants have to be rogue out up to 30 days. Root rot disease may be controlled with soil application *P. fluorescens* or *T. viride*– 2.5 kg/ha + 50 kg of well decomposed FYM or sand at 30 days after sowing; Neem cake @ 150 kg/ha and spot drench Carbendazim @ 1 gm/litre. Grains may be treated with coconut oil or groundnut oil @ 10 ml/ kg of grains meant for household consumption. Seeds may be treated with Neem oil @ 10 ml/ kg of seeds or Malathion 50 EC or Dichlorvas 76 SC @ 10 ml/kg of seed. Pulses may be stored as split dhal for immediate use to avoid egg laying by bruchids.

Economic threshold level (ETL)

- Aphids: 20/2.5 cm shoot length
- Pod borers: 10% of affected pods
- Spotted pod borer: 3/plant
- Stem fly: 10% of affected plants
- Tobacco cut worm: 8 egg masses/100 m^2

Harvest: Greengram simultaneously has green leaves, open flowers, green pods and ripe pods. The mature pods are commonly picked as they ripen. Long season varieties require 3 to 5 pickings but new short season varieties may need only one or two pickings. Hand pickings is laborious and the most expensive single operation in greengram production. The pods are ripe and ready to pick at about three weeks after the flower opens. Pods should be harvested 30 days after the 50% flowering stage, which helps to obtain seeds possessing high germination, vigour and storability. At this stage the colour of the 80% of pods will be brown. The plants are cut when 80% of pods are ripe. The pod moisture content will be 17 to 18%. Stake the plants in the sun for a few days before threshing. Pods picked by hand are dried in the sun and threshed by trampling or placed in a jute bag or in open floor and beaten with a wooden stick. Seed is separated from the hulls by winnowing. Seed is dried to 12% moisture or below before storing. Harvest accounts 25 to 30% of the total production cost and 40 to 50% of the total labour cost. Defoliation of the plants is needed before mechanical harvesting. Ethrel, at 39.5% a.i. and 500 times dilution, will defoliate 90% of foliage without harming seed quality.

Processing: The dehusked and split cotyledons is called '*Dhal*'. The dhal is cooked in water and eaten with rice or other cereals. Commercial operation of dehusking is of two types viz., i) Wet method: the grain is tempered with water for a few hours, coated with earth and sundried to shrink the cotyledon and loosen the hull and ii) Dry method: the seeds are conditioned by spraying with oil and water. The oil tempered seeds are dried and subsequently sprayed with water to soften seed coat before further processing. After the seed coat is loosened, the seeds are dehusked through roller mills or absrasive dehullers.

Utilization: Immature pods and seeds are boiled and eaten. Sprouted greengram seeds are used in salad. Mature seeds are boiled and used in soups. Greengram flour is used in making cakes, desserts, noodles. The plant parts remaining after the pods have been picked are used as animal feed. It is utilized as green manure crop for soil improvement.

REFERENCES

GOI. 2006. Post-harvest profile of green gram. Department of Agriculture and Cooperation, Directorate of Marketing and Inspection. Nagpur – 440001

Sharma, O.P., Bambawale, O.M., Gopali, J.B., Someshwar Bhagat, Suhas Yelshetty, Singh, S.K., Rajesh Anand, and Om Pal Singh. 2011. Field Guide Mungbean & Urdbean. pp 40. National Centre for Intergrated Pest Management, LBS Building, IARI Campus, New Delhi 110 012.

Singh, B.B., Dixit, G.P., and P.K. Katiyar. 2010. *Vigna* Research in India (25 Years of Research Achievements). All India Coordinated Research Project on MULLaRP, Indian Institute of Pulse Research, Kanpur-208 024

http://en.wikipedia.org/wiki/Mung_bean

CHAPTER

23

Cowpea–*Vigna unquiculata* (L.) Walp. (2n=22, 24)

Synonyms: *Vigna sinensis* L. Savi., *Vigna catjang* Wall.

Family: Fabaceae (Leguminoceae)

Vernacular names: Cowpea, Black eye pea, Black eye bean, China pea, Kaffir pea, Marble pea, Yard long bean, Asparagus bean, Bodi bean, Snake bean (English), *Lobia, Mattri* (Hindi), *chola, choli* (Gujarati), *chavali* (Marathi), *alasandulu* (Telugu), *alasande* (Kannada), and *karamani, Thatapayaru* (Tamil), *Chavli* (Marathi) and *Lobhia* (Persia).

Importance: Cowpea is grown for the dry seeds and for its long immature pods. It is tasty, nutritious and capable of increasing milk production. It is a fast growing crop which curbs erosion by covering the ground, fixes atmospheric nitrogen and its decaying residues contribute to soil fertility.

Origin: Cowpea is native of Africa with Nigeria being a major centre of diversity.

Distribution: Cowpea is cultivated mainly in Nigeria, Niger, Brazil, Haiti, India, Myanmar, Sri Lanka, Australia and the United States. In India, it is grown in Maharashtra, Andhra Pradesh, Karnataka and Tamil Nadu.

Botany: Cowpea is an annual, twining herb; stem slightly ridged almost glabrous; leaves, alternate; axillary raceme with flowers, flowers showy, white or yellow or pink, fruit a legume cylindrical; seed is non-endospermous, reniform or globular. Anthesis takes place early in morning between 6.30 and 9.00 a.m. The process of opening corolla takes 45-60 minutes. Dehiscence of anthers is much earlier and it varies from 10.0 p.m. to 00.45 a.m. Stigma becomes receptive from 12 hours before blooming to 6 hours after anthesis. Cowpea pods are smooth, 6 to 10 in. long, cylindrical and generally somewhat curved. As the seeds approach the green-mature stage for use as a vegetable, pod color may be distinctive, most commonly green. yellow or purple. As the seeds dry, pod color of the green and yellow types becomes tan or brown.

The cultivated types of cowpea have been classified into 3 groups:

(i) *Vigna sinensis* (cowpea): These are erect/trailing, early maturing and annual types. Pods are 20-30 cm long containing 0.6-0.9 cm long seed. On drying, seeds are neither flabby nor inflated. It is mainly grown for fodder purposes.

(ii) *Vigna sinensis* sub. sp. *catjang* (Indian cowpea): The pods are 7.5-12.5 cm long, erect ascending when green, spreading or deflexed when dry. Seeds are 0.5-0.6 cm long and nearly as thick as broad. It is mainly grown for fodder purpose.

(iii) *Vigna sinensis* sub. sp. *sesquipedalis* (Yardlong or asparagus bean): They are spreading annual types with 30-60 cm long and pendent pods. Seeds are 0.8-1.2 cm long and are mainly used as vegetable.

Climate: Cowpea is grown between 30°N and 30°S latitude up to an altitude of 1500 m from MSL. Cowpea can be grown in areas receiving mean annual rainfall of 600 to 1500 mm. Cowpea exhibits more tolerance to heavy rainfall than other pulses. Short day, day neutral and long day types of cowpea varieties are available. The cardinal minimum, optimum and maximum temperature for cowpea is 12°C, 21 to 35°C and 40°C respectively. The optimum temperature for nodulation is 24 to 33°C. Cowpea can not tolerate frost at 5 to 10°C for period as short as 24 hours. Maximum dry matter production occurs in cowpea at 27°C day temperature and 22°C night temperature. Cowpea yield is liable to be reduced above 35°C because of flower and pod shedding.

Soil: Cowpea can be grown over a wide range of soil types. It prefers well-drained, highly acid to neutral soils, but can grow well in a range of soil types, including soils with low fertility. Clay loamy soils with high fertility usually result in high yields of hay, but poor seed yields. In light sandy soils, heavy infestation with nematodes is liable to occur. A soil pH of 5.5 to 6.5 is preferred and it can tolerate upto a pH of 9. The plant is very drought resistant and does not survive flooded conditions. It cannot tolerate water logging.

Season: Cowpea can be grown during June-August and September-October under rainfed condition while it can be raised as irrigated crop during February-March.

Land preparation: Prepare the land to fine tilth and form beds and channels. Land is prepared to a fine tilth by 2-3 ploughing and harrowing.

Varieties: Cowpea varieties in different growing zones of India are presented in Table 1 and Table 2.

Table 1 Cowpea varieties recommended for various agro-climatic zones of India

Zone	**Varieties**		
	Grain types	**Vegetable types**	**Dual types**
North-western zone (Punjab, Haryana, Western Uttar Pradesh, Uttarakhand, Rajasthan, Himachal Pradesh, Jammu & Kashmir)	T2, JC 5, JC 10, RS 9, RC 29, V 16, Cowpea 74, Pusa 152, Pusa Sawani (TS 269), Pusa Sampada, Rambha	Pusa Rituraj, Pusa Phalguni, Pusa Dofasli, Pusa Barsati	FS 68, Swarna (V 38), Gomti, Pusa Komal, Arka Garima
North-eastern zone (Eastern Uttar Pradesh, Bihar, West Bengal, Orissa, Assam)	FGC 1, T2, V 16, RC 19, Pusa 152, Pusa Sawani, Cowpea 74	-----	-----
Central zone (Madhya Pradesh, Gujarat, Maharashtra)	Pusa 152, V 240, V 16, Gujarat cowpea 1, Gujarat cowpea 2, K 11, K 14, GC 3	-----	-----
Peninsular zone (Andhra Pradesh, Tamil Nadu, Karnataka, Kerala)	Pusa 152*, Krishnamani (PTB 2), Kanakmani (PTB 1), Co 1, Co 2, Co 3, Co 4, V16 (Amba), V 240, S 228, S 448, JC 5, SU 88, KM 1	-----	-----

Table 2 Particulars of cowpea varieties in different zones/states of India

Variety	Area of adoption Zone/State	Seed yield (Q/ha)	Days of maturity	Remarks
Gujarat Cowpea -3	CZ (MP, Maharashtra & Gujarat)	12-14	65-85	Seed bold, amber colour
V-240	All Zones	14.0	80	Tall, Indeterminate, seed red
Vamban -1	Tamil Nadu	9.5	65	Erect, dwarf, seed white
Gujarat Cowpea -4	Gujarat	8-5	80-90	Seed bold, amber colour
KBC-2	Karnataka	9.5	95-105	Semi-determinate, seed light brown
RC-101	Rajasthan	8.5	85-90	Early, Determinate, seed white
CO-2	Tamil Nadu	13	90	100 seed weight 12.5 g
CO-3	Tamil Nadu	8.3	80	100 seed weight 10 g
CO-4	Tamil Nadu	9.6	85	100 seed weight 11.5 g
CO-6	Tamil Nadu	14.0	85-90	Early, bold seeded
KM 1	Tamil Nadu	9.0	65	100 seed weight 7 g
Paiyur 1	Tamil Nadu	9.0	90	100 seed weight 9.9 g
V 578 (Pusa sampada)	Delhi	12		Early, Resistant to yellow mosaic virus
CL-367	Punjab	12	95-100	Tolerant to YMV
RCP -27 (FTC -27)	Rajasthan	6-13	69-79	Resistant to YMV
UPC 622	Uttarakhand Assam,U.P., M.P., J & K, H.P., Punjab, Raj., Har., WB., Odisha, Bihar, and Jharkhand	4-5	145-150	Tolerant to YMV, Anthracnose, root/collar rot and bacterial leaf blight, Aphids, leaf Miner, flea beetle, pod borer /bugs and root knot nematode & bruchids.
Khalleshwari	Chhattisgarh	6-7		RRF in rabi with restricted irrigations and rainfed upland in kharif season
Swarna Harita (IC285143)	Assam, U.P., M.P., Kerala, A.P., Punjab, Raj., WB., Odisha, Jharkhand , CG., and TN.	60-150 (pods)	75-90	Resistant to rust and mosaic viral disease & tolerant to pod borer.
Kashi Kanchan (VRCP 4)	Punjab, UP, Bihar, Jharkhand, Odisha, CG, MP, AP	150-175 (pods)	50-55	Resistant to golden mosaic virus, *Pseudo-cercospora cruenta diseases,*
UPC 628	Punjab, UP, Bihar, Jharkhand, Odisha, CG, MP, WB,MS	3.5-4.0	145-150	Irrigated summer, and rainfed condition, Medium late variety
IT- 38956 -1	Karnataka	10-12	80-85	Rainfed areas of eastern dry region

Hisar Cowpea 46 (HC 98-46)	Haryana	10	65-70	Resistant to YMV
Pant Lobia -1	Uttarakhand, UP	20	130-135	Moderately resistant to Aphids, Thrips, Bruchids & other field pests. Suitable for spring, summer and *Kharif* season
UPC 628	Uttarakhand, HP, J&K, Punjab, Harya.,Raj., UP,MP,CG, Bihar, Jharkhand, WB, Odisha, MS	350-400 (pods)	145-150	Tolerant to drought and other abiotic stresses, resistant to YMV, Anthracnose/leaf blight, Aphids, Semilooper, Flea Beetle/Defoliators, Pod borer/ bugs & Root knot nematode tolerant to storage weevil
HIDRUDAYA	Kerela	10-11	50-55	Tolerant to leaf rust, Aphids, Pod borer & American Serpentine leaf minor, summer season
C519 (Himachal Lobia 11)	Himachal Pradesh	15-16	80-85	Resistant to Cercospora leaf spot, YMV, Low hills, Sub-tropical zone under rainfed condition in kharif
PKB 4	Karnataka	11-13	80-85	Resistant to Bacterial leaf blight, Rust & Pod borer, suitable for early kharif season
PKB 6	Karnataka	10-12	80-85	Resistant to Bacterial leaf blight, Rust & Pod borer, suitable for early kharif and summer season

Seeds and sowing: Seed is treated 24 hours before sowing with carbendazim or thiram @ 2 g per kg of seed to protect seedlings from seed borne pathogens. After 24 hours of fungicide treatment with seeds, three pockets of rhizobial strains (200 g/pocket) are to be mixed with seeds required for one ha. Use rice gruel as binder. Dry the bacterial culture treated seeds in shade for 15 minutes before sowing. Nodule activity in cowpea starts at three weeks and ends within eight weeks from sowing. Soak cowpea seed in aquous solution of $ZnSO_4$ at 100 ppm (10 g/100 lit of water) at 1/3 volume of seed for 4 hours and quickly air dry in shade to original weight to induce seed hardening. Seeds can be sown in lines or broadcasted or by dibbling. Seed rate for pure crop and intercrop is 20 and 10 kg per ha respectively. Seeds of bushy varieties are dibbled at a spacing of 30 × 15 cm with 1-2 seeds per hole. For semi-trailing varieties provide a spacing of 45 × 30 cm. During rainy season, seeds are sown at the above spacing in raised beds of 90 cm width. Trailing varieties are sown at the above spacing in raised beds of 90 cm width. Trailing varieties are sown in pits of 45-60 cm diameter and 30-45 cm depth at a spacing of 2 × 2m with 3 plants/pit for trailing on bower. Trailing varieties are also grown on trellis by sowing seeds in channels at 1.50 × 0.45m spacing.

Cropping system: cowpea is intercropped with cotton, sorghum or pearl millet. Growing one or two rows of cowpea in widely spaced crops and incorporating of their biomass after picking pods can increase soil fertility and yield of companion crop.

(a) For grain/vegetable purpose: Cowpea-Wheat-Mung/Cheena, Cowpea-Potato-urd/bean, Maize/Rice-Wheat-Cowpea, Maize-Toria-Wheat-Cowpea, Rice-Rice-Cowpea, Rice-Cowpea, and Rice-Mustard-Cowpea.

(b) Fodder purpose: Sorghum+cowpea-berseem-maize+cowpea, Maize-berseem/oat- maize+cowpea, Sudan grass-berseem /oat-maize+cowpea and Cowpea-berseem-maize+cowpea.

Weed management: Fluchloralin may be applied to the irrigated crop @ 1.5 lit/ha or pendimethalin @ 2.0 lit/ha as pre-emergence. Spray the herbicide 3 days after sowing followed by one hand weeding on 30 days after sowing. If herbicide is not applied, two hand weeding are given on 15 and 30 DAS.

Mannures and fertilizer application: FYM @ 12.5 t/ha with 12.5- 25.0-0 and 25-50-0 kg of N, P_2O_5 and K_2O per ha have to be applied to the rainfed and irrigated condition respectively. Spray DAP @ 2% at the time of first appearance of flowers and then second spray 15 days later. In acidic soil, lime pelleting of seed is beneficial along with rhizobium inoculation. Add finely powdered (300 mesh) calcium carbonate to moist freshly Rhizobium treated seeds and mix for 1-3 minutes until each seed is uniformly pelleted. Lime requirement for pelleting varies from 0.05 kg to 1 kg/10 kg seed depending on seed size.

Growth regulator: Spray NAA (Planofix) @ 40 ppm first round at the first appearance of flowering and a second round after a fortnight.

Water management: Irrigate immediately after sowing followed by life irrigation on the third day. Subsequent irrigations are given at intervals of 10 to 15 days depending upon the soil and climatic conditions. Water requirement of cowpea is 400 to 530 mm. The critical stages for irrigation are flowering and pod filling. For wetland bunds, pot water daily for a week after sowing. Cowpea is a drought resistant crop.

Crop protection: Pea aphids, serpentine leaf miner, stem fly, thrips, pod borers, leaf roller, hairy caterpillar etc. are major pests and fusarium wilt, rhizoctonia wilt, anthracnose, powdery mildew and mosaic are serious diseases affecting cowpea.

Harvest: Pull out the plants when 80% of the pods are mature and thresh after drying. Harvest the pods as they turn light straw colour and the seeds within turn brown or mottled in colour. At this stage, the moisture content of seeds will be about 18%. Harvest by hand picking at interval of 3 to 4 days is also practiced. Air dry the pods for 2 days and sun dry until they become brittle and easily break when beat with bamboo stick. At threshing, seed moisture should be about 12%. For grains, the crop may be harvested in about 90-125 days after sowing. The crop should be cut with sickle when it attains the age of 40-45 days.

Yield: A good crop of cowpea may yield about 1.2-1.5 t/ha of grain and 2.5 t/ha of stover/haulms.

Utilization: Cowpea is consumed in many forms: the young leaves, green pods, and green seeds are used as vegetables; dry seeds are used in various food preparations. The mature bean is often dried, stored, and later cooked as a pulse or used as bean sprouts by soaking in water. It is used as forage, hay, and silage. When used as forage, it should only be lightly grazed after flowering. If there are several buds left after defoliation, the plant will regenerate. The haulms are fed to livestock as nutritious supplement to cereal fodder. The large violet-blue flowers and draping pods make yard long bean a useful ornamental in city parks, office buildings, and around homes.

REFERENCES

Allen, O.N., and E.K. Allen. 1981. The Leguminosae: a source book of characteristics, uses, and nodulation. The University of Wisconsin Press, Madison, WI.

Duke, J.A. 1981. Handbook of Legumes of World Economic Importance. Plenum Press, New York, New York.

Fatokun, C.A., S.A. Tarawali, B.B. Singh, P.M. Kormawa, and M. Tamò. 2002. Challenges and opportunities for enhancing sustainable cowpea production. Proceedings of the World Cowpea Conference III held at the International Institute of Tropical Agriculture (IITA), Ibadan, Nigeria, 4–8 September 2000. IITA, Ibadan, Nigeria.

Fery, F.L. 2002. New opportunities in *Vigna*. p. 424–428. In: J. Janick and A. Whipkey (eds.), Trends in new crops and new uses. ASHS Press, Alexandria, VA.

Food and Agriculture Organization (FAO). 2012. Grassland species index. *Vigna unguiculata* http://www.fao.org/ag/AGP/AGPC/doc/Gbase/data/pf000090.htm(accessed 6 Jun. 2012)

Murphy, W.J. 1993.Tables for weights and measurement: crops. Publication # G4020. Univ. of Missouri Extension.

http://extension.missouri.edu/publications/DisplayPub.aspx?P=G4020 (accessed 6 Jun. 2012)

Sheahan, C.M. 2012. Plant guide for cowpea (*Vigna unguiculata)*. USDA-Natural Resources Conservation Service, Cape May Plant Materials Center, Cape May, NJ.

Singh, B.B., Mohan Raj, D.R., Dashiell, K.E. and Jackai, L.E.N. (eds) (1997) *Advances in cowpea research*. Co-publication of International Institute of Tropical Agriculture (IITA) and Japan International Research Center for Agricultural Sciences (JIRCAS). Ibadan, Nigeria.

Stephens, James M. 1994. *Bean, yard-long Vigna unguiculata subsp. sesquipedalis (L.) Verde*. Fact Sheet HS-562. Hort. Sci. Dept., Inst. Of Food and Agri. Sci. University of Florida. Gainesville, Florida.

Yamaguchi, M. 1983. World vegetables. AVI Publishing Company, Inc., Westport, Connecticut. 415 pp.

Borget, M., 1992. Food Legumes. Macmillan Press Ltd., London.

http://www.tropicalforages.info/key/Forages/Media/Html/Vigna_unguiculata.htm

http://dpd.dacnet.nic.in/

http://www.tropicalforages.info/key/Forages/Media/Html/Vigna_unguiculata.htm

http://msucares.com/crops/forages/legumes/warm/cowpea.html

http://www.sare.org/Learning-Center/Books/Managing-Cover-Crops-Profitably-3rd-Edition/Text-Version/Legume-Cover-Crops/Cowpeas

CHAPTER

24

Pea–*Pisum sativum* L. (2n-14)

Vernacular name: Field pea, Garden pea, Spring pea, English pea, Common pea (English), *Patani* (Tamil), *Matar* (Hindi).

Importance: Peas is grown as a vegetable crop for both fresh and dried seed. The fresh green seeds are cooked and eaten. The ripe dried seeds, either whole, split or as flour, are used for human and livestock feed. The plants and haulms are used for forage, hay and silage and also as a green manure. Pea is cool season pulse crop of five month duration.

Origin: This species originated in areas comprising Central Asia, the Near East, Abyssinia and the Mediterranean region.

Distribution It is cultivated in Europe, India, Ethiopia, Uganda, Rwanda, United Kingdom, Morocco, USA, France, Italy, Hungary, Australia, Russia, Netherlands, Canada, New Zealand, Belgium, etc. In India it is grown in Uttar Pradesh and Indo-Gangetic plains.

Botany: Pea is an herbaceous annual tendril-bearing, climbing or trailing plant with tap root system. All peas are vine plants which creep onto low bushes, rocks, or nearby poles for support as well as protection. The pea can be long-vine indeterminate (climbing, late maturing) or short-vine determinate (bush or dwarf, early maturing). Stem is hollow, upright, slender and usually single. Leaves are pinnately compound with the rachis terminating in a single or branched tendril. There are large stipules at base of leaf. Inflorescence is a raceme arising from axils of leaves. It has white or purple coloured flowers on long stalked axillary racemes with one to three flowers. The flowers of field peas are smaller than those of garden pea and are coloured. The colour of the standard is being pale lilac, the wings purplish and the keel yellowish white. The flowers of garden pea are white in colour. Peas are generally self-fertilised but cross fertilization may also occur. Gynoecium is monocarpellary with ovules (up to 13) alternately attached to placenta. Style bends at right angle to ovary and stigma is sticky. Pods are straight or curved. Pods are smooth, inflated, a little flattened or almost cylindrical; green, yellow or violet in colour. The pod contains 4 to 9 seeds. The length of the pod is 5-9 cm and shape is inflated or almost cylindrical. Seeds two to ten in a pod. The seeds are round, smooth or wrinkled and can be green, yellow, beige, brown, red-orange, blue-red, dark violet to almost black, or spotted. Seeds shape varies from round to angular to very rough. Hundred seed weight is 23 g.

There are two types of cultivated peas recognized viz., i) garden pea (*Pisum sativum ssp. hortense*): green seeds are used mostly as vegetable and for canning purposes (*Pachcha Pattani*). It is also called 'table pea'. Seeds are bold and wrinkled. The plants are generally white flowered. Leaf axils are generally green. Seeds are yellowish, whitish or bluish green and ii) field pea (*Pisum sativum ssp. arvense*): mature seeds are used as the pulse *(Pattani or Kerao)*. It is also known as 'dry peas'. This second type is distinguished by its smooth seed surface. The ripe, mature seeds are used as pulse (dal). These are also grown as forage or green manure crop. Field peas are hardy plants which are able to withstand frost. Leaf axils are often pigmented. These have coloured flowers. Seeds are round and little angular. Seeds are grayish green, grayish brown or grayish yellow.

Table 1 Differences between garden pea and field pea

Features	Garden pea - *P.sativum* var. *hortense*	Field pea - *P. sativum* var. *arvense*
Leaf axils	Generally green	Often pigmented
Stem parts	Tender	Little hardy
Flowers	Generally white	Flowers are often purplish
Seed shape	Globular	Angular and rounded
Seed colour	Yellowish, whitish or bluish green	Greyish green, greyish brown, greyish yellow or speckled.

Growth stages: The growth stages particulars are furnished in Table 2.

Table 2 Growth stages in Pea

Crop stage	Fresh	Flower
Establishment	10-25 days	10-25 days
Vegetative	25-30	25-30
Flowering (including pod set)	15-20	15-20
Yield formation (pod development and bean pod)	15-20	20-25
Ripening	0-5	15-20
Total	65-100 days	85-120 days

Climate: Pea is grown in temperate regions. It is a cool-season crop in the sub tropics and at higher altitudes in the tropics. It is grown in areas receiving mean annual rainfall of 800 to 1000 mm. It is grown up to an altitude of 1200 to 1500 m from MSL. Pea is a winter crop. It is well adapted to climates with mild winters. Peas are short day plants. Peas require a cool, relatively humid climate with temperatures of 13 to 18ºC. The cardinal minimum, optimum and maximum temperature of pea are 4ºC, 13 to 18ºC and 25ºC respectively. Optimum temperature for seed germination is 22ºC. Pea tolerates frost to -2ºC in the seedling stage, although top growth may be affected at -6ºC. Winter hardy peas can withstand -10ºC, and with snow cover protection. The optimum temperature levels for the vegetative and reproductive periods of peas are 21 and 16ºC, and 16 and 10ºC (day and night) respectively. A hot spell is more damaging to peas than a light frost. Temperatures above 27ºC shorten the growing period and adversely affect pollination. High air temperatures over 25ºC reduce the number of pods/plant and the number of seeds/pod. As temperature increases the maturity is hastened and yield is reduced. Quality of pods produced is also low at high temperature due to conversion of sugars to hemicellulose and starch. Cool weather and high air humidity of more than 70% are favorable during ripening and extend the harvest period. High temperature and humidity cause early ripening, making grains rather hard in a few days. When soil temperature is below 10ºC, nodulation is scarce affecting productivity. Vernalization at seedling stages accelerates development and stimulates early flowering. The plant can tolerate frost in the vegetative stage, but frost at flowering can cause heavy pod losses and at pod set is liable to produce deformed and discoloured seed.

Soil: Peas can be cultivated over a wide range of soils. Drainage is essential. It cannot tolerate acid soils or water logging. Peas die after 24 to 48 hours in a water-logged condition. The crop does best on loamy to clay loamy or sandy loams overlying clay. The soil pH should be 5.5 to 6.5 although some cultivars can tolerate a pH of 6.9 to 7.5. It can tolerate alkaline soils and does not in acidic soils.

Land preparation: Soil is prepared to a fine tilth with disc or mould board plough which is followed by two to three harrowing and plankings should be given to prepare a well pulverized seed bed. Avoid powdery seedbed. Field should be well leveled and should be prepared after pre-sowing irrigation to ensure adequate moisture at the time of sowing. Form beds and channels or raised beds.

Season: Sowing in mid-October is the best for high yield. Crop duration is 90 to 160 days for field pea, 60 to 85 days for green peas and 75 days for forage crops. In plains of North India, pea is sown from beginning of October to middle of November. Yield is drastically reduced when crop is sown after first week of December. Crop sown in September will be susceptible to wilt disease. In hills, pea is sown in March for summer crop and in May for autumn crop.

Varieties: Field pea varieties grown in different zones and states of India are presented in Table 3. The pea varieties are distinguished by determinate or indeterminate flowering. Determinate varieties mature in 80 to 90 days, while indeterminate varieties mature in 90 to 100 days. Field pea varieties are classified based on maturity period as i) early types varieties produce green pods in 65 days after sowing, ii) mid-season types produce pods in 85-90 days after sowing and iii) late main season types produce pods in 110 days after sowing.

Table 3 Field pea varieties grown in different states of India

Variety	Area of adoption Zone/state	Seed yield (Q/ha)	Days of maturity	Remarks
JP-885	CZ (MP, Maharashtra & Gujarat)	21.0	120-140	Resistant to Powdery mildew PM.
KFP-103 (Shikha)	NWPZ (Punjab, Haryana, Delhi, West UP & North Rajasthan)	15-20	130-140	Resistant to PM.
DMR -7 (Alankar)	NWPZ	20-25	115-135	Resistant to PM.
Uttra (HFP-8909)	NWPZ	20-25	120-140	Resistant to PM., dwarf
Sapna (KPMR-1441)	Uttar Pradesh	20-25	120-130	Resistant to PM., dwarf
Jayanti HFP-8712	Haryana	20-25	120-140	Resistant to PM., Bold Seeded
Swati (KFPD -24)	Uttar Pradesh	25-30	110-125	Resistant to PM. & tolerant to rust Dwarf, escapes leaf minor

Malviya Matar-15 (HUDP-15)	NEPZ (East UP, Bihar, West Bengal). NHZ	25-30	110-130	Resistant to PM., rust and leaf miner
DDR-23 (Pusa Prabhat)	NEPZ	15.0	95-115	Extra early, Resistant to PM
Ambika	CZ	15-20	100-125	Resistant to PM, Tall Plants
DDR -27 (Pusa Panna)	NWPZ	18.0	100-115	Very early, Resistant to PM
Indra (KPMR -400)	CZ	20.0	105-115	Dwarf type, Resistant to PM
Shubhra (LM -9101)	Chhattisgarh	15-20	90-95	Resistant to PM.
Jay (KPMR -522)	NWPZ	23.0	120-140	Dwarf type, Resistant to PM
Adarsh (IPF 99-25)	CZ	23	110-115	Resistant to Powdery Mildew
Vikas (IPFD 99-13)	H.P., Maharashtra, C.G., Gujarat & Bundel khand region of U.P.	23	102	Resistant to PM and tolerant to rust
Prakash (IPFD -1-10)	M.P., Maharashtra . Gujarat, Bundkhand region of UP, J &K, H.P. and Uttarakhand	21	94-121	Resistant to PM and tolerant to rust
Paras	Chhattisgarh	18-24	92-119	Resistant to Powdery Mildew
Pant P -14	Uttarakhand	15-22		Resistant to rust and powdery mildew
VL – Matar -42	Eastern U.P., Bihar, Jharkhand, East Bengal, Assam	20	108-155	Resistant to PM, Moderate resistant to rust
Hariyal (HFP -9907B)	Punjab, Haryana, Rajasthan, Delhi, Western U.P.	17-20	128	Resistant to PM and tolerant to rust
Pant Pea -25	Uttarakhand	18-22	125-128	Resistant to PM & Mod. Resistant to rust
HFP -9426	Irrigated areas of Haryana	20	135	Res. to PM and tolerant to root rot. Mod. Resistant to nematodes.
Pant Pea -42	Western UP, Northern Rajasthan, Punjab, Haryana and plains of Uttarakhand	22	113-149	Resistant to powdery mildew and mod. Resistant to pod borer and stem fly
Swarna Tripti	Jharkhand, Bihar, & WB.	25	65-70	Resistant to rust and powdery mildew. Tolerant to pod borer
Vivek Matar-10 (VP 101)	Uttar Pradesh & Uttarakhand	72-98 (pods)	120-130	Mod. Resistant to PM, white rot, wilt & leaf blight. Less incidence of pod borer

Pant p 13	Western UP, Rajasthan	24-26	110-115	Resistant to powdery mildew
GOMATI (TRCP -8)	Uttarakhand Hills, Jammu & Kashmir and North eastern states	22-24	87-97	Suitable for late sown condition resistant to PM. Tolerant to pod borer and stem fly
Aman (IPF 5-19)	Punjab, Haryana, Plains of Uttarakhand west UP, Delhi and Parts of Rajasthan	22	124-137	Res. to PM and tolerant to rust. Mod. resistant to pod borer and stem fly
IPF 4-9	Suitable to irrigated areas	17	129	Resistant to powdery mildew and mod. resistant to pod borer and stem fly
VL Matar-47 (VL47)	Uttarakhand	14.0	142-162	Resistant to wilt, Rust and powdery mildew
Dantiwada Field pea 1 (SKNP 04 -09)	Uttar Pradesh, Bihar, Jharkhand, and West Bengal.	17.0	98-123	Resistant to powdery mildew

Note: CZ-Central Zone, SZ –South Zone, NEPZ – North Eastern Plane Zone, NWPZ- North Western Plain Zone

Seeds and sowing: Seed rate is 75 to 100 kg/ha for small seeded cultivars and 130 to 150 kg/ha for large seeded cultivars. Seed treatment with fungicides like Thiram / Captan /Carbendazim @ 3.0 g/kg seed should be done before sowing. This is followed by seed treatment with rhizobium culture @ one packet per 10 kg seed and Trichoderma @ 4.0g/kg seed may also be done before sowing. Spacing is 30 × 15 cm. Sowing depth is 3 to 5 cm.

Cropping systems: Peas are grown alone or with cereals for silage and green fodder. Peas can also be grazed while in the field. Peas are grown as green manures and cover crops because they grow quickly and contribute nitrogen to the soil.

Inter-cultivation: When plants are 15 cm high, tall varieties should be stacked with wooden sticks or twigs for trailing. A single row of stakes fixed in middle of raised bed will support both rows of plants in each bed. Earthing up and hoeing is also important operations in peas and helps in root development and growth of plants.

Weed management: Peas grows relatively quickly and forms a canopy which helps the suppression of weeds. Lasso (alachlor) @ 0.75 kg a.i. or tribunal @ 1.5 kg a.i./ha or pendemethalin @ 0.5 kg a.i./ha as pre emergence spray along with one hand weeding at 25-45 days after sowing is recommended to control weeds.

Manures and fertilizers application: FYM at 20 t/ha, 50 kg N, 20 kg P_2O_5, 40 kg K_2O and 15 kg $ZnSO_4$ per ha are applied as basal dressing. High dose of N have adverse effect on nodule formation and N fixation. Fertilizer should be applied in bands at 7-8 cm away and 2.5 cm deeper from seeds. Application of sodium molybdate @ 40 kg/ha either as pre- or post-emergence spray is reported to increase yield and collar rot resistance in peas.

Water management: Water requirement of pea is 350 to 500 mm. Maximum peas yield can be obtained when soil moisture is kept at 60% of field capacity during the period from emergence to just before flowering and at least 90% during flowering. Pea responds to two irrigations given at pre-flowering and flowering stages (40 and 60 DAS). Excessive irrigation in earlier stages increases vegetative growth. Light irrigations at 10-15 days intervals are given for pea. Flowering, fruit set and grain filling periods are critical stages and care should be taken to irrigate crop at these stages.

Crop protection: Stem fly, pea aphid, leaf miner and pod borer are major pests and wilt and root-rot, powdery mildew, rust, *Ascochyta* blight and pod rot are major diseases of pea. Timely sowing is done to avoid powdey mildew and rust, seed treatment with carbendazim + thiram (1+2 g/kg seed), or Trichoderma (4g) + carboxin (1g/kg) seed. Foliar spray of endosulfan 0.07%, foliar spray of wettable sulphur (0.2 -0.3%) against rust and powdery mildew, carbendazim (1g/1 of water) or dinocap, karathane 48 EC (1 ml/1 lit of water) are also used against powdery mildew, removal of diseased plant debris is to be done.

Harvest: Dry peas are harvested when the leaves turn to yellow and the lower pods begin to wrinkle and the peas harden. Generally, plants mature from the bottom up. Crops are nearing maturity when the bottom 30% of pods are ripe, the middle 40% of pods and vines are yellow-coloured, and the upper 30% of pods are turning yellow. This is the proper stage for crops to be swathed or desiccated. Dry peas can be harvested when seed moisture is less than 15%. Complete dry pods along with the plants are harvested and kept for 6 to 8 days for drying. Ideal threshing and separation occurs when the crop is below 13% seed moisture. Garden or green peas are harvested before the seed is mature for the fresh or fresh-pack market. Peas for fresh market are harvested when they are well filled and when colour changes from dark green to light green. When the pea pods are swollen (appear round) and turn light green in colour they are ready to be picked. Pick a few pods every day or two near harvest time to determine when the peas are at the proper stage for eating. Peas are of the best quality when they are fully expanded but immature, before they become hard and starchy. Since tender peas with high sugar content fetch premium price in market, care should be taken to harvest pods at correct maturity. The picking of green pods should be done by giving a simple jerk to the pedicel with minimum possible disturbance to the plant. During maturity, sugar content decreases and polysaccharides and insoluble nitrogen compounds like protein increases. Calcium migrates to seed coat and becomes tougher during ripening. Toughness of seeds is determined using Tendrometer, especially for processing purposes. Peas with low tendrometer reading is offered high price. If harvesting is delayed when attains maturity, the peas lose colour and become brittle, which reduces their market value. The seed losses due to shelling can be avoided by harvesting the crop in the early morning when the pods are more turgid. Green peas are harvested at the immature stage when the pods are well filled but the peas are sweet and soft. For marketing the pods of garden peas, pods are picked by hand in 7-8 pickings at weekly interval which spread over 5-7 weeks.

Yield: Green pod yield varies with duration of variety and is 2.5 to 4.0 t/ha for early varieties, 6.0-7.5 t/ha for mid-season varieties and 8.0-10.0 t/ha for late varieties. Shelling percentage ranges from 35-50. Seed yield varies from 2.0 to 2.5 t/ha. Peas after harvesting are packed in gunny bags or crates. Fresh unshelled peas can be stored for two weeks at 10°C and 90-95% RH. Hundred seed weight is 15 to 25 g. Moisture levels up to 16% and temperatures below 15°C are considered for safe seed storage.

Storage: Unshelled pea pods can be kept better than shelled pea. At room temperature the pods can be kept for two to three days provided they are frequently sprayed with fresh water but in cold storage they can be kept for 15 to 20 days at 0°C with 85 to 90% relative humidity.

Utilization: The peas are used as the food or feed. The pea is consumed as fresh peas. It is also used as canned or frozen or dry peas. Dried pea is used for human consumption as whole, split as *dhal* or roasted or parched or ground peas. Tender seeds are used in soups. Canned, frozen and dehydrated peas are used during off-season. Pea contains 15 to 35% protein and high concentrations of the essential amino acids lysine and tryptophan.

REFERENCES

CPG. 2012. Crop Production Guide. Department of Agriculture, Government of Tamil Nadu, Chennai and Tamil Nadu Agricultural University, Coimbatore.

Hancock, J.F. 2004. Plant evolution and the origin of crop species. CABI Publishing, Wallingford, UK and Cambridge, MA.

Kaplan, Lawrence. 2000. Beans, Peas, and Lentils. The Cambridge World History of Food. Eds. Kenneth F. Kiple, Kriemhild Conee Ornelas. 2 vols. Cambridge University Press, New York.

NRAA. 2011. Technology for increasing production of Rabi Crops in Bundelkhand. Technical Bulletin No.1. National Rainfed Area Authority, New Delhi, India: 66p.

Pavek, P.L.S. 2012. Plant guide for pea (*Pisum sativum* L.). USDA-Natural Resources Conservation Service, Pullman, WA.

CHAPTER

25

Grass Pea / Kesari Bean–*Lathyrus sativus* L. (2n=28)

Vernacular name: Grass pea, Chickling pea, Lathyrus pea, Chickling vetch (English), *Chatri, Matri, Khesari, Khesari* (Hindi), *Lang* (Gujarati) and *Kesari paruppu* (Tamil).

Importance: Grass pea is widely cultivated as a food crop. There are two forms of Grass pea/*kesari* beans available viz., the small seed types are known as *lakhodi* and big seed types are known as *lakh or teora*. Large white seeds are good for human consumption, but only after careful boiling.

Origin: Grass pea is a native of southern Europe and Western Asia (Balkan peninsula).

Distribution: It is grown in India, Bangladesh, Pakistan, Nepal, Ethiopia, Europe and parts of South America. In India, it is grown in West Bengal, Uttar Pradesh, Madhya Pradesh and Gujarat.

Botany: Lathrus is an herbaceous annual, straggling or climbing by means of leaf-tendrils, plants reaching 1 to 1.25 m in height; stem angular, soft. Leaves alternate; pinnately compound; flowers solitary; fruit a flat pod, 2.5 to 5.0 cm long. Grass pea types with blue flowers and speckled (coloured) seed coat are common in the Indian subcontinent whereas white flowers with white seed coat are common in the Mediterranean region and in many European lines. Anthesis takes place between 10 a.m. and 4 p.m. Normally self-pollination takes place. However, natural cross pollination ranges from 1 to 13%. Pods are oblong, flat, slightly bulging over the seeds, about 2.5-4.5 cm in length, 0.6-1.0 cm in width and slightly curved. The dorsal part of the pod is 2-winged, shortly beaked and contains 3-5 small seeds. Seeds are 4-7 mm in diameter, angled and wedge-shaped. Seed colour is white, brownish-grey or yellow, although spotted or mottled forms also exist. Hilum is elliptic and cotyledons are yellow to pinkish yellow. The test weight of 100 seeds is 5 to 6 g.

Climate: Grass pea is a cold season crop. The cardinal minimum, optimum and maximum temperature is 10°C, 20 to 25°C and 30°C respectively. It is grown in areas with annual rainfall of 400 to 650 mm. It tolerates excess rains and flood. It is highly drought tolerant. It grows up to an altitude of 1200 m from MSL.

Soil: It can be grown on a wide range of soils.

Season: The crop is sown in October-November as a winter crop. Early maturing grass pea varieties takes 60 days to 50% flowering and 120 days to maturity while late maturing varieties takes 150 to 180 days to attain maturity.

Varieties: Lathyrus pea varieties grown in different states of India

Table 1. Lathyrus pea varieties grown in different states of India

Variety	Area of adoption Zone/state	Seed yield (Q/ha)	Days of maturity	Remarks
Bio-L-212 (Ratan)	NEPZ (Uttar Pradesh, Bihar, West Bengal)	15.0	108-116	Tolerant to stress, Low ODAP, Bold seed, Blue flower.
Prateek	Madhya Pradesh	6-9 (Utera) 11-15 (sole)	110-115	Tolerant to downy mildew & moderately resistant to powdery mildew.
MahaTeora	Chhattisgarh	15	94	Tolerant to nematode & thirps, mod. Resistant to PM

Land preparation: Grass pea is grown as rice fallow crop utilizing the residual soil moisture.

Seed and sowing: Seed rate of 45 and 30 kg/ha is recommended for pure crop and intercrop respectively. Seeds are broadcasted in standing crop of rice. Seeds are also drilled. Germination is hypogeal. Spacing is 30 × 5 cm. Seeds are treated with fungicides and rhizobial culture.

Manures and fertilizer application: Apply 20 kg of P_2O_5 per ha as basal dressing.

Cropping systems: It is largely grown as catch crop. It is sown 15 days before the harvest of rice which is known as '*Paira*' or '*Utera*' cultivation in Uttar Pradesh and Bihar.

Harvest: The pods are harvested as soon as the leaves turn yellow and the pods turn grey when mature. When fully ripe, the pods dehisce and there is considerable seed loss. The plants are either pulled out by hand or cut with a sickle near the base or uprooted. The plants are spread out on the threshing floor and beaten with sticks and then left to dry for approximately 7 days before being threshed, winnowed, dry in the field or on the threshing floor for 7-8 days and stored similarly to other grain legumes. Pod splitting, which leads to premature shattering of seeds, is common in small type grass pea.

Yield: Seed yield of 1000 and 450 kg/ha can be obtained under irrigated and rainfed conditions respectively. Hay yield is 1200 to 1400 kg/ha. Hundred seed weight is 7 to 16 g.

Utilization: The grass pea is grown for its grain, fodder or green manure. The vegetative types are utilized in the production of fodder or forage for animals. The seeds are boiled and eaten and made into chapattis, paste balls and curries. The seeds are fed to livestock. The leaves are eaten as pot herb. The seeds eaten over a long period can cause lathyrism, a paralysis of the lower limbs in man. In India, the grains are sometimes boiled whole, but are most often processed through a *dal* mill to obtain split *dal*. In Nepal the dried grains are split either in a stone grinder on a home scale or milled to make *dal* which is consumed with rice. Lathyrus dhal is also known as Lattri, Lang and Teora.

Lathyrism or Lathyrismus: It is a serious neurological disease which causes paralysis of the lower limbs on both man and animals. The toxin ß-N-oxalyl-L-a-ß-diaminopropionoc acid (ODAP also known as BOAA) has been identified as the causative principle for lathyrism and is present in all parts of the plant. The toxin is reduced to 90% by steeping the seeds or dhal in water followed by sun drying or parboiling similarly to rice. Lathyrism may set in many ways. Usually a few days prior to actual paralysis, one experiences a sudden agonizing pain in the muscles. Later, it is difficult to stretch legs. There is bending at the knee joints and heaviness in the limbs. This stiffness of muscle goes on increasing till the individual is unable to walk. 75% of the victims are landless labourers who paid their wages in the form of food grains. Lathyrism is high during famine and drought years.

REFERENCES

Ahmed, N. 1985. *Lathyrus*: Guide book of pulses in Bangladesh. FAO/UNDP. Dhaka.

Deshpande, S.S. and S. Damodaran. 1990. Food legumes: Chemistry and technology. Adv. Cereal Sci. Technol. 10:147.

Godt, M.J.W. and J.L. Hamrick. 1991. Genetic variation in *Lathyrus latifolius* (Leguminosae). Am. J. Bot. 78:1163-1171

Jackson, M.T. and A.G. Yunus. 1984. Variation in the grass pea (*Lathyrus sativus* L.) and wild species. Euphytica 33:549-559.

Jeswani, L.M., B.M. Lal and Shiv. Prakesh. 1970. Studies on the development of low neurotoxin (N-N- oxalyl amino alanine) lines in *Lathyrus sativus* (Khesari). Curr. Sci. 22:518.

Kislev, M.E. 1989. Origins of the cultivation of *Lathyrus sativus* and *L. cicera* (Fabaceae). Econ. Bot. 43:262-270.

Kuo, Y., J.K. Khan and F. Lambein. 1994. Biosynthesis of the neurotoxin b-ODAP in developing pods of *Lathyrus sativus*. Phytochem. 35:911-913.

Leakey, C. 1979. Khesari Dhal - The poisonous pea. Appropriate Technology 6:15-16.

Nerkar, Y.S. 1972. Induced variation and response to selection for low neurotoxin content in *Lathyrus sativus*. Ind. J. Genet. and Plant Breed. 32(2):175-180.

Prakesh, S., B.K. Misra, R.N. Adsule and G.K. Barart. 1977. Distribution of b-N-oxalyl-L.ab diaminopropionic acid in different tissues of aging *Lathyrus sativus* plant. Biochem. Physiol. Pflanzen. 171:369-374.

Prassad, A.B. and A.K. Das. 1980a. Relative sensitivity of some varieties of *Lathyrus sativus* to gamma irradiation. Ind. J. Cytol. and Genet. 15(1):156-165.

Prassad, A.B. and A.K. Das. 1980b. Morphological variants in Khesari. Ind. J. Genet. and Plant Breed. 40(1):172-175.

Sinha, S.K. 1977. Food Legumes: distribution, adaptability and biology of yield. Food and Agriculture Organization of the United Nations, Rome.

Smartt, J. 1984. Evolution of Grain Legumes. I. Mediterranean Pulses. Exp. Agric. 20:275-296.

Spencer, S. P. and H.H. Schaumburg. 1983. Lathyrism: A neurotoxic disease. Neurobehav. Toxicol. Teratol. 5(6):625-629.

Vedna Kumari, R.B. Mehra, D.B. Raju and K. Himabindu, 1993. Genetic basis of flower colour production in grasspea. Lathyrus and Lathyrism : 5 (1); pp 10.

CHAPTER

26

Lentil–*Lens esculenta* Moench (2n=14)

Synonyms: *Lens culinaris* Medik.

Family: Fabaceae (Leguminoceae)

Vernacular name: *Masur* (Hindi), *Masur paruppu* (Tamil), *Masooru bele* (Kannada), *Mausuri* (Bengali).

Importance: Lentil is an edible pulse. Lentil is not a high yielding grain legume. Small plant size and small seeds restrict attempts to boost yields. Lentil is inherently a less water demanding (drought tolerant) plant and is often a preferred crop in the water deficient areas. The young pods are used as vegetable in India. The split seeds (dhal) are used in soups. The percentage recovery of the dhal is 65 to 80%. The husks, bran, and dried haulms provide fodder for livestock. In India, lentils soaked in water and sprouted lentils are offered to gods in many temples. Lentil has the anti-nutritional factors such as trypsin inhibitors and relatively high phytate content. The phytates can be reduced by soaking the lentils in warm water overnight. Lentils are a good source of iron.

Origin: Lentil is native of Asia Minor or near east i.e., Israel, Syria, Turkey, Iraq and Iran.

Distribution: Lentil is cultivated in USA, India, Mexico, Chile, Peru, Argentina and Columbia. India is the second largest lentil producer in the world after Canada. In India, lentil is mainly grown in Uttar Pradesh, Madhya Pradesh and Bihar while in West Bengal, Rajasthan, Assam, Haryana and Punjab on a limited scale.

Botany: Lentil is a herbaceous annual, bushy, not exceeding 60 cm in height; leaves pinnately compound, leaflets five to seven pairs, flowers carried on axillary slender peduncles, fruit a legume or pod, short, flattened; seeds generally two per pod. The crop is generally self-pollinated crop. *Lens culinaris* is divided into two sub species. The characteristic features are presented in Table 1.

Table 1 Differences between two lentil species

Characters	Macrosperma	Microsperma
Region	Mediterranean	Indian subcontinent
Plant height (cm)	30-40	20-25
Seed size	Large seeded	Small seeded
Hundred seed weight (g)	5.4	3.5-4.5
Colour	Yellow cotyledon	Orange or yellow cotyledon
Pigmentation	Little or no in flowers and vegetative structures	Presents in pods, leaf later
Common trade name	Masur, Malka masur	Musri

Climate: Lentil is a long day plant while some cultivars are day neutral. The cardinal minimum, optimum and maximum temperature for growth of lentil is 15°C, 18 to 25°C and 30°C respectively. It is grown in areas receiving mean annual rainfall of 750 to 1000 mm. It is grown upto an altitude of 1500 to 2100 m from MSL.

Soil: Lentil can be grown on a wide range of soil types such as sandy loam soils, alluviums, black cotton soils and soil texture ranging from heavy clayey soils to low lying sandy soils. In India, good yields are obtained on light loamy, alluvial soils and on a moderately deep black cotton soils. Lentils can be grown in moderately alkaline or saline soils with a pH of 6.0 to 8.0. Water-logging during seed germination and initial seeding growth phase is a constraint for pulses grown in rice-fallows.

Season: In India, Lentils are grown in the winter season following the *kharif* rice. Seeds are sown during October and harvested in April. The optimum time of sowing is from second fortnight of October to first fortnight of November. The crop duration of early and late maturing lentil varieties is 80 to 110 days and 125 to 130 days, respectively.

Varieties: Lentil varieties with their characteristics in different states of India are furnished in Table 2.

Table 2 Lentil varieties grown in different state of India

Variety	Area of adoption Zone/ State	Seed yield (Q/ha)	Days of maturity	Remarks
JL 1	MP	8.0	120-125	Early, Tolerant to wilt, Seed bold
Sapana (LH 84-8)	NWPZ (Punjab, Haryana, Delhi, West UP)	15.0	135-140	Tolerant to Rust & Bold Seeded
VL Masoor 4	Uttaranchal	12.5	168	Tolerant to wilt & Rust, Small seeded & black.
Pant lentil -4 (PL -81-17)	NWPZ (Punjab, Haryana, Delhi, West UP, North Rajasthan)	16.0	140-145	Resistant to Rust & tolerant to wilt
Lens- 4076	NWPZ (Punjab, Haryana, Delhi, UP) CZ (MP, Maharashtra)	14.0	130-135	Tolerant to wilt & Rust, seed bold
DPL -15 (Priya)	NWPZ (Punjab, Haryana, Delhi, West UP)	15-18	130-135	Tolerant to wilt & Rust, bold seeded,

Pusa Vaibhav (L-4147)	NWPZ (Punjab, Haryana, Delhi, West UP)	20-24	130-135	Resistant to Rust & Tolerant to wilt, small seeded.
Garima (LH – 84-6)	Haryana	15-20	135-140	Tolerant to Rust, wilt & Blight. Bold
Narendra Masoor-1	Uttar Pradesh.	14.0	125-130	Resistant to Rust & Tol. to wilt
DPL -62 (Sheri)	NWPZ (Punjab, Haryana, Delhi, West UP)	17.0	130-135	Resistant to Rust & wilt, bold seeded.
Subrata	West Bengal	12-18	120-125	Tolerant to Rust, bold seeded.
JL-3	CZ (MP, Maharashtra)	15-19	115-120	Tolerant to wilt, bold seeded.
VL Masoor 103	Uttaranchal	12-14	1645	Tolerant to Rust, small seeded.
Noori (IPL -81)	CZ (MP, Maharahtra)	17-18	110-120	Tolerant to Rust, wilt, bold seeded
Pant Lentil -5	Uttaranchal	15-18	135	Resistant to Rust, bold seeded.
Malaviya Vishwanath (HUL 57)	Eastern and Central U.P., Bihar, Jharkhand, West Bengal and Assam	14.0	130	Resistant to rust & wilt, small seeded.
KLS 218	NEPZ (East Uttar Pradesh, Bihar, West Bengal).	14-15	125-130	Tolerant to Rust, wilt, small seeded
VL- Masoor - 507	J & K, H.P., Uttaranchal, North Eastern Hills	10-12	140-209	Resistant to wilt
Haryana Masoor -1 (LH – 89-48)	Haryana	14	138	Moderate resistant to all disease
VL Masoor 125	Uttarakhand	19-20	115-117	Resistant to wilt
VL Masoor 125 (VL -126)	Uttarakhand, H.P., J & K and North Eastern Hills	12-13	126-212	Resistant to GM and Moderately resistant to wilt and rust
IPL -406 (Angoori)	Punjab, Haryana, North Rajasthan, Plains of Uttarakhand and Western UP	17	120-155	Resistant to rust and wilt
Pusa Masoor 5(L -45994)	Delhi	17-18	120-128	Resistant to rust moderately resistant to pod borer
Moitree WBL 77	East UP, Bihar, Jharkhand, Assam & WB	15	117	Resistant to wilt and grey mould
Shekhar Masoor 2 (KLB -303)	Uttar Pradesh	14	128	Moderately resistant to wilt and rust
Shekhar Masoor3 (KLB -320)	Uttar Pradesh	14	128	Moderately resistant to wilt and rust

Pant Lentil 7 (PL 024)	Punjab, Harayana, UP	15	147	Resistant to wilt to rust & Pod borer
Pant Lentil 8 (PL 063)	Punjab, Harayana,Planis of Uttarakhand, Western UP, Delhi and Rajasthan	15	135	Moderately resistant to rust and wilt, resistant to pod borer
Pant Lentil-6 (PL -02)	Uttarakhand	11	125-145	Resistant to rust, Wilt, Ascochyta Blight and Tolerant to pod borer
VL Masoor-129	Uttarakhand	9.0	151	Resistant to Wilt and root rot and no infestation of pod borer
VL Masoor-133 (VL 133)	Uttarakhand	11	150	Resistant to wilt, root rot and rust
VL Masoor-514 (VL 514)	Uttarakhand	10	149-159	Moderately resistant to wilt and root rot disease. Tolerant to pod borer
LL 931	Punjab	12-13	146-147	Resistant to lentil, rust. Tolerant to pod borer

Land preparation: Two to three ploughings are given to prepare the land. Beds and channels are formed under irrigated condition.

Seeds and sowing: Seed rate is 25 to 30, 50 to 60 and 60 kg/ha for small, bold seeded varieties, for *utera* crop respectively while 12 kg/ha for intercrop with barely, mustard and castor. Treat the seed with benomyl or agrosan GN @ 2 g/kg of seed before sowing. The lentil seed should be treated with *Rhizobium* culture before sowing. Spacing is 30 x 22.5 cm. Sowing depth is 2.5 cm.

Cropping Systems: Generally lentil is grown after the harvest of *kharif* crops or as the sole crop in *rabi* season. Rice-lentil, Groundnut–lentil, Sorghum–lentil, Pearlmillet–lentil, Maize–Lentil, Cotton–lentil, etc., are the most common rotations. Lentil is grown mixed with barley, *toria*, rape and mustard crops. Intercropping of linseed+lentil (2:1), lentil+mustard (4-6:1) can be practiced.

Weed management: Critical weed free day for lentil is 40 days from sowing. Fluchloralin (pre-plant incorporation) and pendimethalin (pre-emergence) @ 0.75-1.0 kg/ha or Prometryn @ 1.0 kg/ha can be used for control weeds.

Manures and fertilizers: 20-40-10-15 kg of N, P_2O_5, K_2O and $ZnSO_4$ per ha are applied as basal dressing. The 'nitrogen hunger' phase, which is often experienced when crops are seeded early into cool, wet soil before significant symbiotic dinitrogen fixation begins, can be avoided by the application of a small starter dose of 10 to 25 kg/ha inorganic nitrogen fertilizer placed adjacent to, but not in contact with, the seeds. The zinc deficiency can be rectified by spraying 0.5% $ZnSO_4$ and 0.25% lime.

Water management: Lentil is generally raised as rainfed crop utilizing winter rains. Two to three irrigations are given at four to six leaf, flower and pod formation stages to obtain high yields. The crop requires 200 mm of irrigation water after one pre-sowing irrigation.

Crop protection: Lentil crop is affected by insect pests such as *Sitona* weevil, cutworms, aphids, thrips, bud weevil, podborer, root aphids, leafminer, and stink bugs, alfalfa weevil, semilooper and armyworms. During storage, seed beetles such as *Bruchus* spp. and Callosobruchus spp., can cause severe damage. Some common diseases are collar rot, fusarium wilt, dry root rot, stunt viral disease, Ascochyta blight, rust, etc.

Harvest: Harvesting takes place when the pods are a golden yellow colour and the lower ones are still firm. It cannot be delayed further since seeds in pods become over ripe and shatter, resulting in a considerable yield loss. In many areas, the plant is cut down to the ground level manually and left to dry for about 10 days before being threshed and winnowed. The seeds should be dried to moisture content of 12%.

Yield: Seed yield of 900 to 1200 kg/ha can be obtained under irrigated conditions while, 400 to 700 kg/ha under rainfed conditions. A well-managed crop yields about 1.8-2.0 t/ha of grain and 3.0-4.0 t/ha of straw. Seed size is 3 to 9 mm. Thousand seed weight of small seeds is 40 g whereas grain weight of large seed is 82 g.

Utilization: Lentil is used as '*dhal*' and as an ingredient in soups. Flour prepared from the ground seeds can be used with cereal flours in cakes or bread. It is used in the preparation of invalid and baby foods. In Uttar Pradesh, the whole seed is often eaten salted and fried. Seeds are used as a source of commercial starch for use in the textile and printing industries. The protein content in dry lentils is around 26%. The young immature pods may be eaten as a vegetable. Lentils are also grown as green manure or forage crop. The residue left after starch extraction can be used as an animal feeding stuff. Lentil is considered to be the most easily digested grain legume in the presence of heat. Seeds contain trypsin inhibitor and phytohaemoglutinin.

REFERENCES

Erskine, W., Muehlbauer, F.J., Sarker, A. and B. Sharma. 2009. The Lentil, Botany, Production and Uses. CABI International publication, UK.

Giller, KE. 2001. Nitrogen Fixation in Tropical Cropping Systems. CAB International, Wallingford.

Graham, P.H. and C.P. Vance. 2003. Legumes: Importance and constraints to greater utilization. Plant Physiology 131: 872-877.

Herridge, D., Peoples, M. and R.M. Boddey. 2008. Global inputs of biological nitrogen fixation in agricultural systems. Plant Soil 311: 1-18.

Kay, D. 1979. Food pulses. Tropical Development and Research Institute. TPI Crop and Product Digest No. 3. p.48-71. UK.

Muehlbauer, F. J., Kaiser, W. J., Clement, S. L. and R. J. Summerfield. 1995. Production and Breeding of Lentil. Advances in Agronomy, *54*, 283-332.

CHAPTER

27

Hyacinth Bean–*Lablab purpureus* (L.) Sweet (2n=22)

Synonyms: *Dolichos lablab* L., *Lablab niger* Medikus., *Lablab vulgaris* Savi

Family: Fabaceae (Leguminoceae)

Vernacular name: Hyacinth bean, poor man's bean, Tonga bean (England), lubia (the Sudan), batao (Philippines), hyacinth bean (Brazil). The name '*Lablab*' is an Arabic or Egyptian name describing the dull rattle of the seeds inside the dry-pod.

***Dolichos lablab* var. *typicus* Prain:** Garden bean, Hyacinth bean, Lablab bean, Bonavist bean, Seim bean, Lubia bean, Egyptian bean, Indian butter bean, (English), *Sem* (Hindi) *Shim* (Bengali), *Val* (Gujarathi), *Pavta* (Marathi), *Chikkudu* (Telugu), *Avarai, Pandal avarai* (Tamil), Chapparadavare (Kannada) and *Avara* (Malayalam).

***Dolichos lablab* var. *lignosus* Prain:** Field Bean, Australian pea (English), *Ballar* (Hindi), *Val* (Gujarathi), *Anumulu* (Telugu), *Mochai* (Tamil), *Avare* (Kannada) and *Mochakotta* (Malayalam).

Importance: Pod/grain of *Lablab purpureus* var *lignosus* is boiled and eaten while *Lablab purpureus* var *typicus*.is mainly used as vegetable. It is also grown as a green manure and cover crop. The haulms, either green or as hay or silage, are used as livestock fodder and the dried seeds are also fed to livestock. It is a drought tolerant crop.

Origin: *Lablab purpureus* is native of India or Southeast Asia.

Distribution It is grown throughout the tropics and sub-tropics. In India it is grown mainly in Karnataka, Tamil Nadu and Madhya Pradesh.

Botany: Two ecotypes or sub species of hyacinth bean are viz, i) Garden bean: *Lablab purpureus var typicus* Prain (Syn: *Lablab purpureus bengalensis* Jacf.) is short lived, perennial, twining herb, usually treated as annual. Pods are longer and more tapering. The long axis of seeds parallel to the suture.It does not have oil glands and smell. Entire pod is edible. It is grown for green pods and used as vegetable and ii) Field bean: *Lablab purpureus* var *lignosus* Prain is a long lived, semi erect, bushy, perennial, but is usually treated as an annual. The pods are shorter. Pod has 4 to 6, almost round seeded. Seeds are arranged vertical to the suture. The plants have unpleasant smell. It is grown as a field crop, mainly for ripe seeds and fodder. It is known as Australian pea. Difference between garden bean and field bean is presented in Table 1.

Table 1 Distinguishing features of garden bean and field bean

Characters	Garden bean (var *typicus*)	Field bean (var *lignosus*)
Habit	Perennial twining herb requires support for normal performance	Semi erect, bushy, perennial, cultivated as annual.
Plant parts	No odour / smell	odour / smell due to an oily secretion
Flower	White or yellowish or purple	Usually white
Pod	Whole immature pod used as vegetable. Matured green seeds are also used as vegetable, Slender or soft pericarp	Green immature seeds alone as vegetable, pericarp tough, parchment like. Dried seed pulse
Seed arrangement	Parallel to the length of the suture	Vertical to the length of the suture
Photosensitivity	Photosensitive	Photosensitive
Edible part	Whole fruit edible	Generally seed alone

Climate: Lablab is adaptable to arid, semi-arid, sub-tropical and humid regions. Hyacinth bean is grown up to an altitude 1800 to 2100 m from msl. It is a hardy, drought resistant and grown as rainfed crop in areas with annual rainfall 600 to 1000 mm. It is a drought tolerant crop and grows well in dry lands with limited rainfall. It requires adequate moisture during the early stages of growth, after which its deep root system enables it to sustain growth on residual soil moisture. Both short day, long-day and day neutral cultivars are available. It starts flowering in short days (11-11.5 hour's day length). Short day cultivars take 6 to 7 weeks to flower according to sowing date. The cardinal minimum, optimum and maximum temperature requirement is 10°C, 14 to 28°C and 35°C respectively.

Soil: Hyacinth bean is grown on a wide variety of soils. Well drained sandy loamy soil is preferable. It can tolerate pH ranges of 5 to 8.5. It cannot tolerate waterlogging or brackish soil.

Land preparation: Prepare the field to fine tilth. Form beds and channels for bushy types and form pits of one cubic foot for trailing types.

Season: Hyacinth bean is grown during June-July and September-October under rainfed conditions whereas it is raised during February-March under irrigated condition.

Varieties: The varieties of lablab with their features are furnished in Table 2 to Table 5.

Table 2 Varietal particulars of ***Lablab*** in IARI, New Delhi

Variety	Sailent features
Pusa Early Prolific	It bears early, medium-sized, thin, stringless pods in clusters.
Pusa Sem-2	Its pods are long, dark-green stringless and semi-round in shape. It is high yielding, tolerant to viruses and insect pod borer, etc. Flowers appear on separate spikes, above the plant canopy.
Pusa Sem-3	The pods are green, meaty, very tender, stringless and flat in shape. It is high yielding; flowers appear on nodes in clusters under the plant canopy. It is tolerant to viruses.

Table 3 Varietal particulars of field bean (var ***lignosus***) in Tamil Nadu

Characters	Duration (days)	Days to 50 % flowering	100 seed weight (g)	Pod yield (kg/ha)	
				irrigated	rainfed
CO 1	140	75-90	24.4	-----	1600
CO 2	105	30-45	20.0	1400	900

Table 4 Varietal particulars garden bean (var typicus) in Tamil Nadu

Characters	Duration (days)	Days to 50 % flowering	100 seed weight (g)	Pod yield (t/ha) (irrigated)
CO 3	240	120	----	10.3
CO 4	240	120	----	13.5
CO 5	120	120	----	6.0
CO 6	120	45	24.5	4.5
CO 7	120	45	25.9	4.0
CO 8	120	45	24.4	4.75
CO 9	120	45	32.5	7.5
CO 10	120	45	22.5	7.2
CO 11	95-100	40	35.9	9.9
CO 12	100-110	40	38.4	9.7
CO 13	110-120	40	35.2	10.0

Table 5 Varietal particulars of ***Lablab*** in Karnataka and Maharashtra

Characters	Duration (days)	Days to 50 % flowering	Sailent features
HA-1	90-100	75-90	non-season bound variety, seeds have a good cooking quality
HA-2	105	30-45	It is bushy, determinate with no tendrils
HA-3	135		It is bushy, determinate with no tendrils. It is photo-insensitive and determinate type with yields between 1.2-1.5 t/ha.
HA-4	95-100	40-45	Its green pods can be harvested in 65- 70 days. It produce 1.0 to 1.2 t/ha of dry seeds and 4.5 to 5.0 t/ha of green pods.
Konkan Bhushan	60	40	It is a bush type plant, 60–75 cm tall. The pods are tender and stringless. A plant produces 125–180 pods. Pod yield is 8.8–13.6 t/ha.

Seeds and sowing of field bean: Crop is raised through seed. Germination is epigeal and normally takes about 5 days. Seed is reported to remain viable for two to three years. The seed rate for pure and intercrop is 20 and 10 kg/ha for rainfed crop while it is 25 and 12.5 kg/ha respectively for irrigated crop. Seed treatment can be done with Talc formulation of *Trichoderma viride* @ 4g or *Pseudomonas fluorescens* @ 10 g/kg seed or Carbendazim @ 2 g/kg or Thiram @ 4 g/kg seed. Then seed treatment with rhizobial culture is to be carried out. Dibble the seeds with spacing of 90 × 90 cm and 45 cm × 15 cm depending on varieties.

Weed management: Pre emergence application of Pendimethalin 2.5 lit/ha on 3 days after sowing with flat fan nozzle using 500 litres of water for spraying one ha. After this, one hand weeding on 40-45 days after sowing gives weed free environment throughout the crop period. If herbicide is applied give two hand weeding are given on 20 and 40 days after sowing.

Manures and fertilizer application: FYM @ 12.5 t/ha with 12.5– 25-0 and 25-50-0 kg of N, P_2O_5 and K_2O per ha is recommended to the rainfed and irrigated condition respectively. Spray 2% DAP at the time of first appearance of flowers and then second spray 15 days later. For pandal garden bean varieties, 115 g each in NH_4SO_4 and super phosphate may be applied basally. Soil application of 25 kg $ZnSO_4$ per ha under irrigated condition. However, a dose of 15 kg N + 40-60 kg P_20_5 and 20-25 kg K_2O per ha can be applied for forage and seed production.

Water management: Irrigate immediately after sowing followed by life irrigation on the third day and later at intervals of 15 to 20 days according to soil and climatic conditions. Flowering and pod formation stages are the critical periods when irrigation is most essential.

Growth regulators: In general about 10-20% of the flowers only develop into mature pods. Application of calcium chloride (0.1%) + NAA (100 ppm) when the flowers begin to open in the first inflorescence is reported to improve fruit–set and consequently yield.

Pruning technique in garden bean: Spacing of 300 x 120 cm is recommended. One healthy seedling is allowed per pit. The vine is propped with stick. When the vine reaches the pandal, the terminal bud is nipped. Allow the branches to trail over the pandal. Each branch may be pruned at three feet length so that the pandal is covered with vines. Branches arising on the main vine below the pandal are removed. When flowering starts, prune the tip of the branches bearing inflorescence and having three nodes from the productive axil. Continue this procedure throughout the reproductive phase.

Crop protection: The pests and diseases affecting *Lablab* are aphids, pod borer, lablab bug, mites, Powdery mildew, rust, etc. Spraying of pyrethrum at 1:800 dilution or nicotine sulphate as 1: 600 or 0.1% lindane or parathion 0.25% is recommended to control the aphids. If fruit formation had not taken place, spraying with BHC 50% wettable powder at 2 gm in one litre of water or malathion 50% at 3 ml in 5 liters of water may be done. Pods treated with pesticides should not be harvested or consumed for about a week. Spraying DDT or Pyrethrum is recommended. If the pest occurs just at the time of fruit formation, DDT 50% wettable powder at 1 kg in 300 liters of water can be sprayed to control pod borers. Hand picking of the caterpillars in early stage of attack helps in reducing the intensity of the infestation. Ploughing the fields after harvest of the crop would expose the pupae which would be destroyed by birds. Dusting DDT or spraying 50% DDT wettable powder at 1 kg in 300 liters of water is found to control the bugs. Spraying the crop with 0.05% dicofol or 0.05% monocrotophos or 0.05% quinolphos is effective to control mites. The crop may also be dusted with 200-mesh sulphur at the rate of 25-30 kg per hectare at 10-15 days interval or spray wettable sulphur at 1 kg in 500 liters water to control powdery mildew disease. Treat seeds for half an hour in 0.125% solution of ceresan or 1% organo-mercurial fungicide (Seedex). Dust seedex or captan at 3 g per kg of seed and spray Bordeaux mixture 5:5:50 or a copper fungicide at 1 kg in 250 litres of water as soon as the symptoms appear to control anthracnose. Dusting sulphur at the rate of 25-30 kg/ha protects the crop. Spraying Bordeaux mixture in the early stages prevents spread of rust disease.

Harvest: Harvest the pods as they turned straw yellow in colour or pick the pods when they are completely dry for grain purposes. Pods are picked at intervals of 3 to 4 days. Thresh the pods and clean the beans. The seed moisture content at the stage will be about 15% and the green colour of the seed coat will turn to chocolate brown colour. Dry the pods to 15-18% moisture content. Dry the seeds to 8 -10% moisture content. Pick the tender pods or green mature pods once in a week for vegetable purpose. Three harvests are possible from annual types, but can not stand heavy grazing of stems. For green manure, the crop should be cut before flower initiation. As a forage, the crop should be utilized before flowering. The plant has a peculiar buggy smell (fragrance).

Yield: Seed yield is around 1.2 to 1.5 t/ha in sole crop whereas the seed yield is around 0.4 to 0.5 t/ha in intercrop condition or when grown on trellises as sole crop and 0.5 t/ha as mixed crop. In garden types, the green pod yield is 2.6 to 4.5 t/ha. It yields 3 to 5 t/ha of green matter which can be used as fodder or green manure.

Utilization: The young pods and tender beans are popular vegetables. The ripe and dried seeds are consumed as a split pulse. The beans are also sprouted, soaked in water, shelled, boiled and smashed into paste, which is fried with spices. It is palatable to all types of livestock and used for hay as well as for grazing purpose. It gives good silage with sorghum. Dolichos is also used as green manure and cover crop for soil protection against erosion.

REFERENCES

Cameron, D. G. 1988. Tropical and subtropical pasture legumes. Queensland Agricultural Journal. March-April: 110-113.

Hendricksen, R. E. and Minson, D. J. 1985. Lablab purpureus - A Review. Herbage Abstracts. 55:215-227.

Kay, D. E. 1979 Hyacinth Bean - Food Legumes. Crop and Product Digest No. 3. Tropical Products Institute. xvi:184-196.

Luck, P. E. 1965. Dolichos lablab - A valuable grazing crop. Queensland Agricultural Journal May:308-309.

Schaaffhausen, R. V. 1963. Dolichos lablab or Hyacinth Bean; Its uses for feed, food and soil improvement. Economic Botany. 17:146-153.

Wilson, G. P. and Murtagh, G. J. 1962. Lablab - New forage crop for the north coast. New South Wales Agricultural Gazette. 73:460-462.

Wood, I. M. 1983. Lablab bean (*Lablab purpureus*) for grain and forage production in the Ord River Irrigation Area. Australian Journal of Experimental Agriculture and Animal Husbandry 23:162-171.

http://www.tropicalforages.info/key/Forages/Media/Html/Lablab_purpureus.htm

CHAPTER

28

Horsegram–*Dolichos uniflorus* Lam. (2n=24)

Synonyms: *Dolichos biflorus* L., *Macrotyloma uniflorus*

Family: Fabaceae / Legumininosae

Vernacular name: Horse gram, Madras gram (English), *Kulthi, Kulatha* (Hindi), *Kulattha (Sanskrit), Kollu* (Tamil), *Ulavalu* (Telugu), *Hurali* (Kannada), *Muthira* (Malayalam).

Importance: Horsegram derives its name from the fact that it is important as animal feed and as such fed to horses. It is a hardy, drought resistant tropical grain legume. Horsegram is the cheapest of pulse and is hence the 'poor man's pulse' in southern India where the seeds are parched and then eaten after boiling or frying, either whole or as a meal. The seeds are an important food for cattle and horses and are usually fed after boiling. The stems, leaves and husks are used as fodder. It is sometimes grown in pure stands on new land before cereal crops. It is a good cover crop on undulating, eroded hilly slopes and red laterite soils. It is also grown as catch crop after sesame or cereals in Tamil Nadu. Horsegram is also grown as a green manure crop.

Origin: Horsegram is originated in South East Asia and in Africa.

Distribution: It is cultivated in India, Myanmar, tropical Africa, West Indies and Australia. In India, it is grown in Maharashtra, Karnataka, Andhra Pradesh, Tamil Nadu, Madhya Pradesh, hilly slopes of Himachala Pradesh, Uttar Pradesh and Orissa.

Botany: Horsegram is an erect, sub-erect or trailing, densely hairy annual herb, 0.3 to 0.5 m height, profusely branched at the base. The stem and petioles are mostly purple pigmented. Leaves are compound, alternate, trifoliolate, stipules lanceolate petiole 1-7 cm long. Leaflet is ovate, elliptical apex rounded to acute base rounded lateral leaflets a symmetric hairy to glabrescent on both surface. Flower is short with 6-12 mm in length. The flower is cream yellow with purple spot in auxiliary racemes with 2 appendages at base. Flower is zygomorphic and bisexual. Flowers are always purple. Self-pollination is the rule in horsegram. Pods are formed from the ground to the tips of the vines. Fruit is a linear oblong pod 3-8 cm × 4-8 mm, up curved towards apex acuminate, densely hairy. The pods are either erect or dropping from the leaf axil with 5-10 seeds. It has compressed seed. Seed size ranges from 6-8 mm long and 3-4mm broad smooth of which 100 seed weight is recorded 4 gm. Seed trapezoidal oblong or somewhat rounded, pale to dark reddish brown speckled or mottled with black and orange brown or all black.

Climate: Horse gram is a habitat of tropics and subtropics. It grows well up to an altitude of 1500 m from MSL. It is a drought resistant crop and is grown in areas with mean annual rainfall 500 to 1100 mm. It cannot withstand water stagnation. It is a short day plant. It grows well with temperature range of 20 to 30°C. Best growth is produced during hot moist weather with temperatures of 25 to 35°C. The growth rate is declining markedly below 20°C. It is sensitive to frost.

Soil: It grows on a wide range of soil types provided they are well drained and not highly alkaline. It thrives well in light sandy soils, red loamy and black cotton soil. It can be grown down to 5.5 and up to about 8.0.

Season: Horsegram is grown during June-July and October–November seasons.

Varieties: Horsegram varieties cultivated in India are presented in Table 1.

Table 1 Horsegram (Kulthi) varieties in different states of India

Variety	Area of adoption Zone/State	Seed yield (Q/ha)	Days of maturity	Remarks
KS 2	Rajasthan	6-7	80-85	Early maturing, seed brown
Palem 1	A.P	10-12	80-85	Early maturing, Semi-spreading
Palem 2	A.P.	8-9	100-105	Med. Maturing
Arja Kulthi 21 (AK -21)	Rainfed areas of NW parts	8-9	70-105	Early maturing
Paiyur 2	SZ (KN, AP, Odisha,TN)	8-9	100-106	For Sept-Oct sowing
PHG 9	SZ (KN, AP, Odisha,TN)	7-9	100-105	Semi spreading thick foliage
Pratap Kulthi -1 (AK 42)	Raj., Guj., M.P. Haryana	10-12	83-87	Protein 30% lush green foliage with wax deposition
VL Gahat -8	Uttarakhand	12	92-106	Resistant to anthracnose and stem root.
VL Gahat -10	Uttarakhand	10	110-115	Resistant to YMV & root rot and leaf spot.
GPM 6	Karnataka	8-9	120-130	Resistant to YMV, mod, res. to Rhizoctonia root rot
VL Gahat 15	Northern India	5-6	95-105	Resistant to Anthracnose and leaf spot
VL Gahat 19	North Zone	5	88-94	Multiple disease res. to important disease
CRIDA 1-18 R	Karnataka, AP and TN	8	72-102	Tolerant to YMV, Powdery mildew, leaf blight, and root rot
CRIDA LATHA (RHG 4)	South Zone	8.0	72-110	Tolerant to YMV, Powdery mildew, leaf blight, and root rot & mites

Indira Kulthi 1 (IKGH 01-01)	Chhattisgarh	7.0	92	Up lands under rainfed condition with sowing time of august 15 onwards
Gujarat Dantiwada Horsegram -1 (GHG -5)	Gujarat, Raj. Uttarakhand, Jharkhand, UP & Maharashtra	5-6	89-100	Resistant to root rot, moderately resistant to PM, Collar rot, Cercopsora leaf spot and leaf blight.
CO 1	Tamil Nadu	5.6	110	Hundred seed weight is 4.6 g
Paiyur 1	Tamil Nadu	6.5	110	Hundred seed weight is 3.4 g
Paiyur 2	Tamil Nadu	8.7	100-105	Hundred seed weight is 3.5 g

Note: CZ-Central Zone, SZ –South Zone, NEPZ – North Eastern Plane Zone, NWPZ - North Western Plain Zone, Res. = Resistant , Tol. = Tolerant, Mod. Moderately, YMV = Yellow Moasaic Virus, PM = Powdery Mildew.

Field preparation: Land is prepared well to have fine tilth.

Seeds and sowing: Seed rate is 25 kg/ha is recommended. The spacing is 30 x 10 cm. Dibble the seeds or broadcast the seeds and cover it with ploughing or sow the seeds behind the plough. The seeds are sown to a depth of 2.5 cm. Seeds are treated with carbendazim or Thiram @ 2 g/kg of seed. Then treat the seeds with three packets of rhizobial culture which are sufficient for treating the seeds required for one ha. Use rice gruel as binder. Dry the bacterial culture treated seeds in shade for 15 to 30 minutes before sowing.

Weed management: Give one weeding and hoeing between 20 and 25 days after sowing.

Fertilizer recommendation: Apply 12.5 tonnes of FYM or compost, 12.5 kg N and 25 kg P_2O_5 per ha as basal dressing.

Water management: It is mostly raised as rainfed crop. It is drought resistant but cannot withstand water logging.

Crop protection: This plant is mostly pest-free though occasionally some rust and leaf spot will appear but they do no significant damage.

Harvest: When the crop attains physiological maturity when the pods turn yellowish brown in colour, the plants can be uprooted, dried in the field or under cover and threshed to remove the pods. The plants are usually uprooted when the pods become light brown and begin to shrivel. The leaves begin to dry and fall-off when the crop attains maturity. The harvested plants are stacked and left to dry for about 7 days. Then thresh the pods and extract the seeds. When harvested for seed multiplication, harvest the pods when 75 to 80% of the pods have matured. Then seeds are dried to 10% moisture before storage. Seed yield of 500 to 600 kg per ha can be obtained. When grown as fodder; it is harvested on 45 DAS with green fodder yield of 2 to 5 t/ha.

Utilization: The seed is eaten either boiled or fried. Horsegram seed is not usually split to produce dhal. The seed is used as concentrate for cattle or horses, usually after being boiled or slightly parched. The stems, leaves and pods left after harvesting may be used for animal feed. It makes excellent hay and is suitable as green manure.The majority of anti-oxidant properties are confined to the seed coat and its removal would not do any good. Seed is rich in polyphenols, flavonoids and proteins, the major anti-oxidants present in fruits and other food materials. It is famous for its medicinal use because different part of the plants are used for the treatment of heart disease, jaundice, asthma, bronchitis urinary discharges and for treatment of kidney stones.

REFERENCES

Blumenthal, M.J. and Staples, I.B. 1993. Origin, evaluation and use of *Macrotyloma* as forage. A Review. Tropical Grasslands 29, 16-29.

Kanaka, K. and Durga. 2012. Variability and divergence in horsegram (*Dolichos uniflorus*). Journal of Arid Land. 4 (1): 71–76

Dobhal, V. K. and R.C. Rana. 1994. Genetic analysis for yield and its components in horsegram (*Macrotyloma uniflorum*). Legume Research, 17: 179–182.

http://www.fao.org/ag/AGP/AGPC/doc/Gbase/data/Pf000052.HTM

CHAPTER

29

French Bean–*Phaseolus vulgaris* L. (2n=22)

Family: Fabaceae (Leguminoceae)

Vernacular name: French bean, Common bean, Kidney bean, Haricot bean, Salad bean, Snap bean, Sting bean, Runner bean (English) and *Rajmash* (Hindi).

Importance: French bean is grown for their immature edible pods and for the dry ripe seeds and to a lesser extent for green shelled beans. The leaves are used as a pot herb in tropics. The straw is used as forage. It is consumed both as fresh vegetables and for dry seeds in India.

Origin: French bean is originated in the western Mexico-Guotemala area, Central America. The small seeded french beans are native to Central America and the large seeded types to South America.

Distribution: French bean is largely grown in Brazil, Mexico, Argentina, Chile, Columbia, Honduras, Nicaragua, Africa, USA, Canada, Philippines, Japan, Ethiopia, Malawi, Mozambique. France, Spain, Italy, Egypt, India, the Netherlands, Belgium. In India, it is grown in Maharashtra, Jammu and Kashmir, Himachal Pradesh, Uttar Pradesh hills, Punjab and Bihar.

Botany: French bean is classified into four main growth habit types on the basis of determinacy, node production after flowering and growth habit such as plant height and climbing tendency (Table 1). The growth stages particulars are furnished in Table 2.

Table 1 Classification of French bean based on growth habits

Attributes	Determinate		Indeterminate	
	Bush (dwarf type)	Bush	semi-climbing	climbing (pole beans)
Plant height (cm)	44	92	103	160
Main stem nodes at flowering	8.3	13.3	143	14.8

Main stem at maturity	8.6	17.0	18.9	22.7
Days to flowering	34	39	38	40
Duration of flowering (days)	22	25	28	29
Leaf area duration (days)	114	130	164	261
Seed weight (mg/seed)	336	236	257	294
Yield (t/ha)	2.4	2.7	3.2	3.7

Table 2 Growth stages

Growth stages	**Green bean**	**Dry bean**
Establishment	10-15 (days)	10-15 (days)
Vegetative (up to first flower)	20-25	20-25
Flowering (including pod setting)	15-25	15-25
Yield formation (pod development and bean filling)	15-20	25-30
Ripening	0-5	20-25
Crop duration	60-90	90-120

Climate: French bean is grown in latitude 20ºS to 20ºN with an altitude up to 500 to 1800 m from msl. In India, it is grown in an altitude upto 1200 to 2500 m from msl in Jammu and Kashmir, Himachal Pradesh and Uttar Pradesh. The cardinal minimum, optimum and maximum temperatures for bean are 10ºC, 16 to 24ºC and 30ºC respectively. Germination requires a soil temperature of 15°C or more, and at 18°C germination takes about 12 days, and at 25°C about 7 days. Flowering drops above 30ºC temperature and failure of seed set above 35ºC. Abortion of flowers and failure of seed set at the temperature above 30/25ºC day/night temperature. French bean is a short day plant and day neutral types exist. It is grown in areas receiving mean annual rainfall of 500 to 1500 mm. Rainfall of 300 to 400 mm during crop growth, 125 to 175 mm between planting and first flowering, 25 to 75 mm during flowering and 150 mm from flowering to pod filling and thereafter, the drier is better. Relative humidity should be 50% during flowering. Hail cause severe defoliation. It can be grown in area where there is a frost-free period of 105 to 120 days. Hot dry winds during flowering can cause severe bloom drop resulting in lower seed yields and uneven ripening of the seed.

Soil: French bean can be grown successfully on most soil types, from light sands to heavy clays, but a friable, deep, well drained soil is preferred. When grown as a rainfed crop, the soil should be 60 to100 cm deep, so that the roots may draw water from moisture reserves. Soil pH of 6.0 to 6.8 is preferable. Below pH 5.2, manganese toxicity symptoms (stunting, chlorosis and puckering of the leaves) may be apparent but above pH 7.0, manganese deficiency can occur, causing retardation of growth and also chlorosis of leaves. The leaves show zinc deficiency (poor pod set) when it is grown on calcareous soils. It is sensitive to soil salinity. The yield decrease at different levels of ECe is 0% at ECe 1.0, 10% at 1.5, 25% at 2.3, 50% at 3.6 and 100% at ECe 6.5 mmhos/cm.

Land preparation: Prepare the land to the fine tilth and form ridges and furrows.

Season: It is a *kharif* (June to September) crop. Best time of sowing is second fortnight of June in hilly tracts and in plains, it is grown as *rabi* crop during October to December.

Varieties: The length of the total growing period is 60 to 90 days for green bean and 90 to 120 days for dry bean. Rajmash varieties in different growing zones of India are presented in Table 3.

Table 3 Rajmash varieties grown in different zones/states of India

Variety	Area of adoption Zone/State	Seed yield (Q/ha)	Days of maturity	Remarks
HUR -137 (Malviya Rajmash -137)	NEPZ (East Uttar Pradesh, Bihar, West Bengal)	18-22	112-120	Erect semi dwarf, Red.
HPR -35	Maharashtra	14-15	73	Seed red with purple strips.
Varun (ACPR 94040)	Maharashtra	14-16	66-68	Tolerant to Anthracnose
IPR 96-4 (Amber)	NEPZ (East Uttar Pradesh, Bihar, West Bengal)	15-16	139	Resistant to BCMV & leaf crinckle. 100 seed 39 g
Ankur (RSJ-178)	Rajasthan	12	110-120	Moderately resistant to root rot, leaf crinkle and leaf spot dry root rot.
Gujarat Rajma - 1	Gujarat	20	30-35	Moderate resistant to bean common mosaic virus
VL Rajma 125	Uttarakhand	14-15	82-85	Resistant to root rot, mod. Resistant to Anthracnose, angular leaf spot & rust
VL Bean 2	Uttarakhand	14-15	82	Resistant to root rot, mod. Resistant to Anthracnose, angular leaf spot & rust
Arka Anup	Karnataka	18-20	43-45	Suitable for eastern dry zone of Karnataka in both kharif and rabi season
PDR 14 (Uday)	NEPZ	16-20	----	Tolerant to BCMV, 100 seed wt.44g, white red variegated seeds
HUR 15 (Malviya Rajmash 17)	NEPZ	16-20	----	Moderately susceptible to BCMV. Erect determinate white seeded
IPR 98-5 (Utkarsh)	NEPZ	17	----	Tolerant to BCMV. Gulf red seed
IPR 98-3-1 (Arun)	CZ	24-25	----	Attractive dark red seed, medium seed size (38 g/100 seeds)

Note: CZ-Central Zone, SZ –South Zone, NEPZ – North Eastern Plane Zone, NWPZ- North Western Plain Zone

Seeds and sowing: The seed germination is epigeal. The seeds remain viable for 2 years. The seed rate for bold seeded varieties is 45 to 50 kg/ha and for medium seeded varieties is 35 to 40 kg/ha. 4 to 6 seeds are sown per hill and later thinned to 3 to 4 plants.

Spacing of 30 x 15 cm is recommended. Seeds are sown to a depth of 3 cm in heavy soil and 5 cm in light soils.

Manures and fertilizer application: 20 kg N and 50 kg P_2O_5 per ha are applied as basal dressing at sowing. 25 kg N and 25 kg of K/ha are applied between 20-25 days after sowing and application of remaining 25 kg of N is done between 40-45 days.

Water management: Furrow or overhead irrigation is preferable to flooding. Water requirement is 32 to 64 mm/ha/week depending upon the water holding capacity of the soils, the run off and the evaporation rate of the soil and the crop. Moisture stress at the flowering period is critical and can reduce yields by 20%. The available soil moisture should be maintained above 50% during the flowering period. For a 75-day crop, total evaporation is 220 mm, Crop evaporation is 3.2, 3.2 and 1.7 mm per day or 0.62, 0.77 and 0.38 of pan evaporation (EP) from germination to flowering, flowering to pod development and pod development to maturity, respectively.

Crop protection: Crop is affected by pests like stem fly, thrips, mites, bean beetle, bean weevil, aphids etc. Yellow mosaic, anthracnose, powdery mildew, rust, root rot and wilt and leaf spot are common diseases. Seed treatment is done with carbendazim + thiram (1+2 g/kg seed), or Trichoderma (4g) + carboxin (1g/kg) seed. Foliar spray of metasystox or monocrotophos (0.04%) for aphids that are vectors of bean common mosaic virus, foliar spray of carbendazim (1g/1 of water) against stem blight, removal and destruction of stem blight affected plants, avoiding stagnation of water in the field for limiting the buildup of stem blight and root rot diseases.

Harvest: Picking begins 2 to 4 weeks after the first flowers and 7 to 8 weeks after sowing. Picking is carried out in every 3 to 4 days. In bush varieties, 2-3 harvests and in pole types 3-5 harvests are made. Dry beans are harvested as soon as a large percentage of pods are fully mature and have turned yellow. The whole plants are usually pulled after which they are dried and threshed. Seed yield of 1.5 to 2.0 t/ha can be obtained. Seed with 13% moisture stored at 25^0 C. It may retain its viability and vigour up to 13 months and with 10% moisture for up to 3 years.

Yield: Yield of tender pods varies from 8-10 t/ha in bush varieties and 12-15 t/ha in pole types. Dry beans are harvested when majority of pods are fully ripe and colour turns yellow. Seed yield varies from 1250 to 1500 kg / ha. Store the seeds at 8 -10% moisture.

Utilization: Mature, dry french beans are widely used as human food-stuff in many parts of the world. It is used as tender bean, green shelled bean, and dry bean and as cattle feed.

REFERENCES

CPG. 2012. Crop Production Guide. Department of Agriculture, Government of Tamil Nadu, Chennai and Tamil Nadu Agricultural University, Coimbatore.

CPG. 2013. Crop Production Techniques of Horticultural Crops. Horticultural College and Research Institute, Tamil Nadu Agricultural University, Coimbatore – 641 003

CHAPTER

30

Moth Bean–*Vigna aconitifolia* (Jacq.) Verdc (2n=22)

Synonyms: *Phaseolus aconitifolicus* (Jacq) Marechal

Vernacular name: Moth bean, Moth, Haricot beans, Dewgram, Dewbean, Matbean, Phillipesara (English), *Kallu payaru, narippayir, Pani payaru* (Tamil), *Tulka payaru* (Telugu) and *Matki* (H).

Importance: Moth bean is a source of food, feed, fodder and green manure. The low lying soil cover the crop creates helps to prevent soil erosion by preventing moisture loss. It is a potent source of several confectionary items. It is used as daily snacks as well as dal/dhal. Green pods are delicious source of vegetables. The seeds contain approximately 24–26% protein. It is known for higher proportion of albumin and glutamine fraction of protein along with a good source of lysine and leucine amino acids.

Origin: The moth bean is native to India, Pakistan, Myanmar and Sri Lanka where it is found growing wild and under cultivation.

Distribution: It is cultivated in the USA, Australia, Thailand and other parts of Asia. In India, it is grown in Rajasthan, Gujarat, Maharashtra and Tamil Nadu.

Botany: Moth bean is an herbaceous creeping annual which grows to approximately 30-40 cm tall. Leaf has deeply lobed leaflets. Terminal leaflet has 5 acuminate lobes and lateral leaflet has 4 lobes. Flower is yellow in colour. The pod length is 2.5 to 5 cm which holds 4 to 9 seeds. The pod colour is brown or pale grey when mature. Seed colour is yellow, yellow-brown, whitish green and mottled with black.

Climate: Moth bean is an important arid legume, is known for its tolerance to both drought and heat. It grows up to an altitude of 1200 m from MSL. It is grown in area with annual average rainfall of 500 to 750 mm. It cannot stand heavy rain and is not suited to the subhumid or humid areas of the tropics. It has been known for high degree of adaptation in rainfed arid situation due to tolerance to drought and high temperature. It can produce good harvest under long dry spells (30–40 days) with high ambient temperatures (35–40°C) from seedling emergence to maturity. The optimum temperature for growth is 24–32°C but has been shown to tolerate up to 45 °C during the day. It is susceptible to frost. It is a short day plant. However, day neutral cultivars are available.

Soil: It can be grown on a wide range of soils and will thrive on light, sandy loams or heavy clay loamy soils. It tolerates a wide pH range of 3.5 to 10 and can tolerate slight salinity.

Season: Moth bean can be grown during June-July and September-October. It is frequently planted towards the end of the rainy season and grows mainly on stored soil moisture.

Land preparation: As the seed is small, a well prepared seed bed and a moist friable soil of about 20 cm deep is essential for high yields.

Varieties: Moth bean varieties have different maturity grows in 72-75 days, 65-67, 60-62 and 57-58 days respectively, with semi spreading, semi-erect and erect-upright plant type. These varieties suit to 450-500, 300-450, 150-300 and 130-150 mm rainfall, respectively with reasonable grain yield potential from 500 to 1400 kg/ha depending on rainfall pattern and other agronomic inputs. The moth bean varieties and their characteristics grown in different states of India are presented in Table 1.

Table 1 Particulars of moth bean varieties grown in different states of India

Variety	Area of adoption zone/state	Seed yield (Q/ha)	Days of maturity	Remarks
Maru Bahar (RMO-435)	Rajasthan, Gujarat, Maharashtra	6-6.5	65-67	Early maturing
CAZRI Moth 2	Rainfed areas	5-7	70-72	Drought hardy
CAZRI Moth 3	Rainfed areas	6-5	62-64	Erect, upright growth lush green foliage
RMO-423	Rajasthan	5-6	67-70	Tolerant to insect pests & diseases
RMO -257	Rajasthan	6-7	63-65	Semi erect
CAZRI Moth -3	Rajasthan	6	60-88	Resistant to YMV and dry root rot
Jadia	Rajasthan	5-8	90	seeds test weight 25-29 g
Jwala	Rajasthan	7	90	Resistant to YMV; seeds test weight 30 g
TMV (Mb) 1	Tamil Nadu	9-10	65-70	Moderately resistant to white fly, pod borer, YMV
Rajasthan moth (RMO 257)	Rajasthan	5-6	66	Tolerant to YMV, seed test weight 29-32 g
Baleshwar-12	Gujarat	5	110-115	susceptible to YMV
GAU Moth-1	Gujarat	5	110-115	seeds of chocolate colour.

Seeds and sowing: Seed rate is 10 and 5 kg/ha for pure crop and intercrop respectively. Spacing is 30 × 10 cm for seed and 30 × 5 cm for forage purposes. Seeds are some times broadcasted. Seeds are sown at a depth of 3 cm.

Cropping systems: It is rotated with mustard, gram, linseed and barley under rainfed conditions. Under irrigated conditions, the important rotations are mothbean-potato-wheat, mothbean-*toria*-potato, mothbean-*toria*-wheat-greengram, mothbean-radish-wheat. It is intercropped with pearl millet + mothbean (2:1) and maize +mothbean (4:4).

Weed management: Weeding during early stages of growth is essential. One hand weeding at 30 DAS + pre plant incorporation of fluchloralin (Basalin) @ 0.5 to 1 kg a.i./ha effectively controlled the weeds in mothbean.

Manures and fertilizer application: On light soils application of 20 kg N and 20 kg P_2O_5 per ha are recommended for rainfed crop. 10 kg/ha $ZnSO_4$ Besides their N-fixing capacity they have greater power for absorbing less soluble form of 'P'. Roots have greater CEC and hence, capable of absorbing divalent cautious like Ca^{++} and Mg^{++} but can not compete with cereals for mono valiant K^+. FYM @ 12.5 t/ha is recommended for improving physical condition and improving water holding capacity of soil along with 10 kg N + 40 kg P_2O_5/ha as basal at the time of sowing or last preparation.

Water management: It is mostly grown as rainfed crop. It is a drought-resistant crop.

Crop protection: The common pests affecting moth bean are jassids, white fly, thrips, aphid and mite while the diseases are mung bean yellow mosaic virus, root rot, anthracnose, and seedling blight besides, the root knot nematode. The pests are controlled with application of phorate or aldicarb @ 1.25 kg a.i. effective upto 4 week or spray with monocrotophos @ 25 kg a.i./ha or dimethoate @ 0.15 kg a.i./ha. The white grub is controlled with soil application of phorate @ 1.25 a.i./ha before sowing or applied in the furrows. Termite can be controlled with soil application of phorate 10 G @ 25 kg/ha or before sowing or apply 5% aldrin dust @ 25 kg/ha at the time of soil preparation. The root knot nematode is controlled with carbofuran @ 2 kg a.i./ha. The diseases are controlled with seed treatment with thiram@ 3 gm/kg of seed or spraying with Dithane M 45 @ 2.5 gm/litre of water.

Harvest: Crop may be harvested when pods get mature and turn brown. When grown for grain purpose, the plants are usually cut with a sickle and dried for about one week before being threshed and winnowed.

Yield: Seed yield is 600-750 kg/ha while the stover yield is 1.0-1.5 t/ha. Green forage yield is 2 to 3 t/ha.

Utilization: Moth bean is eaten whole, usually after frying or split and used for the preparation of *dhal*. The ripe seeds, whole or split, are cooked and eaten. Seeds are sprouted and eaten with or without salt, or fried and salted. The seed is used for preparing soup. The ripe seeds are utilized as a feed for fattening oxen and horses. Moth bean is not usually fed to cattle as it prevents the flow of milk. It is used as a feed for sheep. It is grown for forage purpose and to prevent soil erosion. Immature pods are boiled and eaten as vegetable.

REFERENCES

Jain, H.K. and K.L. Mehra. 1980. Evolution, Adaptation, Relationships, and Uses of the Species of Vigna cultivated in India. In "Advances in Legume Science" eds.R.J.Summerfield and A.H.Bunting. Kew Royal Botanic Gardens. pp.459-468.

Khokhar, S. and B.M. Chauhan. 1986. Antinutritional factors in moth bean (Vigna aconitifolia): varietal differences and effects of methods of domestic processing and cooking. Journal of Food Science, 51(3), 591–594.

Purseglove, J.W. 1974. *Phaseolus aconitifolia.* In: Tropical Crops: Dicotyledons. London. Longman. pp.290-294.

Sathe, S. K. and M. Venkatachalam. 2007. Fractionation and biochemical characterization of moth bean (Vigna aconitifolia) proteins. LWT-Food Science and Technology, 40(4), 600-610.

Whyte, R.O., Nilsson-Leissner, G. and H.C. Trumble. 1953. Legumes in Agriculture, F.A.O. Agricultural Studies 21, F.A.O. Rome, Italy.

Yogeesh, L. N., Viswanatha, K. P., Ravi, B. A. and S. Gangaprasad. 2012. Genetic variability studies in moth bean germplasm for seed yield and its attributing characters. Electronic Journal of Plant Breeding, 3(1): 671-675.

www.prota4u.info

CHAPTER

31

Sword Bean–*Canavalia gladiata* Jacq. (2n=22, 44)

Vernacular name: Sword Bean, Jack bean, Scimitar bean, Jamaican horse bean (English), Khadsampal, Badi sem (Hindi), Tebi (Manipuri), Segapputampattai (Tamil), Valpayar, Valaringha (Malayalam), Tamma (Telugu), Tumbekonti, Sembi (Kannada), Makhan shim (Bengali) and Mahasimbi, Asisimbi (Sanskrit).

Importance: It is a vegetable cum grain crop. It is grown as intercrop, border crop, forage crop and as anti-erosion crop. Seeds are a nutritious foodstuff in Asia. It is eaten as whole, usually after frying, or split and used for preparation of 'dhal'. Immature pods are boiled and eaten as a vegetable. Ripe seed is also used for livestock feeding. Sword bean is grown as forage crop and anti-erosion crop.

Origin: Sword bean is originated in Asia.

Distribution: It is cultivated on a limited scale throughout Asia, West Indies, Africa and South America and Australia.

Botany: Sword bean is a vigorous perennial climber but often cultivated as an annual. Leaves are trifoliate, with large pubescent leaflets (10- 18 × 6- 14 em) which are acuminate with a short point at apex. The petioles are shorter than the leaflets with a groove above and stout with large pulvinus at base and at the base of each leaflet. Inflorescence is. an axillary long stalked raceme bearing several flowers in succession. The flowers are inverted, the standard being at the bottom. Flowers 3.5 to 4 em long and are white or pinkish. The seed pods are usually broad and curved with strongly developed ridges. They are about 20-40 cm long and 3. 5-5 cm broad, containing on average 8 to 16 seeds. Seeds are 2.5 -3.5 cm long, white or red in color with a dark brown hilum extending more than one-half length of the seed. The seed has a tough thick coat. Hundred seed weight is 132 g. Germination of the seed is epigeal. It contains 25.9 % protein.

Climate: Sword bean grows well with 15 to 30ºC. It grows in areas receiving mean annual rainfall of 900 to 1500 mm. It is fairly drought resistant crop. It grows up to an altitude of 900 m from MSL.

Soil: It requires fertile soil and is susceptible to water logging.

Season: The crop is raised as *kharif* and *rabi* crop under rainfed condition during June-July and in October respectively. It is also grown in summer season during February-March under irrigated condition.

Growth stages: The crop duration is 120 days. It takes 45 to 50 days to 50% flowering.

Land preparation: Prepare the land to fine tilth and form beds and channels.

Seeds and sowing: Seed rate is 10 to 15 kg for pure crop, 2.5 to 4.5 kg for intercrop and 20 to 25 kg per ha for forage crop. The spacing for sword bean is 45 x 30 cm under irrigated condition while 30 x 20 cm under rainfed conditions respectively. Seed treatment with fungicides and rhizobial culture are to be carried out.

Manures and fertilizers: 25 kg N and 50 kg P_2O_5 per ha are applied as basal dressing.

Water management: It gives good response to irrigation.

Harvest: The plants are cut with a sickle, dried for about one week, before being threshed and winnowed when grown as a grain legume. Tender pods are ready for harvest 75 days after sowing. The moisture content of the dry whole seed is 11.2%.

Yield: Sword bean gives a grain yield of 1360 kg/ha and green pod yield of 7500 kg/ha.

Utilization: The young pods are extensively utilized in Asia as a green vegetable. The mature dry beans may be cooked and eaten as food. Canavanine is a toxic arginine anti-metabolite and canaline, its primary metabolite also a toxic non protein amino acid, are found in seeds.

REFERENCES

Kay, E. Daisy. 1979. Food legumes. TPI crop and Product digest No.3. London: Tropical Products Institute. XVI + 435 pp.

Purseglove, J. W. 1968. Canavalia gladiata (Jacq.) DC. Tropical Crops: Dicotyledons 1, 245. London: Longmans, Green and Co Ltd, 332 p.

Smartt, J. 1976. *Canavalia gladiata* (Jecq.) D. C. (Sword bean). Tropical Pulses. p. 58. London: Longman Group Ltd, 348 pp.

CHAPTER

32

Ricebean–*Vigna umbellata* Thunb var. *umbellata* (2n=22)

Synonyms: *Phaseolus calcaratus* Roxb., *Vigna calcarata* (Roxb.) Kurz., *Dolichos umbellatus*

Family: Fabaceae

Vernacular names: Rice bean, Climbing Mountain Bean, Mambi Bean, Oriental Bean (English); Sutari (Hindi); Bejiamah (Assam); Nagamah (Arunachal Pradesh); Bete (Mizoram); Jami, Agukzungken (Nagaland) and Chak hawai (Manipur).

Importance: Ricebean is a multi-purpose grain legume crop, cultivated for grain, fodder and green manure. It is grown for its dried edible seeds.

Origin: The rice bean is native of Southeast Asia, specifically India, Myanmar, China, Vietnam, Laos and Thailand.

Distribution: It is cultivated in India, Nepal, Bangladesh, Malaysia, Thailand, Vietnam, China, Honduras, Brazil, Mexico, Egypt, Fiji, USA and Australia. In India, it is grown in Assam, Meghalaya, Manipur, Nagaland, Mizoram, Arunachal Pradesh, West Bengal, Sikkim, Odisha, Himachal Pradesh, Uttaranchal, Chhattisgarh, Jharkhand, Madhya Pradesh, Kerala, Tamil Nadu, Gujarat, Punjab, Maharashtra, Haryana and Karnataka.

Botany: It is an annual climbing herb with an erect to semi-erect vine. It grows up to 2.0 to 3.0 m height. The stem is grooved, usually clothed with fine, hairs with taproot system. Leaves are alternate, trifoliolate; stipules lanceolate; leaflets broadly ovate to ovate-lanceolate and almost glabrous. Inflorescence is a raceme, with flowers usually in pairs. Flowers are bisexual and corolla is bright yellow in colour. Flowers are self-compatible, but cross-pollination also occurs. Fruit is a linear-cylindrical pod, glabrous, green when young, black-brown at maturity. The number of pods per plant is approximately 50. The number of seeds per pod is 6 to 8. Seeds are oblong, yellow, green, dark red, brown, black, speckled or mottled. The 100 seed weight is 5 g. The yield per plant is approximately 10 to 15 g.

Growth stages: The early maturing type takes 70 days to flowering and 95 days to harvest whereas late maturing type takes 100 to 115 days to flowering and 130 to 150 days to maturity.

Climate: It is mainly cultivated in humid tropical and sub-tropical climatic region. It is grown up to an altitude of 1400 m from MSL. An average annual rainfall of 1000 to 1500 mm is required. It is grown in areas having the temperature range of 18 to 37ºC. The optimum temperature is 25-35ºC. It is tolerant to drought and high temperature. It is tolerant to some degree of waterlogging, although the young plants appear to be susceptible. It is a short-day plant. It is susceptible to frost. It does not perform well under shade.

Soils: It is grown on well-drained sandy loam to loamy soils. Sandy, waterlogged or saline-alkaline soils are not suitable for its cultivation. It is also known to be tolerant to acidic soils. The optimum soil pH is 6.8–7.5. The wild rice bean types are found in open locations and on roadsides. It thrives well in marginal lands, rainfed tars, drought-prone areas and exhausted soils.

Season: It is cultivated both as *kharif* and *rabi* season crop. The sowing time is July-August and September-October.

Varieties: SRBS-50, SRBS-113, SRBS-368, RBL-35, RBL-50, RBL-99, RBL-202, KRB-14, KRB-16, KRB-18, Bidhan-1, Bidhan Rice Bean 2 (KRB 4) and Beziamah.

Land preparation: Fine tilth is essential for obtaining a good plant stand. The land is to be ploughed 2-3 times followed by leveling.

Seeds and sowing: Seed rate is 25 to 30 kg/ha for broadcasting and 15 kg/ha for dibbling method. Soak the seeds in water overnight for 10 to 12 hours before sowing to promote seed germination. Seeds are either broadcasted or sown by dibbling. It is sown with spacing of 45 × 15 cm or 30 × 15 cm. Sowing may be done with seed drill. It is also sown on rice bunds and in kitchen gardens.

Weed management: Two to three weeding can be taken at about six weeks after sowing seeds and before flowering.

Intercultural operations: During the rapid vegetative growth, terminal clipping or cutting off the tip of the plant is practiced to promote lateral branching and high grain yield. In sole cropping, stakes are provided to save pods from rodents and fungal attack.

Manures and fertilizer application: FYM @ 12.5 t/ha and 25-40-20 kg of N, P_2O_5 and K_2O kg per ha as basal dressing is recommended. The lime is to be applied after soil test in acid soils in order to bring soil pH around 6.0.

Water management: It is grown as rainfed crop in high rainfall areas where drainage is required instead of irrigation. Irrigation is given as and when required during monsoon period in case of failure of rain. It is susceptible to water-logging or excess moisture conditions, which ultimately kill the plants. Hence, it requires proper drainage facilities for its healthy growth and development.

Cropping systems: It is sown as sole or mixed/ intercrop and also on residual moisture after rice. It is grown predominantly under rainfed conditions in mixed farming systems, under shifting cultivation (Jhum) on hillsides, on rice bunds or in kitchen gardens and backyards. The promising crop sequences are ricebean-wheat (late sown), ricebean-potato (spring), ricebean-sunflower, ricebean-moong (summer), ricebean-summer fodders.

Insect pests and diseases management: The pod borer, hairy caterpillar, blister beetle, aphids, pod sucking bugs, pod weevils, green stink bugs and leaf folder are the major insect pests while rust, cercospora leaf spot, powdery mildew, rhizoctonia blight and bacterial blight are the diseases occur in ricebean. Spray Maneb @ 3 g/litre of water to control rust and cercospora leaf spot. Spray Triadimefun-Bayletan 25% EC @ 0.03% and Carbendazim @ 0.5 g/litre water to control powdery mildew. Seed treatment is done with Bavistin @ 1 g/kg of seed to control Rhizoctonia blight. Spraying Margosom, Nemarin @ 1.5 ml per litre of water to control pod borer. Dimethoate @ 0.03% controls aphids. Jassids/Flea beetle/Hairy caterpillar can be controlled with spray of malathion 50 EC @ 1.0-1.5 lit/ha in 500-700 lit of water or dusting with Malathion 5% dust @ 20-25 kg/ha.

Harvest: Harvest the crop when 75% of the pods mature indicating full darkish pod, brittle on slight pressure. Harvesting is done during morning hours to avoid shattering. It is advisable to harvest the pods in 2-3 or more pickings to obtain a good yield. The grain is threshed out by beating the dried pods with a bamboo stick or threshing is done by trampling of draft animals. Seeds are dried for 3 to 4 days before storage. If grown for fodder, rice bean should be harvested when the pods are immature, since the leaves drop easily when the plant reaches maturity. As a green manure, rice bean can be ploughed *in situ* at about 45 to 60 days after sowing.

Yield: The grain yield is 1.0 to 1.5 t/ha. The leaf-stem ratio is 0.75 to 0.8. The green fodder yield is 15 t/ha.

Utilization: The matured dry rice bean seeds are eaten as a pulse. The seeds are boiled and eaten. It is also used as a cover crop, green manure and living hedge.

REFERENCES

Arora, R.K., Chandel, K.P.S., Joshi, B.S. and K.C. Pant. 1980. Rice bean: tribal pulse of eastern India. Economic Botany. 34 (3): 260–263.

Carvalho, N.M. de and R. D. Vieria. 1996. Rice bean [*Vigna umbellata* (Thunb.) Ohwi and Ohashi]. In: E. Nkowolo and J. Smartt (eds). Legumes and oilseeds in nutrition. Chapman and Hall. pp: 222-228.

Chandel, K.P.S., Arora, R.K. and K.C. Pant. 1988. Rice bean - a potential grain legume. NBPGR Scientific Monograph No. 12. NBPGR, New Delhi.

Chatterjee, B.N. and A.K. Mukherjee. 1979. A new rice bean for fodder and grain pulse. Indian Farming 29, 5, 29-31

Duke, J.A. 1981. Handbook of legumes of world economic importance. Plenum Press, New York, London.

Khadka, K. and B.D. Acharya. 2009. Cultivation practices of ricebean. Local Initiatives for Biodiversity, Research and Development (LI-BIRD), Pokhara, Nepal.

National Academy of Sciences, 1979. Tropical legumes: resources for the future. National Academy of Sciences, Washington, D.C., United States. 331 p.

http://uses.plantnet-project.org/en/Vigna_umbellata_(PROTA)

CHAPTER

33

Lima Bean–*Phaseolus lunatus* var. *lunatus* L. (2n=22)

Synonyms: *Phaseolus limensis* Macfad, *Phaseolus viridis* Piper, *Phaseolus falcatus* Benth, *Phaseolus macrocarpus* Moench.

Family: Fabaceae

Vernacular names: Lima bean, Madagascar bean, Butter bean, Sieva bean, Sugar bean, Java bean (English). Bush types are small-seeded lima beans which are called Butter beans, Sieva beans, Burma beans, Madagascar beans, Carolina beans, and Baby limas. Vine types are large-seeded lima beans which are sometimes called Potato lima beans.

Importance: Lima bean is a minor grain legume. It is grown for its edible seeds.

Origin: Lima bean has two centres of domestication viz., i) Central America (Mexico, Guatemala) for the small-seeded types and ii) South America (Peru) for the large-seeded types.

Distribution: It is grown in India, Brazil, Myanmar, China, USA, Mexico, Costa Rica, Guatemala, El Salvador, Honduras, Nicaragua, Panama, Venezuela, Peru, Bolivia, Paraguay, Madagascar, Mauritius and Philippines.

Botany: Lima bean is a perennial herb grown as an annual. It has thin roots. The climbing or trailing plant produces twining stems. The stems scramble over the ground or twine into the surrounding vegetation. The leaves are alternate and trifoliate with ovate leaflets. Inflorescence is an axillary raceme which is 15 cm long and bears 24 white or violet bisexual flowers. Pods are curved and sometimes shaped like a hook on top. The pods are pale green in colour. The pod length varies from 7.5 to 12.5 cm depending upon variety. Pods contain 2 to 4 seeds. Seeds are kidney shaped, rhomboid, globose or crescent-oval-shaped. Seed colour is uniform or speckled or mottled, ranging from white, cream, green, yellow, brown, red, black and purple. The hundred seed weight differs based on growth habit. The hundred seed weight of small seeded types of lima bean ranges from 35 to 50 g while the large seeded types range from 50 to 100 g.

Growth sages: Seed germination occurs in 4 to 10 days after sowing. Vegetative growth accelerates after one month. Flowers appear in 35 to 70 days and ripe pods in 80 to

120 days after sowing with short day length. Of buds, flowers and young pods, 75 to 85% are shed under field conditions. It has two distinct growth habits namely, an indeterminate growth habit (prostrate or climbing; with axillary flowering only) and a pseudo-determinate growth habit (dwarf or bush plants; with terminal and axillary flowering). The vegetative cycle of pseudo-determinate growth types is shorter than that of indeterminate ones. Bush types grow to a height of 60 to 90 cm and have small seeds while the Vine types grow to a height of 2.0 to 3.5 m and have large seeds. The bush cultivars mature within 90 to 110 days whereas the climbing types require 180 to 240 days to reach maturity. In climbing types, flowering and fruiting may extend throughout the wet season. The growth habit of perennial wild types is always indeterminate. The short duration bush type lima bean varieties are ready for harvest from 60 to 80 days from sowing; while the vine bean varieties are ready for harvest in 85 to 90 days. Plants can commence producing their edible seed pods in 85 to 155 days from sowing. The pods can be harvested over a period of several months.

Climate: Lima bean is a tropical and subtropical legume. It is found in humid, sub-humid and semi-arid tropical climates as well as warm temperate climates. The small-seeded wild form is found distributed from Mexico to Argentina, generally below 1600 meters above sea level, while the large-seeded wild form is found distributed in the north of Peru, between 320 and 2030 meters above sea level. It is grown in areas receiving mean annual rainfall of 500 to 1500 mm. It is drought resistant due to their deep, well-developed root system. Seed will not germinate at temperatures below 16°C. The optimum temperature is from 16 to 21°C. Lima beans will not set pods in temperatures above 27°C or in cold or wet weather. It does not tolerate frost.

Soils: It prefers well-drained loam soils with a pH of 4.4 to 7.0.

Propagation: Propagation of Lima bean is by seed only. Seed weight varies between 30 g and 300 g per 100 seeds. The normal seed rate varies between 60 to 80 kg/ha for small-seeded cultivars and 120 to 150 kg/ha for large-seeded types. Spacing for bush types are usually spaced 60 cm × 30 cm or 90 cm × 30 cm while climbing types may be planted on hills in groups of 3 to 4 plants with spacing of 90 cm × 45 cm or 100 cm × 60 cm. Soak the seeds in warm water for 12 hours and sow *in situ* about 2.5 to 4 cm depth. Set poles, stakes, trellis or supports in place at planting time to provide support to climbing/ twining stem.

Weed management: Two to three weeding is necessary during initial growth.

Manures and fertilizer application: Lima bean can fix nitrogen by symbiosis with *Bradyrhizobium* bacteria. If fertilizer is applied, this is often done at planting, in bands below and adjacent to the seeds. Supplemental nitrogen and phosphorus may be side-dressed at the early bud stage and during fruit development.

Water management: Light irrigations at 10-15 days interval are given for pea. Flowering, fruit set and grain filling periods are critical stages and care should be taken to irrigate crop at these stages.

Cropping systems: It is mostly cultivated in home gardens or intercropped with cereals, root and tuber crops or other crops. Sole cropping is more common in drier areas.

Harvest: Bush lima beans will be ready for harvest at 60 to 80 days after sowing while the climbing or pole beans will be ready for harvest at 85 to 90 days after sowing. Green and matured pods of the climbing lima bean types are usually picked manually over a period of 4 to 6 weeks. The harvest in bush lima beans is with 2 or 3 pickings in a season. In drier areas, whole plants are cut and left to dry in the field before the pods are removed and the stems are fed to livestock.

Yield: In the tropics, climbing types yield 3 to 4 t/ha dry seeds whereas bushy types yield 2 to 2.5 t/ha when grown in pure stand. However, the average yield of dry seeds is 300 to 600 kg/ha in intercropping and 1.0 to 1.5 t/ha in sole cropping system. The yield of green fodder is 15 t/ha.

Utilization: The green, immature seeds, pods and leaves are eaten as a vegetable. It is often grown as a cover crop and for green manure. The vines are used as cattle feed. The immature and dry seeds are boiled, fried or baked and are eaten. The seeds are used in soups and stews. The sprouted seeds are used as a vegetable. The young pods are sometimes steamed and eaten. The young leaves can also be steamed and eaten but have a bitter flavour. The shoots and young plants can be cooked and eaten. The leaves and stems of butter bean may be turned into hay or silage. The seed contains hydrocyanic acid, a toxic principle which should not be fed raw to livestock. The toxic cyanide compound is deactivated upon thoroughly cooking the seed. The juice from the leaves is used in nasal instillations against headache and as eardrops. The seeds are used as a diet to relieve fever. It is used as a soil improver since it fixes atmospheric nitrogen and sheds its leaves which are valuable for restoring soil fertility.

REFERENCES

Baudet, J.C., 1977. The taxonomic status of the cultivated types of Lima bean (*Phaseolus lunatus* L.). Tropical Grain Legume Bulletin 7: 29–30.

Duke, J.A., 1981. Handbook of legumes of world economic importance. Plenum Press, New York, United States, and London, United Kingdom. 345 p.

ILDIS, 2002. World database of Legumes, Version 6,05. International Legume Database & Information Service. [Internet] http://www.ildis.org/. Accessed October 2003.

Kay, D.E., 1979. Food legumes. Crops and Product Digest No 3. Tropical Products Institute, London, United Kingdom. 435 p.

http://www.bioversityinternational.org/

http://www.prota4u.org

CHAPTER

34

Winged Bean–*Psophocarpustetragonolobus* (L.) DC. (2n=18)

Family: Fabaceae

Vernacular names: Winged bean, Winged pea, Asparagus bean, Asparagus pea, Four-angled bean, Four-cornered bean, Goa bean, Manila bean, Mauritius bean, Dragon bean, Haricot dragon bean, Princess pea, Princess bean (English); *Chaudhaari-phali, Clzougula sem* (Hindi); *Chara konisem, Lakar-sem (Bengali); Clzaudari sem* (Marathi); *Rakki aravat* (Oriya); *Chathura payar, Kattu avar, Kachang bmbing, Kachang botol, Kachang botor* (Malayalam); *Marisu avara, Morisuavarai, Murukavari* (Tamil).

Importance: It can be grown as a grain legume, green vegetable, tuber-crop or a forage and cover-crop. The genus name gets from the Greek word '*Psophocarpus*' which means 'noisy fruit'. It is so called because the mature pods make a loud popping noise when they split open.

Origin: The origin of Winged bean is Madagascar, Mauritius, Papua New Guinea and Indonesian islands.

Distribution: The winged bean is cultivated in India, Sri Lanka, Bangladesh, Myanmar, Malaysia, China, Vietnam, Taiwan, Thailand, Laos, Cambodia, Papua New Guinea, Indonesia, Philippines, Pacific Islands, Ghana, Nigeria, USA and in West Indies. In India, it is primarily confined to Tripura, Manipur, Mizoram, Nagaland, Assam, Bihar, Kerala, Karnataka and Maharashtra.

Botany: The winged bean is a climbing perennial vine with twining stems. If given support, it can reach a height of 3 to 4 m. The fibrous roots are running horizontally near the soil surface; after a few months roots become thickened and tuberous, near the base of the plant. The roots are normally nodulated. The leaves are trifoliate, on long, stiff petioles; the leaflets are ovate, 7.5 to 15 cm long with the terminal one usually longer than the laterals and attached to the petiole by a marked pulvinus. Stems are usually green, but some varieties have stems with shadings of purple, pink, or brown. The inflorescence is borne on an axillary raceme, up to 15 cm in length, with 2-10 flowers, which may be blue, white, bluish white, purple or lilac. The winged bean is largely self-pollinating, but the bees cause 20% cross pollination to improve pod set. Pods are four-cornered and rectangular or square in cross section, with wings at each corner. Wings may be wavy, serrated, or lobed. Pods are 15 to 22 cm long. The pods take about 20 days to grow its ultimate size and take about 45 days to harden seeds and the surrounding pod shrivels, hardens, and dries out. Pod colour is pale green, green or purple along its entire length. Fresh pods have a chewy texture and taste quite sweet. When mature, the pods turn brown and split open to release the seeds. The number of seeds per pod is 8 to 12. Seeds are mottled and vary in colour from white, yellow and brown to black. The hundred seed weight is 25 to 35 g. The tuber weight is 0.75 to 1.50 kg.

Climate: It grows well in hot, humid equatorial tropical countries. It is cultivated between latitudes of 20°N and 32°S. The plant is grown up to 1500 m from MSL. Winged beans need short days to initiate flowering while long days can inhibit tuber initiation. It is traditionally cultivated in areas with well distributed annual rainfall of 700 to 2500 mm. It is drought-tolerant crop. The winged bean withstands high temperatures but almost never survives frost. Continuous day temperatures higher than 32°C or lower than 18°C inhibit flowering even under short-day conditions.

Soils: It is grown on a wide range of soil conditions with good drainage. It grows well in sandy loam or clay loam soils. Its ideal pH limits are as low as 5.5 and 4.8. In acid soils, the plant is susceptible to aluminium toxicity. However, it thrives in soils with a pH of 8.0. It does not withstand water logging especially at seedling stage.

Season: The crop is sown during July-August and September-October. It can be grown as a dry season crop with adequate irrigation.

Growth stages: The seeds take 8 to 10 days to sprout. After 8 to 10 weeks, the creeper will bear flowers, violet in colour. Tender beans will grow within 4 weeks. Pod-bearing will last for 2 to 3 months. Production of pods starts within 2 months and continues for several weeks. However, when grown for tubers, harvesting is done normally 4 to 8 months after sowing.

Land preparation: Prepare the land to fine tilth and form beds and channels. Land is prepared to a fine tilth by 2-3 ploughing and harrowing. Ridges and furrows are formed.

Varieties: Revathy, PT-62, PT-16, PT-49 and PT-2.

Seeds and sowing: Seed rate is 15-20 kg/ha. Spacing is 90 cm x 60 cm or 45 cm x 15 cm. Seeds are sown at 2 to 2.5 cm depth.

After cultivation practices: Hoeing and weeding should be done as and when required. The twining vines (3-4 m long) require support from the beginning itself which can be trailed over pandal, stakes, trellis or Y-shaped trellis. The bumble bees (*Bombus* sp) are used to enhance pollination of the crop.

Manures and fertilizer application: FYM @ 12.5 t/ha and a fertilizer dose of 40–50-25 kg of N, P_2O_5 and K_2O per ha is recommended.

Water management: Irrigation is required weekly until the plants are large and have deep enough to resist short droughts.

Insect pests and diseases management: There are no serious pests or diseases on winged bean. *Heliothis armigera* causes pod damage. The pod borer, aphids, red spider mite infest the crop to a little extent.

Harvest: Pods can be harvested as a green vegetable in early stages, soft, green, unripe seeds or hard, mature seeds in later stages. Dried pods have a tendency to split open unexpectedly and drop the seeds on the ground. The root tubers are normally harvested when they reach 2.5-5 cm in diameter and 7.5-12 cm in length. The tubers are lifted usually by fork, after pod ripening. The tuber deterioration is rapid under normal tropical conditions, but storage for a few weeks is possible at low temperature and high humidity.

Yield: The yield of fresh pods is from 10 to 15 t/ha. One creeper will bear 4 to 5 kg of beans. The root tuber yield is 2 to 6 t/ha.

Utilization: The economic plant parts are leaves, flowers, pods, green seeds, dried matured seeds, and edible tuberous roots. The young shoots and leaves of winged bean are eaten raw or cooked as green vegetables. The top three sets of leaflets are the tender ones and taste slightly sweet. The flower is eaten raw, fried or steamed. Young pods are good for eating. The half-ripe seeds can be removed from the pod and cooked. The roasted tuber is also edible. Winged bean contains 20% protein and 15 to 20% edible oil. It fixes atmospheric nitrogen in soil.

REFERENCES

Chandel, K.P.S., Pant, K.C. and R.K. Arora. 1984. Winged bean in India. National Bureau of Plant Genetic Resources, New Delhi.

Kay, D. E. 1979. TPI Crop and Product Digest, No. 3. Food Legumes. London: Tropical Products Institute, xvi+435 p.

Sanjive Kumar Singh, Senjam Jinus Singh and N. Reemi Devi. 2013. The Winged Bean: A Vegetable Crop of Amazing Potential. Annals of Horticulture. 6 (1): 159-160

CHAPTER

35

Green Manure and Green Leaf Manure Crops–an Introduction

Green manures refer to a plant material incorporated into the soil while green or soon after maturity, for improving the soil. Green manure crop refers to a crop grown purposely for improving soil productivity and then ploughed in situ. The crop preferred for this purpose is usually a legume, though non leguminous crops are also grown under certain conditions. Green leaf manure refers to the application of green matter consisting of leaves, twigs and loppings from selected plants, shrubs and trees, grown elsewhere to the fields and ploughed in.

Desirable features of green manure crops: The green manure crops should be multipurpose, short duration, fast growing, high nutrient accumulation ability, tolerance for shade, flood, drought and adverse temperature, wide ecological adaptability, efficiency in use of water, early onset of biological nitrogen fixation, high nitrogen accumulation rates, high nitrogen sink in underground plant parts, timely release of nutrients, photoperiod insensitivity, high seed production, high seed viability, ease in incorporation, ability to cross inoculate or responsive to inoculation and pest and disease resistant.

Objectives of green manuring: The objectives of green manuring are i) supply of plant nutrients especially nitrogen and ii) addition of organic matter to soil for improving soil productivity and its sustainability.

Advantages and disadvantages of green manuring: The advantages of green manuring are i) green manure species are potential nitrogen sources with relatively high efficiency, ii) green manure crops withdraw plant nutrients from the lower layers and concentrates them in the surface soil for the use of succeeding crop, iii) incorporation of organic material improves/increases biological activity in the soil and reduces soil erosion, iv) green manure cropping in fallow periods facilitate conservation of nutrients released during weathering process of soil and v) green manure crops require low cash or purchased inputs. The disadvantages of green manure crop are i) a green manure either as a sole crop or inter-crop can compete for time, labour, land and water with the main crop, ii) the cost must be balanced against the cost of inorganic fertilizer or the revenue from a crop, iii) seeds of green manure crop may be costly and scarce, iv) incorporation of a green manure crop may be costly and difficult, v) a green manure crop may attract pests and diseases which have an adverse effect on succeeding crop and vi) control of the quantity and timing of nutrients is more complex and unpredictable.

Green manuring techniques in rice growing region: The green manure crops are grown for 6 to 8 weeks before planting rice. Green manures, when buried *in-situ* just before transplanting rice act as slow release fertilizers and also create reducing conditions which help in mobilizing several other nutrient elements. Green leaf manure is incorporated before or at sowing in semidry rice in Vellore, Cuddalore, Thirivallur, and Chingleput districts, Tamil Nadu. The field is irrigated from tanks after these are full due to monsoon rains. However, the decomposition of incorporated green matter is not uniform and often incomplete.

Sowing of dhaincha or sunhemp with the rice and the green manure crop is pulled out and trampled in about 2 months after sowing when rice is irrigated. Cluster bean is grown as a vegetable cum green manure crop. At 60 DAS, cluster bean yielded 600 kg green pods per ha and 20 tonnes green matter per ha. The material contains 3.86% N on dry weight basis and decomposes easily. In Godavari and Cauvery deltaic areas, cowpea and green gram green pods are harvested for vegetables on 50 to 60 DAS. The crop residues have been used for green manuring. Green manure acts as fixer of atmospheric nitrogen and scavengers of soil nitrogen. The green manures move up to the soil nitrogen released through mineralization following the harvest of the previous crops. This soil nitrogen would otherwise be lost through leaching. The green manure responds to fertilizer phosphorus by showing increasing fresh weight, nitrogen content as well as increased rice yield. The response of rice to green manure depends on the time of its application. Because of succulence and a narrow C/N ratio, a large part of green manure nitrogen is released as nitrates and tends to leach down with water in the course of planting rice.

Dhaincha in reclamation of saline and alkaline soils: Green manuring practice in sodic soil has a unique importance since it helps in the reclamation process besides improving the fertility status of the soil. Usually the fertility status of sodic soils is very poor because of its high pH and exchangeable sodium percentage. The soil organic matter content a measure of available nitrogen, is very low from 0.1 to 0.5 % in sodic soils as sodium carbonate salts solution dissolves the humus. Further the available nitrogen is much lower in the sub soil layers of the sodic soils.

Reclamation of alkali soils basically involves replacing sodium (Na) ions on the exchange complex with more favouable calcium (Ca) ions. The solubility of lime, which is nearly always present in alkali soils in significant amounts, is very low because the pH of alkali soils is high . There is an intimate relationship between the soil pH, partial pressure of CO_2 and Ca ion activity in the calcareous alkali soils. Increase of CO_2 production in the soil enables the soluble Ca status of the soils, that in turn replaces exchangeable sodium (Na), resulting in the improvement of alkali soils.

Soil incorporation of easily decomposable plant material has resulted in increased and rapid production of CO_2. For this reason green manuring has been suggested as an important management practice for the reclamation of alkali soils. Dhaincha (*Sesbania acculeata*) and *Delonix regia* are very effective green manures used for reclamation of sodic soils. Dhaincha is preferred for reclamation of sodic soils. Dhaincha is highly resistant to drought, water stagnation, salinity and alkalinity. It can be grown in soils with pH 4.5 to 9.5. It produces a green matter of 20 t/ha in 90 days. The green matter yield of dhaincha is not affected upto an ESP of 45. Dhaincha nodulates in soils up to an ESP of 70. Dhaincha contains 2.86% N and 3.4% Ca on dry weight basis which helps to replace Na from sodic soil. The acid juice with a pH 4.0 and high seed protein content of 58% seem to be the cause of its resistance to sodicity stresses.

During the reclamation of sodic soils 50% of gypsum requirement has to be spread uniformly over the field. The surface soil is to be ploughed to mix the gypsum in the sodic soils. Irrigate the field with 10 to 15 cm depth of water and maintain the water depth for 3 to 4 days. At this stage the sodium content in clay particles are replaced by the calcium ions from the gypsum, allowing the sodium to wash out the field as leachate. Likewise the field has to be kept with stagnant water for 3 to 4 times after each drainage process. Apply the *Delonix regia* tree looping or dhaincha plants at a rate of 6.25 t/ha. After four to five days of incorporation of green leaves, the field crops like rice can be cultivated without allowing the soil to dry.

Green leaf manuring: Green leaf manuring is a cost-effective alternative to farmyard manure application, especially in areas where cattle populations have declined. Incorporation of fresh leaves harvested from several species of shrubs and trees is an important soil fertility management practice for paddy farmers. Trees are cultivated on field bunds or along the field margins. Branches and leaves are then cut and ploughed into the soil before transplanting rice seedlings.

Effects of green manuring: Green manuring effects have three distinctive effects on physical, chemical and biological properties of soil.

(a) Physical effects: Green manuring improves the crumb structure in soil by aggregation, increase the absorptive capacity and the water holding capacity of the soil, reduce the runoff and increase the percolation of water through the soil.

(b) Chemical effects: Green manuring increases the nitrogen and organic matter content of soil, transfer nutrients from subsoil to the soil surface, increase the availability of soil nutrients during the decomposition of green matter and addition of nitrogen to the soil through atmospheric nitrogen fixation.

(c) Biological effects: Green manuring promotes the activity of soil organisms, depresses the availability of plant nutrients at early stages of decomposition and supplies complex substances like vitamins and hormones produced during decomposition which are helpful for plant growth.

Reasons for decline in the use of green manure: The factors responsible for the decline in green manure use are i) low cost of inorganic nitrogen fertilizers relative to the use of green manuring, ii) increased availability of inorganic fertilizers and relative ease in handling inorganic fertilizers, iii) more profitable alternate uses for land, iv) low seed viability and high seed costs, v) difficulties in incorporating green manure biomass, vi) labour shortages at incorporation time, vii) very low emphasis on organic manure research and extension and viii) often substantial establishment costs of green manure crop.

Current opportunities to increase green manure use: The soil fertility maintenance for sustainable production is enhanced by a judicious combination of green manure and inorganic fertilizers. This approach is more critical in areas where land holdings are small, the resource base is weak, production and productivity have reached a plateau and soil nutrient deficiencies or imbalances have emerged. If green manure is to become popular in those areas, it is essential that it can be fitted into local cropping patterns without competing with other economic enterprises. The technology will be viable only if unit costs of green manure productivity including capital and labour are low. It is essential to maintain production stability and to minimize risk of green manure crop loss.

Green manuring is not practicable in dry lands since the green manure crop will utilize the limited stock of moisture in the lower layers of the soil. The decomposition of the green manure will make further demands in the soil moisture and the succeeding crop will certainly be affected for want of moisture during its cropping period. Green manuring in irrigated garden lands is practiced only to a limited extent.

Green manure crops like sunhemp, daincha are raised as intercrops in sugarcane, banana while calapogonium, sunhemp are raised under coconut plantations. These are incorporated on 60 DAS to improve soil fertility and productivity. Green manuring is particularly suitable for wetlands. Rice responds to green leaf better than to cattle manure or any other manure locally available. Suitable green manure crops can be raised during the fallow period in the single rice crop practicing wetland regions which receive rains in summer season. In double cropped rice areas also a green manure crop can be raised and incorporated *in-situ*.

CHAPTER

36

Green Manure Crops

1. Dhaincha–*Sesbania aculeata* (Willd.) Pers. (2n=12)

Synonyms: *Sesbania bispinosa* (Jacq.) W.Wight

Origin: Dhaincha is native of India, Pakistan, China, Sri Lanka, and tropical Africa.

Botany: It is a soft stemmed shrub with or without prickles. The stems become woody after about 3 months of growth. It grows 1.5 to 2 m in height. The initial growth is slow. However it picks up fast and has vigorous growth later. It makes good growth in 3 to 6 months. The stem is fairly thick, glabrous, branched from the base but soft and pithy; leaves up to 38 cm long, pinnate, leaflets 18–55 pairs, 1.2–2.5 cm long, 0.3 cm wide, glabrous, glaucous; inflorescence 2–8-flowered, 2.5–7.5 cm long; flowers yellow and purple-spotted; pods up to 25 cm long, 0.3 cm thick, curved, cylindrical with 25 to 30 seeds/pod; seeds oblong, grayish brown in colour.

Climate: It is found in subtropical moist through tropical dry to moist forest zones. It grows at an altitude of 1200 m from MSL. It is cultivated in areas receiving rainfall of 500 to 2000mm, annual mean temperature of 19.9 to 27.3°C. Stem produces a spongy mass of aerenchyma under waterlogged conditions.

Soil: It is grown on a variety of soils with a soil pH of 4.5 to 9.5. It comes up well in loamy and clayey soils. It is highly resistant to drought, water stagnation, salinity and alkalinity. The acid juice (pH 4.0) and high protein (2.86 % N) content of dhaincha seem to be the cause of its resistance to sodicity stress. It produces nodules in soils up to an ESP of 70%. It is adapted to wet areas and heavy soils. It thrives in streams, in open wetlands or often as a weed in rice paddy fields.

Seeds and sowing: Seed rate is 25 kg/ha. Spacing for seed crop is 45 x 20 cm. Seeds are usually broadcasted when grown for green manure purpose.

Water management: Irrigation is given immediately after sowing, life irrigation on the third day and subsequently once in 7 to 10 days interval.

Yield: It produces biomass of 8 to 10 t/ha under rainfed condition and 15 t/ha under irrigated condition on 50 DAS. The green matter yield of dhaincha is not affected upto an

ESP of 45. Pods can be harvested on 100 DAS. Seed yield is 500 to 600 kg per ha. Green matter is incorporated on 45 to 60 DAS. Its rapid growth results in accumulation of 80 kg N per ha in 30 days and 230 kg in 60 days after sowing. It contains 2.86% N and 34% Ca which helps to replace sodium from sodic soils. The foliage degrades rapidly and release 50% of their nitrogen within four weeks of incorporation. Seed meal after gum extraction contains 58% protein and 15 % oil.

2. Manila agathi–*Sesbania rostrata* Bremek. & Oberm. (2n=12)

Origin: Manila agathi is native of Sahel region of Africa.

Distribution: It is cultivated as a potential green-manure plant for lowland rice in Senegal, Japan and Philippines due to its vigorous growth, flood and salt tolerance.

Importance: Manila agathi is an excellent green manure crop for lowland rice. The green biomass is to be incorporated on 45 DAS or at flowering stage. Flowering occurs in 4 to 6 weeks after sowing of seeds. It differs from other legumes in having nitrogen fixing nodules also on the stem and branches.

Botany: It is an erect, robust, softly woody, annual or short-lived perennial, grows to a height of 2 to 3 metre on 60 to 90 DAS. Stem is pithy, sparsely pilose, glabrescent and submerged portions clothed with matted fibrous roots. Leaves are paripinnate, 7-25 cm long; petiole 3-8 mm long, pilose; rachis up to 19 cm long, sparsely pilose; stipels present at most petioles; leaflets opposite, in 12 to -24 pairs, oblong, 0.9-3.5 cm x 2-10 mm. Inflorescence an axillary raceme, 1-6 cm long, 3 to 12 flowered; yellow or orange, speckled dark purple or reddish. Pod is 3.5 cm long, thicker at the center than at the sutures, up to 50-seeded. Seed is brown, greenish or dark reddish brown. The nodules are regularly arranged in three or four vertical lines along the main stem and on lateral branches in addition to the presence of root nodules.

Climate: It is grown at an altitude of 1600 m. It grows well with relatively long days and high temperatures. It is a short-day plant with a critical photoperiod of 12-12.5 hours.

Soil: It occurs naturally in marshes, floodplains, on muddy river banks and the edges of pools, open savanna. It tolerates waterlogged soils and flooding to over one metre deep. Its cultivation is always associated with wet rice. It is not suited for salt affected soils.

Season: Seeds can be sown or seedlings may be planted throughout the year. Sowing in February and May records high green matter yield. If grown in winter when days are shorter, it flowers early which results in lesser vegetative growth. It produces nodules sparsely in winter season. It is thus unsuitable as a green manure crop for use in winter season.

Seeds and sowing: Seed rate is 30 kg per ha. Spacing of 45 x 20 cm is used for seed production while broadcasting is recommended for green manure purpose. Treat the seeds with Conc H_2SO_4 acid @ 100 ml/kg of seeds for 3 minutes. Subsequently wash the acid treated seeds with water for 4 to 5 times till the seed is free of acid or with hot water at 80ºC to produce 90% germination. Dry the seeds in shade then sow the seeds. Stem cuttings of one cm thickness can also be used for propagation. Thirty days old seedlings may be transplanted with 1.5 to 2.0 m spacing in 12:1 ratio in rice fields.

Nitrogen fixation: *S. rostrata* is nodulated with fast growing Rhizobium strains and slow growing Bradyrhizobium. Stem nodulating strains nodulate both stems and roots. Root nodualting strains nodulate in roots only. Flowering occurs 4 to 6 weeks after seeding. It fixes 25 to 30 g N/m^2. It fixes nitrogen even in waterlogged conditions because of stem nodualtion but had poor root nodulation. Nitrogen fixation is high on 35 DAS. Nitrogen accumulation is linear between 35 and 53 DAS. The advantages of stem nodulating legumes is their ability to nodualte and fix nitrogen in the presence of combined nitrogen. The nitrogen content is 1.62 g per plant. A population of 250,000 plants per hectare when grown for 55 days can fix about 120-150 kg nitrogen per hectare.

Fertilizer application: Application of 25 kg P_2O_5 per ha as basal dressing is recommended.

Water management: First irrigation is given immediately after sowing, life irrigation on the third day and subsequently once in 10 to 15 days.

Harvest: It is allowed to grow for 45-65 days for green manure purpose depending on its growth rate. If it is left to grow longer than about 55 days, the lignin content increases which decreases the decomposition rate of plant biomass. The green manure crop is ploughed in just before the rice crop is sown or transplanted. Initial decomposition is rapid, with 30-45% of the leaf material decomposing in 10 days after incorporation. The decomposition then slows down considerably, reaching 50% after 35 days, while the half-life of stems and root-stubble is about 110 days.

Yield: Manila agathi produces biomass yield of 12 to 15 t/ha under irrigated condition and 6 tonnes per ha under rainfed condition.

Seed production: Clipping of terminal buds on 60 DAS enhances branching and pod production. Matured pods can be harvested manually from 100 DAS with one week interval. Seed yield is 500 to 600 kg per ha.

3. Sesbania–*Sesbania speciosa* Tam

Origin: This crop is native of South Africa.

Botany: The plant has a grayish appearance with soft grayish tomentose hairs on stems and leaves. It grows to a height of 3 to 4 metres in four months. Its growth is very slow during first 30 to 40 days and then it grows rapidly. The stem is pithy. However, it becomes woody if allowed to grow for more than four or five months and making it difficult to be pulled out or even to be harvested with a sickle. Leaves compound, long with numerous leaflets; leaflets near the pulvinus the longest and gradually reduced in size towards the tip of the leaf; fruiting profuse, fruits long, flat-quadrangular ending in a tapering tip and with 40 to 55 seeds per pod. Seeds are yellowish brown in colour. Each plant gives about 400 to 600 grams of seeds.

Seeds and sowing: Seed rate is 40 kg per ha. The crop comes to maturity in 150 days.

Yield: It produces green matter yield of 12, 20, 50 and 72.5 tonnes per ha on 60, 90, 120 and 150 DAS.

Seed production: Each pod consists of 40 to 55 seeds. Each plant produces 400 to 600 g of seeds. 100 plants in field borders are able to meet seed requirement for sowing one acre.

4. Sithagathi–*Sesbania aegyptiaca* Poiret (2n = 12)

Synonyms: *Sesbania sesban* (L.) Merrill

Vernacular names: Egyptian sesban, Egyptian rattle pod, Sesban, Sesbania (English); sesaban (Arabic); Jainti, Jayant (Bengali).

Importance: Sithagathi is used as a wind break crop in sugarcane and as shade crop in turmeric. *S. sesban* is a fast growing nitrogen-fixing leguminous tree species which has the capacity of rapid decomposition when incorporated into soil serving as a green manure in alley cropping.

Botany: *S. sesban* is a narrow-crowned, deep-rooting single or multiple stem shrub or short-lived tree, which may grow up to 8 m and up to 20 cm stem diameter. The plant is fast growing and it grows 4.5 to 6.0 m high in one year; and normally flowers and produces ripe pods within the first year after planting. If the trees planted are widely spaced they usually develop many side branches. The many branches give the tree a shrubby appearance. Leaves are paripinnate, long (compound 12 to 18 cm long) and narrow; leaflets in many pairs (made up of 6 to 27 pairs of leaflets), rounded or oblong, usually asymmetric at the base, often glaucous and stipules are minute or absent. It has up to 20 flowers which are yellow with purple or brown streaks on the corolla. Flowers are attractive, yellow, red, purplish, variegated or streaked, seldom white, large or small on slender pedicels, solitary or paired in short axillary racemes, usually unpleasantly scented; all petals are long clawed, standard orbicular or obovate. Pods are pale yellow and linear or slightly curved with 10 to 20 cm long and 5 mm wide containing up to 50 seeds. Seeds are oblong or sub-quadrate, brown or dark green mottled with black. There are 55-80 seeds/g.

Climate: It is cultivated in areas with monsoonal, semi-arid to sub-humid regions with mean annual rainfall of 500 to 2,000 mm. It is tolerant to cool temperatures and can grow in the higher elevation regions of the tropics at an altitude of 100 to 2300 m from MSL, but it cannot bear heavy frosts. It grows with the cardinal minimum, optimum and maximum temperature of 7°C to 10°C, 18 to 23°C and 40°C. It has good tolerance to low temperatures. It has moderate shade tolerance.

Soils: It is grown on a wide variety of soils from loose sandy soils to heavy clays. It is tolerant of saline, alkaline and acidic soils as well as of low P levels. It can withstand waterlogging, except during first stages of seedling. It is adapted to a wide variety of soil types, ranging from loose sandy soils to heavy clays. Furthermore, it has an excellent tolerance to waterlogging and flooding as well as saline, acidic, alkaline soils. It grows on stream banks and swamp edges.

Season: It is sown during June-July and September-October months.

Land preparation: Two to three ploughings are given to get a fine tilth. Beds and channels are formed.

Seeds and sowing: Seed rate is 2 kg per ha. Spacing is 60 cm x 60 cm for pure crop; 1.0 m for bund crop and 90 cm × 90 cm for shade crop in turmeric fields.

Weed management: Two hand weeding are given on 20 and 35 days after sowing.

Manures and fertilizer application: 12.5 tonnes of FYM per ha with 20-25-20 kg of N, P_2O_5 and K_2O per ha are applied basally.

Water management: First irrigation is given at the time of sowing, life irrigation on third day of sowing and subsequent irrigations are given once in 12 days.

Harvest: The plants are harvested at one-meter height. The cutting frequency is once in 60 days. The green fodder yield is 35 to 40 tonnes per ha per year for pure crop and 5 to 7 tonnes per ha per year as bund crop.

5. Sunnhemp–*Crotalaria juncea* L.

Vernacular names: Indian hemp, Madras hemp, brown hemp

Origin: Sunnhemp is native of India.

Importance: Sunnhemp is an effective crop in smothering weeds. It is hardy and drought resistant crop. It cannot tolerate water logging condition.

Botany: It is a vigorous and fast growing crop. It grows to a height of 200 to 300 cm.

Climate: It is grown throughout the tropics. Its vegetative growth is favoured by long days. The seed setting is enhanced by short days.

Season: Sunhemp can be sown throughout the year as green manure crop. However it can be grown as seed and fibre crop during March to April.

Seeds and Sowing: Seed rate is 30 kg/ha. Spacing recommended for seed production is 75 × 20 cm.

Water management: Irrigation is given immediately after sowing, life irrigation on the third day and subsequently once in 10 to 15 days.

Harvest: The crop duration ranges from 75 to 150 days, but usually harvested in 70 to 75 days after sowing.

Yield: The green matter yield is 10 to 12 t/ha. Seed yield is 400 to 500 kg/ha. Fibre yield is 350 to 500 kg/ha.

Utilization: Green manure is trampled from 45 to 60 days after sowing.

6. Aeschynomene–*Aeschynomene afraspera* J.Léonard (2n=80)

Vernacular names: Attuneddi (Tamil), Sola, Laugauni (Hindi),

Botany: Aeschynomene grows to about 150 cm tall on 50 to 60 DAS. Plants grow 2 m tall at 3 months. The stem nodules are distributed at random along the principal and lateral branches.

Climate: It is adaptable to warm and humid weather. It tolerates shade conditions and resistant to high temperatures. Seeds normally germinate at 20°C and the plant grows rapidly when temperature is above 25°C.

Soil: It can be grown on lowland with shallow water.

Seeds and sowing: It can be propagated by seeds and stem cuttings. Seed rate recommended is 20 kg/ha.

Nitrogen fixation: Root nodulation and atmospheric nitrogen fixing capacity is strongly inhibited while stem nodulation and related nitrogenous activity are unaffected by increasing amounts of mineral nitrogen. 25 to 35% of the total nitrogen accumulated by aeschnomene could be transferred to the rice crop.

Yield: The green matter yield is 12 to 15 t/ha within six to eight weeks after sowing.

7. Wild indigo–*Tephrosia purpurea* (L.) Pers. (2n=22)

Family: Fabaceae

Vernacular names: Wild indigo, Fish poison, Purple tephrosia (English); Sarphank, Sharpunkha (Hindi); Masa (Rajasthani); Kolingi, Kattu-Koalingi, Apavali, Mollukkay (Tamil); Vempali (Telugu); Sharpunkha (Sanskrit), Kattamari, Kozhinjil (Malayalam), Untoali, Unhali (Marathi), Unnali (Gujarati), Sarpankho, Jhojro (Punjabi),Sarphoka, Satawar (Urdu).

Origin: It is native of Australia, China, India and Sri Lanka.

Distribution: Wild indigo is found in Africa, Southeast Asia and Australia. It is grown throughout India and Sri Lanka.

Importance: It is cultivated as green manure crop. It is suited for summer fallows. When added to the soil as green manure it increases humus content and induces the formation of large, stable soil aggregates. It produces ample seed and builds up a large seed bank in the soil. It is very hardy and drought resistant. It has self-sown capability. It makes slow growth and stands in field for 6 to 8 months. In ayurvedic literature *Tephrosia purpurea* has been described 'Sarva warnavishapak' means it has property to healing all types of wounds.

Botany: It is a small shrub that grows up to 1.5 meters tall. It has bi-pinnate leaves with 7 to 15 leaflets, the terminal leaflet being solitary. The leaflets are 10 to 32 mm long and 5 to 11 mm wide. The pea like flowers is white to purple and arranged in inflorescences that are up to 25 cm long. The individual flowers have corolla parts that are between 2 to 3 mm long. The pods are straight and somewhat upcurved at the terminal end and may range from 20 to 45 mm in length and 3 to 5 mm wide. When dry, the pods split along two valves to reveal 2 to 9 black rectangular seeds 2.5 to 5 mm long and 1.8 to 3 mm wide with about five seeds.

Climate: It occurs naturally in grassy fields, waste places and thickets, on ridges, and along roadsides and seashore. It is grown up to an altitude of 400 to 1300 m from MSL.

Soil: It prefers dry, gravelly or rocky and sandy soils, but it grows well on loamy soils. It is tolerant of saline and alkaline soil conditions.

Season: It is generally sown in March-May for seed production and thoughout the year for green manure purpose.

Seeds and sowing: Seed rate is 15 to 20 kg/ha. Spacing is 30 x 10 cm for seed crop. The seed have a waxy, impermeable hard seed coat and do not quickly germinate. To hasten germination, the seeds are to be abraded with sand or steeped in hot water at 55°C for two to three minutes. The seeds can also be treated in conc H_2SO_4 @ 100 ml per kg of seeds for 3 minutes or with hot water of 80ºC for 3 minutes to promote germination.

Harvest: It is incorporated on 45 to 60 DAS.

Yield: It produces green matter yield of 5 to 8 t/ha. Seed yield is 400 to 500 kg/ha.

Utilization: The leaves and seeds contain tephrosin, a fish poison which paralyzes fish. *Tephrosia purpurea* is used as green manure for vegetables, rice, coconut and banana, especially in India and Sri Lanka, and on a more limited scale in Indonesia, Malaysia and southern China. When grown as a green manure on saline-sodic soils in Rajasthan, it is most successful in reducing soil salinity and lowering the pH. The genus *Tephrosia* is well known for its richness in prenylated flavonoids and considered to possess insect repellant, larvicidal, piscicidal, antimicrobial, anticancer and antioxidant properties.

8. Bengal indigo–*Indigofera tinctoria* Linn (2n = 16)

Synonyms: *Indigofera sumatrana*

Family: Fabaceae

Vernacular names: Indian Indigo (English); Neelamari, Neelamar Amar (Malayalam); Nilini (Sanskrit); Averi, Neeli (Tamil); Nili (Hindi); Karunili (Kannada); Nili Chettu, Nili (Telugu).

Importance: A galactomannan, composed of galactose and mannose in molar ratio of 1:1.52 is isolated from seeds.

Origin: The origin of Bengal indigo is India.

Distribution: It occurs wild or naturalized in most countries of Africa, in Asia from Arabia to South-East Asia and in Australia. It is introduced in tropical America. It is found throughout and cultivated in many parts of India.

Botany: It is a perennial shrub growing up to 2 m. Stems and branches are thin with reddish tinge; leaflets which are elliptic-oblong pale green or bluish. A silvery-cauescent appearance is seen on leaves and the plant especially in the young branches. Leaves are pinnately compound with seven to eleven leaflets. Flowers are small, rose-coloured and are borne in axillary racemes. Pods are long and cylindrical, more or less curved, pale greenish grey when young and dark brown on ripening having 6 to 12 seeded.

Climate: The plant requires good sunlight and grows well in plains as well as hilly areas. It can withstand temperature up to 40°C.

Soils: Sandy loams soils are the best. Clayey soil where water logging is likely is unsuitable.

Season: The best time for planting is August to October.

Land preparation: The field is ploughed well. Manures are applied at the time of final ploughing.

Seeds and sowing: This is usually propagated by seeds. Seed rate is 3 kg/ha. Seeds are mixed with sand and ground gently to break the seed coat. An alternate method for enhancing germination is dipping the seeds in boiling water for a second. Seed treatment with conc H_2SO_4 @ 100 ml per kg of seeds for 3 minutes improves seed germination. Broadcast the seeds preferably mixed with sand 2 or 3 times its volume to ensure uniform coverage. The seedbeds should be covered with straw and irrigated. Seeds germinate within a week. Transplant the seedlings to polybags at 3-4 leaf stage. Two months old poly bagged seedlings are planted in pits at a spacing of 60 × 60 cm.

Weed management: Weeding has to be done at 3 weeks and 6 weeks after sowing.

Manures and fertilizers application: FYM @ 10 t/ha, bone meal 100 kg/ha, neem cake 250 kg/ha, azospirillum and azotobactor @ 2.5 kg each per ha along with cow dung is recommended.

Pest and diseases management: Psyllids are often a problem in neelamari crop. These suck sap from tender leaves and tips. Leaves turn yellow, falls and plant gradually dries up. Spray 3% neem seed extract or 2% garlic neem oil emulsion to control psyllids.

Harvest: Ripe pods are to be harvested in the early morning to prevent loss of seeds by shattering during harvest. Leaves are mainly used in hair oil preparations. Leaves are harvested from 3-4 months of planting. It is done by cutting the plants at a height of about 10 cm from ground level. The crop raised for leaf harvest should not be allowed to flower. Subsequent harvests can be done at 1.5-2 months interval. Harvested leaves are marketed fresh; cut leaves must be used within 8 hours. Four to five cuttings can be taken in a year depending on the growth. After 3-4 years of leaf harvest, the plants are uprooted and the roots are cut with 15 cm stem portion, cleaned, dried and stored in plastic lined jute bags.

Yield: The green matter yield is 8 to 10 t/ha/year. Average leaf yield is 1250 kg/ha/year. The root yield is 800 to 1000 kg/ha.

Utilization: It is used in constipation, liver disease, heart palpitation and gout. It is used chiefly a source of dye, indigo also has been used as a nematicide and treatment for a range of ills including scorpion bites and ovarian and stomach cancer. The leaves are dried in shade, powdered and given in dose of 1 to 5 grams for three times a day for any type of toxicity (herbal, metal or poison of any living creature), fever due to derranged *vatham*, *kamalai* (jaundice), *mantham* (indigestion), fever, etc. The leaf juice is given in the dose of 10-20ml along with honey twice daily for jaundice, inflammation of liver, etc. The root is crushed and prepared into decoction, and given for gunmam (abdominal disorders), vellai (leucorrhoea), all types of toxicities, etc. For poisonous bites, the whole plant is ground and applied as a paste over the bitten area while the leaf juice is given internally to the patient.

9. Pillipesara–*Phaselous trilobus* Ait. (2n = 22)

Synonyms: *Vigna trilobata* L.

Vernacular name: Wild gram (English); Mugvan, Vanmung (Hindi); Naripayaru (Tamil).

Importance: It is a fodder cum green manure crop. It is capable of being cut twice or thrice before being ploughed in the field. Seeds are also used as a minor pulse. The crop is able to withstand drought and excessive soil moisture.

Botany: Pillipesara is a herbaceous creeping annual, growing into a short dense cover when grown thick. The stem is trailing to a length of 25 to 50 cm. It has a strong taproot. Leaves are trifoliately compound which found arranged alternately. Flowers are in a close deltoid head, on a peduncle that usually overtops the leaves. Fruit is a cylindrical pod glabrous and with many seeds. The pods are 2.5 to 5.0 cm long with six to twelve seeds. Seeds are about 0.3 cm long, square and with a prominent white hilum.

Season: It can be sown throughout the year but preference is March-April.

Seeds and sowing: Seed rate is 15 to 20 kg/ha.

Cropping system: It is intersown in the standing rice crop 10 days before the harvest of crop.

Yield: The green matter yield is 10 t/ha. Seed yield is 200 kg/ha.

10. Calopo–*Calopogonium mucunoides* Desv. (2n=24)

Vernacular name: Calopo (E)

Origin: It is native to Mexico, Central America, the Caribbean and tropical South America.

Distribution: It is grown in Belize, Costa Rica, El Salvador, Guatemala, Honduras, Nicaragua, Panama, French Guiana, Guyana, Surinam, Venezuela, Colombia, Ecuador, Peru, Malaya and Indonesia. It has been introduced in 1930 as a cover crop in India. It is well suited to Kerala coasts.

Botany: It is an annual creeper of 0.3 to 0.5 m height. Stem is succulent, covered with long brown hairs. Its lower part becomes creeper while upper part turns to twiner. Leaves are trifoliate and hairy on both surfaces with oval leaflets of 5 cm length. Flowers are small pale blue. The pod is hairy, yellowish-brown, 4 to 5 cm long. It contains 4 to 8 seeds. There are 73,000 seeds in 1 kg.

Climate: It prefers hot, humid tropics with an annual rainfall in excess of 1525 mm. It is moderately drought and shade tolerant.

Soils: It is found on a wide range of soils with 4.5 to 6.5 pH.

Land preparation: One deep ploughing with disc plough followed by two discings and planking are sufficient.

Seeds and sowing: It is sown at the beginning of rainy season through broadcasting the seeds @ 6-8 kg/ha and then covered with soil but in line sowing the distance should be 1.2 to 1.8 m. It can also be propagated by planting the rooted slips. To break the dormancy the seeds are to be treated through any of the following methods viz., i) Conc sulphuric acid (S.G. 1.8) for 20 minutes or with 24 or 36 N sulphuric acid for 7 min. then wash and dry and ii) Scarify with sand.

Manures and fertilizer application: Fertilizer dose of 20-60-30 kg N,P_2O_5 and K_2O kg/ha is recommended.

Harvest: It gives full cover in 2-3 months and can be harvested at 8 weeks intervals after establishment. Grazing can also be done rotationally at interval of 8-12 weeks.

Yield: The forage yield of calopo is 56 t/ha (green) and 14.5 *t/ha* (dry) in 3 to 4 cuttings. Seed yield is 600-700 kg/ha.

Utilization: It is used as a fodder crop and is less palatable at early stage but after flowering it becomes more palatable. Being fast grower it is good for soil conservation purposes. In new mixed pasture, the Calopo can be sown after the grasses get established. It is frost and drought susceptible but is an excellent flood tolerant and grows well in acid soils.

CHAPTER

37

Green Leaf Manure Crops

1. Agathi–*Sesbania grandiflora* (Syn: *S. formosa*)

Botany: Agathi is a short lived tree with life span of 20 years.

Climate: It is best adapted to moist tropics in areas receiving annual rainfall of 1000 mm.

Importance: It provides forage, food and green manure. Seed is the richest source of protein which is more than 40% by weight of all legume seeds. The young leaves, tender pods and giant flowers of agathi are favourite asian vegetables which are used in curries and soups or fried, lightly steamed or boiled.

Botany: The plant has fast growth rate especially during first 3 or 4 years after planting. It attains 4.3 to 5.5 m height in one year and 8 meters in 3 years. The pod length is 60 cm and contains about 50 small seeds.

Seeds and sowing: The species is very easy to propagate by cuttings or seedlings. Over 3000 stems cuttings per ha have been used for planting.

Yield: Agathi produces the green matter of 55 t/ha in 6 to 7 months. The leaves contain 36% crude protein.

2. Sesbania–*Sesbania speciosa* Tam

Origin: This crop is native of South Africa.

Botany: The plant has a grayish appearance with soft grayish tomentose hairs on stems and leaves. It grows to a height of 3 to 4 metres in four months. Its growth is very slow during first 30 to 40 days and then it grows rapidly. The stem is pithy. However, it becomes woody if allowed to grow for more than four or five months and making it difficult to be pulled out or even to be harvested with a sickle. Leaves are compound, long with numerous leaflets; leaflets near the pulvinus are the longest and gradually reduced in size towards the tip of the leaf; fruiting profuse, fruits long, flat-quadrangular ending in a tapering tip and with 40 to 55 seeds/pod. Seeds are yellowish brown in colour. Each plant gives about 400 to 600 g of seeds.

Seeds and sowing: Seed rate is 40 kg/ha. The crop comes to maturity in 150 days.

Yield: *Sesbania speciosa* produces green matter yield of 12, 20, 50 and 72.5 t/ha on 60, 90, 120 and 150 DAS.

Seed production: Each pod consists of 40 to 55 seeds. Each plant produces 400 to 600 g of seeds. 100 plants in field borders are able to meet seed requirement for sowing one acre.

3. Gliricidia–*Gliricidia maculata* (Kunth) ((2n = 52, 104)

Synymous: *Gliricidia sepium* (Jacq.) Kunth)

Vernacular names: Gliricidia (English), Mata raton (Spanish), Cacao de nance, Cacahnanance, Madriado (Honduras), Kakawate (Philippines), Madre cacao (Guatemala). The generic name '*Gliricidia*' refers to 'mouse killer' in Latin, and the species '*sepium*' is named from the Latin saepes meaning hedge. Spanish colonists adapted the local vernacular in naming the *Gliricidia as* 'madre de cacao' (mother of cocoa) to describe its use as a cocoa shade tree.

Importance: It is a fast growing, tropical, leguminous tree. It is used for fuel wood, live fences, animal forage and green manure, soil stabilization and as an ornamental tree. It is generally used for shade in tea, coffee and cocoa plantation. It is good for cultivation in backyards and along fence lines, field bunds and on road sides. It is a strong light demander, drought resistant and has a good coppice.

Origin: It is native of Central America to northern South America (Mexico).

Distribution: It is cultivated in many parts of Central America, Mexico, West Africa, Caribbean, the Philippines, India, Sri Lanka, West Indies, South Asia and tropical Americas.

Botany: It is a small to medium sized, thorn less tree which usually attains a height of 10-12 m. Branching is frequently from the base with basal diameters reaching to 50-70 cm. The bark is smooth but can vary in colour from whitish grey to deep red-brown. The stem and branches are commonly flecked with small white lenticels. Trees display spreading crowns. Leaves are odd pinnate, usually alternate, subopposite or opposite, to approximately 30 cm long, leaflets 5-20, ovate or elliptic, 2-7 cm long, 1-3 cm wide. Leaflet midrib and rachis are occasionally striped red. Inflorescences appear as clustered racemes on distal parts on new and old wood, 5-15 cm long, flowers borne singly with 20-40 per raceme. Flowers are bright pink to lilac, tinged with white, usually with a diffuse pale yellow spot at the base of the standard petal, calyx glabrous, green and often tinged red. Fruits is green, sometimes tinged reddish purple when unripe, light yellow-brown when mature, narrow, 10-18 cm long, 2 cm wide, valves twisting in dehiscence; seeds 4-10, yellow-brown to brown, nearly round.

Climate: It is grown under tropical and sub-tropical climate conditions. The plant grows best in warm, wet conditions up to an altitude of 1600 m from MSL. It is grown in areas receiving rainfall of 800 to 2300 mm. The plant grows well with optimal temperatures of 22-30°C It is susceptible to frost condition and temperature below 15°C. It tolerates fire and the trees quickly re-sprout with the onset of rains. It does not tolerate long period of waterlogging.

Soil: It occurs naturally in coastal sand dunes, river banks, deep alluvial lake-bed deposits, floodplains and fallow land. It establishes well on steep slopes (40% gradient). It adapts very well in a wide range of soils ranging from eroded acidic soils, fertile sandy soils, heavy clay, calcareous limestone and alkaline soils with pH 4.5 to 9.5. It is tolerant to water logging and to a wide range of poorly fertile soil tolerating low pH provided that this is not associated with high aluminium saturation. It can thrive in dry moist, acidic soils or even poor degraded soils under rainfed conditions.

Season: Gliricidia can be grown during July–August and September–October months.

Propagation: Seeds are shed from pods through explosive dehiscence with seed dispersal distances up to 40 m. Seeds are soaked in water for 8-10 hours, preferably overnight. The soaked seeds are sown in small polythene bags or plastic sachets filled with a mixture of red soil, sand, and farmyard manure (1:1:1) and watered regularly. Generally, 3 to 4 month old seedlings can be planted on bunds in the rainy season. Seed propagation method is more convenient for establishing a large number of plants. The 3 to 4 month old seedlings are usually cut back, as 'stumps' prior to planting. It can also be propagated through stem cuttings. *Gliricidia* cuttings are taken from stems of at least one-year-old plants. These should be from brownish green in bark colour of mature branches and should measure 2-6 cm in diameter and 30-100 cm in length. The stem cutting is normally cut obliquely at both ends, discarding the younger tips and the base is inserted 20-50 cm into the soil depending on the length of the cutting. Matured thick stem of 1-2 m length is revoked in cow dung solution and then planted with 2m × 2m spacing.

Planting technique: Cuttings for live fences may be up to 200 cm long whilst those for hedgerows may be 30-50 cm in length. For hedges, cuttings are planted closely at 50 cm spacing. The cuttings should be planted on bunds at 50 to 75 cm spacing in the rainy season immediately after these are cut from the stems. For steep slopes, closer plant spacings of <20 cm is recommended for better soil erosion control. *Gliricidia* plants from stem cuttings grow faster than those grown from seeds. Propagation from stakes is suitable mainly for situations where only a few trees are to be-established.

Intercultural operations: Pruning at 0.3-1.5 m will stimulate leaf production. Pollarding at 2 m or above is recommended for optimal wood biomass production. Coppicing is used for fuelwood production. It tolerates lopping and browsing. It stands repeated lopping. It has no root effect on the crops grown by the side.

Crop protection: It is resistant to the defoliating psyllid (*Heteropsylla cubana*).

Harvest: It is tolerant of regular browsing, lopping or trimming to manageable heights. The optimum frequency of lopping for leaf production depends on the local climate. Trees can be lopped frequently in the wet in comparison to the dry season. In general, total annual biomass yield increases with less frequent cutting. The hedges can be periodically pruned to provide fodder, green manure, firewood, or stakes for new fences. It can be lopped at about 7 months of plant growth from cuttings and 14 months of plant growth from planted seedling and often every 2-3 months during the rainy season and every 3-4 months during the dry season. It can be kept in the form of hedge of 2 to 4 feet height and lopped at an interval of 6 to 8 weeks. The cutting interval is 3 months. One year after planting, harvesting can be started by lopping the plants at 75 cm to 1 m above the ground for green leaf manure. For good management, plants should be pruned at appropriate time. Pruning should be done at least thrice during the year; i.e., in June (before sowing of the rainy season crop), in November (before sowing of the post rainy season crop), and in March (before sowing of the summer crop). After pruning the loppings, they are made into smaller pieces and can be transported to nearby fields. Leaf material is applied on the surface of ploughed soil and mixed into soil before crop planting. The trees re-sprout vigorously after lopping and have potential to tolerate repeated cutting.

Yield: Leaf biomass (dry matter) production depends on various management practices such as establishment methods (seedlings versus stakes of various sizes), plant spacing, lopping height and lopping frequency under a range of climatic conditions. It produces 3 tonnes of dry leaf material per month from 10,000 trees per ha. Gliricidia planted 2 m apart along the bunds can produce 4 tons leaf dry matter per ha. The foliage has crude protein content of 20 to 30 per cent.

Seed production: *Gliricidia* is largely out crossing, needs to be isolated from other trees of the same or related species to prevent cross-pollination. It should be planted in blocks containing at least 30 trees and isolated by at least 200 m. A border row should be established around the block and seed should not be collected from this row. The periodicity of pod ripening is dependent upon the climatic conditions and typically takes 45-60 days. The seed yields of *Gliricidia is* up to 90 g per tree per year. Seeds are shed from pods through explosive dehiscence with seed dispersal distances of up to 40 meters.

Seed collection and storage: Pods are collected in the month of February to March before they dehisce on the tree and dried in the sun for 3 to 4 days to separate the seeds. Viability is retained for one year. Cutting gliricidia to 0.5 m after a seed harvest reduces flowering and seed yield in the following 2 years relative to an uncut tree.

Utilization: The leaves can be used as fodder which are rich in protein and highly digestible, and low in fibre and tannin. It is a good fodder for goat and cattle. It is used as insecticide, repellant and rodenticide. Gliricidia contributes 20 kg of N per tree per year to the soil.

4. Gold mohur–*Delonix regia* (Hook.) Raf

Family: Fabaceae

Vernacular names: Flamboyant flame tree, Gold mohur, Flame tree ,Julu tree, Peacock flower, Flame of the forest, Gulmohr, Flamboyant, Royal Poinciana (English); Kattikayi, Peddaturyl, Gulmohr, Shima Sunkesula (Hindi); Vadhamadakki (Tamil).

Importance: It is the most observed flowering tree in tropical and subtropical climates. Showy and bountiful flowers are etched in the memories of travelers and residents alike.

Origin: It is native of Madagascar.

Distribution: It is grown in Brazil, Burkina Faso, Cyprus, Egypt, Eritrea, Ethiopia, India, Jamaica, Kenya, Mexico, Niger, Nigeria, Puerto Rico, Singapore, South Africa, Sri Lanka, Sudan, Tanzania, Uganda, United States of America.

Botany: The compound leaves have a feathery appearance and are a characteristic light, bright green. Leaves are even-pinnate and alternate. Each leaf is 30 to 50 cm long and has 20 to 40 pairs of primary leaflets or pinnate on it, and each of these is further divided into 10-20 pairs of secondary oblong leaflets or pinnules. The tree starts flowering in its 4th or 5th years. The flowers emit a characteristic Gold mohur tree scent. The flowers are large, with four spreading scarlet or orange-red or yellow petals, standard of red splashed with burgundy spots, arranged on racemes up to 10 inches long. Pods are dark brown and can be up to 60 cm long and 5 cm wide; the individual seeds, however, are small, weighing around 0.4 g on average.

Climate: It is widespread in most tropical and subtropical areas of the world. It is found up to an altitude of 2000 m from MSL. It grows in areas with mean annual rainfall of 700 to 1200 mm and mean annual temperature of 14 to 26°C. The tree demands light and grows weakly and sparsely under shade. It is drought tolerant tree.

Soils: It is grown in various types of soils from clay, clay loam to sandy, but it prefers sandy soils. It tolerates acid, saline and alkaline soils.

Propagation: It is propagated by seeds. Seeds are collected, soaked in warm water for at least 24 hours, and planted in warm, moist soil in a semi-shaded, sheltered position. Seeds can also be 'knicked' or 'pinched' with a small scissors or nail clipper and used for sowing. The seedlings grow rapidly and can reach 30cm in a few weeks under ideal conditions.

Tree management: It is fast growing tree. Careful pruning will achieve good crown form. The trees have shallow root systems and the wood is weak; they are therefore liable to being uprooted during strong storms and broken by strong winds.

Utilization: It is planted as a shade tree in dairy farms, tea plantations and residential street tree. It can be planted as live fence posts. Flowers are reputed to produce bee forage.

5. Cassia–*Cassia fistula* Linn.

Family: Caesalpiniaceae

Vernacular names: Indian Laburnum, Purging Fistula, Cassia; Golden Shower, Amulthus (English); Sonhali, Amultus (Hindi); Garmala (Gujarati); Kakkemara (Kannada); Shrakkonnai, Konai, Irjviruttam (Tamil); Kondrakayi, Raelachettu, Aragvadhamu, Koelapenna (Telugu); Sunaari (Oriya).

Importance: *Cassia fistula* is used as an ornamental and shade tree around houses; on the edges of roads; and in the streets, parks, and gardens of towns. It is the national tree of Thailand, and its flower is Thailand's national flower. It is also state flower of Kerala in India. It is used as an herbal medicine.

Origin: The species is native to the Indian Subcontinent and adjacent regions of Southeast Asia. It ranges from southern Pakistan eastward throughout India to Myanmar and Thailand and south to Sri Lanka.

Distribution: It is grown in India, Indochina, Malaysia Australia, Egypt, Ghana, Mexico, Pakistan, Zimbabwe. In India, it is grown in Maharashtra, Kerala, Deccan and Konkan.

Botany: Cassia fistula is a fast-growing, deciduous tree of up to 15 m in height and 60 cm d.b.h. The trunk is straight, and the open crown consists of horizontal branches with sparse foliage. The leaves are paripinnate, made up of 8 to 16 ovate to lanceolate leaflets, and 8 to 20 cm long. Pod is cylindrical, 40-70 cm long and 20-27mm in diameter. The long pods which are green, when unripe, turn black on ripening after flowers shed. Fruits remain hanging on the tree for 2 or 3 months after ripening. Gradually, they fall to the ground. Seeds broadly ovate, 8 mm long, slightly less in breadth, and 5 mm thick. Seeds average 5,500 per kg.

Climate: It is a tree of deciduous forests ranging from tropics to moist through subtropical thorn to moist forest zones. It grows up to an altitude of 1300 m from MSL. It grows in areas with mean annual rainfall of 500 to 2700 mm, annual temperature of 17 to 37°C. The tree will bloom better where there is pronounced difference between summer and winter temperatures. The tree can withstand moderate amount of shade, is drought tolerant, but not frost hardy.

Soils: Growth for this tree is best in full sun on well-drained soil; it is relatively drought tolerant and slightly salt tolerant. The tree prefers soils of pH 5.5-8.7

Propagation: The seeds can be soaked in boiling water for 5 minutes before planting to stimulate germination because the seed coat is hard. Seeds can be stored at room temperature and remain viable for one year.

6. Avaram–*Cassia auriculata* Linn

Family: Cesalpinaceae

Vernacular name: Tanner's cassia (English); Avaram (Tamil).

Importance: It has attractive yellow flowers which are used in the treatment of skin disorders and body odour. It is widely used in traditional medicine for rheumatism, conjunctivitis and diabetes. Its bark is used as an astringent, leaves and fruits anthelminthic, seeds used to treat eye troubles and root employed in skin diseases. It is also used for the treatment of ulcers, leprosy and liver disease.

Botany: It is a shrub. The leaves are alternate, stipulate, paripinnate compound, rachis 8 to 12 cm long. Its flowers are irregular, bisexual, bright yellow and large, the pedicels glabrous and 2.5 cm long. The racemes are few-flowered, short, erect, crowded in axils of upper leaves. Pod is 7.5 to 11 cm long, 1.5 cm broad, oblong and pale brown. There are 12 to 20 seeds per pod.

Utilization: It is one of the principle constituents of 'Avaarai panchaga chooranam'-an Indian herbal formulation used in the treatment of diabetes to control the blood sugar level. The oral administration of 0.45 g/kg body weight of the aqueous extract of the *Cassia auriculata* flower for 30 days reduces glucose in blood and an increase in plasma insulin.

7. Neem–*Azadirachta indica* A. Juss. (2n=28)

Family: Meliaceae

Vernacular names: Indian Lilac, Margosa tree, Neem tree (English); limbdo (Gujarati); nimb, nim (Hindi); bevu, bevina mana, olle (Kannada); veppu, vepe (Malyalam); nimbay, limbo (Marathi); vembu, veppam (Tamil); vepa, yepa (Telugu); Arista, Nimba, Nimbah, Picumarda (Sanskrit).

Importance: It is an ever green and deciduous tree in dry areas. Neem tree leaves and small twigs are used as mulch and green manure. Farmers in India use neem cake as an organic manure and soil amendment. It is believed to enhance the efficiency of nitrogen fertilizers by reducing the rate of nitrification and inhibiting soil pests including nematodes, fungi, and insects. Neem is extensively used in ayurveda, unani and homoeopathic medicine. The Sanskrit name of the neem tree is 'Arishtha' meaning 'reliever of sickness'. The Latin name of neem, *Azadirachta indica*, is derived from the Persian word 'Azad' means 'free'; dirakht means 'tree' and ; the specific name 'indica' obviously means of Indian origin.

Origin: It is native to India and Myanmar.

Distribution: It is widely distributed in India, Nepal, Pakistan, Bangladesh, Sri Lanka, Myanmar, Thailand, Malaysia, Indonesia, Caribbean and several Central American countries including México, Singapore, Philippines, Australia, Saudi Arabia and Iran. In India, neem is grown in Uttar Pradesh, Tamil Nadu, Karnataka, Madhya Pradesh, Maharashtra, Andhra Pradesh and Gujarat.

Botany: It has straight trunk and long spreading branches forming a broad round crown and hence grown as Avenue tree. Bark is moderately thick, height 15 to 20 m and girth 1 to 2 m. The shiny dark green pinnately compound leaves are up to 30 cm long. Each leaf has 10–12 serrated leaflets that are 7 cm long by 2.5 cm wide. The tree starts flowering and fruiting at the age of 4-5 years, but high seed yield is produced only after 10-12 years. Each panicle has 12–15 small branches and total of at least 100 flowers. The flowers are frequented by bees, flies, ants, butterflies and thrips. Flowers occur during February–March in South and March–April in North India. Flowers are small, white in clusters with fragrance. The time of anthesis is 4 pm to 6 pm. Pollination is by insects such as honeybees. Fruits ripen in about 12 weeks from anthesis and are eaten by bats and birds, which disseminate the seed. Tree can live for over 200 years. Fruits mature in May-June in South and June-July in North India. Fruits are green, turning yellow to brown on ripening, aromatic with garlic like odour. Ripe fruits can be collected when the drupes turn yellowish green on the trees. A single tree of 10 to 12 years may produce 5 to 8 kg of seeds annually, while fully grown trees 20 years or older produce 30 kg of seeds. There are about 2000 to 3000 fruits and 3300 to 6300 seeds per kg.

Climate: The tree is adapted to hot and dry climates. It grows up to 1500 m from MSL. It is grown in areas receiving annual mean rainfall of 500 to 1500 mm and having temperature of 15 to 48°C.Adult neem trees tolerate some frost, but seedlings are more sensitive. It does not tolerate waterlogging.

Soils: It can thrive on a wide range of soils. It grows well only on well-drained soils. It grows on dry, stony, clayey, and shallow soils having a pH range of 5 to 10.

Propagation: It is propagated by seed, air layering, root and shoot cuttings, grafting, marcotting and tissue culture. The seeds germinate and emerge in 14 to 21 days. Neem seeds should be planted fresh, within four weeks of harvest. Direct sowing is done either by dibbling in bushes, broadcast sowing, line sowing, sowing on mounds or ridges, sowing in trenches in sunken beds in circular saucers or by aerial sowing. Seed is normally sown in the nursery in lines at 15 to 20 cm × 2.5 to 5 cm at a depth of 2.5 cm. It is covered lightly with soil or mulch. 12 week old seedlings are planted out in the field. Propagation by root suckers and stem cuttings can be done using 1000 ppm indole-3-butyric acid (IBA) and indole-3-acetic acid (IAA), and by air layering using IBA or naphtalene-1-acetic acid (NAA). Seedlings can be produced in polyethylene bags filled with FYM and soil in a 1:1 ratio. Each bag is seeded with two seeds or one germinated seedling pricked out from a seedling bed. Transplanting stumps prepared from 2-year-old seedlings can be done. One year-old seedlings are uprooted along with a ball of soil around the roots can also be used for planting. It is grown for fuel wood with spacing of 1.3 × 1.3 m or 2.4 × 2.4 m. Then it is thinned to a spacing of 5 × 10 m. At this stage intercropping can be practised.

Tree management: Trees coppice freely, and early growth from coppice is faster than growth from seedlings. It withstands pollarding well, but seed production is adversely affected when trees are lopped for fodder. Pollarding is usually done on 5 to 10 year old trees.

Harvest: Leaves may be collected at any time. In natural conditions, the fruit falls to the ground during the rainy season. Fruits are also beaten down from the tree on to cleaned ground and swept up. Harvesting of wood can be taken from 5 to 7 years onwards. Lopping of tree can be done for fodder purpose at the age of five year.

Yield: Neem tree leaves can be used as an emergency fodder for sheep, goat and cattle. Tree can produce up to 50-100 kg green leaves per tree per year. A fully-grown tree yields about 50 kg fruits, 10 to 15 kg of seed and about 350 kg leaves per year.

Seed production: Only fruits at the yellow green colour stage are pricked from the branches by hand or by using ladder. Fruit collection is also done within 1 to 2 days of natural dropping from trees. After collection the fruits are de pulped immediately. The fruits are soaked in water for 1 to 2 days to remove the fleshy part of the drupe. Fruits are then rubbed and floated in water to separate seed from pulp. The stone is washed in clean water, spread in one layer, and dried in shade for 5 to 10 days. The moisture content of dried seed is 9 to 15% as against 40 –50% moisture at the time of collection of fruits. The seeds are stored in a cool well-ventilated place in cloth or gunny bags. Mature seeds germinate within a week. Germination rate of fresh seeds is 75 to 90% which drops to 40% in 30 days and less than 5% in 60 days. Seed remains viable for 4 to 8 weeks only, but storage of cleaned and dried seeds at 15°C will prolong this period up to 4 to 6 months. Seeds with endocarps show 42% germination after 5 years of storage at 4°C. Kernels (depulped fruits) stored for more than 10 years is done at 4% moisture content at 20°C. Storage of seed in earthen pot containing wet sand (30% moisture) helps to retain viability up to 60% at the end of 3 months. On an average 5000 seeds weigh one kilogram.

Utilization: One tonne of neem seed gives 1.5 kg of Azadirchtin, 200 kgs of neem oil and 780 kg of neem cake. Neem cake is used as good manure and pest repellent.

8. Pungam–*Pongamia pinnata* (L.) Pierre (2n=22)

Family: Fabaceae

Synonyms: *Pongamia glabra* Vent., *Derris indica* (Lam.) Bennett

Vernacular names: Pongam Tree, Indian Beech tree (English); Karanj, Karanja (Hindi, Bengali, Marathi and Gujarati), Pungu, Gaanuga (Telugu), Pungam, Punku, Ponga, Pongam (Tamil), Pungu, Pongam (Malayalam), Koranjo (Oriya), Sukhehein, Karanj, Paphri (Punjabi), Karchuw (Assamese).

Importance: Pungam is often planted in homesteads as an ornamental tree and as avenue plantings in roadsides, stream and canal banks. The tree is frequently found in pastures, waste lands, cultivated lands, roadsides, lawns and in planted forests. It is grown as a windbreak for tea plantation in Sri Lanka. The name Derris, derived from Greek, means 'leather covering or skin'; the specific name 'indica' obviously means of Indian origin.

Origin: Pongamia is native to a number of countries including India, Nepal, Thailand, Malaysia, Indonesia, Taiwan, Bangladesh, Sri Lanka and Myanmar.

Distribution: It has also been naturalised in Pakistan, India, Nepal, Malaysia, China, Sri Lanka, eastern Africa, Australia, New Zealand, Papua New Guinea, Egypt, Fiji, Indonesia, Japan, Mauritius, Philippines, Solomon Islands, Sudan, United States of America.

Botany: It is a small-sized evergreen or semi-deciduous tree 15-25 m high with spreading crown. Tree, branches spreading, twigs glabrous. Leaves alternate, imparipinnate with long slender leafstalk, hairless, pinkish-red when young, glossy dark green above and dull green with prominent veins beneath when mature. Leaflets 5-9, paired except at end, short-stalked, ovate elliptical or oblong, 5-25 × 2.5-15 cm. Inflorescence raceme-like, axillary, 6-27 cm long, bearing pairs of strongly fragrant flowers. Flower clusters at base and shorter than leaves, to 15 cm long, slender and drooping. Flowers are 2-4 together, short-stalked and pea-shaped, purplish or pinkish white in colour. Fruit is flattened, woody pod, 2.5 to 5.0 cm long, oblong, indehiscent and usually one seeded. Seed compressed ovoid or elliptical, bean-like, 1.5 × 2 × 0.8 cm, with a brittle coat long, flattened, dark brown, oily. Pod production starts 5-7 years after sowing. They do not open naturally and must decay before seeds can germinate. There are 1500 to 1700 seeds/kg.

Climate: It is native to humid and sub-tropic environments; found along waterways or seashores, lowland forest on limestone and rocky coral outcrops on the coastal areas, with its roots in fresh or saltwater, along the edges of mangrove forests, tidal streams and rivers. It is found up to an altitude of 1200 m from MSL. It is found in areas with rainfall from 500 to 2500 mm a year. It is tolerant to drought for a period of 2-6 months. It grows well in both full sun and partial shade. It is hardy and can survive in temperatures from 5 to 50°C. It grows in areas having the maximum temperature ranges from 27 to 38ºC and the minimum 1 to16ºC. Mature trees can withstand waterlogging and slight frost.

Soils: It can be grown on most soil types ranging from stony to sandy to clayey soils. It is very tolerant of saline conditions and alkalinity with soil pH range for growth is 6 to 9.

Propagation: It can easily be propagated by seed, cuttings or root suckers. Seed remains viable for a long time. Seedlings reach a height of 60 cm about 1.5 years after sowing and are easy to transplant. In the nursery, seeds are sown with spacing of 7.5 cm ×

15 cm. Germination takes 7 to 30 days of sowing. Germination is hypogeal and the radicle develops quickly before the plumule emerges. Two or three months old seedlings attained a height of 30 to 60 cm can be used for planting. The spacing adopted in avenue planting is about 8 m between plants. In block plantings, the spacing can range from 2 × 2 m to 5 × 5 m. The branches stuck in moist soil develop roots readily. The tree produces root suckers profusely which can be used for further propagation.

Tree management: The tree attains growth of 1.3 m in height and 0.4 cm in diameter in 13 months. Trees coppice well and can also be pollarded. When planted as a shade or ornamental tree, pruning may be necessary to obtain a trunk of appropriate height.

Harvest: Pod production starts 5 to 7 years after sowing. Ripe pods are collected from April-June and are subsequently dried in the sun. Seeds are easily extracted by light hammering or by splitting the pod with a knife along the sutures and winnowing out of the husks.

Yield: Individual trees yield 9-90 kg of pods annually, while mature trees yield 8-24 kg of seeds annually.

Seed production: Seeds are collected from ripen fruits. Seeds remain viable for about a year when stored in air-tight containers. There is about 1500 to 1700 seeds/kg.

Utilization: Pongamia wood has been used for stove top fuels, poles, and ornamental carvings. It can produce up to 40% oil per seed. Flowers are considered good sources of pollen for honey bees, and have been described as having anti-diabetic properties. Leaves have been used as fodder for cattle and goats and as a source of poison used by Australian Aborigines for fish spears. Seed cake leftover after oil extraction has been used as 'green manure' as it is rich in nitrogen.

9. Puvarasu–*Thespesia populnea* (L.) Sol. ex Correa. (2n=24)

Family: Malvaceae

Vernacular names: Portia tree, Pacific Rosewood, Umbrella tree, Indian Tulip Tree, Seaside mahoe, Milo (English); Paras-Pipal, Gajadanda, Pahari Pipal, Parsipu, Porush (Hindi); Buguri mara (Kannada); Gangaraavi (Telugu), Poovarasu, Poovarasam (Tamil, Malayalam), Pakur (Bengali) and Plaksa (Sanskrit).

Importance: *Thespesia populnea* is a sacred tree and often planted near temples and is used in traditional ceremonies. It is also planted for ornament and shade. It is a multipurpose tree suitable for dry and saline conditions. The wood is mainly used for carving. It is suitable for coastal erosion control. It is a host of several cotton pests. The generic name is based on the Greek word 'thespesios'-divine, supposedly because *T. populnea* is frequently planted round temples.

Origin: It is native of Australia, China and India.

Distribution: It is naturally found in Asian tropics or from the coasts of the Pacific and Indian Oceans. It is grown in India, West Indies, USA, Bahamas, Cambodia, Cuba, Fiji, Ghana, Indonesia, Jamaica, Malaysia, Mauritius, Mozambique, New Zealand, Philippines, Puerto Rico, Samoa, Solomon Islands, Sri Lanka, Thailand and Venezuela.

Botany: It is an evergreen shrub or small tree reaches a height of 6 to 10 m and a trunk diameter of 20 to 30 cm, bole branchless for up to 3 m, often twisted or bent, becoming hollow with age, up to 60 cm in diameter. The twigs densely covered with minute brown to silvery scales, glabrescent. Bark gray or light brown, smoothish or slightly fissured, becoming thick and rough. Inner bark yellowish, tough and fibrous. It has long pointed shiny heart-shaped, copper-coloured leaves. Leaves arranged spirally, simple and entire;

stipules lanceolate to subulate, 3 to 10 mm long. Flowers are showy, hibiscus-like, single at upper leaf axils, corolla yellow with a red center, turning maroon by nightfall; stamens united into a column shorter than petals. Flowers are opening and closing the same day, petals withering and turning to purple or pink. It has an indehiscent fruit. Fruit is a leathery, flattened globose, 5-parted capsule, 4 cm wide with a dehiscing outer layer, yellowish to brownish green when mature, exuding a bright yellow gum when cut, many-seeded. Seeds ovoid, 8–15 mm × 6–9 mm, slightly angular, covered by closely matted silky hairs. The 1000-seed weight is 140 to 285 g.

Growth and development: Germination begins 8 days after sowing and may extend for as long as 10 weeks. Seedling is with epigeal germination; hypocotyl elongated; cotyledons leafy; all leaves arranged spirally. Growth in height is rapid in the first few years (0.5–1.5 m per year), but slows down at 7–10 years of age. Stem diameter growth is 1–3 cm/year. Flowering starts when the tree is only 1–2 years old. Flowers and fruits are produced nearly year-round. The pale yellow flowers open at about 10 a.m., turn reddish orange in the afternoon, fade to pink on the tree and do not fall off for several days. Pollination is probably by birds. Fruits and seeds buoyant/ float, adapted to long-distance dispersal by tides and ocean currents.

Climate: It is a tree of tropical and warm subtropical climates. It grows at elevations from sea level to 275 m in areas that receive 500 to 1600 mm of annual rainfall. The mean annual temperature may range from 20–26°C; mean maximum temperature of hottest month range from 27–28°C and mean minimum temperature of the coldest month range from 18–26°C. It can withstand temperatures as low as 4°C. The tree grows best in full sunlight and does not grow well in the shade of other trees. It tolerates wind and salt spray. It is suitable for dry locations because it develops a long taproot in porous soils; it may tolerate a dry season of up to 8 month.

Soils: It grows in a wide range of soil types, preferably on sandy coastal soils in coastal environments, including soils derived from quartz (sand), limestone, and basalt; it favors neutral soils pH of 6–7.4. It tolerates heavier soils, salinity, alkaline and occasional inundation, but does not grow on permanently inundated soils. It occurs at the inland edges of mangrove swamps and along tidal waters.

Propagation: It is usually propagated by seed, but propagation by stem or root cuttings or air-layering is also possible. Seed germination is difficult due to the hard seed coat. It is improved by scarification with a knife, sandpaper or with conc H_2SO_4 for 3 to 6 minutes. Direct sowing is generally practised. Seedlings may be raised in pots. The plants are normally ready for planting in 12 to 16 weeks, but trees up to 3.5 m tall have been planted out from containers. Stump planting involves cutting back the stem to about 1 cm above the root collar before transplanting, thus allowing the roots to recover before new leaves develop. Wild seedlings are also collected and transplanted. For vegetative propagation small cuttings should be rooted in a nursery before planting out. The cuttings up to 2 m long have also been successfully planted directly in the field.

Tree management: Weeding is important until the tree has become established. It tolerates heavy pruning and the regrowth is slow. The tree may take 25 to 40 years to produce usable timber, although branches as little as 5 cm thick are used for carving.

Utilization: Leaves are used for green manure. The leaves are lopped for fodder purpose. The leaves are a good source of protein, calcium and phosphorus for livestock. Wood chippings are also used as a green manure. This tree is suitable for coastal erosion control because of its tolerance of saline conditions. The ethanolic extract of *Thespesia populnea* unripe fruits prevents diabetic neuropathic pain. The fine-grained, strong, hard and durable wood is used for light construction, flooring moulds, musical instruments, utensils and vehicle bodies. As it is very durable under water, it is popular for boat building. The wood is light to medium in weight. The heartwood is dark red and smooth. It is easy to saw and work despite its wavy grain. It is used for horse-drawn carts and wheelbarrows, to carve canoe paddles, bowls, plates and utensils. It is resistant to insect attack. The cooked fruit crushed in coconut oil provides a salve, which, if applied to the hair, will kill lice. The sap of the leaves and decoctions of most parts of the plant are used externally to treat various skin diseases. Leaf and bark decoctions are taken for high blood pressure. Leaf tea is taken for rheumatism and urinary retention. Seeds are purgative. It has light brown sapwood and reddish brown to chocolate brown heartwood, takes a fine polish. It is easy to work. It is classed as resistant to attack from dry wood termites and is used elsewhere in boatbuilding and cabinetwork.

10. Purasu–*Butea monosperma* (Lam.) Taub. (2n=18)

Family: Fabaceae

Vernacular names: Flame of the forest, Bastard Teak (English); Dhak, Palas (Hindi); Elaiporasu, Purasu, Kattumurukhua, Murukamaram (Tamil); Mooduga, Palasamu, Paladulu (Telugu); Khakharo (Guajarati); Muttuga (Kannada); Brahmavriksham, Kimshukam (Malyalam); Palasha (Sanskrit).

Importance: It is sacred tree to both Hindus and Buddhists. The offering of the palash flowers during Saraswati puja is Bengali customs. Without it, the puja is considered incomplete. The flower is also offered to Goddess Kali. When it is in full bloom it gives the appearance of flame, in the forests. Hence it is known as flame of the forests.

Origin: It is native of South and Southeast Asia.

Distribution: It is widely distributed throughout India, Nepal, Myanmar, Sri Lanka, China, Thailand, Vietnam, Cambodia, Indonesia, Japan, and Papua New Guinea.

Botany: It is an erect, small to medium-sized, 10 to 15 m high, up to 43 cm dbh, deciduous tree with a crooked trunk and irregular branches. It grows slowly and attains a height of about 5 to 8 m and diameter of about 20 to 40 cm when it matures at the age of about 50 years or so. Its wood is greenish white in colour, soft and weighs about 14 to 15 kg per cubic foot. The bark is ash colour. Leaves are trifoliate; petiole 7.5-20 cm long with small stipules and leaflets more or less leathery. Flowers are in racemes, 5-40 cm long, near the top on usually leafless branchlets. The bright yellow colour of the flower is attributed to the presence of chakones and aurones. The flowers on the upper portion of the tree form the appearance of a flame from a distance. Fruit is an indehiscent pod, 17-24 × 4-6 cm, stalked, covered with short brown hairs, pale yellowish-brown or grey when ripe, in the lower part flat, with a single seed near the apex. Young pods have a lot of hair, a velvety cover and mature pods hang down. The seeds are flat from 2.5 to 4.0 cm long, 1.5 to 2.5 cm wide, and 1.5 to 2 mm thick. The seed-coat is reddish -brown in colour, glossy, and wrinkled, and encloses two large, leafy yellowish cotyledons. The number of seeds per kg is about 1000.

Growth and development: Leaves are shed during the dry season. At the beginning of the rainy season, the leafless tree flowers abundantly and is very conspicuous in the forest. At the end of the flowering period, new leaves develop, which are initially a pale bronze-tinged green. Birds are the chief pollinators.

Climate: It is a tree of tropical and subtropical climate. It is distributed in Himalayas up to 900 m and in peninsular India up to 1,200 m from MSL. It is grown in areas receiving mean annual rainfall of 450 to 2500 mm. It is a characteristic tree of the plains, often forming pure patches in grazing grounds, wastelands and other open places. It is resistant to browsing and its ability to reproduce from seed and root suckers. It is moderate in its demand for light. Although it can withstand some shade, dense shade suppresses its growth. The tree is very drought resistant and frost hardy.

Soils: It grows on a wide variety of soils including shallow, gravelly sites, black cotton soil, clay loams, and even saline or waterlogged soils. Seedlings thrive best on a rich loamy soil with pH 6-7.

Propagation: Seeds are extracted from the pod. The seeds are sown in 30 x 45 cm size polybags to produce seedlings. Germination takes up to 15 days. One year old seedlings are used for planting. The seedlings attain 3 feet in 6 months. Clonal propagation by air layering helps in the quick establishment of plantation of this tree.

Cultural practices: It is produced with seeds and suckers. Direct seeding in line sowings under irrigated or rainfed condition and lightly covered with sandy soil. The plants are raised at 10 x 10 cm distance in the nursery by sowing the ripe seeds before the rainy season. The seedlings are transplanted 25-30 cm apart during the rainy season. The stem is pruned to 5 cm thickness. Regular watering and weeding is carried out.

Tree management: The trees pollard and coppice well and produce root suckers freely. Tree can also withstand heavy annual lopping. It is well suited for silvipasture at wide spacing of 10 to 15 m in grassland. It is managed by repeated coppicing on a roughly 5-year rotation. Coppice shoots are also cropped in intermediate years for the larger leaves. Under dry land conditions, coppice management yields roughly 100 kg/tree of air-dry fuel wood for every 5 years. If allowed to grow, trees attain a height of 3 to 5 m and dbh of 15-20 cm in 10 years. Plantations can be established on irrigated as well as rainfed land.

Uses: The young leaves are good fodder, eaten mainly by buffaloes. It is used in Ayurveda, Unani and Homeopathic medicine. Bark yield red juice known as 'Butea gum' or 'Bengalkino'. The flowers are widely used in treatment of hepatic disorders, viral hepatitis, diarrhea, depurative and tonic. The seeds are pounded with lemon juice and applied to the skin. The dried flowers are used as a diuretic. The gum obtained from the tree is called Bengal Kino. It contains tannin and is used in the treatment of diarrhoea. The seeds have anti-helminthic properties and are used in the treatment of worms. The wood is soft and durable and is used for making boats. A yellow dye, obtained from the fresh palash flowers is used during the Holi festival.

11. Ipomea–*Ipomea carnea* Jacq. (2n=30)

Family: Convolvulaceae

Vernacular names: Morning glory (English); Neyveli kattamankku, Kodipoovarasu (Tamil); Besharam, Behaya (Hindi); Thootikada, Lottapeeshu (Telugu); Behayo (Oriya); Beshram (Marathi, Bengali).

Importance: It is used as ornamental and hedge plant along the banks of irrigation and drainage canals, lakes /tanks, ponds, puddles and wet places in most parts of India. It is used as a fencing plant by farmers and a tree guard for highways. The plant is originally used for making fence for the road side fields, but due to its massive growth and rapid propagation it has grown rapidly in barren waste lands.

Origin: It is native of South America.

Distribution: It is found in Argentina, Brazil and Bolivia, USA, Pakistan, Paraguay, Egypt, Nairobi, Kenya, India, Pakistan, Sri Lanka, Indonesia and Taiwan,

Botany: It is a hardy shrub with milky juice. It grows to a height of 5 m on terrestrial land, but acquires a shorter height in the aquatic habitats. The stem is thick and develops into a solid trunk with many branches from base. The stem is erect, woody, hairy, and more or less cylindrical in shape and greenish in colour. It is a very quick growing, weak stemmed straggler and profusely branching. The internodes measure 3.5 - 6.0 cm in length. It has heart shaped or somewhat lanceolate leaves. The leaf is simple and petiolate, petiole is cylindrical, attains 4.0–7.5 cm length and 2.5–3.0 mm diameter. The upper surface of leaf is dull green and the lower surface is paler. The leaves which receive lesser sunlight may grow larger than the leaves which receive full sunlight. The difference is more in aquatic condition. It gives abundant green leaf material in a short time. The flowers are pale rose, pink or light violet in lax, dichotomously branched and pedunculate cymes. Ipomea takes 8 to 9 hours for the anthesis, 1 to 2 days for the development or abortion of fruits and 4-5 weeks for the maturation of the seeds. Flowers and fruits are throughout the year. Fruits have a glabrous capsule. Fruits dehisce during winter by the splitting of the dry fruit-wall. Seed is silky.

Ecology: It grows on dry rocks as well as on the banks of lakes and rivers, in water up to 2 m deep. It can survive in all seasons and adverse conditions. It can also survive with or without water for several months in the tanks. It grows in xeric and hydric conditions. It is highly drought resistant and well adopted to dry areas.

Soils: It grows well in sandy, silt and clayey loam soils.

Propagation: It is multiplied through mature stem cuttings. The horizontal branches rapidly root along the downward side in contact with the ground and give rise to new plants separately from the main plant. At the tip of the horizontal branch a secondary shrub (i.e. ramet) develops. During floods and other natural calamities, plants are swept off river beds and embankments and become established in downstream habitats. It can also be produced with seeds. The hairy seeds are dispersed by wind as well as water. The seed do not germinate immediately because of a hard seed coat which is impervious to water. Seeds usually germinate in light within the first three days of contact with water. The germination rate of fully developed seeds attained 17-26%. Seed takes 14-20 days to fully develop the cotyledons.

Ipomoea management: It can be controlled in tanks and channels with foliar application of 2,4-D Na salt 8 g + urea 20 g + soap oil 2ml/litre of water and then removal and burning of dried weeds (or) manual / mechanical removal of grown up plants in channels during summer season.

Utilization: If cut plant materials of Ipomea are applied as green leaf manure, its hard wood twigs sprout easily in rice fields and become weeds. The remedy lies in cutting the shrubs for manure before the twigs turn woody from pure green to a greyish tinge. This is not possible when shrubs are cut every year when farmers need the leaves. The outlet is to cut the shrubs frequently at shorter intervals of 40 to 60 days in rainy or moist conditions or 90 days in dry conditions, each time cut at the lowest node. The leaf contains 2.3% N. The leaves are toxic to cattle. Its leaves contain a polysaccharide-ipomus, one glucoside anthracene a gum-gelapin and saponin. Out of the two materials, one is soluble in water and the other is in ether. Both polysaccharide and anthracene present in *I. carnea* are water soluble poisons when enter into the central nervous system, it damages the respiratory track.

12. Calotropis–*Calotropis gigantea* (L.) W.T.Aiton (2n = 22)

Family: Asclepiadaceae (APG: Apocynaceae)

Vernacular names: Crown flower, crown plant, giant milkweed, gigantic swallow-wort, rubber bush, bow-string hemp (English), Erukku (Tamil).

Origin: *Calotropis gigantea* is native to continental Asia and South-East Asia.

Distribution: It is grown in Pacific Islands, Australia, India, Central and northern South America and Africa.

Botany: There are two species of Calotropis viz., *Calotropis gigantea* (Madar fibre) and *Calotropis procera* (French cotton or Akund). It is a fast growing large shrub or small tree grows up to 2 to 4 m tall, much-branched at base, stems erect, up to 20 cm in diameter; bark pale grey, and latex in all parts. Leaves are opposite, blade broadly ovate to oblong-obovate, 9.5–20 cm × 6–12.5 cm and short-hairy beneath. Inflorescence is an axillary, umbellate to almost corymbose cyme up to 12.5 cm in diameter. Flowers are bisexual, white, cream, lilac or purple; corolla 2.5 to 4 cm in diameter. It flowers throughout the year, but especially during the hot season. It is primarily pollinated by bees, butterflies and wasps. Fruit is a pair of follicles, each follicle ovoid, boat-shaped, inflated, 6.5–10 cm × 3–5 cm, many-seeded. Seeds ovoid, 5–6 mm long, with 2–3 cm long coma at one end.

Climate: *Calotropis gigantea* grows in dry uncultivated land, open waste land, along roadsides and railways, up to 1000 m altitude.

Soils: It grows on a variety of soils, but prefers sandy soils.

Propagation: It can be propagated by seed or stem cuttings. The seeds are dispersed by wind and water. The seedlings can be planted with a spacing of 0.5 × 0.5 m or 2 × 2 m.

Harvest: The leaves, flowers and roots of *Calotropis gigantea* are harvested throughout the year.

Yield: *Calotropis gigantea* reaches a maximum height of 166 cm in 1 year at a spacing of 0.5 m × 0.5 m, producing 7.3 t fresh leaves (1.1 t dry leaves) and 56 kg latex per hectare.

Utilization: The leaves can be used for green manuring of rice fields, mulching and for binding sandy soil. The long-lasting flowers of *Calotropis gigantea* are used in various floral arrangements in temples and in rosaries. It is also widely planted as an ornamental hedge plant. All parts of *Calotropis gigantea* are toxic, due to the presence of several cardiac glycosides (cardenolides). The latex contains the cardenolides calotropin, calotoxin and uscharin as well as the proteinase calotropain. Calotropin is a quick-acting heart stimulant, and is known to be 15–20 times more poisonous than strychnine: minute amounts can cause death. The lethal dose for calotropin is 0.12 mg/kg. The whole plant or its parts is used medicines.

REFERENCES

Ahmed, K.K.M., Rana, A.C. and Dixit, V.K., 2005. *Calotropis* species (Ascelpediaceae) - a comprehensive review. Pharmacognosy Magazine 1(2): 48–52.

Anthony. 2009 Agroforestree Database: a tree reference and selection guide version 4.0 http://www.worldagroforestry.org/resources/databases/agroforestree

Apte Madhavi. 2013. An overview of *Butea monosperma* (flame of forest). World Journal of Pharmacy and Pharmaceutical Sciences. 3 (1): 307-319

Arulprakash, R. and Veeravel, R., 2007. Effect of milkweed plant, *Calotropis gigantea* R. Br. on biochemical constituents of some important storage pests. Journal of Plant Protection and Environment 4(2): 47–50.

Chaudhuri, H., Ramaprabhu T. and Ramachandran V. 1994. *Ipomoea carnea* Jacq. A new aquatic weed problem in India. Journal of Aquatic Plant Management, 32: 37-38.

Chitme, H.R, Chandra, R. and Kaushik, S., 2006. Evaluation of analgesic activities of *Calotropis gigantea* extract in vivo. Asia Pacific Journal of Pharmacology 16(3/4): 157–162.

Chintu R, Mafongoya PL, Chirwa TS, Kuntashula E, Phiri D, and Matibini J. 2004. Propagation and management of *Gliricidia sepium* planted fallows in sub-humid eastern Zambia: Experimental Agriculture. 40:341-253.

Daniel JN. 1997. *Pongamia pinnata* - a nitrogen fixing tree for oilseed. A Publication of the Forest, Farm, and Community Tree Network (FACT Net). Winrock International.

Dev, U., Devakumar, C., Agarwal, P.C, Mohan, J., Joshi, K.D. and Rani, I., 2002. Antifungal effect of *Vitex negundo*, *Calotropis* spp. and other plant extracts against seed-borne fungi. Pesticide Research Journal 14(2): 229–233.

Firdaus Rana and Mazumder Avijit, 2012. Review on *Butea monosperma*. International Journal of Research in Pharmacy and Chemistry. 2(4): 1035-1039

Frey, R. 1995. *Ipomoea carnea* ssp. *fistulosa* (Martius ex Choisy) Austin: taxonomy, biology and ecology reviewed and inquired. Tropical Ecology journal, 36 (1): 21 - 48.

Imam Hashmat, Hussain Azad and Ajij Ahmed. 2012. Neem (*Azadirachta indica* A. Juss) - A Nature's Drugstore: An overview. International Research Journal of Biological Sciences. 1(6), 76-79

Joy, V., Paul John Peter, M., Yesu Raj, J. and Ramesh. 2012. Medicinal values of avaram (*Cassia auriculata* Linn.): a review. International Journal of Current Pharmaceutical Research. 4(3): 1-3

Kausik Biswas, Ishita Chattopadhyay, Ranajit K. Banerjee and Uday Bandyopadhyay. 2002. Biological activities and medicinal properties of neem (*Azadirachta indica*). Current Science. 82 (11): 1336-1345

Laxmappa, B., Srinivasulu, P. and Mahender, J. 2014. Status of *Ipomoea carnea* weed infestation in Inland Water bodies and its effects on Fisheries in Mahabubnagar district, A.P, India. International Journal of Fisheries and Aquatic Studies 2014; 1(6): 12-19

MacDicken, G.K. 1994. Selection and management of nitrogen fixing trees. Winrock International and Bangkok: FAO.

Mannetje, L. and Jones, R.M. 1992. Plant Resources of South-East Asia. No. 4: Forages. Pudoc Scientific Publishers, Wageningen.

Mazumder, P.M., Das, M.K. and Das S. 2011. *Butea monosperma* (LAM.) Kuntze- A comprehensive Review. International Journal of Pharmaceutical sciences and technology, 4(2), 2011, 1390-1393.

NAS.1980. Firewood crops: shrub and tree species for energy production, volume 1. National Academy of Sciences, Washington, D.C. 237 p.

National Academy of Sciences. 1980. Firewood crops. National Academy Press. Washington D.C.

National Academy of Sciences. 1981. Fuelwood crops, shrub and tree species for energy production. National Academy Press. Washington D.C.

National Research Council. 1992. Neem: tree for solving global problems, National Academy Press, Washington D.C.

Natanam R, Kadirvel, R, and Chandrasekaran D. Chemical composition of karanja (*Pongamia glabra* Vent [*P.pinnata*]) kernel and cake as animal feed. Indian Journal of Animal Nutrition 1989;6:270-273.

Neem Foundation. All about neem. Mumbai: Neem Foundation; 2012. [Online] Available http://www.neemfoundation.org/

Nyirongo J, Wolf J de. 2007. How to establish *Gliricidia sepium* trees in a gliricidia-maize intercrop. Lilongwe, Malawi: ICRAF. 2p.

Nyirongo J, Wolf J de. 2007. How to raise *Gliricidia sepium* seedlings in the nursery. Lilongwe, Malawi: ICRAF. 2p.

Orwa C, Mutua A , Kindt R , Jamnadass R, and Simons A. 2009. Agroforestry Database: a tree reference and selection guide version 4.0

http://www.worldagroforestry.org/resources/databases/agroforestree

Patil, S.V., Salunke, B.K. and Bhat, J.A., 2003. Herbal rennet from *Calotropis gigantea*. Journal of Medicinal and Aromatic Plant Sciences 25(2): 392–396.

Rahman, M.A. and Wilcock, C.C., 1991. A taxonomic revision of *Calotropis* (Asclepiadaceae). Nordic Journal of Botany 11(3): 301–308.

Saravanapriya, B. and Sivakumar, M., 2005. Management of root knot nematode *Meloidogyne incognita* on tomato with botanicals. Natural Product Radiance 4(3): 158–161.

Savita Sangwan, Rao, D.V. and R.A. Sharma. 2010. A Review on *Pongamia pinnata* (L.) Pierre: A Great Versatile Leguminous Plant. Nature and Science. 8(11): 130-139.

Scott, PT., Pregelj, L., Chen, N., Hadler, JS., Djordjevic, M.J. and Gresshoff, P.M. 2008. *Pongamia pinnata*: an untapped resource for the biofuels industry of the future. Bioenergy Research 1: 2-11.

Shaltout, K.H., Al-Sodany, Y.M. and E.M. Eid 2006. The biology of Egyptian woody perennials: *Ipomoea carnea* Jacq. Ass. Univ. Bull. Environ. Res. Vol. 9 No. 1, March 2006

Sharma, A. and R. K. Bachheti. 2013. A review on *Ipomoea carnea*. International Journal of Pharma and Bio Sciences. 4(4): (P) 363 - 377

Sharmna N, and Garg, V. 2009. Antihyperglycemic and antioxidative potential of hydroalcoholic extract of *Butea monosperma* Lam flowers in alloxan-induced diabetic mice. Indian J Exp Biol. 2009 Jul; 47(7):571-576.

Shilpkar, P., Shah, M. and Chaudhary, D.R., 2007. An alternate use of *Calotropis gigantea*: biomethanation. Current Science 92(4): 435–437.

Sindhia V.R., Bairwa R. Plant Review: *Butea monosperma*. International Journal of Pharmaceutical and Clinical Research 2(2), 2010, 90-94.

Stewart JL, Allison GE, and Simons AJ. 1996. *Gliricidia sepium* genetic resources for farmers: Tropical Forestry Papers No 33. Oxford Forestry Institute. Oxford University Press.

Tewari, D.N. 1992. Monograph on Neem (*Azadirachta indica* A. Juss.). Indian Council of Forestry Research and Education (ICFRE), International Book Distributors, Dehra Dun, India. 279p.

Verma M., Shukla Y.N., Jain S.P. and Kumar S., Chemistry and biology of the Indian dhak tree *Butea monosperma*. Journal of Medicinal and aromatic plant sciences, 1998, 20, 85-92.

http://agroforestry.org/free-publications/traditional-tree-profiles

http://www.unep.org/wed/tree-a-day/neem.asp.

http://www.prota4u.org/protav8.asp?p=Azadirachta+indica

CHAPTER

38

Forage Crops, Importance and their Classification

Forage crops

The common feeds and forages include green fodders, dry fodders and concentrates. Feed stuffs may be defined as any product of plant or animal origin which the animals are capable (without risk to their health) of utilizing its organic and/or inorganic components. The term 'forage' is defined as vegetative plant materials (root, leaves, stem, flower), primarily grasses and legumes used for feeding of domestic animals. It includes wild as well as cultivated plants that are used as livestock feed. The green forages wilted to 40-50 % dry matter are often referred as 'haylage'. The green forage after harvesting fed to animals as such or after chopping is called 'soilage'. Fodder crops are cultivated forage crops which are cut and stall fed to livestocks. The term 'fodder' is referred to the cultivated forage crops like cereals and legumes, which are primarily grain crops but are also raised as soiling crops (green forage crops that are cut and fed in fresh conditions to the animals).

Animal feeds, including crop residues, agro industrial by products and non conventional feeds, provide a link between crop and animals. The 'feed' is general term that includes vegetative and non-vegetative plants parts (grains, fruits) fed to the animals. The word conventional feed, means the feeds or forages such as maize, sorghum, oats, lucerne, berseem, cowpea, cluster bean, etc., used customary or in practice to the animals by the every individual farmer. Non-conventional feed resources are defined as those feeds that are traditionally not used in animal feeding and or are not normally used in commercially produced rations for livestock. The traditional feeds of crop origin tend to be mainly from annual crops whereas the non-conventional feed resources commonly include a variety of feeds from perennial crops and feeds of animal and industrial origin. Non-conventional feeds are of both animal and plant origin. The main categories of feeds used for ruminants (horses, asses, mules, buffaloes, cows, cattle, goats and sheep) and non ruminants (pigs, ducks, poultry) are viz., cereal fodders (maize, oats, barely, teosinte, sorghum); grass fodders (hybrid grasses like

NB 21, BN 2, CO 1, guinea grass, hamil grass and para grass); fodder legumes (cowpea, horsegram, lucerne, hedge lucerne, berseem, desmodium); pasture grasses (cenchrus grass, rhodes grass, cynodon grass, dinanath grass); pasture legumes (siratro, stylo, centro, pea blue); crop residues (straw, haulms, vines, cane tops) and agroindustrial byproducts such as molasses, single cell proteins, poultry excreta, feather, urea, blood meal. Livestock feeds can be generally divided into two classes viz. roughages and concentrates (Table 1).

Table 1 Classification of feed

S.No.	Forage group	Dry matter content (%)	Level of nutrient per unit weight	Crude fibre content (%)
1	Roughage	<60	Low	>18
2	Concentrate	>60	High	<18

Roughage have low weight per unit volume are often coarse with low readily available carbohydrates. Roughage are divided into two groups viz., succulent (green) and dry based on the moisture content. The succulent (green) roughage feeds contains 60 to 90 % moisture. The green forages that are cut and fed in fresh condition to the animals are termed as soiling of crops. The succulent roughage includes pasture, cultivated fodder cereals, grasess and legumes,tree leaves, weeds, silage, etc,. Dry roughage contains 10 to 15 % moisture. Dry roughage includes stover, hay, straw and haulm or bhusa. Stovers of maize, sorghum and pearl millet refer to matured crop where the grains are not separated and hence more nutritious than their respective straws. Hay refers to grasses or legumes that are harvested at appropriate growth stage dried and stored. Straw refers to the left over portion after threshing of grains in cereals and millets while haulms or bhusa refers to the legume crop residues and are found to be more nutritious. Concentrates refers to the feed materials contain relatively high total digestible nutrients (TDN > 60 % and crude fibre less than 18 %). Energy rich concentrates include grains, seeds, milling by-products, molasses, roots and tubers. The common grains used in cattle feed include maize, barley, oats, sorghum and pearl millet. Protein rich concentrates include oilseed meals or cakes such as groundnut, linseed oil cake, mustard cake, cotton seed cake, coconut oil cake, sesame oil cake, pulse proteins and brewer's grains.

Importance

India has the largest livestock wealth while the animal performance is one of the lowest in the world. The success of an efficient dairy, sheep, goat, piggery, poultry and other livestock industry revolves around the supply nutritious forage and feeds. Besides green forages are also required for the maintenance of huge draught power in the country. The main reasons for poor performance of cattle and buffaloes in India are inadequate supply of nutritious forage and feeds and lower production potential of the animals. Only 6.9 million or 4.4 % of the country's cropped area is under fodder crops and there is hardly any scope of expansion because of pressure on agricultural land for food and cash crop. The forest grazing resources are also dwindling roughly 1.5 millions of forests every year. The monsoonal grasslands of India are also impoverished, over grazed and infested with bushes. The grazing intensity is very high viz., 2.6 adult cattle unit per ha against 0.8 adult cattle unit per ha in the developed countries. This underlines the need to rejuvenate the natural grass lands, pastures etc., and also to increase the productivity of fodder crops. There is a lack of sufficient amount of good quality fodder (both green and dry) and the concentrates. In the absence of appropriate feeding schedule even the best animals fail to express their inherent productivity. Year round supply of forage is very important in order to stabilize animal production specially in the milkshed areas and also for small farmers who maintain dairy animals as a regular source of income. The solution lies in maximizing forage production in space and time, identifying new forage resources, increasing forage production within the existing farming systems and utilizing marginal, submarginal drylands and problem soils for developing feed and fodder resources. Moreover, the green forages form the cheap source of needed nutrients. This in turn governs the overall economics of livestock farming since feeding alone accounts 66 percent of the total cost of animal production.

Feeding systems of livestock: Animal production is a function of i) genetic potential, ii) nutrition iii) husbandry management and iv) health. It has been estimated that milk production can be increased by 20-25 % if adequate care is taken to properly feed the animal. Feeding systems in developed or in western countries are the grain based intensive production system, where complete ration is commonly adopted and that is why the animal efficiencies are very high. In India, low input-low output, crop residue approach is widely practiced. The roughage will constitute $^2/_3$ of the animal's ration and concentrates will account for $^1/_3$ of the ration. For example an animal weighing 400 kg and yielding milk of 8 to 10 lit/day requires 10 kg of dry matter, 3 kg from concentrates and 7 kg from roughage. Generally 2/3rd of roughage requirement should be met through green fodder. A 3:1 grass / cereal and legume mixture seem to be more ideal. Such balanced feeding provides sufficient minerals and vitamins to the animal and favours growth and production. Animals yielding milk upto 8 to 10 lit/day can be easily maintained without feeding concentrates if calculated quantities of legumes are included in the daily ration. This will reduce the cost of milk production by 30 %, which is an additional income to the farmer. The common feeding practices adopted in most part of country include grazing alone (> 10 hours per day), grazing + supplementing with green fodder, dry or preserved forage/crop residues, feed ingredients and stall feeding of green and dry fodder plus concentrates / feed ingredients. Young cattle and buffaloes are mainly (exclusively) maintained through grazing for 7 to 8 hours. The lactating cattle and buffaloes are maintained by grazing for 5 to 6 hours and supplemented with green and / or dry fodder. In the organized farms, concentrates are fed at the rate of 40 % of production plus 1 kg for maintenance. For example animal yielding 10 kg milk per day is fed with 5 kg of concentrates (4 kg for 10 kg milk + 1 kg for maintenance). The relation between milk yield and feeding variations between cows and buffaloes is furnished in Table 2.

Table 2 Milk yield and feeding variations between cows and buffaloes

Particulars	Cows	Buffaloes
Milk yield (kg / day)	4.1–6.3	4.0–5.0
Concentrate (kg / day)	2.3–2.8	1.3–2.7
Green fodder (kg / day)	1.6–8.7	2.8–7.8
Grazing duration (hours)	7.0–9.0	7.0–9.0

Cows are fed with more concentrates compared to buffaloes. The feeding cost is more in cows than buffaloes. 62-65 % of the total cost of milk production is due to the feeding of animals. When the animal is fed on fodder based ration, the cost structure could be brought down even to 40 %, but on the concentrates oriented feeding system the cost structure raises to 80 % of the production cost. For a production of 3 tonne of milk, 8 tonne of straw, 16 tonne of green fodder and 1 tonne of concentrates are required. Feeding cost can be minimized through the following measures: avoid over feeding, prevent/reduce wastage, process and preserve the seasonal excess, include locally available and cheaper non-conventional feed ingredients, fortify the poor quality roughage, include more quantity of green fodder round the year, preferably with one third quantity legume and avoid fodder containing any toxic principles.

Condition of the animals in relation to forage production

(i) Semi starvation is the rule
(ii) Emaciated in appearance
(iii) Prolonged dry periods compared to even distribution of rainfall in temperate climate
(iv) Poor milk yield (200 kg compared to 4000 kg per year in temperate countries
(v) Poor health except higher resistance to disease
(vi) Unproductive cattle are competing with productive farm animals and human beings.
(vii) Poor quality forage in protein carbohydrate, energy, fat and minerals.
(viii) Pastures are not being maintained due to lack of even distribution of rain and heavy pressure of land because of population explosion.

Calculation of fodder requirement: Fodder requirement for a farm or state or country can be calculated with animal population, body weight of animals, fodder requirement of animal type, fodder yield of crops, crop residues, fodder available from forest and other grazing lands. Average quantities of daily feed offered to milk animals are 6.3 kg dry fodder, 5.3 kg green fodder and 0.25 kg concentrates for cows, while a milking buffalo is offered 8.3 kg dry fodder, 2.25 kg green fodder and 0.1 kg of concentrates. Some of the data for calculation of fodder requirement are furnished in Table 3 and Table 4.

Table 3 Feed and fodder requirement for cattle (kg/day/animal).

S.No.	Animal description	Concentrate	Green fodder	Dry fodder
1	Calves of 0-6 months old	0.5	5-10	--
2	Calves of 6-12 months old	1.5	10-15	--
3	Pregnant cow	3.5	25	2.5
4	Millk cows with 5 litres milk	2.5	25	2.5
5	Millk cows with 5-8 litres milk	3.5	30	2.5
6	Millk cows with 10-15 litres milk	6.5	35	2.5
7	Dry cows	1.5	1.5	5
8	Bullocks	2.5-3.5	5.0	6-8

Table 4 Rural feeding practices for animals during lactation and dry period

Particulars	During lactation (kg/animal/day)		During dry period (kg/animal/day)	
	Buffalo	Crossbred cow	Buffalo	Crossbred cow
Green fodder	20	20	10	10
Dry fodder	6	4	7	5
Concentrate	3	4	1	2

Normally the dry matter intake will be 2.5 to 3.5 % of body weight and the green fodder intake will be 10 to 12 % of body weight. Average body weight of adult animal is around 250 kg. Average body weight of young animals below three years is about 150 kg.

Classification of forage crops

A. Grasses

I. Stall feed grasses

(a) **Annuals**

1. Sorghum – *Sorghum glaucum*
2. Maize – *Zea mays*
3. Teosinte – *Euchlaena mexicana*
4. Pearl millet – *Pennisetum typhoides*
5. Oats – *Avena sativa*

(b) **Perennials**

1. Guinea grass – *Panicum maximum* (Hamil grass)
2. Para grass – *Brachiaria mutica*
3. Napier grass – *Pennisetum purpureum*
4. Hybrid Napier grasses – NB21, BN2, CO1, CO2, CO3, KKM1

II. Pastures

(a) **Indigenous pasture grasses**

1. Kangayam tract (Tamilnadu)
   Anjan grass or
   White Kolukattai – *Cenchrus ciliaris*
   Black kolukattai – *Cenchrus setigerous*
2. Hosur tract (Tamilnadu)
   Spear grass – *Heteropogan contortus*
3. Ongole tract (Andhra Pradesh)
   Thengai pul – *Iseima laxum*
   – *Sehima nervosum*
4. Chhota Nagpur (Bihar)
   Dina grass – *Pennisetum pedicellatum*
   (also native of Africa)

(b) **Exotic pasture grasses**

1. Australia
   Blue panic grass – *Panicum antidotale*
   Blou buffel – *Cenchrus glaucus*
2. South America
   Dallis grass, Golden crown – *Paspalum dilatatum*
3. Africa
   Giant star grass – *Cynodan plectostachyon*
   Kikiyu grass – *Pennisetum clandestinum*
   Rhodes grass – *Chloris gayana*
   Pangola grass – *Digitaria decumbens*
   Molasses grass – *Melinis minutiflorus*
   Signal grass – *Brachiaria decumbens*
   Palisade grass – *Brachiaria brizantha*
   Setaria grass – *Setaria sphacelata*
   Hariyali grass or Doob grass – *Cynodan dactylon*

B. Legumes

I. Stall feeders

(a) Annuals

Cowpea	–	*Vigna unquiculata*
Cluster bean	–	*Cyamopsis tetragonaloba*
Velvet bean	–	*Styzolobium deeringianum*
Rice bean	–	*Phaseolus calcaratus*
Fenugreek	–	*Triagonella foenum graecum*
Sweet clover (Senji)	–	*Melilotus indica*
Sunhemp	–	Crotalaria juncea
Pillipesara	–	*Phaseolus trilobus*
Dew gram	–	*Phaselus acontifolius*
Field bean	–	*Dolichos lablab* var. *lignosus*
Horse gram	–	*Dolichos biflours*
Berseen	–	*Trifolium alexandrium*
Pea	–	*Pisum sativum var arvense*
Persian clover (Shaftal)	–	*Trifolium suaveolens (I.resupinatum)*

(b) Perennials

Lucerne	–	*Medicago sativa*
Subabul	–	*Leucaena leucocephala*
Hedge Lucerne	–	*Desmanthes virgatus*
Agathi	–	*Sesbania grandiflora*
Sithagathi	–	*Sesbania aegyptica*

II. Pasture

Siratro	–	Macroptilium atropurpureun
Centro	–	*Centrosema pubescence*
Calopo	–	*Calopogonium intortum*
Green leaf desmodium	–	*Desmodium intortum*
Stylo	–	*Stylosanthes guyanensis; S.humilis S.hameta; S.scabra*
Soybean	–	*Glycine wightii*
Tropical kudzu	–	*Purereraria phaseoloides*
Pea blue	–	*Clitoria ternatea*
Subabul	–	*Leucaena leucocephala*

C. Miscellaneous

Buckwheat	–	*Fagopyrum esculentum*
Sweet potato	–	*Ipomea batatas*
Sugarcane toppings	–	*Saccharum sp.*

CHAPTER

39

Fodder Maize–*Zea mays* L.

Importance: Fodder maize is grown throughout the country. It grows well under irrigated condition. It is a C_4 plant having high fodder production capacity in shorter durations. It is quick growing, high yielding and palatable fodder crop. The chaffed green maize fodder is eaten fully by all kinds of livestock.

Origin: Maize is native of Mexico.

Distribution: Maize is cultivated in USA, China, Europe, Russia, Africa, India.

Botany: Maize has a very distinct growth form; the lower leaves being like broad flags, 50–100 cm long and 5–10 cm wide; the stems are erect, conventionally 2–3 metres in height, with many nodes, casting off flag-leaves at every node. Internodes can reach 20–30 centimetres. Under these leaves and close to the stem grow the ears. The ears are female inflorescences, tightly covered over by several layers of leaves; they do not show themselves easily until the emergence of the pale yellow silks from the leaf whorl at the end of the ear. The silks are elongated stigmas that look like tufts of hair, at first green, and later red or yellow. Corn hybrid develops on average about 20-21 total leaves, silks about 60-65 days after emergence, and matures about 110 days after emergence.

Climate: Maize is generally grown in the areas with rainfall of 50 cm, but higher yields are achieved in 120–150 cm rainfall areas. It is susceptible to water logging. In the early stage upto 35 days after sowing, the crop is drought tolerant. Fodder maize can be grown in areas receiving an annual rainfall of 600 to 1000 mm. Frost damages the crop.

Soils: Maize grows best in fertile, well drained loamy soils. It is grown in alluviual and black cotton soils. Maize grows well with a soil pH 5 to 8 and tolerates salt affected soil.

Composites / hybrid: Fodder maize varieties recommended in different states of India are presented in Table 1.

Table 1 Fodder maize varieties recommended in different states of India

Varieties	Areas of cultivation	Green fodder (t/ha)
African Tall	Entire country	55-70

Vijai, Moti and Jawahar composite	Entire country	35-47
J-I006	Punjab	45-55
A-de cuba	North East	25-44
VL-54	Hilly areas	30-45
J-1006	Punjab	45-55
APFM-8	Andhra Pradesh	35
Pratap Makka Chari 6	Punjab, Haryana, Rajasthan, Uttar Pradesh and Uttarakhand.	45-50

Season: Maize is grown throughout year under irrigated condition and June-July as well as September-October seasons under rainfed condition.

Land preparation: Two to three ploughings are given to obtain good tilth. Beds and channels are formed.

Seeds and sowing: A seed rate of 40 kg per ha for line sowing and 75 kg per ha for broadcasting is recommended. Spacing for line sowing is 30 cm x 15 cm. Sowing depth is 3 to 5 cm. Seed treatment with 3 packet (600 g) of azosprillum is to be carried out.

Cropping system: Maize intercropped with cowpea in 1:1 ratio gives high yield and quality fodder.

Weed management: One hand weeding may be given on 20 DAS. Atrazine/simazine @ 1 kg/ha (pre-emergence) effectively controls the weeds in maize.

Manures and fertilizers application: 12.5 tonnes FYM per ha and 10 packets of azosprillum is to be applied basally. A fertilizer dose of 60-40-20 of N, P_2O_5 and K_2O per ha is recommended. Half the dose of nitrogen and full dose of phosphorus and potassium before sowing while the remaining dose of nitrogen is applied on 35 DAS.

Irrigation: First irrigation on immediately after sowing, life irrigation on third day and subsequently irrigation is given once in 10 days interval.

Crop protection: Diseases such as rust, stalk rot (and leaf blight while insect pests such as shoot fly, cutworm, cyst nematode occur in fodder maize. Downey mildew, which can easily be identified with symptoms like narrow, chlorotic or yellowish stripe later developing in brown lesions. The disease is seed borne and may be escaped by seed treatment with Thiram@ 2g/kg seed. The bacterial stalk rot in which plant shows rotting from base of stalks upward or from the top downward. The soil trenching with bleaching powder@ 3g/10L water controls the disease. The brown spot in which lower parts of plants appear slightly bleached is caused due to water stagnation, drain of excess water controls the disease.

Harvest: The seed to seed maturity is 90–95 days during *kharif* and 105–110 days in winter. It is done when the cob is in milk stage or after emergence of tassels or 50% flowering or 65 to 70 days after sowing.

Yield: It provides 35 t/ha green and 7.5 t/ha dry fodder yield. The crude protein content is 10%.

REFERENCES

http://agridr.in/expert_system/cattlebuffalo/Fodder%20Production.html

CHAPTER

40

Fodder Sorghum–*Sorghum bicolor* (L.) Moench

Importance: Forage sorghums are typically taller, more leafy, and late maturing than grain sorghum hybrids. Many of the forage sorghums have a 'sweet stalk' making them more palatable to livestock when utilized for grazing or for hay. The stalks of forage sorghum tend to be large and succulent, making them less palatable for grazing and sometimes slow in drying down for hay production. Sorghum is grown for grain, green fodder, silage and hay. Sorghum fodder should not be used before flowering stage due to presence of poisonous hydrocyanic acid at early growth stage.

Origin: Abyssinia, East central Africa or near Ethiopia or Sudan in Africa are the places of origin of sorghum.

Distribution: Sorghum is grown in America, West India, Australia, Africa, India and Europe.

Botany: The plants are annual or short-lived perennial without rhizomes; culms 0.3–4 m tall often robust, the nodes glabrous or pubescent. Leaf-blades are variable, often large, 5–75 cm long, 5–70 mm wide. Panicle is linear to broadly spreading, 10–60 cm long; primary branches compound, ultimately bearing racemes of 2–7 spikelet pairs. Sessile spikelet lanceolate to narrowly ovate, 4–9 mm long, glabrescent to white pubescent, sometimes tomentose or fulvously pubescent, awnless or more often with an awn 5–30 mm long. Pedicelled spikelet is linear to lanceolate, male or barren, smaller than the sessile.

Climate: A well distributed rainfall of 250 to 300 mm is sufficient during the crop growing period for rainfed crop. It is resistant to drought and not to water logging.

Soils: Sorghum is grown in alluvial and red sandy loam soils but black cotton soil is preferable. It tolerates salinity and alkalinity. It grows well with a pH of 5.6 to 9.5.

Season: Fodder sorghum is raised during June-July and September-October under rainfed condition while throughout the year under irrigated condition.

Varieties: The particulars of the fodder sorghum varieties grown in different states of India are furnished in Table 1 and Table 2.

Single cut varieties: K1, K3 (Irungu cholam), K-tall, K7, K10, K11, CO 11, IS 3541, CO 18 (Muthuvellai cholam), CO 19 (Thalaivirichan cholam), CO 27, UP chari 1, UP Chari 2, CSV 21 SF (Surat), HJ 513 (Hisar), Pant Chari 5 (UP), Haryana Chari-308 (Hisar).

Sorghum (Multi-cut type) varieties: CO 27, CO FS-29 (TNAU), Pusa Chari Hybrid (109) -PCH 109 (IARI), CSH – 20 MF Hybrid (UP), Pusa Chari 615 (IARI), Pant Chari 6 (UP), CSH – 24 MF Hybrid (UP), Methisudam, Pusa chari - 23, Pusa chari - 6, Pusa chari 9, Hara sona, Hc 308, Haryana Chari - 136, 171 and 260.

Table 1 Fodder varieties of sorghum for different areas

Variety	Region
Single cut	
PC-6, PC-9, PC-23, HC-171, HC 260, HC 308, Pro-Agro-chari, Jawahar Chari 6	For whole country (early medium duration)
HC-136, Raj chari-1, Raj chari-2, Haryana Chari, HJ-523, MFSH-3	Whole country (late varieties)
Sl-44, MP Chari, UP chari-2	North India
Pant chari-3, UP chari-1, Haryana chari- 6, JS-3	Uttar Pradesh, Andhra Pradesh, Tamil Nadu, Maharashtra
Double cut	
Co-27	Tamil Nadu
AS-16, GFS-1	Gujarat
Multicut	
PCH-106, Punjab Sudex, Pusa chari-23, Jawahar chari-69, Hara sona, Pioneer-988, SSG-59-3 (Meethi Sudan), MFSH-3, SPV-932	Whole country
Dual purpose (grain & fodder)	
CSV 15, CSH 13	As single cut variety for north-west India Single cut hybrid superior to HC 6

Table 2 Particulars of fodder sorghum varieties in India

Variety	50 % flowering (days)	Duration (days)	Fodder yield (qtl/ha)	
			Green	Dry
Pusa Chari-6	85-90	135-140	440	165
HC-136	90	140	550	175
Jawahar Chari-6 (JC-6)	80	130-135	412	120
U.P. Chari-1	75-80	115-120	330	90
U.P. Chari-2	75-80	110	380	125
Pusa Chari-9	80-85	120	425	135
Rajasthan Chari-1 (RC-1)	80-85	110-115	450	125
Rajasthan Chari-1 (RC-2)	65-70	150	330	100
HC 171		-	410	122
HC 260	55-60	90	450	140

Haryana Chari - 6 (HC-308)	-	-	415	128
CSV 15	65-70	112	440	128
CSH 13	65-70	113	480	145
M.P.Chari	65-70	110	300	95
Meethi Sudam	55-60	95-110	570	138
Hara Sona	60-65	110	630	147
Punjab Sudex	60-65	110	592	170
PCH 106	60-65	110	640	180

Land preparation: Two to three ploughings with iron plough or country plough should be given. Form beds and channels. Field preparation can be done utilising summer showers.

Seeds and sowing: A seed rate of 40 kg per ha for line sowing under irrigated condition and 75 kg per ha for broadcasting under rainfed condition is recommended. Line sowing is done with spacing of 30 cm × 15 cm. Sowing depth is 3 cm. Seed hardening with 2% KCl or KH_2PO_4 for 6 hours and dried to original moisture in shade drying. Seed treatment with 3 packets (600 g) of Azospirillum is to be carried out.

Cropping System: Mixed cropping of sorghum + cowpea in 1:1 ratio will improve yield and quality of fodder and soil fertility.

Weed management: Hand weeding is given on 20 DAS and 35 DAS. Thinning is done with country plough to have a control on sorghum population and weeds.

Manures and fertilizer application: 12.5 tonnes of FYM per ha is applied basally in the ploughed field along with 10 packets of Azospirillum. Blanket recommendation of 60-40-20 of N, P_2O_5 and K_2O kg per ha is recommended under irrigated condition as basal dressing. Apply 30 kg N per ha on 30 DAS as top dressing. A fertilizer dose of 30-20-20 kg N, P_3O_5 and K_2O per ha is recommended under rainfed condition as basal dressing.

Water management: Irrigate immediately after sowing, life irrigation is given on the third day and subsequently irrigation is given once in seven to ten days interval. Under drought conditions, sorghum leaves tend to fold rather than roll. A heavy white wax (bloom) usually covers sorghum leaf blades and sheaths, protecting them against water loss under hot, dry conditions.

Crop protection: Diseases such as downy mildew and anthracnose while insect pests such as shoot fly, stem borer and cyst nematode occur in fodder sorghum.

Harvest: It is done when earheads are fully emerged and up to grain is milk or early dough stage or 60 to 65 days after sowing (50% flowering). If it is a single cut, it should be harvested on 60 to 65 DAS (50 % flowering). If it is a multicut variety, the first cut is on 50 days after sowing and a second cut on 40 to 45 days after the first cut.

Yield: Green fodder yield of 35 to 45 tonnes per ha in first cut and 25 tonnes per ha in second cut under irrigated condition while 20 to 25 tonnes per ha can be obtained under rainfed condition.

Management of HCN poisoning: Hydrocyanic acid (HCN) or prussic acid or Dhurrin is maximum at early stages of crop growth from 35 to 40 DAS. It decreases gradually with the growth of the crop. When the crop is cut and field-cured, or is ensiled, the hydrocyanic acid degrades (2 to 3 weeks after ensiling), greatly reducing the toxicity. The HCN in excess of 200 ppm is toxic to animals. HCN content increases under moisture stress. In most of sorghum varieties HCN decreases below toxic level after 40 days of the crop growth. In summer, crop should be irrigated 2 to 3 days before harvesting or else it is safer to harvest crop after flowering.

Nitrate poisoning: Sorghum can accumulate nitrates (NO_3) during any weather condition that interferes with normal plant growth, however drought is the most common cause. This NO_3 is converted to nitrite (NO_2) in the rumen, which diffuses out into the bloodstream and binds to haemoglobin. This prevents the transport of oxygen (O_2) causing the animal to die from oxygen deprivation. Most NO_3 accumulate in the stem or lower portion of the plant. If NO_3-N exceeds 0.35% it should either be disregarded or diluted with safe feed (preferably grain). Unlike HCN, NO_3 will not leach out by the sun, however ensiling the forage can lower the NO_3 by approximately 50% or combined with other feeds low in nitrate to reduce daily nitrate intake.

REFERENCES

Casler MD. 2000. Breeding forage crops for increased nutritional value. Advances in Agronomy 71:51–107.

Chapman SR and Carter LP. 1976. Crop production, principle and practices. San Francisco, California, USA: WH Freeman. 566 pp.

Gourley LM and Lusk JW. 1978. Genetic parameters related to sorghum silage quality. Journal of Dairy Science 61:1821–1827.

ICAR. 1989. Technology for increasing forage production in India. Indian Council of Agricultural Research, New Delhi, 40 pp.

CHAPTER

41

Fodder pearl millet–*Pennisetum typhoides* (Burm. f.) Stapf & C.E. Hubb. *Pennisetum glaucum* L.R. Br.

Importance: Pearl millet green or dry fodder (*karvi*) is fed to the cattle. It is the most drought resistant fodder crop.

Botany: Annual, in tufted clumps, the culms slender, 15–75 cm high; leaf blades linear or linear lanceolate, 5–30 cm long, 3–10 mm broad, glabrous or with some long white hairs towards base on upper surface; spike erect, cylindrical, golden-brown in colour, 1–15 cm long, 6–12 mm broad; spikelets broadly oblong, 3–3.5 mm long, the upper lemma rugose; spikelets subtended by 4–12 bristles in each involucre, these are 3–10 mm long, finely antrorsely scabrous. It grows to a height of 3 metres. Two to three cuts can be taken.

Climate: Pearl millet requires warm climate with annual mean rainfall of 350 mm to 800 mm for its proper growth. The optimum temperature for the growth is between 20 and 28ºC.

Soils: Well drained sandy loam soil is preferable. Water logging condition is detrimental to crop growth.

Season: Pearl millet is raised during June-July and September-October season as rainfed crop and throughout the year as irrigated crop.

Land preparation: Two to three ploughings can be given to have good tilth. In heavy soils, harrowing rather than ploughing may be done to have fine tilth. Beds and channels are formed.

Varieties: Pearl millet varieties recommended for fodder production in different states are given in Table 1.

Table 1 Pearl millet varieties recommended for fodder production

State	Varieties/Hybrids
Uttar Pradesh, Punjab, Haryana, Madhya Pradesh and Rajasthan	Pusa Moti, UPFB 1, T 55, S 530, A 1/30, Rajko, AVKB-19, Giant Bajra, PCB-164, FBC-16, Avika Bajra Chari (AVKB-19), Narendra Chara Bajra-2 (NDFB- 2), Hybrids: NB 3, NB 17, NB 18, NB 21, NB 25, PHB 12, MH 30, BJ 105, Composite 6, K 674, K 677, L 72, L 74, Anand S 11, Raj Bajra Chari 2, Haryana, Composite 10, Nandi 32, MH-564, Nandi-8, PB 106, RHB 90, DRSB-2, proagro-1, Pusa 605, Pusa 415, MLHB-44, HAB-9, Pusa Composite 334.
Maharashtra and Gujarat	Malbandro, G 2, G 5 (Drought resistant), GFB-1
Tamil Nadu	CO 1, CO 2, CO 8, Nad Kumbu, TNSC1 (Cumbu)
Andhra Pradesh	APFB-2
Karnataka	B 247
For saline soils	DL 454, DL 532, DL 36
Rainfed: Entire bajra tract, Rajasthan and Gujarat	Giant Rajko

Seeds and sowing: Seed treatment with fungicides is done 24 hours before sowing. Seed rate is 10 kg per ha for broadcasting and 4 kg per ha for line sowing. Spacing is 30 × 10 cm. Removal of ergot and sclerotia affected seeds by soaking in salt solution (2.5 kg salt in 12.5 litre of water).

Cropping system: Fodder pearl millet crop is intercropped with cowpea. This can be harvested together to feed balanced nutritious fodder.

Weed management: Atrazine @ 0.5 kg per ha is applied as pre- emergence under irrigated condition which is followed with one hand weeding on 30 DAS or two hand weedings are given on 20 and 35 DAS.

Manures and fertilizer application: Apply 12.5 tonnes per ha FYM or compost along with 25-20-12.5 kg of N, P_2O_5 and K_2O per ha as basal dressing during last ploughing. 25 kg of N per ha is applied on 30 DAS and after each cutting for good harvest.

Water management: Pearl millet is mostly raised as rainfed crop for forage purposes. Under irrigated condition, first irrigation is given immediately after sowing, life irrigation is third day and subsequently irrigation once in 7 to 10 days.

Crop protection: Diseases such as downy mildew, ergot, smut, rust and leaf blast while insect pests such as shoot fly, jassids and thrips occur in pearl millet. A seed treatment with Apron 35 SD @ 2 g ai/kg seeds is followed by Ridomil 25 WP (1,000 ppm) spray 20-25 days later will effectively check the disease. Rotation of different varieties and hybrids in alternate years is also effective in arresting spread of downy mildew. Seed treatment with *neem* oil 5 ml/kg seed + spray of 5% (*neem*-seed-kernel extract (NSKE) at 50% flowering was found effective in controlling pests. White-grub infestation is managed by mixing of Phorate 10G or Quinalophos 5G @ 12 kg /ha with seed and applying in furrows at sowing.

Four varieties, MH 1336, MH 1364, MH 1392 and Pusa 383 are found to be resistant to smut ergot and blast.

Harvest: Crop is harvested on 45 to 50 DAS or at boot leaf stage or immediately after a few plants has flowered. Subsequently harvest is done once in 40 days. Harvest should not be delayed beyond 50 % flowering stage.

Yield: Green and dry fodder yield of 15 to 20 t/ha and 3.5 to 5.0 can be obtained and 25 to 30 t/ha and 6 to 7.5 t/ha respectively under rainfed and irrigated condition.

CHAPTER

42

Fodder barley–*Hordeum vulgare* L.

Family: Poaceae

Vernacular names: Barley (English), ogre, escourgeon (French), cebada (Spanish).

Importance: It is a valuable source of winter forages.

Distribution: Barley forage is grown in North America, Europe and the Mediterranean area.

Botany: The barley plant is an annual, erect and tufted grass, up to 50 to 120 cm high. It is a leafy species. The leaves are linear and lanceolate, up to 25 cm long, placed opposite their neighbours along the stem.

Soils: It is adapted to well-drained, loamy soils and should not be planted on poor, sandy, or wet soils. It withstands moderate droughts and tolerant to saline and alkaline soils. It is often used as soil reclaimer as it dilutes and excludes salt.

Varieties: The particulars of the barley varieties grown in different states of India are furnished in Table 1.

Table 1 Barley varieties grown in different states of India

Variety	Source	Suitability	Major characteristics
Ratna	IARI, New Delhi	Uttar Pradesh, Bihar and West Bengal	matures in 125-130 days, good tillering ability
Azad (K.125)	C.S.A. Univ of Agrl and Tech, Kanpur	Uttar Pradesh, Bihar and West Bengal	matures in 115-120 days,
Vijaya	C.S.A. Univ of Agrl and Tech, Kanpur	Haryana, Punjab, Uttar Pradesh, and Madhya Pradesh	matures in 120-125 days
Amber	C.S.A. Univ of Agrl and Tech, Kanpur	Uttar Pradesh	matures in 130-135 days
RS-6	Rajasthan	Rajasthan	matures in 130-135 days

Jyoti	C.S.A. Univ of Agrl and Tech, Kanpur	Haryana, Punjab, Uttar Pradesh. Rajasthan, Bihar and West Bengal	matures in 120-125 days
Clipper	Australia		matures in 135-140 days, susceptible to yellow rust
PL-56	Punjab	Punjab	moderately resistant to yellow rust
Ranjit (DL- 70)	Punjab	Punjab	heavy tillering variety
Karan 201, 231 and 264		Madhya Pradesh, Rajasthan, Haryana	huskless varieties, alkaline and saline soils
C-164			resistant to yellow rust
Kailash		Himachal Pradesh	matures in 145-150 days, resistant to yellow rust
Dolma		Himachal Pradesh and Uttar Pradesh hills	matures in 145-150 days, resistant to yellow rust
Himani	Simla	Himachal Pradesh and Uttar Pradesh hills	matures in 135-140 days
LSB-2		Himachal Pradesh and Uttar Pradesh hills	matures in 145-150 days
RDB-1	Rajasthan	Rajasthan	suitable for rust free areas of Rajasthan
BG-25	HAU, Hissar	Haryana	matures in 120-130 days
BG-108	HAU, Hissar	Haryana	matures in 120-125 days, resistant to yellow rust
Neelam	IARI, New Delhi	Punjab, Haryana, Uttar Pradesh, and Bihar	profuse tillering, resistant to yellow rust,

Manures and ferilizers management: After each cut the crop should be fertilized to ensure good quality regrowth.

Water management: It is mainly cultivated under rainfed areas.

Harvest: The crop should be mowed or grazed for about a six-week period until the first node appears on the crops. It is harvested at the soft-dough stage for silage purpose. It can be cut and dried to make hay. The crop should be cut or grazed at 30-40 cm height for best quality feed and good regrowth. This should be well prior to the development of awns which will reduce the palatability of grazing.

Yield: The green fodder yield is 15 t/ha while the dry matter yield is 3 to 8 t/ha.

Utilization: Barley forage can be grazed, cut for hay or silage while green or cut after grain harvest as straw. When barley forage is cut for hay at milky stage, NO_3-N level is not higher than 0.14% which is a safe level for livestock.

CHAPTER

43

Fodder Oats–*Avena sativa* L. (2n=42)

Vernacular name: Oat, common oat, (English), *avoine, avoine commune* (French), *hafer* (German), *avena* (Spanish), *aveia* (Portuguese).

Importance: Oat as a dual purpose crop grown for grain and forage purpose. The crop is widely cultivated for use as food, feed and fodder. Its grain makes a good feed particularly for horses, sheep and poultry. The oat crop may be used as winter cover to protect soil from erosion. Oat is one of the most important cereal fodder crops of *rabi* season in North, Central and West Zone of the country.

Origin: The center of origin of oats is Mediterranean region/Asia Minor. The countries bordering Mediterranean, Turkey, Iran, Iraq and parts of the Russia are rich in the diversity of oat species.

Distribution: Oats are presently grown in temperate parts of the world including USA, Canada, Europe as spring-sown cultivars. In the tropical countries and higher altitude region it is grown as a winter / *rabi* season annual crop. In India, it is grown in Uttar Pradesh, Madhya Pradesh, Haryana, Punjab, Himachal Pradesh, Rajasthan, Bihar, Gujarat, Andhra Pradesh and hilly tracts of southern plateau.

Botany: The oat plant reaches up to 1.2 m in height. It has a fibrous root system. Primary tillers arise in the axils of the older basal leaves of the main stem. Secondary tillers arise in the leaf axils of primary tillers. Its culms are erect, smooth, and hollow. The leaves are linear, alternate, and acuminate. Leaf blades are 15-40 cm long and 0.6-1.5 cm broad. The inflorescence is a terminal, loose, curved, and branched panicle that bears solitary and pendulous spikelets. The spikelets have generally two overlapping husks (glumes). At anthesis, the anthers dehisce and ripe pollen is shed on to the feathery stigma. The florets may open due to swelling of the lodicules at base so that the anthers are exerted to hang outside the glume. Oat is a self-pollinated crop. Natural cross pollination by wind occurs occasionally and varies from 0.4 to 1.3%. The fruit is a hairy cylindrical and slightly ridged caryopsis, 0.6 to 0.8 cm in length, enclosed in hulls which are difficult to remove.

Growth stages: Plant height ranges from 75 to 150 cm. Tiller number is 15 to 20. Number of leaves is 7.5 cm. Leaf length is >50 cm while leaf width is >2.5cm. The days to 50 % flowering is <80 days for the crop duration for maturity is 120 or < 120 days. The days to 50 % flowering is >125 days for the crop duration for maturity is 150 or > 150 days. The green fodder yield at 50 % flowering stage is >450g while the dry fodder yield at 50 % flowering stage is >75 g. Total green fodder yield is >2500g and total dry matter yield is >300g under multicut system.

Climate: Oats flourish in cool and moist climates. Oat production is concentrated between latitudes 35–65ºN and 20–46ºS. The crop can be grown at temperatures varying between 5 to 30ºC, however, optimum temperature required is 15-25ºC. At low temperature, its germination is delayed, while grain production is hampered by hot, dry weather, especially from heading through the grain filling period. Oats are mostly found between 45°-65° North and 20-46° South latitude. Oats grow on a wide range of temperatures ranging from 5 to 26°C and in regions with rainfall over 500 mm. It can tolerate frost up to some extent but its fodder yield and quality is reduced due to hot and dry conditions.

Soils: Oat is grown in loam to clay loam soil with adequate drainage. It tolerates acidic and saline soil with pH ranging from 4.5 to 8.6.

Varieties: The particulars of the oats varieties grown in different states of India are furnished in Table 1.

Table 1 Oats varieties grown in different states of India

Variety	Source	Green fodder yield (t/ha)	Suitability	Major characteristics
Palampur-1		50	Himachal Pradesh	Flowering in 145 days and matures190 days. plant height of 115 cm at 50% flowering, profuse tillering and resistant to lodging.
Kent	USA	45-50	entire country	Flowering in 125 days and matures180 days. plant height of 100 to 125 cm at 50% flowering, moderate tillering, resistant to rust, blight and lodging
OS-6	CCSHAU, Hissar	50-54	Haryana, suited to temperate and sub-tropical regions	plant height of 120 to 130 cm at 50% flowering
OS-7	CCSHAU, Hissar	55-60	Haryana, suited to temperate and sub-tropical regions of the country	suitable for single cut
OL-9	PAU, Ludhiana	40-50	north, north-west and south hills	
OL-125	PAU, Ludhiana	50-58	north-west and central zone	suitable for single cut/multi cut
JHO-822	IGFRI, Jhansi		central part of India	plant height of 120 to 130 cm at 50% flowering
UPO 2005-1	GBPUA&T, Pantnagar			plant height of 130 to 140 cm at 50% flowering

JO 2003-78				plant height of 120 to 130 cm at 50% flowering
FOS-1/29		40-45	Punjab, Haryana, Delhi and Uttar Pradesh	resistant to drought, raised under rainfed conditions, profuse tillering, green fodder yield is 40–45 t/ha in 140 days
HFO-114 (Haryana Javi - 114)		50-55	entire country	two cuts and have good tillering, resistant to lodging and diseases
Algerian		40-45		plant height is 100–125 cm,
Brunker-10		40-45	Punjab, Haryana, Delhi and Uttar Pradesh	resistant to drought, raised under rainfed conditions, profuse tillering, green fodder yield is 40–45 t/ha in 140 days,
Weston 11		50	Punjab	plant height is 155 cm, days to flowering is 110 days and maturity is 160 days,
Bundel Jai-822	IGFRI, Jhansi	50	central zone	takes 95–100 days for flowering and matures in 125–130 days,
UPO-94	GBPUA&T, Pantnagar	40-45	north and central India	plant height of 75–80 cm at 50% flowering, resistant to rust, blight and lodging
UPO–212	GBPUA&T, Pantnagar	60	north and central India	flowers in 140–150 days,
Haryana Javi-8 (HJ -8)	CCS HAU Haryana	60-65	Haryana	suitable for two cuts.
Sabzaar (SKO-7)	SKUA&T, Srinagar	35-40	Jammu & Kashmir	profuse tillering, leafy and suitable for dual purpose
Bundel Jai-851	IGFRI, Jhansi	45-47		high tillering, multicut variety, takes 110 - 115 days for flowering and 140– 145 days for seed setting,
Bundel Jai 992 (JHO 99-2)	IGFRI, Jhansi	50	north-east and north-west zone	flowering in 100–105 days with maturity at 140–145 days,
Bundel Jai 2004 (JHO 2000-4)	IGFRI, Jhansi	50	north-east and north-west zone	single cut variety, tolerant to root rot, crown rust, leaf blight and powdery mildew
JO-1	JNKVV, Jabalpur		central zone	
Harita (RO -19)	MPKV, Rahuri	50	Maharashtra	multicut variety, high leaf to stem ratio and is resistant to leaf blight disease
Bundel Jai 991 (JHO 99-1)	IGFRI, Jhansi	30	hilly zone	single cut system, resistant for grasshoppers and aphids, flowering in 120–125 days with maturity at 150–155 days,
Bundel Jai 2001-3 (JHO 2001-3)	IGFRI, Jhansi	50	North- West and Southern zones	single cut variety,

Season: It is mostly grown in rabi season. October (first week) to December (second week) is considered as the best sowing period. The crop should be sown in October for forage purpose.

Seeds and sowing: Seed rate varies from 70-80 kg (seed crop) to 90-100 kg (fodder crop) per hectare. The seed is broadcasted in case of fodder crop purpose. Low tillering varieties should be sown with 20-25 cm row spacing while higher tillering type should be sown 30 cm apart for grain purpose. Oats are sown at a soil depth of 2 to 4 cm.

Weed management: Oat is infested with winter season grassy and broad-leaved weeds mostly found as in wheat. Effective control of weeds in oats can be obtained with weeder cum mulcher at 4 week crop stage followed by application of 2, 4-D @ 0.37 kg a.i./ha at 6 weeks crop stage.

Manures and fertilizers application: Application of FYM @ 20-25 t/ha before 10-15 days of sowing along with the application of 80 kg N, 40 kg P_2O_5 per ha to single cut and a dose of 120 kg N, 40 kg K_2O/ha to multicut varieties attains good crop growth. In double and multicut varieties, top-dressing of 40 kg N/ha after first cut and two equal split doses of 40 kg N/ha after first and second cut should be done respectively.

Water management: Oats require 4 to 5 irrigations including the pre-sowing irrigation. If soil is dry, first irrigation is given before preparing the seedbed. Subsequent irrigations are given at intervals of about one month mostly after each cut. The tillering and heading stages are critical for irrigation.

Harvest: The oat matures within 120 to 150 days after sowing for seed production. For fodder production, the harvesting of single cut oat varieties is done at 50% flowering (about 50-55 days of sowing). In double cut varieties, first cut should be taken at 60 days followed by second cut at 50% flowering stage. However, in multicut varieties, the first cut is taken at 50-60 days after sowing and the subsequent cuts at 30 days interval.

Yield: The fodder yield ranges from 30 to 50 t/ha green and 12 t/ha dry. The grain and straw yield is 1.6 to 2.0 t/ha and 2.5 to 3.0 t/ha respectively.

Uses: Oats both as forage and grain are good source of protein, fiber and minerals. It is used as green crop, hay and silage for animals. It is principally fed to dairy cattle, horses, mules and turkeys, with lesser quantities fed to hogs, beef cattle and sheep. It may be grazed, cut-and-carried, made into hay, silage or balage.

CHAPTER

44

Teosinte–*Euchlaena mexicana* *Schrad.* (2n-20) Family: Poaceae

Vernacular name: Teosinte (English), *Makchari* (Hindi).

Importance: Teosinte is a multicut fodder. It resembles maize in many ways, notably their tassel (male inflorescence) morphology while it differs from maize by their numerous branches each bearing bunches of distinctive, small female inflorescences. It produces tillers profusely and has dark green narrow leaves than maize and remains green for a longer period. It is suitable as green fodder, silage and hay.

Origin: Teosinte is a native of Mexico, Guatemala, Honduras, and Nicaragua.

Distribution: It is grown in several areas of Central America, Mexico, Guatemala, Honduras, Nicaragua, Columbia, Argentina, Philippines, Indonesia, Japan, Burma, Australia and India.

Botany: Mexican teosinte is an annual, warm-season grass which resembles maize in general vegetative appearance. It is a monoecious with tassel and cob. It produces tillers. It grows to a height of 2 to 4.5 m. It is coarse, branches at the base, and the leaf blades are sword-shaped and succulent. Leaf length is 90 cm with width is 5 cm. Clusters of slender 'ears' (seed pods) are produced in each of the 5 to 7 uppermost leaf axils. The seeds of teosinte are triangular, dark brown to black in colour, and much smaller than maize seeds. The teosinte female ears bear approximately 5 to 12 kernels, whereas the maize ears may contain 500 or more kernels. There are a few physical differences between maize and teosinte. Teosinte plants are branched and produce many ears, while maize plants produce a single upright stem with ears. Maize ears have 8-12 rows of seeds, while teosinte only has two rows. Each segment of the teosinte fruit can produce only one seed because one of the paired potential seeds (spikelet primordia) is aborted. Both of these spikelets are fertile in maize. The seed surrounded by a hard 'fruit case' in teosinte, while the fruit case is reduced and develops into part of the maize cob.

Climate: It is suited for warm humid region with a mean annual rainfall of 1000 mm. It requires 165 to 170 frost free days from seeding to maturity.

Soil: Well drained loamy soils are best for its growth. Saline, alkaline, sandy and swampy soils should be avoided.

Land preparation: Two to Three ploughings are necessary to have good tilth.

Season: It can be grown during June-July and February-March under irrigated condition and in June-July under rainfed condition. The crop duration is 110 to 120 days.

Varieties: Sirsa, TL 1, Teosinte improved.

Seeds and sowing: Seed rate is 40-45 kg per ha. Line sowing is done with a spacing of 30 cm x 15 cm. Sowing depth is 2.5 cm. Broadcasting method of sowing can be also practised.

Weed management: Hand weeding may be given on 20 and 35 DAS or apply atrazine or simazine @ 1.25 kg per ha as pre-emergence which is followed by one hand weeding on 35 DAS under irrigated condition.

Manures and fertilizer application: 12.5 tonnes per ha of FYM or compost along with 60-30-30 of N, P_2O_5 and K_2O are applied as basal dressing during last ploughing. Apply 30 kg N per ha after each harvest.

Water management: Teosinte requires more water than maize. First irrigation is given immediately after sowing. Life irrigation is given on third day and subsequently once in 7 to 10 days.

Harvest: First cut can be taken up on 90 DAS or at 50 % flowering stage. Subsequent harvests are taken on 6 to 7 weeks interval. The best stage for harvesting is 7 to 10 days before the appearance of tassel.

Yield: Fodder yield of 35 to 40 tonnes per ha under irrigated condition whereas 15 to 20 tonnes per ha under rainfed condition can be obtained. Seed yield is 600 to 1000 kg per ha.

REFERENCES

Adriana Natividad Avendaño López. 2011. Seed Dormancy in Mexican Teosinte. Crop Science, 51: 2056–2066.

Collins , G. N. 1921 . Teosinte in México. *The Journal of Heredity* 12 : 339–350 .

John Doebley, Adrian Stec and Charles Gustus. 1995. *teosinte branched1* and the Origin of Maize: Evidence for Epistasis and the Evolution of Dominance. Genetics 141: 333–346

Wilkes, H. Garrison. 1967. Teosinte: The Closest Relative of Maize. Bussey Inst., Harvard Univ.: Cambridge. 159 pp

Wilkes , H. G. 1985 . Teosinte: The closest relative of maize revisited. *Maydica* 30 : 209–223 .

http://en.wikipedia.org/wiki/Teosinte_(disambiguation)

http://maize.uga.edu/index.php?loc=ancestors

http://plants.usda.gov/factsheet/pdf/fs_zeme.pdf

CHAPTER

45

Guinea Grass–*Panicum maximum* Jacg (2n=18, 36, 48)

Family: Poaceae (Gramineae)

Importance: Guinea grass is an important multicut forage grass, because of its ease of propagation, fast growth and high quality forage during the rainy season. It can be used as a cut and carry forage or as a green chop feed. It is a high yielding, and drought resistant grass. Guinea grass is a shade tolerant grass and is suited to grow in coconut plantations. It is a valuable grass for grazing, silage and hay making.

Origin: Guinea grass is native of Africa and later spread to countries like Australia, Philippines, India and USA.

Distribution: Guinea grass is cultivated in Uganda, Kenya, Tanzania, West Indies, Trinidad, Brazil, Australia, Malaysia, Sri Lanka, India and Philippines. It is cultivated in Haryana, Punjab and Himachal Pradesh. It has also wide adaptability in humid tracts of eastern and southern India.

Botany: Guinea grass vigorous, coarse, tufted, perennial and bunch grass. It grows to a height of 3 to 4 metre. Roots are fibrous. Culm is erect, 50 to 100 cm long and 1 to 2.5 cm wide. Inflorescence is a loose spreading panicle. Seeds are 3mm long and one mm wide. There are three cultivars / ecotypes in guinea grass.

(i) Green panic grass *(Panicum maximum* var *trichoglume*): A small leafy, perennial bunch grass; 0.6 to 1.3 m height, yellowish green in colour; leaves 30 cm long, 1.4 cm wide; panicles 20 to 35 cm long.

(ii) Hamil grass *(Panicum maximum* var *hamil)*: Hamil grass grows to a height of 3 to 4 m. It is robust, erect, smoother than green panic grass. Foliage bluer, leaf blades softly hairy and basal leaf sheath is with stiff hairs. Seed sets freely.

(iii) Colonial grass (*Panicum maximum* var *colonial)*: It is cultivated in Brazil. It grows to a height of 3 to 4 m. It has very smooth foliage and thick stems. Foliage is bluish green in colour.

Climate: Guinea is one of the major pasture grass of the tropics and subtropics. The crop performs best in warm and moist climates of tropics. It can be grown up to an elevation of 1,800 m. It also comes up well under shade and therefore suitable for intercropping in orchards and plantations especially in coconut. It can withstand drought because of its deep root system. Green panic grass with annual rainfall of 550 mm to 1750 mm and it is drought resistant. It grows rapidly in warm weather following rains. It can be planted throughout the year provided the temperatures do not fall below 15°C. It is frost sensitive.

Soils: Guinea grass is adapted to a wide range of soils. It needs good drainage. It does not come up well in heavy clay soils or flooded or waterlogged conditions

Season: Guinea grass can be grown throughout the year under irrigated condition. It can be grown during June-July and September -October under rainfed condition.

Land Preparation: The land is ploughed 2 to 3 times to obtain a good tilth. Ridges and furrows are formed with 50 cm row spacing.

Varieties: Guinea grass varieties recommended in different states of India are presented in Table 1.

Table 1 Guinea grass varieties recommended in different states of India

Variety	Suitable for	Green fodder yield (t/ha/year)
Macueni, Riversdale, Haritha, Marathakam, Harthasree.	Kerala for rainfed conditions	60-80
CO 1, CO 2	Tamil Nadu	150-200
Hamil, Bundel Guinea-1 (JHGG-96-5), Bundel Guinea-2 (JHGG 04 –01)	South, north east, east and central India	90-130
PGG-4, PGG-9	Hills, North-west and central India	85-130
PGG-13, PGG-14	Central India and Hills	95-140
Punjab Guinea Grass 1 (PGG-1), PGG-19, PGG 101, PGG-518, PGG 616	Punjab	90-130

Seeds sowing / Setts or slips planting: Guinea grass can invade and naturalize lands without subsequent sowing or planting. Stool forming, spreading somewhat by short rhizomes can be propagated from seed or crown divisions. Seed viability is often poor. Seeds can be broadcasted or drill-sown. Seed rate is about 2 to 3 kg per ha. Seeds have to be stored for more than six months before sowing for breaking seed dormancy. Sow the seeds in the main field or raise the seedlings in the nursery and transplant 20 to 25 days after sowing. Vegetative propagation can be accomplished by dividing the basal crown into segments and planting the divisions in furrows or on the side of ridges when there is excess moisture during rainy season.

Green panic: Seed rate is 2.5 kg per ha. One kg seed contains 1.1 million seeds. Seeds can be broadcasted or drill sown (shallowly). Rooted slips of 66000 are required for one hectare. Slips are planted with spacing of 50 cm x 50 cm.

Hamil grass: Seed rate is 2.5 kg per ha. One kg seed contains 1.1 million seeds. Rooted slips required are 66,000 per ha. Slips are planted with spacing of 50 cm x 50 cm.

Colonial grass: It is propagated by seed or crown divisions. Rooted slips required are 66,000 per ha. Slips are planted with spacing of 50 cm × 50 cm.

Cropping system: Guinea grass can be intercropped with centro, siratro and hedge lucerne. Legumes should be combined with guinea grass in 1:3 ratio. For pastures under coconuts, green panic is probably more suitable than the hamil or colonial varieties. A major strength of guinea grass and its relatives are shade tolerant.

Weed management: First hand weeding is given on 30 DAS. Subsequent weedings are taken whenever necessary. Earthing up is done once for each harvest. Pre-emergence application of atrazine @ 1 kg/ha followed by 2,4-D @ 1 kg/ha on 30-35 days after sowing is recommended.

Manures and fertilizers application: 12.5 tonnes of FYM per ha along with 50-50-40 of N, P_2O_5 and K_2O are applied as basal dressing and 50 kg N per ha after each harvest.

Water management: First irrigation immediately after sowing of seeds or planting of slips, life irrigation is on the third day and subsequently irrigations are given once in 10 days.

Harvest: First cut on 60 to 65 days after sowing or 45 days after planting of slips. Subsequent cuts can be taken at an interval of 50 days. Quartering is done after 2 to 3 years of planting *i.e.,* three fourth of clumb is removed and one-fourth is retained. A thumb rule is to graze it to a height of about 15 to 20 cm. Too heavy grazing can lead to injury to the growing points of the grass and a ventual death of the grass may result. Rotational grazing will help to maintain a uniform height for the pasture.

Yield: The annual green fodder yield ranges from 80-120 tonnes/ha in 7-9 cuts ha with average dry matter yield from 25 to 35 t/ha under irrigated condition to 50-60 t/ha in 4-6 cuts under rainfed condition. The green fodder yields can be realized in 10-12 cuts under sewage irrigation up to 200 t/ha. The fodder contains 9-13% crude protein and 30-35% fibre with an in vitro dry matter digestibility of about 50%.

CHAPTER

46

Napier grass–*Pennisetum purpureum* Schum. (2n=28)

Family: Poaceae

Vernacular name: Elephant grass, Uganda grass

Importance: Napier grass is drought tolerant. Young leafy fodder is palatable. Mature grass has large proportion of stem and becomes rough. The name Elephant grass derives from it being a favourite food of elephants.

Origin: Napier grass is native of Rhodesia in South Africa. The name napier grass is given in the honour of Col. Napier, who first drew the attention of the Rhodesian Department of Agriculture in 1909 to the fodder value of this grass. It was introduced in India in 1912 from South Africa.

Distribution: Napier grass is grown in Uganda, USA, Australia, South America, Srilanka, Malaysia, Indonesia, Philippines, India, Burma, Bangaladesh, and Pakistan.

Botany: This is a perennial grass. It has stoloniferous with a creeping rhizome. It is a deep rooted grass. Culms usually 180–360 cm high, branched upwards. It grows to a height of 3.0 to 4.5 meter. Leaf-sheaths glabrous or with tubercle-based hairs; leaf blade length is 30 to 90 cm and width is up to 2.5 cm, margins thickened and shiny. Inflorescence a bristly false spike up to 30 cm long, dense, usually yellow-brown in colour, more rarely purplish

Climate: The elephant grass grows well at sea level and upto 2000 m. It grows well in areas receiving rainfall of 800 mm to 1500 mm/year. It is best adapted to humid condition. It can survive moderate drought because of its deep root system. The optimum temperature is 28°C. It is most susceptible to frost.

Soil: It is adapted to a wide range of soils. However, it grows best in well drained loamy and fertile soils. The desirable soil pH is 6.5 to 8.0 . It can not withstand water logging and high rainfall.

Season: Napier grass is grown throughout the year under irrigated condition.

Varieties: Pusa Giant Napier has been developed by IARI, New Delhi. It provides high yield (250–300 t/ha/year) of green matter under irrigated condition. It contains 25%

and 12% more protein and sugar respectively than common napier grass. It is less fibrous, juicy and palatable. It is susceptible to *Helminthosporium* sp.

Land preparation: Two to three ploughings are given to obtain good tilth. Ridges and furrows are formed using ridge former with 50 cm spacing.

Seeds and sowing: 40000 rooted slips or stem cuttings with 2-3 nodes per ha are required. Slips are planted with spacing of 50 cm x 50 cm.

Weed management: First weeding is given on 20 days after planting and subsequently whenever necessary.

Manures and fertilizer application: FYM 12.5 tonnes per ha and 50-50-40 kg of N, P_2O_5 and K_2O per ha are applied as basal dressing and 100 kg N per ha after each cutting.

Water management: First irrigation is given at planting of slips, life irrigation is given on the third day and subsequently once in 7 to 10 days.

Harvest: First cut is carried out on 65 to 70 days after planting and subsequently once in 45 days. Totally 7 to 10 cuts per year can be obtained. Quartering is done after 2-3 years of planting i.e., three fourth of clump is removed and one fourth is retained.

Yield: Green fodder yield of 30 to 35 tonnes per ha in summer and 45 to 50 tonnes per ha in rainy season can be obtained at each harvest.

CHAPTER

47

Napier × Bajra Hybrid (NBH) Grasses

Pennisetum purpureum × P. glaucum

Importance: Napier x Bajra Hybrid (NBH) is an interspecific cross between *bajra* x napier grass which is known as king grass. The perennial and heavy tillering characteristics of napier grass have been combined with leafyness of *bajra*, and the Napier x bajra hybrid has been evolved. Among the grasses, Napier-bajra hybrids are the highest green forage yielder in an unit time and space. It is vigourous, nutritious, succulent, palatable and responds to heavy nitrogen fertilizer application. Young leafy fodder is highly palatable and of fairly good quality. However the mature grass has large proportion of stem. Hybrid grasses are vegetatively propagated. These are cultivated mostly under irrigated condition. Hybrid napier is a triploid grass, so does not produce any seeds. It produces high number of tillers and numerous leaves. It grows fast and produces high herbage but the stems are hard and the plants less persistent.

Origin: Napier grass is native of Rhodesia in South Africa.

Distribution: It is widely distributed in tropical and sub-tropical regions of Asia, Africa, southern Europe and America. In India, it is cultivated in the states of Tamil Nadu, Punjab, Haryana, Uttar Pradesh, Bihar, Madhya Pradesh, Orissa, Gujarat, West Bengal, Assam, and Andhra Pradesh. This grass does not survive in heavy rainfall and frost prone areas.

Climate: Warm tropical climate is desirable for cultivation of hybrid grasses. Napier *bajra* hybrid performs well in areas having temperatures above 15°C. The optimum temperature is about 31°C. Light showers alternated with bright sunshine are very congenial to the crop. Total water requirement of the grass is about 800–1000 mm. It can withstand drought for a short spell and regenerates with rains but is susceptible to frost.

Soils: Hybrid grass can be grown in all type of soils. However well drained loamy soil is preferred. It can grow on a variety of soils. Light loams and sandy soils are preferred to heavy soils. The grass does not thrive well on waterlogged and flood prone lands. Phenomenal yields are obtained from very deep fertile soil rich in organic matter. It tolerates pH ranging from 5 to 8.

Land preparation: Two to three ploughings are given to have good tilth. Ridges and furrows are formed with 50 cm spacing.

Season: Hybrid grass has been grown throughout the year under irrigated conditons.

Napier x Bajra Hybrids: The Napier x Bajra Hybrids particulars cultivated in different states of India is furnished in Table 1 Table 3.

Table 1 The Napier x Bajra Hybrids particulars cultivated in different states of India

Hybrids	Characteristics
IGFRI-3, IGFRI-5, IGFRI-6	Suitable for central India, north-east and northern hills. It yields fresh green fodder and dry matter yield of 114 t/ha and 37 t/ha respectively
IGFRI-7	It is suitable to hilly, sub-humid and sub-temperate areas of India. It yields 140-170 t/ha of green fodder.
IGFRI-10	It is suitable for whole country with a green fodder yield of 150-180 t/ha. Suitable for acidic and saline soils
PBN-83	Suitable for Punjab with a green fodder yield of 125-170 t/ha.
Yashwant (RBN-9)	Suitable for Maharashtra. It has low oxalic acid content (2.46%) and high crude protein content (10.15%). It produces 150 t/ha green fodder.
PBN-16	Suitable for Punjab, Maharashtra, Karnataka
PBN 233	375 t/ha green fodder in seven cuttings in a year. It is photosensitive and flowers only in winter.
APBN-1	Suitable for Andhra Pradesh, Karnataka, Tamil Nadu with yield of 200-250 t/ha. The plants are 70 cm tall with 50 tillers/plant and leaf stem ratio 13.62.
Sampoorna (DHN 6)	Suitable for Karnataka, oxalic acid content (1.9%) and yields 120–150 t/ha green fodder in 6–8 cuts.
Hybrid Napier-3 (Swetika)	Suitable for north and central zone of the country. It is tolerant to frost and low temperature and is suitable for low pH, and have field resistance to *Helminthosporium* blight. It gives 70–80 t/ha green fodder and 18 t/ha dry fodder.
NB-37	suitable for sub-tropic pastures. It is drought tolerant and has low oxalates (2–3%). It produces 35–40 t/ha of green fodder and has crude protein of 9-10%.
Suguna	suitable for Kerala. It is a semi-perennial, multicut and yields 260 t/ha/ year green fodder yield.
Supriya	suitable for Kerala. It yields 270 t/ha green fodder.

Table 2 The Napier x Bajra Hybrids particulars cultivated in Tamil Nadu

Hybrids	BN 2	NB 21	CO_1	CO_2
Green fodder yield (t/ha)	170	170	180	180
Dry matter yield (t/ha)	41.25	37.8	40.6	55.0
Crude protein (%)	10.00	10.31	9.38	9.38
Crude protein yield (t/ha)	4.13	3.90	4.58	6.00
Oxalate (%)	2.92	2.87	2.97	2.92
Source	CCSHAU, Hissar	KAU, West Bengal	TNAU, Coimbatore	TNAU, Coimbatore

Table 3 The Napier x Bajra Hybrids particulars cultivated in Tamil Nadu

Hybrids	KKM 1	CO 3	CO (CN) 4
Parentage	Interspecific hybrid between Cumbu IP 15507 x FD 429	Interspecific hybrid between Cumbu PT 1697 x *Penneisetum purpureum*	Interspecific hybrid between Cumbu CO 8 x FD 461
Duration (Days)	Perennial	Perennial	Perennial
Green fodder yield (t/ha/yr)	288	350 (7harvests)	375-400 (7harvests)
Plant height (cm)	155-160	300 – 360	400-500
No. of leaves per clump	165-170	400-450	400-450
No. of tillers per clump	10-15	30 – 40	30 – 40
Leaf stem ratio	-	0.70	0.71
Leaf length (cm)	110-115	80 – 95	110-115
Leaf width (cm)	4.5-5.0	3.0 4.2	4.0-5.0
Dry matter yield (t/ha/yr)	47.23	65.12	79.87
Crude protein yield (t/ha)	4.65	5.40	8.71
Dry matter (%)	16.4	17.0	21.3
Crude protein (%)	9.85	10.5	10.71
Oxalate (%)	1.50	2.51	2.48
Source	TNAU, Coimbatore	TNAU, Coimbatore	TNAU, Coimbatore

Planting: Hybrid grass is raised with rooted slips and stem cuttings (setts). Slips @ 40,000 per ha are planted with the spacing of 50 cm x 50 cm at a depth of 3-5 cm on the side of ridge, and half the distance from the top of the ridges. Half part of the slip / sett is allowed to remain in the air while the rest of portion is buried in soil. The planting is done at 45° slanting position on the ridges so that the lower end may not go deeper in the soil.

Cropping system: Three rows of hybrid grass and one row of desmanthes can be raised to increase the nutrient value of fodder crops. A combination of napier grass with berseem, lucerne or cowpea provides good quality palatable fodder for cattle. The important fodder based cropping systems for all the year round fodder production are furnished in Table 4.

Table 4 Fodder based cropping systems for all the year round fodder production

Location	Cropping system
Palampur (Himachal Pradesh)	NBH + velvetbean-berseem-sarson
Pantnagar (Uttarakhand)	NBH + subabul
Hisar (Haryana)	NBH + berseem / lucerne
Jhansi (Uttar Pradesh)	NHB + cowpea-cowpea-berseem
Anand (Gujarat)	NBH + *guar*-lucerne
Coimbatore (Tamil Nadu)	NBH + hedge lucerne

Weed management: First weeding can be taken on 20-25 days after sowing. Subsequent weeding is carried out wherever necessary. When broad leaf weeds pose a serious problem, application of 2,4-D @ 1 kg/ha is recommended.

Manures and Fertilizer application: Farm Yard Manure @ 12.5 tonnes per ha along with 50-50-40 of N, P_2O_5 and K_2O kg per ha are applied as basal dressing before planting and 100 kg N per ha as top dressing after each harvest.

Irrigation: First irrigation is given at the time of planting, life irrigation on the third day and subsequently once in 10 days. Sewage or waste water can be used for irrigation.

Crop protection: Lleaf blight damage caused by *Helminthosporium sacqhri* in some varieties.

Quartering: Quartering has to be done every year from third year onwards or whenever the clumps become closer and large. Quartering means three fourth of clumb is removed and one fourth is retained.

Harvest: The optimum stage for first cut is 60 to 65 days from the time of planting where maximum digestible dry matter can be obtained. In a year, 8-10 cuts are possible. The grass is to be harvested at a height of 15 cms from the ground level to avoid damage on growing points at the base of the clumb. Subsequent harvests are done at interval of 50 days. The hybrid once planted supplies fodder continuously and regularly for a period of three years. The cost of production is almost half that of single-cut crops.

Yield: A good crop yields about 150-300 t green fodder/ha/year.

Quality: The crude protein and fibre content of NBH fodder is 10 and 30% respectively. The fodder is rich in Ca (0.5%) and P (0.4%). The oxalate content of some of the varieties may be high. Ill effects of oxalates in this grass can be over come by (i) feeding 5 kg of leguminous fodder per day per animal along with these grasses, (ii) providing calcium, bone meal or mineral mixture to animals, or (iii) giving half litre of supernatant clear lime water daily along with the drinking water or sprinkling lime water on the feed or it can also be mitigated if harvested at longer intervals (45 to 60 days).

Utilization: The grass leaves are large and green, the sheaths are softer and the margins less serrated and hence the herbage is palatable. It is juicer and succulent at all stages of growth. It is less fibrous and more acceptable. This grass is coarse for hay making, but can be used for silage making. It is considered as a soil-restoring crop, as grass leaves the soil richer in organic matter.

CHAPTER

48

Para grass–*Brachiaria mutica* (Forssk.) Stapf

Family: Poaceae

Synonyms: *Brachiaria mutica* (Forsk.) Stapf, *B. purpurascens* (Raddi) Henr., *Panicum muticum* Forsk., *P. purpurascens* Raddi., *P. barbinode* Trin., *P. guadaloupense* Spreng. ex Steud., *P. spectabile.*

Vernacular name: Para grass, Water grass, Buffalo grass, Mauritius signal grass, California grass, Giant couch grass, Dutch grass, Bancroft grass, Scotch grass, African wonder grass. (English) and Erumaipul (Tamil).

Importance: Para grass is a semi-aquatic plant. It grows luxuriantly in swampy and nearly waterlogged lands. It is suited for sewage farm. It can be used for browse, silage and hay. It will not withstand heavy grazing. It is useful to reduce soil erosion along the banks of waterways. This grass has become the backbone of the beef industry in South and Central America, Australia, Philippines and Cuba.

Origin: Para grass is native to flood plains of sub-Saharan tropical Africa and Brazil.

Distribution: Para grass is manly cultivated in Africa, Central and South America, USA, India, Cambodia, Myanmar, Australia. In India, it is grown in Maharastra, Tamil Nadu, Karnataka, Bihar and Assam.

Botany: Para grass is a perennial stoloniferous (has horizontal stems, called stolons) grass. It grows to a height of 1 m. Culms are long, long-decumbent and rooting at the lower nodes, vertical height 90 to 200 cm; nodes villous. The lower portions of the stems have roots at swollen, hairy nodes and spread horizontally and then upwards. Lower sheaths with papillose-based hairs, more dense distally, margins ciliate; collars pubescent; ligules 1 to 1.5 mm; leaves / leaf blades 7.5 to 35 cm long, 4 to 20 mm wide, glabrous or sparsely pilose on both surfaces, margins scabrous. A row of short, stiff hairs called the ligule occurs at the junction of the leaf sheath and blade. The inflorescence grows as a primary axis and is arranged alltternate with numerous spikelets that are about 3 mm long with a purple tint. Panicles are 10 to 25 cm long, 5 to 10 cm wide, pyramidal, glabrous or with a few papillose-based hairs. It is a poor seed producer. This grass reproduces from seeds and from broken segments of stem (stolons) that can produce new roots at their nodes.. Caryopses 1.8–2 mm.

Climate: This grass is adapted to warm moist tropics and sub-tropics. It can be grown at elevations ranging from sea level to 1500 m but can persist at elevations up to 1200 m. Para grass does not persist on dry land in arid or semi-arid areas. It is grown in areas receiving an annual mean rainfall of 1000 mm to 2500 mm. It prefers naturally open, shallow, tropical freshwater wetlands and associated floodplains, swampy areas, coastal areas including the margins of rivers, creeks and lakes. It has stems that float over the water surface when growing along the banks of deep waterways. It can survive seasonal dry conditions by relying on the residual moisture of the marshy habitats. It is a shade tolerant grass and therefore suited to grow under coconut plantation. It has evolved hollow stems and large aerenchyma in its roots, which allows the roots to develop under low soil oxygen levels. This is presumably an adaptation to seasonal flooding. It cannot survive in permanent water deeper than 0.3 m–0.6 m. This grass prefers an annual temperature range of 18.7 C to 27.4°C. A mean annual temperature of 21°C is optimal for growth. The grass growth occurs with minimum temperature of 15.5°C. It is extremely frost intolerant but can regrow after. It does not tolerate temperatures below 8°C.

Soils: It has ability to grow in waterlogged or nearly saturated soils. It can be grown in slightly brackish water or sewage water. It prefers soil pH ranges from 4.4 to 5.5. However, it tolerates saline and alkaline soils.

Season: It can be planted throughout the year under irrigated condition.

Planting: This grass can reproduce sexually from seed, but more important is its ability to reproduce vegetatively by means of its above-ground creeping stolons. It is propagated by stem cutting or rooted slips. 40000 rooted slips are required for planting in one hectare. The slips are planted with a spacing of 50 cm × 50 cm.

Weed management: Weed free period for first two months of planting is required. Two-hand weeding are given on 20 and 35 days after planting. Subsequent weeding is carried out whenever necessary.

Manures and fertilizer application: FYM 12.5 tonnes per ha and 50-50-40 of N, P_2O_5 and K_2O kg per ha are applied as basal dressing. Subsequently, 50 kg N per ha is applied after each harvest.

Water management: First irrigation is given immediately after planting, life irrigation on third day after planting and subsequently once in 10 days. Para grass should be grown in high rainfall areas or in areas where irrigation water is available plenty.

Harvest: First cut is carried out on 75 days of planting and subsequently once in 50 days interval.

Yield: Green fodder yield of 75 to100 t/ha per year can be obtained. Seed production can be up to 30 kg/ha. A low germination rate recorded for fresh seeds is attributed to seed dormancy. Germination of the fresh seed improves after 6–8 months storage. Germination rates after 2 months storage range from 51–57%.

REFERENCES

Bogdan, AV 1977, Tropical Pasture and Fodder Plants, Longman Inc., New York, United States.

Daehler, C 2005, Urochloa mutica—risk assessment, http://www.hear.org/pier/wra/pacific/Urochloa_mutica_htmlwra.htm

Dhar S, Gupta SD, Singh A and Arya RL 2001, Performance of grasses with cutting management under seasonal waterlogged conditions. Indian Journal of Agricultural Sciences, vol. 71, pp. 698–700.

Global Biodiversity Information Facility (GBIF). 2011. Urochloa mutica (Forsk.) T.Q. Nguyen, http://www.gbif.org/

Grof, B. 1969, 'Viability of para grass (Brachiaria mutica) seed and the effect of fertilisér nitrogen on seed yield', Queensland Journal of Agricultural and Animal Sciences, vol. 26, pp.271–276.

PIER, 2006. 'Urochloa mutica', Pacific Island Ecosystems at Risk, viewed 17 May 2006, <http://www.hear.org/pier/species/Urochloa_mutica.htm>.

Martin Hannan-Jones and Steve Csurhes. 2012. Para grass. Queensland Department of Agriculture,Fisheries and ForestryGPO Box 46, Brisbane 4001.

CHAPTER

49

Congo Signal Grass–*Brachiaria brizantha* (Hochst ex A. Rich) Stapf (2n=18)

Family: Poaceae

Vernacular names: Congo signal grass, Palisade grass, Ceylon sheep grass (E).

Origin: It is native to East Africa.

Distribution: It is an exotic grass, and has been introduced in India in 1950 from Australia. In India it is found in coastal regions.

Botany: It is tufted, semi-erect, spreading, 1.20 m high, drought-resistant and perennial grass. Leaves are deep green. The seeds remain viable for about 3 years.

Climate: It requires warm and humid climate.

Soils: This grass thrives well on loamy soils.

Seeds and sowing: The rooted slips with 2-3 nodes are transplanted. The pasture can also be established by direct seedlings in lines at 30 cm apart at the onset of monsoon season in dryland conditions and any time in the irrigated conditions as well as in heavy rainfall regions.

Weed management: Weed free period for first two months of planting is required. Two-hand weeding is given on 20 and 35 DAS. Subsequent weeding is carried out whenever necessary.

Manures and fertilizer application: FYM 12.5 tonnes per ha and 50-50-40 of N, P_2O_5 and K_2O kg per ha are applied as basal dressing. Subsequently, 50 kg N per ha is applied after each harvest.

Harvest: It can be heavily grazed if used as a monospecific sward and regularly fertilized with nitrogen. If grown with legumes, the grazing system must favour the legume and adequate phosphorus must be maintained.

Yield: Green fodder yield of 75 t/ha per year can be obtained.

Utilization: The grass remains green for major part of the year even under the rainfed conditions. It is much relished and is quite palatable.

CHAPTER

50

Cenchrus Grass (2n=36)

Cenchrus ciliaris L.	–	Buffel grass (E)
Cenchrus setigerous Vahl	–	Bird wood grass (E)
Cenchrus glaucus	–	Blou buffel (E)

Vernacular names:

(i) *Cenchrus ciliaris* is known as buffel grass, foxtail buffalo grass, blue buffalo grass, Rhodesian foxtail grass, African foxtail grass (English); White kolukattai (Tamil); cenchrus cilié (French); Büffelgras (German); zacate buffel (Spanish - Mexico); capim-búfel (Portuguese - Brazil), buffel grass (Australia), African foxtail (U.S., Kenya) Dhaman grass and Anjan grass (India).

(ii) *Cenchrus setigerus* is known as bird wood grass, cow sandbur (English); *Dhaman* (Hindi), *Black Kolukattai* (Tamil).

(iii) *Cenchrus glaucus* is known as Neela kolukattai (Tamil), Blou buffel (E).

Importance: Cenchrus is an excellent grazing grass grown in dry areas in tropics and sub-tropics. It is more palatable than other grasses of the desert regions and maintains animal acceptance even at maturity. It is a drought resistant grass. It is dormant during summer months and quickly regenerates and produces enough canopy to cover the soil on receipt of monsoon rainfall. It is a good soil binder and used for soil conservation. Buffel grass contains oxalates which can cause a calcium deficiency in stock. It is preferentially grazed when green and actively growing. It is suited for hay making.

Origin: Cenchrus grass is a native of tropical and subtropical Africa, India and Indonesia. However, blou buffel grass is a native of Australia.

Distribution: It is widely naturalised in sub-humid and semi-arid tropics and subtropics. It is widely distributed in Australia, South America, Africa, Arabia and India. It is found in open bush and grassland in its natural habitat. It is widely distributed in the plains of Rajasthan, Haryana, Gujarat, Punjab, western Uttar Pradesh, Maharashtra, Andhra Pradesh and Tamil Nadu extending up to foot hills of Jammu up to an altitude of 400 m.

Botany: Cenchrus is a tufted, tussock forming, erect, annual or perennial grass. The culm grows to a height of 60 to 120 cm1–2 mm in diameter, wiry or sometimes almost woody at base, sometimes many-branched from lower or basal nodes. Plants are rhizomatous in nature which helps in drought tolerance. The sheaths are keeled, glabrous or sparsely pilose; ligule 0.5–2.5 mm long; blades 3–25 cm long, 2–13 mm wide, apex caudate. The inflorescences are paniculate, gray, purple, or yellowish, densely cylindrical to ovoid, 2–14 cm long, 1–2.6 cm wide, burs 6-16 m long, spikelets 1–4 per bur, 2–5.5 mm long, dorsally compressed and lanceoloid. Caryopsis is turgid, ovoid, 1.4–1.9 mm long and 1 mm in diameter.

Climate: It is suited for arid and semiarid regions. It is grown from about 33ºS to 35ºN. It is cultivated from sea level to 2,500 m from MSL. It can be grown in areas with annual mean rainfall of 300 mm to 1250 mm. It is highly drought tolerant. It does not survive prolonged waterlogging, particularly in cold season, but can stand up to 5 days of flooding with negligible adverse effect. The losses of 15-70% occur after 20 days of flooding. It has maximum growth at 30°C/25°C (day/night temperature). The growth rate falls sharply when the temperature is below 18°C/13°C, with slow growth below 15°C/10°C. In general, it performs best in areas where mean minimum winter temperatures are >5ºC. It withstands high temperatures.

Soils: It can be grown over a wide range of soils light-textured sandy soils, gravelly areas, alluvial flats, heavy black clays and calcareous soil. It is tolerant to salinity and alkalinity.

Land preparation: Two to three ploughings may be given to ensure good tilth.

Season: It is sown during June-July and September-October seasons under rainfed conditions.

Varieties:

Cenchrus ciliaris varieties: CO 1, Mallapo, Buttel, IGFRI No.3108, 3132, CAZRI-358, Marwar Anjan (CAZRI-75), Pusa giant, Australian S-3106, Australian S-3108, Cagri-75, Molopo, Gyandah, Bundel Anjan-1, Bundel Anjan-3 (IGFRI- 727).

Cenchrus setigerus varieties: Marwar Dhaman (CAZRI-76), Marwar Dhaman (CAZRI-175), Pusa yellow - Anjan, 175 and 415.

Seeds and Sowing: Seed rate is 2 to 3 per ha. Spacing is 45 cm between rows with continuous line sowing or broadcasting. Sowing depth is one cm. Care to be taken for seeds not to be blown away by wind while sowing. Seeds are viable in the soil for up to 4 years. Fresh seed often has high level of dormancy. Germination rate can be improved with storage of 6-18 months after harvest or by separating caryopses. Total live seed content is commonly 30-50%. Six weeks old seedlings/rooted slips can also be transplanted at 50 cm row spacing and 30 cm plant to plant. Thus about 33,000 seedlings or rooted slips are required for one hectare area with 2 seedlings at each spot.

Cropping system: Cenchrus is mixed with centro and siratro, in 3:1 ratio. It grows well in a subabul, hedge lucerne alleys. For quality forage after every two rows of grass spaced at 50 cm, one row of legume viz., Siratro or Caribbean stylo can also be planted.

Weed management: First weeding may be given after 30 DAS and subsequently as and when necessary since it is a perennial crop.

Manures and fertilizer application: Apply FYM @ 12.5 tonnes per ha, 25-40-20 kg of N, P_2O_5 and K_2O per ha as basal dressing and 25 N per ha as top dressing after each harvest.

Water management: Cenchrus grass is mainly grown as rainfed crop

Harvest: The grass growth is slow to establish and the animals are allowed in the field for grazing is to be delayed 4-6 months after sowing, and up to 9-12 months, depending on establishment conditions. Since quality declines rapidly with age, should be cut or grazed at least every 8 weeks. Once established, it can withstand heavy grazing. It can carry up to 1 steer or 6 sheep/ha, depending on rainfall and soil fertility. The pasture is more productive during 2nd to 4th year, afterwards for its sustained production the pasture should be burnt moderately or ploughed. The grass should be allowed to seed every two to three years to thicken the stand. Generally first cut is given on 70 to 75 DAS and subsequently 4 to 6 cuts after 30-45 days intervals depending on the rainfall distribution. Leafiness is maintained by low cutting at about 7-10 cm.

Yield: Green forage yield of 9 to 10 t/ha and dry matter yield 4–5 t/ha per year can be obtained under rainfed pasture condition. It matures in eight weeks and seeds heavily. Seed yield is around 125 kg seeds/ha. The oxalate levels > 2% can cause 'big head' (*Osteodystrophia fibrosa*) in horses and sheep. 1000 seed weight is 2.18g.

REFERENCES

Winkworth, R.E. 1971. Longevity of buffel grass seed sown in an arid Australian range. Journal of Range Management 24(2):141-144

Venter, P.S., and N.F. Rethman. 1992. Germination of fresh seed of thirty *Cenchrus ciliaris* ecotypes as influenced by seed treatment. Journal of Grassland Society of South Africa 9(4):181-182

Ward, J.P., S.E. Smith, and M.P. McClaran. 2006. Water requirements for emergence of buffelgrass (*Pennisetum ciliare*). Weed Science 54(4):720-725

Rao, A.S., Singh, K.C. and J.R. Wight. 1996. Productivity of *Cenchrus ciliaris* in relation to rainfall and fertilization. Journal of Range Management 49:143-146

Sharif-Zadeh, F. and A.J. Murdoch. 2000. The effects of different maturation conditions on seed dormancy and germination of *Cenchrus ciliaris*. Seed Science Research 10:447-457

Franklin, K.A., K. Lyons, P.L. Nagler, D. Lampkin, E.P. Glenn, F. Molina-Freaner, T. Markow, and A.R. Huete. 2006. Buffelgrass (*Pennisetum ciliare*) land conversion and productivity in the plains of Sonora, Mexico. Biological Conservation 127:62-71

http://www.pi.csiro.au/ahpc/

https://hort.purdue.edu/newcrop/duke_energy/Cenchrus_ciliaris.html

http://www.buffelgrass.org/

http://www.tropicalforages.info/key/Forages/Media/Html/Cenchrus_ciliaris.htm

CHAPTER

51

Spear Grass–*Heteropogon contortus* (Linn.) P. Beauv. (2n=20)

Vernacular names: Spear grass, Tangle head grass, Black spear grass, Bunch spear grass, Tangle grass, Assegai grass, Pili grass (English), Lampa, parwa (Hindi), Orsi Pillu (Tamil), Paretu Mullu gaddi (Telugu), Sankari hull (Kannada), Sukli (Gujarati).

Importance: Spear grass is a good forage grass when young. Its young herbage can be conserved as hay or silage. It is planted for erosion control and revegetation of degraded habitats. It is used as thatch for huts and is commonly woven into mats. The cellulose content of the grass is quite high and it has been examined for pulping for paper manufacture.

Origin: It is native to the tropics and subtropics of Africa, southern Asia, northern Australia and parts of Oceania.

Distribution: It is grown in Australia, India, South America and the United States. In tropical Africa it is commonly found from Cape Verde, Mauritania and Senegal eastwards to Eritrea and Somalia and southwards to Angola, Mozambique, South Africa, and in the Indian Ocean islands. It is indigenous to India and occurs in all arid and semi-arid regions and upto an elevation of 2000 m from Himalaya to Cape Camorin and in the grasslands of east to west and whole of the south of the country.

Botany: It is densely tufted, perennial and highly palatable, 0.9 to 1.0 m tall, erect or decumbent grass. It is leafy mainly at base. Leaves are firm, linear up to 60 cm long and 3.7 mm broad, often hairy with bulbose base. Racemes are terminal, erect, 4.8 cm long with prominent dark brown awns (3-12 cm long) which are jointly twisted together to form a bundle at maturity. Sessile spikelet 7 mm long, hidden by the pedicelled spikelets. Caryopsis is cylindrical, 3.5 to 4.5 mm long, grooved, whitish, 60,000 caryopses/kg; şeeds with awns 140,000/kg, de-awned 510,000 seeds/kg.

Climate: It is found all over the world in Tropics and subtropics. Its natural distribution lies between 30ºN and 30ºS. It is grown up to an altitude of 1,300 m from MSL. It grows well in arid and semi-arid conditions in the rainfall zones ranging from 600 to 1200 mm. It does not tolerate long periods of flooding, but can tolerate a few days' inundation on otherwise well-drained soils. It does not tolerate waterlogging. The optimal temperature for germination lies between 20 and 35°C.

Soils: It is highly variable and adaptable to all types of soils ranging from pure gravelly sand to sandy loam. In drier areas it dominates even by suppressing other grasses especially on poor and rocky soil.

Land preparation: Being hardy it does not require well prepared land and even comes in the land once ploughed.

Seeds and sowing: The seeds are broadcasted @ 5.0 kg/ha just before the monsoon. It can be planted from seeds or vegetatively. Seed should be sown no deeper than about 1 cm, preferably into a fine, firm, clean seedbed, and rolled after sowing.

Fertilizer application: Application of 60 kg N/ha as basal dose increases its productivity.

Harvest: First harvest can be taken after three months from pastures raised through seeds and subsequently at an interval of 60 days and 15 cm cutting height for quality forage in pastures.

Yield: The green fodder yield of about 15 to 20 t/ha of can be obtained with 6 cuttings per year under rainfed conditions. The dry matter yields is up to 4 to 5 t/ha/year.

Utilization: It is very good fodder grass due to presence of sharp awns or spears at maturity. It is grazed accordingly to its carrying capacity or may be cut at pre-flowering stage for hay or silage. It is also used in soil conservation programme.

REFERENCES

Cooksley, D.G., Butler, K.L., Prinsen, J.H. and Paton C.J. 1988. Influence of soil type on *Heteropogon contortus-Bothriochloa bladhii* dominant native pasture in south-eastern Queensland. Australian Journal of Experimental Agriculture 28, 587–591.

Grice, A.C. and Mcintyre, S. 1995. Speargrass (*Heteropogon contortus*) in Australia: Dynamics of species and community. The Rangeland Journal, 17, 3–25.

Tothill, J. C. 1966. Phenological variation in *Heteropogon contortus* and its relation to climate. Australian Journal of Botany 14, 35–47.

Tothill, J. C. and Hacker, J. B. 1976. Polyploidy, flowering phenology and climatic adaptation in *Heteropogon contortus* (Gramineae) Austral Ecology 4, 213–222.

CHAPTER

52

Saen Grass–*Sehima nervosum* (Rottl.) Stapf. (2n=20, 34, 40)

Vernacular names: Rat's tail grass, white grass, Saen grass (E), *Sain, Poona and Suekai* (H), *Sheda* (Marathi), *Karaitoi, Vennai Pillu* (Tamil), *Mendra gaddi* (Telugu) and *Sinnaisphadai hullu* (Kannada).

Origin: The generic name is derived from the Arabic 'saehim', local name of the type species in its native Yemen.

Distribution: It is widely spread in south-east Asia, Sudan, China, Australia and East Africa. In India it is cultivated in Madhya Pradesh, Uttar Pradesh, Gujarat, Rajasthan, Maharashtra, Karnataka, Andhra Pradesh, Tamil Nadu and from sea level to 1830 m elevation.

Botany: It is a perennial grass forming dense tufts with numerous tillers, upto 1 m and above in height. Stems/culms are erect, hollow, slender, pale straw yellowish and bright on ripening. It has abundant and soft foliage. Leaves are 15-40 cm long, 0.8 to 1.5 cm wide with linear leaf blade. Racemes are solitary 7.5 to 15.0 cm long and erect or slightly flexed. Both sessile and pedicelled spikelets are awned. The awns are slender and twisted at base.

Climate: It is cultivated up to an altitude of 2750 m from MSL. It prefers hot and dry climate in the rainfall zones of 500 to 1500 mm with an optimum up to 1000 mm.

Soils: It grows on lava and on black, seasonally waterlogged clays in Africa, and on lateritic red earths in northern Australia. The suitable soil type is brown, dark gray gravelly with good drainage. It grows well on loamy sands with a pH of 6.5 in India, but grows best on black soils. It is also seen on rock crevices of undulating topography and on hill slops. The soil pH of its habitats is 6.5.

Variety: Bundel sain Ghas-1 (IGS 9901)

Seeds and sowing: Seed rate is 6 kg/ha in lines at 50 cm apart or broadcasting or transplanting of seedlings/rooted slips at the onset of monsoon. Transplanting of seedlings is successful and about 1.33 lakhs seedlings are required for 1 hectare.

Weed management: Weeding is required at least twice in first year during growing season and this can be done in the form of interculturing between rows of grasses.

Manure and fertilizer application: FYM @ 12.5 tonnes per ha and 40-20-20 kg per ha N, P_2O_5 and K_2O are applied as basal dressing under rainfed conditions and subsequently 25 kg of N per ha after each harvest.

Cropping systems: Saen grass is compatible with *Clitoria,* Siratro and Carribean stylo. For every two rows of grasses one row of legume can be maintained in mixed stand.

Harvest: It is totally a rainfed crop. In first year only one cut should be taken. Meanwhile seed collection should be carried out. In subsequent years two cuttings may be taken depending upon the pattern of rainfall distribution.

Yield: It gives average production of 10 to 15 t/ha green fodder and 4.0 t/ha dry matter. Seed yield is 110 kg/ha.

Utilization: It is a good forage grass and may be utilized for grazing as well as for hay too.

CHAPTER

53

Deenanath Grass–*Pennisetum pedicellatum* Trin

Family: Poaceae

Vernacular names: Kaysuwa (E), Deenanath, Dinanath grass (H), Kyasuwa grass (Nigeria), desho, desho grass (India).

Importance: Deenanath grass is very palatable to livestock at all stages of its growth. The grass has good ability to spread naturally by self-seed set; regenerating each year and therefore has a tendency to become a notorious weed in certain areas. The grass is sometimes used as cut and carry fodder. It is suited for silage and hay making.

Origin: It is a native of Ethiopia.

Distribution: It is grown in India, Malaysia, Philippines, Thailand, Australia, Fiji and USA. Its habitat is drier sites, savannahs and woodland margins, a weed in croplands, grasslands, waste places. In India, it is cultivated in Bihar, Orissa, West Bengal, Haryana, Punjab, Madhya Pradesh and Uttar Pradesh.

Botany: Deenanath grass is an erect or geniculate, annual grass of 0.7-0.9 m height with a hollow stem. Culm is bright with light reddish at base. It tillers profusely with long leaf. Leaves are 45-60 cm long and light to dark green in colour. Inflorescence is pink in beginning but becomes white at maturity. It sets seeds profusely. Deenanath grass is not a rhizomatous plant.

Climate: It is grown in semi-arid and sub-tropics. It prefers warm climate and is found in regions of rainfall ranging from 800 to 1250 mm. Optimum rainfall for grass growth is 500-650 mm but can grow well and produce seed in even less rainfall. The optimum temperature for growth is 30-35°C. However, it has little frost tolerance.

Soil: It can be sown in all types of soil with good drainage. It grows well on fertile loam soil but can be grown in sandy soils. The grass can tolerate both acidic and alkaline soils. It does not come up well on heavy clay soil or flooded or waterlogged conditions.

Season: The crop can be mostly raised in June-July and October under rainfed conditions.

Land preparation: The field should be disced and ploughed two to three times to obtain good tilth and form beds and channels.

Varieties: Pusa Deenanath grass, Pusa 3, COD-1 grass, PS-3, IGFRI-4-5-1, IGFRI-4-3-1, Bundle Dinanath-1, Bundle Dinanath-2, Jawahar Pennisetum-12.

Seeds and sowing: The crop is generally sown with the onset of rain. Seed rate is 2.5 kg per ha. Break the seed dormancy by mechanical scarification in a defluffer followed by soaking in a mixture of GA3 (200 ppm) and KNO_3 (0.25%) (1:1) for 16 hours. Pellet the seed with DAP @ 60 g/kg and arappu leaf (*Albizzia amara*) powder @ 500g/kg of seed to enable easy handling of seed during sowing and also for good establishment. Slurry treatment of the seeds can be done with carbendazim @ 4g kg^{-1} of seed. The seeds can be sown with 30 × 10 cm or row spacing of 30 cm and continuous line sowing or drilled at 1 cm in rows 40-50 cm apart or can be broadcasted. Before sowing the soil should be mixed with seed. Sowing is usually done just before rainy season when one or two good showers are received. The grass comes every year by self-seeding. The pasture may be established by transplanting six weeks old seedlings at 50 cm distance from row to row and same for plant to plant. Thus 33000 seedlings are needed for one hectare, planting 2 seedlings per hill.

Weed management: Hoeing and weeding on 30th day after sowing. The grass should be cut in bloom stage to save the field from weeds. During first year one or two weeding in growing season help in better establishment of pasture.

Manure and fertilizer application: FYM @ 12.5 tonnes per ha and 40-20-20 kg per ha N, P_2O_5 and K_2O are applied as basal dressing under rainfed conditions and subsequently 20 kg of N per ha is top-dressed at one-month crop stage or after each harvest.

Water management: The crop is sown in pasture or forest lands as rainfed crop. It is usually sown during the onset of rain. Two or three protective irrigations are required depending upon the rainfall distribution. Irrigation is once in ten days or depending on soil condition.

Harvest: Harvest is done on 55-60th day after sowing. The grass comes to flowering stage in 90 to 100 DAS. The maturity period is 120–130 days. It should be cut before flowering or otherwise it will set seeds and fall over field. It can withstand several cuts per year for green fodder production. The grass is quite agreeable to the cattle both in green and dry stages.

Yield: Deenanath grass gives green forage yield of 15 to 20 tonnes per ha per year under rainfed conditions and 25-30 t/ha under irrigated condition. It has been found to be quite agreeable to the cattle both in green and dry stages. The average leaf stem ratio is 1.27. It is an annual crop but being profuse seeder. The seed yield is up to 300 to 400 kg/ ha. The germination is higher for the seeds from the first and second formed tillers. The delayed harvesting resulted in seed shattering loss. The middle and proximal portions of the spike produce high quality seeds. Store the seeds in gunny or cloth bags for short term storage (8-9 months) with seed moisture content of 8 to 10%. Store the seeds in polylined gunny bag for medium term storage (12- 15 months) with seed moisture content of 8 to 9 %. Store the seeds in 700 gauge polythene bag for long term storage (more than15 months) with seed moisture content of less than 8%.

Utilization: It is used as fodder crop and is relished by all kind of livestock for grazing.

REFERENCES

http://www.cabi.org/isc/datasheet/39769

http://agritech.tnau.ac.in/agriculture/foragecrops_deenanath.html

http://www.fao.org/ag/agp/AGPC/doc/gbase/DATA/pf000299.htm

CHAPTER

54

Marvel Grass–*Dichanthium annulatum* (Forsk) Stapf (2n=20)

Vernacular names: Marvel grass, Blue stem grass, Delhi grass, (English), *Kail, Kared, Apang* (Hindi), *Chhijhavo* (Gujarati), Sheda grass (Australia), Kleberg blue-stem (United States).

Importance: Marvel grass can be used in pastures, in cut-and-carry system or for hay-making or silage making if it is cut before flowering. It can stand very heavy grazing and support 7 sheep per ha in India.

Origin: Marvel grass is native of India and Africa.

Distribution: It is distributed from tropical Africa to south-east Asia, New Guinea, Australia. India, China, Myanmar, Thailand, Indonesia, Malaysia, Nepal, Pakistan, Kenya, Tanzania, Uganda, Ethiopia, Somalia, Algeria, Egypt, Libya, Morocco, Tunisia, Botswana, Namibia, South Africa, Mali, Senegal, Cape Verde, Afghanistan, Iran, Iraq, Israel, Jordan, Turkey, Bahrain and Saudi Arabia. It grows on the plains and up to 900 m on hills in India except in northern mountains.

Botany: Marvel grass is an erect tufted, fine stemmed, perennial grass and 60-100 cm high. It has deep root system to a depth of one metre. Culms are purplish red or bluish in colour, distinct rings of whitish hairs at each node. Leaves are green to bluish green, 23-45 cm long. Inflorescence is a compound raceme, made of a cluster (2-8) of purplish false spikes, arising nearly from the terminal tip of the culms. Each false spike is a raceme of paired spikelets, one sessile and the other pedicelled. The absence of pits on the glums is observed in its spikelets. The seed is a 2 mm long, oblong-obovate caryopsis.

Climate: This grass is mainly found within 8-28°N latitude at elevations between sea level and 1500 m in dry to moist subtropical and tropical areas. It grows well in the areas of 350 to 2000 mm rainfall of arid and semi-arid regions. Optimal growth conditions are annual rainfall ranging from 500-900. The optimum temperature for germination is 32°C, although germination can be achieved between 15 and 40°C.

Soils: It can grow on a wide range of soils but moist, well drained, medium black or red alluvial soils are preferred. It can tolerate a fair degree of drought, salinity and alkaline but does not thrive on acidic soils.

Season: It grows during the wet season from June to November in India and after harvest in November for hay. It provides spring growth from February to March, but this growth is stemmy.

Varieties: IGFRI 495-1, Marvel 7, Marvel-8, Marvel 93, GMG-1 (Gujarat Marvel Grass-1), CAZRI 490 and CAZRI 485.

Land preparation: Two to three ploughings with iron plough or country plough should be given. Form beds and channels. Field preparation can be done utilising summer showers.

Seeds and Sowing: Seed rate is 3 to 4 kg per ha. The seeds can be sown with spacing of 45 cm between rows and continuous line spacing or can be broadcasted. Five weeks old seedlings/rooted slips can also be transplanted in a drizzly day at 50 cm row spacing and 30 cm plant to plant.

Weed management: One or two interculturing operations, depending on weed infestations are essential during first year after sowing.

Manure and fertilizer application: FYM @ 12.5 tonnes per ha and 40-20-20 kg per ha N, P_2O_5 and K_2O are applied as basal dressing under rainfed conditions and subsequently 25 kg of N per ha after each harvest.

Water management: It is mostly raised as rainfed crop.

Cropping systems: Siratro or stylos are the appropriate legumes and may be maintained in 2:1 ratio i.e. two rows of grass and one row of legume.

Harvest: In first year it should be harvested once after seed shedding while in subsequent years 3 to 4 cuttings can be taken at an interval of 60 days and 15 cm cutting height for quality forage.

Yield: This grass gives high green fodder yield even under rainfed condition and dry matter yield ranges from 2.5 to 4.0 t/ha in single cut while 20 to 30 t/ha/year. It can give 3 times more forage under irrigation. Nutritive value: Its crude protein content varies from 5 to 7%.

Utilization: It is one of the best grasses for forage and is utilized as hay as well as for grazing purpose. This is also good for soil conservation.

REFERENCES

http://www.fao.org/ag/AGP/AGPC/doc/Gbase/data/pf000213.htm
http://www.hear.org/pier/species/dichanthium_annulatum.htm
https://hort.purdue.edu/newcrop/duke_energy/Dichanthium_annulatum.html

CHAPTER

55

Bermuda Grass–*Cynodon dactylon* Pers.

Family: Poaceae

Vernacular name: Bermuda grass, Devil grass, Star grass (E), Doob, Hariyali (Hindi), Durba (Sanskrit), Arugampul (Tamil), Garike hullu (Kanada), Dhoorva (Marathi), Garicha gaddi, Garike, Thella gariki (Telulgu), Darodi (Gujarati).

Importance: It is of high nutritional value, excellent palatability and withstands drought. It is a high-yielding, sod-forming grass that is well-suited for grazing or hay production. It is a worst type of farm weed.

Origin: *Cynodon dactylon* is native of India while Cynodon *plectostacycon* is a native of Africa.

Distribution: It is distributed throughout the world in tropics and sub-tropics.

Botany: Cynodon grass is a fine leafed perennial grass forming a dense mat. It is a stoloniferous and rhizomatous perennial grass with C4 metabolism. It grows to height of 30 to 70 cm.

Climate: It is well suited for semiarid tropics. It grows well with mean annual rainfall of 500 mm to 1500 mm. Growth is reduced at temperatures less than 10°C. If seeded, this grass requires daily mean temperatures above 18°C for proper germination. It is a drought tolerant grass.

Soil: It is well suited to all types of soils. It is tolerant to saline and alkaline soils.

Season: Seeds are sown in June-July and September–October seasons under rainfed conditions and throughout the year under irrigated conditions.

Seeds and Sowing: Seed rate is 3 to 4 kg per ha. The seeds can be sown with spacing of 45 cm between rows and continuous line spacing or can be broadcasted. It is also propagated by the freshly dug sprigs (stolons and rhizomes). Mature top growth (clippings or tops or culms) can be used to establish stands. The sprigs are to be planted at a rate of 1.30 to 1.73 m^3 ha^1 (0.035 cubic meters = 1.25 cubic feet = 1 bushel) for proper establishment. If not broadcast, sprigs should be planted every 45 to 60 cm in rows 90 to 120 cm apart.

Weed management: One hand weeding may be given on 25 days after sowing.

Manure and fertilizer application: FYM @ 12.5 tonnes per ha and 40-20-20 kg per ha N, P_2O_5 and K_2O are applied as basal dressing under rainfed conditions and subsequently 25 kg of N per ha after each harvest.

Water management: It is mostly raised as rainfed crop.

Harvest: Generally the animals are allowed for grazing. This grass is well adapted to grazing because of its prolific rhizomes and stolons and can withstand frequent defoliation. Improved varieties should not be grazed any lower than five cm and until height is at least 15 cm. Cuttings every four to six weeks during the growing season are adequate.

Yield: Green forage yield of 7 to 10 tonnes per ha per year can be obtained under rainfed conditions.

CHAPTER

56

Blue Panic Grass–*Panicum antidotale* Retz. (2n=18, 36)

Family: Poaceae

Vernacular names: Blue panic grass, giant panic grass, perennial sudan grass, Gramna (E), Bansi (H).

Origin: It is native to temperate and tropical Asia (in the Middle East to India, Afghanistan, Iran, Yemen, India, and Pakistan.

Distribution: It is widely distributed in arid and semi-arid regions of Afghanistan and Persia. It has been introduced into the United States from India, via Australia. It grows on sand dunes and dry river beds in Pakistan, Afghanistan, and Iran. It is grown in Rajasthan and Uttar Pradesh in India.

Botany: The whole plant gives a bluish appearance. It is a thin stemmed, tufted perennial grass with smooth and solid culms. It has fibrous roots growing to 45 cm deep, with thick knotty, bulbous rhizomes. Rhizomes are robust, coarse, short, and bulbous. Leaves are 45 cm long. Inflorescence is a panicle which is erect, with branches widely spreading and flexuous, mostly open to contracted, ovate to elliptic, 13-30 cm long, dense clusters of spikelets from branch base to tip; spikelets 1-2 per node. Inflorescence is panicoid panicle, terminal, loose and pyramidal. It is profuse seeder and seeds mature and shed easily in acropital manner.

Climate: The grass is highly adapted and grows well in dry areas to low hills. It is grown in areas receiving rainfall up to 1000 mm is quite favourable for its growth. It tolerates temporary flooding. A germination percentage of 70% can be observed with average minimum temperatures of 17°C and average maximum temperatures of 46°C.

Soils: It is grown on a variety of soils especially light sandy soil and heavy clay loam soil. It can be grown saline and alkaline soils within a pH range of 4.0 to 11.5.

Seeds and sowing: It is propagated by seed, either drilled in rows or broadcast or by line sowing of seeds. Sow just before the expected rainy season at 6-7 kg/ha broadcast, or 1.25 kg/ha in rows 1 m apart. It is sown in rows 45 cm or 50 cm apart. When drilled, cover not more than 1 cm; when broadcast, sow on surface and, if possible, give a light cover. Under irrigated conditions rooted slips may be planted in furrows at 50 x 50 cm apart.

Weed management: One or two weeding are enough for enhancing yield. Monthly irrigation is required for irrigated pastures.

Manures and fertilizer application: Application of 12.5 t of FYM along with 30 kg P_2O_5 as basal dose and 20 kg N/ha after each cutting is recommended for optimum forage production.

Harvest: Under rainfed conditions first cut of the grass can be taken in four months after seed sowing and subsequent cutting after 2 months, thus 4-6 cuttings in a year could be harvested, while under irrigation, fortnightly cutting is possible.

Yield: Under rainfed it gives 5.0 t/ha green forage in 4-6 cuttings. Seed yield is 100 kg/ha under rainfed and 250 kg/ha under irrigated condition.

Utilization: The grass should be cut and fed for forage but it can also be used as pasturage. It is very fast in regrowth.

REFERENCES

Kumar, A., and I.P. Abrol. 1982. Relative tolerance of grasses to sodic soils. Indian Farming 32:41-43.

Parihar, S.S., P. Agarwal, and V. Shankar. 1999. Seed production and seed germination in blue panic grass (*Panicum antidotale* Retz.). Tropical Ecology 40(1):75-78

Pathak, P.S., P. Rai, and R. Debroy. 1983. Comparative growth analysis of *Panicum antidotale* Retz. cultivars. Indian Journal of Ecology 10(1):141-143

Trew, E.M. 1954. Blue panic grass. Texas Agricultural Experiment Station Bulletin B-245

Uvalde Research and Extension Center. 2000. Blue panicum (*Panicum antidotale* Retz.). Texas A&M University System, Research and Extension Center Virtual Herbarium. Website: http://uvalde.tamu.edu/herbarium/paan.htm

Wright, L.N., and A.K. Dobrenz. 1970. Water use in relation to management of blue panicgrass (*Panicum antidotale* Retz.). Journal of Range Management 23:193-196.

http://www.halophyte.org/pdfs/drkhan_pdfs/202.pdf

CHAPTER

57

Golden Timothy Grass–*Setaria sphacelata* Stapf. ex Hubb (2n=18)

Vernacular names: Setaria grass, Golden timothy grass, Golden bristle grass, golden bristle grass, golden millet, African pigeon grass, Rhodesian grass (E) and *Nandi* (H).

Importance: Setaria grass is used in medium rainfall subtropical regions as a pasture grass for or cut and carry or grazing of dairy and beef cattle, as well as for silage and hay. It forms stable groundcover for soil conservation. The name 'Setaria' is derived from Latin 'Seta' which means the numerous bristles on the inflorescence.

Origin: It is native to tropical and southern Africa.

Distribution: It is cultivated in Australia, Queensland and USA. In India, it is grown in Himachal Pradesh and Uttaranchal.

Botany: It is an erect, densely-tufted, bunchy perennial grass growing to a height of 1.2 to 1.5 m with flattened culms. Leaf blades are flat, 30-45 cm long, 6-10 mm wide, linear and lanceolate. Infloresecence is terminal, compressed panicle about 15 cm long, appearing as a dense cylindrical spike and orange to purple in colour. It is largely cross-pollinated crop. Seeds count 1.4 to 1.7 million.

Climate: It is grown in tropical and subtropical countries from 29°N to 27°S latitude. It is found in its native environment from sea level to 3,300 m, most commonly between 600 and 2,700 m asl. It is grown in areas receiving yearly rainfall of 750 to 1800 mm. The optimum temperature for growth and yield ranges from 18 to 25ºC. It can also survive long, hot and dry seasons. It is more tolerant of waterlogging and flooding than many tropical grasses. It is drought and cold tolerant.

Soils: It thrives well on soils with texture ranging from sand to clay loam and light clay, but will grow on heavy clay. It is adapted to alkaline or acid soils with pH of 5.5-6.5.

Season: The crop can be taken during March-December. It provides fodder during the lean periods viz., April-June and October-November.

Varieties: Setaria 92, PSS 1 (Golden timothy), Nandi, Kazungula, Narok.

Land preparation: Two to three ploughings with iron plough or country plough should be given. Field preparation can be done utilising summer showers.

Seeds and sowing: Seed rate is 1.5 kg/ha for mixed pasture. It can be established by planting seedlings/rooted slips in furrows of 50 cm apart and 30 or 50 cm distance from plant to plant at the onset of monsoon.

Weed management: One or two interculturing operations, depending on weed infestations are essential during first year after sowing.

Manure and fertilizer application: FYM @ 12.5 tonnes per ha and 40-20-20 kg per ha N, P_2O_5 and K_2O are applied as basal dressing under rainfed conditions and subsequently 25 kg of N per ha after every year.

Water management: It is mostly raised as rainfed crop.

Cropping systems: It sustains the forage supply round the year if intercropped with legumes such as stylos and siratro.

Harvest: It is ready for first cutting after 3 months of planting (sowing) and subsequently it can be harvested at monthly intervals in a irrigated and properly manured pasture. The cutting height is 5 to 10 cm from ground level. It remains green for 9-10 months in a year and provides 3–4 cuttings.

Yield: The average green fodder yield and dry matter yield is 20 to 30 t/ha and 7.5 t/ha respectively under rainfed condition but two protective irrigation increased the forage yield up to 60 t/ha with nearly 30% dry matter yield. Seed yield is 100 to 300 kg/ha.

Utilization: The grass can be used for soiling, hay or grazing.

Toxicity: The setarias contain oxalates which can poison cattle. The amount of oxalate varies with the cultivar and stage of growth. Young plants contain more than older plants and strains highest in nitrogen are also highest in oxalate. The amount of oxalates ranges from 3.7 to 7.8%. Lactating cows, cattle and horses are affected with oxalate poison which shows a staggering gait and diarrhea. The animals die within three weeks of eating the grass. Death results from a build-up of calcium oxalate crystals in the kidney, which brings on acute hypocalcaemia.

REFERENCES

Bogdan, A.V. 1977. Tropical Pasture and Fodder Plants: Grasses and Legumes. pp. 249-260. Longman: London and New York.

CHAPTER

58

Job's Tears–*Coix lacryma-jobi* Linn (2n=20)

Family: Poaceae

Vernacular names: Job's tears (E), Adlay millet, Kahado (G). It is known by different names in local languages in different States such as *giral, kara, koa sangti, gurgur, kesai, kasi, golugu,* etc.

Importance: Job's-tears is a minor cereal. It is also popularly known as 'Adlay' millet. It is also used for making beverages. Job's-tears is cultivated as a fodder crop.

Origin: Job's-tears or *Coix lacryma-jobi* L. is one of the cereal plants native to South-East Asia.

Distribution: It is cultivated in the tropical countries of Africa and America. It is extensively cultivated in Philippine Islands, Indochina, Thailand, Burma, and Sri Lanka. In India it is abundantly found in hotter and swampy parts especially in Northeastern Hill regions. In India, it occurs in Rajasthan, Madhya Pradesh, Uttar Pradesh, Bihar, Jharkhand, West Bengal, Orissa, Assam, Meghalaya, Andhra Pradesh, Tamil Nadu, Kerala and Karnataka.

Botany: It is an annual/perennial, 1-2m tall, erect or floating plants. Stem is erect, stout with brace-roots from the lower nodes. It is a leafy grass. Leaves are 10-45 cm long and 3-5 cm wide with cordate base and smooth sheath. Ligules are very short. The inflorescence is of a fascicled axillary and terminal spaciform raceme. It is usually formed of one female spikelet completely enclosed in a globose or; ovoid basal bract through which the rachis grows out and bears the male spikelets above. The bract eventually becomes the outer covering of the seed or "the fruit-case". It may become very hard, strong and polished or may be thin, soft and coarse. Sometimes it is striated. Male spikelet are imbricate in pairs- or in threes and one of them, the central, is pedicelled. Each male spikelet has two florets, both of which are stamina'te. Sometimes the upper floret is empty. The grain is subglobose or ellipsoidal and enclosed in the bract. The seeds are yellow, purple, white or brown.

Climate: It is a species of humid climate and requires high annual rainfall in excess of 1500 mm.

Soils: It grows in swampy places near streams and requires fertile soils for best growth. It is reported to tolerate laterite, low pH, photoperiodic latitude, poor soil, slope, virus, and waterlogging.

Season: The crop can be mostly raised in June-July and October under rainfed conditions.

Varieties: KCA 3, KCA 4

Land preparation: The field should be disced and ploughed two to three times to obtain good tilth and form beds and channels.

Seeds and sowing: The crop is generally sown with the onset of rain. Seed rate is 6 to 10 kg/ha. The crop is raised by direct seed sowing (dibbling or broadcast) at a depth of 2.5 cm in rainy season with a spacing of 40-60 cm between plant to plant and 75-100 cm from row to row. Seeds are sown at spacing of 60 x 60 cm under rainfed condition. It requires sufficient rains in the early stage of growth and dry period for seed setting.

Fertilizer application: Job's tears gives good response with the liberal application of organic manures. Application of inorganic fertilizers is recommended after the analysis of soil in respect of nutrients status.

Crop protection: Coix is hardy and remarkably free from diseases and insects pests.

Harvest: It is a crop of 140-160 days. Plants are cut off at base and grain separated by threshing.

Yield: Its average green forage yield is about 13.9 t/ha. Normally the seed production is 1500 kg/ha. Nutritive value: Green material contains about 29.9% dry matter, 8.5% crude protein, 27.9% crude fiber and 8.9% ash. The husked grain gave 10.8% moisture, 13.6% protein, 61% fat, 58.5% carbohydrate, 8.4% fiber and 2.6% ash.

Utilization: The green material is very palatable and is utilized as food, fodder and both. The foliage may be used as fodder for cattle and can be turned into ensilage. It is also used for preparing the light beer. The seeds are used in medicines and also as livestock feed. The bran may be used as substitute for wheat bran in feeding the poultry.

REFERENCES

Venkateswarlu, J. and Raju S.K. Chaganti. 1973. Job's-Tears (*Coix lacryma-jobi* L.). Indian Council of Agricultural Research, New Delhi.

CHAPTER

59

Karnal Grass–*Diplachne fusca* (Linn.) P. Beauv (2n=20)

Synonyms: *Leptochloa fusca*

Vernacular names: Karnal grass, Kallar grass, salt grass, Brown beetle grass (E), *Dhaner, Choti gendar, Harri* (H) and *Nandi pillu* (T).

Distribution: It is distributed in upper Gangetic plains, Bengal, Orissa, Andhra Pradesh, Tamil Nadu and Kerala along coasts.

Botany: Karnal grass is a perennial, tuft and erect grass growing to a height of 0.9 to 1.5 m. Culms are hollow and roots come upwards from lower nodes. Leaves are 45 cm long and linear. Panicles are 10-30 cm long and margins of lemmas and paleas are hairy. It is a relatively succulent and palatable grass.

Climate: It grows successfully in tropical and sub-tropical areas. It prefers hot and dry season and even continuous submerged conditions. High humidity and wet soil moisture conditions are also favourable for its growth. It occurs in 300-450 mm rainfall regions and up to 1000 mm or more in coastal areas.

Soils: It gives good growth in alkali and saline soils.

Land preparation: One deep tractor ploughing should be followed by 2 to 3 ploughings with country plough. In case of tractor one ploughing with mould board plough followed by two disc harrowing are sufficient.

Seeds and sowing: Sowing/planting (rooted slips or stem cuttings with 3 nodes) should be performed a few days prior to or with the onset of monsoon so that germination is ensured. Under irrigated condition it can be planted any time during summer but mid-June to mid-July is the optimum time for its planting. Row to row and plant to plant distance of 25 and 20 cm should be maintained respectively. One irrigation is essential, immediately after the planting.

Manures and fertilizer application: Karnal grass grows well even without fertilizer. To obtain higher yield and to maintain nutrition and palatability of the grass 20 kg N/ ha is recommended after each cut. Since, alkali soils are rich in available phosphorus and potassium, these are not recommended. These soils are generally deficient in zinc, therefore if organic manure has not been added, zinc sulphate @ 25 kg/ha in first year might be beneficial for the growth of grass.

Harvest: When planted in the month of July, 2 to 3 cuts can be obtained up to November in first year. While under good management (fertilized and irrigated) 5 to 8 cuttings may be taken from 1st year onwards. It is recommended that the grass should be cut at 5 cm height from ground level. In the monsoon the grass may be harvested at an interval of 25 to 30 days while in other months at an interval of 30 to 45 days.

Yield: In highly alkali soil this grass gives 30 to 40 tones green forage in one hectare without fertilizer but with the application of recommended dose of fertilizer the green forage yield reaches upto 60 t/ha. The yield levels are generally higher in rainy season than others. 'Nutritive value: The nutritive value of the grass is quite satisfactory and it contains 8.5% crude protein.

Utilization: Animals relish this grass in young stage but compared with other fodder crops, it is less palatable. Alkali soils are reclaimed generally by growing the Karnal grass.

CHAPTER

60

Rhodes Grass–*Chloris gayana* Kunth (2n=20)

Vernacular name: Rhodes grass (Australia, United States, Africa); *chloris, herbe de rhodes* (French); *rhodasgras* (German); *pasto rhodes* (Peru).

Importance: It can be used for grazing or cut for hay and its vigorous spreading nature makes it well suited for use in erosion control.

Origin: It is a native of South Africa.

Distribution: Rhodes grass is a vigorous, perennial grass with a strong root system that gives it good drought tolerance. It spreads quickly, forming good ground cover and grows to 1.5 m. It is named after the famous Cecil Rhodes, who has popularised it. It is cultivated in Africa, Australia, India, Japan and South America.

Botany: It is fine stemmed, leafy, perennial, erect, rhizomatous or spreading, stoloniferous grass. Culms are 0.6 to 1.2 m tall with long and stout internodes. Leaf blades are 15-30 cm long and 3-5 mm wide, tapering to fine pointed tips. Inflorescence is spreading with 10-15 cm long, spikelets crowded, straw coloured on ripening. There is a flush of flowering when day length falls below 12 hours. It produces profuse amount of seeds. Seed count is 7250000 to 9500000 per kg.

Climate: It grows well in warm-moist conditions. It occurs from sea level to 2,000 m in the tropics, and sea level to >1,000 m from MSL in the subtropics. It is seasonal grown in areas receiving rainfall of 500 to 1500 mm per year. It tolerates seasonal waterlogging and up to 15 days' flooding. The optimum growing temperature ranges from 20 to 37ºC with extremes of 5º and 45ºC. This grass prefers high temperatures with maximum growth at 30°C/25°C (day/night temperature). Growth is reduced greatly below 18°C/13°C and there is negligible growth when the average daily temperature is below 8°C.

Soils: It can be grown in a variety of soil conditions. It prefers loamy to sandy loams and can grow even on a fair degree of salinity but cannot withstand stiff clayey or waterlogging conditions. It shows good salt tolerance. It grows best in soils with pH between 5.5 and 7.5, but it can be grown in soils with pH of 4.5 and up to 10.

Season: Fodder sorghum is raised during June-July and September-October under rainfed condition while throughout the year under irrigated condition.

Land preparation: Two to three ploughings with iron plough or country plough should be given. Form beds and channels. Field preparation can be done utilising summer showers.

Seeds and sowing: It can be propagated by seeds as well as by rooted slips. The establishment by seed is cheaper and for this a firm seed bed is prepared on well ploughed land and seeds are broadcasted @ 1.5 to 3.0 kg/ha at the onset of monsoon by mixing the moist soil. Seed is best sown with spacing of 45 cm between rows and continuous line spacing or drilled or can be broadcasted at a depth of 2 cm in a well-prepared seedbed, followed by rolling. Otherwise, it is important not to cover seed with more than 5mm of soil. The seed germinates from 3 to 7 days. In high rainfall zones or under irrigated condition, rooted slips can be transplanted in the lines at a distance of 50 x 50 cm. Nearly 40000 slips are required for one hectare. Planting material can be obtained by breaking up larger clumps into pieces, or using the small tussocks along the stolons that establish readily.

Weed management: One or two weeding can be taken depending on weed infestations.

Manure and fertilizer application: FYM @ 12.5 tonnes per ha and 40-20-20 kg per ha N, P_2O_5 and K_2O are applied as basal dressing under rainfed conditions and subsequently 25 kg of N per ha after each harvest.

Water management: It is mostly raised as rainfed crop. Its roots are able to extract water at a depth of 4.25 meters.

Cropping systems: It grows well with Stylo and glycine.

Harvest: In pastures raised through seeds, first clipping can be taken after three months, while that from rooted slips after two months and subsequently at an interval of 60 days and 15 cm cutting height for quality forage in pastures. In first year it should be harvested once after seed seeding while in subsequent years 3 to 4 cuttings can be taken at an interval of 60 days and 15 cm cutting height for quality forage. It is tolerant of heavy grazing, but production is reduced by very frequent defoliation. It makes good hay if cut at or just before very early flowering.

Yield: An average green fodder yield of about 17.0 t/ha of can be obtained with 6 cuttings per year under rainfed conditions. The seeds are collected in October-November. Seed yield is 100 kg/ha. The mature seed may have dormancy. It remains viable for up to 4 years depending on environmental conditions. Seed reaches maximum germination for 3-6 months (sometimes up to 18 months) after harvest.

Yield: Rhodes grass seed matures 23–25 days after flowering. Yields up to 350 kg seed per hectare can be harvested. Seed can remain viable in storage for up to 4 years.

Utilization: The grass is suitable for pasturage, silage and hay but it is generally used for soiling.

REFERENCES

Bogdan, A.V. 1969. Review article: Rhodes grass. Herbage Abstracts, 39, 1-13.
http://www.fao.org/ag/AGP/AGPC/doc/Gbase/data/pf000199.htm
https://hort.purdue.edu/newcrop/duke_energy/Chloris_gayana.html
http://ecocrop.fao.org/ecocrop/srv/en/home
http://www.tropicalforages.info/key/Forages/Media/Html/Chloris_gayana.htm

CHAPTER

61

Sewan grass–*Lasiurus scindicus* Henrard (2n=48)

Vernacular name: Sewan grass, *Gorkha* (Rajasthan, India), *Karera* (Pakistan).

Importance: Sewan grass is a palatable perennial used for pasture, hay and fodder for livestock. It does not withstand heavy grazing and disappears when overgrazed. It is used to stabilize desert sandy dunes. It is drought resistant and xeric in habit.

Origin: It is a native grass of India and sudano-sahelian Africa.

Distribution: It is grown in Arabia, Africa, Mall, Niger, Ethiopia, Egypt, Mali, Niger, Ethiopia, Iraq, India and Pakistan. In India, it is cultivated in arid zones of Rajasthan, extending to the parts of Haryana and Punjab.

Botany: It is an erect, tufted and branched perennial grass and attains a height of about 1.2 m. The stem is stout, and smooth. Leaves are alternate with a thin leaf-blade. Leaves are 20 to 45 cm long with setaceous tip. The inflorescence is a silky, 10 cm long raceme bearing hairy spikelets. Out of there are three spikelets at each node, two spikelets are sessile and one is pedicelled. The fruit is a caryopsis

Climate: Sewan grass is found between 25-27°N latitude in dry open plains. The dry arid climate suits to it with an annual rainfall from 125-250 mm. It is highly tolerant of drought resistant grass but protected from wind during establishment of the grass in sandy desert areas using wind breaks.

Soils: It can be grown on rocky ground, gravelly soils, sandy alluvial soils or alluvial soils or sand dunes, interdunal sandy areas, interdunal clayey area and in saline soils with a pH of 8.5.

Varieties: CAZRI-30-5, CAZRI-317, 318, 319, 351.

Land preparation: Two to three ploughings with iron plough or country plough should be given. Field preparation can be done utilising summer showers.

Seeds and sowing: Seed rate is 5 to 7 kg/ha. It is best sown in rows 30 to 50 cm apart at 1.5 cm sowing depth or broadcasted. Seed treatment is made 5 mm diameter pellets containing one or two spikelets with a mixture of fine silt and cow dung. It is also established through transplanting of rooted slips/seedlings at the distance of 2x2 or 2x3 m. Sowing or transplanting is done always during the rains or with 24 hours of rains.

Weed management: One or two interculturing operations, depending on weed infestations are essential during first year after sowing. The plants are protected from winds till they get establish.

Manure and fertilizer application: FYM @ 12.5 tonnes per ha and 40-20-20 kg per ha N, P_2O_5 and K_2O are applied as basal dressing under rainfed conditions and subsequently 25 kg of N per ha after every year.

Water management: It is mostly raised as rainfed crop.

Cropping systems: It performs well with Tropical Kudzu.

Harvest: It is harvested once at 10-15 cm height during establishment but from 2nd year onwards it may be harvested 2-3 times depending on rainfall. The pasture may be maintained for longer period through rotational grazing. It can also withstand excessive grazing pressure.

Yield: The green fodder yield is 2.5 to 3.5 t/ha while dry-matter yields of 0.5 to 1.0 t/ha. The seed yield is 150-200 kg/ha in normal conditions. The harvest is done in 30-day cutting interval at 15 cm height gives the best green fodder or dry matter yields.

Utilization: It is an extremely palatable grass and is used for hay, silage and grazing. It also establishes sand dunes in desert areas.

CHAPTER

62

Fodder Cowpea–*Vigna unguiculata* L. (2n=22)

Family: Fabaceae

Vernacular name: Cowpea, asparagus bean, black-eyed pea, catjang, catjang cowpea, Chinese long bean, clay pea, cream pea, crowder pea, pea bean, purple-hull pea, southern pea, yard-long bean (English), Southern Pea, Black eye pea, Lobia (Hindi), Thattai payaru (Tamil).

Importance: Cowpea is grown for seed, vegetables and fodder purpose. The green vine with pods serves as excellent forage to livestock.

Origin: Africa (Nigeria) and China are considered as the centre of origin.

Distribution: Cowpea is grown throughout the lowland tropics of Africa, India, south eastern Asia, Australia and coastal areas of South and Central America. In India, cowpea is grown only in some parts of Rajasthan, Gujarat, Maharashtra, Karnataka and Tamil Nadu.

Botany: Cowpea includes four cultigroups viz., i) *unguiculata*- major group, ii) *biflora orcatiang* with small erect pods grown in southeast Asia, iii) *sesquipedalis*-yard long bean grown in Asia, iv) *textiles*-grown in west Africa, for textile fibres obtained from peduncles. It may be climbing, erect as well as prostrate and creeping depending on the cultivar. It can grow up to 80 cm and up to 2 m for climbing cultivars. The leaves are trifoliate with oval leaflets, 6-15 cm long and 4-11 cm broad. The inflorescence is axially raceme with flowers congested at the top of the peduncle often in alternate pairs. Flowers are snowy white, yellow, pink or violet, bracteate, bracteolate, bracteoles two, shortly pedicellate, bisexual, hypogynous, zygomorphic and complete. Cowpea is highly self-pollinated crop, although significant out crossing can also occur due to large bees like bumble bees. Pods are cylindrical, 6 to 2 cm long and 3-12 mm broad and contain 8 to 20 seeds. Seeds can be white, pink, brown or black.

Climate: Cowpea is suited for tropics and subtropics. It is widespread throughout the tropics and in most tropical areas between 40°N to 30°S and below 2000 m altitude. Cowpea grows in savannah vegetation. It withstands moderate drought and heavy rainfall. It is well adapted to a wide precipitation range of 650 to 2000 mm. For forage, annual rainfall regimes of 750 to 1100 mm are preferable. It is often grown in annual rainfall regimes as low as 400 mm. The extended water logging or poor drainage should be avoided. It is sensitive to cold climate. The optimum temperature for its growth is 25 to 35°C, and the minimum is 15°C. It requires average soil temperature >19°C for 3 days from sowing to emergence, that limits its cultivation in north during winter. It is a short day plant requiring a day length >12.5 hours for flowering. It is tolerant of shading.

Soil: Loamy soil with a pH range of 5.0 to 6.5 is the best suited for cowpea. It can be grown in salt affected soil.

Season: It can be grown in February-March and June-July under irrigated condition while in June–July months under rainfed condition.

Varieties: Cowpea varieties recommended for different regions and states of India are furnished in Table 1.

Table 1 Cowpea varieties recommended for different regions and states of India

Region/States	Varieties
Whole country	GFC-1, GFC-2, GFC-3 (*kharif*), GFC-4 (summer) with 25-35 t/ha of fodder yield; Bundel lobia-1 (32-35 t/ha), UPC-287, UPC-5286 (30-45 t/ha), Bundel Lobia-1 (IFC - 8401), UPC- 625,
Haryana, Punjab and Delhi	FOS-1, FOS-10, K-395, K-585, IGFRI-S-450 (Kohinoor), C-88 (25-35 t/ha for Punjab). HFC-42-1 (Hara lobia) HFC-128, Cowpea-74, HFC-42-1 (Hara Lobia), EC 4216, Cowpea-88, CS – 88 (Haryana Lobia - 88), CL-367,
North India	UPC-5287 (30-45 t/ha), Russian Giant (30-35 t/ha), Pusa Rituraj, Pusa Sampada
Uttar Pradesh	Russian Giant, IGFRI-S-978;IGFRI-S-985, DFC-1, C-152, UPC-5286, UPC-5287, UPC – 287, UPC-4200, UPC- 8705, UPC- 621
North, West and Central India	EC-4216 (35-40 t/ha), Bundel lobia-2 (30-50 t/ha for north-west) and UPC-4200 (34-45 t/ha for north east India), UPC-607, UPC-610, UPC-612, UPC-616, Pusa Barsati, UPC 9202, UPC 618, Kohinoor, UPC- 9202, UPC 607, UPC – 618, UPC-622,
Gujarat	Chharodi 14-20, Chharodi 26-28, GFC-1 (Gujarat Forage Cowpea-1), GFC-2, GFC-3, GFC-4,
Southern states and West Bengal	Co-1, Russian Giant, EC-4216, TNFC-9901
Himachal Pradesh	PC-1, 3, 12, 14, 16 with resistant to collar rot
Tamil Nadu	Co-5 with 60-65 days duration, green fodder yield is 18-20 t/ha; Co (FC) 8 with 60-65 days duration, green fodder yield is 18-22 t/ha

Land preparation: Two to three ploughings are given to obtain good tilth. Beds and channels are formed

Seeds and sowing: Seed rate is 35 to 40 kg per ha. Spacing is 30 cm x 10 cm for line sowing method. Seeds are also broadcasted. Seed treatment is done with 3 packets of rhizobium culture.

Cropping system: Cowpea is rarely grown as sole crop and more often intercropped with maize+cowpea and sorghum+cowpea. The intercropping makes a balanced nutritious and palatable feed for livestock.

Weed management: Two hand weeding on 20 and 35 days after sowing. Use of trifluralin @ 0.5 kg/ha (pre-plant incorporation) in sole cowpea and alachlor @ 1 kg/ha in maize + cowpea mixed cropping has been recommended.

Manures and fertilizer application: FYM @ 12.5 tonnes per ha and 25-40-20 kg per ha of N, P_2O_5 and K_2O kg per ha are applied as basal dressing before sowing.

Water management: First irrigation is given immediately after sowing, life irrigation on third day and subsequent irrigations once in 7 days. It is also raised as rainfed crop.

Crop protection: Diseases such as root rot, collar rot, seedling blight, wilt, rust and mosaic virus disease while insect pests such as jassids, aphids, hairy caterpillar, stem fly, root knot nematode and spiral nematode occur in cowpea.

Harvest: Harvest is to be done on 55-60 days after sowing or during 50 % flowering to pod formation stage.

Yield: Green fodder yield is 18 to 20 t/ha. The dry mater fodder yields of cowpea ranges from 0.5 to 4.0 t/ha under rainfed conditions. Cowpea fodder contains 20 % crude protein. The haulms, which are the crop residues of seed production, contain about 45-65% stems and 35-50% leaves. Cowpea pod husks obtained after threshing are also used to feed livestock.

CHAPTER

63

Cluster bean–*Cyamopsis tetragonoloba* L. Taub

Family: Fabaceae

Vernacular name: Guar (H), *Kothavari* (T)

Importance: Cluster bean is a legume crop of arid and semiarid regions cultivated for feed, fodder and manure. It is a drought and salt tolerant crop. Since the seed of Cluster bean contains about 30-33% gum (galactomannon gum) in the endosperm, it is considered as an important cash crop for industrial gum production in India. It is contributing 80% to guar seed production in the world. Guar gum is an important ingredient in producing food emulsifier, food additive, food thickener and other Guar gum products. Cluster bean is a drought and high temperature tolerant deep rooted summer annual legume. India is the largest producer of Guar and contributes 80% of total production in the world. This spherical-shaped endosperm contains significant amounts of galactomannan gum (19 to 43% of the whole seed), which forms a viscous gel in cold water.

Origin: Cluster bean is originated in Southeast Asia and specifically in India.

Distribution: It is commercially grown mainly in India, Pakistan, USA, South Africa, Malawi, Zaire and Sudan, Australia and Brazil. In India, it is being grown in arid and semi-arid areas of north-western states mainly in Rajasthan, Gujarat, Haryana, Punjab, Madhya Pradesh, Uttar Pradesh, Maharashtra, Chhattisgarh, Andhra Pradesh, Karnataka and Tamil Nadu.

Botany: Cluster bean plant is an erect, robust annual with a height of 50-100 cm. It bears 4 to 10 branches (branch types). However, non-branch type-varieties have main stem only. It has well developed tap root system. The leaves are trifoliate and toothed. The flowers are white to purple in colour and borne on short axillary racemes. The pods are flattened, fleshy, beaked, 2.5–12 cm long containing 5 to 8 seeds and are borne in clusters. Seed is hard, flinty, flattened, ovoid and about 5 mm long. Seeds are white, grey or black in colour. The seed consists of three parts: the seed coat (14-17%), the endosperm (35-42%), and the germ (43-47%). This spherical-shaped endosperm contains galactomannan gum (19 to 43%

of the whole seed), which forms a viscous gel in cold water. Germination is epigeal type. It is an indeterminate crop.

Climate: Cluster bean is a tropical plant. It requires warm growing season. It is grown up to an altitude of 900 m from MSL. Cluster bean is a sun loving plant. It is a day neutral plant. Long day period favours vegetative growth while the short day period for flowering and pod formation. It performs well under arid and semi-arid condition. In arid condition, guar grows as rain fed crop which requires 300 to 400 mm rainfall in 3 to 4 spells. It is a drought tolerant crop. It is grown in areas receiving mean annual rainfall of 500 to 750 mm. Dry weather is essential once the pods have set seed since the rain or humidity can cause blackening of the pods and seeds. Soil temperature ideally should be around 20 to 25°C for proper germination. The crop requires 25 to 35°C temperatures at the sowing time for proper germination and 32 to 38°C temperature encourages good vegetative growth, but high temperature at flowering stage can result in pre-mature flower drop. It can tolerate temperature as high as 45-46°C. The optimum temperature is 25-30°C. Frost free period of 110 to 130 days required for growing cluster bean crop.

Soil: It can be grown in different type of soils but light textured sandy soils are more suitable for the crop. It performs well on medium-fertile, medium textured; un-logged conditions with neutral pH and well-drained subsoil. Soils with pH 7.0 is desirable. It is a moderately saline and alkaline tolerant crop up to pH of 9.0. The crop has the ability to fix nitrogen to the tune of 30-40 kg/ha. It cannot withstand acidic and waterlogging condition.

Land preparation: Land is ploughed to a depth of 20 cm to give a reasonable tilth and cleared of weeds.

Season: Cluster bean can be grown during June - August and September-October under rainfed condition while it can be raised as irrigated crop during February-March. In India, it is cultivated mainly during *kharif* season. The seed is normally sown in second week of July after rainfall starts. Sowing may get extended upto to August depending on rainfall and is harvested during October and November. It is usually a 90 to 110 days crop. The crop is cultivated in *kharif* as well as in summer season.

Varieties: The particulars of the Cluster bean varieties grown in different states of India are furnished in Table 1.

Table 1 Cluster bean varieties grown in different states of India

Variety	Source	Green fodder yield (t/ha)	Suitability	Major characteristics
Guara-80	cross between FS 277 x No. 119	20-30	Punjab and Haryana	resistant to bacterial blight, alternaria leaf spot and stem breakage
FS-277	-----	20	Punjab and Haryana	It is non-shattering variety susceptible to bacterial blight. It is tolerant to drought and *Alternaria* leaf spot
HFG-119	-----	22	entire country, except Punjab	drought tolerant, non-shattering and *Alternaria* blight resistant
Durgajay	ARS, Durgapura	27	Rajasthan	dual type for fodder and seed. seed yield is 12.6 q/ha

Durgapura Safed	ARS, Durgapura	25	Rajasthan	dual type variety suitable for late sown conditions
Agaita Guara-111	PAU, Ludhiana	23	Punjab	-----
Agaita Guara-112	PAU, Ludhiana	30	Punjab	Early maturing type, plants are hairy, unbranched, high number of pods.
Guara-80	PAU, Ludhiana	25-27	north western zone	15 q/ha seed yield
FS-277	CCSHAU, Hisar	-----	entire country	-----
HFG-119	CCSHAU, Hisar	25-30	entire country	drought tolerant, non-shattering and resistant to Alternaria leaf spot. Duration 130–135 days
HG-75	CCSHAU, Hisar	25	entire country	It is more preferred for seed. 20 q/ha seed yield
HG-182	CCSHAU, Hisar	25	entire country	Duration 110-125 days
HFG-156	CCSHAU, Hisar	35	Haryana	It is a tall, branched variety
Maru Guar (2470/12)	CAZRI, Jodhpur	22.5	Rajasthan	seed yield is 9.5 q/ha
Bundel Guar- 1	IGFRI, Jhansi	35	entire country	6.5 t/ha of dry fodder, moderately resistant to leaf blight, lodging resistant, drought tolerant and non-shattering character
Bundel Guar- 2	IGFRI, Jhansi	25-30	entire country	Used for grain / gum production, moderately resistant to bacterial blight, tolerance to lodging, drought and shattering, 5-6 t/ha of dry fodder
Bundel Guar- 3	IGFRI, Jhansi	35-40	entire country	maturity is 50–55 days, moderately resistant to bacterial blight and powdery mildew, tolerant to shattering and resistant to drought.
Guar Kranti	ARS, Durgapura	34	Rajasthan	Seed yield 14.6 q/ha

Growth period: Cluster bean takes 80 days for vegetables, 110 - 130 days for seed and 60-90 days for green manure. It is a three-four months crop. It takes about 90 to 110 days from sowing to harvest. The seeds are known to germinate only after sufficient imbibition, approximately twice their original weight. It is not able to germinate well under high saline and submergence conditions. Pre-soaking of seed in good quality water for 2 hours followed by half an hour shade drying may enormously increase germination percentage. Germination takes place in 4 to 6 days of sowing. It is an indeterminate plant showing continuous flowering and poding even upto 80 to 85 days of sowing, if soil moisture is available. However, in general flowering stage starts after 40 to 60 days of sowing (DAS). The pod formation takes place after 50 to 70 days from the date of sowing. Pod matures in 80 to 90 DAS. The harvesting of the crop begins when 90 percent pods are matured roughly 90 to 110 DAS (depending on the variety, soil and climatic conditions). Bolder seeds give higher germination compared to small or medium sized seeds. Bolder seeds may give better shoot length and higher dry matter production. Maximum germination occurs at 25 to 30°C and it takes about 6 days for complete germination.

Seed and sowing: Seed rate is 20 to 25 kg per ha for seed and vegetables, 35 to 40 kg per ha for fodder and green manure. Spacing is 45 x 15 cm for seed and 30 x 10 cm for fodder purpose. Seed should be treated with either *Trichoderma* @ 4 g/kg of seed or mancozeb or with carbendazim @ 2 g/kg of seed followed by chloropyriphos @ 2 ml/kg of seed. The seed should be inoculated with suitable rhizobium culture @ 600 gm/15 kg of seed. Three packets (200 gm each) of the rhizobium strain should be mixed with a solution of jaggery by mixing 250 gm in one litre of water. After having uniform coating of slurry over the seeds, it should be dried for 30 minutes in shade. Dried seeds should be sown within 24 hours of inoculation.

Cropping systems: Clusterbean can be successfully intercropped with pearl millet in 2:1 row proportion of clusterbean and pearl millet. Clusterbean based crop rotation can be practiced as follows:

Clusterbean-pearlmillet (two year crop rotation in rainfed condition)
Clusterbean-wheat (one year rotation for irrigated condition)
Clusterbean-cumin (one year rotation for irrigated condition)
Clusterbean-wheat-clusterbean-cumin (two year rotation)
Clusterbean-wheat-mung bean-mustard (two year crop rotation)
Clusterbean-cumin-pearlmillet-mustard (two year crop rotation)
Clusterbean-wheat-pearlmillet-cumin (two year crop rotation)

Weed management: The crop should be weed free at least for initial 30 to 35 days after sowing. Two manual weeding given at 25 and 45 days after sowing are sufficient to keep the crop weed free. However, sometimes due to non-availability of labour, herbicides like pendimethalin @ 2.5 to 3.30 lit/ha can be applied by mixing with 500 lit of water as pre-emergence application (within 2 days of sowing). After that one manual weeding at 30 DAS or post emergence application of imazythypr @ 400 gm/ha mixing with 500 litre water at 20-25 days after sowing can be applied for controlling weeds of clusterbean field.

Manures and fertilizer application: FYM @ 12.5 t/ha with 25-50–25 kg of N, P_2O_5 and K_2O per ha are applied as basal dressing under irrigated condition. Subsequently apply 25 kg N per ha on 30 DAS.

Water management: Cluster bean is usually grown as rainfed crop in arid and semi-arid condition. Irrigate immediately after sowing followed by life irrigation on the third day. Subsequent irrigations are given at intervals of 7-10 days. Spray of 0.1% thiourea solution at 25 and 45 DAS also improves the seed yield during moisture stress condition.

Crop protection: Diseases such as alternaria leaf spot, bacterial leaf blight, powdery mildew, anthracnose and while insect pests such as jassids, aphids and root knot nematode and reniform nematode occur in clusterbean

Harvest: The green immature pods are picked by hand while the plants are usually uprooted or cut a few inches above ground level for seed production. It should be harvested when pods become brown and dry for seed purpose. The plants are stacked to dry and threshed by hand. Seeds should be dried to less than 14 % moisture.

Yield: Green pod yield is 6200 to 8300 kg per ha. Dry seed yield is 600 to 900 kg per ha. Hundred seed weight is 4 to 6 gm. Seed contain 33.3 % protein and 40 % carbohydrate.

Utilization: The green immature pods are used as vegetable. The young pods are sweet and often cooked but become bitter as pod mature. Cluster bean is used as fodder crop, green manure crop and to reclaim salt affected soil. The seed flour has high viscosities at low concentrations, possessing 5-8 times the thickening powder of ordinary starch. The principal mucilaginous substance is mannogalacton. It is used to improve the strength of certain grades of paper industry and as stabilizer and thickener in food products such as ice cream, bakery mixes and salad dressings. Cluster bean is a source of vegetable gum. The seed of clusterbean contains about 30-33% gum in the endosperm. Cluster bean is also raised as a green manure and cover crop. Being a leguminous crop, it enriches the soil fertility by fixing the atmospheric nitrogen.

REFERENCES

Gipson, A. and Balakrishnan, R. 1992. Genetic diversity in cluster bean. Indian J. Hort. 49(1):70–74.

NRAA, 2014, Potential of Rainfed Guar (Cluster beans) Cultivation, Processing and Export in India. Policy paper No.3 National Rainfed Area Authority, NASC Complex. DPS Marg, New Delhi-110012, India: 109p.

Raj Singh. 2014. Improved Cultivation Practices for Cluster bean in *Kharif* and Summer Season. Central Arid Zone Research Institute (Indian Council of Agricultural Research) Jodhpur - 342 003 (Rajasthan)

http://www.igfri.res.in/Default.aspx

http://www.guargum.co.in

CHAPTER

64

Fenugreek–*Trigonella foenum-graecum* L. (2n=16)

Family: Fabaceae

Vernacular names: Fenugreek, bird's foot, (English); Chandrika, Gandhabiijaa, Kairava, Medhika, Methika, Sarapunkha (Sanskrit); Mathai, Uluvaarisi, Vendayum, Ventayam (Tamil); Methi, Shanbalid (Urdu); Methi, Metha (Hindi); Venthiam (Malayalam);

Importance: It is cultivated as a leafy vegetable, spice, condiment and fodder crop. It has a high medicinal value as it prevents constipation, removes indigestion, stimulates spleen and liver and is appetizing and diuretic. Seeds are exported to Saudi Arabia, Japan, Sri Lanka, Korea and United Kingdom.

Origin: Fenugreek, a native of south Eastern Europe, west Asia (Turkey) and India.

Distribution: It is grown in Europe, north Africa, western Asia, India, China, Indonesia, Ethiopia and Russia. The major states growing fenugreek in India is Rajasthan, Madhya Pradesh, Gujarat, Uttar Pradesh, Maharashtra and Punjab.

Botany: It is an annual herb reaching a height of about 40 to 80 cm. The plant has compound leaves, 7–12 cm long. The flowers are white or purplish blue. The flowers are hermaphrodite (have both male and female organs) and are pollinated by insects. The fruits occur as pods of 2–10 cm, long, thin and pointed and contain 10–20 seeds. The plant and seeds have a characteristic strong odour. There are two species of the genus Trigonella which are of economic important viz., *T. foenum graecum*, the 'common methi' and *T.corniculata*, the 'kasuri methi' or 'Champa methi'. These two differ in their growth habit and yield. The latter one is a slow growing type and remains in rosette condition during most of its vegetative growth period.

Climate: It has wide adaptability and is cultivated both in the tropics as well as temperature regions. It is suitable for areas with moderate or low rainfall of 500 to 700 mm per year. It is fairly drought resistant. It is a sun-loving crop. It tolerates frost to -15°C and freezing weather. It cannot be grown in the shade. The seed crop requires warm dry weather for ripening and harvest.

Soils: It prefers well-drained fertile loams, clayey loam or sandy loams, but grows fairly well on gravelly and sandy soils. It is not adapted to heavy clay soils. It is moderately tolerant to saline soils. The optimum pH range is 5.8 to 8.2 for its growth and development.

Season: In warm climates it is usually grown as a cool season crop, while in temperate areas it is grown as a summer crop. It is cultivated both under irrigation and as a rainfed crop. It is grown as both irrigated and as a rainfed crop. Sowing in the plains is generally taken up in Sep-Nov while in the hills, it is grown from March. The October plantings are harvested in February–March, while January plantings are harvested in April for forage purpose.

Varieties:

Variety	Parentage	Special Characters	Duration	Yield per ha
CO 1 (TNAU,CBE)	Reselection from TG 2336	Dual purpose quick growing, suited for intercropping, high seed protein	90	650 to 700 kg grain, 4.5 t of green
Rajendra Kanti (Dholi, RAU)	Mass selection	High yield, medium height, bushy suited for pure as well as intercropping. Seed protein 9.5%	120	1200 to 1400 kg grain
RMt – 1 (Jobner, RAU)	Pure line selection from Nagpur type	High yield, moderately branched, moderately tolerant to root rot and powdery mildew, Seed protein 21%	145	1500 kg grain
Lam Sel. 1 (Guntur, APAU)	A selection from Germplasm	High yield, bushy plant type. Seed protein 53%	70	700 to 750 kg grain
T-8 (AAU, Anand)		recommended for cultivation in Gujarat	-----	-----
ML-150 (PAU, Ludhiana)	hybridization of Type 8 × Type 36	medium to late variety with red leaf margins and yellow seeds.	-----	-----
Pusa Early Bunching		Bushy variety. 115 to 125 days, seeds are bold.	-----	-----

Land preparation: Land is ploughed thrice and beds of uniform size are prepared. The seedbed should be free from weeds.

Seeds and sowing: Broad casting the seed in the bed and raking the surface to cover the seeds is normally followed. Seeds dibbled in lines with 60 x 20 cm at 1 to 2 cm depth which facilitates the intercultural operations. Seed rate is 25 kg/ha. Seeds take about 4 to 6 days to complete its germination.

Intercultivation: Thinning may be done on 20-25 days to keep the distance between the plants at 10-15 cm and to retain 1-2 plants per hill.

Weed management: Hoeing and weeding during the early stages of plant growth are required to encourage proper growth.

Manures and fertilizer application: FYM @ 12.5 t/ha may be incorporated during the last ploughing. A fertilizer dose of 30 kg N, 25 kg P_2O_5 and 40 kg K_2O per ha is recommended as basal dressing and 20 kg N at 30 days after sowing. To obtain more successful leafy growth, nitrogen should be applied after each cutting.

Water management: The crop is grown as rainfed in heavy moisture retentive soils. Crop requires 3 to 6 irrigations. First irrigation is given immediately after sowing and subsequent irrigation is applied at 7 to 10 days interval.

Cropping systems: It is intercropped with coriander, sesame or chickpea.

Plant protection: Root rot is a serious disease and can be controlled by drenching carbendazim 0.05% first at the onset of the disease and another one month.

Harvest: It is an annual and ripens 90 to 150 days after planting. The plant flowers in 30 to -40 days after sowing. Pods mature within 60 to 90 days after sowing. If it is grown as a leafy vegetable, leaves are clipped off 10 cm above ground level in about 25 to 30 days and subsequent cuttings of leaves may be taken after 15 days. Afterwards 4-5 cuttings are taken. For a grain crop, 1-2 cuttings are taken before the crop is allowed for flowering and fruiting. The plants are pulled out for seed 3 to 5 months after planting by uprooting when the pods are dried in the open sunlight on threshing floor for 2-3 days and seeds are threshed by beating with stick or by rubbing with hands. Seeds are winnowed, cleaned and dried.

Yield: Green forage yield is about 4 to 5 t/ha while the seed yield is 0.5 to 0.9 t/ha may be obtained in crops grown for both the purposes under rainfed condition. However, a seed yield of 1.2 to 1.3 t per ha can be obtained under irrigated conditions.

Utilization: It is used as green feed or hay. It is not suited to grazing. It is highly palatable to livestock. The fresh tender leaves and stem are consumed as curried vegetable and the seeds are mainly used as spice for flavoring almost all dishes. It has a high medicinal value as it prevents constipation, removes indigestion, stimulates spleen and liver and is appetizing and diuretic.

CHAPTER

65

White Clover–*Trifolium repens* L. (2n=32)

Common names: White clover, Dutch clover, Ladino clover, White Dutch clover. The genus name, *Trifolium*, derives from the Latin *tres*, 'three', and *folium*, 'leaf', so called from the characteristic form of the leaf, which has three leaflets (trifoliate); hence the popular name 'trefoil'. The species name, *repens*, is Latin for 'creeping'.

Importance: White clover is the most common cultivated clover species worldwide. It is a widely used forage legume, grown alone or in mixed stands with grasses, in rainfed or irrigated stands. It plays an important role in soil conservation, soil improvement and crop rotations. Three types of white clover are grown according to leaf size. Small white clover, generally called white 'Dutch', originated in Holland. The intermediate white clovers are larger than white Dutch. Large white clover, or Ladino, is much larger and produces three to five times as much growth as white Dutch clover.

Origin: White clover probably originated from the eastern Mediterranean region of Asia Minor and Europe.

Distribution: It is cultivated in Europe, North America, South America, Central Asia, Morocco, Tunisia, Papua New Guinea, Kenya, China, India and Australia. It is cultivated in sub-temperate and temperate rangelands of Himachal Pradesh and Jammu & Kashmir.

Botany: White clover is a creeping, herbaceous, short-lived perennial legume that spreads by means of a branched network of stolons. It can live for 3 to 5 years. It is a leafy plant that often grows 20–30 cm tall and spreads by stolons (above ground runners) and forms shallow roots at nodes. It develops a taproot that dies after the first year and is replaced by a secondary, mostly shallow root-system that develops from the stolons at the nodes. The stolons are creeping, 10-40 cm long, and can produce roots, leafy branches or inflorescence stalks. Leaves are petiolated, alternate and palmately trifoliate but they can vary widely in form and size, depending on cultivar or type. Leaves are non-hairy and usually marked with a white 'V'. Leaflets are ovate, broad, solid dull green or occasionally marked with a white 'V' and sometimes with dark red flecks. It bears globular racemes at the end of long peduncles that arise from the stolons leaf axils. The inflorescence has 20-40 white fragrant flowers. Flowers white or sometimes yellowish or pinkish, with more than 20 flowers and flowers are clustered into heads. It is cross pollinated by honey bees and bumble bees. Once pollinated, the flowers develop into linear sessile pods containing 3-4 heart-shaped, smooth, bright yellow to yellowish brown seeds. Seed maturity occurs 3-4 weeks after pollination. Seeds are extremely small with over 700,000 per pound.

Growth stages: Seedlings develop a slender taproot and have a rosette growth habit with very little primary stem elongation. The seedling tap root persists for 1 to 2 years only. Within six to eight weeks of germination, stolons develop and adventitious roots form at each node along the stolon to serve as the primary root system. The stolons are indeterminate in growth habit. The plant spreads and survives by the development of stolons. Roots are shallow and are usually less than 20 cm in length but sometimes are as long as one m. Every three to four days new leaves are being formed. Old leaves fall off after about 40 days.

Climate: It is distributed in temperate and subtropical climates in the northern and southern hemispheres. It grows best in humid areas of the temperate zones during cool, moist seasons. Flowering is a photo period response >13-16 hours. It is found in areas with mean annual temperatures situated between 4.3°C and 21.8°C. Its optimal temperature for growth is 20°C to 25°C. It is as winter hardy as alfalfa and it can withstand frost. It grows best in humid areas of the temperate zone during cool, moist seasons. It is not tolerant to drought. It grows in areas with annual rainfall of 750–1300 mm and conditions are cool and moist. The plant has some tolerance of shade, heat and drought. It has poor tolerance of flooding.

Soils: It can grow on a wide range of soils but does better on clay or loam soils than on sandy soils. The soil pH range for white clover is between 5.5 and 7.0. It is very sensitive to high soil aluminum levels which can occur when soil pH drops below 5.5. It grows in damp and swampy soils and also on lawns and grassy places. It is tolerant of moderately acidic to moderately alkaline soils, wet soils, and flooding but does not grow well under saline conditions or heat and drought.

Varieties: Palampur Composite 1

Seeds and sowing: The ideal seeding depth is approximately 0.05 to 1.0 cm.

Weed management: Two hand weeding are given on 20 and 35 DAS.

Manures and fertilizer application: 12.5 tonnes FYM or compost per ha along with 25-40-20 kg per ha of of N, P_2O_5 and K_2O kg per ha are applied as basal dressing.

Cropping systems: It is most often grown in association with cool-season grasses.

Harvest: White clover tolerates grazing very well; however, close continuous grazing can weaken a stand. By subdividing pastures into three or more paddocks, rotational grazing can be used to increase stand life and improve forage utilization. Short rest periods of two to three weeks allow the clover to renew its vigor. Rotational grazing also makes it easier to mow for weed control or to spread manure piles. The crop is best suited for grazing, it can be used for haylage, hay, soil improvement and land reclamation. However, the green forage yield is less as compared to red clover and Lucerne. White clover provides high quality grazing, is an excellent nitrogen-fixing perennial legume, and can play an important role in soil conservation, soil improvement, and crop rotations.

Yield: It gives 3–4 cuttings in a year, if harvested for green fodder or hay. It is used for grazing purpose. It gives yield of 4.0 to 5.0 t/ha of green fodder.

Quality: White clover can cause bloat in ruminants, when they graze clover-rich swards or pure clover. It can be prevented by gradually introducing livestock to the pasture, and by supplementing them with hay or straw. In extreme situations, anti-foaming agents can be administered directly into the rumen. White clover may contain low levels of phytoestrogens (coumestrol) (0.02-0.06 % of DM) compared to red clover (*Trifolium pratense*) (1-2.5 %). Coumestrol is sometimes the main active agent. However, the phytoestrogen level is too low to have adverse effects on reproduction. Hydrogen cyanide poisoning occurs with white clover. Some varieties rich in HCN are not recommended, especially for lactating animals.

CHAPTER

66

Senji–*Melilotus indicus* (L.) All. (2n=16)

Synonyms: *Melilotus parviflorus* Desf., *Melilotus tommasinii* Jord., *Trifolium indicum* L.

Vernacular names: Sweet clover, Sour clover, Indian sweet-clover, King Island melilot, King Island clover, Bokhara clover, Yellow sweet clover, Californian lucerne (English), *Senji* (Hindi).

Importance: Sweet clover is a drought resistant crop.

Origin: *Melilotus indicus* originated from the Mediterranean, south-western Europe and from India.

Distribution: The plant is found in Asia, Europe and throughout Arabia. In India, it is grown in Punjab and Haryana.

Botany: It is annual or biennial, erect or prostrate herb, branching from base, 10 to 50 cm high sometimes to 80-100cm in favorable conditions. Leaves compound, aromatic, trifoliate, petioles up to 40 mm long, stipulate; leaflets 8-22mm, 3-12mm wide, obovate, obtuse with denticulate margin. Inflorescence terminal or axillary racemes with long axis up to 45 mm long, to 50 to 60 mm wide in fruit; flowers small 3.5-5mm., yellow in colour. Fruit legume, pod round or oval wrinkled, 3mm long, with one or rarely two seeds; seed smooth,ovoid-elliptical,1.5-3 mm long, yellow or greenish yellow, sometimes with purple spots.

Climate: It is adapted to the boreal moist subtropical dry to moist forest life zones. It persists up to an altitude of 2,000 m from MSL.

Soil: It requires loams to heavy soils for its propagation. It is a salt-tolerant wild leguminous herb with high potential for use as a forage crop in salt-affected soils.

Season: The sowing time of the crop is between September to the end of October of *rabi* season.

Varieties: Senji Safed-76, FOS-1, YSL-106, PC-5, HFWS-55.

Seeds and sowing: Seed rate is 20 to 25 kg/ha. The ideal seeding depth is approximately 0.05 to 1.0 cm.

Weed management: Two hand weeding are given on 20 and 35 DAS.

Manures and fertilizer application: 12.5 tonnes FYM or compost per ha along with 25-40-20 kg per ha of N, P_2O_5 and K_2O kg per ha are applied as basal dressing.

Water management: First irrigation is given immediately after sowing, life irrigation on the third day and subsequently once in 7 to 10 days depending on soil and climatic conditions. Thus, about 12-15 irrigations will be needed during the entire crop season. Normally the crop should be irrigated after each cutting. Total water requirement is 1000 mm per year.

Harvest: The crop is of long duration and matures in 172 days.

Yield: It provides 45-50 t/ha green fodder and 8-10 dry fodder and seed yield is 12.0-15.0 q/ha.

CHAPTER

67

Berseem–*Trifolium alexandrium* L. (2n=16)

Family: Fabaceae

Vernacular name: Berseem, Berseem Clover, Egyptian clover, Alexandria clover (English).

Importance: The name berseem is derived from Arabic name 'Bersym' or 'Berzum'. It is highly nutritious forage. It can be used for silage and hay making. It is a soil builder as it fixes atmospheric nitrogen. It remains soft and succulent at all stages of growth. The merit of the crop lies in its multicut nature (4–6 cuts), long duration of green fodder availability (November to April), high green fodder yield (85 t/ha), good forage quality (20% crude protein), and digestibility (up to 65%) and high palatability. The fodder is available for fairly long period during winter, spring and early summer seasons. It has a cosmopolitan adaptability and diverse quality. It is therefore aptly called as king of fodder crops for irrigated regions of north India.

Origin: Berseem is a native of Asia Minor and Egypt.

Distribution: Berseem is grown in Egypt, India, Pakistan, Israel, Syria, Turkey, Russia, Mediterranean countries, Cyprus, Greece, Spain, Romania, Portugal, Europe, Balkan Peninsula, Italy, South Africa, South America, USA and Australia. In India, it is grown in Himachal Pradesh, Punjab, Haryana, Rajasthan, Uttar Pradesh, Jammu and Kashmir, Bihar, Madhya Pradesh, Maharastra, West Bengal, Assam, Odisha and Kerala.

Botany: Berseem is an annual, sparsely hairy, erect forage legume, 60 to 80 cm high. It has a shallow taproot. The roots are mainly restricted to the top 60 cm of soil. Its stems are hollow, branching at the base, with alternate leaves bearing 4-5 cm long x 2-3 cm broad leaflets. Flowers are yellowish-white and form dense, elliptical clustered heads about 2 cm in diameter. The flowers must be cross pollinated by honey bees to produce seeds. The fruit is a pod containing one single white to purplish-red seed. Seed is small about 2 mm and yellow in colour. On the average there are about 440 seeds/gram.

Climate: Berseem is grown in temperate, subtropical and the humid boreal region climate. It can be grown in areas below 1700 m altitude with irrigation facilities. Berseem can grow in areas where annual rainfall ranges between 550 mm and 750 mm. The optimum temperature for germination and flowering is 25 to 30ºC and 35 to 37ºC. Freezing temperature of less than 4^0C to 5^0C kills the plants. Its regenerative growth is retarded during severe cold or frosty period or at temperature above 40°C. Berseem clover does not tolerate shading conditions.

Soils: Well drained loam to clay loam soil is preferable. It is grown in mild acidity to slightly alkaline soils. It will tolerate short periods of flooding. Waterlogged soils should be avoided.

Land preparation: The seeds being very small, berseem requires a fine seedbed. Two to three ploughing are given to have good tilth. Beds and channels are formed.

Season: Berseem is a *rabi* season crop. It is sown from second fortnight of September to first week of December in North West to Eastern and Central India. The time of sowing berseem is ideal when mean day temperature is 25°C, which is recorded mostly in the first to third week of October in north India.

Varieties: Pusa giant, Mescavi, IGFRI-S-99-1, UPB-110, Wardan (S-99-1), Mescavi, Wardan, BL-1, BL-10, BL-22, JB-2, BL-2, BL-42, BL – 180 (PAU), UPB-110, Bundel Berseem –2 (JHB 146), Bundel Berseem –3 (JHTB 96-4), HFB 478. Hisar Berseem 1, Jawahar Berseem 5, Bundle Berseem-2, Bundle Berseem-3, Wardan (S-99-1),

Table 1 Berseem varieties grown in different states of India

Variety	Source	Green fodder yield (t/ha)	Suitability	Major characteristics
Pusa Giant	IARI, New Delhi	55-60		High winter hardiness and frost resistance
Meseavi		50-55		10-12.5 t/ha of dry matter yields in five cuttings
BL-1	PAU, Ludhiana	60		12 t/ha of dry matter yields in five cuttings
Berseem Ludhiana-1 (BL-22)	PAU, Ludhiana	70-75	temperate and north west zone	13.5 t/ha of dry matter yield
Jawahar Berseem-1 (JB-1)	JNKVV, Jabalpur	70-75	Entire country, central and north western zones	12-15 t/ha of dry matter yield
Wardan	IGFRI, Jhansi	70-75	central zone	12-15 t/ha of dry matter yield; days to 50% flowering is 150–165 days, days to maturity is 175–190 days,
UPB-10	GBPUA&T, Pantnagar	70-75	Uttar Pradesh	matures in 200–210 days.

Bundel Berseem-2 (JHB-146)	IGFRI, Jhansi	90-100	central zone	flowers in 150–160 days and matures in 180–190; tolerant to acidic conditions
Bundel Berseem-3	IGFRI, Jhansi	50-55	north east zone, Bihar Odisha, WB and eastern UP.	50% flowering in 155–170 days and maturity in 175–185 days. moderately resistant to stem rot and root rot diseases
JB-5	JNKVV, Jabalpur	45-50	MP, UP, Maharashtra, Gujrat and Chattisgarh	matures in 185-195 days
Hisar Berseem-1 (HFB-600)	CCSHAU, Hisar	75	suitable for late sowing in hill areas of the country	maturity of 205–210 days.
BL-180	PAU, Ludhiana	60-65	Punjab, Haryana, Uttranchal, J&K and Himachal Pradesh	matures in 260–265 days. 3–4 t/ha of seed yield.

Seeds and sowing: Seed rate is 4 to 6 kg per ha. Sowing depth is 3 to 5 cm. For elimination of chicory weed (*kasani*), the seed should be poured in 1% common salt. Floating chicory seed should be taken out and remaining seed of berseem should be sown. Seed treatment with *Rhizobium* culture can be done to enhance biological nitrogen fixation. Seed treatment with fungicides is to be carried out. Then treat the seeds with rhizobium culture after 24 hours. There are two methods for sowing of berseem i.e. dry and wet bed. Seeds are sown either by broadcasting or line sowing with a spacing of 30 cm x 5 cm. Berseem clover can be sown into a conventional seed bed, or direct drilled to a depth of not more than 10 mm. For satisfactory germination and good plant stand, wet method is better. Seed should be sown in beds of convenient size by broadcast method after flooding the beds with 5-6 cm deep water. Before sowing seeds, the water in the beds should be stirred thoroughly with the help of puddler or rake so as to break the clods and capillary to avoid leaching during successive irrigations. The crop should be re-irrigated after 5-6 days of sowing when germination is complete.

Weed management: Two hand weeding are given on 20 and 35 DAS. Application of Fluchloralin @ 1.2 kg a.i./ha at pre planting stage controls the weeds.

Manures and fertilizer application: 12.5 tonnes FYM or compost per ha along with 25-40-20 kg per ha of N, P_2O_5 and K_2O kg per ha are applied as basal dressing.

Water management: First irrigation is given immediately after sowing, life irrigation on the third day and subsequently once in 7 to 10 days depending on soil and climatic conditions. Thus, about 12-15 irrigations will be needed during the entire crop season. Normally the crop should be irrigated after each cutting. Total water requirement is 1000 mm per year.

Harvest: The first cutting is taken at 65 to 70 days after sowing and subsequent cuts once in 50 days interval. Berseem clover is better suited to fodder conservation than grazing, as it has high growing points that can be easily damaged by inappropriate grazing. A rotational grazing strategy should be used if berseem clover is to be used for grazing purposes.

Yield: Green fodder yield is 50 to 60 tonnes per ha per year. Seed yield is 250 to 300 kg per ha. The crude protein percent is 18 to 22%. Berseem contains osterogens sex hormones which affect fertility of animals.

REFERENCES

Carmen Dragomir, Cristina Oproi, Nicoleta Dragomir, Sebastian Toth, 2010. Egyptian clover (*Trifolium Alexandrinum L.*) contribution to yield increase in temporary pastures. Animal Science and Biotechnologies. 43 (2): 156-158

Roy, A. K., Malaviya, D. R., Kaushal, P., Chandra, A. and U. P. Singh. 2009. Descriptors for Tropical Forage Legume: Egyptian clover / Berseem (*Trifolium alexandrinum* L.). Indian Grassland and Fodder Research Institute, Jhansi 284 003

http://www.dpi.nsw.gov.au/aboutus/resources/factsheets

CHAPTER

68

Shaftal–*Trifolium resupinatum* L.

Vernacular names: Persian clover, shaftal clover, bird eye clover, reversed clover (English).

Importance: The young shoots are used as spinach in Pakistan and Afghanistan. It is grown as a fodder often under irrigation, for hay in its traditional areas. In warmer places, Persian clover is preferred to alfalfa.

Origin: It is native to native of Europe, Middle East, western Asia and northern Africa.

Distribution: It is traditionally cultivated in cold regions of Iran, Afghanistan, Pakistan, India and other Asian areas with cold winters. It is cultivated in cool temperate climate of Himachal Pradesh and Punjab.

Botany: Shaftal is an annual, prostrate or semi-erect branched legume, up to 20-60 cm high. It forms dense swards and has a rosette growth habit under grazing. The stems are hollow, branching from the lower part. Leaves are trifoliate with 1 to 3 cm long, oval-oblong leaflets. Flowers are pink to violet and mature to white woolly seed heads, with a resupinate corolla, hence the name *resupinatum*. Fruits are dehiscent single-seeded pods. It has pale brown seeds. Seed is ovoid, brown, about 1.2 mm long. One kg contains about 1480000 seeds.

Climate: It is grown within 40°N and the Sahara at an altitude of 2,500 m from MSL Afghan uplands, and 2750 m in the Himalayas where winter temperatures fall below -12°C; under such conditions the plant is dormant and snow-covered for months, but grows rapidly when temperatures rise in spring. It is usually cultivated as a winter annual, sown in Autumn but produces little before spring; it can be grown as a summer forage. It is a long-day plant.

Soils: It can be grown on a wide range of soils. It thrives on wet; heavy soils in low areas and can be found on roadsides, in fields, waste places, humid grassy plains and mountains. It prefers neutral to alkaline soils with pH range of 5.5 to 9.0. It is tolerant to waterlogging.

Season: It is usually cultivated as a winter season, sown in autumn but produces little before spring. It is grown both as an overwintering annual and as a summer catch-crop.

Varieties: Shaftal-48, SH-48, SH 69.

Seeds and sowing: It is broadcasted @ 5 to 10 kg/ha in autumn into a firm level seedbed.

Harvest: It can be cut 2 or 3 times with cutting intervals of 6-9 weeks in the growing season. Seed is often taken from part of the last cut set aside. The crop is mown, dried on a threshing floor and hand-threshed. It is usually the crop is windrowed before threshing.

Yield: The green fodder yield is 50 to 60 t/ha. Shaftal is a free seeder. Seed yield is 500 to 800 kg/ha.

CHAPTER

69

Red Clover–*Trifolium pratense* L. (2n=14)

Vernacular name: Red Clover, Beebread, Cow Clover, Cow Grass, Meadow Clover, Purple Clover, Wild Clover.

Importance: Red clover is used for forage in the form of hay, silage, or pasture. Red clover is primarily used for hay, pasture, silage and soil conservation. The botanical name *Trifolium,* is Latin, which means *tres* 'three' and *folium* 'leaf' while pratense is Latin for 'found in meadows', which is very much true. It is the national flower of Denmark.

Origin: It is native of Southeastern Europe and Asia Minor.

Distribution: Red clover is one of the leading forage legumes in Canada, United States and northern and eastern Europe.

Botany: Red clover is a fast starting, highly productive, but short lived perennial clover for 2 to 4 years. It has mostly upright growth habit. Primary stem produces many axillary branches at lower nodes. Therefore, unlike white clover, red clover is unable to spread through the pasture or hay field and colonize vacant sites. The basal buds produce upright, pubescent (hairy), hollow stems, 60–80 cm in length, oblong to wedge-shaped leaflets. The root system is mostly concentrated in the top 30 cm of soil. Inflorescence is globose (spherical) head (umbel), borne on long peduncles with up to 125 florets. Flowers are rose, purple, or magenta in colour. Corolla tubes are 9.0-10.5 mm in length. It is cross-pollinated mainly by bumble bees and honey bees. Red clover seeds are short and mitten shaped, 2-3 mm long, varying from pure yellow to purple.

Climate: It is widely adapted throughout the humid, temperate regions of the world. It is adapted where summer temperatures are moderately cool to warm and moisture is sufficient throughout the growing season. It requires minimum annual precipitation of approximately 380 mm in temperate regions. Growth initiates at 7°C and stops at 40°C. The optimum temperature for growth is 20-25°C, with a growth range of 7-32°C. It requires at least a 14 hour photoperiod to flower.

Soils: It grows well on loams, silt loams, and sandy loams as well as on clayey soils. It grows best on well drained soils. It is more tolerant of wet acid soils than alfalfa. Red clover will usually give higher yields than alfalfa at pH 5.5 to 6.5 in wetter areas. The soil pH for optimum growth is 5.0-7.6. It will grow with soil salinity from 0 to 4 mmhos/cm.

Variety: PRC-3

Seeds and sowing: Seeding rate is 15 kg/ha. Sowing depth should be 1.0 to 1.5 cm.

Manures and fertilizer application: 12.5 tonnes FYM or compost per ha along with 25-40-20 kg per ha of N, P_2O_5 and K_2O kg per ha are applied as basal dressing.

Weed management: Two hand weeding are given on 20 and 35 DAS. Application of Fluchloralin @ 1.2 kg a.i./ha at pre planting stage controls the weeds.

Harvest: It is productive for two cuts a year at 3 or 4 week cutting and /or grazing intervals throughout the growing season. It is harvested when 25% of the blossoms are showing colour. The upright growth habit restricts its use to cutting and/or rotational grazing.

Yield: It provides 40–60 t/ha green fodder.

Quality: There are two anti-quality factors of concern viz., bloat and oestrogenic compounds.

Uses: Red clover makes excellent hay and silage and increases the quality of grass pastures. It is also used as green manure with vegetable crops. It is one of the richest sources of isoflavones, water-soluble chemicals that act like estrogens. It is therefore used for osteoporosis, menopausal symptoms, and high cholesterol.

REFERENCES

Smith, R.R., N.L. Taylor, and S.R. Bowley. 1985. Clover Science and Technology: Red Clover. Agronomy Monograph No. 25. ASA-CSSA-SSSA. Madison, WI. 457-470.

Taylor, N.L., and R.R. Smith. 1995. Forages: An Introduction to Grassland Agriculture: Red Clover. 5th Ed. Iowa State Univ. Press. Ames, Iowa. 17:217-226.

CHAPTER 70

Lucerne–*Medicago sativa* L. (2n=32)

Family: Fabaceae (Leguminoseae)

Vernacular name: Alfalfa, Lucerne (English), Kudiraimasal (Tamil), Snail clover, Purple medick, purple medic, chilean clover. Lucerne originates from French word 'luzerne'. The name alfalfa is widely used in North America and Australia while the name 'lucerne' is used in the United Kingdom, South Africa and New Zealand. Lucerne is commonly called as rijka in northern India.

Importance: Lucerne is popularly known as the 'queen of forage crops'. Alfalfa is a perennial forage legume which normally lives four to eight years, but can live more than 20 years, depending on variety and climate. Lucerene is a highly palatable, high yielder and protein rich fodder used for horses, cattle and poultry. It can be used as hay conveniently. It is grown for fresh produce or for hay. It contains 23% crude protein. The leaf and stem ratio is 48:52. Alfalfa can be used as an important break crop in the rotation and most crops can follow alfalfa because of the high amount of root residue left in the soil.

Origin and distribution: Lucerne is originated in south-central Asia (Iran). Lucerne is grown in Europe, France, Italy, Turkistan, Siberia, Australia, USA, Canada, Latin America, South Africa and India. In India, lucerne is cultivated mostly in irrigated areas of Punjab, Haryana, Uttar Pradesh, West Bengal, Gujarat, Maharashtra and Tamil Nadu.

Botany: Lucerne is an herbaceous perennial legume with trifoliate leaves. The leaves are trifoliate and the middle leaflet possesses a short petiole. It has 2-4 m deep roots and about 60-70% of total root mass is in the upper 15 cm of soil surface. Deep penetrating roots make alfalfa quite drought resistant. Flowers are in clusters with 20 to 30. Flower is a modified raceme. The flowers are usually purple, but it may be blue, yellow or white. They are fertilized by insects, especially bees. It produces loosely coiled pods. Pods are glabrous or pubescent. Pods turn from green to brown as they mature. Pods contain 2 to 6 oval or kidney shaped seeds. Seeds are light and dark brown to greenish yellow in colour.

Growth stages: The details of crop growth stages of lucerne are furnished in Table 1.

Table 1 Growth stages of lucerne

Stage 0	*Early vegetative:* stem length up to 15 cm ; obviously vegetative (no visible buds, flowers or seed pods) ; axillary buds not easily seen.
Stage 1	*Mid vegetative:* stem length 15-30 cm ; obviously vegetative ; axillary buds developing (with 1 or 2 leaves), especially at the mid stem.
Stage 2	*Late vegetative:* stem longer than 30 cm but still vegetative ; axillary buds beginning to elongate ; inflorescence buds at apex enclosed by young leaves beginning to develop.
Stage 3	*Early bud:* one or two nodes with developing buds near apex on main axis or on branches ; no flowers or pods.
Stage 4	*Late bud:* three or more nodes with visible buds ; no flowers or pods ; clear separation of flower buds in raceme.
Stage 5	*Early flower:* one node with an open flower ; no seed pods.
Stage 6	*Late flower:* two or more nodes with open flowers ; no seed pods ; nodes with flowers spread around mid portion of stem
Stage 7	*Early seed pod:* one to three nodes with green pods usually on inflorescences at lower nodes initially.
Stage 8	*Late seed pod:* four or more nodes with green seed pods ; older stems highly branched ; leaves falling off.
Stage 9	*Ripe seed pod:* most pods brown and mature ; stem thick and fibrous ; seed ready to harvest.

Source: Kalu and Fick, 1981.

Climate: Lucerne grows up to an altitude of about 2500 m. It performs well in cooler and dry climate than cloudy, humid and wet conditions. In warm climates, the production is higher as compared to humid conditions. Lucerne is a long-day plant. The cardinal minimum, optimum and maximum temperature for lucerne are 8ºC, 25 to 27ºC and 41ºC respectively. At the second trifoliate leaf stage (15 DAS), seedlings become more susceptible to cold injury and may be killed by 4 or more hours at 26°F or lower temperatures. It needs at least 6 weeks growth after germination to survive the winter. The plant will generally survive if it develops a crown before a killing frost. The crown allows the plant to store root reserves for winter survival and spring regrowth.

Soil: It is successfully grown on a wide variety of soils, with deep, medium textured and well drained soils with rich in lime being preferred. The crop is moderately sensitive to soil salinity. This crop does not withstand waterlogging and sodic soil. The desirable soil pH is 6.5 to 8.0. Yield decrease related to electrical conductivity (ECe of extraction saturated paste in mmhos/cm) of soil is: 0% at ECe 2.0 mmhos/cm, 10% at 3.4, 25% at 5.4, 50% at 8.8, and 100% at ECe 15.5 mmhos/ cm. Avoid planting alfalfa on sites with poor surface drainage or where a high water table (within 2 1/2 to 3 feet of the soil surface) exists for most of the growing season.

Land preparation: Three to four ploughings are given to break the clods and to obtain fine tilth. Bed and channels are formed.

Season: Lucerne is grown throughout the year under irrigated condition but its growth is good during July-December.

Varieties: There are also some promising lucerne varieties like Moopa, IGFRI S-54, IGFRI S-244, IFGRI 112 (Suitable for all areas), Nimach 1, Nimach 2, Composite 3, T8 and T15. The particulars of lucerne varieties grown in different states of India are furnished in Table 2.

Table 2 Lucerne varieties grown in different states of India

Variety	Source	Green fodder yield (t/ha/year)	Suitability	Major characteristics
Sirsa 8	Sirsa, Haryana	35-40	Punjab, Haryana, Delhi and Uttar Pradesh	seed yield is 0.2-0.3 t/ha
Sirsa-9	Sirsa, Haryana	30-40	North India	quick growing, DMY is 8-12 t/ha/year, seed yield is 0.25 to 0.43 t/ha
Anand-2 (GAUL-1)	AAU, Anand	40-50	Gujarat, Rajasthan and Madhya Pradesh	Seed yield 0.2-0.3 t/ha
GAUL-2 (SS-627)	GAU, Banaskantha	50-60	Gujarat	10 to 12 cuttings in a year
Anand-3	AAU, Anand	40-50	Himachal Pradesh and Gujarat	tillers vary from 10-15/plant, seed yield 0.30 t/ha
Anand Lucerne-3 (AL-3)	AAU, Anand	50-60		profuse tillers (47/plant),
Chetak (S-244)	IGFRI, Jhansi	40-50	Punjab, Haryana, Uttar Pradesh and Gujarat	resistance to aphids
Lucerne No. 9-L	PAU, Ludhiana	50-60	Punjab	It grows well for a period of 5-6 years
LL composite 3	PAU, Ludhiana	35-40	Entire country	seed yield is 0.3 t/ha, resistant to lodging and frost
LL composite 5	PAU, Ludhiana	50-60	Punjab	seed yield is 0.3-0.5 t/ha
Rambler	Canada	50-60	hilly areas of the country	tolerant to very low temperatures
RL-88	MPKV, Rahuri	50-60	Maharashtra, Madhya Pradesh and Uttar Pradesh	resistant to major diseases and pests
NDRI Selection No. 1	NDRI, Karnal	60	Haryana	It grows well for a period of 5-6 years
Co-1	TNAU, Coimbatore	60	Tamil Nadu	

Seeds and sowing: Seed rate is 10 kg per ha. Seeds are sown with row spacing 30 cm with continuous line spacing or broadcasted. The optimum depth for seeding varies with soil texture. Planting depth should be 1.0 to 1.5 cm in heavy soils and 1.5 to 2.5 cm in lighter soils. Seeding too deep inhibits seedling emergence. Seed inoculation with *Rhizobium meliloti* is promising for crop performance especially in soils where lucerne is being cultivated for the first time.

Cropping systems: It is usually raised as a sole crop. The common lucerne based crop rotations adopted in north India are viz., maize-lucerne, paddy-lucerne, sorghum-lucerne, greengram-lucerne, soyabean-lucerne, cowpea + maize (fodder)-lucerne and sorghum (grain)-lucerne-maize (fodder).

Weed management: Hand weeding is done on 20 days after sowing and subsequently once after each harvest. Cuscuta (dodder) is a seed borne total stem parasite for which pure seeds are used. Remove the lucerne plant along with cuscuta parasite and burn it. It should not be cultivated in cuscuta infested field at least for three years. Pendimethalin 1-2 kg /ha (pre-emergence) or diquat @ 6-10 kg/ha 5-10 days after sowing controls *Cuscuta*. Pre-sowing application of diuron @ 2.0 kg/ha or fluchloralin @ 1 kg/ha or EPTC @ 3.0 kg/ha or MCPB @ 0.75 kg/ha after 3 DAS or pronamide @ 1.0 kg/ha just after sowing is recommended to control the weeds.

Manures and fertilizer application: 12.5 tonnes FYM or ha is applied as basal dressing before ploughing. 25-40-20 kg of N, P_2O_5 and K_2O kg per ha are applied before sowing. Subsequently, top dressing of 25 kg N per ha is given after each cutting. It is capable of fixing atmospheric nitrogen which meets its requirements for high yields. Boron deficiency is generally noticed in leached and coarse textured soils. The leaves develop numerous pale-yellow spots leading to disorder known as Lucerne yellow. Spray of 0.2% borax can overcome this deficiency. Iron deficiency, leading to chlorosis, is fairly common in poorly drained alkaline soils. Liming the soil well in advance of sowing is helpful in areas where soil is acidic. Application of 20 kg/ha each of S and Zn along with 2 kg/ha of Mo may enhance the effectiveness of biological nitrogen fixation.

Water management: It can be raised both as rainfed or irrigated crop in high water table areas. Its roots can extend 2 to 4 m deep in most soils. Being a deep rooted crop, it extracts water from the deeper zone of the soil. If the crop is adequately irrigated, it will get most of its water from the upper 1.5 to 2.0 m soil layer. An adequate level of soil moisture is defined as more than 50% of the available water capacity of the soil. It is relatively drought tolerant, but its growth does depend on available soil water. A good rule of thumb to estimate water needs is that each ton of cured hay consumes approximately 15 cm of water. First irrigation is given immediately after sowing. Life irrigation is given on the third day and subsequently once in 7-10 days. Crop water requirements are between 800 and 1600 mm during its growth period depending on climate and length of growing period. The kc value is about 0.4 just after cutting, increasing to 1.05 to 1.2 just prior to the next cutting with a mean value of 0.85 to 1.05. For seed production, the kc value is equal to 1.05 to 1.2 during full cover until the middle of flowering, after which the kc value is reduced sharply. Plant stress can occur when available soil moisture falls below 50%. Border, corrugation (furrow), controlled flooding, and sprinkler irrigation can be adopted.

Crop protection: The most common diseases are rust and leaf spot. Dithane M-45 or 0.1% Chlorothalonil is recommended to control the diseases.

Harvest: In climates with mild winters, it is grown for 3 to 4 years continuously, but in continental climates with cold winters it is grown for 6 to 9 years, with a dormant period in winter. The crop is also grown as a short season annual crop. It can supply green fodder continuously for 3 to 4 years from the same crop stand. First harvest is taken at 60 to 75 days when sparse (few) flowering occur. Subsequent harvest can be taken up once in 50 to 55 days. If harvesting for maximum persistence, cut alfalfa between first flower and 25% flower. This is approximately 35 to 40 days between cuttings. Take the second cutting 28 to 33 days after the first cut and take subsequent cuttings at 38- to 55-day intervals or at 10 to 25% bloom.

Yield: Green forage yield per cut for a given location varies over the year due to soil and climatic differences. Green fodder yield of 50 to 70 tonnes per ha can be obtained in 8 to 10 cuts. It produces its highest yields during the second year of growth. Seed yield is 200 to 250 kg per ha. Thousand seed weight is 2 grams. Crude protein content is 20 to 24 %.

Seed Production: The seed crop should be sown in rows 50 cm apart. Foliar spray of 0.5% borax at pre-floweirng stage is done to enhance seed production. The harvesting of mature crop should not be delayed to avoid the shedding of pods. Harvest the crop when two-thirds of the pods become dry. The seed yields usually vary from 0.2-0.3 t/ha.

Good hay-making practices: Fresh plants contain about 80% moisture. Rake the plants at 40–50% for hay making. Bale hay at 18–20% moisture content. Hay can be stored under cover to protect from rain, sun, etc. To speed drying, a drying agent such as sodium or potassium carbonate can be used as it is cut in addition to mechanical conditioning. They will shorten drying time by 5 to 24 hours. Drying agents do not work on grasses. Preservatives such as propionic acid, ammonium propionate, acetic acid, etc., are used to bale the hay at higher moisture contents than can normally be stored.

Good silage practices: Fresh plants are chopped and ensiled at 30–50% dry matter content to optimize fermentation. Silo is sealed for at least 14 days to allow complete fermentation. The deteriorated silage is discarded. Lactic acid forming bacteria, *Lactobacillus buchneri* is uniformly mixed throughout the forage for good silage fermentation. Total losses from cutting to feeding are 20% to 30% of the standing crop dry matter in typical hay and silage systems. In hay-making, most of the losses result from mechanical handling and weather damage in the field. In silage-making, most losses occur during storage and feed out.

Utilization: Livestock fed exclusively with lucerne suffer from bloat due to the presence of saponin glucoside and rich in protein content (20 to 24 %). The green lucerne is to be chafed and mixed with grass in 1:2 ratio. Lucerne can be given to animal up to 10 kg in a day.

Autotoxicity: Lucerne leaves produce medicarpin, a toxin that accumulates in the soil and increases in potency during the life of a stand. The compounds impair development of the seedling taproot by causing the root tips to swell and by reducing the number of root hairs. This limits the ability of the seedling to take up water and nutrients and increases the plant's susceptibility to other stress factors. This toxin decreases germination and growth of new lucerne seedlings. This phenomenon is known as autotoxicity. Medicarpin dissipates under warm, moist soil conditions. Removing all top growth of the old stand before plowing it also reduces the potential for autotoxicity.

REFERENCES

Christian, K. R. 1977. Effects of the environment on the growth of alfalfa. Advances in Agronomy 29: 183-227.

Duke, J.A. 1981. Handbook of legumes of world economic importance. Plenum Press. NewYork.

Kalu, B. A.; Fick, G. W. 1981. Quantifying morphological development of alfalfa for studies of herbage quality. Crop Science 21: 267-271.

Smith, D.H. Beck, G.K. Peairs, F.B. and W.M. Brown. Alfalfa: Production and Management no. 0.703. Colorado State University Cooperative Extension. 10/99.

www.colostate.edu/Depts/CoopExt

http://en.wikipedia.org/wiki/Alfalfa

http://www.fao.org/ag/AGP/AGPC/doc/Gbase/DATA/Pf000346.htm

http://www.soilcropandmore.info/crops/alfalfa/A3681_Alfalfa_Germination_and_Growth.pdf

http://learningstore.uwex.edu/assets/pdfs/A3681.PDF

https://www.agronomy.org/files/publications/alfalfa-management-guide.pdf

http://alfalfa.ucdavis.edu/IrrigatedAlfalfa/pdfs/UCAlfalfa8289GrowthDev_free.pdf

http://www2.ca.uky.edu/agc/pubs/agr/agr76/agr76.htm

http://www.hort.purdue.edu/newcrop/duke_energy/Medicago_sativa.html

https://www.agronomy.org/files/publications/alfalfa-management-guide.pdf

CHAPTER

71

Hedge Lucerne–*Desmanthus virgatus* (L.) Willd. (2n=28)

Synonyms:

Desmanthus virgatus (L.) Willd. var. *strictus* (Bertol.) Griseb.
Desmanthus depressus Humb. & Willd.
Desmanthus pernambucanus (L.) Thellung
Desmanthus diffusus Willd. *Desmanthus strictus* Bertol.

Family: Fabaceae

Vernacular name: Desmanthus, slender mimosa, donkey bean, hedge lucerne, virgate mimosa, wild tantan (English); *Vellimasal* (Tamil); *Little mimosa, anil, jureminha* (Brazil), Bundle flower (USA), *Leucaena mini* (Indonesia); *adormidera, brusca prieta, dwarf koa, frijolillo, ground tamarind, guajillo* (Mexico); *guashillo, huarangillo, langalet, petit acacia, petit cassie, petit mimosa, virgate mimosa, dwarf koa* (Hawaii);

Importance: The word *Desmanthes* originates from the Greek word '*desmos*' meaning bundle and 'anthos' meaning 'flower'. Desmanthes is an evergreen shrub. Leaves are highly palatable. It grows in moist roadside situations, ditches, and abandoned pastures; also common in coastal thickets and along the edges of marshes. It is an ideal forage crop for saline and alkaline soils. Hedge lucerne can be pruned and trained as a hedge plant and the prunings can be fed to animals. It withstands heavy grazing pressure and recovers well from fire. The seeds are dispersed by ruminant animals.

Origin: Hedge lucerne is a native of Mexico, Central America, South America, and the Caribbean.

Distribution: It is grown in Australia, Africa, south-east Asia, USA (Florida), Indonesia, Queensland, Mauritius, Hawaii, Malaysia, India, Thailand, Guyana, Surinam, Brazil, Caribbean, Anguilla, Antigua, Barbados, Bermuda, the Bahamas, Cuba, Grenada, Guadeloupe, Hispaniola, Martinique, Montserrat, Puerto Rico, Tobago, Trinidad, the Virgin Islands, Pacific and Indian Ocean Islands,.

Botany: Hedge Lucerne is a perennial shrub. It grows to a height of 2 to 3 metre. The plant is erect or diffuse. It has pithy stem and tap root. Leaves are bipinnate, 2.4–8 cm long with 6-8 pinnae 2–9 mm long persistent stipules. Flowers are white in colour and are borne is a globose head. Flowering commences in 45 to 50 DAS. Pods are flat, long and slender with acute tip. Pod colour is reddish to golden brown. Pod contains 9 to 27 seeds and 5–8 cm long. Seed is 2.1-2.9 mm long, 1.4-2.7 mm wide, obliquely inserted, ovate, flattened, red- or golden-brown. Seed weighed an average of 0.0005 g/seed or 244,000 seeds/kg.

Climate: Hedge lucerne grows well in tropical and subtropical environments. The species grows from near sea level up to 1900 m in elevation. It is a drought tolerant crop and cultivated mostly in semiarid regions receiving annual rainfall of 500 to 1600 mm. It is moderately tolerant of poor drainage. The green tissue killed by frost, but regrows from old wood and to some extent from the base.

Soils: Hedge lucerne grows well in deep, well to moderately well-drained silts, clays, loams, and sands with pH's from about 5 to about 8, formed over both igneous and sedimentary rocks. It is seldom seen in shallow, rocky sites. It grows fairly well in sandy loam and loamy soils. However it grows on a wide range of soils including acidic, saline and alkaline with a soil pH range of 5.0 to 9.5.

Season: The crop is raised during June-December under rainfed condition utilizing the rainfall of both southwest and northeast monsoon seasons and throughout the year under irrigated condition.

Varieties: CO1.

Land preparation: Two to three ploughings are given for pure crop. Ridges and furrows are formed at a spacing of 50 cm.

Seeds and sowing: Seed rate is 2 and 3 kg per hectare for irrigated and rainfed conditions respectively. Seeds are soaked in hot water at 80ºC for 5 minutes. Boiling water removed from flame and kept for four minutes to obtain temperature of 80ºC. Seeds can also be treated with conc H_2SO_4 for 8 minutes followed by thorough washing with water is recommended to enhance germination. Seeds are sown with a row spacing of 50 cm on both side of the ridges closely in lines under irrigated condition at a depth of 1–2 cm. However, the seeds are sown in lines with a row spacing of 30 cm under rainfed condition. It is easy to plant with stem cuttings.

Cropping system: The crop can either be grown as sole crop or as mixed crop with grasses. It can be grown in hedgerows. Alley cropping of hedge lucerne with perennial hybrid grasses like NB 21, BN 2, CO 1 , CO 2, CO 3 and guinea grass under irrigated condition and with pasture grasses like cenchrus under rainfed condition in 1:3 ratio can be raised to provide a balanced feed to livestock.

Weed management: One or two weeding is required during establishment.

Manures and fertilizer application: Generally 12.5 tonnes FYM per ha and 100-60-30 kg of N, P_2O_5 and K_2O per ~~hectare~~ are applied basally before sowing and 50 kg N per ha after each cutting. 15 grams of super phosphate and 2-3 hand full of farm yard manure (FYM) per plant in bund cropping for every two years.

Water management: First irrigation is given at the time of sowing. Life irrigation is given on the third day. Subsequently irrigations are given once in a week. Pot watering is done for bund crops if necessary. Necessary drainage has to be provided for Hedge lucerne in sodic soil. Continuous water stagnation for a week causes iron deficiency showing yellowing and whitening of leaflets. This can be overcome by foliar spraying of one per cent ferrous sulphate solution.

Harvest: The optimum stage for harvest is at the early pod stage and can be harvested 4 times/year at 10 to 15 cm height above the ground. Harvest can be taken up when plant grows 50 cm height. First harvest can be taken up on 90 DAS and subsequently at an interval of 60 days at 45–60 cm height with six cuts per year. It produces flowers 45–50 days after cutting. There is vigourous development of crown with a number of branches after each harvest favours a good fodder yield. The crop can be kept in field for a period of three years after which it has to be re-sown in fresh area.

Yield: Green fodder yield is 50 and 20 tonnes per ha per year under irrigated and rainfed condition respectively. Green fodder yield of 35 tonnes per hectare per year is in sodic soils having a pH of 9.5, EC 3 millimhos per cm and ESP 30 under irrigated condition.

Utilization: *Desmanthus* spp. do not cause bloat in ruminants because they contain 2-3% (of total DM as tannic acid equivalent) condensed tannins. Leaves contain 22% and stems contain 7.1% crude protein respectively. The whole plant contains 12% crude protein.

REFERENCES

Burt, R.L. 1993. *Desmanthus*: a tropical and subtropical forage legume. Part 1. General Review. Herbage Abstracts, 63, 401–413.

Department of Primary Industries. 2002. Legumes for the tropics and subtropics: Desmanthus (*Desmanthus virgatus*). Queensland Government, Brisbane, Australia. http://dpi.qld.gov.au/pastures/4490.html. 2 p.

Hopkinson, J.M. and English, B.H. 2004. Germination and hardseededness in desmanthus. Tropical Grasslands. 38: 1-16.

Institute of Systematic Botany. 2002. Atlas of Florida plants: *Desmanthus virgatus*. University of Florida, Gainesville, FL.
http://www.plantatlas.usf.edu/maps.asp?plantID=1474. 1 p.

Jones, R.M. and Brandon, N.J. 1998. Persistence and productivity of eight accessions of Desmanthus virgatus under a range of grazing pressures in tropical Queensland. Tropical Grasslands, 32, 145–152.

Luckow, M. 1993. Monograph of *Desmanthus* (*Leguminosae-Mimosoideae*). Systematic Botany Monographs, Vol. 38. The American Society of Plant Taxonomists.

MacRoberts, M.H. and B.R. MacRoberts. 2011. *Desmanthus virgatus* (Fabaceae): New to Louisiana. Phytoneuron. 54: 1–3.

Pengelly, B.C. and Liu, C.J. 2001. Genetic relationships and variations in the tropical mimosoid legume *Desmanthus* assessed by random amplified polymorphic DNA.Genetic Resources and Crop Evaluation, 48: 91–99.

Skerman, P.J., D.G. Cameron, and F. Riveros. 1988. Tropical forage legumes. 2nd. Ed. Food and Agriculture Organization of the United Nations, Rome. 692 p.

Turner, B.L. 1950. Texas species of *Desmanthus* (Leguminosae). Field and Laboratory, 18: 54–65.

Internet links

http://www.hear.org/pier/species/desmanthus_virgatus.htm

http://www.fao.org/ag/AGP/AGPC/doc/Gbase/data/pf000151.htm

http://www.tropicalforages.info/key/Forages/Media/Html/Desmanthus_virgatus.htm

CHAPTER

72

Pea Blue–*Clitoria ternatea* L. (2n=16)

Family: Fabaceae (alt. Leguminosae)

Vernacular names: Pea blue creeper, blue-pea, bluebellvine, butterfly-pea, cordofan-pea, kordofan pea, Darwin-pea, Conch flower creeper, Mussel-sheel, (English), Sangu pushpam, Kokkattan, Kuvalai (Tamil), Sankhu Poolu / Sankham Poolu (Telugu), Sankhulpushpi (Malayalam), Aparajita (Hindi, Bengali), Shankha Pushpa (Konkani), Gorani (Gujarat), Gokarna (Marathi) and Buzrula (Arabic), Honte (French); Blaue Klitorie (German) and Cunha (Brazil).

Importance: Pea blue is known as 'tropical alfalfa'. It is a drought resistant, hardy perennial pasture legume. It is highly palatable. It is susceptible to continuous heavy grazing. It persists best when grazed lightly during the wet season. It combines well with buffel grass, pangola grass or with native pastures to grow as mixed pastures. It can be grown all along the fence or as a border crop. The animals like the soft stem and leaves. It is grown for ornamental uses. The flowers are chunk shaped and are used in religious rituals. The root and the seeds are of medicinal value. The pea blue contains the pentacyclic triterpenoids such as taraxerol and taraxerone.

Origin: It is a native of India and Latin America. It is naturalized in all the semi-arid and sub-humid tropics of Asia, Africa and Australia.

Distribution: It is native to tropical America and widely grown in warmer parts of the world.

Botany: The butterfly pea is a vigorous, trailing, scrambling or climbing tropical legume. It is an indeterminate, vine creeper and fast in its growth. Its sparsely pubescent stems are sub-erect and woody at the base and may be up to 5 m long. They root only at the tips. The leaves are pinnate, bearing 5-7 elliptical, 3-5 cm long leaflets. The flowers are solitary or paired, deep blue or pure white, about 4 cm broad. It has an infundibular calyx, a standard lacking a spur, wings longer than the keel petals, a geniculate style. The flowers of this vine have the shape of human female genitals, hence the Latin name of the genus

'*Clitoria*' from 'clitoris'. Pods are flat, linear, beaked, 6-12 cm long, 0.7-1.2 mm wide and slightly pubescent with up to 8-10 seeds. The seeds are olive, brown or black in colour, often mottled, 4.5-7 mm long and 3-4 mm wide.

Climate: Pea blue grows within 20°N and 24°S and from sea level up to an altitude of 1600-1800 m. It grows in rainfall ranging from 400 to 1500 mm. It has also some drought tolerance. It does well under irrigation but has only a low tolerance of flooding or waterlogging. It can survive a 5-6 month drought in the drier tropics. It can tolerate shade to some extent. It grows well with average temperature of about 19-28°C. However, it tolerates temperatures as low as 15°C and even some frost as it may regrow from the stems or from the plant base, provided it is already woody when the frosting occurs.

Soils: It is adapted to a wide range of soil conditions from sandy to deep alluvial loams and heavy black cracking clays. It has tolerance to moderately saline, alkaline and calcareous soils with pH 5.5-8.9.

Season: Pea blue is sown during June-July under rainfed conditions.

Varieties: White and blue petal as well as broad petal types

Land preparation: Plough the field two to three times to obtain a good tilth.

Seeds and sowing: Seed rate is 15 to 20 kg per ha. Seeds require conc sulphuric acid of S.G. 1.8 treatment for 20 minutes to break the dormancy. Spacing is 45 cm x 10 cm. Seeds are sown at a depth of 2-3 cm. The plants may be grown with support crops (or) staked with bamboo to facilitate hand picking of the pods.

Cropping systems: It is hardy perennial highly suitable for mixed cropping with cereal or grass fodder. It can be grown as mixed crop with perennial grasses like guinea grass, para grass, blou buffel grass, etc.

Weed management: One or two weeding is required during establishment.

Manures and fertilizer application: 12.5 tonnes of FYM and 30-40-20 of N, P_2O_5 and K_2O kg per ha is recommended under rainfed condition.

Water management: Pea blue is mostly raised as a rainfed crop.

Harvest: A fast summer-growing legume, it can cover the soil within no more than 30-40 days after sowing and yield mature pods within 110 to 150 days. In the first year, it gives only one cut but in subsequent years depending upon the amount of rainfall and its distribution two or more cuts may be taken. First cut is given on 75 to 80 days after sowing and subsequent cuts once in 50 days based on growth. It produces a large amount of seeds and will readily self-seed, when the dry pod shatters.

Yield: Green fodder yield of 30 and 12 tonnes per ha per year can be obtained in irrigated and rainfed condition respectively. The dry matter yield varies from 1.1 to 3.3 t/ ha in first year under rainfed condition while under irrigated condition it yields around 13.3 tonnes dry matter per hectare. Seed yield is 50 to 75 kg/ha.

Utilization: Butterfly pea is used in fences and in trellises as an ornamental for its showy flowers, valuable for dyeing and in ethno-medicine. It is the most palatable fodder for all types of cattle. It should be grazed lightly and in rotations to preserve the pasture for longer period. It is also used for cut-and-carry feeding systems and cut for hay and silage. As nitrogen fixing legume, it is used as a ley legume or as green manure. It is a valuable cover crop in rubber and coconut plantations. The young pods are edible and used as vegetables.

REFERENCES

Conway, M.J., McCosker, K., Osten, V., Coaker, S. and Pengelly, B.C. 2001. Butterfly Pea - A Legume Success Story in Cropping Lands of Central Queensland. In; Rowe, B., Mendham, N. and Donaghy, D. (eds) Proceedings of the 10th Australian Agronomy Conference, Hobart.

Jones, R.M., Bishop, H.G., Clem, R.L., Conway, M.J., Cook, B.G., Moore, K. and Pengelly, B.C. 2000. Measurements of nutritive value of a range of tropical legumes and their use in legume evaluation. Tropical Grasslands, 34, 78-90.

Michael Gomez, S. and A. Kalamani. 2003. Butterfly Pea (*Clitoria ternatea*): A Nutritive Multipurpose Forage Legume for the Tropics - An Overview. Pakistan Journal of Nutrition 2 (6): 374-379

Pengelly, B.C. and Conway, M.J. 2000. Pastures on Cropping Soils: Which Tropical Pasture Legume to Use? Tropical Grasslands, 34, 162-168.

Reid, R. and Sinclair, D.F. 1980. An Evaluation of a Collection of Clitoria ternatea for Forage and Grain Production. Genetic Resources Communication, 1, 1-8.

Staples, I.P. 1992. Clitoria ternatea L. In: 't Mannetje, L. and Jones, R.M. (eds) Plant Resources of South-East Asia No. 4. Forages. pp. 94-96. (Pudoc Scientific Publishers, Wageningen, the Netherlands).

Fantz, P.R. 1991. Ethnobotany of *Clitoria* (Leguminosae). Econ. Bot. 45:511–520.

Fantz, P.R. 1993. Revision of cultivated *Centrosema* and *Clitoria* in the United States. HortScience 28:674–676.

Banerjee, S.K. and Chakravarti, R. N.1963. Taraxerol from *Clitoria ternatea,* Bull Calcutta School Trop Med, 11: 106-107.

Hall, T. J. 1985. Adaptation and Agronomy of *Clitoria ternatea* L. in Northern Australia, Tropical Grasslands, 19(4): 156-163.

Reid, R. and Sinclair, D.F. 1980. An evaluation of a collection of *Clitoria ternatea* for forage and grain production. Genetic Resources Communication, 1: 1-8.

https://en.wikipedia.org/wiki/Clitoria_ternatea

http://www.fao.org/ag/AGP/AGPC/doc/Gbase/data/pf000021.htm

http://www.tropicalforages.info/key/Forages/Media/Html/Clitoria_ternatea.htm

CHAPTER

73

Siratro–*Macroptilium atropurpureum* (DC.) Urb. (2n=22)

Family: Fabaceae

Synonym: *Phaseoulus atropurpureus* DC.

Vernacular name: Siratro, Atro, purple bush-bean (English), Siratharai (Tamil), Conchito (Spanish), Purpurbohne (German).

Importance: Siratro is a tropical, herbaceous, twining legume and long trailing stems. Siratro is drought resistant pasture legume. It tolerates heavy grazing. It is a shade tolerant crop. It withstands high summer temperature. It grows for 4 to 5 years. The protein content is 15%. It is commonly used for cattle pastures intercropped with grass, used in hay, or as a ground cover to prevent soil erosion and to improve soil quality. It is well suited for grazing when intercropped in fields with grass. It is very adaptable to soil, drought resistant, has good N fixation, and has a high palatability to domesticated animals. Sowing siratro on a weed infested field will reduce and kill weed populations, and improve the soil at the same time.

Origin: Siratro is a native of Central and South America with specific region of Peru, Brazil and Caribbean.

Distribution: It is grown in Mexico, Brazil, Argentina, Colombia, Ecuador, French Guiana, Peru, El Salvador, Guatemala, Honduras, Nicaragua, Panama, Bahamas, Cuba, Puerto Rico, USA, Australia and India.

Botany: Siratro is a deep rooted perennial with trailing, hairy stems and rooting at the nodes. It rapidly develops dense dark hairy green vines approximately 5 mm in diameter, with dark purple flower buds. The vines contain trifoliate dark green leaves on the upper surface, and silvery and very hairy on the lower surface. Stem produces root at the nodes. Leaves are approximately 2-7 cm long with smooth hairs on the underside. Leaflets are oval shaped. The petiole is 0.5 to 5 cm long; stipules 4 to 5 mm long, ovate-acuminate, strongly veined. Inflorescence is a raceme with 6 to 12 deep purple flowers with a velvety sheen, reddish at the base. It is self-pollinated crop. Pods are straight, about 75 mm long, cylindrical shattering and sharp pointed. It contains 5 to 12 seeds per pods. Seeds are flattened ovoid shape with brown to black in colour. The seeds are small brown peas with a white spot. The long hirsute pods shatter when ripe. The seed count is 75000 per kg.

Climate: Siratro is one of the major tropical forage legumes in tropics and subtropics. It is found in areas from 30°N to 22°S up to an altitude of 2900 m from MSL. It prefers hot and humid climate. It can be grown in areas receiving annual rainfall of 650 mm to 1500 mm. It is tolerant to drought because of its taproot and pubescent leaves that reduce evaporation. It cannot tolerate waterlogging. The optimum temperature for growth is with day/night temperature of 27-30/22-25°C while the growth is greatly reduced with temperature below 18/13°C day/night and ceases at 14°C. Leaves are burnt by light frost. More severe frosts kill the plant back to the crowns, however, plants recover with the onset of warm, moist conditions.

Soil: It comes up in all soils with good drainage. Sandy loam soil is preferred. It can be grown in soils with a pH level from 5.0 to 9.0.

Season: Seeds may be sown in June-July and September-October. It needs warm, subtropical climate. It is dormant in winter and slow to start growing again in spring.

Land preparation: Two to three ploughings are given and clods are broken to have fine tilth. Beds and channels are formed.

Seeds and sowing: Seed rate is 5 to 6 kg per ha. Seeds are sown in a row spacing of 45 cm with continuous line sowing or broadcasted. The seeds sowing depth is 1 to 2 cm. Siratro is propagated naturally by legumes bursting and disbursing seeds forcefully into the ground. The nodes from the stems close to ground level may root in favourable weather and soil conditions.

Cropping system: Siratro can be intercropped with cenchrus grass in 1:3 ratios under rainfed condition.

Weed management: Two hand weedings are given on 20 and 35 DAS. Subsequent weedings are given depending upon the necessity.

Manures and fertilizer application: FYM 10 tonnes per ha, and 20-40-20 kg of N, P_2O_5 and K_2O per ha are applied as basal dressing. Subsequently 20 kg of nitrogen per ha is applied as top dressing after each cut.

Water management: It is mostly grown as pasture crop under rainfed condition. It cannot tolerate waterlogging.

Cropping systems: It can be grown along with grasses like Rhodes, Setaria, Green panic and Guinea grass.

Harvest: First harvest is taken up at 120 days after sowing and subsequent harvests are taken once in 60 days. Pods mature unevenly. Seeds shatter at maturity. Pods are collected in the early morning. It has a high palatability to domesticated animals.

Grazing management: It is to be lightly grazed at all times leaving at least 15 cm of stubble with green leaves after grazing. In thinning stands, it should be allowed to set seed to improve the recruitment of new plants. Individual plants are not long-lived and die out after three to four years and need to be replaced by natural seedling regeneration.

Yield: Green fodder and dry fodder yield is 20 to 25 and 3 to 5 t per ha.

Seed production: It produces the seeds two times in a year i.e. October-November and April-May. The ripe pods should be picked early in the day because as the day progresses, ripe pods dry out and dehisce violently with little stimulation. Seed yield varies from 100 to 300 kg/ha. Hundred seed weight is 1.25 grams.

Utilization: It is grown as a cover crop for soil conservation, fallow crop or as a forage crop. It is mainly used for permanent and short-term pastures. It is suited to grazing and for cut-and-carry or conserved as hay. Its value as a protein bank for dry season feeding is reduced by its tendency to drop leaf under very dry conditions.

REFERENCES

Bray, R.A. and Woodroffe, T.D. 1994. *Macroptilium atropurpureum* (DC.) Urban (atro) cv. Aztec. Australian Journal of Experimental Agriculture, 34, 121.

Cameron, D.G. 1985. Tropical and subtropical pasture legumes 5.Siratro (*Macroptilium atropurpureum*): the most widely planted subtropical legume. Queensland Agricultural Journal, 111, 45-49.

Cook, S. and Ratcliff, D. 1985. Effect of fertilizer, root and shoot competition on the growth of Siratro (*Macroptilium atropurpureum*) and Green Panic (*Panicum maximum* var. *trichoglume*). Australian Journal of Agricultural Research, 36, 233-245.

Costa, N de L., Paula, J.R.F. de, Fernandes, R.N. and Jacques, A.V.A. 1985. Siratro: Production and Management. Lavoura Arrozeira, 38, 20.

English, B.H. 1999. *Macroptilium atropurpureum* in Australia. In: Loch D.S. and Ferguson J.E. (eds) Forage seed production. Volume 2: Tropical and subtropical species. pp. 407-412. CABI Publishing, Wallingford, UK.

Hutton, E.M. 1962. Siratro - a tropical pasture legume bred from *Phaseolus* DC. Australian Journal of Experimental Agriculture and Animal Husbandry, 2, 117-125.

Hutton, E.M., Williams, W.T. and Andrew, C.S. 1978. Differential tolerance to manganese in introduced and bred lines of *Macroptilium atropurpureum*. Australian Journal of Agricultural Research, 29, 67-79.

Hutton, E.M., Beall, L.B. and Williams, W.T. 1978. Evaluation of bred lines of *Macroptilium atropurpureum*. Australian Journal of Experimental Agriculture and Animal Husbandry, 18, 702-707.

Hutton, E.M., Williams, W.T. and Beall, L.B. 1957. Reactions of lines of *Phaseolus atropurpureus* to four species of root-knot nematode. Australian Journal of Agricultural Research, 23, 623-632.

Jones, R.M., Bishop, H.G., Clem, R.L., Conway, M.J., Cook, B.G., Moore, K. and Pengelly, B.C. 2000. Measurements of nutritive value of a range of tropical legumes and their use in legume evaluation. Tropical Grasslands, 34, 78-90.

Jones, R.M. and Bunch, G.A. 1987. The effect of stocking rate on the population dynamics of Siratro in Siratro-Setaria pastures in south-east Queensland. II Seed Set, Soil Seed Reserves, Seedling Recruitment and Seedling Survival. Australian Journal of Agricultural Research, 39, 221-234.

Jones, R.M. and Bunch, G.A. 1988. The Effect of Stocking Rate on the population dynamics of Siratro in Siratro-Setaria pastures in south-east Queensland. I.Survival of plants and stolons. Australian Journal of Agricultural Research, 39, 209-219.

Jones, R.M. and Bunch, G.A. 2003. Experiences with farm pastures at the former CSIRO Samford Research Station, south-east Queensland, and how these relate to results from 40 years of research. Tropical Grasslands, 37, 151-164.

Jones, R.M. and Jones, R.J. 2003. Effect of stocking rates on animal gain, pasture yield and composition, and soil properties from setaria-nitrogen and setaria-legume pastures in coastal south-east Queensland. Tropical Grasslands, 37, 65-83.

Jones, R.M. and 't Mannetje, L. (1992) *Macroptilium atropurpureum* (DC.) Urban. In: 't Mannetje, L. and Jones, R.M. (eds) Plant Resources of South-East Asia No. 4. Forages. pp. 155-157. (Pudoc Scientific Publishers, Wageningen, the Netherlands).

Kretschmer, A.E. Jr., Sonoda, R.M., Bullock, R.C., Snyder, G.H., Wilson, T.C., Reid, R. and Brolman, J.B. 1985. Diversity in *Macroptilium atropurpureum* (DC.) Urb. In: Proceedings of the XV International Grassland Congress, Kyoto, Japan. 24-31 August 1985. pp. 155-157. (Science Council of Japan and Japanese Society of Grassland Science, Nishi-nasuno, Japan).

Lenne, J.M. and Sonada, R.M. 1985. Diseases of *Macroptilium atropurpureum* - A review. Tropical Grasslands, 19, 28-34.

Loch, D.S. and Ferguson, J.E. (eds) 1999. Forage Seed Production Volume 2: Tropical and Subtropical Species. (CAB International, Oxon., UK).

Mannetje, L. 't and Jones, R.M. 1990. Pasture and animal productivity of buffel grass with Siratro, lucerne or nitrogen fertilizer. Tropical Grasslands, 24, 269-281.

Moore, G, Sanford, P and Wiley, T. 2006. Perennial pastures for Western Australia, Department of Agriculture and Food Western Australia, Bulletin 4690, Perth.

Walker B. 1977. Productivity of *Macroptilium atropurpureum* cv. *Siratro* pastures. Tropical Grasslands, 11, 79-86.

Internet links

http://www.fao.org/ag/agp/AGPC/doc/gbase/data/pf000049.htm
http://www.fao.org/ag/aga/agap/frg/afris/default.htm
http://www.pi.csiro.au/ahpc/legumes/pdf/siratro.pdf
http://www.fao.org/ag/agp/AGPC/doc/gbase/data/pf000049.htm
http://www.tropicalforages.info/key/Forages/Media/Html/Macroptilium_atropurpureum.htm
https://www.daf.qld.gov.au/__data/assets/pdf_file/0003/65289/IPA-Siratro-PP93.pdf
http://plants.usda.gov/core/profile?symbol=MAAT80
http://www.totoagriculture.org/PDFs/Wikipedia/1/24/24_1135.pdf
https://en.wikipedia.org/wiki/Macroptilium_atropurpureum

CHAPTER

74

Centro–*Centrosema pubescens* Benth (2n=20)

Family: *Fabaceae* (alt. *Leguminosae*)

Vernacular names: Centro, Butterfly pea, Spurred butterfly pea (English), Jetirana (Argentina, Brazil), Campanilla (Colombia), Senthiri (Tamil).

Importance: Centro is a pasture legume. It is a shade tolerant crop. It tolerates poor drainage and withstands seasonal flooding. It has a reasonable drought tolerance due to its deep root system. It has quick regeneration capacity with vigorous growth. It is a good cover crop and controls soil erosion. It is palatable and withstands heavy grazing.

Origin: It is native to tropical Central and South America.

Distribution: Centro is grown in India, Australia, Brazil, Nigeria, South-East Asia, Indonesia, Africa and Pacific Islands. In India, it is well suited to west coast of the peninsula.

Botany: Centro is a perennial vine of trailing habit. It forms a dense cover up to a height of 45 cm. Stem is dark green and remains succulent for 18 months. The root system can reach up to 30 cm in depth. Stems do not become woody until about 18 months after planting. Leaves are trifoliate, with elliptical leaflets approximately 2.5 cm long, dark-green and glabrous above but whitish and densely tomentose below. Flowers are generally pale violet with darker violet veins, born in axillary racemes. Fruit is a flat, long, dark brown pod 7.5–12 cm long, containing up to 20 seeds. Seeds are spherical, about 4–mm (0.16–inches) in diameter, dark brown when ripe. The number of seeds per kg is 40,000.

Climate: Centro grows well in hot humid tropical climate. It is cultivated in regions from 22°N and 22°S. It is widespread below 600 m but can be grown from sea level to 915 m from MSL. The rainfall requirements range between 750 to 1700 mm. It withstands 3 to 4 month dry periods. It survives stagnant water for at least two months. It is a short day plant, that is, the plant needs short days to flower.

Soil: It will grow on a wide range of soils, from sandy loams to clay loams. It thrives well in alluvial to medium fertile soils and is moderately tolerant to poor drainage. It grows best in a soil pH from 4.0 to 9.5. It yields better at pH levels between 6.1 and 6.4 and it grows well in sandy loam soils.

Season: Seeds may be sown in June-July and September-October.

Seeds and sowing: Seeds of *Centro* have a mechanical dormancy that has to be broken by soaking the seeds for 3–5 minutes in water at 85°C. After this seed inoculation should be carried out for sowing. Seed rate is 4 to 5 kg/ha for pure stand and 2 to 2.5 kg/ha for mixed pasture. For green manure, it can be sown up to 8 kg/ha. Seeds are sown at a row spacing of 45 cm with continuous line spacing in lines. The seed can be sown to depths of 2.5 to 5 cm. Seed are also broadcasted. It takes about four months for establishment, but once established, it can be maintained up to 10 to 15 years.

Land preparation: Two to three ploughings are given to prepare a good seed bed. Clods are broken and fine seed bed is prepared. Beds and channels are formed.

Cropping system: Centro can be mixed with guinea grass and signal grass in pastures. It is grown as cover crop in coconut and rubber plantations.

Weed management: Two hand weeding are given on 20 and 35 days after sowing.

Manures and fertilizer application: FYM 12.5 tonnes per ha and 20-40-0 kg of N, P_2O_5 and K_2O per ha are applied as basal dressing.

Water management: Centro can be grown as pasture crop under rainfed condition. It can be raised under coconut plantations as irrigated crop. First irrigation is given immediately after sowing, life irrigation on the third day and subsequently once in 7 to 10 days.

Harvest: The growing season for centro ranges between 4 to 8 months, but the seeds typically mature within 4 to 6 months. Seed harvest is usually performed by hand. Mechanical harvesting is difficult due to the plant architecture. First harvest is taken up at 120 days after sowing and subsequent harvests are taken once in 60 days.

Yield: Green fodder yield is 10 to 15 tonnes per hectare per year. The number of seeds per kg is 39600. Seed is harvested by hand. It is difficult to harvest mechanically because of uneven ripening. It is best grown on trellises or fence-lines to facilitate hand-picking. The seed yield is 200 to 275 kg/ha. It produces about 1250 seeds/m^2 collected on a fence-line.

Utilization: It is mainly used as cover crop in plantations and utilized as fodder crop. Older plants fix nitrogen effectively, increases soil nitrogen levels and the crude protein of associated grasses. Being deep rooted it is drought and flood tolerant.

REFERENCES

Fantz, P.R. 1996. Taxonomic notes on the *Centrosema pubescens* Bentham complex in Central America (Leguminosae: Phaseoleae: Clitoriinae). Sida 17:321–332.

http://www.tropicalforages.info/key/Forages/Media/Html/Centrosema_pubescens.htm

CHAPTER

75

Stylo–*Stylosanthes guyanensis* (2n=20)

Vernacular name: Stylo, Brazilian stylo, Brazilian lucerne (English), *Brasilianische luzerne* (German), *Muyal masal* (Tamil).

Importance: Stylo is a perennial pasture legume for the humid to semi-arid tropics. It is a fodder cum leguminous cover crop, which is suited for cultivation as sole or as intercrop. The crop is suitable for grazing as the crown of growing point is near the soil surface. It is used in pastoral, agropastoral and silvipastoral systems for animal production.

Origin: Stylo is native of south and Central America and the Caribbean islands.

Distribution: Stylo is grown in Malaysia, Brazil, USA, Australia, China, Colombia, India, Costa Rica, Mexico, Nicaragua, Panama, Cuba, Venezuela and Africa. In India, it is cultivated in Maharashtra, Odisha, Andhra Pradesh, Karnataka, Tamil Nadu and Kerala.

Botany: Stylo is a perennial pasture legume. It is a short-lived, erect or semi-erect perennial legume that can reach 1-1.5 m in height. It is strongly taprooted and nodulated. The stems are many-branched and may be woody at the base. It is a leafy species that remains green under dry conditions. The leaves are trifoliolate with elliptical to lanceolate leaflets, 0.5-45 mm long x 20 mm broad. The inflorescence is a densely flowered spike, up to 40 flowers/head. Flowers are yellow to orange with black or red stripes. The fruit is a single-seeded pod, 2-3 mm x 1.5-2.5 mm wide. The seeds are very small, pale brown or purple in colour. The different types of stylosanthes suited for Indian conditions are:

(i) Brazilian stylo (*Stylosanthes guianensis*): The plants are erect to semi-erect and are not profusely branched at the base and can attain a height up to 1.5 m. The stem is coarse and hairy. The trifoliate leaves are long, narrow and pointed which can be sticky also. Flowers are small, terminal and yellow, borne in cluster on capitate spike. Pods are flattened, single seeded with small and coiled peak. Seeds are kidney shaped, yellowish brown and 1.7 mm long. It does not tolerate shade and can grow very well in areas receiving 900 to 4000 mm of rainfall.

(ii) Townsville stylo (*Stylosanthes humilis*): This is an annual type and attains an average height of 0.7 m. The stem is much branched and fibrous with short white hairs along one side of the stem. Leaves are trifoliate. The leaflets are narrow, pointed, lanceolate and without hairs. These are 15 mm long and 3.5 mm wide. The flowers are small and yellow borne in cluster of 5–15. Pod is hairy. Seed is small, yellowish to brown.

(iii) Caribbean stylo (*Stylosanthes hamata* cv. Verano): This is a short-lived perennial legume similar to Townsville stylo. It is slow growing and develops a flat crown under grazing. Erect stem may grow up to 80 cm. The stems are smooth with a line of very fine, short white hairs on one side only.

(iv) Shrubby stylo (*Stylosanthes scabra*): This is an erect and woody perennial. It is dark green in colour and slightly sticky. It attains a height of 1.0–1.5 m. Stems are hairy and rough. Leaves are trifoliate and leaflets are small and broad. It flowers late and flowers are small and yellow. The pod is 4.4 cm long and single seeded.

(v) Caatinga stylo (*Stylosanthes seabrana*): It is an erect perennial of 0.9-1.0 m height, with triplicate leaves varying in shape. Presence of rudimentary axis at the nose of the pod is the identifying character.

Climate: Stylo is suited for growing in warm, humid tropical climate. It is fairly drought and shade tolerant. It is found from 20°N to 32°S and from sea level up to an altitude of 2000 m. It can grow in places where annual rainfall ranges from 450 to 2500 mm. Stylo is a warm season growing legume that thrives in places where annual temperature are in the range of 23 to 33C. It cannot tolerate cold temperatures and may be killed by frost. It can remain productive down to 15°C.

Soil: It is adapted to a wide range of soils. It can be grown in acid soils, gravelly sandy soils, sandy loams, clay loam and clay soils. It grows well on coastal sandy and shallow rocky soils. It is grown in soils with pH of 4 to 8.3. It has tolerance of high levels of available Al, Mn and tolerance of salinity

Land preparation: Plough the field two to three times to obtain good tilth. Beds and channels are formed.

Season: Seed can be sown during June-July and September–October seasons.

Seeds and sowing: Seed rate is 2 to 3 kg/ha for line sowing and 5 to 6 kg/ha for broadcasting. Spacing is 30 cm x 5 cm. Stylo seeds possess hard seed coat. Seeds are treated with conc H_2SO_4 @ 100 ml per kg of seeds for 3 minutes. Wash the seeds with tap water. Acid scarified seeds are again to be presoaked in cold water overnight. Germination of seed can be improved by mechanical scarification or hot water with 80ºC for 10-15 minutes, then cool and dry. Scarified seeds are again to be presoaked in cold water overnight before sowing. Seed treatment with rhizobium culture can be done before sowing.

Cropping system: Intercropping of cenchrus and stylo in 3:1 ratio is recommended under rainfed conditions to provide balanced feed to livestock.

Weed management: One or two weeding is given, first at four weeks after sowing and again after eight weeks is required during establishment.

Manures and Fertilizer application: 12.5 tonnes FYM per ha and 20-60-15 kg of N, P_2O_5 and K_2O per ha are applied as basal dressing during last ploughing. Top dressing of 20 kg nitrogen per ha after each harvest can to be done.

Water management: Stylo is grown as rainfed crop.

Harvest: First harvest on 75 DAS or at flowering stage and subsequent harvests at 8 to 12 weeks intervals depending upon the growth to ensure survival of the crop. Stylo is moderately palatable for livestock. When leaves are dry especially at later growth stages, it can be hayed off. It can withstand moderately heavy grazing. It is ready for green forage harvest in 65 to 70 days and 90 to100 days for seed.

Yield: Green fodder yield is 25 to 30 tonnes per ha per year. Dry matter yields range from as little as 5 t/ha to a high of 7 t/ha depending on soil and climatic conditions. The stylo pasture can support upto 2.5 beasts/ha, although 0.5-1 beasts/ha is more realistic. The average live weight gain of 0.3-0.5 kg/head/day is achievable. Seed yield is 400 to 500 kg/ha. The leaves contain 12 to 18 % crude protein. One kilogram contains 3,50,000 seeds.

Utilization: Stylo is used as cut and carry feed and for grazing cattle, buffalo, sheep, goats and pigs. It is used to conserve *in situ* moisture, as a nurse crop, intercropping with rainfed crops, revegetation of mine spoil sites.

REFERENCES

Amaresh Chandra, P. S. Pathak and R. K. Bhatt. 2006. *Stylosanthes* research in India: Prospects and challenges ahead. Current Science, 90 (7): 915-921.

Ramesh, C. R., Bhag Mal, Hazara, C. R., Sukanya, D. H., Ramamurthy, V. and Chakraborty, S. 1997. Status of *Stylosanthes* development in other countries. III. *Stylosanthes* development and utilization in India. *Trop. Grassland.* 31: 467–475.

CHAPTER

76

Sainfoin–*Onobrychis viciifolia* Scop. (2n=28)

Family: Fabaceae

Vernacular names: Sainfoin, cock's head, holy clover, esparcet, (English). Its English name 'sainfoin' is derived from the French 'sainfoin', which means 'healthy hay'.

Importance: It is highly palatable legume forage crop. It's very high quality hay was used to feed the heavy working horses and the aftermath grazing is preferred for fattening lambs. It is a non-bloating forage. It has the potential to persist for 3 to 4 years.

Origin: It is native to South Central Asia.

Distribution: It is cropped in USA, Canada, England, New Zealand, Italy, Spain, Iran, Turkey Australia and New Zealand.

Botany: Sainfoin is a deep-rooted perennial legume arising from a branching root crown. It is an erect or sub-erect plant. It grows to a height of 40 to 90 cm. Its large, deep tap root also makes fairly drought tolerant. Each stem has pinnate leaves formed with 10 to 28 leaflets grouped in pairs on long petioles and with a terminal leaflet. The stipules are broad and finely pointed. A plant may produce 5 to 40 tillers, each having 3 to 5 inflorescences. It produces seeds on an inflorescence consisting of 5 to 80 flowers. Each flower has the potential to produce single seed but at best only 55% of the flowers that are pollinated produce seed. Flowers are showy and pink, white or purple and tightly arranged in a compact raceme with 20 to 50 flowers per head. It blossoms produce copious amounts of nectar and are highly attractive to pollinating insects, particularly honey bees. Each flower can produce a kidney-shaped seed contained in a brown pod. The fruit is either spiny or spineless. The seeding unit is a single-seeded pod. Seeds are large with only 18,500 (pre-husked) seeds per pound. The size of the true seeds is variable from 2.5 to 4.5mm long, 2 to 3.5mm broad and 1.5 to 2 mm thick. Sainfoin seeds are available in two forms. The first, termed 'unmilled seed' or 'seed in husk', consists of the whole one-seeded fruits. The other form, termed 'milled seed', is comprised of only the true seed with the dry husk removed. The weight per thousand unmilled seed and milled seed is 24 g and 15 g, respectively.

Sainfoin has been divided into two agricultural types termed 'Giant Sainfoin' and 'Common Sainfoin'.There are a number of key differences between the two types. The common type (*Onobrychis sativa* var. *communis* (Ahlefed)) is from central Europe. Its growth habit remains prostrate in the year of sowing and regrowth after the first spring cut is slow and vegetative. The aftermath is normally grazed. It is also named single-cut sainfoin due to this limitation in terms of regrowth. The Giant type or double-cut sainfoin (*Onobrychis sativa* var. *bifera* Hort.) is from the Middle East. It grows more quickly into an erect habit during the first year of growth. It has the ability to re-flower after being cut. It can be cut more than once per year, but unlike common types, which will persist for 10 years or more, the giant types will not normally survive beyond three years.

Pollination: Honey bees (*Apis mellifera*) and leafcutting bees (*Megachile rotundata*) are recommended for pollination of Sainfoin as they are efficient pollinators. Two to three colonies of honey bees per ha may be installed to enhance pollination.

Climate: It can be grown in areas receiving rainfall of 400 mm and greater precipitation areas. The optimal temperature for germination is between 10 and 20°C; and advised that 5°C was the minimum temperature to sow. It is extremely tolerant of frosts. It is tolerant to flooding.

Soils: Sainfoin prefers to grow on calcareous soil or limestone soil with a pH of 6.0 to 8.0. It grows well on soils that are low in phosphorus. It is adapted to neutral and alkaline soils of pH 6 or above. It is cultivated in sagebrush, pinyon-juniper and mountain shrub areas in deep, calcium based soils. It does not do well in sites with high water tables or wet soils and so the soil needs to be well drained. However, it withstands wet or saturated soils for approximately one week. It is best adapted to soils at least 45 cm deep.

Season: Sainfoin can be planted in the spring or fall. It is traditionally sown as a spring crop (March to May).

Seeds and sowing: The seeding rate and row spacing from 7 kg ha^{-1} and 60 cm for seed production to 40 kg ha^{-1} and 15 cm for irrigated hay production. Seed 2 to 5 kg/ha as mixed crop in rangeland. The seeds can be drilled to a depth of 1 to 2 cm or broadcasted. It may be sown in monocultures at 50 kg/ha of milled seed, and up to 120 kg/ha of unhulled seed. Seeds are inoculated with the appropriate rhizobium prior to planting.

Weed management: The crop growth is slow to start with but it produces enough ground cover to compete against weeds once it is well established.

Intercultural operations: It needs topping 2 to 3 times a season, although this will vary considerably with season and soil type.

Manures and fertilizer application: 12.5 tonnes FYM or compost per ha along with 25-40-20 kg per ha of of N, P_2O_5 and K_2O kg per ha are applied as basal dressing.

Weed management: Two hand weeding are given on 20 and 35 DAS. Application of Fluchloralin @ 1.2 kga.i./ha at pre planting stage controls the weeds.

Crop protection: It is susceptible to crown rot, which can reduce the persistence of the stand. It is resistant to many pests especially weevils or root and crown rot diseases due to wet conditions.

Harvest: It is harvested at 50 to 100% flower bloom without a large loss in quality. The best stage of growth for horse hay is mid-flower. Sainfoin maintains its quality into full bloom. Care must be taken to avoid leaf shatter in field handling and dust in the hay. Do not allow to graze for two seasons after planting. Stands should be allowed to naturally reseed every 2 to 3 years for reestablishment. Stands will persist 3 to 6 years under irrigation but will last longer if root and crown rot diseases are controlled. It is best to avoid heavy grazing and poaching which can reduce the population of viable tap roots. It should not be planted next to shelterbelts because trees and shrubs will suffer from browsing by deer and elk. Seed should be harvested when seed moisture is <40% or when seed is beginning to shatter. Seed shatter is a problem with this species and seed can be harvested by swathing followed by combining after 2 to 5 days of drying.

Yield: The green fodder and dry matter yield is 20 to 25 t/ha and 5 to 7 t/ha under irrigated and rainfed respectively. Seed yields is 500 and 200 kg/ha under irrigated and rainfed conditions respectively. However, the seed yield up to 500 to 1000 kg/ha can be obtained with some cultivars.

Utilization: It is a good quality feed for race horses. It is highly palatable to sheep and cattle and is preferred over alfalfa. It can be used as hay, or grazed in pastures. It is well suited to haying due to its upright growth habit. It has an average digestibility of 63% and crude protein of 18% during early bloom. It does not cause bloat in livestock. The high protein, high palatability and its non-bloat characteristic make it a good choice for range improvement for livestock or wildlife. The foliage is readily eaten by elk, deer and sage grouse, and the seed is eaten by many other birds and rodents. The condensed tannin content of sainfoin appears to restrict protein breakdown not only in the rumen but also in the clamp as well.

REFERENCES

Bolger, T.P., and A.G. Matches. 1990. Water-use efficiency and yield of sainfoin and alfalfa. Crop Science. 30:143–148.

Kallenbach, R.L., Matches, A.G. and Mahan, J.R. 1996. Sainfoin regrowth declines as metabolic rate increases with temperature. Crop Sci.36:91-97

Morrill, W. L. and R. L. Ditterline. 1998. Insect pests and associated root pathogens of sainfoin in western USA. Field Crops Research. 59(2): 129-134.

Mowrey, D.P., and A.G. Matches. 1991. Persistence of sainfoin under different grazing regimes. Agron. J. 83:714-716.

Peel, M.D., K.A. Asay, D.A. Johnson, and B.L. Waldron. 2004. Forage production of sainfoin across an irrigation gradient. Crop Science 44:614–619.

Stannard, M. 2002. Sainfoin. USDA-NRCS Plant Fact Sheet. Pullman, Washington. 2p.

CHAPTER

77

Desmodium

Family: Fabaceae (alt. Leguminosae)

Vernacular names of Greenleaf desmodium (*Desmodium intortum* (Mill.) Urb.): greenleaf desmodium, beggarlice (English); *grünes desmodium* (German); *pega pega, amor seco, desmodio verde* (Spanish); desmodie (French); *karikuy-ritkuk* (Philippines); *thua kleen leap* (Thailand), Greenleaf desmodium (Australia), *pega-pega* (Philippines, Costa Rica, Venezuela), *kuru vine* (Zimbabwe), *amor seco* (Colombia).

Vernacular names of Silverleaf desmodium (*Desmodium uncinatum* (Jacq.) DC): Silverleaf desmodium, Spanish tick-clover (English); spanischer Klee (German); desmodio plateado, pega pega (Spanish).

Importance: Desmodium, in general, is known as 'Alfalfa of tropics' or 'Tropical medic'. There are two economic tropical species in desmodium for forage purpose namely greenleaf desmodium and silverleaf desmodium. Silver leaf desmodium is known as 'Japanese clover'. Desmodium is palatable and make nutritious feed. Livestock relish both leaves and young branches. It can be grazed as a long-term pasture, cut and offered fresh in cut-and-carry systems, or cut from irrigated pastures for conservation as hay or silage. It is a valuable ground cover providing abundant leaf material. Desmodium has self-seeding ability.

Origin: Desmodium is a native of Central and South America.

Distribution: It is mainly grown in Argentina, Brazil, Venezuela, Costa Rica, El Salvador, Guatemala, Honduras, Mexico, Nicaragua, Panama, Caribbean, Jamaica, Puerto Rico, Colombia, Ecuador, Peru, Venezuela, Brazil and India.

Botany: There are two economic tropical species in desmodium for forage purpose namely greenleaf desmodium (*Desmodium intortum*) and silverleaf desmodium (*Desmodium uncinatum*). Green leaf desmodium is a trailing, creeping, herbaceous, and perennial legume. Green leaf desmodium grows to height of 2.0 to 2.5 metre. It has a strong taproot and the long trailing stems can root at the nodes if it contact with moist soil. These stems are grooved. The <u>trifoliate</u> leaflets, to 7 cm x 5 cm, have deep green, leaflets often have a reddish-brown flecking on the upper surface with very fine gray hairs while lower surface is grayish green with many white silky hairs. Short day induces flowering. Flowers are small, deep lilac to pink in colour. Flowers are in compact, terminal and axillary racemes. The small narrow and recurved seed pods are segmented and have hooked hairs that cause them to adhere to clothing and animals. Each pod has 8 to 12 seeds. Silver leaf desmodium is a trailing perennial tropical legume with thin, ovate, hairy leaflets which have a broad irregular silver band along the mid rib. Moistness induces rooting along the pubescent stems, and swollen as well as fibrous roots are produced. The cylindrical or angular stems are covered with short, hooked hairs that stick to hair or clothing. Stems can root at the nodes if they touch moist soil. Leaves are coarse and stems are hairy. Flowering occurs in short days. Flowers are pink to bluish in colour. The fruits are light brown hairy pods that break readily into 4 to 8 segments at maturity. Pods are in chain like form. Pods mature from top to bottom. Pods stick to animals and clothing. The seeds are olive-green in colour, triangle or oval shaped, 3 mm long x 2 mm wide. The seeds are flat and larger than those of green leaf desmodium. Under adequate moisture, the fallen seeds germinate and a new crop comes up densely. Once mature woody runners have developed and leaves start falling.

Climate: Greenleaf desmodium grows best under warm subtropical conditions. It can be found within 30°S and 19°N at altitudes ranging from sea level to 2000 m in the subtropics and from 500 m to 2400 m in the tropics. It grows well on slopes. It can be grown in areas where annual rainfall is above 900 mm and up to 3000 mm. It can withstand waterlogging but does not withstand drought. It has better tolerance of flooding and waterlogging than silverleaf desmodium. Green leaf grows well in moist elevated areas. The greenleaf desmodium has good shade tolerance. It has a sensitive short-day control of flowering in the southern hemisphere. However, temperature as well as daylength controls flowering in greenleaf desmodium. It is tolerant of shade and can be grown in coffee plantations. The optimum temperature for growth and yield is from 25 to 30°C. It is not tolerant of heavy frosts or fire. Silverleaf desmodium is a warm season legume but it is also one of the most cold tolerant tropical legumes grown between 10°S and 10°N up to an elevation of 1000 m. It can survive frost down to -10°C. Though the leaves may be killed by frost, the plant recovers quickly in warm weather and it is one of the earliest tropical legumes to grow in spring season. Optimal growth is obtained at average temperatures of 25 to 30°C. Silver leaf is not hardy as green leaf and thrives only in moist coastal areas where annual rainfall is 500 to 1600 mm or even more. Heavy rainfall of 3000 mm is deleterious to its growth. Silver leaf desmodium is less drought resistant and tolerant to water logging than green leaf desmodium. It has poor tolerance of medium to heavy shade.

Soil: It is adapted to a wide range of soil conditions. It tolerates acid, saline and alkaline soils and is adapted to some poorly drained or water logged situations. Greenleaf desmodium will grow on a wider range of soils than will Silverleaf desmodium from sands, light loams to medium clays. Green leaf thrives on a variety of soils in the coastal areas. Silverleaf desmodium grows on a wide range of soils from sands to clay loams, but does better on light, friable soils. It can withstand moderate acidity (down to pH = 5) and soils with high Al and Mn contents. It is somehow tolerant to flooding and drought (up to 3 months). Silverleaf desmodium may recover from fire.

Season: Desmodium can be grown throughout the year under irrigated condition. However, sowing during June–July is preferable.

Land preparation: Two to three ploughings are given to have a fine tilth. Beds and channels are formed.

Seeds and sowing: Seed rate is 2 kg/ha. There are about 7,50,000 seeds per kg. It establishes well after broadcasting into a cool ash seedbed in periods of reliable rainfall. If seeds are not available, stem cuttings can be used.

Cropping system: Desmodium is a shade tolerant legume. It comes well under the shade of trees such as coconut, tamarind and eucalyptus. It will not affect the coconut yield. It is suited to raise in pasture with grasses and in silvipasture systems.

Weed management: One hand weeding may be given on 20 DAS. Once established, desmodium spreads vigourously and can smother weeds under adequate management.

Manures and fertilizer application: FYM @ 12.5 tonnes per ha and 25-40-20 kg per ha of N, P_2O_5 and K_2O kg per ha are applied as basal dressing before sowing. Top dressing of 25 kg of N per ha is done after each harvest.

Water management: First irrigation is given immediately after sowing, life irrigation on third day and subsequent irrigations once in 7 days. It is also raised as rainfed crop.

Harvest: The crop is harvested at flowering stage. Four cuttings are taken in a year. Grazing is needed during the establishment phase to provide good cover. This legume has considerable tolerance to heavy grazing pressure.

Yield: Desmodium green fodder yield is 20 to 25 tonnes per ha per year under shade environment. Commercial seed yields of 100 to 120 and 200 to 300 kg per ha can be obtained from green leaf and silver leaf desmodium respectively under irrigation during the dry season. Desmodium fodder contains 27 % dry matter and 20.9 % crude protein.

REFERENCES

Sweeney, F.C. and Hopkinson, J.M. 1975. Vegetative growth of nineteen tropical and sub-tropical pasture grasses and legumes in relation to temperature. Tropical grasslands 9: 209-217.

http://www.regional.org.au/au/asa/2003/p/12/gardiner.htm

http://www.fao.org/ag/agp/AGPC/doc/Gbase/data/pf000030.htm

http://www.tropicalforages.info/key/Forages/Media/Html/Desmodium_uncinatum.htm

http://www.fao.org/ag/AGP/AGPC/doc/Gbase/data/pf000026.htm

CHAPTER

78

Buckwheat–*Fagopyrum esculentum* Moench (2n=16, 32)

Family: Polygonaceae

Vernacular names: Buckwheat, beech wheat (English), Kotu, Phaphra (Hindi), *ogal* (Kumaon, India), *mite phapar* (Nepal), *jare* (Bhutan), *soba* (Japan). The name 'buckwheat' comes from the Anglo-saxon word *boc* (beech) and *whoet* (wheat) because the seed resembles a small beech nut.

Importance: Buckwheat is one of the minor grain crops. It is used a green manure, smothering crop for weed control, winter cover crop which aids in erosion control and forms a cover to catch the snow. Buckwheat plants are not grasses, but the seeds (strictly, achenes) are usually classified among the cereal grains because of similar use. Buckwheat grain is used for feeding livestock and poultry and as food for humans. The protein of the buckwheat is an excellent quality. Buckwheat flour is used in pancake mixes and flaked breakfast foods.

Origin: The origin of buckwheat is Nepal, Kashmir of India, China or Siberia.

Distribution: The buckwheat is mainly cultivated in China, Russia, Ukraine, Kazakhstan, Japan, Korea, Poland, Brazil, USA, Canada, France, India, Pakistan, Nepal and Bhutan. In India, the crop is widely grown in Jammu and Kashmir, Himachal Pradesh, Sikkim, Assam, Nagaland, Manipur, Arunachal Pradesh and Tamil Nadu.

Botany: Buckwheat plants are herbaceous, erect, annuals, with single main stems that usually bear several braches. It grows to a height of 0.6-1.2 m. It is coarse, strongly grooved, succulent stalks usually produce several branches. It has indeterminate growth. The number of leaves varied from 10 to 45. Leaf width ranged from 2.8 to 8.0 cm while leaf length ranged from 2.1 to 8.9 cm. It begins to bloom 5-6 weeks after seeding. The plants produce flowers daily for about 4 weeks. Flowers are white to pink. It is naturally cross-pollinated. Fruit is 02-04 inches long with keeled edges varies in colour form silvery grey to brown or black. The seed are actually one-seeded fruits called achenes. It has triangular seeds with black soft hull, light green to white kernel. The 100-seed weight varied from 1.2 to 5.0 g while yield per plant ranged from 2.3 to 20.0 g. In general, the crop takes 24 to 78 days to flower and 75-150 days to maturity.

Flowers of cross-pollinating species of buckwheat are attractive to insects because of the nectar secreted by the glands at the base of the ovary. Bees and other insects contribute to the distribution of pollen. The glands secrete nectar only in the morning and early afternoon and therefore if honey bees are introduced to increase seed set they must be forced to work the buckwheat. An acre of buckwheat flowers might yield enough nectar to produce 100 to 150 lbs of honey.

Climate: Buckwheat is a short-duration crop of 3 to 4 months and requires a moist and cool temperate climate to grow. It is cultivated in the Himalayas at an altitude of 500 to 2500 m from MSL. It needs about 70 mm of rainfall from germination to full bloom, and about 20 mm from full bloom to maturity. The optimal humidity is 60-80%. Relative humidity should not fall below 50 to 60 % for best growth during and after pollination in buckwheat. It is generally considered a short-day plant. It has good growth and yield at light intensity of 18 klux or more. Air temperatures of 17ºC to 19ºC are mostly satisfactory during blossoming. When the plants are in bloom, the crop is very sensitive to high temperatures and hot dry weather. Under these conditions flowers are likely to blast without forming seed. Flowering at temperatures above 30°C is accompanied by desiccation of the fruit and lowering of yield. Common buckwheat has little tolerance to frost.

Soil: It can be grown on variety of soils, but thrives well on well-drained sandy soils. It has good tolerance to acid soils.

Season: It is a quick growing crop. In Northern India, *Fagopyrum esculentum* is usually a rainy season crop. It is sown in July and harvested in October. In Nilgiris, it is generally sown in April and harvested in August.

Land preparation: The seedbed is prepared in much the same way as for other small grains.

Seeds and sowing: Seed rate is 3 to 4 kg per ha. If buckwheat is grown as smothering crop for weed control, it is desirable to seed at a heavy rate of 4 to 5 kg per ha.

Weed management: One hand weeding may be given on 20 DAS..

Manures and fertilizer application: FYM @ 12.5 tonnes per ha and 25-40-20 kg of N, P_2O_5 and K_2O per ha are applied as basal dressing before sowing. Top dressing of 25 kg of N per ha is done after each harvest.

Water management: First irrigation is given immediately after sowing, life irrigation on third day and subsequent irrigations once in 7 days depending on climatic condition.

Harvest: Due to its indeterminate flowering habit, it has flowers, green seeds and mature seeds are present on the plant at the same time. The crop is harvested when the maximum number of seeds are ripe. Shattering losses may be reduced by harvesting when the dew is on or on a damp day. If harvest is delayed, seeds can shatter due to wind. The crop is left in the fields until dry as it is full of moisture at the time of harvest. The seeds are generally dried to 16% or less moisture content before being stored. If the seeds are artificially dried the drying temperature should not exceed 43°C. The seeds are stored with the hulls on and are dehulled shortly before use to prevent development of rancidity.

Utilization: The buckwheat grain is generally used as human food and as animal or poultry feed. The small leaves and shoots are used as leafy vegetables, the flowers and green leaves are used for rutin extraction for use in medicine. Buckwheat is a source of rutin, a flavonol glucoside used in the treatment of capillary fragility associated with hypertension, or high blood pressure, in man.

REFERENCES

Campbell, C.G. 1995. Buckwheat. pp. 409-412 *in* Evolution of Crop Plants. Second edition. J. Smartt and N.W. Simmonds, eds.). Longman, London.

Campbell, C.G. 1997. Buckwheat *Fagopyrum esculentum* Moench. International Plant Genetic resources Institute, 25-27.

IPGRI. 1994. Descriptors for buckwheat (*Fagopyrum* spp.). International Plant Genetic Resources Institute, Rome, Italy.

Joshi, B.D. and R.S. Paroda. 1991. Buckwheat in India. National Bureau of Plant Genetic Resources, New Delhi.

Joshi, B.D. and R.S. Rana. 1995. Buckwheat (*Fagopyrum esculentum*). pp. 85-127 *in* Underutilized Crops. Cereals and Pseudocereals (J.T. Williams, ed.). Chapman & Hall, London.

Joshi, B.D. 1999. Status of Buckwheat in India. Fagopyrum, 16, 7–11.

Parminder Ratan, Preeti Kothiyal, 2011. *Fagopyrum esculentum* Moench (common buckwheat) edible plant of Himalayas: A Review. Asian Journal of Pharmacy and Life Science. 1 (4): 426-442

Tahir, I., Farooq, S. 1988. Review article on buckwheat. Fagopyrum, 8, 33-53.

CHAPTER

79

Fodder Trees and Shrubs

Forage crops are mostly available to livestock during rainy seasons and not in the summer season. Tree tops come to the rescue of the livestock during summer season. The parts of trees that are commonly used as feed include leaves, tender shoots or twigs, fruits, pods and seeds. Leaves also contain more crude protein on average than pods but the latter were found with higher organic matter and digestibility. The young leafy, succulent material, highly nutritive and rich in crude protein and minerals, serves as a concentrate even if fed in small quantities along with other dried grasses and crop residues. Some trees that produce gum are Subabul, Sithagathi and Daincha. The gum content in the seeds of the two species of Sesbania is of superior quality and has a property to reduce the cholesterol content in the blood. These trees can be planted on the boundaries of the fields, in the cattle yards, etc. to serve as shade cum fodder cum gum producing plants. Trees are planted in the pastures as companion species with grasses. This practice of growing fodder cum fuel trees in association with the grasses and legumes is popularly known as the silvipastural system. The important fodder trees and shrubs are viz., Subabul (*Leucaena leucocephala*), Hedge Lucerne (*Desmanthus virgatus*), Agathi (*Sesbania grandiflora*), Seemai vel (*Prospis juliflora*), Kalyana murungai (*Erythrina indica*), Kodukapuli (*Pithacellobium dulce*), Israeli babool (*Acacia tortilis*), Australian babool (*Acacia legulata*), Babool (*Acacia arabica*), Acha (*Hardwickia binata*), Siris (*Albizia lebeck*), *Albizia amara*, Pipal (*Ficus religiosa*), Mulberry (*Morus alba*), etc,. The most browsed species is observed according to their voluntary intake and subsequent performance of browsers on different fodder trees and shrubs. Tree leaf feeds are particularly valuable for goats.

Characteristic features of Tree fodder

(i) Leaves and pods should have a high nutritive value, which means that they contain a lot of protein.

(ii) Trees should produce many leaves and regrow easily after frequent pruning.

(iii) Edible parts of the tree should not contain (too much) toxins.

(iv) Tree leaves need to have a high palatability, which means that the animals like to eat them and can digest them well.

(v) Trees must preferably be tolerant to drought, pests and diseases.

(vi) Trees should not compete too much with other crops. For example, good fodder trees should form deep roots in order to avoid competition with shallow rooted crops for water.

Harvesting techniques of leaves from the fodder trees

(i) Pruning and Lopping: These are methods in which the side branches of a tree are cut. Pruning is different from lopping in that the branches are cut from the base. Lopping is not always done starting from the lower part, but can be done more haphazardly.

(ii) Pollarding: If all branches and the crown of a tree are cut off but the whole stem is left (about 2 meter), this is known as pollarding.

(iii) Coppicing: Many species of trees and shrubs have the ability to resprout after the whole tree has been cut. If this ability is used for regene-ration of the tree the practice is known as coppicing.

Feeding fodder trees leaves to the animals

(i) **Browsing or 'cut-and-carry':** Fodder tree leaves can be browsed or 'cut and carried' to stall-fed (zero-grazed) animals. Browsing is possible but it should be done in a rotational manner. This means that the trees are browsed for a couple of days and left for six or more weeks to recover. Overgrazing should be avoided otherwise fodder trees will die! The cut-and-carry system is also practiced. Here the fodder is cut and carried to animal stalls. For most tree species, the fodder leaves should be fed to the animals within an hour of harvesting. Pruning, coppicing and pollarding is common in the sub-humid zones.

(ii) **Live fence or boundary systems:** Single or double rows of fodder trees are planted along farm boundaries to provide fodder and serve as live fence. Thorny species are planted as thick hedges to prevent livestock from straying into crop plots and also to fence them off from wild animals.

(iii) **Alley cropping system**: It is also practiced where fast growing and coppicing trees are planted in a single or multiple rows making a dense hedge rows. This system is mostly used in high rainfall areas on specially sloppy lands to obtain leaf fodder and fuel wood through lopping frequently. Species like *Leucaena leucocephala*, *Gliricidia sepium*, *Sesbania sesban,* etc. can be established as fodder hedges by planting closely in rows and managed as bushes by cutting the main stem at a height of 0.5 to 2.0 m at intervals of 30-60 days depending on species, season, soil fertility and moisture availability. The spaces in between hedge rows are planted with pasture grasses. As in fodder banks, herbage may be cut and carried to animal feeding stalls. The more common practice is to let the animals forage on the cut tree branches and pasture grasses.

(iv) **Tree plantation + animal grazing systems:** The understorey of tree plantations is utilized as grazing area for cattle, sheep and goats. The plantation may be of forest trees, fruit trees, coconuts, oil palms or rubber. The livestock are allowed to graze freely on improved pasture grasses planted under trees.

Silvicultural practices of Fodder trees

1. Agathi - *Sesbania grandiflora* (L.) Pers. (2n = 52, 104) (Syn: *Sesbania formosa*)

Vernacular names: Agati, scarlet wistaria tree, vegetable hummingbird, West Indian pea, Corkwood Tree, West Indian Pea (English), *Agati* (Bengali), *Turibaum* (Gujarati.), *Agasti, Basna* (Hindi), *Agasti, Agastya, Agati* (Sanskrit), *Agathi, Peragathi, Akkati* (Tamil), *Khae baan* (Thailand), *Turi, Tuwi* (Indonesia), *Agastya* (Kannada), *Akatti* (Marathi), *Agasti* (Nepalese).

Importance: It provides forage, food and green manure. It is raised as shade crop in turmeric, supporting crop in betelvine, border crop in sugarcane and bund crop of paddy fields. Leaves and flowers are used for human consumption. It regenerates quickly after subsequent cuttings. Leaves are readily eaten by cattle and goats. Leaves contain 30 % crude protein and 9000 i.u. of vitamin A. The young leaves, tender pods and giant flowers of agathi are favorite vegetables which are used in curries and soups or fried, lightly steamed or boiled.

Origin: It is native of Southeast Asian countries (India, Indonesia).

Distribution: This tree is found in all the tropical to sub-tropical parts of India. Best adapted to regions with annual rainfall of 2,000-4,000 mm, but has been grown successfully in semi-arid areas with 800 mm annual rainfall and up to 9 months dry season. It is grown Karnataka, Maharashtra, Chhattisgarh, Bihar, West Bengal, Uttar Pradesh and Madhya Pradesh.

Botany: This is very fast growing tree attains 4.3 to 5.5 m height in one year and 8 meters in three years. Its lifespan is about 20 years. Its roots are heavily nodulated and some floating roots may develop in waterlogged conditions. The trunk is straight with few branches. The leaves, up to 30 cm long, are pinnately compound with 20 to 50 oblong leaflets, 1-4 cm long and 0.5-1.5 cm broad. The flowers are white, yellowish, pink or red and borne in axillary racemes. The pods are 50 to 60 cm long, glabrous and indehiscent and hang vertically. Pod contains 15 to 25 dark brown seeds, 5 mm long and 2.5-3 mm broad. The number of seeds is 17,000 to 30,000 seeds per kg.

Climate: It is well adapted to hot, humid environments at an altitude from sea level to 800-1000 m. It is best adapted to moist tropics with annual rainfall of 750 to 2000 mm. It is tolerant to flooding over short periods of times and waterlogging environments. During waterlogging and floods, it develops floating adventitious roots and protects their stems, roots and nodules with spongy aerenchyma tissue. It does not thrive in temperatures below 10°C. The optimum temperature for growth is from 22 to 30°C. It is intolerant of high winds that can break stems and branches. It is frost sensitive and intolerant of extended periods of cool temperatures.

Soil: Agathi is grown in all types of soils. It withstands acidic soils. It is tolerant to saline and alkaline soil conditions with pH ranging from 4.5 to 9.5. It can be grown on heavy clay as well as poorly drained soils.

Season: It is sown during June-July and September-October months.

Land preparation: Two to three ploughings are given to have a good tilth.

Seeds and sowing: Seed rate is 3 to 4 kg per ha. Spacing is 60 cm x 60 cm for pure crop; 1.0 m for bund crop and 90 cm x 90 cm for shade crop in betelvine and turmeric field. It is not hard-seeded and usually germinates well without scarification. It is easy to propagate by cuttings or seedlings raised in polythene bags. Over 3000 stems cuttings per ha have been used for planting to produce pole timber, or sparsely planted to produce dry-season forage and fuel wood. It is also planted as individual trees or in rows, spaced 1 to 2 m apart along fence lines, field borders and the bunds.

Manures and fertilizer application: FYM @ 12.5 t/ha with 20-25-20 kg of N, P_2O_5 and K_2O per ha are applied basally.

Water management: First irrigation is given at the time of sowing, life irrigation on the third day of sowing and subsequent irrigations are given as and when necessary. Pot watering may be given for bund crops.

Harvest: The cutting heights is 2 metres. Harvest is done once in 30 days. The side branches are cut for feeding, leaving the trees to develop tall poles. Side branches can be harvested from the third month onwards and the main branch can be harvested after six months. After the tree has reached a height of 3 m or more, the leading branch can be cut back above 1.5 m height. The young tree is not tolerant of severe and regular pruning. Large trees are heavily pruned during the long dry season without significant mortality. The tree has a potential lifespan of 20 years.

Yield: The green fodder yield is 70 and 30 tonnes per ha per year under irrigated and rainfed condition respectively. It produces 10 tonnes of green fodder per ha per year as a bund crop. The annual yield of 27 kg of green leaf/tree can be achieved by harvesting side branches every 3 to 4 months.

Seed collection and storage: It is able to produce ripe pods 9 months after planting. Seeds are collected from matured pods in month of May and June. Collected seeds are stored after proper drying. Seed is immediately germinable without any scarification and viability deteriorates rapidly.

Utilization: Tree leaves contains 25-30% crude protein. Its low tolerance to defoliation makes it badly suited to direct grazing. The tree is not generally directly grazed by livestock as high plant mortality occurs. Tree leaves used as cut-and-carry forage integrated into cropping systems. Tree leaves used as fodder particularly for dry season feeding of cattle and goats. Leaves are highly palatable to ruminant animals. The leaves contain 36% crude protein. Its sparse canopy casts relatively little shade which is suitable to shade-loving crops and gardens. It grows fast enough to be used as an annual green manure crop. The leaves, flowers and pods are eaten as a vegetable in Southeast Asia.

2. Subabul - *Leucaena leucocephala* (Lam.) de Wit (2n=104)

Family: Fabaceae

Vernacular name: Savundal (English), Leucaena (Australia, United States), Ipil ipil (Philippines), Lamtoro (Indonesia), Katin (Thailand), Subabul (India), Koa haole (Hawaii), Yin ho huan (China), Guaje (Mexico).

Importance: This fast growing multipurpose evergreen leguminous forage tree has a vast potential use of making as forage, round wood (poles, fence posts), firewood and charcoal, timber and other construction materials and pulp for paper and rayon manufacture not to speak of it value in the environmental development. It is used as hedges, shade trees around home and roadside as ornamentals and wind breaks. Subabul coppices readily. Stumps from plants of almost any age and variety quickly resprout new shoots. Coppice regrowth is even more vigourous than seedling growth because the new shoots are served by a developed root system. This fast growing multipurpose nitrogen fixing evergreen leguminous forage tree has to do a potential role in afforestation programme in India

Origin: It is originated in Central America and Mexico.

Distribution: Of the ten leucaena species, only *Leucaena leucocephala* has been widely distributed throughout the tropics. It is found naturalised on the rocky coralline terraces of Pacific island countries.

Botany: It is a thornless long-lived shrub or tree which may grow to heights of 7-18 m. Leaves are bipinnate with 6-8 pairs of pinnate bearing 11-23 pairs of leaflets 8-16 mm long. The inflorescence is a cream coloured globular shape. Flowers are small, fluffy, white balls. It is self-pollinated. It produces a cluster of flat brown pods of 13-18 mm long containing 15-30 seeds. The seed weight ranges from 10 mg to 70 mg.

Climate: It is cultivated up to an altitude of 500 to 1500 meter from MSL in most part of the tropics. It grows best where the annual rainfall is 650 to 2500 mm with even distribution of 100 to 125 mm per month. It is drought tolerant and withstands long dry seasons (5-7 months) from the second year of its establishment. It recovers rapidly during rainy seasons. It grows well in high light intensities, tolerates partial shade and its growth is slow under heavy shade. This is grown in areas with temperature range of 22 to 41°C. It requires temperature of 25-30°C for optimum growth. Heavy frosts will kill all above ground growth.

Soil: It can be cultivated on deep, well drained, acid soils (pH > 5), neutral to calcareous soils. It is tolerant to salinity and alkalinity. It grows well in neutral or alkaline (pH 9) soils. Its growth is poor in acidic and calcium deficient laterite soils with a pH below 5 and high levels of aluminum saturation (85%). These factors inhibit deep penetration of roots and favours only superficial roots.

Season: It is sown during June-July and September-October under rainfed condition and throughout the year under irrigated condition. It does not tolerate waterlogging and must be lopped aggressively to avoid shading.

Varieties: Hawaian giant, Ipil Ipil, K8, FD 1423 and CO 1.

Establishment for forage and plantations: Propagation is by seed. Since the seeds have hard seed coat, mechanical scarification is required for good germination. Seeds stored for three years germinate well and require less scarification than fresh seeds. A more rapid and uniform emergence of seedlings could be obtained when the seeds are allowed to imbibe water for 72 hours. Under field conditions, hot water treatment of seeds at 80ºC for 5 minutes or (boiling water for 4 seconds) followed by soaking cold water overnight gives more than 80 percent germination. Boiling water reaches 80^0C in ambient air temperature after 4 minutes. Seed treatment is done with conc H_2SO_4 @ 100 ml per kg of seeds for 3 minutes which is followed by washing of seeds with fresh water 4 to 5 times. This will enhance seed germination. Direct seeding is recommended for forage production. Normal seeding rate is 10 kg per ha for fodder green manure purpose while 1.25 kg per ha for fuel purpose. Seeds sown at 3 cm depth recorded better germination than those broadcasted on the surface. Germination occurs in 5 to 8 days. Generally seeds are slow to establish in the field. The best forage production appears with the population of 1,00,000 to 1,50,000 plants per hectare. Populations higher than 1,30,000 per hectare are effective in controlling weeds. Row spacing of 50 to 100 cm with 20 cm between plants is recommended for forage purposes under irrigated conditions. A spacing of 30 cm x 10 cm is suggested for the best forage productions under rainfed conditions. As transplanting of subabul seedlings is expensive, it is recommended only under adverse conditions and for the plantation programmes but not for forage production. Seedlings grown in polythene bags usually establish better than the bare-rooted seedlings from the nursery. Seedlings are transplanted with a spacing of 1 m x 1 m for poles and 4 m x 4 m for fuel wood and timber.

Cropping systems: Subabul growth is compatible with guinea grass, pangola grass and paragrass. Alley cropping of subabul with food as well as forage crops is also feasible.

Soil fertility management: Subabul fixes 90 to 130 kg of atmospheric nitrogen per hectare annually. Two strains of rhizobium cul 81 and NGR 8 are widely used for acidic and alkaline soils respectively. The association of mycorhiza, *Glomus fasciculatus* with roots aids in the uptake of phosphorus, particularly in soils with low phosphorus levels. The small size leaflets which contain 4 per cent nitrogen decay within two weeks to form humus under tropical environments. By this about 8.5 tonnes of dry litter or about 10 kg nitrogen is added annually in a stand of 10,000 trees per hectare. High forage yields can be obtained with the application of 20 kg nitrogen, 60 kg phosphorus and 30 kg potassium per ha annually. Gypsum is recommended over lime and dolomite since the calcium in gypusm has a better downward movement than the rest in acid soils.

Water management: A water requirement of 54 mm per tonne of dry matter and a total of 1100 mm rainfall or equivalent irrigation is required to produce 50 tonnes of dry matter per hectare annually. Subabul can withstand several months of dry weather. Irrigate the crop once in 6 weeks. The bund should be adequately moist up to first 5-6 months from sowing.

Harvest: Subabul plants are harvested at 50, 75 or 100 cm height in fields. First harvest is done 6 months after sowing. Initial cutting should not be done until the trunk has attained atleast 3 cm diameter. Subsequent harvest may be taken on 60 to 70 DAS depending upon growth and season. The optimum stage for the harvest of forage occurs when the stem starts turning gray from green, and or with a stem thickness around 10 mm. Maximum forage yield can be obtained when the plant is cut at 90 to 100 cm height under irrigated condition. The cutting frequency is at an interval of 6 to 8 weeks. Eight cuts can be given per year under irrigated conditions. Yield can be maximised when 25 percent of foliage is left on plants during harvest.

Yield: The green fodder yield of 80 to 100 tonnes per ha per year inclusive of leaves and tender stems can be obtained which accounts 18 tonnes dry weight annually. The forage yield of 35 tonnes per hectare in the first year and 52.5 tonnes per hectare during the second year has been recorded in a soil with pH of 9.5, EC 0.300 milli mhos per cm and ESP 35, and with the well water pH of 9.0, EC 1.6 millimhos per cm and RSC 6-8 (Residual sodium carbonate). The wood yield is 75 tonnes per hectare as poles in 2.5 years with 1 m x 1 m spacing. The cutting interval of 4 to 6 years are being used commercially to produce wood for pulp and timber. The wood yield would be 750 tonnes per hectare in five years with 4 m x 4 m spacing. The fuel wood of 2 to 4 years has a heating value of 4620 kcal per kg which is 70% heating value of fuel oil. Subabul is one among the best tropical hardwoods for paper and rayon manufacture. Wood pulp yield is 50 to 52%.

Seed production: Ripe pods should be collected before they split and dried in the sun for 3 to 4 days. The pods then split, seeds then can be gathered by sieving. Seeds are viable for three to four years. Seeds come out of pods which grow in clusters, mostly from self-pollinated flowers, which look like fluffy white ball. The seed has a waxy white coat, and needs to be treated. One kilogram of giant types seed contains about 20,000 seeds. In general there are 30000 seeds in a kg of 100% purity with about 6% moisture. Spraying maleic hydrozide, a growth regulant at 500 ppm accelerates flowering, pod formation and seed maturation. It increases seed yield by increasing pod number, pod length and seed weight. Generally the trees set flowers and pods twice a year. Seed yield of 4 to 5 kg can be collected from a single tree.

Nutritional management of subabul and animals system: Young or mature, green or dry the foliage is eagerly sought by livestock particularly when green feeds are scarce during dry months. This is a common feature in the seasonally dry tropics. Animals can be fed on the standing bushes or the tree foliage can be harvested and carried to the animals or dried into leaf meal or fermented to silage. For ruminant animals such as cattle, water buffalos and goats, this forage is palatable, digestible and nutritious. The aminoacid mimosine causes weight loss and ill health in non-ruminants such as horses, rabbits, and poultry. Mimosine content in the forage is 0.87 to 2.39% on fresh weight basis which is equivalent to 3.5 to 9.23% on dry matter basis. The subabul leaves contain as high as 25 to 30% protein, even after correcting mimosine problem. The foliage and thin stem should be fed to a maximum of 30 percent in ruminants diet and 5 to 10 percent in non-ruminants diet due to the high protein and mimosine content. It is also a rich source of carotene and vitamins. Its provitamin-A content (Carotene) is among the highest ever recorded in plant specimens which increases yellowness in the fat of subabul fed cattle, skin of goats and egg yolk of chickens. The leaves are also a good source of calcium, phosphorus and other dietary mineral nutrients except sodium and iodine. The digestibility rate is 50 to 70 per cent for both leaves and stems and 70 percent of leaves alone.

3. Acacia fodder trees (2n = 52, 104)

Vernacular name: Babul, Babul acacia, Umbrella thorn (E), Velamaram, Karuvel, Vellaivel, Kodaivel (T).

Importance: The pods and leaves are excellent fodder. The tree provides fuel wood. Acacia acts as soil improver for binding sand dunes or controlling soil erosion. It is used for wood and fuel.

Botany: Three important species of of acacia are grown in semiarid tropics viz., (i) Israeli babul (*Acacia tortilis*) is a fast growing tree and resistant to drought,(ii) Australian babul (*Acacia legulata*) is drought resistant trees having ornamental value with dropping branches and (iii) Kikar (*Acacia arabica*) is a medium growing tree, suitable for dry areas. *Acacia nilotica* flowers at a relatively young age, around 3–4-years old under ideal conditions.

Climate: It is suited for semiarid tropics. It can tolerate floods.

Soils: It can be grown in all types of soils. However it grows well in black cotton soils. Tree tolerates a wide range of soil types, thriving in alluvial and heavy clay soils with pH 5.0 – 9.0. When used in land reclamation, *Acacia nilotica* can be planted onto degraded saline/alkaline soils with a soluble salt content below 3%. This plant can tolerate the poor soil.

Land preparation: Pits are taken to a size of 1' x 1' x 1'. Seedlings are planted during onset of monsoon rains.

Season: It is planted during June-July and September-October months.

Propagation: Seed rate is 500 gm per ha. Soaking seed in water for 12-24 hours enhance germination. Seeds can also be nicked. More mature seeds can be immersed in conc sulphuric acid for 3 minutes or dropped in boiling water for 5 seconds. Seedlings are raised in polythene bags and then setting the seedlings in the pit with a spacing of 25' x 25'. *Acacia tortilis* seeds should be treated with hot water for 5-7 minutes before sowing in a polythene bag. *Acacia arabica* seeds are treated with conc H_2SO_4 for 3 minutes for better germination. *A. senegal* is usually raised in the nursery in polyethylene pots, 2-4 seeds per pot, thinned to one seedling after 4-6 weeks. Direct seeding (5-8 seeds in 30 x 30 x 30 cm pits or larger) can also be used. Protection of young trees from livestock browsing is essential.

Weed management: One or two weeding may be given during establishment period.

Manures and fertilizer application: Two to three handful of farm yard manure mixture with 500 grams of super phosphate per tree seedling have to be applied.

Water management: Pot watering may be done during establishment period.

Harvest: The lower shoots can be cut and used as fodder to feed goats while the pods are collected and fed to goats and even cattles. *Acacia nilotica* is not much suitable for loping purposes. It is mostly used for browsing in arid area of Rajasthan. Tree can be looped for browsing by small animals. An ideal stocking density at 30 years is 100 trees/ ha for 28 m tall trees and 400 trees/ha for 14 m tall ones. Spot sowings with 10–15 seeds per hole did not require weeding, but needed early thinning and then thinning at 5-yearly intervals. For 5-year-old stands a density of 1200 trees/ha is recommended. Trees pollard well and can be freely lopped for thorn fences; however heavy lopping reduces tree growth.

Yield: Acacia produces pod yield of 750 to1000 kg per ha per year after fifth year onwards. Fuel wood yield of 300 tonnes per ha can be obtained after 15 years.

Utilization: The leaves and pods are widely used as fodder and, in arid regions of India, constitutes the chief diet for goat and sheep. Leaves, pods and shoots are used as forage for camels, sheep and goats, especially in arid area. It is considered as chief diet for goats and sheep, and seeds are a valuable cattle food.

4. Jack tree - *Artocarpus heterophyllus* Lam. (2n = 52, 104)

Family: Moraceae

Vernacular names: jack, jack tree, jackfruit, jak, jakfruit (English); *chakki*, *kanthal*, *kathal*, *kathar*, *panos, halasu, kathar, alasa* (Hindi); panasa (Sanskrit); kanthal (Bengali); *palla, palapalam* (Tamil); kos (Sinhala); nangka, nongko (Indonesian, Javanese); *jackfrutchbaum* (German); *jacquier* (French); *langka*, *nancas* (Filipino).

Importance: Jack is an important tree in tropical home gardens. It is also known for its remarkable, durable timber, with ages to an orange or red-brown colour. The leaves and fruit waste provide valuable fodder for cattle, pigs and goats. The generic name comes from the Greek words 'artos' (bread) and 'karpos' (fruit); the fruits are eaten and are commonly called breadfruit. The specific name, 'heterophyllus', is Latin for various leaved, or with leaves of different sizes and shapes; it is from the Greek word 'heteros' (different).

Origin: The tree is native to the rainforests of Malaysia and the Western Ghats of India.

Distribution: Jackfruit is cultivated in Southeast Asia, especially in India, Bangladesh, Myanmar, China, Sri Lanka, Malaysia, Indonesia, Jamaica, Thailand, the Philippines, Africa, Brazil, Suriname, the Caribbean, Fiji and Australia. This tree is mainly found in sub-Himalayan tract from Nepal eastwards to West Bengal, Assam and Tripura. It is also found in Western Ghats, Andaman and Nicobar Islands.

Botany: It is a fairly fast-growing, large evergreen tree with a large, dense crown. It is a small to medium-sized evergreen tree reaching a height of 8 to 30 m and a trunk diameter of 30 to 80 cm or even up to 200 cm. It usually has a straight stem, rough, scaly bark, branches near the base. Canopy is dense, dome shaped or rarely pyramidal. Inflorescence is solitary, axillary, and cauliforous, with ramiflorous, short, leafy shoots. Inflorescence occurs from November through January. Male heads are sessile or on short peduncled receptacles, sometimes borne on the ultimate twigs; female heads are on oblong ovoid receptacles with simple spatulate styles exerted to 1.5 mm, syncarpous, 30 to 100 cm by 25 to 30 cm, cylindrical or somewhat clavate. It is cross pollinated crop. The fruits are borne on the trunk and on the main branches. They are oval or oblong, 30 to 60 cm in length and 15 to 30 cm in diameter, and sometimes weigh as much as 10 to 50 kg. The skin is studded with short spikes, is pale green when immature, and becomes greenish-yellow to brown when ripe. The ripened fruit is sweet, fleshy, and has a crisp pericarp. Each fruit can contain between 100 and 500 seeds, and each seed is enclosed in a yellowish, juicy sheath with a strong flavor. The seed is kidney-shaped and has thin, white, coriaceous, smooth testa.

Growth stages: Germination begins in 1–3 weeks, or longer (up to 6 weeks) if seeds were stored for long days after collection. It grows moderately rapidly in early years, up to 1.5 m per year in height, slowing to about 0.5 m per year as trees reach maturity. In 5 years, the tree height reaches 7 m and the canopy diameter 4 m; trees of 20 years old are about 18 m, as tree growth slows down with age. Jack begins fruiting after 2 to 3 years after planting but mostly after 8 years. However, the 'Singapore' (or 'Ceylon') cultivar begins bearing fruit 18 to 30 months after transplanting. Trees of 2 years old produce about 25 flowers and 3 fruits; trees of 5 years old bear as many as 840 flowers, and trees of 6 years old 1500 flowers. However, only 15-18 fruits develop due to the low production of female spikes (about 0.6-5% of the total number of inflorescences). Young trees bear more male than female flowers at a ratio of 4:1; production of female flower increases with age. A male-to-female ratio of 2:1 produces 250 fruits per tree, and as the trees ages, fruit productivity declines. Trees bear fruits and flowers throughout the year in favourable environments, but in areas with distinct dry and wet seasons, flowering occurs in the wet season. Fruits are usually borne on branches in young trees and on trunks and roots in older trees. The tree is wind and insect pollinated. Insects normally visit the scented male flowers, which release pollen that is carried to female flowers by the wind. Wilting and drying stigmas are the best indicators of fruit set. Fruits mature in 80-160 days after pollination and a sweet and strong aroma indicates that the fruit is ripe. The unripe fruits used as vegetables can be harvested after 3 to 4 months. The fruiting season lasts about 4 months. The tree lives up to 100 years, but productive up to 20 years.

Climate: It is best grown in tropical, near tropical and subtropical regions. It is found at the altitude of 450-1600 m in elevation. It fruits up to latitudes 30°N and 30°S in frost-free areas and bears good crops between 25°N and 25°S. It is grown in areas with evenly distributed rainfall of 750 to 2400 mm. The tree will not tolerate flooding. If the root touches water, the tree will not bear fruit and might even die. It tolerates 3 to 4 months of drought. It does best with even and continuous soil moisture. Seedlings are best grown in 30–50% sunlight, with sun exposure increasing to 100% as the tree matures. The tree prefers full sun at maturity. It is cultivated in areas having the mean annual temperature of 24–28°C, mean maximum temperature of hottest month of 32–35°C and mean minimum temperature of coldest month of 16–20°C. It can also withstand lower temperatures and frost. It tolerates with minimum temperature of -3 to 0°C. At 0°C, the leaves may be damaged, and at –2°C (28°F), branches or the whole tree may die. It grows in well humid climate with temperature of 16 to 35°C.

Soils: It grows best in well drained, deep soils. It thrives in gravelly or lateritic soils, limestone, deep soils, sandy or stony soils, alluvial, sandy-loam or clay loam soils with a pH of 5-7.5. It exhibits moderate tolerance to saline soils. The tree does not tolerate water stagnation or poor drainage. If the roots touch stagnant water, the tree fails to bear fruit, or it may die.

Propagation method: The most common and simplest method of raising jackfruit trees is from seeds. Inarching and grafting is known for propagation of jackfruit in Malaysia as a means of perpetuating desirable clonal stocks. Jack seedlings are very easy to grow. Seedlings develop very quickly, reaching 25 cm in height within 3 to 4 months. Seeds are not true-to-type since the crop is cross-pollinated. So grafting of known varieties onto rootstocks is often done, especially for commercial production. The root systems are sensitive to damage during transplanting of seedlings. Field sown seedlings can be top-worked (grafted) with select varieties once they are established. Propagation by vegetative means such as cuttings and air-layering is also possible. There is more than 80% rooting success using a low-cost, non-mist polythene propagator when cuttings are taken from 3-month-old seedlings and treated with 0.4 or 0.8% IBA.

Nursery management: Seed soaking in water or a dilute gibberellic acid solution for 24 hours prior to sowing hastens germination. Hot water treatment has been used successfully to stimulate germination. A well-drained medium is recommended, such as 50% peat moss, 25% perlite, 25% vermiculite amended with a little compost, dolomite lime, gypsum, and a 14-14-14 slow-release or an organic fertilizer. In the nursery, 2 to 4 litre root-training containers work well. Germination requires 3 to 8 weeks which is expedited by soaking in water for 24 hours. The seeds are stored in light or darkness at ambient temperatures or at 6°C germinated 1 to 4 days after storage. After 22 days, germination ranged from 80 to 86%. The seeds germinate satisfactorily on coconut husks containing enough soil to cover the seed, and are out planted with the husks. Seeds are sown at a depth of 2–3 cm and can be laid flat or planted with the hilum pointing down. Germination begins in 1 to 3 weeks, or longer (up to 6 weeks) if seeds were stored for long period after collection. Daily watering is often necessary once seeds germinate. Seedlings are ready for transplanting when attains 20 cm height and have a stem diameter of 9 mm. This takes about 3 to 4 months in good growing conditions. The seedlings should not be allowed to root through the container into the underlying substrate, as the roots would have to be cut or broken for transplanting.

Spacing: For fruit production, trees are planted 7.5 to 12 m apart. Closer in-row spacing can be used for slower growing or more compact clonally reproduced cultivars (usually grafted clones) with in-row spacing of 4.6–7.6 m and between-row spacing of 6.1–7.6 m. For timber production, closer spacing should be used to inhibit side branching by shading and promote long, straight trunks. Spacing for timber of 2 x 3 m or 3 x 3 m is suitable. Dig pits or holes of 60-80 x 40-50 cm size for planting.

Cultural practices: Trees should eventually be thinned to a spacing of 7.5 to 12 m and lack of thinning may lead to die-back. Dead branches should be removed from the interior of the tree so that sufficient light is obtained for the developing fruit and to check the spread of pests.

Water management: It is good practice to water trees during the dry season, but the soil at the base of the plant should be raised, and drainage pathways need to be constructed to avoid waterlogging.

Manures and fertilizer application: It is recommended that fertilizer be applied twice yearly – at the onset and the end of the rainy season.

Cropping systems: It is used as a shade tree for coffee, pepper, betel nut, and cardamom with spacing such as 15 x 15 m.

Harvest: At an age of 2 to 3 years, the trees can be topped at 3 to 5 m height to encourage lateral growth for fruit production at an accessible height. Top pruning will have to be done throughout the life of the tree to avoid branches breaking off due to wind or the weight of the fruit. Pruning is advised to remove damaged branches, especially on the lower interior of the tree. Jackfruit is heavy and cumbersome to transport and should be harvested when mature only 3 to 5 days from ripening, so it is best to have a market close by if selling fresh fruit. This tree can be lopped two times in a year depending on growth of tree and feed requirement of farm animals. About 130 kg green leaves can be harvested from a tree in a year. Lopping should be done from the age of ten years. Livestock readily eat jackfruit foliage, so young trees would not survive exposure to grazing animals. However, livestock can be pastured among mature trees. Fallen fruits are readily eaten by livestock and make an excellent contribution to their diet. Cattle, goats, and other small ruminants relish the leaves. Cattle and pigs also readily eat fallen fruit. The waste after removing the pulp from fruits is considered good fodder for cattle. Leaves are cropped in India for fodder, and over ripe, immature or fallen fruits are fed to hogs and cattle.

Fruits yield: Average yield of mature trees are 70–100 kg of fruit/tree/yr depending on variety, cultural practice, and environmental factors. Potential yield is 100 to 200 fruits per tree per year. The fruits may weigh 10 to 50 kg each with an average between 20 to 30 kg. Each fruit contains 100 to 500 seeds; there is no correlation between fruit size and the number of seeds it contains. There are about 50 to 90 seeds per kg.

Seed processing: Open the fruit with a large knife. Then the seeds are separated from the fleshy sheaths that enclose the seeds. The thin, slimy coating around the seed (perianth lobe) should be removed and the seeds are rinsed in water to remove any remaining pulp juice or sugary residue. The large seeds should be used for high germination and produce good seedlings. Seeds may be air-dried in partial shade for about an hour for ease of handling. Seeds should not be stored for more than one month before sowing. Germination for seed sown within a few days of harvesting is 90%.

Seed storage: Seeds are recalcitrant, i.e., they do not retain viability when dried or stored for extended periods. They should be planted immediately for best germination and seedling vigor. Seeds can be stored moist in a plastic container in the refrigerator for a few weeks. Stored seeds germinate more slowly than fresh seeds. Seeds stored at 20°C retain viability for 32 weeks with 48% germination. Seed viability is maintained for two years in moist storage at 15°C, seeds kept in polythene bags filled with perlite at 6°C.

Utilization: The fruit is edible. The tender young leaves are cooked and eaten as a vegetable. Young male flower spikes can be grated or smashed and eaten with salt and vinegar as a vegetable, or pickled. The young fruit is cooked as a vegetable, pickled or canned while the mature fruit is eaten fresh or made into chutney, jam, jelly and paste, or preserved as candies. The fruit pulp is also used to flavour ice-cream and beverages. The seeds are edible when boiled, roasted or dried and salted. The tree can be used as shade tree for coffee and for reforestation. The wood is durable and used for tool handles, guitars, ukuleles, timber and fuel. Leaves can be used as fodder. Cattle, goats, and other small ruminants relish the leaves. Cattle and pigs also readily eat fallen fruit. The waste after removing the pulp from fruits (rags) is considered good fodder for cattle and pigs.

5. Junglee kikar - *Prosopis juliflora* (Sw) DC (2n=28, 56)

Vernacular names: Mesquite (Arabic); Ganda-babul, Vilayati babul, Vilayati khejra, Vilayati kikar (Hindi).

Importance: It is a multipurpose leguminous tree. It is a pioneer species that rapidly colonizes denuded, abandoned ravines in dry lands of India. It serves as one of the main sources of fuel for the rural and urban poor in the country. It provides >90% of the fuel wood because its wood has excellent burning qualities. The wood is used for parquet floors, furniture, and turnery items, fencepost and pilings. Toasted seeds are added to coffee. In Tamil Nadu, Haryana, Gujarat, and Karnataka farmers raise Prosopis hedges along field boundaries to meet fuel wood requirements.

Origin: It is native of Peru, Chile and Argentina

Distribution: This tree is grown in Argentina, Brazil, Uruguay, Chile, Peru, Bolivia, Chile, Colombia, Costa Rica, Ecuador, El Salvador, Guatemala, Honduras, Mexico, Nicaragua,

Panama, Peru, United States of America, Iran, India, Mexico, Venezuela, West Indies, Puerto Rico, Jamaica, Nicaragua, El Salvador, South Africa and Australia.

Botany: It is an evergreen tree with a large crown and an open canopy, growing to a height of 5 to 10 m. It has a twisted stem, flexible branches with long strong thorns. It has a deep taproot. Leaves are compound; leaflets in 13-25 pairs, oblong and dark green, bipinnate with 1 or 2 pairs of rachis, almost pendulous. Flowers are tubular and light greenish-yellow in colour. Fruits are a non-dehiscent pod, straight, linear, falcate to annular, with a coraceous mesocarp. Seeds are compressed, ovoid, hard, dark brown, with mucilaginous endosperm surrounding the embryo; cotyledons flat, rounded, epigenous when germinating. It bears fruits from 2 to 3 years of age.

Climate: It is a tree typical of arid and semi-arid regions. It is a xerophytic plant. It grows from the coast to altitudes of 100 m, preferring lower quality, arid, rocky and dry lands. It is successfully grown with an average annual rainfall of about 250 to 1500 mm. It remains green and bears fruit even in years of drought.

Soil: It is adapted to various soil types. It tolerates acid, saline and alkaline soils. It tolerates seasonal water logging.

Season: It is planted at the beginning of the rainy period. Water the seedling immediately after it is planted in order to insure sufficient moisture and hence plant survival; watering is not necessary if the soil is sufficiently moist, right after a rain.

Propagation methods: Propagation is by seed, root cuttings or grafting. Seeds are embedded in indehiscent fruits of hard endocarp of the pods. Seeds can be extracted with lengthwise cut on the pods and crosswise between the seeds using a small knife. This method is very time-consuming when the amount of seeds to be withdrawn from the pods is large. Seed is extracted with feeding the whole pod to the animals; the seeds are liberated from the pods during digestion and come out clean embedded in the droppings, in suitable condition for sowing. Seeds may also be extracted with the digestion of the fruits by a 5% HCL solution at a temperature of 30°C. Seed treatment is required either by soaking for 15-20 minutes in sulphuric acid followed by thorough washing in cold water or by mechanical scarification. About 80-90 % of the seeds germinate in 4-6 days. Germination is only 20% when sowing seeds with mesocarp since the moisture to penetrate the hard cover is slow. Seeds can be soaked in water for six hours to enhance germination in 24 hours. The seeds are rubbed with sand in a mortar for 10, 15 and 20 minutes; germination rates recorded are 25%, 81%, 68%, 86% and 100% respectively. With this process, the seeds germinated between 5 and 10 days. Seeds can be sown in beds. The beds must be constantly and abundantly watered, because if there is a lack of moisture, the seeds will dry out, affecting the germination rate, this is due to the fact that the germination process has already begun, and should not be interrupted. When the seedlings reach a height of 15 to 25 cm, they are ready for planting at the definitive place. The stump sprout material can be obtained with stakes 15 cm long and 2.5 mm to 4.5 mm in diameter, clipped from the top crown branches. The stakes can be treated with 2,000 ppm indolbutiric acid (IBA) to promote rooting.

Planting technique: Spacing is 2 × 2 m or 3 × 3 m for timber production, 5 × 5 m, for stake production, 6 × 6 m for fodder tree plantations, 15 × 15 m or 15 × 10 m or 5 x 5-10 m for pod production. If the pastures have bushy grasses, the distance should be 10 m. The spacing is 20 × 15 meters or 15 × 15 m whenever associated to buffel grass depending on soil type.

Cultural practices: Young plants benefit from weeding around the stem and need protection from grazing animals. Thinning and pruning are needed to keep the plantation accessible.

Harvest: Direct browsing of leaves and pods by cattle, sheep in the field, constitutes a practical and economical way of feeding animals. Naturally, the animals ingest both ripe and unripe pods, snatching them before they drop to the ground or picking up the dry ones from the ground. The pod harvest is easy but costly operation. It consists of manually collecting the pods directly from the plants, together with those naturally fallen to the ground. Lopping stage depends on growth of plant which is varied with availability of rain and soil condition. In India, tree can be lopped in scarcity of fodder. Lopped trees bear lush green leafy fodder in the subsequent year.

Yield: It bears pods in summer and winter seasons. The pods can be collected in May/June and September/October. Peak pod production occurs at 15-20 years of age. Mesquite starts fruiting at 3-4 years of age; 10 year-old plants may yield up to 90 kg pods annually, however, annual pod yield ranges up to 100 kg/tree. A high yield of 170 kg/tree/year is also observed. A good plantation may produce up to 6 to 10 tons of pods per ha considering a stocking rate of 100 trees per hectare.

Seed collection and storage: Seed is difficult to extract from the pods Seed storage behaviour is orthodox, 60% germination following 50 years storage, viability of seed can be maintained for several years in airtight storage at 10°C with 5-9% moisture. There are 20000-26000 seeds/kg. The pods can be stored stratified in sand. The process consists of placing a fine sand layer on the ground, and on top of it a layer of 20 to 25 cm thick of well-dried pods. On top of this, a new layer of sand is deposited. Then come new layers of pods, alternated with sand layers. Pods can be conserved for 14 months fumigating three times with phosphine during the storage period.

Utilization: Leaves used as fodder purposes for different categories of animals. The seeds obtained from pods are also fed to dairy cows. Pods have high palatability, nutritive value and without any adverse effects on their performance eaten by cattle, sheep and goat. The common uses Prosopis is the production of hard wood for mosaics, boards and sleepers; good-quality firewood; coal; stakes for fences; pods for animal and human consumption. The flour may make up 40-60% of concentrate rations. It is fed unmixed to sheep. Ripe pods contain 12-14% crude protein. The short-fibred parts are also suitable for pigs and poultry.

6. Khejari - *Prosopis cineraria* (L.) Druce ($2n = 28$)

Vernacular names: Jand, Jandi, Musquit bean, Screw bean (English); *Ghaf*, (Arabic), *Shami* (Bengali), *Kamra, Khijado, Sami, Semru, Sumri* (Gujarati); *Banni, Chaunkra, Chonksa, Jand, Janti, Khejiri, Sangri, Shami* (Hindi); *Jhand, Jhind* (Sanskrit); *Jambu, Perumbay, Vanni* (Tamil).

Importance: The pods are used as vegetable in the dried and green form in many parts of the Thar desert in India. It is grown for fuel wood and provides high-quality charcoal. Wood is used for boat frames, houses, posts, and tool handles. The tree yields a pale to amber coloured gum with properties similar of gum acacias. The pods and leaves are used as animal food or feed. The trees are planted for sand dune stabilization.

Origin: It is native of India.

Distribution: It is grown in Punjab, Rajasthan, Gujarat, Uttar Pradesh, in dry parts of central and southern India, extending into Pakistan, Afghanistan, Iran, and Arabia. It is a characteristic tree of secondary dry deciduous forest, desert thorn forest, ravine thorn forest, and desert dune scrub.

Botany: Khejari is a tree of 6.5 m in height, prickles inter nodal, scattered, straight, conical with broad bases. Tap root is 3 m long. Leaves are glabrous or puberulous; petiole and rachis 0.5-4 cm long, the pinnae 2-7 cm long. Fruits are slender, elongate, 8-19 cm long, 4-7 mm in diameter; seeds are ovate, 6 mm long, 10-15 in a pod, brown in colour.

Climate: The suitable climate is dry to arid characterized by extremes of temperature ranging from 48 to 52°C with annual rainfall varies from 250 to 1500 mm.

Soils: It grows on a variety of soils including coarse sandy alluvial soils. Good growth is obtained on deep sandy loam soil with adequate availability of moisture in lower layers. It is tolerant to saline and alkaline soils.

Propagation method: It is propagated with root suckers and seeds. The seeds should be soaked for 24 hours before sowing. Spacing is 2 x 2 m.

Weed management: Tree should be weeded until well established.

Lopping stage: Lopping stage depends on growth of plant which is varied with availability of rain and soil condition. Tree can be lopped in scarcity of fodder. The lopped trees bear lush green leafy fodder in the subsequent year.

Fodder production potential: The leaves are known as loong, locally and available for 4-5 months (June-October). It is used as dry fodder for animals and sometimes mixed with animal feed also.

Seed collection and storage: Ripe pods are collected by lopping or shaking the branches. The pods are dried in the sun, beaten and winnowed to separate clean seed. Seeds weighing a kg contain 25000-27000 numbers of seed. Ripe pods are collected by lopping or shaking the branches. The pods are dried in the sun, beaten and winnowed to separate clean seed. Seeds weigh 25 000-27 000 per kg. The 1000 seed weight range 37 to 40 g.

7. Calliandra - *Calliandra calothyrsus* Meisn (2*n* = 22)

Family: Fabaceae (alt. Leguminosae)

Vernacular names: calliandra, red calliandra (English); kalliandra, kalliandra merah (Indonesia); barba de gato, barbillo, barba de chivo, barbe jolote, barbe sol, cabello de angel, carboncillo, clavellino, pelo de angel (Spanish).

Origin: It is native of humid/sub-humid regions of Central America and Mexico.

Distribution: It is grown in Mexico, Belize, Guatemala, Honduras, El Salvador, Nicaragua and Costa Rica, Panama, Indonesia and east Africa.

Botany: It is a small, perennial, thornless leguminous tree growing to a height of 2-12 m. Trunk diameter is up to 30 cm with white-reddish brown bark. Leaves are bipinnate and alternate; there are 19-60 pairs of leaflets; leaflets are linear, oblong and acute. Inflorescences are particulate with flowers in umbelliform clusters, 10-30 cm long. Flower sepals and petals are green. Flowering will cease over the dry season where greater than 4 months dry occurs. Flowers are andromonoecious, bearing both hermaphrodite (bisexual) and staminate (male) flowers. Fruits are broadly linear, flattened, 8-11 cm x 1.0 cm with thickened and raised margins, finely pubescent or glabrous, brown dehiscent, 8 to 12 seeded. Seeds are ellipsoid , flattened, 5-7 mm long and mottled dark brown.

Climate: It is grown in areas receiving annual rainfall of 700-3,000 mm with 3-6 dry months. It is poor tolerance of inundation. It can withstand dry periods, particularly in riverine environments, or where a perched water table is present. It grows best at 250-1,850 m altitude. It is adapted to the areas with mean monthly maximum temperatures of 24-28°C, and mean minimum temperatures of 18-24°C. It is frost susceptible but possesses considerable cool tolerance for a tropical species.

Soils: It grows well on a wide range of soil types ranging from deep volcanic loams to more acidic metamorphic sandy clays. It does not tolerate waterlogged conditions. It does not grow well on poorly drained calcareous soils.

Propagation: Seed requires scarification. Seeds are soaked in cold water for 48 hours. Mechanical scarified seed is sown at 1-3cm depth or seedlings raised in nurseries when the plants are 20 to 50 cm tall. Seedlings can be planted 0.5-1.0 m apart in hedgerows spaced 3 to 4 m apart, or in fodder banks spaced 0.5-1.0 m apart in a grid pattern. It does not establish well from cuttings.

Harvest: Defoliation depends on use. The tree provides fodder in 8-12 months after sowing as first cut. For maximum leaf production, cut to a height of 0.5-1.0 m every 2-3 months. It is generally recommended for cut-and-carry feeding rather than for use in direct grazing systems.

Yield: It produces yields of 7 to 10 t/ha/year of dry matter.

Seed production: Pod ripening is basipetal and takes 90 to 120 days. Seed dispersal is through explosive apical dehiscence of the pods. Seed ripens sequentially along the inflorescence, starting from the base. Single harvest result in seed losses from early dehiscing of pods, or varying quantities of unripe seed. Tarpaulins can be laid under seed trees to collect seed as it falls.

Utilization: It is grown primarily for forage for ruminant livestock. The fodder can be given to all types of ruminants and fulfils 40-60% of their needs. The freshly cut forage within 4-6 hours has higher digestibility value of 60-80%. It is also used for the provision of green manure, shade for coffee and tea, land rehabilitation, erosion control. It is an excellent fuelwood for cooking and small fires. The wood dries very quickly and burns well with a smokeless fire.

8. Anjan - *Hardwickia binata* Roxb

Vernacular names: Anjan (English, Sanskrit); Anjan, *Karachi, Alti, Kamra, Yana, Katt-Udugu, Parsid, Ura* (Hindi); Yepi (Telugu); Kamara (Marathi, Kannada); Aacha (Malayalam:); *Katudugu, Acha, Calam, Ura, Karacha* (Tamil).

Importance: The tree is generally planted as an ornamental tree and the fiber from the bark is used for making ropes.

Origin: It is native of India.

Distribution: It is grown in Afghanistan, Bangladesh, Brunei, Cambodia, India, Indonesia, Iran, Laos, Malaysia, Myanmar, Nepal, Pakistan, Papua New Guinea, Philippines, Thailand and Vietnam. The tree is found in the Deccan peninsula, Central India, Uttar Pradesh and Bihar.

Botany: It is a medium or large deciduous tree, growing up to 30 m in height and 3 m in girth, with a clean cylindrical bole 15 m in height. The leaves are small, alternate, kidney-shaped, pinnate, grayish green in colour and measure about 2-6 cm in length, and 2-3 cm in width. Flowers are borne in lax panicles. Flowers are small in size, pale yellowish-green in colour. The pod is one seeded. Seeds have a hard testa, pointed at one end and rounded at the other end. Seed count is 3000 to 5000 per kg.

Climate: It is found in teak forests, dry savannah and degraded dry deciduous forests. It grows in dry and hot climate, characterized by a long period of drought, low to moderate rainfall and intense heat during summer season. The tree occurs up to an altitude of 760 m from MSL. Young shoots are sensitive to fire, but recovery is good. It is grown in areas with mean annual temperature of 22 to 34°C in areas receiving mean annual rainfall of 250 to 1500 mm.

Soils: The tree grows well on soils of sandy loam or reddish gravelly sand. It tolerates acidic to neutral soils.

Propagation method: It is propagated normally by the seeds, but the root suckers also produce seedlings during the rainy season. The nursery beds should be prepared in the shade, covered lightly with hay and burnt. Seeds should be soaked in water for 24 hours. Seed germinates within 10 to 35 days. The seedlings can be transplanted into the fields after 12 months during the rainy season. The stumps can be made from one year old seedlings. These stumps should contain 4 cm of shoot and 13- 15cm of roots. The stumps should be planted in polythene bags or nursery beds. The sprouted stumps can be transplanted into the fields after 12 months during the rainy season.

Harvest: The tree pollards well even up to a comparatively advanced age and old pollards when re-pollarded almost invariably produce abundant new shoots. The tree coppices poorly. Old trees, which send out vigorous pollard shoots if cut 1 or 2 m above the ground, produce no coppice shoots if cut flush with the ground; old pollards when felled at ground level never coppice.

Seed collection and storage: The seed is flat about 2 cm long and 0.75 cm wide. The light-winged pods are best collected from the tree since fallen pods are often damaged by insects or rodents. The optimum period for collection is when the colour of the pods turns brown with the moisture content of about 10-12%. Seeds can tolerate 4 to 5% moisture content. The seeds can be stored for more than 5 years if stored at freezing temperature of 0 to -20°C. The seed is capable of retaining its viability at ambient temperature.

Utilization: Leaves contain about 9% crude protein, but the amount varies with age of the leaves. Leaves are used as cattle fodder and green manure. It contains about 9-11% crude protein. The species is regarded promising for fodder production in arid zone areas. The wood is used as timber. The bark yields a fibre used for well ropes.

9. Ardu - *Ailanthus excelsa* Roxb.

Family: Simaroubacea

Vernacular names: Indian Tree of Heaven, Coramandel ailanto (English); Maharukha, Limbado, Maharuk, Mahanimb, Ghodakaranj (Hindi); Aralu (Assamese); Moto ardusa, Adusa, Arduri, Arlabo, Moto adusa (Gujarati); Bende, Dodabevu, Dodda, Doddamaru, Hemaraheera mara, Dodumani (Kannada); Perumaram, Mattipongilyam (Malayalam); Mahanimb, Maharukh (Marathi); Pimaram, Pinari, and Perumaram (Tamil); Pedu, Pey, Pedda, Peddamamanu, Putta (Telugu).

Importance: It is grown as a shade and avenue tree. It is a relatively salt-tolerant species. Ailanthus is derived from 'ailanto', an Ambonese word probably meaning 'tree of the gods' or 'tree of heaven' while in Latin 'excelsa' means 'tall'.

Origin: It is native of India and Sri Lanka.

Distribution: It occurs throughout the tropical and subtropical regions of India, especially in the dry districts of Gujarat, Rajasthan, Haryana, Punjab, Uttar Pradesh, Bihar, Odisha and the Deccan plateau. It is not found in the high rainfall regions of the West Coast. It also occurs in coastal areas of Andhra Pradesh and in dry belts of Tamil Nadu, West Bengal and Karnataka.

Botany: It is a large deciduous tree, 18 to 25 m tall, trunk straight, 60 to 80 cm in diameter; bark light grey and smooth, aromatic, slightly bitter. Leaves alternate, pinnately compound, large, 30-60 cm or more in length; leaflets 8-14 or more pairs, long stalked, ovate,

6-10 cm long, 3-5 cm wide, hairy gland; edges coarsely toothed and often lobed. Flower clusters droop at leaf bases, shorter than leaves, much branched; flowers many, mostly male and female on different trees, greenish-yellow in colour. Fruit is one seeded samara, lance shaped, flat, pointed at ends, 5 cm long, 1 cm wide, copper red, strongly veined, twisted at the base. There are about 9500 seeds/kg.

Climate: It is commonly found in mixed deciduous forests and sal forests. It grows well in semi-arid and semi-moist regions, both in the plains and the hills. The tree can be seen growing up to an elevation of 900 m. It has been found suitable for planting in dry areas with annual rainfall of about 400 to 2500 mm. It is a strong light demander. It is susceptible to frost and prolonged drought. It is grown in areas with average mean temperature from 10ºC to 40°C.

Soil: It grows in a wide variety of soils, but thrives best in porous sandy loams. It avoids clay soils with poor drainage and waterlogged areas. Water logging and poor drainage cause high seedling mortality.

Propagation methods: Seeds are sown in line and lightly covered with a thin layer of fine sandy soil of 5 mm thick. Germination starts in 8 to 10 days and is completed in about 40 to 45 days. From one kg of seeds about 1425 healthy seedlings can be obtained. About 15 gm of seeds are required for sowing one sq. m of bed. The seedlings which attain height of 50-100 cm are suitable for planting. The root shoot ratio of 1:2 is considered good for stump planting. It coppices well. Natural regeneration through coppice and root suckers is also possible. Seedlings of 6 to 10 months are used for planting in pits at a spacing of 3m x 3m in block plantations and 6 m x 6 m for Agro-forestry.

Harvest: Mature leaves lopped twice or thrice in a year. Lopping of leaves can be started from fourth year onwards. Mostly lopping is done during October to February and May-July.

Yield: An average tree yields about 5 to 7 quintals of green leaves twice a year. The tree leaves yields about 11.9 to 82.2 kg DM per tree. Mature leaves are highly palatable and nutritious, usually fed to sheep and goats. Some trees are lopped for green leaves while leaves from others can be lopped, dried and stored for feeding during scarcity period. Wood makes good timber and used as firewood.

Seed collection and storage: Greenish yellow flowers in large panicles appear in January-February. Fruits ripen in April-May. The brownish fruit bunches at the end of the branches are cut because of the possibility of blowing away by wind. Then they are dried in the sun, beaten and winnowed to separate the seeds. Seeds lose viability rapidly and hence to be used immediately. The seed loses viability fast but under proper storage conditions they can remain viable for up to 240 days otherwise the normal viability is 4-5 months. The number of seeds in one kg is 8000-10,000.

10. Kala siris - *Albizia amara* (Roxb.) Boiv.

Vernacular names: *Lallei, Veracchi, Nella, Wunja, Renga* (India); *Tugli* (Kanara); *Arad, Arrad* (Arabic); *Krishnasirisha, Moto Sirisio* (Gujarati); *Chigara, Krishnasirisha, Moto Sirisio, Sikkai, Tugal, Tugli, Unjal* (Hindi); *Munja, Sherkam, Suranji, Thuringi, Unjal, Woomjai, Wunja* (Tamil).

Origin: It is native of Boivin, Kenya, Ethiopia and India.

Distribution: It is grown in Botswana, Malawi, Mozambique, Zambia, Zimbabwe, Kenya, Tanzania, Uganda, Sudan, Sri Lanka and India. The tree is found in Himalayan tracts, West Bengal, Assam, Gujarat, Tamil Nadu, Andhra Pradesh, Karnataka and Andamans.

The tree occurs chiefly in moist places along river banks, in swampy lands and low lying savannahs.

Botany: It is a small to moderate-sized, much-branched deciduous tree with smooth, dark green, scaly bark. Its root system is shallow and spreading. The leaves are pinnately compound, with 15-24 pairs of small, linear leaflets, on 6-15 pairs of pinnae. Flowers develop when the tree is almost leafless. The flowers are yellow, fragrant, in 12-20 globose heads in clusters. Fruits are oblong pods, about 10-28 x 2-5 cm, light brown, puberulous, thin, and 6-8 seeded; seeds flattened, 8-13 x 7-8 mm. It produces root suckers when the stem is mutilated or the tree becomes old.

Climate: Tree survives up to an altitude of 1500 m from MSL with mean annual temperature of 10 to 47°C and mean annual rainfall of 400 to 1000 mm. It has good adaptability for growing in moist as well as fairly dry conditions. It is fairly drought resistant. It is a strong light-demander, although it withstands moderate shade in the pole stage. It is sometimes raised as a shade tree in tea gardens.

Soil: It tolerates clays and is often found streams where it can reach to more water.

Propagation methods: It is propagated with seed in areas protected from fire and grazing. Seed pretreatment involves immersion in boiling water for 5 minutes followed by soaking for 12 hours. The treated seed can be then sown and will germinate within 7 to 10 days. Germination is about 80%. It reproduces very freely from coppice; it produces coppice shoots as many as 50 to 100 shoots.

Planting technique: Spacing generally adopted is 9 to 10 m apart along contour lines; plants are thinned when 2-3 m tall in the first year and 5-8 m in the third or fourth year depending on the rate of growth. Young seedlings should be protected from fire and grazing.

Harvest: This tree is not suitable for lopping but leaves can be harvested during scarcity. After attaining the age of ten years plant can be used for fodder purposes.

Yield: The leaves make excellent fodder. Ten years old plant is able to produce 20-22 kg fresh leaves in a year.

Seed collection and storage: Fruits are oblong pods, about 10-28 x 2-5 cm and light brown in colour. Pod contains 6-8 seeds. Seeds are flattened, 8-13 x7-8 m. The seeds can be stored up to 2.5 years without losing viability. They are best stored in mud pots with wood ash or in sealed tins or gunny bags.

11. Indian coral tree - *Erythrina indica* Lam. ($2n = 42$)

Synonym: *Erythrina variegata* L.

Family: Fabaceae.

Vernacular names: Indian coral tree, tropical coral tree, tiger's-claw, moochy wood tree, variegated coral tree (English), Paribhadraka, Kantakimsuka, Paribhadra (Sanskrit), Pharahada, Pangara, Ferrud, Dadap, Pharad (Hindi), badisa, bodita (Telugu), murukku, mulmurukku (Malyali), varjipe, harivana (Kannada), palidhar, Palitamadar, palitu-mudar (Bengali), kalyana murangai (Tamil), panarawas, pararoo, Panervo (Gujarati), pangara (Marati). It is known as Pangrah in unani system of medicine. Erythrina comes from the Greek word 'eruthros' meaning red, alluding to the showy red flowers of the Erythrina species.

Importance: It is used *as* fodder, light timber and more recently, pulp for the paper industry. Erythrina comes from the Greek word '*eruthros*' meaning red, alluding to the showy red flowers of the Erythrina species.

Origin: It is a native of India and Malaysia.

Distribution: It is grown in Malaysia, Myanmar, Thailand, Australia, Bangladesh,

Brunei, Cambodia, Ethiopia, Fiji, Guam, India, Indonesia, Laos, Madagascar, Malaysia, Myanmar, Nigeria, Papua New Guinea, Philippines, Solomon Islands, Sri Lanka, Taiwan, Province of China, Thailand, Vietnam, Nepal, United States of America. In India, it is distributed in coast forests from Bombay to Malabar and from the sundarbans along the coast and in the Andaman's and Nicobars much planted for ornamental purpose.

Botany: It is a medium-sized, spiny, deciduous tree normally growing to 6-10 m tall and 60 cm dbh. Young stems and branches are thickly armed with stout conical spines up to 8 mm long, which fall off after 2-4 years; rarely, a few spines persist and are retained with the corky bark. Bark smooth and green when young, exfoliating in papery flakes, becoming thick, corky and deeply fissured with age. Leaves trifoliate, alternate, bright emerald-green, on long petioles 6-15 cm, rachis 5-30 cm long, prickly; leaflets smooth, shiny, broader than long, 8-20 by 5-15 cm, ovate to acuminate with an obtusely pointed end. Leaf petiole and rachis are spiny. Flowers are bright pink to scarlet erect terminal racemes 15-20 cm long; stamens slightly protruding from the flower. Fruit a cylindrical torulose pod, green, turning black and wrinkly as they ripen, thin-walled and constricted around the seeds. There are 5–10 seeds per pod. Seeds are kidney-shaped, dark purple to red, and 1–1.5 cm in length. There are 1450–5000 seeds/kg.

Growth stages: The tree is grown from cuttings or seed. Sapling growth is rapid, and a 1-year-old sapling can reach 3 m in height. Trees of 3 or 4 years old can start flowering. It typically reaches 3 m in height in a year, and 15–20 m in 20 to 25 years. On favorable sites, the stem can reach a diameter at breast height (dbh) of 50–60 cm in 15 to 20 years. It can live to about 100 years. In general, rooting is superficial, with most roots in the upper 30 cm of the soil; older trees, however, root deeper. The fruits are ripe from October to November in the Southern Hemisphere and March to April in the Northern Hemisphere.

Climate: It is found in coastal forest up to 1500 m in elevation. The tree is found in the humid tropics and subtropics and can tolerate a wide variety of climates within this zone. It does particularly well in monsoonal climates that have a wet summer and a dry winter, and it requires little water during the winter dry season, because it drops its leaves at that time. It grows in areas receiving mean annual rainfall of 500–1500 mm. It is somewhat frost-sensitive, tender shoots dying back but quickly re-grows under favourable conditions. It is grown in areas with mean annual temperature of 20 to 32°C.

Soils: The tree occurs in a wide range of soils frequently on sandy loams, alluvial loams, silts and clays. On gravelly skeletal soils, its growth is normally stunted. It is also tolerant of a wide range of soil pH, ranging from 4.5 to 8.0. It is typically found on sandy soil in littoral forest, and sometimes in coastal forest. It tolerates seasonally waterlogged soils.

Propagation: Propagation is by two common methods, cuttings and seeds. This tree is most commonly propagated vegetatively for live fences, windbreaks, and establishment in areas where livestock is present (which could eat shoots from small seedlings). Large-size branch cuttings at least 1.5 m long are used, usually 2–3 m in length and 5–10 cm in diameter. Smaller cuttings may be used, a minimum of 30 cm in length and a diameter of 4–5 cm. It is best to retain the terminal bud of branch cuttings to ensure fast new top growth. Cuttings can be taken any time of year usually at the onset of the rainy season. Cuttings are the only way to propagate clonal varieties, as seeds are not true to type.

Storage of cuttings: Cuttings are stood upright in shady, dry, and cool conditions for a minimum of 24 hours and a maximum of 2 weeks. This standing time allows the cuttings to dry slightly and helps prevent rotting and fungal problems. For larger stakes 2–2.5 m tall, the lower portion of the cutting is buried 20–40 cm deep. For smaller cuttings, generally about 20% of the cutting's length should be underground.

Spacing: A spacing of 8-10 m is used for shade in coffee plantations; spacing of live stakes for betel and pepper is 2-3 m x 2 m.

Cropping systems: It is planted as a shade tree in coffee, cinchona plants and cacao plantations and as a trellis plant with betel nut, black pepper, vanilla, and yam. It is used as a support crop for the betel vine. It should be planted 3-4months before the main crop is planted out. When used as a shade tree for coffee it should be pollarded at 2-3m yearly.

Harvest: When trees are used to support black pepper vines, side branches are lopped at interval of 6-8 weeks, the foliage being used as green manure or fodder. When planted for shade, lower branches are removed immediately after establishment and the trees are pollarded once per year in the middle of the rainy season. The trees are used as support for betel vines yield 15-50 kg fodder per year; shade trees in coffee plantations produce about 100 kg fodder and 25-40 kg wood per year.

Yield: It yields 15 to 50 kg of fodder per tree per year.

Seed storage: Seed germination is 60-75% for fresh seeds. They can be stored for a long time if kept in cool, dry and insect free conditions. There are about 4500-6250 seeds/kg.

Utilization: The leaves are used as green vegetable and cattle fodder. It is the most preferred plant by goats. The leaves are used for treatment of fever, inflammation and joint pain. The juice of the leaves is used in ear ache, tooth ache, constipation and also known to stimulate lactation and menstruation. Juice from the leaves is mixed with honey and ingested to kill tapeworm, roundworm and threadworm. Women take this juice to stimulate lactation and menstruation. The bark is used for fever, hepatosis, malaria, rheumatism, toothache, also for boils and fractures.

12. Kardhai–*Pithecellobium dulce* (Roxb.) Benth. (2n=26)

Vernacular names: Blackbead tree, Bread and cheese tree, Madras thorn, Manila tamarind, Sweet Tamarind, sweet Inga (English); *Jungle Jalebi* (Hindi); *Simachinta* (Telugu); *Amil, Balati, Dekhani babul* (Bengali); *Jangal Jalebi, Kodukapuli* (Tamil).

Importance: *P. dulce* is a hardy, nitrogen-fixing tree that tolerates harsh sites, heat and drought, and heavy cutting. It is also able to withstand poor and saline soils.

Origin: It originates from Central America (Mexico).

Distribution: It is grown in Argentina, Bolivia, Brazil, Colombia, Ecuador, French Guiana, Guyana, Mexico, Paraguay, Peru, Surinam, United States of America, Uruguay, Venezuela, Philippines, India, Malaysia and Thailand. In India this tree is found in drier parts of Rajasthan, Gujarat, Uttar Pradesh and Madhya Pradesh.

Botany: Trees are evergreen. It grows to a height of 10 to 15 m, but ranges from 5 to 18 m. The crown of the tree is broad spreading with irregular branches. The bark is grey, becoming rough, furrowed, and eventually peeling. Leaves are bipinnate, with two pairs of two kidney-shaped leaflets. Thin spines (2 to 15 mm in length) are in pairs at the base of leaves. The flowers are in small white heads. Pods are 10-15 x 1.5 cm. The pod is spiral and pinkish or reddish-brown in colour as they ripen. Each pod contains 5 to 10 shiny black seeds up to 2 cm long. The pods are grey in colour and tightly coiled. Seeds are 7-8 seeds, black in colour, orbicular with white aril.

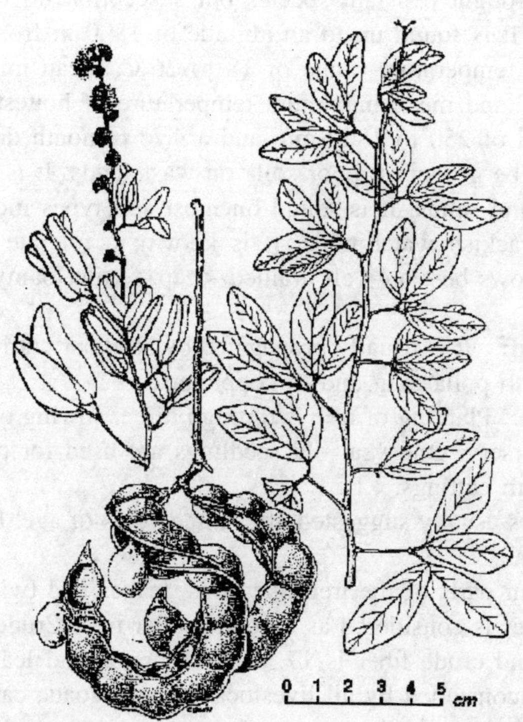

Phenology: On most sites growth rates of <1 m height per year are normal and it may take 40 years to reach full size. Trees start to flower and fruit at 4-8 years of age. Flowering and fruiting can occur throughout the year as moisture permits, but usually flowering peaks during the dry season with pods ripening 2-4 months later. Trees are generally evergreen or almost so, even in seasonally dry climates, as leaf fall and flush of new leaves often overlap. Trees flower from 4-8 years old, over an extended period of the year, with a peak in the dry season and pods ripen 2-4 months after flowering. Flowers are pollinated by a wide range of generalist insect pollinators, including large and small bees. The seeds remain attached to the pods after they open. Pods often ripen over a long period of the year rather than synchronously over a shorter period. There are approximately 10 seeds in each pod and 9,000 to 26,000 seeds/kg. Seeds are dispersed by birds attracted by the red pods, and sweet fleshy white, pale-pink, or occasionally red arils which persist after the pods open. Seeds germinate within 1-2 days and germination of 20-70% can be expected.

Climate: It is a drought resistant species but susceptible to frost and growing well in areas of low rainfall. It is found up to an altitude of 1800 m from MSL. It is grown in areas with mean annual temperature range of 18 to 26°C, mean minimum temperature of coldest month 8 to 20°C and mean maximum temperature of hottest month of 32 to 48°C and mean annual rainfall of 250 to 1800 mm and a 4 to 6 month dry season.

Soil: The tree can be grown on poor soils on wastelands. It is also found along river banks on alluvial soils and sands. It is found on most soil types including clay, limestone and wet sand with a brackish water table. It is known to tolerate moderate salinity and impoverished soils. It grows best on well drained, deep, fertile loamy soils, including black cotton vertisols.

Propagation method: It is usually propagated by seeds or cuttings. It suckers are well and responds positively to pollarding and to coppicing.

Planting technique: Planting of tree is through direct sowing of seeds. Seedlings may also be grown in the nursery. One year old seedlings are used for planting. It can also be easily propagated by stem cuttings.

Harvest: Pruning is usually suggested after three years of age. Round the year lopping is done after seven years.

Yield: The different parts of the tree like pods, leaves and twigs are good fodder for herbivores and the species is considered as a good fodder tree. Crude protein content in the foliage is up to 29 % and crude fiber is 17.5 %. The pods and leaves are collected from hedge clippings and are consumed by all livestock, horses, goats, camels, cattle and sheep. The press cake residue from seed oil extraction may be used as stock feed. This fodder tree is actively managed in Rajasthan.

Seed collection and storage: Seeds are collected from ripen fruits and stored after proper drying. Seed storage behaviour is recalcitrant. The seed count is around 6400 to 6700 in a kg. Seeds extracted from mature fruits remain viable for about six months.

Utilization: *P. dulce* is most often cultivated as an ornamental, shade or street tree planted on roadsides, and in backyards and hedge. *P. dulce* is perhaps best-known for its sweet edible aril, which is eaten fresh, as an infusion, or macerated in water to make a lemonade-like beverage. The pods are also relished by livestock and chickens. The leaves contain 29% crude protein and the young shoots are used for livestock fodder in some areas, either browsed directly or by lopping branches and allowing the leaflets to dry and drop off. Hedge trimmings are often used in this way as fodder for goats in parts of India.

REFERENCES

Burkart, A. 1976. A monograph of the genus *Prosopis* (Leguminosae subfam. Mimosoideae). J. Arn. Arb. 57(3/4):219–249; 450–525.

Chamberlain, J.R. (ed.) 2001. *Calliandra calothyrsus*: an agroforestry tree for the humid tropics. Tropical Forestry Papers No. 40. Oxford Forestry Institute, Oxford, UK.

Chawla HM and Sharma SK. 1993. Erythritol, a new isoquinoline alkaloid form *Erythrina variegata* flowers. Fitoterapia. 64(1): 15-17.

Evans, D.O. 1996. Proceedings of the International Workshop on the Genus *Calliandra*. Forest, Farm and Community Tree Research Reports (Special Issue). Winrock International, Morrilton, Arkansas, USA.

Foster, A.H. and Blight, G.W. 1983. Use of *Leucaena leucocephala* to supplement yearling and two year old cattle grazing speargrass in Southeast Queensland. Tropical Grasslands 17, 170-178.

Fukuda, N. and Hidaka T. 1990. Isolation and characterization of a lectin from *Erythrina variegata* var. *orientalis* seed. Agricultural and Biological Chemistry. 54(2): 413-418.

Hamilton, R.I., Donaldson, L.E. and Lambourne, L.J. 1971. *Leucaena leucocephala* as a feed for dairy cows: direct effect on reproduction and residual effects on the calf and lactation. Australian Journal of Agricultural Research 22, 681-692.

Hendro Sunarjono H, Coronel RE, 1991. *Pithecellobium dulce*. In: Verheij EWM, Coronel RE eds. Plant Resources of South-East Asia No. 2. Edible Fruits and Nuts. Wageningen, Netherlands: Pudoc, 256-257.

Huang KF and Yhen YF. 1997. Constituents of *Erythrina variegata* (II). Chinese Pharmaceutical Journal. 49(1): 21-29.

Hughes, C.E. 1998. *Leucaena*: a genetic resources handbook. Tropical Forestry Papers No 37. Oxford Forestry Institute, Oxford, UK.

Larbi, A., Kurdi, I.O., Said, A.N. and Hanson, J. 1996. Classification of *Erythrina* provenances by rumen degradation characteristics of dry matter and nitrogen. Agroforestry Systems, 33: 153-163.

National Academy of Sciences, 1980. Firewood Crops: Shrub and Tree Species for Energy Production. Washington DC, USA; National Academy of Sciences.

Orwa C, A Mutua, Kindt R , Jamnadass R, S Anthony. 2009 Agroforestree Database:a tree reference and selection guide version 4.0

http://www.worldagroforestry.org/resources/databases/agroforestree

Palmer, B. and Jones, R.J. 2000. The effect of PEG addition in vitro on dry matter and nitrogen digestibility of *Calliandra calothyrsus* and *Leucaena leucocephala* leaf. Animal Feed Science and Technology, 85, 259-268.

Paterson, R.T. Kiruiro, E. and Arimi, H.K.1999. *Calliandra calothyrsus* as a supplement for milk production in the Kenya Highlands. Tropical Animal Health and Production, 31, 115-126.

Pravin, G., Priti, G., Shaikh, A., Sindha, S. and Khan M.S. 2012. *Sesbania sesban* Linn: A Review on Its Ethnobotany, Phytochemical and Pharmacological Profile. Asian J. Biomed. Pharmaceut. Sci. 2(12):11-14

Ghosh, P.K. et al., 2014. Fodder tree in India: A potential feed resource during drought and lean period. Indian Grassland and Fodder Research Institute, Jhansi, UP, India. pp 170.

Sharma, I.K. 1981. Ecological and economic importance of *Prosopis juliflora* in the Indian Thar Desert. J. Econ. Taxon. Bot. 2(0):245–8

Shelton, H.M. 2000. Tropical forage tree legumes in agroforestry systems. Unasylva, 51: 25-32.

Srinivasa Rao, Ch., Venkateswarlu, B., Dinesh Babu, M., Wani, S.P., Dixit, S, Sahrawat, K.L. and Sumanta Kundu. 2011. Soil Health Improvement with *Gliricidia* Green Leaf Manuring in Rainfed Agriculture, On Farm Experiences. Central Research Institute for Dryland Agriculture, Santoshnagar, P.O. Saidabad, Hyderabad 500 059, Andhra Pradesh, p:16.

Troup RS, Joshi HB, 1983. Troup's The Silviculture of Indian Trees. Vol IV. Leguminosae. Delhi, India; Controller of Publications.

Troup RS, Joshi HB, 1983. Troup's The silviculture of Indian Trees. Volume IV. Leguminosae. Govenment of India Press, Nasik, India.

Weber, J.C., Larwanou, M., Abasse, T.A. and Kalinganire. A. 2008. Growth and survival of *Prosopis africana* provenances related to rainfall gradients in the West African Sahel. Forest Ecology and Management 256, 585-592.

Wesley HD. 1987. Bird activity and seed productivity in the coral tree, *Erythrina indica.* Indian Forester. 113(9): 640-647.

Wesley HD and Wesley SV. 1990. The relative fertility of an isolated coral tree, Erythrina indica. Indian Forester. 116(4): 292-295.

Wiersum, K.F and Rika, I.K. 1992. *Calliandra calothyrsus* Meissn. In: 't Mannetje, L. and Jones, R.M. (eds) Plant Resources of South-East Asia No. 4. Forages. pp. 68-70. (Pudoc Scientific Publishers, Wageningen, the Netherlands).

Wildin, J.H. 1986. Tree leucaena - top feed, shade and fertilizer too. Bulletin Series RQR86003. Queensland Department of Primary Industries, Brisbane, 12 pp.

Wong, C.C. and Devendra C. 1983. Research on leucaena forage production in Malaysia. In: Leucaena Research in the Asian-Pacific Region. IDRC, Ottawa pp. 55-60.

Zerihun Nigussie and Getachew Alemayehu. 2013. *Sesbania sesban* (L.) Merrill: Potential uses of an underutilized multipurpose tree in Ethiopia. African Journal of Plant Science. 7(10): 468-475.

http://www.cabi.org/isc/datasheet/41187

https://hort.purdue.edu/newcrop/duke_energy/Calliandra_calothyrsus.html

http://www.prota4u.org/

CHAPTER

80

Grasslands and Pastures: Concepts, Establishment, Management and Grazing Systems

Grazing resources

Out of the total geographical area of 328 m ha in India, nearly 122 m ha area is occupied by uncultivable land, forest outskirts, village panchayat purampokes, current fallows and cultivable wastelands. These habitats are earmarked as grazing areas, the land under forest, permanent pastures, land under miscellaneous tree crops and rooves, wastelands and fallow lands occupied 22.0% , 3.9% , 1.1% , 5.0% and 8.0% of land respectively. Out of the total area of country, about 35.5 % land is utilized for grazing purposes. The main source of forage for livestock is from permanent pasture and grazing lands. Pastures and grasslands naturally formed from degradation and destruction of forests until savannas develop.

Grasslands: Grassland refers to land used to grow grasses and is used for grazing. Natural grasslands have a few limitations. There are (i) low quality forage leads to inadequate nutrient to animals, (ii) lower protein content in forage grasses. Most grasses contain 3.6% crude protein. A good quality grass should contain 7% crude protein, (iii) natural pasture/ grasslands provide green forage during rainy season for 4 to 6 months only, (iv) legumes absent in natural grasslands, (v) poor response of native grasses to fertilizer application which prevent improvement in feed value or carrying capacity and (vi) frequent overgrazing of grasslands suppresses good quality productive perennial grasses and allows the undesirable perennial and annual grasses to dominate. The grassland can be improved through (i) use of supplements or concentrates, (ii) use of fertilizers especially nitrogen, and (iii) use of legumes with grasses to increase the protein content. The vegetation of overgrazed natural range lands can be improved through seeding of desirable species of grasses and legumes that are natural component of the original range vegetation. However, the natural range species have limitations of herbage production and its nutritional quality. Therefore, sown pastures with high yielding and better nutritional quality grasses and legumes are raised. These sown pastures can be used for hay or grazing by livestock.

Dabadghao and Shankaranarayan (1973) have classified grasslands of India into five types.

(i) ***Sehima - Dichanthium*** grasslands are spread over the Central Indian plateau, Chota Nagpur plateau and Aravallis, covering an area of 1,740,000 km^2. The elevation ranges between 300-1200 m. There are 24 species of perennial grasses, 89 species of annual grasses, and 129 species of dicots including 56 legumes.

(ii) ***Dichanthium - Cenchrus - Lasiurus* grasslands** are spread over an area of 436,000 km^2, including northern parts of Gujarat, Rajasthan, Aravalli ranges, south-western Uttar Pradesh, Delhi and Punjab. The elevation ranges between 150-300 m. There are 11 perennial grass species, 43 annual grass species, and 45 dicots with 19 legumes.

(iii) ***Phragmites - Saccharum - Imperata* grasslands** cover an area of 2,800,000 km^2 in the Gangetic plains, the Brahamputra Valley and the plains of Punjab. The elevation ranges between 300-500 m. There are 10 perennial grasses, 26 annual grasses, and 56 herbaceous species including 16 legumes.

(iv) ***Themeda - Arundinella* grasslands** cover over 230,400 km^2 and include the States of Manipur, Assam, West Bengal, Uttar Pradesh, Himachal Pradesh and Jammu and Kashmir. The elevation ranges between 350-1200 m and there are 37 major perennial grasses, 32 annual grasses, and 34 dicots with 9 legumes.

(v) ***Temperate - Alpine grasslands*** are spread across altitudes higher than 2100 m and include the temperate and cold arid areas of Jammu and Kashmir, Himachal Pradesh, Uttar Pradesh, West Bengal and the north-eastern states. There are 47 perennial grasses, 5 annual grasses and 68 dicots including 6 legumes

Pastures: Pasture is land where grasses and other plants including legumes are grown and used for grazing of livestock.

Kinds of pasture

1. **Permanent pasture** These pastures are established with perennial or annual or both types of grasses, legumes which are maintained for a period of at least 5 years or more. For example, pasture once established with buffel grass + stylo/ siratro/ centro/ pea blue serve the grazing need for a long time.
2. **Ley farming or Rotational pasture:** Ley is an English term designating pasture portion of a rotation that includes cultivated crops. Ley farming is a temporary grazing land systems in which pastures are grown in rotation with crops. The benefit crops receive in a ley is increased nutrient availability. Efficient use of crop residues is possible in the ley farming. Deep rooting grasses bring up nitrogen and phosphorus from deeper layers of the soil and make these nutrients more available to shallow rooted crops. Ley farming has a beneficial effect on soil physical structure. This has been measured in terms of improved infiltration rate, increased content of water table crumbs, and decreased energy inputs necessary for cultivation. These effects are fairly temporary in the tropics, and the main benefits are recorded in the first crop following pasture ploughing. A reduction in erosion losses occurs under ley farming, since a well-managed pasture crop is protective. Pasture may also provide a break in the crop sequence to help control of weeds, pests and diseases. Ley farming is established as an integral part of the cropping

cycle with perennial or self seeding annual grasses to run for 3 to 5 years and after this pasture is ploughed up and crops are sown in the same land. The sequence is repeated over the years. For example, siratro / pea blue / stylo + buffel grass pastures in 1:3 ratio are grazed for 3 to 5 years and then food crops are raised on the same land for one or two years. Then the sequence is repeated subsequently. In Kangayam tract of Tami Nadu, pasture are maintained for 2 to 3 years and on 4th or 5th year the land allotted to cultivate field crops. *Cenchrus ciliaris* and *Cenchrus setigerus* are grown for 2 to 3 years and then cotton or finger millet is raised in the field for one or two years.

3. **Temporary or complementary pasture** is the miscellaneous vegetated land used for grazing.
4. **Summer pasture** is temporary or complementary pasture used for grazing during summer season.
5. **Winter pasture** are temporary or complementary pasture used for grazing during the winter season.
6. **Annual pasture** are established every year on the whole or a part of the permanent pastures.
7. **Mixed pasture:** Grasses and legumes are grown together to improve the nutritional quality and yield of forage crops. For example, buffel grass + siratro/ stylo/pea blue/centro in 3:1 ratio.
8. **Silvipasture** is defined as growing ideal combination of grasses, legumes and trees for optimising land productivity, conserving plants, soils and nutrients and producing green fodder, timber and firewood on a sustainable basis.
9. **Range** lands are defined as extensive area mostly under natural vegetation, unfenced where wild and domesticated animals graze.
10. **Agroforestry** is defined as growing of trees with field crops, or pasture crops or both in an effort to optimize use of accessible resources to satisfy the objectives of the producer in a sustainable way.

Establishment of pasture:

A. Selection of crops and varieties: The crops and varieties are selected for pasture based on climatic and land features which helps in soil and water conservations. The selected species for improved pasture should have the following features viz., (i) high productivity, (ii) high nutritional values, (iii) ease in establishment, (iv) quick growth, (v) competitive ability, (vi) high seed production, (vii) high germinability and establishment, (viii) tolerance to grazing / defoliation, (ix) drought tolerance, (x) high regeneration and (xi) high soil binding capacity. Crops and varieties suitable for problem soils are as follows

(i) **Acid soil:** oats, deenanath grass, maize, setaria grass, hybrid grasses, guinea grass, cowpea, rice bean.

(ii) **Calcareous soil:** sorghum, pearl millet, hybrid grasses, teosinte, lablab bean, lucerne, cenchrus grass.

(iii) **Saline soils:** sudangrass, sorghum, oats, teosinte, hybrid grasses, lucerne, hedge lucerne, paragrass, rhodes grass.

(iv) **Coastal saline:** Signal / congo grass, guinea grass, hamil grass, oats, paragrass, barley.

(v) **Sodic soil:** berseem, shaftal, sorghum, oats, karnal grass, cenchurs grass, hybrid grasses, hedge lucerne, subabul, blue panic grass, paragrass, rhodes grass.

(vi) **Marshy / wetland**: paragrass, teosinte, congo / signal grass, rhodes grass.

(vii) **Degraded forests and ravines:** deenanath grass, guinea grass, hamil grass, stylo, subabul, prosophis, *Acacia nilotica.*

(viii) **Desert and arid soils:** *Cenchrus ciliaris, Cenchrus setigerus* and *Cenchrus glaucus.*

(ix) **Rocky gravelly sites:** *Acacia tortilis, Albizia amara, Hardwickia binnata,* Cenchrus ciliaris, siratro, etc.

10. Rainfed red soil

Trees +	Grasses +	Legumes
Albizia amara	Blue panic grass	Stylo, Pea blue,
Acacia nilotica	Deenanath grass	Desmodium
Hardwickia binnata	*Cenchrus ciliaris*	Siratro, Centro

11. Rainfed black cotton soils:

Trees +	Grasses +	Legumes
Albizia amara	*Cenchrus setigerus*	Siratro
Acacia nilotica	*Cenchrus ciliaris*	Stylo
Hardwikia binnata	*Cenchrus glaucus*	Pea blue

12. Normal, medium to deep soils

Trees +	Grasses +	Legumes
Subabul	Guinea grass	Siratro
Agathi	Blue panic grass	Stylo
Albizia lebbeck	Cenchrus grass	Pea blue
Albizia amara		Centro

13. Moist sites

Trees +	Grasses +	Legumes
Subabul	Paragrass	Desmodium
Agathi	*Paspalum dilatatum*	*Glycine javanica*
Acacia nilotica		

B. Practices in establishment of pasture

1. Season: The most reliable time of land preparation, and sowing seed of grass/legume mixture or introducing legume seed into the existing pasture is just before the rainy season or shortly after the first rain have fallen during June–July months.

2. Land preparation: The pasture seedbed must provide favourable conditions for rapid seed germination and seedling emergence. The competition from unwanted plants must be reduced or if possible removed. Unwanted plants and bushes must be removed. Burning of native plants followed by ploughings and harrowing is essential.

3. Seeds and sowing: High yielding forage crops should be selected and sown depending on soil and climatic conditions. Upgrading deteriorated pasture can be carried out by seeding legumes like siratro, pea blue, stylo, etc., is a low cost operation, with a higher benefit / cost ratio. Multipurpose or fodder tree saplings such as Subabul, Acacio, etc., can be planted with a spacing of 6 m to provide passage of farm machineries. This is better than developing a new complete seedbed and sowing of grass / legume mixture. Holding a freshly harvested seeds in storage will overcome dormancy problems in many grass seeds. For example, *Cenchrus ciliaris* requires 3 to 12 months to overcome dormancy. Drying the seeds for 10 days at 40ºC is effective on new seed of *Cenchrus ciliaris*. Seed treatment with hot water at 80ºC for four minutes for subabul; at 75ºC for 6 minutes for siratro and at 55ºC for 20 minutes for stylo. Seeds should be dried and preferably sown immediately after treatment. Seed pelleting is done to protect the seed from ants. Germination of seeds will be better when seed pelleting is carried out by mixing grass seeds, cow dung, tank silt or clay and sand in 1:1:3:1 ratio. Seed pelleting with lime is also practiced for subabul, and lucerne. The ingredients needed are 30 to 50 kg finely ground lime, 2.5 kg gum arabic, 600 grams rhizobial peat innoculant and 5 litre rain water for treating the seeds for sowing one hectare. Grass seeds have to be innoculated with azospirillum while legumes with rhizobium culture. Choice of seed rate is determined by viable seeds, degree of seedbed preparation, climatic risks, habitat of plant (runner or bunch type), extent of weed problem, etc. High seeding rates improve the growth of pasture, suppression of weeds, avoidance of erosion, amount of nitrogen fixed by sown legumes and the grazing returns are greater for high sowing rates atleast in moist environments. The alternative of low seeding rate is only successful if land preparation and weed control are faultless and if grazing is delayed. This permits plant stand thickening through natural seed drop, seeding regeneration, plant crown extension and runnering. Sowing can be done either through broadcasting or drilling. Drilling seed is better than broadcasting since the seed can be placed positively at the desired depth in the soil and correct degree of coverage made. Optimum sowing depth can be deeper in sandy soils than in heavy soils. Sowing depth is 2.5 to 3.0 cm for small seeded crops and 3 to 5 cm for large seeded crops. Closer spacing will give more cover, 50 cm row spacing produce good growth regarding plant height, number of tillers and forage production.

4. Weed management: Two hand weeding are carried out on 20 and 40 DAS. Weed growth is controlled in pasture through regular grazing.

5. After cultivation practices: Management of tree components in silvipasture is primarily required for providing necessary light interception to ground flora. An agroforesty system meets the fodder and firewood requirements of the farmer at proper intervals besides optimizing the total productivity of the agroforesty systems. Many trees and shrub species have the capacity to regenerate new growth from stumps, roots or branches after being cut. Several harvesting methods allow the trees or shrubs to regenerate through sprouting are coppicing, pollarding, lopping, prunning and thinning. The management techniques enable a sustainable yield of wood or fodder prone trees over a long period of time.

(i) **Coppicing**: Trees are cut at 15 to 75 cm above ground level. New shoots develop from the stumps. Two to three sprouts should be allowed for pole production. Subabul, Gliricidia, Calliandra are good examples for good coppices.

(ii) **Pollarding**: All branches including top of the trees are removed at 2 to 3 m above ground level. *Azadiracta indica, Eucalyptus, Giricidia sepicum*, subabul trees require pollarding operation.

(iii) **Lopping:** is a form of harvesting in which most of the branches are removed. The branches of lopped trees are mainly used for fuel wood and fodder. *Prosopis juliflora, Acacia tortilis, Albizia amara, Albizia lebbeck, Hardwickia binnata* tree loppings are used for fodder purposes.

(iv) **Prunning** refers to the removal of smaller branches and stems which constitute a major source of fodder, fuel, and mulch for tree crops. Pruning is done in *Albizia saman,* subabul, gliricidia, acacia , calliandra tree crops.

(v) **Thinning** is done for fodder, fuel wood or pole production depending on plant density. Stands should be thinned after a few years of establishment. Thinning eliminates poor and diseased trees and to improve the stand by reducing competition for light and nutrients.

6. Manures and fertilizer application: Apply FYM or compost or pressmud at the rate of 12.5 tonnes per ha to increase water holding capacity of soil and to increase the soil fertility. In acid soils apply 2.5 tonnes lime per ha. In alkaline soils apply gypsum @ 2 tonnes per ha. Apply 40-40-20 kg N, P_2O_5 , K_2O per ha every year at the start of rainy season.

7. Water management: Pasture are raised in rainfed condition. The grasses and legumes grow and produce herbage utilizing the rainfall occured during both monsoons. In summer season, the crops are dry and in dormant condition. On receipt of rains the rhizomatous stems rejuvenates well.

C. Grazing systems: There are four natural grazing systems based on periodic / seasonal migrations in the country viz.,

(i) **Total nomadism** covers systems in which the animal owners do not have a permanent place of residence. They move from place to place with their herd in search of forage and water. They do not practice regular cultivation. At present total nomadic are present in Kashmir, Uttar Pradesh, Madhya Pradesh, Rajasthan, and Andhra Pradesh.

(ii) **Seminomadism** is a grazing system where owners have a permanent place or residence near which supplementary cultivation is practiced. However, for greater part of the year, they travel with their herds to distant grazing areas. Seminomadic pastoral tribes are available in Himachal Pradesh, Punjab, Haryana, Rajasthan, Maharastra, Gujarat, Andhra Pradesh and Karnataka.

(iii) **Transhumance** is practiced by the farmers with a permanent place of residence who send their herds tended by herdsman for long periods of time to distant grazing areas. The herdsmen take the animals of rich farmers to the grazing areas of other states also. Such practice is observed in Rajasthan, Kashmir, Madhya Pradesh and Ladakh regions.

(iv) **Partial nomadism** is a grazing system where the farmers have permanent settlements and who keep herds in the vicinity. Village grazing system may also be put in this category in which the grazing animals remain in the vicinity. Partial nomadism is practiced in Uttar Pradesh, Madhya pradesh, Rajasthan, Bihar, Gujarat, Maharastra, Nilgiris of Tamil Nadu and Orissa.

D. Grazing management practices: Grazing management is necessary (i) to maintain a high production of good quality forage for the longest possible period, (ii) to maintain a favourable balance between herbage species, (iii) to achieve efficient utilization of the forages produced and (vi) high animal production. There are different man-made grazing management practices namely, continuous grazing, deferred grazing, deferred rotational grazing or strip grazing. Farmers should adopt a grazing plan that is suitable to their particular conditions.

(i) **Continuous grazing** is an extensive system of grazing in which the stock remains on the same grazing area for prolonged period. This systems of grazing is the normal practice in India. After a long period of continuous grazing, the fields have low herbage yield and show sign of deterioration with an increase in the percentage of weed species. Normally pastures are under grazing during the rainy seasons and over-grazing during the dry season which leads to the deterioration of the grazing land, continuous grazing should be adopted for the highest livestock return.

(ii) **Rotational grazing** is an intensive system of grazing. It overcomes the disadvantages of continuous grazing. Rotational grazing involves the use of dividing the entire grazing area in to eight or more paddocks (field) and the animals are moved systematically from one to another in rotation. Each paddock (field) is grazed for a period of about 7 days after which the stocks are moved out to next paddock while the first is rested. By the time the last paddock in sequence has been grazed, the first paddock becomes ready for grazing for the second time. The length of grazing period should depend on stocking rate and herbage availability. The advantages of rotational grazing are (i) The vegetation gets a chance to make better growth, (ii) Young and nutritious grass becomes continuously available during the grazing period. Persistancy and productivity of legumes are maintained, (iii) the grazing period with better quality herbage is prologned, (vi) the livestock maintain better condition and high stocking rate can be maintained, (v) Soil erosion is reduced and (vi) a roational system of grazing management will be beneficial when dry condition persit for several months. The disadvantage is that the rotational grazing system does not ensure seed formation which is a vital consideration in improvement of grassland through grazing management.

(iii) **Deferred grazing** implies delay in grazing until the most important forage plants have set seed. By allowing the plants to mature before grazing, their vigour is built up, root system are allowed to develop and self sown seedlings to become established. For example, the grazing area is divided into different subunits and grazing is allowed one by one sub unit in notation which is delayed in the one subunit till plants have not set seeds for regeneration in pasture. Deferred grazing practice is used to improve denuned grasslands.

(iv) **Deferred rotational grazing** consists of dividing the grazing land into three fenced block of convenient size and the animals are allowed for grazing alternatively into two while protecting the third compartment during the growing season so as to permit the palatable species to recoup their vigour. The animals are then allowed to graze on the protected block before the grass stems harden. Protection of one compartment once in three years maintains the grasslands in high state of production. This system is mainly recommended for grassland in which perennial species are dominant.

(v) **Strip grazing** (Close folding, ration grazing) is a more intensive method of rotational grazing based on the use of the electric fence. A movable electronic fence is placed across grazing paddock (field) and is moved forward once or twice each day. Strip grazing technique can be practiced in highly productive nutrients pastures. In general the rotational and deferred rotational grazing systems provides timed rest periods and opportunities for natural reseeding in pastures. The biomass production of herbage is higher in deferred rotational grazing than controlled continuous grazing.

Productivity of animals: On the basis of grazing on one site, the animal weight gains could be 75 kg per ha per year at the rate of one adult cow unit per ha. The annual gain in the animal population is of the order of 12 lambs and 12 kids per year or one animal per ha per year. Silvipasture systems on degraded grazing lands have enhanced biomass by up to 7 to 15 tonnes/ha/year.

Carrying capacity: The most influenced single factor in management is the number of animals carried per hectare. Carrying capacity is estimated as the number of animal to one ha on year basis. Carrying capacity can be increased at 2.5 animals per ha per year of pasture. In semi-arid areas, the present stocking rates are 1 to 51 ACU/ha against the carrying capacity of 1 adult cattle units (ACU)/ha while in the arid areas, the stocking rates are 1 to 4 ACU/ha against the carrying capacity of 0.2–0.5 ACU/ha. The appropriate stocking rates are 25–30, 20, 17, 12 and 6 ACU/100 ha for the management of excellent, good, fair, poor and very poor classes of rangelands, respectively.

Mosaic farming: *Mosaic farming* creates agricultural landscapes made up of "patches" of annual crops and pastures interspersed with deep-rooted perennial vegetation such as lucerne and trees. Each vegetation type is located so that its requirements are matched to landscape, vegetation and soil characteristics. Such deployment of deep-rooting perennials in carefully targeted locations may provide environmental benefits and profitable opportunities, even though they may not be economically viable for use on a wider scale. A key challenge in implementing a mosaic farming system is determining how to match the spatial location of the various enterprises of the mosaic with landscape position and soil attributes within a farm in a multiple-paddock or subpaddock context.

The list of forage grasses, legumes, shrubs and trees for grassland/grazing land on agro-ecological basis is furnished in Table 3.

Table 3 List of forage grasses, legumes, shrubs and trees for grassland/grazing land on agro-ecological basis

Agro-eco Regions	Grasses	Legumes	Shrubs/Trees
Western Himalaya, cold arid with shallow skeletal soils	*Agrostis spp., Poa alpina, Trisetum spicatum*	*Medicago sativa/subsp sativa, M. sativa, subsp faslculata*	*Hippophae rhamonides*
Western plains and Kaccha Penisula, hot arid with desert and saline soils	*Cenchrus ciliaris, C. setigerus (Sandy plains), Lasiurus scindicus (Sandy interdunal plains), Panicum turgidum (Sand dunes) Chloris gayana, Sporobolus marginatus (salt affected lands)*	*Cassia rotundifolia*	*Acacia nilotica, A. tortilis, Albizia lebbeck, Ailanthes excelsa, Dichrostachys cinerea, Prosopis cineraria, Ziziphus nummularia, Prosopis juliflora, Salvadora oleoides, S. persica (Saline soil)*
Deccan Plateau, hot arid with red and black soils	*Andropogon gayanus, Chrysopogon fulvus (Red soil), Dichanthium annulatum, Bothriochloa intermedia (Black soil)*	*Clitoria ternatea, Stylosanthes hamata, S. scabra*	*Acacia nilotica, Albizia amara, A.lebbeck, Desmanthus virgatus, Leucaena leucocephala, Tamarindus indica*

Northern plains and central highlands includling Aravallis, hot semi-arid with Alluvium	*Bothriochloa intermedia, Cenchrus ciliaris, Chrysopogon fulvus, Dichanthium annulatum, Sehima neroosum*	*Macroptilium atropurpureum, Stylosanthes hamata, S. scabra*	*Acacia nilotica, A. holosericea, Albizia amara, A.lebbeck, A. procera, Azadirachta indica, Dichrostachys cinerea, Hardwickia binata, Leucaena leucocephala, Sesbania grandiflora, Sesbania sesban*
Central (Malwa) highlands, Gujarat plains & Kathiawar Peninsula, hot semi-arid with red loamy soils	*Bothrichloa intermedia, Chloris gayana, Cynodon dactylon, Dichanthium annulatum, Pancium maximum*	*Arachis hagenbackii, Clitoria ternatea, Stylosanthes hamata, S.scabra*	*Albizia lebbeck, Artocarpus lackoocha, Dendrocalamus strictus, Gliricidia sepium, Faidherbia albida, Holoptelia integrifolia, Pithecellobium dulce*
Deccan Plateau, hot semi-arid with shallow and medium black soils	*Bothriochloa intermedia, Brachiaria decumbens, Cenchrus setigerus, Dichanthium annulatum, Pennisetum pedicellatum, Panicum maximum*	*Arachis hagenbackii, Stylosanthes hamata, S. scabra*	*Acacia nilotica, Albizia procera, Anogeissus pendula, Bauhinia variegata, B. purpurea, Leucaena leucocephala, Moringa oleifera, Pterocarpus marsupium, Sesbania sesban, Terminalia arjuna*
Deccan Plateau (Telangana)	*Andropogon gayanus, Bothriochloa intermedia*	*Atylosia scrabaeoides,*	*Albizia lebbeck, Gliricida sepium, Faidherbia albida, Holopteaia*

Silvipasture systems with fodder trees such as *Acacia, Leucaena, Artocarpus, Albizia, Melia, Bauhinia, Dalbergia, Dendrocalamus, Ficus, Grewia, Terminalia,* and *Toona* with understorey grasses *Cenchrus ,Chrysopogon, Panicum, Pennisetum, Dichanthium* and legumes *Stylosanthes hamata* and *Macroptilium* are becoming popular with the farmers.

REFERENCES

Dabadghao, P.M. and Shankarnarayan, K.A. 1973. The Grass Cover of India. ICAR, New Delhi.

Hazra, C.R. 1995. Improved Cultivars of Forage Crops for Different Agro- Environments. In: R.P. Singh (ed.) Forage Production and Utilization. IGFRI, Jhansi (India), pp. 326–335.

Kaul, R.N and Ganguli, B.N. 1963. Fodder potential of *Zizyphus* in the shrub grazing lands of arid zones. *Indian Forester*, 39, 623–630.

Misra, R. 1983. Indian Savannas. In: F. Bourliere (ed.) *Tropical Savannas*. Elsevier, Amsterdam, pp. 155–166.

Pathak, P.S and Roy, M.M. 1995. Agrosilvipastoral farming systems for optimizing forage and energy resources in rainfed areas. In: R.P. Singh (ed.) *Forage Production and Utilization*. IGFRI, Jhansi (India), pp. 154–178.

http://www.fao.org/ag/AGP/AGPC/doc/counprof/india.htm

CHAPTER

81

Silage and Hay Making

Silage Making

The term 'Silage' is term used for the preserved, succulent fodder formed when any green plant material is put in an air tight pit or place where it can ferment in the absence of air. The major changes which occur during ensiling are the fermentation of sugars to form acids and the breakdown of some of the Forage proteins to simpler compounds, including ammonia. This fermentation occurs during the first two to three month, after that the silage remains practically unchanged for another 12 to 18 months. The process of making silage is called ensiling and the container for keeping silage as Silo. The advantages of silage making are as follows i) crops can be ensiled when the weather does not permit curing them into hay or dry fodder, ii) The use of silage generally makes it possible to keep more animals on a certain area of land, iii) silage furnishes high quality succulent feed for any season of the year at a low expense, iv) silage can be produced from weed crops which would make poor hay. v) the ensiling process kill many of weeds seeds, vi) crop from a given area can be stored in less space as silage than as dry fodder. Silage contains 230 kg of dry matter per m^3 as against 66 kg/m^3 in hay vii) staining forage crops when converted into silage become soft and are better utilized by the stock and viii) many undesirable things present in a fresh crop are eliminated after ensiling.

Characteristics and classification of silage: Silage quality is a measure of ensiling process, the amount of nutrient loss and relative palatability. Well prepared silage is yellowish green in colour with pleasant vinegar sweet. Silage should be palatable to livestock. It should be fed carefully to cows to prevent off flavour in milk. Silage containing stubble and foreign matters or fodder cut at too late a stage will naturally be less palatable and nutritious. Silage having moisture content more than 75% reduces the feed value than silage having low moisture. If juice runs freely when silage is squeezed in the hand, it indicates that the material has high moisture content. The pH of silage should be less than 4.2. In general, butyric acid concentration of good silage is less than 0.2 per cent and the ammonical nitrogen content less than 11 per cent of the total nitrogen.

Table 1 Classification of Silage

Grade	Crude protein (%)	Source
Grade I	> 15	Young grass and legumes (no flower)
Grade II	10-12	Grass and legumes at flowering /ear emerging stage
Grade III	< 10	Grasses at seed development stage.

Steps for making good silage: Avoid rainy days at the time of harvest of fodder to avoid wetting of fodder as it is susceptible to fungal infection. Assess the quantity of fodder crop to be harvested depending upon the number and kind of animals to be fed, length of feeding period and the number and size of silo pits. Walls of silo pits should be air proof to prevent the growth of moulds which spoil the silage. Growth stage of fodder crop should be neither dry nor have high water content for preparing good silage. Forage crop with solid stems should be selected so that only a small amount of air will remain in the mass after it has settled. If hollow-stemmed crops are used, the cut forage must be trampled with special care to force out as much as air as possible. Milky growth stage is recommended for making silage from maize, oats, etc., whereas flowering stage for sorghum. Pre-wilt the fodder if needed. The practice of wilting the crop (Table 2) in the field for a few hours before ensiling is valuable for young succulent grass or legumes. The moisture content should not be more than 70 to 75 per cent. It encourages butyric acid type of fermentation and further, the amount of liquid effluent from the silage is reduced.

Table 2 Simple rules for ensilage

Crop	Silage packing	Air control	Wilting process	Use of molasses
Short leafy growth (vegetative stage)	Very compact	Build shallow layers and allow each one in heat to about 38^0C.	May be an advantage as it will avoid tight packing and produce loss of effluent	Must be used, 30 to 40 kg per tonne of crop
Early flowering stage, it will have some stems.	Satisfactory	Low layers in silo should be allowed to heat; fill quickly thereafter.	Only for the bottom layer.	Not necessary. 20 kg per tonne may be added if desired.
Full flowering stages, stemmy	Lossely	Must be consolidated thoroughly, especially top layers or the over heated.	Wilting should be avoided; water may have to be added	Not needed.

For proper consolidation the fodder crop should be chopped into short lengths of 4 to 5 cm lengths before ensiling. Good silage can be made from a mixture of legumes such as lucerne, berseem, soybean, cowpea with a suitable proportion of green maize, sorghum, paddy straw, etc which may contain more sugar. The mixture can be in the following ratio: sorghum + lucerne 1:3 ratio, maize + cowpea is 1:3, paddy straw + berseem ratio is 1:5 , paddy straw + water hyacinth ratio is 1:4. Forage containing more than 10 % water soluble carbohydrates, for example maize, millets, oats and grasses are very suitable for ensiling than lucerne and berseem in view of their low content of soluble carbohydrates. Addition of molasses to leguminous forage crops improve the quality of silage. Salt adds taste to the silage. Unwilted leguminous fodders may be ensiled after mixing with dry roughage in 4:1 ratio. To ensure good silage, the chopped fodder material is distributed and trampled uniformly throughout the Silo especially near the wall. It will be better if leveling, trampling, addition of additives, etc., are carried out after every 30 cm layer of packing. Water should be added if forage is dry. Molasses is applied @ 40 to 50 kg per tonnes of legume fodder and 80 to 100 kg per tonnes of chopped maize, barley, oats or grass fodder. To check the growth of undesirable organisms and to increase the growth of lactic acid producing bacteria, preservatives like common salt @ 18 to 20 kg or Sodium metabisulphite @ 5 kg or diluted acetic acid @ 10 liters or phosphoric acid @ 6 kg per tonne of chaffed fodder may be added. Once the pit is full it should be covered with wet straw, saw dust, etc., and plastered with 10 to 30 cm thick layer of soil. After covering, weights such as slabs, concrete posts, concrete cylinders and wooden logs should be kept for better compression. Silos should be kept off the rain. It is advantage to check the seal from time to time, flatten it down and seal any cracks. There should be small door near the surface of the silage to allow the CO_2 to escape. Nitrogen dioxide gas produced from nitrate in the forage is favored in silage soon after filling the silo pit.

Different types of silos

Pit silo: It is cylindrical or rectangular. It extends below the ground and not subjected to water logging. Provision should be made for effluent to escape. A pit of average width of 4 m and with silage settled to a depth of 2 m will hold 1.5 tonnes of silage for each 30 cm length. Pit silo is very economical. It has a large capacity, less power is required for filling pit silo, silage settle uniformly and retain the juice. However, it is inconvenient to take out the feed. It is difficult to compress the silage uniformly. There is 20% loss of silage.

Trench silo: It has greater length in relation to breadth. Horse, cows or tractors can be used to pack the silage. The power required for filling the trench silo is less. Silage settle well into the trench below ground level. Unloading and carrying of silage are much easier. However, more silage is spoiled in trench silo method.

Tower silo: It is round, cylindrical and is placed above the ground level. The height is varying from 6 to 10 m with 3 to 6m diameter. The materials used include wood, reinforced concrete or sheet metal. The silage material can be well preserved with no side waste due to air leakage. Since silage is at kept at height, the mass of silage itself apply pressure and acts as an air seal to the lower layer. Hence no need for sealing of silo. The loss of silage is only 10 per cent. However, tower silo is very expensive to make silage. Chopper blower is needed for filling up of the silo. Emptying tower silo is very laborious. The silage gets dehydrated in dry hot place.

Nutritive value and feeding of silage: Good silage retains vitamin A better than hay, apart from its nutrient content. It is more palatable than dry roughage. Silage feeding is especially suited to milch cow which would produce more milk on ration of silage and concentrates than on straw and concentrates. A milch cow can easily consume 13 to 15 kg of silage in a day.

Hay making: Hay is any leafy forage crop cut before it is fully ripe and dried for storage. It is pliable, green leafy and free from mould, weeds and dust and has a pleasant characteristic smell and aroma. Hay is more nutritious and palatable than straw. Hay is the main source of feed for cattle during lean months. Hay is less expensive to prepare. At low expense it furnishes high quality feed. The crop from a given area can be stored in less space as hay after being baled (66 kg m^3). Many undesirable things present in a fresh crop are eliminated after it is converted into hay.

Characteristics of good hay: Hay must be leafy and green, and have soft and pliable stems. It should be free from mustiness or mould and be palatable. Fresh crop contains 70-80 per cent of moisture when young. Good quality of hay should contain about 20 per cent moisture.

Table 3 Classification of Hay

Classification	Digestible crude protein (%)	Crude fiber % on DM basis	Starch equivalent
Very good	10-11	25-27	250-400
Medium	6-7	30-31	380-410
Poor	<= 5	35	<= 300

Choose right crop and growth stage: Thin stemmed and leafy forage plants are best suited for hay making. The grasses *Cynodon dactylon*, *Cenchrus ciliaris*, Oats, etc., remain sufficiently nutritious and palatable after drying. Among legumes, lucerne, berseem, pea, cowpea are suited for hay making. The forage crops should be in general harvested at the pre-flowering stage for maximum nutrient content. The feeding value of hay depends largely on the stage of the crop growth when it is cut. The right growth stage at which the crops should be harvested is furnished in Table 4.

Table 4 Crops and stage of harvest for hay making

Name of the crop	Stage of harvest
Cenchrus ciliaris, Cynodon dactylon, Dicanthium annulatum	Flowering stage- At the end of Southwest monsoon, after taking first cutting for green feeding, when half of flower heads have emerged.
Deenanath grass, Napier grass, Oats	Early flowering stage or milk dough stage.
Clover, Lucerne, Berseem	At half bloom stage; only second and third cuts are best used for hay making.
Pea, soybean	Full boom stage. 10-20 % flowering stage
Cowpea, mungbean, rice bean	At full bloom stage.

The leaves of the legume hay dry out much sooner than the stem. The leaves shatter badly when raked or otherwise handled if they become too dry or brittle. Since leaves are richer in nutrients (e.g., 2 to 3 times higher protein) than stems, measures should be taken to prevent the loss of leaves during drying, transportation and storage. Leaves also contain most of the carotene, and other vitamins, are much richer in minerals and have much less fibre than stems. Crushing of the cut forage crop hastens drying. The drying time is reduced by 30 to 50 percent by this technique. It involves the use of rollers which is expensive. Fodder is dried with ventilation to possess bright green colour and pleasant aroma. Such hay has been greatly relished, both by cattle and buffaloes. The Indian Grassland and Fodder Research Institute, Jhansi, has devised a new technique. In IGFRI method, galvanized iron woven wire fencing material, along with angle iron post, are utilized in fabricating a gable-shaped drying frame. This device provides a well ventilated slopping support for the fodder being dried, which completely prevents the loss of leaves. Such frames may also be made of bamboo to suit an average farmer. This method is suitable for making hay from green fodders like berseem, lucerne, peas and graminaceous fodders.

Hay making process: Hay making is one of the methods of preserving green fodder, harvested at early bloom stage and stored after reducing water by different process to avoid much loss of nutrients. The moisture content of hay should be 15 to 20 %. Make hay while sun shines.

Conventional method / Natural method: On the first day, start harvests either by sickle or mower as soon as the dew has dried off the grass say 8.30 to 9.00 a.m. and cut for about 4 hours. Lease cutting for the day. At 4.00 p.m. turn the grass with the side rake with small, fluffy wind rows. On the second day, turn the first day's moving for second with ventilation. Watch it state of drying. If needed dry it for one more day. On the third day, turn the first days moving once again for drying with ventilation. At about 4.00 to 5.00 p.m., the first day's moving should be ready for stacking or fit to be baled straight from the wind row. Hay is packed not so tightly into small bundles when it becomes free from mould at moisture content of 20 to 22%. The other hay making methods are conventional or natural methods, tripod method, tetrapod method, fence method, herbage poles method and hay racks method. During season when it is likely to rain, good hay is made by tripod stand or tetrapod stand or by fencing method. Tripod will hold about 250 to 300 kg of herbage which becomes 125 to 150 kg of finished hay. In tetrapod's 300 to 400 kg of herbage can be held up. The hay should be clear off the ground and the central ventilator kept free while the heaps are in the field to promote quick drying. Herbage is also dried in fence type racks and herbage poles.

Silage versus Hay: When hay and silage are made in the ordinary way, considerable losses of starch equivalent and digestible protein take place. In this respect, ensiling is superior to hay making. Further silage making is less dependent on weather conditions than hay making. Expenses and spoilage (fungal infection) are higher in silage than hay. Silage making suits to large agricultural and dairy farms since it needs exact technical requirements. Silage making is difficult for average Indian farmer in view of exact technical requirements.

References for Further Reading

Chatterjee, B.N. and Bhattacharya, K.K. 1986. Principles and Practices of Grain Legumes Production. Oxford and IBH Publishing Co. Pvt. Ltd., New Delhi.

Chatterjee, B.N. and Das, P.K. 1989. Forage Crops Production: Principles and Practices. Oxford and IBH Publishing Co. Pvt. Ltd., New Delhi.

Cobley, L.S. 1957. The Botany of Tropical Crops. Longmans Green and Co., London.

CPG. 2012. Crop Production Guide. Department of Agriculture, Government of Tamil Nadu, Chennai and Tamil Nadu Agricultural University, Coimbatore.

Crane, J.C. 1947. Kenaf: Fibre plant rival of Jute. Economic Botany. 1: 334-350

Daisy E. Kay. 1979. Food legumes. Food Tropical Products Institute, London. 435p

De Datta , S.K. 1981. Principles and Practices of Rice Production. John Wiley & Sons, New York.

Fageric, N.K. 1992. Maximizing Crop Yields. Marcel Dekker Inc., New York.

Fehr, W.R. and Caviness, C.E. 1977. Stages of Soybean Development. Spec. Rept. 80. Iowa State University of Science and Technology, Ames, Iowa, U.S.A.

Gozazzina, E., Gething, P.A., and Mazzali, E. 1997. Fertilizing for High Yield Maize. IPI Bulletin No.5, International Potash Institute, Bern, Switcherland.

Gooding, H.J. 1982. The Agronomic aspects of Pigeon pea. Field Crop Abstract 15: 1-5

Hunter, 1952. The Barley Crop. Crossby Lockwood Ltd., London.

ICAR. 1985. Rice Research in India. Indian Council of Agricultural Research, New Delhi.

ICRISAT, 1975. International Workshop on Grain Legumes. Hyderabad, India. 349p

ICRISAT,1981. International Workshop on Pigeon pea. Vol 2. ICRISAT, Patencheru, India. 451p

Jaswani, L.M. and Baldev, B. 1992. Advances in Pulse Production Technology. ICAR, New Delhi.

Kemmler, G. 1983. Modern Aspects of Wheat Manuring. IPI Bulletin No.1, International Potash Institute, Bera, Switcherland.

Kirby, E.J.M. and Appleyard, M. 1984. Cereal plant development and its relation to crop management. pp 161-173. In: Gallagher, E.J (ed) Cereal Production. Butterworth in association with Royal Dublin Society, London.

Maiti, R.K. and Bidinger, F.R. 1981. Growth and development of the Pearl millet plant. Research Bulletin No.6, ICRISAT, Patancheru, India.

Mortan, J.F. 1976. The Pigeon Pea: A high protein tropical lush legume. Horticultural Science 11: 11-19

Padwick, G.W. 1979. Growth phases in plants and their bearing on Agronomy. Experimental Agriculture. 15: 15-26

Purseglove, J.W. 1974. Tropical Crops: Dicotyledons. English Language Book Society, Longman. 719p

Rachie, K.J. and Roberts, L.M. 1974. Grain Legumes of the Lowland Tropics. Advances in Agronomy. 26: 1-132

Singh, N.N. 1997. Kharif Maize. Directorate of Maize Research, Indian Agricultural Research Institute, New Delhi-110 012.

Singh, N.N. 1998. Winter (Rabi) Maize in India. Directorate of Maize Research, Indian Agricultural Research Institute, New Delhi-110 012.

Smart, J. 1976. Tropical Pulses. Longman Group Ltd., London. 348p

Srivatsava, H.C., Bhaskaran, S., Menon, K.K.G., Ramajayam, S. and Rao, M.V. 1984. Pulse Production - Constraints & Opportunities. Oxford and IBH Publishing Co. Pvt. Ltd., New Delhi.

Summerfield, R.J. and Bunting, A.H. 1980. Advances in Legume Science. Royal Botanic Garden. Kew, England.

Thakur, C. 1981. Scientific Crop Production. Vol II. Metropolitan Book Pvt. Ltd., New Delhi.

Yoshida, S. 1981. Fundamentals of Rice Crop Science. International Rice Research Institute, Los Banos, Philippines.

http://www.ipmcenters.org

FIELD CROPS
Production and Management

Volume II

Srinivasan Jeyaraman, M.Sc.(Ag), Ph.D.
Former Dean,
Anbil Dharmalingam Agricultural College and Research Institute (TNAU),
Tiruchirapalli, Tamil Nadu
Former Director
Centre for Soil and Crop Management Studies,
Tamil Nadu Agricultural University, Coimbatore, Tamil Nadu
Former Principal
Institute of Agriculture (TNAU), Kudumiyanmalai,
Pudukkotai district, Tamil Nadu

OXFORD & IBH PUBLISHING CO. PVT. LTD.
New Delhi

FIELD CROPS : PRODUCTION AND MANAGEMENT

Oxford & IBH Publishing Company Pvt. Ltd.
113-B Shahpur Jat,
Asian Games Village Side
New Delhi 110 049, India

Fax: (011) 4151 7559
Email: oxford@oxford-ibh.in

ISBN 978-81-204-1795-3

Printed at Chaman Enterprises, New Delhi.

1-R16-07

Preface

Crop cultivation is a dynamic process of taking different decisions at different time as the cultivation practices changes with change in agro climatic conditions, soil types and locations. It is necessary to achieve 'more crop per drop' through adoption of improved crop/water management systems. The science of crop production embraces the knowledge to perform the various operations at the farm in a skillful manner to enhance input and output efficiency.

The open global trade of agricultural produces leads to competitive agriculture throughout the world. The crop products produced in one part of the world are made available to the other part of the world. Fair market price for agricultural produce can be achieved only through yield maximization per unit area with high quality. This can be done through adopting scientific principles in crop production and management for maximum profit with minimum production costs. Documentation of basic principles and practices for sustainable agriculture is necessary to enrich the knowledge of the students of agriculture. This book will be a ready reckoner for under graduate students of agriculture, extension officers and personnel involved in agro based industries. This book is mainly a compilation of scientific information available from different journals, papers, bulletins, books and reports. The new chapters on hybrid rice, System of Rice Intensification, basmati rice, rice cultivation in salt affected soils, rice ratooning management practices, coconut, oil palm, jatropha, potato, chinese potato, sweet potato, aroids, tapioca, tea, turmeric, betel vine, aromatic and medicinal plants are included in this book to meet the competitive examinations at national level in agronomy subject since these crops are included in syllabus of different State Agricultural Universities in India. Each chapter covers with standard headings such as importance, origin, distribution, climate, soils, varieties, growth stages, seeds and sowing, intercultivation, weed management, manures and fertilizers application, water management, harvest, yield and utilization. I hope that this book will be helpful to the students and extension officers in fulfilling the objective of holistic knowledge of crops production i.e., soil, climatic requirement, crop production and crop protection technologies for increasing the income and sustainability to the production system.

I am thankful to Prof Dr. R. Rajagopal, former Professor and Head, Department of Plant Breeding and Genetics; Prof Dr. K. Annadurai, Agronomist, Prof Dr. S. Avudaithai, Professor and Head, Department of Agronomy; Prof. Dr. R. Kavimani, Professor and Head,

Cotton Research Station, Veppanthattai; Dr. R. Arulmozhian, Professor (Horticulture); Dr. N. Thavaprakaash, Dr. P.M. Shanmugam, Dr. T. Ramesh, Dr. S. Somasundaram, Dr. S. Anandha Krishnaveni, Dr. S. Rathika, Assistant Professors (Agronomy); Dr. R. Neelavathi, Assistant Professor (Horticulture); Dr. M. Shanmuganathan, Assistant Professor (Plant Breeding and Genetics) and Dr. M. Surulirajan, Assistant Professor (Plant Pathology) for their kind help in writing the book. The author is thankful to fellow agronomists and other scientists of Tamil Nadu Agricultural University who have helped me during the course of preparation of the manuscript. The author is thankful to the Publisher, Oxford & IBH Publishing Co. Pvt. Ltd, New Delhi for bring out this book in time.

I am highly grateful to the beloved Prof. Dr. K. Ramasamy, Vice Chancellor, of Tamil Nadu Agricultural University, Coimbatore who has kindly given the foreword to this book.

S. Jeyaraman

TAMIL NADU AGRICULTURAL UNIVERSITY

Dr. K. Ramasamy, Ph.D.,
Vice - Chancellor

Coimbatore - 641 003
Tamil Nadu, India

FOREWORD

Food, feed and fibre production have to be increased to meet the requirement of ever increasing population. The land area under cultivation is dwindling due to urbanization in cultivable lands. In view of the burgeoning human population, horizontal expansion of crop cultivation is impossible. Therefore, there is a very limited scope for extension of cultivated area but it is necessary to produce more food, feed, fodder, fuel and fiber to fulfill the future requirements from the existing land area with the available resources. This challenging scenario, demands to increase the productivity, increase in yield per hectare per crop, increase in the number of crops per hectare per year, reduction of post-harvest losses besides focusing on ecological approaches for sustainable agriculture. The water availability for crop production is shrinking due to global warming coupled with low rainfall and increase in temperature. Under such conditions, the crop productivity can be achieved with the use of short duration high yielding varieties which are resistant to pest and diseases and other abiotic stresses such as drought, flood, salinity and alkalinity; diversity of crops and varieties to suit a wide array of the soil and climatic conditions, timely operations and appropriate techniques on efficient use of inputs such as seed, bio-fertilizers, green manures, green leaf manures, balanced application of organic manure and fertilizers, irrigation water, herbicides and pesticides based on requirement at various crop growth stages without any detrimental effects on soil quality. Use of low cost technology, eco-friendly inputs and non-monetary inputs such as cultivar, time of sowing, maintenance of plant population, timely inter-cultivation, crop rotation play an important role in cost reduction in crop production which really benefit the farmers. Good agricultural practices include choice of crops and varieties, time of sowing/planting, input management based on crop growth stages, cropping systems involving legumes, bio-fertilizers green manures, green leaf manures, shade management, balanced application of organic and mineral fertilizers and other agro-chemicals, water-saving measures, method and scheduling of irrigation, micro irrigation and fertigation, integrated pest and disease management resulting in safe and healthy food and non-food agricultural products, while taking into account economic, social and environmental sustainability.

• **Tel: Off.: +91 422 2431788** • **Res.: +91 422 2430887** • **Fax: +91 422 2431672** • **Email: vctnau@tnau.ac.in**

Vice - Chancellor
Tamil Nadu Agricultural University

Coimbatore - 641 003
Tamil Nadu, India.

This book on 'Field Crops: production and management' is planned in two volumes to cover the above scientific principles and practices on different field crops. The Volume-I deals with special reference to food crops under cereals, millets, pulses, forage crops, green manure and green leaf manure crops. The second volume deals with special reference to commercial crops under oilseeds, fibre crops, sugar and starch crops, spices and condiments, narcotics and beverages, tuber crops, aromatic and medicinal crops. Each crop is dealt in sub-titles *viz*., vernacular name, importance, history, origin, distribution, botany, climate, soil, varieties, growth stages of crops, season, land preparation, seeds and sowing, after cultivation, weed management, manures and fertilizer application, water management, cropping system, plant protection, harvest, post-harvest technology, yield and quality.

Prof.S.Jeyaraman, a reputed agronomist, has worked in different field crops and has vast practical knowledge on crop production with farming background. He has taught B.Sc.(Ag.) and B.V.Sc. Courses in Agronomy. Besides, the author has wide experience in different capacities as Dean, Anbil Dharmalingam Agricultural College and Research Institute (TNAU), Tiruchirapalli, Tamil Nadu; Director, Centre for Soil and Crop Management Studies, Tamil Nadu Agricultural University, Coimbatore, Tamil Nadu and Principal, Institute of Agriculture (TNAU), Kudumiyanmalai, Pudukkotai District, Tamil Nadu. The author has included special chapters/topics covering the emerging issues in crop production such as basmati rice, hybrid rice cultivation, rice cultivation in salt affected soils, rice ratoon management, system of rice intensification, sustainable sugarcane initiative, aroids, aromatic and medicinal crops, tea husbandry and biofuel crops such as jatropha. This book has been prepared based on syllabus of different State Agricultural Universities, ICAR Junior Research Fellowship and ARS–NET Examinations which will be very useful to the students to prepare for Competitive Examinations. I am confident that this text book in two volumes would be a valuable resource for the students of under graduate and post graduate degree programmes in the field of Agriculture and Horticulture, Diploma in Agriculture courses and Extension officers. I congratulate Prof.S.Jeyaraman, a dedicated Agronomist, for bringing out an excellent educational treatise for the benefit of students and farming community.

13/1

(K.RAMASAMY)

Place : Coimbatore – 641 003.
Date : 13.01.2016

Contents

CHAPTER

1

Oilseeds

Oilseed is one of the most important sources of vegetable oil. The vegetable oils are classified under two groups namely, (a) fixed or fatty oils and (b) the essential or volatile oils. The fixed oils do not evaporate under normal conditions of temperatures and exposure. The essential oil volatilizes or evaporates in the atmospheric air on exposure.

The oil is glycerides of fatty acids i.e., glycerol + fatty acids. The oil is mostly in liquid at room temperature while fat is solid at room temperature. The wax is the fatty ester (fatty acid + alcohol) which is mostly solid rather than liquid at room temperature. The fat provides 9 kcal/g while both the carbohydrates and protein provide 4 kcal/g. The oils (liquid) and fats (solid) are essential in the diet of humans which comes from seeds except for olive and avocado from fruits or from animals. Fat is an essential component of the diet, which provides energy and essential fatty acids (EFAs) to meet the body's metabolic requirements and facilitates the absorption of fat-soluble vitamins. If one calculates the average fat needs of Indians by considering different age groups and physical activity, fat intake of 29 g per head per day is adequate to meet the nutritional needs, which translates into annual vegetable oils requirement of 10.585 kg per person. The fat portion is supplied by oils, which give the necessary energy for human metabolism, besides adding taste to the food.

Vegetable oils and fats are located as small insoluble droplets in plant cells either in vascular or found along the cell wall. Oil occurs mostly in seeds (endosperm, cotyledon, mesocarp, embryo) and less frequently in root, stem and foliage. The occurrence of oil in different plant parts is presented in Table 1.

Oilseeds are important both for consumption and for industrial purposes including biodiesel. The use of oil for medicinal purposes is also well known. The oils usage is considerable in the preparation of soaps, cosmetics and lubrication in industry. Castor oil and coconut oil are very important industrial oils, which figure in the export trade. India is the fifth largest vegetable oil economy in the world, next only to USA, China, Brazil and Argentina. Vegetable oils, fats, and waxes are used for foods and for industrial purposes.

Table 1 Occurrence of oil in different plant parts

Endosperm	Cotyledon	Mesocarp	Embryo	Root, stem and foliage
Sesame Castor Coconut Oil palm	Groundnut Sunflower Safflower Soybean Niger Linseed	Olive Oil palm	Cereals Corn oil	Sandal wood Eucalyptus Pepper mint Lemon grass

Classification of oilseed crops based on the nature of oil

(i) **Edible oilseed crops:** The important sources of supply of edible oils are the seeds known as edible oilseeds. The crops belong to this category are known as edible oilseed crops. Edible vegetable oil is composed of glycerides of fatty acids. They may contain small amounts of other lipids such as phosphatides, of unsaponifiable constituents and of free fatty acids naturally present in the fat or oil. Examples are groundnut, rapeseed-mustard, soybean, sunflower, sesame, safflower and niger.

(ii) **Non-edible oil seed crops:** The most important source of supply of non-edible oils are the seeds known as non-edible oilseeds and the crops belong to this category are known as non-edible oilseed crops. Examples are castor, linseed, etc.

Classification of vegetable oil based on sources

(i) Primary Sources

- **Edible group:** groundnut, rapeseed & mustard (toria, mustard, sarson), soybean, sunflower, sesamum, safflower and niger.
- **Non-edible group:** castor and linseed.

(ii) Secondary sources

- **Edible group:** coconut, oil palm, cotton seed, rice bran, corn oil, oil bearing tree crops such as sal seed, mahua, mango-kernel, cheura / phulwara, kokum, dhupa, simarouba.
- **Non-edible group:**
  Seasonal crops: mesta seed, tobacco seed
  Plantation crops: rubber seed
  Other tree borne oilseeds: Neem, karanj, pilu or khakan, palash, nahor, undi, pisa, wild-apricot, rattan-jyot, maroti, jojoba

Classification of oils based on drying properties

(i) **Drying oils:** It absorbs oxygen on exposure to atmosphere and form into thin elastic films. These are used for manufacture of varnishes and paints, soaps, food, etc. E.g. linseed oil, niger seed oil, safflower oil, tung (*Aleurites fordii*), buffalo gourd (*Cucurbita foetidissima*), hempseed, poppy and soybean oil.

(ii) **Semidrying oils:** It dries up slowly on continued exposure to the atmosphere. The semidrying oils are used in the manufacture of soaps and detergents. Examples are sesame oil, rapeseed and mustard oil, sunflower oil, cottonseed oil and corn oil.

(iii) **Non-drying oil:** Non-drying oils remain liquid and do not dry up at ordinary room air temperature. These are used for soap manufacture even under high temperature.

Castor is used for lubrication purposes. Examples are castor oil, groundnut oil, palm oil, olive oil, almond oil and grape oil.

(iv) **Fats:** Fats remain solid or semi-solid at ordinary temperatures. These are used in food and manufacture of soap and candle manufacture. Examples are mahua butter (Madhuca longifolia var. latifolia) and cocoa butter.

Classification of vegetable oils based on fatty acid type

(i) Polyunsaturated fatty acid

- linseed oil (*Linum usitatissimum* - seeds)
- tung oil (*Aleurites fordii* – seeds)

(ii) Unsaturated fatty acid

- safflower (*Carthamus* – 1-seeded fruits)
- soybean (*Glycine max* – seeds)
- sunflower (*Helianthus annuus* – 1-seeded fruits)
- corn oil (*Zea mays* – germ)
- sesame oil (*Sesamum indicum* – seeds)
- cottonseed oil (*Gossypium* – seeds)
- canola oil (*Brassica* – seeds)

(i) Moderately saturated fatty acid

- peanut oil (*Arachis hypogaea* – seeds)
- olive oil (*Olea europea* – fruit pulp)

Oil extraction methods

(i) Mechanical oil extraction: Oil is extracted normally by mechanical pressure of either heated or unheated seed. Pressure is applied gradually and the extracted oil is refined to improve the quality by removing impurities such as water, dirt and vegetable matter. The residue remaining after the extraction process has been completed is known as the seed cake or oilcake, which is rich in protein. Many seed cakes are used as feed to cattle and some as nitrogenous fertilizer.

(ii) Solvent extraction: This process involves a solvent to leach out the oil. By this method, even the last traces of oil from press cake or plant tissues can be extracted. Soybean oil is extracted using this method and oils are freed from solvents by fractional distillation. This process is expensive but extraction is perfect.

Classification of oilseed crops based on processing method:

(i) Virgin oils are obtained, without altering the nature of the oil, by mechanical procedures, e.g. expelling or pressing, and the application of heat only. They may have been purified by washing with water, settling, filtering and centrifuging only. Example is Virgin coconut oil is obtained from fresh and mature kernel of the coconut by mechanical or natural means with or without the application of heat, which does not lead to alteration of the nature of the oil.

(ii) Cold pressed oils are obtained, without altering the oil, by mechanical procedures only, e.g. expelling or pressing, without the application of heat. They may have been purified by washing with water, settling, filtering and centrifuging only.

Keeping quality of oils: Most fats and oils are naturally unstable: when kept for considerable lengths of time, they become rancid due to breakdown of the glycerides into various lesser products. The ratio of oleic to linoleic acid also affects the storage ability of edible oil and hence affects the nutritional quality. It should be greater than 1.6 for longer shelf life. The sunflower and safflower oils cannot be stored for longer periods. Soybean oil loses its original flavour after once deep frying. Groundnut oil has very good stability due to the presence of tocopherols (vitamin E) and can be stored at room temperature even upto 18 months without any quality deterioration. Sesamum oil is highly stable due to the presence of anti-oxidants such as sesamin and sesamol. Mustard oil is also rich in tocopherols (vitamin E) and has good stability even at high temperature.

CHAPTER

2

Groundnut–*Arachis hypogaea* L. (2n=40)

Family: Fabaceae

Vernacular name: Groundnut, Peanut, Monkeynut (English), *Mungphali* (Hindi), *Nilakadalai* (Tamil), *Chenabadam* (Bengali), *Neelashanga* (Telugu), *Phuimug* (Marathi), *Magafuli* (Gujarati) and *Kalale kayi* (Kannada).

Importance: Groundnut is a vegetable cum oilseed leguminous crop. Groundnut is known as 'King of vegetable oilseeds'. Globally, the groundnut is used for extraction of oil and food as raw or processed as snack. In India, it is used for oil extraction, seed, direct consumption and export trade. Groundnut haulm is used as nutritious fodder for livestock. The botanical name *Arachis hypogaea* for groundnut has been derived from two Greek words, '*Arachis*' meaning legume and '*hypogaea*' meaning underground, referring to the formation of pod in the soil.

Origin: The cultivated groundnut, *Arachis hypogaea* is believed to have originated in northern Argentina and southern Bolivia and the centre of diversity of the genus *Arachis* is the Mato Grasso, Brazil wherein majority of the species are found. The features of different cultivated forms of groundnut subspecies are furnished in Table 1.

Table 1 Features of different groundnut subspecies

Subspecies	Cultivar	Primary area of origin	Branching pattern	Growth habit	Seed/ pod
Arachis hypogaea ssp. *hypogaea* var. *hypogaea*	Virginia (bunch, runner)	Southern Bolivia and Northern Argentina	Alternate, less hairs and short branches	prostrate to semi erect	2-3
Arachis hypogaea ssp. *hypogaea* var. *hirsuta*	Peruvian runner (spreading)	Peru	Alternate, more hairs and long branches	prostrate	2-4

Archis hypogaea ssp. *fastigiata* var. *fastigiata*	Valencia (bunch)	Peru, Brazil and Paraguay	Sequential, less branches	erect	3-5
Archis hypogaea ssp. *fastigiata* var. *fastigiata*	Spanish (bunch)	Paraguay, Uruguay, and Brazil	Sequential, more branches	erect	2

Distribution: Groundnut is cultivated in China, India, Nigeria, USA, Sudan, Senegal, Myanmar, Indonesia, Brazil, Argentina, Pakistan, Malaysia and Sri Lanka. India is the largest producer of groundnut in the world. Groundnut production in India is mostly concentrated in six states, viz., Gujarat, Andhra Pradesh, Tamil Nadu, Karnataka, Maharashtra and Rajasthan. These five states account for about 90% of the total groundnut area. The remaining areas are scattered in the states of Madhya Pradesh, Uttar Pradesh, Punjab, Odisha, and West Bengal. Groundnut is cultivated in India over an area of 58.812 lakh ha with a production of 69.815 lakh tones and productivity of 1187 kg/ha. India accounts for about 25% of global area and contributes 19% to world groundnut production. Based on rainfall pattern, soil factors, diseases and pest situations, groundnut growing areas in India have been divided into five agro-climatic zones (Table 2).

Table 2. Groundnut zones in India and their characteristics

Zone	**Rainfall (mm)**	**Soil Type**	**Temp (°C)**		**RH (%)**	**Crop duration (Days)**	**Pest and diseases**
			Min	Max			
Zone I: Rajasthan, Haryana, Punjab, Uttar Pradesh	466-478	Sandy to sandy loam (saline-alkaline soils in some cases)	12	38	57-88	120-150	White grub, termites, collar and stem rots, leaf spots
Zone II: Gujarat, Southern Rajasthan	547–866	Medium black (calcareous; low soil depth)	19	35	76–94	115-130	Thrips, jassids *Spodoptera*, collar and stem rots, leaf spots
Zone III: Northern Maharashtra, Madhya Pradesh	610–939	Medium black to loamy (neutral soil with good depth)	16	37	65-90	100-110	Rust, leafspots, *Spodoptera*, *Helicoverpa*, thrips, jassids
Zone IV: Jharkhand, West Bengal, Odisha, North Eastern Hill region	747–1268	Alluvium, sandy loam to clay-loam (mostly acidic)	14	34	78-92	110- 120	Aphids, thrips, rust and leaf-spots, pod rots
Zone V: Southern Maharashtra, Andhra Pradesh, Karnataka, Tamil Nadu	455-900	Red-lateritic to sandy- loam (high alumina content forms hard crusts)	15	31	68-95	95 -120	Rust, leaf spots, root rot, stem rot, leaf miner, *Spodoptera*, termites

Source: Singh, 2014.

Botany: Groundnut is an herbaceous annual crop. The tap root system generally penetrates to a depth of 35 cm and root spread is confined to a radius of 12 to 15 cm. Root nodules make their first appearance on 15 days old plants. Stems are solid, but become slightly hollow when the plant matures. There are three distinct growth habits in groundnut namely spreading (runner, trailing), procumbent and prostrate; erect (upright, erect bunch and bunch) and an intermediate semi-spreading form (spreading bunch, bunch runner and runner bunch) are recognized. Erect and semi-spreading types are mainly grown for seed purpose while spreading type (runner, trailing) is grown for fodder purpose. Leaflets on the main stem vary from those of lateral branches. Groundnut is a self pollinated crop.

Flowers are borne on axils of leaves on primary or secondary branches, spike-like, simple or compound and each node has up to five flowers. However, three flowers per inflorescence are most common. Only one flower per inflorescence opens at any given time. The style is contained within the calyx tube and both calyx tube and style elongate rapidly up to 5-7 cm in 24 hours prior to anthesis. Groundnut produces more flowers than the plant can sustain to develop in to pods. About 40% of flowers fail to develop into mature pods, while another 40% produce only pegs. Less than 20% of flowers only produce mature fruits. Depending on photoperiod, temperature and genotype, flowering starts at about 25 days after emergence. The number of days require to first flowering reduces from 38 to 24 days when the daily mean temperature rises from 20 to 30°C in spreading and semi-spreading types while it drops from 35 to 24 days in Spanish and Valencia types. The dehiscence of anthers takes place one hour before flower opening. Similarly the stigma is receptive only a few hours before anthesis. The pollen tube grows at the rate of about one cm per hour resulting in fertilization, 5-6 hours after pollination. The pollen matures 6 to 8 hours before anthesis. The self-pollination occurs because the stigma and anthers are enclosed by the keel. Cross pollination (ranging from 0 to 6.16%) also occurs through bees. Pollination takes place at or near the time of anthesis. The ovary is unilocular, and has 1 to 3 ovules, superior with the calyx tube attached to the base of the ovary.

The groundnut flower opens from 6 a.m. to 8 a.m. and fertilization is complete within 6 hours after pollination or before midday. After fertilization, the flower drops, the corolla closes, the calyx tube bends down and the flower withers away in 3 days. On fertilization, the thalamus portion below the ovary begins to grow into a peg or gynophore. The peg grows towards the soil i.e., positively geotropic, to begin with, a lignified tissues forms a protective cap at the tip for the fertilized ovary. The developing ovary pierces through the floral parts to reveal an elongating peg, a green stalk like structure which is botanically a 'carophore' or 'gynophore' and starts growing (elongating) geotropically. The peg carries the fertilized ovule at its tip. The peg carrying the ovary pushes itself into the soil. The peg enters the soil in 2 to 8 days after initiation of the gynophore development in the bunch types and in about 5 to 10 days in the semi spreading and spreading types. Elongation of the peg stops after it has penetrated the soil to about 5 cm. Usually 5 to 6 days after the peg enter the soil, development of pod commences. The gynophore is positively geotropic until it has penetrated the soil to a maximum depth of 7 cm. Then carpel develops into pod attaining a diageotropic horizontal position. If the peg fails to contact and enter the soil after it expands to 15 cm, it usually withers away. The peg thickness is 2 mm. The pod development is initiated when the peg ceases to elongate. Pod development takes about 60 days from the time of fertilization to full maturity. The normal pod zone is located 4 to 7 cm below the soil surface. Pod set percentage (the ratio of the number of mature pods to the total number of flowers) ranged from 8 to 17. Only 40 to 60% of the pegs form the pods. Groundnut pods are elongated with varying degrees of reticulation on the surface.

They contain two to five seeds although differences exist among members of the subsp. *hypogaea* and *fastigiata.* The groundnut seed contains two large cotyledons, a stem axis, leaf primordial, hypocotyl and primary roots. Germination is epigeal, hypocotyl is white and very prominent in the early stages of growth but becomes indistinguishable from the roots as the plant matures. Seeds (kernels) may be oval, round or elliptical, have pointed or flattened ends, vary in seed coat colour from off-white to deep purple and may be monochrome or variegated. Seed size ranges from 0.15 to more than 1.3 g/seed while the seeds of the wild species weigh as low as 0.047 g/seed.

Groundnut has been classified on the basis of growth habit, branching pattern, inflorescence, pod and seed characters, seed dormancy, etc. The cultivated groundnut, *A. hypogaea*, an annual herb of indeterminate growth habit, is divided into the following two subspecies (subsp.):

1. Subsp. hypogaea: absence of flowers on the central or main axis (stem) and regular alternation of vegetative and reproductive branches on the laterals (alternate ramification), long life cycle.
2. Subsp. fastigiata: presence of flowers on the central or main axis (stem) and no specific order of vegetative and reproductive branches on the laterals (sequential ramification), shorter life cycle.

Subsp. hypogaea is divided into two botanical varieties (var.):

(i) Var. hypogaea: leaflets with glabrous dorsal surface or with some hair along the midrib, short central or main axis, prostrate to erect growth habit, simple inflorescence, 2- 3 seeded pods. The var. hypogaea is also known as Virginia type (large-seeded) or Runner type (small-seeded).

(ii) Var. hirsuta: leaflets with entire dorsal surface hairy (1-2 mm), prostrate growth habit, 2-4 seeded pods. The var. hirsuta is also known as Peruvian runner.

Subsp. fastigiata is divided into four botanical varieties as follows:

(i) Var. fastigiata: Leaflets with glabrous dorsal surface and hair only on the midrib, 3-5 seeded pods, erect growth habit. The var. fastigiata is also known as Valencia type.

(ii) Var. peruviana: Leaflets with glabrous dorsal surface and hair only on the midrib, pods with very marked reticulation and with prominent longitudinal ribs.

(iii) Var. aequatoriana: Leaflets with entire dorsal surface hairy (1-2 mm), purple stems, more branched, erect growth habit.

(iv) Var. vulgaris: Pods mostly 2-seeded, bunched fruits pointing to the base of the plant, erect growth habit, more branched. The var. vulgaris is also known as Spanish type; Spanish types are likely to have originated due to hybridization between Virginia × Valencia types.

The difference between spreading and bunch groundnut are furnished in Table 3.

Table 3 Difference between spreading and bunch groundnut

Spreading (Virginia/Peruvian)	Bunch (Spanish/Valencia)
Central axis erect, lateral branches prostrate. Lateral branches exceed the length of central axis	Central axis erect, laterals also erect, but, it will not exceed the length of axis
Main axis is vegetative	Nodes on main axis above primary laterals are productive for the first 6 nodes

Alternate branching in the laterals i.e., vegetative branch followed by reproductive branch. More or less indeterminate in growth habit	Sequential branching. Nodes of primary laterals usually reproductive for first 6 nodes followed by sterile axis
Indeterminate growth habit	Determinate growth habit
Comparatively high oil content	Comparatively less oil
Longer in duration (120–130 days)	Short duration (90–105 days)
large and bold seeded	small seeded type
Seed dormancy present	No dormancy
Leaf dark green, small	Pale green, larger
Perennial tendency	Annual

Climate: Groundnut is essentially a tropical plant. It is mostly cultivated in tropical and subtropical regions. It is grown from 40°N to 40°S latitude. It is commercially grown upto an altitude of 1,250 m from MSL. Rainfall of 500 to 1200 mm will allow commercial production of groundnut while a well-distributed rainfall of at least 500 mm is essential during the crop-growing season. A sum of 80 to 120 mm rainfall in summer months just preceding the sowing to facilitate preparatory cultivation, 150 to 120 mm at sowing, 200 mm from commencement of flowering to peg penetration and 200 mm from early pod development to pod maturity are desirable for rainfed groundnut crop. Groundnut cannot withstand waterlogging. It produces high seed yield in dry season with adequate soil moisture. It is highly resistant to drought. Groundnut is considered a day-neutral plant. The optimum light intensity for groundnut ranges from 21.5 to 86.1 klux. The optimum temperature for different critical stages ranged from 25 to 35°C. Its growth ceases below 15°C and slows down very much above 42°C. For germination, the optimum temperature range is from 25-30°C. It ceases below 18°C and slows down above 35°C. The optimum temperature for vegetative growth is between 25 to 30°C and for reproductive growth is 23 to 30°C with a most suitable temperature of 27°C. The optimum temperature for nitrogen fixation is 25°C. The ideal temperature for pod development is warm day with 23-27°C followed by cool night with 23-25°C. Night temperature below 10°C delays maturation. Decreased fruit-set at high temperatures is mainly due to poor pollen viability, reduced pollen production and poor pollen tube growth, all of which lead to poor fertilization of flowers. Increase in daytime temperature from 26–30 to 34–36°C will reduce the number of subterranean pegs and pods, seed size and seed yield by 30–50%. Groundnut cannot withstand frost.

Soil: Well drained, light coloured, loose friable sandy loam soil well supplied with calcium and a moderate amount of organic matter is an ideal soil for groundnut. Clayey or heavy soils are not suitable for this crop as they interfere the penetration of pegs and their development and make harvesting quite difficult. The field should have adequate drainage facility since waterlogging for periods as short as 24 hours is injurious to plants. Soil pH 5.0 to 6.4 is desirable. Alkali soil is undesirable. When soil pH is 7.5 to 8.5, yellowing of leaves and blackening of parts of pods occur. Groundnut requires EC of 4.0 dSm^{-1} of saturation extract of soil and irrigation water; ESP less than 5%; $CaCO_3$ equivalent less than 4%; pH less than 8 and RSC of irrigation water less than 2 mg per litre.

Season: In India, groundnut is grown in four seasons namely *kharif* (85% area), *rabi* (10% area), summer (4% area), and spring ($\leq$ 1% area). In India, groundnut is cultivated largely in *kharif* season (June-July to September-October) under rain-fed conditions with low inputs and if available, with a few protective irrigations. In *kharif*, the pressure of diseases, insect-pests and weeds is high and hence the productivity is low. In *rabi* season

(October-November to February-March), the crop is grown under residual moisture of rice fallows or river beds under minimal irrigation situations where winter is not severe and night temperature do not go below 15°C and also in summer season (January-February to April-May) as an irrigated crop. The spring groundnut, grown during March-April to July-August after the harvest of potato/toria, also gives high productivity. In general, the sowing date has to be chosen in such a way that favourable soil and air temperature prevail during the crop period. The minimum temperature in the top 10 cm soil should not be less than 18°C.

Varieties: The particulars of groundnut varieties grown in different states of India in Table 4 and Table 5.

Table 4. Groundnut varieties in India

Variety	Yield (kg/ha)	Oil content (%)	State /region	Specific features
VRI (Gn) 5	2133 (R) 2384(I)	50	Tamil Nadu	Resistant to rust and Late Leaf spot (LLS); suitable for both *kharif* and *rabi*-summer seasons
Co (Gn) 4 (TNAU 269)	1500 (R), 1950(I)	53	Tamil Nadu	Resistant to rust and LLS; suitable for all seasons
GG 7 (J-38)	2149	49	Gujarat & Rajasthan	Tolerant to LLS; suitable for *kharif* season
AK 159	1606	51	Maharashtra and Madhya Pradesh	Early (105-110 days) maturity; recommended for *kharif* season
Kalahasti (TCGS 320)	3764	52	Andhra Pradesh	Tolerant to Peanut Bud Necrosis Disease (PBND) and Jassids; suitable for *rabi* season in kalahasti malady endemic areas and also for *kharif* in north coastal and north Telangana regions
Narayani (TCGS 29)	3764	48	Andhra Pradesh	Tolerant to mid-season moisture stress conditions; recommended for both *kharif* and*rabi*-summer seasons
Sneha	-	-	Kerala	Recommended for *kharif* season
Snigdha	-	-	Kerala	Recommended for *kharif* season
GG 6	2782	50	Gujarat	Suitable for cultivation in summer
GG 14 (JSP 28)	2159	52	Rajasthan, Punjab, Haryana, Uttar Pradesh	Tolerant to thrips, *Spodoptera and* leaf miner; recommended for *kharif* season
TPG 41	2008	49	All India	Moderately resistant to rust; bold-seeded (HSM > 60g); high O/L ratio; recommended for *kharif* season
TG 37A	1900	48	All India	Tolerant to collar rot, rust and late leaf spot; suitable for both *kharif* and *rabi*-summer seasons; possess fresh seed dormancy up to 15 days

Vikas (GPBD 4)	1900-2200	49	All India	Resistant to LLS and rust; recommended for *kharif* season
TLG 45	1506	51	Maharashtra	Large-seeded (HSM = 59 g); medium maturity (114 days); recommended for *kharif* season
SG 99	2501	52	Punjab	Tolerant to bud necrosis disease; possess long fresh seed dormancy (30days); suitable for summer season
Phule Unap (JL 286)	2231	49	Maharashtra	Tolerant to LLS, rust and stem rot; also tolerant to thrips, leaf miner and *Spodoptera*
Prutha (Dh 86)	4022	48	All India	Tolerant of LLS and sucking pests; suitable for *rabi*-summer season
Kadiri 5	1800-2200	48	Andhra Pradesh	Tolerant to leaf spots and drought; recommended for *kharif* season
Kadiri 6 (K 1240)	1800-2400	49	Andhra Pradesh	Tolerant to leaf spots; recommended for *kharif* season
Pratap Mugphali 2 (ICUG 92195)	1800-2800	49	Rajasthan	Tolerant to Early Leaf Spot (ELS), LLS and PBND; tolerant to *Spotdoptera*,, leaf miner and thrips; recommended for *kharif* season
Pratap Mugphali 1 (ICUG 92035)	2500.-3000	49	Rajasthan	Moderately resistant to ELS, LLS and PBND; moderately resistant to *S. litura*, leaf miner and thrips; recommended for *kharif* season
Co (GN) 5	1585	54	Tamil Nadu	Tolerant to rust, PBND; tolerant to leaf minerand *Spodoptera;* recommended for *kharif* season
Ratneshwar (LGN 1)	1487	51	Maharashtra	Moderately resistant to LLS, stem rot, rust and PBND; tolerant to sucking pests; recommended for *kharif* season
Utkarsh (CSMG 9510)	21.92	49	Uttar Pradesh, Punjab, Northern Rajasthan	Resistant to rust, possess fresh seed dormancy up to 40-45 days; recommended for *kharif* season
GG 21 (JSSP 15)	1843	53	Uttar Pradesh, Punjab, Rajasthan	Recommended for *kharif* season
Durga (RG 382)	2203	55	Rajasthan	Suitable for sandy and loamy soils; recommended for *kharif* season
TMV (Gn) 13	2580	50	Tamil Nadu	Tolerant to early and mid season moisture deficit stress conditions; recommended for *kharif* season

GG 8 (J 53)	1716	46	Northern Maharashtra and Madhya Pradesh	Moderately tolerant to PBND, collor rot and stem rot diseases
TG 38B (TG 38)	2768	48	Odisha, West Bengal and north eastern states	Tolerant to stem rot; suitable for *rabi*-summer season
Prasuna (TCGS 341)	2000-2500 R) 4000 -4500 (I)	50	Andhra Pradesh	Tolerant to *kalahasti* malady; recommended for *kharif* season
Abhaya (TPT 25)	2300 (R) 3756 (I)	52	Andhra Pradesh	Tolerant of LLS, jassids and thrips and *Spodoptera*, tolerant to early and mid season moisture deficit stress; suitable for both *kharif* and *rabi*-summer seasons
GG 16 (JSP 39)	2058	46	Tamil Nadu, Andhra Pradesh, Kerala, Maharashtra	Tolerant to bud necrosis, root rot diseases; tolerant to *thrips*, *Spodoptera*, leaf miner; recommended for *kharif* season
Vasundhara (Dh 101)	2877	50	West Bengal, Odisha, Jharkhand and Assam	Tolerant to stem rot and PBND; tolerant to *thrips* and *Spodoptera;* suitable for *rabi*-summer season
ICGV 91114	2000	48	Andhra Pradesh	Tolerant to rust and LLS; early maturity (100 days); tolerant to drought; recommended for *kharif* season
AK 265	1903	47	Maharashtra, Karnataka, Andhra Pradesh, Tamil Nadu	Resistant to foliar diseases; drought tolerant; recommended for *kharif* season
M 548	2185	51	Punjab	Tolerant to leaf spot and collar rot; recommended for *kharif* season
AK 303	2100	49	Maharashtra	Bold seeded (HSM = 80g); recommended for *kharif* season
TG 51	2675	49	West Bengal, Odisha, Jharkhand and Assam	Tolerant to stem rot and root rot; suitable for *rabi*-summer season.
Ajeya (R 2001-3)	2440	46-48	Maharashtra, Karnataka, Andhra Pradesh, Tamil Nadu	Resistant to PBND; drought tolerant; recommended for *kharif* season
Girnar 2 (PBS-24030)	2907	51	Uttar Pradesh, Punjab, northern Rajasthan	Virginia bunch type with 'stay green' leaves and bold seeded (HSM =62g); tolerant to rust, LLS PSND; recommended for *kharif* season

ICGV 00348	2013	47	Southern Maharashtra, Karnataka, Andhra Pradesh, Tamil Nadu	Tolerant to late leaf spot and rust; recommended for *kharif* season
VRI (Gn) 7	1865	48	Tamil Nadu	Moderately resistant to leaf miner, LLS and rust; recommended for *kharif* season
VL-Moongphali-1	1943	-	Uttarakhand	Resistant to late leaf spot and root rot; recommended for *kharif* season
VRI (Gn) 6 (VG 9816)	2259	47	Maharashtra, Karnataka, Andhra Pradesh, Tamil Nadu	Tolerant to LLS, rust, PBND; recommended for *kharif* and *rabi*-summer seasons
Jawahar Groundnut 23 (JGN 23)	1631	49	Madhya Pradesh	Tolerant to ELS and LLS; drought tolerant; recommended for *kharif* season
Kadiri 9	2500-3000	52	Andhra Pradesh	Tolerant of thrips, jassids and nematodes. Tolerant to late leaf spot, rust, dry root rot and collar rot. Recommended for *kharif* season
Greeshma	2000-2500 (R); 4000-4700 (I)	49	Andhra Pradesh	Tolerant to LLS, drought, high temperature and aflatoxin; recommended for *kharif* and *rabi*-summer season
Kadiri 7	1643	47	Andhra Pradesh	Tolerant to sucking pests and leaf spots; bold seeded (HSM =65-75 g); recommended for *kharif* season
Kadiri 8	1523	47	Andhra Pradesh	Tolerant to sucking pests and leaf spots; bold seeded (HSM = 65-75 g)
Mallika (ICHG 00440)	2579	48	All India	Resistant to collar rot and PBND; bold seeded (HSM=73g), recommended for *kharif* season
TGLPS 3 (TDG-39)	2500-3000	-	Karnataka	-
JSP-39	3000	49	AP, Karnataka, TN, Maharashtra	Tolerant to foliar diseases and root rot.
JL 501	1661	48	Gujarat and Rajasthan	Suitable for early as well as late sown rainfed condition
Vijetha (R 2001-2)	1600	47	West Bengal, Odisha, Jharkhand Maharashtra, Karnataka, Andhra Pradesh, Tamil Nadu	Resistant to PBND; recommended for *rabi*-summer season

HNG 69	2800	50	Uttar Pradesh, Punjab, northern Rajasthan	Tolerant to collar rot, stem rot and ELS; recommended for *kharif* season
Girnar 3 (PBS 12160)	1520	45	West Bengal, Odisha, Manipur	Tolerant to leaf miner and thrips; recommended for *kharif* season
Kadiri Haritandhra (K 1319)	3728	48	Karnataka and Maharashtra	Multiple diseases and insect pests resistant, possess fresh seed dormancy upto 20 days; recommended for *rabi*-summer season
VL-Moongphali-1	1940	42.2	Uttarakhand (Kharif)	Resistant to LLS and root rot diseases. (State release)
GPBD-5	1500	46.0	Jharkhanad and Manipur (K)	Resistant to LLS and rust.
GJG-HPS-1 (JSP-HPS-44)	2120	49.0	Gujarat (Kharif)	Rose colour seed.
Phule vyas (JL-220)	2000	52	Maharashtra	Early maturing, High oil content.
GJG-9	1700	48	Gujarat	Suitable for *Kharif*, tolerant to stem rot.
GJG-31	3500	49	Gujarat	Free from PBND, Suitable for Summer.
Bheema	3500-5000	45	Andhra Pradesh	Suited to *Kharif* and rabi regions
Rohini	3700-4000	50	Andhra Pradesh	Suited to *Kharif* and rabi areas. Tolerant to mid and end season.
Pratap Raj Mungphalli	1600-2200	48	Rajasthan	Moderately tolerant to ELS, LLS and PBND, Suited for *Kharif* and Summer
ALG-06-320	3500	49	Vidharbha & Southern M.P.	Suitable for rabi/summer
RG-510	2600	49	Rajasthan & Punjab	Resistant to collar rot, stem rot, early leaf spot, rust and stem necrosis.
RG 425	1800-3600	48	Rajasthan	Resistant to collar rot and tolerant to drought. Suitable for *Kharif*.
RHRG-6021	3800	51	Western Maharashtra	Resistant to rust, LIS and stem rot and spodoptera
Divya (CSMG-2003-19)	3000	49	Uttar Pradesh	Resistant to leaf spots and tolerant to BND.
HNG-123	3000	49	Rajasthan & Punjab	Virginia bunch variety
GJG-17	1800	49	Gujarat	Suitable for *Kharif*, spreading groundnut area. Tolerant to stem rot
GJG-22	1800	52	Saurashtra & South Gujarat	Suited to *Kharif*, semi spreading groundnut area. Tolerant to collar rot.
ICGV-00350	3000-4400	48	Tamil Nadu & Andhra Pradesh	Tolerant to LLS, rust, stem rot, High fodder value.

Source: http://oilseeds.dacnet.nic.in/crop.html

Table 5. Groundnut varietal particulars of Tamil Nadu

Variety	Growth habit	Duration (days)	Pod yield (kg/ha)	Shelling (%)	Oil content (%)	100 seed weight (g)
TMV 2	----	100-105	1250	76	49.4	36
TMV 7	Bunch	100-105	1400	74	49.6	36
TMV 10	Semi-spreading	120-130	1650	77	51.4	43
TMV 12		100-105	1650	72	51.1	38
TMVGn 13	Bunch	100-105	1600 (R) 2580 (I)	71	50.0	44
CO 1	----	100-105	1675	74	50.4	34
CO 2	----	100-105	1650	77	51.4	41
CO 3	Bunch	115-120	1750 (R) 2150 (I)	71	49.2	65
COGn 4	Bunch	115-120	1500 (R) 1950 (I)	70	52.7	60
COGn 5	Semi-spreading	125-130	1600	70	51.0	47
TNAU Co 6	Semi-spreading	125-130	1910	74	51.0	47
JL 24		95-105	1650	75	50.1	46
VRI 1		105	1500	70	47.0	30
VRI 2		105	2500	72	48.0	40
VRI 3		90	2500	71	48.0	35
VRI 4		105-110	2200	72	47.0	35
ALR 1		115-120	1840	70	49.0	33
ALR 3		110-115	1680	69	50.0	35

Growth stages: The length of growing period is 90 to 115 days for the sequential, branched varieties and 120 to 140 days for the alternately branched varieties. Groundnut seed absorbs moisture very rapidly at temperatures of 12°C and emergence occurs after 4-6 days. Temperature of 15-20°C are more favourable than lower ones and will ensure a rapid and uniform emergence. The particulars on growth stage of groundnut is presented in Table 6.

Table 6. Growth stages of groundnut

Stage	Length of growing period (days)			
Germination	4-6	4-6	4-7	4-7
Vegetative stage	6-25	6-30	7-30	30
Panicle initiation	25-30	28-30	28-30	30-35
Flowering stage	28-35	30-35	35-40	45
Aerial peg	36-42	40-42	40-45	55
Peg soil penetration	43-50	45-52	50-55	70-75

Pod filling stage	50-55	55-60	65-70	80-90
Soft dough stage	55-70	70-75	75-80	130
Hard dough stage	70-85	85-90	90-95	150
Maturity	95-100	100-105	105-110	160
Harvest	100-105	120	130	165-170

Growing ecosystems:

(i) Rainfed and irrigated *kharif* Ecosystem: Gujarat, Andhra Pradesh, Tamil Nadu, Karnataka, Maharashtra, Rajasthan, Madhya Pradesh and Uttar Pradesh.

(ii) Summer Irrigated Ecosystem: Gujarat, Andhra Pradesh, Tamil Nadu, Karnataka and Odisha.

(iii) Rice-Fallow Ecosystem: Andhra Pradesh, Tamil Nadu, Karnataka and Odisha

(iv) Riverbed Ecosystem: Andhra Pradesh, Tamil Nadu, Karnataka and Odisha

(v) Spring Irrigated Ecosystem: Uttar Pradesh

Season: Groundnut is sown during *kharif* season, i.e., June 15th to first week of July and during December-January under irrigated condition. In India, groundnut is cultivated largely in *kharif* season (June to October) usually under rainfed conditions with low input use. In *kharif,* the pressure of insect pests and diseases including weeds is high and hence, the productivity is low. In *rabi* season (October to March), the crop is grown on residual moisture in rice fallow lands or river beds under minimal irrigation situations and also in summer season (January-February to April-May) as an irrigated crop. The spring groundnut is grown during March-April to July-August after the harvest of potato/toria.

Land preparation: A good tilth to a depth of 15 to 20 cm is adequate. Three to four ploughings are necessary to have a good tilth. Groundnut pegs containing fertilized ovary have to penetrate the solid soil surface 2 cm deep before the development of pods occur. This pegging process is vulnerable to hard surface soil. Deep ploughing is not recommended, though groundnut is a deep rooted crop, as it tends to encourage forming of pods in deeper layers, rendering harvest more difficult.

Seeds and sowing: Seed treatment is done to protect the young seedlings from root-rot and collar rot infection and to enhance biological nitrogen fixation. Biocontrol agents are compatible with biofertilizers. Fungicides and biocontrol agents are incompatible. First treat the seeds with biocontrol agents and then with *Rhizobium* and *Phosphobacteria.* Treat the seeds with *Trichoderma viride* @ 10 g/kg seed or it can be applied @ 10 kg/ha as soil application or *Pseudomonas fluorescens* @ 10 g/kg seed just before sowing (or) Treat the seeds with chloropyriphos @ 12.5-25 ml/kg seed or Thiram or Mancozeb @ 4 g/kg of seed or Carboxin or Carbendazim at 2g/kg of seed. Then treat the seeds with 3 packets of Rhizobium @ 600 g ha^{-1} + Phosphobacterium @ 600 g ha^{-1} using rice gruel as binder. If the seed treatment is not carried out, apply 10 packets/ha (2000g) with 25 kg of FYM and 25 kg of soil before sowing. Seed treatment with fungicide and insecticide can be done just 1 or 2 days prior to sowing. The seed treatment with recommended fungicide and insecticide should be carried out gently so as to avoid injury to seed radicle. Seed treatment ensures good germination and establishment of healthy seedlings, which leads to higher productivity in the crop. Spreading and semi-spreading types of groundnut varieties have dormant seeds just after harvest which usually requires a resting period of 60-70 days. Dormancy can be broken by exposing seeds to ethrel solution of 250 ppm. Bunch type varieties can be used immediately after harvesting for sowing.

The seed rate depends on the recommended plant spacing for the region and 100-seed weight of the variety under multiplication.

Varieties	Seed rate (kernels in kg/ha)	Spacing (cm)	Population (lakh/ha)
Bunch	100-110	30 × 10	3.33
Runner	95-100	30 × 15	2.22

The recommended spacing for bunch type varieties is 30 × 10 cm while for runner type varieties the recommended spacing is 45 × 10 cm or 30 × 10 cm. The groundnut is sown in flat beds. Some of the improved methods to get higher yield over the conventional method are Criss-cross sowing, Broad-bed and furrow method (BBF) and Ridge and furrow method. Seeds are sown either by hand dibbling, dropping seed with hand in furrow formed by the country plough or seed drill sowing. Seeds are sown at a depth of 5 to 7 cm. The isolation distance for certified seed is a minimum 15 m between two groundnut plots. For foundation seed the minimum isolation distance required is 30 m since there is a chance of 0.25 to 6.16% natural cross pollination. In India where natural cross pollination is almost negligible, an isolation distance of 3 m between varieties is required for all classes of certified seeds.

Cropping system: Groundnut can be grown in different cropping systems like mono, sequential, multiple, and intercropping with an array of field crops (sorghum, maize, bajra, pigeon pea, green gram, black gram and commercial/ plantation crops (cotton, sugarcane, coconut). The mono cropping of groundnut accounts for 60% of the total groundnut area in the country while multiple (inter and sequential) cropping occupies the rest of the area. Groundnut is grown in rotation with pearlmillet/sorghum/foxtail millet and cotton / gingelly. It is cultivated as intercrop with cotton, cowpea, pigeonpea, blackgram, castor and sesame.Groundnut is usually rotated with wheat, mustard and grams. The groundnut based intercropping systems practised in different states are presented in Table 7.

Table 7 State wise important Intercropping Systems

State	Intercropping system	Ratio to base crop (groundnut)
Andhra Pradesh	Groundnut + Pearlmillet	3 : 1
	Groundnut + Cowpea	6 : 1
	Groundnut + Redgram	6 :1 or 8 : 1 or 10 : 2
	Groundnut + Castor	5 : 1 or 7 : 1
Gujarat	Groundnut + Sesame	1 : 1
	Groundnut + Sunflower	1 : 1
	Groundnut + Redgram	3 :1
	Groundnut + Castor	1 : 1
Karnataka	Groundnut + Redgram	4 : 1
	Groundnut + Cotton	3 : 1 or 5 : 1
	Groundnut + Sorghum/Ragi	6 : 1
Maharashtra	Groundnut + Sorghum	4 : 1 or 6 : 2
	Groundnut + Redgram	6 : 1 or 10 : 2

Madhya Pradesh	Groundnut + Redgram	8 : 2 or 10 : 2
	Groundnut + Soybean	4 : 1 or 6 : 1
	Groundnut + Sesame	4 : 1
Tamil Nadu	Groundnut + Blackgram/Greengram	6 : 1
	Groundnut + Cotton	5: 1
	Groundnut + Castor	7 : 1
	Groundnut + Sesame	6 : 1
Rajasthan	Groundnut + Pearlmillet	4 : 1
	Groundnut + Sesame	4 : 1

Intercropping in Tamil Nadu

(i) Raise one row of cowpea for every five rows of groundnut wherever red hairy caterpillar is endemic.

(ii) Raise intercrops like redgram, blackgram, sunflower, gingelly or other pulses.

(iii) Groundnut + Gingelly or Groundnut + Blackgram in the ratio of 4:1 or Groundnut + Cowpea at 6:1 ratio and Groundnut + Sunflower at 6:2 ratio may be raised.

The groundnut based sequential cropping systems practiced in different states are presented in Table 8.

Table 8 State wise important Sequential cropping systems involving groundnut

State	Rainfed (Monocropping in two years)	Residual moisture (Double cropping in one year)	Irrigated (Double or triple cropping in one year
Andhra Pradesh	Groundnut-Sorghum	Groundnut-Bengalgram	Groundnut-Maize
	Groundnut-Millet	Groundnut-Safflower	Groundnut-Wheat
	Groundnut-Tobacco	Groundnut-Sesame	Groundnut-Onion
Gujarat	Groundnut-Sesame	Groundnut-Fodder Sorghum	Groundnut-Mustard-Greengram
		Groundnut-Mustard	Groundnut-Wheat-Greengram
Karnataka	Groundnut-Sorghum	Groundnut-Safflower	Groundnut-Wheat
			Groundnut-Maize
			Groundnut-Sunflower
Maharashtra	Groundnut-Sorghum	Groundnut-Safflower	Rice-Potato-Groundnut
		Groundnut-Fodder Maize	Groundnut-*Rabi* Sorghum
			Groundnut-Safflower
Tamil Nadu	Groundnut-Sesame	Groundnut-Sesame	Rice-Rice- Groundnut
	Groundnut-Cotton		Groundnut-Rice-Greengram
			Groundnut-Maize
Rajasthan	Groundnut-Pearlmillet	Groundnut-Barley	Groundnut-Wheat-Greengram
		Groundnut-Mustard	Groundnut-Wheat
Madhya Pradesh	Groundnut-Sorghum	Groundnut-Safflower	Groundnut-Wheat/ Mustard
Odisha	Groundnut-Sorghum/ Pearlmillet	Groundnut-Bengalgram	Groundnut-Rice/Ragi
		Groundnut-Sesame	Groundnut-Coriander/ Cumin

Interculitivation: Intercultivation usually starts around 10 days after emergence and continues up to 35 DAS at 7 to 10 days interval till pegging begins. Earthing up is done after second hand weeding. Earthing up provides medium for the peg. Do not disturb the soil after the 45th day of sowing as it will affect pod formation adversely.

Roguing: Minimum two (preferably three) roguings should be carried out before harvest to remove off-type groundnut plants in the seed production field. At the seedling stage, weak, distorted, variegated, diseased, and out of the row alignment seedlings should be removed and destroyed. At the flowering stage, variants, not conforming to flower morphology, branching pattern, growth habit, and other diagnostic characteristics of the variety under seed multiplication should be removed from the field. Similarly, at the podding stage, based on peg morphology and other vegetative characters, the remaining off-types including late flowering plants should be removed. The last roguing is done on the harvested plants to remove plants with diseased pods and off-types based on pod and seed characteristics.

Weed control: The critical weed free period for groundnut is 45 days or 5 to 6 weeks after sowing. Weeds cause 30 to 50% yield reduction. Two mechanical weeding each at 15-20 days and 30-35 days after sowing to keep the crop weed free for 6 weeks. Apply Pendimethalin 30% EC @ 1.0-2.0 kg a.i./ha as pre-emergence or Oxyflourfen 23.5% EC @ 425-850 g/500 lit. of water or Alachlor 50% EC @ 1.5 -2.5 kg/250-500 lit. of water or Alachlor 10% GR @15-25 kg/ha or Nitrofen (Tok-E-25) @ 5 lit/ha as pre-emergence or Fluchloralin 45% EC @ 1.5 lit/ha as pre-emergence after sowing but before emergence of weeds. Use of post- emergence weedicides like Fluzifopbutyl 9% EC @ 0.25 kg ai/ha or Imazethapyr 10% SL @ 1.0-1.5 lit/500-700 lit. of water or Quizalofop ethyl 10% EC @ 750-1000 g/500 lit. of water for grass and broad leaf weeds are recommented.

Manures and fertilizer application: For every one tonne of pod yield and two tonne of haulm yield, groundnut crop removes 63 kg N, 11 kg P_2O_5, 46 kg K_2O, 27 kg CaO and 14 kg MgO from the soil. To obtain high yield well decomposed farm yard manure @ 12.5 t/ha should be applied at least 21 days before sowing of crop. Apply lime @ 2 tonnes per ha along with FYM or composted coir pith @ 12.5 tonnes per ha to tide over the surface crusting. When converted coir pith into compost by inoculating with *Pleurotus* is applied, it serves as a good source of nutrient. Application of 17-34-54 and 10-10-45 kg of N, P_2O_5 and K_2O per ha are recommended under irrigated and rainfed conditions respectively. 50% N, entire P_2O_5 and K_2O are applied as basal dressing. The remaining half quantity of nitrogen is applied on 30 DAS i.e., during peak flowering stage. Single super phosphate (SSP) is the best source of phosphorus as it contains 15% P, Ca 17.5% and S 12.5%. Application of phosphorus to groundnut promotes root growth, multiplication of rhizobium and helps the crop to overcome moisture stress. The recommended fertilizer doses for groundnut in different states are furnished in Table 9 and Table 10.

Table 9 Recommended doses of NPK for different states

State	Situation	N- P- K (kg/ha)
Andhra Pradesh	Rainfed	20 - 40 - 20
	Irrigated	30 - 60 - 45
Gujarat	Rainfed	12.5 -25 - 0
	Irrigated	25 - 50 - 0
Karnataka	Rainfed	15 - 30 - 25
	Irrigated	25 - 75 - 25
Madhya Pradesh	Rainfed	20 - 40 - 20

Punjab	Irrigated	15 - 40 - 25
Rajasthan	Rainfed	20 - 60 - 0
	Irrigated	20 60 0
Maharashtra	Irrigated	20 - 40 - 0
Uttar Pradesh	Rainfed	15 - 30 - 45
West Bengal	Irrigated	15 - 30 - 45
Tamil Nadu	Rainfed	11 - 22 - 33
	Irrigated	22 - 44 - 66

Source: DOD, Hyderabad

Table 10. Schedule for controlling micronutrient deficiencies

Micronutrient	Form and rate of allocation to soil	Spray schedule
Boron	Borax 5-20 kg/ha	0.2% borax
Copper	Copper sulphate 5-10 kg/ha	0.1% copper sulphate + 0.05% lime
Manganese	Manganese sulphate 10-25 kg/ha	0.6% manganese sulphate +0.3% lime
Zinc	Zinc sulphate 25 kg/ha	0.5% zinc sulphate + 0.2 % lime
Molybdenum	Sodium or ammonium molybdate 0.5-1.0 kg/ha	0.07-0.1% ammonium molybdate
Iron	Ferrous sulphate 10 kg/ha	0.5% ferrous sulphate + 0.02% citric acid

Source: DOD, Hyderabad

Gypsum application: There must be enough Ca in the top 8 to 12 cm of soil after the peg enters it. Excessive K in the podding zone interferes with Ca uptake and results in pod rot and pops (unfilled pods). Calcium increases ovule fertilization, pods per plant, filling of pods and seed yield. Calcium is required to produce high quality kernels. Calcium deficiency causes unfilled pods (POPS), darkening of the plumule of the seed embryo and reduced pod development. Pops is the condition of occurrence of unfilled or partially filled pods with shriveled kernels. It has been found to be caused by calcium deficiency in the soil. The occurrence of pops in groundnut has been first observed in 1885 by Jones and reported that unless the soil contains a good percentage of lime in some form in an available state no land will produce a crop of pods, although it may yield large luxuriant vines. The problems of pops become severe in heavily leached soils which have low pH, usually in high rainfall areas. Pops are more area specific than season specific. The conditions are favourable for pops when the calcium levels reach below 1.0 milliequivalent Ca. Pops may also be caused by unfavorable conditions such as drought during pod filling. Even if there is enough calcium, lack of moisture during pod filling stage results in pops. Sulphur is essential in the biosynthesis of oil and increases nodulation in the plant. As Ca is relatively immobile in plant tissues and is not translocated in sufficient quantities from the roots to meet the needs of developing pods, Ca and S have to be made available in adequate quantities in the pod zone. Supply of Ca and S to groundnut is done through gypsum. Gypsum is applied in soil on 30 DAS i.e., early flowering stage at the pegging zone. Gypsum is the cheapest source of calcium (25%) and sulphur (18.6%). Gypsum is to be broadcasted and incorporated in the pegging zone of 7 to 9 cm. Gypsum is to be applied on the soil surface as close to the base of the plant as possible when it is in peak flowering stage. Application of gypsum is recommended in the sandy loam soils @ 500 kg per ha in 2 split doses, half at planting and the other half at the peak flowering stage. Generally 2 to 4 t lime per ha depending upon the degree of acidity should be applied one month before sowing.

Micronutrients: Application of $ZnSO_4$ to the soil @ 25 kg per ha, once in 3 years has been recommended for groundnut. Foliar application of 0.2 % $ZnSO_4$ is recommended. Boron deficiency is corrected by either soil application of Borax @ 5 to 10 kg per ha along with N and P fertilizers or of 0.1 ppm boric acid (300 mg boric acid in 500 litres of water). Foliar spray of 0.1 % borax is recommended on 30 and 50 days after sowing depending upon the severity of deficiency. Iron chlorosis in groundnut is prevalent in the calcareous black soils which can be corrected by spraying 1.0 % ferrous sulphate, 0.1% ammonium citrate on 25, 40 and 75 DAS. Spanish bunch varieties JL 24 and GG 2 are tolerant to iron chlorosis.

Combined nutrient spray: Pod filling is a major problem especially in the bold seed varieties. To improve pod filling spraying of nutrient solution is to be given. This can be prepared by soaking DAP 2.5 kg, Ammonium sulphate 1 kg and borax 0.5 kg in 37.5 lit of water overnight. The next day morning it can be filtered and about 32 litre of mixture can be obtained and it may be diluted with 468 lit of water so as to make up to 500 litre to spray for one ha. Planofix at the rate of 350 ml. can also be mixed while spraying. This can be sprayed on 25th and 35th day after sowing.

Nutritional disorders:

Nitrogen deficiency: Leaves show general chlorosis. N accumulation in leaves and stem ceases after the onset of pod development, i.e., on 80 DAS. Spray 4% urea at pegging stage is desirable.

Phosphorus deficiency: The older leaves turn dark green and then become orange, yellow, brittle and are finally shed.

Potassium deficiency: Interveinal chlorosis followed by yellowing of edges of leaves.

Calcium deficiency: Necrotic spots on both leaf surfaces which give the leaves a bronze colour, basal stem black and dieback of shoots. Lime induced iron chlorosis occurs in basic and calcareous soils.

Magnesium deficiency: Leaves show interveinal chlorosis which starts from the margins and advances towards the midrib. N, P, Ca and B deficiencies increase Mg content in plants.

Zinc deficiency: Zinc deficiency symptoms are characterized by yellowing from midribs of young leaves and red spots in the old leaves. Light yellow stripes along the veins of leaf blade under acute condition-veinly chlorosis and cessation of growth of terminal bud

Iron deficiency: Interveinal chlorosis, depression on growth of aerial parts and roots and stunted growth. For correction of iron deficiency, spray 1.0 % $FeSO_4$ on 30, 40 and 50 days after sowing.

Boron deficiency: Growth of young leaves restricted giving a rosette effect. The pod development is affected resulting in the production of 'pop' pods. Apply borax @10 kg per ha and gypsum 200 kg per ha at 45th day after sowing for boron deficient soil.

Sulphur deficiency: Stunted growth, uniformly chlorotic plants, thin stemmed and spindle appearance is symptoms of sulphur deficiency. Magnesium and boron deficiency decreases sulphur concentration in the plant tissue.

Copper deficiency: Leaves show chlorotic with yellowish white spots.

Water management: Groundnut crop requires water on an average of 600 to 700 mm for light soil and 500 to 600 mm for heavy soils. Groundnut requires 11 to 12 irrigations in summer season. Groundnut crop requires 11 to 12 irrigations, one irrigation at 0-12 days, one at 10-30 days, two at 30-50 days, three at 50-80 days, one at 80-90 days, two at 90-105 days and one at 105-120 days after sowing are found to be most ideal. It requires 5 to 7 mm per day during its peak growth. Water requirement reaches a maximum during flowering and continues up to pod formation. It requires 5 to 6 irrigations in *rabi* season. A light soil moisture stress during the vegetative phase (up to 3-4 weeks after sowing) does not affect yield but it can reduce vine growth enough to minimize diseases later. However, moisture stress during flowering can delay or inhibit flower formation. After flowering, peg penetration requires adequate moisture. Once active pegging and pod formation have begun (about 50-60 days after sowing), the pegging zone (the top 8-12 cm of soil) should be kept moist. This can be achieved by applying frequent but small amount of irrigation. The presence of moisture in the podding zone improves Ca uptake, which is essential for proper pod and seed development. A lack of water in this zone during pod addition and development results in more pops, more one-seeded pods, a less mature crop, and a lower Ca content in the seed, which in turn affects germination and seed quality. The critical growth stages for moisture stress are flowering (30–50 days), pegging (50-60 days) and pod development (70-80 days) since most of the dry matter is accumulated during the periods. Groundnut is irrigated with IW/CPE ratio of 0.60 from sowing to flowering (0–30 DAS); 0.60 from effective pegging (30–45 DAS) and 0.90 from pod formation to maturity (45 DAS to maturity). Irrigation is to be given at 50% of soil available moisture. Limits for use of irrigation water are with EC less than 4 dSm^2 and residual sodium carbonate (RSC) less than 2 meq/litre. Spraying 0.5% potassium chloride during flowering and pod development stages will aid to mitigate the ill effects of water stress. Sprinkler irrigation will save water to the tune of about 30%. Composted coir pith increases moisture availability and good drainage in heavy textured soil. The crop is usually irrigated by check basin method. Border strip irrigation is recommended in command areas for light textured soils. Sprinkler irrigation is ideal for the crop on sandy soils. Drip irrigation is gaining popularity among the groundnut farmers.

Plant protection:

Crop stage-wise pests:

(i) **Pre-sowing stage:** White grubs, Nematodes, Red hairy caterpillar, Bud necrosis, Soil borne pathogens (stem rot, collar rot, dry rot, *Aspergillus flavus*)

(ii) **Seed and seedling stage:** Aphids, Jassids, Leaf miner, Tobacco caterpillar, White grubs and Termites, Red hairy caterpillar (RHC) and Tobacco caterpillar; Bud necrosis, Collar rot, Stem rot, Seed rots/ collar rots and root rot.

(iii) **Vegetative stage:** Aphids, Jassids, Leaf miner, Tobacco caterpillar, Red hairy caterpillar, Bihar hairy caterpillar, Leaf spot, Rust, Stem rot, Bud necrosis clump leaf spots and rust.

(iv) **Peg and pod formation stage:** *Spodoptera*, Leaf miner, Rust, Leaf spot, Stem rot; leaf spots and rust.

(v) **Maturity stage:** Pod rot, Leaf spots, Rust, Pod borers, Termites, Rodents, Aflatoxins

(vi) **Harvest and storage stage:** Pod infection by Alpha-root (*Aspergillus flavus*), Bruchids infection in stored produce.

Pests management:

(i) Deep ploughing during April-May to expose pupae to sunlight and predatory birds.
(ii) Clean cultivation by rouging out weed hosts and self-sown plants.
(iii) Growing of resistant varieties like, BR 2, ICGV 87160, ICGV 86031, ICGV 86699 (Leaf Miner), ICGV 86590 (*Spodoptera*), BG 2, Girnar 1 (aphids), Girnar 1, Co-1, Dh-3-30, ICGS 11, MH 1, POL 2, S 206 (Leafhoppers) and Girnar 1 (Thrips).
(iv) Early sowing escapes the damage caused by Leaf Miner and White Grubs.
(v) Intercropping with Soybean (Leaf Miner), Castor (*Spodoptera*), Cowpea (Hairy Caterpillars, Aphid and Leafhopper) and Pearlmillet (Thrips).
(vi) Set up the petromax light traps @ 1-2/ha to attract and kill the moths during June-August.
(vii) Install pheromone traps @ 10 traps/ha for *Spodoptera* and *Helicoverpa* and 25 traps/ha for leaf miner.
(viii) Spray neem oil @5ml/lt water alongwith suitable surfactant like soap powder @ 1g/lt or NSKE 5% as it acts as oviposition deterrent.
(ix) Erect bird perches @ 10-12/ha.
(x) Conserve the natural enemies like, Coccinellids, Spiders, Hymenopteran and Dipteran Parasitoids.
(xi) Release *Trichogramma chilonis* @ 50000/ha, two times at 7-10 days interval followed by release of *Bracon hebetor* @ 5000/ha two times at 7-10 days against Leaf Miner and Defoliators.
(xii) Spray commercial formulation of Nuclear Polyhedrosis Virus (NPV) for the management of *Spodoptera* and *Helicoverpa* @ 250 LE (6x109/LE /ha) and @ 200 LE for Hairy Caterpillars.
(xiii) Spray *Bacillus thuringiensis* @ 1-1.5 kg/ha against Hairy Caterpillars, *Spodoptera* and *Helicoverpa*.
(xiv) Spray entomo-pathogenic fungus like, *Nomuraea rileyi* and *Beauveria bassiana* @ 2g/lt of water for lepidopteran Caterpillars and *Verticillium lecanii* for sucking pests.
(xv) Shoot webber may be controlled with foliar spraying of Monocrotophos 36WS 625ml/ha or Neem seed kernel extract 5% or Neem oil 2% or or dusting of Malathion 5D 25kg/ha at 25 and 50 DAS.
(xvi) Papaya mealy bug can be controlled with release of *Acerophagus papayae* parasitoid @ 100 per village.

Diseases management:

(i) Deep burial of surface organic matter and crop debris.
(ii) Use good quality seeds of resistant/tolerant varieties.
(iii) Seed treatment with commercial formulation of *Trichoderma harzianum* or *T. viride* or *Pseudomonas fluorescens* @ 10g/kg seed or soil application of *Trichoderma viride* @ 2.5 kg/ha mixed with 100 kg FYM. or Thiram or Carbendazim or Captan or Mancozeb @ 3-4g/kg seed or Tebuconazole (Raxil 2 % DS) @ 1.25g/kg.
(iv) Avoidance of deep sowing and injury to the seedling.
(v) Crop rotation with wheat and gram, mixed cropping with mothbean.
(vi) Soil application of neem cake or castor cake @ 500kg/ha or neem seed kernel powder @ 3-5%.
(vii) Foliar application of Carbendazim (0.025%) + Mancozeb (0.2%) at 2-3 weeks

interval, 2 or 3 alternate spray of Mancozeb (0.2 %), Carbendazim (0.02 %) and Mancozeb (0.2 %) or three sprays of Chlorothalonil (0.2 %) or Hexaconazole (0.005 %) or Difenoconazole 25% EC @ 2ml/L at 30, 50 and 70 DAS effectively reduces the early leaf spot and late leaf spot severity.

(viii) Spray Mancozeb (0.2 %) or Copper Oxychloride (0.2 %) and destroy the collateral weeds and self-sown plants.

Nematode management practices:

(i) Crop rotation with poor or immune host crops like cereals.

(ii) Deep summer ploughing.

(iii) Soil solarization by a transparent polythene sheet (25-50 μm) for 15 days during summer also helps to control nematodes. Soil amendments such as neem cake or castor cakes @ 1 tonnes/ha preferably seven days prior to sowing has been found to reduce nematode population. Their combination with seed treatment, with Carbosulfan (25 DS) @ 3% a.i. (W/W) further improves efficacy in reducing the nematode population and enhancing yield significantly.

(iv) Use resistant varieties like Tirupathi-2 and 3 for the management of Kalahasti malady disease.

(v) Farmers of south Saurashtra region of Gujarat are advised to sow groundnut with castor as an intercrop (row ratio 2:1) along with soil

(vi) Application of Carbofuron @ 1kg a.i/ha to reduce the population of root-knot nematodes infecting groundnut

Resistant varieties:

(i) Sucking insects such as aphids, thrips and jassids: Girnar 1

(ii) Early leafspot, late leafspot, Rust: ALR 1, ALR 2, ICGV 86590, ICGV 87160, Girnar 1, CSMG-84 -1, GG 52-1, DH 8.

(iii) Peanut bud necrosis: Kadril 3, R8808, ICGV 87160.

(iv) Stemrot, collar rot, Aspergillus flavus: J-11, OG 52-1 and JCC 88.

Post harvest losses management:

(i) The moisture content in the stored produce should be less than 10% which prevents aflatoxin production and contamination.

(ii) Spray Malathion 1.25 % or Deltamethrin 0.04% on the walls, floor and roof of the Warehouses or Godowns before storage and use Aluminium Phosphide @ 3-5 tablets/ tonnes of pods for the management of bruchids.

Aflatoxin management in groundnut:

What is Aflatoxin?

(i) Aflatoxins are toxic and produced as secondary metabolites by two fungi called *Aspergillus flavus* and *Aspergillus parasiticus* on variety of food products. *A. flavus* is the predominant species present in the Indian soils.

(ii) Mould contamination can occur in the field as well as during harvest, processing, transportation and storage.

(iii) Aflatoxins can cause acute disease manifested by kidney or liver failure or chronic disease including carcinoma, birth defects, skin irritation, neurotoxicity, and death.

Reasons for aflatoxin contamination in groundnut at pre-harvest stages:

(i) Soil population of *Aspergillus flavus* depending on soil types and crop rotations
(ii) Development of cracks during pods growth
(iii) Mechanical injury to pods during intercultural operations
(iv) Infestation of insect-pests (termites, pod borers and wire worm) causing damage to pods
(v) Death of plants caused by diseases (stem, root and pod rots) at pod maturity stages
(vi) Nematode damage to the pod

Strategies to reduce aflatoxin content in groundnut are:

(i) Deep ploughing (8-10 inches) to expose the soil to sun for 2-3 weeks.
(ii) Apply neem/ castor cake @ 500 kg/ha in furrow at the time of sowing or enriched with 2.5 kg of commercial formulation of *Trichoderma* sp.
(iii) Harvest the crop at right maturity (blackening of inner surface of shell).

Strategies to reduce aflatoxin content in groundnut at the time of harvesting and post-harvestings are:

(i) Dry the plants till the leaf/pegs become brittle (6-7 days).
(ii) Pick the immature pods first and do not mix them with the main lot of mature pods.
(iii) Remove all the pods showing mechanical or insect damage.
(iv) Keep the storage space free from any kind of seepage or leakage water that may lead to build up of moisture.
(v) Prevent insect damage to the pods in storage by fumigating with phosphene (use 3-5 aluminum phsophide tablets for every 100 kg of pods for 7-8 days).

Harvest: As groundnut is an indeterminate crop, hence synchronous maturity of its pods cannot be obtained. When the crop is mature, there will be drying and falling of older leaves and yellowing of the top leaves. The mature pods become reticulated and within it, seed is separated from the shell of the pod and the inside of the shell becomes dark in colour. Pull out a few plants at random and shell the pods. If the inside of the shell is brownish black and not white, then the crop has attained maturity. When high percentage of pods (about 75%) matures, harvest can be commenced. Well-filled pods with pink seed coat indicate maturity in large-seeded Virginia type. At the time of harvesting, pods usually have moisture content in excess of 40% on wet basis. Irrigate prior to harvest, if the soil is dry as this will facilitate for easy harvesting. If there is enough moisture in the soil, there is no need for irrigation for harvesting. If water is not available for irrigating the field prior to harvest, work a mould board plough or work a country plough, so that the plants are uprooted. Engage labour to search pods left out in the soil if necessary. When harvesting is done at ideal condition, about 5 to 10% pods are left in the soil.

Drying: The groundnut pods generally contain 35-60% moisture at the time of harvest. Until the moisture is reduced to <10%, the pods are prone to mould attacks. Plants are allowed to remain in the windrows until whole pod moisture drops down to 18-20% for mechanical threshing and to 15% for hand threshing. Do not keep the pulled out plants in heaps, when they are wet especially the bunch varieties as the pods will start sprouting. Dry the uprooted plants along with the pods in small heaps by keeping them up-side-down *i.e.* foliage towards ground and pods upwards under shade. The shade could be provided either by a big tree or a house where free air movement exists. In the summer season, due to the natural movement of hot air, the drying of the pods will take place quickly. This facilitates rapid drying and thus shortens the risk-period of invasion by the fungi. Dry the plants till the leaf/pegs become brittle (6-7 days). Repeat drying for 2 or 3 days more after an interval of 2 or 3 days to ensure complete drying. When temperature is very high, avoid direct sun drying. When the bundles are dried, the pods may be detached from the plants and spread in a thin layer under shade for further drying. Exposure to high temperatures of above 45°C may affect the seed quality. The moisture content in unshelled groundnut should be brought down to 8% (in case of seed 6%) for storage.

Storage: Farmers dispose off their groundnut pods within 3 to 4 weeks after harvesting, although some rich farmers store for 3 to 4 months expecting favourable market prices. Pods are stored for 7 to 8 months for seed purpose and stored till the commencement of the next harvesting season for edible purposes. Collect the pods in gunnies and store on the ground over a layer of sand to avoid any moisture coming in contact with dry pods. Pods are dried to bring the moisture to 10% for safe storage. Best storage conditions for normal dry bulk storage of unshelled groundnuts is about 8.0% kernel moisture content at 10°C and 65% relative humidity i.e. moisture in the surrounding air. The expected storage life will be longer with the lower temperature. Temperatures below 13°C inactivate most insects and arrest growth and influence of other seed quality deteriorating factors. The relative humidity (RH) should be between 65% and 70%. Mould growth is encouraged at higher RH levels. However, at RH below 65%, the groundnut pods lose weight and the seeds become brittle and split during handling. Groundnuts always should be stored as pods rather than as kernels. Prevent insect damage to the pods in storage by fumigating with phosphine (use 3-5 Aluminium phosphide tablets (Celphos) for every 100 kg of pods for 7 to 8 days). Do not open the doors of the warehouse for at least a week after fumigation.

Processing: There are traditional and modern technologies in groundnut processing.

1. Village ghani: The *ghani* is driven by bullocks on the principle of a pestle rotating in a mortar. Oil is extracted within the mortar as a result of the friction caused by the revolving pestle. It takes 2 to 3 hours to crush 10 to 25 kg of kernel at a time and extract oil. Oil yielded from *ghani* is generally 5% of the seed weight crushed.

2. **Oil mills**: There will be an expeller which consists of a fight steel worms on a shaft revolving inside a steel cage. The worms are so arranged to produce a gradually increasing pressure on seeds. Oil is expelled through perforations in the steel cage and the cake comes out from the opposite end. The expelled cake has an oil content of about 7%. The usual capacity of a mill is 10 to 15 kg at a time and will be crushed within 30 minutes.

3. **Solvent extraction plants**: In a solvent extraction, plant oil bearing material is brought into contact with a solvent (hexane) to release the entrapped oil. The solvent is percolated several times to achieve complete extraction. This process is extremely useful for oil cakes having very low oil content of 7 to 8%.

Utilization: Groundnut seed contains 40% oil and 26% protein. Groundnut seeds are used for salad purpose. Groundnut seed can be consumed raw, boiled or roasted or crushed for edible oil. Groundnut cake is a valuable cattle feed and can be used as manure. The dried green parts of plants (haulm) are used as fodder. Its haulms are used as animal feed and shells that constitute about 25% of the total pod mass are used as fuel, filler in the feed and fertilizer industries and in manufacture of particle boards, etc. There are four major by-products of groundnut viz., shell, groundnut skins (peels), cakes and haulms. Oil extracted from runner type has best quality due to high concentration of Tocopherol.

REFERENCES

Baldwin John, A and Lee R. Dewey. 1990. Producing high quality seed peanuts. The Cooperative Extension Service Bulletin 1037, The University of Georgia College of Agriculture, Athens, USA. pp 11.

Basu, M.S. and P.S. Reddy. 1989. Technology for increasing groundnut production. National Research Centre for Groundnut (NRCG), Indian Council of Agricultural Research (ICAR), P.O. Timbawadi, Junagadh 362 015, Gujarat. Publications & Information Division, ICAR, New Delhi. pp 18.

Boote, K.J. 1982. Growth stages of Peanut (*Arachis hypogaea* L.). Peanut Science, 9: 34-40

Bunting, A.H., R.W. Gibbons, and J.C. Wynne. 1985. Groundnut (*Arachis hypogaea* L.). In Grain Legume Crops. R.J. Summerfield and E.H. Roberts, eds. pp. 747-800. London: Collins.

Coffelt, T.A. 1989. Peanut. In: Oil Crops of the World: Their Breeding and Utilization. G. Robbelen, R.K. Downey, and A. Ashri, eds. pp. 319-338. New York: McGraw-Hill Publishing Company.

DOD. 2012. Recent Varieties and Hybrids of Annual Oilseeds Recommended for Different States. ICAR & Directorate of Oilseeds Development, Himayatnagar, Hyderabad

Faujdar Singh and D.L. Oswalt. 1995. Groundnut Production Practices. International Crops Research Institute for the Semi-Arid Tropics. Patancheru

Gibbons, R.W., Buntings, A.H. and Smartt, J. 1972. The classification of varieties of Groundnut (*Arachis hypogaea* L.). Euphytica, 21: 78-85.

IBPGR and ICRISAT. 1992. Descriptors for groundnut. International Board for Plant Genetics Resources, Rome, Italy: International Crops Research Institute for the Semi-Arid Tropics, Patancheru, India, pp 125.

Knauft, DA and Wynne, JC. 1995. Peanut breeding and genetics. Advances in Agronomy 55: 393-445

Nigam, S.N., Giri, D.Y. and A.G.S. Reddy. 2004. Groundnut seed production manual. Patancheru 502 324, Andhra Pradesh, India: International Crops Research Institute for the Semi-Arid Tropics. 32 pp.

Pattee, H.E. and Stalker, H.T. 1995. Advances in Peanut Science. American Peanut Research and Education Society, Inc., Stillwater, OK, USA.

Singh, R. P. 2014. Status Paper on Oilseed Crops. Directorate of Oilseeds Development, Hyderabad.

Sivakumar, M.V.K. and Virmani, S.M. 1986. Agroclimatology of Groundnut. International Crop Research Center for Semi-Arid Tropics, Patancheru, India.

Smartt, J. 1994. The Groundnut Crop: A Scientific Basis for Improvement. Chapman and Hall, London, UK.

Syamasonta, M.B. 1990 A Review of 'Pops' Research in Zambia. Fourth Regional Groundnut Workshop for Southern Africa, Arusha, Tanzania, 1990, pp. 17-21. ICRISAT.

Weiss, E.A. 2000 Oilseed Crops. London: Blackwell Science.

Wynne, J.C. and Gregory, W.C. 1981. Peanut breeding. Advances in Agronomy 34: 39-72.

http://www.fao.org/nr/water/cropinfo_groundnut.html

http://pmil.caes.uga.edu/documents/GroundnutAtAGlanceFinalVersion6Jan2015.pdf

CHAPTER

3

Sesame–*Sesamum indicum* L. (2n=26)

Synonyms: *Sesamum orientale* L.

Family: Pedaliaceae

Vernacular name: Gingelly, Sesame, Sesamum, Simsim or Benniseed (English), *Til* (Hindi), *Tili* (Punjabi), *Ellu* (Tamil, Malayalam, Kannada), *Tal* (Gujarati, Punjabi), *Nuvvulu, Manchi Nuvvulu* (Telugu), *Tila / Pitratarpana* (Sanskrit), *Rasi* (Odia), *Huma* (Chinese), *Goma* (Japanese), *Gergelim* (Portuguese) and *Ajonjoli* (Spanish).

Importance: Sesame is called as 'Queen of oilseeds' because of its excellent qualities of the seed, oil and meal. Sesame seeds represent a symbol of immortality in Hindu legends. Oil cake is very good cattle feed. Seed contains 50% oil and 25% protein. The oil contains 47% oleic acid and 39% linoleic acid. Seed oilcake contains 34 to 50% protein. India is the largest exporter of Sesame seed in the World accounting for a share of 23%. The total global exports of Sesame seed were at a level of 1.31 million tonnes during 2012-13(Source: Oil World 2013). China is the largest importer of sesame seed followed by Japan.

Origin: Sesame is native of India and Africa.

Distribution: Sesame is cultivated in India, Pakistan, Myanmar, China, Korea, Thailand, Japan, USA, Sudan, Uganda, Sri Lanka, Egypt, and Ethiopia. In India, it is grown in Uttar Pradesh, Madhya Pradesh, Rajasthan, Odisha, Karnataka, Gujarat and Tamil Nadu.

Botany: Sesame is an herbaceous annual growing 50 to 100 cm height. The stem is erect and quadrangular. The basal leaf is opposite while the upper leaf is alternate. The plant has an unpleasant odour. The extra floral nectar gland is present on peduncle base. The flowers are borne in the axils of leaves. The flowers are tubular, 3 to 5 cm long, with a four-lobed mouth. The flowers may vary in colour with some being white, blue or purple. Sesame is mainly a self-pollinated crop, though cross-pollination takes place to a limited extent of 5 to 6%. The fruit is oblong or ovoid capsule, containing 50 to 100 seeds which are compressed, oblong or obovate. The seeds mature 4 to 6 weeks after fertilization. The fruit naturally splits open (dehisces) to release the seeds by splitting along the septa from top to bottom depending on the varietal cultivar. Each plant may bear 15-20 fruits and each fruit contains 50-100 seeds. The sesame seeds are pear-shaped, overate, and small flatted.

The dimensions of sesame seed are 2.80 mm in length, 1.69 nm in width and 0.82 mm in thickness. Seeds colour varies from pure white, various shades of brown, yellow, red, grey and black. Seed coat may be rough or smooth.

Climate: Sesame is cultivated in regions from 40°N to 40°S latitude up to an attitude of 1500 m from MSL. It is grown in tropical and subtropical regions. It can be grown in areas receiving rainfall of 500 to 650 mm during its growing season. It cannot tolerate heavy rainfall and high humidity. The crop is very sensitive to excessive water in the field. Stagnation of water for long period in the standing crop will completely affect the crop. Alternate spells of wet and dry weather in the pre flowering and flowering phases are beneficial to sesame crop. The growth of sesame roots is very poor. Sesame is a poor utilizer of stored soil moisture. Sesame is basically is a short day plant, but long day types also exist. Critical day length is 12 hour or less. The cardinal minimum, optimum and maximum temperature is 12°C, 25°C to 35°C and 40°C respectively. The minimum temperature for germination is around 12°C. Growth is inhibited from 18 to 20°C. Warm conditions above 23°C favour growth and yield. High temperatures particularly during night, promote stem growth and leaf production. Low temperature at flowering can cause pollen sterility and flower drop. Temperature of 40°C or above at flowering can seriously affect fertilization and number of capsules set. Temperature higher than 45°C completely kills the pollens of sesame. Pollination and the formation of capsules is inhibited during heat wave periods above 40°C. On the other hand, if temperature falls below 20°C for a length of time, germination and seedling growth are delayed and these processes are inhibited at temperatures below 10°C. Sesame needs a constant high temperature, the optimum range for growth, blossoms and fruit ripeness is 25-30°C. If the temperature goes beyond 45°C or less than 15°C there is a severe reduction in yield. If the temperature is more than 40°C with hot winds the oil content reduces. Sesame is frost-susceptible. A killing freeze will terminate the crop and typically dry down the plants in 7 to 10 days. Usually a freeze after physiological maturity does not damage the crop. Commercial varieties of sesame require 90 to 120 frost free days.

Soils: Sesame prefers well drained loamy soil, though it comes up on wide range of soils. The desirable soil pH is 5.5 to 8.0. It does not grow well in acidic, saline or alkaline soils. It is mostly grown under rainfed conditions on marginal and sub-marginal soils.

Seasons: Sesame is cultivated during *kharif* and summer seasons in Peninsular India and only during *kharif* in Northern India. In Tamil Nadu, sesame is grown during June–July and October-November under rainfed condition and during February-March under irrigated condition.

Varieties: The particulars of sesame hybrids/varieties grown in different states of India are presented in Table 1.

Table 1 Sesame hybrids/varieties grown in different states of India

State/Variety	Seed Yield (kg/ha)	Oil content (%)	Duration (days)	Salient characters
Gujarat				
Guj-Til-1	650-700	48-52	86-92	White seed, higher market price
Guj-Til-2	750-800	48-52	88-92	White seed, High seed yield, export quality, medium maturing, higher market price

Guj-Til-3	750-800	48-52	84-88	White and bold seed, high seed yield, higher market price, suitable for export purpose
Guj-Til-4	750-800	48-52	79-83	White seed, early maturing, glabrous, narrow oblong capsules
Guj-Til-10	750-800	48-52	88-92	Black seed, high seed yield
Rajasthan				
RT-46	700-750	48-50	82-85	White seed, medium maturity, export quality, higher market price
RT-54	700-800	43-46	78-80	Light brown seed, high seed yield,
RT-103, RT-125	700-800	48-50	83-88	White seed, medium maturity, high seed yield
RT-127	750-850	50-52	82-86	White seed with high oil content, drought hardy, medium maturing, suitable for export purpose, higher market price
RT-346, RT-351	750-850	48-51	82-86	White seed with high oil content, suitable for export purpose, higher market price
Maharashtra				
Phule Til.1	600-700	49-51	90-95	White seed, tolerant to macrophomina
N-8	600-650	50-51	120-125	Brown tinge seed, resistant to powdery mildew, tolerant to macrophomina and alternaria leaf spot
Tapi (JLT-7)	600-700	48-52	85-90	White bold seed, tolerant to alternaria leaf spot
Padma (JLT-26)	700-750	48-50	82-86	Light brown seed, tolerant to Alternaria and Cercospora
AKT-64	700-750	47-48	85-90	White seed, medium tall, tolerant to macrophomina and phytophthora blight
AKT-101	750-800	48-49	88-90	Oxalic acid below 1 % and FFA below 2 %, tolerant to phyllody, macrophomina and bacterial blight
PKV-NT-11	800-850	48-49	82-92	White seed, tolerant to phyllody, macrophomina and bacterial blight
JLT-408	700-800	51-53	80-85	White bold seed, low FFA content, tolerant to powdery mildew
Madhya Pradesh / Chhattisgarh				
Kanchan (JT-7)	600-700	50-53	84-88	White seed, medium size seed
TKG-21 (JT-21)	650-700	52-54	85-90	White seed, tolerant to bacterial leaf spot and alternaria leaf spot
TKG-22	650-700	50-54	82-85	White seed, tolerant to phytophthora blight
TKG-55 (JT-55)	650-700	50-53	82-85	White seed, tolerant to phytophthora blight
JTS-8	650-700	50-53	82-85	White seed, capsules alternate, non-hairy; flowers hairy, tolerant to macrophomina, alternaria leaf spot and phytophthora blight

Jawahar Til-11 (PKDS-11-Venkat)	650-700	46-50	82-85	Dark brown seed, tolerant to macrophomina
TKG-306	700-800	49-52	86-90	White seed, alternate capsule, leaves alternate, flower blue white hairy, capsules medium hairy and tolerant to phytophthora, phyllody, macrophomina, cercospora, powdery mildew, and alternaria leaf spot
Jawahar Til –12 (PKDS-12)	700-750	48-52	82-85	White seed, tolerant to macrophomina
TKG-308	700-750	46-50	85-90	White seed, tolerant to Phytophthora, macrophomina, cercospora, powdery mildew and alternaria leaf spot
Jawahar Til -14 (PKDS-8)	700-750	50-53	85-88	Black seed, tolerant to capsule borer
Uttar Pradesh/ Uttaranchal				
T-12	650-700	46-50	85-88	White seed, tolerant to phyllody and leaf curl
T-13	600-700	48-52	82-88	White seed, tolerant to lodging
T-78	650-700	46-50	85-90	White seed, tolerant to lodging and leaf curl disease, suitable for export purpose, higher market price
Sekhar	700-800	50-52	85-90	White seed, tolerant to leaf curl, powdery mildew, macrophomina and phytopthora blight
Pragati (MT-75)	700-750	48-52	85-90	White seed, tolerant to leaf curl, powdery mildew, macrophomina and phytopthora blight, good for export, higher market price
Tarun	700-800	52-53	90-95	White seed, moderately resistant to diseases
West Bengal				
Tilottama (B-67)	900-1000 (Summer)	42-44	80-85	Blackish brown seed, tolerant to macrophomina
Rama (Imp.Sel.5)	1000 (Summer)	46-48	85-90	Reddish brown seed, tolerant to macrophomina
SWB-32-10-1 (Savitri)	1100-1400 (Summer)	48-50	84-88	Light brown seed, erect branching type, tolerant to lodging, leaves deep green, flower light pink colour, tetra locular capsules, tolerant to macrophomina
Haryana				
Haryana Til-1	700-750	48-50	85-90	White seed, early maturing, dark green, long and thick leaves, tolerant to leaf curl
Haryana Til-1	700-750	48-50	85-90	White seed, tolerant to phyllody and leaf curl

Andhra Pradesh				
Gauri	650-700	46-48	85-90	Dark brown seed, suitable for early *Kharif and summer*
Madhavi	650-700	46-48	78-82	Light brown seed
Rajeshwari	700-750	48-50	85-90	White seed, tolerant to stem rot and powdery mildew
Varaha (Yel.1)	800-850	50-53	82-85	Dark brown seed, uniform maturity, good for export purpose, higher market price
Gautama (Yel.2)	750-800	50-52	76-80	Light brown seed, uniform maturity, tolerant to alternaria leaf spot
Swetha Til	750-800	50-52	82-86	White seed, determinate, tolerant to powdery mildew, stem rot, leaf curl and macrophomina
Chandana (JCS-94)	800-850	45-48	84-88	Brown seed, tolerant to bacterial blight
Hima	800-850	48-50	80-85	Shiny white seed, long capsules, early in duration, field tolerant to alternaria leaf spot, good for export purpose, higher market price
Odisha				
Kanak	600-700	46-48	85-90	Light brown seed, tolerant to lodging
Kalika	600-700	45-48	85-90	Light brown seed, tolerant to macrophomina
Vinayak	600-650	43-46	85-90	Light brown seed, tolerant to alternaria leaf spot
Uma (OMT-11-6-3)	750-850	42-46	75-80	Pale white seed, tolerant to Macrophomina and phyllody
Usha (OMT-11-6-5)	700-750	43-46	85-90	Light brown seed, tolerant to alternaria alternaria leaf spot
Nirmala (OS-Sel-164)	800-900	42-44	80-85	White seed, tolerant to bacterial leaf spot, powdery mildew and alternaria leaf spot
Prachi (ORM-17)	800-900	42-45	85-90	Black seed, tolerant to cercospora, powdery mildew
Amrit [OSC-24(95)2-1-3]	800-900	43-46	82-85	Light brown seed, tolerant to powdery mildew and alternaria leaf spot
Shubhra	800-900	48-52	78-84	White seed, delayed shattering moderately tolerant to macrophomina and alternaria leaf spot, good for export purpose, higher market price
Smarak	800-900	48-52	80-85	Golden yellow bold seed, delayed shattering, Synchronous maturity, tolerant to macrophomina and alternaria leaf spot
Tamil Nadu				
TMV- 3	650-700	50-52	80-85	Black seed, suitable for all seasons
TMV-4	700-850	48-50	85-90	Brown seed, four loculed capsule
TMV-6	700-950	52-54	85-90	Brown seed, tolerant to drought

CO -1	650-750	50-52	85-90	Black seed, tolerant to macrophomina
Paiyur-1	750-850	50-52	85-90	Black seed, four loculed capsule, , good for export, higher market price
TSS-6 (SVPR-1)	750-850	50-54	75-80	White seed, four loculed capsule, tolerant to alternaria leaf spot
VRI (SV)-1	600-700	50-52	72-75	Dark brown seed, short, four loculed capsule, early maturing
VRI (SV)-2	700-800	50-53	80-85	Reddish brown seed, glabrous, mixed in phyllotaxy, basal branch habit, profuse branching, tolerant to macrophomina
TMV(SV)-7	800-900	48-50	80-85	Brown seed, tolerant to root rot, high protein content (24.5 %), suitable for value addition
Bihar/Jharkhand				
Krishna	700-750	45-48	88-95	Black seed, tolerant to alternaria leaf spot
Jawahar Til-11 (PKDS-11-Venkat)	600-700	46-50	82-85	Dark brown, tolerant to macrophomina
Karnataka				
DS-1	400-500	48-50	95-100	White seed, tolerant to bacterial leaf blight
DSS-9	550-600	49-50	85-90	White bold seeded, early maturing
DS-5	600-700	50-52	90-95	White bold seeded, tall 4-5 branched with long capsules, tolerant to bacterial leaf blight
Kerala				
Thilothama	600-650	48-50	85-90	Brown bold seed, multicapsuled, shy branching
Soma	600-650	44-48	85-95	White seed, tolerant to alternaria leaf spot
Surya	600-650	44-48	85-95	Black seed, tolerant to phyllody
Kayamkulam-1	600-650	48-50	80-85	Brownish black seed, tolerant to drought
Thilak	600-650	48-50	85-90	Blackish brown seed, suitable for both rice fallow and rabi upland, tolerant to drought
Thilathara	600-650	48-52	84-88	Blackish brown seed, tall, shy branching, resistant to powdery mildew
Thilarani	650-750	46-50	82-86	Dark brown seed, semi tall, compact capsule packing, resistant to powdery mildew
Punjab				
Punjab Til-1	650-700	48-52	80-85	White seed
TC-25	700-800	50-52	80-85	White seed
TC-289	700-800	48-52	84-88	White seed, tolerant to macrophomina
Himachal Pradesh				
Brijeshwari (LTK-4)	800-850	48-52	85-90	White bold seed, medium tall with spreading branches, good for export purpose, higher market price

- TKG-21 variety of Madhya Pradesh and Uma variety of Odisha are recommended for Assam.
- Sekhar variety of Uttar Pradesh and Amrit variety of Odisha are recommended for Chhatisgarh.
- PKDS-11 variety of Madhya Pradesh is recommended for Jharkhand.
- Brijeshwari variety of Himachal Pradesh is recommended for Jammu and Kashmir.
- TKG-21, TKG-22 and TKG-55 varieties of Madhya Pradesh, Tilottama variety of West Bengal, GT-10 of Gujarat and Uma variety of Odisha are recommended for North Eastern States.
- RT-46 and RT-125 are recommended for Punjab, Haryana, and Uttar Pradesh besides Rajasthan state.
- RT-54 and RT-103 are also recommended for Gujarat, Maharashtra and Telangana region of Andhra Pradesh.
- RT-346 is also recommended for Haryana, Punjab, Himachal Pradesh, Gujarat and adjoining areas of western Uttar Pradesh, Maharashtra and Karnataka besides Rajasthan state.
- RT-351 is also recommended for Gujarat, Uttar Pradesh, Maharashtra, Haryana, Punjab and Himachal Pradesh besides Rajasthan state.
- TKG-21, TKG-22 and TKG-55 are also recommended for Eastern UP, Bihar, Odisha, West Bengal and North Eastern states besides Madhya Pradesh state.
- Uma is also recommended for Uttar Pradesh, Bihar, West Bengal and North Eastern States, besides Odisha state.
- Amrit is also recommended for West Bengal besides Odisha state.
- Sekhar is also recommended for Haryana and Bihar besides Uttar Pradesh state.
- PKDS-11 is also recommended for Punjab, Haryana, Uttar Pradesh, Himachal Pradesh and West Bengal besides Madhya Pradesh state.
- JTS-8 is also recommended for Uttar Pradesh, Maharashtra and adjoining parts of Andhra Pradesh besides Madhya Pradesh state.
- TMV-3 is also recommended for Karnataka besides Tamil Nadu state.
- TC-25 is also recommended for Vidarbha region of Maharashtra besides Punjab state.

Growth stages: The particulars on growth phases and stages of sesame are furnished in Table 2.

Table 2 Growth phases and stages of sesame

Stage/Phase	**End point of stage**	**DAS**	
Vegetative phase			
Germination	Emergence	0–5	0–5
Seedling	3rd pair true leaf length = 2nd	6–20	6–25
Vegetative stage	First buds	21–32	26-39
Reproductive phase			
Flower bud	First bud	32-35	40-44
Flowering	50% open flowers	38	45-52
Early bloom	5 node pairs of capsules	45	45–52

Mid bloom	Branches/minor plants stop flowering	53	53–81
Late bloom	90% of plants with no open flowers	82	82–90
Ripening	Physiological maturity (PM)	82-91	91–106
Full maturity	All seed mature and dry capsules	91–106	127–146

Growth habit is generally indeterminate, but determinate cultivars are available. Flowers arise in leaf axils on the upper stem and branches, and the node number on the main shoot at which the first flower is produced is a highly heritable cultivar characteristic. Most flowers open at 5–7 a.m., wilt after midday, and are shed at 4–6 p.m. Pollen is released shortly after the flowers open; the interval between flower opening and pollen release is a cultivar characteristic. The stigma is receptive one day before flower opening and remains receptive for another day. Under natural conditions, pollen remains viable for 24 hours. Flowers are mostly self-pollinated, but cross-pollination is possible and may reach 50%. Depending on cultivar, the crop matures in 75–150 days after sowing. Capsules near the stem base normally ripen first, those nearest the tip ripen last.

Land preparation: Land is ploughed well to bring the soil into a fine tilth to facilitate quick germination as the seeds are small.

Seeds and sowing: Seed rate is 5 kg per ha. Seeds are sown at a depth of 3 cm with spacing of 30 × 30 cm. Treat the seeds with Thiram or Carbendazim @ 2 g/kg of seeds before sowing. Seeds are also treated with Trichoderma @ 4 g/kg or *Pseudomonas fluorescens* @ 10g/kg of seeds. This can be done just before sowing. It is compatible with biofertilizers. Such seeds should not be treated with fungicides. Soil application of *Trichoderma viride* @ 2.5 kg/ha (mixed with 100 kg FYM) can be done as basal dressing. The optimum time of sowing and spacing for sesame in different states/ regions are furnished in Table 3.

Table 3 Optimum time of sowing and spacing in different states/ regions.

State	Season	Sowing time	Spacing (cm)
Andhra Pradesh Coastal Telangana	*Kharif* Summer *Kharif*	Second fortnight of May Second fortnight of January Second fortnight of July	30 × 15 30 × 15 30 × 10-15
Gujarat	*Kharif* Semi-*rabi* Summer	Last week of June to second fortnight of July Mid September January-February	45 × 10 45 × 10 45 × 15
Madhya Pradesh/ Chhattisgarh	*Kharif* Semi-*rabi* Summer	First week of July Late August-Early September Second to last week of February	30 × 10-15 30 × 15 30 × 15
Maharashtra	*Kharif* Semi-*rabi* Summer	Second fortnight of June to July Early September February	30 × 15 30 × 15 45 × 15
Rajasthan	*Kharif*	Late June-Early July	30 × 15
Odisha	*Kharif* *Rabi* Summer	June-July September-October February	30 × 15 30 × 15 30 × 15
Uttar Pradesh/ Uttaranchal	*Kharif*	Second fortnight of July	30-45 × 15

Bihar/ Jharkhand	*Kharif*	July	30 × 15
West Bengal	Summer	February-March	30 × 15
Tamil Nadu	*Kharif* *Rabi* Summer	Second fortnight of May to Second fortnight of June November-December Second fortnight of January to March	22.5 × 22.5 22.5 × 22.5 30 × 10
Karnataka North South	 *Kharif* Early *Kharif*	 June-July April-May	 30 × 15 30 × 15
Assam	*Kharif*	July-August	30 × 10-15
Punjab/Haryana	*Kharif*	Second fortnight of July	30 × 10-15
Kerala	*Kharif* Summer	August December	30 × 10-15 30 × 15

After cultivation: Thinning is carried out to a spacing of 15 cm between plants on 15 DAS and 30 cm on 30 DAS.

Weed management: The critical crop weed competition period in sesame is up to 40 DAS. Pre-plant incorporation of Fluchloralin @ 1 kg a.i./ha or pre-emergence application of Pendimethalin @ 1 kg a.i./ha or Alachlor @ 1.25 kg per ha on 3 DAS which is followed by one hand weeding and hoeing at 30 days after sowing are recommended to control weed growth. Subsequently one hand weeding or mechanical weeding can be given on 30 DAS. Otherwise two hand weedings are given on 15 DAS and 30 DAS.

Growth regulators: Spray salcylic acid @ 2% (20 g/100 litres of water) on 30 and 50 DAS to increase seed yield.

Manures and fertilizer application: Apply FYM or compost coir pith @ 12.5 tonnes per ha as basal dressing. Fertilizer dose of 23-13-13 and 35-25-25 kg of N, P_2O_5 and K_2O per ha are recommended for rainfed and irrigated conditions respectively. Foliar spray of 2 % urea, DAP at flowering and capsule development stage is also recommended. Apply 5 kg of $MnSO_4$ per ha at the time of sowing. 25% of nitrogen can be substituted by 3 packets of azosprillium by seed treatment and 10 packets (2 kg) of azosprillium per ha as soil application. The recommended fertilizer doses for sesame in different states are furnished in Table 4.

Table 4 Recommended fertilizers doses in different states

State/ Situation	N:P:K (kg/ha)	Specific recommendation
Andhra Pradesh Coastal region Telangana region	 40-40-20 30-30-20	 ------
Gujarat *Kharif* Semi-*rabi*	 30-25-0 25-25-0	 Apply sulphur 20-40 kg/ha
Madhya Pradesh/ Chhattisgarh Rainfed Summer	 40-30-20 60-40-20	 Apply 25 kg/ha zinc sulphate once in three years in zinc deficient soils.
Maharashtra	50-0-0	Half N at 3 weeks after sowing and remaining half 6 weeks thereafter.

Rajasthan Heavy soils Light soils	 20-20-0 40-25-0	 For areas with less than 350 mm rainfall For areas with more than 350 mm rainfall.
Odisha	30-20-30	-----
Tamil Nadu Irrigated Rainfed	 35-23-23 25-15-15	 Apply full dose of N, P_2O_5, K_2O as basal. Seed may be treated with *Azospirillum*.
Uttar Pradesh/ Uttaranchal	20-10-0	-----
Haryana	30-0-0	------
Bihar/ Jharkhand	40-40-0	------
West Bengal Irrigated Rainfed	 50-25-25 25-13-13	 No fertilizer if sown after potato.
Assam	30-30-20	Entire as basal
Kerala	30-15-30	N may be applied 75% as basal + 25 % as foliar 2% urea, 30-35 days after sowing.
Karnataka	37.5-25-25	Half N + full P_2O_5 and K_2O as basal Remaining half N at 30-35 DAS

Water management: Sesame requires 300 mm of water. Beds and channel method of irrigation has been found to be the best method to provide uniform moisture availability throughout the field. Irrigation immediately after sowing followed by another at flowering and then at pod formation, consumes only 150 mm of water. With one rain during the growing period, the total water requirement does not exceed 200 mm. The critical stages of crop water requirement are flowering (30 to 35 days) and pod development (45 to 65 days). Stress should be avoided during these periods. Otherwise there will be reduction in yield. Sesame is highly sensitive for water stagnation and excessive irrigation is harmful to the crop. It is drought tolerant, requiring 1/4 the water for corn, 1/3 the water for sorghum and 1/2 the water for cotton. Approximately 80-90 percent of sesame grown in the country is in dry land and 10-20 percent has supplemental irrigation.

Cropping systems: The sequence cropping and intercropping systems with sesame in different states of India are presented in Table 5 and Table 6. Sesame is a short duration crop and fits well into a number of multiple cropping systems either as a catch crop or a sequence crop.

Table 5 Sequence cropping with sesame in different states of India

State	Crop sequence
Andhra Pradesh	Rice-Groundnut-Sesame, Sesame-Horse gram, Ragi/Sorghum/Horse gram (Early)-Sesame, Sesame- Upland Rice.
Bihar	Early Rice -Potato-Summer Sesame/Moong bean *Kharif* Sesame-Maize/Pigeon pea/Rabi gram Wheat-Summer Sesame/Moong bean
Gujarat	Sesame-Wheat/Mustard
Karnataka	Sesame-Horse gram/Chickpea
Madhya Pradesh	Cotton-Sesame-Wheat, Rice -Summer Sesame, Sesame-Wheat

Maharashtra	Sesame (Early)-Rabi Sorghum/Safflower Cotton-Sesame-Wheat.
Odisha	Rice/Potato-Sesame, *kharif* Sesame-Maize/Pigeon pea/ Rabi Gram
Rajasthan	Sesame-Wheat/Moong bean/Barley
Tamil Nadu	Rice/Groundnut-Sesame, Sesame-Urd bean, Sesame-Rabi Sorghum, Sesame-Moong bean, Cowpea-Sesame.
Uttar Pradesh	Sesame (Early)-Gram/Rapeseed-Mustard/Lentil/Pea
West Bengal	Potato-Sesame (Late Jan./Early Feb) Rice – Sesame

Table 6 Inter cropping systems for different states.

State	**Intercropping system**
Gujarat	Sesame+Groundnut / Urd bean (3:3) Sesame+Pearl millet / Cotton (3:1)
Karnataka	Sesame+Groundnut (1:4)
Madhya Pradesh	Sesame+Moong bean / Urd bean (2:2 or 3:3) Sesame+Soybean (2:1 or 2:2)
Maharastra	Sesame+Pearl millet / Urd bean (3:1)
Odisha	Sesame+ Summer Groundnut (2:3) Sesame+Moong bean/Urd bean (2:2)
Rajasthan	Sesame+Pearl millet / Moth bean (1:1)
Tamil Nadu	Sesame+Moong bean / Urd bean (3:3) Sesame+Pigeonpea (3:1), Sesame+Groundnut (2:4)
Uttar Pradesh	Sesame+Moong bean (1:1), Sesame+ Pigeon pea (3:1)
West Bengal	Sesame+ Groundnut (1:3 or 2:2)

Crop protection: The pest management practice of sesame is presented in Table 7.

Table 7 Pest management practices of sesame

Pests	**Management strategies**
Shoot and Leaf webber *Antigastra catalaunalis*	Two sprayings with neem formulation 0.03% Spray any one of the following :
Pod borer *Elasmolomus (= Aphanus) sordidus*	Carbaryl 50 WP 1000 g/ha in 500 litre of water Neem seed kernels extract (5%). Neem oil 2% (two rounds)
Gall fly *Asphondylia ricini*	Use alternate insecticides each time and avoid the usage of same insecticide every time.
Leaf webber *Antigastra catalaunalis*	Dust any one of the following : Carbaryl 10% DP @ 25 kg/ha
Whitefly Leaf hopper	Spray any one of the following : Methyl demeton 25% EC @ 1200 m l/ha Quinalphos 25%EC @ 2000 ml/ha
Storage pests *Triboilum castaneum* *Corcyra cephalonica*	Dust any one of the following on gunny : Malathion 5 D, Carbaryl 10 D Mix one kg of activated clay with 100 kg of seeds after adequate drying of seeds

Seed treatment is done with any one of the following *P. fluorescens* @ 10g/kg of seed, *T. viride* @ 4g/kg of seed, Thiram @ 4g/kg of seed, Carbendazim @ 2g/kg of seed. The diseases management practices of sesame is presented in Table 8.

Table 8 Diseases management practices of sesame

Disease	Management
Powdery mildew: *Erysiphe cichoracearum*	Apply any one of the following Sulphur dust 25 kg/ha Wettable sulphur 25 kg/ha
Alternaria blight: *Alternaria sesami*	Spray Mancozeb 1000g/ha
Cercospora leaf spot: *Cercospora sesami*	Spray Mancozeb 1000g/ha
Root rot: *Macrophomina phaseolina (Rhizoctonia bataticola)*	Soil application of *P. fluorescens* or *T. viride* – 2.5 Kg / ha + 50 Kg of well decomposed FYM or sand at 30 days after sowing. Spot drench Carbendazim – 1 gm/ litre
Phyllody: Phytoplasma Vector: *Orosius albicinctus*	Remove and destroy infected plants. To control vector, spray Monocrotophos 36 or Dimethoate 30 EC 500 ml/ha combined with intercropping of Sesamum + Redgram (6 : 1)

Harvest: Symptoms of maturity in sesame show 25% of bottom leaves shed and top leaves turn yellow. The capsule colour turns yellow upto the middle. Harvesting can be done before the bottom capsule turn brown. Open the tenth capsule from the bottom, if the seeds turn black, harvest can be taken up for black seeded varieties. If harvest is delayed, the capsules will shatter the seeds resulting in yield reduction. Cut the plants at the bottom with sickle. Stack in the open areas one over the other in a circle with the stems pointing out and the top portion pointing inside. Cover the top with straw so that moisture and temperature increase. Cure it for 3 days. Shake the plants. All the mature seeds will fall off. If necessary turn the sun dried plants upside down and beat the plants to shed seeds. Winnow the seeds and dry in the sun for 3 days. Stir once in 3 hours to give uniform drying. Collect the seeds and store it in gunnies. In sesame, harvest index varies from 15 to 20% and possibilities exist to double with improved plant types.

Utilization: Sesame seeds approximately contain 50% oil and 25% protein. The sesame oil contains about 47% oleic and 39% linoleic acid. Sesame is rich in sulfur containing amino acids and limited in lysine and contains oxalic (2.5%) and phytic (5%) acids. Sesame seeds have the highest total phytosterol content (400-413 mg/100g). Sesame seeds are used to make cakes, wine and brandy, bread stick, bread, cakes, confectionery, sesame seed buns, seed paste (tahini), chips, cracker, salad, cooking oil, roasted seed, soup, spice and seed oil. The oil having 85% unsaturated fatty acids has reducing effect on cholesterol and prevent coronary heart diseases. In general, the paler varieties of sesame seem to be more valued in West and Middle East, while the black varieties are prized in the Far East. The oil extracted from black seeds is best suited for medicinal purposes. Sesame oil is used as a solvent, oleaginous vehicle for drugs, skin softener, and used in the manufacture of margarine and soap. Sesame oil is exceptionally resistant to rancidity. The oil fraction shows a remarkable stability to oxidation. This could be attributed to endogenous antioxidants namely lignins and tocopherols. The stability to oxidation can be attributed to its endogenous anti-oxidant lignans along with tocopherols. The sesamin, episesamin, sesaminol and sesamolin are major lignans (a non-fat constituent) constituents of sesame oil and all have chemically methylenedioxyphenyl group. Sesamin and sesamol are responsible for high stability of oil at room and frying temperatures. Sesamol has been found to be an antioxidant which inhibit the development of rancidity in the oil and thereby prevents the spoilage of oils. Sesamol and sesaminol substances belong to lignans and g -tocopherol have been shown to possess cholesterol-lowering effect in humans and to prevent high blood pressure and increase vitamin E supplies in animals. Sesame lignans such as sesamin, episesamin, and sesamolin and g -tocopherol play important roles in plant defense, such as antifeedant as well as potent antioxidants and insecticides. Many health benefits of sesame may be attributed to its lignans, especially sesamin. The refined sesame oil is rich with antioxidant components like lignans allowing for greater shelf-life of foods plus improving their flavour and taste. Sesamin has been found to protect the liver from oxidative damage. The oil has been used for healing wounds for thousands of years. It is naturally antibacterial for common skin pathogens such as *Staphylococcus* and *Streptococcus* as well as common skin fungi such as athlete's foot fungus. It is anti-viral and anti-inflammatory. Sesame oil is used after exposure to wind or sun to calm the burns. It nourishes and feeds the scalp to control dry scalp dandruff and to kill dandruff causing bacteria. It has been successfully used in the children's hair to kill lice infestations. It protects the skin from the effects of chlorine in swimming pool water. Sesame oil helps joints keep their flexibility. It keeps the skin supple and soft. It heals and protects areas of mild scrapes, cuts and abrasions. It helps tighten facial skin, particularly around the nose and controls the usual enlargement of pores as skin ages.

REFERENCES

Bedigian, D., 2003. Evolution of sesame revisited: domestication, diversity and prospects. Genetic Resources and Crop Evolution 50(7): 773–778

Bedigian, D., 2004. History and lore of sesame in Southwest Asia. Economic Botany 58(3): 329–353.

Bedigian, D., Seigler, D.S., and Harlan, J.R. 1985. Sesamin, sesamolin and the origin of sesame. Biochemical Systematics and Ecology 13: 133–139.

Bedigian, D., Smyth, C.A. and Harlan, J.R. 1986. Patterns of morphological variation in sesame. Economic Botany 40: 353–365

Joshi, A.B. 1961. Sesamum. The Indian Central Oilseeds Committee, New Delhi, India. p. 109.

Nayar, N.M. and K.L. Mehra. 1970. Sesame: Its uses; botany, cytogenetics, and origin. Econ. Bot. 24:20-31.

Ranganatha, A.R.G. 2013. Improved Technology for Maximizing Production of Sesame. All India Coordinated Research Project on Sesame and Niger, Indian Council of Agricultural Research, JNKVV Campus, Jabalpur-482004

Singh, R. P. 2014. Status Paper on Oilseed Crops. Directorate of Oilseeds Development. Hyderabad. pp 118-135.

https://www.hort.purdue.edu/newcrop/proceedings1990/V1-225.html

CHAPTER

4

Rapeseed and Mustard

Family: Cruciferae / Brassicaceae

Vernacular name: Rapeseed/canola is known as *Toria, Sarson*, *Rai* (Hindi), *Ava* (Telugu) and *Sarisha* (Bengali) while Mustard is known as *Kalisarason* (Hindi), *Kadugu* (Tamil), *Ava* (Telegu), *Kalimohri* (Marathi), *Kari sasava rai* (Kannada) and *Rai* (Oriya). Rapeseed is also known as canola throughout the world. Canola is the term used to describe edible oil cultivars that have low erucic acid oil and low ucosinolate meal. In common Indian language, 'Raya' refers to mustard while sarson, toria and Taramira are rapeseed. The word 'rape' comes from the Latin word 'rapum', means 'turnip'. On the other hand, the word 'mustard' is derived from Latin word mustum' or 'must', which denotes 'expressed juice of grapes and 'ardens' means 'hot and burning'.

Importance: Rapeseed (*Brassica napus* L.) is one of the most important oilseed crops in the world. India is the third largest rapeseed-mustard producer in the world after China and Canada. India and China are the two major consumers of Rape/Mustard seed. *B. rapa* (rapeseed) is grown as leafy vegetable (China and Japan), oilseed (India) and as turnip-rape (Europe). Rapeseed and mustard is one of the most important winter season (*rabi*) oilseed crops of the country. Rapeseed and mustard come next importance to groundnut in India. The average oil content ranged from 40 to 49% depending on cultivars. Canola oil is widely used as a cooking oil, salad oil, and in making margarine. It is appealing to health-conscious consumers because it has the lowest saturated fat content of all major edible vegetable oils. The alkaloid 'Glucosinolate' give rise to the hot principle and pungent taste of mustard. The presence of anti-nutritional factor 'Glucosinolate' in the mustard cake render it unsuitable as source of human protein and is used as a manure and cattle feed. The refined oil is called 'Colza' is used in Europe. Rapeseed can be divided into two types viz., industrial rapeseed or canola. Visually, the seeds of the two types are identical. The distinguishing difference between the two types is their individual chemical or fatty acid profiles. Generally, canola refers to the edible oil crop that is characterized by low erucic acid (less than 2%). 'Industrial rapeseed' refers to any rapeseed with a high content (at least 45 percent) of erucic acid in the oil. Canola is Canadian Oil Low Acid, which has eurucic acid levels (unpalatable for food) of less than 2%. The term 'canola' is a registered trademark of the Canadian Canola Association. The '**Can**' part stands for **Can**ada and '**ola**' refers to oil and continue to claim it stands for **Can**(ada) + **o**(il) + **l**(ow) + **a**(cid). Canola cultivars normally produce oils that contain less than 4% linolenic acid (18:3) and/or greater than 70% oleic acid (18:1). Canola is produced extensively in Europe, Canada, Asia, Australia, and to a more limited extent here in the United States. Canola oil is widely used as a cooking oil, salad oil, and in making margarine. It is appealing to health-conscious consumers because it has the lowest saturated fat content of all major edible vegetable oils. Canola oil (or rapeseed oil) contains both Omega-6 and Omega-3 fatty acids in ratio of 2:1 and is only second to flax oil in Omega-3 fatty acid. It is one of the most heart-healthy oils and has been reported to reduce cholesterol levels, lower serum tri-glyceride levels, and keep platelets from sticking together. 'Canola' quality varieties are commonly developed from either of two species of *Brassica napus* and *Brassica campestris.* The characteristics of canola Include i) higher yield with < 2% erucic acid, ii) relatively shorter duration of the crop, iii) perceived as a healthy cooking medium having less than 2 per cent erucic acid, iv) demand as livestock feed as oil meal contains less than 30 micromoles glucosinolates per gram of defatted meal, v) used as salad oil for it's light colour and texture and vi) used in baking industry (reduces the saturated fatty acid intake, modifies the texture of baked product by making it more moist and softer).

Origin: The different types of mustard- rapeseed have different centres of origins as follows:

(i) *Brassica campestris* var. *brown sarson* (Brown sarson) has originated from Eastern Afghanistan, together with the adjoining north-western India and Pakistan.

(ii) *Brassica campestris* (Yellow sarson) has originated from eastern part of India.

(iii) *Brassica juncea* L. (Indian mustard or Rai) has originated from China to India via North Eastern India and spread to Afghanistan via Punjab.

(iv) *Brassica napus* (Gobhi sarson) has originated in the South-West Europe and Mediterranean region.

(v) *Brassica carinata* has originated in East Africa.

(vi) *Brassica nigra* (Black or true mustard or *Banarsi rai*) is native of Eurasia.

(vii) *Eruca sativa* (Taramira) is native to South Europe and North Africa.

Distribution: Rapeseed and mustard are grown in India, Pakistan, Bangladesh, Australia, Canada, China, Northern Europe, and the United States. In India, it is cultivated in Uttar Pradesh, Rajasthan, Madhya Pradesh, Haryana, Punjab, Odisha, Assam, West Bengal, Bihar and Gujarat. Yellow sarson is mainly grown in Assam, Bihar, Northeastern states, Odisha, Eastern Uttar Pradesh and West Bengal while brown sarson is cultivated in limited areas of Eastern Uttar Pradesh. Toria is a short duration crop cultivated largely in Assam, Bihar, Odisha and West Bengal in the east mainly as winter crop. Taramira is grown in the drier parts of north-west India comprising the states of Rajasthan, Haryana and Uttar Pradesh. Gobhi sarson is a long duration crop confined to Haryana. Himachala Pradesh and Punjab. Rajasthan, Uttar Pradesh, Haryana and Madhya Pradesh are the major rapeseed-mustard growing States contributing about 77% and 82% of the total rapeseed-mustard area and production of the country. More than 44% cultivated area under rapeseed-mustard is under rainfed condition.

Botany: Rapeseed and mustard are tall herbaceous annuals, growing 1.0 to 1.5 metres; stem glabrous or glaucous; leaves of 2 types-the lower ones known as the stem leaves and the upper arising on the axis of the inflorescence known as floral leaves. Inflorescence is a long raceme, corymbose fruit a siliqua; the seeds are either black or white depending on the varieties. Mustard (*B. juncea*) is self-fertile while Rapeseed (*B. campestris*) is cross fertile. The cross pollination is through the agency of bees. Indian rape seed (*B. campestris*, 2n=20) is self sterile in nature. It is an important oilseed crop of North India. There are three cultivated types viz., *B. campestris* var. *brown sarson* – (Brown sarson), *B.campestris* var. *yellow sarson* - (yellow sarson) and *B.campestris* var. *toria* – *(Toria)*. European rape seed (*B. napus,* 2n=38) is a self-fertile. The fruit is a siliqua. The pods are two-valved, three-valved or four-valved, depending upon the number of carpels in the ovary. The flowers begin to open from 8 a.m. and continue up to 12 noon. Oil is pungent due to presence of glucoside sinigrin mostly used as condiment. Leaves are used in herbal medicines. White mustard or *Ujli sarson* (*B. alba,* 2n=24) young seedlings are used as salad. Seeds are yellowish in colour and contain 30% oil. The classification of rapeseed and mustard is presented in Table 1.

Table 1 Classification of Rapeseed and Mustard

S. No.	Botanical name	Common name	Local name	Chromosome number (2n)	Characteristics of seeds
1	*Brassica juncea* (L.) Czern & Coss.	Indian mustard or Brown sarson	*Rai, laha, Raya,* ryada, lahta , sasve, herbo,	36	Seeds are medium sized, round and dark brown or black in colour
2	*Brassica nigra* (L.) Koch.	Black mustard	*Banarasi Rai*	16	-------
3	*Brassica alba* (L.) Rabenh	White mustard	*Ujli sarson*	24	-------
4	*Brassica campestris* L. *oleifera* var *toria*	Indian rape seed	*Toria* tori, lahi	20	Seeds are dark brown, bold and large sized
4.1	*Brassica campestris* L. spp *oleifera* var *yellow sarson* Prain	Turnip rapeseed; Indian colza, colza, rape	*Yellow sarson, Pilli sarson*	20	-------
4.2	*Brassica campestris oleifera* var *brown sarson*	Turnip rapeseed	*Brown sarson*	20	-------
5	*Brassica rapa*	Canola/ rapeseed	*Sarsan,* Brown sarson,	20	Seeds are light reddish in colour , bold, large sized
6	*Brassica napus ssp oleifera var annua*	Gobhi sarson	Gobhi sarson	38	Seeds are brownish black and large sized
7	*Brassica carinata A. Br.*	Abyssinian mustard/ Ethiopian mustard	Karan rai	-------	Seeds are small, round and in reddish brown

8	*Eruca vesicaria* L. spp. *Sativa (*Mill.) Thell Syn: *Brassica eruca* L. / *Eruca sativa*	Rocket Cress	*Taramira or tara*	22	Seeds are light reddish brown coloured and ovoid shape

Climate: Rapeseed and mustard grow well in a wide range of climatic conditions. These are basically temperate crops and require somewhat cool weather for satisfactory growth. They require cool and dry weather with plentiful bright sunshine. The cardinal minimum, optimum and maximum temperature requirements for growth and yield of rapeseed and mustard are 10°C, 23-30°C and 35°C respectively. High temperature at maturity reduces the seed yield. Mustard suffers more to frost than rapeseed. Both are long day plants. Both are sensitive to drought condition and water logging. Rapeseed and mustard are cultivated in areas receiving mean annual rainfall of 700 to 1000 mm. Sarson and toria are preferred in low-rainfall areas, whereas raya and toria are grown in medium and high-rainfall areas, respectively.

Soils: Well drained, fertile sandy loamy to clay loamy soils having neutral pH is suitable for rapeseed and mustard cultivation. These tolerate to moderate salinity. The rapeseed and mustard thrive best in light to heavy loams. Excessive alkaline or acidic soils are not suitable for cultivation of these crops. Raya may be grown on all types of soil, but toria does best in loam to heavy loam. Sarson is suited to light-loam soil and taramira is mostly grown on very light soils.

Land preparation: A fine seed-bed is required to ensure good germination. In irrigated areas, the first ploughing is done with a medium sized soil-turning plough, followed by two to four ploughing with a desi plough or a cultivator. Sohaga (Planking) is given after every ploughing. In rainfed areas, one to two ploughings with a desi plough or a cultivator, each followed by planking, may be given. Toria, in particular, requires a fairly moist seed-bed for good germination, but excessive moisture should be avoided. Gypsum should be added before sowing in salt affected areas.

Varieties: The varieties of Rapeseed-Mustard are furnished in Tables 2 to Table 10.

Table 2 Indian mustard (*Brassica juncea* L)

Variety/ hybrid	Yield potential (kg/ha)	Oil content (%)	Recommended states/ region/ situations	Specific features/traits
Aravali Mustard	1265	42	Haryana, Punjab and Rajasthan	Suitable for rainfed conditions
RCC 4	1000-1200	39-40	Himachal Pradesh	Compact plant type
Basanti	1179-1226	40-42	Uttar Pradesh	Yellow seeded, white rust resistant
NDRE 4	800-1000	40	Uttar Pradesh	Early maturing
Urvashi	1600-1700	39	Uttar Pradesh	Tolerant to high temperature at seedling stage
Geeta (RB 9901)	1773	40-42	Delhi, Haryana, Punjab and South Rajasthan	Suitable for rainfed conditions

Kanti	1458-2068	40-42	Uttar Pradesh	Tolerant to high temperature at seedling stage
Maya (RK 9902)	2500-2900	39-40	Madhya Pradesh and Uttar Pradesh	Suitable for irrigated conditions, resistant to white rust
RGN 13	2200	41-43	Rajasthan	Suitable for normal sown irrigated conditions
Swaran Jyoti (RH 9801)	1377	39-43	Haryana, Madhya Pradesh, South Rajasthan and Uttar Pradesh	Suitable for irrigated late sown and frost conditions
Vasundhra (RH 9304)	2109	38-40	Haryana, Madhya Pradesh, South Rajasthan, Uttar Pradesh and Uttranchal	Suitable for normal sown irrigated conditions
CS 54 (CS 614-4-1-4)	1932	39-41	Haryana, MP, Rajasthan, UP and Gujarat	For salt affected soils
Ashirwad (RK-01-03)	1450-2358	31- 41	Madhya Pradesh, Rajasthan, Uttar Pradesh, and Uttaranchal	Moderately resistant at leaf and pod stage for *alternaria*blight and resistant for White rust. Suitable for irrigated late sown conditions
Jawahar Mustard 2	1632-2588	38-41	Madhya Pradesh and Chhattisgarh	White rust resistant
Jawahar Mustard 3	1500-2500	40	Madhya Pradesh	Tolerant to A*lternaria* blight
Narendra Swarna Rai 8	1681-2211	36-45	Uttar pradesh	Yellow seeded, high oil content
Pusa Karishma	1731-2506	37-38	Delhi	Low erucic acid, yellow seeded
Pusa Mahak (JD-6)	597-1049	39-44	Orissa, WB, Bihar, Jharkhand, Chhattisgarh, Assam	An early maturing, for rainfed areas
Shivani	653-930	37-41	Jharkhand	Suitable for rainfed
Gujarat Mustard 3	Rainfed :901-1362 Irrigated : 1673-2317	35-40	Gujarat	Tolerant to high temperature and salinity
RGN-48	1692-2924	39- 41	Haryana, Punjab and Rajasthan	Suitable for rainfed and frost conditions
Navgold (YRN-6)	1253-1803	39-40	Punjab, Haryana	Yellow seeded and suitable for late sown condition

RRN 505	1200-1400	40	Rajasthan	Suitable for late sown conditions
PBR 210	2080-2532	38-39	Punjab	Early maturity in Punjab conditions
RLC 1(ELM 079)	1600-2000	38	Punjab	Low erucic acid
NRCDR-2	2213	39	Haryana, Punjab, Jammu, parts of Rajasthan & Delhi	Suitable for Irrigated conditions
Pusa Mustard-21 (LES 1-27)	2111	34.0 – 40.0	Delhi, Haryana, Jammu & Kashmir (Plains) Punjab, Rajasthan, Western U.P.	Low erucic acid (<2%)
RGN-73	2006	40	Uttar Pradesh, Uttaranchal, Madhya Pradesh and parts of Rajasthan	Suitable for Irrigated, frost conditions
Shatabdi (ACN 9)	468-1291	32-40	Maharashtra	Suitable for timely and late sown conditions
Pusa mustard 22 (LET 17)	2007	35.5	Haryana, Punjab, Jammu, parts of Rajasthan and Delhi	Suitable for Irrigated conditions, low erucic acid variety
TPM 1	1127-1682	34-39	Maharashtra	Yellow seeded
Pusa Vijay (NPJ 93)	1870-2715	35-41	Delhi	High temperature tolerant at seedling stage and salinity
CS 56 (CS-234-2)	1170-1423	34.2-38.0	Haryana, Punjab and parts of Rajasthan	Suitable for late sown conditions, salt tolerant, 1000 seed weight more than 6 g.
Pusa mustard-24 (LET-18)	1241-2904	32.0-39.7	Haryana, Punjab, New Delhi and parts of Rajasthan	Low erucic acid (<2%)
Dhara mustard hybrid 1 (DMH 1) Hybrid	1782-2249	38 – 42	Delhi, Haryana, Punjab, J & K and Rajasthan	High pod density, resistant to white rust
NRC HB 101	1382-1491	35- 42	Madhaya Pradesh, Uttar Pradesh, Uttarakhand and Eastern Rajasthan	Suitable for late sown irrigated conditions
NRCHB506(Hybrid)	1550-2542	39- 43	Madhaya Pradesh, Uttar Pradesh, Uttarakhand and Eastern Rajasthan	High adaptation
RB 50	846-2425	39- 40	Delhi ,Haryana,Punjab, Jammu& Kashmir, Rajasthan	Suitable for rainfed conditions
RGN 145	1448-1640	35- 39	Delhi ,Haryana,Punjab, Jammu& Kashmir, Rajasthan	Suitable for late sown irrigated conditions

Pusa mustard 25 (NPJ 112)	1324-1654	36-41	Delhi ,Haryana,Punjab, Jammu& Kashmir, Rajasthan, Western Uttar Pradesh	Suitable for early sown irrigated conditions, high temperature tolerance at juvenile stage
Pusa Tarak	1852-1996	38-42	Delhi	High temperature tolerant and bold seeded
Coral PAC 432 (Hybrid)	1831-2581	40-42	Uttar Pradesh, Uttarakhand and Rajasthan	Hybrid
NRCDR 601 (DMR 601)	1939-2626	39-42	Delhi, Haryana, Punjab, Jammu & Kashmir, Rajasthan	Tolerant to salinity and high temperature at sowing time
Pusa Mustard 26 (NPJ 113)	1481-1895	30-41	J&K, Punjab, Haryana, Rajasthan, Delhi & U.P.	Suitable for late sown irrigated conditions in rabi season.
Pusa Mustard 27 (EJ 17)	1437-1659	40-45	U.P., M.P., Uttarakhand & Rajasthan	Suitable for early sown irrigated conditions & for multiple cropping

Table 3 Karan rai (*Brassica carinata*)

Variety/ hybrid	**Yield potential (kg/ha)**	**Oil content (%)**	**Recommended states/ region/ situations**	**Specific features/ traits**
Pusa Swarnim (IGC-01)	1567	42-43	Delhi, Haryana, Himachal Pradesh and Punjab	Suitable for irrigated and rainfed conditions
Pusa Aditya	1400	38.2-41.2	Delhi	Suitable for rainfed conditions

Table 4 Taramira (*Eruca sativa*)

Variety/ hybrid	**Yield potential (kg/ha)**	**Oil content (%)**	**Recommended states/region/ situations**	**Specific features/ traits**
Karan Tara (RTM-314)	1050	34-35	Harayana, Punjab, Rajasthan and Uttar Pradesh	For rainfed areas
Narendra Tara (RTM-2002)	1001-1115	37-38	Rajasthan	For rainfed areas
Vallabha Taramira1 (PUT 93-11)	616-1133	38-40	Uttar Pradesh	Moderately resistant to alternate blight and aphid
Vallabha Taramira2 (PUT 93-1)	500-1283	38-39	Uttar Pradesh	Resistant to white rust, moderately resistant to alternate blight and aphid

Table 5 Brown sarson (*Brassica rapa* var brown sarson)

Variety/ hybrid	Yield potential (kg/ha)	Oil content (%)	Recommended states/ region/ situations	Specific features/traits
KBS 3	900-1200	44	Himachal Pradesh	Resistant to white rust and tolerant to cold

Table 6 Toria (*Brassica rapa*)

Variety	Yield potential (kg/ha)	Oil content (%)	Recommended states/ regions/situations	Special features/traits
Parbati	1380	42	Orissa	For rainfed areas
Anuradha	1460	44	Orissa	For rainfed areas
VL Toria-3	769-1106	39- 41	Uttarakhand Hills	-
Uttara	1000	42	Uttarakhand plains	Moderately resistant to white rust, downy and powdery mildew

Table 7 Yellow Sarson (*Brassica rapa* var yellow sarson)

Variety	Yield potential (kg/ha)	Oil content (%)	Recommended states/regions/ situations	Special features/traits
Ragini (MYSL-203)	1567	44	Suitable for irrigated conditions	Bihar, Eastern Uttar Pradesh and West Bengal
NRCYS 05-02	1239-1715	38.2-46.5	Early maturity, medium height and high oil content	Yellow sarson growing areas of the country
YSH 0401	1273-1651	43-45	Bold seeded	Yellow sarson growing areas of the country
PYS-1 (Pant Pili Sarson)	1050-1163	42-44	For irrigated areas	Uttaranchal

Table 8 Gobhi Sarson (*Brassica napus*)

Variety	Yield potential (kg/ha)	Oil content (%)	Recommended states	Special features/traits
Hyola-401 (Hybrid)	1200-1640	42	Haryana, Himachal Pradesh and Punjab	Double low (low erucic acid, 0.8 % and low glucosinolate content, 15 micromoles/g defatted seed meal)
Neelam	1570	43	Himachal Pradesh	Suitable for late sown conditions

GSC-5	1719-2390	37-43	Punjab	Low erucic (< 2%) and low glucosinolate (26-41 micromoles/g defatted seed meal)
TERI –Uttam-Jawahar [TERI (00) R 9903]	1619-2685	43-45	Madhya Pradesh	Low erucic acid (< 2%) and low glucosinolate (12.2 micromoles/g defatted seed meal)
GSC-6 (OCN-3)	1795	39.0	Punjab, Haryana and Jammu	Irrigated areas
NUDB-26-11	984-1339	38-42	Himachal Pradesh, Jammu &Kashmir	Low erucic and low glucosinolate content, suitable for irrigated conditions.
Him Sarson 1 (ONK 1)	693-1789	38-42	Himachal Pradesh, Jammu &Kashmir	Low incidence of *Alternaria* leaf blight and *Sclerotinia* stem rot, resistance to white rust and moderate resistance to downy and powdery mildew

Table 9 Black mustard (*Brassica nigra*)

Variety	Yield potential (kg/ha)	Oil content (%)	Recommended states	Special features/traits
Surya (LBM-428)	1000-1200	40-41	Andhra Pradesh	-------

Table 10 Indian mustard varieties for specific environmental conditions

Stress/situation/condition	Varieties Recommended
Salinity	Indian mustard : CS-52, CS-54,CS 234-4, Narendra Rai -1
High temperature tolerant	Indian mustard: Kanti, Pusa Agrani, RGN-13, Urvashi
High Oil Content	Indian mustard: Narendra Swarna Rai 8
Earliness	Indian mustard: Kanti, Narendra Ageti Rai 4, Pusa Agrani, Pusa Mahak.
Intercropping	Indian mustard: RH-30, RH781, Vardan.
Non Traditional Areas	Indian mustard: Pusa Agrani, Pusa Jaikisan, Gujarat Mustard 2.
Late Sown	Indian mustard: Ashirwad, RLM 619, Swaran Jyoti, Vardan, Navgold
Frost Tolerant	RGN13, RH-781, Swaran Jyoti
Drought (Rainfed)	Indian mustard RH-819, RH-781, GM1, Pusa Bahar, Pusa Bold, Aravali Mustard, Sej-2, JD-6, Geeta, RGN-48, RL-99-27, Shivani, PBR-97
Low erucic acid / glucosinolate	Indian mustard: Pusa Karishma, Pusa Mustard 21, LET-17, LET-18 Gobhi Sarson: Hyola 401, GSC 5, GSC 6, NUDB 26-11, Teri Uttam Jawahar.
White rust resistance	Indian mustard: Basanti, JM 1, JM 2, Maya

Growth stages: The particulars on growth stages of Rapeseed-Mustard are presented in Table 11.

Table 11 Growth stages of Rapeseed-Mustard

Phenological events	Crop duration (days)				
Start emergence	7	7	7	7	8
Complete emergence	14	14	14	14	14
Start flowering	55	60	63	66	80
50% flowering	64	65	67	81	99
Early siliqua development	72	76	80	86	106
Pod filling	82	90	95	100	116
Physiological Maturity	115	125	129	137	149

Season: The first half of September is best for sowing toria (if wheat is to follow, it should be sown by the end of August), 25th September to 15th October for sarson, 30th September to 15th October for raya and taramira is sown throughout October.

Seeds and sowing: Rapeseed and mustard are cultivated during September–October to February–March. Seed rate is 5 and 2 kg/ha for sole and intercrop, respectively. In White Rust and Downy Mildew endemic areas, seed treatment with Apron SD 35 @ 6 g/kg of seed is advised. For other seedling diseases, seed treatment with Carbendazim, Thiram or Captan @ 2 g/kg of seed is recommended. Seeds are sown with spacing of 30 × 15 cm for mustard and 45 × 15 cm for Rapeseed. Seeds are sown at a depth of 3 cm. Thinning is done on 15 DAS.

Weed management: The critical period of crop weed competition is the initial 45-60 days after sowing. one hand weeding on 25 DAS under rainfed conditions while two hand weedings 25 and 40 DAS are necessary under irrigated conditions. Pre-plant incorporation of fluchloralin @ 1.00 kg/ha or pre-emergence application of pendimethalin @ 1.00 kg/ha are quite effective in controlling weeds. Nitrofen @ 1.5 kg/ha or Isoproturon @ 1 kg/ha can be applied as pre-emergence which is followed by one hand weeding on 35 DAS.

Manures and fertilizer application: Apply 12.5 tonnes of FYM or compost per ha as basal dressing. Fertilizer dose of and 60-60-40 and 40:20:20 kg of N, P_2O_5 and K_2O per ha are recommended for irrigated and rainfed conditions respectively. Half of nitrogen and full dose of P_2O_5 and K_2O are applied as basal dressing while the remaining 50% of N is applied on 35 DAS. Apply borax @ 5 kg per ha are recommended for boron deficient soils to increase seed yield. Application of 20-25 kg/ha at sowing time of zinc sulphate overcomes zinc deficiency. Zinc deficiency can also be corrected by foliar spray of zinc sulphate @ 0.5% solution along with 0.25% slaked lime. Spray 0.2% copper sulphate solution 2-3 times at weekly intervals to control copper deficiency. Spray 0.5% ferrous sulphate solution 3-4 times at weekly intervals to control iron chlorosis. Spray manganese sulphate @ 0.2-0.3 % two to three times at weekly intervals to control Manganese chlorosis. Apply gypsum@ 250 kg/ha or Sulphur @ 20 kg/ha in soil to overcome sulphur deficiency. Use Sodium Molybdate 37 to 39% to control molybdenum deficiency.

Water management: Rapeseed and mustard are mainly raised as rainfed crops. If irrigation facilities are available, two irrigations at pre-bloom and seed filling stages are desirable. Mustard requires about 200 to 300 mm of water. Due to its low water requirement, rapeseed-mustard crops fit well in the rainfed cropping system.

Cropping system: Rapeseed and mustard are grown as mixed crops and in crop rotations. Intercropping or mixed cropping systems as Potato + Mustard (3:1), Field pea + Mustard (3:1), Wheat + mustard 9:1, Gram + mustard 4:1 and sugarcane + mustard 1:1 are widely practiced in mustard growing areas. Mustard + gram, Mustard + potato intercropping are also practiced in Assam, West Bengal, Odisha, Bihar, Uttar Pradesh, Madhya Pradesh, Haryana, Rajasthan and Gujarat. The common crop rotations/ cropping sequence practiced are Toria-spring Greengram-Rice; Toria-Wheat-Maize; Toria-Wheat-Cowpea; Toria-Wheat-Blackgram/Greengram; Toria-spring Greengram/Cowpea (fodder); Mustard /Brown Sarson/ Yellow Sarson-Groundnut; Toria-Rice-Rice and Toria-Pumpkin/ Watermelon/Cucumber/ Sponge gourd-early Rice. Suitable cropping systems for Bundelkhand region are namely, sorghum/ maize /pearl millet- rapeseed mustard, Black gram/ cowpea/ sorghum fodder – rapeseed mustard, green gram (green manuring after first picking) – rapeseed mustard, pearl millet+cowpea (fodder)–gram+rapeseed mustard, rice-mustard and sesame-mustard. Suitable intercropping systems for Bundelkhand region are viz., Mustard + Gram (1:4), Mustard + Lentil (5:1) and Mustard + Pea (1:3).

Plant protection: The major pests and diseases of rapeseed and mustard are given in Table 12.

Table 12 The major pests and diseases of rapeseed and mustard

Pests	**Crop stage attacked**	**Period of activity**
1. Insect pests		
Mustard Aphid (Lipaphis erysimi)	Vegetative/flowering and pod formation	December-March
Painted Bug (Bagrada hilaris)	Leaves	August-October
Tobacco Caterpillar (Spodoptera litura)	i. Seedling ii Maturity stage	i. October-November ii. March-April
Mustard Sawfly (Athalia proxima)	Vegetative	October-December
Leafminer (Chromatomyia horticola)	Reproductive	February-March
2. Diseases		
White rust (Albugo candida)	i. Vegetative ii. Reproductive	i. November ii February-March
Alternaria Leaf Spot (Alternaria brassicae)	Throughout crop growth	February-March
Powdery Mildew (Erysiphe cruciferarum)	Reproductive	February-March
Sclerotia rot (Sclerotia sclerotiarum)	i. Vegetative ii. Reproductive	i. October-November ii. February-March

Source: GOI, 2014.

Seed treatment with biocontrol agents viz., *T. viride,* G. *virens* or botanicals like *Allium sativum* bulb extract (1 % w/v) or carbendazim @ 0.1% a.i. or mixture of carbendazim with Apron 35 SD (6 g/ kg) or or Thiophanate Methyl against seedling diseases and Imidacloprid @ 5g/kg of seeds against insect pests. There is a need for mixture of fungicides for avoiding resistance development in pathogens to fungicides. If the white rust disease severity is more than 3%, apply ridomil MZ 72 WP @ 3g/l. If the alternaria blight disease severity is more than 3%, spraying of Mancozeb 50 WP @ 2g/l needs to be taken up at 50 and 70 days after sowing. Disease affected plants should be uprooted and destroyed by burning at a distance from the crop field. If Powdery Mildew disease is observed at a later stage of the crop, there is no need to worry. However, if the disease appears before or at flowering, dusting of Sulphur @ 1.5 kg/ha or spraying of Sulfex 2 g/l may be done. Spray Chlorpyrifos 20 % EC @ 200 ml in 200-400 l of water/acre or Dimethoate 30% EC @ 264 ml in 200-400 l of water/acre or Malathion 50% EC @ 400 ml in 200-400 l of water/acre or Phosphamidon 40% SL @ 200 ml in 200 l of water/acre or Thiamethoxam 25% WG @ 20-40 g in 200-400 l of water/acre to control aphids. Spray 2% Neem oil and 5% Neem Seed Kernel Extract (NSKE) effective against the mustard aphid. Pests and diseases resistant/tolerant varieties of mustard/rapeseed can also be grown as furnished in Table 12.

Table 12 Resistant/tolerant varieties of mustard/rapeseed

Pest	**Tolerant/ resistant varieties**
White Rust	Pusa Karishma (LES-39), Jawahar Mustard-1, PBR 91
Alternaria blight and aphids	Coral-432, NRCHB 5-6, NPJ 112 (Pusa Mustard 25), NRCDR 601, RYSKS-2 and DMH-I
Alternaria blight	Saurabh
White rust, Alternaria blight, Powdery mildew	Indian mustard-91, Toria-16, Yellow sarson-11; Gobhi sarson-11; Brown Sarson-3; Karan Rai-4; Taramira-5 and Black mustard-1
Aphids	RH-7846, RH-7847, RH-9020 and RWAR-842

Harvest: Canola is an indeterminate plant, which means it flowers until limited by temperature, water stress or nutrients. As a result, pod development can last over 3 to 5 weeks, with lower pods maturing before higher ones. Physiological maturity occurs when the seed moisture content reaches 35 to 45%. Rapeseed is ripe when plants turn a straw color and seeds become a dark brown. Because shattering is a potential problem, it is recommended that the crop be swathed when 25% of the seeds have turned from green to brown. In addition, the seed should be firm and not break when rolled between the thumb and forefinger. Seed moisture at average maturity stage is about 35 percent. Harvesting can be done as soon as the leaves begin to turn yellow at maturity. Toria takes 75 to 90 days to mature. The time for direct harvest occurs when all the pods are dry and rattle when shaken. Although tops of plants may contain a few green pods, do not wait for them to ripen as the rest of the crop will begin to shatter. Begin harvest only when the moisture content has fallen to eight per cent. Ensure moisture meter is properly calibrated before harvest. Do not use the general colour of the crop as a guide, use seed moisture content. Rapeseed shatters seed lesser than mustard. So care is to be taken to harvest mustard just before the pods open in order to avoid losses due to shattering of seeds. The plants are uprooted or harvested with sickles. Threshing is done with help of sticks. Threshing is done by bullocks or by running a tractor over the dried plants. Seed are cleared and then dried for a couple of days. Seeds can be stored at the moisture content of less than 8%.

Yield: Rapeseed produces a seed yield of 1400 to 2000 kg/ha. Mustard produces a seed yield of 2000 to 2500 kg/ha.

Utilization: The oil of rapeseed and mustard possesses a sizable amount of Erucic acid (41 to 57%), together with linoleic acid up to 12 to 18 % . The oleic and linoleic acid together constitute only about 22 to 34%. A lower proportion of erucic acid will make the oil more palatable, nutritive besides reducing metabolic disorders. The protein contained in rapeseed and mustard normally ranges between 24 and 30%. The oil is used for the manufacture of margarine, cooking fats and special lubricants for jet engines. The residual cake is used for livestock feeding.

REFERENCES

Arvind Kumar, Pankaj Sharma, Lijo Thomas, Abha Agnihotri and S.S. Banga. 2009. Canola cultivation in India: scenario and future strategy. 16th Australian Research Assembly on Brassicas. Ballarat Victoria.

GOI. 2014. Status paper on Oilseeds. Ministry of Agriculture, Government of India. New Delhi.

Kirk, J.T.O., and R.N. Oram. 1981. Isolation of erucic acid-free lines of Brassica juncea: Indian mustard now a potential oilseed crop in Australia. J. Aust. Inst. Agr. Sci. 47:51–52.

NIPHM. 2014. AESA based IPM – Mustard/Rapeseed. National Institute of Plant Health Management, Rajendranagar, Hyderabad

Prakash, S. 1980. Cruciferous oilseeds in India. p. 151–163. In: S. Tsunoda, K. Hinata, and C. Gomez-Campo (eds.), *Brassica* crops and wild allies. Biology and breeding. Japan Scient. Soc. Press, Tokyo.

Raymer, P.L., D.L. Auld, and K.A. Mahler. 1990. Agronomy of canola in the United States. p. 25–35. In: F. Shahidi (ed.), Canola and rapeseed: Production, chemistry, nutrition, and processing technology. Van Nostrand Rhienhold, New York.

Singh, R. P. 2014. Status Paper on Oilseed Crops. Directorate of Oilseeds Development. Hyderabad. pp 118-135.

Vaughn, J.G., MacLeod, A.J. and B.M.G. Jones (eds.), 1976. The biology and chemistry of Cruciferae. Academic Press, New York.

http://oilseeds.dacnet.nic.in/oilseeds/mustard/MUSTARD%20AND%20RAPESEED.htm

http://oilseeds.dacnet.nic.in/ProdTech.html

CHAPTER

5

Sunflower–*Helianthus annuus* L. (2n=34)

Family: Compositeae / Asteraceae

Vernacular name: Sunflower (English), *Surjamukhi* (Hindi), *Suriyakanthi* (Tamil), *Suryaphul* (Marathi), *Suryamukhi* (Oriya).

Importance: The name 'Helianthus' is derived from 'Helios' meaning 'sun' and 'anthos' meaning 'flower'. It is known as sunflower as it follows the sun by day, always turning towards its direct rays. Sunflower is described as 'drenched with sun-vitality' because the head follows the sun, ending up facing the west 'to absorb the few last rays of the dying sun'. Sunflower is grown as ornamental cum oilseed crop. It is tolerant to drought, salinity and alkalinity. It is a short duration, photo-insensitive and thermo insensitive crop. So it is grown throughout the year.

Origin: Sunflower is native of Mexico and USA.

Distribution: Sunflower is grown in Russia, Ukraine, Argentina, China, Romania, France, India, Hungary, USA, Spain, Yugoslavia, Bulgaria, Canada, and Mexico. In India, it is grown in Karnataka, Maharashtra, Tamil Nadu, Andhra Pradesh, Punjab, Haryana, Uttar Pradesh, Madhya Pradesh, Bihar and Rajasthan. Nearly 40% of sunflower area is found in *kharif* and the rest in *rabi/summer* and spring seasons. In Northern India, the crop is grown mainly in *spring/zaid* season. Sunflower is grown in 15 agro-regions of the country. Among them, Deccan plateau, hot semiarid ecoregion contributes the highest area and production of sunflower closely followed by Deccan plateau, hot arid ecosystem. The productivity of sunflower is highest in sub-humid ecosystem (1039 kg/ha) while it is lowest in arid ecosystem (500 kg/ha). Among different eco-regions, Western Himalayas, warm sub-humid (moist) to humid (inclusion of per-humid), ecoregion (1400 kg/ha), Eastern plain, hot sub-humid (moist) ecoregion (1384 kg/ha) and Northern plain, hot sub-humid ecoregion (1370 kg/ha) have high sunflower productivity.

Botany: Sunflower is an annual with rather stout, erect, herbaceous stem, 2.5 to 7.5 cm in diameter and 1.5 to 6.0 metres in height. Inflorescence is a head, consisting of pistillate or sterile ray florets at the periphery and central hermaphrodite, disc florets. Ray flowers are unisexual while disc flowers are bisexual. Head is 10 to 15 cm in diameter with 40 to 80 yellow rays and brown or nearly black discs. Sunflower is protandrous in which the male and female elements mature at different times. Flower opening starts from outer whorl and proceeds towards centre of head. The head bloom within 5-10 days. The pollen grains are viable for 12 hours. Anthesis take place at 5-8 a.m. Self-incompatibility operates in sunflower. It is essentially a cross pollinated plant. Pollinating agent is honey bees. The disc florets mature into what are normally called 'sunflower seeds', actually the fruit (an *achene*) of the plant, with the true seeds encased in an inedible husk.

Climate: Sunflower is basically a temperate crop. Sunflower is grown from 40°S to 55°N, but greater production from 20°N to 50°N and from 20°S to 40°S. The crop can be grown up to an altitude of 2500 m. However, the highest amount of oil per hectare is obtained below 1500 meters from MSL. It is a day neutral crop. It is a lover of warmth and sunshine. It comes under heliophyte plant group. Sunflower head always face east because of heliotropic response. The crop requires a cool weather during germination and seedling growth, warm weather from seedling up to flowering and non-cloudy sunny days during flowering to maturity. Sunflower is a photo-insensitive and thermo-insensitive crop. A well distributed annual rainfall of 500 to 1000 mm is required for growth and yield of sunflower. The growth stage from the formation of flower cluster to flowering must be warm. The cardinal minimum, optimum, maximum air temperature for the growth and yield of sunflower are 6°C, 20 to 28°C and 40°C respectively. Seeds are not affected by vernalization (cold) in the early germination stages. Seedlings in the cotyledon stage have survived temperatures down to -5°C. At later stages freezing temperatures may injure the crop. Temperatures less than -2°C are required to kill maturing sunflower plants. Temperatures above 38-40°C during flowering period cause desiccation of pollen and drying of stigma, resulting in poor seed set and yields. Extremely high temperatures have been shown to lower oil percentage, seed fill and germination. Heat stress (>35°C) decrease grain weight, oil content and oleic/ linoleic and increase percent of pericarp. Storage temperatures of >48°C, severely affect germination

Soils: Sandy loam soils are preferred than clayey soils. Sunflower is sensitive to water logging. Good drainage is essential. It grows well on fertile, well-drained neutral soils. Sunflower can profitably be cultivated in moisture retentive soils like Vertisols. The ideal pH is around 6.5-8.5. It can tolerate slight alkaline conditions but not acidity.

Season: Sunflower is grown in *kharif, rabi* and summer seasons in India. The oil content of sunflower varieties in the rainy season varied from 38.6 to 42.0% while for the winter crop the variation is from 44.8 to 47.3%. The winter season sunflower plants produce larger heads than the *kharif* season plants. Sowing time should be decided in such a way that the flowering and seed filling stages of the crop do not coincide with continuous rainy period or high temperatures above 38°C. Sunflower is grown in traditional areas during *kharif* (rainy season), second fortnight of June to mid-July in light soils and up to second fortnight of August in heavy soils. In *rabi* (winter season), the crop can be sown from September to first fortnight of October. The sowing time for sunflower in different states of India are furnished in Table 1.

Table 1 Sowing time for sunflower in different states of India

Centre	*Kharif*	*Rabi*	Summer	Spring
Maharashtra	July 1st week	1st FN of October	Last week of January to First week of February.	------
Karnataka,	July	2nd FN of September to 1st FN of October	January to 1st FN of February.	------
Tamil Nadu	July 15	November 1	February 15	------
Andhra Preadesh	July 10 to August 20	2nd FN of November	------	------
Punjab, Haryana, Uttar Pradesh, Bihar, West Bengal, Odisha and Chhatishgarh	------	------	------	January to February end

Varieties: Four varieties of Russian origin viz., EC 68413, EC 68414, EC 68415 and EC 69874 have been introduced to India in 1969. Sunflower varieties grown in different states of India are presented in Table 2 to Table 3.

Table 2 Sunflower varieties/hybrids recommended in different states of India

States	Cultivars	Varieties/hybrids
Karnataka	Hybrids	BSH-1, MSFH-8, KBSH-1, MSFH-17, Jwalamukhi, Sungene-85, PAC-36, PAC-1091, DSH-1, MLSFH-47, KBSH-41, KBSH-42, KBSH 44, Pro.Sun 09, RSFH-1, SH-416, KBSH-53, DRSH-1
	Varieties	Morden, TNAUSUF-7, DRSF-108, DRSF-113
Maharashtra	Hybrids	BSH-1, MSFH-8, KBSH-1, MSFH-17, LSH-1, LSH-3, PKVSH-27, Sungene-85, PAC-36, PAC-1091, MLSFH-47, KBSH-44, Pro Sun 09, SH-416, DRSH-1, LSFH-35
	Varieties	Morden, TNAUSUF-7, Surya, SS-56, LS-11 (Sidheswar), DRSF-108, DRSF-113, TAS-82, LSF-8, PKVSF-9, Phule Raviraj
Andhra Pradesh	Hybrids	BSH-1, APSH-11, MSFH-8, KBSH-1, MSFH-17, Jwalamukhi, Sungene-85, PAC-36, PAC-1091, MLSFH-47, KBSH-44, Pro. Sun.09, NDSH-1, SH-416, DRSH-1
	Varieties	Morden, TNAUSUF-7, DRSF-108, DRSF-113
Tamil Nadu	Hybrids	BSH-1, MSFH-8, KBSH-1, MSFH-17, Jwalamukhi, Sungene-85, PAC-36, PAC-1091, TCSH-1, MLSFH-47, KBSH-44, Pro. Sun.09, SH.416, DRSH-1
	Varieties	Morden, TNAUSUF-7, CO-1, CO-2, DRSF-108, DRSF-113, COSFV-5
Punjab	Hybrids	BSH-1, KBSH-1, PSFH-67, Jwalamukhi, Sungene-85, PAC-36, PSFH-118, KBSH-44, DRSH-1, PSFH-118, PSFH 569
	Varieties	Morden, DRSF-108, DRSF-113

Haryana	Hybrids	BSH-1, KBSH-1, Jwalamukhi, Sungene-85, PAC-36, KBSH-44, Pro.Sun.09, HSFH-848, DRSH-1
	Varieties	Morden, DRSF-108, DRSF-113
Gujarat	Hybrids	BSH-1, KBSH-1, Jwalamukhi, Sungene-85, PAC-36, PAC-1091, MLSFH-47, KBSH-44, SH-41, DRSH-1.
	Varieties	GAUSUF-15, Morden, TNAUSUF-7, DRSF-108, DRSF-113
Other states	Hybrids	BSH-1, KBSH-1, Jwalamukhi, Sungene-85, PAC-36, PAC-1091, KBSH-44, Pro.Sun.09, DRSH-1
	Varieties	Morden, TNAUSUF-7, DRSF-108, DRSF-113

Source: Directorate of Oilseeds Research (ICAR), Hyderabad

Table 3 Sunflower varieties/hybrids particulars in different states of India

Variety/ hybrid	Duration (days)	Seed yield (kg/ka)	Oil (%)	State	Salient features
DRSF-108 (V)	------	900-1800	36-39	All India	High oil
CO-5 (V)	------	1000-1700	39-42	Tamil Nadu	------
TAS-82 (V)	------	800-1200	40-42	Maharashtra	Black seeded variety
LSF-8 (V)	------	1000-1400	36-39	Maharashtra	Tolerant to downy mildew, rust and *Alternaria*
DRSF-113 (V)	------	1000-1500	36-39	All India	High yield
Phule Raviraj (V)	------	1795	34	Western Maharashtra	Tolerant to Necrosis, *Alternaria* and Capitulum borer.
SS-56 (V)	60	700-800	36-38	Maharashtra.	Short duration.
Gujarat Sunflower-1	90-95	800-1200	38-42	Gujarat.	Medium duration. Tall type.
TNAU-SUF-7 (V)	95-100	800 -1200	38-42	Tamil Nadu.	Medium duration
TNAU-SUF-10 (V)	85-90	1250 (K, RF) 1390 (R, IR)	38.2	Tamil Nadu, Karnataka, Maharashtra.	Suitable for both *Kharif* and *rabi*. Moderately resistant to pests like leaf hopper, head borer and ash weevil and diseases like Alternaria and rust.
PKVSF-9 (AKSF9) (V)	80-85	1200 -1300	38-40	Vidarbha Region of Maharashtra.	Early maturing. Suitable for all seasons, Moderately resistant to Alternaria and downy mildew.
Sidheshwar (LS-11) (V)	85	1313 (*kharif*)	36.1-39.4	Maharashtra (rainfed situation)	Moderately resistant to downy mildw.

KBSH-41 (H)	------	1300-1500	39-41	Karnataka	Tolerant to moisture stress
KBSH-42 (H)	------	1300-1500	38-41	Karnataka	Tolerant to moisture stress
PSFH-118 (H)	------	1400	40	Punjab	Resistant to stem and head rot.
KBSH-44 (H)	------	1400-1600	36-38	All India	Resistant to downy mildew
LSFH-35 (Maruti) (H)	------	1400-1500	39-41	Maharashtra	Resistant to downy mildew
NDSH-1 (NDSH-15) (H)	------	1400	40	Andhra Pradesh	Early maturing hybrid
RSFH-1 (H)	------	1300-1500	39-41	Karnataka	High oleic hybrid
HSFH-848 (H)	------	1800-2400	41-42	Haryana	Dwarf hybrid
DRSH 1 (H)	------	1300-1600	42-44	All India	High oil
KBSH-53 (H)	------	1700-2700	42-44	Karnataka	Resistant to powdery mildew
PSFH-569 (H)	------	2232	40	Punjab	High oil, early hybrid
Suryamukhi (H)	------	2000-2200	40	Punjab	------
LDMRSH-1 (H) CMS 338 A × MRH A2)	85-90	900-1200	37-39	Maharashtra.	Resistant to downy mildew. Low yields in seed production.
LDMRSH-3 (H) CMS 207 A × MRHA-2)	95-100	1000-1500	38-40	Maharashtra.	Resistant to downy mildew. Low yields in seed production.
KBSH-1 (H) CMS 234A X6D-1)	90-95	1200-1500	42-44	Wide adaptability. (All sunflower growing states)	Wide adaptability.(All sunflower growing areas of the country) High yield and oil content.
PSFH-67 (H)	90-95	1000-1500	38-42	Punjab	Spring cultivation
PKVSH-27 (H) (CMS2A × AKHR)	80-85	1300-1600	38-40	Vidarbha Region of Maharashtra.	Early maturing. Suitable for all seasons. Moderately resistant to Alternaria and downy mildew.
Jwalamukhi (PSCL-5015)(H)	95-100	1500 -1800(RF) 2400 -2800(IR)	40-42	All sunflower growing areas of the country.	Medium maturity. Suitable for all seasons.
SUNGENE-85 (H)	80-85	1000-1500(RF) 2250-3000(IR)	40-43		Resistant to downy mildew and tolerant to Alternaria and rust.

PAC-1091 (H)	95-100	1680	37.4	All kharif sunflower growing areas.	Medium maturity and height. Tolerant to loding.
PAC-36 (H)	103	1688	38.8	All rabi / summer sunflowr growing areas.	Tolerant to downy mildew.
TCSH-1 (H) (CMS 234A x RHA 272)	85	1600	39	Tamil Nadu	

Note: (K)=*Kharif*; (R)=*Rabi*; (RF)=Rainfed; (IR)=Irrigated; (H) =Hybrid; (V)=Variety.

Growth stages: The particulars on growth stages of sunflower are furnished in Table 4.

Table 4 Growth stages of sunflower

Phenological events	Crop duration (days)					
Emergence	4-5	4-5	6-8	21	11	20
4 leaf stage	-----	-----	16	36	28	31
6 leaf stage	-----	-----	21	39	38	41
8 leaf stage	-----	-----	28	42	45	50
12-14 leaf stage	-----	-----	35	53	63	67
Star stage	-----	-----	50	66	83	86
Bud formation start	30	40	54	69	86	89
Bud formation completed	35	45	59	74	91	91
Start flowering	40	50	63	78	93	92
Complete flowering	45	55	74	91	111	108
50% Grain filling	55	65	81-90	105	118	116
Physiological Maturity	80	90	105	120	136	152

Land preparation: Plough the land once with tractor or twice with iron plough or three to four times with country plough till all the clods are broken and a fine tilth is obtained.

Seeds and sowing: Seed rate is 15 kg per ha for varieties and 5 to 6 kg per ha for hybrids. Spacing is 45 cm × 30 cm. Sowing depth is 3 to 5 cm. Seeds are soaked in 2% $ZnSO_4$ for 12 hours with a seed to solution ratio of 1:0.06 and are dried back to their original moisture content of 8-9% in shade drying for rainfed sowing. For quick germination and better stand establishment in dryland conditions, soak the seed in fresh water (1:1 W/V) for about 10 hours and shade dry. Fresh seeds of sunflower exhibit physiological dormancy which could be broken by soaking the seeds in 300 ppm ethrel for 8 hours or 0.5% KNO_3 for 16 hours. Moist hydration of seed with water for 24 hours followed by dry dressing with thiram or bavistin @ 2g/kg of seed to protect from seed-borne diseases. Seed treatment with metalaxyl @ 6 g/kg can protect the crop against downy mildew disease. Treat the seed with imidacloprid @ 5 g/kg before sowing against insect vectors for the necrosis management. Apply azospirillum @ 2 kg per ha (10 packets) mixed with 25 kg of FYM and 25 kg of sand before sowing.

Weed management: Critical weed free period is up to 45 DAS. Two hand weeding are given on 20 and 35 DAS. Use of fluchloralin @ 1.0 kg per ha, pendimethalin @ 1 kg per ha, Alachlor @ 1 kg per ha as pre-emergence followed by one hand weeding on 30 DAS is found to be effective. Pendimethalin @ 0.75 kg a.i./ha in black soils or @ 1.0 kg a.i./ha in shallow red soils as pre-emergent application is found to be effective for sunflower + groundnut intercropping system.

Growth regulator: At head opening stage, spray DAP @ 2% and NAA (Napthelene Acetic Acid)) / Planofix @ 20 ppm (280 gm in 625 lit of water) on 30 and 60 DAS to increase seed setting.

After cultivation: Gap filling is carried out on 10 days after sowing. Thinning is done at the formation of the second pair of leaves or at 10-15 days after germination to retain single healthy plant per hill which is essential for obtaining higher yields.

Manures and fertilizer application: Sunflower is a heavy feeder and a tonne of crop removes as much as 63.3 kg N, 19.1 kg $P_2 0_5$, 126.2 kg K_2O, 11.7 kg S, 68.3 kg Ca, 26.7 kg Mg, 47g Zn and 1075 g Fe. Apply 12.5 tonnes of FYM or compost per ha as basal dressing. Fertilizer dose of 40-20-20 for varieties and 60-40-40 for hybrids are recommended for irrigated conditions while 40-20-20 kg of N, P_2O_5 and K_2O per ha is recommended for rainfed conditions respectively. Nitrogen is applied in three splits viz., 50 % at sowing, 25% at button stage (30 DAS) and 25% at flowering stage (40 to 45 DAS). The entire P_2O_5 and K_2O is applied at the time of sowing. Soil application of sulphur @ 40 kg/ha through gypsum, ammonium sulphate or single super phosphate once in the cropping system enhances yield besides maintaining soil fertility. Soil application of 25 kg $ZnSO_4$ + 15 kg borax ha^{-1} is recommended for irrigated sunflower crop. Soil application of micronutrients mixture @ 12.5 kg with sand 37.5 kg per ha is also recommended. Spray $ZnSO_4$ @ 0.5% and $MnSO_4$ @ 0.5% on 30, 40 and 50 DAS to overcome deficiencies. Direct spray of borax @ 0.2 % concentration or dusting 2 to 3 kg borax at capitulam or ray floret opening stage increased the seed yield to 20 to 25% apart from test weight, seed filling percentage, oil content, germination and vigour index. Fertiliser recommendations for sunflower in different states are furnished in Table 5.

Table 5 Fertiliser recommendations in different states of India

State	Ecosystem	Fertilizer (kg/ha)		
		N	P_2O_5	K_20
Andhra Pradesh	Rainfed (Rabi)	60	60	30
	Irrigated	60	90	30
Karnataka	Rainfed (Rabi)	50	25	25
	Irrigated	60	75	60
Maharashtra	Rainfed	40	60	0
	Irrigated	60	95	60
Tamil Nadu	Rainfed	40	20	20
	Irrigated	60	90	60
Uttar Pradesh, Punjab, Bihar and Haryana	Rice fallow situation	80	60	40
	Rainfed	80	60	40
	Irrigated	40	40	20
West Bengal, Bihar, Odisha, Chhattisgarh, Madhya Pradesh	Irrigated	80	60	40

Water management: The consumptive use of sunflower is 500 to 600 mm. Sunflower root system grows to a depth of 2 meters. Sunflower is irrigated at 40% available soil moisture in the top 30 cm soil. The growth stages in sunflower are seedling, bud initiation, button stage, flowering and seed development stage. At all growth stages, it is better to keep soil moisture at 50% of field capacity, except at flowering, when 70% of field capacity gives highest yield and oil content. The three critical stages are bud initiation (30 to 40 DAS) flower opening (40 to 60 DAS) and seed filling (50 to 80 DAS) stages. Sunflower is an excellent indicator plant to figure out the soil moisture status. First irrigation is given immediately after sowing. Life irrigation is given on 3 DAS. Five to seven irrigations in black soil and 10 to 12 irrigations in light soil are required.

Artificial pollination: Since sunflower is self-incompatible, it depends on insect, mainly honey bees, for cross pollination and seed set. The insects/honey bees act as pollinators in the field for pollen movement and seed set. Lack of adequate pollinators results in chaffy and partially filled seeds. Keeping beehives in the field @ 5 per ha improves seed setting and yield. Sunflower head blooms with in 5 to 10 days depending on size and season. Anthesis takes place between 5 to 8 a.m. The pollen grains are viable for 12 hours. The stigma continues receptiveness for 2 to 3 days. Hand pollination may also be resorted to induce cross pollination. Hand pollination is done at mid-flowering stage. Mild rubbing of the capitulam with the hand covered with palm or muslin cloth or soft cloth or rubbing two neighbouring flowers with each other, face to face gently on alternate days for two weeks to enhance pollination. In hybrids, the palm is first gently rubbed on the male parent flowers and then on the female line to transfer the pollen. Mid flowering occurs on 58 to 60 DAS for long duration varieties and 45 to 48 DAS for short duration varieties. Hand pollination is done from 7 to 11 a.m. for 2 weeks when pollen shedding is high. Hand pollination increases the seed yield to about 20 to 25%.

Cropping systems: It is possible to have double cropping involving sunflower in drylands in areas receiving more than 750 mm rainfall with a soil moisture storage capacity of more than 20 cm of available water. Crop sequences involving sunflower under rainfed conditions for different states are Sunflower-finger millet, Mungbean-sunflower, Sunflower-safflower, Sorghum-sunflower, *kharif* pulses-sunflower, Soybean-sunflower, Sunflower-chickpea. Crop sequences involving sunflower under irrigated conditions for different states are Groundnut-sunflower, Sunflower-groundnut, Mungbean-rice-sunflower, Sunflower-rice-sesame, Sunflower-rice, Rice-sunflower–mungbean, Rice-sunflower, Sesame-sunflower–groundnut, Sorghum-sunflower, Cotton-sunflower. Sunflower after/before groundnut in Alfisols and sunflower after soybean in Vertisols performs well. Under rice-fallow situation, sowing of sunflower should be done as early as possible i.e., a week after the harvest of rice to realise higher yield of sunflower. Growing sunflower after sunflower should be avoided as it increases pest and disease complex and depletes the soil nutrients resulting in low yield. Legumes are preferred in sequence with sunflower for better productivity and sustainability.

Intercropping of Sunflower + groundnut in 2:4 ratio, sunflower + groundnut in 2:6 ratio and sunflower + finger millet in 3:6 ratio is commonly practiced. It is possible to have double cropping involving sunflower in drylands in areas receiving more than 750 mm rainfall with a soil moisture storage capacity of more than 20 cm of available water. The remunerative intercropping systems and crop sequences involving sunflower are identified for various regions/states are furnished in Table 6 and Table 7.

Table 6 Intercropping systems identified in different agro-climatic zones

State	Soil type	Efficient intercropping	Row ratio
Karnataka	Alfisols	Groundnut+sunflower	4:2, 3:1
	Alfisols	Pigeonpea+sunflower	1 :2/1:1
	Vertisols	Pigeonpea+sunflower	3:1
	Alfisols	Fingermillet+sunflower	4:2
Maharashtra	Vertisols	Pigeonpea+sunflower	3:3
	Vertisols	Soybean+sunflower	2:1
	Vertisols	Groundnut+sunflower	6:2
Andhra Pradesh	Alfisols	Groundnut+sunflower	4:2
	Alfisols	Pigeonpea+sunflower	1 :2
	Alfisols	Castor+sunflower	1: 1
Tamil Nadu	Alfisols	Groundnut+sunflower	3:1
	Alfisols	Castor+sunflower	1:1
Gujarat	Alfisols	Groundnut+sunflower	1:1
	Vertisols (irrigated)	Castor+sunflower	1:1
Non-traditional areas	Inceptisols	Urdbean/mungbean+ sunflower	4:2/3: 1

Table 7 Suggested crop sequences involving sunflower for different states/regions

S. No.	State	Situation	Crop Sequence
1	Karnataka -Southern region	Rainfed	Sunflower-finger millet
		Irrigated	Rice-sunflower, Groundnut-sunflower, Sorghum-soybean-sunflower, Cotton-sunflower, Maize-sunflower –groundnut, Sunflower-groundnut
2	Karnataka -North region	Rainfed	Fallow-sunflower, Mungbean-sunflower, Sunflower-chickpea, Sunflower-safflower
		Irrigated	Cotton-sunflower, Rice-sunflower, Soybean-sunflower, Maize-sunflower, Sunflower-chickpea, Sunflower-groundnut
3	Maharashtra -Vidarbha	Rainfed	Sorghum-sunflower, *Kharif* pulses-sunflower, Sunflower-safflower, Sunflower-chickpea
		Irrigated	Cotton-sunflower, Sorghum-sunflower, Groundnut-sunflower
4	Maharashtra -Marathwada	Rainfed	Soybean-sunflower, *Kharif* pulses-sunflower, Sunflower-chickpea
		Irrigated	Ground nut-sunflower-sesame, Cotton-sunflower Sorghum, Pigeon pea-sunflower, Pigeonpea-sunflower, Sorghum-sunflower, Maize-cowpea (F), Groundnut-sunflower, Soybean-sunflower
5	Western Maharashtra	Rainfed	*Kharif* pulses-sunflower, Soybean-sunflower
		Irrigated	Cotton-sunflower ,Sunflower-groundnut

6	Andhra Pradesh and Telangana	Rainfed	*Kharif* pulses-sunflower, Millets-sunflower, Groundnut-sunflower, Sorghum-sunflower
		Irrigated	Rice-sunflower –mungbean, Rice-sunflower, Sesame-sunflower –groundnut, Groundnut-sunflower, Sorghum-sunflower
7	Tamil Nadu	Irrigated	Groundnut-sunflower, Mungbean-rice-sunflower, Sunflower-rice-sesame, Sunflower-rice
8	Punjab, Haryana, Uttar Pradesh, Bihar	Irrigated	Maize-potato-sunflower, Cotton-sunflower, Fodder-potato-sunflower, Fodder-potato-sunflower, Fodder -mustard-sunflower, Urdbean-mustard –sunflower, Groundnut -mustard-sunflower
9	Odisha, Chhattisgarh, West Bengal	Irrigated	Rice-sunflower, Soybean-sunflower, Rice-potato-sunflower

Crop protection: The major insect pests of sunflower are jassids and white fly, tobacco caterpillar / Bihar hairy caterpillar / semi loopers, gram pod borer while the important diseases are alternaria blight and leaf spot, rust, downy mildew, necrosis, etc. The important insect pests, diseases and their management methods are presented in Table 8 and Table 9. Seed treatment can be done with any one of the following fungicides/ bio-control agents: *T. viride* @ 4g/kg of seed,Thiram @ 4g/kg of seed, Carbendazim @ 2g/kg of seed.

Table 8 Pest management strategies

Pests	**Management strategies**
Weevil *Myllocerus* spp. Tobacco cut worm *Spodoptera litura* Gram pod borer *Helicoverpa armigera*	Hand pick the *Helicoverpa* larvae and destroy. Spray any of the following insecticides : Dichlorvos 76EC 500 ml/ha Phosalone 35 EC 1000 ml/ha Dust any one of the following : Phosalone 4 D @ 25 kg/ha Insecticidal application at the time of bee visit is toxic to honey bees. So, apply the insecticides after 4 pm when the bee activity is minimum. Do not spray insecticides on the same day when NAA is sprayed
Leaf hopper *Amrasca devastans*	Seed treatment with imidacloprid 70 WS at 7 g/kg protected the sunflower plants from leaf hopper upto 7 weeks. Spray Imidacloprid 70% WS 490 ml/ha Imidacloprid 17.8%SL 100 ml/ha
Whitefly	Spray Imidacloprid 70% WS 490 ml/ha Imidacloprid 17.8%SL 100 ml/ha Malathion 50 EC 500 ml/ha
Thrips	Spray Imidacloprid 17.8%SL 100 ml/ha

Table 9 Disease management strategies

Alternaria leaf spot: *Alternaria helianthi*	Spray Mancozeb 1000g/ha
Rust: *Puccinia helianthi*	Spray Mancozeb 1000g/ha
Charcoal rot: *Macrophomina phaseolina (Rhizoctonia bataticola)*	Soil application of *P. fluorescens* or *T. viride* – 2.5 kg / ha + 50 Kg of well decomposed FYM or sand at 30 days after sowing. Spot drenching with Carbendazim @ 1 gm/ litre
Head rot: *Rhizopus sp*	Spray Mancozeb 1000g/ha in case of intermittent rainfall at the head stage, directing the spray to cover the capitulum. Repeat fungicidal application after 10 days if humid weather continues
Necrosis virus disease: *Tobacco streak virus (Ilarvirus)* Vector: Thrips	Raise sorghum as border crop (One month prior to sunflower sowing). Imidacloprid seed treatment 2g/kg and Imidacloprid foliar spray at 30 and 45 DAS.

Harvest: Harvesting the crop at physiological maturity i.e., when back of capitulam or Head turns lemon yellow is ideal to minimize losses due to lodging. The heads are cut with sickles and allowed for drying. Seeds are separated from the heads by beating with stick. Seed moisture at harvest will be 25% and is to be dried to 15 to 16%. Seed yield is 700 to 1000 kg/ha for varieties and 1000 to 1500 kg/ha for hybrids.

Seed storage: The seeds lose their quality during storage due to deterioration and pest infestation, when the germination falls below 5-10 % of the required standard the seeds are imposed with midstorage correction, where the seeds are soaked in double the volume of 10^{-4} M solution of potassium dihydrogen phosphate @ 3.6mg/lit of water for 6 hours and the seeds are dried back to original moisture content of 8-9%.

Seed Dormancy: Seed dormancy period is 30 to 50 days. However, dehusked seed shows 75% germination after 11 days of harvest. Seeds should be soaked for 6 hours in Ethrel 25 ppm solution and dried in shade or can be sown directly. Spraying ethrel @ 150 ppm on 20 to 25 days after seed filling releases dormancy to the extent of 30 to 35% by the first week after harvest.

Utilization: Sunflower seed contains about 48 to 53% edible oil. Sunflower oil is a rich source of linoleic acid (64%) which is good for heart patients. Linoleic acid helps in washing out cholesterol deposition in the coronary arteries of the heart. Sunflower is also a source of lecithin, tocopherols and furfural. It is used as nutritious meal for birds and animals. It is also used in the preparation of cosmetics and pharmaceuticals. Sunflower oilcakes used as cattle feed. The refined oil is used as salad oil and cooking oil. The oil has non-cholesterol and anti-cholesterol properties. The oil is used for culinary purposes, in preparation of vanaspati and in the manufacture of soaps and cosmetics. Oil cake is rich in high quality protein (40–44%) and used as cattle and poultry feed.

REFERENCES

CPG. 2012. Crop Production Guide. Department of Agriculture, Government of Tamil Nadu, Chennai and Tamil Nadu Agricultural University, Coimbatore.

Singh, R. P. 2014. Status Paper on Oilseed Crops. Directorate of Oilseeds Development, Hyderabad. http://oilseeds.dacnet.nic.in/crop.html

CHAPTER

6

Safflower–*Carthamus tinctorius* L. (2n=24)

Family: Compositae/Asteraceae

Vernacular names: *Kusum* (Hindi), *Kusumbha* (Tamil), *Kardi* (Marathi), *Kosambi* (Gujarati) and *Kusuma* (Telugu).

Importance: Safflower is an ancient oilseed crop. The crop is cultivated, primarily for its seeds which yield oil. The oil content varies from 29 to 36%, depending on the variety, soil, climate and other conditions. The seeds are edible and are eaten after roasting. It is ideal for growing in scanty rainfall regions. It is a spiny crop. However, the entire production of safflower in China is under spineless cultivars. It is also cultivated for its orange red dye (Carthamin) extracted from its coloured florets. Safflower petals are of great medicinal value and find use in curing skin diseases, hyper tension, muscular arthritis and joint pains, jaundice and viral fever, etc.

Origin: Safflower has origin over a vast region of China, India, Nile valley and Ethiopia. It is believed to be originated in an area bounded by the eastern Mediterranean and Persian Gulf, encompassing southern parts of Russia, Afghanistan, Western Iran, Iraq, Syria, Southern Turkey, Jordan and Israel.

Distribution: Safflower is grown primarily in semiarid parts of North Africa, India, China, Thailand, USA, Canada, Australia, Spain, Portugal, Egypt, Ethiopia, Afghanistan, Argentina and Mexico. India is the largest producer of safflower. In India, it grows in Madhya Pradesh, Maharashtra, Andhra Pradesh, Karnataka, Bihar, West Bengal, Uttar Pradesh, Tamil Nadu and Rajasthan.

Botany: Safflower is an annual herb and has a thistle-like appearance. The plants may be glabrous or pubescent. Plant height ranges from 30-150 cm. Leaves are spiny or non-spiny, alternate and serrate. Inflorescence is broad, flat or slightly curved and densely bristled owing to the presence of numerous floral bracts. Head is homogamous, florets tubular, varying from light yellow to orange yellow or red in colour. The total flowering period varies from 3 to 4 weeks. Safflower is self-pollinated crop. However, cross-pollination takes place in nature.

Natural crossing is about 15 to 16%. Bees' increases 5% seed yield. However, intense bees' activity is observed due to coloured florets and presence of nectar. The flowers are yellow to red, containing 20 individual florets, each of which produces a seed. It produces white, shiny, and smooth seeds (fruits) called achenes, each weighing from 0.01 to 0.1 g. Seed is exendospermous. The cotyledons have 29 to 32% oil.

Climate: Safflower is a crop of cool and bright climate. Safflower is adapted to a wide range of climatic conditions. It is mostly cultivated between 14 and 22ºN and 73.5 and 79ºE. Safflower grows well in areas receiving mean annual rainfall of 300 to 600 mm. It comes up better in relatively drier areas. It does not stand waterlogged conditions. Frequent and prolonged rains and heavy dew at flowering stage adversely affect pollination, seed development. Rain after flowering causes dis-colouration of the seed; reduce oil content and leads to quiescent germination. The cultivated safflower is day neutral but a day length or 12 to 14 hours is considered essential for flowering and fruit set. It is thermosensitive crop. The cardinal minimum, optimum and maximum air temperatures are 10°C, 24 to 30°C and 40°C respectively. There is no germination below 2°C. The optimum temperature for germination is 15°C. The crop is tolerant to frost at seedling and vegetative stages but sensitive at elongation, flowering and post-flowering stages. The optimum temperature at flowering is between 24 and 32°C, however adequate soil moisture reduces the adverse effect of high temperature. It is sensitive to high humidity at flowering period. It does not favour extremes of either heat or cold. Emerging plants need cool temperatures for root growth and rosette development (mean daily temperatures 15 to 20°C) and high temperatures during stem growth, flowering and yield formation periods (20 to 30°C).

Soils: Safflower is grown on a wide range of soils from sandy loam to clay loam. Well drained, moisture retentive and deep soil is desirable. Safflower is grown in medium and deep black soils. It is highly salt tolerant. It tolerates up to ECe upto 7 ds/m with a slight reduction in yield of about 10%. In terms of salinity, it is intermediate between barley and sorghum. It is vulnerable to saline water in early stages.

Seasons: Safflower is sown in September-October and is harvested in March-April. Crops can also be grown with late receipt of rains in early November. The optimum time of sowing for different safflower growing areas taking advantage of monsoon rains is furnished in Table 1.

Table 1 Sowing time of Safflower in different states of India

Region	Recommended sowing time
Andhra Pradesh -Coastal and Rayalseema areas	October
Telangana	Late September or early October
Maharashtra	
Drought prone areas of Western Maharashtra and adjoining areas	Second fortnight of September
Assured rainfall areas of Western Maharashtra and Marathwada	Last week of September to mid October
Vidarbha Rainfed	Last week of Sept to first week of October
Irrigated	First and second fortnight of October
Karnataka - Rainfed	Second fortnight of September.
Irrigated	Mid September to early November

Gujarat	Last week of Oct to first week of Nov
Madhya Pradesh	Last week of Sept to second week of Nov
Chhattisgarh	Second week of Oct to second week of Nov
Odisha	Second week of September to mid October
Uttar Pradesh Rainfed	Second to third week of October
Irrigated	Mid October to first week of November
Rajasthan and Tamil Nadu	Early to mid October
Bihar	Second week of October

Varieties: Safflower varieties/hybrids with salient features/ traits in different states of India in Table 2 and Table 3.

Table 2 State-wise safflower hybrids/varieties under cultivation

State	Cultivars	Varieties/Hybrids
Andhra Pradesh	Hybrids	DSH-129, MKH-11, NARI-NH-1, NARI-H-15, MRSA- 521
	Varieties	Manjira, Sagar Muthyalu, NARI-6, Parbhani Kusum, Phule Kusuma, PBNS-40, NARI-38, SSF-658
Karnataka	Hybrids	DSH-129, MKH-11, NARI-NH-1, NARI-H-15, MRSA- 521
	Varieties	S-144, A-300, A-1, A-2, NARI-6, Parbhani Kusum, Phule Kusuma, PBNS-40, NARI-38, SSF-658
Maharashtra	Hybrids	DSH-129, MKH-11, NARI-NH-1, NARI-H-15, MRSA- 521
	Varieties	Bhima (Western Maharashtra and Vidarbha), Nira (irrigated areas), NARI-6, Girna (Khandesh area), Sharda (Marathwada region), Parbhani Kusum, Phule Kusuma, PBNS-40, AKS-207 (Vidarbha region), NARI-38, SSF-658
Madhya Pradesh	Hybrids	DSH-129, MKH-11, NARI-NH-1, NARI-H-15, MRSA- 521
	Varieties	JSF-1, JSI-7, NARI-6, JSI-73, Parbhani Kusum, Phule Kusuma, JSI-97, JSI-99, PBNS-40, NARI-38, SSF-658
Orissa	Hybrids	DSH-129, MKH-11, NARI-NH-1, NARI-H-15, MRSA- 521
	Varieties	NARI-6, Annigeri-1, Parbhani Kusum, Phule Kusuma, PBNS-40, NARI-38, SSF-658
Bihar	Hybrids	DSH-129, MKH-11, NARI-NH-1, NARI-H-15, MRSA- 521
	Varieties	S-144, A-300, Annigeri-1, NARI-6, Sagara Muthyalu, Parbhani Kusum, Phule Kusuma, PBNS-40, NARI-38, SSF-658
Chhattisgarh	Hybrids	DSH-129, MKH-11, NARI-NH-1, NARI-H-15, MRSA- 521
	Varieties	JSF-1, JSI-7, NARI-6, Parbhani Kusum, Phule Kusuma, PBNS-40, NARI-38, SSF-658
Uttar Pradesh	Hybrids	DSH-129, MKH-11, NARI-NH-1, NARI-H-15, MRSA- 521
	Varieties	Malaviya Kusum, NARI-6, Type-65, Parbhani Kusum, Phule Kusuma, PBNS-40, NARI-38
Tamil Nadu	Hybrids	DSH-129, MKH-11, NARI-NH-1, NARI-H-15, MRSA- 521
	Varieties	K-1, CO-1, NARI-6, Parbhani Kusum, Phule Kusuma, PBNS-40, SSF-658
Rajasthan	Hybrids	DSH-129, MKH-11, NARI-NH-1, NARI-H-15, MRSA- 521
	Varieties	Annigeri-1, NARI-6, Parbhani Kusum, Phule Kusuma, PBNS-40, NARI-38, SSF-658

Table 3 Safflower varieties with salient features/ traits in different states of India

Cultivars	Yield potential (kg/ ha)	Oil content (%)	Recommended states / regions situations	Salient features/ traits
Varieties				
Parbhani Kusum (PBNS-12)	1900	29	All India	Spiny
Phule Kusuma (JLSF-414)	1200-1500 (R) 2000-2200 (I)	29	All India	Spiny
JSF-97	1500-1600	30	Madhya Pradesh	Non-spiny, Moderately tolerant to *Alternaria,* wilt and aphids
JSF-99	1100-1200	28-29	Madhya Pradesh	Semi-spiny, Moderately tolerant to *Alternaria,* wilt and aphids
PBNS-40	1500	27	Madhya Pradesh	Non spiny
AKS-207	1200-1400	27	Maharashtra	-
NARI-38	2038	28	All India	Resistant to wilt
SSF-658	1430	28	All India	Tolerant to wilt and aphids
Hybrids				
NARI-NH-1 (PH-6)	1936	31	All India	Non spiny hybrid, moderately resistant to *Cercospora* leaf spot, wilt; tolerant to *Alternaria* and aphids
NARI-H-15	2200	29	All India	-
MRSA-521	1000-1500 (R) 2200-2500 (I)	27	All India	Resistant to wilt; tolerant to*Alternaria* and aphids

Growth stages: Initial growth after the germination of seed is slow in safflower. During the slow growing period, called the rosette stage, several leaves are produced at the stem base. The duration of the rosette stage in safflower varies from 20 to 35 days. After this stage, the stem elongates quickly and branches profusely. It has a taproot system that elongates to 2 to 3 m in soils with adequate depth. The deep root system in safflower helps to extract the water and nutrients from deep layers of soil and thus makes it an ideal plant for rain-fed cropping systems. The flowering period in safflower lasts for a month. The flower starts in the capitula on primary branches flower first, followed by flowering of the capitula on secondaries and tertiaries. Flowering in a capitulum begins in the outermost whorl of florets and proceeds centripetally over 3 to 5 days. Safflower crop duration is 150 to 180 days in north India and 115 to 140 days in south India. The particulars on growth stages of safflower are presented in Table 4.

Table 4 Growth stages of safflower

Crop stage	Days
Establishment	4–10
Early vegetative (rosette development)	25
Late vegetative (elongation and branching)	60
Flowering	30
Yield formation (seed filling)	25
Ripening	10
Total	150–160

Seeds and sowing: Seed rate is 10 to 15 kg/ha. Seeds should be treated with Thiram, Captan or Carbendizim @ 3 g/kg seed before sowing to prevent losses from seed and soil borne diseases. In Chhattisgarh, presoak the seeds in pure water for 24 to 48 hours and shade dry for about 4 hours. Thereafter, treat the seeds with fungicide before sowing. Seeds are sown with a spacing of 45 × 15 cm. Seed rate and spacing for safflower in different states of India is furnished in Table 5.

Table 5 Seed rate and spacing for safflower in different states of India

Region	Ecosystem	Seed rate (kg/ha)	Spacing (cm)
Andhra Pradesh		7.5–10	45 × 20
Maharashtra	*Kharif* fallows	10–12	45 × 20
	Double cropped areas	15	45 × 20
Karnataka		7.5	60 × 30
Gujarat		10–12	45 × 20
Madhya Pradesh		20	45× 20–25
Chhattisgarh	Rice fallows	10–15	45 × 20
Odisha	Rice fallows	20	30 × 15
Uttar Pradesh and Bihar	Adequate moisture	12–15	45 × 20
	Scanty moisture	15–20	45 × 20
Tamil Nadu		10	45 × 20

After cultivation: Thinning should be done on 10 to 15 DAS. This ensures a desirable population density.

Weed management: Two hand weeding are given on 20 and 45 DAS.

Manures and fertilizers application: Apply 12.5 tonnes of FYM or compost per ha as basal dressing. Fertilizer dose of 40-20-30 and 20-20-0 kg of N, P_2O_5 and K_2O per ha are recommended for irrigated and rainfed conditions, respectively. The recommended levels of manures and fertilizers for different regions are given in Table 6.

Table 6 Manures and fertilizers recommended for safflower for different regions of India

Region	Ecosystem	FYM (t/ha)	Fertilizer (kg/ha)		
			N	P_2O_5	K_2O
Andhra Pradesh	Rainfed	-	40	25	-
Western Maharashtra	Rainfed	-	50	25	-
	Irrigated	5-10	50	50	-
	Scarcity zone	-	25	12.5	-
Marathwada- Maharashtra	Rainfed	-	40	20	-
	Irrigated	-	60	40	-
Vidarbha- Maharashtra	Rainfed	5-10	25	25	-
	Irrigated	5-10	50	50	-
Karnataka	Rainfed	6	35	50	-25
	Irrigated	6	75	75	35
Gujarat	Rainfed	-	25	10	-
Odisha	Rainfed	6	25	25	-
Uttar Pradesh	Rainfed	-	40	30	20
	Irrigated	-	60	30	20
Madhya Pradesh Malwa tract	Rainfed	-	40	40	20
	Irrigated	-	60	40	30
Chhattisgarh	Rainfed	-	20-30	15-20	10-15
	Irrigated	-	50-60	20-30	20
All other areas	Rainfed	-	30	20	-

Growth regulators: Spray cycocel @ 500 ppm at flower initiation to get 8 to 10% higher seed yield and returns from safflower. Use about 300 litres of spray solution per hectare.

Water management: Safflower is mostly raised as a rainfed crop. Flowering and grain filling are critical growth stages to moisture stress condition. It is sensitive to excessive moisture / water logging condition. It is a good soil drier because of its root characteristics and hence suited to humid regions. In poorly-drained soils like heavy black soils, improper surface irrigation may lead to prolonged waterlogged condtions which leads to the associated problems of wilts and root rots. Give a light pre-sowing irrigation if the soil moisture in the seed zone is not adequate for germination. Later, depending on the soil moisture status, give one irrigation at 35 days after planting in early elongation stage and another at 65 to 70 days during flowering. After this, no irrigations are generally required unless soils are too light with low water holding capacity. In light soils about 5-6 irrigations are needed including a preplanting irrigation. Under irrigated conditions, it is advisable to plant the crop on broad beds with furrows laid at intervals of 1.35 to 1.8 m or preferably on flat bed and then form irrigation furrows after every 2 rows at the time of first irrigation. Such a system of planting would greatly help to minimise contact of water with above-ground parts. In soils that crack, apply irrigation well before cracks develop for better control of water.

Cropping system: Safflower is intercropped with linseed/coriander/gram or wheat. Safflower based cropping sequences are cowpea / greengram / groundnut-safflower and sesame/sorghum/setaria-safflower. The suggested crop sequence and intercropping systems in different states of India is furnished in Table 7 and Table 8.

Table 7 Safflower based crop sequence in different states of India

State	Suggested crop sequence
Karnataka (Dharwad, Belgaum and adjoining areas)	Mungbean – safflower, Soybean – safflower, Groundnut – safflower
Scanty rainfall areas of Karnataka (Medium deep black soils of Bijapur & western parts of Bellary)	Hybrid Sorghum- Safflower Mungbean-Safflower
Northern Telangana of Andhra Pradesh (parts of Ranga Reddy and Mahaboobnagar district, Adilabad, Medak and Nizamabad)	Mungbean – safflower, Maize-safflower, Hybrid sorghum – safflower, Sesame – safflower
Assured moisture areas of Maharashtra (Khandesh tract, parts of Marathwada and Vidarbha)	Mungbean – safflower, Urdbean – safflower,Hybrid sorghum – safflower, Groundnut – safflower, Sesame – safflower, Sunflower – safflower
Drought prone areas of Maharashtra	Mungbean –safflower Urdbean – safflower
Malwa plateau of Madhya Pradesh	Soybean – safflower, Maize – safflower, Groundnut –safflower
Chhattisgarh	Upland rice – safflower
Bundelkhand region of Uttar Pradesh	Soybean – safflower, Cowpea (fodder) – safflower, Mungbean – safflower, Hybrid sorghum – safflower
Eastern Uttar Pradesh	Upland rice – safflower, Hybrid pearlmillet – safflower, Mungbean – safflower, Urdbean – safflower, Sesame – safflower
Jharkhand	Upland rice – safflower Maize – safflower
Medium and uplands of Odisha	Upland rice – safflower
South-eastern Rajasthan (Udaipur and adjoining areas)	Mungbean – safflower Urdbean – safflower Maize (fodder) – safflower Sorghum (fodder) – safflower Cowpea (fodder) – safflower Cowpea (vegetable) – safflower

Table 8 Safflower based intercropping systems in different states of India

State/ Region	Suggested intercropping system	Row proportion (Other crop : Safflower)
Maharashtra	Chickpea + safflower	3:1 or 2:1
	Wheat + safflower	3:1 or 2:1
	Linseed + safflower	3:1 or 4:2
	Coriander + safflower	3:1
Karnataka	Chickpea + safflower	3:1
	Coriander + safflower	3:1 or 2:1
	Wheat + safflower	3:1 or 5:1

Andhra Pradesh	Chickpea + safflower	3:1 or 2:1
	Wheat + safflower	3:1 or 2:1
	Coriander + safflower	3:1 or 2:1
	Linseed + safflower	3:1 or 2:1
Madhya Pradesh	Mustard + safflower	6:2
	Toria + safflower	6:2
	Chickpea + safflower	2:1 or 6:2 or 4:2
	Linseed + safflower	2:1 or 6:2
	Amaranthus + safflower	6:2
Chhatisgarh	Chickpea + safflower	2:1 or 6:2 or 4:2
	Linseed + safflower	2:1 or 6:2
	Mustard + safflower	6:2
Eastern Uttar Pradesh and Bundelkhand region	Linseed + safflower	3:1
	Chickpea + safflower	3:1
	Barley + safflower	6:2
	Toria + safflower	1:2

Crop protection: Safflower is affected by number of insect pests and diseases although only a few of them are of economic importance. The management practices for insect pests and diseases of safflower are presented in Table 9.

Table 9 Management practices of insect pests and diseases of safflower

Insect pest	**Management practices**
Aphids (*Uroleucon compositae* Theobald)	Spray Dimethoate (0.05%) or Methyl parathion (0.05%) or Monocrotophos (0.05%) or Chlorpyriphos or Endosulfan (0.05%) or alternatively dust Quinalphos (1.5%) or Methyl parathion (2. 5%) or Malathion (5%) or Endosulfan (4%) at 40 and 60 DAS. Use 500 litres of spray mixture and 20 kg dust formulation/ha. One spray of NSKE (5%) a week after first incidence followed by the spray of recommended insecticide 15 days later gives good control of aphid.
Capsule borer (*Helicoverpa armigera* Hubner)	Spray Dimethoate (0.07%) or Endosulfan (0.07%) at the rate of 500 litres of spray mixture/ha.
Disease	
Alternaria **leaf spot** (*Alternaria carthami* Chowdhary)	Spray Mancozeb (0.25%) immediately after disease is noticed and repeat the spray 15 days later depending on the intensity of disease.
Ramularia leaf spot (*Ramularia carthami* Zaprometov)	Spray Mancozeb (0.25%) immediately after disease is noticed and repeat the spray 15 days later depending upon the intensity of disease.
Rust (*Puccinia carthami* Corda)	One or two sprays of Calixin (0.050) or Dithane M-45 (0.25%) at 15 days interval.
Wilt (*Fusarium oxysporum* f.sp.*carthami*)	Treat the seed with Carbendazim @ 0.1-0.2%
Root rot (*Rhizoctonia bataticola*)	Treat the seed with Thiram or Dithane M-45 @ 0.2%.

Birds scaring: Birds pose a threat to the crop during seed setting and harvesting. They could be effectively kept at bay through proper bird scaring in the morning and evening when the crop approaches maturity.

Harvest: The crop will be ready for harvest when the leaves and most of bracteoles, except a few on last formed flower, heads dry and become brown. The crop should be harvested in the early hours in order to avoid shattering of seeds. The spines will also be relatively soft, and it will make harvesting easier. Seeds are winnowed before storage. Whole plants are either cut or pulled out and stacked in the fields in small well-pressed heaps till they are fully dry. Threshing is done by beating with sticks or bullock drawn stone rollers or tractor. The resulting produce is then cleaned by windowing. Threshing and cleaning operations can be done with power threshers used for other crops such as wheat. Combine harvesters used in case of wheat can also be used for harvesting and threshing of safflower.

Yield: A well managed sole crop of safflower can yield between 800 and 1,200 kg/ha in rainfed conditions. In areas with minimal irrigation facilities, it can produce seed yield of 2500 kg/ha. The oil may be extracted from whole seed, or from dehulled seed. Safflower varieties released for commercial production in India in general possess low oil content of 28 to 32%, while HUS-305, NARI-6, and non-spiny hybrid NARI-NH-1 contains 35% oil.

Utilization: Safflower is grown for its orange red dye (*Carthamin*), which is used as a source of dye for clothing and food. The cold pressed oil is golden yellow and largely used for cooking and lighting and in manufacture of soap, paints, varnishes, etc. The oil is used for cooking as a foodstuff, for making food dressings and for lighting. It is processed to make margarine. The cake produced by processing oil from unhulled seed is an excellent feedstuff for ruminant livestock. Roasted safflower seeds are used for consumption. The cold-pressed oil is golden yellow and is used for culinary purposes, or for making soap. The oil obtained by dry hot distillation is black and sticky and is used only for greasing well ropes and leather goods exposed to water. Safflower oil has also good drying properties and is, therefore, used in the manufacture of paints, varnishes and linoleum. It can be mixed with white paint without any after-yellowing effects. The cake, particularly from decorticated seed, is used as a concentrated cattle feed, and that from undecorticated seed is sometimes used as a manure. The safflower oil is rich in poly unsaturated fatty acids (Linoleic acid 78%) which play an important role in reducing the blood cholesterol level.

REFERENCES

Anjani, K. 2000. Components of seed yield in safflower (*Carthamus tinctorius*). Indian J. Agric. Sci. 70: 873–875.

Anonymous. 2002. Safflower Research in India. Directorate of Oilseeds Research, Hyderabad, 96 pp.

Ashri, A. and P.F. Knowles. 1960. Cytogenetics of safflower (*Carthamus* L.) species and their hybrids. Agron. J. 52: 11–17.

CPG. 2012. Crop Production Guide. Department of Agriculture, Government of Tamil Nadu, Chennai and Tamil Nadu Agricultural University, Coimbatore.

Deshpande, R.B. 1952. Wild safflower (*Carthamus oxyacantha* Bieb.): a possible oil seed crop for the desert and arid regions. Indian J. Genet. 12: 10–14.

Knowles, P.F. 1969. Centers of plant diversity and conservation of crop germplasm: safflower. Econ. Bot. 23: 324–329.

Nagaraj, G. 1995. Quality and Utility of Oilseeds. Directorate of Oilseeds Research, Hyderabad, India, 70 pp.

Prakash, K.S. and B.G. Prakash. 1993. Yield structure analysis of oil yield in safflower (*Carethamus tinctorius* L.). Oleagineux 48: 83–89.

Ramachandram, M. 1985. Genetic improvement of oil yield in safflower: problems and prospects. J. Oilseeds Res. 2: 1–9.

http://oilseeds.dacnet.nic.in/crop.html

CHAPTER

7

Castor–*Ricinus commnuis* L. (2n=20)

Family: Euphorbiaceae

Vernacular name: Castor Bean, Palma Christi, Wonder Tree (English), *Rehri* (Hindi), *Erandi* (Marathi), *Amanakku* (Tamil), *Amudhulu* (Telegu) and *Rehri* (Bengali).

Importance: Castor is an industrial oilseed crop. Seed consists of 20% husk or shell and 80 % of soft kernal. Castor soft kernel contains 58 to 66% oil while the unshelled or undecorticated seed contains 40 to 53% of oil. The small seeded castor varieties contain about 7 to 8% more oil than the large seeded varieties. Castor seed also contain lipase which split the oil into glycerin and fatty acids and is prepared for industrial uses. The oil contains 85% of recinoleic fatty acid. Castor oil is a non drying oil. The castor oil cake contains an alkaloid called '*Ricin*'. It is a poison and act as blood coagulant. Alkaloids are vegetable bases containing nitrogen. They have much value in medicine and drugs. Cold drawn castor oil is free from ricin. The demand for castor oil is more in industrialized nation. The major importers of castor oil in the world market are European Union, USA and Japan. India is the major supplier of castor oil to all these destinations. India dominates the castor oil exports market as both China and Brazil have a large domestic demand for castor oil and as such their domestic production goes to own industry. India consumes only a small percentage of its castor oil production and exports the maximum part to the countries like USA, European Union and Japan, which are totally dependent on castor oil imports. The world demand for castor oil is increasing at the rate of about 3-5% per annum.

Origin: Castor is native of Africa (Ethopia) and India. It is also reported as of polyphyletic origin with four centres of variability viz., Afghanistan, Russia, Palestine-West Asia, Indo-China and Arabian Peninsula.

History: The genus name '*Ricinus*' is derived from Latin term meaning 'dog tick' because of its seed resemblance to the common pest of dog. *Communis* means common in Latin. The name 'Castor' has said to be coined by English traders who confused it with the oil of another shrub, *Vitex agnus-castus,* which the Spanish and Portuguese in Jamaica called 'agno-casto.' The expressions 'castor bean' and 'castor seed' are used interchangeably in referring to the seed of *Ricinus communis* L, the source of castor oil. Although it is commonly known as the castor bean plant, the seed is really not a true bean and it is not related to the Bean or Legume Family (Fabaceae).

Distribution: Castor is grown in Brazil, India, Russia, Argentina, China, Thailand, USA, Egypt and Sudan. India ranks first in area and second in production to Brazil. India accounts 30% of world production. In India, the major Castor growing states are Gujarat and Andhra Pradesh accounting for 85% in area and 90% in production. The castor is grown in limited areas in Rajasthan, Tamil Nadu, Odisha, Karnataka, Maharashtra, Madhya Pradesh, Assam and Bihar.

Botany: Castor is a tall glabrous annual or perennial, growing to a shrub. The stem is erect, circular in section, partially hollow, smooth, glabrous with good branching. The stem is marked by a number of well-defined nodes, from each of which a leaf arises. The lower internodes are shorter and their length increases with the height. It has a well-developed root system, going to a depth of two metres and more. It has also surface root spread with numerous roots forming mat between 15 and 25 cm depth of soil. Leaves are alternate, large, palmate with 5-11 lobes, acuminate, margins notched, serrate or indented. They are carried on long stout petioles. Castor has ashy coating on the leaves and stem of the plant. The inflorescences are borne terminally on the main and lateral branches. Flowers are large, in terminal sub-panelled racemes. It is monoecious (unisexual) condition with male flower at the bottom of the peduncle and female at the top. Flowering occurs from the base upwards. The female flowers are reddish brown in colour. The proportion of male and female flowers varies from 95 to 5% and is greatly influenced by environment. Temperature, growth regulator, nitrogen and chemicals like silver and cobalt nitrate also alter sex ratio. A fairly high percentage of self-pollination occurs. A natural crossing of 5 to 14% occurs in castor. The fruits are egg-shaped capsules, about 2.5 cm long, thickly covered with soft flexible spines. Four distinct size groups of fruits *viz*., very small, small, medium and large are recognized. Very small fruits are found in ornamental types and in some of the wild perennial types. Each capsule contains three seeds that look like fat swollen dog ticks and are deadly poisonous. Seeds are albuminous, anatropous, broad, oval, compressed with a marked carnnele and longitudinal raphe. Medium and small seeded types are preferred for cultivation since they have a fairly high oil content varying from 45 to 57%. Big seeds have generally low oil content of less than 40%. Very small seeds are preferred for medicinal purposes. The seed colour ranges from white to grey, deep chocolate purple and red. Mottling is also much varying. The hilum almost concealed under the caruncle; presence of thin leaf like cotyledon. The smallest seed is 5 mm to 8 mm while the largest seed is about 13 mm to 21 mm. The testa is thin, brittle, varying in colour and mottling. Below the testa is the thin tegmen, covering the whitish oily endosperm containing the embryo. Four year old seeds have recorded 60% viability. Germination is quicker if the seed coat is removed.

Growth stages: Seedlings of castor emerge 10 to 20 days after sowing. The growth is sympodial. The successive formation of branches and inflorescences continues throughout the plant's life. In annual cultivars, the first inflorescence is the largest one and may account for up to 80% of the seed yield. In perennial cultivars, flowering is more diffuse. The first flowers may open 40 to 70 days after sowing. Pollen is mainly shed in the morning and pollination is by wind. Ripening of fruits within an infructescence is uneven, the lower fruits maturing before the upper ones. In wild types, the period of maturation between the first and the last fruits within a given infructescence may be several weeks. In cultivars grown as annuals, the period from emergence to maturation varies from 140 to 170 days. Under favourable conditions, castor has a high rate of photosynthesis, which has been attributed to the high chlorophyll content in the leaves.

Climate: Castor is grown in tropics, sub-tropics and temperate zones. It is cultivated mostly in the arid and semi- arid regions in the world. It has been commercially cultivated from 40°S to 52°N, from sea-level to 2000 m altitude at the equator, with an optimum at 300–1500 m. Castor is considered to be a drought resistant crop. Castor can be grown in areas receiving a well distributed mean rainfall of 500-750 mm. Castor can withstand dry arid climates, but also heavy rains and short flooding. Castor can tolerate water stress because of its deep root system, but is sensitive to excess of water and humidity. In heavy rainfall areas, the crop puts on excessive vegetative growth and assumes a perennial habit. It can withstand long dry spells as well as heavy rains, but is highly susceptible to water logging. Cloudy or humid days irrespective of temperature will reduce yields and castor is highly susceptible to frost. Humid and cloudy weather during flowering period will promote fungal diseases (Botrytis) of the inflorescence resulting in total loss of spike. The warmer temperatures at the time of flower initiation promote more male flowers while cooler temperatures produce more female flowers within a given genotype. High rainfall and high temperature at flowering results in poor seed set. Castor beans are grown best where temperatures remain fairly high throughout the growing season of 140 to 180 days. Castor is sensitive to frost. The cardinal minimum, optimum and maximum temperature for growth and yield of castor is 10°C, 20 to 32°C and 40°C respectively. Moderate temperature of 32°C promotes female flowers while high temperature promotes male flowers. If the temperature increases beyond 32°C, the production of male flower increases which leads to seed yield reduction. The seed may fail to set, however, if the temperature stays above 38°C for an extended period.

Varieties and hybrids in Castor: The varieties and hybrids in Castor grown in different states are furnished in Table 1 to Table 4.

Table 1 State-wise farmers preferred hybrids/varieties of Castor

State		Recommended varieties/hybrids
Andhra Pradesh	Varieties	Jyothi, Kranti, Kiran, Haritha, 48-1(Jwala)
	Hybrids	GCH-4, DCH-32, DCH-177, PCH-1, DCH 519
Gujarat	Varieties	VI-9, GAUC-1, SKI-73 (GC 2), 48-1, GC-3
	Hybrids	GAUCH-1, GCH-2, GCH-4, GCH-5, GCH-6, GCH-7, DCH-519
Karnataka	Varieties	RC-8, Jyothi, 48-1
	Hybrids	GCH-4, DCH-32, DCH-177, DCH-519
Maharashtra	Varieties	Jyothi, AKC-1, 48-1
	Hybrids	GCH-4, DCH-177, DCH-32, DCH-519
Rajasthan	Varieties	Jyothi, 48-1
	Hybrids	GCH-4, GCH-5, DCH-32, RHC-1, DCH-177, DCH-32, DCH-519
Tamil Nadu	Varieties	SA-2, TMV-5, TMV-6, Jyothi, Co-1, 48-1
	Hybrids	GCH-4, DCH-32, TMVCH-1, DCH-177, DCH 519, YRCH-1
Uttar Pradesh	Varieties	T3, T4, 48-1, Kalpi-6, Chandra Prabha
Haryana & Punjab	Varieties	CH-1, Jyothi, 48-1
	Hybrids	GAUCH-1, GCH-2, GCH-4, GCH-5, DCH-32, DCH-177, DCH-519

Source: Singh, 2014

Table 2 Castor varieties particulars grown in different states of India

Variety	Year of release	Releasing Centre / Duration	Yield potential (kg/ha)	Oil content (%)	Recommended states/regions situations	Salient features/ traits
Haritha (PCS-124)	2002	ANGRAU, Palem	1400-1600 (R)	49	Light soils of Southern Telangana, Rayalaseema and Prakasam District.	Resistant to wilt
Kiran (PCS 136)	2002	ANGRAU, Palem	1200-1500 (R)	51	Rainfed areas of A.P. and also late sown*kharif* conditions with one or two irrigations.	Tolerant to Botrytis
Jwala (48-1)	2007	DOR, Hyderabad	1000 (R) 1800 (I)	50	All castor growing areas under both rainfed and irrigated	Resistant to *Fusarium* wilt; tolerant to *Botrytis*, salinity

GC-3	2009	JAU, Junagadh	2340 (I)	49	Irrigated areas of Gujarat	Resistant to wilt
Chandra prabha	2009	CSAU&T Kanpur			Uttar Pradesh	Suitable for intercropping
DCS-107	2010	DOR Hyderabad	1762 (I)	46	Identified for both rainfed and irrigated areas of the country	Resistant to wilt
AKC-1	----	120-150	1200 (R)	48	Rainfed areas of Maharashtra	Medium height, red stem, high yielding.
Jyothi (DCS-9) (REC-9)	----	90-150	1028 (R)	49	Rainfed areas of Andhra Pradesh, Karnataka and Tamil Nadu	Superior to Aruna for wilt resistance.
Gujarat Castor-2 (SK1-73)	----	120-180	1190 (R) 2770 (I)	49	Irrigated areas of Gujarat	Red stem, early maturity, triple bloom. Susceptible to wilt.
TMV 6	----	160-170	1166 (R)	52	Rainfed areas of Tamil Nadu	Red stem, medium height elongated inter-nodes and cat green leaves with brown tinge, double bloom
Kranti (PCS-4)	----	----	1200 (R)	48-50	Rainfed areas of A.P.	Medium height, double bloom, red stem, flat leaf, mostly female spiny capsules, moderately tolerant to wilt.

Table 3 Castor Hybrids in India

Hybrid	Year of release	Releasing Centre / Duration	Yield potential (kg/ha)	Oil content (%)	Recommended states/regions situations	Salient features/ traits
PCH-1	2001	ANGRAU Palem	1500 (R) 2000 (I)	-	Rainfed areas of A.P.	Tolerant to wilt resistant to Jassids
RCH-1	2002	RAU, Mandor	1600(R) 2000 (I)	49	Rajasthan	

DCH-519	2006	DOR Hyderabad	1500 (R) 2200 (I)	49	Both rainfed and irrigated castor growing areas of the country	Green, triple bloom, resistant to *Fusarium* wilt
GCH-5	----	180-240	1740 (R) 2820 (I)	50	Gujarat, Rajasthan	Resistant to root rot and wilt.
GCH-6	----	120-140	1400 (R) 2300 (I)	48	Gujarat, Rajasthan, Maharastra	Resistant to root rot, tolerant to wilt, whitefly, jassids, thrips,
GCH-7	2006	SK Nagar Gujarat	2450 (I)	49	Irrigated areas of Gujarat	Resistant to nematode-wilt complex
YRCH-1	2009	Yethapur, TN	1860 (R) 3000 (I)	49	For Tamil Nadu	----
TMVCH-1	----	Tindivanam, TN 160-170	600 (R) 1300 (I)	52	For Tamil Nadu	moderately resistant to wilt.
Deepti	---	90-180	1800 (R) 2800 (I)	49	Gujarat, Rajasthan, Maharashtra, Karnataka, Tamil Nadu, Andhra Pradesh	Wilt tolerant
DCH-519	----	----	1500 (R) 2200 (I)	48	All states	Resistant to fusarium wilt

Table 4 Castor varieties and hybrids particulars in Tamil Nadu

Particulars	**CO 1**	**TMV 5**	**TMV 6**	**Hybrid TMVCH 1**	**Hybrid YRCH 1**
Parentage	Pureline selection from Anamalai	Derivative of SA 2 X S 248/2	Derivative of VP 1 X RC 962	LRES 17 X TMV 5	DPC 9 X TMV 5
Duration (days)	perennial	120	160	160 -170	150-160
Yield (kg/ha)					
Rainfed (mixed crop)	----	----	500	600	----
Rainfed (pure crop)	2.5 kg / tree/ year	850	950	1300	2000
Irrigated (pure crop)	----	----	----	----	3000
Oil content (%)	57	50	51.9	51.7	49

Special features					
Stem colour	Pinkish green	Rose	Red	Red	Light red
Bloom (waxy coat)	No bloom	Triple	Double	Triple	Triple
Receme/ capsule	Bold, sparse setting, non dehiscent	Spiny, non dehiscent, resistant to leaf hopper	Medium, lengthy, spiny capsule	Semi compact, spiny capsule	Spiny, non dehiscent, resistant to leaf hopper
Suitability	Bund crop and fit for raising in vacant areas	Pure and mixed crop	Pure and mixed crop	Pure and mixed crop	Pure and mixed crop

Soil: Deep red loam and alluvial loam soils are suitable for castor. It can withstand clayey soils of rice fields. Castor crop tolerates acid soil, saline soil up to 8 ds/m and alkaline / sodic soil up to 30 ESP as long as the subsoil is permeable and there is good drainage.

Season: Indian castor is produced under two contrasting environments *viz*., irrigated intensive cultivation with high productivity in Gujarat and Rajasthan; and rainfed culture coupled with poor management with very low productivity in Andhra Pradesh, Karnataka, Tamil Nadu, Odisha, etc., which requires location specific technologies different for these two situations for sustainability. Castor is sown in June-July both under irrigated and rainfed conditions in Tamil Nadu. It is also grown in November-December season under irrigated condition. If it is sown in July/August, the harvesting commences around December/January. The crop duration is 4-5 months. The most ideal time to sow *kharif* castor in drylands is immediately after the receipt of first rains from south-west monsoon. Optimum seeding time for rainfed castor is second fortnight of June in Andhra Pradesh and other southern states and first fortnight of July in Gujarat and Rajasthan. Information generated hitherto suggests that September-October is ideal period for *rabi* castor, while January is optimum for sowing summer castor. In case the other *kharif* crops could not be sown due to delayed monsoon, castor can form a better contingent crop even for sowing up to 15th August in rainfed areas of Andhra Pradesh.

Land Preparation: Plough the land with tractor twice with iron plough or three to four times with country plough till all the clods are broken to have a friable soil tilth.

Seeds and sowing: Seed rate is 6 to 12 kg per ha depending on seed size. Seeds are sown with the spacing of 60 × 45 cm and 90 × 45 cm in rainfed and irrigated conditions respectively. A spacing of 120 × 90 cm with seed rate of 5 kg per ha is recommended for hybrid castor such as YRCH-1. Castor seedling takes around 7-8 days to emerge and complete germination by 10-12 days in *kharif* and 15 days in winter depending on the prevailing temperatures. Seeds may be treated with Thiram or Captan @ 3g/kg seed or Carbendazim 2 g/kg to protect plants from seed borne diseases like *Alternaria* Leaf Blight, Seedling Blight and Wilt. Treat the seed with *Trichoderma viride* @ 10 g/kg seed and soil application of 2.5 kg incubated in 125 kg FYM/ha for managing wilt. Germination can also be improved by pre-soaking with KNO_3 @ 500 ppm followed by $ZnSO_4$ at 100 ppm. Soaking the seed in 1% sodium chloride (common salt) for 3 hours before sowing imparts tolerance to sodicity wherever the problem exists. Seed treatment can also be done with *Azospirillum* or phosphorus solubilizing bacteria (PSB) @ 50 g/kg seed. The spacing recommended for castor in different regions and situations is given in Table 5.

Table 5 State-wise recommendation of spacing for castor

Region	Situation	Spacing (cm)
Andhra Pradesh	Rainfed	90 × 60
	Delayed	60 × 30
Karnataka	Rainfed	60 × 30
	Irrigated	90 × 60
Tamil Nadu	Rainfed	90 × 60
Maharashtra	Rainfed	90 × 60
Delayed sowings	Rainfed	60 × 30/90 × 20
Gujarat-North-west	Rainfed	90 × 60
	Irrigated	120 × 60
Saurashtra	Rainfed	120 × 60
Rajasthan	Irrigated	90 × 60

After cultivation: Nipping of axiliary buds is carried out soon after emergence of spike to facilitate the growth and development of main spike and induces early maturity from 30 to 50 days.

Weed management: Two hand weeding are given on 20 and 40 DAS. Generally, the crop may need 2 or 3 hand weeding at intervals of 15-20 days in order to keep weeds under check. Pre-plant application of herbicides such as Fluchloralin or Trifluralin @ 1.0 kg a.i./ha or pre-emergence application of Alachlor @ 1.25 kg a.i./ha is equally effective under irrigation.

Manures and fertilizer application: Apply 12.5 tonnes of FYM or compost per ha as basal dressing. Fertilizer dose of 60-30-30 and 30-15-15 kg of N, P_2O_5 and K_2O per ha is recommended for irrigated and rainfed conditions respectively. 50% of N and entire P_2O_5 and K_2O are applied basally while the remaining 50% of N is applied on 60 DAS. The State wise fertilizer recommendations are given in Table 6.

Table 6 State wise fertilizer recommendations for Castor

Region	Situation	N	P_2O_5	K_2O	S
Andhra Pradesh	Rainfed	60	40	30	-----
Tamil Nadu	Rainfed	30	15	15	-----
Karnataka	Irrigated	75	50	25	-----
	Rainfed	40	40	20	-----
Maharashtra	Rainfed	60	30	0	
Rajasthan	Irrigated	80	50	0	20
North Gujarat	Irrigated	120	25	0	-----
	Rabi castor	80	50	0	-----
Saurashtra (Gujarat)	Irrigated	120	50	0	20
	Rainfed	100	50	0	20

Source: Singh, 2014

For irrigated castor, nitrogen can be given in 5 splits, 40 kg each at sowing, 40, 70, 100 and 130 days after sowing. In soils deficient in Zn and Fe, 10 kg ZnS04 and 30 kg FeS04/ha, respectively has to be applied to obtain enhanced yields. In saline/sodic soils, spot application of 12.5 t/ha FYM and sowing the crop on the side of the ridge has been found beneficial. Besides, adoption of salinity tolerant varieties like GCH-5, 48-1, GC-2 also helps in combating salinity/sodicity problem.

Water management: Castor is grown as rainfed crop. Castor being a deep rooted crop can extract water from considerable depth in the soil. The medium and long duration hybrids/ varieties require 5 to 7 irrigations. The first irrigation should be applied at around 55-75 days or around full flowering of primary spike. The subsequent irrigations may be given at intervals of 20 days after first irrigation, so as to ensure availability of adequate moisture for the development of different spike orders. Since, castor is a deep rooted crop, sufficient water to wet at least 30-40 cm soil profile should be applied. Application of irrigation at 0.8 IW/CPE is recommended. Castor require 20 acre inch of water. The critical growth stages for moisture stress in castor are flowering and fruit set. Castor is a highly drought resistant crop.

Cropping system: Intercropping of castor in groundnut, sesame, sunflower, turmeric, ginger and sugarcane and mixed cropping with finger millet, groundnut, cotton, tobacco and horsegram is commonly practiced in Tamil Nadu. Farmers rotate castor with crops such as sorghum, pearl millet, redgram in Andhra Pradesh and sorghum, pearl millet, redgram, cowpea, guar, wheat and mustard in Gujarat. Avoid continuous cropping of castor year after year to avoid wilt and root rot. The different types of castor based intercropping practices are as follows:

Table 7 State-wise castor based intercropping systems

State	Intercropping	Row proportion
Andhra Pradesh	Castor + Pigeonpea	1:1
	Castor + Cowpea	1:2
	Castor + Urdbean	1:2
	Castor + Mungbean	1:2
	Castor + Clustebean (vegetable)	1:1
	Castor + Groundnut	1:5-7
	Castor + Turmeric/Ginger	1:5
	Castor + Horsegram (relay)	1:6-8
Gujarat	Castor + Pigeonpea/Clusterbean	1:1
	Castor + Cowpea	1:2
	Castor + Urdbean	1:2
	Castor + Mungbean	1:1/1:2
	Castor + Bunch groundnut	1:3
	Castor + Pearlmillet	1:2
	Castor + Sesame	1:1

Karnataka	Castor + Groundnut	1:5-7
	Castor + Horsegram (relay)	1.6-8
	Castor + Pigeonpea	2:1
Tamil Nadu	Castor + Groundnut	1:5-7
	Castor + Chillies	1:8
	Castor + Urdbean	1:2
Bihar	Castor + Soybean	1:1
	Castor + Lathyrus	1:5
	Castor + Turmeric/Ginger	1:5
	Castor + Chillies	1:8
Rajasthan	Castor + Mungbean	1:2
	Castor + Mothbean	1:2

Table 8 State-wise castor based crop sequence systems

State	Crop sequence	One year/two year	Irrigated/ rainfed
Andhra Pradesh	Castor-groundnut	One year	Irrigated
	Castor-sunflower	One year	Irrigated
	Castor-pearl millet	Two year	Rainfed
	Castor-finger millet	Two year	Rainfed
	Castor – pigeonpea	Two year	Rainfed
	Castor – groundnut	Two year	Rainfed
	Castor-sorghum	Two year	Rainfed
Gujarat	Pearlmillet-castor	One year	Irrigated
	Sorghum-castor	One year	Irrigated
	Castor-mungbean	One year	Irrigated
	Castor-groundnut	One year	Irrigated
	Castor-sesame	One year	Irrigated
	Castor-sunflower	One year	Irrigated
	Mustard-castor	Two year	Irrigated
	Castor-pearl millet-mungbean	Two year	Irrigated
	Castor-pearl millet-chickpea	Two year	Irrigated
Karnataka	Castor-finger millet	Two year	Rainfed
	Castor-groundnut	Two year	Rainfed
Tamil Nadu	Castor-finger millet	Two year	Rainfed
	Castor-groundnut	Two year	Rainfed
Rajasthan	Castor-pearl millet	Two year	Irrigated

Plant protection: Insect-pests are castor semi-looper, capsule borer, tobacco caterpillar, jassids, white flies, mites and thrips. As soon as the egg masses of Castor semi-looper are first detected over the second leaf from the top of the main stem and secondary shoots, spray Monocrotophos (0.05%) or Quinalphos (0.05%) or Dimethoate (0.05%) or Endosulfan (0.05%). For effective control, direct the spray on the lower surface of foliage more particularly the second leaf from the top of different order of branches/shoots. Hand pick and destroy semi-looper larvae as well as Tobacco caterpillar. Direct the spray of Monocrotophos (0.05%) or Quinalphos (0.05%) or Dimethoate (0.05%) on the primary and secondary spikes in order to check further spread of Capsule borer to higher order racemes and Tobacco caterpillar. Spray can be taken up timely with Dimethoate (0.05%) or Monocrotophos (0.05%) to control sucking pests such as jassids, white flies, mites and thrips when observed on the lower side of foliage for their presence. The castor diseases are seedling blight, alternaria leaf blight, wilt and root rot. Seed treatment with Thiram or Captan @ 3 g per kg of seed is recommended to control seedling blight and alternaria leaf blight. For wilt and root rot management, avoid water logged conditions. Avoid planting castor immediately after sorghum if the latter has been infected with charcoal rot. Uproot and destroy affected plants as and when they are detected.

Harvest: Castor is a multiple branching type which produces 4 to 5 sequential order spikes over a span of 180 to 240 days, one each at an interval of about 30 days. The main spike is ready for harvest within 90-120 days after planting. The subsequent pickings can be taken up at intervals of 30 days. Physiological maturity in castor is attained when some of the capsules in a spike turn brown in colour. Harvesting is done when capsule turn yellowish and start drying. However, all spikes do not mature at the same time. The matured spikes are harvested first and the remaining spikes later. Two to three picking are required. Spikes dried in hot sun for four to five days. Threshing is usually done by either beating the capsules with sticks or alternatively by trampling with bullocks. It is essential to dry the seeds completely before storage. Castor beans should be stored at less than 6% moisture.

Yield: Castor produces a seed yield of 2.0 to 2.5 t/ha. The average yield of castor has been recorded as 1451 kg/ha with highest in Gujarat (2006 kg/ha) followed by Rajasthan (1299 kg/ha). Similarly, the average yield recorded in the world is 1140 kg/ha and highest yield is in India (1451 kg/ha) followed by Paraguay (1157 kg/ha) and Thailand (876 kg/ha).

Utilization: Castor oilcake contains 30 to 45% protein. Castor cake is not fed to livestock but used as valuable manure. Castor oil cake contains 5.5 % N, 1.8 % P and 1.1 % K. Castor oil is used as lubricant in high-speed engines and aeroplanes, in the manufacture of soaps, transparent paper, printing-inks, varnishes, linoleum and plasticizers. It is also used for medicinal and lighting purposes. The plant stalks are used as fuel or as thatching material or for preparing paper-pulp. In the silk-producing areas, leaves are fed to the silkworms.

REFERENCES

Brigham, R.D. 1967. Natural outcrossing in dwarf-internode castor, *Ricinus communis* L. Crop Sci. 7:353-355.

Duke, J.A. 1998. *Ricinus communis*. From Purdue University New Crop Resource Online Program. http:// https://hort.purdue.edu/newcrop/duke_energy/Ricinus_communis.html

Hegde, D.M., Sujatha, M. and N.B. Singh. 2003. Castor in India. Directorate of Oilseeds Research, Hyderabad, India.

Kadambi, K. and S.N. Dabral. 1955. The silviculture of *Ricinus communis* Linn. Indian Forester 81(1): 53-58.

Meinders, H,C, and M.D. Jones. 1950. Pollen shedding and dispersal in the castor plant Ricinus communis L. J Agron 4:206–209

Singh, R. P. 2014. Status Paper on Oilseed Crops. Directorate of Oilseeds Development. Hyderabad. pp 118–135.

Zimmerman, L.H. 1958. Castorbeans: a new crop for mechanized production. Adv. Agron. X: 257-288

http://oilseeds.dacnet.nic.in/crop.html

http://ntbg.org/plants/

http://waynesword.palomar.edu/plmar99.htm#castor

CHAPTER

8

Linseed/Flax–*Linnum usitatissimum* L. (2n=30)

Family: Linaceae

Vernacular name: Flax, Linseed (English), Lin (French), Lein, Flachs (German), Lino (Spanish), *Alsi* (Hindi), *Alivithai* (Tamil), *Avise* (Telugu), *Javas* (Marathi) and *Tishi* (Bengali)

Importance: Flax is one of the oldest cultivated plants known to mankind. Two types of flax are grown, seed flax for the oil in its seed, and fiber flax for the fiber in its stem. It is grown chiefly for its valuable fibres and principally for its seeds in southwest Asia. Linseed is known as flax when grown for fibre purpose. Flax is grown commercially for two products, fiber and seed. Linseed or flax seed are different names for the same seed. Flax seeds produce a vegetable oil known as flaxseed or linseed oil. For a long time flax has been cultivated as a dual-purpose crop, but nowadays fiber flax and linseed represent different gene pools. The cultivated groups are grown for fiber flax and linseed. Fiber flax has a long unbranched growth habit, whereas linseed (oil flax) is much shorter and highly branched.

Origin: Linseed is originated in Mediterranean region as primary centre while (i) Ethiopia (Somali Land), (ii) Asia minor, (iii) Punjab, Kashmir and Afghanistan as subcentres.

Distribution: Fibre flax is grown in Canada, Russia, Belorussia, Ukraine, France, Belgium, Italy, Germany, United Kingdom, Holland, Belgium, Argentina, China, Egypt while Linseed is grown in Canada, USA, Argentina, Chile, Poland, Netherlands, Czechoslovakia, Rumania, Hungary, Yugoslavia, Chile, Mexico, Australia, India, Ethopia, Afghanistan, Turkey, Pakistan, Bangladesh and Japan. In India, it is grown mainly for the production of oil in Maharashtra, Rajasthan, Punjab, Karnataka, Haryana, Madhya Pradesh, Andhra Pradesh and Odisha. It is grown for fibre in some temperate hill regions such as Himachal Pradesh, Jammu and Kashmir and West Bengal.

There are four agro-climatic zones for linseed cultivation in India *viz.*,

(i) Zone-I: Himachal Pradesh, Punjab, Haryana, Jammu & Kashmir

(ii) Zone-II: Uttar Pradesh excluding Bundelkhand, Bihar, Jharkhand, West Bengal, Assam and Nagaland.

(iii) Zone-III: Bundelkhand area of Uttar Pradesh, Madhya Pradesh and Rajasthan
(iv) Zone-IV: Chattisgarh, Odisha, Maharashtra and Karnataka.

Botany: Linseed is an erect herbaceous annual with slender stem, glabrous, grayish-green stem, which grows to a height of about 90 to 120 cm, depending upon variety, soil conditions and length of growing season. The tap root could reach to a depth of 80 to 130 cm. The varieties differ in their branching habits; those grown for fibre are almost unbranched except near the top of the stem, whereas the oil-seed types are bushy in character and often bear two or more basal branches arising just above the surface of the soil. The leaves are narrow, lanceolate, alternate and sessile. The leaf size varies from 2 to 5 cm in length and 5 to 10 mm in width. Flowers found in terminal and axillary raceme. Most flax varieties have blue flowers while some varieties have white flowers. The flowering is indeterminate and continues until the end of the growth period. It is a self-pollinated crop. However 0.3 to 3.4% out crossing may occur under normal circumstances. Insects such as honey bees are the primary agents of out crossing. The pollen is viable for only a few hours, from the time of anther dehiscence until about the time the petals dehisce between 4 and 7 hours. As the flower opens, the anthers come together and form a cap over the stigma. The mature fruit of the flax plant is a dry boll or capsule. Ripening of the boll begins 20 to 25 days after flowering. The boll has five segments which are divided by a wall (septum). Each segment produces two seeds separated by a low partition called a 'false septum', whose margin may be hairy or smooth, depending on the variety. With complete seed set, the boll contains ten seeds, with an average of six to eight seeds per boll. Flax seeds are flat, oval, and are pointed at one end. A thousand seeds weigh from about 5 to 7 g depending on variety and growing conditions. Seed contains 30 to 40% oil. Seed of different varieties ranges in colour from light to dark reddish brown or yellow. The yellow flax called Linola or Solin. Solin oil contains less than 5% linolenic acid compared to more than 50% in flax oil, producing a light oil suitable for cooking. Flax and solin are grown using the same agronomic practices. There are two distinct morphological types *viz.*, 'flax' and 'linseed' which are recognized in this cultivated species. The flax types are primarily grown for extraction of fibre which is tall growing with straight culms and less number of secondary branches. The varieties of flax that is grown for seed are much bushier and more branched than those grown for fibre. Flax is a soft bast fibre. The linseed types which are predominantly grown in India which are meant for extraction of oil. Varieties grown for seeds grow to a height of 30 to 45 cm and generally assume a bushy character with lateral branches. The linseed grown in India can be further classified into the Indo-Gangetic and the peninsular groups. The Indo-Gangetic group has tall plants with small seeds and low oil content. The peninsular group has shorter plants with bold seeds and high oil content.

- Cultivated linseed can be charactrised as an annul, glabrous having one to many stems, 20-110 cm tall, leaves alternate, sessile, linear-lanceaolate, about 40 mm long and 4.7 to 7.5 mm wide, acuminate and ciliate on the inner sides, pollen three colpate, styles almost free to the base, stigma clavate, capsules globose, mostly indehiscent, seeds ovate and brown or yellow in colour.
- Annual, herbaceous winter crop with short and slender tap root that grows vertically downwards in to numerous lateral branches.
- Two distinct morphological types viz., seed type and flax/dual purpose type are recognized.
- Linseed in India mostly utilized for extraction of oil.
- Self-pollinated crop although 0.43-3.4% natural out-crossing has been reported.

Growth stages:

(i) Germination: 4 to 6 days
(ii) Vegetative period: 45-60 days
(iii) Flowering period: 15-25 days
(iv) Maturation period: 30 to 40 days

Climate: It is a cool/winter season crop and requires moderate cool temperature during the growing season. Flax requires humid and cloudy weather during the growing season. It is grown up to an altitude of 800 m from MSL. Fibre flax is essentially a crop of temperate climate free from heavy rains and frosts. It grows in regions receiving mean annual rainfall of 480 mm to 760 mm. Linseed can be sown when soil temperature is >7-9°C. It requires a minimum and maximum temperature of 10°C and 38°C, respectively. The optimum temperature for seed germination is 18 to 25°C, vegetative and reproductive development is 21-27°C and for seed formation is 15 to 20°C. Moderate air temperatures (18-20°C) and the accompanying cloud cover promote high yields. Flax may be damaged or destroyed at temperatures between -8 and -3°C in the seedling stage, and a light freeze of -1°C may cause injury in the blossom or green boll stage, but between these stages the plants may survive temperature of -9°C or even lower. High temperature of >32°C accompanies with moisture stress during flowering stage reduced the seed yield, oil content and quality of oil.

Soils: Linseed makes its best growth on well drained, fertile medium heavy soils, especially silt loam, clay loam and silt clays. Light soil is unsuited for the seed crop particularly in the regions of deficient rainfall. Linseed has the best growth at 5.0 to 7.0 pH. It is sensitive to salt affected soils. Alkaline soils are unfavorable for its economic production.

Season: Linseed is grown during *rabi* season during October-November to March-April.

Varieties: The particulars of Flax / Linseed varieties grown in different states of India are furnished in Table 1.

Table 1 Flax / Linseed varieties grown in different states of India

Variety	Yield (kg/ha)	Oil content (%)	Specific features	Recommended states /regions/situations
Parvati	1600(S), 1020(F)	42.00	Blue fowered, brown seeded moderately resistant to rust and PM	U.P.(Excl. Bundelkhand), Bihar, West Bengal & Rajesthan (Kota)
Sheela	1379 (R)	41.00	Erect deep blue flower, shiny brown seeded, resistant to rust, wilt and moderately resistant to AB and BF	Himanchal Pradesh, Punjab, Haryana and J &K
Shekhar	1555 (I), 920 (R)	43.00	Violet blue flower, shining brown seeded, resistant to PM, rust, wilt and moderately resistant to AB and BF	U.P.(Excl. Bundelkhand), Bihar, West Bengal & Assam
NL-97	641 (R)	42.00	Blue flowered, early maturing, moderately resistant to bud fly, AB, PM & tolerant to wilt.	Bidarbha Region of Maharashtra

RL-914	1617 (I)	41.10	Tall in height, brown seeded, resistant to rust and wilt	Kota Command Area of Rajasthan
Suyog (SLS -27)	1509 (I)	41.43	Medium in height, white flower, light brown seeded, moderately resistant to rust, PM and BF.	Rajasthan.,U.P., M.P., Maharastra,CG, Odisha, A.P. & Karnnataka.
Binwa (KL-210)	1142 (I)	40.00	Medium in height, blue flower, yellow seeded, resistant to rust.	Haryana, Punjab, Himanchal Pradesh and J & K.
Baner (KL-224)	511 (U)	39.70	Medium in height, purple flower, brown seeded, resistant to rust.	Haryana, Punjab, Himanchal Pradesh and J & K.
Kartika (RLC-76)	1078 (R)	42.93	Dwarf in height, light brown seeded, moderately resistant to wilt, PM and BF.	Rainfed areas of Chhatisgarh
Indira Alsi-32	780 (R)	39.18	Dwarf in height, blue flower, dark brown seeded, resistant to PM.	Maharastra, Karnataka & Odisha
Sharda (LMS-4-27)	762 (R)	41.32	Dwarf, early duration, white flower, brown seeded, moderately resistant to wilt, PM & BF	Maharastra, Kernataka, Andhra Pradesh & Odisha,
Pratap Alsi-1 (RLU-6)	1997(S), 834 (F)	41.08	Tall in height, erect, funnel shaped, white flowered, light brown seed, moderately resistant to wilt, PM, AB & BF, tolerant to rust.	Rajasthan Kota command areas
Deepika (RLC 78)	1272 (I)	41.39	Medium in height & early, blue flower, brown seeded, resistant to PM	Partially irrigated as well as *Utera* situation of CG state
LC-2063	1200 (I)	41.00	Tall plant, medium maturity, blue flower, dark brown seed, moderately resistant to BF	Irrigated areas of Punjab State
Azad Alsi-1 (LMS 9-2K)	1610 (I)	39.92	Disk shape violet blue flower, dark brown seed, resistant to rust, wilt , PM and bud fly.	Bundelkhand of UP, MP & Rajasthan
Himani (KL-214)	583 (U)	36.40	Blue fowered, small brown, seeded moderately resistant to rust and PM	HP, J&K, Haryana and Punjab
PKV NL 260	963 (R)	37.67	Medium height, early, light violet funnel shaped flower, moderately resisatnt to AB, PM, BF	Maharashtra
SLS 67 (Shival)	1252 (R)	40.16	Dwarf star shaped white flower, early maturity, light brown seed, moderately resistant to PM & rust	Bundelkhand of UP, MP & Rajasthan

Bhagsu (KL-215)	428(U)	36.38	-------	Himachal Pradesh, J&K, Uttaranchal, Punjab, Haryana
Ruchi (LCK-5021)	1366(S) 1055(F)	39.84	-------	UP (Except BKD), Bihar, JKD, WB, Assam & NEH Region
Jawahar Linseed-41, (PKDL-41)	1600 (I)	40.00	-------	M.P limited irrigation facility
Jawahar Linseed-66, (SLS-66)	1200 (R)	42.80	-------	Rainfed areas of M.P
NDL 2004-05	1800 (I & R)	42.00	-------	U.P.
NDL-2002	1800 (I)	43.00	-------	U.P.

Source: http://oilseeds.dacnet.nic.in/crop.html

Table 2 Agronomic practices for flax/linseed in different crop growing situations

State	Situation	Recommended Varieties	Optimum time sowing	Spacing (cm)	Seed rate (kg/ha)	Fertilizers N:P (kg/ha)
Assam	Rainfed	Shekhar, Sweta, Shubhra, T397	1st fortnight of October	25	30	40:20
	Irrigated	Shekhar, Garima, Shubhra	2nd fortnight of October	25	25-30	60-80:30
	DP	Rashmi, Meera, Shikha, Gaurav, Parvati, Ruchi	1st week of November	20	45	80:30
Bihar	Rainfed	Shekhar, Sweta, Shubhra, T397	Mid October	25	30	40:20
	Irrigated	Shekhar, Garima, Shubhra	21nd fortnight of October	25	25-30	60-80:30
	DP	Rashmi, Meera, Shikha, Gaurav, Parvati, Ruchi	2nd fortnight of October	20	45	80:30
Chhattisgarh	Rainfed	Indira Alsi-2, Sharda, Deepika, Padmini,Mau Azad Alsi2	1st fortnight of October	25	25-30	30:15
	Utera	R552, Kartika, Padmini, Kiran, T397	1st to 3rd week of Oct.	Broad cast	35-40	10-20:00
	Irrigated	Suyog, JLS-9, J23 and T397, RLC92	Mid October	25	25-30	60-80:30

Haryana	Irrigated	Binwa, Himalini, LC-54	2nd fortnight of Oct. to Ist week of Nove.	25	20-25	60-80:30
	DP	Nagarkot, Jeevan	2nd fortnight of Oct. to Ist week of Nove.	20	45	80:30
H.P. and J&K	Rainfed	Surabhi, Janki, Sheela	2nd fortnight of October	25	25-30	40:20
	Utera	Baner, Surabhi, Himalini, Bhagsu	October	Broad cast	50-60	20:00
	Irrigated	Binwa, Janki, Himalini, LC-54	2nd week of Oct. to Ist fortnight of Nove.	25	25-30	60-80:30
	DP	Nagarkot, Jeevan	Mid October	20	65-75	90:30
Jharkhand	Rainfed	Shekhar, Sweta, Shubhra, T397	Mid October	25	30	40:20
	Irrigated	Shekhar, Garima, Shubhra, T397	1st fortnight of October	25	25-30	60-80:30
	DP	Rashmi, Meera, Shikha, Gaurav Parvati, Ruchi	1st fortnight of October	20	45	80:30
Karnataka	Rainfed	Indira Alsi-2, Padmini, Kiran, Sharda, Mau Azad Alsi2	Up to Ist week of Oct.	25	25-30	40:20
	Irrigated	Suyog, J-23	Up to Mid Oct.	25	25-30	60-80:30
MP	Rainfed	Padmini, Kiran, T-397, JLS9, SLS67, SLS73, SLS66	1st fortnight of October	25	25-30	30:15
	Utera	R-552, Sweta	1st to 3rd week of Oct.	Broad-cast	35-40	10-20:00
	Irrigated	Suyog, J-23 and T-397, Azad Alsi 1, SLS 41	Mid Oct.	25	25-30	60-80:30
Uttar Pradesh	Rainfed	Shekhar, Sweta, Shubhra, T397, SLS67, SLS73	Mid October	25	30	40:20
	Irrigated	Shekhar, Garima, Shubhra, T397, Azad Alsi 1	1st fortnight of October	25	25-30	60-80:30
	DP	Rashmi, Meera, Shikha, Gaurav Parvati, Ruchi	1st fortnight of October	20	45	80:30

Source: Directorate of Oilseeds Development, Hyderabad

The desirable characteristic features of flax cultivars are *viz.*,

(i) Resistance to pathogen complex
(ii) Resistance to abiotic stresses (drought, high temperature)
(iii) Middle vegetation period
(iv) Yielding potential of unretted stem (7-8 $t.ha^{-1}$)
(v) Yielding potential of seeds (1.10-1.30 $t.ha^{-1}$)
(vi) Long fibre content potential (22-25 %)
(vii) Total fibre content potential (39-41 %)
(viii) Long fibre yielding potential (1.25-1.40 $t.ha^{-1}$)
(ix) Total fibre yielding potential (2.50-3.50 $t.ha^{-1}$)

Seeds and sowing: Seed rate is 25 to 30 kg/ha in Peninsular India and 47 to 50 kg/ha in Indo-Gangetic region, respectively. Spacing is 30 × 10 cm for seed purpose and 20 × 20 cm for fibre purpose. High plant population density gives high stem yields and fibre quality, suppressing stem branching. Seeds are sown either by broadcast or drilling at a depth of 3 to 4.5 cm.

Weed management: The linseed is a poor competitor with weeds because of its very low photosynthetic surface. The common broad leaf weeds are *Chenopodium album, Convolvulus arvensis, Fumaria parviflora, Cathamus oxycantha, Melilotus* sp., *Medicago* sp, and the grassy weeds are *Phalaris minor*, *Avena ludoviciana* and *Lolium tenulentum. Cuscuta* (a parasitic weed) is the problematic weed in Flax. Two hand weeding are generally recommended on 20 DAS and 40 DAS. Post emergence application of isoproturon @ 1.00 kg/ha and 2,4-D @0.5 kg/ha can manage weeds effectively

Manures and fertilizer application: Apply 12.5 tonnes of FYM or compost per ha as basal dressing. Fertilizer dose of 60-20-20 and 20-20-00 kg of N, P_2O_5 and K_2O per ha is recommended for irrigated and rainfed conditions respectively. The NPK ratio recommended for Flax is of 1:2:3 or 1:2:4 ratio. The NPK ratio recommended for linseed is 1:0.8:2.5 ratio. Hence, fertilizer recommendations can be made based on soil test values for high yields. Fertilizer dose of 75-40-25 kg of N, P_2O_5 and K_2O per ha is recommended for fibre purpose. Nitrogen fertilization requirements of 10-20 kg/ha if the humus resources of the soil > 2%, 40 kg/ha if humus 1 to 2 % and 60 kg/ha if the humus < 1% N. Nitrogen has a critical impact on i) fibre content, ii) fibre length and iii) stem diameter. Nitrogen is necessary for growth, but excessive doses cause thickening of stems and reduce fibre strength. Excessive doses of nitrogen, however, lead to flax lodging particularly in high precipitation conditions and large plant density. Phosphorus has impact on proper length of straw and number of fibre bundles in each stem. Excessive doses of phosphorus, however, lead to shortening and branching of the stem, which reduces the fibre's tensile strength. Potassium has a beneficial influence on i) fibre strength, ii) fibre elasticity and iii) dew retting process. The beneficial effect of potassium is only revealed when nitrogen is correctly applied. $ZnSO_4$ can be applied @ 25 kg/ha to correct Zn deficiency. Soil pH has a significant effect on zinc assimilation. The more acidic the soil reaction, the better the uptake of zinc. Zinc can be a yield limiting nutrient and symptoms may include stunting, yellowing and terminal die-back. Excessive amounts of phosphorus may induce zinc deficiencies. MgO deficiency causes leaf chlorosis and stem shortening. When growing flax on soils with low magnesium content, it is recommended to apply magnesium fertilizers at 40-80 kg MgO per ha. Copper plays an essential role in chlorophyll formation and in seed formation. Copper is essential for proper enzyme activity. When growing flax on soils with low copper content, it is recommended to apply copper fertilizers at 6-10 kg Cu per ha. (300 g/ha Cu in post emergent application).

Water management: Linseed is generally raised as rainfed crop. However, the crop responds to irrigation. Fibre flax should be irrigated frequently during the vegetative growth period before flowering and to a lesser extent during flowering. The seed yield of linseed requires an adequate water supply just prior to flowering and after flowering stages. Under irrigated conditions, the small seeded and shallow rooted varieties are cultivated. Frequent light irrigations to maintain enough moisture in the root zone are more beneficial and economical than heavy irrigation at long intervals. Three irrigations are given on 35, 55 and 75 DAS. The consumptive use of linseed is 950 to 1100 mm for total biomass production and 2500 to 3000 mm for seed production.

Cropping systems: Linseed crop is under cultivation in three ecosystems namely *utera*, rainfed and irrigated. Flax should not be grown on the same field more than once in 7 years. This time is required for the soil to clean itself of *Fusarium* pathogens which will ensure high yields and pathogen free plants. Linseed is grown as mixed crop in rotation with cereals and legumes. The common linseed based cropping systems are *viz*., maize - linseed+wheat-sorghum, maize-linseed+gram, maize-linseed-cowpea-wheat, rice-linseed, barley+linseed, linseed+wheat, linseed+chickpea, linseed+lentil/safflower and linseed+potato. The Linseed based intercropping systems particulars are furnished in Table 3. The crop rotation with soybean-linseed in Madhya Pradesh and Chhattisgarh whereas, black gram–linseed followed by paddy-linseed in Jharkhand are remunerative cropping systems under rainfed situation. Crop rotation with maize-linseed at Himachala Pradesh and paddy-linseed at Uttar Pradesh are remunerative crop sequences under irrigated situation. Linseed crop may be intercropped with cereals, pulses and oilseeds of *rabi* season very well. When linseed is intercropped with chickpea the incidence of wilt and pod borer in chickpea is reduced.

Table 3 Linseed based intercropping systems

State	Situation	Intercropping system
Uttar Pradesh (Excluding Bundelkhand)	Rainfed	Linseed + Chickpea/Lentil (3 : 1 or I : 3)
	Irrigated	Linseed + Wheat (4:2),
		Linseed + M ustard (5:1)
		Linseed + Potato (3:3)
Bundelkhand of U.P.	Rainfed	Linseed + Chickpea/Lentil (3:1 or 1:3)
		Linseed + Wheat (1:3)
Madhya Pradesh and Chhattisgarh	Rainfed	Linseed + Chickpea (3: I)
Bihar and Jharkhand	Rainfed	Linseed + Chickpea (3: 1
		Linseed + Mustard (5:1)
West Bengal	Irrigated	Linseed + Mustard (5: 1)
		Linseed + Potato (3:3)
Maharashtra and Karnataka	Rainfed	Linseed + Chickpea (3:1 or 1:3)
		Linseed + Safflower (Different row ratio)
Punjab and H.P.	Rainfed	Linseed + Chickpea (3:1)
	Irrigated	Linseed + Mustard (5:1)
		Linseed + Wheat (4:2)

Paira or *Utera* cropping is a practice for utilizing the residual moisture in rice fields where tillage is difficult. About 25% of the linseed area (0.25 million hectares) is under utera cropping. In this practice, linseed is broadcast in the standing rice fields when the rice crop is between flowering and dough stages. Growing linseed in *utera* system is the predominant practice of regions of Bihar, Jharkhand, Odisha, Maharashtra, Madhya Pradesh, Uttar Pradesh and Himachal Pradesh. High yielding varieties are especially suited to *utera* system of linseed cultivation namely R552, Baner and Surabhi.

Plant Protection: The important diseases are rust, powdery mildew, alternaria blight, wilt and rust which causes loss. Seed treatment with Thiram or Topsin or Agrosan GN @ 2.5 g/kg of seed is done to manage wilt and alternaria blight diseases. Rust can be controlled with spray of Sulfex @ 0.05% or Calaxin @ 0.05% or Dithane M-45 @ 0.25%. If Dithane M-45 is to be applied, spray twice at 15-day interval. Powdery Mildew can be controlled with spray of Sulfex or wettable sulphur (0.3%) or Karathane (0.2%) at 15-day interval, depending on the intensity of the disease. Alternaria blight can be controlled with spray of Dithane M-45 (0.25%) @ 2.5 kg/ha at 15-day intervals, depending on the intensity of the disease. Cultivation of diseases resistant varieties is one of the approaches to control diseases (Table 4). The pests such as budfly, semilooper, wireworm, cutworm and termites infest the crop and cause considerable damage which can be controlled with soil application of 5% Aldrin dust @ 25-30 kg/ha with the last harrowing would ensure protection against these pests. For other pests, the following control measures are quite effective. Two fortnightly sprays of spinosad 45 SC (0.015%), Phosphamidon 85 EC @ 250 ml or Oxydemetonmethyl 25 EC @ 600ml or Monocrotophos 36 SL (0.04%) or Endosulphan 35 EC @ 1.2 litres/ha reduces bud fly infestation. Two application of neem based commercial formulation containing azadirachtin 300 ppm reduces bud infestation. Semilooper and caterpillar can be controlled by dusting with 2% Methyl parathion @ 25 kg/ha. Thrips can be controlled with spray of Monocrotophos 36 SL @ 0.04% or Dimethoate 30 EC @ 0.03%.

Table 4 List of disease resistant varieties of Linseed

Name of disease	Resistant varieties
Wilt	Kiran, R 552, J-23, JLS-9, Padmini, T-397
Alternaria blight	Sheela, Meera, Parvati
Powdery Mildew	Kiran, J-23, Jeevan, Padmini, R-552
Rust	Padmini, Jawahar-23, LC 54, JLS-9

Harvest: Flax can be harvested as an early crop for fiber, with attached immature seeds, or harvested as a late, mature crop, for both seed and fiber production. The crop requires 70 to 100 days to reach maturity. The harvesting operation is carried out when the lower two-thirds of the stem have turned yellow and the leaves have fallen off. This takes place about one month after the appearance of the first bloom on the plant. Linseed crop is ready for harvesting when the bolls are ripe. Maturity symptom shows 90 % brownish bolls, the leaves are yellow in colour while stems may remain green after the bolls are ready to harvest. Lower portion of leaves wither away. Seeds have become shiny. Cut the plants close to the ground using sickle or pull the whole plant. Flax plants are pulled from the field instead of being cut. Pulling is considered better than cutting as the cut straw tends to deteriorate at the cut end. Pulling by hand is still practised in some regions, but is an expensive and slow process. The harvested crop is left in the field for a few days to ensure the drying of the plants and facilitates easy recovery of seeds. The separation of seeds is performed by beating the plants with sticks or by treating under the feet of the cattle. Winnowing is done to clean the seed from the chaff.

Processing: The harvested plants are generally left in the field to dry for two to three days. The preparation of flax is much more laborious and costly process than that of cotton. The process of extracting the fibres involves four steps *viz.*, rippling, steeping or 'retting', scutching and hackling. Dried flax is stripped-off all leaves and capsules with least possible damage to the stem and fibre, a process termed 'rippling'. The denuded stems, in small bunches, are then subjected to 'retting' by immersing them in a dam (dam retting), or in concrete tanks (tank retting) in 30 cm deep in water for 10 to 16 days. In retting tank luke warm water at 32 to 34°C is used. Retting is the name given to the action of bacteria / fungi, which break down the tissue holding the fibres. Retting is the process of stem preparation for fibre extraction through decomposition of pectin (fibre bundles are cemented to wood by pectin).

(i) Dew-retting: The most common method of retting is dew retting or field retting. Dew-retting is an art that depends upon the removal of matrix materials from the cellulosic fibers before cellulolysis, and therefore weakening of the fibers occurs. In dew-retting, flax plants are pulled from the soil and flax straw is exposed to the action of dew and rain (dew retting) for four to eight weeks in fields for selective attack by the fungi under natural conditions. The aerobic fungal consortia produces the plant cell-wall degrading enzymes which decompose fiber-bundle matrix releasing the bast fibers from each other and from the woody core. Polysaccharide degrading enzymes such as pectinases, xylanases, hemicellulases, produced by the fungi, are primarily responsible for fibre separation. The dew-retting (ground or field retting) is mainly caused by fungi (*Alternaria spp., Aspergillus, Cladosporium, Mucor, Rhisopus*) and bacteria (*Bacillus spp., Bacterium coli*). *Bacillus subtilis*, an isolated bacteria has been identified for microbial retting of flax. Retting duration is also reduced on bio-inculation. Retting is carried out with the morning dew for a period of 14 to 21 days. *Alternaria alternata* fungus hastens the retting process. The enzymes removes pectins, gums and liberates fibre. Then the stems are dried in the sun. Afterwards the stems are cleaned by maceration. Subsequently fibres are separated and combed. The dew-retting is weather dependent, requiring a settled spell of dry weather during the 3-4 week retting time, but enough moisture to activate the fungi. If conditions are too wet, over retting can occur which reduces the fibre strength. On the other hand, if the dry condition exists, then there is not enough moisture to facilitate retting and under-retting is a problem. Under-retting means that it is difficult to get good separation of the bast fibres from the core, which is a serious problem for processing. The dew-retting process depends on particular geographical regions that have the appropriate moisture and temperature ranges for retting. Dew-retting fibres are coarser and lower quality than water retting, poor consistency in fiber characteristics, and occupation of agricultural fields for several weeks. Further, dew-retting results in a heavily contaminated fibre that is dusty and particularly problematic in textile mills areas. Dew-retting supplies the linen used in high quality textiles.

(ii) Water retting: Water retting is a more rapid process and in general yields a high quality fibre, but involves more labour. Water retting uses bacteria under anaerobic conditions, but it does have potential problems with water pollution. Tank retting under controlled temperature, however, is known to produce superior fibre over stream or dew retting, and it also takes less time. The long unbroken fibres are called line fibres. The water retting is essentially a bacteriological decomposition while the dew retting is predominantly a fungal activity.

(iii) Enzyme-retting method: Enzyme-retting has been developed to replace dew-retting. The pectinase-rich enzyme mixtures and / oxalic acid are required for retting. Enzyme-retting produced fibers having the fineness, strength, colour and waxiness comparable to the best water-retted fibre. The advantages of the enzyme method are: i) time saving of 4 to 5 days, ii) increased yield of Ca 2% over water-retting and iii) fibre consistency. With enzyme-retting, field dew-retting is not required and the time of land occupation is shorter so farmers can be assured of the availability of the land for planting other crops. Enzyme retting produces a fewer pollutants. After retting the straw is baled using a round or square baler and stored before collection by the processor. It is important that the straw is to be dried (less than 18% moisture) before baling.

Yield: Flax can be harvested safely at moisture levels up to 18%, but seed is safe to store at 8% with 11% moisture being the upper end. Higher moisture will result in heating and mould formation. Linseed produces a seed yield of 2000 to 2500 and 500 to 800 kg/ha under irrigated and rainfed conditions respectively. Seed contains 42% oil and 23% protein. The oil cake contains 32% protein and 10% oil. The fibre yield is approximately 20 to 25% by stem weight. The flax fibre is made of 64% cellulose, 17% hemicelluloses and 2% lignin.

Utilization: The flax plant supplies both linseed oil and bast fibre used to produce textiles, composites, and paper/pulp. Flax fibres are known for their fineness, beauty, durability, flexibility, heat conductivity, moisture absorbency and great strength, especially when wet. Fibres are pale yellow or cream in colour and lustrous because of about 1% wax which also imparts suppleness to the fibre. The flax fibre is stronger than cotton, rayon or wool, but weaker than ramie. Owing to its length, flax is suitable for strong yarns which are used for sewing threads. The best grades of flax fibre are used for linen fabrics such as damasks, cambrics, lace and sheetings and for linen thread. The coarser grades are used for twines, canvas, bags, etc. The yarn and fabric made from flax fibres are called *linen.* Flax fibre has ability to absorb up to 12% of its own weight of water and its strength increases by 20% when wet. It also dries quickly. Fax fibre is twice as strong as those of cotton and five times as strong as those of wool. Hand-kerchiefs, dresses, cushion covers, hand towels, etc, are made from flax fibre. High grade writing paper and cigarette paper are made out of flax fibre.

Seeds contain 35-45 % oil and 20-25 % protein. Linseed oil is used as a drying oil. Linseed contains cyanogenic glucoside linamarin which on hydrolysis produces toxic hydrocynic acid. Brown flax, can be eaten however, it is grown for the commercial linseed oil, paint, and solvents industries. It is also used for making linoleum, oil cloth, printers ink, soap, etc. The linseed oilcake contains 3% oil and 36% protein which can be fed for the livestock. The golden flax seed called Omega for its high omega-3,6 fatty acid content and for human consumption and is preferred for the food market because of its nutty-buttery flavour. The cooler climate and longer summertime daylight hours during the growing season help to acheive a higher Omega-3 content in the flax seed. Low linolenic-acid flax cultivars are called linola or solin flax.

Oilcake is used as manure to prevent soil from unwanted microbes due to its germicidal property.

REFERENCES

Akin, D., W. Morrison, G. Gamble, L. Rigsby, G. Henriksson, and K. Eriksson. 1997. Effect of retting enzymes on the structure and composition of flax cell walls. Textile Res. J. 67:279–287.

Akin, D., R. Dodd, W. Perkins, G. Henriksson, and K. Eriksson. 2000. Spray enzymatic retting: A new method for processing flax fibers. Textile Res. J. 70:486–494.

Akin, D., H. Epps, D. Archibald, and H. Sharma. 2000. Color measurement of flax retted by various means. Textile Res. J. 70:852–858.

Akin, D., J. Foulk, and R. Dodd. 2002. Influence on flax fiber of components in enzyme-retting formulations. Textile Res. J. 72:510–514.

FCC. 2005. Growing Flax: production, management, and diagnostic guide. Flax Council of Canada. 64p

Jurat Repeckien and Zofija Jankauskien. 2009. Application of Fungal Complexes to Improve Flax Dew-Retting. Lžuu Mokslo Darbai. 83 (36): 63-70.

Parks, C., J. Frederick, P. Porter, and E. Murdock. 1993. Growing flax in South Carolina. Clemson Univ.Cooperative Ext. Serv.

UPOV. 2010. Guidelines for the conduct of tests for distinctness, uniformity and stability. International Union for the Protection of New Varieties of Plants, Geneva.

Van Dam, J., G. van Vilsteren, F. Zomers, W. Shannon, and I. Hamilton. 1994. Industrial fibre crops. ATO, DLO, Wageningen, The Netherlands.

http://oilseeds.dacnet.nic.in/crop.html

http://oilseeds.dacnet.nic.in/oilseeds/linseed/LINSEED.htm

http://www.fibrafp7.net/Portals/0/02_Jankauskiene.pdf

http://flaxcouncil.ca/

http://www.worldjute.com/flax.html

http://www.cs.arizona.edu/patterns/weaving/books/bf_flax.pdf

https://www.ag.ndsu.edu/agnic/flax/index.html

https://www.ag.ndsu.edu/agnic/flax/productioni.htm

http://www.dakotaflax.com/#!why-dakota-flax/c1bs6

http://flaxcouncil.ca/38.htm

CHAPTER

9

Niger–*Guizotiaa abyssinica* L. (2n=30)

Family: Compositae

Vernacular name: *Ramtil, Jagni* or *Jatangi* (Hindi), *Ramtal* (Gujarati), *Karale* (Marathi), *karale* or *khurasani* (Marathi), *uhechellu* (Kannada), *Payellu* (Tamil), *Verrinuvvulu* (Telugu), *Alashi* (Oriya), *sarguza* (Bengali), *Sorguja* (Assamese), *Ramtil* (Punjabi) and *Sorguja* (Assamese).

Importance: Niger is a minor oilseed crop. India ranks first in area, production and export of niger in the world. It is tolerant to drought. It resists water logging due to presence of aeranchyma. Seed contains 32 to 40% oil with 18 to 24% protein. Niger oil is used as cooking oil, lubircations, cosmetics and base oil in perfume industry. The oil is readily subject to oxidative rancidation rendering its keeping quality poor due to high content of unsaturated fatty acid (oleic acid 38% and linoleic acid 51.6%). The oil is used in paints, soaps, etc. Oil cake is a good cattle feed.

Origin: Niger is origin of Ethiopia.

Distribution: Niger is cultivated in Germany, Switzerland, France, Russia, China, Nepal, Myanmar, India, Ethiopia, Sudan, Uganda, Tanzania, Malawi, West Indies, Brazil and Mexico. It is also cultivated to some extent in hilly areas of Andhra Pradesh, Bihar, Jharkhand, Odisha, Gujarat, Maharashtra, Madhya Pradesh, Uttar Pradesh, Rajasthan, Tamil Nadu, Karnataka, West Bengal, Assam and Arunachal Pradesh.

Botany: Niger is an herbaceous annual, erect and smooth stem; growing about a metre in height. The number of branches per plant varies from five to twelve. Leaves are sessile, linear or lanceolate obtuse with serrated margin. First leaf is paired, small and successive leaves are larger. The leaf margin varies from pointed to smooth and leaf colour varies from light green to dark green. Inflorescence is heterogamous head or capitulum, 1.5 to 2.5 cm in diameter. The hermaphrodite disk florets are arranged in three whorls. The niger flower is yellow and rarely, slightly green. The long flowering period of 45-80 days, make Niger the ideal crop for bee keeping. Fruit is an achene. The head produces about 40 fruits. Seed is glabrous, dorsally compressed, tip rounded, black and ex-endospermous.

Germination is epigeal and seedlings have pale green to brownish hypocotyls and cotyledons.

Climate: It can be grown in tropical and temperate zones. It is grown mainly in lower altitude of 500-1600 m, mid altitude and highland areas of 1600- 2200 m from MSL. It requires a rainfall of 1000 to 1250 mm. It is not suited for high rainfall areas. It is a short day plant. It is a drought tolerant crop. The cardinal minimum, optimum and maximum temperature is 10ºC, 20 to 30ºC and 38ºC respectively. The optimum soil temperature is 17ºC to 25ºC. High wind velocity and hail storm lead to seed shattering.

Soil: Niger is adapted to a wide range of soil types from clay loam to sandy loam, sandy and gravely soil. However, it thrives best on well drained, loamy soils of good depth and texture with pH range of 5.2 to 7.3. It can withstand slight alkalinity and salinity also. Heavy clay and black cotton soils are not suitable for high yield.

Season: Niger is a *kharif* season crop and is harvested in December. The crop flowers on 90 DAS and matures in 35 to 45 days after flowering. The optimum sowing time is middle of July and continue to early August for *kharif* crop and September for semi *rabi* crop. The sowing time of Niger in different states of India is presented in Table 1.

Table 1 Sowing time of Niger in different states of India

State	Optimum time of sowing
Madhya Pradesh/ Chhattisgarh	Third week of July to second week of August
Maharashtra	July and continue to early September
Odisha	Second fortnight of Aug to first week of September
Bihar/ Jharkhand	Second fortnight of Aug to first week of September
Andhra Pradesh	Second week of August
Gujarat	July-August
Karnataka	July-August

Varieties: Niger varieties and their characteristic features grown in different states of India are presented in Table 2.

Table 2 Niger varieties and their characteristic features grown in different states of India

Variety	Seed yield (kg/ha)	Oil content (%)	Days to maturity	Salient features	Recommended State
JNC-6	650-700	37-38	95-100	Shining dark black seed	Madhya Pradesh, Maharashtra, Bihar, Karnatka and Rajasthan
JNC-1 (Jawahar Niger Composite-1)	650-700	38-40	90-100	Black seed	Madhya Pradesh, Bihar, Maharashtra, Karnatka and Rajasthan
JNS-9 (Jawahar Niger Composite-9)	650-700	38-40	95-100	Black seed, tolerant to moisture stress	Odisha
GA-10	600-650	39-41	115-120	Tall, dark black seed	Odisha

Utkal Niger-150	650-700	38-40	105-110	Black seed, tolerant to Alternaria and Cercospora leaf spot	Maharashtra
IGP-76 (Sahyadri)	500-550	35-38	95-100	Black Seed	All niger growing states
N-5	500-600	36-39	95-100	Small black and sickle shaped seed	Maharashtra and Bihar
IGPN-2004-1, (Phule Karala-1)	650-700	39-41	98-105	Shining black seed, Tolerant to Alternaria leaf spot, powdery mildew.	Maharashtra Karnataka, Bihar/ Jharkhand
Birsa Niger-1	550-600	36-38	95-100	Light black seed	Bihar/ Jharkhand and Madhya Pradesh
Birsa Niger-2 (BNS-8)	600-650	35-38	95-100	Black seed	All niger growing states
BNS-10 (Pooja-1)	650-700	36-38	95-100	Shining black seed	All niger growing states
RCR-317	500-550	35-38	90-95	Black seed	Karnataka
RCR-18	400-450	34-35	100-110	Light black seed and robust growth habit	Karnataka
KBN-1	550-650	36-38	85-95	Black seed	Karnataka
DNS-4	500-600	39-41	90-95	Shining black bold seed	Karnataka
Guj. Niger-1	600-650	35-38	95-100	Black seed	Gujarat; All niger growing states
Guj. Niger-2 (NRS-96-1)	650-700	35-38	90-95	Black seed	Gujarat
Paiyur –1	600-650	35-38	90-95	Black seed	Tamil Nadu

- JNS-9 variety of Madhya Pradesh, BNS-2 and BNS-10 varieties of Jharkhand are recommended for Andhra Pradesh.
- JNC-1, JNC-6 and JNS-9 varieties of Madhya Pradesh are recommended for Rajasthan.
- Birsa Niger-2 and BNS-10 varieties of Jharkhand are recommended for West Bengal.
- JNC-6 variety of Madhya Pradesh and IGP-76 (Sahyadri) variety of Maharashtra are recommended for North Eastern Hill Region.
- JNC-1, JNC-6 and JNS-9 are also recommended to Maharashtra, Bihar, Rajasthan and Karnataka besides Madhya Pradesh.
- Birsa Niger-2 and BNS-10 are also recommended to all niger growing states.
- IGP-76 is also recommended to Odisha, Gujarat, Tripura, Daman besides Maharashtra state.

Seeds and sowing: Seed rate is 4 to 6 kg/ha. Seed treatment with Carbendazim 5 g/kg or *Trichoderma viride* 10 g/kg of seed before sowing to protect the crop from seed and soil borne diseases. Seed treatment with phosphorus solublising bacteria (PSB)/Azotobactor/Azospirillum @ 10g/kg of seed is recommended. Spacing is 30 × 10 cm. Seeds are mixed with ash or sand and sown to a depth of 3 cm. Seeds germinate in 3 to 6 days.

Cropping systems: Niger is sown under rainfed situations in *kharif* and *rabi* seasons as a sole crop or mixed with sorghum, pearlmillet, little millet, cotton, sunflower, groundnut, blackgram, greengram, etc., in different states. The crop sequences and intercropping systems followed in different niger growing states of India are furnished in Table 3 and Table 4.

Table 3 Niger based cropping sequence in different states of India

State	Crop sequence
Madhya Pradesh/ Chhattisgarh	Early Black gram – Niger
Maharashtra	Little millet/ Finger millet – Niger; Horse gram – Niger
Odisha	Common millet (*Panicum miliaceum*) – Niger; Little millet (*Panicum miliare*) – Niger; Early Finger millet – Niger; French bean – Niger.
Bihar/ Jharkhand	Little millet – Niger Early Finger millet – Niger Black gram – Niger Early rice – Niger

Table 4 Niger based intercropping systems in different states of India

State	Intercropping	Row ratio
Madhya Pradesh	Niger + Kodo/ Kutki/ Pearl millet/ Green gram	2:2
	Niger + Groundnut	4:2 or 6:2
Maharashtra	Niger + Finger millet / Horse gram/ Rice bean	2:2 or 4:2
	Niger + Finger millet/Little millet	2:4 or 3:6
	Niger + Pearl millet/ Groundnut	3:3
	Niger + Groundnut	2:6 or 3:6
Odisha	Niger + Finger millet / Black gram	2:2 or 4:2
	Niger + Ricebean/ Cowpea/ French bean	4:2
Bihar	Niger + Finger millet / Black gram	2:2
	Niger + Red gram	3:2
	Niger + Rice bean / Groundnut	4:2
Andhra Pradesh	Niger + Cow pea	4:2
Karnataka	Niger + Groundnut	6:3
	Niger + Finger millet	1:1

After cultivation: Thinning is done on 15 DAS or when the seedlings attain 8-10 cm height.

Weed management: One or two hand weedings are given on 20 and 35 DAS to control weeds. Pre emergence application of Pendimethalin @ 1 kg a.i./ha *is* recommended followed by one hand weeding on 30-35 DAS. Cuscuta (*Cuscuta hyalina/ C. chinensis*) infestation has become a major problem in niger. Seed should be obtained from *Cuscuta* free areas. If the *Cuscuta* seed is found mixed with niger seed, sowing should be done after separation by sieving with a 1 mm sieve. The niger seed infested with *Cuscuta* may be treated with 10 % brine solution (Table salt) to obtain a *Cuscuta* free crop.

Manures and fertilizer application: Fertilizer dose of 20–40-10 kg of N, P_2O_5 and K_2O kg/ha is applied as basal dressing. Top dressing of 10 kg of N per ha on 35 DAS is recommended. Application of sulphur @ 20-30 kg/ha is recommended to increase seed yield and oil content. The nutrient management of Niger in different states of India is presented in Table 5.

Table 5 Nutrient management of Niger in different states of India

State	Recommended dose of fertilizer
Madhya Pradesh	20:20:10, N:P:K (kg/ha) at the time of sowing as basal and remaining 20 kg N/ha at 30-35 days after sowing
Maharashtra	Four tonnes of FYM and 20:20, N:P (kg/ha) at the time of sowing and 20 kg N/ha 30 days after sowing (top dressing)
Odisha	20:40, N:P (kg/ha) at sowing and remaining 20 kg N/ha at 30 days after sowing
Bihar/ Jharkhand	20:20:20:15, N:P:K:S (kg/ha) as basal
Andhra Pradesh	Five tonnes of FYM and 10 kg N/ha at sowing
Karnataka	20:20:10, N:P:K (kg/ha) at sowing

Water management: Niger is mostly raised as rainfed crop.

Crop protection: The major insect pests are Niger caterpillar (*Condica conducta*), Cutworm (*Agrotis ipsilon*), Bihar Hairy caterpillar (*Spilosoma obliqua*), Surface grasshopper (*Chrotogonus sp.*), Aphids (*Uroleucon carthami*), Semilooper (*Plusia orichalcea*) and Niger capsule fly (*Dioxyma sarorcula*) while the diseases are Cercospora leaf spot (*Cercospora guizoticola*), Alternaria leaf spot (*Alternaria sp.*), Powdery mildew (*Sphaerotheca* sp.) and Stem/root rot (*Macrophomina phaseolina*). The caterpillar (green with purple markings) feeds on leaves and defoliates the plants. Proper weeding of fields reduces hiding places. Dusting with 5% Carbaryl or 5% parathion @ 20-25 kg/ha or spray Thiodon 35 EC @ 0.07%, Carbaryl 0.1% or Monocrotophos 0.05% or Dichlorvos 0.05% to control caterpillar. Cut worm moth hides under dried twigs and leaves at day and lays eggs on leaves. Larvae attack the crop and eat plants at ground level. Keeping grass bundles in fields for hiding the caterpillars during day and killing them by dusting 5% Carbaryl or 5% Parathion dust. Spraying with Carbaryl 0.2% or Monocrotophos 0.05% or Dichlorovos 0.05% to control cut worm. The hairy caterpiller remain gregarious underneath leaves in early stages and cause serious loss in yield at 3rd and 4th instar. Collection and destruction of egg masses and early instars of caterpillers. Spraying with Dichlorovos 0.05% or Carbaryl 0.2% or Monocrotophos 0.05%. Dusting with 5% Carbaryl or Parathion @ 20-25 kg/ha to control hairy caterpillar.

Powdery mildew disease symptoms occur as small cottony spots on leaves which spreads on the lamina followed by defoliation of affected plants in severe cases. When the disease is noticed spray 0.3% solution of wetsulf or sulphur or Bavistin (0.1%) or Karathane (0.1%) and repeat after 15 days. Cercospora leaf spot symptoms occur as small straw to brown colour spots on leaves. When severe, entire leaf may be covered followed by defoliation. Alternaria leaf spot occur as concentric rings, brown with gray center, oval to circular or irregular in shape. Seed rot (*Rhizoctonia bataticola* and *Sclerotium rolfsii*) infected seeds become discoloured. The seedlings from such seeds die. Seed treatment with 0.3% Thiram + Bavistin (0.1%) before sowing or two sprays with 0.2% Zineb or Bavistin (0.1%) + Dithane M 45 (0.25%) @ 15 days is recommended to control the diseases. Application of 2.5 kg/ha *Trichoderma viride* mixed with 50 kg FYM in the field before sowing to control diseases.

Harvest: Niger usually matures in 95-105 days after sowing. The crop should be harvested when the leaves dry up and the capitula turns brownish / blackish in colour. The crop is harvested in the base when leaves fall on ground with sickle and stacked for a week in threshing floor. The stem turns yellow and leaves shed at maturity. The produce is dried to reduce moisture content up to 8% and then stored properly.

Yield: The average seed yield of 300 to 400 kg/ha can be obtained. However, it has the yield potential of 800-1000 kg/ha under optimum growing conditions. Thousand seed weight is 3 to 5 grams.

Utilization: Seed contains 34-36% oil and 18-20% protein. The oil has good keeping quality and has 70% unsaturated fatty acids free from toxins. The oil has 75% linoleic acid and a small amount of lignoceric acid 2%. Niger oil seed cake has 14% fibre and 30% protein. Niger flour is used for bread making. Niger oil is used for birth control and for the treatment of syphilis. Niger sprouts mixed with garlic and 'tej' are used to treat coughs.

REFERENCES

CPG. 2012. Crop Production Guide. Department of Agriculture, Government of Tamil Nadu, Chennai and Tamil Nadu Agricultural University, Coimbatore.

GOI. 2014. Status paper on Oilseeds. Ministry of Agriculture, Government of India. New Delhi.

Ranganatha, A.R.G. 2013. Improved Technology for Maximizing Production of Niger. All India Coordinated Research Project on Sesame and Niger, Indian Council of Agricultural Research, JNKVV Campus, Jabalpur-482004

http://oilseeds.dacnet.nic.in/oilseeds/niger/NIGER.htm

CHAPTER

10

Soybean–*Glycine max* (L.) Merr. (2n=40)

Family: Fabaceae

Vernacular name: Soybean, Soyabean (English), *Soya payaru* (Tamil), *Soyachikkudu* (Telugu), Soybean (Marathi).

Importance: Soybean is known as the 'golden bean'. It is the world's most important seed legumes which contribute 25% to the global edible oil, about two third of the world protein concentrate for livestock feeding and is a valuable ingredient in formulated feeds for poultry and fish. It contains 43% protein and 21% oil. It produces roughly 2.5 times more seed and protein yield than other pulses. Soybean has remained a key foreign exchange earner due to export of soybean de-oil cake. Japan, China, Indonesia, Philippines, and European countries are importing soybean to supplement their domestic requirement for human consumption and cattle feed. Vegetable soybean is popular in Japan, Korea, China and Taiwan.

Origin: Soybean is originated in northeastern China between 35ºN and 45ºN.

Distribution: Soybean is primarily confined to temperate region. It is now cultivated in the tropics and sub tropics. USA and Brazil are the major soybean growing countries followed by Argentina, China and India. The area under soybean is mainly spread in Madhya Pradesh, Maharashtra, Rajasthan, Chhattisgarh, Andhra Pradesh and Karnataka. Madhya Pradesh and Maharashtra are the two major soybean growing states contributing about 86% of total soybean area and production.

Botany: It is a herbaceous, annual, erect, much branched; pubescent throughout the plant body, 0.6 to 1.8m height, rarely creeping or twinning habit. Both determinate and indeterminate types are available. Leaves are trifoliate, leaflets ovate acute. Infloresence is a highly condensed raceme. Flowers are small, style curved, glabrous with capitate stigma and monodelphous stamens. Flowers up to twenty on short racemes arising in the axils of the leaves, white or purple; pods linear-oblong, three to seven seeded, seed varying from white, red, yellow, brown to black. Yellow seeded types contain high oil and less protein while black seeded types contain high protein and low oil. Soybean is normally a self-pollinated crop. Cross-pollination is usually less than 1%.

Climate: Soybean is a habitat of sub-tropical and temperate climate. Soybean is grown up to an altitude of 2000 m from MSL within an latitude of 0°N and 20°N and up to an altitude of 1000 m from 20°N to 40°N and with warm sea currents from 40° N to 60°N. Soybean is a short day plant. All cultivars flower quickly with 14 to 16 hours of darkness. A well distributed annual rainfall of 500 to 750 mm is required for growth and yield of soybean. The minimum temperature for germination and emergence is 10°C. Soybean requires relatively high soil water for germination to occur in 5 to 8 days at 20°C. The cardinal minimum, optimum and maximum temperature for soybean is 10°C, 20 to 30°C and 40°C. The thermal optimum for CO_2 exchange is 35°C. Seed germination rate is maximum 25 to 30°C. Days to flowering of soybean are shortest at 30°C and an increase in days to flowering at 25°C, 35°C and 20°C has been observed. Flower formation is inhibited at temperature of less than 15°C and pod set is low above 40°C. Temperature requirements at different stages of soybean development are presented in Table 1.

Table 1 Temperature requirements at different growth stages of soybean

Growth stage	Temperature °C		
	Biological minimum	**Sufficient**	**Optimum**
Germination	6-7	12-14	20-24
Emergence	8-10	15-18	20-22
Flower initiation	16-17	18-19	21-23
Flowering	17-18	19-20	22-25
Seed formation	13-14	18-19	21-23
Ripening	8-9	14-16	19-20

Soil: Soybean can be grown in soils with a wide range of textural classes. Well drained loamy soil with a pH of 6.0 to 7.5 is ideal for soybean crop. Soybean is predominantly grown as rainfed crop in *Vertisols* and associated soils with an average crop season rainfall of 900 mm. Soil with excessive salts/ sodium and poorly drained conditions are not suitable for soybean cultivation. Soils having pH below 4.5 and above 9.0 are not suitable for soybean cultivation. Drained *Histosols* can be physically quite satisfactory but mineral soils that can reduce emergence as found on the Indo-Gangetic plains. Roots and nodules develop well in the shallow layer of moist aerated soil. Rhizobial populations are low in sandy and clay loamy than in sandy loam or loamy soil.

Season: The crop can be grown throughout the year if water is available. Soybean is grown during June-July and September-October under rainfed conditions whereas it is raised during February-March under irrigated condition. It is also raised as rice fallow crop during January.

Varieties: The soybean varieties grown in different states of India are furnished in Table 2.

Table 2 State wise recommended soyabean varieties

S.No.	State	Suitable Varieties
1	Andhra Pradesh	LSb-1, Pratikar (MAUS 61), Pooja (MAUS 2), MACS 450, Pant Soybean 1029, MACS 124, Monetta and Bragg
2	Arunachal Pradesh & Assam	Ahilya 1 (NRC 2), JS 80-21, Samrudhi (MAUS 71), Pratap Soya (RAUS 5), Bragg, Indira Soya 9
3	Bihar	PK 416, Pusa 16, Pusa, 24, Pant Soybean 564, Pant Soybean 1024, Pant Soybean 1042, Bragg and SL 525
4	Chhattisgarh	Ahilya 1 (NRC 2), Ahilya 3 (NRC 7), Ahilya 2 (NRC 12), Ahilya 4 (NRC 37), JS 71-05, JS 335, JS 80-21, JS 75-46, MACS 58, JS 9041, Indira Soya 9, JS 93-05, Kalitur, Parbhani Sona (MACS 47), Pratishta (MAUS 61-2), Monetta, Punjab-1, PK 472, Shakti (MAUS 81), Samrudhi (MAUS 71) and Pratap Soya (RAUS 5)
5	Delhi	PK 416, Pusa 9712, Pant Soybean 564, Pant Soybean 1024, Pant Soybean 1042, Bragg and SL 525
6	Gujarat	Gujarat Soybean 1, Gujarat Soybean 2, JS 93-05, JS 335, JS 80-21, JS 75-46, MACS 58, Monetta, Parbhani Sona (MAUS 47), Pratishta (MAUS 61-2), Shakti (MAUS 81) and PK 472
7	Haryan	Punjab-1, Pk 416, Pusa 16, Pant Soybean 564, Pant Soybean 1024, Pant Soybean 1042 and SL 525
8	Himachal Pradesh	VL Soya 2, VL Soya 47, Shilajeet, Shivalik, Bragg, Pusa 16, Hara Soya and Palam Soya
9	Jharkhand	Birsa Soybean 1, Ahilya 1 (NRC 2), JS 80-21, Samrudhi (MAUS 71), Pratap Soya (RAUS 5) and Bragg
10	Karnataka	Hardee, Bragg, Sneh (KB 79), KHSb-2, Pratikar (MAUS 61), Pant Soybean 1029, MACS 124, MACS 450 and Pooja (MAUS 2)
11	Madhya Pradesh	Ahilya 1 (NRC 2), Ahilya 3 (NRC 7), Ahilya 2 (NRC 12), Ahilya 4 (NRC 37), JS 71-05, JS 335, JS 80-21, JS 75-46, MACS 58, JS 90-41, Indira Soy 9, JS 93-05, Kalitur, Parbhani Sona (MAUS 47), Pratishta (MAUS 61-2),Monetta, Punjab-1, PK 472 and Shakti (MAUS 81)
12	Vidharbha and Marathwada region of Maharashtra	Ahilya 1 (NRC 2), JS 335, JS 93-05, JS 80-21, MACS 58, Parbhani Sona (MAUS 47), Pratishta (MAUS 61-2), Shakti (MAUS 81), MACS 13, Monetta, Prasad (MAUS 32) PK 472, Shakti (MAUS 81), TAMS-38 and Phule Kalyani (DS-228)
12.1.	Southern Maharashtra	MACS 124, MACS 450, Pant Soybean 1029, Pooja (MAUS 2), Pratikar (MAUS 61), Prasad (MAUS 32), MACS 13, Monetta and Phule Kalyani (DS-228)
13	Manipur, Meghalaya and Nagaland	Ahilya 1 (NRC 2), JS 80-21, Samrudhi (MAUS 71), Pratap Soya (RAUS 5), Bragg, Indira Soya 9
14	Odisha	JS 80-21, Pusa 24, Indira Soya 9, Ahilya 1 (NRC 2), Ahilya 2 (NRC 12), Ahilya 3 (NRC 7), Ahilya 4 (NRC 37) and Pusa 16
15	Punjab	PK 416, Pusa 16, Pant Soybean 564, Pant Soybean 1024, Pant Soybean 1042, SL 295, Bragg and SL 525

16	Rajasthan	Pratap Soya (RAUS), Bragg, Punjab-1, PK 472, MACS 58, JS 80-21, JS 335, Ahilya 4 (NRC 37), Parbhani Sona (MAUS 47), JS 93-05, Pratishta (MAUS 61-2) and Shakti (MAUS 81)
17	Sikkim	NRC 2, JS 80-21, MAUS 124, MAUS 71, RAUS 5, Pusa 16, Bragg and Indira Soya 9
18	Tamil Nadu	Co 1, Co Soya 2, ADT-1, MACS 124, MACS, 450, Pooja (MAUS 2), Pratikar (MAUS 61) Hardee, Pant Soybean 1029, and Bragg
19	Tripura and West Bengal	Ahilya 1 (NRC 2), JS 80-21, Samrudhi (MAUS 71), Pratap Soya (RAUS 5), Bragg and Indira Soy 9
20	Uttar Pradesh	Pusa 16, Pant Soybean 1092, Pant Soybean 1042, Pant Soybean 1024, Pant soybean 564, PK 472, PK 472, PK 416, Pratishta (MAUS 61-2), JS 93-05, Ahilya 4 (NRC 37), JS 335, SL 525, PS 1241, PK 262 and PK 327
21	Uttrakhand	Hara Soya, Palam Soya, Punjab-1, Pusa 16, PS 1241, VL Soya 1, VL Soya 2, VL Soya 21, VL Soya 47, Pant Soyabean 1092 and Shilajeet.

Varieties with special characters:

Early (< 95 days): JS 71-05, JS 90-41, JS 93-05, MAUS 47, Monetta, Ahilya 2, Ahilya 3, PK 564

Medium (96-100 days): Indira soya 9, JS 80-21, JS 335, MACS 450, Ahilya 1, Ahilya 4, PK 472, PS 1024, PS 1029

Late (> 100 days): Bragg, MAUS 61-2, PK 416

Good germinability : JS 335, Ahilya-1, Pusa-16, MACS 450, Ahilya-4

Bold seeded : JS 71-05, Ahilya-3, PK 472

Resistance to shattering : PK 472, PK 416, JS 335, Ahilya-4, Ahilya-4

Useful varieties in delayed sowing : JS 335, Ahilya-1, PK 472, Pb-1 etc. Increase seed rate and reduce row to row distance

High Protein content (> 40%): ADT 1, MACS 58

High oil (> 20%): NRC 7, VLS 1, PK 416

Tofu quality: Pb 1, Hardee, PK 472

Resistance to lodging: JS 71-05, Pusa-16, Ahilya-1

Resistance to stem fly: JS 335, Ahilya 2, MACS 124, JS 71-05

Resistance to defoliator larvae: JS 80-21, PK 416, PK564, Ahilya-3

Tolerant to rust: PK 1029, PK 1024, Indira soya-9, JS 80-21

Resistance to YMV: SL-295, PK 416, PK 564, PK 308

Growth stages: Soybean takes 5 days for germination, 30 days for first flowering, 60 days for complete flowering, 30 to 70 days for seed development and 90 days for attaining maturity. The growth stages of soybean are furnished in Table 3.

Table 3 Growth phases and stages of soybean

Stage	**Description**	**DAS**
Vegetative phase		
VE – Emergence	Cotyledons above the soil surface	5
VC - Cotyledon	Unifoliate leaves	7
V1 – First node	Fully developed unifoliate leaves	10

V2 – Second node	Fully developed trifoliate leaves at node above the unifoliate nodes	14
V3 – Third node	Three nodes on the main stem	17
V4 – Fourth node	Four nodes on the main stem	20
V5 – Fifth node	Five nodes on the main stem	24
V6 – Sixth node	Six nodes on the main stem	27
Reproductive phase		
R1 - Beginning bloom	One open flower	30
R2 – Full bloom	90 % open flower	35
R3 - Beginning pod	Pod 5 mm long	35-40
R4 – Full pod	Pod 2 cm long	50
R5 - Beginning seed	Seed 3 mm long in pod	58
R6 – Full seed	Pod containing a green seed that fills the pod cavity	75
R7 - Beginning maturity	One normal pod has reached its mature pod colour	85
R8 – Full maturity	95% of the pods have reached their mature pod colour.	90-95

In general, soybean takes 30 to 45 days from sowing to flowering depending on cultivars while the flowering duration is 14-20 days. Soybean flowers are white or purple, very small and borne in short clusters. Only about 25 to 60% of the flowers actually produce pods, which become prominent one to two weeks after the flowers appear. Pod setting lasts two to several weeks, longer for the indeterminate cultivars. When conditions are not limiting, a pod produces three to four seeds and they fill in about one month. Most roots are located in the upper 30 cm of soil, but prolonged dry periods cause roots to proliferate more in the deeper soil layers.

Land preparation: The land should be prepared well to get good tilth. The ridges and furrows are formed at an interval of 60cm and sowing can be done on both sides of ridges or on the beds and channels (check basin) of convenient size are formed depending upon irrigation source.

Seeds and sowing: Seed rate is 75 kg/ha for small seeded varieties and 100 kg/ha for bold seeded varieties. Half the quantity of seed required for sole cropping of soybean is sufficient for the intercrop. Seeds can be sown in lines with 30 × 5 cm spacing. Dibbling has been found to be best method of sowing over broadcasting for rice fallow crop. It is also useful as bund crop in rice fields. There should be 67 plants/m^2 for high seed yield. Sowing depth of 2 to 3 cm is preferred to ensure good germination. The seeds should be treated with fungicides like carbendazim or Thiram or Brassicol and the insecticide Carbofuran at the rate of 2 g/kg of seeds. The seeds should be treated at least about 24 hours before sowing. Seed treatment will protect the seedlings from seed borne pests and diseases in the early growth stages. The chemically treated seeds are to be again treated with peat based *Bradyrhizobium japanicum* culture (G3 and 61 A 76) or *Rhizobium fredii* culture and phospho-bacterium @ one packet (200 gm) per 10 kg of seed with equal quantity of rice gruel to enable better seed coating. The treated seeds should be dried under shade for 15 minutes and sown within 24 hours. Nodules begin to develop early from third week and remain active for 6-7 weeks in the root system. Zone wise seeds and sowing technology recommended are furnished in Table 4.

Table 4 Zone wise seeds and sowing technology

Zone	**Sowing time**	**Seed rate**	**Spacing**
North Hill Zone	Last week of May to June end	Bold seeded- 80-90 kg/ha Medium seeded-70-75 kg/ha Small seeded-55-60 kg/ha	45 × 5 cm
North Plain Zone	Mid of June to First week of July	Bold seeded- 80-90 kg/ha Medium seeded-70-75 kg/ha Small seeded-55-60 kg/ha	45–60 × 5 cm
Central Zone	Middle of June to Middle of July	Bold seeded-80-90 kg/ha Medium seeded-70-75kg/ha Small seeded-55-60 kg/ha	30–45 × 5–8 cm
Southern Zone	(i) For Kharif-Middle of June to end of July (ii) For Rabi- First week of October to December (iii) For Summer- Second fortnight of January	Bold seeded- 80-90Kg/ha Medium seeded-70-75kg/ha Small seeded-55-60 kg/ha	30–45 × 5 cm
Eastern Zone	Middle of June to middle of July	Bold seeded- 80-90 kg/ha Medium seeded-70-75 kg/ha Small seeded-55-60 kg/ha	30–45 × 5 cm

Source: http://www.nrcsoya.nic.in/

After cultivation: In indeterminate soybean varieties, nipping is done either mechanically or chemically through 20 ppm TIBA spray during early flowering stage which enhances seed set and seed yield by 10 to 15%.

Weed management: The field should be kept weed free for 30 to 40 days. Once the plants cover the inter row space weed growth will be suppressed. Spray Fluchloralin @ 2.0 lit/ha immediately after sowing as pre emergence for irrigated crop and then the field is irrigated immediately. Hand weeding should be given on 30-35 days after sowing to control voluntary crops and perennial weeds or two hand weeding one at 15-20 DAS and the other at 30-35 DAS are carried out for efficient weed management. Herbicides recommended for soybean is presented in Table 5.

Table 5 Herbicides recommended for soybean

Chemical name	**Mode of application**	**Dose kg ai/ha**
Fluchloralin	PPI	1.00
Trifluralin	PPI	1.00
Alachlor	PE	2.00
Metolachlor	PE	1.00
Clomozone	PE	1.50
Pendimethalin	PE	1.00
Imazethapyr	POE	0.10

Note: PPI: Pre-plant incorporation, PE: Pre-emergence, POE: Post emergence

Manures and fertilizers application: Apply 12.5 t/ha of FYM or compost along with 20–80-40-40 kg of N, P_2O_5 K_2O and S as basal dressing. Sulphur is applied as gypsum 220 kg/ha. If nodulation is not adequate a top dressing or foliar application of nitrogen may be applied. In case of sandy soil, nitrogen can be split into two halves. Apply first half as basal dressing and the rest as top dressing on 30 days after sowing preferably by placement. Spray 2% DAP solution on 40 DAS to supply N and P to meet N requirement if there is poor nodulation and P to increase seed yield. Apply $ZnSO_4$ @ 5 kg/ha and $MnSO_4$ @ 5 kg/ha with 50 kg FYM as basal dressing. Spray of 1% $MnSO_4$ on 25 and 40 DAS is recommended. Zone wise Manures and fertilizers recommendation is furnished in Table 6.

Table 6 Zone wise Manures and fertilizers recommendation

Zone.	**Manures and fertilizers**
North Hill Zone	10 t FYM/ha + 20:80: 20: 20 N:P_2O_5: K_2O:S kg/ha
North Plain Zone	10 t FYM/ha + 20:60: 20: 20 N:P_2O_5: K_2O:S kg/ha
Central Zone	10 t FYM/ha + 20:60: 20: 20 N:P_2O_5: K_2O:S kg/ha
Southern Zone	10 t FYM/ha + 20:80: 20: 20 N:P_2O_5: K_2O:S kg/ha
Eastern Zone	10 t FYM/ha + 20:80: 20: 20 N:P_2O_5: K_2O:S kg/ha

Source: http://www.nrcsoya.nic.in/

Water management: First irrigation is given immediately after sowing and the life irrigation on the 3rd day. Subsequent irrigation should follow at an interval of 10 to 15 days depending on soil and weather conditions. Soybean can tolerate drought to certain extent in the vegetative phase. Since 60-75% of most flowers abort on soybean, any stress that increases this abortion will greatly influence yield. Half of most flowers are lost before pods begin developing and the other half are due to pod abortion. If moisture stress persists during flowering and pod filling stages the seed yield will be affected adversely. To alleviate moisture stress, spray either kaolin @ 3% or liquid paraffin @ 1% on the foliage. Since soybean is sensitive to excess moisture, drainage facilities should be provided to harvest a good crop. Excess moisture severely affects germination and early growth of soybean. However soybean tolerates flooding or waterlogging up to 7 days, but yield can be reduced by more than 40 percent if prolonged flooding occurs at floral initiation or beginning of the seed filling stage. In addition to being detrimental to root activities, flooding reduces nodulation. In India, soybean evapotranspiration (ET) is reported to be around 450 mm. Water used by closed canopies (near full bloom to beginning of pod filling) may be 7.5 to 9 mm/day. Water requirement for soybean is 5 to 8 mm/day during seed filling. The transpiration ratio for soybean is 580 g of water to produce one g of dry matter.

Cropping system: Soybean can be intercropped with corn, pigeon pea, sorghum, cotton, pearl millet, ground nut, sugarcane, wheat, roselle, etc. The intercropping followed in different agro climatic zones in India is given below:

Northern Hill Zone: Maize + Soybean
Northern Plain Zone: Maize + Soybean, Arhar + Soybean,
Rice + Soybean (on bunds)
Central Zone: Cotton + Soybean, Arhar + Soybean,
Sorghum + Soybean

South Zone: Arhar + Soybean, Ragi + Soybean, Sorghum + Soybean, Cotton + Soybean, Sugarcane + Soybean, Maize + Soybean

Plant protection: Insect management can be done with furrow application of Phorate 10G @ 10 kg/ha, followed by 1 or 2 sprays of Quinalphos 25 EC (1.5 lit/ha) or Endosulfan 35 EC @ (1.5 lit./ha) or Monocrotophos 36 SC (0.8 lit./ha) or Triazophos 40 EC (0.8 lit./ha) or Methomil 40 SP (1 kg/ha) or Chlorpyriphos 20 EC (1.5 lit./ha) or Ethofenprox 10 EC (1 lit./ha), depending on the insect pest and intensity of damage. The spray volume has to be 750 to 800 lit for one ha. Two spray of carbendazim 50 WP or thiophanatemethyl 70 WP @ 0.5 kg in 1000 l water/ha at 35 and 50 DAS can be done for control of foliar diseases like Myrothecium, Cercospora leaf spot and Rhizoctonia aeria blight. Spray of Copper oxychloride 2 kg + Streptocycline 200g /1000 l water at the appearance of the disease for control of bacterial pastule. Spray of thiomethoxam 25 WG @ 100 g/ha or methyl deinaton @ 0.8 l/ha is done to control yellow mosaic. Two to three sprays of hexaconazole or propiconazole or triadimefon or ocycarboxin @ 0.1% can be done at 15 days interval for control of rust. The first spray is done at the time of appearance of rust followed by subsequent sprays at 15 days control. For hot spot areas one prophylectic spray of any of above fungicide can be given at 35 to 40 days after sowing. The major insect pests and diseases are given in Table 7. Some insect tolerant or less susceptible varieties are furnished in Table 8.

Table 7 Major insect pests and diseases in Soybean

Pests	Region
1. Insect-pests	
1. Stem fly (*Melanagromyza sojae*)	All soybean growing areas
2. Gridle beetle (*Obereopsis brevis*)	MP, Rajasthan, Uttaranchal
3. Tobacco Cattter pillar (*Spodoptera litura*)	MP, Rajasthan, Maharashtra
4. Green semi loopers (*Chrysodexis acuta* and *Diachrysia orichalcea*)	MP, Uttaranchal, Maharashtra
5. Blue beetle (*Cneorane* spp.)	MP, Maharashtra
6. Leaf miner *(Aproaerema modicella)*	Maharashtra, Karnataka
7. Bihar hairy caterpillar *(Spilosoma obliqua)*	Uttaranchal
8. White fly *(Bemisia tabaci)*	Delhi, Punjab, Uttaranchal
9. Cotton gray weevil *(Myllocerous* spp.)	Delhi,Punjab,Rajasthan, Karnataka
10. Pod borer*(Helicoverpa armigera, Cydia ptychora)*	All soybean growing areas
2. Diseases	
1. Soybean rust *(Phakopsora pachyrhizi)*	Parts of MP, Southern, Maharashtra and Northern Karnataka
2. Yellow mosaic (Mung bean yellow mosaic virus)	Punjab, Uttaranchal, Pockets in western MP
3. Collar rot *(Sclerotium rolfsi)*	All soybean growing areas
4. Bacterial pustule (*Xanthomonas campestris* pv. *glycines)*	All soybean growing areas
5. Leaf spot *(Myrothecium roridum)*	MP
6. Soybean mosaic (Soybean mosaic virus)	Uttaranchal, Punjab, Karnataka

Table 8 Some insect tolerant or less susceptible varieties are listed below:

Insect-pest/ disease	Resistant varieties
Stem fly	JS 335, PK 262, NRC 12, MACS 124
Defoliators	NRC 7, NRC 37, JS 80-21, Pusa 16, Pusa 20, Pusa 24, PS 564, PK 472
Girdle beetle	JS 71-05
Soybean rust	JS 80-21, PK 1029, PK 1024, Indira Soya 9
Collar rot	PK 262, PK 416, PK 472, PK 1042, NRC 37
Myrothecium leaf spot	Bragg, JS 71-05
Bacterial pustule	PK 416, PK 472, PS 564, Bragg
Soybean mosaic	Ankur, PK 327, PK 416, PS 564
Yellow mosaic	PK 416, PK 472, PS 564, PK 1024, PK 1029, PS 1042, PS 1092, SL 295

For the management of seed and seedling diseases (collar rot), seed treatment with thirum + carbendazim (2: 1 ratio) i.e. 2 g + 1 g/kg seed should be done. Alternatively, seed treatment with *Trichoderma viride* (@3-4 g/kg seed) can also be done where charcoal rot and collor rot are perpetual problems. In yellow mosaic virus prone areas, seed treatment with thiomethoxam 70 WS @ 3 g/kg seed is recommended. To prevent early seedling mortality due to stem fly and blue beetle infestation, soil application of phorate 10 G @ 10 kg/ha, should be done. Regular field scouting and pest monitoring should be done to facilitate mechanical removal of plant parts/plants infested with girdle beetle, or gregarious phases of tobacco caterpillar or Bihar hairy caterpillar or plants infected by yellow mosaic virus disease. This also helps in assessing the pest population with respect to economic threshold level (ETL) as furnished in Table 9.

Table 9 Economic threshold level (ETL) of pests in soybean

Insect pests	Crop stage	ETL
Blue beetle	Seedling	4 beetle/m row
Green semiloopers	Flowering, Podding	4 larvae/m row 3 larvae/m row
Bihar hairy caterpillar	Pre-flowering	10 larvae/m row
Tobacco caterpillar	Pre-flowering	10 larvae/m row
Pod borer	Podding	5 larvae/m row

Based on the ETLs, one foliar spray of any one of the insecticides Triazophos 40EC @ 0.8 l/ha, Chlorpyrifos 20EC @ 1.5 l/ha, Quinalphos 25EC @ 1.5 l/ha, Ethion 50EC @ 1.5 l/ha, Methomyl 12.5L @ 2.0 l/ha, Ethofenprox 10 EC @ 1.0 l/ha is to be given preferably at the time of flowering. One spray of any of the microbial-pesticides (Dipel *(Bacillus thuringiensis)* @ 1 l/ha or Biobit *(Bacillus thuringiensis)* @ 1 kg/ha) or Dispel *(Beauveria bassiana)* @ 1 l/ha) should be given 15 days after the spray of chemical insecticide for the control of defoliators. In rust prone areas, prophylactic spray of Hexaconazol, Propiconazol,

Triadimefon @ 0.8 kg/ha is recommended. For the management of foliar diseases *viz. Myrothecium Cercospora* and *Altemaria* leaf spot diseases and *Rhizoctonia* aerial blight, two sprays of carbendazim or thiophenate methyl @ 0.5 kg/ha at 35 and 50 days after sowing may be given. For the control of bacterial pustule disease, crop may be sprayed with mixture of copper oxychloride (2 kg) + streptocyclin (200 g/ha) at the time of appearance of disease. For the control of yellow mosaic disease, spray of methyl dematon 25EC @ 0.8 l/ha or thiomethoxam 25WG @ 100 g/ha is recommended for the control of vectors. In order to build up and conserve naturally occurring bio-control agents' fauna *viz.* Coccinellid beetles, *Chrysoperla* etc., biodiversity should be created in the form of intercropping. In rainfed areas, intercropping soybean with maize, sorghum or short duration pigeon pea (in 4:2 row ratio) is beneficial.

Harvest: The symptoms of maturity is yellowing and shedding of leaves. Seed has 17% moisture at maturity. When all the pods have turned yellow, cut the entire plant with sickle. Over maturity will cause shattering of seeds causing losses in yield. Dry the pods adequately in the sun and threshing is done by beating the dried plants with wooden sticks. Hand threshing is preferable for seed purpose. Thresh at about 14% grain moisture using thresher. Maximum possible total dry matter in soybean is 10.2 t/ha with associated seed yield of 2.8 t/ha. The harvest index is 27%. The mean seed yield is 2.0 t/ha. The seed moisture content at physiological maturity is approximately 50 to 60% which coincides with one pod on the main stem has reached its yellow colour. Store in moisture proof bags at seed moisture of about 9-10%. Commercially grown soybean varieties have 20 to 25% oil and 40 to 45% protein. Vegetable soybean harvesting is done when 80% of the pods have reached physiological maturity stage. It may take 65 to 75 days after germination for vegetable soybeans to be ready for harvest depending upon variety, temperature and weather conditions. The pods are still green. The green pod yield is 7–10 t/ha while the green bean yield is 4–7 t/ha.

Storage: The moisture content of seed is normally 14% during threshing. The seeds should be dried adequately to reduce the moisture level to 10% for storage. Soybean seeds have a relatively short storage life. In order to obtain a maximum life, soybeans should mature during dry weather. Pods should be permitted to dry as much as possible in the field before harvest. For seed purpose, mix one kg of carbaryl @ 10% or Malathian @ 5% or activated kaolin for every 100 kg of seeds. Store the seeds in gunny bags. Seed stored for 13 and 18 months showed 76 and 30% emergence respectively. Seeds obtained from drought and stored for one and six months gave 73 and 42% emergence respectively.

Utilization: Soya milk is recommended for infants where mother milk cannot be fed at the time of birth. Tofu is the soybean curd which can be converted into a variety of value added products. Soya nuggets products have >50% in the form of the granules. Noodles and vermicelli are produced from soybean. Soya flour is used for fortifying wheat flour and in preparation of bakery products. Soya Lecithin can be obtained from degumming of soya oil.

REFERENCES

CPG. 2012. Crop Production Guide. Department of Agriculture, Government of Tamil Nadu, Chennai and Tamil Nadu Agricultural University, Coimbatore.

GOI. 2014. Status paper on Oilseeds. Ministry of Agriculture, Government of India. New Delhi.

Kanemasu, E. T. 1981. Irrigation water requirements and water stress. In: Proceedings irrigated soybean production in arid and semi-arid regions. Ed. W.H. Judy and J.A. Jackobs., 31 August–6 September 1979. International Soybean Programme Series No.20.

Mayaki, W.C., Teare, L. D. and Stone, L. R. 1976. Top and root growth of irrigated and non-irrigated soybeans. Crop Science 16:92-94.

McWilliams, D.A., Berglund, D.R. and G.J. Endres. 1999. Soybean growth and management. North Dakota State University, Fargo, North Dakota 58105.

Specht, J.E., Hume, D.J. and S.V. Kumudini. 1999. Soybean yield potential – a genetic and physiological perspective. Crop Science. 39: 1560-1570.

http://www.nrcsoya.nic.in

http://oilseeds.dacnet.nic.in/crop.html

CHAPTER

11

Coconut–*Cocus nucifera* (L) (2n=32)

Family: Arecaceae / Palmae

Vernacular name: Coconut (English), Gola, Narial (Hindi), *Kalpavriksha* (Sanskrit), *Tenga, Thenna* (Malayalam), *Tengu* (Kannada), *Kobbari, Tenkaya, Narikadam* (Telugu), *Narikel* (Bengali), *Thennai* (Tamil), Nardiya (Oriya) and *Jhada, Naliyer, Nariel, Nariera* (Gujarati). The word coconut is derived from the Portuguese and Spanish word 'coco' which means 'monkey/grotesque face' or 'head' or 'skull', from the three indentations or 'eyes', on the hairy coconut shell that resemble facial features or the nut resembles a 'monkey's face' and nucifera referred to the 'nut-bearing' property of the palm

Importance: The coconut palm is one of the most beautiful and useful palm in the world. It provides a variety of useful products like food, fuel and timber. In Sanskrit, coconut is called '*Kalpavriksha*' which provides humans with everything. It is also known as the 'tree of heaven', 'tree of life', 'king of the tropical forest' and 'One of the Nature's greatest gifts to man'. Coconut oil is widely used in the manufacture of soaps, hair oil, cosmetics and other industrial products, its husk is a source of fibre which supports a sizable coir industry. The tender nut supplies coconut water, a popular thirst quencher of health and hygienic value. Coconut is grown in more than 90 countries of the world.

Origin: Coconut is originated in Southeast Asia, especially India, Malaysia, Indonesia and Philippines.

Distribution: Coconut is cultivated in Fiji, India, Indonesia, Malaysia, Papua New Guinea, Philippines, Solomon Islands, Sri Lanka, Thailand, Vietnam, Kenya and Jamaica. Philippines, Indonesia, India, Sri Lanka, Thailand and Malaysia together accounted for 80.65% of the total area under coconut cultivation and about 82% of world production. The major coconut crop acreage is concentrated on the West Coast region of the country comprising the states of Kerala, Karnataka and Maharashtra, followed by East Coast of Tamil Nadu, Andhra Pradesh, Odisha and Pondicherry. The coconut cultivation areas also traditionally located in the coastal region of Gujarat, Goa, West Bengal, Islands of Andaman & Nicobar and Lakshadweep. About 90% of the area of coconut cultivation and equally the same per cent of production of coconut are from the four Southern states, viz. Kerala, Karnataka, Tamil Nadu and Andhra Pradesh.

Botany: Coconut palm is a tall tree which grows to a height of 15 to 30 m. The stem develops from the single terminal bud called the 'cabbage' which is the palm's only vegetative growing point. The palm has characteristic inclined trunk (leaning towards the light at the edge of plantations and along the direction of the wind) marked by prominent ring like leaf scars. In a growing seedling, the trunk begins to form after the emergence of 12 to 18 leaves. Under favourable conditions, the foundation of the trunk of a young palm reaches full development within 3-4 years. The main stem is normally unbranched with a thick swollen base (blunt bottom) surrounded by a mass of adventitious roots. The rate of trunk elongation in the tall coconut is from its early peak of greater than 1 to 1.5 m/year to 10-15 cm per year at about the 40th year and 5 cm/year at 60 years of age. The diameter of the stem is about 30 to 40 cm under normal situations. The diameter may reach 0.8 to 1 m at the base which diminishes only by 30% over the period. The coconut has an adventitious root system. In an adult palm, 3600 to 8000 roots are observed depending on the soil conditions. The direction of the main roots is more or less horizontal. The normal length of the roots of a mature tree is about 5 m in firm soil and 7m in sand. Roots of 6 to–10 mm diameter can extend 5–7 m outward from the base of the palm. Over 80% of the active roots lie in the soil cylinder of 2 m radius around the palm. Vertical extension down the profile is normally 1 to 15 m. In a soil with 30% clay, cored to a sandy one, root depth is much reduced. Primary roots have first order (major) root branches of 4–5 mm diameter. Second (2 mm) and third (0.5–1 mm) order branches fill the role as feeder roots. Coconut roots have no root hairs and no nutrient scavenging mycorrhiza. Roots have small protrusions called pneumatophores which act as 'breathing organs' for the exchange of gases. There are small, whitish, pointed organs known as pneumatophores on the upper surface of primary roots, which evidently evolved to maintain oxygen supply to the root tip during diurnal submersion in the water table.

The leaf primordial is differentiated about 30 months before it emerges as a 'sword leaf'. Generally, a normal adult palm produces 12-16 new leaves annually at the rate of every 25 days, each bearing a corresponding flower cluster (inflorescence or spadix). Tall coconuts produce 18-20 leaves (called 'fronds') per year. Dwarf coconuts produce 20-22 leaves per year. As leaves senescence about 2.5 years after unfolding, tall coconuts have 30-35 leaves in their crown at any given time. The matured frond (leaf) length of the tall coconut is around 6 m, compared to the 4 m of the dwarf coconut, resulting in a much larger crown of the tall form. The frond has 200-250 leaflets. Leaves are arranged in the crown in whorls. A leaf (frond) remains on the palm for about 2 to 3 years and thereafter, shed leaving a permanent scar on the trunk. On the lower trunk of the palm, up to 5 m in height, the average vertical spacing (interval between leaf scars) of fronds has a maximum value of 7 cm. The spacing diminishes to around 4 cm in 25 years, causing the upsweeping attitude seen in all the fronds of a young palm. The number of scars on the stem, divided by 13, gives the approximate age of the palm in years.

The inflorescence primordium can be detected about 4 months after the leaf primordium is differentiated. The opening of the fully grown inflorescence occurs 1 year later. The coconut inflorescence is enclosed in a double sheath or 'spathe', the whole structure known as a 'spadix' which is borne singly in the axil of each leaf. The palm is monoecious, i.e. its inflorescence carries both male and female flowers on the same inflorescence. The rate of inflorescence production is one in each leaf axil. Male flowers ranges from 200 to 300 which borne singly or in twos or threes in the upper part of the floral axis, while the female flowers are relatively few in number and are located, usually singly, at the base of the inflorescence branches. Coconut palm keeps on flowering all the year round. Male flowers open first, within

15 days of opening of the inflorescence. The period between the opening of the first male flower and the last male flower lasts about 10 to 22 days while for female flower lasts for 4-10 days. A female flower remains receptive for 1 to 3 days. The spathe opening lasts 3-5 days in tall palms and about 8-15 days in dwarfs. A normal inflorescence may have 10-50 female flowers. With natural pollination, 50-70% usually abort and fall off, especially those which emerge during severe dry weather. The remaining flowers develop into fruits, which take about 11 to 12 months to develop into a mature nut. It is cross pollinated crop. Both wind and insects help in cross pollination up to 80 to 100%. Flower takes about 11 to 12 months after fertilization to develop into large ovoid mature fruit which is popularly known as 'nut'. The nut is made up of an outer exocarp, a thick fibrous fruit coat known as husk; underneath lies the hard protective endocarp or shell. Lining the shell is a white albuminous endosperm or 'coconut meat' and the inner cavity is filled with a clear sweet refreshing liquid called 'coconut water'. The mature fruit is a fibrous drupe, weighing 1.2 to 2.0 kg. Around 50 to 70% of the female flowers may fall during the first two months (button-stage) due to the failure of fertilization or some other causes. A few immature (developing) fruits may fall after three to four months. The number of fruits, which reach maturity, is generally between three and seven per inflorescence.

There are three botanical varieties of *Cocos nucifera* based on the stature or height viz.,

(i) **Tall coconut varieties** (*Cocos nucifera* var. *typica* Nar.): This is a hardy type with the life span of 80 to 90 years but the economic life is about 60-70 years. Hardy palms tolerate a wide variation of soil type and climate. Stem is broad with mean girth of 80 cm at 150 cm above ground. The stem attains a height of about 18 m. The leaves are long with mean length 5.5 m. Coconut palms attain bearing in 5-6 years after planting and steady bearing in about 12 to 15 years. It has long and stout trunk growing to a height of about 15 to 18 meters. The nuts mature within a period of 12 months after pollination. Tall coconuts produce about 50 to 80 fruits per year. Tall coconuts produce economic yield up to 25 years and live more than 100 years. The nuts are medium to big in size with colour varying from green, greenish yellow to brown. About 4000 to 5200 nuts are required to produce one ton of copra. The copra, oil and fibre are of good quality as compared to dwarf cultivars. This variety is cross pollinated because male flowers open first and female flowers are receptive one week after male flowers open. e.g., West coast tall (WCT), East coast tall (ECT), Kappadam, Andaman Giant (AG), Laccadive Ordinary (LO).

(ii) **Dwarf coconut varieties** (*Cocos nucifera* var. *nana* Griff): It grows to the height of 5 to 7 m. It grows with an average life span of 40 to 50 years but the economic life is 30 to 40 years. It thrives well on fertile soils with a well distributed rainfall. It suffers adversely from drought. Stem is narrow with mean girth of 55 cm at 150 cm above ground. It attains a height of about 10 m. Leaves are short with mean length of 4 m. They are normally self-pollinating and therefore considered to be homozygous. It is early flowering and start bearing from 3 to 4 years after planting. The nuts are small and the copra is soft, leathery and low in oil content. About 9,000 to 12,000 nuts are required to produce one ton of copra. The endosperm is not suitable for oil extraction. The nuts of dwarf cultivars have orange, yellow or green colours. The tall palms are not true to type as they are cross-pollinated, whereas dwarf palms are self-pollinated in nature because of overlapping of male and female phases and hence true to type. e.g., Chowghat dwarf, Gangabondam. It suffers adversely from drought; susceptible to pests and diseases.

(iii) **Intermediate coconut varieties** (*Cocos nucifera* var. *aurantiaca* Liy): The economic life period is 30 to 40 years. Stem is semi-broad with mean girth of 70 cm at 150 cm above ground. It attains a height of about 12 m. Leaves are short with mean length 4.3 m. Flowering is late and occur from 6 to 8 years after planting. Flower production is seasonal; predominantly in-breeding as male and female reproductive phases of inflorescence overlap. The nuts are medium size and about 8000 nuts are required to produce one ton of copra. The epicarp of nut orange is in colour, endosperm thin and of little value for copra production. Tender king coconut (*Thembili*), with its bright orange epicarp, and high sucrose, sweet nut water, is a very popular beverage coconut. Palms are very susceptible to drought, pests and diseases. It thrives well on fertile soils with a high water table and a well distributed rainfall.

Growth stages:

Vegetative phase (up to three to five years)

(i) Seedling stage is from 9 to 12 months

(ii) Vegetative growth stage is from three to five years

Reproductive phase (after three to five years)

(i) Initiation of flowers and fruit set is for three months

(ii) Period of fruit enlargement is four months

(iii) Period of fruit ripening is two months

Climate: It is essentially a tropical tree, growing mostly between 20ºN and 20ºS latitudes and up to an elevation of about 1000 m from MSL. A well distributed mean annual rainfall of 1000 mm to 2500 mm with distribution of 150 mm per month and relative humidity more than 60-90% is the best for proper growth and maximum yield of coconut. It needs solar radiation incidence of 300-900 W m^{-2}. It requires 120 hours sunshine or more per month for good yields. Day length is not critical for coconut. It grows in areas with a cardinal minimum, optimum and maximum temperature of 13ºC, 22 to 32ºC and 40ºC respectively. A diurnal range of 6-7ºC is ideal for coconut growth and yield. It has poor tolerance of frost. Severe frost is fatal to seedlings and young palms. It does not tolerate waterlogging within one m of surface, and will not survive more than two weeks of surface waterlogging.

Soil: Coconut is adaptable to a wide range of soil conditions, from light sandy soils to clay loam with a pH ranging from 5.0 to 8.6. The best soils are deep, friable and loamy soils. In heavier soils, it requires good drainage. Red sandy loam, laterite, and alluvial soils are suitable. Select site with deep (not less than 1.5 m depth) well drained soil. Avoid shallow soil with underlying hard rocks, low-lying areas subject to water stagnation and heavy clayey soils.

Varieties: Coconut palms are broadly classified into three groups, the tall varieties, dwarf varieties and hybrids. Selection of right variety and quality of seed material is important as the quality of planting material has got a direct bearing on production and productivity of coconut. The important coconut varieties and hybrids with salient features grown in India are presented in Table 1, Table 2 and Table 3.

Table 1 Coconut varieties and hybrids grown in India

State	Tall varieties	Dwarf varieties	Hybrids
Tamil Nadu	West Coast Tall, Chandrakalpa or Lakshadweep ordinary (LCT), VPM – 3 (Andaman Ordinary), East coast tall, Aliyar Nagar 1, Kera Chandra (Philippines ordinary)	Chowghat Orange Dwarf (COD), Chowghat Green Dwarf (CGD), Chowghat Yellow Dwarf (CYD), Gangabandom, Malaysian Dwarf Yellow, Strait Settlement Dwarf Green, Kalpa Raksha	Kerasankara (WCT × COD), Chandrasankara (COD × WCT), Kerasoubhagya (WCT × SSAT), VHC 1 (ECT × MGD), VHC 2 (ECT × MYD), VHC 3 (ECT × MOD)
Kerala	West coast tall, Chandrakalpa or Lakshadweep ordinary (LCT), Java, Cochin China, Philippines Ordinary (Kerachandra), VPM-(Andaman ordinary), Andaman ordinary, Kappadam, Lakshaganga (lakshadweep ordinary), gangabondam	Chowghat Orange Dwarf (COD), Chowghat Green Dwarf (CGD)	Kerasankara (WCT × COD), Chandrasankara (COD × WCT), Chandralaksha (LCT × COD), Keraganga (WCT × GBGD), Lakshaganga (LCT × GBGD), Anandaganga (ADOT × GBGD), Kerasree (WCT × MYD), Kerasoubhagya (WCT × SSAT)
Karnataka	West Coast Tall, Tiptur Tall (TPT), Chandrakalpa or Lakshadweep ordinary (LCT), VPM – 3 (Andaman Ordinary), Kera Chandra (Philippines Ordinary)	Chowghat Orange Dwarf (COD)	Kerasankara (WCT × COD), Chandrasankara (COD × WCT), Chandralaksha (LCT × COD), Kerasoubhagya (WCT × SSAT),

Table 2 Salient features of coconut varieties/ hybrids

S. No	Cultivars	Area for which recommended	Annual nut yield/ palm	Copra (g/nut)	Oil content (%)
Varieties					
1	Chandrakalpa	Kerala, Karnataka, TN	97	195	70
2	Kerachandra	AP, Maharashtra	110	198	66
3	Chowghat Orange Dwarf	All coconut growing regions	Tender nut variety		
4	KalpaPratibha	West Coast region and peninsular India	91	256	67
5	KalpaDhenu	West Coast region and Andaman and Nicobar Islands	86	242	65.5

6	KalpaMitra	West Coast region and West Bengal	80	241	66.5
7	Kalpatharu	Kerala, Karnataka, TN	116	176	68
8	Kalparaksha	West Coast region and root (wilt) diseases tracts of Kerala	65	215	65.5
9	Kalpasree	West Coast region and root (wilt) diseases tracts of Kerala	90	96.3	66.5
10	Pratap	Konkan region	150	152	59
11	VPM-3	Tamil Nadu	77	191	66
12	ALR 1	Tamil Nadu	126	131	64
13	Kamrupa	Assam	101	162	64
14	KeraSagara	Kerala	99	203	67.8
15	KeraKeralam	Kerala, Tamilnadu and West Bengal;	109	186	67.8
16	KeraBastar	Andhra Pradesh, Konkanregion in Maharashtra and Tamilnadu	117	151	----
17	Kalyani coconut-1	West Bengal	80	154	
18	Gauthami Ganga	Andhra Pradesh	90	157	68
Hybrids					
1	Chandra Sankara	Kerala, Karnataka, Tamil Nadu	110	208	68
2	Kera Sankara	Kerala, Karnataka, Maharashtra, Andhra Pradesh	106	198	68
3	Chandra Laksha	Kerala, Karnataka	109	195	69
4	Kalpa Sankara	West Coast region and root (wilt) disease tracts of Kerala	84	170	67.5
5	Kalpa Samrudhi	Kerala and Assam	117	214	69
6	Laksha Ganga	Kerala, Tamil Nadu	108	195	70
7	Kera Ganga	Kerala	100	201	69
8	Kerasree	Kerala	112	216	66
9	Kera Sowbhagya	Kerala	130	195	65
10	Ananda Ganga	Kerala	95	216	68
11	Godavari Ganga	Andhra Pradesh	140	150	68
12	VHC-1	Tamil Nadu	98	135	70
13	VHC-2	Tamil Nadu	107	152	69
14	VHC-3	Tamil Nadu	156	161	64.5
15	KonkanBhatye Coconut Hybrid-1	Konkan Region, Maharashtra	122	180	67.1

Source: Central Plantation Crops Research Institute, Kasaragod

Table 3 Salient features of indigenous and exotic coconut varieties

S.No.	Cultivars	General characteristics
A	Tall varieties	
1	West Coast Tall (WCT)	Cultivated in the West Coast region. Life span 75 years. Regular bearer producing 12 inflorescences per year. Start bearing in about 6-7 years. Average annual yield 80 to 100 nuts per tree. Average copra content 176 g. Oil content 68%.
2	East Coast Tall (ECT)	Cultivated in the East Coast region. It is morphologically similar to WCT. The palm takes 6-8 years to start bearing. The average annual yield is 100 to 120 nuts per palm. It has mean copra content of 125 g per nut and oil content of 64%.
3	Tipton Tall	This is popular tall cultivar of Karnataka. Morphologically similar to WCT. Average yield is 86 nuts per palm per year. Mean copra content of 178 g per nut with an oil content of 68%.
4	Benaulim Tall (Pratap)	Cultivated in Goa, Konkan region of costal Maharashtra. Morphologically similar in appearance to WCT but nuts are smaller and round and closely grow in heavy bunches. The variety starts bearing in about 7-8 years. Average annual yield is 150 nuts per palm with copra content of 152 g per nut and 64% oil content. The variety is cultivated by name Pratap in Maharashtra.
5	Lakshadweep Ordinary (Chandra Kalpa)	The variety is indigenous to Lakshadweep Islands and resembles to WCT except in nut size which is smaller with prominent three ridges on triangular nuts. The palms are good source for tapping toddy. Average yield of 100 nuts per palm, copra content 176 grams with 72% oil content. The variety is cultivated in Kerala, Karnataka and Andhra Pradesh.
6	Andaman Ordinary	The variety is largely grown in Andaman & Nicobar Islands. The palms are tall massive and more vigorous than WCT in vegetative growth. The average yield is 94 nuts per palm per annum. Copra content is 169 g per nut with 66% oil content.
7	Sevvelanir	This is a tall cultivar found in Pondicherry, tender nut water has high medicinal value. The nuts are green in colour, tender nut water is sweet in taste. The mean annual yield is 40 to 45 nuts per palm.
8	Kamrupa	The variety is known as Assam Green Tall, a selection from the local germplasms, one of the most promising cultivars of this region. The yield is 106 nuts per palm per year. Copra yield is 16.34 kg per palm per year, with 64.50% oil content. In tender nut the water content is 253 ml.
B	Exotic cultivars	
1	Philippines Ordinary	It is an exotic cultivars from Philippines. Palms grow up to the height of 10-12 m. Annual average yield is 110 nuts, with copra content 183 g per nut and oil content 66%. The variety is found suitable for cultivation in the West Coast, Konkan region of Maharashtra, Andhra Pradesh and West Bengal.
2	Philippines laguna	Annual average yield 88 nuts, per palm, per year, with copra content of 258.9 g per nut and oil content 66.5%.
3	Fiji longtongan	Annual yield of 104 nuts per palm per year. Copra content is 210.5 g per nut and oil content 66%.

4	Fiji tall	Annual yield of 106 nuts per palm per year, copra content 199.1 g per nut and oil content 65.2%.
5	S.S. Green	Annual yield of 108 nuts per palm per year, copra content 186 g per nuts and oil content 67%.
6	Sanramon	Annual yield of 64 nuts per palm per year, copra content 349.6 g per nut and oil content 68%.
C	Indigenous Dwarf Cultivars	
1	Chowghat Orange Dwarf	The variety is an indigenous dwarf cultivar in Chavakkad area of Trissur district in Kerala. It is also known as Gaurigathram or Chenthengu and Kenthali in Karnataka. The palm has thin stem, small and compact crown with orange coloured leaf, petioles, inflorescence and nuts. The Palm starts bearing from 3-4 years. Average annual yield is 65 nuts per palm, mean copra content 150 g. with 66% oil content. The variety is ideal for Tender coconut water purpose. It can be cultivated in Kerala and Karnataka.
2	Chowghat Green Dwarf	These dwarf cultivars bear nuts with orange, yellow and green colours as in their names. It is cultivated for tender nuts, ornamental value and for production of hybrids.
3	Malayan Yellow Dwarf	
4	Malayan Orange Dwarf	
5	Malayan Green Dwarf	
6	Ganga Bondam	
7	Kentholi Orange Dwarf	This cultivar is from Karnataka State, having high medicinal value for its tender nut water.
8	Chitta Gangapani	These dwarf cultivars are found in Arsikere and Tiptur area in Karnataka with green, medium sized round fruits. Tender nut water is very sweet.
9	Udha Gangapani	

Nursery practices

(i) Selection of mother palms: The need for collecting seed materials from high yielding coconut palms can hardly be over emphasized in a perennial crop like coconut. The following points may be remembered. Select seed gardens, which contain larger proportion of high yielding trees with consistence yield. High yielding mother palms yield about 100 nuts/palm/year should be chosen for collecting seed nuts. Alternate bearers should be avoided. The age of the palm chosen be middle age i.e., from 25 to 40 years. Even trees with 12 years age can be selected, if it is high yielding and has stabilized yield. A good regular bearing mother palm produces on an average one leaf and an inflorescence in its axil every month. So, there will be twelve bunches of varying stages of maturity at any one time. Harvest seed coconuts during the months of March–August to ensure good quality seedlings. The seed nuts should be round in shape and when tapped by finger should produce metallic sound. Fully ripe nuts develop twelve months after fertilization. Harvest the coconut bunches intended for seed nut by lowering them to the ground using a rope to avoid injury to seed nuts.

Select mother palms having the characters, viz.,

(i) 30-40 fully opened leaves with short strong petioles and wide leaf base firmly attached to the stem,

(ii) bearing at latest 12 bunches of nuts with strong bunch stalks,
(iii) bearing nuts of medium size and oblong shape.
(iv) husked nuts should weigh not less than 600 g,
(v) mean copra content of 150 g per nut or more.

Avoid mother palms which are having the following characters viz.,

(i) have long thin and pendulous inflorescence stalks,
(ii) produce long, narrow small sized or barren nuts
(iii) show shedding of immature nuts in large numbers and
(iv) growing under favourable environmental conditions.

(ii) Collection and storage of seed nuts: Collect mature nuts (above 11 months old) during the period from December to May. Seed nuts stored in shade for a minimum period of 60 days prior to sowing in the nursery. During storage, the seed nuts are arranged with the stalk end up over an 8 cm layer of sand in a shed and cover with sand to prevent drying of nut water. Nuts can be arranged one over the other up to five layers. The nuts can also be stored in plots provided the soil is sandy and the ground is sufficiently shaded. In the case of nuts harvested in May, heap them in partial shade, till husk is well dried and then sow them in the nursery.

(iii) Nursery management: Seed nuts may be preserved in sand in shade for a month preceded by a week of shade drying. Not more than four layers of seed nut may be stacked in storage. Sprinkle water, if there is excess temperature due to summer. Seed nuts may be dipped in Lindane solution before planting. 400 g of Lindane wettable powder may be dissolved in 100 litres of water for this purpose. Plant the seed nuts in raised sand bed with a spacing of 1' × 1'. Five rows of nuts may be planted in each bed accommodating 25 to 50 nuts per row. Irrigate the nursery beds every alternate day. Provide shade to the nursery by raising shade trees such as *sesbania* or *leucaena* on the sides of beds. The seed nuts start germination 6 to 8 weeks after planting and germination continues up to six months. Remove seed nuts, which do not germinate within 6 months after sowing as well as dead sprouts. Make a rigorous selection and select only good quality seedlings (9-12 months old) based on the characteristics such as early germination, rapid growth and seedling vigour, having 4 leaves for 9 months old seedlings and minimum of 6 to 8 leaves for 10-12 months old seedlings, 10 to 12 cm girth at collar and early splitting of leaves. However, 18 to 24 months old seedlings are preferred for planting in water logged areas. Do not select the seed nuts, which have just germinated. The recovery of quality seedling will be about 60 to 65%. Since early germination is one of the criteria for selection of seedling, the storing and sowing of seed nuts should be in lots rather than in a staggered manner. Never allow to lift the seedlings from the soil by pulling on the leaves or stem. Remove seedlings from the nursery by lifting with spade and cutting the roots. Keep the seedlings in shade and do not expose to sun. Plant the seedlings as early as possible after removal from nursery. Husk burial can be done for moisture conservation.

Season: The best planting seasons suitable for coconut are from June to July and from December to January.

Land preparation: Size of the pit depends on the soil type and water table. In laterite soils large pits of the size 1.2 m × 1.2 m × 1.2 m may be dug and filled up with loose soil, powdered cow dung and ash up to a depth of 60 cm before planting. In loamy soils, pits of size 1m × 1m × 1m filled with top soil to height of 50 cm are recommended. While filling the pits, two layers of coconut husk can be arranged at the bottom of the pit with concave surface facing upwards for moisture conservation. After arranging each layer, BHC 10% DP should be sprinkled on the husk to prevent termite attack. In laterite soils, common salt @ 2 kg per pit may be applied, six months prior, on the floor of the pit to soften the hard pans.

Coconut planting: Adopt spacing of 25 feet × 25 feet with 175 palm per ha. For planting in field border as a single row, 20 feet spacing between palms may be adopted. In general, the spacing depends upon the planting system, soil type etc. The spacing of 7.6 m for triangular; 7.6 m × 7.6 m, 8 m × 8 m, 9 m × 9 m for square and 6.5 to 6.5 m in rows and 9 m between pairs of rows for double hedge planting system are recommended in sandy and laterite soils.

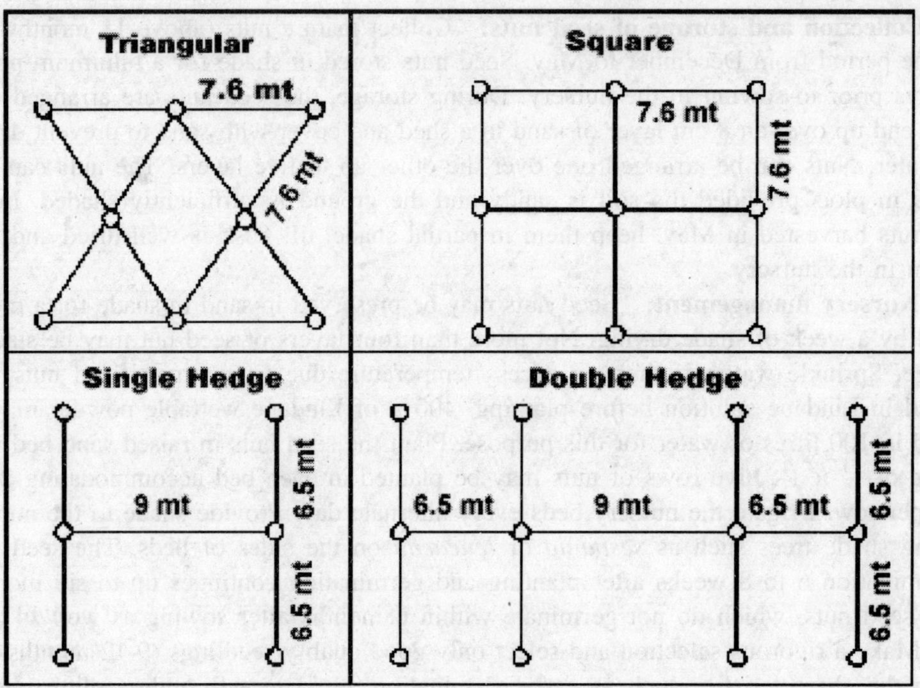

Pit size should be 3' × 3' × 3'. In the pits, Lindane 10% dust may be dusted to prevent white ant damage. The pit should be filled to a height of one foot with FYM, red earth and sand mixed in equal proportions. At the center of pit, the seedling should be planted after removing all the roots. The soil around the nut should be pressed well and the seedling should be provided with shade by using plaited coconut leaves or palmyra leaves.

After cultivation: The interspace in the coconut garden has to be ploughed twice in a year in June–July and December–January. Intercultural operation is essential to keep weed population under check, to ensure the utilisation of the applied plant nutrients by the coconut trees, to facilitate proper aeration to the roots of coconut and to induce fresh root growth.

Weed management: For the control of broad-leaved weeds, pre-emergence spraying of atrazine @1.0 kg a.i./ ha is recommended. Post emergence spraying of glyphosate @ 10 ml and 20 g ammonium sulphate/litre of water is recommended for the control of grasses and sedges.

Manures and fertilizer application: The quantity of manures, fertilizers and micronutrients to be applied to coconut are furnished in Table 4 and Table 5 respectively.

Table 4 Requirements of organic manures and major nutrients for coconut palm.

Year	FYM or green manure (kg/ha)	Quantity of fertilizer (g/palm)			Neem cake (kg/palm)	Gypsum (kg/palm)
		N	P_2O_5	K_2O		
First year	10	150	80	300	1.5	0.5
Second year	20	300	160	600	2.5	1.0
Third year	30	450	240	900	3.75	1.5
Fourth year onwards	40	600	320	1200	5	2.0

Table 5 Requirements of micronutrients to be applied from fourth year old coconut palm

Micronutrients	Quantity (g/palm)	Micronutrients	Quantity (g/palm)
$MgSO_4$	500	$CuSO_4$	50
$ZnSO_4$	250	$MnSO_4$	50
$FeSO_4$	250	Ammonium molybdate (or) Sodium molybdate	10
Borax	250	----	----

Apply the manures and fertilizers with the pit in the first year; 1.0 m radius basin in the second year; 1.5 m radious basin in the third year and 2.0 m radious basin from the 4th year onwards. Irrigate immediately after application of the fertilizers. The fertilizers may be applied in two doses, first in June-July and the second in December-January. Basal application of FYM @ 10 kg + top dressing of NaCl @ one kg 3 months after planting nuts or FYM / composted coir pith @10 kg or both as basal application is effective for the good growth of seedlings. Any one of the green manure crops such as sunhemp, wild indigo, calapagonium or daincha may be sown and ploughed in-situ or applied in trenches at the time of flowering as a substitute to compost. The root activity is maximum around a radius of 1.5 m to 2 m from the base of the palm. Application of fertilizer to the entire area around the palm is recommended and the fertilizer is forked in. Sufficient moisture should be present when manuring. Mix 50 g of *Azospirillum*, 50 g of Phosphobacteria or 100 g *Azophos* and 50 g of VAM in sufficient quantity of compost or FYM and apply near feeding roots once in 6 months / palm starting from planting. Don't mix with chemical fertilizers and pesticides

Water management: Coconut is irrigated @ 800 litres/palm of water once in seven days or 5 cm depth of water in 2 metres radius basins around the palm. Summer irrigation @ 40 litres per palm per week increases the yield of nuts by 50%. Under basin irrigation, 200 litres per palm once in four days is recommended. The optimum water requirement is found to be equal to the potential evaporation rate. The total transpiration of coconut is 90 litres per day. The IW/CPE ratio of 1.0 is found to be the best for irrigation scheduling. Coconut palms suffer due to water logging especially during rainy season and in wetland areas where rice fields are converted to coconut gardens. Shallow trenches are formed to drain the excess water from the root zone during the rainy season. In the first and second year, irrigation should be given twice a week and afterwards once in 10 days. During summer months and also whenever there is no rain, irrigation is a must, depending upon soil moisture. Application of the coconut husks at about 30 cm depth around the coconut trees at a radius of 1 m and covering it up with earth will conserve soil moisture in light textured soil. Use of coir waste @ 50 kg/palm as soil mulch around the tree to a thickness of about 15 cm is also advantageous to conserve soil moisture especially under water scarcity condition. The husk can also be buried either in linear trenches taken 3 m away from the trunk between rows of palms or in circular trenches taken around the palm at a distance of 2 m from the trunk. The trenches may be 0.5 m wide and as deep. The husks are to be placed in layers with concave surface facing upwards and covered with soil. The beneficial effect of husk burial will last for about 5 to 7 years. In areas where water is scarce drip irrigation system can be adopted. Coconut requires about 100 litres/day/palm through drip irrigation for matured plantation. The quantity of water recommended for drip irrigation in coconut is 66% of the open pan evaporation. Pitcher irrigation is given under severe water scarce condition. Four pitchers per tree can be followed. In water scarcity areas, pots with 20 litres capacity of water are fixed at a distance of 75 cm from the seedling on both sides and filling them periodically with water is recommended. Supply of water through drip system at 100% Eo level (Pan evaporation) with addition of super phosphate (0.950 kg), muriate of potash (2.0 kg) every year per palm enhances the nut yield. Occurrence of drought 15 to 16 months before opening of spadices lead to abortion of spadices. This is the reason for occurrence of abortion of spadices sometime in the rainy season. Moisture stress at flower primordial initiations one year before the opening of spathe affects the nut yield for two years later. Moisture stress at flower initiations and fruit set affects the number of nuts. Moisture stress at period of coconut fruit enlargement affects the nut size and copra content.

Cropping systems: Schedule for inter-mixed cropping may be drawn up based on the canopy size and orientation of palms. Intercropping can be practiced with crops like pineapple, banana, elephant foot yam, groundnut, chillies, sweet potato, tapioca, turmeric, etc. up to five years. Avoid crops like paddy and sugarcane, banana in first five years. Annual crops such as groundnut, bhendi, turmeric, tapioca, sweet potato, sirukizhangu, elephant foot yam, ginger, pineapple and biennial crops such as banana varieties viz., poovan and monthan are suitable as intercropping above 20 years of age (20 years of age has to be adjusted based on the sunlight transmission of above 50% inside the canopy). Green manure crops and fodder crops (Napier grass and guinea grass) can also be cultivated. Green manure and cover crops will help to increase the organic matter content of the soil and also will prevent soil erosion in coconut gardens. The green manure / cover crops such as *Crotalaria juncea*, *Tephrosia purpurea,* *Gliricidia maculata,* *Calapagonium muconoides* and *Mimosa invisa* are recommended for cultivation in coconut gardens. Sow the green manure / cover crops during April-May with the onset of pre-monsoon showers. The green manure crops should be ploughed in and incorporated in the soil during August-September. In the shade of the well grown up plantation, cocoa, pineapple, banana and forage crops like desmodium and desmanthus can be raised. In multistoried cropping system, banana and pineapple combination with coconut gives higher net returns per unit area. A multi-storey cropping system is developed to accommodate two or more intercrops of different heights, canopy patterns and rooting systems, to maximize the use of available sunlight, nutrients, moisture and land area under coconut. The fundamental objective is to increase the productivity of coconut land. The coconut palm serves as the 'top floor', whereas perennials such as coffee, bananas, papaya etc., form the mid-storey crops, and short-growing crops such as taro, vegetables, pineapples etc., form the ground floor. A crop combination of coconut + pepper + cocoa or cinnamon+ pineapple met the general criteria of a multi-storey cropping system. While coconut formed the top floor, pepper grown along the coconut trunk formed the second floor, cinnamon grown to 2 or more metres high formed the first floor, and pineapple with its shallow root systems represented the ground floor. Multi-storey cropping systems require high-level management. At the conventional spacing of 7 × 7 m to 9 × 9 m, compatible crop mixtures may be introduced, but each crop species has to be individually managed. One of the promising crop mixtures in a multi-storey cropping system was coconut + pineapple + papaya + coffee + jackfruit. The cropping sequence involved planting pineapple under coconut first, at a spacing of 30 × 100 cm, followed by two rows of papaya spaced at 3 × 3 m within the inter-rows and jackfruit between the coconut. Then within the rows of papaya, coffee is planted at the centre of each coconut block of four palms. The early-maturing papaya is harvested at the end of the first year, continuing until the end of the third year. Pineapple is harvested in the second year and allowed to ratoon thereafter. By the time papaya and pineapple have been abandoned, coffee and jackfruit are expected to provide continuous returns.

Crop protection

Rejuvenation of existing garden: The low yield in vast majority of gardens is due to thick population, lack of manuring and irrigation. These gardens could be improved through thinning of thickly populated gardens. In the farmer's holdings, 41% of the trees give a yield of less than 20 nuts/palm/year. After removal of low yielding trees, the population should be maintained at 175 to 200 palms/ha. Adequate cultural operations, manuring and irrigation especially during summer season, the yield of nuts per palm can be increased.

Button shedding: Shedding of buttons and premature nuts may be due to excess acidity or alkalinity, lack of drainage, severe drought, genetic causes, lack of nutrients, lack of pollination, hormone deficiency, pests and diseases. The remedial measures suggested are:

(i) Rectification of soil pH: Excess acidity or alkalinity of soil may cause button shedding. If the soil pH is less than 5.5, it is an indication of excess acidity. This could be rectified with lime application. Increase in alkalinity is indicated by soil pH higher than 8.5. This can be rectified with gypsum application.

(ii) Providing adequate drainage facilities: Lack of drainage results in the roots of coconut trees getting suffocated for want of aeration. Shedding of buttons occur under such condition. Drainage channels have to be dug along the contours to drain the excess water during rainy season.

(iii) To rectify the drought situation, coconut husks may be buried @ 100 husks with concave surface facing upwards or 25 kg of coir waste in semi circular trenches, dug to one foot width and two feet depth at 1.5 metres radius. This may be applied at the bottom and the usual manures and fertilizers applied above this layer when there is moisture in the soil. The monsoon rains are preserved by soaking of the coconut husk or coir waste as the case may be. Besides decomposition of these materials, these provide addition of potash to the coconut.

(iv) Genetic causes: In some trees button shedding may persist even after ensuring adequate crop pest and disease management. This is an indication of inherent defect of the mother palm from which the seed material is obtained. This underlines the need for proper choice of superior mother palm for harvesting seed coconut to ensure uniformly good yielding trees.

(v) Lack of nutrition: Button shedding occurs due to inadequate or lack of manuring. The recommended dose of manurial schedules and proper time of application are important to minimise the button shedding. Apply extra 2 kg of potash with 200 g of borax per palm over and above the usual dosage of fertilizer to correct the barren nuts in coconut.

(vi) Lack of pollination: Button shedding also occurs due to lack of pollination. Setting up of beehives at 15 units per ha may increase the cross pollination in the garden.

(vii) Pests: Button shedding occurs due to the bug infestation. Spraying of systemic insecticides like methyl demeton 0.025% or Dimethoate 0.03% may reduce the occurrence.

(viii) Diseases: Button shedding also occurs due to disease incidence such as Thanjavur wilt. Aureofungin @ 2 g + copper sulphate @ 1 g + Tridemorph @ 2 ml dissolved in 100 ml water may be applied as root feeding. 40 litres of 1.0 % Bordeaux mixture per palm should be applied as soil drench around the trunk in a radius of 1.5 metre. This will control the spread of diseases and prevent shedding of buttons.

Integrated pest management practices

(i) Set up pheromone trap for rhinoceros beetle @ 1 trap/100 ha by fixing it to the plant at 0.6 to 1 m height to trap and kill the beetles. Release of *Baculovirus oryctes* inoculated adult rhinoceros beetle @ 6 beetles/acre reduces the leaf and crown damage caused by this beetle. Soak castor cake at 1 kg in 5 liter of water in small mud pots and keep them in the coconut gardens to attract and kill the adults. Apply mixture of either neem seed powder + sand (1: 2) @ 150 g/palm or neem seed kernel powder + sand (1: 2) @ 150 g/palm in the base of the 3 inner most leaves in the crown. Apply Phorate 10 G @ 5 g in perforated sachets in two inner most leaf axils for 2 times at 6 months intervals to control rhinoceros beetle.

(ii) Set up pheromone trap for red palm weevil @ 1 trap/100 ha by fixing it to the plant at 0.6 to 1 m height to trap and kill the Red palm weevil beetles. Setting up of attractant traps (mud pots) containing sugarcane molasses 2½ kg or toddy 2½ lit (or pineapple or sugarcane activated with yeast or molasses) + acetic acid 5 ml + yeast 5 g + longitudinally split tender coconut stem/logs of green petiole of leaves of 30 numbers in one acre to trap adult red palm weevils. In red palm weevil attacked palms, observe for the bore- holes and seal them except the top most one. Through the top most hole, pour 1% carbaryl (20gm/litre) or 0.2% trichlorphon using a funnel. Then plug this hole also. If needed repeat after one week. Fill the crown and the axils of top most three leaves with a mixture of fine sand and neem seed powder or neem seed kernel powder (2:1) or lindane 1.3 D (1:1 by volume) once in three months to prevent the attack of rhinoceros beetle damage in which the red palm weevil lays eggs.

(iii) Bark weevil can be controlled with stem injection through a stove wick soaked in 0.2% fenthion or 0.2% dichlorvos and plugging the hole and repeating the treatment using the same wick and hole a month after.

(iv) A slanting hole is drilled one meter above the ground level and 10 ml of Monocrotophos 36 WSC is taken in a syringe and the needle is guided to the hole. After the insecticide has been absorbed the hole has to be plugged with clay mixed with copper oxy chloride or select pencil size roots and make a sharp cut at an angle and insert the root in the insecticidal solution containing monocrotophos 36% WSC @ 10 ml + water 10 ml in a 7 cm × 10 cm polythene bag to control leaf eating caterpillar / black headed caterpillar. Harvest is done prior to root feeding or 45 days after root feeding with pesticide.

(v) Coconut eriophyid mite is controlled with spray of Fenpyroximate 5% EC @ 10 ml/l. Root Feeding of monocrotophos 36 WSC @ 15 ml or triazophos 40 EC @ 15 ml or carbosulfan 25 EC @ 15 ml / 15 ml of water. Next harvest should be done 45 days later after root feeding.

(vi) Spray neem oil 5% (50 ml/l) once on the base and up to 2 m height of the trunk for effective control of termites.

(vii) Rats attack tender nuts which results in fall of immature nuts. Rats can be controlled with poison baiting with zinc phosphide or warfarin; fumigate the hiding places using Aluminium phosphide tablets; place wax blocks containing the poison Bromodioline @ 30 blocks per ha (each weighing 10g) on 5 palms two times at 12 days interval.

(viii) Spray copper oxychloride 50% WP @ 1 kg in 300-400 l of water/acre on the crown of the neighbouring palms as a prophylactic measure before the onset of monsoon to control bud rot.

(ix) Tanjore wilt / Basal stem rot / Ganoderma wilt can be controlled with application of *Pseudomonas fluorescens* @ 200g/palm + *Trichoderma viride* @ 200g / palm / year. Aureofungin-sol 2 g + 1 g Copper sulphate in 100 ml water or 2 ml of Tridemorph in 100 ml water applied as root feeding. (The active absorbing root of pencil thickness must be selected and a slanting cut is made. The solution to be taken in a polythene bag or bottle and the cut end of the root should be dipped in the solution). Trunk injection / root feeding with Calixin 3 ml/tree. Forty litres of 1% Bordeaux mixture should be applied as soil drench around the trunk in a radius of 1.5 m.

(x) Stem bleeding disease is controlled with application of Tridemorph @ 25 ml in 25 litre of water as soil drenching once in four months. Chisel out completely the affected tissues and paint the wound with Tridemorph 5% or Bordeaux paste. Apply coal tar after 1-2 days on the treated portion and burn off chiseled pieces. Root feeding with tridemorph 5% (5% Calixin) 5 ml in 100 ml water, thrice a year during April-May, September-October and January-February is recommended to prevent further spread of lesions/disease onto upper portion of trunk.

Harvest: Coconuts become mature in about 12 months after the opening of the spathe. Tall varieties start bearing coconuts from 5 to 7 years after planting. Economic life of the coconut palm is about 60 years. Tall varieties normally produce 80 to 100 nuts per palm per year. Dwarf varieties start bearing nuts from 3 to 4 years after planting. It produces economic yields up to 35 years. It requires high water and does not tolerate drought. Incidence of pests and diseases are high in dwarf varieties. Hybrid coconut starts bearing nuts from 3½ to 5 years after planting. It gives 30 to 50% more yield than tall varieties. It cannot tolerate drought, pests and diseases. Coconuts are harvested at varying intervals in a year. In well maintained and high yielding gardens, bunches are produced regularly and harvesting is done once a month. Nuts which are eleven months old give fibre of good quality and can be harvested in the tracts where green husks are required for the manufacture of coir fibre.

Yield: Coconut produces 80 to 100 nuts/palm/year depending on variety. The dwarf varieties yield about 70 to 80 nuts/palm/year while the tall varieties produce about 80 to 100 nuts/palm/year. The hybrid coconut yields about 130 to 150 nuts/palm/year. Copra is the dried endosperm from which oil is expelled. In ripe nuts, the endosperm contains about 50% water and 30 to 40% oil.

Utilization: Coconut industry is mainly confined to traditional activities such as copra making, oil extraction, coir manufacture and toddy tapping. Coconut products such as virgin coconut oil, desiccated coconut, coconut water based vinegar, tender coconut water, coconut oilcake are also made. Coconut oil is processed through two methods viz., the refining, bleaching, and deodorizing (RBD) processes and wet method. During the RBD process, heating process is applied especially during deodorization process, which is carried out at high temperature between 204 and 245°C. The coconut oil produced through the wet method is known as virgin coconut oil (VCO). The virgin coconut oil refers to an oil that is obtained from fresh, mature kernel of the coconut by mechanical or natural means, with or without the use of heat and without undergoing chemical refining. It is a pure form of coconut oil, crystal white in colour, with natural vitamin E (Tocopheorl) contents. It has a mild scent of fresh coconut. Virgin coconut oil differs from traditionally produced copra-derived coconut oil, which has to undergo chemical processing, bleaching and dehydration process to make it suitable for human consumption. The refined bleached coconut oil is yellow in colour, odourless and tasteless. It does not contain vitamin E, as it gets lost in the process of thermal and chemical processing. Virgin coconut oil can be produced either from the fresh comminuted coconut meat, coconut milk or coconut residue. Tapping in coconut is performed in the inflorescence before the flowers fully develop. The fresh tapped juice is the *neera* (*padhaneer*), and when allowed to ferment it is the toddy and when fresh juice is evaporated, it yields unrefined sugar known as jaggery. The shell as such is used for fuel purpose, shell gasifier as an alternate source of heat energy, making handicrafts, ice-cream cups and other commercial products like shell powder, shell charcoal and activated carbon. The husk yields fibres, which is converted into coir and coir products *viz.*, coir carpets, coir geo-textile, coir composite, coir safety belts, coir boards, coir asbestos and coir pith. Coir pith, a secondary by product obtained during defibring process is used as soil conditioner and mending all types of soils. The spongy nature of pith helps in disintegration of clay soil and allows free drainage. Its sponginess helps to retain water and oxygen and also prevents loss of vital nutrients from soil.

REFERENCES

Batugal, P., V. Ramanatha Rao and J. Oliver. (eds). 2005. Coconut Genetic Resources. International Plant Genetic Resources Institute – Regional Office for Asia, the Pacific and Oceania (IPGRI-APO), Serdang, Selangor DE, Malaysia.

Child, R. 1974. Coconuts. Longman, Green & Co., London. 335 p.

CPG. 2012. Crop Production Guide. Department of Agriculture, Government of Tamil Nadu, Chennai and Tamil Nadu Agricultural University, Coimbatore.

Kenneth, R.S.P. 2005. A guide to intercropping coconut. Upland Development Programme in Southern Mindanao. Philippines.

Liyanage, D. V. 1958. Varieties and phenotypes of coconut palms grown in Ceylon. Ceylon Coconut Quarterly. 9: 1–10

Mahindapala, R. and J.L.J.G. Pinto. 1991. Coconut cultivation. Coconut Research Institute, Bandirippuwa Estate, Lunuwila, Sri Lanka. 156 p.

Manciot, R.M., Ollagnier and R. Ochs. 1979. Mineral nutrition and fertilization of the coconut around the world. Oleagineux. 34 (12): 563-580.

Menon, K.P.V. and K.M. Pandalai. 1958. The coconut palm, A monograph. Indian Central Committee. Times of India Press, Bombay, India.

Ohler, J.G. 1984. Coconut, tree of life. Plant Production and Protection Paper No.57. FAO, Rome.

Satyagopal, K., S.N. Sushil, P. Jeyakumar, G. Shankar, O.P. Sharma, D.R. Boina, S.K. Sain, M.N.

Reddy, N.S. Rao, B.S. Sunanda, Ram Asre, K.S. Kapoor, Sanjay Arya, Subhash Kumar, C.S. Patni, S. Gangopadhyay, R. Mesta, Venkateshalu, S.D. Ekabote and K. Rajashekarappa, 2014. AESA based IPM package for Coconut. pp 38. National Institute of Plant Health Management, Rajendranagar, Hyderabad – 500 030

Singh, H.P., Hari Prasad and B.D.Sherkar. 2008. Production and marketing of coconut in India. Government of India, Directorate of Marketing and Inspection. Nagpur. pp 1-229.

Subburaj and R.K. Singh. 2003. Marketing of Coconut, Disposal Strategies of farmers. Indian Coconut Journal : 1-7 March, 2003 Vol. XXXIII No. 11, ISSN 0970-0579.

Thomas, M., Anand, B.L and M.K. Sukumaran Nair. 2004. The Mite and Coconut Economy of India. Indian Coconut Journal, Vol. XXXIV No. 9 January 2004, 6-14.

Wahid, P.A., Salam, M.A. and R.R. Nair. 1993. A Farmer's Primer on Coconut Cultivation. Kerala Agricultural University, Vellanikkara, Thrissur -680 654

https://www.nabard.org/english/plant_coconut3.aspx

http://agmarknet.nic.in/productioncoconut.pdf

http://www.worldagroforestry.org/treedb/AFTPDFS/Cocos_nucifera.PDF

http://www.bioversityinternational.org/fileadmin/bioversity/publications/Web_version/108/begin.htm

http://agridr.in/expert_system/coconut/coconut/pdf/faq.pdf

http://agritech.tnau.ac.in/farm_innovations/pdf/Modified_CPCRI.pdf

http://agritech.tnau.ac.in/horticulture/horti_plantation%20crops_coconut.html

http://agmarknet.nic.in/productioncoconut.pdf

CHAPTER

12

Oil Palm–*Elaeis guineesis* Jacg. (2n=32)

Family: Arecaceae / Palmae

Vernacular name: Oil palm, African oil palm, (English), *palmier a huile* (French), *Kelapa sawit* (Indonesia), *Ennai pannai* (Tamil).

Importance: Oil palm is the highest edible oil yielding crop, which yields an average of 4 to 6 tons of palm oil per hectare per year. However, it has a potential oil yield of 17 tons per hectare per year. Oil palm produces two distinct oils, i.e., palm oil from the fleshy mesocarp of the fruit, which contains 45 to 55% of oil and palm kernel oil, is obtained from the stony seed which contains 50% of oil.

History: Dutch has introduced oil palm to Brazil during 15th century. Oil palm has been planted in Botanic Gardens of Java during 1848 by East Indies. Oil palm has been introduced as ornamental palm in Sumatra islands during 1900. M. Adrien Hallet has established first commercial oil palm plantation in Sumatra during 1911. Later in 1915, Indonesia has taken up commercial planting of oil palm. In 1917, first commercial planting has taken up in Malaysia at Tennarmaran Estate, Selangor. In 1981, pollinating weevil *Elaeidobius kamerunicus* has been introduced to Asia. Oil palm has been first brought to India as a botanical collection at the National Botanical Garden, Calcutta during 1886. Subsequently, Maharashtra Association for Cultivation of Sciences (MACS) has introduced African durra palms during 1945-47. Though oil palm research has originally started in 1960 at Thodupuzha, Kerala, a systematic approach has been made only from 1976 at Central Plantation Crops Research Institute (CPCRI) Research Centre, Palode (Kerala). Subsequently, four research centers one each in Andhra Pradesh (Vijayarai), Karnataka (Gangavathy), Maharashtra (Mulde) and Tamilnadu (Aduthurai) have been started form 1987-88 under All India Coordinated Research Project on Palms. During 1993, cold and drought tolerant material have been collected by CPCRI from Guinea, Bissau, Cameroon, Zambia and Tanzania. Recognizing the vast potential of oil palm cultivation, the ICAR has established National Research Centre for Oil palm at Pedavegi on 19-2-1995 to cater the research needs for oil palm development.

Origin: Oil palm is originated in the tropical rain forest region of West Africa which spread over southern latitudes of Cameroon, Ghana, Liberia, Nigeria, Sierra Leone, Togo and into the equatorial region of Angola and the Congo.

Distribution: The oil palm is mainly distributed in Brazil, Malaysia, Indonesia, Columbia, Ecuador, Panama, Costa Rica, the Salmon Island, Papua New Guinea, Thailand, Philippines and India. Four seed gardens of 20 ha were set up one each in Andhra Pradesh (Rajahmundry, Lakshimipuram), Karnataka (Taraka) and Kerala (Thodupuzha). The potential areas for oil palm cultivation in India is presented in Table 1. In Cauvery delta of Tamil Nadu, it is presently cultivated in Tiruchirapalli, Karur, Perambalur, Ariyalur, Thanjavur, Thiruvarur and Nagapattinam districts in about 3500 acres. Tamilnadu ranks fourth after Andhra Pradesh, Karnataka and Kerala in growing oil palm in India. It is grown as rainfed crop in Kerala, Andaman and Nicobar islands.

Table 1 Potential areas for oil palm cultivation in India

S. No.	State	Potential area identified in ha	Area under oil palm in ha as on March, 2012
1	Andhra Pradesh	400000	128160.21
2	Chhattisgarh	40000	0.00
3	Goa	2000	863.24
4	Gujarat	90000	3437.04
5	Karnataka	250000	29386.00
6	Kerala	6500	5366.85
7	Maharashtra	0	280.00
8	Mizoram	61000	14306.45
9	Odisha	25000	10569.00
10	Tamil Nadu	162000	15292.05
11	Tripura	0	31.00
	Total	1036500	207691.84

Botany: The genus *Elaeis* consists of three species *viz*., the African *E. guineensis*; and two species indigenous to South and Central America (*E. oleifera* and *E. odora*). The longevity of the oil palm is believed to be 200 years. Leaves are produced @ one leaf per month up to six months. The leaf consists of leaflets (each with lamina and mid rib), a central rachis (to which leaflets are attached), a petiole (stalk between the lowest leaflet and the trunk, and a leaf sheath (normally remnants are visible). The matured leaf is pinnate with linear pinnate (leaflets) on either side of the leaf stalk (rachis). At normal planting density an adult palm may have about 48 to 50 leaves. Leaf arrangement in oil palm is spiral. The matured leaf is pinnate with linear leaflets on each side. Each leaf will be 8 m long and 3 m broad with 50-300 leaflets. The rate of leaf production may be about 35 to 40 leaves per annum at 5-6 years of age, thereafter it declines to 20-25 leaves per annum. Palm stem has one growing point. Stem apex is conical in shape. The average increase in height of the stem may be 0.35-0.75 m per year depending upon the growing conditions and genotypes. The width of the stem unclothed by leaf bases varies from 20-75 cm. Oil palm has typical adventious root system. Individual primary roots travel up to 90 m from the stem. There are two types of primary roots *viz*., descending and radiating primaries are there. In adult palm total quantity of absorbing roots increases at least to a radius of 4.5 meter. Bulks of the absorbing roots are seen in the top 15-30 cm of the soil. Oil palm root spread to a radius of one, two and three metres during the first, second and third year old oil palm plantation. Majority of the active root system lies in between 5 and 55 cm depth. Thus, most of the feeder roots are located at shallow depth and comparatively closer to the tree trunk. It has been estimated that roots of fully-grown oil palm extend beyond 4.5 m, though most of the roots get concentrated around the bole to a radius of 2-3 meters. Vertically, roots get concentrated to a depth of 40-50 cm though some of the roots are extended up to 5 meters. Even then, about 65 per cent of land area can be available in grownup oil palm for exploitation by other crops. Roots of oil palm have a tendency of accumulation in areas where water and fertilizers are applied.

Oil Palm is monoecious wherein male and female inflorescences are produced separately on the same palm in cycles. Female inflorescence has 110 numbers of spikelets. There will be over 4000 flowers. Male inflorescence is borne on a longer peduncle than the female. There are 600 to 1200 male flowers which have distinct aroma and matures from the bottom. Anthesis in a normal inflorescence usually lasts for 36-40 hours, but may be extended to a week in *E. guineensis.* In *E. oleifera,* the period of anthesis is much more erratic than *E. guineensis* and may last for 3-4 weeks or have two peaks. Male inflorescence is borne on a longer peduncle than the female. Male and female inflorescences are produced in cycles. Inflorescence takes 3 days for complete flowering. Flower opens from bottom to top and completes in 3 days. An inflorescence may produce 25-50g of fresh pollen grains. Maximum pollen shedding takes place in 2-3 days following the anthesis and production ceases within 5 days. However, viability of late produced pollen is low. Oil palm is a cross pollinated crop and pollination takes place through insects, the weevils (*Elaeidobius kamerunicus*). Stigma is receptive for 3-5 days. Oil palm is naturally a cross pollinated crop. Oil Palm fruit is a sessile drupe, varying in shape from nearly spherical to ovoid or elongate and are 2-5 cm long and weigh about 3-30 g. Oil Palm seed is a nut consisting of shell and kernel. The pericarp of the fruit consists of the outer exocarp (skin), mesocarp (pulp rich in oil) and endocarp (hard stony shell surrounding kernel). The kernel has testa (skin), a solid endosperm (oil rich) and embryo. Oil palm seed is a nut consisting of the shell and one, two or three kernels. Seed weight ranges from 1.0 to 13.0 g. There are three types of oil palm fruits namely,

(i) Tenera: This is the commercial planting material which is the cross between dura (female parent) and pisifera (male parent). Tenera fruits have thin shell (0.5-4.0 mm) with medium to high mesocarp content (60-95%). There is prominent fibre ring in its mesocarp.

(ii) Dura: Dura palms have fruits with thick shell (2-8 mm in thickness), low to medium mesocarp content (33-55%) and are not commercially grown. These are used as female parent (mother palms) in developing tenera hybrid.

(iii) Pisifera: Pisifera palms are female sterile, shell-less and used as male (pollen) parent in development of tenera hybrids.

Climate: The natural habitat of oil palm is in swamps and along the river banks. It is mostly grown in the humid tropical region between 13°N and 13°S in the forest lands under rainfed condition. Oil palm is ideally suitable in areas with an annual rainfall of 2000 mm or more without a defined dry season, maximum temperature of 29 to 33°C minimum temperatures of 22-24° C and a constant sunlight of at least 5 hours per day or 1800 to 2000 sunlight hours annually. Solar radiation below 350 cals cm^{-2} day^{-1} affects the growth and yield of oil palm. Oil palm is a sun loving plant. It needs at least 157 kcal cm^{-2} per year.

The minimum, optimum and maximum temperature for growth and yield of oil palm are 18°C, 24 to 33°C and 40°C respectively. Growth is arrested below 15°C. Oil palm requires a mean daily temperature of 20°C during growing period of 270 days or more. Oil palm requires an average annual rainfall of 2000 mm with evenly distribution of 150 mm per month. For every deficit of 200 mm rainfall below the normal annual rainfall requirement of 2000 mm, there will be yield reduction of 5 tonnes of fresh fruit bunch per hectare.

Table 2 Climatic conditions prevailing in oil palm growing states in India

State	Temperature (°C)		Relative humidity (%)	Total rainfall (mm)
	Min	Max		
Andhra Pradesh	23.7	47.0	65 to 76	1017
Karnataka	11.9-23.9	41.0	--	723 to 1528
Tamil Nadu	17.3-26.8	27.7-37.1	61 to 76 during May-August and 73 to 89 during September-April	1010
Gujarat	10.5-26.5	28.7-41.7	51.1 to 95.4	776 to 1046
Odisha	15.8-26.5	28.7-37.1	84 to 92	1500
West Bengal	10.1-27.4	23.4-36.3	> 80	1750

Soils: Oil palm can be grown on a wide range of soils. Well drained deep loam to clay loam soils with moderate permeability are found to be suitable for oil palm. The soil depth of more than 75cm, slope of less than 12 %, pH of 4 to 6 are favourable for oil palm cultivation. Soil types prevailing in oil palm growing states of India is presented in Table 3. Oil palm is grown in India with pH ranging from 3.5 to 8.5. Soils having pH above 8.5 are to be avoided. Avoid highly alkaline, highly saline, waterlogged and coastal sandy soils. Soils with poor drainage, laterite soil containing concretionary iron stone, sandy coastal soils and deep peat soil must be avoided. Oil palm tolerates periods of flooding provided it is not prolonged.

Table 3 Soil types prevailing in oil palm growing states of India

State	Soil type
Andhra Pradesh	Red sandy, red loamy, red loamy with clay base, deltaic alluvium pH of 6.5 to 9.0
Assam	Tea growing soils and river alluvium
Goa	Red loam and laterite soil
Gujarat	Black soils and alluvial soils are pre-dominant. Some are coastal alluvium and laterite soils
Karnataka	Deep red brown sandy loamy and black clay loamy soil with 6.5 to 8.3 pH. River alluvium soil with 7.5 to 8.5 pH. Upper Krishna project and Thungabadra – 60% black cotton soils with 7.8 to 9.5 pH and 40% red soil with 7.4 to 9.5 pH.
Kerala	Laterite soil
Odisha	Red and yellow soils, red loamy soils, river alluvial, coastal alluvial and brown forest soils.

Tamil Nadu	Alluvium with coastal sands. Laterite/gravelly laterite on a few patches. River alluvium and red sandy soils.
Tripura	Deep sandy loam and river alluvium soils
West Bengal	Sandy loam with acidic pH.

Season: Best season for planting is June-December i.e., during monsoon. In case of planting during summer, adequate irrigation, mulching and growing cover crops like sunhemp in the basin would help in avoiding hot winds during summer season.

Varieties: Tenera hybrids like Deli.Avros, Deli.Ghana, Deli.Ekona and Deli.Lame are widely cultivated. Tenera hybrid oil palm a cross between thick shelled durra and shell less pisifera. The characteristic features of oil palm varieties are presented in Table 4.

Table 4 Characteristic features of oil palm varieties

Fruit type	Mesocarp (%)	Shell (mm)	Kernel (mm)
Durra	20 to 65	2.0 to 8.0	4.0 to 20.0
Pisifera	92 to 99	Nil	3.0 to 8.0
Tenera	60 to 96	0.5 to 4.0	3.0 to 15.0

Nursery management: The fruits are separated from the bunch and seeds are extracted by scrapping the exocarp and the mesocarp with a knife or by retting in water. The seeds are dried by spreading them on a concrete or wooden floors under the shade for two days. Such seeds can be stored for 3 to 9 months at about 27^{0}C without much reduction in viability. Seeds are soaked in water with changing the water daily. Thereafter the seeds are spread out for drying for 24 hours. The dried seeds are placed in polythene bags and put in the germinator maintaining a temperature at 40°C for 80 days. The seeds are removed from the polythene bags, soaked in water for 5 days changing water daily and dried in the shade for 2 hours. The seeds are put back into the bag and kept in a cool place in order to maintain the moisture content. Germination commences in about 10 to 12 days. The percentage of germination obtained by this method is 90 to 95%. Oil palm seedlings are raised in the nursery for 12 to 14 months. The nursery area can be calculated keeping in view of having 19900 polybags/ha for main nursery, if spaced at 0.76 m triangle spacing or 13800 seedling if 0.91 m triangle. There are 2 types of polybag nurseries viz., i) Raising of seedlings in beds or small poly bags (250 gauge and 23 × 13 cm size) up to five leaf stage and transplanting into bigger bags (500 gauge and 40 × 45 cm in size) and ii) Transplanting the sprouts directly into large poly bags and raising the nursery up to the stage of transplanting. In India, a mixture of topsoil, sand and well decomposed cattle manure 1:1:1 ratio is commonly used. The bags should be well watered but not water logged. They should be placed in the shade to avoid drying out and watering is given every two days. Seeds whose plumules (shoots) and radicles (roots) have fully differentiated should be selected for planting. The optimum stage of seed development for planting is when the radicle is in between 5 and 10 mm. Shading is given up to two leaf stage. Palm leaves can be removed when the seedlings are 2 to 3 months old. Oil palm seedlings up to 5 leaf stages or 8 weeks old seedlings are maintained in primary nursery. The fertilizer doses applied in nursery in Costa Rica is furnished in Table 5.

Table 5 Fertilizer schedule is recommended for oil palm nursery in Costa Rica

Age of the plant (months)	Type of fertilizer used	Amount (g/plant)
2	18-46-0	3.0
3	18-46-0	5.0
4	18-46-0	8.0
5	18-46-0	12
6	15-15-15 + $MgSO_4$	16+0.5
7	15-15-15 + Borax (11% B)	20+0.5
8	15-15-15 + $MgSO_4$	25+1.0
9	15-15-15 + Borax	30+1.0
10	15-15-15 + $MgSO_4$	35+1.5
11	15-15-15 + Borax	40+1.5

This has to be applied 6 to 8 cm away from the seedlings during the first application, 10 to 12 cm away during the second and 15 to 20 cm away during the third application in primary nursery.

Secondary nursery and single stage polybag nursery: Black polythene bags of 500 gauge with dimensions of 40 × 45 cm size are used. The bags are provided with perforations in the lower half of the bag at an interval of 7.5 cm to facilitate drainage. Each bag can carry approximately 16 kg of the mixture. Shade has to be provided up to 2 leaf stage. Water requirement of oil palm seedling at different stages has been estimated as 0 to 2 months @ 4mm/day, 2 to 5 months @ 5 mm/day, 4 to 6 months @ 7 mm/day and 6 to 8 months @ 10mm/day. Normally planting in nurseries is initiated with the onset of the rainy season. A first round plant selection should be done when the seedling are two months. A second selection round should be done when the seedlings are 11 months old (1 month before transplanting). Seedlings should have 13 leaves, 1 to 1.5 metres height at 12 to 14 months old.

Growth and development: Oil palm seed has dormancy period. Germination takes place between 3 to 6 months. Pre-heating the seed (40ºC) for 80 days, followed by cooling at a higher moisture content induces the germination. Seedlings are usually kept in poly-bag nurseries for about 12-14 months and planted in the field when they have 18-24 leaves. The stem has a single growing point, from which a leaf primordium develops about every second week. Rate of leaf production is up to 40 per year in the first two years, dropping to a rate of 18-24 per year from year 8 onwards. From leaf primordium to fully expanded leaf (2-10 m^2) takes about 2 years. The normal photosynthetically active life of a leaf is about 2 years. In plantations, the leaf number is usually kept at about 40 per palm. The fruit development takes place from 15th to 19th day after anthesis while the time taken between anthesis of female flowers and fruit ripeness is 4-5 months. Fruit ripening on the bunch proceeds simultaneously from top to bottom and from outer to inner fruits. Ripe fruits become detached. Coconut kernel is found nearer to the stalk whereas oil palm kernel is away from the stalk. The plumule emerges after the radicle reaches 1.0 cm in length. First adventitious roots are produced in a ring just above radicle/ hypocotyle junction which gives rise to secondary roots before the emergence of first foliage leaf. First leaf emergence takes place after one month of germination. After this, leaves are produced @ one leaf/month up to six months. The first few leaves are lanceolate. In later leaves splitting is formed resulting in bifurcate leaves and subsequently, the leaflets become separate. There are three growth stages in oil palm *viz.*, seedlings stage (12 to 14 months), establishment stage (1 to 4 years) and productive stage (5 to 25 years).

Planting methods: Oil palm seedlings of 10 to 14 months old are planted in triangular planting with a spacing of 9 m × 9 m × 9 m to accommodate 143 palms per hectare. Planting can be done during June to December.

Cropping system: Oil palm is a widely spaced perennial crop with a long juvenile period of three years. The average canopy size was 13.19, 47.86 and 63.58 sq.m/palm during the first three years of oil palm. The availability of space mainly depends on the management condition of the plantation. 81% and 35.2% space would be available for growing intercrops at the end of first and second years, respectively. At the end of third year, only 9% of space will be available for growing intercrops. In 4 to 12 years plantation, the canopy of palm completely covers the ground space available in between and will make inter cropping difficult due to crisscrossing of leaves and short trunk, which is hardly a meter in height. There are a few management practices under inter/mixed cropping systems. There must be a basin around oil palm, when inter cropping is practiced in juvenile phase. A minimum of 3 m radius has to be kept free around each palm to avoid any competition. Care should be taken to avoid sowing/planting of inter/mixed crops in basins of oil palm to avoid the competition for nutrients and water. Ploughing of basin area close to the stem should be avoided as it destroys the root system of oil palm and affect absorption of water and nutrients. Tying of leaves/fronds should not be done as it affects the growth of plant due to reduction of total photosynthetic area. Wherever inter/mixed cropping is practiced in oil palm, it is better to have separate irrigation facility for the main crop and inter/mixed crops. Avoid severe pruning of leaf fronds for intercropping purpose since it affects yield of oilpalm bunches. All the crops should be managed as per the recommended package of practices. Crops like banana (dwarf cavendish, *rasthali*), tobacco, chillies, turmeric, blackgram, greengram, cowpea, groundnut, gingelly, cotton, vegetables, flowers, etc., can be grown as intercrops during the first three years to raise the revenue of farmers. Subsequently cover crops like calopogonium or green manure crops such as *dhaincha*, sunhemp, etc., can be grown which can later be incorporated in situ. Inter/mixed crops grown in oil palm in different states of India are presented in Table 6.

Table 6 Inter/mixed crops grown in oil palm in different states of India

State	Juvenile phase	Mature Phase
Andhra Pradesh	Tobacco, maize, banana, oil seeds like groundnut, sunflower, sesamum vegetables-bhendi, chillies, brinjal, tomato, yam, tapioca, cucurbits, turmeric, pulses-black gram, green gram, horse gram, fodder crops, cotton drumstick, etc.	Cocoa, banana, black pepper, long pepper, elephant foot yam, pine apple, etc.
Karnataka	Cereals-ragi, maize and sorghum, vegetables-onion, brinjal, cucurbits, chillies, tomato, cole crops, oil seeds-ground nut and sunflower, flowers-marigold and china aster, fruits-banana and fig, sugar cane, tobacco, cotton, red gram, turmeric, ginger, drumstick, fodder crops, etc.	Banana, coffee, vanilla, medicinal and aromatic plants, arecanut, Annona, etc
Tamil Nadu	Sugarcane, banana, maize, vegetables-bhendi, tomato, brinjal, chillies, cucurbits, groundnut and flowers-crossandra, tuberose, marigold etc.	Banana

Odisha	Maize, sunflower, ground nut, banana, cotton, chillies, tomato, brinjal, etc.	Banana, turmeric, arrow root and pine apple, etc.
Gujarat	Paddy, bajra, groundnut, sugarcane, banana, brinjal, etc. ,	Banana
Mizoram	Paddy, banana, pine apple, ginger, chillies, cucurbits, cowpea, beans, mustard, maize, soybean, etc.	--
Goa	Vegetables-bhendi, chillies, brinjal, cowpea, cluster beans, cucurbits, banana, groundnut, tapioca, cowpea, fruits-banana, papaya, pine apple etc.	Arecanut

Weed management: Cover crop and intercrop minimize weed growth. Herbicide mixtures of paraquat @ 2.0 kg a.i. per ha and atrazine @ 1.0 kg a.i. per ha can be sprayed in ground twice a year. Herbicides like 2,4 –D, 2,4,5-T, Dalapon and TCA should not be used since these cause abnormalities in oil palm. Herbicide mixtures of 2 kg a.i. of Paraquat with 3-4 kg Atrazine, Monuron and Diuron per ha sprayed/ground applied twice a year has been found to give control of weeds.

After cultivation

(i) **Leaf pruning**: It is done with clean cut to the petiole as close to the stem with sharp chisel. Only the dried and diseased leaves must be pruned. One leaf frond is allowed below the fruit bunch at the lower side of the palm. A minimum of 40 leaves per palm is to be maintained in oil palm for high photosynthetic activity and sustainable growth and yield of oil palm.

(ii) **Abalation**: Oil palm comes to flowering from 14 to 18 months after planting. Flowers produced are removed by hand pulling up to 3 years old palms. This operation is known as *abalation*.

(iii) **Pollination:** Oil palm is a monoecious crop, producing male and female inflorescence separately on the same palm. Inflorescence generally contains 100 spikelets with over 4000 flowers. A single inflorescence produces up to 50 gram pollen which is liberated over a period of 2 to 3 days. It is a cross pollinated crop. Pollination is assisted by wind and insects. Pollinating weevil congregates and multiplies on the inflorescence during flower opening. The weevils also visit the female flowers and pollinate them effectively. Pollinating weevil, *Elaeidobius kamerunicus* activity and production of male inflorescence are high during January to April. So the pollinating weevil can be released during these months (January–April) after third year to increase the number and weight of fruit. The pollinating weevil activity is high during morning hours from 7.30 a.m. to 11.30 a.m. Release of this weevil after 2½ year of planting is advisable. If the plants are not having good girth and vigour, release the weevils after 3 years.

Manures and fertilizer application: Both organic and inorganic fertilizers have to be applied to oil palm for increased yield and quality of fresh fruit bunches. Potassium, magnesium and boron deficiencies were observed in the oil palm plantation. Fertilizer schedule containing N, P_2O_5, K_2O, $MgSO_4$ and Borax at the level of 400–200–400–125–50 during the first year, 800–400–800–250–100 during the second year, 1200–600–1200 –500–150 during the third year and 1500–750–1500–500–200 in grams per palm from fourth year onwards is recommended. Fertilizers can be applied either in two or four splits. If applied in two splits, fertilizers can be applied during June–July and January–February. When fertilizers are applied in four splits, it may be done during the January, April, July and October months. For adult palms, fertilizers are applied to a circular area within a radius of 2 m around the base of individual palms.

Nutritional disorders

(i) Nitrogen deficiency shows lemon yellow discolouration. In potassium deficiency leaves colour changes from pale greenish to yellowish rectangular spots to orange necrotic spots. Bronzing and tapering of the crown is observed. Mg deficiency gives chrome yellow to deep orange discolouration in upper plane of old leaves.

(ii) Boron deficiency shows hook leaf, white strips little leaf. Copper deficiency shows orange discoloration of inner whorl leaves followed by die back of spear. Foliar spray of copper sulphate @ 200 ppm controls copper deficiency.

(iii) Phosphorous deficiency shows dull pale olive green and pre-mature dessication of old leaves. Bunch failure occurs due to death of fruit bunches.

(iv) Boron deficiency shows little leaf at initial stage and fish boron leaf or hook leaf symptom at advanced stage.

Water management: The water requirement for oil palm is 150 to 200 liters of water for palms of < 3 years plantation and 400 litres per palm per day for > 3 years plantation. Irrigation of the oil palm at 1.0 IW/CPE ratio is optimum for the yield of fresh fruit bunches. The production of male inflorescence increases when oil palm is irrigated at 0.6 to 0.8 IW/CPE ratio. If water is plenty, basin irrigation is recommended. Basin of one metre radius is to be taken during the first year around the palm removing the soil from inside so that soil will not accumulate at the collar region. Basins have to be widened to 2 m radius during second and 3 m radius from third year onwards. This is done as the oil palm roots spread to a radius of one, two and three metres from palm base during the first, second year and third year respectively. Basin area of the oil palm represents its active root zone. Drip irrigation is advocated if water is limited for irrigation. Apply 50 to 100 kg of farm yard manure or green manure and 100 grams in each of azosprillum and phosphobacterium per palm per year. Apply fertilizers 50 cm away from the palm base and incorporate by forking. Drip jet system with pressure of 1.5 kg per cm^2 emitting 30 litre per hour is recommended for oil palm. Two emitters discharge water at 35 lps hour and release 300 litres of water in five hours @ 30 litres per two emitters per hour.

Crop protection: The pests such as rhinoceros beetle, red palm weevil, case worm and diseases such as stem wet rot, bud rot disease and basal stem rot are affecting oilpalm. Rhinoceros beetle adults can be trapped with fermented castor cake or pheromone bait. Use bio-agents like virus (*Baculovirus oryctes*) and fungi (*Metarrihizium anisopliae*). Treat the compost pit with Carbaryl or Quinalphos @ 0.025 % to kill the young stages of pest. Red palm weevil damages and causes rotting of bunches. Red palm weevil is controlled with pasting of tar to the wounds and cuts on the stem portion to avoid egg laying. Trap the adult beetles using pheromone baits. Root feeding of Monocrotophos (10ml of insecticide in 10 ml of water) is done to control the weevil and case worm. Stem wet rot is controlled with removal of infected portion and protective covering with Carbendazim (1%) + Monocrotophos (1ml) paste. The affected crown due to bud rot disease should be removed and drenched with Carbendazim or Thiram @ 0.1 %. Basal stem rot affected plants have to be removed and then apply 5 kg of Neem cake per year per tree. Root feeding with Calixin 10 ml or 10 g Aureofungin sol in 100 ml of water per trce per year to control basal stem rot.

Harvest: The optimum stage for harvest is reddish orange in colour of fresh fruit bunch (FFB) or detachment of 25% or more fruits per bunch or falling of 10 or more fruits per bunch from young palms and 5 fruits per bunch from adult palms. Harvesting of fruits can be carried out once in 15 to 30 days interval. Harvesting is done with a 6 to 9 cm wide chisel attached to 1.5 m long G.I. hollow pipe.

Yield: The number of bunches produced per palm per year decreases with age i.e., from 15 to 25 bunches in the fourth year to about 10 bunches in the twelth year. After that the number of bunches decline slowly. The bunch weight increases with the age of palm from 5 kg in the fourth year to 20 to 25 kg in the 14th year. The fruit to bunch ratio is 55 to 70%. Mesocarp to fruit ratio is 75 to 90%. Palm oil is found in the fleshy portion of the fruit (mesocarp) whereas palm kernel oil is found in the kernel or the seed of the fruit. The oil to mesocarp ratio is 45 to 55 %. The palm oil to bunch ratio is 18 to 20 %. The kernel to fruit ratio is 7 to 15%. Oil palm of 6 years plantation produce the fresh fruit bunch yield of 16 tons per ha per year while ten years old plantation produced 19 tons per ha per year. One tonne of FFB comprises of 220 kg of empty fruit bunch, 500 kg of Palm Oil Mill Effluent (POME), 120 kg of fibre and 50 kg of shell.

FFB grading standards

(i) **Unripe bunch:** Unripe bunch will have a black or purplish black fruits and the outer layer fruitlets mesocarp is yellowish orange in colour. This bunch does not have any fresh sockets of detached fruitlets at the time of inspection at the processing unit.

(ii) **Ripe bunch:** Ripe bunch will have reddish orange colour and the outer layer fruitlets mesocarp is orange in colour. The bunch has atleast 10 fresh sockets of detached fruitlets and more than 50% of the fruits still attached to the bunch at the time of inspection at the processing unit. It should be brought to the processing unit within 24 hours from the harvesting time.

(iii) **Under ripe bunch:** Under ripe bunch has reddish orange or purplish red in colour and the outer layer fruitlets mesocarp is yellowish orange in colour. This bunch has leas than 10 fruit sockets detached fruitlets at the time of inspection at the processing unit. The FFB should be brought within 24 hours from the time of harvesting.

(iv) **Over ripe bunch:** Over ripe bunch will have darkish red coloured frits and has more than 50% of detached fruitlets but with atleast 10% of the fruits still attached to the bunch at the time of inspection at the processing unit. The bunch and the loose fruits are sent to the processing unit within 24 hours after harvesting.

(v) **Loose fruit:** Fruits detached from a fresh fruit bunch because of ripeness and is reddish orange in colour. All good loose fruits have to be sent to mill within 24 hours after harvesting.

Utilization: Fibres from fronds and empty bunches are used for production of fibre boards and chipboards. Fibres from trunk of the empty bunch can be utilized as medium for mushroom growth. The mesocarp fibre comes out of the mill is found to contain 10 to 12% of oil. If one million ha of land is brought under oil palm cultivation, it would be possible to produce 3.5 to 4.0 million tones of oil and thereby there is scope for releasing some of the land locked up with annual oilseeds which can go for any other food or cash crop.

The crude oil is used for soap making, perfumes, etc. Traditionally, about 80% of palm oil is for edible use and 20% for non-edible use such as oleochemical manufacture. Palm oil contains 500-700 ppm □-carotene (pro-vitamin A) which gives it its red colour and 600-1000 ppm vitamin E. Vitamin E is present as tocotrienols (70%) rather than tocopherols (30%). It confers on the oil a natural stability against oxidation and a longer shelf-life as well as a potent ability to reduce LDL (bad) cholesterol and anti-cancer properties. Palm oil also contains 250-620 ppm sterols.

Oil production from oil palm: A small scale oil palm factory capacity is 10 tons/hour. Palm oil is extracted from the fleshy mesocarp of the oil palm fruit which is highly perishable in nature. To obtain good quality of palm oil, the fruits have to be processed within 24 hours after harvesting. The free fatty acid (FFA) content should be normally 3.5% but in any case should not exceed 5% oil at harvest. Oil having more than 5% FFA is not used for refining as the refining losses would be more. Processing of oil palm fruits starts with sterilization, stripping, digestion, oil extraction, clarification, oil purification and separation of the nuts from the fibre.

(i) Sterilization: After harvesting, the fresh fruit bunches are sterilized with steam for one hour to inactivate lipase enzyme and to loosen fruits for better stripping and to soften the tissues for better digestion. Bunches are cooked for 70 minutes at 140 to 160°C through steam. Steam pressure of 3 kg/cm^2 for 40 to 60 minutes is required. Then the containers containing the cooked bunches are taken out. The fruits are very soft and easy to remove.

(ii) Stripping: After sterilization, the bunches must be stripped. This process involves the separation of the fruits from the bunches. The fruits are separated into mesocarp and kernel through rotating crusher drum. The rotary drum stripper is rotated at 20 to 22 rpm. The empty bunches are taken out through other point.

(iii) Digestion: It is done to disrupt the mesocarp to break up the maximum quality of oil bearing cells to facilitate oil release. This process frees the nut from fibre. Digester is used to separate the oil from the mesocarp. The sterilized fruits are agitated in the digestion with beater arms rotating at about 25 rpm. The temperature is maintained at 90 to 95°C during digestion by jacket steam and open steam. The crude oil is taken to the clarifier where oil floats over sludge and water. Purifier removes oil from water, sludge and other impurities. The crude oil is then stored in tanker.

(iv) Oil extraction: The digested fruits mass is taken to the crusher and pressed. The oil and water mixture is expelled through perforations while the press cake is expelled.

(v) Clarification: Crude palm oil from screw press contains about 50% water and non-oily solids which will be removed by means of clarification. The crude oil is further diluted with hot water (Oil to water in 1:2 ratio). Separation of oil from water and other impurities is activated by the difference in density of oil and water at 90 to 95°C. Oil being lighter in density rises to the top. The oil will be separated at high level opening. The sludge, water and impurities remain at the bottom which is removed at the lower level opening.

(vi) Oil purification and drying: Oil from clarifier tank contains 0.4 to 0.6 % moisture and 0.1 to 0.2 % sludge and other impurities. Moisture is responsible for the increase of free fatty acids while sludge and impurities increases the oxidation, Bulk of water and impurities are removed by a high speed centrifuge with 8500 rpm. Further the moisture is reduced to the optimum level of 0.15 to 0.10% by vacuum drying. The vacuumized oil is pumped to the oil storage tank. There will be 19% oil recovery in factory.

(vii) Separation of nuts from fibre: The press cake is passed through steam at 75^0C and hot air is blown. This separates fibres and nuts.

REFERENCES

Corley, R.H.V., Hardon. J.J. and Wood, B.J. (eds), 1976. Oil palm research. E1-sevier Scientific Publishing Co., Amsterdam. 532 pp.

CPG. 2012. Crop Production Guide. Department of Agriculture, Government of Tamil Nadu, Chennai and Tamil Nadu Agricultural University, Coimbatore.

Hardon, J.J., Rao, V. and Rajanaidu, N., 1985. A review of oil-palm breeding. In: Russell. G.E. (Editor). Progress in plant breeding 1. Butterworths, London. p. 139-163.

Hartley, C.W.S., 1977. The oil palm. Longman, London. 806 pp.

Jose Kallarackal; Jeyakumar, P. and Suman Jacob George. 2004. Water use of irrigated oil palm at three different arid locations in Peninsular India. Journal of Oil Palm Research. 16(1): 45-53

Rethinam, P. 1994. Oil palm development in India. Tree World, 3:1-4

Rethinam, P. and Nagendra Rao, T. (eds). 1997. Nutrient Management of Oil Palm Plantations in South India. National Research Centre for Oil Palm. Pedavegi, India.

Rethinam, P. and Suresh, K. (eds) 1998. Oilpalm Research and Development. National Research Centre for Oil Palm. Pedavegi. India.

Thampan, P.K. 1992. Comments on oil palm development in India. Tree World, 1(7): 1-3

CHAPTER

13

Jatropha–*Jatropha curcas* L. (2n=22)

Synonymous: *Curcas purgans* Medic.

Family: Euphorbiaceae

Vernacular names: Jatropha, Physic nut, Pig nut, Fig nut, Purging nut (English), *Purgerbuske* (Swedish), *Purgiernuß, Brechnuß* (German), *Piñoncillo* (Mexico), *Mundubi-Assu* (Brazil), *Kattamanakku* (Tamil), *Ratanjyot, Bagbherenda, Jangliarandi, Safed Arand, Bagaranda* (Hindi), *Kampma ramda, Kanananaeranda, Parvataranda* (Sanskrit), *Bagbherenda, Erandagachh* (Bengali), *Mogalierenda, Ranayerandi* (Marathi), *Jamalgota, Ratanjota* (Gujarati), *Nepalamu, Peddanepalamu, Adaviamidamu* (Telugu), *Adaluharal, Bettadaharalu, Maraharalu, Karnocchi* (Kannada), *Kattavanakka, Kadalavanakka* (Malayalam), *Jahazigaba* (Orya), *Bongalibhotora* (Assam).

Importance: *Jatropha curcas* is a most promising 'biodiesel crop' or 'biofuel crop' suited for growing in the wasted lands / degraded lands. Jatropha plants are not browsed by animals. It is also suitable for preventing soil erosion and shifting of sand dunes. It can also be planted as fence crop. The oil cake can be used as organic fertilizers.

Origin: Jatropha is native of South America (Mexico, Brazil) and Central America.

Distribution: Jatropha is believed to have been spread by Portuguese seafarers from its centre of origin in Central America and Mexico via Cape Verde and Guinea Bissau to other countries in Africa and Asia. It is now widespread throughout the tropics and sub-tropics. In India, it is grown in Gujarat, Chhattisgarh, Tamil Nadu, Uttar Pradesh, Maharashtra, Uttrakhand, Haryana, Andhra Pradesh, Karnataka, Arunachal Pradesh, Manipur, Mizoram and Nagaland.

Botany: *Jatropha curcus* L. is a shrub. The genus name Jatropha derives from the Greek *jatrós* (doctor), *trophé* (food), which implies medicinal uses. The *curcas* is the common name for physic nut in Malabar, India. Jatropha has smooth grey bark, which exudes a whitish colored, watery, latex when cut. Normally, it grows between three and five meters in height, but can attain a height of up to eight or ten meters under favourable conditions. The plants reach a height of three meters within three years. Leaves are large, green to pale-green in colour, alternate to sub-opposite, three-to five-lobed with a spiral phyllotaxis. The petiole length ranges between 6 to 23 mm. Jatropha is monoecious, meaning it carries separate male and female flowers on the same plant. The inflorescence is formed in the leaf axil. Flowers are formed terminally, individually, with female flowers usually slightly larger and occur in the hot seasons. In conditions where continuous growth occurs, an unbalance of pistillate or staminate flower production results in a higher number of female flowers. The ratio of male to female flowers averages 29:1 but this is highly variable and may range from 25-93 male flowers to 1-5 female flowers produced on each inflorescence. Generally flowers between September and January and second flowering in June is also reported. The inflorescence, once it begins flowering, flowers daily, and the flowering lasts for 11 days. One inflorescence will normally produce 10 or more fruits. Fruit set generally results from cross pollination with other individual plants through insects, because the male flowers shed pollen before the female flowers on the same plant are receptive. In the absence of pollen arriving from other trees, jatropha has the ability to self-pollinate. Fruit is known as capsule. The fruits are ellipsoidal, green and fleshy, turning yellow and then brown as they age. Fruits are mature and ready to harvest around 90 days after flowering. Each fruit contains two or three black seeds, around 2 cm × 1 cm in size. Seeds resemble to castor seeds. The exocarp remains fleshy until the seeds are mature. Fruits are produced in winter when the shrub is leafless. The seeds mature about 3 to 4 months after flowering. A three, bi-valved cocci is formed after the seeds mature and the fleshy exocarp dries. The seeds become mature when the capsule changes from green to yellow, after two to four months from fertilization. The seeds are black and the seed weight per 1000 is about 727 g, there are 1375 seeds/kg on an average. The seeds are 18 mm long (11–30) and 10 mm wide (7–11). The seeds are oblong and resemble small castor seeds. It generally takes four to five years to reach maturity. The life-span of the *Jatropha curcas* plant is more than 50 years.

Growth and Development: Seedlings produce flowers 9 months after sowing. However plants established through cuttings, produce flowers from the 6 month onwards. Wherever *Jatropha* is cultivated under irrigated condition, the flowering is throughout the year. The flowers are yellowish green in loose panicles. The flowering occurs twice in a year *i.e.*, in March-April and in September-October. Each inflorescence yields a bunch of approximately 10 or more ovoid fruits. The seeds are ripe about 90 days after the flowers have been pollinated and the fruit turns yellow and hardens. After this the fruit turns brown, and becomes black and opens when it is fully dried. The ripe fruits are about 2-5 cm large. The seeds resemble with castor seed in shape either ovoid or oblong. The average number of pods per plant ranges from 90 to 100 with two to three seeds per pod. The mean hundred seed weight is 65 to 68 g.

Climate: *Jatropha* grows well under tropical and subtropical climates, with cultivation limits at 30ºN and 35ºS. It is found from sea level up to 3000 m altitude. It is usually found at lower elevations (below 500 m). The minimal rainfall requirement to survive is 300 mm ha^{-1} y^{-1} while the minimum rainfall to produce fruits is 600 mm ha^{-1} y^{-1}. The optimal rainfall requirement is 1000 to 1500 mm ha^{-1} y^{-1}. The short periods of drought (one month) will induce blossoming. Jatropha is a drought resistant crop but the plant sheds its leaves and does not produce flowers or fruit in dry seasons. Where rainfall is above 1,000 mm, it does better in hot rather than temperate climates. It thrives well in low rainfall regions and problem soils. A dry climate has been found to greatly improve the oil yields of the seeds. The plant is well adapted to conditions of high light intensity and is unsuited to growing in shade. It can tolerate extremes of temperature from 0 to 5°C but not the frost and also the temperature below 0°C. The plant requires average temperatures between 18 and 28°C. There is no fruiting or grain filling during summer season in most of the states where temperature goes above 43°C. Its water requirement is extremely low and it can stand long periods of drought by shedding most of its leaves to reduce transpiration loss.

Soil: Jatropha grows well in aerated sands and loams of at least 45 cm depth. It can also be grown on gravelly, sandy and saline soils and alkaline soils having pH ranging from 6.0 to 8.5. It can thrive on the poorest stony soil. It can grow even in the crevices of rocks. Heavy clay soils are less suitable and should be avoided, particularly where drainage is impaired, as jatropha is intolerant of waterlogged conditions. It does not thrive in wetland conditions. The leaves shed during the winter months form mulch around the base of the plant. The organic matter from shed leaves enhances earthworm activity in the soil around the root-zone of the plants, which improves the fertility of the soil.

Season: The monsoon seasons (June-July, October-November) may be preferred for better establishment of seedlings.

Land preparation: The land should be ploughed once or twice depending upon the nature of soil. A disc/deep ploughing will help to improve the soil drainage and proper water holding capacity. After land preparation, pits of size 30x 30 x30 cm may be dug and filled with soil and Farm Yard Manure @ 5 kg/pit + 150 g Neem cake+100 g super phosphate+ 20g VAM 10 g + biofertilizers each of azospirillum and phosphobacteria before planting.

Varieties: High yielding types collected from African countries (Madagascar, Zimbabwe and Cape Verde) are ideal for cultivation. Jatropha variety is TNMC-6. The Cape Verde variety is spread all over the world.

Propogation practices: Jatropha can be grown easily from seeds. However, commercially it can be propagated by three different ways such as by seeds, nursery and stem cutting.

(i) **By sowing seeds in polybags:** *Jatropha* is normally propagated through seeds. Fully matured seeds should be selected for sowing. For one-hectare plantation about 5 to 7.5 kg seeds are required. Pre-soaking of seeds in water / cowdung slurry for 12 hours before sowing is advised to get vigorous seedlings. Soaked seeds are generally sown in poly bags of 10 × 20 cm size filled with soil, sand and FYM (Farm yard manure) in the ratio of 1:2:1 respectively. Germination is generally noticed after 4-5 days and continues up to 15 days and seedlings are ready for planting after 4–6 months of maintenance in the nursery. If the seedlings are to be retained by 5-6 months before transplantation, then bigger poly bags (15×25 cm) should be taken.

(ii) **By sowing in nursery beds:** Well developed plumpy seeds are selected for sowing in the raised nursery beds. Before sowing, seeds are soaked in cow dung solution for 12 hours and kept under the wet gunny bags for 12 hours. Treated seeds are to be sown in the raised mother beds of 10 × 1 m size. Germination starts from the third day and one week old seedlings can be pricked out and transplanted in polybags filled with red soil, sand and organic in the ratio of 1:1:1 respectively. Shifting of the polybags has to be carried out once in every 30 days coupled with root pruning. Adequate watering has to be done taking care to avoid water stagnation. The seedlings can be kept in the nursery for a period ranging from two months to six months.

(iii) **By stem cutting:** From the selected tree, collect the branches having 2-3 cm diameter, put them in a bucket of water. Cut them into 15-20 cm long pieces with a knife or a mini hand-saw. Dip the cuttings into a true rooting hormone mixture (may be commercial seradix or growth hormones like IBA or NAA of 100 ppm concentration. Put the cuttings in a poly bagged rooting media (Soil + Sand in 1:1) by inserting the basal region (about 3 cm). Place the poly bags inside a closed polythene chamber or mist chamber to avoid the drying of cuttings. Allow the cuttings inside the chamber for a period of 2-3 weeks.

Planting method: Plants raised from the seeds take 2 to 3 years to yield seeds. For commercial plantations, the crop can be grown with a spacing of about 2 m × 2 m or 3 m × 3 m. It is planted 2 m × 2 m with 2500 plants in one hectare under rainfed condition while 3 m × 3 m with 1110 plants per ha under irrigated conditions. Seedling age is 4 to 6 months. At least 20 % of the plants have to be replaced. Cuttings of 60 cm length may be planted 25 cm apart and 20 cm into the ground as a live fence hedge.

Weed management: Two to three times weeding per year is recommended.

After cultivation: For early flowering, GA @100 ppm may be sprayed. It also helps better pod development and yield.

Manures and fertilizer application: Organic manures such as FYM @ 5 kg per plant has to be applied. Inorganic fertilizers such as 20 g of urea, 120 g super phosphate and 16 g muriate of potash and biofertilizres such as azospirillum 10 g and VAM 50 g are applied to the individual plant / shrub. Mycorrhyzal soil fungi are generally known to improve a plant's ability to absorb mineral nutrients and water from the soil, and to increase drought and disease resistance. From 4th year onwards, 150g super phosphate is recommended over and above the regular dose. From second year onwards fertilizers are applied @ 50-100-50 kg of N,P_2O_5 and K_2O ka/ha and micronutrient mixture @ 12.5 kg/ha along with Azophos 10 kg/ha + VAM 12.5 kg/ha.

Water management: Jatropha is deciduous, a mechanism that limits water loss and minimizes the effect drought, which is manifest by the abscission of the leaves. Jatropha is mostly raised under rainfed condition with proper and well distributed rainfall. Additional irrigation is given depending on situation. Three to four irrigations a month during summer will help in improving the growth of the plants. The crop requires a life irrigation on the third day after field planting. After this, irrigation at an interval of 15 to 20 days may be done, based upon need. Foliar spray of Brassinolides @ 0.25 ppm at pre flowering stage increase the tolerance to drought and pod yield per plant.

Canopy management: The terminal growing twig is to be pinched to induce secondary branches. Likewise the secondary and tertiary branches are to be pinched or pruned at the end of first year to induce a minimum of twenty five branches at the end of second year. Early pruning at 6 to 8 months after planting or just before flowering is advisable. First pruning must be at 45 cm from ground level which will enhance more number of auxillary branches. Pruning has to be taken up for the first three years to increase number of fruiting branches. Second pruning must be made in the newly developed shoots by cutting two thirds of the new flush leaving one third in the plant. Similar procedure has to be followed on the new flush upto the completion of the third year. The pruning operation can be adapted and modified according to the location and also growth and development of the plant. Avoid flowering and seed setting during the first two years in order to develop a broader canopy. Once in ten years, the plant may be cut leaving one meter height from ground level for rejuvenation. Pruning will be useful to induce new growth and yield stabilization.

Cropping systems: Intercrops can be raised in between the rows for the first two years in Jatropha plantation. Crops like tomato, sunflower, ash gourd, bitter gourd, pumpkin, cucumber, cowpea and blackgram can be grown profitably. Cowpea is the best intercrop under Jatropha during the initial years.

Crop protection: Seed treatment with *Pseudomonas fluorescens* @10g/kg of seed to control seed borne pathogens. Soil application of *Trichoderma viride* or *Pseudomonas fluorescens* @ 2.5 kg/ha mixed with 50 kg of FYM, before planting to control soil borne pathogens. Spray Bt @ 1g/litre or neem oil (2%) or Monocrotophos36 WSC @1.25 ml/ litre or Profenophos 50 EC (1ml/litre) for leaf webber. Spray Methyl Parathion 25 EC @ 2ml/litre or Dimethoate 30 EC @2ml/litre or Monocrotophos 36 WSC @ 1.25 ml/litre for thrips. Scutellerid bugs can be controlled by spraying carbosulfan 25 EC @ 1ml/litre. Damping off or Root rot may become a problem in the beginning and can be controlled by application of 1% of Bordeaux drenching or application of neem cake @ 150 kg /ha. The insects and diseases control measures in Jatropha are furnished in Table 1.

Table 1 Diseases control measures in Jatropha

S. No.	Name of disease	Control
1	Damping off	Spray of Keptan 50% @ 0.2%
2	Collar rot	Drenching of 1% Bourdeax mixture
3	Root rot	Spray of Thiram @ 0.2%
4	Leaf spot	Spray of Blitox @ 0.2%
	Name of Insects	
1	Leaf minor	Spray of 1.5 ml/litre of water of Metasystox 25 CC
2	Blue Bug	Spray of Phosphomedin/ Dimethoate @ 2.0 ml/3 lt. of water
3	Green bug	Spray of 2 ml/3 lt. water of Phosphomedin

Harvest: The fruits are harvested after change of colour of capsule from yellow to brown and shrinking of fruit capsules for easy dehusking. Seeds are ready for harvesting around 90 days after flowering when the fruits have changed from green to yellow-brown. Threshing can be done by manual separation by beating and mechanical separation using decorticator. Seeds can be shade dried and stored in gunny bags at room temperature.

Yield: Seed yield starts from first year. However, the harvestable yield starts from third year while the economic seed yield is from 5th year. Seed yield is 500 g to 5 kg under irrigated condition and 200 g to 2.5 kg under rainfed condition. Generally, seed production varies from 0.4 tonnes initially to ten tonnes per hectare per year, as the bushes grow larger. Each plant can yield up to 1.5-2.0 kilos of seed depending on plant condition. In relatively poor soils, the yields have been reported to be 1 kg per plant while in lateritic soils, the seed yields have been reported between 0.75-1.00 kg per tree. If planted in hedges, the reported productivity of Jatropha is from 0.8–1.0 kg of seed per meter per year of live fence. This is equivalent to seed production of between 2.25 t/ha and 5 t/ha, depending upon whether the soils are poor or average for plantations and between 2.5 ton per ha per year and 3.5 ton per ha per year for hedges. Assuming a square plot 400 m^2, a fence around it will have a length of 400 m and a production of 0.4 tons of seed. A hedge along one hectare will be equal to 0.1 hectare of block plantation. The dried pods are collected and seeds are separated either manually or mechanically. The dried pods are collected and seeds are separated either manually or mechanically. Seeds are dried in the sun for 4-5 days to reduce moisture level to 10% before packing. If kept dry and ventilated, the seeds may be stored up to 12 months without loss of germination or oil content. The shelling % is 59.6 and 100 seed weight is 45 g. Seeds are dried under sunlight for four days until the moisture is brought to 6-10% before oil extraction. The seeds contain 35% oil while the kernels contain 55-60% oil. Assuming oil content of 35 % and 94 % extraction, one hectare of plantation will give 1.6 tons of oil if the soil is average, 0.75 tons if the soil is lateritic soil. Plantation per hectare on poorer soils will give 0.9 tons of oil.

Utilization: The jatropha plant contain toxins such as phorbol esters, curcins and trypsin inhibitors. When it is detoxified, it can be applied as a supplement to animal food, because it consists of 56% protein. The chemical composition of the Jatropha seed shows moisture 6.20 %, protein 18.00 %, fat 38.00 %, carbohydrates 17.00 %, fiber 15.50 % and ash 5.30 %. The oil contains 21 % saturated fatty acids and 79% unsaturated fatty acids. The latex of Jatropha contains an alkaloid known as 'jatrophine' which is believed to have anti-cancerous properties. It is also used as an external application for skin diseases and rheumatism and for sores on domestic livestock. The tender twigs of the plant are used for cleaning teeth, treating, toothache gum inflammation, gum bleeding, pyorrhea while the juice of the leaf is used as an external application for piles. Bark is used as fish poison. The roots are used as an antidote for snake-bites. The bark of Jatropha yields a dark blue dye which is used for colouring cloth, fishing nets and lines. Jatropha leaves are used as food for the tusser silkworm. The seeds are considered anthelimintic in Brazil, and the leaves are used for fumigating houses against bed-bugs. Seeds are used to treat arthritis, gout, jaundice and contraceptives. The ether extract shows antibiotic activity against S*typhylococcus aureus* and *Escherichia coli.* The toxicity of the seeds is due to the toxic protein (curcin) and diterpene esters, making it unsuitable for animal feed. However, it does have potential as a manure. Plant extract is used for treating dermatomucosal disease, allergies, burns, cuts, wound inflammation, wound healing, leprosy, leucoderma and smallpox.

REFERENCES

Dahmer, N., Schifino-Wittmann, M.T. and L.A.S. Dias. 2009. Chromosome numbers of *Jatropha curcas* L.: an important agrofuel plant. Crop Breed. Appl. Biotechnol. 9: 386-389.

Dias, L.A.S., Missio, R.F. and D.C.F.S. Dias. 2012. Antiquity, botany, origin and domestication of *Jatropha curcas* (Euphorbiaceae), a plant species with potential for biodiesel production. Genetics and Molecular Research 11 (3): 2719-2728

Divakara, B.N., Upadhyaya, H.D., Wani, S.P. and C.L. Laxmipathi Gowda. 2009. Biology and genetic improvement of *Jatropha curcas* L.: A review. Applied Energy.30: 1-11

Freitas, R.G., Missio, R.F., Matos, F.S. and M.D. Resende. 2011. Genetic evaluation of *Jatropha curcas*: an important oilseed for biodiesel production. Genet. Mol. Res. 10: 1490-1498.

Gour, V. K. 2006. Production practices including post-harvest management of *Jatropha curcas*. In: Singh, B., Swaminathan, R. and V. Ponraj. (eds). Proceedings of the biodiesel conference toward energy independence – focus of Jatropha, Hyderabad, India, June 9–10. New Delhi, Rashtrapati Bhawan, 2006: 223–251.

Gubitz, G.M., Mittelbach, M. and M. Trabi. 1999. Exploitation of the tropical oil seed plant Jatropha curcas L. Bioresource Technology, 67; 73–82.

Heller, J. 1996. Physic nut, *Jatropha curcas*. Promoting the Conservation and Use of Underutilized and Neglected Crops. International Plant Genetic Resources Institute (IPGRI), Rome, Italy.

Jones, N. and J. H. Miller. 1992. *Jatropha curcas*: A multipurpose Species for Problematic Sites. The World Bank, Washington DC., USA.

Kureel, R. S., Singh, C.B., Gupta, A.K. and Ashutosh Pandey. 2007. JATROPHA: An alternate source for biodiesel. National Oilseeds & Vegetable Oils Development Board, Ministry of Agriculture, Government of India, 86, Sector-18, Institutional Area, Gurgaon – 122015.

Makkar, H.P.S., Becker, K., Sporer, F. and M. Wink. 1997. Studies on nutritive potential and toxic constituents of different provenances of *Jatropha curcas*. J. Agric. Food Chem. 45, 3152-3157

Nair Dahmer, Maria Teresa Schifino Wittmann, and Luiz Antônio dos Santos Dias. 2009. Chromosome numbers of *Jatropha curcas* L.: an important agrofuel plant. Crop Breeding and Applied Biotechnology 9: 386-389

NOVOD. 2007. Jatropha – An Alternate source for Biodiesel. New Delhi: National Oil Seeds and Vegetable Oils Development Board (available at http://www.novodboard.com).

Openshaw, K. 2000. A review of Jatropha curcas: an oil plant of unfulfilled promise. *Biomass and Bioenergy.* 19:1–15.

Ratha Krishnan, P. and M. Paramathma. 2009. Potentials and *Jatropha* species wealth of India. Current Science, 97 (7): 1000-1004.

Shivani Sharma, Hitesh K. Dhamija and Bharat Parashar. 2012. *Jatropha curcas:* A Review. Asian J. Res. Pharm. Sci. 2 (3): 107-111

www.Jatropha.org

CHAPTER

14

Fibre Crops

The basic unit in any clothing or textile industry is a fibre. The word fibre has many meanings. Botanically, 'fibre' consists of very long narrow cells. The average length of the fibre cell is 1-3 mm but the ramie fibres are among the longest cells in higher plants being up to 55 cm in length. Fibre cells consist of cellulose with hemi-cellulose, lignin or any other substances. The fibres are obtained from the sclerenchymatous cells found in the plant body. Chemically, vegetable fibres are mainly composed of cellulose (64 to 94%) that constitutes the structural framework of the cell wall. Commercially, the term fibre refers to thin and slender substances, which can be spun or made use of as fine stuffing material. Fibres are a class of hair-like material that are continuous filaments or are in discrete elongated pieces, similar to pieces of thread. They can be spun into filaments, thread, or rope. Clothing is the most important in the needs of man, next to food, which is obtained from fibres.

Fibre definitions

(i) Botanical definition: Fibre is a long narrow tapering cell, dead and hollow at maturity, thick cell wall composed mostly of cellulose and lignin, rigid, for support, found mainly in vascular tissue.

(ii) Commercial definition: Fibre is a long narrow flexible material, may be animal (hair, wool), mineral (asbestos), synthetic (nylon, dacron), or plant.

(iii) Nutritional definition: Fibre is a indigestible material in food

The textile fibres are broadly classified as natural fibres and man-made (synthetic polymer fibers) fibres. Natural fibres refer to fibres that occur within nature, and are found in vegetables respectively plants (cellulose fibres), animals (protein fibres) and minerals (asbestos). Vegetable fibres are generally composed of three structural polymers (the polysaccharides cellulose, and hemicelluloses and the aromatic polymer lignin) as well as by some minor non-structural components (i.e. proteins, extractives, minerals). On the other hand, animal fibres such as wool and the hairs of various other animals are composed mainly of proteins. Man-made fibres are those that are not present in nature, although they may be composed of naturally-occurring materials. They are classified into three main groups: those made by transformation of natural polymers (regenerated fibres), those made from synthetic polymers (synthetic fibres), and those made from inorganic materials (fibres made of metal, ceramics, and carbon or glass).

Classification of natural fibres based on origin: Fibres that are used for industrial purposes may be derived from any of three sources viz., i) obtained from plants, ii) derived from animals (wool and other animal hairs, silk from insects and the feathers of various birds) and iii) mineral origin (asbestos, glass filaments, drawn metal threads and various coal tar derivatives, of which nylon is the best known).

(i) **Plant fibres** or **Vegetable fibres** are obtained from various parts of plants, such as the seeds (cotton, kapok, milkweed, poplar, calotropis), stems (flax, jute, hemp, ramie, kenaf, isora, nettle, bamboo), and leaves (sisal, manila, abaca, pineapple, banana, opuntia, paja, jukka, palm, curaua, ananas), fruit (coconut coir, luffa), grass fibres (bamboo, totora) and other woods (hard wood, soft wood- examples are Pine wood, Baobab bark- *Adansonia digitata*). Fibres from these plants can be considered to be totally renewable and biodegradable. These vegetable fibers are generally comprised mainly of cellulose.

(ii) **Animal fibres** includes silk from cocoons of silkworms and wools from fleece (hair) of sheep or lambs. The animal fibres are known as protein fibres. The term 'wool' refers to the fibres of the fleece of the sheep, or of animals whose hair is generally assimilated to wool (sheep, alpaca, llama, vicuna, yak, camel, horse, cashmere goat, mohair goat, cashgora goat, angora rabbit). The Avian fibres are from birds, e.g. feathers and feather fibre.

- Merino sheep hair: Cashmere (pashmina) is considered as one of the finest and softest animal fibres which is highly suitable for comfortable garments and baby wear. Wool (also known as sheep hair) mainly comes from fleece of a sheep, mainly Merino sheep.
- Mohair or Angora goat hair: Mohair comes from fleece of the Angora goat and more than 50% of the world's production of mohair is produced by South Africa. Angora goats that produce mohair are naturally white in colour which makes their fleece to be also naturally white.
- Cashgora comes from fleece of the goats. Cashgora is produced in Kazakstan, Russia and other central Asian countries.
- Angora fibre is produced by Angora rabbits. Angora fibre production is the third largest animal fibre industry in the world after wool and mohair.
- Horse hair interlining fabric is regarded as having the highest resilience to bending. Single horse tail hairs are used as warps.
- Reindeer hair has enjoyed intermittent success in fashion fabrics to produce novelty effects. The red deer is found in New Zealand.
- Shahtoosh fibre comes from the highly endangered Tibetan antelope (*Pantholops hodgsoni*) found mainly in the Tibet region of China. Shahtoosh is used to make the famous shahtoosh shawl which is produced in Kashmir state of India.
- Camel hair from Bactrian camels is longer than that obtained from Dromedaries (one-humped camels).
- Qiviut is the downy fibre harvested from the Musk-ox (*Ovibos moschatus*), a distant relative of the cashmere goat. Musk-ox are found in Greenland, Canada, Norway and Alaska.

(iii) Mineral fibres are asbestos, glass, mineral wool, basalt, ceramic, aluminium, borate, silicate, carbon, etc. Mineral fibers are naturally occurring fiber or slightly modified fiber procured from minerals. These can be categorized into the following categories:

- Asbestos: The only naturally occurring mineral fiber. Variations are serpentine (chrysotile) and amphiboles (amosite, crocidolite, tremolite, actinolite, and anthophyllite).
- Ceramic fibers: Glass fibers (Glass wool and Quartz), aluminum oxide, silicon carbide, and boron carbide.
- Metal fibers: Aluminum fibers.

Fibres of plant origin are known as plant fibres or vegetable fibres. Nearly 90% of the world production of fibres is from the natural vegetable fibres. Vegetable fibres are often grouped into three types according to their botanical/ anatomical origin viz., i) surface fibres, ii) soft, stem or bast fibres and iii) hard, leaf or structural fibres.

(i) Surface fibres: These are borne on the surface of stems, leaves, fruits and seeds. These fibres arise as single-celled outgrowths from the seeds or inner walls of the fruits. Since these fibres are found on the surface, these are known as surface fibres. Examples are cotton, kapok and coconut coir fibre. Cotton fibres still represent the world's most important natural fibres.

(ii) Soft or Bast fibres: These are found associated with the pholem, pericycle and cortex are referred to as stem, bast or soft fibres. They are sclerenchymatous tissues associated with the phloem and pericycle found in the stem bark. These are obtained by retting. The plants producing bast fibres are flax, jute, sunnhemp, kenaf, roselle and ramie.

(iii) Hard or leaf fibres: Structural fibres are strands of small, short lignified cells ensheathing both xylem and phloem (fibrovascular bundle) and are primarily found scattered in the leaves of monocotyledonous plants. Leaf fibres are sclerenchymatous and vascular tissues found in the leaves. These cells are lignified and generally hard, hence known as hard fibres. Examples are Agave, Manila hemp or Abaca (*Musa textiles*).

Commercial classification of fibres based on use:

Fibres are classified into the following five categories depending upon the use.

(i) Textile fibres: Kenaf, rosselle, sunnhemp, flax, hemp, jute and cotton are more commonly used, but sometimes flax and hard fibres are also used. All fibres used for the manufacture of fabrics, netting and cordage or ropes. The textile fibres of economic importance are cotton, flax, kenaf, hemp, roselle hemp, deccan hemp, sunhemp, etc. Flax, ramie and hemp are used for superior quality fabrics while jute is used for inferior grade fabrics such as clothing bags, caves, burlap, bagging, sacking, etc.

(ii) Cordage fibres: Cordage fibres are used for making twines or ropes, cables and hawsers from coconut, agave, manila hemp.

(iii) Brush fibres or braiding fibers: Twigs, leaves and bark are used for making brushes and brooms. Eg. palmyra leaves, coconut, agave, roots of grasses, surface fibres of palm leaves and stems, broomcorn.

(iv) Plaiting and rough weaving fibres: More elastic flat strands or strips (plaits) are roughly woven into hats, sandals, hammocks, baskets, plaited walls, roof thatching, chair seats, matting and thatched roofs of houses. Bamboo strips are used in the manufacture of fishing rods, furniture, baskets, etc. *Cyperus* sp. is used for mats weaving while vettiver grass is used for curtains. These weaving fibres include hard fibres like Palmyra leaves, Coconut coir, Agave fibres.

(v) Wicker works: Bamboo, canepalans, willows, etc., are used for baskets and chairs.

(vi) Filling fibres: Fibres used for stuffing matresses, pillows, cushions, life belts, packing materials, etc. The chief stuffing fibres are kapok, cotton, calotropis floss, coconut coir, jute, hemp, sunhemp, inflorescence of *Aerva tomentosa*, spanish moss, several hard fibres and innumerable grasses.

(vii) Paper-making fibres: wood fibres, textile fibres, various grasses and sedges are much used for making paper and paper products.

Plant-based natural fibres can very well be used as reinforcement in polymer composites, replacing to some extent more expensive and non-renewable synthetic fibres such as glass. The maximum tensile, impact and flexural strengths for natural fibre reinforced plastic (NFRP) composites reported so far are 104.0 MN/m^2 (jute-epoxy), 22.0 kJ/ m^2 (jute-polyester) and 64.0 MN/m2 (banana-polyester), respectively. The properties of some of the natural fibres are compared in Table 1. There are many examples of the use of cellulosic fibres in their native condition like sisal, coir, jute, banana, palm, flax, cotton, and paper for reinforcement of different thermoplastic and thermosetting materials like phenol formaldehyde, unsaturated polyester, epoxy, polyethylene, cement, natural rubber, etc.

Table 1 Properties of selected natural fibres

Property	Jute	Banana	Sisal	Pineapple	Coconut coir fibre
Width or Diameter (mm)	-	80-250	50-200	20-80	100-450
Density (gms./cc)	1.3	1.35	1.45	1.44	1.15
Volume Resistivity at 100 volts (W cm × 105)	-	6.5-7.0	0.4-0.5	0.7-0.8	9-14
Micro-Fibrillar Angle (degree)	8.1	11	10-22	14-18	30-49
Cellulose/Lignin Content (%)	61 /12	65 /5	67 /12	81 /12	43 /45
Elastic Modulus (GN/m^2)	-	8-20	9-16	34-82	4-6
Tenacity (MN/m^2)	440-533	529-754	568-640	413-1627	131-175
Elongation (%)	1-1.2	1.0-3.5	3-7	0.8-1.6	15-40

Synthetic or man-made fibres: Man-made fibres are not damaged by microorganisms like natural fibers. Man-made fibres can deteriorate in bright sunlight and melt at a lower temperature than natural fibers. Synthetic fibres are made from different types of chemicals which have the ability to form very long chains of molecules called polymers. Therefore, polymers form the basis of man-made fibres. Man-made fibres are generally divided into two broad groups, namely:

(i) **Regenerated fibres** (derived from cellulose):

(a) Rayon is the most common of this type of fiber. It can imitate (duplicate) natural fibers, but it is stronger.

(b) Celenese is cellulose chemically combined with acetate and is often found in carpets.

(c) Polyamide nylon is cellulose combined with three acetate units, is breathable, lightweight, and used in performance clothing.

(ii) **Synthetic Polymer fibres** (derived from petrochemical sources): Petroleum is the basis for these fibres, and they have very different characteristics from other fibres.

(a) Polymer fibres are non-cellulose

(b) Monomers in large vats are joined together to form polymers. The fibres produced are spun together into yarns.

Examples of synthetic polymer fibers:

- Polyester is found in 'polar fleece,' wrinkle-resistant, and not easily broken down by light or concentrated acid; added to natural fibers for strength.
- Nylon is easily broken down by light and concentrated acid; otherwise similar to polyester.
- Acrylic is inexpensive, tends to 'ball' easily, and used as an artificial wool or fur.
- Olefins is high performance, quick drying, and resistant to wear.

CHAPTER

15

Cotton

Vernacular name: *Kapas* (Hindi), *Paruthi* (Tamil), *Hatti* (Kannada), *Pratti* (Telugu). The English word 'cotton' comes from the Arabic word '*al qutun*' or '*kutun*. The same root is also to found in many other languages; *Katoen* (Dutch), *Coton* (French), *Cottone* (Italian) and *Algodon* (Spanish). The botanical term *Gossypium* seems to have risen from the word "Gossypines" in Tylos for cotton. The sanskrit words kurpasa or kurpusum denoting cotton and cotton cloth are mentioned in the sacred writing of Manu (3000 BC) and the word kapas have been derived from them.

Importance: Cotton is grown commercially in the temperate and tropical regions of more than 100 countries. It is known as 'King of fibres' and 'White Gold'. The scientific term *Gossypium,*is derived from the Arab, Persian and Afghan words *Goz*, *Gozah* and *Gozeh* respectively. Cotton produces the world's most important vegetable fibre. It has four cultivated species. The cultivated species have spinnable seed fibres called lint. The wild cotton species has only short seed fuzz or smooth seeds. The oil content in cotton seed is 20% which is extracted from the cotyledons. It is semi drying oil. Gossypol content in seed is associated with resistance to pests. Technology Mission on Cotton has been launched in February, 2000. Cotton Corporation of India Limited (CCI) has been established on 31st July 1970 as a Government Company registered under the Companies Act 1956.

World cotton scenario: The major cultivators of cotton in the world are India, China, Egypt, Pakistan, USA, Uzbekistan, Turkey, Australia, Greece, and Brazil. The four major cotton producing countries are China, India, the USA and Pakistan, which accounted for approximately three-quarters of world's cotton production. India has the largest area under cotton production. China is the largest producer of cotton in the world, whereas India is the second largest. Interestingly, China with almost half the area under cotton production compared to India, but produces more than 2½ times yield (kg/ha) of cotton as compared to India.

Indian cotton scenario: India is recognised as the cradle of cotton industry for over 3000 years (1500 BC to 1700 AD). India produces finest and beautiful cotton fabrics since time immemorial. There are more than 1,500 spinning mills, 250 composite mills in India having an installed capacity of approximately 35 million spindles and more than 1 lakh handlooms. The cotton cultivation sector not only engages around 6 million farmers, but also involved another about 40 to 50 million people relating to cotton cultivation, cotton trade and its processing. There are four major species of cultivated cotton, of which two are diploid (*Gossypium arboreum* and *G herbaceum*) and the other two tetraploid (*G. hirsutum* and *G. barbadense*). India is the only country to grow all four species of cultivated cotton. In addition, hybrid cotton, which is produced from crossing tetraploid species *G. hirsutum* are also cultivated in the central and southern zones. The diploid species referred to as the 'Desi' cotton, having resistance to insects, pests, diseases and drought, low productivity and low quality cotton, contributes 25 to 30% of the country production. The tetraploids variety contributes remaining 70% of the cotton production in India. These varieties have fine quality fibre, and are normally used by the textile industry. India is the second largest exporter of cotton behind the US. Major export destinations are Bangladesh, Pakistan, China (Mainland) and other Far-east countries. But today India not only becomes self-sufficient for own cotton requirement, but also becomes a leading exporter of cotton globally.

History: The chronological events in cotton developments are as follows:

1917: Indian Cotton Committee set up under the chairmanship of J. MacKenna.

1921: Central Cotton Committee was established at Mumbai.

1923: Central Cotton Committee became a statutory body for promoting agricultural and technological research in cotton.

1924: Central Cotton Committee set up Technological Laboratory (presently Central Research Institute for Research on Cotton technology, Mumbai) at Bombay with Dr. A. J. Turner as its first Director.

1924–1937: Central Cotton Committee provided entire financial assistance to cotton research and development in India.

1937: First National Conference of Cotton Workers was organized by Central Cotton Committee at Bombay.

1966: Central Cotton Committee was wound up and the functions were transferred to ICAR, New Delhi.

1967: All India Coordinated Cotton Improvement Project was established.

1970: First commercial cotton hybrid of the world (Hybrid 4) was released from Surat by the noted breeder Dr. C. T. Patel.

1972: First commercial interspecific cotton hybrid of the world (Varalaxmi) was released from Dharwad by the noted breeder Dr. Katarki

1976: Infrastructure of cotton research got strengthened with the establishment of Central Institute for Cotton Research at Nagpur.

1995: Department of Biotechnology, Govt. of India permitted Mahyco to import a small stock of Bollgard® (Bt cotton) seeds from Monsanto Company, USA.

1999: Technology Mission on Cotton was launched.

2000: Based on the recommendation of RCGM (Review Committee for Genetic Modification), the GEAC (Genetic Engineering Approval Committee), Ministry of Environment & Forests, Govt. of India, gave approval for Mahyco to conduct large scale field trials in 85 ha and also undertake seed production in 150 ha.

2002: Transgenic Bt Cotton hybrid was approved for commercial cultivation in India.

2005: Transgenic interspecific cotton hybrid (MRC 6918) released for cultivation.

Origin: The origin of different cultivated cotton species is presented in Table 1.

Table 1 Origin and chromosome number of different cotton species

Species	Chromosome (2n number)	Geographic origin
A. Old world species (Asia, Africa, Australia) *G.herbaceum* L. *(Uppam cotton)*	26	Afghanistan, Africa
G.arboreum L. (Karunganni cotton)	26	Indo-Myanmar, China, Arab
B. New world species *G.hirsutum* L. *(American upland cotton,* Combodia cotton)	52	Mexico and Central America
G.barbadense L. *(Egyptian cotton, Sea island cotton, Pima cotton)*	52	South America (Peru, Colombia)

Distribution: Cotton growing regions in world and in India are presented in Table 2 and Table 3. Cotton is cultivated mainly in USA, Russia, China, India, Brazil, Mexico, Pakistan, Turkey and United Arab Emirates. In India, it is grown in Maharashtra, Gujarat, Madhya Pradesh, Punjab, Haryana, Rajasthan, Andhra Pradesh, Karnataka, Tamil Nadu and Odisha. In the world, *G. hirsutum* cotton is predominant with 92-93% area and production, *G. barbadense* is grown for nearly 4-5% and the diploid cottons *G. arboreum* and *G. herbaceum* accounted for less than 2%. In India, *G. hirsutum* (hirsutum × hirsutum hybrid cotton) is cultivated in 85% of total area which is followed by *G. arboreum* (Karunganni cotton) in 9.43%, *G. herbaceum* (Uppam cotton) in 4.72% and *G.barbadense* 0.85%. *G. hirsutum* is best adapted to areas with irrigation. *G. arboreum* and *G. herbaceum* are adopted to regions with uncertain and irregular rainfall pattern. *G.barbadense* is grown in Tamil Nadu, Andhra Pradesh and Karnataka.

Table 2 Cotton growing regions in World

Species	Staple length	Region / Country	Mode of cultivation
G.arboreum *G.herbaceum*	Short staple (< 21 mm)	India, Pakistan, China	Rainfed
G.hirstum	Medium staple (21-25 mm)	USA, India, Pakistan	Rainfed (20% of world production)
G.hirsutum	Medium long staple (26-28 mm)	USA, India, Russia, Turkey, Mexico, Brazil, Pakistan	Rainfed (75% of world production)
G. hirsutum *G. barbadense*	Long staple (29-34 mm)	USA, Egypt, Greece, Sudan, Russia, Spain, Peru, Turkey, Uganda, Brazil.	Irrigated / rainfed (9% world production)
G.barbadense	Extra long staple (>= 35 mm)	Egypt, India, Sudan, Peru, USA, Russia,	Irrigated (6% world production)

Table 3 Indian cotton production zones

Zones/ Area in m ha.	States	% of total Cotton area	% of contribution in total production	Irrigated/ Species grown	Rainfed / Species grown
North Zone 1.459	Punjab, Haryana and Rajasthan.	16	18.5	100 % *G. hirsutum, arboreum* *Intra hirsutum* hybrids and Diploid hybrids	-
Central Zone 6.180	Maharashtra, Madhya Pradesh and Gujarat.	68	60	23 % *Intra hirsutum* hybrids *G. hirsutum*	77 % *G. herbaceum, G. arboreum, G. hirsutum, Intra hirsutum* hybrids, Diploid hybrids
South Zone 1.398	Andhra Pradesh, Karnataka and Tamil Nadu.	15.3	16.3	40 % *Intra hirsutum* hybrids *Inter specific* hybrids (H × B) *G. hirsutum* *G.barbedense*	60 % *G. herbaceum* *G. arboreum* *Intra hirsutum* hybrids *Inter specific* hybrids (H × B) Diploid hybrids
Eastern region	Assam, Tripura, Manipur	------	------	------	*G. arboreum*

Botany: Cotton is a tropical perennial shrub which is grown as an annual crop. Cotton is a hypocotyl plant. Plant height ranges from 60 to 180 cm. Cotton has a tap root system. Two thirds of root system by a weight are commonly found in the top 30 cm of the soil. The main stem arises from the terminal growing point while the lateral branches from the terminal axils of the leaves of main stem. The latter group consists two types viz., vegetative and fruiting. Axillary buds develop into *monopodial* or vegetative branch and the extra axillary buds grow into fruiting (*sympodial*) branches. The vegetative branches are more vertical and ascending while the fruiting ones are nearly horizontal. Sympodial branch has a flower bud at each node. Thus, the main stem and the monopodia do not bear flowers directly but produce fruiting branches. The leaf is palmatifid. The hairiness of the leaf is heritable. The adult leaf has 3 to 5 lobes either half cut or fully cut leaf. In general the leaves of *G. barbadense* are more deeply divided them those of *G. hirsutum* while *G. arboreum* is more deeply divided than *G. herbaceum.* The distinguishing characters of *Gossypium* species are presented in Table 4. The flowers are appearing from leaf axils but from its sides. The flower bud is surrounded by three green bracts or bracteoles and joined at their bases. The bracteoles are heart shaped forming a pointed canopy over the bud, often referred to a '*square*'. If a bud or young boll is punctured by boll worm, the bracts flare outwards leaving the bud exposed and start to lose green colour. These flared squares are useful in indicator of bollworm attack in the crop. The corolla / petals is white or slightly cream coloured in *G. hirsutum*, bright yellow in *G.barbadense* and *G.arboreum*. *G. herbacium* have a red or purple spot on the petals near the base. There are approximately 10,000 pollen grains in a flower. Under normal conditions, the pollen grains are viable up to 24 hours. The petals turn pink following anthesis and on the next day or two, the petals wither to a purple colour and drop off to reveal the young green boll.

Table 4 Distinguishing characters of *Gossypium* species

Character	G. arboreum	G. herbaceum	G. hirsutum	G. barbadense
Chromosome Number	2n : 26	2n : 26	2n : 52	2n : 52
Ploidy	Diploid	Diploid	Tetraploid	Tetraploid
Origin	Africa	Indo-China	Central America, Mexico	South America
Popular names	Asiatic cotton, Desi cotton	Asiatic cotton, Desi cotton	American cotton, Upland cotton, Cambodia cotton,	Egyptian cotton, Sea Island cotton, Pima cotton
Habit	Annual or perennial shrub	Annual shrub	Annual, small shrub	Perennial or annual tall shrub
Branches	Sub-erect	Branched	Branched	Branched
Leaves	5 to 7 deeply lobed' light green	3 to 5 lobed, green	2 to 5 lobed large, dark green	3 to 5 lobed large, very dark green
Bracteoles epicalyx	3, closely enclosing the bud and flowers, entire 3-4 teeth, nectar absent	3, flaring widely from the flower with 6-8 broadly triangular teeth nectar absent	3, longer bracteoles 8-12 teeth, nectar present	3, almost as broad as a long cordates with 10-15 acuminate teeth, nectar present
Flower	Yellow, petal spot present	Yellow, petal spot present	Pale yellow, petal spot absent	Bright yellow, petal spot present
Bolls	Tapering and pitted with prominent oil glands, 3-4 locules, 6-17 seeds/locule, pale green	Rounded beaked with a few oil glands, 3 loculi, pale green smooth to moderately pitted.	Medium rounded with a few oil glands, 3-4 locules, 11 seeds.locule, Non-pitted, green	Long capsule with 3-4 locules, broad at the base tapering to acute tip, rough surface oil glants at the bottom, 5-8 seeds/ locule profusely pitted dark green
Seeds	With short fuzz hairs	With fuzz and lint	Bearing lint hairs and thick coat of fuzzy hairs	Bearing even coat of lint with full coat at the end or absent of fuzz
Lint	Short staple	Short staple	Medium staple	Long staple

Cotton is a self-pollinated crop. Cross pollination occurs up to 20% due to the activities of honey bees (*A.dorsata, A.florea, A.indica*), bumble bees (*Bombus* sp.), leaf cutting bees (Hymenoptera megachilidae). Cotton fruit is known as boll, a leathery capsule which is splitted into 3 to 5 locs or loculi each containing about 8 seeds (Table 5). Each seed surrounded by downy fibers called lint.

Table 5 Locs and seeds in different cotton species

Species	Locs/boll	Seeds/locs
G. arboreum	3, rarely 4	6-17
G. herbaceum	3-4	6-11
G. hirsutam	4, sometimes 5	5-11
G. barbadense	3, some times 4	5-8

Cotton boll is a dehiscent schizocarp. Boll diameter size increases one mm per day which attains full growth on 25 day and the dehiscence on 48th day. The boll period or the ripening period of fruit is 50 to 70 days. The boll wall is pitted black and deeply sunken in the outside with gossypol glands. The gossypol glands are little in in *G. hirsutum* but high in others. '*Motes*' refer to undeveloped seeds or aborted ovules due to non-fertilization, hereditary and environment. The cotton seeds are generally lesser than the number of ovules. Seeds may be naked (without hairs) or bear short hairs called 'fuzz'. All cultivated cotton bear long fibres named lint and a majority of them have also fuzz on the same seed. The lint is removed by ginning while the fuzz remains attached to the seed. Hundred seed weight is 5 to 10 gm.

Bt Cotton: A genotype or individual which is developed by the techniques of genetic engineering is referred to as transgenic. In other words, genetically engineered organisms are called transgenics. A transgenic may be a plant, an animal or a microbe. Transgenic plants contain foreign gene or genetically modified gene of the same species. The foreign gene may be from a distantly related species, closely related species or unrelated species or from micro-organisms such as fungi, bacteria and viruses. The Bt cotton refers to transgenic cotton which contains endotoxin protein inducing gene from soil bacterium *Bacillus thuringiensis*. Bt cotton, a transgenic plant, produces an insect controlling protein Cry1A(c), the gene for which has been derived from the naturally occurring bacterium, *Bacillus thuringiensis* sub sp. *kurstaki* (B.t.k.). The cotton hybrids containing Bt gene produces its own toxin for bollworm attack, thus significantly reducing chemical insecticide use and providing a major benefit to cotton growers and the environment.

In India, five genes/events for insect resistance have been approved so far. These include:

(i) cry1Ac (MON 531) by Mahyco-Monsanto
(ii) cry1Ac + cry2Ab (MON 15985) by Mahyco-Monsanto
(iii) cry 1Ab+cry 1Ac " (GFM event) by Nath Seeds
(iv) cry1 Ac gene (Event-1) by J.K. Agri Genetics
(v) cry1Ac gene (CICR) by Central Institute of Cotton Research
(vi) cry1 C by Meta-Helix Life Sciences

In addition, research and field trials have been undertaken/are underway for the following insect resistance and herbicide tolerant genes:

(i) vip3A+cry1 Ab by Syngenta
(ii) cry1 Ac+ cry 1F by Dow Agrosciences
(iii) cry1 Aa3, cry1 F, cry1 Ia5, cry1 Ab by ICAR institutions
(iv) cry1 Ec by J.K. Agri Genetics
(v) cp4epsp gene by MAHYCO

The commercially cultivated Bt cottons are presented in Table 6.

Table 6 Commercially cultivated Bt cotton

Biotech Cotton	**Active Gene(s)**	**Countries Bt cotton cultivated**
BXN[TH]	Nitrolase (Mon 1445/1698)	USA
Round Ready	cp4 epsps (Mon 88913)	Australia, USA, South Africa
Liberty Link	bar (LL Cotton 25)	Australia, USA
Bollgard	cry 1 Ac (Mon 531)	Argentina, Australia, Brazil, China, Colombia, India, Indonesia, Mexico, USA, Africa
Bollgard II	cry 1 Ba (Mon 531) + cry 2 Ab (Mon 15985)15	Australia, Combodia, USA, South Africa, India
Wide Strike [TH]	cry 1 Ac + cry 1F (Event 3006-210-23 + Event 281-24-236)	USA
Guokang	cry 1A + CpTI	China, India
Round Ready + Bollgard	cp4 epsps (Mon 1445/1698) + cry 1Ac (Mon 531)	Australia, Combodia, USA, South Africa, Mexico
Round Ready Flex + Bollgard II	cp4 epsps (Mon 88913) + cry 1Ac (Mon 531) + cry 2Ab (Mon 15985) 15	Australia, Combodia, USA, South Africa
Event 1	cry 1Ac	India
Wide Strike [TH] + Round Ready	cry 1Ac + cry 1F (Event 3006-210-23 + Event 281-24-236)	USA
Wide Strike [TH] + Roundup Ready Flex	cry 1Ac + cry 1F (Event 3006-210-23 + Event 281-24-236) + cp4 epsps (Mon 88913)	USA
Liberty Link + Bollgard II	bar + cry 1Ac (Mon 531 + cry 2Ab (Mon 15985) 15	USA

Source: Tripathi et al, 2011

Climate: Cotton is cultivated from 30°S to 47°N in old world and 32°S to 37°N in new world up to an altitude of 500 m from MSL. Cotton is adapted to the humid tropics and subtropical climates where the frost free period of 160 to 200 days and summer isotherm about 25°C or more. Cotton crop requires 500 to 625 mm rainfall during its growing period with well distributed rainfall of 175 to 200 mm over the growing season. A mean annual rainfall of 1000 to 1500 mm is required for raising cotton as rainfed crop. Heavy rain at planting time and early growth stages are undesirable as its leads to shallow roots development. Heavy rainfall during flowering causes flower buds and young bolls to fall. Continuous rain during flowering and boll opening will impair pollination and reduce fibre quality. Late in the season, rains stimulate top growth, delay maturity, interfere with picking and discolour lint. Cotton is relatively resistant to short periods of water logging and heavy rainfall.

Cotton is sun loving plant. Solar radiation of 400 to 500 cal cm^{-2} day^{-1} is ideal for cotton. Cotton is a short-day plant but day-neutral varieties exist. Perennial cotton is short to medium day plant. The wild ancestors of cotton are short day plants. However the effect of daylength on flowering is influenced by temperature. It cannot tolerate shade particularly in the seedling stage. The optimum light intensity required for growth and yield of cotton is 32.3 to 86.1 klux. Cool nights and low daytime temperatures result in vegetative growth with few fruiting branches but cool days and hot night result mainly fruiting branches. Cotton prefers relatively dry and cool condition at harvest. Reduced light intensity due to prolonged over cast weather, shading from with planted crops or too dense stand of cotton, retard flowering and fruiting increase boll shedding. The optimum temperature is 17 to 27°C, 24 to 30°C and 32°C for germination, seedling growth and for continuous growth respectively. The optimum temperature for Asiatic cotton is 21 to 27°C for vegetative growth and 27 to 32° C for fruiting with cool nights. The optimum temperature for American cotton is 32 to 34°C. The optimum air temperature for germination is 25 to 30°C with soil moisture content of 90% of field capacity. The seeds will not germinate below 15°C and above 40°C. The optimum soil temperature is 20°C for germination. Delayed germination exposes seeds to fungus infections in the soil. The cardinal minimum, optimum and maximum temperature for American cotton is presented in Table 7. Cotton is very sensitive to frost and a minimum of 200 frost-free days is required. Night temperature around 16ºC facilitates Red leaf disease in cotton. The heat unit for cotton is 3000 to 4000 above with a base temperature of 10ºC. Strong and/or cold winds seriously affect the delicate young seedlings and at maturity will blow away fibre from opened bolls and cause soiling of the fibre with dust and vegetative matter.

Table 7 Cardinal temperature for American cotton

Growth phase / stages	**Cardinal air temperature (°C)**		
	Minimum	Optimum	Maximum
Germination	15	25-30	40
Vegetative phase	15	25-30	35-40
Reproductive phase (flowering, boll formation and maturation)	21	25-32	40

Soil: Cotton is grown on a wide range of soils, but medium and heavy textured, deep soils with good water holding characteristics are preferred. The soils under grassland or well drained, alluvial flood plains are best. Heavy clays tend to delay maturity and result in undesirable greater vegetative growth since conducive for insect damage. Cotton grows well in black cotton soil which is typically a clayey soil rich in organic matter, $CaCO_3$ and $MgCO_3$ It cannot withstand water logging. Water logging, shallow soils, hard pan, *etc.*, restricts root growth of cotton. Acid or dense subsoils limit root penetration. Cotton tolerates 3000 to 6000 ppm of salts and grows best with soil pH of 6.5 to 7.5. Black alkali (sodic) soil is more destructive to cotton than white alkali (saline) soils. Cotton is grown on soils with pH values ranging from to 5.5 to 8.5. The crop is tolerant to soil salinity. Yield decreases at different ECe values are: 0% at ECe 7.7 mhos/cm; 10% at 9.6, 25% at 13, 50% at 17 and 100% at ECe 27 mmhos/cm. The desirable bulk density for cotton root growth is 1.8 to 1.9 g/cc.

Agronomy of Irrigated Cotton

Desi cotton varieties: *Gossypium arboreum* and *G. herbaceum* are grouped into desi cotton or Asiatic cotton or Old world cotton. *G. arboreum* is locally known as 'karunkanni' cotton. Eg: K2, K5, K6, K7, K8, K9, K10 while *G. herbaceum* is known as Uppam cotton in Andhra Pradesh and Broach cotton in Maharashtra. Eg. Hagari 1.

Exotic cotton varieties: *Gossypium hirsutum* and *G. barbadense* are grouped as exotic cotton or American cotton or Sea island cotton or New world cotton. Examples for varieties of *G. hirsutum* are MCU 5, MCU 7, MCU 9, MCU 10, PRS 72 and *G. barbadense* are Sujatha, Suvin, TNB 1, Varalakshmi and Jayalakshmi.

The cotton cultivated zones, seasons and varieties/ hybrids in India and Tamil Nadu are presented in Table 8, Table 9, Table 10 and Table 11.

Table 8 List of cotton varieties/hybrids released in India

S.No.	Name	Species	Year of Release	AICCIP centre	Remarks/Special characteristics
1	HD 123	*G.arboreum*	2000	CCSHAU, Hisar	High yielding desi variety for Irrigated tracts of Haryana
2	RS 810	*G.hirsutum*	2000	RAU, Sriganganagar	Resistant to cotton leaf curl virus
3	L 603 (H)	*G.hirsutum*	2000	ANGRAU, Hyderabad	Suitable for rainfed and also irrigated tracts of A.P.
4	L 604 (H)	*G.hirsutum*	2000	ANGRAU, Hyderabad	Suitable for rainfed and also irrigated tracts of A.P.
5	Aravinda	*G.arboreum*	2000	ANGRAU, Hyderabad	Drought tolerant desi variety
6	LAHH-4	*G. hir* × *G. hir*	2000	ANGRAU, Hyderabad	Hybrid suitable for rainfed and irrigated tracts of A.P.
7	MCU 12	*G. hirsutum*	2000	TNAU, Coimbatore	Suitable for 50s count and moderate sucking pest tolerance
8	SVPR-3	*G. hirsutum*	2000	TNAU, Coimbatore	Suitable for rice fallows of delta districts in TN
9	PHH-316 (Ganga)	*G. hir* × *G. hir*	2000	MAU, Parbhani	Drought tolerant hybrid for Marathwada
10	RHB-0388	*G. hirsutum* x *G.barbadense*	2000	MPKV, Rahuri	Quality lint interspecific hybrid
11	RG 18	*G.arboreum*	2001	RAU, Sriganganagar	Leaf curl virus tolerant high yielding desi variety
12	Vagad Kalyan	*G. hirsutum*	2001	MPUA&T, Banswara	Rainfed tracts of south Rajasthan
13	G.Cot.18	*G. hirsutum*	2001	GAU, Surat	Suitable for Gujarat
14	G.Cot.21	*G. herbaceum*	2001	GAU, Surat	Suitable for coastal saline tracts
15	AKA 7	*G.arboreum*	2001	Dr. PDKV, Akola	Suitable for poor rainfall areas in Maharashtra
16	PKV HY 4	*G.hir* × *G.hir*	2001	Dr. PDKV, Akola	hybrid for rainfed conditions
17	Pratima	*G. hirsutum*	2001	CICR, Nagpur	Suitable for south zone

18	Sahana	*G. hirsutum*	2001	UAS, Dharwad	Yields better both under rainfed and irrigated situations
19	RAMPBS 155	*G. hirsutum*	2001	UAS, Dharwad	Robust genotype responsive to inputs
20	Sumangala	*G. hirsutum*	2001	CICR, Coimbatore	High yielding hirsutum for south zone
21	Bunny	*G.hir* × *G.hir*	2001	M/s Nuziveedu seeds	Popular private sector hybrid tested through AICCIP
22	VagadKalyan	*G.hirsutum*	2001	MPUAT, Banswara	Suitable for rainfed situations in south Rajasthan
23	H 1117	*G.hirsutum*	2002	CCSHAU, Hisar	Suitable for Haryana
24	HHH 223	*G.hir* × *G.hir*	2002	CCSHAU, Hisar	Intrahirsutum hybrid for north zone
25	RS 2013	*G.hirsutum*	2002	RAU, Sriganganagar	Irrigated tracts of North zone
26	JK 4	*G.hirsutum*	2002	JNKVV, Khandwa	Suitable for M.P.
27	G. Cot.23	*G.herbaceum*	2002	GAU, Surat	Suitable for coastal marginal areas
28	Phule 492	*G.hir* × *G.hir*	2002	MPKV, Rahuri	Deccan canal areas of Maharashtra
29	Phule 388	*G.hir* × *G.bar*	2002	MPKV, Rahuri	Fine quality interspecific hybrid
30	VICH 5	*H* × *H*	2002	Vikram Seeds	Private sector hybrid evaluated through AICCIP
31	LAHH 5	*G.hir* × *G.hir*	2002	ANGRAU, Guntur	Suitable for cotton growing areas of A.P.
32	F 1861	*G.hirsutum*	2003	PAU, Faridkot	Moderately high oil content
33	PA 402	*G.arboreum*	2003	MAU, Nanded	Suitable for less rainfall areas
34	DLSA 17	*G.arboreum*	2003	UAS,Dharwad	Long linted desi variety
35	Swadeshi	*G.arb* × *G.arb*	2003	M/s Ankur Seeds	Private sector hybrid evaluated through AICCIP
36	Mallika	*G.hir* × *G.hir*	2003	M/s Nuziveedu Seeds	Private sector hybrid evaluated through AICCIP
37	G.Cot.19	*G.arboreum*	2003	NAU, Surat	Suitable for rainfed tracts of Gujarat
38	G.Cot.MDH 11	*G.hir* × *G.hir*	2003	NAU, Surat	High yielding hybrid for rainfed areas
39	G.Cot.Hy.102	*G.hir* × *G.hir*	2003	NAU, Surat	High yielding intra-hirsutum hybrid
40	Veena	*G.arboreum*	2004	ANGRAU, Mudhol	Good drought tolerance
41	CISAA-2 (CICR-2)	*G.arb* × *G.arb*	2004	CICR, Sirsa	Gms based arboretum hybrid for North zone

42	PKV DH ? 1	*G.arb* × *G.arb*	2004	Dr.PDKV, Akola	Desi hybrid for rainfed farming
43	PKV Hy.5	*G.hir* × *G.hir*	2004	Dr.PDKV, Akola	Cms based hybrid
44	NH.545	*G.hirsutum*	2004	MAU, Nanded	Drought tolerant variety
45	ParbhaniTurab (PA 255)	*G.arboreum*	2004	MAU, Parbhani	Long linteddesi cotton
46	CSHH-198 (Shresth)	*G.hir* × *G.hir*	2005	CICR, Sirsa	High yielding intra-hirsutum hybrid for north zone
47	PratapKapi (RBDV.7)	*G.herbaceum*	2005	MPKVV, Banswara	Desi variety for rainfed tracts of south Rajasthan
48	JLA 794	*G.arboreum*	2005	MPKV, Jalgaon	Suitable for rainfed tracts in khandesh region of Maharashtra
49	HD 324	*G. arboreum*	2005	HAU, Hisar	High yielding arboretum variety for irrigated conditions of Haryana
50	MCU 13	*G. hirsutum*	2005	TNAU, Coimbatore	Early duration for winter irrigated tracts of T.N. and has 50s spinning potential
51	Mahabeej 106	*G.hir* × *G.hir*	2005	Mahabeej, Maharashtra	Intrahirsutum hybrid for rainfed situation
52	Mahabeej DH 986	*G.arb* × *G.arb*	2005	Mahabeej, Maharashtra	Desi hybrid for rainfed conditions
53	HHH 287	*G.hir* × *G.hir*	2005	CCSHAU, Hisar	Gms based intra-hirsutum hybrid
54	PH 348 (Yamuna)	*G.hirsutum*	2005	MAU, Parbhani	Suitable for Rainfed
55	PA 402 (Vinayak)	*G.arboreum*	2005	MAU, Parbhani	Drought tolerant
56	Raj DH 9	*G.arb* × *G. arb*	2006	RAU, Sriganganagar	Good tolerance to CLCuV disease
57	CISA 310 (CICR 1)	*G. arboreum*	2006	CICR (RS), Sirsa	Desi variety for irrigated north zone conditions
58	NACH 6 (Navinya)	*G. hir* × *G. hir*	2006	M/s Nirmal Seeds	Private sector hybrid evaluated through AICCIP
59	Vasant (Navkar 5)	*G. hir* × *G. hir*	2006	MsNavkar Seeds	Private sector hybrid evaluated through AICCIP
60	Ajeet II (AHH 90-1)	*G. hir* × *G. hir*	2006	MsAjeet Seeds	Private sector hybrid evaluated through AICCIP
61	Ajeet 33 (AHH 90-2)	*G. hir* × *G. hir*	2006	MsAjeet Seeds	Private sector hybrid evaluated through AICCIP
62	G Cot Hy 12	*G.hir* × *G.hir*	2006	NAU Surat	High yielding hybrid for irrigated situation in Gujarat

64	NDLHH 240 (Sona)	*G.hir × G.hir*	2006	ANGRAU, Nandyal	Suitable for Rayalaseema region of A.P.
65	NDLA 2463 (Srinandi)	*G. arboreum*	2006	ANGRAU, Nandyal	Suitable for Rayalaseema region of A.P.
66	NH 615 (Anusuya)	*G.hirsutum*	2007	MAU, Nanded	Drought tolerant
67	JK 5	*G.hirsutum*	2007	JNKVV Khandwa	Suitable for M.P.
68	CSHH 238 (Kalyan)	*G. hir × G. hir*	2007	CICR (RS), Sirsa	High yielding intra-hirsutum hybrid for north zone
69	Dhruv	*G. hir × G. hir*	2007	MsZuari Seeds	Private sector hybrid evaluated through AICCIP
70	LAHH 7	*G.hir × G.hir*	2007	LAM, Guntur	Suitable for rainfed tracts of coastal A.P.
71	KC 3	*G.hirsutum*	2007	TNAU, Kovilpatti	Well suited for low rainfall areas
72	AKA 8	*G. arboreum*	2007	PDKV Akola	Drought tolerant
73	Moti	*G.arboreum × G.arboreum*	2007	PAU, Ludhiana	Desi hybrid for Punjab
74	H 1226	*G.hirsutum*	2007	CCS HAU, Hisar	Irrigated conditions of Haryana
75	AKH 8828	*G.hirsutum*	2007	Dr.PDKV, Akola	Suitable for rainfed situation
76	Suraj	*G.hirsutum*	2008	CICR, Coimbatore	Jassid tolerant long staple variety
77	RAHH 98	*G. hir × G. hir*	2008	UAS Dharwad	Intra-hirsutum hybrid for south zone
78	G Cot 20	*G.hirsutum*	2008	NAU Surat	Irrigated tracts of Gujarat
79	NDLH 1755 (Sivanandi)	*G.hir × G.hir*	2008	ANGRAU, Nandyal	Well suited for rainfed conditions
80	CICR 1 (CISA 310)	*G.arboreum*	2008	CICR RS, Sirsa	Short staple desi genotype
81	CSHH 243 (Simran)	*G.hirxG.hir*	2008	CICR,RS,Sirsa	High yielding intra-hirsutum hybrid with good tolerance to CLCuV
82	G.Cot.20	*G.hirsutum*	2008	NAU, Surat	Irrigated tracts of Gujarat
83	RAS 299-1	*G.hirsutum*	2009	UAS, Dharwad	Rainfed farming
84	DDhc-11	*G.herbaceum*	2009	UAS, Dharwad	Desi genotype suitable marginal soils
85	Suvidha (DHH 543)	*G. hir × G.hir*	2009	UAS, Dharwad	High yielding hybrid for Karnataka
86	CNHO 12	*G.hirsutum*	2009	CICR, Nagpur	Suitable for denim purposes
87	CISA 614	*G.arboreum*	2009	CICR, RS, SIRSA	Non-spinnable desi with good yield potential

Source: http://aiccip.cicr.org.in

Table 9 List of varieties/hybrids released by CICR, Nagpur

Varieties/ hybrids	Year of release	Spinning potential (counts)	Area of adaptability	Remarks
G. hirsutum				
MCU5VT	1982	60s	Verticillium wilt prone tracts of Tamil Nadu	Verticillium wilt tolerant
LRA 5166	1982	30s-40s	Rainfed and irrigated tracts of southern cotton zone &Vidarbha (Maharashtra)	Drought tolerant & adaptable to different agro-climatic conditions
Supriya	1984	40s-50s	Irrigated tracts of southern cotton zone	Whitefly tolerant
Kanchana	1988	40s	Whitefly prone area of southern cotton zone	Whitefly tolerant
Anjali (LRK 516)	1992	40s	Rainfed and irrigated conditions of Maharashtra, Gujarat and south Rajasthan	Early maturing, compact &semidwarf, suited for closer spacing
CNH 36	1993	40s	Irrigated areas of western Maharashtra and southern and middle Gujarat	Dwarf early maturing
Arogya	1995	12s	Rainfed areas of central Zone	Bacterial blight immune
Surabhi	1997	50s - 60s	South Zone	Verticillium tolerant extra long staple high yielding than MCU 5 VT
Sumangala	2000	30s- 40s	South Zone	
CNH 120 MB	2001	30s	Irrigated areas of south zone, also suitable for rainfed condition	compact, early maturing, medium staple with high fiber strength
G.arboreum				
CISA 310	2006		Irrigated Area of North Zone	Tolerant to fusarium wilt and root-rot
Intraspecific hybrid - Intra-*hirsutum*				
Savita	1987	60s	Irrigated tracts of southern cotton zone	Intra-*hirsutum* **hybrids of MCU 5 quality**
Kirti (CICR HH1)	1992	40s	Rainfed areas of Marathwada region of Maharashtra	Early maturing hybrid
Surya (TM 1312)	1994	50s-60s	Southern cotton zone	Presence of genetic marker character
OM- Shankar	1997	30s-40s	Northern cotton belt	Early maturing, high yielding hybrid

CSHH 198	2004	50s	Northern cotton belt	CLCuV resistant
CSHH 238	2006	40s	Irrigated Area of North Zone	CLCuV resistant
CSHH 243	2007	50s	Irrigated Area of North Zone	CLCuV resistant
Interspecific hybrid - ***G. hirsutum* × *G. barbadense***				
HB 224	1989	80s	Irrigated tracts of southern cotton zone	Extra long staple Hybrid
Shruthi	1997	80s	Southern Cotton zone	Compact, short duration hybrid
Intra-***arboreum* hybrid**				
CISSA 2	2004	10s	Northern cotton zone	GMS based hybrid

Source: http://www.cicr.org.in

Table 10 Bt Cotton Hybrids approved by Genetic Engineering Approval committee

Private companies	**Bt cotton hybrids released**
Mahyco	MECH 184 Bt, MECH 162 Bt, MECH 12 Bt, MRC 6322 Bt, MRC 6918 Bt, MRC 7351 BG II, MRC 7201 BGII, MRC 7160 BGII, MRC 7347 BG II
Rasi Seeds	RCH 2 Bt, RCH 20 Bt (Srinithi Bt), RCH 368 Bt, RCH 371 Bt, RCH 111 Bt (ACE Bt), RCHB 708 Bt (Rasi XL Bt), RCH 530 BGII, RCH 533 BGII, RCH 2 BG II,
Nuziveedu seeds	Bunny Bt, Mallika Bt, NCS 913 Bt, NCS 950 Bt, NCS 954 Bt, NCS 929 Bt, NCHB 990 Bt, NCHB 992 Bt, NCS 145 Bt2,
Ganga Kaveri Seeds	GK 209 Bt, GK 207 Bt
Emergent Genetics	Brahma Bt
Nath Seeds	NCEH 3R Bt, NCEH 2R Bt, NFHB -109 Bt (Kashinath)
JK Seeds	JK Durga Bt, JKCH 99 Bt, JKCH 634 Bt (JK Iswar Bt)
Prabhat Ag. Biotech	PCH 2270 Bt, PCH 930 Bt, PCH 207 Bt, PCH 205 Bt, PCH 115 Bt
Ajeet Seeds	ACH -33-1 Bt, ACH 33 -2 BGII, ACH 21 Bt
Krishidhan Seeds	KDCHH 9632 Bt, KDCHH 621 BG II, KDCHB 407 Bt
Vikram Seeds	Vikram Seeds
Tulasi Seeds	Tulasi 117 Bt, Tulasi 4 Bt
Pravardhan Seeds	Rudra Bt
Nandi Seeds	NSPL 405 Bt, NSPL 36 Bt, NSPL 999 Bt
Ankur Seeds	Jai Bt, Akka Bt
Vikki Agro Tech	VCH 111 Bt
Bioseeds Research India Pvt. Ltd.	BCHH340 Bt, BCHB 6188 Bt
Amar Biotech	ABCH 1165 Bt, ABCH 1220 Bt
Zuari Seeds	ZCH 50064 (Dhruv Bt)
Vibha Seeds	Sigma Bt, Dyna Bt, Ole Bt
Bayer Bioscience	SP 504 B1
Kaveri seeds	KCH 135 Bt, KCH 707 Bt

Table 11. Cotton varieties/hybrids in Tamil Nadu

Variety	Year of release	Species	Duration (days)	Yield (kg/ha)	Mean fibre length (mm)	Ginning Out turn (%)	Highest Standard Count	Special attributes
I Winter irrigated								
MCU 5	1968	G. *hirsutum*	165-170	1900	29.0	34.0	70s	versatile and highly adaptable variety
MCU 9	1978	G. *hirsutum*	160-165	2010	29.0	36.0	70s	higher ginner and good combiner
MCU 11	1988	*G.hirsutum*	150-155	2200	27.7	34.6	50s to 60s	
MCU 12	1998	G. *hirsutum*	150-155	1574	28.2	34.8	60s	high yielding and pest tolerant
TCHB 213	1990	Interspecific hybrid between G. *hirsutum x* G. *barbadense*	165-175	2500	32.8	32.0	60s	high yielding with superior quality and better blending properties
II Winter rainfed								
MCU 10	1982	*G.hirsutum*	150-160	746	25.0	37.0	40s	drought tolerant
K 10	1985	G. *arboreum*	140-145	726	23.9	38.0	30s	drought tolerant
K 11	1993	G. *arboreum*	130-135	1100-1200	23.6	34.7	30s	drought tolerant
KC 2	1998	G. *hirsutum*	150	772	24.4	37.5	40s	good ginning out turn and jassid resistance
Paiyur-1	1991	G. *hirsutum*	145-150	1173	26.3	37.1	40s	drought tolerant
III summer irrigated								
SVPR 2	1996	G. *hirsutum*	150-160	1658	24.1	36.4	40s	Multiple pest tolerant and high adaptability
SVPR 3	2000	G. *hirsutum*	135-140	1300	26.2	35.2	40s	tolerant to stem weevil and leaf hopper and alternaria leaf spot.
IV Rice fallow								
SVPR 3	2000	G. *hirsutum*	135-140	1300	26.2	35.2	40s	tolerant to stem weevil and hopper and alternate leaf spot.

MCU 7	1972	G. *hirsutum*	130	1200	25.0	33.2	40s	A versatile, short duration variety suitable for rice fallows
ADT 1	1992	G. *hirsutum*	120-125	1263	23.6	33.7	40s	short duration variety suitable for rice fallows

Growth stages: The length of the total growing period is about 150 to 190 days. The growth stages of cotton are presented in Table 12. Cotton grows from seed. The seed/embryo then begins to swell as it absorbs moisture. Under favourable conditions, the radicle (root tip) emerges within 2-3 days from the seed and newly germinated seedlings emerge above the soil 5-6 days after emergence of the radicle. The root may reach a depth of 15 cm or more in soil when the time the cotyledons unfold. The roots had reached a depth of 1 to 1.5 m when the young plant is 20 to 25 cm height while the final depth at maturity is 3 m. If plant grows in wet soil or soil with a compacted zone near the surface, most of the roots will be concentrated within the top 15 to 20 cm of soil. The first cotton leaf appears 10-12 days after emergence and leaf development reaches its peak about three weeks after the first buds are formed. The first square (flower-bud) appears on the lowest fruiting branch 35-45 days after emergence of the seedling, depending upon prevailing temperatures. The time taken between the appearance of first flower bud and opening of the flower may be between 25-30 days. The flower then opens at dawn and withers before the evening of the same day. The interval between successive flowers in the spiral is approximately 3 days. The interval between flowers on the same fruiting branch is approximately 6 days. Self-pollination normally takes place. Cross-fertilization takes place owing to the visits of insects to the flowers. Cross pollination by air movement is little. During first 25 days after fertilization the fibres attain full length after which they start to increase in thickness by deposition of cellulose on the inside of the primary wall. Complete maturity of fibre take places in 45 to 50 days. The ovary, or young boll, develops rapidly and reaches full size in about 21 days in upland varieties. An additional 20 to 50 days elapse before the boll is mature and ready to open. The time from flowering to open boll ranges from 40 to 70 days. Usually from 110 to 200 bolls are required to make a kilogram of seed cotton. The hundred seed weight fluctuates between 6 and 17 g/100 seeds.

Table 12 Growth stages in Cotton

S.No.	Stage	Days after sowing (DAS)
1	Emergence	2 to 3 (Good seeds) 4 to 6 (Weak seeds)
2	Seedling stage	6 to 20
3	Seedling to early square – vegetative stage	21 to 50
4	Squaring – 4 true leaves stage	40
5	Early square to flower opening stage	51 to 75
6	First bloom stage	65
7	Peak bloom stage	90
8	Early boll filling stage	100
9	First open boll stage	110

10	20% open boll stage	115
11	50% open boll stage	130
12	95% mature stage	140
13	Harvest stage	160

Land preparation: Prepare the field to get fine tilth to have a good seed bed. Chisel the soils having hard pan formation with chisel plough at 50 cm interval once in three years. The field should be ploughed in deep and keeping the stubbles intact and with big clods. Ridges and furrows are formed with required spacing depending on variety.

Cropping systems: The cropping systems adopted in cotton growing states are presented in Table 13.

Table 13. Cotton based cropping systems in India

State	Cropping System
Punjab, Haryana and Rajasthan	Cotton-wheat, cotton-mustard, cotton-berseem
Madhya Pradesh, Maharashtra and Gujarat	Cotton (monocrop), cotton-jowar (two year rotation), cotton-wheat, cotton intercropped with blackgram, greengram, soyabean, groundnut and redgram
Tamil Nadu	Cotton (monocrop), rice-cotton, rice-rice-cotton, cotton-jowar, cotton-pulse-jowar, cotton intercropped with onion, blackgram, groundnut
Andhra Pradesh	Cotton (monocrop), cotton-rice (1 year), cotton-chilli or cotton-tobacco (two year rotation)
Karnataka	Cotton (monocrop), cotton-wheat, cotton intercropped with chilli, blackgram, greengram, groundnut

Seeds and sowing: The seed rate and spacing for cotton are presented in Table 14 and Table 15.

Table 14 Seed rate and spacing for cotton varieties and hybrids

Varieties/ Hybrids	Quantity of seeds (kg/ha)			Spacing for cotton (cm)	
	Seeds with fuzz	Delinted seeds	Naked seeds	Pure crop	Intercrop
MCU5, MCU 5 VT, MCU 9, MCU 7, MCU 11	15.0	7.5	---	75 × 30	(60+90) × 30
Suvin	---	---	6.0	90 × 45	(80+100) × 45
Jayalakshmi, HB 224, TCHB 213	3.75	2.50	---	120 × 60	(100+140) × 60
SVPR I	15	7.5	---	60 × 30	---
MCU 7, ADT 1	18	9.0	---	60 × 30	---
LRA 5166	25	15	---	45 × 15	---

Table 15 Seed rate and spacings of cotton species in different zones of India

Species	Growing conditions	Cotton zone	Seed rate (kg/ ha)	Spacing (cm)
G.hirsutum	Irrigated	Northern Southern Central Southern	20–22 10–15 18–20 18–20	75 × 15 75 × 30 75 × 45 60 × 30 60 × 30
G.arboreum	Irrigated	Northern Central	10–12 10–12	60 × 30 60 × 30
G.herbaceum	Rain fed	Central Southern	12–15 12–15	45 × 30 60 × 30
G.barbadense	Irrigated	Southern	8–10 12–15	90 × 30 75 × 30
Hybrids	Irrigated	Southern	2–3	90 × 30 45 × 60 90 × 60
Hybrids	Irrigated	Central	2–3.5	120 × 40 120 × 60
Hybrids	Irrigated	Northern Central	3–3.5 3–3.5	67.5 × 67.5 150 × 60
Hybrids	Rain fed	Southern	2.5–3	120 × 60

Seed treatment of cotton seeds

(a) Acid delinting of seeds: Choose either plastic bucket or enamel bucket for acid delinting of seeds. Do not use earthen wares, metal vessels, porcelain wares, or wooden drums for acid delinting as conc. H_2SO_4 will corrode them. Put the required quantity of seeds in the container and add commercial conc. H_2SO_4 @ 100 ml per kg of fuzzy seed. Stir vigorously and continuously with a wooden stick for 2 to 3 minutes till the fuzz sticking to the seeds is completely digested and the seed coat attains a dark brown colour of coffee powder. Add water to fill the container. Drain the acid water and repeat the washing 4 or 5 times to remove any trace of acid. Remove the floating; ill filled, damaged, diseased and insect attached seeds white retaining the healthy and good seeds which remain at the bottom. Acid delinting in cotton eliminates some externally seed borne pathogenic organisms; kills eggs, larva and pupae of pink bollworm; helps to remove immature, ill filled, damaged seeds; makes seed dressing more effective and facilitates easy sowing.

(b) Local method: Red earth and cow dung at the ratio of 2:1 is mixed to form a slurry. Treat the fuzy cotton seeds in the slurry and shade dry it before sowing.

(c) Seed treatment with fungicides and bio-fertilisers: Treat the seeds with carbendazim 2 g or carboxin 2 g or captan 2 g or thiram 2 g/kg of seed. Then treat 15 to 20 kg of seeds with 3 pockets (600 g) of azospirillum innoculant along with 250 ml of jaggery solution or rice gruel. Mix this with delinted seeds thoroughly and dry it under shade. Ten packets of azosprillum is mixed with 25 kg each of soil or farm yard manure and is applied to the soil at sowing over an area of one hectare.

(d) Film coating of cotton seeds with polymers: Seed deterioration can be prevented up to 18 months of storage and can retain the viability of 77% when pre cleaned seeds are coated with seed polymer polykote @ 3ml/kg of seed diluted with 5 ml water combined with carbendazim (Bavistin) @ 2 g /kg and the seeds are stored in cloth bag under ambient condition. Viability can also be retained to 76% by coating the seeds with polykote @ 3ml/kg of seed diluted with 5 ml water combined with carbendazim @ 2 g /kg and Imidacloprid @ 7g /kg when seeds are stored in polythene bags.

Cotton nursery: With the introduction of hybrid cotton for large scale cultivation, attempts are made to transplant the cotton through seedlings so that less seed required and at the same time better establishment is maintained. The tap root system of cotton is the bottleneck in transplanting. To obviate this difficulty, attempts were made to arrest the tap root growth temporarily and to induce the fibrous root system in the seedlings stage by providing an impervious layer below the rootings medium and the following practices are formulated and recommended for adoption. Input requirement for producing seedlings in *pai* nursery to plant one hectare of main field is given in Table 16.

Table 16 Input required for variety and hybrid to produce seedlings

Inputs required	Variety	Hybrid
nursery area (m^2/ha)	25	12.5
polythene sheet (m^2)	30	15
gunny bags (m^2)	30	15
FYM: sand 1:1 mixture (kg)	125	100
Acid delinted fuzz seeds	9 kg/ha	2 kg/ha
	350 gm/m^2	160 gm/m^2
Diammonium phosphate	1.25 kg/ha	0.625 kg/ha
	50 gm/m^2	50 gm/m^2
Furadon granules	125 gm/ha	62.5 gm/ha
	5 gm/m^2	5 gm/m^2
age of seedling (days)	30	35

Form raised bed in a leveled ground under direct sunlight area. Spread polythene sheet over the raised bed on which jute gunny bits spread. Over the gunny bits spread, form small bund (5 cm × 3 cm) all around so as to give protection to the medium. Spread the medium (FYM: sand @ 1:1) to a thickness of 4 cm. Apply diammonium phosphate at 50 g/m^2 and mix it. Apply carbofuraon at 5 g/m^2 by uniformly spreading it over the area. To avoid damping off disease, apply phytolon at 5 g/m^2 by mixing it with the medium. Dibble the seeds (delinted) @ 250-300 g/m^2 for varieties and 150-160 g/m^2 for hybrids. Compact the seed in the medium by giving slight press with the hand. Cover the exposed seeds with thin spread of the medium. Cover the nursery with straw mulch. Water the nursery on the third day with rosecan. Remove the mulch next day. Irrigate the nursery with the rosecan twice a day, sufficient enough to keep the medium moist. The seedlings can be transplanted within 24-40 days. Plant one seedling per hill. The advantage of raising pai nursery seedlings helps in reduction of field duration, avoidance of late sowing, reduction in water use, plant protection and increased yield.

Direct sowing: In hybrids, two fuzzy seeds and one delinted seed per hole and in varieties three fuzzy seeds and two delinted seeds per hole have to be sown. Dibble the seeds at a depth of 3 to 4 cm and cover with soil.

Gap filling: To take up gap filling, dibble 3-4 seeds of cotton varieties in each gap and pot water on 10 DAS. For hybrid cotton, raise seedlings in polythene bags of size 15 × 10 cm. Fill the bags with FYM and soil in the ratio of 1:3. Dibble one seed per bag on the same day when sowing is taken up in the main field. Pot water and maintain. On 10th day of sowing, plant the seedlings maintained in the polythene bags, one in each of the gaps in the field by cutting open the polythene bag and planting the seedling along with the soil intact and then pot water.

After cultivation: Thinning of seedlings is carried out on 15 DAS or when plants are about 15 cm tall. Allow only one seedling per hole in case of fertile soil whereas two seedlings per hole in poor fertile soil. Rectification of ridges and furrows are carried out on 40 to 45 DAS after first top dressing of fertilisers in such a way that the plants are on the top of ridges and well supported by soil.

Topping: Topping or nipping in cotton is done to intermediate types/varieties. Topping breaks apical dominance. Topping is done to prevent lodging, reduce boll rot, increase flowering and fruiting, branches and thereby increases boll production. Topping is done between 75 and 90 days. This corresponds to square to flowering stage (75th day) or flowering to half boll stage (90th day). The terminal bud of the main stem, 1-2 inches from the top is pinched off by hand or clipped with a pair of scissors. 3 or 4 women can cover an acre per day. For MCU5, MCU9 and MCU11, rip the terminal portion of the main stem beyond the 15th node (70 to 80 DAS) and for Suvin, Jayalakshmi, TCHB 213 beyond 20th node (90 DAS).

Weed management: Pre-emergence application of Fluchloralin @ 2.0 lit/ha, pendimethalin @ 3.3 lit/ha, alachlor 50% EC @ 4-5 lit/ha or alachlor 10% GR @ 20-25 kg/ha or diuron 80% WP @ 1-2.2 kg/ha followed by one hand weeding on 35 to 40 days after sowing can be done. Use a hand operated sprayer with a deflecting or jar type nozzle. Sufficient moisture should be present in the soil at the time of herbicide application or irrigate immediately after herbicide application. If herbicide is not applied at the time of sowing, hand weeding on 18 to 20 and 35 to 40 days after sowing can be done. At post-emergence stage (15 to 30 DAS) paraquat dichloride 24% SL @ 1.25-2.0 lit/ha may be applied as direct spray and give good control of weeds in later stages. Pyrithiobac sodium 10% EC @ 625-750 ml/ha or quizalofop-ethyl 5% EC @ 1000 ml/ha or glufosinate ammonium 13.5% SL (15% w/v) @ 2.5-3.0 l/ha or fenoxaprop-p-ethyl 9.3% w/w EC (9% w/v) @ 750ml/ha (20 -25 DAS) can also be opted

Manures and fertilizer application: Apply 12.5 tonnes of FYM or compost per ha uniformly before last ploughing. Fertilizers are applied 5 cm to the side and 5 cm below seed. Band placement of P is found superior to broadcast method due to less contact with soil which results in less P fixation. Normally NPK should be applied to cotton in a ratio of 2:1:1 or 3:1:1. Potassium is to be applied in the proportion of 1/2 to 1/3 dose of nitrogen. Magnesium deficiency may occur with excess of calcium or in acid soils, leading to reddening of leaves and reduced photosynthetic activity. Nitrogen deficiency shows a pallid or yellowish green colour in the leaves which first appears on older leaves and reduced size of young leaves compared with a vivid green where nitrogen is plentiful. Severe nitrogen deficiency causes reddening of the leaves. Phosphorous deficiency shows a bronzing of the margin of older leaves. Phosphorous deficiency reflects in a dark green colouration of the foliage. Acute potassium hunger is expressed in cotton as rust or browning. The fertilizer requirements for cotton crop yielding 336 kg of fibre per ha are 150-96-161 kg of N, P_2O_5 and K_2O per ha. Sulphur @ 10 to 15 kg/ha is adequate for cotton. Use of super phosphate and NH_4SO_4 fertilizers avoids sulphur deficiency. Boron deficiency is common in alkaline soils. It induces shedding of flowers and young bolls, reduced branch tip growth and shortens internodes, giving bushy appearance. Borax @ 5 kg/ha can be recommended to correct the deficiency. Liming materials will be needed when the pH value is below 6. Two-thirds of the nutrients are taken up during the first 60 days of the growing period. Nitrogen should be readily available at the start of the growing season. Nitrogen is applied in three splits: first after sowing, second at thinning of the plants and second at the start of flowering. The demand for N at flowering and boll formation is very high. Deficiency of N at this stage results in bud and boll shedding. Therefore, application of N through 2 to 3 foliar sprays of urea or DAP @ 2% at flowering 15 to 20 days later proves beneficial. Hybrids show better response to nitrogen than varieties. It is necessary to apply the whole of the phosphorus and potassium as basal dressing before the crop is sown, either before ploughing or between ploughing and the surface cultivations. The particulars of fertilizer doses and method of application for cotton in different zones of India are furnished in Table 17 to Table 20.

Table 17 Fertilizer recommendations for cotton in different zones of India

Cultural Practices	Northern cotton zone	Central cotton zone	Southern cotton zone
Fertilizers	N - 60 - 100 kg / ha. P and K dose as per soil test. No P need be applied if previous wheat received / recommended P does 5.5 kg.Zn / ha. as ZnSo4 once in two cotton - wheat cycles	N : P : K 40 - 20 – 20; 50 - 25 – 25; 80 - 40 - 40 for varieties 100 - 50 – 50; 160 - 80 – 80; 240 - 120 - 120, for hybrids	N : P : K 40 - 20 – 20; 60 - 30 – 30; 90 - 45 - 45 for varieties 100 - 50 - 50, 120 - 60 - 60, 150 - 60 - 60 for hybrids
Method of fertilizer application	(i) Half N at thinning and remaining at first flowering; (ii) Half N at sowing time in late sown crop; (iii) Foliar application of N if needed; P to be drilled at sowing.	N in three splits at sowing, squaring and peak flowering stages. (P & K according to soil test). Application of 2 % urea or DAP at flowering and early boll development.	N at squaring and peak flowering, P & K at sowing. In Karnataka entire NPK at planting (rainfed cotton), Half N and entire P & K planting, remaining N at flowering (irrigated cotton) N in 4 splits in irrigated hybrid cotton.
Bio-fertilizer		Seed treatment with azotobacter is beneficial.	Seed and soil treatment with Azospirillium in Tamil Nadu.

Table 18 Fertigation schedule for hybrid cotton with WSF

100% dose @120:60:60 kg NPK/ha					Spacing = 150cm × 60cm	
Stage (days)	Fertilizer	Fertilizer grade			Dose kg/ha/ day	Total qty kg/ha
	Form	N	P	K		
0-30	MAP	12	61	0	2.00	60
	Urea	46	0	0	3.33	100

30-60	MKP	0	52	34	1.17	35
	Urea	46	0	0	3.50	105
60-75	PF	19	19	19	2.00	30
	Multi K	13	0	45	5.00	75
75-90	Multi K	13	0	45	1.67	25
	SOP	0	0	50	4.80	72

Table 19 Fertigation schedule for cotton with normal fertilizers

Stage (days)	**Fertilizer**	**Fertilizer grade**			**Dose (kg/ha/day)**	**Total qty (kg/ha)**
	Form	N	P	K		
10 to 30	Urea	46	0	0	0.75	15
	Potash	0	0	60	0.25	5
31-60	Urea	46	0	0	1.00	30
	Potash	0	0	60	0.33	10
61-90	Urea	46	0	0	2.33	70
	Potash	0	0	60	0.83	25
90-110	Urea	46	0	0	0.75	15
	Potash	0	0	60	0.50	10

Note: Entire P (30 kg/ha) may be applied as basal with super phosphate.
Dose @ 60:30:30 kg NPK/ha with spacing =75cm × 30cm

Table 20 Fertilizers suitable for fertigation

Fertilizers	**Grade**	**Solubility (g /lit)**
Conventional fertilizers		
Potassium sulphate (white)	0 – 0 – 50	110
Potassium chloride (red)	0 – 0 – 60	347
Urea	46 – 0 – 0	1100
Ammonium sulphate	21 – 0 – 0	760
Water soluble fertilizers		
Mono Ammonium Phosphate MAP)	12 – 61 – 0	282
Poly feed (PF)	19 – 19 – 19	-
Mono Potassium Phosphate (MKP)	0 – 52 – 34	230
Potassium Nitrate (Multi K)	13 – 0 – 46	316
Sulphate of Potash	0 – 0 – 50	-
Ortho Phosphoric Acid	0 – 52 – 0	47

Water management: Cotton requires 6000 to 9000 m^3 of water per ha for its growth cycle to take care of ET, exclusive of 25 to 30% water to meet water losses like conveyance, drainage, etc,. Seasonal water requirement of cotton is 650 mm in heavy soil and 700 mm in light soil. In cotton, 60% of water is extracted from top 60 cm of soil and 75% from top 75 cm soil. Small amounts are used from 120 to 180 cm of soil. In Tamil Nadu, the winter sown cotton crop gets North East Monsoon rains and is usually given supplementary irrigation, once in 15 to 20 days depending upon the climate, soil characters and crop growth. The summer irrigated crop gets frequent irrigations at intervals of 8 to 12 days while the short duration crop raised under rice follow receives only 4 to 5 irrigations. Flowering initiation stage, peak flowering stage and boll development stage are the three critical periods for moisture stress. In relation to reference evapotranspiration (ETo), the crop coefficient (kc) for the different development stages is: for the initial stage 0.4 -0.5 (20 to 30 days), the development stage 0.7-0.8 (40 to 50 days), the mid-season stage 1.05-1.25 (50 to 60 days), the late-season stage 0.8-0.9 (40 to 55 days), and at harvest 0.65-0.7.

Physiology of cotton

Red leaf in cotton

(i) Symptoms: Reddening of leaves in cotton is also referred as red leaf disease, red blight and yellow-red disease. Varieties of *G.hirsutum* and *G.barbedense* as well as hybrids manifest reddening of leaves. The development of deep pink to red colour in leaves commences when the crop age is about 50 to 60 days. It is spread by jassids. The reddening leaves have low chlorophyll and high anthocyanin. Leaf reddening is initially seen in mature leaves and gradually spreads throughout the canopy. Reddening begin with the leaf margins turn yellow, red colour is developed on the fringes of the leaves or patches or intervascular portions. Later red pigmentation is formed over the whole leaf area. The affected leaves start drying from the edges and ultimately prematurely shed. A change in colour from green to red may also occur without yellowing. Red leaf generally appeared during flowering or early boll filling stage of growth and arrested further development of bolls, which cracked prematurely. As the red leaf affected crops ceased to grow further, reduction in yield occurs. If reddening occurs at early stages (50 to 60 days) of growth and not corrected immediately, there will be severe loss of yield due to stunted growth and cracking of bolls. Reddening at later stage depending on its duration will not affect the yield. Reddening at later stage proceeds from bottom, largely due to senescence and may be sometimes desirable. Leaves at this stage turns red and drop off, thus providing good aeration and light is needed for proper boll burst.

(ii) Causes

(i) The causes of reddening in leaves are due to nitrogen deficiency, night temperature around 16°C, high wind velocity, water logging, water stress and magnesium deficiency. Even if the Mg in soil is high, its uptake is interfered with high soil K and Ca. Generally 1.5-2.0% nitrogen is considered as the critical level. Low Nitrogen (N) level in the leaves could be due to low soil N availability, impaired nitrogen uptake (water logging/moisture stress), diversion of leaf N to the developing bolls or synchronized boll development- high boll N demand. Nitrogen deficiency in leaf occurs if leaf nitrogen is < 2-2.5%N.

(ii) Deficiency of P and K also hastens leaf reddening.

(iii) Low night temperature: when night temperature falls below 15°C, it stimulates the formation of anthocyanin pigment in the leaf and the appearance of red colour of the foliage.

(iv) Deficiency of micronutrients particularly Zn.

(v) Severe leaf hopper infestation.

(vi) High wind velocity leading to desiccation injury. Hot desiccating wind during the fruiting period leads to rapid maturation of the crop. This leads to rapid depletion of N and photo- assimilates from the leaves.

(vii) Moisture stress or low moisture level in leaf tissue brings in adverse chemical reactions leading to degradation of chlorophyll and formation of anthocyanin pigment in the leaf.

(viii) High water table and soil compaction causing low oxygen in the rhizosphere.

(iii) Leaf reddening management

(i) Timely correction of N status either by optimum supply in the soil or through foliar application (DAP 2 % or Urea 1-2 %) during boll development stage

(ii) Preventing water logging, since this result is non-availability of magnesium and other nutrients

(iii) Providing protective irrigation to avoid stress and maintain relative water content of the leaf above 55-60%.

(iv) Soil application of 20 to 25 kg $MgSO_4$ per ha to control reddening or Spray $MgSO_4$ @ 5% + urea @ 1% twice at 15 days interval (50 g of $MgSO_4$+10 g urea in 10 litres of water as soon as the reddening symptoms appear in leaf reduces this disorder

(v) A combination of $MgSO_4$ @ 5%, urea @ 1.0% and $ZnSO_4$ @ 0.1% as foliar sprays on 50 and 80 DAS correct and prevents reddening of leaf.

(vi) Leaf hopper management with recommended pesticides

(vii) Foliar application of urea (1-2%) with 15-20 ppm chlormequat chloride and 0.10 % citric acid, 2-3 times at weekly intervals

(viii) Spray ascorbic acid (500 ppm) + 10 ppm Phenyl Mercuric Acetate (PMA) (AA increases leaf respiration and leaf N).

(ix) Soil moisture conservation and water harvesting/recycling to minimize soil moisture stress during boll formation

Para-wilt/New wilt

(i) Symptoms

(i) Leaves show wilt like drooping, became chlorotic and turned bronze or red followed by drying and premature abscission of leaves and fruiting parts.

(ii) Squares and young bolls are shed and immature bolls are forcefully opened.

(iii) Some of the wilted plants gradually recover and produce new flushes; however their contribution to yield is negligible.

(iv) Plants at grand growth phase *i.e.* squaring, flowering and early boll development are more sensitive to wilt.

(v) Para-wilt is sporadic (random) in distribution.

(ii) Causes

(i) Environmental conditions like high temperature, bright sunlight followed by heavy rainfall are found to favour the occurrence of wilt.

(ii) Plants with large canopy and heavy boll load are more prone to wilting.

(iii) Prolonged dry spell with high temperature and sunlight followed by soil saturation due to heavy downpour or irrigation

(iv) Wilt incidence is high in heavy clayey and deep soils

(v) Incidence is more in ill drained soils as compared to well drained soils

(iii) Para-wilt management

(i) Planting of wilt tolerant genotypes: Wilting is not seen in *G.arboreum* and *G.herbaceum* genotypes. Hybrids like JKHY 1, DCH 32, NHH 44 are wilt sensitive. Varieties such as LRA 5166, LRK 516 (Anjali), SRT1, MCU 5 VT, AKH 4, G 27 and Jayadhar are relatively tolerant to wilt

(ii) Provision of adequate drainage to avoid waterlogging of the fields to maintain adequate oxygen content of the soil.

(iii) Irrigation if available may be provided during grand growth phase to avoid prolonged exposure of plants to dry condition

(iv) Excessive use of farm yard manure and fertilizers may be avoided in heavy soils

(v) Cobalt chloride spray at 10 ppm within 24-48 hours of symptom appearance

Boll abcission in cotton/ bud and boll shedding in cotton: Premature shedding of bud , flower and boll is a serious problem in cotton. In all cases shedding is caused by an abscission layer one cell thick near base of peduncle. Most of the cotton varieties produce 100 to 150 squares (floral buds) whereas only 22 to 27% of these develop into harvestable bolls. A general shedding of 60% squares, 8% flowers and 5% bolls is common. Flowers mostly produced during vegetative phase or before the commencement of reproductive phase exhibit maximum shedding. The causes for bud, flower and boll shedding in cotton are due to copious water supply, prolonged drought and moisture stress; inadequate nutrient supply; poor drainage or flooding or inadequate moisture during and after flowering; high humidity during reproductive phase; zinc and boron deficiency; high temperature; heavy rainfall during reproductive phase causing boll rot; insect, disease and nematode attack of cotton roots; partial sterility of pollen or un successful pollination; cloudy weather and shading affect the availability of sunlight to the cotton crop. Two foliar sprays of Planofix (NAA) at 40 ppm (40 mg NAA in one litres of water) at one-month interval at square and peak flowering stages prevent the shedding of squares and flowers. Spray of Gibberllins @ 100 ppm at the opening of flowers and young bolls helps in retention of bolls with plants (GA 61 g per ha). Spray of CCC @ 40 ppm twice in cotton is recommended.

Defoliation in cotton: Defoliation in cotton is done for easy mechanical harvesting. Defoliation is done just prior to harvest, which helps in good aeration and sunlight, which are important for normal boll opening. Admixture of leaves, dusts and trash and staining can be reduced at harvest. Thus quality of cotton is good. Gramaxone @ 2.5 lit per ha can be recommended for defoliation in cotton.

Herbicide injury in cotton: Cotton is sensitive to 2, 4-D herbicide (2, 4 - dichlorophenoxy acetic acid). Characteristic 2, 4 -D injury is the leaf growth of bracts and finger like expansion of young leaves. Once 2, 4 -D injury sets in, there is no chance of recovery. Water spray can be given to minimize the effect of 2, 4-D. Urea spray (1 %) is advisable for quick production of new flushes. Foliar application of calcium carbonate (1.5%) or Gibberellic acid (50 ppm) led to the recovery of plants from the effects of low concentration of 2, 4-D.

Plant protection: Cotton is attacked by a number of insect pests such as jassids, aphids, thrips, whiteflies and bollworms (Pink, Spotted and American) and Spodoptera. The cotton pest management particulars are furnished in Table 21 to Table 26.

Table 21 Insect pests of cotton

S.No.	Insect pests	Scientific Name	Crop age (days)
	Sucking pests		
1	Jassids	*Amarasca bigutulla*	1-50
2	Aphids	*Aphis gossypii*	1-50
3	Thrips	*Thrips tabaci*	1-60
4	White fly	*Bemesia tabaci*	35-110
	Boll feeders		
5	Red cotton bug	*Dysdercus cingulatus* *Dysdercus koengnii*	50-110
6	Dusky cotton bug	*Oxycarenus laetus*	45-110
7	Spotted boll worm	*Earias vitella* *Earias insulana*	35-110
8	Pink boll worm	*Pectinophora gossypiella*	65-100
9	American boll worm	*Helicoverpa armigera*	65-110

Table 22 Economic threshold level for chemical application in cotton

Insect Pests	Economic threshold level (ETL)
Jassids	2.0 nymphs/leaf
Aphids	10 aphids/leaf
Thrips	10 thirps/leaf
Whitefly	5 whiteflies/leaf
American bollworm	1.0 larvae/plant
Spotted bollworm	5.0% damage
Pink bollworm	10 % damage

Table 23 Insect pests tolerant cotton varieties

Biotic Stress	Variety	Zone
White fly	Abadhita, LK 861, Kanchana, Supriya	South zone
Boll worm	Abadhita	South zone
Cotton leaf curl virus disease	LHH 144, CSHH 198,RS 810, RS 875, RS 2013, F 1861, H1117	North zone
Fusarium wilt	G Cot 13, Eknath, Rohini	Central zone
Bacterial Blight	Arogya	Central zone
Verticillium wilt	MCU 5 VT, Surabhi	South zone

Source: AICCIP

Seed treatment with Imidacloprid (Gauho) @ 7.5 g or Carbosulfan 50 g / kg of seed helps in delaying the first spray for hybrids in protecting the crop against sucking pests such as jassids up to 40-60 days. Spray Acetamiprid 20 SP @ 15 g a.i./ha, Confidor 200 SL, Thiomethoxam 25 g a.i./ha or Methly demeton @ 2 MVI are used for control of sucking pests namely aphids and jassids. For the management of whiteflies, Thimethoxam 25 WG at higher dose of 50 g.a.i./ha, neem product @ 2.5 l/ha) or neern seed kernel extracts (5%) may be sprayed. In case of severe infestation, Triaophos @, 600-800 ml/ha should be sprayed. During flowering period, Endosulfan, chlorpyriphos or Quinalphos @ 2.5 l/ha, Spinosad 48 EC @ 50-75 g a.i/ha, Indoxacarb 15 EC at 75 g a.i./ha or Decis tablet 25% @ 12.5 g a.i./ha should be sprayed for the management of bollworms. Insect growth regulator Rimon 10 EC (Novuluron) @ 100 g a.i./ha and Match (Lufenuron) 5% EC @ 60 g a.i./ha are used for control of bollworms in cotton. Safer chemicals like Imidacloprid and Rimon and Spinosad need to be incorporated in the IPM module to reduce environment pollution and health hazards.

Cotton suffers from a number of diseases. The important ones are bacterial blight, leaf spots caused by Alternaria and *Myrothecium*, grey mildew, boll rots, Fusarium wilt, *Verticillium* wilt, root rot and leaf curl virus. In the areas where the bacterial blight appears in severe form acid delinted seeds should be treated with Streptocycline (0.01%) before sowing. Later a foliar spray of a combination of Streptocycline (0.01%) and Copper oxychloride (0.25%) should be given when the disease starts appearing. Copper oxychloride is effective against fungal leaf spots too. Grey mildew is managed with the use of wettable sulphur (0.2%) or carbendazim (0.1%). For the management of boll rots, Copper oxychloride or carbendazim should be sprayed along with the insecticides. *Fusarium* and *Verticillium* wilts are best managed by growing recommended resistant varieties in the wilt affected areas. In an area where root rot occurs regularly, seeds should be treated with carbendazim @ 2-3 g/kg seed before sowing. When the root rot symptoms start appearing, drenching of affected plant as well as nearby plant with 0.1% carbendazim should be done. Intercropping with moth bean is also recommended in such area. Application of zinc sulphate @ 24 kg/ha also helps in reducing the diseases. Leal curl disease caused by Gemini virus has become prominent in the northern cotton zone. The virus is transmitted by whitefly. Therefore, the crop should be free from the whitefly attack. The insecticides, which control whitefly should be sprayed. Desi cottons are not affected by this disease. Some varieties like LRK-516, LRA-5166 , RS-875,LHH-144 are resistant to this disease. These varieties should invariably be grown in leaf curl prone areas.

Cotton leaf curl virus (CLCUV) disease caused by a Gemini virus and transmitted by whitefly has been a concern to farmers of north zone since last 4-5 years. Spray any of the following insecticides for the control of whitefly viz., Oxydemetan methyl (metasystox) 25 EC @ 400 ml, Dimethoate (Rogor) 30 EC @ 350 ml, Triazophos (Hostathion) 40 EC @ 600 ml, Ethion (Formite) 50 EC @ 800 ml or Acephate (Starthene) 75 SP @ 400 g per acre in 200 litre of water. Incidence of Grey mildew and Alternaria blight were severe during January 1st to February 11th where the maximum temperature ranged between 22.1. to 32.0°C, minimum temperature between 16.9 to 21.9°C and humidity ranged between 87 to 90% in the morning and 39 to 62% in the evening. Spray of carbendazim (0.1%) or Propiconazole (0.1%) or Prochloraz (550g ai/ha) to control of grey mildew. Apply FYM @ 12.5 t/ha seed treatment with *Trichoderma viride* MNT–7 mutant or application of neem cake @ 150 kg/ha + *T.viride* MNT-7 mutant to control of root rot.

Seed treatment with carbosulfan 25% SP at 2% of W/W and soil application of carbofuran 3G at (1 kg ai/ha) 30 kg/ha, 25 days after sowing, is found to give good control of this nematode. A bionematicide, namely 'Sincocin' containing cytokinins and fatty acids derived from plant extracts applied at 2.5 litres/ha in two split doses, the first at 15 DAS as foliar spray and the second at 45 DAS as soil drench has increased the yield of cotton by suppressing the multiplication of the reniform nematodes.

Table 24 Recommended chemicals to manage foliar diseases

Disease	**Recommended fungicide/bactericide**
Root rot	Carbendazim or vitavax as seed treament
Bacterial blight	Streptomycin sulphate (100 ppm) + Copper oxychloride (0.3%) as foliar spray twice at an interval of 10 days
Alternaria leaf spot	Dithane M 45 or copper oxychloride as foliar spray twice at an interval of 10 days
Grey mildew	Carbendazim @ 250 g/ha as foliar spray
Cotton is long lasting if well looked after *Myrothecium* leaf spot	Carbendazim/copper oxychloride as foliar spray

Source: AICCIP

Pest: Defender ratio (P: D ratio): The P: D ratio can vary depending on the feeding potential of natural enemy as well as the type of pest. The natural enemies of cotton pests can be divided into 3 categories 1. parasitoids; 2. predators; and 3. pathogens. The general rule to be adopted for management decisions relying on the P: D ratio is 2:1. However, some of the parasitoids and predators will be able to control more than 2 pests. Wherever specific P: D ratios are not found, it is safer to adopt the 2: 1, as P: D ratio. Whenever the P: D ratio is found to be favourable, there is no need for adoption of other management strategies. In cases where the P: D ratio is found to be unfavourable, the farmers can be advised to resort to inundative release of parasitoids/predators depending upon the type of pest. In addition to inundative release of parasitoids and predators, the usage of microbial biopesticides and biochemical biopesticides such as insect growth regulators, botanicals etc. can be relied upon before resorting to synthetic chemical pesticides.

Table 25 Predators/ Parasitoids feeding potential/ Egg laying capacity

Lady bird beetle	**Predatory rate of adult coccinellid on aphids is 50 aphids per day**
Green lacewing	Each grub can consume 100 aphids, 329 pupa of whitefly and 288 nymphs of jassids
Hover fly	1st instar larva can consume 15-19 aphids/day 2nd instar larva can consume 45-52 aphids/day 3rd instar larva can consume 80-90 aphids/day In total life cycle they can consume approx. 400 aphids
Spider	5 big larvae/day
Predatory mite	Predatory rate of adult is 20-35 phytophagous mites/female/day
Bracon hebetor	Egg laying capacity is 100-200 eggs/female. 1-8 eggs/larva
Trichogramma sp.	Egg laying capacity is 20-200 eggs/female

Source: Mohan *et al*, 2014

Rodent management practices: Application of 0.005% bromadiolone in ready to use form (wax blocks) or loose bait in packets near rodent burrow. Apply 2% zinc phosphide poison baits when the rodent infestation is very high. Practice pre-baiting incase of ZNP poison baiting. Don't apply ZNP poisons more than one time in a crop season.

Table 26 Stage wise IPM Practices for management of cotton pests

S. No.	Crop stage/pest	Stage-wise IPM Practices
1	Pre-sowing	Deep ploughing in summer for removal of weeds as well as towards destruction of insect stages Clean up of the fields free of weeds and alternate host plants including vegetable crops Adopt crop rotation with cereals (sorghum) or pulses (soybean) or green manure crops (sun hemp or dhaincha) at least once in two to three years
2	At sowing	
2.1	Soil & seed borne diseases	Select tolerant/resistant cultivars. Acid delinting treatment for seeds. Seed treatment with Thiram 75% WS @ 2.5-3.0 gm/kg seeds. Seed borne infection can be eliminated by soaking the seeds in 40 to 50 ppm solution of Streptomycin Sulphate 90% + Tetracycline Hydrochloride 10% SP for a period of two hours
2.2	Sucking pests	Timely sowing of sucking pest tolerant genotypes- Immediate to receipt of monsoon keeping the fields ready for sowing after the receipt of first rains, and taking up dry sowing. Growing refugia (for Bt cotton). Two border rows of non-Bt along with Bt cultivars. Seed treatment with insecticide Imidacloprid 48% FS or Imidacloprid 70% WS @ 500 to 1000 g per 100 kg seeds
3	Vegetative growth stage (20-50 days)	
3.1.	Sucking pests	Monitoring pest and natural enemy population on attractant/ trap & inter crops. Inoculative release of *Chrysoperla* grubs @ 10,000/ ha. Spray of neem based insecticides as initial sprays (Azadirachtin 0.03% (300 ppm) Neem Oil Based WSP @ 2.5-0.5 l/ha). Spray recommended insecticides when pest crosses ETL
	Whitefly	Fix yellow sticky traps for monitoring population
	Spotted & spiny bollworm	Crushing of larvae in the shoots mechanically
	Bollworms	Set up pheromone traps @ 5 traps/ha for monitoring
	Stem weevil	Soil application of carbofuran 3%CG @ 33300 g/ha
	Root rot & wilt	Remove & destroy root rot/wilt affected plants
4	Early fruiting stage (50-80 days)	
	Sucking pests	Release *Chrysoperla* @ 10,000 /ha
	Whitefly	Use yellow sticky traps for monitoring population Spray recommended insecticides
	Bollworms	Use pheromone traps and change lures Management of population in trap crops, release of *Trichogramma* @80,000/ha. Set up bird perches
	CLCuD Disease	Destroy affected plants
	Parawilt	Foliar application of 10ppm cobalt chloride on infected plants
5	Peak flowering & fruiting stage (80-120 days)	
	Whitefly	Use yellow sticky traps for monitoring population Spray recommended insecticides

	Bollworms	Use pheromone traps @ 5 traps/ha Physical collection & destruction of grown up larvae Use of *Ha*NPV 0.43% AS @ 2700 ml/ha Removal of terminals (topping) to be done at times of high oviposition by *Helicoverpa*
	Spodoptera	Use pheromone traps@ 5 traps/ha Sowing castor seeds at field borders serves as an indicator cum trap crop Hand collection & destruction of egg masses & early instar gregarious larvae Spray recommended insecticides
	Black arm disease	Spray recommended chemicals (Streptomycin Sulphate 90% + Tetracyline Hydrocloride 10% SP). Streptocycline 25-40 ppm to be sprayed thrice – Before flowering, after flowering and twenty days after second spray
	Leaf reddening	Foliar application of urea (1-2 %) with 15-20 ppm chlormequat chloride and 0.10 % citric acid, 2-3 times at weekly intervals.
	Parawilt	Foliar application of 10 ppm cobalt chloride on infected plants
6	Boll opening stage (120-150 days)	
	Whitefly	Use yellow sticky trap for monitoring population
	Bollworm complex	Need-based application of recommended insecticides Do not extend the crop period Use pheromone traps for monitoring of *Helicoverpa, Spodoptera* and pink bollworm Spray recommended insecticides keeping IRM strategies in focus
7	After last picking of cotton	
	Bollworms and mealybugs	Allow grazing by animals (cow, buffalo, sheep, goat, etc.) immediate to final picking. Avoid staking of the cotton stalks near the fields. Pulled out stalks should be burnt off *in situ* before ploughing the field. Shredding and incorporation of crop residues.

Harvest: Cotton is harvested by hand picking. Large bolls and bolls that flare back on ripening are easy to pick by hand. Uniform ripening makes picking more economical. A compact plant with bolls spaced along the main stem and set high off the ground is best suited for machine harvesting. Cotton should be picked clean and free from pieces of leaves, bolls, bracteoles and turgs. 3-4 pickings are usually made. Pick kapas from well burst bolls only. Harvest at less than 7 days interval between each harvest. Harvest in the morning hours up to 10 to 11 a.m. only. When there is moisture due to dew, the leaves and bracts, do not stick to the kapas and lower the market value. Daily picking of cotton is 30 to 60 kg per day per labour. Both hands should be used for picking and only two or three bolls picked at one time before being bed into the bag. As kapas is picked, sort out good puffy ones and keep separately. Keep strained, discoloured and insect attacked kapas separately.

Yield: The yield of a cotton plant is determined by the number of bolls, size of the bolls and percentage of lint. Every ton of cotton picked in the field yields about 600 kg of seed and 350 kg of cotton lint, but the value of the lint is about seven times that of the seed. A large seed set is desirable since the lint is produced on the surface of seed. The density of lint on the seed also affects the total lint production. Lint percentage is referred to as ginning percent or gin out turn. It is the weight of the lint expressed as a percentage of the weight of the seed cotton (seed and lint). Large seeded varieties normally have a low lint percent and small seeded varieties have high lint percent.

Utilization: Cotton seed contains 94% cellulose on dry weight basis. Cotton seed has 16-20% protein, 18-24% oil. Cotton seed contains 0.4 to 2.0% a poly phenolic poisonous compound located in the pigment gland. This substance adds undesirable flavour, colour, and physiologically detrimental property of foods unless removed. Glandless cotton seed can be used for flours, protein concentration and edible nuts prepared by roasting, toasting or frying.

Rice Fallow Cotton

In Tamil Nadu rice fallow cotton cultivation confined mainly to Thanjavur district. The rice fallow cotton crops occupy the land from January to February onwards which is followed by rice cultivation from June to July.

Land preparation: If the soil is in waxy conditions, there is no need for any preparatory cultivation. The seed can be dibbled directly. If the soil is dry and not in condition to take up sowing, let in water and then allow the soil to come to waxy condition. At the lower level of the field, dig a trench 15 cm wide and connect this trench to the outside channel to drain off the excess water.

Seeds and sowing: The particulars of varieties, seed rate, spacing and sowing are furnished in Table 27.

Table 27 Seeds and sowing rice fallow cotton

Particulars	Varieties		
	ADT 1	MCU 7	LRA 5166
Seed rate (kg/ha)			
i. Fuzzy seed	18	15	25
ii. Acid delinted	9	7.5	15
Spacing (cm)	60 × 30	60 × 30	45 × 15
Number of seeds per hole			
i. Fuzzy seeds	4	4	4
ii. Acid delinted seeds	2	2	2
Depth of sowing (cm)	3	3	3

Pre-treatment of acid delinted seeds is carried out with fungicides and later with azospirillum.

After cultivation: Gap filling is done on 10 DAS. Dibble 2 to 3 acid delinted seeds or 4 to 5 fuzzy seeds in the gap in the case of MCU 7 and ADT 1. Thinning of seedlings is carried out on the 20 DAS. Leave only one healthy and vigorous seedling per hill. Spray 40 ppm of NAA (40mg of NAA dissolved in one litre of water) at 40 to 45th day and thereafter 15 days of first spraying. Arrest the terminal growth by nipping the terminal 10 to 12th node for controlling excessive vegetative growth.

Weed management: Pre-emergence application of Fluchloralin 50EC @ 2.5 litres per ha or Pendimethalin @ 3.3 litres per ha ensures weed free condition for 40 to 45 days. This should be followed by one hand weeding and earthing up during 40 to 45 DAS. If herbicide is not applied, take up hoeing and weeding on 20 and 40 to 45 DAS.

Manures and fertilizer application: Application of fertilizers at the rate of 60-30-30 N, P_2O_5 and K_2O kg per ha is recommended. Apply half the dose of N and full dose of

P_2O_5 and K_2O on the 30th to 35th day of sowing along the rows of cotton plants. Form ridges and incorporate the fertilizer in the soil around the plants between the 30th and 35th day of sowing. Apply the remaining N on 50 to 55th day of sowing. In case of zinc deficient soils, application of $ZnSO_4$ at 50 kg per ha is recommended.

Water management: Water required for rice fallow cotton is 600 to 650 mm. Irrigate the crop depending on weather conditions and receipt of rains. Irrigation is given IW/CPE ratio of 0.75. The particulars of water management in rice fallow cotton are presented in Table 28.

Table 28 Water management in rice fallow cotton

Growth stage	Duration	Nos. of irrigation	
		Old delta (clay loam)	New delta (sandy loam)
Seedling stage	0 – 20	Residual soil moisture	Residual soil moisture
Vegetative stage (regulate irrigation)	21 – 50	1	2
Flowering stage (more frequent irrigation)	51 – 100	5	10
Maturity stage (control irrigation)	101 – 150	2	3

Stop irrigation from the 113rd day onwards. Skip furrow method of irrigation is recommended to prevent excessive irrigation.

Rainfed Cotton

Table 29 Seasons, varieties and region of Tamil Nadu

Seasons	Varieties	Region
G.hirsutum		
September to first week of October	MCU 10, LRA 5166	Thirumangalam, Sattur, Kovilpatti, Aruppukkottai
G.arboreum		
October to November	K 8, K 9, K 10	Dindugul, Madurai, Tirunelveli, Ramnad
	Paiyur 1	Dharmapuri, Salem

Preparatory cultivation: Soon after the harvest of standing crop in March to April, the field should be ploughed deeply keeping the stubbles intact and with big clods. This will enable to receive subsequent rains, reduce erosion, evaporation as well as control the weed population. The first ploughing should be across the slope and along the contour, which will ensure each furrow to serve as micro water shed. During July-August rains, the fields should be ploughed again and brought to fine tilth to have a good seed bed. Leveling is advisable to ensure uniform placement of seeds and subsequent establishment. Form broadbed cum furrow method of land preparation with 150 cm spacing to provide adequate drainage and keep the seed row area under friable conditions.

Table 30 Seeds and sowing

Varieties	Quantity of seeds			
	Fuzzy seeds	Delinted seeds	Intercrop	
			Blackgram	Cowpea
K 9, K 10	20	-	10	7.5
MCU 10, LRA 5166, Paiyur 1	20	15	10	7.5

Spacing for normal row planting is 45 × 15 cm while for paired row planting is (30 + 60) × 15 cm. One row of pulse with 30 × 10 cm spaced plants within row in 60 cm space between paired rows. APK 1 black gram is best suited for intercropping. Acid delinting of fuzzy cotton seeds has to be carried out. Pre-treatment of seeds with fungicides is carried out to control seed borne diseases. Seed hardening is done with Cycocel (CCC) @ 500 ppm for inducing drought tolerance. For this, 10 ml of Cycocel has to be dissolved in 20 litres of water and 15 kg of cotton seeds per ha is soaked in this solution for 12 hours. If Cycocel is not available KCl 2% can be used for this purpose (400 g of KCl in 20 litres of water). To fully utilize the monsoon rains taking of sowing to the second week of September under dry conditions anticipating the onset of monsoon. The soaking rains occur in the third week of September to the second week of October. The pre-monsoon sowings can be taken only in deep or medium deep black soils and not suited for light red soils. Two cotton seeds per hole and one pulse seeds per hole is sown at a depth of 5 cm for cotton and 3 cm for blackgram.

After cultivation: Gap filling is done by dibbling 3 to 4 seeds per hill on 10 days after sowing. Thinning of seedlings is done on 15 DAS allowing two seedlings per hill.

Weed management: Two hand weedings are taken up on 20 days and 35 to 40 days after sowing. Pre-emergence application of Fluchloralin @ 2.5 litres per ha or Pendimethalin @ 3.3 litres per ha or Thiobencarb @ 3.0 litres per ha followed by one hand weeding on 40 days after crop emergence to control weeds.

Manures and fertilizer application: Rainfed soils are not only thirsty but also hungry. Rainfed soils should be adequately manured. Apply 12.5 tonnes of FYM or compost or composted coir pith per ha uniformly on the unploughed soil. Apply 10 packets of Azospirillium innoculant plus 25 kg FYM plus 25 kg soil.

Varieties	Quantity of fertilizers (kg/ha)		
	N	P_2O_5	K_2O
K9, K10	20	0	0
MCU 10, LRA 5166, Paiyur 1	40	20	20

For cotton crop, 40 kg can be split into two i.e., first 20 kg of N is applied as basal and the remaining 20 kg N is top dressed 30 days after germination when adequate soil moisture is available. Apply 12.5 kg of micronutrients and 37.5 kg of sand uniformly over the furrows after sowing and cover the seeds. Do not incorporate in the soil. Enriched FYM @ 750 kg/ha can be applied (750 kg of FYM is mixed with 125 kg of super phosphate and cover it with mud. After one month mix it with required quantity of nitrogen of the crop that is grown in the field.

Quality Characters in Cotton

Immediately after picking, dry the kapas is shade. If it is not dried immediately, the colour will change which will lower the market value. Do not dry the kapas under direct sun as the fibre strength and buster will be lost. Spread a thin layer of dry sand in the ground and keep the kapas over it. The dry sand will absorb moisture and prevent it from counting in contact with the kapas as moisture will strain the kapas and lower its valve.

Seed cotton or kapas yield: Cotton along with seed removed from locks of the boll. Seed cotton yield = weight of lint + weight of seeds.

Grade: Grade is composed of three factors viz., colour, leaf and preparation. American cotton is graded in steps from ordinary through middling to fair while American pimal grades are numbered from 1-9 grade, 1 being the best. Colour and trash content can be measured in the laboratory, but there is no standard test for preparation, a general term covering faults introduced during storage and handling. Colour test involves two measurements, reflectance and yellowness obtained on the Nickerson-Hunter cotton colorimeter. The amount of trash is measured by Shirley analyzer.

Cotton fibres: Fibres are epidermal prolongation of seed coat cells. Cotton seeds are covered with one to two layers of unicellular convolated hairs. Lint arises from the epidermal layer and fuzz arises from the sub-epidermal layer. 'Lint' is the long outgrowths of epidermal cells while 'fuzz' is the short outgrowth of epidermal cells. In diploid cotton and upland cotton both lint and fuzz are present whereas in Sea Island cotton (*G. barbadense*) only lint is present and fuzz is absent, such seeds are called naked seeds. During first 25 days after fertilization, the fibres attain full length after which they start to increase in thickness by deposition of cellulose on the inside of the primary wall. Complete maturity of fibres takes places in 45 to 50 days. Lint is removable or separable by ginning. Fuzz fibre is short in nature. Fuzz is practically unseparable or not removable from the seed by ginning. Acid defuzzing is done to destroy pathogens before sowing. The desirable level of fibre properties for processing with spinning technologies are furnished as follows:

Table 31 Desirable level of fibre properties for processing of cotton

Fibre properties	Desired level
Fibre fineness	120 to 150 m tex
Micronaire value	3.0 to 3.8
Percent mature fibre	80 or more
Bundle strength	28 g/tex or higher
Elongation at break	7 % or more
Fibre length	25-28 mm for rotor and friction spinning. 30 mm and higher for air-jet spinning.
Trash content	Less than 2% with less level of microdust.

Ginning: A cotton gin is a machine that quickly and easily separates cotton fibers from their seeds. Ginning is the process of separating the cotton fibers from the cotton seeds. A badly set gin can lead to an apparent drop in ginning percentage by failing to remove all the lint from the seed. Drying and cleaning machinery in the ginnary reduces the ginning percentage by removing moisture, dust, trash and broken particles of lint during the various operations. Loss during ginning may range from 2 to over 7%. Ginning percentage (GP) or lint percentage (LP) or ginning out-turn (GoT) is the ratio between lint and seed and is expressed in percentage.

$$\text{Ginning percentage (GP)} = \frac{\text{Weight of lint}}{\text{Weight of seed cotton}} \times 100$$

GP is usually around one third. Ginning percentage also varies from season to season. GP depends primarily on seed weight and lint weight. Seed weight is determined by seed volume and specific gravity, while lint weight varies according to the number of fibres length, thickness and specific gravity of fibre walls. Smaller seeds raise the ginning out turn, but it does not necessary increase the lint production. Ginning percentage of *G. barbadense* is 28 to 30% and *G. hirsutum* is 34 to 42% and for desi cotton is 36 to 42% respectively.

Staple length or Fibre length or Halo length: Lint fibre length is generally measured on a sample of lint after ginning. Early measurement of lint can be done by combing out the lint in the form of holo or butterfly and measuring the length with the transplant protractor which has a hole at the centre to fit over the seed. A steel comb is used to trace out and straighten the fibres which are then pressed flation a black velvet covered board. The lint length is measured to the nearest mm. Staple length is divided into 5 classes as shown below:

Table 32 Staple length classification adopted in India

Description	CIRCOT, Mumbai, Range of 2.5% Span Length (mm)	Bureau of Indian Standards, Range of 2.5% Span Length (mm)
Short	≤ 20.0	≤ 20.0
Medium	20.5-24.5	20.5-25.5
Medium Long	25.0-27.0	26.0-27.5
Long	27.5-32.0	28.0-33.5
Extra Long	≥ 32.5	≥ 34.0

Source: Sundaram, 1976 and BIS, 1996.

Table 33 Staple length classification adopted by USDA

Description	Staple length mm
Short	≤ 21.0
Medium	22.0-25.0
Medium Long	26.0-28.0
Long	29.0-34.0
Extra Long	≥ 34.0

Source: Bradow and Davidonis, 2000.

Staple length is the most important among fibre properties. This character is determined in a number of ways viz., pulling method, professional estimate based on eye and hand judgement, combing the fibre with brass comb and measuring the combed fibre by celluloid disc scale known as Halodisc / Halo– butterfly, use of fibre sorting instruments like Ball' sorter or Baer's sorter, use of photoelectric instruments namely Fibro graph. Staple length or Holo length is the overall length of the lint without the fibre taken out of seed.

Fibre fineness: It is measured through the principle that the resistance to the flow of air through a bundle of fibre varies in proportion to the surface area of the fibres. Micronaire is a measure of airflow through a compressed specimen of fibre. Micronaire (fibronaire) reading is determined by an airflow instrument. It is a measure of maturity and cell wall development. It relates to fibre fineness and/or maturity. It is expressed in terms of weight per unit length of fibre i.e., 10^{-6} g/cm or millitex. Millitex is the weight in mg of a km length of fibre. Micronaire is equal to the average weight of the fibre in micrograms and the fineness in cotton is graded as:

Table 34 Fibre fineness

Description	**Micronaire (μ g)**
Very fine	< 3.0
Fine	3.0 – 3.9
Average	4.0 – 4.9
Coarse	5.0 – 5.9
Very coarse	> 6.0

Fibre maturity: The most reliable method of measuring maturity is that of counting mature fibres (N), immature fibres (I) and dead fibres (D) in a sample treated with 18% caustic soda. As a measure of maturity, maturity ratio (M) is calculated as follows:

M= {(% N-%D) / 200} + 0.70 where N = fully matured hairs and D = dead hairs

Fibre strength: The inherent strength of individual cotton fibre is an important factor in the strength of the thread spun from them and strong fibres usually give less trouble in processing. A sample of cotton lint is combed into a flat bundle of fibres and then gripped by a pair of clamps in a machine called 'Pressely tester' or 'Stelometer' to measure fibre strength. Weight is read on the scale which indicates the strength in lbs/inch^2 of fibre or kg/cm^2. The different grading system in fibre strength classes are given below.

Table 35 Fibre Strength

Description	**Medium staple**	**Extra long staple**
	(zero gauge 000' psi)	
Very low	70-75	93-96
Low	77-83	97-100
Average	84-90	101-104
High	91-97	105-108
Very high	98-104	109-112

Lint index: Lint index is the weight of lint from 100 seeds. Lint index represents the absolute weight of lint produced per seed and it more useful in breeding than ginning percentage.

$$\text{Lint \%} = \frac{\text{Lint index}}{\text{Lint index + seed index}}$$

$$\text{Lint index} = \frac{\text{Weight of 100 seeds}}{100 - \text{Ginning percentage (GP)}} \times \text{GP}$$

Seed index: It is a test weight or the weight of 100 seeds expressed in grams. The average seed index (weight of 100 seeds) of different cotton species are as follows:

G. arboreum = 4.8 to 5.0 g

G. herbaceum = 5.5 to 6.0 g

G. hirsutum = 8.0 to 11.0 g

G. barbadense = 9.0 to 11.0 g

High standard counts (HSC): HSC is expressed as hanks which is equal to 840 yards. 40 counts (40s) means a pound of cotton gives 40 hanks of 840 yards. The metric yarn counts is the number of km making up 1 kg of yarn. This metric counts is 1.693 times the English count.

Earliness is given by Bartletts index (BI)

$$BI = \frac{(P_1) + (P_1 + P_2) + ... (P_1 + P_2 ... P_n)}{n\,(P_1 + P_2 ... P_n)}$$

where $P_1, P_2 ....\ P_n$ are the weight of seed cotton collected in 1st, 2nd and nth picking.

Earliness (E) can also be given as

$$E = \frac{\text{Yield of first picking}}{\text{Yield of total picking}} \times 100$$

Fruiting efficiency or Fruiting index (FI)

$$FI = \frac{\text{Dry weight of open bolls}}{\text{Dry weight of above ground plant (biomass)}}$$

Cotton seed oil: Cotton seed contains hull and kernel. The hull produces fibre and linters. The kernel contains oil, protein, carbohydrate and other constituents such as vitamins, minerals, lecithin, sterols etc. Cotton seed oil is extracted from cottonseed kernel. Cotton seed oil, also termed as 'Heart Oil' is among the most unsaturated edible oils. In India, about 68 lakh tonnes of cottonseeds are available as byproduct of cotton and utilized for extraction of oil. Cotton seed is one of the important byproducts of cotton, the use of which has not been exploited to the fullest extent. The seed collected after ginning is used for extraction of oil. The ginned seed is covered with small fuzzy fibres known as linters. The small fuzzy fibres are removed by delinting machines. Thereafter, the seed coat (cortex) is removed by dehulling to get kernels, which contain mostly storage proteins, oil and a toxic pigment, gossypol. The approximate constituents of cottonseed belonging to different species are given in Table 1.

Table 36 Constituents of cotton seed of different species

Species	Whole seed			Kernel		
	Linter	Kernel	Hulls	Seed oil	Kernel oil	Protein
G. arboreum	5.9	52.0	43.0	19.6	32.5	33.8
G. herbaceum	4.3	53.0	45.0	19.1	31.3	34.6
G. hirsutum	10.5	55.0	35.0	21.3	33.7	35.5
G. barbadense	-----	61.0	39.0	21.8	30.8	36.6

Source: www.cicr.org.in

Cotton seed contains about 19-22% of oil depending on the species. The oil content in kernel varies from 30.8 to 33.7% depending on the species. Crude cottonseed oil which contains varying amounts of free gossypol and gossypol like pigments which are toxic to human-beings and have to be removed by refining to make it suitable for edible purposes. Refining is generally done by neutralization with caustic alkali, followed by bleaching and deodorization, etc. Refined cotton seed oil is one of the few oils which is in 'Ok food' list of American Heart Association (AHA). Cotton seed oil is cholesterol free. It contains 50% linoleic acid which is essential fatty acids. It does not allow blockening of coronary arteries by forming hard pellets of cholesterol. In view of above two qualities, it can be made compulsory to be used in the form of blend either with other oil or with hydrogenated vanaspati or it should be supplied in the form of encapsule. The keeping-quality of the oil is comparable to groundnut and safflower oils and its nutritional value is around 9 k cal/g. The average digestibility of cotton seed oil is 97% and could be compared with that of soybean, safflower and sunflower oils. Cottonseed oil with practically no gossypol is pale yellow in colour, rich in vitamin E, and can be used directly as a cooking medium and also for the manufacture of vanaspati, soap, etc.

Gossypol: Gossypol is the most important pigment present in the cotton seed and create enormous problem of seed processing and utilisation of cotton seed as by-product. Gossypol is located all over in plant. It gives undesirable colour to the oil and reacts with protein to reduce the nutritive value of cotton seed product. Ruminant animals can tolerate the Gossypol, but it is toxic to non-ruminants.

Table 37 Gossypol content in seed and kernel of different cotton species

Species	Seed (%)	Kernel (%)
G. arboreum	0.69 (0.30-1.25)	1.31 (0.65- 2.38)
G. herbaceum	0.77 (0.43-1.09)	1.44 (0.82-1.96)
G. hirsutum	0.77 (0.42-1.25)	1.39 (0.73-2.35)
G. barbadense	1.11 (0.73-1.49)	1.78 (1.22- 2.35)

Source: www.cicr.org.in

Naturally coloured cotton: Cotton with naturally coloured lint, other than white, is commonly referred as coloured cotton. In India, brown linted varieties of tree cotton (*G.arboreum* L.) namely Cocanada 1, Cocanada 2 and Red Northerns are under commercial cultivation mainly on black soils under rainfed condition in parts of Andhra Pradesh. Coloured linted varieties could not remain popular with growers, mainly because of low productivity per unit area, poor fibre characteristics and non-uniformity of colours. The awareness about the toxicity and pollution caused by synthetic dyes has revived the interest in cultivation of organic coloured cotton in view of its ecofriendly character. The lint colour of cotton under commercial cultivation is often white. In the cultivated species, brown and green colours are most common. Some of the genotypes in germplasm collection of USA and Russia are reported to have coloured lint with shades of pink, red, blue, green and also black. Lint colour is a genetically controlled character. Accumulation of pigments in the lumen of lint starts before boll bursting. In upland cotton (*G.hirsutum*), pigmentation starts appearing in the developing lint 32 days after fertilization and it takes nearly six days to develop colour. In Asiatic cotton (*G.arboreum*) colour pigments observed 46-47 days after fertilization which take 5-6 days for colour development. Naturally coloured cottons are low yield potential, poor fibre properties, limited colours, instability of colours, contamination of white cotton, low market demand, and lack of marketing facilities.

British standard yarn measures:

- 1 thread = 55 inches (about 137 cm)
- 1 skein or rap = 80 threads (120 yards or about 109 m)
- 1 hank = 7 skeins (840 yards or about 768 m)
- 1 spindle = 18 hanks (15,120 yards or about 13.826 km)

REFERENCES

Anupam Barik.2010. Cotton Statistics at a Glance. Directorate of Cotton Development, Ministry of Agriculture, Government of India, Mumbai and Director, National Centre for Integrated Pest Management, New Delhi. pp.148

Bhatt, J. G. 1996. Cotton Physiology. Indian Society for Cotton Improvement, Bombay.

BIS. 1996. Indian Standard Textiles: Grading of Cotton. ICS 59.060.10. Bureau of Indian Standards. New Delhi -110002

Bradow, J.M. and G. H. Davidonis. 2000. Quantitation of Fiber Quality and the Cotton Production-Processing Interface: A Physiologist's Perspective. The Journal of Cotton Science 4:34-64

Fryxell, P.A. 1984. Taxonomy and Germplasm resources in Cotton. Agronomy monograph ASA-CSSA-SSSA. pp 27-29. 677 South Segoe, Road Madison, WI 53711, USA.

Gulati, A.N. 1947. The cotton of Dacca Muslins, Textile Digest 8(4): 10-11

Meredith, W.R.J. and Bridge, R.R. 1973. Natural crossing in cotton (*Gossypium hirsutum* L.). Crop Science. 13: 551-552.

Mohan, S., D. Monga, Rishi Kumar, V. Nagrare, Nandini Gokte-Narkhedkar, S. Vennila, R. K. Tanwar, O. P. Sharma, Someshwar Bhagat, Meenu Agarwal, C. Chattopadhyay, Rakesh Kumar, Ajanta Birah, N. Amaresan, Amar Singh, S. N. Sushil, Ram Asre, K. S. Kapoor, P. Jeyakumar and K. Satyagopal. 2014. Integrated Pest Management Package for Cotton. National Centre for Integrated Pest Management, Faridabad, Haryana -121 001. p. 84

Narayanan, S.S. Singh, V.V. and Kothandaraman R. 1990. Cotton Genetic Resources in India; In: Cotton Scenario in India A Souvenir, pub: Central Institute for Cotton Research (ICAR), Nagpur. pp.9-20.

Santhanam, V. and Sundaram, V. 1997. Agri history of cotton India - An overview. Asian Agri-History. 1, 4: 235-251.

Sethi, B.L., 1960. History of cotton. In: Cotton in India. A Monograph. Vol. I. Indian Central Cotton Committee, Bombay, India. pp. 1-39.

Sundaram V. 1974. Antiquity of cotton, In: 50 Years of Research at Cotton Technological Research Laboratory; Indian Council of Agricultural Research, New Delhi, India, p. 212.

Sundaram, V. 1976. Revision of Staple Length Classification of Indian Cottons. CTRL Publication No. 69. Cotton Technological Research Laboratory, Matunga, Bombay - 400 019

Tripathi, K.K., Govila, O.P., Rajini Warrier and Vibha Ahuja. 2011.Biology of *Gossypium* spp. Department of Biotechnology, Ministry of Science & Technology and Ministry of Environment and Forests, Government of India. New Delhi. 41 p.

http://wwf.panda.org/what_we_do/footprint/agriculture/
http://bettercotton.org/resources/
https://www.icac.org/
http://www.fao.org/agriculture/crops/core-themes/theme/pests/ipm/en/
http://pan-international.org/resources/
http://www.indiantextilejournal.com/
http://ffymag.com/admin/issuepdf/Cotton-Oct09.pdf
http://www.pcotexport.com/cottonvarieties.html
http://cotcorp.gov.in/index.aspx
http://www.indiaonestop.com/cotton/cotton.htm
http://www.mahacot.com/cotton.html
http://texmin.nic.in/
http://www.cottoninc.com/corporate/Market-Data/MonthlyEconomicLetter/
http://dcd.dacnet.nic.in/NPT.asp?name=pop1.htm
http://www.cicr.org.in/pdf/TAP_for_cotton.pdf
http://agmarknet.nic.in/cotton_profile.pdf

CHAPTER

16

Jute-*Corchorus olitorius* L. (2n=14) *Corchorus capsularis* L. (2n=14)

Family: Tiliaceae

Vernacular name: Jute, Jews mallow (English), *Tossa pat, Sadha Pat* (Bengali), *Pat* (Hindi), *Marapat* (Assamese), *Jhota* (Oriya), *Sanal* (Tamil),), *Joot* (Marathi), *Senabu* (Kannada), *Janumu* (Telugu), *Chanambu* (Malayalam), *Moti* (Gujarati). The term jute is probably derived from the Oriya word '*jhout*' or '*jhut*'. *The genus name 'Corchorus'* is believed to have been derived from the word '*Korkhores*' used by the Greeks for the 'Pot herb'. In ancient time, Jute was termed 'Jew's mallow' (*i.e.* food for people) and would be used as 'Pot herb' and as well as fiber. The green leaves of *C. olitorius* L are used as vegetables. The leaves of *C. capsularis* contain a bitter glucoside called 'corchorin' and taste bitter on chewing. Hence it is known 'Tita (bitter) pat', whereas leaves of *C. olitorius* L. are tasteless, so it is called 'Mitta (sweet) pat'. The fiber of *C. capsularis* is whitish and of *C. olitorius* is yellowish and accordingly they are called white jute and Tossa jute, respectively. *C. capsularis* jute is also known as *Guti pat* while *C. olitorius* jute is also known as Jews mallow, *Desi pat*. The term 'pat' is used in India for jute and seems to be an abbreviation of the Sanskrit word '*patta*' which is used to describe a wide range of fibres.

Importance: Jute is known as the 'Golden fibre of Bengal'. Jute was a monopoly crop for India until the country was divided in 1947 while India retained all the jute mills, most of the area producing fine quality jute went to Bangladesh. Raw jute and jute products together form one of the main sources of earning foreign currency for India. Jute is a bast fibre or phloem fibre. The fibre is developed in the outer portion or bark of the stem. Jute is used to manufacture sacks and bags, twins, carpet yarns and in paper manufacture. Globally, India is the largest producer and second largest exporter of jute goods. There are 83 jute mills in the country. Of these, 64 are in West Bengal, 7 in Andhra Pradesh, 3 each in Bihar and Uttar Pradesh, two each in Assam and Chattisgarh, one each in Odisha and Tripura. As on 01-01-2012, total number of looms installed in jute industry stood at 49, 529 consisting of 21122 Hessian looms, 26663 sacking looms, 903 C.B.C looms and others at 841. Jute industry supports nearly 4 million farm families, besides providing direct employment to about 2.6 lakh industrial workers and livelihood to another 1.4 lakh people in the tertiary sector and allied activities. Jute trade is currently centered around the Indian subcontinent. Bangladesh is the largest exporter of raw jute, and India is the largest producer as well as largest consumer of jute products in the world. The local price of Jute Goods in India is the international price. Nearly 75% of Jute goods are used as packaging materials, burlap (Hessian) and sacks.

History: Indian Central Jute Committee (ICJC) has been stated in 1936. Indian Jute Industries Research Association [IJIRA] has been established in 1937. The Institute of Jute Technology (IJT) has been established in 1947. Jute Agricultural Research Institute (JARI) has been started at Barrackpore in 1953. Jute Corporation of India (JCI) Ltd, Kolkata is set up in 1971. National Jute Manufactures Corporation Ltd. (NJMC), Kolkata has been started on 3rd. June, 1980. Jute Packaging Material (Compulsory Use in Packaging Commodities) Act, 1987 (JPM Act) has been enacted to provide for the compulsory use of jute packaging material in the supply and distribution of certain commodities. This is followed in 90% food grains and 4% sugar packing compulsorily at present. Jute Agricultural Research Institute (JARI) is renamed as Central Research Institute for Jute and Allied Fibres (CRIJAF) in 1990. The Indian Government has announced the National Jute Policy 2005 of India. The Jute Technology Mission has been started on 2-6-2006. Jute Technological Research Laboratories (JTRL) has been opened on 3-1-1939 which later renamed as National Institute for Research in Jute and Fibre Technology (NIRJFT). The International Jute Study Group (IJSG) is an intergovernmental body set up with Headquarters in Dhaka, Bangladesh on 27 April 2002.

Origin: *Corchorus olitorius* is native of Africa and Indo-Miyanmar region while *Corchorus capsularis* is originated in Indo-Miyanmar including South China.

Distribution: Jute is cultivated in India, Bangladesh, China, Thailand, Nepal, Taiwan, Malaysia, Java, West Indies, Mexico, Brazil, and Africa. In India, it is grown in West Bengal, Assam, Bihar, Uttar Pradesh, Odisha, Andhra Pradesh, Megalaya and Tripura.

Botany: *C. olitorius* is a tall growing herbaceous annual, growing up to five metres. It is an upland species. Stem is green or reddish. Branches develop more than *C.capsularis*. Leaves are glabrous and flowers born axillary. Fruit is an elongated capsule. Seeds are smaller than that of the *C.capsularis*. Seeds are blackish to bluish green in colour. The fibre of *C. olitorius* is a yellowish, reddish or grayish colour depending upon the nature of the retting water. Fibre is frequently finer, softer, stronger and more lustrous than that of *C. capsularis*. *C. capsularis* is a herbaceous annual growing from 1.5 to 4 metres, stem slender, cylindrical, grayish. Branches sparsely produced, leaves simple, alternate, stipulate, ovate to oblong, coarsely toothed, flowers small and yellow. White jute (Corchorus capsularis) is a predominately self pollinated and tossa jute (Corchorus olitorius) is a partially cross pollinated species. Fruit is a capsule, wrinkled fairly toothed, five loculed seeds, smaller and brownish in colour. The comparison of *Corchorus* species is presented in Table 1.

Table 1 Characteristic features of capsularis and olitorius jute species

Features	White jute - Corchorus capsularis L.	Tossa jute - Corchorus olitorius L.
General	5-12 feet tall, grown in both low and high lands. It can withstand waterlogging is later stages. 3-5 months duration depending on time of sowing. Early sown crop *i.e.*, middle of February	5-15 feet tall, cannot withstand waterlogging. 4-5 months duration depending on time of sowing. Late sown crop middle of April
Stem	Cylinderical, colour of stems varies from dark green to red colour with different shades	Cylinderical, colour of stems varies from green to light or deep red colour
Leaves	Leaves are straight, relatively smaller in size, dark green in colour with deep serration. Leaves contain a bitter glycoside, corchorin. Hence it is called '*Tita*' (bitter) pat. Non bitter types of C. *capsularis* are also available	Leaves are drooping, bigger in size, yellowish green with less serration. Leaves shining in upper surface and a rougher under surface. Leaf is almost tasteless. Hence it is called '*Mitha*' (sweet) pat
Flowers	Flowers are smaller in size 0.3-0.5 cm in length and 0.5-0.6 cm in width. Flowering in sequence of flush for 4-6 weeks	Flowers are longer in size (2 to 2 ½ times of those of *capsularis* jute). Flowering is distributed in periodic flushes for about 10 weeks
Anthesis	1-2 hours after sunrise	An hour or less before sunrise
Capsule / pod	Capsule 1 to 1.5 cm diameter , roughly globular, rounded or wrinkled	Capsule 6 to 10 cm long, cylindrical or elongated
Seeds	Seeds are larger and chocolate brown colour, 300 seeds/g, 35 to 50 seeds/ capsule	Seeds smaller than capsularis. 500 seeds/g, 140 to 200 seeds/ capsules
Root:shoot ratio	Root:shoot ratio is less (1:7)	Root:shoot ratio is more (1:12)
Roots in clay loam soils	25 to 30 cm	40 to 50 cm
Roots in sandy loam soils	50 to 70 cm	60 to 80 cm
Fibre	Fibre is white	Fibre is yellowish, reddish, or grayish depending on nature of retting water. C. *oiltorious* fiber is finer, softer, stronger, silkier and more lustrous than that of C. *capsularis* fibre.
Seed coat	Dull black	green

Climate: Jute thrives well under a warm and humid climate. Jute is grown during the rainy season. It thrives best with an annual rainfall of about 1000 to 1500 mm with summer rainfall of 250 mm. Both species are short day plants. The critical light period is 12 ½ hours above which flowering is retarded. Jute requires a cardinal minimum, optimum and maximum temperature of 15°C, 25 to 30°C and 37°C, respectively. Temperatures below 15ºC and above 43ºC during growth are not suitable for jute crop. Soil temperature needs to be 15°C or above for favourable seedling growth. At a temperature below 10ºC, no germination occurs in both the species. C. capsularis can withstand higher temperature at germination (up to 32ºC), while C. olitorius is sensitive to such high temperatures. It requires a relative humidity at 70 to 95%. *C. olitorius* can not withstand water logging, however, *C. capsularis* can withstand water logging, but its fibre quality is impaired with prolonged water stagnation.

Soil: Jute can be grown in all types of soils varying in texture from clay to sandy loam. It thrives well on alluvial soils. The inundated areas near river banks are very suitable for the jute corp. Jute is grown in acidic soils. The acidity increases with increase in rainfall and heaviness of soil. The ideal pH ranges from 6.0 to 7.5.

Varieties: State wise Jute varieties in India is presented in Table 2.

White Jute (*C.capsularis*)**:** KJC-7 (Shresthaa), JBC-5 (Arpita), RRPS-27-C-3 (Monalisa), NDC-2008 (Ankit), JRC-517 (Siddhartha), JRC-532 (Sashi), JRC-80 (Mitali), Bidhan Pat 2, Bidhan Pat I, Bidhan Pat 2, Bidhan Pat 3, JRC-698 (Shrabanti), KTC-1(Rajendrasada pat-I), KC-1 (Joydev), UPC-94 (Reshma), Padma, JRC-4444 (Baldev), JRC-7447 (Shyamali), JRC-212 (Sabuj Sona), JRC-321(Sonali).

Tossa Jute (*C.olitorius*)**:** JRO-2407, JBO-1 (Sudhangsu), CO-58 (Sourav), JBO 2003 H (IRA), AAU-OJ-1 (Tarun), JRO-204 (Suren), S-19 (Subala), JRO 128 (Surya), JRO-8432 (Shakti Tossa), JRO-66 (Golden Jubilee Tossa), KOM-62 (Revati), JRO-3690 (Savitri), TJ-40 (Mahadev), JRO-524 (Navin), JRO-7835 (Basudev), JRO-878 (Chaitali Tossa), JRO-632 (Baisakhi Tossa), Bidhan Rupali Tossa.

Table 2. State wise Jute varieties in India

State	Jute varieties
Assam	AAUOJ-1 (Tarun), JBO-2003H (Ira), JRO-204 (Suren), S-19 (Subala), JRO-8432 (Shakti), CO-58 (Sourav), JBO-1 (Sudhangshu), JRC-698, JRC-80, JBC-5 (Arpita), RRPS-27-C-3 (Monalisa)
Bihar	JBO-2003H (Ira), JRO-204 (Suren), S-19 (Subala), JRO-8432 (Shakti), JRO-128 (Surya), JRO-66 (Golden Jubilee Tossa), CO-58 (Sourav), JBO-1 (Sudhangshu), JRC-698, JBC-5 (Arpita), RRPS-27-C-3 (Monalisa)
Meghalaya	JRO-204 (Suren), S-19 (Subala), JRO-8432 (Shakti), CO-58 (Sourav), JBO-1 (Sudhangshu), JBC-5 (Arpita)
Nagaland	AAUOJ-1 (Tarun), JBO-2003H (Ira), JRO-204 (Suren), S-19 (Subala), JRO-8432 (Shakti), CO-58 (Sourav), JBO-1 (Sudhangshu), JBC-5 (Arpita)
Odisha.	JBO-2003H (Ira), JRO-204 (Suren), S-19 (Subala), JRO-8432 (Shakti), JRO-128 (Surya), JRO-66 (Golden Jubilee Tossa), CO-58 (Sourav), JBO-1 (Sudhangshu), JBC-5 (Arpita), RRPS-27-C-3 (Monalisa)
Tripura	AAUOJ-1 (Tarun), JBO-2003H (Ira), JRO-204 (Suren), S-19 (Subala), JRO-8432 (Shakti), CO-58 (Sourav), JBO-1 (Sudhangshu), JBC-5 (Arpita)
Uttar Pradesh	JBO-2003H (Ira), JRO-204 (Suren), S-19 (Subala), JRO-8432 (Shakti), JRO-128 (Surya), JRO-66 (Golden Jubilee Tossa), JRC-80, CO-58 (Sourav), JBO-1 (Sudhangshu), JBC-5 (Arpita), NDC 2008 (Ankit)
West Bengal	JBO-2003H (Ira), JRO-204 (Suren), S-19 (Subala), JRO-8432 (Shakti), JRO-128 (Surya), JRO-66 (Golden Jubilee Tossa), JRC-80, JRC-698, CO-58 (Sourav), JBO-1 (Sudhangshu), JBC-5 (Arpita), RRPS-27-C-3 (Monalisa)

Source: http://farmer.gov.in/cropstaticsjute.html

The varietals particulars of jute are furnished in Table 3.

Table 3 Varietal particulars of jute.

Varieties	Cultivation systems	Sowing time	Duration (days)	Fibre yield (t/ha)	Region
JRC 212 (Sabuj Sona)	mid to high lands	April	150 (F)	3.0-3.8	West Bengal, Assam, Tripura, Bihar, UP, Odisha.
JRC 321 (Sonali)	lowlands	Feb-March	130 (F)	2.5-3.3	West Bengal
JRC 7447 (Shyamali)	mid to high lands	----	----	3.0-4.0	West Bengal, Assam, UP, Bihar, Odisha.
UPC 94 (Reshma)	lowland	March	130 (M)	2.7	West Bengal, Assam, Tripura, Bihar.
Hybrid C (Padma)	lowand	----	----	2.8	West Bengal, Assam
JRC 4444 (Baldev)	----	----	135 (M)	2.6-3.3	Odisha.
KC 1 (Jaydev)	----	----	----	2.6	Odisha.
KTC 1 (Rajendra Sanda Pat1)	----	----	----	2.1-3.3	Bihar
JRC 698 (Shrabanti White)	----	mid March-mid April	----	3.0-3.5	West Bengal, Assam, Bihar
Bidhan Pat3	----	early March to early July	110	2.5-2.7	Flood prone belt
Bidhan Pat1	----	early March to early August	65	1.3-1.4	Flood prone belt
Bidhan Pat2	----	early March to early July	90-110	2.0-2.3	Flood prone belt
JRC80 (Mitali)	----	Mid March to early April		3.0-3.5	North Bengal, Assam and U.P.
JRC532 (Sashi)	----	Mid April	110 (M)	3.0-3.5	North Bengal, Assam, Bihar, Orissa, U.P.
JRC517 (Sidhartha)	----	mid March-mid April	120 (M)	3.2-3.5	North Bengal, Assam, Bihar, Orissa, U.P.
RRPS27C3 (Monalisa)	----	Mid March	----	3.4	West Bengal, Assam, Bihar and Orissa

NDC 2008 (Ankit)	----	Mid February to mid March	----	2.7	Bihar, Orissa, Assam, W.B. , U.P.
JBC5 (Arpita)	----	Mid March	----	2.8-.0	Entire belt
KJC7 (Shresthaa)	----	Early March to early April	----	2.8-3.0	Entire belt
JRO 878 (Chittali tossa)	----	3rd week of March	135	3.2	West Bengal
JRO 7835 (Basudev)	high lands	3rd week of March to May	----	4.1	West Bengal Assam, Tripura, Bihar, Odisha.
JRO 524 (Navin)	mid to high lands	3rd week of March	120-140	2.5-4.0	West Bengal, Assam, Tripura, Bihar. Orissa
JRO 632 (Baisakthi tossa)	mid to high lands	15th April	----	4.0	West Bengal Assam, Tripura Odisha.
JRO 3690 (Sabitri)	mid to high lands	mid April	----	3.0	West Bengal
TJ 40 (Mahadev)	----	mid April	----	3.5	Odisha.
KOM 62 (Rabati)	----	mid March	----	2.5	Odisha.
JRO 66 (Golden jubilee tossa)	rainfed	mid April	120 (F) 150 (M)	2.5-4.0	West Bengal
JRO 8432 (Shakti tossa)	----	mid-March	----	3.5-4.0	Entire belt
JRO 128 (Surya)	----	mid March-mid April	----	.2-3.8	Entire belt
S19 (Subala)	----	mid-March	----	3.0-3.5	West Bengal, Bihar, Assam and Orissa
JRO204 (Suren)	----	1st week March	----	3.6-3.8	West Bengal, Bihar, Assam and Orissa
AAUOJ1 (Tarun)	----	mid-March	----	3.6	Assam
JBO2003H (Ira)	----	mid-March	----	3.8	Assam, West Bengal, Bihar and Orissa
CO58 (Sourav)	----	mid-March	----	3.4	Entire belt
JBO1 (Sudhangshu)	----	mid March-mid April	100-120 (M)	3.0-3.5	Entire belt
JRO2407 (Sampati)	----	Early March	----	3.5-4.0	Entire belt

Note: M stands for maturity stage and F stands for 50 % flowering stage.

Season: *C. capsularis* jute is sown during i) mid February to early April for fibre (120 days) and ii) first week of July to first week of November for seed (125 days). *C. olitorius* jute is sown during i) early April to early May for fibre (120 days) and ii) first week of October to January for seed (110 days). Jute is predominantly (>85%) grown as rainfed crop. Emergence of seedlings will take place in 3 to 6 days after sowing.

Land preparation: Jute seeds are very small. Therefore, the soil should be thoroughly pulverized and a seed bed with fine tilth obtained. Seed bed is prepared by ploughing and cross ploughing five or six times. Clays soils usually require more ploughing. Beds and Channels are formed depending on water resource.

Seeds and sowing: Seeds can be sown either by broadcasting or by line sowing 3 to 4 seeds/hill can be sown. Seed treatment with 5 gm of Dithane M45 or 2 g of Bavistin per kg of jute seeds will control root rot and stem rot diseases. Inoculation of jute seed with *Azotobacter chroococcum, Azospirillum brasiliense* and *Bacillus megaterium can* can save 25% N. The particulars on seed rate and spacing for jute is presented in Table 4.

Table 4 Seed rate and spacing for jute

Species	Seed rate (kg/ha)		Spacing (cm)		Plants/ m2
	Line sowing	Broadcasting	Fibre	Seed	
C. olitorius jute	5	7	25 × 5	30 × 20	80
C. capsularis jute	7	10	30 × 5	30 × 30	67

After cultivation: Gap filling is done on 6 DAS. First thinning is carried out on 10 DAS, while second thinning is on 20 DAS

Weed management: Hand weeding should be given on 21 and 35 DAS. Weed free condition is necessary till jute crop attains a height of 50 to 60 cm. Application of Fluchloralin @ 1.0 kg a.i./ha, Butachlor 50% EC or 5G @ 1 to 1.5 kg a.i./ha or Pretilachlor 50% EC @ 0.8 to 0.9 kg a.i./ha on 3 days after sowing followed by one irrigation and with one hand weeding on 35 DAS controls the weeds effectively. Quizalofop ethyl @ 50g a.i./ha at 21 DAE effectively controls the dominant grassy weeds.

Manures and fertilizer application: *C. olitorius* jute requires 2.06 kg N, 1.66 kg P and 5.18 kg K and *C. capsularis jute* requires 3.14 kg N, 1.54 kg P and 7.96 kg K for the production of 1 quintal of fibre. The increase in fibre yield is more in *olitorius* jute than *capsularis* jute especially at low application of nitrogen. At higher application of N the fibre yield is more in *capsularis* jute than in *olitorius* jute. Application of 12.5 tonnes of FYM or compost per ha can be carried out as basal dressing. A fertilizer dose of 80-40-40 and 60-30-30 kg of N, P_2O_5 and K_2O are recommended for *capsularis* and *olitorius* jute respectively. Entire P_2O_5 and K_2O fertilizers and ½ N are applied as basal dressing. The remaining ¼ N is applied on 21 DAS and the balance ¼ N is applied on 35 DAS after hand weeding. One fourth of nitrogenous fertilizer (15 kg N/ha) may be applied to jute crop through foliar spray. N may be applied @ 5 kg N/ha at weekly or fortnightly intervals as foliar spray from 25 to 45 days onwards. The nitrogen concentration should be 2 % in high volume sprayer and 15% for low volume sprayer.

Water management: Total water requirement for jute is 500 mm. Jute germinates at 18 to 20% soil moisture. Irrigation can be given at 50% available soil moisture. First irrigation is to be given after sowing and life irrigation is on 4 DAS. Afterwards irrigation can be given once in 15 days. 7 to 10 irrigations are required for jute cultivation. The irrigation requirement may be known when the top shoots dropping during day hours but recoup at night. This is called temporary wilting. Young plants are very sensitive to waterlogging conditions. Grown up *C. capsularis* plants to a height of 3 to 4 feet are able to tolerate waterlogging condition. It is therefore necessary to adjust the sowing time in such a way that the crop may attain a height of 3 to 4 feet before heavy monsoon starts.

Cropping System

Rainfed

Jute- groundnut
Jute- blackgram
Jute- toria

Irrigated

Jute- rice- greengram
Jute- rice- potato
Jute- rice- vegetables

Plant protection: Three to four spray of Endosulphon @ 0.2% control for semilooper, hairy caterpillar, stem weevil and yellow mites. Spraying Kelthane @ 0.2% controls the red mites. Spray of Bavistin @ 0.1% on crop controls the root rot and stem rot diseases. Two spray of endosulfan/ profenofos @2ml/litre of water at 15 days interval can be done to control mites, Bihar hairy caterpillar and semilooper. Seed treatment with carbendazim @ 2gm/ kg of seed and spraying of the same fungicide @ 1.5g/litre of water can control the major diseases of jute particularly Jute root and stem rot. Varieties like JRO-66. JRO-8432, JRO-7835, JRO-524 and S-19 are found to be moderately resistance to the root knot nematode.

Harvest: The ideal stage of harvest is when the plants are in early pod stage between 100 to 140 days after sowing. The harvest index remains more or less constant over 60–105 day period following sowing. The fibre remains weak if it is harvested before flowering. Early harvesting gives finer fibre of good quality. Cutting the crop too early leads to low yield and weakened fibres. On the other hand, late harvesting gives a high fibre yield but a coarse fibre which lacks lustre if harvesting is delayed beyond 120-day age. Harvesting around 120 days makes a compromise between quality and quantity. Harvesting is done by cutting the plants close to the ground level with sharp sickles. The plants are then sorted out according to height and diameter. The assorted plants are tied into bundles. Each bundle weighs 8-10 kg. The bundles are kept standing in the field for 3–4 days for shedding of

http://www.jute.com/web/guest/green-jute/agriculture-raw-jute

Source: http://jute.org/harvesting.html

http://www.jute.com/

leaves/ defoliation. The apical portion of the plants may also be severed and left in the field. When the crop is 50 to 120 cm under water at harvest time, the plants are pulled out by hand and the roots then cut off. Jute attains maturity for seed production at 140 DAS when it is sown in May.

Fibre yield: Jute yields 5 to 10 t of dry matter per acre of land. About 1 t of dry matter is put back to the soil in the form of leaves. About 3 t of roots remain in the soil. The fibre yield is 42 qtl per ha at experimental stations and 35 qtl per ha at national demonstrations. The national average fibre yield is 14.2 qtl per ha. Fibre yield of *olitorius* jute is 25 to 30 qtl per ha and of *capsularis* jute is 20 to 25 qtl per ha. The fibre yield of more than 35 qtl per ha may also be obtained if good management practices are followed and good soil fertility is maintained. One bale of fibre is equal to 180 kg fibre and 1 metric tonne is equal to 1000 kg fibre.

Seed yield: The seed yield of *C. olitorius* jute is 13 to 15 qtl per ha while that of *C. capsularis* jute is 7 qtl. per ha in Tamil Nadu. However, the seed yield of northeastern states is 4 to 5 qtl per ha in *C. capsularis* and 2.5 to 3.0 qtl per ha in *C. olitorius* jute. Seed production in jute is better in Tamil Nadu than West Bengal. Seed moisture is reduced to 7 to 8% during storage.

Retting: Retting is the process through which the fibres in the bark are being separated from the plants with removal of pectins, gums and other mucilaginous substances. This is usually affected by the combined action of water and microorganisms. This is the most important aspect of jute cultivation after harvesting. It determines the quality of the fibre that fetches price to the growers for their labour and to the industry basic raw material.

The important conditions for good retting are:

(i) The water should be non-saline and clear.

(ii) The volume of water should be enough to allow jute bundles to float.

(iii) Bundles, when immersed, should not touch the bottom.

(iv) The same retting tank or ditch should not be used when water becomes dirtier.

Retting is best done in slow moving clear water. Where such facilities are not available, land with a water depth of about 2 metre are quite suitable. The fibre develops various shades of black when retting is done in stagnant pool of muddy water. Generally, higher the volume of water, shorter is the period of retting. The minimum ratio of plant material to water in stagnant water should be 1:20 to develop fibre of good quality. Just after harvesting, the plants are tied into bundles of 20 to 23 cm. Thick and thin plants should be bundle separately. These bundles are kept in the field for 2 to 3 days, which helps in the shedding of leaves on the soil surface for subsequent incorporation in to the soil. The bottom portions of malleted stems are dipped for 3 to 4 minutes in 5% urea solution. The treated stems are then steeped in water for normal retting. Mixed microbial culture suspension may be sprayed on the jute bundles, particularly on the basal portion along with 5% urea before the bundles are placed in the retting water under JTRL method. Thereafter the bundles are to be taken to the source of water and kept standing in 30 to 60 cm deep water for one to four days for softening the hard basal portion, which will later favour uniform retting. The layer of bundle sinks partially from weight. If the water is deep enough it may be possible to have a second layer of bundles over the first and at right angles to it and even a third layer. Each layer consists of 25 bundles. Usually 2 to 3 layers are made. A few stem of *daincha* or sunhemp or *Gliricidia maculata* as activators may be placed on the top layer and is submerged by weighing it down with logs of wood. Banana stem is rarely used since it discolours the finished fibre. Care should be taken to see that the bundle does not touch the bottom of the tank. The bundles should remain in the tank water for 7 to 8 days, and thereafter one reed (stem) should be pulled out from any bundle and examined if the fibre is loose for extraction. If the fibre slips out easily from the wood on pressure from thumb and fingers, retting is considered complete. Over retting results in dazed weak fibre. Retting process lasts for 12 to 30 days depending upon the maturity of the plant, temperature, etc. Normally when the temperature of the retting water is 34^{0}C, the bundles become ready for extraction between 12th and 16th day.

Fibre extraction: Fibre is removed from the bark of the retted jute stalks by hand. A labour who stands two to three feet deep in water, take up in left hand, a handful (six to ten) of the retted stems and gently beats them at the base with a wooden or bamboo mallet. The fibres at the bottom are thereby loosened. He then breaks the woody core, one to two feet from the base. With slight jerks the pieces of broken stems from the base are taken out and thrown on dry ground. The loosened fibre is grasped by the right hand and the remaining portions of the retted stems are jerked backward and forward in the water. Thus the fibre is separated from the sticks. Taking a bundle of fibre, the labourers lashes it on the surface of water and again pulls it towards himself by a few jerks, by which the adherent bits of periderm and wood are got rid of. Finally the labour spreads out the clean fibre on the surface of the water and picks of any bits, if still adhering by hand. Water is then wrung out of the fibres and later is dried in the sun, over a bamboo frame for 3 to 4 days. Then the fibres are tied into bundles for the market. During drying the fibre is bleached by the sun. A labourer can strip, 40 lbs of dry fibre per day working for 6 to 7 hours. A highly skilled labourer can however strip up to 80 lbs of dry fibre during the same period. Jute fibre is marketed in bundles of fibre hanks. A fibre hank is composed of about 10 to 15 fibre reeds obtained from 10 to 15 plants.

Grading: In India with object of enabling the cultivators to get proper price for their fibre, Bureau of Indian Standards (BIS) introduced a standard for grading of raw jute on the basis of its quality. Both white and tossa jute fibres are classified into eight grades, W1-W8, and TD1-TD8 (Table 5). Each grade is assessed by scoring six characters - strength, fineness, defects, root contents, colour and density. The ceiling price of the fibres goes down at every step from 1 to 8, *i.e.,* maximum at 1 and minimum at 8.

Table 5 Grades and grading of White Jute (W) and Tossa Jute (TD) in India

Grade	Characteristics
W1/TD1	Very good strength and colour, very fine heavy bodied fibre, free from major and minor defects. Maximum root content: W1-10%, TD1-5%.
W2/TD2	Good strength and colour, fine heavy bodied fibre, free from major and minor defects. Maximum root content: W2-15%, TD2-10%.
W3/TD3	Fairly good colour and strength well separated medium bodied fibre, free from major and minor defects except a few specks. Maximum root content: W3-20%, TD3-15%.
W4/TD4	Fair, average strength and colour, well separated medium bodied fibre, free from major defects and substantially free from specks and loose sticks. Maximum root content: W4-26 % , TD4-20 % .
W5/TD5	Average strength and colour, fine from major defects. Maximum root content: W5-36%, TD5-26%.
W6/TD6	Average strength, free from centre root and dazed/over-retted fibre and reasonably free from entangled sticks. Maximum root content: W6-46%, TD6 - 35%.
W7/TD7	Weak mixed fibre with maximum root content :W7-57%. TD7-35%.
W8/TD8	Entangled or any other jute not suitable for any of the above grade but of commercial value.

Jute by-products: Jute fibre is a natural, renewable, bio-degradable and eco-friendly product that meets all the standards for safe packaging.

Jute products fall into five classes of manufacture:

(i) Fabrics: Hessian or Burlap, Sacking, Scrim, Carpet Backing Cloth (CBC) and Canvas are the most commonly used fabrics of jute. Fabrics are used in cloth form and in the form of bags.

(ii) Sacking: It is also known as 'heavy goods,' made from lower grades of fiber, loosely woven cloth, in plain or twill weave, weighing from 12-20 ozs. per yard of different widths. It is used for bags of all types.

(iii) Canvas: The finest jute product, closely woven of the best grades of fiber widely used in India for protection from the weather.

(iv) Yarns: The types of jute yarn manufactured can be classified according to the application/use to which they are put *i.e.,* fine yarns, hessian yarns, carpet/special yarns, sacking yarns, etc. These yarns can be further classified into warp and weft yarns, the warp yarns normally being superior to the weft yarns as they have to withstand the cycles of stress during weaving while the weft yarns act more as filler and undergo less strain during weaving process.

(v) Twines, Ropes, Cordage etc.: Twines, ropes, cordages etc. are used for the purpose of sewing, tying, knotting, binding, etc. particularly agricultural commodities. Jute yarns of various dimensions are plied together to make twines as per requirement and use.

Conversion factors

100 yards of hessian = 54 lbs of raw jute

4148 yards of hessian = 1 tonne raw jute (5.55 bales raw jute)

1 tonne of sacking = 1.11 tonne of raw jute (6.17 bales of raw jute)

1 tonne of hessian, sacking etc.=1.05 tonnes raw jute (5.85 bales of raw jute)

REFERENCES

Kundu, B.C., Basak, K.C. and P.B. Sarkar. 1959. *Jute in India,* Monograph, Indian Central Jute Committee, Calcutta.

GOI. 2006. Report of the working group on Textiles & Jute industry for the eleventh five year plan (2007-2012), Government of India, Ministry of Textiles. New Delhi. p 97-106

GOI. 2008. Guidelines for the conduct of test for distinctiveness, uniformity and stability on (*Corchorus olitorius* L. *and Corchorus capsularis* L.). Plant Variety Journal of India. Vol. 2(1 & 2):1-32.

Kirby, R.H. 1963. Vegetable Fibers. Leonard Hill (Books) Ltd., London.

http://djd.dacnet.nic.in/

http://www.jute.com/web/guest/welcome

http://crijaf.org.in/

http://jute.org/plant.html

http://www.plantauthority.gov.in/pdf/jute.pdf

http://farmer.gov.in/cropstaticsjute.html

http://jutecomm.gov.in/

http://www.jutegeotech.com/index.html

CHAPTER

17

Kenaf–*Hibiscus cannabinus* L. (2n=36)

Family: Malvaceae

Vernacular names: Kenaf, Mesta, Ambani jute, Palungi, Jute of Madras, Deccan hemp, Bombay hemp (English), *Bat Patsan, Mesta-pat, Bimli Jute, Bimlihemp, Gambo hemp, Bimlipatum jute, Binbli jute* (India), *Pulichai* (Tamil), *Sah* (Assamese), *Gohu* (Telugu), *Binabli* (Bengali), *Ambari* (Taiwan) and Java jute (Indonesia).

Importance: Kenaf is a backyard crop satisfying to produce leaves as greens and the bark fibre for cordage purposes. Kenaf has a few advantages over Jute. It is adopted to a variety of growing conditions. It will produce a fibre crop quickly with in 90 to 120 days. It needs little care during crop growth. Kenaf is a best fibre next to jute in importance. The kenaf seed contains 16 to 22% of oil and 32% of protein.

Origin: Kenaf is native of Angola in Africa and India.

Distribution: It is grown extensively in India, Thailand, China, Vietnam, Russia, Brazil, Bangladesh, Indonesia, Cuba, USA, Senegal, Nigeria and Mexico. In India, it is grown in Madhya Pradesh, Andhra Pradesh, Orissa and Tamil Nadu.

Botany: Kenaf is an herbaceous annual, tall growing habit with straight stem with green, red or purple colour. It develops a strong tap root and has a long, slender, unbranched, glabrous or prickly stem. Stem may reach a height of 2.5 to 5.0 m or even more in about 5-6 months. Kenaf plants produce two general leaf types, divided and entire. The juvenile or young leaves on all kenaf seedlings are simple, entire, and cordate. The divided (split-leaf) cultivars have deeply lobed leaves with 3, 5, or 7 lobes per leaf. Leaves are with serrated edges on the main stalk (stem) and along the branches. The position of these leaves alternate from side to side on the main stalk and branches. The plants produce flowers with day length of 12.5 hours. Kenaf is hermaphrodite producing large showy, light yellow, creamy colored flowers that are bell-shaped and widely open. The flowers of many cultivars have a deep red or maroon colored center. The flowers are 8 to 13 cm in diameter with 5 petals and are borne singly in the leaf axis along the stalk and branches. Flower production is indeterminate. The flowers open before sunrise and close in the night. It is a self-pollinated crop but cross pollination also occurs which is caused by honey bees. The self-pollination is by the twisting closing movement of the petals. The cross-pollination has ranged from 2% to 4%. Once pollinated, the seeds require 4 to 5 weeks to maturation. Fruit is a capsule with pointed tip. Each capsule contains 5 segments with a total of 18 to 20 seeds per capsule. Seeds are dark grayish-brown, glabrous, flattened triangular shaped, 5 to 6 mm long and 4 mm wide. There are roughly 35,000 to 40,000 number of seeds/kg. The bulk of the fibre is found in the lower portion of the stem. In a plant 2.5 m high, for example, about 75% of the fibre is in the basal 1.25 m of the stem, about 20% being in the first 0.30 m. This is because the fibre percentage in properly grown plants increases from the base upwards for the first 60 to 90 cm and then decreases fairly quickly.

Climate: Kenaf is grows well tropics and subtropics in the latitudes from 30°S to 45°N up to an altitude of 1500 m above MSL. Rainfall of 500 mm to 750 mm in 4 to 5 months periods is essential for the successful production of crop. It cannot withstand water logging or water stagnated condition. A dry period following the rainy season is necessary for the production of seed. If rains occur during the time of seed production, the seeds germinate in the capsule before the harvest. In areas with low rainfall conditions, kenaf is grown as a partially irrigated crop. It is drought tolerant than jute. It can withstand against strong wind and extremely heavy rains. Kenaf is a short day plant and needs 12 hours and 30 minutes of daylight per day to flower. The crop is grown during long days in order to obtain as much rapid vegetative growth as possible, before flowering begins. Kenaf requires a minimum, optimum and maximum temperature of 10°C, 20 to 26°C and 35°C for germination and growth. It is sensitive to cool temperatures and to frost. Wind bents down the plants, which make difficulty in harvesting.

Soils: A well drained sandy loamy, loamy or sandy clay-loamy soil with a pH 6.0 to 7.5 is desirable. The crop is generally grown on marginal soils. A considerable quantity of humus in the soil is important for good fibre production.

Land preparation: Kenaf seeds are relatively small and require good seed-soil contact for germination. Therefore, a fine, firm, well-prepared seedbed is necessary. The soil should be deeply ploughed to provide favourable condition for good germination and crop growth.

Season: Kenaf is planted from April to August for seeds and from May to August for fibres. Sowing during first week of May is desirable for high fibre yield. For maximum yield of fibre per unit area of land, kenaf should be planted at the beginning of the rainy season. For production of seed, planting should be done from 4 to 8 weeks before the length of daily light period becomes shorter than 12.5 hours.

Varieties

State	Mesta varieties
Andhra Pradesh	AMV-5 (Durga), GR-27 (Madhuri) and MT-150 (Nirmal), JRM-5 (Shrestha), AMC 108 (Bhimpli), HC 583
Bihar	GR-27 (Madhuri) and MT-150 (Nirmal), JRM-5 (Shrestha)
Meghalaya	JRM-5 (Shrestha),
Odisha	AMV-5 (Durga), GR-27 (Madhuri) and MT-150 (Nirmal), JRM-5 (Shrestha),
Tripura	GR-27 (Madhuri) and MT-150 (Nirmal), JRM-5 (Shrestha),

Source: http://farmer.gov.in/cropstaticsjute.html

Crop duration: 120 to 130 days for fibre.

Seeds and sowing: The seed rate is 15 kg/ha. The spacing adopted is 30 × 5 cm. Warm, moist soils after danger of a killing frost has passed are the ideal planting conditions. Seed should be planted less than 2.5 cm deep. The plants will emerge in 3 to 6 days after sowing.

After cultivation: Thinning is done on 15 DAS.

Weed management: Two hand weeding and hoeing are needed on 20 and 35 DAS. Application of Fluchloralin @ 2.0 l/ha or Alachlor @ 2 kg per ha can be applied just after sowing (3 days before sowing) preferably in moist soil is recommended in controlling the weeds which is followed by one hand weeding on 35 DAS.

Manures and fertilizer application: Apply 12.5 tonnes of FYM or compost per ha. A fertilizer dose of 20-20-20 kg of N, P_2O_5, K_2O per ha is to be applied as a basal dressing and 20 kg of nitrogen is applied on 35 DAS as top dressing.

Water management: First irrigation immediately after sowing. Life irrigation is given on the third day. Then, irrigate the field once in 7 to 10 days depending on soil and climatic conditions. The plants need 12 mm of water per day for seed germination and initial crop establishment. Irrigation needs 500 to 625 mm rainfall over a period of 5 to 6 months is needed for a suceessful production of kenaf fibre. In general the water stress results in shorter plants with lower LAI, thinner stems and lower stem yields. Kenaf can be grown successfully on a saline soil when the irrigation water has good quality. It cannot withstand water logging or water stagnation. Kenaf is grown under rainfed situations.

Cropping systems: Kenaf can be rotated with non-host nematodes crops such as groundnut, rice, maize and sesame in order to reduce the nematodes populations such as *Meloidogyne incognita, M. javanica* and *M. arenaria*. Kenaf may not be suitable for rotations with cotton and peanut. When kenaf is rotating with a legume (like soybean) the stunt nematode (*Tylenchorhynchus* spp) can be reduced. To reduce the incidence of nematodes in affected areas, kenaf could be planted following maize and sorghum.

Plant protection: Cut worms, leaf miners, and other chewing/sucking insects are potential problems at seedling emergence and during young seedling growth. For all sucking pests, seed treatment with thiomethoxam @ 6g/kg or thiochloprid @ 8g/kg or chlorpyriphos @ 6 ml/kg is highly recommended. Spray Methyl demeton @ 2 ml or Dimethoate @ 2 ml or Monochrotophos @ 1.6 ml or Oxymethyl demeton @ 2 ml or Imidacloprid @ 0.25 ml or Thiomethoxam @ 0.2 gms or Acetamipride @ 0.2 gms /lit. of water to control jassids and aphids. Mealy bug can be effectively controlled by spraying Neem oil 5ml or Pongamia oil 5 ml or Propenophos 2 ml or Triazophos 2 ml or Methyl parathion 2 ml per lit. of water. Add adhesive like Sandovit –triton AE or Teepol @ 1 ml/lit. of spray fluid for effective penetration on the skin of the insect. Spray Thiodicarb @ 1.0 g/l or Indoxacarb @ 1ml/l or Novuluron @ 1.0 ml/l or Phosphomidon 80 SL @ 1 ml/litre in 500 litres of spray fluid/ha to control green semilooper. Anthracnose is perhaps the most serious potential disease problem. Damping-off is a moderate concern during seedling stages. The root-knot nematode and fusarium wilt complex are the most serious constraint to kenaf production. Seed treatment may be done with Mancozeb @ 3 gm or Metalaxyl 3 gm or Metalaxyl mz 2 gm or with cyamoxanil + soil application of *Pseudomonas fluorescens* @ 20 g/kg of seed. Spraying with ridomyl @ 2 gm/lit. when foot and stem rot disease symptoms are noticed. Spray Mancozeb @ 3 gm or Copper oxychloride 3 gms /lit. of water twice with an interval of 7 days to control leaf blight. Spraying with Thiamethoxam @ 0.1g/lit or Imidacloprid @ 0.25 ml/lit. at 50 days after sowing to check the white fly which is vector for the yellow vein mosaic virus disease.

Harvest: The length of cropping period is 150 to 180 days. Its duration varies from 75 to 105 days for early varieties or 105 to 120 days for semi-early varieties. Kenaf can form a dense canopy in 5 weeks after sowing. The plants begin to flower within 60 days of sowing, produces seed at the end of 100 days. The vegetative phase lasts from 120 to 140 days in long duration varieties. The seed requires 60 to 90 days to mature after fertilization. At harvest, leaf materials are left in the field which can return from 50 to 100 kgs of nitrogen/ha.

Harvesting method for fibre is similar to that in jute. For fibre production the plants are cut at the ground level with a sickle or with a mechanical harvester. The best yields and the best quality of fibre are obtained when harvested at the time of flowering and before seed production starts. The plants are uprooted, dried in the sun, tied into bundles of 30 to 40 stalks, topped to remove any fruits that may have formed, of the stalks. Then the bundles are kept in the field for 2 or 3 days for shredding of leaves. Highest quality fibre is obtained when kenaf is harvested during the flowering period. With an increase in the age of the plant, there appears to be an increase in coarseness and a decrease in luster of the fibre. The growing period takes from 110 to 140 days. The best fibre is usually obtained from kenaf that is about 120 days old, about 3 to 3.6 m height, has no branches and is not too thick. For paper pulp production the harvesting technique is different. At maturity the stems are cut by hand, tied in bundles and stocked vertically for sun drying. After 6 weeks, when the stems are dried to a moisture content of about 10%, the top 30 cm is removed and the stems are rebundled for transport to the mills. The best time for seed harvesting is when the seeds at the lower and middle portion of the bearing area are fully mature, since the seed of kenaf matures progressively from the lower toward the upper portion of the plant. If harvesting is delayed until all the capsule reach the maturity, then considerable seed loss would occur due to shattering of seeds, especially from the lower portion of the seed bearing area. The dried plants are beaten with wooden mallet to thresh the seeds. Then the seeds are cleaned through winnowing.

Retting

(a) Whole stalk retting: The harvested plants are steeped horizontally in water for retting. Then these are covered with weeds, grasses, soil or wood to protect them from direct exposure to the sun or change of temperature. The stem bundles are weighted down in two to three layers and submerged nearly 10 cms in water with the help of weights made of cement blocks or stones to facilitate the retting process. Care should be taken not to put any weighed material which releases tannin and iron. For hastening the retting process spraying of 1.25 % urea solution (12.5 gm/lit of water) on mesta sticks is desirable before retting of the bundles. Microbial consortium @ 1 kg/layer diluted in 10 litres of water may be applied to the mesta bundles in the pond. The retting process will be completed with 12-15 days, depending on the temperature of the water and the stage at which the plants are harvested. It is necessary to check regularly after a lapse of 10 days whether the stems are properly retted as over or under-retting deteriorates the fibre quality. Then the fibres are washed and dried in the sun for 4 or 5 days. Subsequently, the well dried fibre strands are made into bundles and carried to the markets or the factories. Fibre forms about 16% of the weight of the dry stalks. The yield of fibre from a good crop varies from 900 to 1350 kg/ha. The fibre will be with golden yellow colour.

(b) Ribbon retting: The power operated bast fibre extractor can extract 25 kg dry fibre ribbons per hour with broken sticks while manually operated Mesta fibre extractor can extract 15 kg dry fibre per hour with unbroken sticks. When kenaf bark material is retted at its ideal temperature of 34°C, dry ribbons of bark may take 70 hours to ret, compared to green, moist ribbons of bark which takes 29 hours. Hand-stripped green bark ribbons and mechanically separated bark material can be chemically retted using 7% and 1% sodium hydroxide, respectively, to produce good textile quality fibers from kenaf. These green ribbons can be treated with the microbial consortium @ 100 ml/10 kg of green ribbons and kept for one hour covering with polythene sheet. Polythene lined retting tank of one m^3 size (1m × 1m × 1m) containing 1000 litres of water can be used for small scale retting. The retting process will be completed after five to seven days.

Yield: The moisture content of actively growing plants at harvest is normally about 75%. Fibre yield is 15 to 30 qtl/ha. The bark portion accounts 25 to 35% of whole stalk. Stalk yields normally range from 11 to 18 t/ha on oven dry weight basis. Bast fibers make up 20 to 25% of the stem on a dry weight basis. The average fibre out turn is about 4% of weight of green stalks or 16% of weight of the dry stalks. Among the coarse fibres, jute fibre contains 11.5% lignin while kenaf/mesta fibre contains 5.9% lignin. Seed yield is 700 to 800 kg/ha. Seed stored at 8% relative humidity remained fully viable for 5.5 years when stored at either –10°, 0°, or 10°C, and fully viable for 5.5 years when stored at –10° or 0°C at 12% relative humidity. Seed contains 20 to 24% oil which is refined and used for cooking purposes. The oil cake usually serves as a stock feed.

Utilization: The per cent of conversion from dry biomass yield to dry stems of kenaf is 28 to 30 %. Kenaf stems accumulate the highest (67.05%) biomass, followed by root (21.15%). The kenaf stalk's average composition is 35% bark and 65% woody core by weight. The bark of the kenaf stalk contains a long fiber called bast fiber, while the woody core contains short core fibers. Whole stalk kenaf (bast and core fibers) has been identified as a promising fiber source for paper pulp. The kenaf fibers, bast and core fibre can be pulped as a whole stalk or separated and pulped individually. The combined (bast and core fibre) bleached fiber yield from chemical pulping is about 46% by weight. Kenaf fibre is used for making twine, rope, gunny-bag, sackcloth, fishing nets and cordage of all kinds. Leaves are used as green. The leaves are rich in protein (15 to 30%) and are used as animal feed. It can be ensiled and fed to young cows. The whole plant is used for making paper pulp. The bast fibre can also be converted to pulp for newsprint, tea bags, and grass mats (biodegradable mats impregnated with grass and/or flower seeds), hydrocarbon free bags, twine, rugs, ropes and textiles. The seeds are roasted and ground into meal for consumption.

REFERENCES

Angelini, L., Macchia, M., Ceccarini, L. and Bonari, E., 1998. Screening of kenaf genotypes for low temperature requirements during germination and evaluation of feasibility of seed production in Italy. Field Crops Research, 59, 73-79.

Campbell, T and White, G., 1982. Production density and planting date effects on kenaf performance. Agron. J. 74: 74-77.

Carberry, P. S. and Abrecht, D.G. 1990. Germination and elongation of the hypocotyls of kenaf in response of temperature. Field Crops Res. 24: 227-240.

Charles L. Webber III, Venita K. Bledsoe, and Robert E. Bledsoe. 2002. Kenaf Harvesting and Processing.pp.340-347. In: Trends in new crops and new uses. J. Janick and A. Whipkey (eds.). ASHS Press, Alexandria, VA.

Charles, L., 2002. Trends in New Crops and New Use, ASHS Press, Alexandria, VA.

Cook, J.G., 1960. Handbook of textile fiber. Merrow publishing, Watford, UK.

Crane, J.C. and J.B. Acuna. 1945. Effect of plant spacing and time of planting on seed yield of kenaf, Hibiscus cannabinus L. J. Am. Soc. Agron. 37:969–77.

Dempsey, J.M. 1975. Fiber crops. The Univ. Presses of Florida, Gainesville.

Hossain, M. D., Hanafi, M. M., Saleh, G., Foroughi, M., Behmaram, R.and Noori. Z. 2012. Growth, photosynthesis and biomass allocation of different kenaf (*Hibiscus cannabinus* L.) accessions grown on sandy soil. Australian Journal of Crop Science. 6 (3): 480-487

Jones, M.D, C. Puentes, and R. Suarez. 1955. Isolation of kenaf for seed increase. Agron. J. 47:256–257.

Killinger, G. B. 1969. Kenaf (*Hibiscus canabinus* L.), a multi-use crop. Agron. J. 61:734-736.

Liu, A.M., 2000. World production and potential utilization of jute, *H.cannabinus* and allied fibers. In: Proceedings of the 2000 International *H.cannabinus* symposium. Hiroshiam, Japan, pp. 30-35.

Mohamed, A., H. Bhardwaj, Hamama, A., and C. Webber, 1995. Chemical composition of kenaf (Hibiscus cannabinus L.) seed oil. Ind. Crops Prod. J. 4:157–165.

Tamargo, M.A. and M.D. Jones. 1954. Agents concerned with natural crossing in kenaf in Cuba. Agron. J. 46:456–459.

Toole, E.H., V.K. Toole, and E.G. Nelson. 1960. Preservation of hemp and kenaf seeds. USDA Tech. Bul. 1215. Washington, DC.

Vikram S. Negi, R.K. Maikhuri, L.S. Rawat, and P.C. Phondani. 2010. Current Status and Future Potential of Fiber Yielding Crop *Hibiscus cannabinus* L. in Mountain Region of Central Himalaya, India. Int. J. Sci. Tech. 5: 87-96

Webber, C.L. 1993. Yield components of five kenaf cultivars. Agron. J. 85:533–535.

Webber, C.L. 1994. Kenaf (*Hibiscus cannabinus* L.) yield components as affected by sewage sludge applications. Nonwood Plant Fibers. Prog. Rpt. 21:19–24.

Webber, C.L. and R.E. Bledsoe. 1993. Kenaf: Production, harvesting, and products. p. 416–421. In: J. Janick and J.E. Simon (eds.), New crops. Wiley, New York.

White, G.A., D.G. Cummins, E.L. Whiteley, W.T. Fike, J.K. Greig, J.A. Martin, G.B. Killinger, J.J. Higgins, and T.F. Clark. 1970. Cultural and harvesting methods for kenaf. USDA Prod. Res. Rpt. 113. Washington, DC.

Wood, I., Muschow, R. and Ratcliff, D. 1983. Effect of sowing date on the growth and yield of kenaf grown under irrigation in tropical Australia II. Stem production. Field Crop Res. 7: 91-102.

http://corn.agronomy.wisc.edu/Crops/Kenaf.aspx

https://hort.purdue.edu/newcrop/ncnu02/v5-327.html

https://hort.purdue.edu/newcrop/afcm/kenaf.html

http://www.westafricanplants.senckenberg.de/root/index.php?page_id=14&id=811

http://jute.org/plant.html

http://www.fibrafp7.net/Portals/0/05_Alexopoulou.pdf

CHAPTER

18

Roselle–*Hibiscus subdariffa* L. (2n=72)

Family: Malvaceae

Vernacular names: Roselle, Rozelle, Rosella, Sorrel, Red Sorrel, Jamaican Sorrel, Indian Sorrel, Guinea Sorrel, Sour-Sour, Jelly okra, Queensland Jelly Plant, Sour Tea, Red Tea, Lemon Bush, Florida cranberry (English), *Gongura, Lalambari, Patwa* (Hindi), *Lal-mista, Chukar* (Bengali), *Lal-ambadi* (Marathi), *Yerra gogu* (Telugu), *Pulichchai kerai* (Tamil), *Pulachakiri, Pundibija* (Kannada), *Polechi, Pulichchai* (Malayalam) and *Chukiar, Tenganora* (Assamese), *L'oiselle* (French), *Jamaica* (Spanish), *Karkade or carcadé* (Arabic), *Rohzelu* (Japanese).

Importance: Roselle is the drought tolerant crop than kenaf. Roselle fibre is a phloem fibre obtained from the stems of the plant. Roselle is very similar in its features to kenaf. It can be used for paper pulp production. The seeds contain about 17% of oil, which is similar in its properties to cotton seed oil. Roselle is grown for its calyces, which are exported from the Sudan, Jamaica, China, Thailand and Mexico to Europe and USA.

Origin: Roselle is native of Sudan, Africa.

Distribution: It is cultivated in India, Bangladesh, Taiwan, Thailand, Malaysia, Indonesia, Sri Lanka, China, Sudan, Mexico, Egypt, Senegal, Tanzania, Mali and Jamaica. Germany and the United States are the main countries importing roselle. In India, it is grown in Uttar Pradesh, Bihar, Madhya Pradesh, West Bengal, Assam, Tripura, Odisha, Maharashtra, Andhra Pradesh, Tamil Nadu and Karnataka.

Botany: Roselle is an erect, glabrous, annual shrub growing from a long taproot. The alternate, glabrous leaves are carried on long petioles, usually palmately divided into three or five lobes with serrated margins. Species grown for their fiber are tall, with fewer branches, sometimes growing to more than 3 to 5 m in height. Culinary varieties are many-branched, bushy, and generally 1 to 2 m tall. The stem colour varies according to the variety. The flowers are showy and are carried singly in the axils of the upper leaves, on very short peduncles. Flowers are red to yellow with a dark center containing short-peduncles. The flowers open before sunrise and close in the night. Roselle is a self-pollinated crop. Natural cross-pollination of 0.20 to 0.68% is observed in roselle. The capsule/pods begin ripening towards the bottom and proceed to the top. Each capsule contains 18 to 20 seeds. Seed is kidney shaped. Two botanical varieties of roselle are recognized viz., *H. sabdariffa* var. *altissima* Wester, is a tall, vigorous, practically unbranched plant, 3.1- 4.9 m high with fibrous, spiny and inedible calyces, and is grown for fibres. *H. sabdariffa* var. *sabdariffa* is a bushy, branched shrub with a red or green stem and red or pale-yellow inflated edible calyces from which refreshing drinks, jellies, sauces, chutneys and preserves are made.

The differences between kenaf and roselle are furnished in Table 1.

Table 1 Difference between *Hibiscus cannabinus* and *Hibiscus subdariffa*.

S. No.	Plant part	Kenaf (*H. cannabinus*)	Roselle (*H. subdariffa*)
1	Apex of bracteole	entire	furrowed
2	Calyx pubescence	white and highly hairy	pubescent, smooth
3	Calyx nectary	swollen	not swollen and without hairs
4	Calyx lobe shape	acuminate	ovate to lanceolate
5	Corolla colour fresh	pale yellow to yellow	pale yellow
5.1	Corolla colour withered	pale yellow to yellow	pale yellow to reddish tinge
6	Stem	prickly	unawned
7	Seed		
7.1.	Seed shape	triangular, angular acute	rounded, hemispherical, angles rounded
7.2.	Seed surface	dull grey, many yellowish brown raised spots	dull grey without spots
7.3.	Seed hilum	yellowish brown, relatively small	brownish red, relatively large

Climate: Roselle grows well in tropics and subtropics. It grows up to an altitude of 700 m from MSL. It needs a warm, more humid weather climate. It is more susceptible to frost, mist and foggy weather. It is a short day plant. It requires 13 hours of sunlight during the first 4-5 months of growth to prevent premature flowering. Hence, the planting time to be fixed based on day length and not according to the rainfall requirements. It requires a minimum of 500 mm rainfall. However, it is grown a well distributed rainfall from 1500 to 2000 mm per annum. The plant needs ample water during its vegetative period, but cannot tolerate stagnant water. Roselle can tolerate shade and can be grown in greenhouse. Roselle thrives is between 18 and 35°C, with an optimum of 25 to 27.5°C. It requires 4-8 months with night time temperatures not below 21°C. It tolerates floods and heavy winds.

Soils: Roselle plants prefer well-drained humus and rich-fertile soils with a pH of 4.5 to 8.0. It can be grown even in heavy soils. Lighter soils or light sandy soils with poor organic matter are not suitable.

Hibiscus calyces

Source: http://natureproducts.net/Forest_Products/Malvaceae/Hibiscus_sabdariffa.html

Varieties: GR-27 (Madhuri), NON BRIS4 (Jaya), AMV-1, AMV-2, HS-4288, HS-7910 (Ujjal).

Variety	Duration (days)	Fibre yield (q/ ha)	Special characteristics
Madhuri (GR-27)		27-30	Green stem with red patches only in nodes
AMV-3 (Surya)	140-150	18.52	Tolerant to drought, and foot and stem rot
AMV-4 (Kalinga)	140-150	20.00	Withstands to drought
AMV-5 (Durga)	135-140	25-30	Good fibre quality, higher fibre yield
AMV-7 (Janardhan)	130-140	30-35	Moderately tolerant to leaf rot

Seeds and sowing: The seed rate is 15 kg/ha. The spacing is 30 × 5 cm.

Weed management: Two hand weeding and hoeing are given on 20 and 35 DAS. Alachlor @ 2 kg/ha can be applied preferably in moist soil just after sowing.

Manures and fertilizer application: Apply 12.5 tonnes of FYM or compost per ha during last ploughing. A fertilizer dose of 20-20-20 kg of N, P_2O_5 and K_2O per ha is applied as a basal dressing and 20 kg of nitrogen is applied on 35 DAS.

Water management: First irrigation immediately after sowing and life irrigation is given on the third day. Then irrigate the field once in 7 to 10 days, depending on soil and climatic conditions.

Harvest: Crop duration is 130 to 150 days.

(i) Harvesting for fibre: It is similar to that in jute. Highest quality fibre is obtained when roselle is harvested during the flowering period. With an increase in the age of the plant, there appears to be an increase in coarseness and a decrease in luster of the fibre. For fibre production the plants are cut at the ground level with a sickle. The stems are tied in bundles of 15 to 25 after removal of branches and leaves. Then the bundles are kept in the field for 2 or 3 days. The harvested plants are piled on the field and are covered with grasses to prevent the loss of humidity. Subsequently, they are carried to the retting tank for fibre extraction. For paper pulp production, the harvesting technique is different. The optimum stage for harvesting is 50% flowering for paper pulp. The stems are cut by hand using sickle, tied in bundles and stocked vertically for sun drying. After 6 weeks, when the stems are dried to a moisture content of about 10%, the top 30 cm is removed and the stems are rebundled for transport to the mills.

(ii) Harvest for calyces: The calyces ripen about three weeks after the start of flowering. The fleshy calyces are harvested after the flower has dropped but before the seed pod has dried and opened. Ripe calyces are harvested with clippers leaving the stalks and immature calyces to ripen in the field. The calyx is separated from the seed-pod by hand, or by pushing a sharp edged metal tool through the fleshy tissue of the calyx separating it from the seedpod. The fleshy calyces are harvested after the flower has dropped but before the seedpod has dried and opened. The more time the capsule remains on the plant after the seeds begin to ripen, the more susceptible the calyx is to sores, sun cracking, and general deterioration in quality. All harvesting is done by hand. On average, each fruit yields about 7 to 10 g of sepals. Drying of the calyces is done in one of two ways: by harvesting the fresh fruit and then sun-drying the calyces, or by leaving the fruit to partially dry on the plants and harvesting the dried fruit, keeping the crop well protected during the process. The fully developed fleshy calyx is peeled from the fruit by hand and dried naturally in shade. Total calyces yield is about half a ton per hectare. For every 100 pounds of fresh calyx, 11 pounds of dry calyx is produced.

(iii) Harvest for seeds: The harvest is timed according to the ripeness of the seed. The fruit ripens progressively from the bottom of the plant to the top. The fruit may be harvested when fully grown and can be easily snapped off by hand / clippers. The fruit is easier to break off in the morning than at the end of the day.

Fibre extraction: In roselle, the process of fibre extraction and criteria of realization vary from one place to another and even within the same region with varying climatic conditions. The bundles of stems are placed in a water tank or reservoir. Then these are covered with weeds, grasses, soil or wood to protect them from direct exposure to the sun or change of temperature. The stem bundles are weighted down in two to three layers and submerged to facilitate the retting process. It takes about 15 to 20 days depending on the maturity, temperature and quality of the water. It is necessary to check regularly after a lapse of 10 days whether the stems are properly retted as over or under-retting deteriorates the fibre quality. Then the fibres are washed and dried in the sum for 4 or 5 days. Subsequently the well dried fibre strands are made into bundles and carried to the markets or the factories.

Yield: Fibre yields is15 to 20 qtls per ha. Seed yield is 1200 to 1300 kg/ha per ha.

Utilization: Roselle is cultivated for stem fibers, paper pulp or edible calyces, leaves and seeds. It is used in food, animal feed, nutraceuticals, cosmeceuticals and pharmaceuticals. Roselle is a fine, silky, soft, shining, light yellow fibre, which makes a useful substitute for jute, particularly in the manufacture of coarse sackings. Fibre is used to manufacture sacks, cordage, rope and fishing nets. The stalks left over after fibre extraction is used as fuel. Roselle plant parts such as flower, leaves, fleshy calyx (sepals) surrounding the fruit (capsules) and corolla are used in beverages in China, Taiwan and Thailand both as a thirst quenching drink and medical purposes for its alleged antihypertensive properties. The extracts of roselle are also used in folk medicine against high blood pressure, liver disease and fever. The calyx juice contains vitamin C, anthocyanins, flavonoids and polyphenols and other antioxidants. The red anthocyanin pigments in the calyces are used as food colouring agents. The young leaves and tender stems of roselle are eaten raw in salads or cooked as greens alone or in combination with other vegetables or with meat or fish. They are also added to curries as seasoning. The juice of the boiled and strained leaves and stems is utilized for the same purposes as the juice extracted from the calyces. The dry calyx is used to produce a flavorsome and healthy drink, tea, jelly, marmalade, ices, ice cream, sorbets, butter, pies, sauces, tarts, and other desserts. The dried calyces can be soaked in water to prepare a colourful cold drink, or may be boiled in water and taken as a hot drink. The best quality is dark red in colour with a sour-fruity taste. The calyces may be merely chopped and added to fruit salads. For making a finer-textured sauce or juice, syrup, jam, marmalade, relish, chutney or jelly, the calyces may be first chopped in a wooden bowl or passed through a meat grinder or the calyces, after cooking, may be pressed through a sieve. The dried calyces are packed in plastic bags. It is calculated that 5 kg of fresh calyces dehydrate to 0.45 kg of dried calyces, which is equal to the fresh for most culinary purposes. The dried calyces is pressed/squeezed into solid cakes or balls weighing 80 kg for export to Europe or USA, where they are utilized to make extracts for flavouring liqueurs. The dried calyces as sold for 'tea' do not yield high colour and flavour if merely steeped; they must be boiled.

REFERENCES

Bahaeldeen Babiker Mohamed, Abdelatif Ahmed Sulaiman and Abdelhafiz Adam Dahab. 2012. Roselle (*Hibiscus sabdariffa* L.) in Sudan, Cultivation and Their Uses. Bull. Environ. Pharmacol. Life Sci. 1 (6): 48-54

Juhi Agarwal and Ela Dedhia. 2014. Current Scenario of Hibiscus Sabdariffa (Mesta) in Maharashtra, India. International Journal of Social Sciences and Humanities Invention. 1(3): 129-135

Kirby, R.H. 1963. Vegetable Fibres (Botany, Cultivation and Utilization). New York: Interscience Publishers,Inc.

Mahadevan, N., Shivali, K.P. 2009. *Hibiscus sabdariffa* Linn: An overview. Natural Product Radiance, 8: 77-83.

Ngamjarus C, Pattanittum P, Somboonporn C.2010. Roselle for hypertension in adults. Cochrane Database of Systematic Reviews.10, Issue 1. Art. No.: CD007894. DOI: 10.1002/14651858. CD007894.pub2.

Plotto, A., Mazaud, F., Röttger, A. and Steffel, K. 2004. Hibiscus: Post-production management for improved market access organisation: Food and Agriculture Organization of the United Nations (FAO), AGST.

Purseglove, J.W. 1974. Tropical crops: Dicotyledons. Longman, London.

Singh, D.P. Mesta- *Hibiscus cannabinus* & *Hibiscus sabdariffa*. N.d. Retrieved on Jan 01, 2013 from http://assamagribusiness.nic.in/mesta.pdf

Vaidya, K.R. 2000. Natural cross-pollination in roselle, *Hibiscus sabdariffa* L. Genetics and Molecular Biology. 23 (3): 667-669

Wahabi, H. A., L. A. Alansary, A. H. Al-Sabban, et al. (2010). The effectiveness of Hibiscus sabdariffa in the treatment of hypertension: A systematic review. Phytomedicine 17: 83-86.

Wilson, F.D. and Menzel, M.Y. 1964. Kenaf (*Hibiscus canabinus*), roselle (*Hibiscus sabdariffa*). Economic Botany. 18(1): 80–90.

Yadong Qi, Kit L. Chin, Fatemah Malekian, Mila Berhane, and Janet Gager. 2005. Biological Characteristics, Nutritional and Medicinal Value of Roselle, *Hibiscus Sabdariffa*. Southern University and A&M College System. Agricultural Research and Extension Center. Baton Rouge, LA 70813.

http://jute.org/plant.html

https://hort.purdue.edu/newcrop/Indices/index_ab.html

https://hort.purdue.edu/newcrop/duke_energy/Hibiscus_sabdariffa.html

https://hort.purdue.edu/newcrop/morton/roselle.html#Food%20Uses

http://natureproducts.net/Forest_Products/Malvaceae/Hibiscus_sabdariffa.html

http://www.trop-hibiscus.com/index.html

http://www.hibiscus.org/toeat.php

http://www.fao.org/3/a-av006e.pdf

http://bepls.com/may2012/10.pdf

CHAPTER

19

Ramie–*Boehmeria nivea* L. Gaud (2n=28)

Family: Urticaceae

Vernacular names: There are two cultivated species in ramie. *Boehmeria nivea* var. nivea is known as the true ramie or 'China grass', 'Chinese silk plant' or 'white ramie' or 'Grass linen' or 'China Linen'. It is the Chinese cultivated plant. *B. nivea* var. *tenacissima* is known as 'green ramie' or nettle fibre or 'rhea' which is originated in the Malay Peninsula.

Importance: Ramie produces unicellular fibre, which is longest, strongest and finest among the vegetable fibers in the world. It is the strongest natural fibre, thus nicknamed as 'steel wire fibre'. The fibre is twice stronger than cotton and has better length: breadth ratio than cotton. This property is being utilized to design bullet proof panels by developing ramie based fibre composites. Ramie possesses the longest filament compared to other fibres obtained from stem. The length varies from 3 to 30 cm with an average of 20 cm, the width is 40-100 μm, and the proportion of length to width of the fibre cell varies from 3000 to 3500. The fibre is produced in the bark as a bast or phloem fibre. The fibre is obtained in the 'bark' ribbon of the stem. The fibre is highly resistant to water. It lacks elasticity and flexibility. It has much higher tensile strength than cotton and this increase on wetting. It is one of the most satisfactory vegetable fibres and can be used for clothing, table-cloths, hand kerchiefs, curtains, twine, thread, fishing nets, gas mantle, shoelaces and parachute barriers. The fibre is spun and the cloth known as grass cloth or Chinese liven is used for clothing, table cloths, mat, etc. Ramie fibre has a tensile strength about eight times that of cotton and seven times that of silk cotton, and its tensile strength increases when wet by 60 to 70%. The fibre are exceptionally long (20 to 40 cm), very durable and have a high lustre.

History: The word 'Ramie' has been derived from the Malayan word, which is anglicized as 'ramie'. In India it is called by different popular names such as Rhea, Popah, KhunKoora, Kurkunda, etc. Ramie fibre finds a reference in the ancient Indian literature i.e. drama "Sakuntala" written by Kalidas about 400 AD and also in Ramayana. Japan has been producing fabrics made of ramie fibre known as 'Echigojfu' and 'Satsumjofu' since ancient time. According to chronicle of Nester (900 B. C), the ships in volga (Russia) were utilizing ramie fibre for the preparation of their sails. Ramie was introduced from Malaysia in 1278 during the regime of 'Ahom' kings in Assam.

Origin: Boehmeria nivea var. nivea, a native of China and Japan while *Boehmeria nivea* var. *tenacissima* Miq. is a native of Malaysia.

Distribution: Ramie is grown in China, Brazil, Indonesia, Philippines, Korea, Vietnam, Japan, India, Taiwan, France, USA, Germany, Italy and Switzerland. China is the major producer of ramie fibre contributing to 96.3% of the global production. In 2011, ramie has cultivated in 72,934 ha area in China, producing 1,24,000 tonnes fibre with an average productivity of 1700 kg/ha. In India, it is grown in West Bengal, Assam, Nagaland, Tripura, Manipur, Mehalaya, Mizoram, Arunachala Pradesh.

Botany: Ramie is a perennial monoecious species belonging to the family Urticaceae. It is a shrubby perennial with unbranched, pubescent, green and slender stems, which are also called 'canes'. The stems grows to a height of 2 to 3 m with stems 8–16 mm thick. The plants have rhizomatous roots which contain storage roots, small fibrous roots and rhizomes. The branches, petioles, leaves are soft and hairy. The leaves are green in upper surface and silvery appearance in the backside. Leaves are simple, alternate, green above and whitish below from the dense covering of felty hairs between the reticulate veins. Inflorescence is a branched panicle. The flowers are greenish white in colour, borne in declinate clusters in the axils of the leaves. Male and female flowers are found on the same stalk. The female flowers are found on the upper part of the stalk. They have one celled, one seeded ovary and a slender style, hairy on one side. The male flowers are arranged on the lower part, have five stamens and a rudimentary ovary. The male flowers open first. Flower is with perianth 3-5 lobed, stamens 3-5, perianth tubular 2-4 toothed; ovary superior with slender style and hairy stigma. It is a wind pollinated crop. Anthesis occurs between 8.00 AM and 10.30 AM but may extend depending on environmental conditions. Fruit is an achene, oblong compressed and hairy. Seeds are very small in size and weigh nearly 7000 seeds per gram. There are two cultivated varieties viz.,(i) *Boehmeri nivea* var. *nivea* (the true ramie, white ramie or China grass) is chiefly cultivated in the subtropical or temperature parts of China, Japan, Taiwan and India and yields fibre of superior grade. The leaves are bright green on upperside and white thick white fell of hairs on the under surface and (ii) *Boehmeri nivea* var. *tenacissima* Miq. (green ramie or rhea) is better suited to the tropical climates of the Malayan Peninsula, Algeria, Africa, Mexico and Brazil, etc. The leaves are small compared to white ramie and are green on both surfaces. It is a bushier plant with leaves that are green on both surfaces, in contrast to 'China grass' which has dark green leaves covered on the under surface by a mat of short, white hairs, giving the leaves a silvery appearance.

Climate: Ramie is cultivated in tropics, subtropics and temperate climates. It grows best in warm moist climate. Ramie is grown between 40°N and 40°S latitude up to an elevation of 500 metre from MSL. Ramie flowers early under short day condition. The favourable temperature for growth ranges from 25 to 31°C with the maximum of 35°C. The relative humidity (RH) should not be below 21% while the optimum RH is 80%. A well distributed annual rainfall of 1500 to 2500 mm is suitable for luxuriant growth of ramie. It is highly sensitive to waterlogging, frost and strong wind. It can tolerate partial shade.

Soil: Well drained loamy soil is preferred. Since it is an exhaustive crop, it requires heavy manuring. Soil pH of 5 to 7 is ideal for ramie cultivation. If soil is acidic (pH range 4-5), lime @ 1 t/ha is to be applied.

Season: Ramie can be planted during February-March and also during September-October. Under irrigated condition, planting can be done throughout the year except the winter periods as the sprouting is not uniform. The planting of ramie under rainfed condition can be done with the onset of monsoon. Adequate drainage is to be provided to prevent waterlogging. Ridge and furrow planting is very effective as it provides adequate drainage.

Varieties: Ramie varieties recommended for commercial cultivation in India are presented in Table 1.

Table 1. Ramie varieties recommended for commercial cultivation in India

Variety	Green Plant (t/ha)	Defoliated Stalk (t/ha)	Plant height (cm)	Fibre % on the basis of total green weight	Dry decorticated fibre (q/ha)	Total gum (%)	Tenacity (gm./tex)
R-67-51	95.28	71.97	122.5	5.38	29.01	24.9	27.6
R-67 -46	100.22	65.64	139.1	3.45	20.22	24.8	32.5
R-67 -36	94.91	62.00	129.3	3.56	20.19	22.6	29.9
R-67-34	73.48	47.16	121.3	4.89	19.39	21.9	26.4
R-67-20	65.09	50.05	104.1	5.19	18.59	27.8	32.2
R-67 -44	89.04	57.49	130.1	3.34	17.26	24.5	33.4
R-67 -43	88.20	57.29	120.5	3.15	15.82	24.9	27.2
R-67-38	84.07	52.4 7	118.3	3.32	15.34	24.1	30.1
R-67-35	77.65	55.69	117.3	3.61	14.52	27.1	26.1
R-67-21	70.39	51.96	107.3	3.76	14.38	27.0	28.1
R-67 -52	51.51	39.43	95.4	5.11	14.10	27.7	32.1
R-1452	48.59	36.50	96.5	5.28	13.30	24.8	32.0
R-67 -45	76.94	51.06.	115.1	3.11	13.02	25.1	27.5
R-67-30	54.32	44.24	99.2	4.24	12.64	29.1	27.4
R-1412	46.18	36.50	97.8	4.94	12.39	25.2	26.3
R-1411	50.81	37.74	88.0	4.45	11.12	26.7	27.4
R-67-40	79.78	50.94	109.7	2.78	10.92	31.1	27.8
R-67 -59	48.72	35.21	98.8	3.79	9.53	27.6	33.5

Source: CRIJAF, Barrackpore.

Propagation: Ramie can be propagated through: i) rhizome cuttings, ii) plantlets/ suckers, iii) stem cuttings, iv) waste stalk (< 2ft in length) planting and v) seed. Plants from seed usually take two years for germination. So it is propagated vegetatively by rhizome cuttings of 10 to 15 cm long and 1.2-2.5 cm diameter (each weighing about 10-15 g) taken from 4 year old plantation are suitable for planting. Seed rhizomes cuttings of 3,750 to 4,000 weighing near about 250 to 275 kg/ha have been found to be sufficient to cover an area of one hectare in a spacing of 60 × 45 cm. The quantity of rhizomes obtained from one hectare of 2 years old plantation can give enough planting material to cover at least 20 hectares of land. Cut rhizomes when exposed to sunlight loose viability rapidly and should be planted preferably within 72 hours for better sprouting. The establishment is faster in case of rhizome planted crop. Ramie is a crop with spreading habit, therefore, a good planting distance should be provided for proper growth and development. The spacing of 60 × 45 cm or 60 × 30 cm is recommended depending upon the soil fertility, soil type and variety used. Ridge and furrow planting is very effective as it provides adequate drainage. It is planted at a depth of 5 to 7.5 cm. The sprouting of plants starts after a week or ten days after planting and continues till 3 to 4 weeks depending mainly on the soil type and the weather conditions.

Plantlets: Plantlets are matured canes (< 1.0 m) along with 2-5 cm underground stem piece from more than one year old crop. The plantlets are to be placed in the furrows opened at 60 cm × 30 cm spacing and covered with soil. Sprouting takes place within 15-30 days depending upon temperature and moisture, and sprouting occurs from 2-5 cm underground stem portion as well as from 5-10 cm basal portion of the stem above ground. One to two irrigations need to be provided if planting is done in dry season.

Stem cutting: The matured stem (2/3rd of the length brownish) unsuitable for fibre extraction (> 1m in length) are to be collected at the time of harvest and cut into small pieces each having 2-3 nodes. The top part of the stem is rejected. The stalks cuttings should be planted in fields at 45° angle at 60 cm × 30 cm spacing. The termite attack can be prevented to certain extent if the cut ends of the stem cuttings are dipped in chlorpyriphos 20 EC solutions (3 ml/l) of systemic insecticide. The cut end at the top may be covered with raw cow dung which enhances the sprouting rate. The sprouting starts within 15-20 days after planting. The young plantlets obtained from the waste stalk should be uprooted from the nursery bed at 50-55 days without disturbing the root system and transferred in the black polythene bags containing mixture of soil, sand and FYM or other well rotted compost (2:1:1) and kept in shaded place for transportation. The plantlets can be kept for 3-4 weeks without loss in viability. About 8-10 q rhizome of ramie is required to cover one hectare area. In case of alternate planting materials (plantlets, stem cutting, waste stalk), about 60,000 planting materials are required per hectare area (inclusive of 5 to 10% mortality rate).

Stage back: The growth of winter crop is retarded and not uniform in growth. Hence the winter crop is cut in end of February to March depending upon temperature, close to ground for uniform growth of subsequent crop. This operation is called 'stage back' and is extremely important for getting good fibre yield in subsequent cuts.

Weed management: Weeding is usually done in the early stages of growth, i.e., in the first and second years. Application of quizalofop ethyl 5% EC @ 60-90 g a.i./ha or atrazine and paraquat @ 1 kg + 2 lit a.i. kg/ha on 15 days after emergence or @ 1 kg + 2 lit a.i. /ha/cutting after each cut can effectively control weeds in ramie. The herbicide has to be sprayed carefully between the rows after each cutting.

Manures and fertilizers application: The nutrient requirement for one quintal of undegummed ramie fibre is 8.1, 2.0 and 9.5 kg of N, P_2O_5 and K_2O, respectively. Application of 3-4 t/ha of hydrated lime at the beginning improves fibre yield of the crop particularly if the soil is acidic and deficient in Ca. Farm yard manure on compost @ 12.5 tonnes per ha is applied on basal dressing every year after 'stage back'. Application of N, P_2O_5, K_2O @ 20-10-10 kg/ha after 2 months of planting is recommended. Subsequently fertilizer is to be applied as N, P_2O_5, K_2O @ 30-15-15 kg/ha/per cut for the 1st and 2nd year. From 3rd year onwards, the dose has to be doubled (60-30-30 kg/ha/cut). Application of Borax @ 15 kg/ha and $ZnSO_4$ @ 20 kg/ha increase in fibre yield and also reduce chlorosis.

Water management: Ramie being a perennial crop, suffers from both drought and waterlogging. Irrigation supplied during pre- and post monsoon periods (1st, 2nd and 5th cut of ramie) at 0.6 to 1.0 IW/CPE ratio. The death of plants is observed when the water stagnation prolonged for 10 days.

Cropping systems: Cultivation of ramie as a sole crop is practiced in Assam. Ramie based multi-tier cropping systems introduced in the non-traditional areas are: ramie + rubber and ramie + cinnamon in Tamil Nadu, ramie + coconut and ramie + coconut + black pepper in Goa and ramie + mango or ramie + mango + black pepper at Ratnagiri, Maharashtra while in Assam the most successful model is ramie + arecanut.

Plant protection: Ramie leaf roller (*Syleptra derogata*) and leaf eating caterpillar (Spodoptera exigua), hairy caterpillar (*Spilosoma obliqua*) incidence are common. Carpospores leaf spot is caused by *Cercospora boehmereae*. The pest management practices are :

(i) Judicious application of fungicides, i.e. spraying of propiconazole or difenconazole @ 0.1% or mancozeb @ 0.25% at 15 days after previous harvest for management of foliar diseases

(ii) For Sclerotium rot, drenching with copper oxychloride @ 0.25% is effective

(iii) Dipping of plantlets or stem cuttings in 0.2% carbendazim solution for 2 minutes before plantation is effective in management of soil borne diseases

(iv) The planting materials should be treated with chlorpyriphos (0.05%) at the time of planting

(v) For the management of insect pests and termite, spraying of chlorpyriphos @ 0.05% at 15 days intervals is effective

Harvest: Crop field duration is 6 years. The plants usually reach their highest yield in the third and fourth years after planting and maintain this until the sixth year at least. The crop persists for 20 years in China. The crop is replanted after 7 to 10 years. Under rainfed condition, usually four cuts may be taken in a year at 50, 45, 45 and 50 days interval respectively. Under irrigated condition five cuts can be taken per year. Harvesting is done manually by cutting the stalks manually near ground level. It is proper to harvest the crop for good yield with the following symptoms viz.,

(i) Lower parts of the stem takes coppery red colour/yellow in colour.

(ii) Lower leaves turn yellow and one-third of the leaves have fallen/shed.

(iii) New sprouts begin to appear above the ground.

(iv) Bark of the stem can easily be peeled out from the stick

(v) Emergence of inflorescence

All the above symptoms are exhibited by the crop when the plants are about 45 to 50 days old in age. Good results have been obtained when the first and fourth crops are harvested around 50 days and second and third crops around 45 days of crop age. Harvesting is done by cutting the canes sufficiently close to the ground with the help of a sickle. When the plant is in active growth, the harvesting could be done once in 75 to 90 days. During cold months, it may extend up to 120 to 135 days. Stems are harvested at 4 to 6 feet height in ten months after planting and subsequently 3 to 5 times per year. After taking each harvest, the underground rhizomes send up new canes, which again provide fresh crop for the subsequent harvests.

Retting and fibre extraction: Fibre is extracted from harvested defoliated canes by a decorticator machine. 1000 kg cane can be decorticated everyday using decorticator machine which produces approximately 35 to 45 kg fibre. The raw fibre is then washed thoroughly in clean water to remove some of its gum and then sun dried for 2 to 3 days. The raw fibre cannot be extracted satisfactorily by the usual retting methods like many stem fibres such as jute, flax, hemp, etc., owing to the nature of the pectinous or gummy matter found in the bark of the stem. The fibres are contained in the secondary phloem, heavily coated with a gummy substance that makes the fibre extraction process quite difficult. The best-cleaned fibrous strands contain at least 10% gummy matter, whilst the poor grades as much as 30%.

Ramie fibres cannot be extracted satisfactorily because of the resistance to microorganisms of the gums binding the bast bundles. The outer thin skin of the bark is peeled off. The outer bark and the adhering fibres are stripped from the woody stem either by hand, as practiced in China and Japan, or by decorticating machines in the United States. These strips are known as 'ribbons'. Decorticated fibre stands are generally termed 'crude ramie fibre'. In either case, the extraneous matter is scraped off from the gummy fibrous strands with the help of a bamboo or bone knife against a bamboo thimble covering the operator's thumb. The fibre is then hung over poles to be dried and bleached by the sun and wind. The ribbons, of partially separated but not yet degummed fibre, are known as 'filasse'. They are then dried and bales into bundles varying from 45 to 225 kg and are exported in this form to Europe and other markets. Most of the fibre is locally spun and woven into coarse cloth, often without degumming. However, it is necessary to remove the plant gums by a degumming process before this crude fibre can be spun.

Degumming: The fibre extracted by the decorticators contains nearly 25 to 30% gum. This fibre must be properly degummed before spinnable fibre is obtained. The most important factor involved in the degumming of ramie fibre is the concentration of the chemicals, temperature and the time of penetration of the degumming solution to all parts of the ribbons. The perfectly degummed fibre is pale creamy in colour. In order to make it even more whiter it is bleached with chlorine or hydrogen peroxide or in combination of both. Generally two types of degumming have been recommended viz., i) Microbial degumming and ii) Chemical degumming.

(i) Microbial degumming: *Bacillus pumillus* bacterium DKS1 having high pectinase activity is used in a mild alkali solution (0.1%) for degumming of ramie fibre. The decorticated fibre is treated with these liquid bacterial cultures. In this process the gums are removed by the bacteria. The degummed fibre is then washed in water.

(ii) Chemical degumming: Ramie fibre is initially boiled in the aqueous alkaline solution and then washed in water. The alkaline solution helps in dissolving the gums which are removed by washing. The steps followed in this process are as follows:

(i) One percent sodium hydroxide is prepared and taken to a vat.

(ii) Raw decorticated ramie fibre is then immersed into this solution. The fibre and liquid ratio may be kept at 1:6 or 1:7. Some suitable wetting agents may be used.

(iii) The fibre-liquid mixture is then boiled at a temperature of 96-98°C for two hours. The boiling may be done by means of steam generated from a baby boiler operated by power.

(iv) The fibre is then washed in water and then dried.

(v) This process helps in reducing the gum content to 5%. In case the boiling is done at a higher temperature of 120-125°C for two hours, the gum content is reduced to even 2%.

(vi) Use of 0.5% of sodium sulphite helps in improving the tenacity of the fibre.

(vii) Bleaching of fibre is done with the help of an oxidising agent such as dilute hypochlorite, hydrogen peroxide or chlorine dioxide. Bleaching is however, optional. It makes the fibre more whiter and soft.

(viii) The bleaching is followed by washing in water & drying perfectly in hot air.

(ix) If required the drying is again done at this stage.

(x) The oiling of the fibre with a sulphonated hydrocarbon as softening agent may be done to facilitate combing and carding.

Degumming is also taken in the mills that spin the fibre. The heavy coating of gum is removed with the treatment of soap solution, lime or 0.25 per cent of sodium hydroxide. After a 24 hours soaking period, the fibres are boiled for one to four hours. After cooking, they are rinsed, neutralized, washed and centrifuged several times. They are then dried either in the sun or over charcoal fire and treated with softeners such as glycerine, wax, soap, etc., to prevent the fibres from becoming brittle. The cleaned fibres are graded and processed for various commercial uses. Individual fibre cell is 13 to 15 cm long and may reach up to 55 cm. Degummed ramie fibre is pure cellulosic (>90%) with little hemicellulose (<3%) and negligible lignin (<0.5%).

Yield: 50 t of green stem/ha/year produces 2.5 t of dried ribbons which in turn yields about one t of degummed fibre. The yield of ribbons or crude fibre varies between 2 to 4% of the weight of the green plant and the yield of degummed fibre is normally about 1% of the weight of the green plant. The air dry stems, after the leaves have been removed, contain from 12 to 15% of crude fibre. The average fibre yield is 1200 to 1800 kg/ha. The average fibre content in ramie is about 3 to 4% of green biomass of the crop. Rhizome yield is 750 to 1250 kg/ha.

A 10 t of green ramie plant will yield 6.0 t of green stems (60%) and 4.0 t of green leaves (40%). On drying, the green leaf yield is reduced to 0.75 t of dry leaves (7.5%). On drying, the green stem yield is reduced 1.25 t of dry stems (12.5%). Dry stems yield about 0.35 t of dry decorticated fibre (3.5%) in the processing of fibre. The decorticated fibre yields 0.20 t of degummed fibre (2.0%). The stem and leaves together contain nearly 80% moisture and thus 10 t of green plants are equal to 2 tons of dry plants which in turn yield about 350 kg of decorticated fibre. This will ultimately give 200 kg of de-gummed fibre.

Renovation and Replanting: Ramie needs replanting after every 5 to 6 years although ramie is a perennial crop plantation can easily stand for 6 to 7 years. The yields, however, again go down after that period. In real sense good and economic yields are obtained only up to third year of plantation with a total of twelve harvests and the decline in gross return starts after fourth production years. Hence it is always advisable to go for replanting after every 12 harvests. This will ensure profitable returns.

Utilization: Ramie fibre is used for clothing, table cloths, hand kerchiefs, towels, napkins and other household items. It is also made into twines and ropes for sail boats and mountain climbing, industrial sewing thread, fishing and mosquito netting, parachute cords, shoelaces, conveyor belts, water hoses, insulation for cables, bags for mortar, hats and upholstery works, and packing material for ship propeller shafts. Short fibres and wastes are used in the manufacture of specialty papers such as cigarette paper and bank notes. Ramie is strongest fibre in the nature and used to make Armour garments, which is used in the war and could not cut by sword and spear. Young ramie leaves and shoots are of high nutritive value and excellent feed for cattle, pigs and poultries. Leaves and tops contain 16 % protein of dry matter and may be fed to livestock.

REFERENCES

Biswas, S.K. and Chakravarty, K. 1986. Ramie – a valuable fibre crop and its properties in India. Tropical Agriculture (Trinidad) 63 (4): 135-140.

Mitra, Sabyasachi, Saha, Suprakash, Guha, Biswajit, Chakrabarti, Krishanu, Satya, Pratik, Sharma, A. K., Gawande, S. P., Kumar, Mukesh and Saha, Monidipta. 2013. Ramie: The Strongest Bast fibre of Nature, Technical Bulletin No. 8, Central Research Institute for Jute and Allied Fibres, ICAR, Barrackpore, Kolkata-120, 38 p.

Saha, M. N., Samaddar, M. and Ghorai, D. 2006. Ramie cultivation and diversified uses. Technical Bulletin 8, CRIJAF, Barrackpore, Kolkata. 27p.

Sarkar, D., Sinha M.K., Kundu, A., Kar, C.S., Saha, A., Kharbikar, L.L. and Mahapatra, B.S. 2010. Why is ramie the strongest but stiffest of bast fibres? Current Science **98**(12): 1570-1572.

Satya, P., Mitra, S., Sharma, A.R., Ray, D.P., Jana, S., Karan, M. and Chakraborty, K. 2013. Genetic variability in flowering behaviour of ramie (Boehmeria nivea). National Seminar on Jute and Allied Fibre in Changing Times: Issues and Strategies, January 3-5, 2013, National Institute of Research on Jute and Allied Fibre Technology, Kolkata, India.

Satya, P., Sarkar, D., Kar, C.S., Mitra, J., Sharma, H.K., Biswas, C., Sinha, M.K., Mahapatra, B.S. and Maiti, R.K. 2010. Possibilities for reducing gum content in ramie, the strongest and finest bast fibre by genetic modification of pectin biosynthesis pathway. International Journal of Agriculture, Environment and Biotechnology **3**: 261-264.

Singh, D.P. 1982. Ramie. Central Research Institute for Jute & Allied Fibres, Barrackpore, Kolkata, West Bengal. pp 1-52

Singh, D.P. 1986. Ramie – a wonder fibre crop. Indian Farmers' Digest. **19** (11-12): 13-6.

Tarafdar, J.C. and Ray A. K. 1987. Effect of land reclamation technology on ramie cultivation – an economic analysis. Jute Developement Journal. **7**(3): 25-29.

http://en.wikipedia.org/wiki/Ramie

http://assamagribusiness.nic.in/ramie.pdf

CHAPTER

20

Sunnhemp–*Crotalaria juncea* L. (2n=16)

Family: Leguminoseae

Vernacular Name: Sunnhemp, Indian hemp, Madras hemp, Brown hemp, Bombay hemp, Banaras hemp (English), *Samn, Sonai* or *San* (Hindi), *Sana* (Sanskrit),*Tag* (Marathi), *Janumu* (Telugu), *Sanappai* (Tamil), *Vakku* (Malayalam), *Saab* (Kannada), *Sanpat* (Oriya), and *Sanpat, Shan* (Bengali). The genus name '*Crotalaria*' means 'rattle' and refers to the noise made by the seeds shaken in the mature pods.

Importance: Sunhemp is a bast fibre cum green manure crop. It follows in importance next to cotton, jute and kenaf. Fibre is used for manufacture of rope, twines, nets, mats, sacks, marine ropes, pulp, cigarette paper, silk and cordage. When it is grain as green manure, the crop is incorporated in field at 6 to 8 weeks old. It adds 40 to 60 kg N per ha in soil after decomposition. The crop is used to improve soil in rotation with cereal crops. It produces large amounts of biomass in a short duration of 60 to 90 days.

Origin: Sunnhemp is native of India and Myanmar.

Distribution: Sunnhemp is cultivated in India, Bangladesh, Pakistan, Brazil, China, Russia, Romania, Hungary, Poland, Turkey, and Chili. In India, it is grown in Uttar Pradesh, Bihar, Madhya Pradesh, Maharashtra, Rajasthan, Odisha and Tamil Nadu.

Botany: Sunhemp is a tall, erect-growing annual, about 1 to 3 m in height with a strong tap root system that penetrates into the soil. Root nodules are freely produced. All the vegetative parts of the plant are covered with short downy hairs. The stem is straight and thin with 1.25 to 2.0 cm in diameter for fibre purpose. The leaves are small, lanceolate, subsessile (5 to 7.5 cm long) and alternately placed. Flowers are small, yellow, borne in axillary racemes. It is a cross pollinated crop. The lowest flowers of the inflorescence open first and remain open for 2 days. The flowers open between 2 and 3 am. The dehiscence of anthers is complete in between 12 noon to 1 pm in about one hour and a half. The stigma remains receptive on the day of the opening of flower. Sunnhemp may be considered as a highly cross-pollinated crop. Self-pollination only takes place after the stigmatic surface is stimulated by the bee or mechanically. The fruit is an inflated hairy pod with a pointed beak and contains kidney shaped dark grey to black seeds that become loose at maturity and produce a rattling sound. Seed production is dependant on bees. A normal pod contains nearly 10 to 15 seeds. Seeds are dark green and about ¼ inch long. There are about 15,000 seed per pound.

Climate: Sunnhemp grows well in tropical and sub tropical climates. It is grown in areas of India range from latitude 17-30°N. It can be grown up to an attitude of 800 m from MSL. It is a short-day plant, but vegetative growth is favoured by long days. Growth period is favoured by the atmospheric temperature of 20-35°C. It requires well distributed rainfall of 400 mm with high relative humidity of 60 to 85% in growing season. It requires warm weather with a frost-free period of 8 to 12 weeks.

Soil: Sunnhemp is adapted to a wide range of soils and performs better on poor sandy soils than most crops. It grows best on well-drained loamy soils with a pH from 5 to 8.4.

Varieties: CO 1, SUN-053 (Swastika), SH-4, K12 Yellow

Crop duration: The crop duration for fibre and green manure are 120 to 135 DAS and 45 to 60 DAS respectively.

Land preparation: Three to four ploughings are required to make the soil friable and weed free condition for sowing. Beds and channels are formed under irrigated condition.

Season: Sunnhemp is grown throughout the year under irrigated conditions. The best growth is obtained from plants sown in April and May. As the vegetative growth period become progressively shorter with later plantings, the plant height becomes progressively shorter.

Seeds and sowing: The seed rate is 20 to 25 kg/ha for fibre and 60 kg/ha for green manure. The spacing is 30 × 5 cm is for fibre and 30 × 10 cm with topping at 30 days after sowing for seed production. Seeds are sown at a depth of 3 to 4 cm. Seedlings will appear within the first week after planting. Sensitive to moist soil conditions, if initial planting is followed by heavy rains within the first few weeks, many seedlings will not survive. Nitrogen fixation and growth will be reduced under cool and shady conditions.

Manures and fertilizer application: A fertilizer doses 40-20-20 kg of N, P_2O_5 and K_2O per ha is applied as basal dressing.

Water management: Crop is irrigated once in 10 to 15 days depending on soil and climatic conditions.

Cropping systems: Sunnhemp could also be used in rotations with kenaf (*Hibiscus cannabinus* L.) for non-wood fiber production. Kenaf is susceptible to root knot nematodes while sunnhemp is resistant. Sunnhemp could be grown after early season vegetables, tobacco and small grains or other winter crops (Canola) in the summer or after maize/corn and prior to winter small grains. The cropping sequence found suitable are *viz.*, sunnhemp-wheat; sunnhemp-potato; sunnhemp-mustard; sunnhemp-paddy-wheat; sunnhemp:-mustard-wheat.

Plant Protection: Sunnhemp is susceptible to diseases such as *Fusarium* wilt and anthracnose; and pests such as the sunnhemp moth, pod borers and stink bugs.

Harvesting: Harvest at pod formation stage to produce good quality fiber. Harvest the plants close to the ground with a sickle. Most of the leaves get dried up after 2 to 3 days of harvest. These plants are shaken to shed the leaves. The plants are tied into bundles having 15 to 20 plants in each bundle. When grown as forage, *C. juncea* can be harvested 6 to 8 weeks after planting and every 4 weeks thereafter. For green manure, incorporate crop stand in situ before it reaches full-bloom stage through plough or disc plough. This stage will occur be between 10–12 weeks after date of planting or when 50% of the flower buds have opened.

Retting and fiber extraction: It is done as in jute crop. The period of retting is 10 to 15 days. Microbial retting of sunnhemp is carried out using various strains of *Aspergillus niger* and *Bacillus subtilis.* The retting time is reduced from 7 or more days in water with naturally occurring water microorganisms to 4 days with known microbes in the distilled water. The optimum temperature for retting is 21 to 27°C. The fibre has more tensile strength and durability under exposure to humidity than jute. One person can strip only the equivalent of 2.7-3.6 kg of dry fibre per day.

Yield: As green manure crop, sunnhemp yields 2 t/ha of high quality dry matter in 6-8 weeks. Total green matter yields average 18-27 t/ha with forage yields ranging from 5-19 t/ha. Sunnhemp produces a fibre yield of 10 to 12 qtl/ha. Sunnhemp has 2 to 4% fiber on green stem weight basis and 8 to 12% on dry weight basis. Pulp yield is 50 to 54% from the whole stem of sunnhemp. Seeds are heart-shaped, with narrow end strongly in-curved, up to 6 mm long, dark brown to black. The seed weight is highly variable, ranging from 18,000 to 35,000 per kg. In optimum condition, the seed yield may range between 18-20 q/ha. The hundred seed weight ranges from 3.4-6.0 g depending on varieties. Seed should be dried to below 10% moisture and stored at 4.5°C. The ultimate fibre cells of sunnhemp are 3 to 5 mm in length. The strength of cordage fibre of sunnhemp is 185 kg as compared to 157 kg for cotton rope, 132 kg for hemp and 102 kg for coir.

Utilization: Sunnhemp is an excellent cover crop (green manure). Root nodulates freely and fix nitrogen of 40 to 60 kg/ha. Plants are erect, fast growing and ploughed in situ after 45 to 60 days. Plants are generally drought resistant and high yielding. The crop can be grown on infertile soil. It is useful for rotation with potato, tobacco, sugarcane, tea, coffee and other crops which are subjected to damages by nematodes. It is used as forage for goats and cattle. *C. juncea* contains toxic pyrrolizidine alkaloids, particularly in the seeds and pods. The pyrrolizidine alkaloid is a liver toxin which is poisonous to most livestock. sunnhemp hay can be safely incorporated at up to 45% level in rations of sheep under an intensive feeding system. Sunnhemp should not be fed to horses, and the intake of hay by cattle should be restricted to about 10% of their diet. The whole dry stems of sunnhemp can be used as raw materials for paper industries. Sunnhemp fibre is used for rope, string, fishing nets, mats, sacking, tarpaulins, soles of shoes, sandals and marine cordage. Sunnhemp fibre is stronger when wet and is also fairly resistant to moisture and micro-organisms in salt water.

REFERENCES

Abdul-Baki, A.S., H.H. Bryan, G.M. Zinati, W. Klassen, M. Codallo, and N. Heckert. 2001. Biomass yield and flower production in sunnhemp - effect of cutting the main stem. J. Veg. Crop Prod. 7:83–104.

Chaudhury, J., Singh, D.P. and S.K. Hazra. 2012. Sunnhemp. Central Research Institute for Jute & Allied Fibres (ICAR), Barrackpore 743, 24-Parganas (North), West Bengal

Chittapur, B.M. and Kulkarni, S.S. 2003. Effect of sowing dates on the performance of sunnhemp. J. Maharashtra Agric. Univ. 28(3):331-331.

Mansoer, Z., D.W. Reeves, and C.W. Wood. 1997. Suitability of sunnhemp as an alternative late-summer legume cover crop. Soil Sci. Soc. Am. J. 61(1): 246–253.

Nang Kyu Kyu Win, Hee–Young Jung and Shoji Ohga. 2011. Characterization of Sunn Hemp Witches' Broom Phytoplasma in Myanmar. J. Fac. Agr., Kyushu Univ., 56 (2), 217–221

Ram H, Singh. G. 2011. Growth and seed yield of sunnhemp genotypes as influenced by different sowing methods and seed rates. World J. Agric. Sci. 7(1):109-112.

Tripathi, M.K., Chaudhary, B., Bhandari, H.R., Harish, E.R. 2012. Effect of varieties, irrigation and nitrogen management on fibre yield of sunnhemp. J. Crop. Weed. 8(1): 84-85.

Ulemale, R.B., Giri, D.G., Shivankar, R.S. 2002. Effect of sowing date, row spacing and phosphate level on biomass studies in sunnhemp. J. Maharashtra Agric. Univ. 26(3): 323-325.

Ulemale, R.B., Giri, D.G., Shivankar, R.S. and Patil, V.N. 2003. Effect of sowing dates, spacing and phosphate levels on the growth and productivity of sunnhemp. Legume Res. 26(2): 121-124

http://plants.usda.gov/plantguide/pdf/pg_crju.pdf

http://edis.ifas.ufl.edu/

http://plants.usda.gov/java/

http://edis.ifas.ufl.edu/pdffiles/NG/NG04300.pdf

http://www.fao.org/ag/agp/AGPC/doc/Gbase/data/pf000475.htm

http://www.hort.purdue.edu/newcrop/proceedings1996/v3-389.html

https://hort.purdue.edu/newcrop/duke_energy/Crotalaria_juncea.html

http://agroecology.ifas.ufl.edu/sunn%20hemp.htm

http://www.sare.org/Learning-Center/Bulletins

http://www.fao.org/ag/AGP/AGPC/doc/Gbase/DATA/PF000475.HTM

http://www.fao.org/ag/AGP/AGPC/doc/Gbase/Default.htm

http://ecocrop.fao.org/ecocrop/srv/en/cropView?id=801

http://www.tropicalforages.info/key/Forages/Media/Html/Crotalaria_juncea.htm

http://www.sare.org/Learning-Center/Books/Managing-Cover-Crops-Profitably-3rd-Edition/Text-Version/Printable-Version

CHAPTER

21

Agave–*Agave sisalana* Perrine (2n=138)

Synonyms: *Agave americana* L

Family: Agavaceae

Vernacular names: Agave, Sisal or Sisal hemp, Century Plant, American aloe, (English), *Maguey* (Spanish), *Maguey* (Mexico), *Rakaspatta* (Hindi), *Semaikathalai* (Tamil), *Seemakathala* (Telegu), *Gayapathi* (Marathi) and *Kathaale* (Kannada). The nickname '*Century Plant*' emerged from the mistaken belief that the tall rosette of pale yellow flowers blossomed only after 100 years. In truth, depending on its environment, the plant's single bloom can appear anytime between 5 and 100 years.

Importance: Agave is a hard fiber crop which is obtained from the leaves or leaf-stalks. It is a monocotyledonous plant. Agave fibre is used extensively to make twines, ropes, sacks, cordage and carpets. The fibres of low quality are used as filling in furniture and in the manufacture of paper. Agave fiber cannot be spun as fine as jute. Agave is grown in containers as ornamental plant of potential in urban areas especially in city apartment balconies. As an ornamental, *Agave americana* is planted in private and public gardens and on roadsides. It is used as a hedge plant and planted along contours for erosion control and for reclamation of denuded and overgrazed land. Agave plants are grown along railway line, road sides, river banks and as a hedge plant in dryland areas throughout the country. It is grown in patches and as border crop in a neglected condition. Agave are often used for fencing in homesteads.

History: Agave is derived from the Greek 'agavos' meaning 'illustrious.' Sisal is native to Mexico. In the 19th century, its cultivation spread to Florida, the Caribbean islands and Brazil, as well as to countries in Africa, notably Tanzania and Kenya, and Asia. The first commercial planting in Brazil has been made in the late 1930's and the first sisal fiber exports from there have made in 1948.

Origin: *Agave sisalana* is native of Central America and Mexico.

Distribution: Agave is cultivated in Brazil, USA, Mexico, East Africa, Tanzania,

Angola, Kenya, Mozambique, South Africa, Venezuela, Haiti, Madagascar, Venezuela, China, Taiwan, Indonesia, Australia, India and South Africa. In India, it is grown in Tamil Nadu, Andhra Pradesh, Maharashtra, West Bengal and Karnataka.

Botany: The genus Agave was established by Linnaeus in 1753. The genus Agave is comprised of about 140 species which occur and are cultivated in semi-arid and arid regions of the world. Important fibre plants in this genus include *Agave tequilana* (AT), *Agave sisalana* (AS) and *Agave americana* (AA). The most common exploitation of AT is in the production of tequila, an alcoholic beverage produced by fermentation of sap from the pinas of the plant. The AS has been utilized over centuries for the production of fibre and subsequent products like rope, cordage and bags. On the other hand, AA has been used in many ways including a source of fibre.

Agave is a robust, perennial shrub. It is a monocotyledonous plant. Agave grows to a height of 3 to 6 m before blooming. The agave plant has a 7 to 10 year life-span. Agave leaves are gray-blue to blue-green with spines at the tips and on the margins. The leaf margin spines are recurved like fishhooks and the tip spines can be more than 2 to 2.5 cm long. The older leaves often gracefully arch down. Leaves are thick, fleshy, covered with a thick cuticle and a waxy coating. The margins of the leaf are prickly and the leaf itself ends in a spiny pointed tip. It bears rosette of long, erect, pointed fleshy leaves at the base up to 4 m wide. The leaves are long, straight, dark green and tip taper into sharp spine. The leaves are born in a broad basal rosette. Leaves are produced in spiral either in clockwise or anticlockwise direction. A plant produces 200 to 250 leaves before flowering. Matured leaf is 120 cm long and 10 to 15 cm wide. The leaf cuticle contains about 20% wax. Leaf production and size are best in *A. sisalana* followed by *A. americana*. It has a tufted, spreading, fibrous root system and no taproot. In favourable soil conditions, the roots will extend horizontally for distances of 1.5 to 3.0 m and even up to 5 m if the plant has free range. Roots concentrate in the upper 30 to 40 cm of soil while a few may go downwards, although rarely deeper than 1.50 m.

The 'Century plant' doesn't really take a century to bloom, but it lives for 10 years or so in warm regions and as much as 60 years in colder climates. Flowers are produced after 10 to 12 years though not a century. Agave blossoms only once during its life time and then dies after blooming (a condition called monocarpic). When the plant blooms, the central inflorescence spike, commonly termed the pole, elongates rapidly at the rate of 20 to 30 cm day^{-1}. The stem apex is transformed into a long flowering shoot or 'pole' to a height of 6 to 9 m, which bears a few scale-like leaves and has branches bearing clusters of white or slightly green flowers. It produces small plants form around the base of the parent plant. The small plants are technically called 'offsets' or 'offshoots' but are commonly called 'pups.' These may be separated from the parent plant for propagation. The plant produces a flower stalk that can be 15 feet tall or more. The plants are hermaphroditic. *Agave americana* produces cluster of yellow, rose, or white tubular or funnel-shaped flowers on the tall stems 6 to 9 m above the plant. Bulbils are small buds that develop in the axils of the flower stalks after flowering has taken place. Bulbils develop into small plants whilst still attached to the parent. These bulbils are then shed from the plant. If the bulbil falls into moist soil under favourable condition, it takes root easily and develops into adult plant. A massive agave pole with many flowering branches will bear from two to three thousand bulbils of various sizes. Bulbils are produced from buds in axils of papery bracteoles. The plant dies after producing flowers and bulbils. Pollination is effected by flying insects, especially bees, which are attracted to the nectar and carry off the pollen or else by the wind. Fruits are triangular capsules with many black seeds inside. Agave plants rarely set seeds. Seeds of about 150 may be produced from an inflorescence. Seeds are used for propagation. Freshly harvested seed has poor germination and viability is not lost during storage for a year. On dry weight basis of the plant, the leaves constitute 70%, stem (hole) 8% and roots 22%. Agave produces rhizomes from buds, which are located on the plant below the surface of the ground. Rhizomes vary in thickness from 1.5 to 3 cm and also in length. They extend horizontally, mostly at depths of 5 to 15 cm in the soil, before growing upwards to give rise to a new plant known as a 'sucker' at the surface. A plant may produce 5 to 10 rhizomes at one time and about 20 during its life span. Suckers are removed during cultivation. The required suckers may be used for propagation.

Physiology: Agave grow reasonably well since its Crassulacean Acid Metabolism (CAM) photosynthetic pathway permits it higher productivity on lands with severely restricted water availability and prolonged droughts. Agave sugar is a rich source of bioethanol for renewable energy. Agave could be an excellent Climate Change Mitigation (CDM) crop for bioethanol as well as for afforestation over poor quality lands with very limited access to water giving both permanent credits for bioenergy and the temporary credits of forestry for carbon sequestration. It causes no threat to food security and places no demand for the scarce water. Crassulacean Acid Metabolism (CAM) is the ideal photosynthetic pathway for places with low moisture availability. The CAM pathway is able to function in low moisture because the plants open their stomata only during night time thus limiting transpiration by a huge extent. In arid and semiarid lands the soil nutrition is also very limited which is, in itself, a consequence of low moisture availability. CAM plants have got over these handicaps by two major adaptations. One is closing stomatal openings during the hot day time and the other having surface roots that are able to access the moisture from the rare rains and the dew along with whatever little nutrition available on the surface soil.

Climate: Agave can be grown in the tropics and the subtropics. Agave is grown up to an altitude of 2000 m from MSL. It is a hardy survivor, tolerating heat, drought, and salty seaside conditions. It is a xerophytic crop. It can withstand long periods of drought. It grows best in areas with annual rainfall of 700 to 1300 mm. It is extremely sensitive to water logging. It can also be grown in areas receiving annual rainfall of as low as 250 to 375 mm with moderate humidity. Excessive rainfall is harmful. It is a hardy tropical plant requiring full sunlight or partial shade. The cardinal minimum, optimum and maximum temperature for agave is 10°C, 22 to 32°C and 38°C, respectively. Night temperature of -4°C causes leaf 'burn' and dieback.

Soil: Agave prefers well aerated soil sandy, sandy loamy or sandy clay loam soil. It can be grown in black cotton soil and in calcareous soils. It can be grown on poor, shallow rocks. It is intolerant of wet or swampy soils. It can be grown in acidic to alkaline soils with a pH between 6 and 8. The plants need a soil depth of at least 75 cm for good growth.

Varieties: Leela

Season: Planting can be carried out during June-July.

Propagation: Agave rarely set seeds. When seeds are produced, it is only used for experimental or breeding purposes. Agave is produced vegetative through suckers appearing at the base, or bulbils/plantlets formed on the flower spike or clones. The grown up suckers can be dug out and planted during rainy months. Bulbils are four inches in length. The bulbils are first sown in mother beds at close spacing of 50 × 25 cm. The bulbils are planted at a depth of 1.5 cm. After 6 months, the seedlings are pulled out and again planted in secondary nursery bed of size 20 × 1 m@ 500 plants/20m^2. In the second stage, it is kept for three months. After 9 months from the date of planting bulbils suckers weighing 1/4 to 1/2 kg and 9 to 12" height are ready for planting.

Main field planting: Bulbils are transplanted to the field after 12 to 18 months in the nursery when they are 25 to 40 cm height and with about 15 matured leaves. The planting depth is 5 to 7.5 cm. Spacing of 2.0 × 0.8 m is practiced @ 5000 plants/ha. The spacing of 2 × 2 m in pits of size 30 cm^3 can also be adopted in the main field.

Weed management: Weeding is carried out during the first two years.

Manures and fertilizer application: A fertilizer dose of 40-20-20 kg of N, P_2O_5 and K_2O per ha as basal dressing is recommended.

Water management: Agave is an xerophytic crop mostly grown under rainfed condition. Water is well for the first few days when transplanting to give them a good start, but continued supplemental watering is unnecessary.

Harvest: Mature leaves can be harvested from 3rd year onwards. When plants are grown on fertile soils without severe prolonged drought, the first harvest of the basal leaves may be made about 2 to 3 years after planting. The agave plant lives for 10 to 12 years. The first harvest of leaves is carried out when plant height is 1.5 m by which time the plant may produced about 100 leaves of which 35 to 40 are of economic value. Generally, the leaves making an angle of 60° or less with the ground level are cut by hand with a machete or a very sharp knife. The cut is made approximately 5 cm from the leaf base. Leaves terminal spines are cut off. Then the leaves are bundled or tied in groups of 30 to 50 for transport to the factory. At least 25 leaves should be left after the first cut and 16 to 18 leaves at subsequent cut. 25 to 50 or even more leaves can be harvested once or twice a year. After harvest, farmers should leave 20 to 25 leaves with a vertical inclination of 40° on the plant to provide sufficient leaf area at the first cutting. This is reduced from 16 to 18 leaves at subsequent cuttings. Subsequently harvest the leaves at intervals of 6 to 12 months. The number of cuts depends on the conditions of growth. A plant produces 200 to 250 leaves during its productive life. Each green leaf weighs between 500 to 700 g. Each leaf contains an average of around 1000 fibers. The fibers account for only about 2 to 5% of the plant by weight. A yield of 900 kg fibre can be obtained from a stand of 5000 plants per hectare. *Agave sisalana* produces a better quality fibre than *Agave americana*. Leaf yield of 75 to 100 t/ha could be harvested from 3rd year onwards. The leaf: stem dry weight is 4:1.

Processing: The fibres are located in the succulent tissue of the leaf, associated with vascular bundles. Retting and mechanical methods is used for fiber extraction. In the retting process, the harvested leaves are immersed in water for about a week. During this period the pithy matter adhering to fibre disintegrates due to bacterial action. The retted leaves are then beaten on stones to remove the remaining extra matter. The separated fibre is washed and dried in the sun and then baled. Instead of traditional hand scraping process now mechanical decorticators are available for the extraction of fibre. In mechanical process, modern decorticator is used. A modern decorticator can treat 25000 leaves weighing 10 to 20 t/hour, for which 100 ha of agave is required to keep it in permanent use. A modern decorticator requires 36000 to 45000 liters of water per hour for processing. Fibre decortication involves stroking and scrapping on rollers of 100 to 150 cm diameter and 30 cm width. The scraping knives of 5 cm length and spaced at 25 cm are located in front of the rollers at a right angle to the direction of rotation. As the leaf passes through the rollers, the pulp gets separated from the fibres, which are then washed with water, cleaned and dried and packed in bales. As a precaution, fibre extraction is done on a bright sunny day and within 2 days of the harvesting of the leaves or else the quality of the fibre will be deteriorated. Processed fibre contains 60% water. It should be dried as early as possible. Sun drying takes 8 to 12 hours in fine weather. Ultra violet rays are used to bleach the fibre. The fibre is dried to 15% moisture. The packed fibre contains 10 to 12% moisture. The fibre colour varies from milky white to golden yellow.

Fibre grades: Subsequently the fibres are dried under the sun, combed and classified. The fibre is classified as Grade A: white fibre, >90 cm long; Grade B: white fibre, 60 to 90 cm long; Grade C: white fibre, 50 and 60 cm long; and Grade D: refuse.

Yields: Agave produces 2000 to 2800 kg fibre/ha/year. Each green sisal leaf weighs on an average about 700 g and contains between 2 to 5% of fibre or about 1,000 individuals fibres. The individual fibre cells are small and 2 to 5 mm in length. Sisal fibre has a high percentage of cellulose of 72.0% and a relatively high proportion of lignin of 14.5%.

Utilization: Agave fibre is widely used for making ropes, cordage, twine, fishing nets, door mats and rugs and the short fibres are used for making mops, brushes. The waste material left after decorticating the leaves is used for making craft paper and paper boards. The fibres also contain about 73 to 78% of lignified form of cellulose. Sap from the inflorescence stalk of *Agave* sp. may be fermented to produce *pulque,* the national drink of Mexico. Agave syrup is a natural fructose sweetener extracted from Mexico's Agave plant. Traditionally, the Native Americans of Mexico collect the sweet juice, 'aguamiel', from several varieties of agave. They also used it as a sweetener, for special celebratory beverages. *A americana* is used to make a detergent, an insecticide, and an ingredient in an alcoholic drink (pulque). Calcium oxalate is highly concentrated in the sap of *A americana.* The calcium oxalate crystals, called '*raphides*' and saponins are found in the sap of the plant can function as a systemic poison that cause acidosis, vascular damage, renal tubule obstruction, erythematous eruption, severe pruritus and burning symptoms. The pina of the *Agave americana* contains up to 25% of inulin per weight basis. The leaf base contains up to 16% of fructans. Pina and leaf base can be used for the commercial production of fructans and long-chain inulin, which can be used as vaccine adjuvant in the pharmaceutical industry. Inulin is a valuable component used widely in the food industry as an additive, sugar substitute and prebiotic agent. Fructans are oligo- or polysaccharides and have value in the health and food arenas.

REFERENCES

Borland, A. M.; H. Griffiths; J. Hartwell; J. A.C. Smith. 2009. Exploiting the potential of plants with crassulacean acid metabolism for bioenergy production on marginal lands. Journal of Experimental Botany. Vol. 60. Issue 10. pp. 2879-2896

Kelly, J., and M. Olsen. 2006. Problems and pests of agave, aloe, cactus and yucca. AZ1399. Tucson: University of Arizona Cooperative Extension.

http://extension.arizona.edu/sites/extension.arizona.edu/files/pubs/az1399.pdf

Nobel, P.S., McDaniel, R.G. 1988. Low temperature tolerances, nocturnal acid accumulation and biomass increases for seven species of Agave. Journal of Arid Environments. 15(2): 147-156.

Pinkie E. Zwane , Michael T. Masarirambi, Nelisiwe T. Magagula, Abednigo M. Dlamini and Evison Bhebhe. 2011. Exploitation of *Agave americana* L plant for food security in Swaziland. Am. J. Food. Nutr, 2011, 1(2): 82-88

Sahu, G.C., Mishra, A. 1994. Morphology, characteristic and classification of soils under sisal (*Agave sisalana*) cultivation. Journal of the Indian Society of Soil Science. 42(1):111-114.

http://www.crijaf.org.in/SideLinks/Research/Varieties.html

http://edis.ifas.ufl.edu.

http://en.hortipedia.com/wiki/Agave_americana

http://.www en.wikipedia.org/wiki/Agave_americana.

http://www.cactus-art.biz/schede/AGAVE/Agave_americana/Agave_americana/Agave_americana.htm

http://www.hardytropicals.co.uk/Suculents/Agave_americana.php

https://ag.arizona.edu/pima/gardening/aridplants/Agave_americana.html

http://www.fao.org/economic/futurefibres/fibres/sisal/en/
http://www.fao.org/agriculture/crops/news-events-bulletins/theme/plant-production-and-climate-change/en/?no_cache=1
http://www.ars-grin.gov/cgi-bin/npgs/html/taxon.pl?1690
http://agritech.tnau.ac.in/agriculture/fibrecrops_agave.html
https://edis.ifas.ufl.edu/pdffiles/EP/EP41900.pdf
http://www.hardytropicals.co.uk/Suculents/Agave_americana.php
http://www.issg.org/database/species/ecology.asp?si=1664&fr=1&sts=&lang=EN
http://.www. Ag.arizon.edu/pina/gardening/aridplants/Agave-americana html
http://scihub.org/AJFN/PDF/2011/2/AJFN-1-2-82-88.pdf

CHAPTER

22

Kapok–*Ceiba pentandra* (L) Gaertn. (2n=72, 84)

Synonyms: *Bombax ceiba* L., *Bombax pentandrum* L., *Bombax malabaricum*

Family: Bombacaceae / Malvaceae

Vernacular name: Kapok, Silk-floss Tree, Floss-silk tree, silk-cotton tree (English), White cotton tree, White silk cotton tree (Mexican), Ceiba (Colombian), *Ceibo* (Spanish), *Ilavum* (Tamil), *Sveta salmali* (Sanskrit), *Safed semal* (Hindi), Pannimaram (Malayalam), *Dudi* (Kannada), *Tellaburga* (Telugu), *Safed simal* (Hindi), *Samali* (Marathi), *Schwet simul* (Bengali), *Kekabu* , *Kabu-kabu* (Malay), *Kapokier* (French), *Kapokbaum* (German). The generic name, 'Ceiba' comes from a local South American word Ceiba (seiβa / Seyba) which is a small municipality in northeast Puerto Rico. The specific name, 'pentandra', is Latin for 'five-stemmed'; from the Greek word 'penta' (five) and 'andron' (male).

Importance: Kapok fibre is obtained from the fruit of the kapok tree. This fibre is the best known tree fibre which follows cotton as the most valuable of the seed fibres. A short fibre, averaging about 19 mm in length, it is silky, lustrous, fluffy and very light, being only about one-sixth of the weight of cotton. Its springiness, resilience and vermin proof properties make kapok ideally suited for stuffing mattresses, pillows, cushions and other upholstery articles. Because of its lightness, water repellency and high degree of buoyancy (about five times that of cork), the fibres are valuable in making life-belts, life-jackets, life-buoys and other naval life-saving appliances. In a compressed state, kapok can support about 36 times its own weight in water. It retains its buoyancy even after prolonged immersion in water, the loss being quite gradual, i.e. only 10% after a month of continued immersion. Because of its low thermal conductivity, it is an excellent material for insulating ice boxes, refrigerators, cold storage plants, offices, theatres and aeroplanes, etc. Kapok is also a good sound absorber and hence is widely used for acoustic insulation in aeroplane cabins, tanks, broadcasting studious, theatres and hospitals. Kapok tree is a national emblem of Guatemala and has also been adopted by Puerto Rico.

Origin: Kapok is originated in tropical South America (southern Mexico to the southern

Amazon). Kapok is considered native to the Bahamas and the Caribbean, Northern Mexico to Northern South America. It is also native in many countries of tropical West Africa from Cape Verde, eastward to Chad and south to Angola. One explanation for its transcontinental nativity is that sea currents might have carried the trees' floating seed capsules to new locations.

Distribution: Kapok is especially abundant in secondary forests in tropical Africa. It is mainly cultivated in Indonesia, Thailand, Myanmar, Jawa, Sri Lanka, Thailand, Indonesia, Cambodia, Colombia, Venezuela, Ecuador, West Indies, Cuba, Jamaica, Trinidad, Tobago East Africa and India. The main kapok producing states in India are West Bengal, Tamil Nadu, Bihar, Assam and Uttar Pradesh.

Botany: Kapok is a tall, deciduous tree of the tropical evergreen forests. It grows to a height of 10 to 15 m and has a broad, tall trunk reaching 1.80 m in diameter, with pronounced buttresses at the base. It bears short, sharp conical spines all along the trunk and branches. The leaves are compound palmate consisting of 5 to 8 leaflets, leathery and smooth lanceolate to elliptic sessile leaflets, with pointed tips. Kapok is considered to be a hermaphroditic species (flowers contain both reproductive parts). The small white or pink flowers are produced in dense clusters up to 20 cm in length. The flowers have nocturnal anthesis, with 1 to 20 flowers opening per inflorescence per night. The five petals, have staminal filaments and pistil are creamy white giving a mostly whitish colour to the flowers and inflorescence, spotted with the gold-yellow of the anthers. Most flowers are inclined or pendant, thus giving a globose appearance to the whole inflorescence. The calyx is bell-shaped, 1 cm long, with 5 to 10 short lobes; corolla 3-3.5 cm long, with 5 lobes, white to rose coloured, covered with silky hairs and five stamens. *Pentandra* means 'with five stamens'. The flowers are pollinated by birds, bats and bees. The fruits are elliptic, pendant, brown capsules that dehisce by five valves. In each capsule there are 66 to 250 ovoid seeds, 4 to 6 mm in diameter, weighing 45 to 65 mg, surrounded by the pale yellow silk cotton, which develops from the endocarp. The seeds have a float structure in their basal portion, indicating that in addition to wind dispersal, they can also be dispersed by water. The fruit is a brown ellipsoidal capsule which splits open into 5 valves. Fruit is 15 to 30 cm long, pointed at both ends, which rarely undergoes dehiscence on the tree. The tree is leafless when the fruits are ripe. The ripe fruits packed with whitish floss-kapok in which the seeds are embedded. The seeds are many, dark brown to black and are embedded in copious white, pale yellow or grey silky floss, which is derived from the epidermal cells of the capsule wall. The seed is pea-like in shape. Seeds cannot be stored for a long period of time. It can be stored in glass or plastic containers at 4°C with 60% relative humidity and 15 to 20% physical moisture, seeds can be preserved up to one year. When they are ripe and fresh, seeds have 90 to 100% viability. Seeds are pretreated by soaking in water for 24 hours and cutting longitudinally without damaging the embryo. Seed germination is good, using a substrate of sand under conditions of 20 to 30°C alternating temperature, with 8 hours of light daily.

Kapok can be differentiated into three types based on the branches, colour of the flower, number of fruits besides length and breadth of the fruits, pod breaking and colour of the fibre viz., i) *Ceiba pentandra* var. *indica*, ii) *Ceiba pentandra* var. *caribea* and iii) *Ceiba pentandra* var. *africa*. Of the above three types, the first two are mainly for fibre. *Ceiba pentandra* var. *caribea* is otherwise called as 'Singapore Kapok'. The local variety is *Ceiba pentandra* var. *indica*. The difference between indica and caribiea varieties are as follows: i) the branches will start from centre of the trees and grown horizontally in *indica* variety whereas in *caribea* variety the branches will go upright. ii) the fruits will be more in Singapore variety. After the age of 8 to 10 years, it will yield 800 to 900 fruits in the Singapore variety whereas in the local variety it is only 450 to 500. iii) the fruits are long and the length of fruits may vary from 25 to 35 cm in Singapore variety whereas in the local variety the length of the fruits is 10 to 15 cm. iv) the pods are not broken even in the fully matured fruits so there is no loss of fibre in Singapore variety whereas in the local variety, the fruits burst in the tree itself resulting in loss of fibre. v) colour of the fibre is pearl white in Singapore variety whereas in the local variety it is only dull pale white colour. vi) the number of seeds per fruits is very low in Singapore variety as compared to local variety.

Source: Elumalai et al, 2012

Kapok flowers

Kapok fruits

Climate: *Ceiba pentandra* grows in the dry, humid, and very humid tropics. It is cultivated widely in the tropics between 16°N and 16°S. Kapok can be grown at elevations below 900 m from MSL. The ideal climate is to have abundant rain during the growing season, followed by a dry period from the time when the flowers appear until the pods are harvested. It requires an annual rainfall of 800 to 1500 mm with abundant rainfall distribution during the vegetative period and a relatively drier period during flowering and fruiting. It is grown in areas with temperature of 18 to 38°C. Fruit setting fails when the night temperature is below 19°C.

Soil: A good, deep, permeable loamy soil is desirable for kapok. Laterite and heavy clay soils should be avoided for this crop. Soil should be free from waterlogging condition. Kapok seeds germinate and tolerate the salinity variant of up to 9 dSm^{-1}, which can be regarded as a LD50 value for severe damages and classified as moderately saline tolerant multipurpose tree species in seedling stage.

Season: Planting is done during rainy season or at any time with irrigation facilities.

Seeds and sowing: Fresh seeds can be used for sowing. The seeds are sown directly in the polythene bags of size 20 × 10 cm and watered daily. Seeds start germinating at 7th day and it will be over after 15 days. Four to six months old seedlings are generally used for planting with spacing of 7 × 7 m in pits with size of 30 × 30 cm. However, it is also easily raised by means of cuttings.

Weeding: Weeding is done based on necessity.

Cropping systems: Pulse crops can be raised as intercrop for initial two years. Bund planting with a spacing of 6 m can be taken up.

Water management: In areas receiving less than 1000 mm rainfall, irrigation is a must especially during summer season at 10 days interval.

Harvest: The tree comes into bearing in the fourth or fifth year, reaching its maximum after seven to ten years. Trees may remain productive for as long as 30 years or even more. The fruits are clipped by hand when the colour changes to brown and the surface becomes wrinkled. A mature tree produces as many as 600 pods in a year.

Yield: The kapok tree produces 500 to 900 fruits per tree per year with each fruit containing 120 to 175 seeds. It produces 15 to 20 kg of fibre floss in one year. Trees remain productive for 30 years. Seeds may be stored up to one year in glass or plastic containers at 4°C and 60% relative humidity.

Processing: The fruits are broken to open with a mallet and the fibre and seeds are removed by hand. The seeds with the fibres are then stirred with a stick and the heavy seeds drop to the bottom. During baling, care is taken not to exert excessive pressure otherwise the elasticity and other qualities of the fibres are altered. The mature pod, on average, contains 44% husk, 32% seeds, 17% floss and 7% placental material and pedicel on weight basis.

Fibre characteristics: Individual fibres are 0.8 to 3.0 cm long, thin walled with a wide air–filled lumen. Kapok fibres are very light, only about one-sixth the weight of cotton. The fibres cannot be spun into yarn because of their short, soft lustrous, untwisted and brittle nature. Kapok fibres have a cellulose content of about 64% and a lignin content of about 13%. The fiber is brittle and inflammable and not suitable for spinning into threads.

Utilization: Kapok fibre stuffing material is used in life preservers as the fibers are water-resistant and trap air. Kapok fibre is used for stuffing pillow, mattresses, lifebuoys, life-jackets, saddles, sleeping bags and protective clothing. Kapok floss in the fruits is used for stuffing and thermal and acoustic insulation. Very young, unripe pods are eaten in Java. It is used for making tennis ball, boxing gloves, shooting suits. Its oil is used for making soap, its cake used as cattle feed. Shell is used for fuel. Kapok seeds contain 20 to 25% non-drying oil, which is fairly similar to cottonseed oil and is used as a lubricant, in soap manufacture and for cooking purposes. Kapok oil, which is extracted from the seeds, is used for the manufacture of the soap and as a substitute for cotton-seed oil. The seeds contain large amounts of oil that tend to go rancid quickly and the viability diminishes rapidly. When the seeds are stored at 10-12 % moisture content in hermetically closed plastic bags at 15°C, they retain viability for 5-6 months. The residual mass, left after the extraction of the oil, is an important stock feed containing 26% protein.

REFERENCES

Anon. 1986. The useful plants of India. Publications & Information Directorate, CSIR, New Delhi, India.

Elumalai, A., Nikhitha Mathangi, Adarsh Didala, Raju Kasarla and Yetcharla Venkatesh. 2012. A Review on *Ceiba pentandra* and its medicinal features. Asian J. Pharm. Tech. 2 (3): 83-86

Murawski, D. A. and Hamrick, J. L. 1992. Mating system and phenology of *Ceiba pentandra* (Bombacaceae) in Central Panama. Journal of Heredity. 83:401-404.

Rex Immanuel, R. and M. Ganapathy. 2007. Growth and physiological attributes of Ceiba pentandra (L.) Gaertn. Seeds and seedlings under salt stress. Journal of Agricultural and Biological Science. 2(6): 12-16.

Rogeârio Gribel, Peter E. Gibbs and Aldenora L. and Queiroâ, Z. 1999. Flowering phenology and pollination biology of *Ceiba pentandra* (Bombacaceae) in Central Amazonia. Journal of Tropical Ecology. 15:247-263

Sankaram, A. 1948. A note on the cultivation of kapok. Indian Forester. 74: 383-385.

Singh, S.P. and Singh, D. 2010. Biodiesel production through the use of different sources and characterization of oils and their esters as the substitute of diesel: A review. Renewable and Sustainable Energy Reviews. 14 (1): 200-216.

Thakur, M.L.; Pillai, S.R.M. 1984. A new defoliator of *Ceiba pentandra* from Tamil Nadu, India. Indian Forester. 110(6): 558-560.

Ceiba Foundation for Tropical Conservation: www.ceiba.org/ceiba.htm

http://www.tropilab.com/ceiba-pen.html

http://www.worldagroforestry.org/treedb/AFTPDFS/Ceiba_pentandra.PDF

http://lee.ifas.ufl.edu/Hort/GardenPubsAZ/Ceiba_pentandra.pdf

http://www.asianpharmaonline.org/AJPT/1_AJPT_2_3_2012.pdf

http://www.worldagroforestry.org/treedb2/speciesprofile.php?Spid=501

http://www.forestrynepal.org/images/simal_present_report3.pdf

CHAPTER

23

Sugars and Starches

Carbohydrates play a major role in human diets, comprising some 40-75% of energy intake which is a universal fuel for all cells, the cheapest source of dietary energy. It is divided into available and unavailable forms. The available carbohydrate is defined as 'starch and soluble sugars' and unavailable form occur as 'mainly hemicellulose and fibre (cellulose)'. The importance of carbohydrates in foods is shown in Table 1.

Table 1 List of important carbohydrates in foods

Monosaccharides	**Glucose, Fructose**
Disaccharides	Sucrose, Lactose
Oligosaccharides	Raffinose, Stachyose, Fructo-oligosaccharides
Polysaccharides	Cellulose, Hemicelluloses, Pectins, b -Glucans, Fructans, Gums, Mucilages, Algal polysaccharides
Sugar alcohols	Sorbitol, Mannitol, Xylitol, Lactitol, Maltitol

The food starches are derived from seed (wheat, maize, rice, barley) and root (potato, cassava, tapioca) sources. Sugar is used to describe purified sucrose as are the terms 'refined sugar' and 'added sugar'. Sugars are used as food sweeteners which are rich sources of energy. Sugar is highly esteemed because it is non-toxic, although its consumption has been linked to dental caries, obesity, etc. Sugar is classified as natural caloric sweeteners and synthetic sweeteners.

(a) Natural caloric sweeteners: These are nutritive sweeteners provide calories in the form of sugars. Some are found naturally in foods like fruit and milk and others are added to foods during preparation or processing, for example, high fructose corn syrup.

(i) Sucrose: Cane sugar can be processed as brown sugar, domestic syrup or treacle. Treacle is a by-product of the industry with about 50% sugar and is used to prepare alcohols, citric acid, yeast, animal feeds, etc. Sucrose is manufactured mainly from sugarcane and sugarbeet.

(ii) Fructose is also known as levulose and fruit sugar. It is the white crystalline reducing sugar that occurs naturally in fruits, flowers, honey and root vegetables. It makes up about half the sugar in sucrose or high fructose corn syrup. It does not increase blood sugars like other sugars but excessive amounts have been shown to increase blood lipids so it is not recommended for diabetics or those with active heart disease.

(iii) Glucose is the main source of fuel for the body and is the only sugar usable by brain cells. The body converts other sugars and carbohydrates to glucose. It is found in nearly all plant foods. Glucose makes up half of the sugar in sucrose and high fructose in corn syrup and is also known as dextrose and blood sugar.

(iv) Galactose is found in milk and other dairy products, galactose is less sweet than glucose. When paired with glucose, galactose forms the sugar, lactose.

(v) Lactose is known as milk sugar. It is a di-saccharide containing galactose and glucose. Since it is less sweet and soluble than other sugars, it is rarely added directly to foods except commercial baby formulas, baked products and stout beer. Some people lose the ability to digest lactose and become lactose intolerant. This sweetener is not recommended for infants less than 1 year of age.

(vi) Maltose is composed of two glucose molecules, maltose is sometimes referred to as malt sugar. It is found in molasses and is used in fermentation. It can be found in beverages, mostly beer, and in foods like cereal, pasta, potatoes and many processed products.

(vii) Polyhydric sugar alcohol: These sugar alcohols such as sorbitol, mannitol, and xylitol which are also considered nutritive sweeteners because they provide energy to consumers along with sweetness. The sorbitol is prepared from dextrose, maltitol prepared from maltose and xylitol prepared from wood waste. These are used mainly in soft ice creams and are thought to cause less tooth decay. Xylitol is suspected of being carcinogenic.

(viii) Starch hydrolysates: Glucose syrups derived from starch which contains dextrose (D-glucose) and other higher saccharides.

Non-Nutritive Sweeteners provide zero or very low calories since the molecules are large and are partially indigestible. These are usually much sweeter than sugar.

(i) AcesulfameK with brand names are 'Sweet One', 'Sunett' and 'Swiss Sweet'.

(ii) Aspartame with brand names are 'Equal' and 'Nutrasweet'. Aspartame has been approved for general distribution in the USA, Canada and Western Europe. It is available as table top sweetener and is used in processed foods and beverages.

(iii) Neotame is manufactured by the Nutrasweet company and labeled as 'Neotame'.

(iv) Saccharin with brand names 'Sweet N'Low', 'Sugar Twin', 'Necta Sweet'. It is under suspicion of being carcinogenic. It is the most widely used synthetic sweetener. It is also the cheapest.

(v) Cyclamate is 30 to 80 times sweeter than sucrose. This product has been prohibited from sale because of being carcinogenic.

(vi) Sucralose with brand name 'Splenda'

(vii) Stevia is also known as Stevioside, and with brand names 'Sweet Leaf' and 'Only Sweet'. Stevia sweetener is derived from the leaves of herbs native to subtropical and topical regions in western North and South America. Its sweetening compound is a diterpene glycoside called stevioside which is 300 times sweeter than sugar, but it provides no calories. It is often found in diabetic foods since it has zero calories, does not raise blood sugars and may improve glucose tolerance. This product can be found in liquid or powder forms.

Sugar crops: Sugar crops are plants that are natural sources of the sweet substance called sugar in large quantities. The main sources of sugar are sugarcane, sugarbeet and the palms namely palmyrah and date. The use of sugarcane for the production of jaggery has been in existence for many centuries. In Europe, Canada and USA, sugarbeet is the source of sugar. This has not come into prominence in the tropical countries because of the sugarcane crop which gives high cane and sugar yield under the tropical conditions. In general, the costs of producing sugar from sugarcane are lower than for sugarbeet. Approximately 80% of the sugar output is sourced from sugar cane, which is largely grown in tropical countries. The remaining 20% is produced from sugar beet, which is grown mostly in the temperate zones of the northern hemisphere.

The Indian Sugar Industry, with an annual productive capacity of over 25 MMT, stands out to be the second largest in the world after Brazil, accounting for around 15% of the global sugar production. The country consumes approximately 22 MT of sugar annually, with Maharashtra contributing over 60% of it while the rest of the output come from states like Tamil Nadu, Karnataka, Uttar Pradesh and Madhya Pradesh. Sugar companies have been established in large sugarcane growing states like Uttar Pradesh, Maharashtra, Karnataka, Gujarat, Tamil Nadu, and Andhra Pradesh and these six states contributing more than 85% of total sugar production in the India while, over 60% of total production is together contributed by Uttar Pradesh and Maharashtra. The largest world markets for sugar are confectionary, bakery products and soft drinks. By-products derived from sugarcane include rum, molasses (which is used to produce syrups for the food industry), ethanol (motor fuel), bagasse (woody cane fiber used as biofuel for mills, pulp for paper industry and building materials) and filter cake (animal feed and fertilizer).

The tapped juice from the palmyrah is generally converted into palm gurs, which form the source of cheaper sugar to the people. The particulars of the sugar crops which are sources of various types of sugar are furnished in Table 2.

Table 2 List of sugar crops that are major sources of various types of sugar

Common name	Scientific name	Family name
Sources of Sucrose		
Sugarcane	*Saccharum officinarum*	Poaceae / Gramineae
Sugarbeet	*Beta vulgaris*	Chenopodiaceae / Amaranthaceae
Sugar maple, Maple tree	*Acer saccharum*	Aceraceae
Black maple, Black sugar maple, Hard maple, Rock maple	*Acer nigrum*	Aceraceae
Sweet sorghum, sorgo	*Sorghum bicolor*	Poaceae / Gramineae
Palms (various names)	various species	Arecaceae / Palmae
Sources of Glucose		
Corn, maize (corn syrup or crystallized corn sugar from starch)	*Zea mays*	Poaceae / Gramineae
Potato, White potato (from starch)	*Solanum tuberosum*	Solanaceae
Sources of Fructose		
Dahlia (from inulin)	*Dahlia pinnata*	Asteraceae

Jerusalem artichoke, sunchoke, sunroot (from inulin)	*Helianthus tuberosus*	Asteraceae
Corn, maize (high-fructose corn syrup from starch)	*Zea mays*	Poaceae / Gramineae
Sources of Galactose		
Flax, Linseed (from flaxseed gum or mucilage)	*Linum usitassitimum*	Linaceae
Sources of Mannose		
Ivory nut, Ivory nut palm (from endosperm)	*Phytelephas macrocarpa*	Arecaceae / Palmae
Manna ash tree, Flowering ash (from juice secreted through the bark)	*Fraxinus ornus*	Oleaceae
Sources of Maltose		
Barley	*Hordeum vulgare*	Poaceae / Gramineae
Rice	*Oryza sativa*	Poaceae / Gramineae

Starch crops: In the Indian diet, the cereals supply the bulk of the starch as in rice, sorghum, maize and other cereals. Starch is also much needed in confectionaries, textile and cosmetics industries. Starchy food is also obtained from root and tuber crops. Tuber crops (stem modifications) are white potato (*Solanum tuberosum*) and cocoyam (*Colocasia esculenta*). Root crops are elephant foot yam, cocoyam or taro, sugarbeet, carrot, yam, sweet potato, tapioca, arrowroot, radish, etc.

FAO defines roots and tubers as plants yielding starchy roots, tubers, rhizomes, corms and stems. They are used mainly for human food, for animal feed and for manufacturing starch, alcohol and fermented beverages including beer. FAO classifies seven primary crops under the heading of root and tuber crop. The crops are Irish potato, sweet potato, tapioca, yautia (*Xanthosoma*), taro (Cocoyam) and Yam (*Dioscorea*).

A tuber is an underground stem that stores food. Potato is a tuber because it has nodes (eyes) which produce new shoots. The examples of tuber crops are yam, jerusalem artichoke, potato, etc. The examples of root crops are beet, carrot, radish and of tuber crops are yam, potato, sweet potato and tapioca. The root and tuber crops have very high yield potential although their protein, mineral and vitamin content are generally low compared to cereals. The orange-fleshed sweet potato is also well known as a rich source of beta-carotene, a precursor of vitamin A. The root and tuber crops are rich in carbohydrate and are commonly used as staple food, livestock feed, or as raw materials for the production of industrial products such as starch and alcohol, or processed into various food products. In tropical countries, cassava, sweet potato, taro or gabi, yams and arrowroot are staple food crops. Potato is common in the temperate and subtropical countries.

The modified root consists of the tuberous and fleshy roots while the modified stem consists of tubers, corms, bulbs, rhizomes and stolons. Tuber is a fleshly modification of a stolen or stem which enlarge rapidly and longitudinally during its growth phase. It has ability to store starches and sugar in specialized parenchymatous cells, structure at or below the soil surface. A corm is a swollen, vertical stem with a papery covering. Rhizomes are underground stems, horizontally-growing and produce shoots and adventitious roots. Stolons are aboveground stems which grow horizontally, produce shoots and adventitious roots.

A tuberous root is a thickened secondary root as in yam bean while the fleshy root is an enlarged primary root as in carrot (*Daucus carota*), ginseng (*Panax* spp.) and sugar beet (Beta vulgaris). In these root crops which produce fleshy roots the upper portion on which roots are absent is a hypocotyl, the first internode of the stem. The portion on which secondary roots develop is a root. However, in radish (*Raphanus sativus*), the fleshy root consists mainly of the hypocotyl. A tuber is an enlarged tip of an underground stem with leaves reduced to scales or scars subtending the auxillary buds. Tubers are the enlarged storage tips of a rhizome. The 'eyes' are actually buds in nodes, arranged in spiral pattern from the base to the apex of the tuber. Aerial tubers, which are common in yams, are called tubercle. A corm is a short, solid, thickened compressed modification of underground stem found in monocots. Small corms are called cormels. The corms are usually flat in shape having numerous roots at the bottom, and a tuft of leaves at the top. They are distinguished from bulbs with their lack of fleshy leaves but having a covering of dry papery leaves. Corms store food in the stem, unlike the bulbs which store food reserves in the leaves. The rhizome is a modified underground stems (e.g. ginger, iris) and bulb (e.g. onions, tulips, daffodils, hyacinths, lilies), but they are not produced for starch. A rhizome or rootstock is an underground, horizontally enlarged shoot, more or less fleshy when fresh. Bulbs, like corms, are modified stems found in monocots. Bulbs are erect underground stems having both fleshy and papery leaves with the fleshy leaves serving as food storage organs. The term 'bulb' is also used commonly to refer to plants having fleshy underground storage organs including the root crops, but only a few are true bulbs.

CHAPTER 24

Sugarcane

Family: Gramineae (Poaceae)

Vernacular name: Sugarcane, Noble cane (English); *Sessal, pundia, paunda* (Hindi); *Karumbu* (Tamil), *Oos* (Marathi), *Kabbu* (Kannada), *Cherukku* (Telugu); *ikshu, khanda, sarkara* (Sanskrit); *Karimbu, Karimpu* (Malayalam).

Importance: The name sugar derived from the sanskrit word 'Sarkara' meaning gravel or sand. The original name has been changed to 'Sukkar' in Arabic, 'Sakharon' in Greek, 'Sucre' in French and finally to sugar in English. Of the two major sugar crops, 62% of world sugar production is achieved through sugarcane while sugarbeet contributes 38%. India is the 2nd largest sugar producer in the world which consistently producing 25 million tons per annum. There are about 526 working sugar factories are located in the country with total crushing capacity of about 242 lakh tonnes. There are 5 million farmers and their families grow 350 million tons of cane. India is the largest consumer of sugar in the world. The highest cane price is paid per ton of sugarcane in the country. The production of molasses as a by-product of the sugar industry and production of alcohol by distilleries also play a very significant role in the Indian Economy. The alcohol is required as a base material for pharmaceutical and in several other industries. In India, 41.9% of cane is utilized for white sugar production while 58.1% used for *gur* and *khandasari* production. The *gur* and *khandasari* produced in India amounts to 8.3 billion tons. Sugarcane is also grown as energy crop.

History: The earliest record on sugarcane is found in a sacred book of the Hindus, the Aharva veda which has been composed between 5000 and 1000 B.C. The first mention of sugar is found in the treatise called 'Preeti Mokhsha' a code of conduct for Buddhists, which is written around 600 B.C. Sugar forms important part of 5 nectars (other being honey, milk, curd and ghee which are commonly used in Hindu rituals). Alexander the Great came to India and his scribe mentions that they have seen local people chewing reed which produce honey without bees. The first sugarcane hybrid has been developed in 1708 by Fair-Child. First sugar factory on a small scale has been established in the year 1824 AD at Aska, Odisha by a Frenchman named Mr. James Frederick Vivian Minchin. Sugarcane research work has been initiated first at Manjari Research Station near Pune in 1892. The Sugarcane Breeding Institute, Coimbatore has been established by the Government of India in 1912. The first Indian commercial sugarcane hybrid variety, Co 205 has been evolved in 1918 at Coimbatore. Indian Sugar Committee has been appointed by the Government of India in 1920. Indian Sugar Mills Association has been established in 1932 at New Delhi. The Government of India *vide* Sugar Industry (Protection) Act 1932, has granted protection to the entrepreneurs who established sugar factories. Indian Institute of Sugarcane Research, Lucknow has been established on 16-02-1952. National Federation of Co-operative Sugar Factories Ltd has been established in 1960 at New Delhi. National Sugar Institute has been established in 1963 at Kalyanpur, Kanpur, Uttar Pradesh. All India Co-ordinated Research Project on Sugarcane is functioning from 1970-71 at IISR, Lucknow. Vasantdala Deccan Sugar Institute has been established in 1975 at Manjari near Pune.

Origin: The origin of different species of sugarcane is furnished in Table 1.

Table 1 Origin of different species of sugarcane

Species	Origin	Chromosome number
S. spontaneum	Southern Asia	2n= 40-128
S. robustum	Papua New Guinea	2n= 60-200
S. officinarum	South Pacific islands, Papua New Guinea, Indonesia, India.	2n= 80
S. barberi	India	2n= 111-120
S. sinense	China and India	2n= 80-124
S. edule	South Pacific islands and Papua New Guinea	2n=60-80 with aneuploid forms

Source: Buzacott 1965; Daniels and Roach 1987.

Distribution

S. officinarum L. Is known as 'Noble cane' because of fine thick stem. It is distributed in Mauritius, Java, South Pacific islands and India. It is spread in tropics and subtropics. It is rich in sugar and has thick stems with soft rinds. They grow well only under favorable conditions. Stem is large barreled. It has low fibre and high sucrose content. It is susceptable to diseases and pests. Sugarcane seed is a caryopsis.

S. sinense Roxb. Is distributed in India, Combodia, China and Taiwan. It has thin stem and narrow leaves. Stems have high fibre content and the juice quality is poor. Stem is hard and small barreled. It contains fair amount of sucrose and is resistant to diseases. It is mainly used as fodder.

S. barberi Jeswiet. Is distributed in India, Egypt, Syria, Cyprus, Spain, Portuguese, Indonesia and Taiwan. It is suited to subtropics and temperate region. It tillers profusely. It is an intermediate between noble cane and wild cane. Internodes are cylindrical, small barreled, spindle shaped. It has high fibre content and resistant to diseases.

S. robustum Brandes and Jeswiet. Is distributed in New Guinea and Hawaii. Its growth is vigorous and luxuriant. Stem is thick. Many tillers are formed and may grow to a height of 10 m. It has low sucrose and resistant to diseases.

S. spontaneum L. Is known as 'Darba grass' or 'wild cane'. It is distributed in India, China, Malaysia, Taiwan, Middle East, New Guinea and Africa. It is vigorous, thin and grassy forms. It does not contain sucrose. It is resistant to drought, flood, pest and diseases.

Brazil, Europe Union, India, China, USA, Australia, Thailand, Mexico, Cuba and Pakistan are some of the largest producers of sugar in the world. Sugarcane cultivation in India is divided in to 5 zones based on area *viz*., peninsular zone, north western zone, north central zone, north east zone, and east coast zones. The zones vary widely in productivity of sugarcane and sugar recovery. The sugarcane growing states are grouped based on productivity and sugar recovery in Table 2 and 3.

Table 2 Sugarcane productivity groups in India

Productivity group	Yield (t/ha)	States
High productivity	> 70	Tamil Nadu, Maharashtra, Andhra Pradesh, Gujarat and Karnataka.
Medium productivity	50 – 70	Uttar Pradesh, Haryana, Punjab, Odisha, West Bengal and Kerala.
Low productivity	< 50	Bihar, Madhya Pradesh, Rajasthan and Assam.

Table 3 Sugar recovery groups in India

Sugar recovery group	Recovery	States
High recovery	> 10%	Maharashtra, Gujarat and Karnataka.
Medium recovery	9-10%	Uttar Pradesh, Haryana, Bihar, Madhya Pradesh and Rajasthan.
Low recovery	< 9%	Andhra Pradesh, Tamil Nadu,West Bengal, Assam, Kerala and Odisha.

Production potential of sugarcane: Theoretical maximum and recorded levels of cane, sucrose % and sugar yield for sugarcane is presented in Table 4.

Table 4 Theoretical maximum and recorded levels of cane, sucrose % and sugar yield

Parameters	Theoretical maximum	Recorded level
Cane yield (t/ha)	339.42	264.0
Sucross (%) in juice	26.00	22.5
Sugar yield (t/ha)	55.90	42.0

Although maximum yields of about 264 t/ha has been recorded in sugarcane crop, the average yield in farmers' fields are still at 49.81 t/ha in subtropical region and 81.75 t/ha in tropical region. However, an average cane yield of 108 t/ha is observed in certain areas of Tamil Nadu. A few farmers have produced 200 t/ha in Tamil Nadu. It is thus evident that the recorded yield might touch the potential yields at least under favourable soil and climatic conditions.

Botany: Sugarcane is a perennial grass growing erect and tall, roots adventitious, fibrous; stems form a big clump, stems or culms more or less cylindrical, solid, jointed with distinct nodes and internodes; inflorescence a silky panicle called 'arrow', each spikelet is subtended by soft silky hairs called the 'callus hairs', fruit a very small caryopsis and used as seed. The root system is fibrous and consists of two types/kinds of roots will develop from a planted seed piece, namely i) sett roots and ii) shoot roots. The root primordia (translucent dots) situated at the base of every cane joint is activated and produces roots.

These roots are known as 'sett roots' and are mostly temporary. The germinating seed piece depends entirely on the sett roots for water and nutrients in the initial stages before shoots form. After the emergence of the primary shoot from the bud, other roots are produced from lower rings of the lower nodes of the shoot/ tillers. These roots produced from shoot are known as 'shoot roots'. The sett roots' function will eventually be taken over by the shoot roots as they develop. The life of the shoot root is also limited. New roots are continuously produced as new tillers are formed. In a ratoon crop of sugarcane, the root system is shallower than that of the main plant crop. Each new tiller (shoot) will develop its own roots that eventually take over the function of the original shoot roots. These are permanent roots and are thick, fleshy and white in colour. New roots are continually produced from tillers. The stalk consists of segments called joints. Each joint is made up of a node and an internode. The node is where the leaf attaches to the stalk and where the buds and root primordia are found. A leaf scar can be found at the node when the leaf drops off the plant. The joints at the base are short and internodal length gradually increases. One bud is present on each node and they alternate between one side of the stalk to the other. When seed-cane is planted, each bud may form a primary shoot. From this shoot, secondary shoots called 'tillers' may form from the underground buds on the primary shoot. In turn, additional tillers may form from the underground secondary shoot buds. All colours of the stalk derive from two basic pigments: the red colour from anthocyanin and the green from chlorophyll. The ratio of the concentration of these two pigments produce colours from green to purple-red to red to almost black colour. Yellow stalks indicate a relative lack of these pigments. The surface of the internode, with the exception of the growth ring, has a distinct waxy coating. The stem of the sugarcane is roughly cylindrical and is divisible into distinct nodes and internodes. There is a distinct root ring called the '*keim ring*' in the stem of sugarcane. This root ring has three or more rows of root primordial. The average length of a cane is about three to four metres and the average weight is 2 to 3.5 kg. The leaf of the sugarcane plant is divided into two parts: sheath and blade, separated by a blade joint. The number of green leaves per stalk is around ten, depending on variety and growing conditions. The blade joint is where two wedge-shaped areas called 'dewlaps' are found. The 'top visible dewlap' leaf is a diagnostic tissue that is frequently used in nutritional studies. When a sugarcane plant has reached a relatively mature stage of development, the growing point ceases forming leaf primordia and starts the production of an inflorescence. The inflorescence, or tassel, of sugarcane is an open-branched panicle. Each tassle consists of several thousand tiny flowers, each capable of producing one seed. Spikelets open from 3 a.m. to 8 a.m. with peak between 6 a.m. to 8 a.m. The seeds are extremely small and weigh approximately 250 per gram or 113,500 per pound. Generally, a day length close to 12.5 hours and night temperatures between 20 and 25°C will induce floral initiation. Flowering or *arrowing* in sugarcane varies with varieties, seasons and environmental conditions. Some varieties do not flower at all, some varieties flower sparsely and some flower is profusely. Extreme drought and ill drained conditions induce flowering; crops planted in February-March in South India arrow by the end of October and beginning of November. Crops planted after June do not arrow ordinarily. Arrowing is less if the crop is well manured and grown under rich nutrition and good irrigation.

Classification of *Saccharum*

The plants under Saccharum genus have been classified into separate allied genera viz. *Erianthus, Sclerostachya* and *Nareaga*. Thus Saccharum now consists of 5 species viz. *S. officinarum, S. barberi, S. sinense, S. sponteneum* Linn. and *S. robustum* Brand and Jesw.

Sugarcane and allied genera

(a) *Saccharum* with spp. *officinarum, spontaneum, barberi, sinense* and *robustum*
(b) *Erianthus* with spp. *vavennae, mungo*, and *arudianceum*
(c) *Nareaga* with spp. *prophy* and *recoma*
(d) *Sclerostachya* with spp. *nitroye* and *ridlelic*
(e) *Miscanthus sinensis*

Generally, *Erianthus, Nareaga* and *Sclerostachya* generas are closer to each other. *Saccharum barberi* species has been put into four groups viz., Mungo, Saretha, Sunnabile and Nargori. The thin canes of north India viz. 'Katha' and 'Saretha' and thicker canes *i.e.*, Mungo, Nargori and Sunnabile are placed in *S. barberi*, while Panashi falls in *S. sinense*. *Erianthus* spp are low sugar grasses and potential donors of diverse abiotic stress tolerances diseases resistance and rationing ability traits. *Miscanthus* is a potentially valuable genetic resource for improving sugarcane which is a source of resistance to downy mildew), smut, nematodes, as well as tolerance to drought and cold. Miscanthus is also named 'miscanes' and also show promise as a highly productive cellulosic biomass crop. *Miscanthus sinensis* is native to eastern Asia covering China, Japan, Taiwan and Korea. It is an herbaceous perennial grass, growing to 0.8 to 2.0 m tall, forming dense clumps from an underground rhizome. The leaves are 18 to 75 cm long and 0.3 to 2.0 cm broad. The flowers are purplish, held above the foliage. The common names of *Miscanthus sinensis* include Chinese silver grass, Japanese silver grass, Silver Feather, Eulalia grass, Maiden grass, Zebra grass, Susuki grass, Eulalia grass, and porcupine grass. It is widely cultivated as an ornamental plant in temperate regions around the world. The characteristics of different species of genus *Saccharum* are presented in Table 5.

Table 5 Characteristics of different species of genus *Saccharum*

Species	Common name	Stem diameter (cm)	Sucrose content (%)	Fibre content (%)	Adaptability
S.officinarum (2n=80)	Noble	Thick >2.2	High 18-25	Low 5-15	Tropical and subtropical
S. sinense (2n=80-124)	Chinese	Medium 1.4-2.2	Medium 12-15	High 10-15	Tropical and subtropical
S. barberi (2n = 111-120)	Indian	Medium 1.7-2.1	Medium 13-16	High 10-15	Tropical and subtropical
S. spontaneum (2n = 40-128)	Wild	Slender 0.5-0.9	Very low 1-4	Very high 25-40	Tropical and subtropical
S. robustum (2n = 60-80)	Wild	Medium 1.1-1.7	Low 3-7	Very high 20-35	Tropical
S.edule (2n= 60-80)	Edible	Medium 1.1-1.8	Low 3-8	-----	Tropical - compacted inflorescence eaten as a vegetable
Erianthus	Related genus	-----	Very low	Very high	Tropical and subtropical

There are three categories of sugarcane varieties according to their physical and chemical characteristics—chewing canes, crystal canes, or syrup canes.

(i) Chewing canes are generally softer and contain fibers that stick together when chewed, making it easier to spit out the pulp once the sugary juice has been consumed. Many chewing canes are also used for syrup production.

(ii) Crystal canes (typically commercial varieties) must contain a high percentage of sucrose since this is the sugar molecule that easily forms into crystals when concentrated during a heating and evaporation process.

(iii) Syrup canes contain less sucrose than crystal canes, but have additional kinds of sugar molecules that are not as readily crystallized, so when the juice is concentrated into syrup, there is a lower likelihood of crystallization as compared with a crystal or commercial cane.

Climate: Sugarcane is cultivated both in tropics and subtropics. It is grown in latitudes from 30°N to 30°S and up to an altitude from sea level to 1000 m. The productivity of sugarcane is higher in tropics than subtropics. The crop cycle in subtropics is generally restricted from 40 to 52 weeks as compared from 52 to 72 weeks (depending on the time of planting) in tropics. Sugarcane's response to weather parameters varies with phenological stages due to its long duration. A mean annual rainfall of 1250 to 2500 mm is considered to be suitable for sugarcane cultivation without supplementary irrigation. It tolerates occasional flooding. Sugarcane is a sun-loving C_4 plant. Saturation light intensity is 64.6 klux. Sugarcane is a short day plant. Night length of more than 10 hours and day length of more than 11 hours is essential for flower initiation. Flowering of cane is photoperiodically controlled with night period of 11 hour 30 minutes to 11 hour 15 minutes. Most *Saccharum* varieties will not flower under day lengths longer than 13 or shorter than 12 hours. Photoperiod sensitivity depends on variety. Sunlight of 12.4 hours and 20 to 25°C night temperature will induce floral initiation. The cardinal minimum, optimum and maximum air temperature for sugarcane growth and yield are 12°C, 24-31°C and 42°C, respectively whereas the cardinal minimum, optimum and maximum soil temperature for sugarcane growth and yield are 19-21°C, 27-28°C and 38°C, respectively. The vegetative growth is checked when the mean daily air temperature drops below 21°C. The cane leaves and meristem tissues are killed at temperature -1° C to 2°C. Standing stalks of sugarcane freeze at -4 to -5.5°C depending on cultivar and length of exposure. Frost causes injury to sugarcane buds. The optimum temperature for different physiological activities are as follows: carbon assimilation, tillering, root growth in soil, shoot growth, sugar synthesis and sugar transport is 30°C, 33.3 to 34.4°C, 35°C, 33 to 36°C, 30°C and 30 to 35°C respectively. The translocation of sugar is to approximately half of that at 24°C. In equatorial regions, high temperature and rainfall, with heavy cloud cover and little difference in the length of the nights, cause vigorous vegetative growth and militate against ripening. On the other hand, in sub-tropical areas cool and long nights immediately before and during harvest favour the accumulation of the sucrose in the stems. For ripening, however, relatively low temperature in the range of 12 to 14°C is desirable. Thus low air temperature is the most effective in inducing ripening and lead to better juice quality while temperature of - 1°C to -2°C will kill the leaves and even meristems. The juice will not freeze and its quality will remain good for several months, provided the ambient temperature remain low. If temperatures falls further from -7°C to -8°C, the juice will freeze and destroy the cells and even at such temperature, sucrose will be hydrolyzed into glucose and fructose. A temperature of 28 to 30°C with humidity of 70 to 85% is most conducive for vegetative growth. Relative humidity of 95% is preferable for proper flowering of tropical varieties. A relative humidity of 55 to 75% during ripening phase is ideal. Relative humidity < 50% during growing season is not suitable for sugarcane cultivation. Cool moist conditions are favourable for flowering which is less in hot, dry, irrigated areas. As sugarcane absorb water through leaves and transmit it into the soil by roots, dew and light showers can make a useful contribution to the plant water balance. Strong wind is a natural hazard for sugarcane. High velocity winds exceeding 60 km/hour are harmful to grown up canes leading to lodging and cane breakage. Wind damage results in termination of growth of stalk, metabolic depletion and reduced assimilation.

Effect of climate on sugarcane yields and sugar recovery in India: The sugarcane productivity and juice quality are profoundly influenced by weather parameters during the various growth phases of cane. Maharashtra records higher sugar recoveries than Uttar Pradesh and Tamil Nadu. In Maharashtra, the period from November to March (the main crushing period in most places), is dry with very less rainfall and low mean humidity of 44 to 64% and lower daily mean temperature of 21.4 to 27.0°C, the nights are cooler and the sunshine hours (9.9 to 10.2 hours) are greater. The minimum temperature during November to March ranges from 13.0 to 21.7°C while the maximum temperature ranges from 27.1 to 35.9°C. The diurnal variation in temperature is 14.2°C, which is also fairly high. These conditions favour higher sugar accumulation. The latitudes between 15°N and 20°S, where ideal climatic conditions are met with record higher sucrose levels than the other latitudinal positions. Maharashtra is situated in this position and thus achieves the highest sugar recovery in the country. Both Tamil Nadu and Uttar Pradesh are not in favourable latitudinal positions. In Tamil Nadu, the mean temperatures throughout the year are more favourable for vegetative growth and thus the sugarcane yields are the highest in the country. The ripening phase has a higher mean air temperature of 24.7 to 28.8°C, the diurnal temperature variations of 8.9 to 11.2°C are narrow, the relative humidity of 75 to 81% is higher and the short or less duration of sunshine of 6.5 to 9.5 hours. These factors are not favourable for high sugar build up. In Uttar Pradesh, weather extremes are observed, very cold winter period follows very hot summer months. The length of growing season is very much restricted largely to south west monsoon and autumn months (June-November). The restricted growth phase results in lower yield levels. Very low temperatures restrict sugar accumulation also. This is why productivity levels are much less in the subtropics. Therefore thermo-insensitive varieties, if could be developed, are highly useful in subtropics.

Soil: Sugarcane can be grown in almost any kind of soil, provided it has sufficient moisture and fertility. In India, it is grown under widely varying soil conditions. Well drained loam or clay loam soils are found to be the most suitable. Sugarcane grows well in neutral soils with pH of 6.5 to 7.5. Sugarcane tolerates a wide range of soil acidity and alkalinity. The pH values of sugarcane soils range from 5.5 to 9.0. Cane yields are good under proper management in most calcareous soils (pH 8.7) since Na is not dominant (ESP 2-7%) and exchangeable Ca is generally high (35-60%). Critical level of salinity (level of salinity at which cane yield is reduced to 50%) ranged from 6.6 to 11.0 EC (m mhos/cm at 25°C) both in subtropical and tropical India. Land slope should be 1.0 to 3.0%. Higher the slope being is more acceptable in heavy soils. Sugarcane grows well in deep soils with minimum of 60 cm soil depth. Moderately heavy and medium deep (1-2 meters) loams to clay loam are better suited than heavy clay or shallow soils. High salinity and compactness of soil leads to diseased roots as well as chlorotic and stunted shoot growth with poor quality cane. Prolonged stagnant water kills the root rapidly and impair plant growth. In general sugarcane requires a well prepared and adequately aerated soil profile to a depth of at least 40 to 60 cm without compaction, hard pan, lime band or salt zone in the subsoil. Cane is very susceptible to soil compaction. The soil would normally have a bulk density below 1.4 g/cc and a pore space of at least 50%. The desirable bulk density values depend on soil types. Suppression of sugarcane rooting at bulk densities of 1.52 in alluvial soils and of

1.05 in hydrol humic laterosols. Ideally the ground water table depth should be more than 1.5 to 2.0 m from the soil surface. Shallow water table is mainly responsible for poor root growth and low cane yield. Ideally, the available water capacity of the soil should be 150 mm per metre or more to ensure adequate reservoir of water available to the plant roots between rainfall and irrigation cycles. The criteria to classify the soils for growing sugarcane are presented in Table 6.

Table 6 Criteria to classify the soils for growing sugarcane

Characteristics	Class			
	Good	**Average**	**Restricted**	**Unfit**
Effective depth	Deep	Medium	Shallow	Too shallow
Soil texture	Clayey	Medium to clayey	Sandy	Too sandy
Relief	Flat	Rolling	Too rolling	Hilly
Fertility	High	Medium or low	Too low	Too low
Drainage	Good	Medium to accentuated or incomplete	Incomplete	Excessive or deficient
Restraints to mechanization	Absent	Medium	Strong	Too strong
Susceptibility to erosion	Low	Medium	High	Too high
Source: Kofeler and Bonzelli, 1987 and http://www.sugarcanecrops.com/soil_requirement/				

Seasons in India: Sugarcane take generally one year to mature in sub-tropical states (Uttar Pradesh, Punjab, Haryana and Bihar) called '*Eksali*' however in some tropical states it matures in 18 months (Karnataka and Maharashtra) called '*Adsali*'. In India, planting seasons of Sugarcane in subtropical regions are September to October (Autumn) and February to March (spring), whereas in tropical regions it is June to August (*Adsali*) and January to February and October to November (*Eksali*). Apart from this in some states like Karnataka and Tamil Nadu sugarcane planting continue throughout the year except few months. The planting and harvesting time of sugarcane in different states of India is furnished in Table 7.

Table 7 Planting and harvesting time of sugarcane in different states of India

State	Time of planting		Harvesting time
Andhra Pradesh	*Adsali*	Aug-Sept	March-April
	Eksali	Jan-Mar	December-April
Assam	*Eksali*	Jan-Mar	November-March
Bihar	*Eksali*	Oct Feb-Mar	November-February Nov- April
Gujarat	*Eksali*	Jan-Feb Oct-Nov	Nov-Dec. Nov-Dec
Haryana	*Eksali*	Oct Feb-Mar	Nov- Dec. Dec- April
Karnataka	*Adsali*	Jul-Aug	Dec- Jan
	Eksali	Oct-Nov Jan-Feb	April-May Jan-March

Kerala	*Eksali*	Oct-Dec	August-March
Madhya Pradesh	*Eksali*	Oct-Nov Jan-Feb	Dec- April Jan-Feb.
Maharashtra	*Adsali*	June-July	Nov- Dec
	Eksali	Oct-Nov Jan-Feb	Jan-Feb. Feb-March
Orissa	*Eksali*	Jan-Mar	Dec-May
Punjab	*Eksali*	Feb-Mar	Nov-March
Rajasthan	*Eksali*	Oct Feb-Mar	Nov-March
Tamil Nadu	*Eksali* –Spring Early *Eksali* –Spring Mid *Eksali* –Spring Late	Dec-Jan Feb - March April-May	Dec-Feb Feb to May April-July
	Adsali - Special	July-Aug	Jul-Sept
Uttar Pradesh	*Eksali*	Sept-Oct Feb-Apr	December Dec- May
West Bengal	*Eksali*	Feb-Apr	Feb-Apr

Sugarcane varieties in India: The sugarcane varieties grown on a large part of the area under sugarcane in different states of India and Tamil Nadu are furnished in Table 8 to Table 11. The promising saline tolerant sugarcane varieties and sugarcane varieties tolerant to waterlogging or drought tolerant are furnished in Table 12 and Table 13 respectively.

Table 8 Sugarcane varieties in India

S.No.	Name of State	Early ripening varieties	Mid/late ripening varietes
1.	Andhra Pradesh	CoC671, Co6907 Co7704, Co8013 Co8014, CoA 890981 CoA 89082, CoA 880881 CoA 89085	Co.8021, Co T 8201 Co 7805, Co V 92101 COV 92102, Co 7219 Co7706, CoR 8001
2.	Karnataka	KHS3296, IC225 Co419, Co6415 Co7219, Co7704	B-37172, Co449 Co740, Co62175
3.	Tamil Nadu	CoC671, CoC771 CoC8001 CoC85061	CoC772, CoC773 CoC774, CoC775 CoC775, CoC778 CoC779, CoC6304 CoC62174, Co 86032
4.	Maharashtra	Co.7219, Co419	Co678, Co740 CoM.7125, Co7527 Co62175
5.	Punjab	Co.J64, CoJ83 CoP221	CoJ79 CoJ84 Co1148, CoS767
6.	Haryana	CoJ58, CoJ64 Co6914, Co 7314 Co7717	CoS 767, Co1148, Co1158

7.	Bihar	BO99, BO102 BO120, BO90	BO109, BO116 CoS 767, BO128 BO91, BO108, BO110, Co1148, BO104
8.	Uttar Pradesh	CoS687, CoS8436 CoS90265, CoJ64 CoS92254 CoS88230	CoS7918, CoS8016 CoS8009, CoS8315 CoS86216, CoS88216 CoS91269, CoS87220

Table 9 Sugarcane varieties particulars in India

Variety	Maturity	Cane yield (t/ha)	Sucrose (%)	State (s) for which recommended	Resistant to
Co 85004	Early	90.5	19.5	Maharashtra, Karnataka, Gujarat, Kerala, Interior TN, AP, MP and Chattisgarh	Smut
Co 86032	Midlate	102.0	20.1	--do--	-do-
Co 87025	Midlate	98.2	18.3	--do--	-do-
Co 87044	Midlate	101.0	18.3	--do--	-do-
Co 8371	Midlate	117.7	18.6	--do--	-do-
CoM 88121	Midlate	88.7	18.6	--do--	-do-
Co 91010	Midlate	116.0	19.1	--do--	Smut
Co 94008	Early	119.8	18.3	--do--	Red rot
Co 99004	Midlate	116.7	18.8	--do--	-do-
Co 2001-13	Midlate	108.6	19.03	--do--	Red rot, Smut, Wilt
Co 2001-15	Midlate	113.0	19.37	--do--	Red rot, Smut
Co 0218	Midlate	103.77	20.79	--do--	Red rot
Co 0403	Early	101.6	18.16	--do--	Red rot, smut
Co 86249	Midlate	104.2	18.7	--do--	-do-
CoC 01061	Early	110.8	17.4	--do--	Red rot
CoS 91230	Midlate	68.2	18.8	Punjab, Haryana, Rajasthan, Central and Western U.P. and Uttarakhand	Red rot
Co Pant 90223	Midlate	73.3	18.5	--do--	Red rot, Smut
CoH 92201	Early	70.0	18.2	--do--	Red rot
CoS 95255	Early	70.5	17.5	--do--	--do--
CoS 94270	Midlate	81.5	17.1	--do--	--do--
CoH 119	Midlate	82.8	17.5	--do--	--do--
Co 9814	Early	76.3	17.6	--do--	--do--

CoS 96268	Early	69.8	17.9	--do--	--do--
Co Pant 97222	Midlate	88.2	18.2	--do--	--do--
CoJ 20193	Midlate	75.9	17.9	--do--	--do--
CoS 96275	Midlate	80.8	17.3	--do--	--do--
Co 0118	Early	78.2	18.45	--do--	Red rot, Smut, Wilt
Co 0238	Early	81.08	17.99	--do--	Red rot, Wilt
Co 0124	Midlate	75.71	18.22	--do--	Red rot
Co 0239	Early	79.23	18.58	--do--	Red rot
Co 87263	Early	66.3	17.4	Eastern U.P., Bihar, West Bengal and Jharkhand	Red rot, Smut
Co 87268	Early	78.9	17.5	--do--	Red rot, Smut
Co 89029	Early	70.6	16.3	--do--	Red rot
BO 128	Midlate	69.2	17.6	--do--	Red rot, Smut
CoSe 95422	Early	67.8	17.7	--do--	Red rot
CoSe 92423	Midlate	70.1	17.5	--do--	--do--
CoSe 96234	Early	64.1	17.9	--do--	--do--
CoSe 96436	Midlate	67.1	17.7	--do--	--do--
CoLK 94184	Early	76.0	18.0	--do--	--do--
Co 0232	Early	67.83	16.51	--do--	Red rot, Smut, Wilt
Co 0233	Midlate	67.77	17.54	--do--	--do--
Co 0232	Early	67.83	16.51	Assam	--do--
Co 0233	Midlate	67.77	17.54	Assam	--do--

Source: DSD, 2013.

Table 10 Sugarcane varieties from Andhra Pradesh

S. No.	Variety	Season	Parentage	Cane yield (t/ha)	Sucrose %
1.	CoA 89085	Early	Co6806 × Co775	105.0	20.00
2.	CoV 94101	Early	Co7704 × Co775	102.0	19.15
3.	CoV95101 (Krishna)	Early	Co 798 × Co 62198	115.0	18.50
4.	Co A 99082	Early	Co T8201 × B38192	105.0	18.50
5.	CoA 92081 (Viswamitra)	Early	Co7704 × CoC 671	125.0	18.50
6	83R23 (Vasudha)	Early	Co 740 × Co 6806	114.0	18.60
7.	CoV 92102	Midlate	CoC 671 × Co 6806	119.2	19.30

Table 11 Sugarcane varieties in Tamil Nadu

Variety	Duration	Yield/ha (t)	CCS %	CCS yield (t/ha)
Early season				
COC 90063	Early	128.0	12.6	16.2
COC 92061	Early	133.5	13.2	17.5
COSi 86071	Early	113.4	12.1	14.0
COC 671	Early	123.5	13.0	16.0
CoC 772	Early	133.0	12.0	20.0
CO 8021	Early	130.0	11.8	15.5
COSi 95071	Early	143.2	12.2	17.5
COC 85061	Early	135.4	12.4	16.8
COSi 96071	Early	145.0	11.9	17.3
CO 86010	Early	146.1	10.7	15.6
COG 95076	Early	108.2	11.5	12.4
CO 8208	Early	136.5	11.1	15.3
COG 94077	Early	133.0	13.3	17.6
COC 771	Early	130.0	12.5	16.3
COC 91061	Early	131.0	12.8	16.8
COC 98061	Early	120.0	11.6	13.8
CO 86249	Early	127.6	11.2	13.8
CoC (Sc) 23	Early	133.0	12.9	17.1
CoSi 6	Early	142.0	13.1	17.6
Mid-late season				
CO 6304	Mid late	135.0	11.5	15.5
CO 86032	Mid late	132.0	13.1	17.2
COG 93076	Mid late	134.0	12.9	17.4
CO 62175	Mid late	125.0	13.2	16.5
CO 85019	Mid late	134.5	12.5	16.8
CO 87025	Mid late	130.0	12.3	15.9
COC 99061	Mid late	130.3	11.9	15.6
CoC (Sc) 22	Mid late	135.9	12.1	16.4
CoG 5	Mid late	120.0	12.7	15.4
COSi 98071	Mid late	144.7	12.3	17.7

Source: CPG, TNAU and DOA, 2012

Table 12 Sugarcane varieties tolerant to salinity:

Co 87002	Co 90010	Co 91002	Co 91005
Co 91011	CoJn 86141	Co 89010	Co 89027
Co 92012	Co 93005	Co 93019	Co 93079
MS 92121	Co 93009	Co 93011	Co 93016
CoG 93076	Co 93021	Co 93015	Co 94004
Co 94005	Co 94008	Co 93018	Co 94010
Co 94011	Co 94012	Co 94015	CoJn 94141
CoG 93079	Co 85019	Co 95003	Co 95006

Source: http://sugarcane-breeding.tn.nic.in/agronomy.htm4

Table 13 Sugarcane varieties tolerant to waterlogging and drought

S.No.	**Variety**	**Zone**	**Reaction to abiotic stresses**
1	Co 98014 (Karan-1)	North West zone	Tolerant to drought
2	Co 99004 (Damodar)	Peninsular zone	Tolerant to drought
3	Co 94012	Peninsular zone	Tolerant to drought
4	Co 2001-13 (Sulabh)	Peninsular zone	Tolerant to drought
5	Co 2001-15	Peninsular zone	Tolerant to drought
6	Co 0118 (Karan-2)	North West zone	Tolerant to drought and waterlogging
7	Co 0232	North Central zone	Tolerant to waterlogging
8	Co 0233	North Central zone	Tolerant to waterlogging
9	Co 0238 (Karan-4)	North West zone	Tolerant to drought and waterlogging
10	Co 0239 (Karan-6)	North West zone	Tolerant to drought and waterlogging
11	Co 0218	Peninsular zone	Tolerant to drought
12	Co 0237	North West zone	Tolerant to drought and waterlogging
13	Co 0403	Peninsular zone	Tolerant to drought

Growth stages: The crop cycle is characterized by the establishment (germination), formative (tillering or shoot buildup), elongation (grand growth) and ripening (sucrose accumulation). Each crop phase has duration of nearly 12 to 13 weeks in subtropics. In tropics, the crop elongation phase is almost twice as compared to subtropics. In sugarcane, there are six phenophases, *viz.* germination and settling expansion or sprouting and early growth, formative phase, grand growth phase, flower primordia development or flowering, ripening/maturity phase and aerial rooting phase. Sugarcane growth phases and stages are presented in Table 14. Under field conditions germination starts from 7 to 10 days and usually lasts for about 30-35 days. Functional life of sett root is for 9-10 weeks. The active roots, i.e., shoot roots come out after four weeks of planting in *S. spontaneum* and 9-10 week in *S. officinarium* and *S. robustum*. Most of the feeding roots occur in the top 0-15 cm soil layer. Setts-roots contribute to sugarcane nutrition appreciably in early stage of growth. Transition of nutrient uptake function from sett-roots to shoot roots takes place around 50 to 55 days after planting. The tiller emergence coincides with the development of 6-8 leaves on the mother shoot. In formative phase (differentiation phase or tillering phase), tillers are initiated in sugarcane when there are two opened leaves on mother shoot. The maximum frequency of tiller production coincides with 6-8 open leaves on the mother-shoot. Grand growth phase prolongs for 90 days. In sugarcane, flower (arrow) emergence takes place after completion of the five stages *viz*., induction, initiation of axis primordium, initiation of branch primordium, spikelet primordium and growth. Each of the former four stages requires at least 14 days while last growth stage requires 21 days for its completion. Thus 77 days are required for visibility of an arrow. In tropical zone, a minimum of around 70 days are required for arrowing. In subtropical India, flowering is erratic in nature. A flower bud takes about two months to develop into tassels present inside the flag leaf. An average period of 44 days is required for flag leaf to complete emergence of arrow (inflorescence). There is an increase in quantity of sucrose in the stalk. A crop matures when it contains a minimum of 15-16% sucrose in juice and 85% and above purity. A crop or stalk ripens when the rise in sucrose has attained a plateau and its juice has 20% brix, 18% pol percent and juice purity reaches 85%.

Table 14 Growth stages in sugarcane

Growth stages	**Cropping period (days)**
Germination stage	0 - 10
Planting to 0.25% full canopy – seedling stage	0 - 30
Formative phase- Tillering stage	31 - 120
0.25 to 0.50% full canopy	31 - 60
0.50 to 0.75% full canopy	61 - 75
0.75 to full canopy	76 - 120
Grand growth phase (internode elongation)	101 - 270
Complete green cover	120 - 300
Stable population	150 - 180
Maturity stage	271 - 365
Ripening - early leaf senescence	300 - 330
Ripening	330 - 365

Land preparation: Deep ploughing with a soil turning plough (chisel plough) to a depth of 45 cm once in every three years induces good root development. Land preparation in sugarcane has two component namely ploughing and harrowing. Ploughing is done to bring the disturbed soil compaction while harrowing is done to bring the disturbed soil to optimum soil structure. Form ridges and furrows with 90 cm wide and 20 cm depth. Open irrigation channels at 10 m interval in problem soil areas provide drainage facilities. In wetlands (heavy soils), no preparatory cultivation is done. After harvest of the paddy crop, form irrigation and drainage channels of 40 cm depth and 30 cm width at an interval of 6 m long, along and across the field and along field borders. Form ridges and furrows with a spacing of 90 cm between rows with spade. Dig with hand hoes in the furrows and allow the soil to weather for 4 to 5 days. Use junior hoe to break the clods and get a fine tilth free of weeds and stubbles. Open ridges and furrows 90 cm apart with the help of victory plough. The depth of furrow must be 20 cm. Open irrigation channels at an interval of 10 m to a depth of 25 cm.

Seed rate and planting: Sugarcane is a highly heterogenous crop. The seeds collected from the flower and sown do not 'breed true'. The stalk of sugarcane has vegetative buds at the nodes which can sprout and develop into sugarcane plant. Hence under commercial cultivation, sugarcane is vegetatively propagated and the seed material of sugarcane is called "setts" which is a small cut sections of sugarcane stalk. Depending upon the number of buds (eyes) present in each sett, they are called as single budded, double budded and three budded setts. It is preferable to select the top $^1/_3$ of the cane for seed purpose. Immature cane is best suited for seed material. Ideal age of seed crop is 6 to 8 months. The cane crop intended for seed should receive 25 kg more N than normal crop. Sometimes separate nursery crops are raised exclusively for seed purpose. This is known as 'short crop' / 'seed crop'. The crop is planted as a normal crop and irrigated at frequent intervals. The seed materials are cut when the crop is about 6 to 8 month old. Normally an area of one tenth hectare of nursery crop will give setts for planting one hectare. Short crop has 80% germination and vigorous seedlings. The seed material should be free from borers, scales, mealy bugs, grassy shoot and smut and they should be resistant to red rot disease. Detrash the cane with hand. Use short knife to prepare setts without any splits. Discard setts with damaged buds, sprouted buds splits, etc.

Sett treatment: Select healthy setts free from borers, scales, grassy shoot, smut and red rot for planting. Seeds showing red colour at the cut end and hollows should be rejected and burnt. Setts should be soaked for 15 mts in carbendazim @ 0.05% or Agallol @ 0.5% or Aretran @ 0.25% or Tridemefon @ 0.05% along with urea @ 0.1% solution. Carbendazim and Tridemefon will eradicate pathogens causing smut and pineapple diseases. Treat the setts with the aerated steam at 50^{0}C for 1 hour to control primary infection of grassy shoot diseases. It is desirable that planting is done immediately after the setts treatment.

Planting: Mostly two budded setts are used. A seed rate of 50000 three budded setts or 75000 two budded setts or 187500 single budded setts per ha is required. Furrow must be in good tilth with loosening of the soil at the bottom of the furrow. Apply Lindane 5% dust and Chloradine 5% dust @ 25 kg/ha which in turn prevent termite attack. Planting is done in furrow in light soil and at the side of furrow in heavy soils. Buds should be planted in lateral position while planting and setts are covered with thickness of about 3 cm. Irrigation is given to form a slurry and then setts pushed inside the soils. Place the setts end to end. Plant extra setts near the channels for gap filling. Avoid exposure of setts to sunlight. Gap filling is done 30 to 45 days after planting with sprouted setts if gaps are more than 30 cm.

Seed cane standard: The seed cane shall be taken from short plant crop only (not ratoon), materially undamaged and reasonably clean. Immature top portion of the cane can be used for planting. If the cane has flowered, the top portion develops piths and fibre and is useless as seed materials. In such cases top portion should be avoided. The best seed material which assumes good germination and healthy growth of the crop is from a 6 to 8 month crop maintained as nursery crop, setts given the heat treatment, good fertilization of NPK, adequate irrigation and free from pests and diseases. Two budded setts are generally recommended for planting. Cutting the cane into the setts is to be done by skilled labourers to avoid damage to the buds and the cut ends. The cut is effected at about the middle of the internode to allow as much nutrient as possible. Seeds should be selected from immature canes having viable buds. The buds on the top portion of the mature canes germinates faster and in greater number than on lower portion. Immature canes rich in nitrogen content with grown fresh buds which makes good seed material and are in vigorous state of growth. Cane grown in water logged and salt affected soils are unsuitable for seed to obtain good seed. It is better to have separate nursery for seed purposes. Each node of seed cane shall bear one bud. The number of nodes without bud shall not exceed 5% (by number) of the total number of buds per seed cane. The number of buds which have swollen up or have projected outward beyond one centimeter from the rind surface shall not exceed 5% (by number) of the total number of buds. Seed canes should not have nodal roots. In waterlogged area, relaxation may be given up to a maximum of 5%. Moisture in seed cane should not be less than 65% on wet weight basis. Germinability of buds should not be less than 85%. Physical purity should be 100% while the genetic purity of seed should be 100%.

Polybag nursery: Select shady places preferably coconut shade with enough sunlight. Bed size is 7 m × 1.5 m. Polybags of size 12.5 cm × 10 cm with two small holes in lower side is required. One kg may contain 1200 to 1400 polybags. Mix tank silt, pressmud and sand in equal proportions. These mix may be filled about 200 to 250 g in a polybag. Single budded setts are taken to prepare the polybag nursery. There should be 2.5 cm length on top of bud and 4 cm below the bud. The setts are placed in polybag such a way that the bud is in the soil surface. The polybags are watered with rosecan. Sometimes trenches are made in the main field. The polybags are arranged in lines in the trenches. The sides are supported with soil and plot irrigated directly with flow of water. In this method adequate moisture is provided with less expense. 13000 seedlings are required for one acre. Optimum age of seedlings is 25 to 30 days. It can be planted up to an age of 60 to 65 days. Polybag seedlings are planted with a spacing of 90 cm × 30 cm. Advantages of ploybag seedlings include low seed cane weight. Generally one metric tonne of seed cane is required to obtain 13,000 single budded setts and 6 to 7 tonnes of seed cane is required to have 40,000 to 45,000 two budded setts per acre. Only 300 to 400 kg of seed cane is required to obtain 13,000 chip buds. Seed multiplication ratio is 1:40 as against 1:7 in conventional method. Promising varieties can be multiplied fast. There is one month age advantage in polybag seedlings. Nursery can be raised one month earlier before harvest of standing crop. Avoid time between two crops to get more returns from an unit area in a year. Take polybag nurseries during summer and keep nursery area free of weeds. The expense of two weeding can be avoided and so also 3 to 4 initial irrigations when compared with actual (usual) planting method. No gaps are found after transplanting polybag seedlings and so high cane yield can be obtained. There is uniformity in growth of polybag seedlings, which is not in ordinary method as there is different age group in the same unit area itself. Early vigour is observed due to use of well established seedlings for planting and it is comparatively less than normal method. Water required for polybag nursery for 25 to 30 days is negligible. Varietal purity can be maintained. Healthy and disease free seedlings can be produced. It is possible to have a successful crop during the acquired one month of hot period as well as during heavy downpour rainy period of one month by this method and take up transplanting after unfavorable periods. It will enhance the tillering ability due to spaced transplanting between plants. When the field is moist soon after harvest of paddy in wetlands and garden lands, trench is formed with spade allowing a thin film of water and straight line planting of polybag seedlings can be taken up using ropes. All the polybag seedlings utilize the residual phosphorus and potassium. This is an added advantage for early vigour and healthy crop. In garden land, thorough land preparation and planting, the seedlings to a depth of 22.5 cm will give better establishment and high yield.

Systems of planting

1. Flat bed planting: The field is thoroughly prepared by repeated ploughing and irrigation. Shallow furrows of 10 to 15 cm deep are formed by country plough. Channels are formed. The long beds are flooded with water and the setts are pressed in lines end to end continuously at a spacing of one metre between the rows.

2. Ridges and furrows planting: Ridges and furrows are formed in the well prepared land at a spacing of 85 to 90 cm and with a furrow depth of 20 to 25 cm. Cross irrigation channels are formed at convenient intervals (of about 10 m or 20 to 25 m). The ridges and furrows are formed before a few days of planting and then left for soil to weather. The bottom of furrows are loosened again for a depth of 15 cm to form a good seed bed. Setts are spread on ridges. Water is let into furrows to soak the soil to the maximum saturation. Labourers walk through the furrows, take the setts from ridges and press it down the wetted soil to a depth of 2.5 cm keeping the buds in lateral position. This is done in January to March when the weather is dry. For *Adsali* planting, there may be monsoon rains. So, the setts are placed at the middle of ridges to promote better germination than planted in the bottom of furrows which result in poor germination due to water stagnation in rainy season.

3. Trench planting: This system is also known as 'Hawaiian method of planting'. Instead of furrows, sometimes deep trenches of about 30 to 45 cm depth and 30 cm width are formed with inter-row spacing of 100 cm and setts are planted at the bottom of the trench. Due to repeated working of the soil and irrigations, the top loose soil fall into the trenches and gets automatically earthed up. Trench planting is beneficial in subtropics and in coastal areas, particularly to prevent lodging.

4. IISR 8626 method: The technique referred to planting of long 'rayungans' vertically in deep, fully fertilized and irrigated trenches. *Rayungan* is a word of Japanese used by Dutch planters to mean a single node piece of cane in which bud has been made to grow into a shoot. A long rayungan is several noded 40 cm top sett having side-shoots at the top. Planting of pre-germinated raynugan in IISR 8626 technique has potential of making up yield loss in summer (late) planting season. (IISR stands for Indian Institute of Sugarcane Research, Lucknow).

5. Spaced Transplanting Technique (STP): STP has been developed at IISR, Lucknow. Spaced transplanting technique method is useful to economise seed rate. Spaced transplanting technique (STP) consists of raising single bud settings (nursery) in 50 m^2 area for one hectare transplanting. An area of 10 × 5 m (50 beds of 1 m^2) is planted at least 30 days in advance before transplanting. A total of 600-800 single bud setts are planted in bed. Thus only 2 tonne seed is sufficient for 1 ha crop. It would be still better if ½ of the upper cane stalk is used for planting. Sprouted seedlings are transplanted in the field after 3 to 4 weeks. A higher germination (up to 95% buds) is achieved in space transplanting. Furrows at 90 cm distance are open and filled with water before transplanting of germinated (3-4 leaf stage) settlings at 60 cm apart. STP method helps for fast multiplication and observation for diseases is also possible at nursery and transplanting stages. The seed multiplication ratio in this technique is 1:40 as against 1:10 in conventional planting. The system is sustainable for seed cane multiplication of new improved varieties to effect varietal replacement in shortest possible time.

6. Ring-pit planting method: Leave 65 cm space around the boundary of the field. Then mark the field at 105 cm both length and width wise. At the cross section of these lines, dig pit of 75 cm diameter and 30 cm depth with pit digger machine which comes 9000 pits per ha. In case, the machine is not available the pits are dug manually. The soil dugged up from the pit is kept in the periphery of the ring in 30 cm space left in between the two pits. Cut the cane stalks in 2 budded setts and dip them in 0.2 per cent solution of carbendazim (2 g in one liter of water). In each pit, apply mixture of 3 kg FYM + 20 g DAP + 8 g urea + 16 g MoP + 2g $ZnSO_4$ and mix it with soil. Twenty-three budded setts are placed horizontally in a circular manner in each pit as of spokes in a cycle wheel. 2.5 cm soil cover is made over the setts. Spray the solution of 5 liters Chlorpyriphos 20 EC dissolved in 1500 liters of water on setts for one ha area and 2-5 cm soil cover is made over the setts. Interconnect each pit with narrow channel manually for irrigating the pits. One irrigation just after planting prompts germination of setts. Thirty days after germination, 16 g urea is applied in each pit and half of the soil remaining at periphery is filled back in the pit at 4th leaf stage (35-40 days after planting). The filling of soil is completed when all the mother shoots have emerged. The crop under ring planting consists mainly of mother shoots, which are thicker and heavier than tillers. Carry out weeding as and when required. Apply Furadan 3 G @ 33 kg per ha. A gap of at least 3-4 days must be kept in application of urea and Furadan. Remove lower dry leaves. Harvest the cane close to the ground level to take good ratoon crop. Ratoon yields are also higher because of deeper planting of plant crop. As the irrigation water is applied in the pits only, more than 40% surface area remains dry. For this reason, with the ring planting 25-30% irrigation water is saved. Higher sugarcane yield and reduced quantity of irrigation water improves 30-40% higher irrigation water use efficiency.

Trench planting is beneficial in sub-tropics and in coastal areas, particularly to prevent lodging. IISR 8626, STP, Pit system of planting helps to sustain greater number of shoots per ha. STP increases shoot multiplication ratio of 1:40 as compared to conventional method of 1:10. The STP is useful to economize seed rate. Ring planting increases cane yields in sub-tropics and increase 25% cane yield in tropics. Planting techniques which appreciably increased sugarcane yield include IISR 8626, spaced transplanting technique (STP), ring system and bud chip method. Use of rayungans, water soaked seed cane and planting polythene bag raised settlings are some of the techniques to improve crop stand under late planted conditions.

Sustainable Sugarcane Initiative (SSI): Sustainable Sugarcane Initiative method of sugarcane production which involves less seeds, less water and optimum utilization of nutrients and soil to improve the cane yield from 100 t ha^{-1} to 300 t ha^{-1}. In conventional system, about 6-8 tons seed cane/ha (nearly 10% of total produce) is used as planting material, which comprises of about 25-30 cm stalk pieces having 2-3 buds. This large mass of planting material poses a great problem in transport and handling of seed cane. The auxiliary buds of cane stalk can be used as seed to plant cane which is popularly known as 'bud chips'. These bud chips are less bulky, easily transportable and more economical seed material. The bud chip technology holds great promise in rapid multiplication of new cane varieties. The bud chip seed material has relatively low food reserves (1.2-1.8 g sugars/bud) compared to conventional three budded seed material (6.0-8.0 g sugars/bud). The highest germination of 80.94% is found under single node cane segments as against 42.50% under 3-bud setts at 40 days after planting. The cane yield obtained under cane node technology is higher by 10.90% over that of 3-bud setts planted crop. Only 1.7-1.8 t/ha or <1.0 t/ha seed cane is required in cane node method as against 6-8 t/ha under conventional method of planting.

Nursery is raised using single budded chips from 7-8 months old healthy canes with internode length of 7 to 8 inches and with10-12 buds per cane. The number of potential buds per cane is 10-12. The single budded chip seed/setts required is 5000 per acre or 12500 per ha. The number of canes required is 500 to 1250 to produce required setts for one acre and one ha respectively. Avoid taking buds from extreme top and 3-4 short internodes at bottom. Avoid canes with disease infestation. The bud along with a portion of the nodal region is chipped off using a bud chipping machine. Large number of buds (about 150/hour) can easily be chipped off.

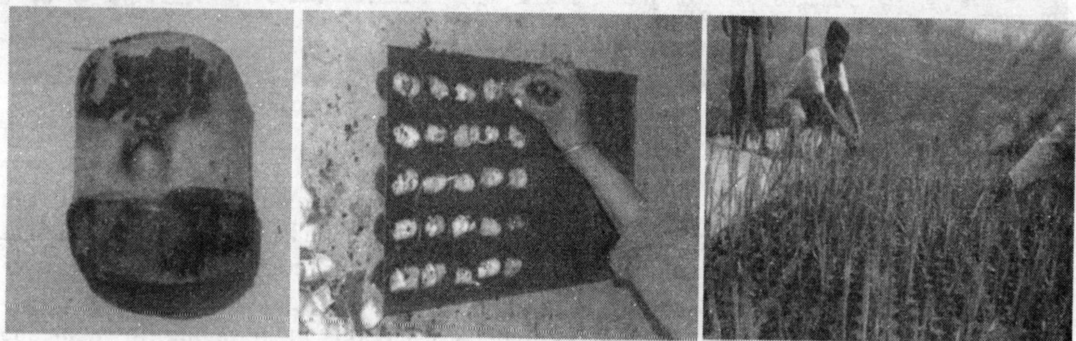

Single bud chips are prepared in such a way that the bottom portion is longer than that of the top portion. The bud chips are treated with fungicides. Then the setts are planted in the raised bed nursery or polythene bags of size 12.5 × 10 cm filled with FYM or pressmud, soil and sand mixture in 1:1:1 proportion. The raised nursery bed is thoroughly soaked with water. The chip bud is pushed vertically into the soil side by side. The eye of the bud should be just touching the soil surface. The longer end of the sett should be pushed down. Over this, a thin layer of dry soil is placed. The bed is watered using a rose cane. The nursery should be allowed to dry. Five to six weeks old seedlings are transplanted in the main field. Sugarcane seedlings can also be produced in green house or shade net house using single bud chips @ 12500 per ha. For this, a quantity of 1.0 to 1.5 tonne of seed cane is required. After taking chip buds, the left over seed cane can be sent for milling.

Cocopith (coconut coir waste) can be used for raising the settlings. Hundred trays and 150 kg of coconut coir pith are required for this purpose. Fill half of each cone in the tray with coconut coir pith. Bud treatment with mixing carbendazim @ 125 g and urea @ 100 g in 250 l of water for soaking the buds for 10-15 minutes. Bud chips are also treated with ethephon (2-chloroethylphosphonic acid) @ 100 mg/l increase the root number about 38% and root weight about 160%. Place the buds flat or in a slightly slanting position bud side faces up. Cover the bud chips in trays completely with coconut coir pith. After filling all the trays, place them one above the other and finally keep an empty tray upside down at the top. About 100 trays (4 sets, each consisting of 25 trays) are to be placed together and wrapped tightly with polythene sheets. Keep it for 5 to 8 days in the same position to create high temperature and humidity. Create artificial warmth through electric bulbs if the climate is too cold (crucial phase). Bud chips, after fungicide treatment, planted in trays are stored at low temperature conditions (10 ±1°C) exhibit about 80% bud germination after 10 days of storage than one stored at room temperature (about 40%).

The germination of chip buds can also be induced when the treated chip buds are placed in a moist gunny bag for five days under air tight dark condition. No watering is done during this period. Within 3-5 days, white roots (primodia) and shoots develop in next 2 to 3 days. Either on 5th or 8th day (based on the climatic conditions), 100% sprouting occurs. Germination of all setts will be completed in 10 days. Trays can be removed and kept on ground. Watering is done with rose cane during evening for 15 days. Shoots will start growing after appearance of two leaves.

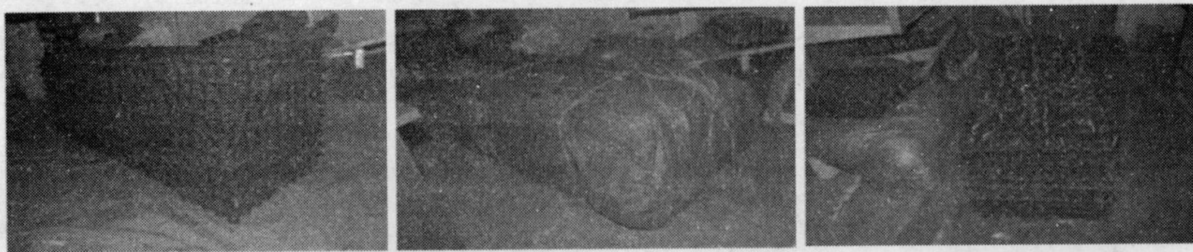

The spacing of 150 cm × 60 cm is practiced in SSI method for easy sunlight penetration and profuse tillering. Transplanting is done with 25-35 days old seedlings (three to four leaf stage). While transplanting in main field, zigzag method of planting (see picture) can be followed to utilize more space and achieve maximum tillers. The number of tillers/plant is 20 to 25 and 9-10 millable canes/plant is produced in SSI method as compared to 10-15 tillers and 4-5 millable canes in conventional cultivation. The single cane weight is 2 to 3 kg. There is high accessibility to air and sunlight. SSI supports intercropping in sugarcane with crops like pulses, vegetables, *etc.*

Apply 20 to 25 t/ha of organic manure (FYM/compost/well-decomposed press mud). Quantity of organic manure could be adjusted in such a way as to supply 300 kg N/ha through one or more sources depending on their N content. *Trichoderma* and *Pseudomonas* (each 2.5 kg/ha) and decomposing cultures can be mixed with the organic manures. This will improve the soil fertility to realize higher yields.

The fertigation schedule for sugarcane is presented in Table 15. The recommended dose of fertilizer for sugarcane is 300-100-200 of N, P_2O_5 and K_2O kg/ha. Fertigation is done once in 10 days. Water requirement is less through drip fertigation. The cane yield is 200 to 250 t/ha under SSI.

Table 15 Fertigation schedule for sugarcane

Stage (days after planting-DAP)	**Fertilizer dose (kg/ha)**		
	N	**P**	**K**
0-30	39.4	0	0
31-60	48.6	26.25	9
61-90	51.4	20.50	13.5
91-120	55.2	16.25	14.6
121-180	57.8	0	40.5
181-210	10.5	0	35.0
Total	275.0	63.0	115.0

Source: http://agridr.in/expert_system/sugar/ssi.html

Sugarcane based cropping systems

Sub-tropical cropping systems

Paddy- Autumn Sugarcane-ratoon-wheat

Greengram- Autumn Sugarcane-ratoon-wheat

Maize- Autumn Sugarcane-ratoon-wheat

Kharif crops-Potato-Spring Sugarcane-ratoon-Wheat

Kharif crops-Mustard-Spring Sugarcane-ratoon-Wheat

Kharif crops-Pea/Coriander-Spring Sugarcane-ratoon-Wheat

Kharif crops-Wheat-late Planted Sugarcane-ratoon-Wheat

Tropical cropping systems

Bajra-Sugarcane(pre-seasonal)-Ratoon- wheat

Paddy-Sugarcane-Ratoon- Finger millet

Paddy-Sugarcane-Ratoon- Wheat

Paddy-Sugarcane-Ratoon- gingelly

Paddy-Sugarcane-Ratoon- urd.

Cotton-Sugarcane-Ratoon–wheat

Sugarcane-Ratoon-*Kharif* rice-*rabi* rice.

Intercropping: In general, a less spreading short-duration crop fits well with sugarcane as a companion crop. In an autumn planted crop, most of the winter (*rabi*) season crops such as cereals, potato, Indian mustard, can be accommodated as a companion crop. In spring-planted sugarcane, onion, greengram, blackgram, soybean and cowpea can be grown as an intercrop.

After cultivation

(i) Interculture and Hoeing: To suppress weed population and to create aeration for optimum tillering 3 to 4 hoeings are recommended. Early tillering phase (40 to 70 days after planting) is the most critical for hoeing. Hoeing during 40 to 70 DAP increases 38% cane yield over no hoeing. Two hoeings, one each at early and mid tillering phases, produced the same results as that of three hoeings, one each during early, mid and late tillering phases.

(ii) Earthing up: Earthing up is also known as 'hilling up'. Good earthing up from 3 to 4 months crop age after completion of tillering stage provides anchorage to sugarcane reducing lodging and facilitates irrigation and drainage. Partial earthing up is done on 45 to 50 days after second top dressing of fertilizers. Full earthing up is done on 120 days after planting coinciding with peak tillering stage. High earthing up helps in good root development and provides support to the plant. At 150 days after planting, earthing up may be done with spade.

(iii) Wet earthing up: It is done in waterlogging areas or flood-affected areas or in excess moisture during monsoon rainy season. The soil is removed from the furrow, is plastered on the ridges. This gives a firm support to the crop against lodging. Excess water during rainy season or flood water drains quickly through furrows and provides a well aerated soil condition.

(iv) Propping: Propping is tying of cane clumps together by wires or by trash leaves to prevent lodging of cane at 7-8 months old crop. Trash-twist propping is advocated to the cane crop. Trash twist propping refers to the twisting of trash (dry leaves and lower green leaves without removing from the cane) to form a sort of rope and cane stalks are tied together.

(v) Wrapping: It is twisting of dry leaves around the cane in such a way that no part of internode is exposed to prevent damage from jackals, etc. To keep the sugarcane crop erect so as to minimize losses in juice quality due to lodging and to avoid damage of rats, jackals, *etc*, tying of the crop should be done when it attains the height of 2 metres.

(vi) Detrashing: A normal cane stalk bears on an average 30 to 35 leaves under good growing conditions. Only 8 to 10 leaves are required for optimum photosynthesis. The bottom leaves usually dry as the crop ages. The bottom green leaves used the food materials, which are used for stalk growth. So it is necessary to remove both the dry and lower green leaves. This operation is known as 'detrashing'. It increases hardness in exposed stalk. Detrashing is carried out at 5 and 7 months old crops. It prevents water drops adhering to bud region, which will induce bud sprouting in standing cane.

Weed management: The sugarcane growth is very slow in the initial stages. It takes about 30 to 45 days to complete germination and another 60-75 days for developing full canopy cover. There is a chance of high weed infestation in initial stages. The critical period for competition with weeds is up to 120 days after planting for a plant-crop and 90 days after ratooning for a ratoon crop. The sugarcane crop is planted at a wider spacing. The initial establishment and growth of crop is very slow. These factors along with frequent irrigation and monsoon rain provide ample opportunity for weeds to grow profusely. Normally hoeing, hand weeding and inter culture methods are employed to control weeds. The effectiveness of these methods depends on field conditions and also on the type of weeds associated with the crop. The suppression of sprouting and subsequent growth of weeds using pre-emergence herbicides are more efficient than allowing weeds to grow and their removal by hand weeding. The weeds remove four times of N and P and 2.5 times of K than of sugarcane even at 35 days. Atrazine @ 1.75 kg per ha, Sencor (metribuzin) @1kg/ha or Oxyflurofen @ 750 ml per ha mixed in 500 liters of water is found to be the best for safer use and maximum weed control efficiency. Pre -emergence application of herbicide is carried out on 3rd day of planting using a deflector or fan type nozzle. Pre-emergence application of Thiobencarb @ 1.25 kg a.i./ha under intercropping system in sugarcane with soybean, blackgram or groundnut gives effective weed control. If pre-emergence spray is not carried out, go in for post-emergence spray of Gramaxone @ 2.5 litre and 2,4 -D sodium salt @ 2.5 kg per ha on 21st day of planting or apply 10% NH_4SO_4 on 45th, 75th and 105th day after planting as direct spray. *Convolvulus arvensis* is a twining perennial. It produces long weak stems, which raise themselves up twinning round sugarcane plant. It propagates by means of both root sockets and seeds. The weed possess a deep root system penetrating upto a depth of 5 to 6 meters in soil. *Ipomoea sepiaria* and *Convolvulus arvensis* are difficult to control through mechanical means mainly because its roots and rhizomes can regenerate from much deeper layer of the soil, which are normally inaccessible with implements. *Striga sp.* is an angiospermic parasite on sugarcane plant. It produces haustoria and connects them with internal tissues of sugarcane roots. *Striga* possess green leaves and is thus capable of producing carbohydrates but for water and mineral nutrients, it depends entirely on sugarcane plants. It is propagated by means of tiny and light seeds produced in large numbers and dispersed through wind. A single plant of *Striga* can produce a half a million of seeds. The genus *Striga* has three important species namely *S.euphrasiodes, S.lutea and S.densiflora.* Cane yield loss due to *striga* infestation is observed in Tamil Nadu, Punjab and Bihar. If the parasitic weed *Striga* is a problem, post-emergence application of 2,4-D sodium salt @ 1.75 kg per ha in 500 lit of water per ha has to be sprayed. 2, 4-D spray should be avoided when neighboring crop is cotton or *bhendi* and in those areas, apply 20% urea for the control of *Striga* as direct spray. Pre-plant application of glyphosate @ 2.0 kg/ha along with 2% ammonium sulphate at 21 days before planting of sugarcane followed by post emergence direct spraying of glyphosate @ 2.0 kg/ha along with 2% ammonium sulphate on 30 DAP suppress the nut sedges (*Cyperus rotundus*). Sempra is used @ 90 g/ha in 500 litre of water to control *Cyperus* sp in plant and ratoon sugarcane as post emergence spray.

Manures and fertilizer application: Sugarcane has a long growing season ranging from 8 to 18 months countrywide and an extensive root system, which favour nutrient uptake, is highly efficient. The nutrient uptake per tonne of cane yield is as follows: macronutrients (kg): N 0.8, P_2O_5 0.30, K_2O 1.32, MgO 0.50, CaO 0.42 and S 0.25; micronutrients (g): Fe 31, Zn 4.5, Mn 11, Cu 2.0, B 2.0 and Mo 0.01. To produce one tonne of cane, it requires 1.7 kg N and 1.0 to 2.5 kg K_2O is required. The recommended NPK ratio of 2:1:4 achieves high fibre and sugar production. Nutrient uptake and utilization are normally slow during the initial period *i.e.*, 1-12 weeks after planting or ratooning (re-growth after harvest). During the grand growth period (12 to 32 weeks after planting), nutrient uptake and plant development are rapid. The maturation period is characterized by low nutrient uptake and low bio mass accumulation, however, during this period, sugar assimilates accumulate. This is referred to as the 'ripening period', which elevates sugar yield. Nutrient needs of sugarcane are best portrayed on a yield basis considering cultivars, soil, nutrient reserves and climatic conditions. The management of crop residues influence sugarcane fertilization needs. Though nitrogen is largely lost, ash remaining in the field supplies soluble potassium and phosphorous. After a number of crops, burning will decrease the soil inorganic reserves added in the sugarcane monoculture.

Nitrogen: It promotes vegetative growth and inhibits the storage of sucrose in the parenchyma of the stems. Nitrogen deficiency in older leaves shows die-back symptom. N deficient plant leaf blades turn uniformly light green to yellow. Stalks become short and slender. The vegetative growth rate is reduced. Tips and margins of older leaves become necrotic prematurely. Consequently, nitrogen fertilizers are applied to ratoon shoots as soon as possible after the previous crop has been reaped and all nitrogen should be applied five months before the cane will be harvested. If the dressing is too heavy or its application unduly delayed, the sucrose content of the juice will be depressed. Excessive nitrogen delays the maturity of the cane which affect the juice quality by increasing non-sugary contents of the juice. Nitrogen is usually applied as a band along the cane row. The sprayed urea entered the leaf in 24-48 hours after spraying.

Phosphorous: It is essential for the formation of healthy root system. Consequently, the growth of cane on phosphate deficient soils is inhibited and the number of tillers (roots which arises from the buds of the root stocks) is limited. P deficiency shows the deep purple discolouration of the leaf tip. Phosphorus fertilizer is placed in the furrows (root zones of the cane) at the time of planting.

Potassium: It is intimately connected with the carbohydrate formation in the leaves and the subsequent translocation of sucrose to the parenchyma of the stems. K deficiency shows necrotic lesions between veins along margins and tips of older leaves. Older leaves may be entirely brown or fired. Young leaves usually remain dark green. Potassium application increases juice quality but not cane yield. It mitigates the ill effects of excess of nitrogen. Nitrogen and potassium has synergetic interaction.

Zinc deficiency: It shows pronounced pale green colour along the major veins, stripped chlorosis, shortening of internodes with stunted growth of the cane. $ZnSO_4$ @ 37.5 kg/ha is applied as basal in Zinc deficient soils. If basal application is not done, 0.5 % $ZnSO_4$ foliar spray is to be given on 90, 105 and 120 days after planting.

Iron deficiency: Immature leaf blades are distorted and necrotic. Interveinal chlorosis occurs from tip to base of leaves. The leaves may become chlorotic in calcareous soils due to iron deficiency. Ratoon plant exhibit iron deficiency when root development is slow. Foliar sprays can be used to correct chlorotic leaf symptoms, but they may not increase yields. Ferrous sulphate can be sprayed at 0.25% (25 g $FeSO_4$ in 10 lit water). If crop is more than 120 days 0.5% $FeSO_4$ has to be sprayed (50 g in 10 lit water). Application of 100 kg $FeSO_4$ per ha is recommended for soils with Fe deficiency.

Biofertilizers: Nitrogen fixing organisms such as *Azospirillum brasilence*, *Acetobacter* (*Acetobacter diazotrophicus)* and phosphorus solublization bacteria such as phosphobacterium (*Bacillus megatherium)* are used in sugarcane. All these biofertilizers can be applied @ 10 kg per ha in two splits on 30 and 60 days after planting cane. Acetobacter is referred as black urea since it reduces nitrogen requirement to 50%.

Time of fertilizer application: Sugarcane crop utilizes the nitrogen in the setts up to 30 to 45 days. The crop requires about 51 % of total N, during tillering phase and about 32 % during elongation period. Hence, it is better to apply nitrogen fertilizers after 30 days after planting in order to avoid leaching losses. Nitrogen is applied by placement in shallow furrows on both sides of rows and covered with soil. Volatilization loss is reduced by placing and covering with soil. Nitrogen is generally applied in two equal splits once at 45th day (start of tillering) and the other at 90th day (start of grand growth period). Three equal splits as top dressing on 30, 60, and 90 days is also adopted for sugarcane. Three splits (30, 60, 90th day) are found to be superior to 2 splits (45, 90th day). Apply nitrogen in 3 splits to crop *i.e.*, 10 percent of planting, 40% after 8 weeks and 50% at the time of earthing up without loss in the yield as compared to 4 splits. Four splits of nitrogen on 30, 60, 90 and 120 days after planting is also recommended for high cane yield. Application of nitrogen beyond 120 days delays maturity by prolonging the vegetative growth period. Half of nitrogen recommended at second split can be sprayed as 5.0 to 7.5% urea twice at 90th and 110 days. The proportion of organic to inorganic nitrogen to be applied in different tracts varies from 1:1 to 2:3. Biofertilizers such as Acetobacter or Azospirillum @ 20 pockets per ha should be well mixed with powdered FYM and uniformly applied in the furrows. Ten pockets can be applied at 30th day while the remaining 10 packets at 60th day. Biofertilizers meet 20 to 25% nitrogen requirement. Potassium is applied in two or three equal splits along with nitrogen. Application of 25% extra N and K on 150th and 180th days after planting increased the cane yield by 13%. The fertilizer recommendation in different regions of India is presented in Table 16.

Table 16 Fertilizer recommendation for sugarcane in different states of India

State	**Fertilizer dose (kg/ha)**		
	N	**P_2O_5**	**K_2O**
Maharashtra – Seasonal/Ratoon	350	140	140
Maharashtra - Pre-seasonal	400	170	170
Maharashtra - *Adsali*	500	200	200
Telangana	250	100	120
Andhra Pradesh -Coastal districts	120-225	100	100
Andhra Pradesh – Inland districts	225	100	120
Karnataka Mandya	250	100	150
Karnataka –Belgaum	250	75	200
Karnataka -Tungbhadra project area	250	75	150

Karnataka -Heavy rainfall area	185	124	124
Karnataka - Ratoon	315	75	200
Odisha	220	100	60
Gujarat	250	125	125
Kerala-Pandalam and Thiruvella area	165	82.5	82.5
Kerala -Chittur area	225	75	75
Kerala -Newly cleared forest areas	115	75	90
West Bengal - Spring	160	60	60
Assam	135	70	60
North Bihar	120	85	30
South Bihar	140	70	30
Rajasthan- Kota and Udaipur	150	50	On soil-test basis
Rajasthan - Sriganganagar	100	50	On soil-test basis
Madhya Pradesh	120	34	24
Haryana - Spring	150	50	50
Haryana - Ratoon	225	50	50
Haryana - Autumn	150	50	50
Tamil Nadu - plant crop (meant for sugar mills)	300	100	200
Tamil Nadu - Ratoon crop (meant for sugar mills)	375	100	200
Tamil Nadu - Jaggery areas	225	62.5	112.5

Source: http://agridr.in/expert_system/sugar/nutrientmanagement.html

Nitrogen application on 45, 90 and 135 days after planting is recommended for plant crop and on 30, 60 and 90 days may be followed for ratoon crop. For jaggery areas, apply 175 kg nitrogen and 112.5 kg K_2O per ha in three equal splits on 30, 60, and 90th days. Subsurface application of nitrogen and potassium at 10 to 15 cm depth by the side of the cane clump through holes is also recommended.

Soil fertility management of problem soils: Salt affected soil can be improved through leaching out salts, incorporation of bulk manures like farmyard manure (12.5 t/ha), pressmud-cake (5-7.5 t/ha) or compost (12.5 t/ha), green manure @ 5 t/ha and application of gypsum (1.7 t/ha) to remove 1 meq Na/100g soil. It may be applied with irrigation water. Crops growing on acid soils suffer due to toxic effects of excessive Al, Fe and Mn due to high phosphate fixation, low rate of nitrification, low water holding capacity and deficiency of Mo, Si, S,K, Ca, B and Mg. These soils may be effectively managed through the application of lime as CaO @ 1.2 to 2 t/ha preferably 10-25% of lime requirement added each year or application of dolomite @ 2.5 to 5.0 t/ha.

Crop logging system: The term 'crop log' is given by Hamilton Agee, who is the Director of the Hawaiian Planters Association Experimental Station at Hawaii. Later on, this system is used in sugarcane production by H.F. Clements (1980) at Hawaii. Crop log is defined as the record of the progress a crop makes from its start until harvest. It provided a running picture of the crop's physiology, which integrates climate and physiology in such a way that alterations in the management of a crop can be made while the crop is growing. The system indicates the general conditions of the plants and suggest changes in the management which are necessary for getting maximum yields. This method is ideally suited to irrigated cane which is cropped for 18 to 24 months. Crop-logging is not efficient for non-irrigated cane. It is not efficient for 12 month cane as it cannot make good use of late corrective application of fertilizer, especially nitrogen. But ripening-log can still be used for 12 month irrigated cane. The crop-log approach is based on critical nutrient concentration (CNC). The single CNC points are customarily located in that portion of the curve where the plant nutrient concentration changes from deficient to adequate. Leaf nitrogen has positive correlation with tillering behaviour of the crop. A unit of 1% leaf nitrogen increases 5.81 tillers per plant. Nutrient concentration in leaf laminae and sheath (3, 4, 5 and 6th) are utilized for diagnosis of N, P, K, Ca, Mg, S and micronutrient deficiencies. This method is sensitive to the age of tissue, the position of leaf, cultivar and other factors, which strongly affect nutrient concentration. Crop log technique is used to reduce nitrogen dose. Nutrient ratios *viz*., N:P and N:K, are negatively associated with sugar content and cane yield. Maintenance of 85% or more of moisture in +3 to +6 leaf sheath in the first four months of a crop (formative phase) and about 1.8 % N in 3 to 6 laminae during the period 150 to 270 days through nitrogen fertilization by three months after planting and frequent irrigation in the formative phase of the crop result in optimum cane and sugar yields. The soil and plant threshold values of nutrients are worked out. The nutritional advances have markedly affected fertilizer use while ensuring greatly improved sugarcane production. About 0.08% phosphorus and 1.99% potassium in 3-4 leaf sheaths on sugar free dry weight basis are found to be optimum for realizing satisfactory cane yields. Sheath moisture is also negatively correlated with yield at ripening phase. Positive correlation is observed between foliar nitrogen and sheath moisture during maturity phase. Clements has used this method for fertilizer and irrigation scheduling in sugarcane crop. Several indices are used to evaluate the crop status, which in turn are used for crop management practices for harvesting potential yield.

i) **Primary index (PI):** It is defined as the total sugar level of the elongating cane sheaths expressed as percent of the dry matter. PI value 10 indicates a balance between sugar production and utilization.

ii) **Nitrogen index (NI):** It refers to leaf nitrogen content as a percentage of the dry weight. NI is used to determine cane yield, sugar yield, sucrose percent, etc.

iii) **Moisture index (MI):** It is used as cultural guide for water use through irrigation. Sheath moisture in formative phase and leaf N in grand growth phase are positively correlated to cane yield, whereas sheath moisture and N content in ripening phase are negatively correlated to juice quality.

iv) **Potassium index**: It is expressed as a percentage of the sugar free dry weight. Cane growth is high with high K index.

Diagnosis and recommendation integrated system (DRIS): DRIS is new approach to interpret leaf or plant analysis. It is developed by Beaufils at the University of Natal, South Africa. It is a comprehensive system which identifies all nutritional factors limiting crop production and so increases the chances of obtaining higher crop yields by improving fertilizer recommendations. DRIS approach can provide slightly more accurate diagnosis of nutrient deficiencies than conventional crop log approach. The method utilizes ratio of tissue nutrient concentration rather than the concentrations itself. DRIS involves ordering nutrient ratios into expressions called DRIS indices. Essentially, a DRIS index is a mean of the deviations of the ratios containing a given nutrient from their respective optimum or norm values. In general, nutrient ratios have been found to be relatively stable with crop age. In order to eliminate possible imbalance among nutrients, the DRIS norms are established for nine nutrients (N, P, K, Ca, Mg, Fe, Mn, Zn and Cu) in the top visible developed leaf lamina of sugarcane. Fertilization according to DRIS increases both cane and sugar yields compared with those obtained when fertilization is guided by foliar analysis using critical nutrient concentration approach or soil testing. Foliar analysis using the DRIS proved to be a better guide for fertilization of sugarcane than the other testing methods currently in use.

Crop lodging: The crop is said to have lodged when the stems of normally upright plants fall over and do not return to their original position. There are two principal types of lodging in plants *viz.*, (i) stem lodging and (ii) root lodging. When lower stem internodes bend or break, the damage is called stem lodging. When the entire plant leaves or falls over because of root failure to provide anchorage the condition is called root lodging. Stem lodging is caused by wind, rain, hail, *etc.* Root lodging results from failures in the root system or unfavourable changes in the soil as a support base or root distribution, length, thickness, number and tensile strength. When stem and root lodging occur, the above ground parts of the plant may fall and the point of breakage or bending is usually at the crown or upper part of the root system. Loss of crop yield caused by lodging is determined by the stage of growth when lodging occurs and the severity of lodging. Lodged plants may partially recover if damage is clearly in the season. Careful monitoring of fertilizer application, especially nitrogen, is an effective way to reduce crop lodging. Crowded or sparsely spaced plants tend to lodge. Development of lodging resistant varieties with stiff and strength solves problem of lodging. Lodging can be prevented by deep trench planting, heavy earthing up, high potassium and low nitrogen application, delayed manuring, mid-crop hardening, removing dried leaves, trash twist single or double line propping, wrapping with leaves round the clumps of cane, supporting cane by bamboos or by tying with wires. Deep ploughing to a depth of 45 cm induces setts root development, which helps to prevent lodging. Lodged plants produce lower yields than upright plants and they are more difficult to harvest.

Water management: The quantity of water required to raise a normal crop in the sub-tropical region is from 350 to 400 ha cm. In the tropical region 600 to 725 ha cm has been found to be optimum. The water requirement of sugarcane is high in tropical region, which varies from 2000 to 3000 mm, and it is 1400-1800 mm in sub-tropical region. Water requirement for sugarcane ranges from 1400 to 2500 mm for 12 month crop. For long duration (18 months) *Adsali* crop, water requirement ranges from 3200 to 3500 mm. About 200 to 250 tonnes of water is required to produce one tonne of cane. In general, 60 to 90 kg of water is required to produce 1 kg of cane depending on climatic conditions, soil type and sugarcane variety. Sugarcane roots are mostly distributed in the upper 60 cm soil layer whereas the roots occur in the upper 30 cm soil are mostly active in water absorption. Irrigation interval and frequency in sandy and loamy soils during different growth stages of sugarcane are presented in Table 17.

Table 17 Irrigation interval and frequency during different growth stages of Sugarcane

Growth stages	Cropping period (days)	Irrigation interval (days) and frequency		
		Sandy soil	**Loamy soil**	**Clay loam soil**
Germination phase (Sprouting of buds)	0-35	5 (7)	7 (5)	10 (4)
Formative phase (Tillering Stage)	36-100	8 (8)	10 (6)	12 (5)

Grand growth phase (Tiller ceaser and internode elongation)	101-270	8 (34)	10 (17)	12 (14)
Ripening phase (Maturity-increase in sucrose)	271 to harvest	10 (9)	12 (8)	15 (6)
Total	12 months	--- (58)	--- (36)	--- (29)
Note: Figures in the bracket indicate the number of irrigations.				

Among the growth stages, tillering and grand growth stages are most sensitive period for irrigation. '*Trash mulching*' can be done in younger crop, which suppresses weed growth, helps in conserving moisture and aggregating soil particles in cases of heavy clayey or excessively sandy soils. Mulch the ridge with cane trash to a thickness of 10 cm uniformly to tide over drought on 21 days after planting. Avoid trash mulching in areas where incidence of termites is noticed. During the maturity and ripening period, irrigation intervals are extended or irrigation is stopped when active leaf area is declining due to leaf senescence and the crop is not able to respond to sunshine. Imposing moisture stress at maturity and ripening periods, increase sucrose content. Frequent irrigation during this period reduces sugar production. Shallow water table is responsible for poor root growth and low cane yield.

The first symptom of change of colour from fresh green to bluish tint may be considered as an index for irrigation in cane and ultimately the leaves become yellow and dry up. The yellowness caused by the deficiency of water is quiet different from that of the deficiency of nitrogen. The latter shows a pale yellow colour whereas the former exhibits a dark bluish yellow colour. The worst affected plants shows wilting by curling of leaves whose margins turn inwards. However, delaying irrigation until the plant appearance points to water deficiency in the soil is a harmful practice. Irrigation at 50% moisture depletion is adequate up to 270 days for sugarcane and 75% depletion from 271 days to harvest. Irrigation can be provided at 0.75, 0.75 and 0.50 1W/CPE ratio during tillering, grand growth and ripening phases respectively. 1W/CPE ratio of 0.9 with 1920 mm water requirement was optimum in Tamil Nadu for maximum cane and sugar yield. Water use efficiency under field conditions is low which ranges between 0.4 and 0.6 t/ha/cm. Ideally, water use efficiency of one tonne of cane/ha/cm of water is desirable. Generally sugarcane crop requires irrigation at weekly or ten days interval during grand growth phase and 15 days interval during ripening phase. Under limited water supply, skip furrow or alternate furrow irrigation system with trash mulch saved 10 to 15% irrigation water. Drip irrigation saves 50% water and increase cane yield from 10 to 25%. Skip furrow, paired row or pit method of planting may be followed in clayey soil when there is shortage of irrigation water. There is 30 to 40% of water saving in skip furrow or alternate furrow irrigation. Sheath moisture indicates the rate of growth or accumulation of sugar. The ideal sheath moisture of the 3 to 6 leaf sheath from the top is 85% for maximizing growth, which must be maintained for 3 to 7 months of planting. Sheath moisture during grand growth phase has a dominant role in cane growth. Gradual withdrawal of sheath moisture from 85% to 73 -74% before harvest hastens maturity and ripening which in turn helps to improve sugar accumulation.

Excess moisture/water management: Prolonged water stagnation may result in death of the root system. Prolonged water stagnation, swampy conditions or constant flooding induce early and uneven maturity with low cane yield, poor juice quality, high ash, low sucrose, low phosphorous uptake, high invert sugar, high gums, high non-proteinous matter and total colloidal contents. A practical problem of subsequence is the difficulty of harvesting sugarcane. The sucrose content in cane increases suddenly after water receding and thereafter, drops again in a few days. So cane should be harvested as early as possible after water recession to get maximum sugar yield. Water logging is associated with pithines, cavity formation, nutrient imbalance and early attainment of ripening, but, peak sucrose is not attained. Sucrose content declining faster with receding flood waters while there is increase in reducing sugars and gums. Shallow water table is mainly responsible for poor root growth and low cane yield. Very shallow water table depth (<60 cm) causes adverse effects on cane growth and yield while a water table depth of 60 to 100 cm can contribute to a total ET. Water table beyond one metre neither contributes to ET nor affects the sugarcane growth and yield adversely. Management practices in water logged areas include opening a number of drains, close planting to increase plant population, planting half way up the ridge, gap filling, application of pressmud, green manuring, wet earthing up, skipping of 25% irrigation, etc., help to sustain cane yield. Pre-harvest spraying of cyocel @ 2500 ppm on waterlogged canes improves 1.52 units more sugar recovery.

Drought management: To alleviate moisture stress, soak the setts in ethrel 200 ppm or lime solution (80 kg lime in 400 litres water for one hour) and plant in deep furrows of 30 cm depth. Then spray potash 2.5% and urea 2.5% during moisture stress period 3 or 4 times at 15 days interval. Spray kaolin @ 3% (30 g in one liter of water) to alleviate the water stress. There is water saving of 30 to 40% in skip furrows and alternate furrow method of irrigation while 50% in drip irrigation. Application of potash @ 125 kg/ha during last irrigation before the closure of canal water is effective in conserving soil moisture and improving the cane yield of 6 t/ha. A variety having thin and more stalks, which maintain relatively higher water content from 105 to 165 days after planting with low membrane injury and lower rates of lipid peroxidation with greater accumulation of proline shows drought tolerance. Cane growing under drought conditions accumulates proline (accumulation being more in drought-tolerant varieties) and contains high soluble nitrogen and reducing sugars.

Measures to mitigate drought / moisture stress effects

Sugarcane production under inadequate irrigation facilities can be brought through selection of drought tolerant varieties, trash mulching, skip-furrow irrigation, trench planting, applying irrigation during most critical period of crop water need and use of moisture absorbent. Organic amendments such as FYM, pressmud, coir waste application improve yield of cane under moisture stress conditions by increasing the moisture availability. Use of absorbents like *jalshakti, stocksorb, alcosorb* have also been found to play an important role under moisture stress conditions. Super absorbents like *jalshakti* have the unique property of absorbing water many times it's weight and making it available to the plants subsequently resulting in overcoming moisture stress. Setts can be dipped in lime solution two hours before planting for drought hardening. The solution is prepared by mixing 8 kg of kiln lime in 400 litres of water. It enhances the germination percentage and also gives the crop the ability to withstand moisture stress. Trash mulching soon after planting combined with urea spray on 90, 105 and 120 days after planting (55 kg N/ha for 3 sprays) overcomes the ill effects of drought and improves cane productivity. Besides these, inducing drought hardiness by soaking setts in lime solution for 2 hours, trench planting, use of antitranspirants (Kaolin @ 6% spray) during formative phase (60-150 days after planting), spray of urea and KCl @ 2.5% each) at 60, 90 and 120 days after planting, and spraying 2% potassium sulphate during hot summer mitigates drought and improves cane yield. Standing crop can also be hardened by withholding irrigation water for 30 days as a result the plant suffers less when it is required to face actual moisture stress during summer months.

Sugarcane maturity and ripening: Maturity means the botanical completion of a stalk suitable for producing new plantlets from each node. The stalk will contain stored sugar to support reproduction but its storage potential has not been realised. Sugarcane plant experiences maturation when the rapid growth process tends to slacken as water and nitrogen resources depleted. The increasingly predominant bulk of the plant becomes a succession of fully elongated joints. Maturity refers to the culmination of diphasic physiological process. The first phase ends at about the time when the aging leaf detaches from its subtending internode, the second phase includes all subsequent events relating to sugar accumulation of about 50% in the fully expanded internodes. Maturity does not refer to age. If water and nitrogen continuously abundant the plant never mature regardless of age. Sugarcane ripening refers to potential or massive sugar or sucrose accumulation in the storage tissue of the stalk. Sugarcane ripening is the physiological senescence, intermediate between rapid growth phase and the ultimate death of the plant. Agronomic concepts of ripening are usually based upon the appearance of internodes no long subtended by green leaves and a parallel accumulation of sucrose in each successive internode towards a common high value. The ripening process is conveniently depicted as a culmination or perfection of maturity. The sucrose content in these plants decreases through the stalk towards the top of the stalk. As the plant matures, a more uniform sucrose content is found throughout the stalk except for the top few internodes and the below-ground stool. The top/bottom ratio of sucrose content approaches one throughout the cane stalk at cane ripening. As a rule each variety will tend to approach a ripening peak after which its sucrose content will decline if the cane remains un-harvested. Ripening can be enhanced with topping, water regulation and fertilizer programming.

Management of burnt cane: Pre-harvest burning of sugarcane practiced in some western cane growing countries *viz.*, Australia, Iran, Jamanica. The burnt cane left standing gained weight during the first day and either gained or maintained additional weight during the ensuing four days. The sucrose decline was accountable to the increased moisture uptake by the burnt stalks whose ET rates were lower than normal. No loss of sugar in standing cane prior to fourth day whereas cut cane loses 3% sugar within 2 days. Most of sugar loss occurs after third day. Cut burnt cane rapidly loses its weight. Burnt standing cane showed slightly greater quality decline than burnt cut cane. Refractometer measures brix (total soluble solids) in cane juice.

Cropping systems: The promising sugarcane based cropping systems in tropical and subtropical regions of India are furnished in Table 18.

Table 18 Sugarcane based cropping systems in India

Sub tropical region	**Tropical region**
Paddy- Autumn Sugarcane-ratoon-wheat	Bajra -Sugarcane-Ratoon -wheat
Greengram- Autumn Sugarcane- ratoon-wheat	Paddy-Sugarcane-Ratoon- Finger millet
Maize- Autumn Sugarcane-ratoon-wheat	Paddy-Sugarcane-Ratoon- Wheat
Kharif Crops-Potato-Spring Sugarcane-ratoon-Wheat	Paddy-Sugarcane-Ratoon- gingelly
Kharif Crops-Mustard-Spring Sugarcane ratoon- Wheat	Paddy-Sugarcane-Ratoon- urd
Kharif Crops-Pea/Coriander-Spring Sugarcane-ratoon-Wheat	Cotton-Sugarcane-Ratoon–wheat
Kharif Crops-Wheat-late planted Sugarcane ratoon-Wheat	Sugarcane-Ratoon-*Kharif* rice- *Rabi* rice; Sugarcane-ratoon-rice-banana

Crop protection: Planting of sugarcane seed from heat treated (moist hot air treatment at 54°C for 2.5 hours at 95-99% RH) crop of a variety moderately resistant to red rot, eradication of diseased plants, especially of smut or GSD and spraying of crop before grand growth phase with fungicides like copper oxychloride (0.25%) against leaf spots effectively reduced incidence of diseases and sustained cane yield and quality. Soil treatment with nematicides, *viz.* Vapam, Nemagon, Dasanit, Carbofuran, has been found effective in reducing the population of nematodes and increasing the cane yield. The economic threshold level (ETL) of sugarcane pests is furnished in Table 19. Chemicals used to control pests of sugarcane are presented in Table 20.

Table 19. Economic Threshold Level (ETL) of sugarcane pests

Name of the pest	Economic threshold level
Early shoot borer	15.0-22.8% for late variety, 16.8% for early variety
Pyrilla	3-5 individuals/leaf or one egg mass per leaf
Stalk borer	17 bored internodes per row of 6 m. length
Internode borer	16.15 to 28.39 bored canes per row of 6 m. length
Top borer	15-22% incidence
White grub	15 beetles/host tree
Rodents	15 live burrows/ha

Table 20 Chemicals used to control pests of sugarcane

Pest	Control measures
Soil & seed borne diseases, insect pests	Application of biocontrol agents like *Pochonia chlamydosporia, Paecilomyces lilacinus* or *Trichoderma viride or Pseudomonas fluorescens* @ 4 kg/acre at the time of planting mixed with moist FYM or cured pressmud and distributed uniformly helps in suppressing the plant parasitic nematodes.
Termites	Chlorantraniliprole 18.5% SC @ 200-250 ml in 400 l of water/acre or clothianidin 50%WDG @100 g in 400 l of water/acre or imidacloprid 70% WS @28-42 g in 40-60 l of water/acre or imidacloprid 17.8% SL @ 140 ml in 750 l of water/acre or chlorpyrifos 20% EC @ 2.5 l/acre
White grubs	Fipronil 40% + imidacloprid 40% WG@175-200 g in 400-500 l of water/ acre or phorate 10% CG @ 10,000 g/acre
Early shoot borer and root borer	Use of pheromone traps @ 4-5/acre for monitoring, application of Phorate10% CG @ 12,000 g/acre, or Carbofuran 3% CG @ 26640 g/ acre or Chlorantraniliprole 0.4% GR @ 7.5 Kg/acre or Phorate10% CG @ 12,000 g/acre and spray of Fipronil 5% SC @ 600-800 ml in 200 l of water/acre, Fipronil 0.3 % GR @ 30-40 ml in 10000-13320 l of water/ acre, Chlorpyrifos 20% EC @ 500-600 ml in 200-400 l of water/ acre, Chlorantraniliprole 18.5% SC@ 150 ml in 400 l of water/acre, or Cypermethrin10% EC @ 260-304 ml in 200-280 l of water/acre, or Quinalphos 5% granule @ 2000 g/acre, or Chlorantraniliprole 0.4% GR @ 7.5 g/acre, Monocrotophos 36% SL @ 600-900 ml in 200-400 l of water/acre or Quinalphos 25 % EC @ 800 ml in 200-400 l of water/acre.

Top borer	Soil application of Carbofuran @ 1.0 kg ai/ha or Phorate @ 3.0 kg ai/ha against 3rd brood
Stalk borer	Application of Monocrotophos @ 0.75 kg a.i/ha
Scale insect and mealy bug	Dipping of setts in 0.1% Malathion/0.06% Dimethoate for 15 min and drench spraying with the above chemicals at 4-5 internode stage of the crop. Monocrotophos 36% SL @ 600 ml in 200-400 l of water/acre
White fly	Spraying of acephate (0.1 %).
Root borer	Soil drenching of imidacloprid @ 0.1 kg a.i. /ha during mid-August / 120 days after planting is effective for management of root borer.
Woolly aphid	Bioagents like *Diphaaphidivora*, *Micromusigorotus*and *Chrysoperlacarnea*may be redistributed in sugarcane fields. Need-based chemical application of imidacloprid 200 SL @ 100 g a.i./ha or chlorpyriphos 20 EC @ 1 kg a.i./ha or oxydemeton methyl 25 EC @ 1 kg a.i./ha or thiamethoxam 25 WG @ 50 g a.i./ha is recommended.
Rodents	Bromadiolone 0.25 % CB @ 0.005% or Bromadiolone 0.005 % RB @ 0.005%

Management of early shoot borer in sugarcane

- **Cultural methods**: Earthing-up (45 days after planting) and trash mulching in moisture stress areas. In termite prone areas, the trash may be treated with insecticide drench before mulching.
- **Insecticidal methods**: If the infestation of the borer is found to exceed 20 percent, spraying the leaf whorls with chlorpyriphos 20EC @ 5ml/lit of water or Chlorantraniliprole 18.5% SC @ 75g a.i./ha followed by irrigation or Fipronil 5% SC or 0.3% GR @ 75 g a.i./ha followed by irrigation.
- **Use of Pheromone lures**: Pheromone lures for shoot borer available with Pest Control India, Bangalore may be used @ 25/ha on 15th day of planting with one lure change after 45 days.
- **Biological methods:** Release of *Trichogramma chilonis* @50,000 eggs /ha bearing trichocards release 6-8 times during the season 30 days after germination.

Harvest: Sugarcane maturity symptom shows the appearance of pale yellow leaves or drying of leaves. When tapped the mature cane gives a sort of metallic sound. There will be glittering of sugar crystals when a mature cane is cut in a slanting way and held against sun. The harvest in sugarcane depends on maturity which in turn depends on the variety, nature of the crop (plant or ratoon), quantity and time of fertilization *i.e.*, late or earlier fertilizer application (particularly nitrogen), season and weather succeeding vegetative growth and control of soil moisture through irrigation during maturity phase of the crop. Harvesting coincides with cool dry period of the year when optimum maturity can be obtained. Prior to cutting the crop, it is customary to give good irrigation for facilitating free extraction of juice. Harvesting in cane includes cutting, stripping, binding, loading and transport to mill yard. The cane for milling is expected to be of high at maturity stage. The harvested cane should be clean and as fresh as possible to maintain a steady but high level of recovery. Cutting immature top portion from the cane stalks is important for good sugar recovery, since this portion is poor in sucrose and rich in nitrogen compounds and starch that add difficulties in clarification and increased final molasses. But, top portion is not usually removed. Depending on a variety and cane maturity, the millable stalks account for 50 to 60% of the total vegetative growth produced above ground. Maturity of cane is assessed using Hand Refractometer brix reading. Brix reading should show 18 to 25% for optimum maturity for cane harvest. Top bottom brix reading should be 1:1. The best is to crush the canes as early as possible after harvest so that sugar output, whether white or gur (jaggery) is high and of higher quality with little processing problems. Staling cane has to be avoided. Likewise, harvest should be clean (free from soil and trash as much as possible) and close to ground (to avoid loss of sugar rich lower stem portions). Harvested canes should be loaded properly for further quick transport. These operations form parts of good harvest for good milling efficiency. Harvested canes are kept in shade and maintain moist condition for a day or two. Otherwise cane loses its weight and juice quality. Harvested cane should be crushed within 24 hours.

Ratoon Management

The word 'ratoon' is seemed to be originated from Latin words 'retonus' means cut down or mown; or 'retono' means to thunder back; resound; the spanish word 'reteno' means fresh root or sprout or the French work 'rejeton' meaning sucker or shoot, scion, descendent, off spring or sprout. When a ratoon is raised from the stubble regrowth of the first ratoon, it is 'second ratoon' and so on. The first crop raised by planting cane setts is referred as 'Plant crop'. Ratooning or stubble cropping is practiced in almost all the sugarcane growing countries. Ratooning is defined as the cultivation of an additional crop from the regrowth of stubbles of previous main crop after its harvest thereby avoiding reseeding or replanting. Farmers treat ratoon as gift crop or free crop. Sugarcane inherits about 7% biomass as stubble from the preceding crop and 3% as root mass. A ratoon crop invariably ripens earlier by at least one to two months due to early development of shoots, maintenance of relatively lesser nitrogen content in the index tissues and rapid run out of nitrogen during grand growth phase and relatively higher inorganic non-sugars. Ratoon yield is normally lower than the plant crop. The yield gap between plant and ratoon crop is 20% in India. But ratoons can yield as much as plant crops if managed properly. Ratoon cane yield is low because of reduced plant population due to failure in sprouting, decline in the soil nutrient status, soil compaction leading to poor physical status of soil and more incidences of pests and diseases.

Management practices for ratoons

1. Variety: Generally early maturing varieties are poor ratooners than mid-late or late varieties. Thin or medium thin varieties give better ratoons than thick varieties. Some varieties yield higher in ratoons than plant crop. Ratooning ability of a variety also differs from place to place.

2. Climate: Moderate temperature of 25 to 30°C is conducive for stubble sprouting.

3. Duration of harvest: The harvest duration of plant crop in a field should not extend beyond a week for ensuring uniformity of the sprouts and growth of ratoon crop.

4. Trash disposal is important soon after the harvest to carry out ratoon operations.

5. Trash burning helps to control insect pests (scales, mealy bugs, termites and rodents), diseases (grassy shoot). Trash burning generates heat which facilitates sprout germination.

6. Trash mulching is useful in extreme weather conditions. Trash mulch raises soil temperature and favours sprouting. Mulching suppresses weeds and conserve moisture wherever water is scarce.

7. Stubble shaving is done to a depth of 4 to 6 cm along the ridges with sharp spade.

The stubbles protruding out of the field are cut to, below the ground level using sharp spade. Stubble shaving facilitates healthy underground bud to sprout and establish deeper root system in the ratoon.

8. Spraying chemicals: Growth promoters like Ethrel @ 500 ppm spray and fungicides like Emissan at 0.25% concentration to the stubbles give better sprouting. Most of the sprouts, which make the millable cane for harvest come from the buds, located on lower half of the stubble. Sprouting of ratoon crop may be improved by spraying certain growth regulators, *viz.*, IBA (100 ppm), TIBA (50 ppm), Ethephon @ 750 ppm or Cycocel @ 8 kg/ha. More sprouting in the February ratoons than December is due to availability of more reducing sugars for respiration.

9. Off barring or shoulder breaking and loosening the interspace: Soil compaction is the major cause for poor growth of ratoon cane. Off barring is an operation where in the ridges are broken or cut on either side. It is also called shoulder breaking. To loosen the soil, the inter spaces between rows are dug. Off-barring increases the water and nutrients absorption.

10. Gap filling: Gap filling to the extent of 20 to 25% is absolutely essential to obtain adequate population. Gap filling the ratoons with pre-germinated setts, raised from single buds in polythene bags, ensured quick establishment with higher tiller and stalk populations. Material for gap filling (sprouted setts) should be at least 30-35 days old. Clumps can be uprooted from spots where excess sprouting is seen and cut into quarters and planted in the gaps.

11. Hoeing and weeding are carried out on 20th and 40th days.

12. Fertilizer application: Entire phosphorous and one fourth of nitrogen and potassium may be applied soon after stubble shaving and off barring on either side of ridge and covered with soil. Top dressing of fertilizers is carried out on 25 days after ratoon (DAR), on 60 DAR after partial earthing up and 85 DAR after final earthing up respectively. Additional 25% extra N and trash incorporation helped to sustain multi-ratoon yields.

13. Chlorosis: Two to three spray of 0.25% $FeSO_4$ plus 0.5% $ZnSO_4$ plus 1% urea solution at weekly intervals for young crop above 60 days old. Chlorosis normally occurs in calcareous and salt affected soils. Because of shallow root system, the ratoons have lesser capacity to withstand soil moisture stress than the plant crop.

14. Detrashing is done on 120 and 180th day

15. Trash twist propping is done at 11th week.

16. Harvest the cane after 11 month since ratoon crop matures one month earlier than plant crop.

Sugarcane Harvest Management: The harvest management includes technical and non-technical aspects. The technical components comprise the harvest at the proper maturity of the crop based on juice quality. The non-technical aspects include adequate labour force for cane harvesting, transport arrangements for moving the cane from the field to factory. The cane harvest technology can be broadly divided into pre-harvest, harvest and post-harvest management.

A. Pre-harvest management

(i) Methods of maturity evaluation: Age is one of the major factors that decide the cane harvest. The sugarcane plant is considered to attain maturity when it contains a minimum of 15 to 16% sucrose in juice and 85% purity. Sugarcane varieties are classified based on juice sucrose and purity which is presented in Table 21.

Table 21 Sucrose and purity content of sugarcane varieties

S.No	Varieties	Sucrose (%)	Purity (0%)	Harvest time
1	Early varieties	16	85	December
2	Mid varieties	18	85	February
3	Late varieties	16	85	April
4	Short duration (8 months)	18	85	----

(ii) Hand Refractometer Brix (HR Brix) in the field: The crop is considered mature enough and fit for harvest when the HR Brix is about 20 and above. Juice is collected with pouch piercer. Sugar recovery can be improved at least by 0.5 to 0.8 units through this practice.

(iii) Top/bottom ratio of sucrose content approaches one throughout the cane stalk nearer to peak maturity (1:1)

(iv) Periodical crusher juice analysis in the laboratory for brix, sucrose and purity, the maturity can be evaluated.

(v) Reducing sugars and invert ratio in relation to maturity: Reducing sugar invert ratio in juice should be low and sucrose will be high at peak maturity.

(vi) Acid invertase at pH 5.4 and neutral invertase ratio at pH 7.0

(vii) Physiological maturity: Maturity refers to the stage where the detachment of aged leaf from internodes and accumulation of 50% sugar in fully expanded internode. Ripening refers to the perfection of maturity where massive or potential sugar or sucrose accumulates in cane storage tissue.

B. Harvest management: Harvesting in cane includes cutting, stripping, loading and transporting to mill yard.

(i) Time for harvest: The harvest of cane crop is decided in the economic or profitable content of sugar in juice. Availability of labour for cane harvest, facility for transporting of cane to factory also influences the time of harvest. If the crop is to be ratooned, the harvest of plant crop is adjusted to be in the best months for ratooning. Harvesting in too early winter or late summer results in poor ratoon sprouts. Harvesting coincides with cool dry period of the year when optimum maturity can be obtained. It is customary to give irrigation prior to harvest or cutting the crop for facilitating free extraction of juice.

(ii) Tools for harvest: Harvesting is generally done by manual labour using the cutting knife of varying size and shapes, while machine harvesting is gaining momentum in India.

(iii) Harvesting technique: Irrigate the crop before harvest. Cane should be harvested at ground level with a cutting knife, since the bottom internodes are rich in sucrose and also helps in good ratooning. Avoid cutting cane into pieces.

(iv) Avoid water shoots, as these add to the weight only and contain very little sucrose.

(v) Removal of binding materials and trash: Clean cane has to be supplied. For every 1% increase in trash and binding material, about 0.2 to 0.5 units drop in sugar recovery will occur since they add only fibre and ash and re-absorb some of the juice.

(vi) Removal of immature top portion: Cutting immature top portion from cane stalks is important for cane recovery since this portion is poor in sucrose and rich in nitrogen compounds and starch which add difficulty in clarification and increased final molasses. But, the top portion is not usually removed. Depending on a variety and cane maturity, the millable stalks account for 50 to 60% of the total vegetative growth produced above the ground.

(vii) Dressing and bundling of cane stalk. Dressing up and bundling of cut cane as well as loading is done by manual labour. Removing leaf, dry or green, from cut cane is done using cutting knife. Cutting off-types may be done and this may add 1 to 2% on weight.

Special harvest management practices

(i) **Burnt cane harvest management:** Burning dry leaves of cane before harvest is a practice due to scarcity of labour. Trash burning has both advantages and disadvantages. If the cane is burnt, it has to be immediately harvested by mobilizing labour for supplying the cane to the mills within 12 to 24 hours after burning. Otherwise, the cane deteriorates if left in the field longer after burning and more so if cut burnt cane is not crushed immediately. Farmers sometimes burn the cane crop to get harvest priority.

(ii) **Cane management under waterlogged condition:** Cane is subject to stagnation of water in low lying areas especially during monsoon period. Under such conditions, the maturity is hastened and leaves turn pale yellow due to physiological drought. Such isolated plots need priority in harvest.

(iii) **Cane management under water stress:** In certain fields, due to drought and inadequate water supply, the crop starts drying and cannot stand in the field any longer. Such crop should be identified and cutting order is issued out of priority to save the crop from further damage.

(iv) **Flowered cane harvest:** Non-flowering varieties are to be planted in March-April. Extra N and irrigation control checks flowering. If 50% flowering takes place after 10 months age, cane harvesting even after 2 months after flowering do not reduce cane yield. Flowered cane with age of less than nine months cannot be given priority in harvest although suffers in yield reduction. Flowering of less than 9 months old crop is not a sign of maturity.

Post-harvest management

(i) **Cane supply for continuous crushing:** The cane for milling is expected to be of high maturity stage, clean and as fresh as possible to maintain a steady but higher level of sugar recovery. One of the problems of the sugar industry is to get the cane into the cane carrier of the mill as early as possible after the harvest without staling. This is possible if the harvest is planned in advance; sufficient labour should be made available to harvest the quantity within the prescribed time. The transport and the roads should be in good condition to facilitate quick movement without any breakdown of vehicles.

(ii) **Post-harvest deterioration of sugarcane:** Harvested cane should be crushed within 24 hours. Harvested cane should be kept in shade and moist to maintain juice quality for a day or two. When cane is harvested and stored, loss occur in cane weight due to moisture loss and bio-deterioration due to *Leuconostoc* bacteria. *Leuconostoc* bacteria can pass up to 20 cm in the stalk from the cut ends within an hour after harvest and causes degradation of sucrose.

(iii) **Loss of cane weight beyond 24 hours of harvest:** Maximum deterioration of harvested cane occurs at 38 to 40°C at 15% moisture loss. The loss in cane weight ranges from 1.5 to 2.5% for every 24 hours storage after harvest.

(iv) **Juice quality and sugar losses:** Loss in sucrose percentage in juice is 0.1 to 0.2 units drop for every 24 hours during the first 48 hours of storage. Rapid deterioration and fall in quality can be observed from 72 hours onwards.

(v) Screening cane varieties for resistances to post-harvest inversion: CoC 671 is comparatively resistant to post-harvest inversion than Co 6304.

(vi) Optimum economic distances for drawl of cane: To avoid post-harvest losses and minimizes transport cost, it is desirable to obtain the cane from short or economic distances only.

(vii) Staling of cane: The terms *stale cane* and *sour cane* are two different stages of cane deterioration after harvest. During the period of deterioration, metabolic conversion of stored sucrose takes place into less economic products through the agency of enzymes and microbes. The *stale cane* is the aging of harvested stalks, which have depleted their sucrose via continuing inversion and respiration whereas *sour cane* is microbiological deterioration of sugarcane stalks by lactic acid bacterium *Leuconostoc mesenteroides* which converts sucrose into organic acids of typical sour odour. However, both types of deterioration seem to operate simultaneously in cane and milled juice. In the prevailing supply system of raw material, *Kill to mill,* delay of 3 to 10 days or more is quite normal which aggravates deterioration process in the harvested cane due to inversion, respiration and development of acid, alcohol and physio-biochemical and microbiological agents has detrimental effect on sugar recovery.

In India, cane is harvested manually, which is mostly unburnt full green cane. There is no planned burning of crop, however, in south Gujarat, irrespective of measures taken by factory, burning of cane is a routine practice and burnt cane up to 80% is observed, especially near the end of crushing season. In some parts of north India, burnt cane is supplied at the end of crushing season, especially when cane is in plenty. Good harvesting practices are followed in western Maharashtra, north Karnataka and parts of south Gujarat, where cane plantation is planned and harvesting is entirely carried out by the factory. Other states depend on farmers for cane supplies, leaving maximum possibilities for irregularities. Appreciable amount of sugar is lost during the time lag between harvesting to milling, even as high as 25 kg/t of cane. The losses are higher in certain areas of Tamil Nadu and Andhra Pradesh where temperature is high. On an average, Indian Sugar mills lose sugar about 10 to 15 kg/ ton of cane crushed. These losses are further escalated when crushing is extended till May/ June or even later. Magnitude of cane deterioration increases if the cane supplies are from problem soil such as saline, alkaline, waterlogged, burnt field, fields receiving excessive application of nitrogenous fertilizer, crop damaged by frost or affected by pests and diseases. The fully mature cane will not deteriorate as rapidly as either immature or over mature cane. Green cane is less susceptible to post-harvest deterioration as compared to the burnt cane. High temperature around 40°C and low atmospheric humidity of 25 to 35% has deleterious effects on juice quality. Freeze also causes considerable deterioration in quality.

Consequences of cane staling: When sugarcane is allowed to stale after harvest, there is deterioration in juice quality. In order to lessen staling of cane, the harvested cane may be covered by dry trash and water is sprinkled. In cool months the deterioration is not so fast as in hot dry summer months. While in transits, the cane wagons may be mist sprayed. Loss in cane weight may be of 1.0 to 1.5% per day of staling. The juice quality of stale cane falls from 72 hours onwards. The sugar recovery of 9.77% in fresh cane is brought to 7.76% at 72 hours after harvest. The stale cane reduce the sugar both white sugar and molasses. Cane starts to lose weight by drying out as soon as it is harvested. The loss in cane is observed after 72 hours of staling. The cane weight loss will be around 7 to 15% under sub-tropical conditions and the weight loss may be as high as 16 to 18% after 120 hours of storage during May and June. Commercial Cane Sugar (CCS) reduces from 13.73

to 10.62 due to cane staling. The delay in cane crushing is normally 3 to 10 days in sub-tropical India and 36 to 48 hours in tropical India as harvesting is totally controlled by the factory. In India, 10 to 15% of sucrose is lost after harvest of cane and its subsequent delay in processing. The loss in cane weight due to staling ranges from 3 to 16% of initial cane weight depending on months of cane harvest. The post-harvest deterioration of sugarcane quality in cane is furnished in Table 22.

Table 22 Post harvest deterioration of sugarcane quality in cane var Co6304 (Big Mill Trial)

Hours After Harvest	Mill Trial juice	Brix	Pol % Sucrose	Purity%	Reducing sugar %	Fibre %	Pol % Cane	Sugar recovery %	Moisture loss
0	Primary	16.79	13.90	82.79	0.77	13.20	12.07	9.77	--
24	Primary	16.94	14.19	83.79	0.93	13.60	12.03	9.73	2.31
48	Primary	16.89	13.78	81.59	1.26	14.00	11.85	9.55	5.64
72	Primary	15.59	11.72	75.18	1.57	14.20	10.06	7.76	7.75
96	Primary	15.69	11.62	74.06	1.71	14.80	9.90	7.60	8.78

Source: SBI, Coimabatore

The most important polysaccharide producing organism is *Leuconostoc* sp, which is responsible for huge sugar losses in the industry, harbours under the leaf sheath. The dextran, which are polymers of glucose are produced directly from sucrose by the bacteria *Leuconostoc mesenteroides* or *L. dextranicum.* Cane from a ratoon crop is prone to dextran formation as compared to canes from a corresponding plant crop. Cane grown in soil, containing high content of K and Mg tend to produce less dextran. Every 0.1% of dextran produced represents sucrose loss of 0.04%. Presence of dextran and other undesirable polysaccharides in raw sugar are adversely affecting the export potential of sugar. Spraying of harvested cane with Benzoic acid @ 100 ppm, Formaldehyde @ 100 ppm, Potassium permanganate (0.1%), Sodium metasilicate (1%) on harvested stored cane covered with trash was found to be much effective in minimizing invertase activity and retaining the juice quality. Dipping both the cut ends in Sucroguard, improves 0.9% sugar recovery and there is 70% reduction in microbial population.

Methods to minimise stale cane losses are i) identification of cane varieties with high sucrose and juice quality maintenance and resistance to post harvest inversion even for long staling period, ii) by pre-harvest maturity survey and planned harvest, mature cane can be supplied, iii) adoption of wireless communication system for drawal of cane from optimum economic distance, iv) foliar spray of polaris improve quality and help to minimize post harvest deterioration, v) cane can be stored in shade covering with trash and sprinkling water minimises stale cane losses and vi) use of biocides. Formalin or Polycide @ 2 ml per litre and bacterinol-100 @ 250 ppm arrest the growth of *Leuconostoc* which helps in juice preservation. Dipping the cut ends immediately after the harvest in polycide @ 2 ml/ per litre and bactrinol–100 @ 1000 ppm solution and spraying the same in stored cane minimizes loss in stale cane juice quality.

Quality improvement in sugarcane

1. Fibre content: Cane quality is considered as the best when it contains optimum fibre of 12 to 13 percent.

2. Balanced NPK application and avoidance of late and heavy N application. Basal application of compost or FYM, green manure, *etc*., improves cane quality. Heavy nitrogen application leads to cane succulence, delayed maturity, low sucrose and high non-sugars. Late nitrogen application reduces juice quality.

3.Irrigation: As per crop logging system of H.F. Clements, the gradual withdrawal of sheath moisture from 84-85% to 73-74% helps in improving the sucrose content and cane quality.

4. Lodging: Lodged canes have low juice quality with high reduced sugars and low purity. The trash twist method of canes has been found to minimize lodging to a certain extent. Lodging can limits cane yield. It should be prevented by heavy earthing up, less use of inorganic nitrogen and applying high potassium, delayed manuring, mid-crop hardening, removing dried leaves, wrapping with leaves round the clumps of cane and supporting the canes by bamboos, use of non-lodging varieties.

5. Flowering: Floral initiation in sugarcane occurs when the plant experiences a day length of 12 hours and 30 minutes. The optimum night temperature to induce flowering is 21 to 28°C. The period from floral intiation and floral emergence is found to range from 75 to 90 days in different cane varieties. In peninsular India, cane varieties flower heavily in October to November whereas in North India (Uttar Pradesh and Bihar), cane flowers heavily in December to January. As a result of flowering, pith formation develops first in upper internodes and later on proliferates downwards, side shoots start developing on the top portion and finally gradual decline in sugar content. In general, ratoon crops flower more than plant crop. Flowering is to be controlled to prevent reduction in sugar yield to maintain it's green top, which can be fed to the cattle at harvest. A rough estimate shows reduction of 25% in yield and 1% in sugar in a profusely flowered cane.

5.1. Methods for suppression of flowering

(i) Mechanical defoliation of top 3 to 4 leaves. Here the labour cost is high.

(ii) Flare to break the continuity of the night. It is a costly operation.

(iii) With holding irrigation during floral initiation period reduces flowering considerably during October-December.

(iv) Late application of extra nitrogen: Well nitrogen fed crop combined with nitrogen stress at floral initiation reduces flowering.

(v) Spraying of Ethrel @ 500 ppm (500 mg in one litre of water) twice at 4 to 5 days interval during floral initiation period check the flowering completely.

(vi) Adjusting planting date: Flowering is considerably reduced in April and May planting. But plant crops ratooned in April to May flowers heavily in October to November. The control of flowering will increase the cane and sugar yield. If a crop is harvested soon after flowering, there can be no loss in sugar yield. The reduction in sugar and cane yield happens when harvested 3 to 4 months after flowering.

6. Chemical ripener: It improves cane juice quality. Polaris @ 5 kg/ha or Sodium metasilicate 4 kg/ha in 200 litres of water sprayed about 6 to 8 weeks before harvest improves 10 to 15% sugar recovery. The other chemical ripeners are Ethrel @ 1.2 kg/ha, CCC @ 2500 ppm, 2,4–D @ 0.5%, Cycocel @ 0.3%, Glyphosate or Polado (phosphorous methylglycine) @ 0.34 kg/ha, Fusidade Super @ 400 ml/ha, *etc*,. can also be used. Maleic hydrazide @ 4% increase sucrose content in 10 to 30 days.

Sugarcane juice extraction, quality assessment, sugar manufacture: Accumulation of sugar in the stalk begins soon after completion of the elongation phase when glucose produced during photosynthesis is not utilized for growth but stored in the stalk as sucrose. When sucrose concentration of juice exceeds 16% and juice purity increases over 85% the cane crop is said to be matured. Maturity of sugarcane can be judged by determining the quality parameters of the cane juice.

Cane juice sampling procedure: At least two whole clumps or 10 canes of all sizes randomly selected should be taken as a sample. The cane samples must be free from extraneous matter. The sample cane is cut into pieces of 45-50 cm, weighted and their juice is extracted. The juice is collected in a previously weighed bucket containing 0.5 g mercuric chloride as preservative. The extracted juice is weighed and the percent of extraction is calculated.

Quality parameters

Juice brix: Juice brix refers to the total solids content present in the juice expressed in percentage. Brix includes sugars as well as non-sugars. Brix is measured using brix hydrometer. Mix the juice sample thoroughly and pass it through a 150 mesh sieve to remove suspended particles. Collect the juice in a 100 ml cylinder and allow it to settle for 15 minutes. Gradually lower the brix spindle and allow it to float. When the brix spindle becomes stationary, take the brix reading at the line with the plain surface of juice. True brix can be obtained with reference temperature. Note the temperature of the juice and correct the brix at 20°C temperature from the chart. Brix can be measured in field using a Hand Refractometer. It is referred as Hand Refractometer Brix or HR Brix.

Sucrose percent or **Pol percent:** POL percent is the actual cane sugar (sucrose) present in the juice. It is measured using a polarimeter or sucrolyser. Transfer 100 ml juice into 250 ml measuring flask with stopper and add minimum quantity of Hornes lead subacetate for clarification. Shake well and filter. Polarize the pol reading is noted. Find out the pol percent of juice from the Schmitz'z table from corrected pol reading and corrected brix based on room temperature.

Purity coefficient or **Purity percent:** It refers to the percentage of sucrose in the total solids in the juice. Purity is a ratio between pol / corrected brix, where juice purity is corrected for the standardised temperature at 27.5°C in tropics.

$$\text{Purity}\,\% = \frac{\text{Sucrose (POL)}}{\text{Brix}} \times 100$$

Pure Obtainable Cane Sugar: {(1.5 × Pol % J0 × [0.95- (Fibre % / 1000)]} – {(0.5 × Brix %) [0.97 – (Fibre % / 100)]}

Reducing sugars: The reducing sugars (RS) refers to the percentage of other sugars (fructose, glucose) in the juice. A lower RS value indicates that much of sugars have been converted into sucrose.

Commercial Cane Sugar percent (CCS %): The CCS % refers to the total recoverable sugar percent in the cane. CCS % = 1.0225 × S – 0.292 × B, where S is the Sucrose % in juice and B is the Brix in juice.

Calculations:

Pol % cane = Pol in juice % cane + pol in bagasse % cane

$$\text{Pol in juice}\,\%\,\text{cane} = \frac{\text{Pol}\,\%\,\text{juice} \times \text{Juice}\,\%\,\text{cane}}{100}$$

$$\text{Pol in bagasse \% cane} = \frac{\text{Pol \% bagasse} \times \text{bagasse \% cane}}{100}$$

Bagasse % cane = 100 – juice % cane

$$\text{Purity of the juice (\%)} = \frac{\text{Pol \% juice} \times 100}{\text{corrected brix \% of juice}}$$

$$\text{CCS (t/ha)} \frac{\text{Yield (t/ha)} \times \text{sugar recovery \%}}{100}$$

$$\text{Sugar recovery (\%)} = \frac{([S - 0.4\,(B - S)]) \times 0.73}{100}$$

where S= sucrose % in juice
B= corrected brix %

As ripening of sugarcane proceeds, the percent of sucrose in stalk gradually increases while the percent of reducing sugar decreases.

Baggasse moisture percent: Take 100 g of well composted sample. Oven dry it at 110°C for 4 to 6 hours till the weight of bagasse becomes constant. Loss in weight gives moisture percent in bagasse.

Bagasse pol percent: The standard Deer baggasse digester is used for analysis. Bagasse is digested with 1.0% sodium carbonate for one hour. After cooling the contents, 200 ml of the extract is treated with standard lead sub acetate. Shake well and polarize the solution in a 400 ml pol tube. The reading directly gives bagasse pol percent. The quantity of bagasse to be taken for analysis for obtaining half normal solution is determined is follows:

$$\frac{w + ax}{x} = \frac{100}{13}$$

where w = water capacity of digester vessel in ml.
a = moisture per unit of bagasse and
x = quantity of bagasse

Sugar production from sugarcane: In India, two types of crystal sugars are produced. One is the plantation white sugar directly from sugarcane for domestic consumption and the another is the raw sugar exclusively meant for export. The plantation white and raw sugar are produced by the vaccum pan process while the '*gur*' and the '*khandsari*' sugar by open pan process. Out of cane production in the country, 49% of cane is utilized for white sugar production, 39% for *gur* and *khadasari* production and 12% for seed, feed and changing purposes. In international sugar trade, *jaggery* (*gur*) and *khandsari* are known as 'non-centrifugal sugar' and also the 'artisan sugar'.

(i) **Harvesting, handling and storage:** Sugarcanes are gathered manually and hand cut and tied in bundles. The harvested cane in the field is transported to the factory by lorries, trucks, tractors and carts. This is weighed in the weigh bridge and unloaded in the cane carrier. Cane has to be processed immediately since it deteriorates rapidly due to cane staling.

(ii) **Cleansing:** The cane is first cut and chopped into pieces by one or two sets of the rotating knives before it is squeezed in the milling tandom.

(iii) **Juice extraction:** The shredded sugarcane travels on a conveyor belt through the milling tandem generally consists of 4 to 5 mills each having a series of 3 heavy duty roller which extract juice from the chopped cane pieces. Cut pieces of cane are squeezed repeatedly when it is passed through pressure loaded rolling mills called milling tandom. Hot water at 70°C is added @ 30 to 35% of cane weight. A maximum amount of 80% of cane juice is extracted in sugar mills. The residue leaving the last tandem is called baggase which is used to produce steam in high pressure boilers and can also be used for pulping in paper mills. The raw juice moves on to be clarified. The last juice to be extracted is of low quality with less than 60% sucrose.

(iv) Evaporation: The filtered juice emerging from clarification station is evaporated to syrup stage and sent to vacuum pans for further concentration.

(v) Purification: The method of clarification is either sulphitation or carbonation for producing plantation white sugar. Very few factories use carbonation for clarification of juice. In sulphitation process the juice is treated by heat, lime and sulphur dioxide. To the cane juice, milk of lime ($Ca\ OH_2$) is added and simultaneously neutralised around 7.2 to 7.4 pH by passing SO_2 gas as clarificants. The SO_2 gas is produced by burning sulphur with compressed air in the sulphur burners. Super phosphate / P_2O_5 is used in sulphitation process to make up the deficiency of phosphate in the raw juice. The treated juice is then heated to boiling and settled in subsiders. The supernatant juice is clear and bright. In carbonation process the juice is treated by a heavy dose of lime and carbon dioxide. Carbon dioxide is bubbled through, which removes excess lime and the resulting juice is filtered. The filtered juice is of higher purity and brightness than in the sulphitation method. The pH 6.5 gives a lighter colour and more turbid juice. The pH 7.5 gives a darker colour and higher clarity. Good clarification may require the addition of phosphate if the level in the juice is too low. Sterilisation of juice by heating is part of the process, but some thermophillic bacteria can survive. The treated juice is heated to 100 to 102^{0}C in juice heaters which is allowed to settle in a continuous settler. The impurities settle leaving a clear supernatant liquor. The sediments settled at the bottom of the settler are filtered in a vacuum filter and the filtrate is taken for process. The sediments/residue settled at the bottom is known as 'filter cake' or 'pressmud' and is disposed as manure for growing crops in the field.

(vi) Crystallisation: The clear supernated liquor is further concentrated in evoporators. The thickened liquor is again sulphited by passing SO_2 gas to form concentrated syrup. The concentrated liquor syrup is further evaporated. On concentration, sugar crystals form in the liquor and these are built up to the required size by further concentration of liquor. The product containing the crystals and its mother liquid is called as 'massecuite'. The mechanical separation of the crystals from the mother liquor is done at the centrifugal machines. The crystals after water wash become white and bagged as commercial white sugar. The mother liqour so separated is sent back repeatedly for process of crystallization to remove all possible sugar. The remaining mother liquor known as 'molasses' or 'treacle' is sent to distilleries. The white sugar has 99.5° or higher pol value, which is used for direct consumption. Raw sugar has 96 to 98° pol value. Brown sugar refers to sugar crystals including any adhering molasses before complete processing as white sugar. Brown sugar is further refined by dissolving in water, filtering through charcoal and bleaching with SO_2.The total sugar losses are also around 2.2% cane. Some of the factories are reporting sugar losses up to 1.6% of cane.

Sugar quality: Sugar varieties *viz.*, very high pol raw sugar, refined sugar, golden granulated sugar, Demarera sugar, Mascovada sugar, *etc.*, are produced in other countries. Indian sugar industry only produces plantation white sugar. The white sugar that is marketed in India contains 99.7% to 99.8% sucrose and is white in appearance. The raw sugar that is exported contains 98.5% sucrose 0.3% moisture and 0.3% mineral matters. The specification of Indian white sugar is as follows: minimum pol % as 99.5, maximum moisture % as 0.08, maximum reducing sugar % as 0.10, maximum sulphur as 70 ppm, maximum calcium-level as 30mg/100 gm, maximum water insoluble matter % as 0.1 and specific conductivity as 100 × 103 mhos. The plantation white sugar is further graded as large, medium and small as per the grain size. The various grain sizes as indicated are in three colour series termed as 29, 30 and 31 colour. The sugar colour is measured in terms of Modulated Reflectance (MR) value of different grades. The MR values are calculated as MR=R × G, where, R=Mean reflectance value of the 4 surfaces of the bottle and G= Grain size of sugar in mm. Another special feature of sugar manufacture in India is the production of bold grains of about 2.0 mm size for which there is a sizable market.

Khandsari sugar: In khandsari, the juice is clarified with lime in three to four stages. The impurities are removed from the top. The clarified cane juice is concentrated in open pan boiling to syrup stage at which it can be stored for some time for crystalisation in small factories called 'khandsari mill'. The crystalline mass is centrifuged and washed with water to produce sugar and molasses. The sugar is dried in the sun and bagged for direct consumption. This is called 'khandsari sugar'. Losses do occur in khandsari plants due to which recovery is poor (5 to 7 %). Even at present, out of total crystal sugar produced about 6-7% appears to be accounted for khandsari. The khandasari units provide an option to the farmers for disposal of the cane wherever sugar factories have inadequate crushing capacity, for crushing the sugarcane grown in their area. The season duration of khandasari units depends upon the availability of sugarcane, employment to a large population of rural area. Khandasari units are mostly equipped with horizontal crushers minimum three rollers and maximum nine rollers. Very few have 12 rollers. The extraction of juice in all these units varies from 50 to 75%. The capacity of these units varies from 50-200 TCD (tones of cane per day). Few units are of capacity up to 500 TCD. These units pay the sugarcane price to the farmers according to the demand and supply. In the lean year of sugar production, these units pay higher price than sugar mills and vice-versa.

Jaggery: Jaggery/Gur is also manufactured in the forms such as solid jaggery (gur), kakavi (liquid jaggery/cane juice concentrate), rab/matsyandi (semi-liquid), shakkar (granular /powdered gur) and vinegar. Jaggery contains carotene (about 160 mg/100 g), thiamin (0.02 mg/100 g), riboflavin (0.05 mg), Vitamin A (900–1200 in), Vitamin B (60-80 in), Vitamin C (0.50 mg/ 100g) and energy (383 kcal). [One international unit (iu/in) is equivalent to 45.5 microgram or 0.0455 milligram]. Gur has been used for direct consumption as well as sweetening agent. The use of jaggery is prescribed as medicine in various diseases *viz.*, anemia, jaundice, cold, cough and breathlessness, kidney problems, *etc.* It has also been frequently observed that the incidence of diabetic patients in urban areas (sugar consuming) is more as compared to rural (jaggery consuming areas). Jaggery is considered as a mild laxative and supposed to have more warming effect and sweetening quality than sugar.

Solid jaggery manufacturing process: Gur / jaggery unit is an unorganized sector of cottage industry. These units mostly use vertical crushers. In southern states mainly, Karnataka and Tamil Nadu, horizontal crushers are also used by jaggery industry. These units are operated by the farmers for production of jaggery for their own use. However, most of these units are now operating for commercial purposes. These units operate at the time of maturity of the sugarcane. The duration of these units depends mainly on the price of jaggery in the market as well as percentage recovery of gur from the sugarcane. By and large, these units operate between October and April. These units also provide employment to the rural work force. Entire family including women are engaged in these units. These units operate day and night. However, the working of these is dependent on the drying of bagasse in open by sunlight. Three main operations involved in the jaggery making process are namely i) extraction of juice, ii) clarification of juice and iii) concentration of juice.

(i) **Extraction of juice**: Sugarcane is crushed in a three to five roller power operated crusher. A maximum of 60% juice is extracted in mechanical crusher used in jaggery industry. Juice obtained after crushing, is passed through multistage filtration and is allowed to settle. Almost clear juice is then pumped into the open pans and heated for boiling and concentration.

(ii) **Clarification of juice**: It is done to remove all objectionable materials present in the juice as well as to prevent formation of new non-sugars during boiling. During this heating and boiling process vegetable/chemical clarificants are added to remove the impurities from juice for making good quality products. Out of various vegetative and chemical clarificants *viz.*, deola, bhendi, phalsa, castor, groundnut, soybean, hydrous (sodium hydro sulphite), lime and sodium carbonate, super phosphate, *deoa*, alum are found to be more effective at juice pH of 6.0 producing superior quality jaggery with high sucrose content and low reducing sugars absorbing minimum moisture and having relatively better storability. Vegetative clarificants such as bhendi green plant is crushed in water. 50 g of mucilage extract is mixed in 100 kgs of juice as clarificants. (Cut 1.25 kg of fresh bhendi shoots in 20 litres of water to get a mucilaginous, colourless liquid. It is added to the juice when it begins to boil). The *bhendi* mucilage removes the maximum possible scum. Deola (*Hibiscus ficuleneus*) @ 150 g with 1.5 lit of sajji water per 400 kg of juice can be added to cane juice. Tapoica flour, soybean flour, *etc.*, are also used as clarificants. The milky liquid obtained when castor or groundnut seed is grinded or decorticated is used as clarificants along the seed with water @ 75 g per 100 kgs of cane juice to improve the gur colour and quality. Chemical clarificants such as hydross (sodium hydrosulphite), lime (calcium oxide), sodium carbonate, sodium bicarbonate, super phosphate, Sajji (50% sodium carbonate, 6.4% calcium sulphate, 4.5% sodium chloride) are used as clarificants to improve the color brightness and gur quality. Lime saturated water is useful to get hard jaggery. Sodium carbonate improves colour temporarily. Super phosphate improves jaggery colour but affects crystalline structure. Hydros (Sodium hydro sulphite) improves colour temporarily. Excess of Hydros should be avoided since SO_2 content of more than 70 ppm in jaggery is injurious to the health of the consuming person. The major drawback with these clarificants is that the chemicals are injurious to health and biological clarificants are either non-available or quite insufficient in quantity throughout the season. Vegetable clarificants are better than chemical clarificants. Cost wise chemical clarificants are cheaper and therefore become popular. Gur quality and storability often depended on effectiveness of juice clarification. The objective of clarification is to make juice clear and light in colour. It also helps in preventing charring and overheating during concentration. Gur made out of clarified juice is light in colour, crystalline in texture, hard less hygroscopic and hygienic. Different vegetative and chemical clarificants are being used to achieve various degrees of jaggery quality. Use of antifoaming agent POEM-Z-001 at 300 ppm during gur making results in saving of 1.3% of bagasse-fuel and increases 2.3% jaggery production. Presence of starch in juice affects jaggery processing. It can be overcome by adjusting pH of the juice to 8.5 and heating the juice at 55°C with 500 ppm of True Floc S-3.

(iii) **Concentration of juice**: After clarification, the juice is boiled vigourously to evaporate most of the water. The temperature of the thick mass of the syrup steadily increases up to 115 to 118°C and becomes a semi-solid material. Addition and stirring of coconut oil @ 20 ml for 100 litres of the juice at this time helps in contained boiling, prevents charring of the syrup and helps in the proper development of crystal size and shape. When the syrup becomes hard mass, the pan is removed from the furnace and cooled by stirring with the help of wooden ladder. It is then transferred to the moulding frames to obtain the jaggery in the desired shapes or sizes. After solidification and cooling, the jaggery is collected and stored. Solid jaggery in particular is being exported to Afghanistan, Angola, Australia, United Arab Emirates, Oman, Bahrain, Bangladesh, Canada, Czech Republic, Germany, Ghana, Greece, Honk Kong, Indonesia, Ireland, Japan, Kenya, Malawi, Mauritius, Philippines, Saudi Arabia, Tanzania, UK and USA.

Granular jaggery /Powder Jaggery / Jaggery powder: In this process the fresh juice with raised pH of 6 to 6.2 by adding lime as clarificant is heated in open pan. The hot concentrating mass of juice is removed at specific striking point temperature (120 to 122^{0}C) from the open pan. Hot mass is removed, allowed to cool down through mixing and transferred from pan to platform and left without stirring for crystal formation. Immediately after solidification, the powder is made manually using wooden scrapers, sieved through 1-3 mm sieves, dried to about 1% moisture content and packed in polyethylene packets.

Factors determining jaggery quality: The characteristics of good quality jaggery are light colour, hardness, good crystalline structure, sweet taste, good flavour and longer storage life. The quality of jaggery prepared depends mainly on the composition of cane juice and also method of boiling and kind of clarificants used.

(i) Variety: Varieties have variations in sucrose content and clarity of the juice. For good clarification, the varieties must have low level of colloidal matter, mineral matter and organic non–sugars such as harmful nitrogen, phenols, aminoacids, starch, gums and a high level of phosphate content in juice. Varieties CoC 671, Co 7704, Co 6304, Co 775, Co 62174, Co 86032, *etc*., can give good quality jaggery.

(ii) Juice composition: The content and ratios of non-sugar in the cane juice are responsible for deciding the jaggery quality.

(iii) Fertilizer application: High nitrogen application increases the organic non-sugars, colloids, gums, pectins and non-protein nitrogen while decreases the phosphorous content of juice which affects the jaggery quality. High nitrogen application affects the clarity leading to dark coloured jaggery and adversely affects purity, invert sugar and moisture content of jaggery and it decreases total organic acid content and pH of jaggery, which affects keeping quality, and poor colour of jaggery. Application of phosphorous improves clarity of the juice and also the quality of jaggery.

(iv) Storage of jaggery: The following methods minimize deterioration in quality of jaggery. Large quantities of jaggery may be stored in godown provided with moisture absorbing agents like calcium chloride and quick lime. Use of sugarcane trash, fly ash, palmyrah leaves, paddy husk, *etc*., between layers of jaggery. Smoking of godown with paddy husk particularly during monsoon period is followed in a few places. Storing of jaggery under low temperature maintains freshness in flavour and no loss in sucrose content. Storing of jaggery can be done in gunny bags lined with black polythene sheets.

Rab (Matsyandi in Sanskrit) is a semi-liquid form of gur and is used as sweetener. The crystals gradually develop upon storage and the remaining parts is molasses which can be used for cattle feed, tobacco curing and alcohol production. Rab crystals are more nutritious than the mill sugar as it contains iron, phosphorous, calcium and reducing sugars as well. Rab can be made from any type of cane particularly when gur does not set well.

Liquid Jaggery: Liquid jaggery is an intermediate product. The cane juice after extraction is filtered through a muslin cloth and collected in a storage tank. The juice is poured in an open pan and is heated to facilitate the coagulation of the suspended particles into a gummy colloidal substance. The nitrogenous impurities present in the juice start coagulating and started floating on the surface when the temperature reaches 85°C in around 115-120 minutes. The impurities are removed with the help of a long handled strainer. At this stage, biological clarificant is added. After addition of the clarificant, a golden coloured substance called scum appears on the surface, which once again was removed. After complete clarification, the boiled juice became clear, transparent and light brownish yellow. Later on, the juice started concentrating. At temperatures 99-100°C, the juice began to froth. To avoid frothing and charring, continuous stirring is done. At striking temperature 108°C, the juice becomes thick and is removed from heating source. The concentrate was cooled. The impurities are removed from the concentrating cane juice at 105-106°C. It is allowed to cool down at ambient temperature and settle. To minimize the crystallization and improve colour, citric acid @ 0.04% or benzoic acid @ 0.5% is added and packed in the sterilized clean bottles in liquid from. The product resembles honey in appearance and colour. Essence of orange, rose etc may be added to give distinct flavour. The concentrate can be stored for 1 year without any deterioration in the quality. The invert sugar content of liquid jaggery is 33.71% which is much more compared to other forms of jaggery (6.28%) and also the white sugar. Invert sugars are quickly and easily assimilated by human digestive system. The orange and grape essences are also added for good palatability and taste. This is preferred by majority of the people and fetches higher prices to the manufacturer. The comparison of solid jaggery, powder jaggery and cane juice concentrate is presented in Table 23.

Table 23 Comparison of solid jaggery, powder jaggery and cane juice concentrate

Properties	Solid jaggery (per 100 g)	Powder jaggery (per 100 g)	Cane juice concentrate (per 100 ml)
Energy (kcal)	383	385	300
Sucrose (%)	85	85	35
Glucose+Fructose (%)	6.28	7.0	33.7
Protein (g)	0.3	0.4	0.14
Fat (g)	0.28	0.1	0.12
Ash (%)	2.63	NA	0.88
Reducing sugars (%)	7-10	6-8	4-8
Calcium (mg)	8.0	9.0	300
Phosphorus (mg)	4.0	4.0	3.0
Iron (mg)	12.27	12	10.23
Copper (mg)	0.32	0.40	2-4
Moisture (%)	5-8	2-4	12-15
Total minerals (g)	0.6-1.0	0.6-1.0	0.75

Vinegar: Vinegar is prepared by inoculating the cane juice with yeast, which accelerates the fermentation of juice. After fermentation is complete, the liquid is filtered and the juice is stored under aerobic conditions. The pH of the juice after few weeks falls to 3.5 leading to the formation of vinegar. Vinegar is produced in 15 to 20 days. This vinegar has high nutritive value and acts as digestive and food flavoring agent as against the commercially available vinegar.

Sugarcane byproducts: The composition of sugarcane and sugarcane juice are presented in Tables 24 and Table 25.

Table 24 Composition of millable sugarcane

Constituents	Contents (%)
Water	73-76 %
Solids	24-27 %
Fibre	11-16 %
Soluble solids	10-16 %

Table 25 Composition of sugarcane juice

Juice constituents	Soluble solids (%)
Sugars	75-92
Sucrose	70-88
Glucose	2-4
Fructose	2-4
Salts	3.0-4.5
Free organic acids	0.5-2.5
Carboxalic acids	0.1-0.5
Amino acids	0.5-2.0
Organic non sugar	
Protein	0.5-0.6
Starch	0.001-0.05
Gums	0.3-0.6
Wax, fat, phosphotides	0.05-0.15
Unidentified non-sugars	3.0-5.0

A typical Indian sugar factory with 100 t of cane produces 10 t sugar, 4 t molasses, 3 t of filter mud, 0.3 t of furnace ash, 120 t flue gases at 180°C and 30 t of bagasse. 7 t bagasse make 1.0 t of paper. 100 t of molasses can produce 4 t of Ethanol which is later blended with petrol to make an excellent biofuel (Gashol). A sugar factory with a capacity of 2500 t of cane per day (TCD) produces daily 5 to 7 t of pressmud. A plant of 2500 TCD would generate 7 to 8 MW surplus power.

Bagasse: Bagasse is fibrous residue of sugarcane stalk left after crushing and extraction of juice. Structurally the cane stalk consists of various fibrous tissues which includes true fibres (55%), pith (40%) and non-fibrous materials (5%). True fibre is cylindrical cells of the rind and vascular tissues. Pith is soft and irregular shaped cells. Pith is found at the centre of cane stalk. The mill wet bagasse contains 48-49% fibre and 49-50% moisture. The composition of dry bagasse is as follows: cellulose (45%), pentosan (28%), lignin (20%), ash (2%) and sugar (5%). Sugarcane bagasse contains two principal types of fibre tissues (i) Cellulose fibre from the rind and vascular tissue (vessels) and (ii) mainly parenchyma cells of the central cane area that is pith. On the basis of dry weight both of them account for nearly 55% of the bagasse residue. One tonne of bagasse is equal to 0.8 tonnes of wood, 0.2 tonnes of furnace oil and 0.4 tonnes of coal. The calorific value of bagasse is 4250 kcal/kg. Bagasse is utilized for various purposes. Cellulose in bagasse is utilized for the generation of steam and power in sugar factories. The sugar mills consume bagasse in the boilers to generate steam to run the prime-movers. On an average, 5-6 % bagasse from cane is available as surplus to be used for co-product industries. It is used for pulping feed stock for paper. Bagasse pulp is used for printing papers, newsprint, compressed panel board and particle board, bedding or litter material for poultry etc. For manufacturing one tonne of paper, 5 tons of bagasse with 50% moisture is required. Mexico is the largest bagasse paper producer. Pentosans in bagasse is used as a raw material for production of 9 to 10% furfural. Furfural is a chemical required in petroleum refineries and in manufacture of furnace resins. Xylitol is a natural sweetner prepared using baggase. Xylitol is used as sugar substitute for diabetic patients.

Molasses: Molasses is the dark viscous liquid discharged from the centrifugal when all the sucrose, which can be recovered from the syrup, is removed. Molasses is the mother liquor in the massecuite, which is separated from the crystals by mechanical means. The mother liquor obtained before washing is termed as 'heavy molasses' and after washing is termed as 'light molasses'. The heavy molasses obtained from the last grade of massecuite is termed as 'final molasses'. Final molasses contain 30 to 36% sucrose, glucose and fructose (30%), water (20 %) and mineral potash (4%). There are three grades of molasses on the basis of percentage of sugar contents in terms of reducing sugar known as Fermentable Sugar. The fermentable sugar is 50%, 47-49.99 % and 44-46.99% in Grade-I, Grade-II and Grade-III molasses respectively. Molasses is utilized for the manufacture of ethanol. The ethanol is utilized for the manufacture of potable liquor and for the manufacture of alcohol based chemicals and other users. Alcohol generally means ethyl alcohol (C_2H_20H) is produced from cane molasses. Alcohol is not only used for potable purposes but also for the manufacture of chemicals like ethylene, acetic acid, acetic anhydride, acetaldehyde, chloral and other organic chemicals. Alcohol is a solvent. Alcohol is used for the preparation of pharmaceuticals and cosmetics. CO_2 and alcohol is produced when molasses is fermented with *Saccharomyces cerevisiae.* The yield of CO_2 is 16% by weight of molasses. Many inorganic and organic chemicals are manufactured from alcohol *viz.* acetaldehyde, acetic anhydride, poly vinyl acetate, monosodium glutamate ethyl amines, acetic acid, citric acid, oxalic acid, glycerol, acetone, Bakers's yeast, *etc*,. Molassess is also used in cattle feed. One tonne of Citric acid can be produced from three tonnes of molasses. Baker's yeast of one tonne can be produced from one tonne of molasses. Molasses is a valuable feed for animals because of it's high carbohydrate. Molasses is diluted with water and directly fed to the cattle.

Pressmud or filter mud or filter cake: On clarification of juice, by adding lime, sulphur dioxide or carbon dioxide, the impurities are precipitated resulting filter cake or filter mud. Press mud is the residue left after extracts of molasses. The percentage of filter mud in the factory using sulphitation process is 3-5% on cane and 7-9% on cane in case of carbonation process. The sulphitation cake is useful and being used as manure due to its nitrogen, phosphate and potash contents. Several sugar factories are producing compost by mixing press mud with distillery spent wash. In the sulphitation, pressmud nitrogen is 1.5%, phosphate 4.0–4.2% and potash 0.7–0.8%. The moisture in cake is around 70 to 80%. The carbonation cake is used for land fills. Pressmud is used as manure in the fields. It can be used as base material in production of wax. The nutrient content of pressmud is furnished in Table 26.

Table 26 Nutrient content of pressmud

N	2.1 %	Cu	130 ppm
P	1.4 %	Zn	430 ppm
K	0.5 %	Mn	450 ppm
Ca	3.2 %	Fe	2580 ppm
Mg	2.0 %	C:N ratio	16 : 1
S	0.5 %	Organic carbon	20 to 24%

Enrichment of pressmud: Fresh pressmud can be spread to one metre width and three metre length (depending upon the quantity) to about 15 cm thicknesses. One layer of 10 cm thickness press mud with sprinkling of urea 0.5% and cowdung slurry 5%. Then pressmud can be enriched with addition of beneficial fungi organisms like *Aspergillus humigatus, Pleurotes,* bacteria like *Bacillus subtilis, Trichoderma viridi* and actinomycetes, such as *Streptomyces vioiaceus* and *Streptomyces flavidovirens*. 5 bottles or 5 packets (each 250 gm) of *Trichoderma viridi* per tonne of press mud is recommended. Then, microbial culture can be sprinked over this layer by mixing them in water. Another layer of pressmud to a thickness of 30 cm can be added over the first layer. Again the mixture of microbial culture, urea and cow dung can be sprinked. This process is repeated until it reaches a height of about one metre. The top layer can be coverd with soil. Water is sprinkled to maintain 50% of water holding capacity with an inoculation period of 45 days. Decomposition will be over within 6 to 8 weeks. Rock phosphate, ferrous sulphate, zinc sulphate, *etc*., can also be added to improve the nutrient contents. The pressmud thus composted is dark in colour with narrow C:N ratio of about 10:1 to 15:1. The well decomposed pressmud can also be used as a source of inoculum @ 1:5 ratio of decomposed pressmud and fresh pressmud for preparation of enriched pressmud. Enriched pressmud is rich in organic carbon, nitrogen, phosphorous and potassium. It can be used for reclamation of saline and alkaline soils.

Bioearth: Bioearth is produced by heaping the fresh pressmud over racks or windrows sprayed with correct proportion of distillery effluent and mixed thoroughly by using aerotillers mounted on tractors. Simultaneously the microbial culture (Fabe-earth inoculum 110) is also sprayed and mixed. The aerotiller thoroughly aerates and agitates the composting mixture and grinds the shredding lumps into a uniform size while traverses windrows. The aerotiller is vital because it ensures that the Fabe-earth inoculm rigidly controlled and enough air is provided for the rapid composting reaction. A distinctive black, loamy, free flowing and dry baggable compost with pleasant earthy smell is produced. It is easy to handle and transport. The experience with this compost has shown a reduction in fertilizer inputs by up to 50 % in Kenya, California, Philippines and Pakistan with simultaneous rise in production of certain crops. This is the best way of restoring organic matter to soil and the most practical and profitable method of disposing effluent. Bioearth is produced and exported to foreign countries profitably at Maharashtra. The composition of bioearth is given in Table 27.

Table 27 Composition of Bioearth

S.No	Composition	Per cent	S.No	Composition	ppm
1	Moisture	30	10	Iron	1250
2	C:N ratio	12.1	11	Manganese	820
3	Nitrogen	2	12	Zinc	21

4	Phosphorus	2	13	Copper	72
5	Potassium	3	14	Boron	20
6	Sulphur	0.6	15	Molybdenum	23
7	Magnesium	0.6	16	Wax	Nil
8	Calcium	4.0	17	pH	7-7.5
9	Sodium	0.15			

25% of chemical fertilizer can be saved by using bioearth or enriched pressmud @ 10 t/ha with *Pleurotus* or *Trichoderma viride* besides improving soil organic matter without affecting soil pH and EC. Application of enriched pressmud or bioearth @ 10 t/ha along with 75% NPK fertilizers is recommended to sustain cane yield.

Sugarcane trash compost: Cane trash is the leafy material at the top of the cane plant (tops), leaves and husky material around the plant that is left over as residue when the cane is harvested green. Cane trash acts as soil opener and as carbohydrate food material for soil bacteria. Trash contains about 0.35 % N, 0.13 % P_2O_5, 0.65 % K_2O, 0.27 % CaO and also appreciable quantity of micronutrients with wider C: N ratio of 60:1. This trash compost can be done in both pit and heap methods. Trash can also be composted along with pressmud. For composting one tonne of sugarcane trash, three spawn bottles of *Pleurotous sajar caju*, three bottles of *Trichoderma viride,* 7.5 kg of urea + 50 to 75 kg of fresh cow dung for every tonne of trash are to be applied to every layer. One tonne of trash is chopped into 8 to 10 cm and divided into ten lots having 100 kg. One lot is spread in 5 m × 3 m area. One bottle *Pleutrotus* is sprinkled on bed. Second lot trash is spread. Apply urea followed by watering. Then third lot of trash is spread. Then one bottle *Trichoderma viride* is to be applied followed by watering. This process is repeated and kept for one month as incubation period for composting. Top layer is covered with silt 250 kg in order to compact the bed. Watering is done once in 4 days. Frequent watering is to be done for maintaining the moisture content. The compost will be ready within 2 to 2½ months for use. After two months, a staple compost of 0.7% N, 0.25% of P and 0.7% of K and C:N ratio of 24:1 is obtained.

Snapshot of Indian sugar industry

Revenue generation of a typical integrated sugar mill: Sugar 89%, Potable liquor/ ENA 5%, Rectified spirit 2%, Fuel ethanol 2 % and co-generation 2 %.

Sugarcane Value Chain: From 1000 kg of sugarcane, 100 kg sugar, 300 kg bagasse, 45 kg molasses and 30 kg press mud can be produced. From 300 kg of bagasse, 660 kg steam and 130 kwH power is generated. From 45 kg molasses, 11.25 litres of alcohol can be produced.

Sugar mills in India

India is the second largest sugar producer in world

Operational sugar mills in India are 526 (695 installed).

Total sugarcane crushing capacity – 2.3 million tons

Total sugar production capacity – over 30 million tons per annum

Average crushing capacity - 4000 tons per day per unit

Out of total operational 526 units:

Integrated units are 30%

Mix of Private/Public/Cooperative units:
Private & Public –309 units (Cane crushing Capacity – 19.1 mln tons/annum)
Cooperative – 217 units (Cane crushing Capacity – 12.7 mln tons/annum)

Integrated sugar mills having Cogen and Distillery are 160 units. Out of 160 units, 147 units make fuel ethanol. Each distillery unit is having a capacity of producing 2 billion lit/yr). Besides, another 168 stand-alone distilleries are functioning in country. The number of co-generation units in sugar mills is 210 with 3200 exportable energy (MW) in India.

Government of India's decisions on fuel ethanol blending

(i) Oct, 2007: Mandatory 5% ethanol blending with petrol with fixed procurement price of INR 21.50 per litre
(ii) 2009: National Biofuel Policy: 20% ethanol blending by 2017
(iii) Aug, 2010: Rs.27 per litre fixed as provisional price ex-factory

Ethanol Distillation capacity in India: There are 328 distilleries in the country including 160 sugar mills. The installed capacity of alcohol production from both stand-alone distilleries and sugar mills is of 6 billion litres. The distillation capacity with sugar industry is 2.3 billion litres. Of this, the capacity for production of fuel ethanol is 2 billion litres. Of this, 1.8 billion litres is with sugar mills and 0.2 billion litres is with stand-alone distilleries. Annual demand of Fuel Ethanol in the country is about 1 billion litres at 5 % blending. 5% blending programme requires 1 billion litres of ethanol. All ethanol currently produced only from molasses. If mandatory blending is increased from 5 % to 10%, the annual demand will double.

Ethanol blending: Advantages in India

(i) Reduces environmental pollution: 'green, renewable fuel'
(ii) Saves foreign exchange, reduces imports
(iii) Fuel mileage of 5% ethanol blended petrol better than pure petrol
(iv) Boosts rural income & directly benefits 5 million cane farmers
(v) Conversion of surplus sugarcane into ethanol: can balance production of sugar

Bagasse based cogeneration of power: At current installed crushing capacity, potential to generate power is said to be 7500 MW

(i) Only 210 mills generating exportable power
(ii) Currently, 3200 MW of power is exported by sugar mills to the grid
(iii) Rates offered by State Government agencies in the range of INR 3.12 - INR 6 per unit
(iv) Some units are trading power at INR 4.5 per unit

Press Mud

(i) India produces about 7.5 million tons of Press mud every year
(ii) Press mud is used as manure and gives best results in phosphorus deficit soils
(iii) Press mud improves biotic conditions, organic content, water holding capacity and reduces salinity
(iv) Average realization from press mud is about INR 150-200 per ton

REFERENCES

Buzacott, J.H. 1965. Cane varieties and breeding. *In* “Manual of cane growing.”, NJ Kim, RW Mungomery, CG Hughes, eds. Sydney, Australia. pp 220-253.

Cox, M., Hogarth, M., and Smith, G. 2000. Cane breeding and improvement. In “Manual of cane growing”, M Hogarth, P Allsopp, eds. Bureau of Sugar Experimental Stations, Indooroopilly, Australia. pp 91-108.

CPG. 2012. Crop Production Guide. Department of Agriculture, Government of Tamil Nadu, Chennai and Tamil Nadu Agricultural University, Coimbatore.

Daniels, J., Roach, B.T. 1987. Taxonomy and evolution. *In* “Sugarcane improvement through breeding”, DJ Heinz, ed Vol 11. Elsevier, Amsterdm, Netherlands. pp 7-84.

DSD. 2013. Status Paper on Sugarcane. Directorate of Sugarcane Development. Govt. of India, Ministry of Agriculture, (Department of Agriculture & Cooperation),8th Floor, KendriyaBhavan, Aliganj, Lucknow.

Fauconnier, R. 1993. Sugar cane. Macmillan Press Ltd, London, UK.

Radha Jain, Solomon, S.Shrivastava, A.K. and A. Chandra. 2010. Sugarcane bud chips: A promising seed material. Sugar Tech. 12(1): 67-69

Duke, J.A. 1978. The quest for tolerant germplasm. p. 1–61. In: ASA Special Symposium 32, Crop tolerance to suboptimal land conditions. Am. Soc. Agron. Madison, WI.

Duke, J.A. 1979. Ecosystematic data on economic plants. Quart. J. Crude Drug Res. 17(3–4):91–110.

MacMillan, H.F. 1925. Tropical gardening and planting. 3rd ed. Times of Ceylon Co., Ltd., Colombo.

Satyagopal, K., S.N. Sushil, P. Jeyakumar, G. Shankar, O.P. Sharma, D.R. Boina, S.K. Sain, M.N. Reddy, N.S. Rao, B.S. Sunanda, Ram Asre, K.S. Kapoor, Sanjay Arya, Subhash Kumar, C.S. Patni, C. Chattopadhyay, M.P. Badgujar, A.K. Choudhary, S.K. Varshney, P.S. Tippannavar, M.K. Basavraj, A.Y. Thakare, A.S. Halepyati, M.B. Patil, A.G. Sreenivas. 2014. AESA based IPM package for sugarcane. National Institute of Plant Health Management, Rajendranagar, Hyderabad – 500 030. pp 56.

http://edis.ifas.ufl.edu/topic_book_sugarcane_handbook

http://ec.europa.eu/agriculture/capreform/sugar/infopack_en.pdf

http://www.ers.usda.gov/Briefing/Brazil/braziliansugar.pdf

http://www.iisr.nic.in/research/varietiesreleased.htm

http://www.upcane.org/sis/en/index.asp

http://www.sugarindia.com/

http://www.vsisugar.com

http://www.staionline.org/

http://www.coopsugar.org/

http://www.indiansugar.com/

http://nsi.gov.in/

https://www.hort.purdue.edu/newcrop/duke_energy/Saccharum_officinarum.html

http://www.indiansugar.com/uploads/PRESIDENT_SPEECH-M-Srinivaasan.pdf

http://www.indiansugar.com/uploads/SPEECH-KVT-79.pdf

http://www.indiansugar.com/uploads/Indonesia_Amit_Bhardwaj.pdf

http://www.indiansugar.com/uploads/Bangkok-_26th_Feb_2014_1_%20(1).pdf

http://agritech.tnau.ac.in/expert_system/sugar/index.html

http://agridr.in/expert_system/sugar/faq.html

http://www.bdbcanegrowers.com.au/images/images/fact_sheets/Cane_Trash_-_Use_and_Management.pdf

https://edis.ifas.ufl.edu/pdffiles/SC/SC03400.pdf

http://www.indiansugar.com/uploads/ISO_London_Abinash_Verma.pdf

http://www.fao.org/nr/water/cropinfo_sugarcane.html

CHAPTER

25

Sugarbeet–*Beta vulgaris* (2n=18)

Family: Chenopodiaceae

Vernacular name: Common Beet, Sugar Beet, Swiss Chard (English), *Palak* (Hindi), *Paleng Sak* (Assamese), *Palanki* (Sanskrit).

Importance: Sugarbeet is a temperate sugar crop. In the 16th century, Olivier de Serres discovered the value of sugarbeets for preparing sugar syrup. Sugarbeet crop has been commercially developed from fodder beet at the end of 18th century. In 1747, a German chemist Andreas Sigismund Marggraf has discovered the sugar in cultivated fodder beets is identical composition to the sugar in sugarcane. In 1784, Marggraf's student Franz Carl Achard has perfected a commercial method of sugar from sugarbeet. It has been introduced in the USA as a sugar crop in early 18th century. Under the patronage of Frederick William III of Prussia, Achard has opened the world's first beet sugar factory in 1801, at Cunern in Silesia. By 1812, Frenchman Jean-Baptiste Queruel, working for the industrialist Benjamin Delessert, has devised a process of sugar extraction suitable for industrial application. Though sugarbeet cultivation in India is initiated in 1914, its commercial cultivation has started in Sriganganagar area of Rajasthan after producing first beet sugar in 1968 by Sriganganagar sugar mill. Systematic research on sugarbeet has been initiated at Indian Institute of Sugarcane Research, Lucknow in 1959. The All India Coordinated Research Project on Sugarbeet (AICRPS) was started in 1971 with its headquarters at GBPUA&T, Pantnagar. Sugarbeet is naturally a biennial plant, even though for sugar production it is managed as an annual crop. However, it is cultivated as 6-7 months old crop and yield as much sugar as 12 month sugarcane crop. The cultivated *Beta vulgaris* includes sugar beet, vegetable beet root and forage beet root. All of them cross freely. The crop is a major source of sugar in temperate countries.

Origin: Sugar beet is believed to have originated in the Mediterranean region covering Middle East, near the Tigris and Euphrates rivers.

Distribution: Sugarbeet is essentially a crop of temperate region grown in 5.49 mha with major share (>72%) of production from Europe. It is mainly grown in USA, Canada, United Kingdom, France, Germany, Denmark, Netherland, Poland, Romania, Czechoslovakia, Italy, Ukraine, Russia, Turkey, China, Egypt, Iran and Japan. Sugarbeet is a crop of the temperate region but its cultivation is now expanding from subtropical tracts to tropical world. Besides its conventional home in the temperate zone, it is now being successfully grown in Iran, Iraq, Algeria, Egypt, Afghanistan and Pakistan. In India, sugarbeet is grown in Uttar Pradesh, Kashmir, Punjab, Himachal Pradesh, parts of Rajasthan and Maharashtra.

Botany: Sugarbeet is a biennial plant. It produces roots in first year and seeds (germs) in second year. During the first growing season, the sugar beet plant has a rich, brilliant green color leaves in rosette and grows to a height of about 35 cm. The leaves are ovate to cordate in shape and grow in a tuft from the crown of the beet, which is usually level with or just above the ground surface. A long conical, white, fleshy taproot with minimal secondary roots and a flat crown develops, prominently swollen at the junction of the stem. The storage organ is called the root has two components viz., tap root occupies 90% and the remaining 10% is the crown which is derived from the hypocotyl. The average weight of sugar beet ranges between 0.5 and 1 kg. It stores up reserves in the root during the first growing season so that it is able to over-winter and produce flowering stalk and seed in the following season. Occasionally, some plants produce a seed stalk in the first year itself is known as 'bolting'. Bolters have smaller roots and thus total root yield is reduced. Bolting is induced by long days and prolonged cool periods, a principle that is utilized in the over-wintering method of seed production. During the second growing season (the reproductive stage), a flowering stalk elongates (bolts) from the root. This angular seed stalk forms an inflorescence and grows approximately 1.2-1.8 metres tall. A large petiolate leaf develops at the base of the stem with small leaves. There are less petiolate leaves and finally sessile leaves developing. At the leaf axils, secondary shoots develop forming a series of indeterminate racemes. These flowers are small, sessile and occur singly or in clusters. Sugar beets produce a perfect flower consisting of a tricarpellate pistil surrounded by five stamens and a perianth of five narrow sepals. Petals are absent and each flower is subtended by a slender green bract. Flowering in sugarbeet is indeterminate and continuous until the plant is harvested. Flowers are produced singly or in dense clusters in the axil of the bract. The flowers are small, cup shaped, without petals and perfect. The ovaries are enclosed by the common receptacle of the flower cluster. The sugarbeet is normally cross pollinated. Wind and insects transport the patterns. Isolation distance is 1-2 km. For seed production, however, an overwintering period of cold temperatures of 4 to 7°C (vernalization) is required for the root to bolt in the next growing season and for the reproductive stage to be initiated. When the fruits formed in aggregate fruit by the fusion of few flowers, the fruit is called 'multiple beet seed'. The aggregate fruit produces a seed ball with 2 or more viable seeds, or germs. If the fruit is formed from a single ovary, it is termed 'single germ beet seed'. The fruits consist of 2-5 seeds which are shiny, having lentil like structure and are about 3 mm long. Seed shape varies from round to kidney-shaped.

Uniqueness of Sugar beet is as follows:

(i) Sugarbeet being a halophytic plant (as well as Na-salts scavenger), its cultivation removes nearly 500 kg of sodium salts per ha. It may be cultivated to reclaim sodic soils. It can tolerate a salinity level of 9.5 m mhos/cm. Thus it may be a most suitable crop to cultivate in uncultivable *usar* lands.

(ii) Sugarbeet is a C_3 plant containing up to 20% sugar on fresh weight basis. It requires vernalization for flowering (for seed production).

(iii) Selective breeding and improved agricultural practices have improved sugar concentration in its roots from 4% to 18% on fresh weight basis and around 75% on dry weight basis, in a relatively shorter period of time.

(iv) It is a temperate biennial plant (for a seed crop) but for the root crop it is cultivated as an annual crop.

Sugarbeet can be an important sugar crop by supplementing sugarcane. This is mainly because of its short-duration (5-7 months as compared to 10-12 months of sugarcane), high sugar content (15-17%), high sugar recovery (12-14%) and high purity (85-90%). It requires a lower water requirement (about 1/3 to 1/2 of the water needed to grow sugarcane) and a slightly higher sugar and ethanol yield per acre (2500 to 3000 liters of ethanol per acre). A standard sugar beet has a sugar content of 16% compared with 12 to 14% in sugar cane. The extraction rate of sugar from sugar beet varies from 40 to 80% (sugar content 16%), while the extraction rate of from sugar cane can vary from 30 to 100% (sugar content 12%). Worldwide, 80% of sugar is made from sugarcane and 20% from sugar beet. Sugar beet is cultivated in countries with temperate climates while Sugarcane is cultivated in tropical and subtropical countries. The salient features of sugarcane and sugarbeet are presented in Table 2.

Table 2 A comparison of sugarcane and sugarbeet

S.No.	Characteristic	Sugarcane	Sugarbeet
1	Global sugar production	80%	20%
2	Countries grown	Tropical and subtropical countries, mainly in Brazil, India, China, Mexico, Cuba, Australia, Thailand, Pakistan and USA	Temperate and mediterranean countries mainly in Europe, Ukraine, USA, Canada, Russia, Ukraine, Turkey, Iran, Egypt, China and Japan
3	Family	Poaceae	Chenopodeaceae
4	Cotyledonous nature	Monocotyledon	Dicotyledon
5	Photosynthetic characteristics	C_4 photosynthesis	C_3 photosynthesis
6	Organ and tissues storing sugar	Parenchymatous cells in the stalk (in the vacuoles and cytoplasm)	Parenchymatous cells in the concentric rings in the roots
7	Propagation	Vegetative propagation, cuttings with single bud or 2 to 3 buds	By true seed
8	Response to salinity	Can grow only on partially reclaimed soil	Grows well and adapted to saline soils
9	Water requirement	1800 to 2000 mm	600 to 800 mm
10	Crop duration	10- 18 months	5-7 months
11	No. of crops /year	One	It is a temperate biennial plant (for a seed crop) but for the root crop it is an annual crop
12	Yield (t/ha/crop)	125-200	40-90
13	Process for extraction of sugar (in juice)	Shredding and application of pressure	Slicing and diffusion
14	Sucrose content in juice	12-25%	12-20%
15	Extraction rate of sugar	30 to 100%	40 to 80%

Growth stages: Sugarbeet takes 5 days for germination. The germinated seedlings have a large pair of healthy cotyledons and about two weeks later a pair of true leaves. At six weeks from emergence, the crop has at least two true leaves. Seedling stage (4 leaves) is up to 35 days. Canopy development stage is for 45 days. Storage root growth predominates for 60 days. The physiological maturity attains in 115 days from sowing. Thus vegetative stage takes about 75-115 days (16 leaves). The crop takes 40-50 days for ripening period after attaining the physiological maturity. Nutrient and water uptake continues and provides for leaf growth and sugar storage until harvest. The roots spread to a radius of 50 cm and most of the roots to a depth of 35 cm. Root development stage is from 35 to 115 days. Ripening stage is from 116 to 165 days (40-50 days). It is a six to seven months old crop.

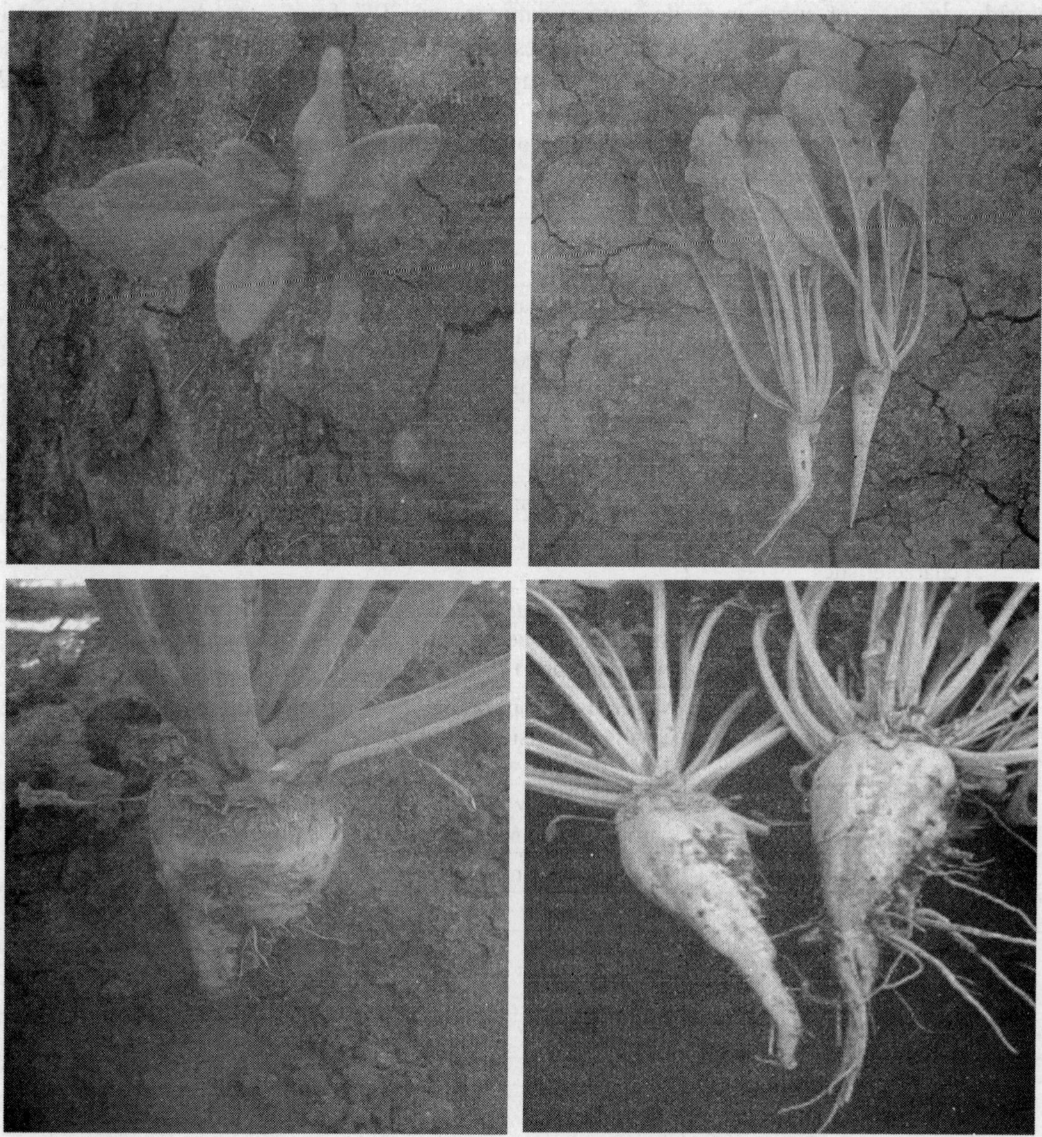

Source: TNAU, Coimbatore

Climate: Sugarbeet is a temperate crop. It is also grown in subtropics. It is grown within the latitude of 33°N and 33°S. Sugarbeet can be grown during October to March with a well distributed rainfall of 300-350 mm across the growing period. High soil moisture or continuous heavy rain may affect development of tuber and synthesis of sugar. It requires cool climate. Sugarbeet is a little less resistant to frost. Vernalization is essential for flowering. For seed production, however, an overwintering period of cold temperatures of 4 to 7°C (vernalization) and long days is required to induce flowering which results in bolting in the next growing season. The cardinal minimum, optimum and maximum temperature for growth and yield are 5°C, 17 to 22°C and 30°C respectively. Temperatures above 21°C favour vegetative growth and temperatures between 4 and 13°C favour initiation of the reproductive phase. For high sugar yield, the vegetative growth must be retarded by low temperature of 20-22°C in the later part or storage phase in order to stimulate translocation. Night temperatures have a strong effect on sucrose concentration in roots while night temperatures from 8 to 10°C favour high root sugar content. The temperature $\geq$ 30°C retards sugar accumulation in sugarbeet. The optimum temperature for seed germination is 15-20°C and for growth and sugar accumulation, it is 20-25°C under temperate and sub-tropical climatic conditions while the optimum temperature for germination is 25°C and for growth and sugar accumulation is 25 to 30°C under tropical climatic condition. Sugar beet is sensitive to cold temperatures and is killed by frost at temperatures below -5°C. The soil temperature of 15°C is ideal for sugarbeet.

Soil: Sugarbeet can be grown in well drained deep soil (45 cm) with stable and porous soil structure and sandy loam to clayey loam soils with good organic matter status. Heavy clayey soils should be avoided. Optimum pH range is from 6.5 to 8.0 but it can also grow in saline and alkaline soil up to a pH of 8.5 to 9.5, but the sucrose content is decreased. It is a halophyte (can tolerate a salinity level of 9.5 m mhos/cm) as well as scavenger of sodium salts.

Season: Sugarcane is grown as a winter (*rabi*) crop. It is sown in 15th September to 15th October and harvested during April-May. In north Indian plains, sugarbeet is raised as a *rabi season* crop in 15 [th] October to 15 [th] November. Delayed sowing adversely affects the yield and quality of beet. Time of sowing for seed production is 15 April in the valley of Himachal Pradesh.

Varieties: Sugarbeet varieties suitable for tropic and subtropics are HI006, H1 0064, Dorotea (Cauvery), Posada (Indus), IISR Comp-1 (Shubra), LS-6, R-06, etc. Sugarbeet varieties are mostly introduced from Germany, Russia, USA, UK, Denmark, Sweden, etc. A few varieties are also released from IISR, Lucknow.

Table 1 The exotic and Indian varieties of sugarbeet recommended for cultivation in India

Exotic varieties	Indian Varieties
Hh-Monitor 212, Maribo maroepoly Maribo magnapoly, Mezzanopoly-A Ertype E, US 75, Bush E, NP poly, AJ poly, USH 6, USH 8, USH 9, US 36, MSH 102, Kawemagapoly Kawegigapoly, Tribel, Ramonskava 06, US 35,Maribo Resistapoly, Triplex, BGW 674, Crimson glob, Detroit Dark Red, GWH 14, GWH 66, VSH 6, VSH 7	**Suitable for all regions** Pant S-1, Pant S-2 , Pant S-3, Pant S 10, Pant composite 6, Pant composite 3 **Certified varieties** Pant composite 1, IIST composite 1 IISR 2, IRSR composite, Rajendra Sakarkand 35, Rajendra Sakarkand 43, Sree Bhadra and Sree Rathna **Suitable for saline soils** Pant S 1, Pant S 10, IISR Composite 1 and IISR 2

Land prepration: Field is prepared with the deep ploughing and three cross harrowings. Proper levelling is necessary so that water stagnation in the field can be avoided.

Seeds and sowing: The seed rate is 4 kg/ha. Seed treatment may be done with Thiram @ 2g/kg or Bavistin @ 1.0 g/kg using bentonite clay as base material and methyl cellulose as sticker have been found effective in containing seed borne and seedling diseases. Seedling and root rot disease can be contained through the application of bio-agents (*Trichoderma viride* or *T. harzianum*) @ 20 kg/ha at the time of planting followed by seed pelleting with bio-agents or drenching with Bavistin @ 0.5% or Thiram @ 1.0% @ at 45 days after sowing. The seeds are soaked in water for 12 hours before sowing. There are 2 common methods of sugarbeet sowing, *viz.* ridge sowing and flat bed sowing. It can be sown in flat bed either by dibbling or drilling with a spacing of 50 cm × 20 cm and to depth of 2.5 cm. Sugarbeet can also be propagated using clonal cuttings. If the central bud of the crown is removed, the mother beet will give rise to a number of seed stalks from buds of old leaf scars. Thus, an abundance of cutting material may be obtained from one plant.

Thinning or gap filling: Sugarbeet seed is a multigerm that produces 4 seedlings / glomerule. To avoid the competition, maintain single healthy plant/glomerule by plucking off the remaining ones when the seedlings develop 3-4 leaves. Thus a uniform spacing of 20 to 25 cm between the plants should be maintained. Gap filling should be done soon after germination.

Weed management: Sugarbeet should be maintained weed free condition for a period of 75 days. Two hand weeding on 20th and 35th days after sowing can be taken up. Pendimethalin @ 1.5 lit / acre can be dissolved in 300 litres of water and sprayed with hand operated sprayer on 3rd day after sowing followed by hand weeding on 25th and 50th day after sowing.

Manures and fertilizers application: A blanket dose of 12.5 tonnes of farm yard manure or compost per ha and 120–60–60 kg of N, P_2O_5 and K_2O kg per ha are recommended to sugarbeet crop. Half dose of nitrogen and full dose of P_2O_5 and K_2O are applied basally during last ploughing. The remaining nitrogen in applied in two splits, the first after thinning on 30 DAS and the second at earthing up on 50-60 DAS. The earthing up operation coincides with top dressing of N fertilizer. In sugarbeet producing minimum beet yields, the tops contain 3% N in dry matter at harvest and the roots contain 0.8% N. Sugarbeet leaves show purple-red colouration when there is phosphorus deficiency. When there is potassium deficiency, the outer margins of the older leaves turns yellowing and then orange colour. Magnesium deficiency shows pale yellow colour on the margin of leaves. The magnesium concentration in leaf dry matter from healthy plants is 0.2 to 0.6% and from deficient plant is 0.1 to 0.2% Mg. Boron deficiency in sugarbeet causes heart rot and dry rot. Heart rot is the term applied when the growing point becomes blackened and dies. Dry rot describes the symptoms on the taproot shoulders, which appear subsequently. Boron deficiency occurs in sandy soil and alkaline soil. Healthy plant has 40 ppm of boron in leaf dry matter. Plants with typical heart rot symptoms contain less than 30 ppm of boron. Application of 45 kg borax/ha is recommended in boron deficient soils. Manganese deficient plant leaves roll inwards with small angular chlorotic spots (speckled yellow symptom). Manganese deficiency is prevalent where soil pH is above 6.5. Sugarbeet leaves with Mn deficiency contain 10 to 12 ppm Mn whereas leaves from healthy plants contain 50 to 200 ppm. Plants show deficiency symptoms of bright yellow leaves when grown in calcareous soil.

Water management: Water requirement for sugarbeet ranges from 550 to 850 mm but usually 700 mm is required for maximum yield of sugar. Water flood in the field should not exceed 24 hours period and drainage has to be provided immediately. Sugarbeet extract 60 to 70% water from the top 60 cm soil layer and an additional 10 to 20% from 60 to 90 cm depth soil layer. The rate of moisture extraction ranged from 1.9 mm per day at the beginning of the growing season to 6.6 to 7.6 mm per day at the peak of the season. . Sugarbeet is particularly sensitive to water deficits at the time of crop emergence and a period of about a month after emergence. Frequent, light irrigations in early vegetative period may retard leaf development and can encourage flowering during the first year (bolting). Water deficits in vegetative and yield formation periods tend to affect sugar yields more strongly when occurring during later periods. Frequent irrigation in the ripening period has an adverse effect on sugar concentration although it may increase the root size. The crop requires about 7 to 10 irrigations depending on soil and climatic conditions. One or two irrigations are before thinning on 30 DAS. Irrigations are subsequently given once in 10-12 days. Irrigation is to be stopped at least 2 to 4 weeks prior to harvest, which helps to improve sugar yield.

Crop protection: Spraying of Quinalphos 20 EC @ 2 ml/l water or two sprays of SlNPV (*Spodoptera litura* nucleopolyhedro viruses) @ 1x109 POBs (500 ml/ha) or use of 10 pheromone traps /ha and change of lures twice at monthly interval can be done to control of army worm (*Spodoptera litura*) and Bihar hairy caterpillar (*Spilosoma oblique*. Methomyl baiting @ 25 g/ha with rice husk + jaggery should be applied at evening time to control rodents. Seed treatment of Thiram 2g/kg or Bavistin 1g/kg of seed is done to control of seed born diseases. Two to three sprayings of wettable sulpher @ 2 kg/ha at 12-15 days interval may be carried out for control of foliar diseases such as powdery mildew. The root rot disease caused by *Sclerotium rolfsii* may be contained through the application of 20 kg *Trichoderma* at the time of planting or drenching with Bavistin 0.5% or Thiram 1% at 45 days after sowing. The sugarbeet crop should not be taken on the same piece of land for at least 3 years to minimize the flare up of root diseases

Harvest: The leaves become yellowish at physiological maturity. Drying up of lower leaves is a good sign of maturity. Hand refractometer less reading is 20 at harvest. One half of leaves may be permitted to wilt before harvest. This period may be 2 to 3 weeks in loamy to clay loamy soil and 7 to 10 days in sandy loam soils during which 3-4% sugar may increase. Then the tops and crown are cut off from the roots. The roots can be lifted through a digger or with a country plough. Sometimes, the roots are hard picked, shacked free from soil and detopped from the base of the lowest leaf of the crown with the help of a sharp blade. Storage of roots beyond 36 hours at high temperature is not fit for processing. Under unavoidable circumstances, roots can be stored with foliage and tops in the field.

Source: TNAU, Coimbatore

Yield: The average sugarbeet yield is 40 to 50 t/ha with sugar yield 7 t/ha. The green top yield is 10 t/ha. Sugarbeet has a potential yield of 100 t/ha of fresh roots. The sucrose (sugar) stored in the roots of sugarbeet ranges from 13 to 18% of the gross weight and 18 to 22% of the root weight depending on variety, cultural and climate conditions in which sugarbeet is grown. The sugar recovery from beet is 10 to 12%. A freshly harvested root of sugarbeet contains 75-76% water, 15-20 % sugars, 2.6% non-sugars and 4-6 % the pulp. Under normal conditions, processing of 1000 kg fresh sugarbeet can give 120-150 kg sucrose, 35 kg molasses (containing 18 kg sugar, 10 kg impurities and 7 kg water), 45 kg dried pulp and varying amounts of filter cake and these products are utilized for various purposes including energy generation. Out of sugar present in fresh sugarbeet roots, nearly 83% is recovered as white sugar, 12.5% is lost in molasses and 4.4 % is lost in some other ways. Beet roots also contain 0.3% betaine, an important osmoregulant.

Seed production: The isolation distance must be maintained for seed production of this cross pollinated crop. It varies from 2 km in commercial to 5-6 km in foundation seed production. Nucleus and breeder seeds are produced in muslin cloth chambers. The seed yield varies from 1.6-2.0 t/ha. The crop needs over wintering for seed production. Heavy snow fall in winter and 10-15 cm rainfall in summer are preferable.

Sugar production from sugarbeet: The harvested sugarbeet roots are transported to factory, weighed and discharged into hopper from which they move on to coarse screens for removal of trash, rocks and dirt matter. Samples are taken for laboratory analysis. The percentage of soluble solids in beets varies from 11 to 20% and sucrose content varies from 10 to 18%. Insoluble matter contains cellulose and hemicellulose. The white beet sugar production from sugarbeet involves the following nine steps:

(i) Washing: The sugar beet is moved to washers fitted with agitator blades to remove soil, weeds and stones.

(ii) Slicing: The washed sugar beet is then put through slicing machines that cut it into thin slices called 'strips'. The beet is sliced into cosettes which are long thin strips of 2 to 3 mm thick and 10 mm long.

(iii) Extraction: The sugar juice is extracted from the strips by diffusion in a long cylinder in which hot water circulates in the opposite direction to the strips. In a process rather like brewing tea in a pot, the sugar from the strips gradually passes into the water. The sugar juice is extracted by diffusion process from the cosettes in the diffuser. Hot water at 80°C is passed through the diffuser to enhance diffusion of juice. To prevent microbiological action, some hydrolysed paraformaldehydehyde is added. The pulp leaving the diffuser has 95% water which is pressed, dried and is sold for cattle feed. Dry pulp is 6 to 8% of the beet sliced.

(iv) Purification: The juice extracted contains all the sugar from the sugar beet as well as impurities (mineral salts), which are removed by adding milk of lime and carbon dioxide and then filtering. Calcium oxide 8 % is added for clarification of beet sugar juice. The amount of lime stone used for carbonating raw juice is 30 kg per tonne of beet. The juice is first limed to a pH of 7.8. Then it is heated to 50°C and juice is carbonated. Passing of CO_2 through the lime juice is the best practice for purifying sugar beet juice. It removes the colouring matter and gums. Sulphitation of sugar beet juice @ 25 to 100 ppm SO_2 is also adopted for further purification of juice to give best white sugar. Phosphatation enables flocculation of impurities in juice. Carbonation enables 40 to 50% colour removal whereas phosphatation removes 30 to 40% impurities. Activated charcoal, bone char, granular carbons and powdered carbon and powdered carbon resin beds are used to decolourise and to remove colloidal materials present in the juice.

(v) Evaporation: The filtered juice contains around 13% sugar and 87% water. It is heated to boiling point and then passed through a series of evaporator pans to convert it to syrup containing 65–70% sucrose. The filtered thick syrup is sent for evaporation at 550^{0}C. All best sugar factories use steam for sugar boiling.

(vi) Crystallization: Tiny sugar crystals are added to the pans to start crystal formation. The mixture of crystals and syrup (or "mother liquor") is known as "massecuite".

(vii) Centrifugal treatment: The massecuite is spun in centrifuges to separate the sugar from the syrup. The sugar settles on the sides of the centrifuge and is then washed with clean hot water to produce white sugar crystals.

(viii) Drying: Still hot and moist, the crystallized white sugar is transferred to hot-air dryers and then cooled. It is then ready for consumption.

(ix) Packaging: After sorting and weighing, the sugar is stored in bulk in huge silos, then bagged or sent for specialized packaging, e.g. as cube, caster or icing sugar, before shipping.

Utilization: Beet pulp accounts for 5 (dry basis) to 30% (wet basis) of total roots, which is a good source of feed (forage/silage) for livestock. One tonne of sugarbeet produces 300 kg of wet pulp and 50 kg of dry pulp. Dry pulp contains 60% carbohydrate and 5% crude protein. Beet pulp makes equally good feed. Filter cake is used as fertilizer. Sugarbeet molasses can also be used as feed. Molasses are combined with beet pulp to provide animal feed, or used as feedstock in the chemical and pharmaceutical industries for fermented products such as citric acid and its esters. The tops contain 10% protein and 60% total digestible nutrients. Since beet-tops contain oxalic acid, fresh-tops must not be fed to cattle. Sun-dried tops 100 kg + 60 g finely ground lime is a good cattle-feed.

REFERENCES

Clarke, M.A. and M.A.Godshall. (eds.). 1988. Chemistry and Processing of Sugarbeet and Sugarcane (Sugar Series 9). Elsevier, Amsterdam, pp. 406.

Cooke, D. A. and R. K. Scott. 1993. The Sugar Beet Crop. Chapman and Hall, Publishers. 675 pp.

CPG, 2012. Crop Production Guide. Department of Agriculture, Government of Tamil Nadu, Chennai and Tamil Nadu Agricultural University, Coimbatore.

Draycott, A.P. 1972. Sugar-beet Nutrition. Applied Science Publishers Ltd., London, pp. 250.

Panella, L. 2010. Sugarbeet as an energy crop. Sugar Tech 12(3-4): 288-393.

Sanjeev Kumar, P.K. Singh, Swapna M. and A.D. Pathak. (eds). 2013. Souvenir on IISR-Industry Interface on Research and Development Initiatives for Sugarbeet in India. Indian Institute of Sugarcane Research and Association of Sugarcane Technologists of India, Lucknow.

Steven Cosyn, Klaas Van der Woude, Xavier Sauvenier and Jean-Noël Evrard. 2011. Sugar beet: A complement to sugar cane for sugar and ethanol production in tropical and subtropical areas. International Sugar Journal. 113 (1345): 120-123

http://www.fao.org/nr/water/cropinfo_sugarbeet.html

https://en.wikipedia.org/wiki/Sugar_beet

https://www.hort.purdue.edu/newcrop/afcm/sugarbeet.html

http://agritech.tnau.ac.in/agriculture/sugarcrops_tropicalsugarbeet.html

http://vikaspedia.in/energy/energy-production/bio-energy/sugarbeet

CHAPTER

26

Sweet Sorghum–(*Sorghum bicolor* (L.) Moench)

Importance: India's growing dependence on petroleum imports exposes its energy needs to external price rise. Therefore, production and use of domestic energy resources including renewable is the high priority to ensure India's energy security. The Government of India's policy to blend ethanol (5-10%) with petrol has necessitated searching alternate feedstock other than sugarcane molasses. Sweet sorghum has emerged as a supplementary crop to sugarcane in dry land pockets for the production of ethanol. Sweet sorghum is similar to grain sorghum except for its juice-rich sweet stalk. The term sweet sorghum is used to distinguish between varieties of sorghum with high concentration of soluble sugars in the plant stalk sap or juice. It is known as the sugarcane of the desert. Its ratooning ability enables multiple harvests per season, a feature that could expand the geographical range of sorghum cultivation. Sweet Sorghum is considered as an important energy crop for the production of bioethanol due to high biomass, drought tolerance, relatively low input requirements and ability to grow under a wide range of environmental conditions. It is rich in sugar and used to produce jaggery, syrup, starch and even ethanol from juice. Traits like plant height, stem diameter, green biomass, stem sugar content, and stem juice extractability are the major contributors for sweet sorghum's economic importance. Sweet sorghum belongs to the same species as grain sorghum and forage sorghum; however, it has been selected to accumulate high levels of sucrose in the stem parenchyma. The juice extracted from sweet sorghum cane contains high levels of sucrose and invert sugar that are easily fermented to produce ethanol.

Distribution: It is cultivated in Maharashtra, Karnataka, Tamil Nadu, Andhra Pradesh, Madhya Pradesh, Uttar Pradesh, Rajasthan and Gujarat.

Climate: It can be grown easily in tropical, subtropical and temperate climates. It can be grown successfully in the semi-arid tropics. It can be grown in areas receiving rainfall of 700 to 1500 mm with 80% relative humidity.

Soils: It can be cultivated in red or black loamy soil with pH of 6.5 to 9.0. Soil with organic matter >0.6%, soil depth >80 cm, bulk density <1.4 g/cc and water holding capacity >20% (weight basis) is preferable.

Season: Sweet sorghum can be cultivated under rainfed condition where growing of sugarcane is not economical.

Varieties: The varietal features of sweet sorghum are furnished in Table 1.

International varieties: Rio, Dale, Brandes, Theis, Roma, Vani, Ramada and Keller.

National (AICSIP) **Cultures**: BJ 248, RSSV 9, NSSV 208, NSSV 255, RSSV 56. NSSH 104, RSSV 56 and PAC 52093, SSV 84, CSV 19, CSH22

Private Hybrids: Hybrid Madhura developed by Nimkar Agricultural Research Institute, Phaltan, Maharashtra. 'Madhura' hybrid is of 110 days duration, yielding 2 tons of grain/ha, 40 tons of cane / ha and 3500 lit/ha of ethanol.

Table 1 Characteristic features of sweet sorghum varieties

Variety	Duration (days)	Cane yield (t/ha)	Brix (%)	Grain yield (t/ha)	Special features
SSV84	120-125	30-35	17-19	1.0-1.2	Turtle grain, Ethanol yield 800- 1000 kl ha^{-1}
VMS 98003	Flowers in 70 to 75 days and matures in 110 to 115 days	45	18.0	2.5	-----
RSSV 9	115-120	35-40	18-20	0.8-1.0	Round creamy grain. Ethanol yield 1000- 1200 kl ha^{-1}
RSSV 16	Flowers in 80 to 85 days and matures in 120 to 125 days	36	16.5	2.3	Juice yield is 16.8 kl ha^{-1}
NSS 104	-----	38	19.0	2.3	-----
ICSA 749 x SSV 74	-----	40	20.0	2.0	-----
ICSA 474 x SSV 74	Flowers in 80 to 85 days and matures in 120 to 125 days	45	16.3	3.0-3.2	Juice yield is 25.4 kl ha^{-1}
ICSA 511 x SSV 74	Flowers in 80 to 85 days and matures in 120 to 125 days	40-45	17.97	4-5	Juice yield is 18.0 kl ha^{-1}
SPV 422	Flowers in 90 to 95 days and matures in 125 to 130 days	40	19	3.0-3.2	Suited for post-rainy season, stalk height of 2.6 to 3.0 m; resistant to leaf diseases; Juice yield is 16.5 kl ha^{-1}
NTJ 2	Flowers in 70 to 75 days and matures in 110 to 115 days	50	18.5	3.5-4.0	Suited for post-rainy season, stalk height of 2.0 to 2.5 m; resistant to leaf diseases

SPV 1411 (*Parbhani Moti*)	Flowers in 80 to 85 days and matures in 120 to 130 days	30	21	3.0-4.0	Suited for post-rainy season, stalk height of 2.0 m; resistant to shoot fly, stem borer and leaf diseases
ICSR 93034	Flowers in 75 to 80 days and matures in 115 to 120 days	40	18	3.0-3.4	Suited for rainy and post-rainy season, stalk height of 2.0 to 2.8 m
ICSV 93046	Flowers in 75 to 80 days and matures in 115 to 120 days	43	15	3.2-3.5	Suited for rainy and post-rainy season, stalk height of 2.0 to 2.5 m; resistant to shoot fly, stem borer and leaf diseases
ICSV 700	Flowers in 80 to 85 days and matures in 120 to 125 days	40	18	3.0-3.2	Suited for rainy and post-rainy season, stalk height of 3.0 to 3.2 m; resistant to stem borer

Growth stages: Sweet sorghum grows rapidly and able to attain maturity in 3 to 5 months. In sweet sorghum, plant height, stem diameter, stem fresh weight, juice yield, brix and stem sugar contents are the most important characteristic features for biofuel production. The above established characters are obtainable only under optimal irrigation conditions. Sweet sorghum varieties differ in their ability to produce and store sugar in stem. Sucrose increases from the pre-boot to anthesis stage. Sucrose concentration in the stalks increases 7 folds between boot and mid-grain filling stages. The sugar concentration of stem juice is approximately 12.5°Brix and 17°Brix at the beginning and at maturity, respectively. During maturity, the juice sugar content ranges from 15 to 25° Brix. Sugar concentration in stalk juice starts to increase during the milk stage to the soft dough stage of the seed and then decreases as the seeds become mature. The particulars on growth stages of sweet sorghum of various duration are presented in Table 2.

Table 2 Growth stages of sweet sorghum

S.No.	Stages	Crop duration (days)		
		110	125	130
1	Days to 50 % emergence	4	4	5
2	Days to 50 % boot stage	80	85	95
3	Days to 50 % flowering stage	90	105	110
4	Days to 50 % milk stage	95	115	120
5	Days to 50 % soft dough stage	105	115	125
6	Days to final harvest stage	110	125	130

Land preparation: Production technology of sweet sorghum is similar to the grain sorghum. The beds and channels or ridges and furrows may be formed. Large scale planting at a time in a given area minimizes shoot fly damage.

Seeds and sowing: Seed rate is 7.5 to 8.0 kg ha^{-1}. Treat the seed with Carbendazim or Thiram @ 2 g per kg of seeds and then with Azospirillum @ 600 g per 10 kg of seeds. Sowing can be done on ridges with a spacing of 45 cm between rows and 15 cm within rows. 3-4 seeds may be dibbled per hill and later eventually thinned at about 12-15 days after sowing to one per hill.

Weed management: Hand weeding is done on 15-20 and 30-35 DAS. Otherwise, pre-emergence spray of Atrazine @ 0.5–1.0 kg ai/ha or Pendimethalin @ 1.5 kg ai/ha immediately after sowing followed by one hand weeding at 45 days after sowing may be practiced to control weeds. *Striga* can be managed with crop rotation with trap crops, *viz.*, cotton, sunflower and groundnut. Hand pulling is done when *Striga* population is less. Spraying of 2,4-D @ 2.0 kg ai/ha is also recommended to control *Striga*.

Manures and fertilizer application: A fertilizer dose of 120-40-40 of N, P_2O_5 and K_2O kg/ha is recommended. Half of N and whole of P and K are applied as basal dressing. The remaining nitrogen is to be top-dressed during 25-30 days after germination, following weeding and intercultivation.

Water management: Protective irrigation if required at germination, seedling stage, flower primordial initiation stage, flowering and dough stage of grain formation. Minimum of 6 to 7 irrigations are required with an interval of 7-10 days. Sweet sorghum of 4 to 5 months duration requires 4000 m^3 of water per crop. It can be grown as rainfed crop.

Harvest: It is a four month duration crop and can be cultivated 2 to 3 times a year. The optimum harvesting time for sweet sorghum is between milk and hard dough stage due to the highest brix, total sugars, reducing sugars and non-reducing sugars; highest level of sugar and juice in the stem. The appropriate stage for crushing is when the crop reaches 18°Brix. The leaves are removed before crushing which is a nutritious fodder. Stalks may be crushed within 24 hours. If grain is required, the stalks can be harvested either along with the grain, or 4-5 weeks after the grain harvest.

Yield: High yielding varieties and hybrids produces green cane yield of 40 to 50 t/ha. However, the average productivity of sweet sorghum stalks (with leaves) is 20.6 t/ha. Sweet sorghum cultivars grain yield is very less as compared to grain sorghum or may not produce grain yield. The sweet sorghum stalks need to be crushed within 8-12 hours of harvesting as sugars start inverting with time delay, which affects juice recovery and fermentable sugar content. Sweet sorghum varieties with a green cane potential of 20 t/ha usually have juice yield of 10,000 lit/ha, TSS (brix) of 18-23 and 43 Mg ha^{-1} per year of juice which contains 11.8% of fermentable sugars in 120-125 days. Thus sweet sorghum can give 4 to 5 t of syrup or 3.0 to 3.5 t of jaggery per ha. The juice and jaggery yield can be increased substantially by improving crushing efficiency to 50 to 60% using three roller crusher and management practices. Ethanol yield from sweet sorghum is 1400 lit/ha. The dissolved solids in sweet sorghum juice contain 70 to 75 % sucrose and 25 to 30% non-sucrose impurities. It is estimated that, under favorable conditions, sweet sorghum can produce around 43 Mg ha^{-1} per year of juice, which contains 11.8% of fermentable sugars.

Post-harvest technology: This juice has been used to produce syrup in the USA and alcohol in Brazil and India.

(a) Syrup production: The sweet sorghum variety with relatively high reducing sugar (glucose, fructose) may be more suitable when TSS reaches above 14^0. The syrup production includes filtering of raw juice, centrifugation and addition of alpha amylase @ 50 ppm to remove impurities and starch, boiling of juice on slow fire in iron pan with zinc layer, addition of okra (*bhendi*) mucilage or okra plant extract @ 1.0 to 1.5 kg for 750 liters of sorghum juice in boiling pan to remove non-sugary matter, removing of scum with perforated ladle on and often, add filtered solution of single super phosphate @ 150 to 200 g made in 5 liters of water in boiling juice in 3 to 4 installments to avoid inversion of sucrose to glucose and for obtaining golden yellow colour of jaggery and transparent syrup. The syrup is formed at 103-104^{0}C temperature of boiling juice. For safe preservation of syrup, mix thoroughly 50 g of sodium benzoate in 100 lit syrup. After cooling the syrup to room temperature, the bottling is done. The bottles/containers should be sterilized in boiling water, filled and sealed. The sorghum syrup may look similar to honey and contain 5% (w/w) sucrose and 65% reducing sugars. Calcium, vitamin C and nicotinic acid content may be 160, 11.5 and 153 mg per 100 g, respectively. It can be used as table syrup, bread spreads and in salad dressing, cakes and biscuits, ice cream topping, etc.

(b) Jaggery production: High TSS and ratio of sucrose to reducing sugars greater than 9 gives good quality jaggery. Proper scum removal is found to be most important parameter in solidification/ Ball formation stage is attained generally at 110-112^{0}C. Further pour semisolid material from the boiling iron pan in cement or porcelain pan, stir slowly but frequently and allow to cool. More vigourous working of the concentrated mass during cooling period gives better colour. The method applied to sugarcane can be used to make both solid and liquid jaggery in sweet sorghum. Addition of alpha-amylase and centrifugation or addition of one gram of crystalline sugar per litre of juice at boiling can give jaggery comparable to that from sugarcane juice. Jaggery prepared from the juice of sweet sorghum contains 78.0% sucrose and 8.8% reducing sugar as compared to 84.2% sucrose and 7.5 % reducing sugar in jaggery produced from sugarcane. The white sugar can also be manufactured from sorghum independently as well as along with sugarcane juice without disturbing manufacturing process. For every unit of non sucrose impurity of juice reduces the sugar recovery by 0.4 units. Removal of maximum amount of starch of 93.7% can be obtained by adjusting the pH of the juice to 8.5 and heating it to 55^{0}C in the presence of Flocculent 53 @ 500 ppm. Addition of amylase enzyme from barley malt in sweet sorghum juice may remove starch to an extent of 96%. The starch can be removed (87%) by centrifugation and addition of alpha amylase @ 50 ppm.

(c) Alcohol (Ethanol): Alcohol is widely used industrial raw material. It can be produced both from the juice of sweet stalk sorghum as well as from grain. The unsterilized juice without addition of any nutrients can be successfully fermented by using *Saccharomyces cerevisiae* strain NCIM 3319. The average fermentation efficiency is 90% and the fermentation will be completed within 48 to 72 hours. The juice containing 10 to 11% (w/w) total fermentable sugar gives 6% (v/v) alcohol yield i.e. 1000-2000 litre per ha per season. The highest yield of ethanol 10.33 (v/v) can be obtained at the end of 72 hours of fermentation using *S. cerevisiae* strain NCIM 3281. The potential ethanol yield from the juice varied between 5,000 and 6,500 litres/ha

Utilization: Sweet sorghum grain is used for food/feed. The stripped leaves and bagasse (the fibrous residue that remains after juice extraction) are used as animal feed. The stalk juice is used for ethanol production. The bagasse of sweet sorghum is highly palatable and intake by livestock is more as compared to normal sorghum stover.

Comparison of sweet sorghum, sugarcane and maize: Sweet sorghum is a multipurpose crop (ethanol / sugar/ grain / fodder) while sugar cane is single purpose (ethanol or sugar). Sweet sorghum takes only 3-5 months from planting to harvest while sugarcane takes 12 months between harvests. The sweet sorghum crop can be established from seed. The grain from sweet sorghum can be used as food, feed or fuel. Sweet sorghum is more efficient in water use than sugarcane. Sweet sorghum water requirement (8000 m^3 over two crops) is 4 times lower than those of sugarcane (12–16 months duration and 36,000 m^3 of water crop^{-1}, respectively). Sweet sorghum stores starch as the principle nonstructural carbohydrate in grain, but primarily stores sucrose in the stems. The sweet sorghum stalk contains 13 to 20% total fermentable sugars that can be easily fermented and thus provides a better source of carbohydrates for the production of fuel ethanol. There is possibility of 2/3 ethanol production compared to sugarcane. Sweet sorghum cultivation cost is also four times lower than that of sugarcane. The particulars on comparison of sweet sorghum, sugarcane and maize are furnished in Table 3.

Table 3 Comparison of sweet sorghum, sugarcane and maize

Particulars	**Sweet sorghum**	**Sugarcane**	**Maize**
Crop duration (months)	3 to 4	12	3.5 to 4.0
Planting material	seed	Cane setts	seed
Water requirement (m^3)	4000	36000	8000
Ethanol source	Juice, Grain, Stillage	Juice, Bagasse	Grain, Stover
Grain yield (t/ha)	2.0	------	3.5
Green stem/stalk yield (t/ha)	35	75	45
Stillage/stover (t/ha)	4	13.3	8
Brix (%)	14 to 19	------	------
Ethanol yield from grain (litre/ha)	760	------	1400
Ethanol from stalk cane (litre/ha)	1400	5600	------
Ethanol from residue (litre/ha)	1000	3325	1816
Total ethanol yield (l/ha)	3160	8925	3220
Corn oil (litre/ha)	------	------	140

Source: ICRISAT, 2008.

REFERENCES

Almodares, A. and Mostafafi, D.S.M. 2006. Effects of planting date and time of nitrogen application on yield ans sugar content of sweet sorghum. J. Environ. Biol. 27:601–605.

Almodares, A. and Hadi, M.R. 2009. Production of bioethanol from sweet sorghum: A review. African Journal of Agricultural Research. 4: 772–780.

Audilakshmi, S., A.K. Mall, M. Swarnalatha, and Seetharama, N. 2010. Inheritance of sugar concentration in stalk (brix), sucrose content, stalk and juice yield in sorghum. Biomass Bioenergy. 34:813–820.

Blummel, M., Rao, S.S., Palaniswami,S., Shah, L. and Reddy, B.V.S. 2009. Evaluation of sweet sorghum (*Sorghum bicolor* L. Moench) used for bio-ethanol production in the context of optimizing whole plant utilization. Anim. Nutr. Feed Techn. 9:1–10.

Broadhead, D.M. 1972. Effect of stalk chopping on leaf removal and juice, quality of Rio sweet sorghum. Agron. J. 64:306–308.

Dayakar Rao, B., Ratnavathi, C.V., Karthikeyan, K., Biswas, P.K., Rao, S.S., Vijay Kumar, B.S. and Seetharama, N. 2004. "Sweet sorghum cane for bio-fuel production: A SWOT analysis in Indian context", National Research Centre for Sorghum, Rajendranagar, Hyderabad, AP 500 030. India. 20 pp.

Dercas, N and Liakatas, A. 2007. Water and radiation effect on sweet sorghum productivity. Water Resources Management, 21:1585-1600.

Djanaguiraman, M., Manickasundaram, P. and Chinnusamy, C. 2005. Response of sweet sorghum to foliar spray of nutrients and plant growth regulators. J. Agric. Resource Management. 4:247–248.

Ganesh Kumar, Fatima, A., Srinivasa Rao, P., Reddy, B.V.S., Rathore, A., Nageswar Rao, R., Khalid, S., Kumar, A.A. and Kamal, A. 2010. Characterization of improved sweet sorghum genotypes for biochemical parameters, sugar yield and its attributes at different phenological stages. Sugar

Tech. 12: 322–328.

Geng, S., Hills, F.J., Johanson, S.S. and Sah, R.N. 1989. Potential yields and on-farm ethanol production cost of corn, sweet sorghum, fodder beet and sugarbeet. J. Agron. Crop Sci., 162 (1): 21-29.

ICRISAT, 2008. Biofuels Market East Africa Conference 17- 18 September, 2008 Dar es salaam, Tanzania.

Kim, M. and Day, D.F. 2011. Composition of sugar cane, energy cane, and sweet sorghum suitable for ethanol production at Louisiana sugar mills. Journal of Industrial Microbiology and Biotechnology, 38: 803-807.

Kumar, C. Ganesh, P. Srinivasa Rao, Soma Gupta, Jayalakshmi Malapaka, and Ahmed Kamal. 2013. Enhancing the shelf life of sweet sorghum [*Sorghum bicolor* (l.) moench] juice through pasteurization while sustaining fermentation efficiency. Sugar Tech. 15(3): 328-337.

Kumar, C.G., Fatima, A., Srinivasa Rao, P., Reddy, B.V.S., Rathore, A., Rao, R.N., Khalid, S., Ashok Kumar, A. and Kamal, A. (2010) Characterization of improved sweet sorghum genotypes for biochemical parameters, sugar yield and its attributes at different phenological stages. Sugar Tech. 12: 322–328.

Lingle, S.E. 1987. Sucrose metabolism in the primary culm of sweet sorghum during development. Crop Sci. 27:1214–1219.

Mastrorilli, M., Katerji, N. and Rana, G. 1999. Productivity and water use efficiency of sweet sorghum as affected by soil water deficit occurring at different vegetative growth stages. European J. Agron., 11: 207-215.

Miller, A.N. and Ottman, M.J. 2010. Irrigation frequency effects on growth and ethanol yield in sweet sorghum. Agron. J. 102:60–70.

Munirathnam, P., Kumar, K. A., and Rao, P. S. 2013. Performance of sweet sorghum varieties and hybrids during post rainy season (maghi) in vertisols of scarce rainfall zone in Andhra Pradesh. Sugar Tech, 15(3): 271-277.

Prasad, S., Anoop Singh, Jain, N. and Joshi, H.C. 2007. Ethanol Production from Sweet Sorghum Syrup for Utilization as Automotive Fuel in India. Energy Fuels. 21:2415–2420.

Pfeiffer, T.W., Bitzer, M.J., Toy, J.J. and J.F. Pedersen. 2010. Heterosis in sweet sorghum and selection of a new sweet sorghum hybrid for use in syrup production in Appalachia. Crop Sci. 50:1788–1794.

Rajendran, C., Ramamoorthy, K. and Backiyarani, S. 2000. Effect of deheading on juice quality characteristics and sugar yield of sweet sorghum. J. Agron. Crop Sci. 185:23–26.

Reddy, B.V.S. and Sanjana Reddy P. 2003. Sweet sorghum: characteristics and potential. International Sorghum and Millets Newsletter. 44:26–28.

Reddy, B.V.S., Ramesh, S., Sanjana Reddy P., Ramaih, B., Salimath, P.M. and Rajashekar Kachapur. 2005. Sweet sorghum-A potential alternative raw material for bioethanol and bioenergy. International Sorghum and Millets. Newsletter. 46:79-86.

Reddy, B.V.S., Ramesh, S., Reddy, P.S., Ashok Kumar,A., Sharma, K.K., Karuppan Chetty, S.M. and Palaniswamy, A.R. 2007. Sweet Sorghum: Food, Feed, Fodder and Fuel Crop. ICRISAT, Pantacheru.

Seetharama, N., Dayakar Rao, B., Ratnavathi, C.V., Shahid Parwez, Md., Binu Mathew, Singh, K. and Singh, B. 2002. Sweet sorghum – an ancillary sugar crop. Indian Farming. 36(4):7–8.

Srinivasarao, P., Prasad, J.V.N.S., Umakanth, A.V. and Reddy, B.V.S. 2011. Sweet sorghum *(Sorghum bicolor* (L.) Moench) - A new generation water use efficient bioenergy crop. Indian J. of Dryland Agriculture. 26: 65-71

Srinivasarao, P., Rao, S.S., Seetharama, N., Umakanth, A.V., Sanjana Reddy, P., Reddy, B.V.S. and Gowda, C.L.L. 2009. Sweet sorghum for biofuel and strategies for its improvement. Information Bulletin No. 77, International Crops Research Institute for the Semi-Arid Tropics, Patancheru 502324, Andhra Pradesh, India. 80 pp.

Subramanian, V.K., Prasada Rao, K.E., Mengesha, M.H. and Jambunathan, R. 1987. Total sugar content in sorghum stalks and grains of selected cultivars from the world germplasm collection. J. Sci. Food Agric. 289–295.

Victor, D., and Miller, F. 1990. Assimilation, partitioning and nonstructural carbohydrates in sweet compared with grain sorghum. Crop Sci. 30:1109-1115.

Wortmann, C.S., Liska, A.J., Ferguson, R.B., Lyon, D.J., Klein, R.M. and Dweikat, I. 2010. Dryland performance of sweet sorghum and grain crops for biofuel. Agronomy Journal. 102: 319–326.

Zhao, Y.L., Dolat, A., Steinberger, Y., Wang, X., Osman, A. and Xie, G.H. 2009. Biomass yield and changes in chemical composition of sweet sorghum cultivars grown for biofuel. Field Crop Res. 111:55–64.

http://agritech.tnau.ac.in/agriculture/sugarcrops_sweetsorghum.html

http://www.ifad.org/events/sorghum/b/Reddy.pdf

https://www.cbd.int/doc/biofuel/Bioversity%20ICRISAT-Report-Biofuels.pdf

http://www2.ca.uky.edu/agc/pubs/agr/agr122/agr122.htm

http://www.uky.edu/Ag/CDBREC/sorghumethanol.pdf

http://edis.ifas.ufl.edu/ag298

http://sseassociation.org/Default.aspx

http://articles.extension.org/pages/26634/sweet-sorghum-for-biofuel-production

CHAPTER

27

Potato–*Solanum tuberosum* L. (2n=48)

Vernacular name: Irish potato, English potato, White Potato (English), *Aaloo* (Hindi), *Urulaikilangu* (Tamil), *Bilati Alu* (Telugu), *Batata* (Marathi) and *Alu* (Bengali).

Importance: The potato in cultivated for it tubers. Potato is a staple food crops of the world. Potato flour is the oldest commercially processed potato product and is utilized by the baking industry. Potatoes may be pulped and fermented to produce alcohol. The leaves and other green parts of the plant consist of poisonous alkaloid '*Solanine*', hence are not suitable as fodder. The glycoalkaloids content of potato peels and flesh is 22.5 to 28.8 and 1.7 to 3.4 mg per 100 g of tissue. Potato tubers contain less than 20 mg glycoalkalids per 100 g fresh weight, which do not cause harmful effects. It can be removed by peeling and boiling of potatoes. The productivity of potato in India is quite low (183.3 q/ha) as compared to Belgium (490 q/ha), New Zealand (450 q/ha), UK (397 q/ha) and USA (383q/ha). This is due to shorter crop duration in India. There is a wide ranging variation in the agro-ecological setting of different parts of the country, which results in wide variations in the productivity levels of different states. The per capita consumption of potato in India is only about 16 kg/year. On the other hand, the per capita consumption in Europe is 121 kg/year and as high as 136 kg/year in Poland. The processing of potatoes constitutes less than 2% of the total annual production in India compared to 60% in USA, 47% in the Netherlands and 22% in China.

Origin: Potato is originated in the high Andean hills of South America. The potato is believed to have originated in the altiplano around Lake Titicaca, at a height of about 3000 m in the Bolivian Andes and the main centre of diversity is in the Andes between 10°N and 20°S at altitudes above 2000 m.

Distribution: Potato is a temperate crop, but it is also grown in mountains of the tropical climate. It is grown in USA, South America, Europe and Russia. India has diverse soil types and agro-climatic conditions. Successful potato cultivation requires night temperatures of 15-20°C with sunny days. Indian sub-tropical plains offer optimum conditions for potato cultivation, where 85-90% of potatoes are grown during short winter days from October to February. The hills account for less than 5% of the total potato production where the crop is grown during long summer days from April to September/October. The plateau regions of South-eastern, central and peninsular India constitutes about 6% area where potato is grown mainly as rainfed or irrigated winter crop. Potato is mainly grown in Uttar Pradesh, West Bengal, Bihar, Punjab, Assam, Karnataka, Madhya Pradesh and Gujarat.

Potato growing zones in India: Based on soil and climate, potato growing areas are grouped into 5 zones.

(i) Western Himalayan zone: It comprises hills of Himachal Pradesh, Jammu and Kashmir and Uttarakhand. In general, the soils are acidic with varying textures and depth of soil. The crop is raised during spring season and in summer season as rainfed crop.

(ii) Plains zone: This zone extends from Punjab to West Bengal. The crop is grown mainly in *rabi* and spring seasons for seed production. The potato is also grown as winter (*rabi*) crop.

(iii) North-eastern hills zone: It comprises of hilly areas of West Bengal, Sikkim and 7 north east states. In this sub-zone, crop is grown from January-August (high hills) and from August-December (low hills).

(iv) Low hills and Plateau zone: It comprises of Gujarat, Maharashtra, Karnataka, Odisha, parts of Madhya Pradesh. Potato is grown as a rainfed (July-September) as well as irrigated (November-February) crop.

(v) Nilgiri and Kodaikanal hills zone (Tamil Nadu): The potato is cultivated all the year round i.e. summer (March-Septemebr), autumn (August-January) and winter (January-May). The first two season crops are rainfed whereas the last is irrigated and grown in valleys.

Botany: Potato is an annual, herbaceous, dicotyledonous and vegetatively propagated plant. It can also be propagated through botanical seed known as 'True Potato Seed' (TPS). Sexual or botanical seed of potato commonly known as 'True Potato Seed' (TPS), a radical alternative to seed tubers for raising a commercial potato crop. The potato tuber is a modified stem developed from underground on a specialized structure called stolon. It contains all the characteristics of a normal stem like dormant buds (eye) and scaly leaves (eyebrow). The aerial part of the stem is erect in the early stages of growth but later it becomes more spreading. It attains a height from 0.5 to 1.5 m depending on the variety. The aerial stem is hollow except at nodes. Stem is robust, hollow and angular with subterraneous tuberiferous stolons below. Tubers develop at the tip of the stolons (part of stem) variable in size. Stem tuber is the apical portion of the stolon. The tuber is a shortened swollen stem. Tuber has the attachment to the stolon, and this end of the tuber is known as a 'stolon end' or 'heel end' while the opposite end of the tuber, which is actually the tip of the branch, is known as the 'rose end' or 'crown end'. The tuber is bearing a group of buds or 'eyes' protected by scale-like leaves which shed and leaves a rudimentary leaf scar (eyebrow) or ridge. The scaly leaves on the tubers are visible only during the growing season. As the tubers form, the scales fall off. Tubers exposed to light develop either a greenish or a purplish colour. The shape of the tubers may be round or oblong or elongated. The leaves are 30 to 60 cm long. The leaflets are opposite and hairy. Flowers are borne in terminal clusters. Flowers are hermaphrodite. Flowers are varying in colour from white to creamy and very often tinges of green or pink or blue or purple are found. The fruit is globular or short oval and goes by the name potato ball, or potato apple or the seed apple. The fruits (seed or potato ball) are spherical with 2.5 cm in diameter. Each berry contains approximately 200 seeds. Potato is a C_3 plant.

Climate: The potato is essentially a 'cool weather crop', with temperature being the main limiting factor. In tropical areas, potato should be grown where the climate is tempered by altitude (1500-4200 m) or at lower altitudes provided the crop is grown during the cool season. Potato is grown in both subtropical and temperate regions. Potato is mostly grown as a rainfed crop in regions receiving a rainfall of 1200 to 2000 mm per annum. A rainfall of 500 to 750 mm, evenly distributed throughout the growing period, is considered essential (approximately 25 mm to 30 mm per week is required). Excessive rainfall lowers tuber growth and yields and high air humidity favours tuber growth. In the tropics, a rainfall up to 150 mm per month can be tolerated. Drought, even for short periods, can have a serious effect upon yields and the quality of the crop. Drought along with high temperatures or drought during the last 9 weeks of growth affects the crop growth, yield and quality. Inadequate or irregular water supply not only results in poor yields, but the tubers are usually malformed, often having very thick peel and a knobby appearance. The importance of altitude is its effect on temperature. Many indigenous South American cultivars can be grown in the tropics at levels of about 2000 m, but at lower altitudes yields are generally poor. Main crop or late-maturing cultivars from temperate climates can usually be grown with moderate success in the tropics at altitudes between about 400 m and 2000 m, and even down to sea level if there is a marked cool season. Day length, light intensity and temperature all interact in their effects on the productivity of potato crops. Early varieties are less sensitive to photoperiod than late season varieties. The longer the photoperiod, the time taken from emergence to tuber initiation will increase. In general, indigenous South American cultivars produce reasonable yields with a day length of 12 to 13 hours, but the early maturing cultivars of temperate regions require a day length of 15 to 16 hours. Varieties get killed at -2°C are called 'frost sensitive'. Potato needs about 25°C at the time of germination, 18 to 20°C for vegetative growth and 17 to 20°C for tuberization and tuber development. Tuber initiation begins at about 34 days after planting at low temperature. The ideal night temperatures are 10 to 14°C with the day temperature of 23 to 25°C for tuber initiation. Tuberization rarely occurs with night air temperatures over 18°C and lateral roots are scarcely produced. At 18°C constant temperature, plants become stunted, chlorotic, and there is no tuber formation. Net assimilation is at a maximum at about 25°C. The optimum soil temperatures are 22 to 25°C for emergence, and 15°C to 20°C for tuber growth. Maximum soil temperature for tuber growth seems to vary from 25 to 30°C. Potato produces higher tuber production under cool condition than in warm conditions. The ideal condition for tuberization is a night temperature of around 16°C, while optimum yields are obtained where mean daily temperatures are in the 18-20°C range. Potato produces high yield of tubers in short days with low night temperature of 17°C. Night temperature of 10°C or more than that during the last 30 days of crop growth improves the quality of the potatoes and produces potatoes with high dry matter and low reducing sugars. Tuber production is ceases at high night temperature of 23°C in short days and at day temperatures above 30°C. Dormancy can be broken by holding the tubers at 20 to 30°C for 30 to 45 days (which is common feature in the tropics). Once the period of natural dormancy has ended, potatoes sprout provided that the temperature, is above 5°C; the higher the temperature the quicker the rate of sprouting. Cut tubers can be stored for up to 30 days if they have been cured at 15 to 21°C and 85% relative humidity for 7 to 10 days.

Soil: The soil should be friable, porous and well drained loam or sandy loam, relatively free from stones. It can be grown on all soil types, except heavy waterlogged clays. The desirable soil pH may range from 4.8 to 8 (optimum 5.5–7). Potatoes are liable to suffer from scab at pH above 6.0. Potatoes are tolerant to soil acidity. Below pH 4.8, however, the crop may fail due to calcium deficiency. Liming may be necessary.

Season: In Europe the potato crop is grown in summer having long photoperiod of up to 14 hours and the crop duration of 140-180 days. The potato in Indian plains is, however, grown in completely contrasting situations. Nearly 85 per cent of the crop is grown during winters having short photoperiod (with about 10-11 hours sunshine) and the crop duration is also limited to 90-100 days because of short and mild winter. The mornings usually have fog, which further reduces the sunshine hours posing severe constraints on photosynthetic activity. Besides, the post-harvest period consists of long hot summer, which creates storage problems. Potato growing season varies depending on regions and physiographic conditions. Potato is sown/planted during i) August –September (Autumn) and Jan-Feb (Spring crop) in the low hill regions up to an altitude of 800 m above MSL, ii) March-April in the Mid hills up to an altitude of 800-1600 m above MSL, iii) March-April in the High hills up to an altitude of (1600-2400 m above MSL): iv) April-early May in very high hills up to an altitude of 2400 m above MSL and above. Planting of seed potato tuber in Punjab, Haryana and North-western plain should be completed in first fortnight of October. Planting in Rajasthan, Western Uttar Pradesh and Central plains may be carried out in second fortnight of October. The planting for seed crop in Eastern Uttar Pradesh, Bihar, West Bengal and Odisha may be taken up in first week of November.

Varieties: The potato varieties cultivated in India is presented in Table 1.

Table 1 Zone-wise popular potato varieties in India

Zone	Season	Varieties
North Western plains	Early	Kufri Ashoka, K. Chandramukhi, K. Pukhraj, K. Gaurav, K. Khyati
	Medium	Kufri Anand, K. Arun, K. Badshah, K. Chipsona-1, K. Sadabahar, K. Chipsona-3, K. Pukhraj, K. Pushkar, K. Surya, K. Jyoti, K. Bahar
North Central Plains	Early	Kufri Chandramukhi, K. Gaurav
	Medium	Kufri Arun, K. Anand, K. Badshah, K. Bahar, K. Chipsona-1, K. Chipsona-3, K. Pukhraj, K. Pushkar, K. Surya, K. Frysona, K. Sadabahar
	Late	KufriSindhuri
North Eastern Plains	Early	KufriAshoka, K. Chandramukhi, K. Gaurav
	Medium	Kufri Arun, K. Chipsona-1, K. Chipsona-3, K. Pushkar, K. Surya, K. Bahar, K. Jyoti, K. Kanchan, K. Chipsona-4
	Late	Kufri Sindhuri
Plateau and Central Plains	Early	Kufri Chandramukhi, K. Lauvkar
	Medium	Kufri Pukhraj, K. Badshah, K. Chipsona-1, K. Chipsona-3, K. Surya, K. Lauvkar, K. Jyoti, K. Chipsona-4

North-Western Hills	Medium	Kufri Giriraj, K. Himalini, K. Jyoti, K. Shailja, K. Himsona, K. Girdhari
North Eastern Hills	Medium	Kufri Giriraj, K. Himalini, K. Jyoti, K. Megha, K. Shailja, K. Girdhari
North Bengal & Sikkim Hills	Medium	Kufri Jyoti, K. Kanchan
Southern Hills	Medium	Kufri Giriraj, K. Himalini, K. Jyoti, K. Shailja, K. Swarna, K. Girdhari

Plains: Early = 70-90 days, Medium = 90-110 days, Late= > 110 days
Hills: Early = 100-110 days, Medium = 110 -120 days, Late = > 120 days
Source: Central Potato Research Institute, Shimla

Growth stages: Potato is propagated vegetatively. The seed tuber bears a number of buds which grow and produce subsequent crop. There are two phases in potato namely growth and development phases and storage phase. The growth and development phase starts with growth of buds into daughter tubers. Storage phase extends from lifting of tubers during harvest to subsequent planting. Pre emergence growth involves the establishment of root and leaf surface from material stored within the mother tuber. The auxiliary seeds on the tuber start to grow. So the number of shoots above the ground is almost always greater than the number of growing sprouts planted. Post emergence growth concerns with growth of hauling and tuber growth, i.e., tuber initiation, tuberization and maturation. Development of haulm of potato plant after emergence provides photosynthetic surface and sites for formation of tubers. In temperate climates, early potato cultivars matures in 3 to 3½ months, medium cultivars in about 4 to 6 months and late cultivars may take up 7 months, depending upon the environmental conditions. Short day-lengths (12–13 hours) lead to earlier cropping and main crop may mature in a short time as 4 months, but with reduced tuber yield. Potato takes 10-15 days for emergence, 16-35 days for vegetative stage, 36-49 days for tuber initiation stage, 50 to 70 days for tuber bulking and 71 to 112 days for maturation stage.

Land preparation: Prepare the field to fine tilth. In hills provide an inward slope of 1.40 in the terraces. Provide drainage channel along the inner edge of the terrace. Form ridges and furrows with a spacing of 45 cm between ridges either by hand hoe or ridger.

Seeds and planting: Seed potato tubers may be 30-80 g in weight depending on variety. Use disease free, well spouted seeds weighing 40 to 50 g per ha. The seed size tubers should have a diameter of 30 to 55 mm and its corresponding weight of 25 to 125 g. Hill seed tubers are slightly larger than the seeds obtained from the plain. Use whole seed tubers having multiple sprouts of about 2-3 cm. Potato tuber seed of 2.5 to 3.0 tonnes per ha is required. Planting of the whole tuber is recommended for disease free quality seed production. Tubers are treated with Ethylene chlorohydrin (1 litre per 20 quintals of seed tubers) to break dormancy of tubers. However cut tubers should be treated with 1% aqueous solution of Thiourea by dipping for one hour (1 kg chemical for 1000 kgs of seed tuber) which is followed with gibberelic acid one ppm for 10 minutes. Use carbon disulphide 30 g per 100 kg of seeds for breaking the dormancy and including sprouting of seeds. Plant the tubers at 60 × 20 cm spacing running across the slope or in ridges and furrows. Tubers are dibbled with *a spade hoe* (a narrow blade tool), 5-7 cm deep on the ridges manually. The seeds are raised in nursery beds like other vegetables and seedlings are transplanted 30 days after sowing. For raising one hectare of crop, a quantity of 100 g seed is required. Direct seeding of Potato true seeds @ 200 g TPS per ha has an emergence of 75% in well-controlled field conditions produce an yield of 28-30 tons of seedling tubers with an average tuber weight of 25-30 g. Raising seedlings in nursery beds and transplanting them to field is the possible alternative of TPS use for potato production. Maximum yields of seedling tubers in nursery beds are obtained at plant densities of 80-100 seedlings per m2 resulting in seedling tubers of 1 to over 40 g with an average size of 10-15 g. Only 750 kg of seed tubers of an average tuber size of 15 g is sufficient to plant one hectare of potatoes, compared with 2-3 tons for 40-60 g seed tubers. This means that 6.0-7.5 g of TPS (0.60-0.75 mg per seed, 80% emergence) and 80 m^2 of nursery bed can produce seedling tubers for 1 hectare potato crop. TPS crop also takes about 20-25 days more for maturity than the conventional potato crop raised from seed tubers.

Cultural practices

(i) **Rouging:** First rouging should be done at 25-30 days after planting immediately before earthing-up to remove off-type and diseased plants. Do the last rouging 3-4 days before haulm cutting. Ensure that all the tubers of diseased and off-type plants are removed.

(ii) **Mulching:** Apply plant material such as paddy straw or farm refuse or pine needles @ 10 t/ha on ridges as mulch. Remove the mulch 20-25 days after planting for inter-cultural and earthing-up operations. Mulch reduces the soil temperature by 3-4^{0}C and also helps to conserve soil moisture and controls the weeds. This practice significantly increases the tuber size and yield and saves irrigation water.

(iii) **Dehaulming:** Dehaulming is done by manually cutting with the sickle close to the ground or by spraying non- selective herbicide like Paraquat @ 0.5 kg/ha. In the plains, if haulms are removed manually, it is preferred to keep the haulms on the ridges to protect exposed tubers from high temperature and direct sunlight.

Weed management: The critical period of weed competition is up to 60 days. Weeding should be performed after full crop emergence (about 4 weeks after planting) and after the plants have reached a height of about 20 cm. Take up the first hoeing on 30th day without disturbing stolons. Second hoeing and earthing up is done on 45th days after planting. The hoeing must not be delayed beyond 30 days in plains and 45 days in the hills after planting to avoid damage to the plant roots, foliage and stolons. Pre-emergence weedicides like Metribuzin @ 0.75 kg/ha, Oxyfluorfen @ 0.15 kg/ha, Linuron @ 0.5 kg/ha, Alachlor @ 1.5kg/ha, Atrazine 50WP @ 1.0 kg/ha and Isoproturon @ 0.75 kg/ha applied 2-3 days after planting are effective. Pre-emergence herbicides are most effective when applied in moist soil. Therefore, if soil is dry apply herbicides after first irrigation as soon as it is possible to enter the field. In case pre-emergence herbicides are not used, spray paraquat @ 0.5 kg/ha at about 5-10% plant emergence of potato provided sufficient weeds have appeared, as it kills only emerged weeds. Spray Grammaxone @ 2.5 lit per ha in 500 lit of water as post–emergence to control weeds. Atrazine spray on potato when intercropped with sunflower crop is not advisable as it affects germination of sunflower. Shallow ridging is done subsequently to prevent the stolons becoming aerials, and to protect tubers against insect pests, disease infection and greening.

Manures and fertilizers application: Apply 15 tonnes of FYM and 120-240-120 kg of N, P_2O_5 and K_2O per ha as basal dose in furrows. FYM application @ 30 t/ha can take care of phosphorus and potassium requirement of potato crop. If FYM is applied at 15t/ha, then half the dose of phosphorus and potassium is to be applied through fertilizers. Nitrogen is applied in 2 equal splits viz., half at planting and half at earthing up (45 DAP) and is band placed (5 cm away and 5 cm deep). Foliar spray of urea 1.5% along with fungicides and insecticides is also recommended. Darkening in potato is caused by an excess of nitrogen in the fertilization. Avoid high or excessive nitrogen dressing as it stimulates haulm growth, delays tuber formation and affects tuber quality (low dry matter content, high reducing sugar content and high protein and nitrate content). Application of phosphorus improves tuber number, size and help in rapid bulking, hastens maturity and counteracts the ill effects of excessive nitrogen. High response to applied phosphorus is observed in acid hill soils and lowest in black soils. Potassium is a quality element and increases dry matter, starch content and better cooking quality. It improves tube yield by increasing their size and helps in efficient utilization of nitrogen. Potassium offers resistance to water stress, frost and diseases. It is applied in furrows in 2 equal splits viz., half at planting and half at earthin up (45 DAP) along with nitrogen. Since potato seed tuber size is small (<75 g), 30 per cent reduction of N and P is recommended for seed crop as against the ware crop. Potash levels are maintained as such both for seed and ware potato. Soil application of $MgSO_4$ @ 60 kg/ ha, $ZnSO_4$, @ 25 kg/ha, $FeSO_4$ @ 25 kg/ha and sodium borate@ 1.0 kg/ha is recommended to correct Mg, Zn, Fe, and B deficiency.

Water management: For best yields, a 120 to 150 day crop requires from 500 to 700 mm of water. First irrigation is given immediately after planting, life irrigation on 5 to 7 days after planting and subsequently once in 7-10 days interval in sandy loam soil and 10-12 days in heavy soil. Avoid flooding over the ridges while irrigating and irrigate as far as possible in morning and evening hours. In a normal seed crop, 6-8 irrigations are required. Irrigation at IW/CPE ratio of 2 gives maximum tuber yield. Irrigation at 20 to 25 mm CPE produces maximum tuber yield in sandy loam soil. Light and frequent irrigations are much better than heavy irrigations given less frequently. Frequent irrigation reduces the occurrence of tuber malformation. Excess moisture makes lenticels prominent due to rupturing and seed tuber quality is impaired. It also promotes certain diseases. Stop irrigation at about 10-12 days before dehaulming in light soil and 15 days in heavy soils. Moisture stress restricts re-growths after dehaulming and hastens curing of peel of seed tubers. For the potato, the critical period for water deficit is during tuber development. The water supply for the potato crop should be regular, especially from the stage of tuber initiation until the end of tuber enlargement. Water deficit in the early phase of yield formation increases the occurrence of spindled tubers and, when followed by irrigation, may result in tuber cracking or tubers with 'hollow hearts'.

Cropping systems: Potatoes are not rotated with tomato and other Solanaceae crops which are susceptible to the same pathogens as potato.

Harvest: The seed crop which is meant for storage should be harvested at full maturity. For most commercial varieties, yellowing of the potato plant's leaves or 2 weeks after foliage death and easy separation of tubers from stolons indicate that the potato crop has reached maturity. Generally the seed crop should be harvested at 15-20 days after haulm cutting and at proper soil moisture. Potatoes are harvested after three and a half to four months after planting in cool temperate areas. Only one harvest in a year is possible in a temperate climate. Only climates with mild winters and light frosts may allow two crops: in spring and in fall. In temperate climates the crop is usually harvested 2 or 3 weeks after the foliage has died down naturally. At the time of harvesting, the field should neither be too wet nor too dry. The harvesting can be done by tractor drawn one or two row potato digger or by the bullock drawn-one row digger or with the help of spade or *hand hoe*. Ideally, the harvest should be carried out in temperatures of between 10-18°C. Do not harvest when tuber pulp temperature is less than 8°C or more than 20°C. Seed crop is dehaulmed and the tubers are lifted 10 to 12 days after halum killing. If the potatoes are to be stored rather than consumed immediately, they are left in the soil to allow their skin to harden since the hard peel also help seed potatoes to resist storage diseases. Ware crop can be harvested after the haulms senescence naturally and tubers peel gets hardened. The tubers are left on the top of the soil for 15 to 60 minutes, depending upon the temperature, in order for the peel to dry and toughen. Early potatoes are frequently harvested by hand, as the immature skins are easily damaged or rubbed off. However, leaving tubers for too long in the ground increases their exposure to the fungal disease black scurf and increases the risk of losing quality and marketable yield. Harvested potato tubers should be surface dried and kept in heaps for 10 to 15 days in shade for curing of skin. Tuber peel is hardened before the soil temperature rises above 27 to 28°C to prevent charcoal not of tubers. Losses can be considerably reduced if the tubers are cured at 12 to 18°C and 85% relative humidity or higher to encourage cell tuberisation, and periderm formation to heal wounds. Tubers can be round, oval or cylindrical, with smooth or somewhat roughened peel which may be white, yellowish or red. The tuber yield of 15 to 20 t/ha can be harvested in 120 days.

Grading: Keep the surface dry tubers in heaps for 10-15 days in shade for curing of peel. Then sort out all cut, cracked, damaged and rotten tubers. Seed tubers are usually graded into four grades, viz. small (<25 g), medium (25-50 g), large (50-100 g) and extra-large (>100 g) to get the good returns and packed in gunny bags. Do not expose the tubers to sun light as far as possible, otherwise they will become green and the inner core will be spoiled. Potato is also graded into i) seed size tubers ii) large size tuber and iii) chats in plain region. The seed size tuber can be used for seed in next year. The last two grades are sold as ware potatoes in markets. In hilly region, tubers are graded in to i) A1 is 62 mm and above ii) A2 is 37 to 62 mm and iii) A3 is below 37 mm.

Storage: In European countries, the potato crop is grown in summer and the main storage season is the cold winter. However, in India, 85% of potato is produced in winter and stored during long hot summer. As the potato is a perishable commodity, the seed potato tubers are packed in bags and stored in cold stores at 4 to 5°C and 90% relative humidity. To avoid sweetening, potato are required to be stored at 10-12°C. Only seed potatoes should be cold stored at 2-4°C. Depending on variety and conditions during growth, at 10-13°C and 93% humidity, potatoes store for 1-3 months; at 8-10°C and 93% humidity for 2-5 months; at 5-8°C and 93% for 4-8 months; at 2-5°C and 93% humidity for 7 months. Ware potatoes should be kept at a temperature of about 6 to 8°C, in a dark, well ventilated environment at 85 to 90% relative humidity. For processing purposes, such as the production of french fries, storage temperatures may range up to 10°C to reduce the risk of increasing sugar levels, which are responsible for a dark colour during frying.

Seed tuber storage: Wash the seed tubers in water and then 1% bleaching solution and again in water. Tuber treatment (spray) with 3% Boric acid before storage is effective in management of black scurf and common scab. Treatment of Boric acid @3% can be done by dipping of tubers for 30 minutes also and the solution once prepared can be used for 20 times dipping. Pack the seed tubers in clean Hessian bags and label them as per standard. Store them immediately in cold storage. If the ambient temperature is high, the seed bags should be kept in pre-cooling chamber or in a cool place and then stored in cold store at 2-3°C and 75-85% relative humidity. The stored bags should be inspected periodically. Seed treated with chemicals should not be used for table purposes. In hills, the treated tubers should be kept in wooden or plastic trays or in baskets or spread in a well ventilated room.

Potato quality requirements: The quality of product is determined by its appearance (i.e., colour), crispiness and taste. Light or light golden yellow colour is desirable, whereas brown or black colour is undesirable in chips and French fries. Potatoes of 40-60mm diameter and above are suitable for preparation of chips and French fries. Round to round-oval potatoes are preferred for preparation of chips whereas long oval potatoes are preferred for preparation of French fries. Small sized potatoes are used for canning. Potatoes contain about 80% water and 20% dry matter. Potatoes with high dry matter are preferred for fried and dehydrated products and with low dry matter are used for canning. For preparing good quality potato chips, the reducing sugars content should be less than 150 mg/100g tuber fresh weight. Specific characteristics of potato varieties for different purposes are listed in Table 2. Tuber characters and processing quality of some Indian potato varieties are furnished in Table 3.

Table 2 Requirement of potato varieties for different purposes

Characters	Use requirements			
	Table potatoes		Processing	
	Boiled	Baking	French fries	Chips
Tuber shape	Long-oval/round	Long-oval/round	Long-oval (>3 inch)	Round (2.5-3.3 inch)
Peel color	White/yellow /red	White/yellow /red	White/ yellow	White/ yellow
Eye depth	Shallow/ medium	Shallow/ medium	Shallow	Shallow
Flesh color	White/yellow	White/yellow	White/ yellow	White/ yellow
Texture	Waxy	Mealy	Mealy	Mealy
Uniformity	High	High	High	High
Defects	Minimum	Minimum	Minimum	Minimum
Dry matter (%)	18-20	>20	>20	>20
Reducing sugars*	-		<200mg	<100mg
Phenols	Less	Less	Less	Less
Glycoalkaloids*	< 15mg	< 15mg	< 15mg	< 15mg
Keeping quality	Good	Good	Good	Good
Damage resistance	High	High	High	High

Note: *mg/100g fresh tuber weight.
Source: Central Potato Research Institute, Shimla

Table 3 Tuber characters and processing quality of some Indian potato varieties

Variety	Shape/Size	Dry matter (%)	Reducing sugars (mg/100g f. wt)	Acrylamide (μg/kg f. wt)
Kufri Chipsona-1	Oval/Large	21-24	45-100	< 100
Kufri Chipsona-2	Round/Large	21-25	44-93	<100
Kufri Chipsona-3	Round-Oval Medium	22-24	30-50	< 200
Kufri Chipsona-4	Round	21-23	60-140	< 200
Kufri Frysona	Oblong/ Large	22-23	<100	< 100
Kufri Himsona	Oval/ Medium	20-25	<50	< 200
Kufri Jyoti	Oval/Large	18-21	106-275	< 800
Kufri Lauvkar	Round/Large	18-20	200-250	< 300
Kufri Chandramukhi	Oval/large	18-20	250-324	< 500

Source: Central Potato Research Institute, Shimla

Utilization: A potato tuber contains 80% water and 20% dry matter consisting of 14% starch, 2% sugar, 2% protein, 1.0% minerals, 0.6% fibre, 0.1% fat, and vitamins B and C in adequate amount. Potato is eaten boiled, roasted, baked or fried and are processed into a very wide range of products, such as canned whole potatoes, frozen french fires or chips, crisps, dehydrated flakes, powder or granules, potato salad, etc. Potato is a good substrate for the growth of microorganisms.

REFERENCES

CIP. 2007. Root and Tubers: The overlooked opportunity, Annual Report. CIP [International Potato Center]. Lima, Peru.

CIP. 2008. Potatoes. Fact sheet. CIP, Lima, Peru.

CIP. 2008. Why potatoes. Fact sheet. CIP, Lima, Peru.

CIP. 2008. Potatoes and the environment. Fact sheet.CIP, Lima, Peru.

CPG. 2013. Crop Production Techniques of Horticultural Crops. Horticultural College and Research Institute, Tamil Nadu Agricultural University, Coimbatore – 641 003

Gopal, J. 2004. True Potato Seed: Breeding for hardiness. Journal of New Seeds, 5(4):57-73

Kumarm, D., Singh, B.P., and Kumar, P. 2004. An overview of the factors affecting sugar content of potatoes. Annals Applied Biol 145: 247-55

Renia, H. 1995. True seed is a commercial reality in USA. Potato Review, 5:48-51

Shekhawat, G.S. and Naik, P.S. 1999. Potato in India. Central Potato Research Institute, Shimla Technical Bulletin No. 1. p 99.

Stark, J.C. and Love, S.L. (eds.). 2003. Potato Production Systems. University of Idaho Extension. USA.

http://agritech.tnau.ac.in/horticulture/horti_vegetables_potato.html http://www.tnaugenomics.com/seedportal/npotato.php

http://agmarknet.nic.in/profile-potato.pdf

http://www.ukia.org/pdfs/water%20management%20for%20potatoes.pdf

http://pdf.usaid.gov/pdf_docs/PNABD861.pdf

http://www.uky.edu/Ag/NewCrops/introsheets/potatoes.pdf

http://migarden.msu.edu/uploads/files/growingPotatoes.pdf

http://www.iivr.org.in/sites/default/files/Technical%20Bulletins/7.%20Vegetable%20Statistics.pdf

http://ir.library.oregonstate.edu/xmlui/bitstream/handle/1957/43803/em8912.pdf

http://ag.arizona.edu/crop/soils/aznpotatoe.pdf

CHAPTER

28

Chinese Potato–*Coleus parviflorus* L.

Family: Lamiaceae (Mint family)

Synonyms: *Plectranthus rotundifolius* (Poir.), *Coleus rotundifolius*, *Solenostemon rotundifolius*, *Plectranthus tuberosus*.

Vernacular names: Chinese Potato, Coleus potato, Hausa potato, Frafra, Zulu potato, Sudan potato, Madagascar potato (English); Arbi (Hindi); Arvi (Marathi); Sambrali, Samagadde (Kannada); Kooka (Konkani); Siru kizhangu, Seppaankizhangu (Tamil:); Koorka kizhangu, Koorka, Koorkka (Malayalam); Cheemadumpa (Telugu).

Origin: It is native of India and of Central or East Africa.

Distribution: It is cultivated for its edible tubers in West Africa, India, Sri Lanka, Malaysia and Indonesia for its edible tubers. In India, it is grown mostly in Kerala, Karnataka and southern Tamil Nadu.

Botany: It is an herb with prostrate or ascending habit and a succulent stem reaching up to 15 to 30 cm in length, forming tubers in clusters around the base of the stem. The stem is decumbent or ascending, quadrangular, with densely pubescent on the angles and roots at the nodes. The leaves are rather thick, juicy, faintly aromatic smell when bruised and arranged oppositely. The leaf-stalk is 1-3 cm long. The blade is ovate to nearly round, measuring 2-6 cm × 1.5-4 cm, wedge-shaped at the base, coarsely toothed at the margins. The inflorescence is a terminal spike, measuring 5-15 cm long, with distant whorls of 4-6 flowers. The flower-stalk is 1-2 mm long. The flower is tubular and 2-lipped, measuring 7-12 mm long, light to dark violet, velvety and gland-dotted. The tube is strongly curved. The tubers are round, dark brown or black and aromatic. The tubers are small and dark brown and produced in clusters at the base of the stem.

Climate: It comes up well in shade with warm humid climate. It requires rainfall of 1000 to 2300mm for its growth and cannot withstand drought conditions. In case rains are not received, irrigation has to be provided for satisfactory growth. The cardinal minimum, optimum and maximum temperature requirement is 18°C, 24 to 28°C and 36°C respectively.

Soils: Red, loamy and well drained soils with soil pH of 5.0 to 7.0 are desirable.

Season: Chinese potato is normally cultivated in Kerala during the months of July and

August. It is cultivated widely in paddy fields and low lying areas where water is available.

Varieties: CO 1, Sree Dhara.

Land preparation of main field: The field is ploughed 4 to 5 times to a fine tilth and form ridges and furrows 60 cm apart.

Nursery management: Select well drained and medium fertile area for planting seed tubers. Raise a nursery 30 to 45 days prior to planting. An area of 500 m^2 is required to produce vines for planting one hectare of land. Cattle manure or compost may be applied @ 1 kg/m^2 and ridges / mounds may be prepared at a closer spacing of 45/60 cm. Healthy tubers that weigh about 15-20 g may be planted at 5cm spacing on the ridges / mounds so as to accommodate 750-1200 kg tubers in 500 m^2 area during March-April. Top dressing is done with urea (5 kg / 500 m^2) at about three weeks after planting to encourage good vine growth. Stem/vine cutting of 15-20cm length from these sprouts also used for planting at a spacing of 15 cm in the raised beds of 60-90 cm width. To enable rapid multiplication of the planting material, single node cuttings can be planted directly in the secondary nursery. Such single node cuttings produce axillary shoots within one week can be planted in the main field.

Planting: Vine cuttings of 10 cm length are taken from the nursery beds after one month and plant in the main field during July-October at a spacing of 30 cm on the ridges either in vertical or horizontal position. Horizontal planting of vines to a depth of 4-5 cm and exposing the terminal bud ensures quick establishment and promote tuber yield. In loose soils having good drainage, planting can also be done on flat beds with provision for drainage. Plant canopy usually spread to cover entire ridges by 10 week after planting.

After cultivation: Two to three weeding and earthing up are taken up to 45 days after planting. Weeds close to plants are rouge out by hand. Cover portion of the vine with soil to promote tuber formation.

Manures and fertilizers application: FYM at 25 t/ha and NPK at 30:60:150 kg/ha are applied as basal dressing. Apply 30 kg N/ha on 30 days after planting at the time of earthing up along with 2 kg of Azospirillum. In case, the soil has eroded from the base of the plant, give one more earthing up at 30 days later to promote tuber formation.

Water management: Irrigation is given at weekly intervals. In dry conditions, it may fail to produce tubers.

Crop protection: Caterpillars and termites are the major pests at the vegetative stage. Termites and millipedes also bore holes into tubers causing tuber rot during storage. The common disease is nematodes which cause leaf curling and stunting at the vegetative stage. Root knot nematode is a serious pest on coleus and the infested plants exhibit serious swellings or galls in the roots resulting in suppressed roots, stunted growth and wilting. Less than a millimetre long, the nematodes are tiny worms that enter the plant roots of the seedling when the plant is most vulnerable. Seed tubers free of nematodes can be used. Deep ploughing of the field immediately after harvest exposes the soil and kills the nematodes. The nematodes are controlled by practising summer fallowing and soil solarization. Cultivation of sweet potato as a preceding crop in May-June enables trapping of root knot nematodes in the soil. To control leaf folding caterpillars and vine borers, dipping the vines in insecticide solution (Dimethoate or Rogar 30 EC ie. 1.7 ml/litre) for 10 minutes prior to planting is helpful. In case, severe damage is noticed in the field, spraying may be adopted with Malathion, Fenthion or Fenitrothion 50 EC 1 ml/litre).

Harvesting: Tubers may be harvested 120 to 150 days from planting. Pull out the plants and dig out the left over tubers in the field. Separate the tubers from the plant and destroy the crop residues by burning.

Yield: Tuber yield of 4 to 7.5 t/ha can be obtained in 120 days.

Utilization: Tubers are mostly boiled and consumed as main meal or snack. It is used to treat stomach pain, nausea, vomiting, diarrhoea, mouth and throat infections. It is also used as purgative, carminatives and as antihelmintics.

REFERENCES

Allemann, J. and Hammes, P.S. 2006. Effect of photoperiod on tuberization in the Livingstone potato (*Plectranthus esculentus* N.E.Br. Lamiaceae). Field Crop Res. 98:76-81.

Allemann, J., Robertse, P.J. and Hammes, PS (2003). Organographic and anatomical evidence that the edible storage organs of *Plectranthus esculentus* N.E.Br. (Lamiaceae) are stem tubers. Field Crop Res. 83:35-39.

Catherine W, Lukhobaa, Monique SJ, Simmonds and Alan, J.Paton. 2006. Plectranthus: A review of ethnobotanical uses. J. Ethnopharmacology, 103(1): 1-24.

Mohankumar, C.R., Nair, G.M., George, J., Ravindran, C.S. and Ravi, V. 2000. Production Technology of Tuber Crops. Central Tuber Crops Research Institute, Sreekariyam, Thiruvananthapuram, Kerala, India. 174p

https://en.wikipedia.org/wiki/Plectranthus_rotundifolius

http://ecocrop.fao.org/ecocrop/srv/en/dataSheet?id=4768

CHAPTER

29

Sweet Potato–*Ipomoea batatas* L. (2n=90)

Family: Convolvulaceae (morning-glory family)

Vernacular names: Sweet potato (English), batata (Spanish).

Importance: Sweet potatoes are true roots and not tubers. It is cultivated for its sweet root tubers. It is mainly used for human food after boiling or steaming, baking or frying and also as animal feed. Certain varieties having yellow flesh are rich in carotene, a precursor of vitamin A. Since roots contain 16% starch and 4% sugar, it is suited for production of industrial starch, syrup and alcohol.

Origin: The sweet potato is native to Peru and tropical Central America and some South Pacific islands.

Distribution: It is an important tuber crop in tropical and sub-tropical countries like China, USA, India, Japan, Indonesia, Philippines, Thailand, Vietnam, Panama, Peru, Mexico, Brazil and Nigeria. In India, it is grown mainly in Andhra Pradesh, Assam, Bihar, Uttar Pradesh, Tamil Nadu and Odisha.

Botany: Sweet potato, the underground part is classified as a storage root, rather than a tuber. It is a perennial herb with trailing vines and with medium sized tubers. For cultivation purpose, it is treated as an annual with duration of 90 to 120 days. It has an extensive fibrous root system both at stem cuttings and at nodes touching soil. Storage roots may be fusiform, spindle or globular in shape and surface is smooth. Skin has white, red or light copper colour. Flesh may be white or with different combinations of orange and red. Leaves are simple, alternate and stipulate. They vary in size and shape, occasionally in same plant. Leaf shape varies from ovate to cordate, hastate or deeply lobed and may change on ageing. Leaf shape is an important character for identifying clones. All clones do not flower and in flowering ones, duration and initiation of flowering vary. Flowers are axillary and borne solitary or in simple cymes. Flowers are bisexual. Corolla is attractive and funnel shaped formed by fusion of five petals. Anthesis starts before dawn and closes by 9 to 11 a.m. The crop is highly cross-pollinated due to self-incompatibility and male sterility. Pollination is entomophilies. Fruit is a capsule with false septa. Seed coat is hard and impervious to water. Hence, scarification is required for promoting germination. Though plants produce viable seeds, highly heterozygous nature of the crop results in a heterogeneous population. Hence stem cuttings are used for propagation purpose

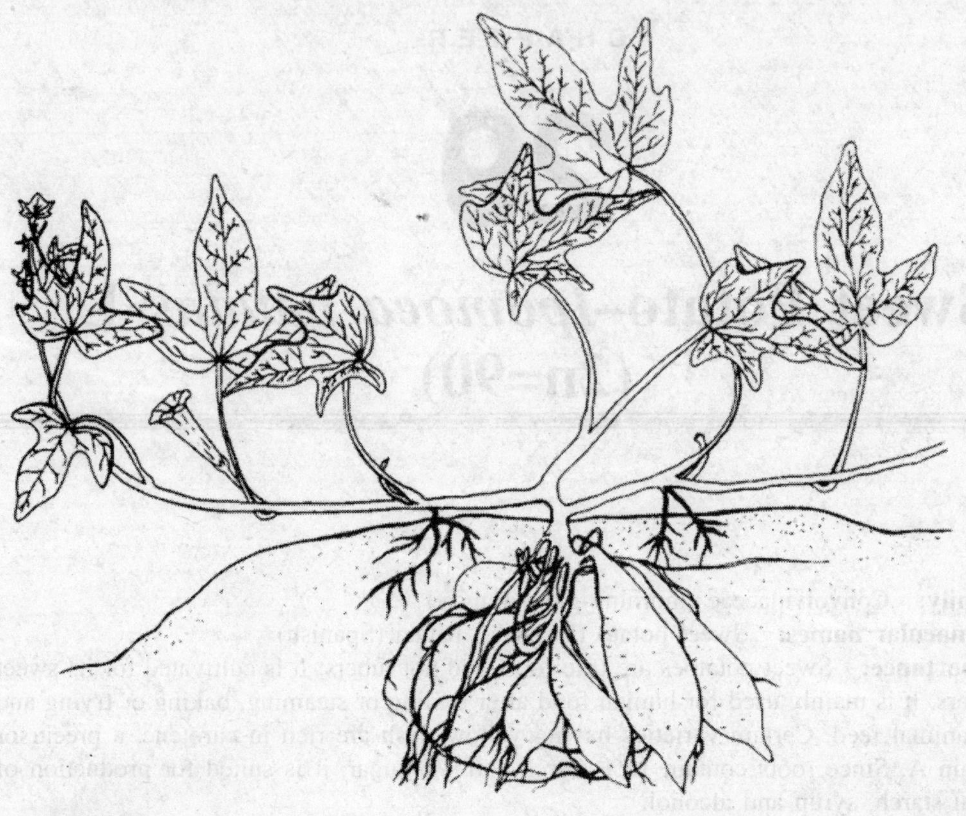

Source: www.adap.hawaii.edu

The skin and flesh of the sweet potato may be white, cream, yellow, orange, pink, or deep purple, although white/cream and yellow-orange flesh depending upon the variety. The intensity of the sweet potato's yellow or orange flesh colour is correlated to its beta-carotene (β-carotene) content which is a precursor of Vitamin A. The purple fleshed sweet potato variety is a source of anthocyanins (especially peonidins and cyanidins). The staple white fleshed sweet potato has high starch and dry matter content.

Growth stages of sweet potato: Sweet potato generally requires a growth season of 4 to 5 months with optimum temperatures of 20°C to 25°C. It can however grow at a wide range of temperatures between 15°C and 35°C. Highest root yields are obtained during day time temperatures of 25 to 30°C and night temperatures of 15 to 20°C. Storage root differentiation may begin as early as two to three weeks after planting, and on average between 4-6 weeks, depending on the variety and the environmental conditions. Early maturing sweet potato varieties can be harvested 3 to 4.5 months after planting. If temperatures are low, the growing period has to be extended to 6-7 months.

Growth phases	Growth stages	Days after planting	Cultivation operations
Establishment phase	Planting and germination	0-6	Planting
	Fast growth of young roots	7-14	Gap filling
	Storage roots start to differentiate	15-21	Avoid stress
	Slow growth of vines	22-28	
Storage root initiation	Initiation of storage roots	28-34	Weeding
	Fast growth of vines	35-42	
	Large increase in leaf area	43 -49	
Storage root bulking	Growth of vines ceases	56	Vine lifting
	Rapid bulking of storage roots	63	
	Reduction of leaf area due to yellowing and falling	70-105	Hilling up
	Harvesting	112	Harvesting

Source: Stathers *et al.*, 2013.

Source: www.google.co.in

Climate: Sweet potato is grown between 32°S and 40°N up to an elevation of 1800 m from MSL. It adapts well to warm climates and grows best during summer season. It is a short day plant. It requires 500 mm rainfall during the growing season and an annual rainfall of 750-1500 mm is considered the best. They can tolerate considerable periods of drought but yields are very much reduced if water shortage occurs at the time of tuber initiation. It is grown under high rainfall frequently produce vigorous vine growth but poor tuber yield. The ideal conditions for high yield are good rains during the period of early growth and dry sunny weather during the period of tuber bulking and maturity. The optimum temperature to achieve the best growth of sweet potatoes is between 21 and 29°C, although they can tolerate temperatures as low as 15°C and as high as 35°C. It is cold and frost sensitive. Hence, it can be grown throughout the tropics, but in temperate regions, the aerial portion dies out during winter season. The warm sunny days and cool nights are very much favourable for storage root formation in sweet potato. The root formation requires cooler mean temperature of 24°C while the weather should be warm temperature for root bulking. The highest root yields are obtained during day time temperatures of 25 to 30°C and night temperatures of 15 to 20°C. Shading of the crop results in reduction of tuber yield.

Soils: It requires well drained fertile soil rich in humus with permeable subsoil. Sandy loam soils with clay sub soil are ideal for tuber production. The heavy clay soil should be avoided as it can retard root development, resulting in reduction in tuber size and formation of irregular shape tubers. The optimum soil pH is 5.6 to 6.7. Liming is necessary if soil pH is below 5.2. The soil pH should be adjusted to about 6.0 by applying lime or dolomite. It is an acid tolerant crop, but it cannot withstand salinity and alkalinity. Salinity reduces growth of stems and roots, causes lateral rolling of leaf lobules, reduction in leaf size and necrosis of older leaves.

Season: The crop is usually raised as a rainfed crop in *kharif* season and as irrigated crop during the rabi season and in summer season.

Varieties: Sweet potato varieties differ in shape, size and colour of leaves, tubers and nature of tuber flesh. There are three categories of sweet potato viz., i) the staple type with white flesh, ii) orange flesh with β-carotene and iii) purple flesh with anthocyanins. The particulars of sweet potato varieties grown in different states of India are presented in Table 1.

Table 1 Sweet potato varieties grown in different states of India

Variety	Duration (days)	Tuber yield (t/ha)	Features
Varsha (CTCRI)	120-125	18-22	semi-spreading hybrid with reddish purple skin, light yellow flesh and excellent cooking quality
Sree Nandini	100-105	20-25	spreading variety with light cream skin, white flesh and good cooking quality
Sree Vardhini	100-105	20-25	semi-spreading variety with purple skin, light orange flesh, tolerant to feathery mottle virus
Sree Ratna	95-105	20-25	spreading variety with purple skin, orange flesh and excellent cooking quality
Sree Bhadra	90-95	20-25	semi-spreading variety with light pink skin and cream flesh; resistant to root knot nematode
Sree Arun	90-100	20-28	spreading variety with pink skin, cream flesh and good cooking quality
Sree Varun	90-100	20-28	spreading variety with cream skin, cream flesh and good cooking quality
Sree Kanaka	75-85	-----	high carotene (8.8 - 10 mg/100 g).
Kanjanghad (KAU)	-----	12	Purple coloured and spindle shaped tubers have yellow soft flesh, tolerant to shade and weevil incidence
Pusa Safed (IARI)	-----	-----	white skinned variety with white flesh
Pusa Sunheri (IARI)	-----	-----	brown skinned variety with yellow flesh rich in carotene; boiled flesh is attractively orange yellow
Cross-4 (APAU, Hyderabad)	90-105	20-30	highly susceptible to weevil infestation
Rajendra Shakarkand-5 (APAU, Hyderabad)	105-120	30	resistant to fusarium wilt and cercospora leaf spot disease
Kalmegh (APAU, Hyderabad)	90	26	short duration variety
CO1 (TNAU)	135	25-30	tubers have pink skin and white flesh, tolerant to root weevil,
CO2 (TNAU)	110-115	32	Vines are dwarf and erect. The tubers are medium sized with light pink skin and white flesh
CO3 (TNAU)	105-110	40-43	tubers are medium sized with red skin and deep orange flesh, tolerant to root weevil
COCIP 1 (TNAU)	95-100	30	tolerant to weevil incidence, tuber has attractive pink skin with yellow flesh

Cultivation practices

Nursery management: Cuttings for planting are multiplied in two nurseries viz., primary and secondary. Tubers in primary nursery are planted three months ahead of planting. A nursery area of 100 m^2 is required to raise vines for planting one hectare. The medium sized weevil free tubers of 125 to 150 g weight each to about 100 kg of planting stock are used for planting one hectare in ridges at 60 cm apart. To ensure quick growth of vines, top dressing with 1.5 kg urea/100 m^2 at 15 days after planting is advisable. The nursery is irrigated on alternate days for the first 10 days and once in three days thereafter. After 40 to 45 days of growth, the vines are cut to a length of 20-30 cm for further multiplication in the secondary nursery. Vines collected from the primary nursery are further multiplied in the secondary nursery in an area of 500 m^2 to produce enough vines for planting one hectare of land. Farm yard manure or compost is applied @ 1 kg m^2 and ridges are formed at a spacing of 60 cm. Vines obtained from the primary nursery or from freshly harvested crop are planted in the secondary nursery at a spacing of 20 cm on the ridges. To ensure enough vegetative growth, five kg of urea is applied in two splits on 15th day and 30th day after planting. For better establishment of vines in the nursery, daily irrigations are provided for the first three days and on alternate days for one week. Thereafter, irrigation may be restricted to once in three days. The vine cuttings of 20-30 cm length with at least 3-4 nodes get ready for transplanting at about 40-45 days after planting. In general, the apical cuttings are the best to secure high yields from sweet potato. Starch content and reducing, non-reducing and total sugars were highest in tubers from plants grown from apical cuttings. Cut vines are stored with intact leaves in shade for two days prior to planting in main field to promote better root initiation, early establishment and higher tuber yield. The optimum root length of cuttings for transplanting is < 1 cm as longer roots are easily broken. The leaves can be removed when the vines are to be transported for off to reduce the bulk.

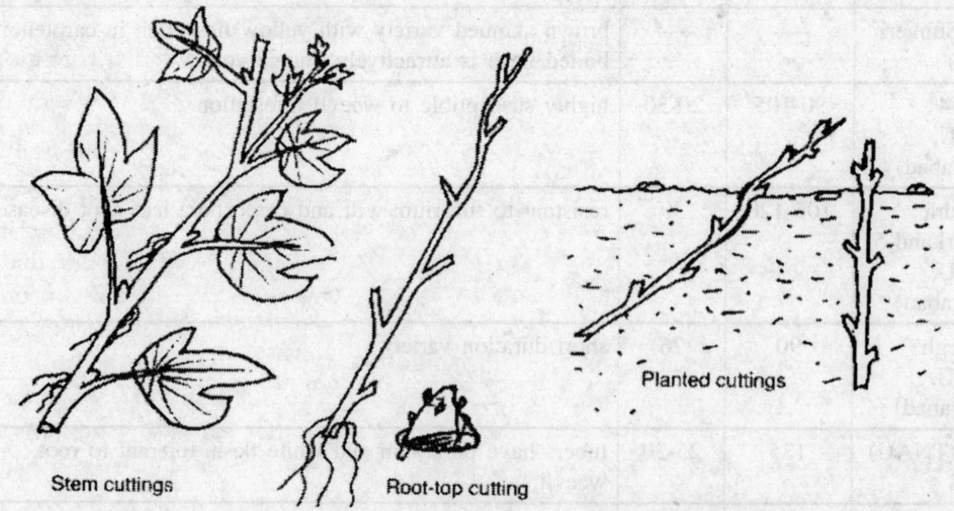

Land preparation of main field: Main field is prepared by making ridges of 25-30 cm high at 60 cm apart after thorough ploughing. The land is ploughed or dug to a depth of 20 cm and harrowed to pulverize the soil. Mounds, ridges and furrows and flat bed methods of planting are being practised in different locations. It is preferable to plant sweet potato on mounds in areas experiencing problems of drainage. Ridges formed across the slope are recommended in slopy lands to prevent soil erosion. Among the different methods of land preparation, the highest tuber yield was obtained when planted on mounds. The higher yield obtained in mound system of planting is attributed to better soil aeration and less tendency for soil compaction.

Planting system: A spacing of 60 × 20 cm has been recommended to accommodate 83,000 plants per hectare on ridges formed 60 cm apart. Vines of 20 to 30 cm length are buried horizontally with 2-3 nodes below soil, leaving the remaining portion above soil. The cuttings of vines can also be planted in the soil with both the ends exposed and the middle portion or three to four nodes buried in the soil. Vines are also planted in an inclined position with half of its length buried in the soil. The number and weight of total and marketable tubers are higher when slips are planted horizontal to the soil with 5 or 6 nodes covered than when planted perpendicular to the soil. Horizontal planting leads to high plant survivals and better development of the root system.

Weed management: Two hand weeding and earthing up operations are carried out about 2 weeks and 5 weeks after planting.

Manures and fertilizers management: Apply 12.5 tonnes of farmyard manure per ha at the time of preparation of ridges. A fertilizer dose of 75:50:75 or 50:25:50 kg N, P_2O_5 and K_2O per ha is recommended depending on soil conditions, of which 50% of N, full P and K to be applied as basal dose while the remaining 50% of N is applied one month after along with weeding and earthing up.

Irrigation: A moist seedbed is required for 4-5 days to ensure proper sprouting and establishment of vines. Irrigation has to be provided daily for the first three days followed by irrigation on alternate days for one week. The tuber initiation phase, which falls around 20 DAP, is very critical period to maintain optimum soil moisture to produce economic yield. During *rabi* and summer seasons, 12-14 irrigations are required at an interval of 7 to 10 days for high tuber yield.

Cropping systems: Under irrigated conditions, sweet potato can be rotated with rice and planted during December-January after harvest of the second crop of rice. As a mixed crop, it can be grown along with colocasia, elephant foot yam etc. Under rainfed conditions, green manure crops such as *kozhinji* (Wild indigo) and sunnhemp can be grown after harvest of the sweet potato and later incorporated into the soil at the time of land preparation for the succeeding crop.

Pests and diseases: Sweet potato weevil is the most important pest causing severe damage to the crop. Adult weevil makes puncturing on vines and tubers. Grubs bore and feed by making tunnels. Even slightly damaged tubers are unsuitable for consumption due to bitterness. Yield loss is up to 100% in severe cases. On an average, 20-55% tuber loss occurs. Dipping vine cuttings in Fenthion or Fenitrothion or Monocrotophos @ 0.05% solution for 10 minutes before planting is recommended to control weevils. Earthing up the crop two months after planting, installing synthetic sex pheromone traps @ 1 trap / 100 m^2 area to collect and kill the male weevils and destroying crop residues after harvest by burning are recommended control measures. Apply *Chromolaena odorata* leaves as mulch @ 3 t ha^{-1} at 30 DAP to control of sweet potato weevil.

Harvest: Maturity is indicated when the leaves turn yellow and begin to fall. The maturity of the tubers can be determined by cutting fresh tubers. The cut surface of the immature tuber gives a dark greenish colour while the cut ends dry clearly in the case of mature tubers. The field is irrigated 2-3 days prior to harvesting to facilitate easy lifting of the tubers. Manual harvesting is done by digging out tuber with pick-axe when leaves turn yellow in colour. The tuber yield per plant also increases if the crop remains in the field longer, but the tubers become less palatable and infestation by sweet potato weevil and/or rotting of tubers may occur. Tubers are cleaned and packed in gunny bags for marketing. Care should be taken to avoid injuries and bruises on tubers. For marketing of fresh tubers, cleaning and grading should be done to get better prices. After harvest, tubers are spread in partial shade for 5-6 days for healing and curing.

Yield: Tuber yield of 6-9 t/ha for rainfed crop and 12-15 t/ha under irrigated condition can be obtained.

Storage: Red skinned varieties store better than white skinned varieties. Tubers are stored better at 15°C and 85% RH. Tubers can also be stored for 2-3 months in a pit covered with straw. Tubers are also stored in a layer of dry sand/soil after curing under ambient conditions.

Utilization: Sweet potato roots can exhibit a variety of colours. Traditionally the market demands an orange-fleshed root, but several other colours, including white, red, yellow and even purple, are grown. Sweet potatoes are nutritious and a good source of carotenoids, vitamin A and vitamin C.

REFERENCES

AVRDC. 1982. Sweet potato: Proceedings of the First International Symposium. Asian Vegetable Research and Development Center. Shanhua, Tainan, p. 481. Taiwan: AVRDC Press.

Bouwkamp, J.C. 1985. Sweet potato products: A natural resource for the tropics, CRC Press. Boca Raton. p. 271.

CPG. 2013. Crop Production Techniques of Horticultural Crops. Horticultural College and Research Institute, Tamil Nadu Agricultural University, Coimbatore – 641 003

Edmond, J. B. and Ammerman, G.R. 1971. Sweet potatoes: Production, processing, marketing, The AVI Publishing Co. Westport, CT, p. 334

IPC. 1987. Exploration, maintenance and utilization of sweet potato genetic resources; Report of the First Sweet Potato Planning Conference, International Potato Center. Lima, Peru. p. 369.

Stathers, T., Low., J., Mwanga, R., Carey, T., David., S., Gibson, R., Namanda, S., McEwan, M., Bechoff., A., Malinga, J., Benjamin, M., Katcher, H., Blakenship, J., Andrade, M., Agili, S., Njoku, J., Sindi, K., Mulongo, G., Tumwegamire, S., Njoku, A., Abidin, E., and Mbabu, A. (2013). Everything You Ever Wanted to Know about Sweet potato: Reaching Agents of Change ToT Manual. International Potato Center, Nairobi, Kenya. pp390

Woolfe, J.A. 1992. Sweet potato: An untapped food resource. Cambridge University Press. Cambridge, UK. p. 389

Yen, D. E. 1974. The sweet potato and Oceania: An essay in ethnobotany, Bishop Museum Press. Honolulu, p. 389.

https://hort.purdue.edu/newcrop/CropFactSheets/sweetpotato.html

http://agridr.in/tnauEAgri/eagri50/HORT281/pdf/lec23.pdf

http://cipotato.org/library/pdfdocs/SW52966.pdf

https://www.hort.purdue.edu/newcrop/duke_energy/Ipomoea_batatas.html

http://www.adap.hawaii.edu/adap/Publications/Ireta_pubs/sweet_planting.pdf

CHAPTER

30

Tapioca–*Manihot esculenta* subsp. *esculenta* Crantz (2n=36)

Synonym: *Manihot ultissima* Phol
Family: Euphorbiaceae
Vernacular name: Cassava, Tapioca, Manihot, Manioc, Brazilian arrowroot, Yuca (English), *Ketalla* (Brazil), *Mandioca* (Brazil), *Mravuli, Kamadi, Ganna* (Hindi), *Maravalli kizhangu* (Tamil), *Maracheeni* (Malayalam), *Simolu aloo* (Assamese), *Sheredi* (Gujarati), *Karrapendalamu* (Telugu), *Maraganasu* (Kannada). In Thailand, tapioca is known as sago, which can lead to confusion with true sago starch obtained from the sago palm, *Metroxylan sagu* Rottb. The English name 'cassava' may have derived from the word 'casabi', which, among the Arawak Indians, signifies 'root', or else came from the word 'cazabe', which is a cake or dry biscuit produced by the indigenous populations of the Amazon Basin. The cassava is called 'madioca' in Brazil; 'manioc' in Africa where French is spoken countries; 'yuca' in Spanish-speaking countries and 'tapioca' in Asia.

Importance: Tapioca is a tropical root crop, grown as famine reserve and against the ravage of migratory locust in Africa. The roots or tubers are the parts used as human food, industrial raw material (starch based industries), animal and poultry feeds. Tapioca starch is used for food and manufacture of adhesives and cosmetics and sizing in textiles, laundering and paper making. Tapioca is used for pudding, biscuits and confectionery. Sweet cassava can be eaten raw after peeling. Bitter cassava is raised in areas where pigs cause serious menace. Flour can be used for propagation of slices, chips etc. The yellow colour of the flesh is due to the presence of carotene, which is the precursor of vitamin A. The carotene content increases with age. The moist starch is made into globules by shaking and the globules are surface gelatinized. Sago is a product obtained from cassava starch. Sago production is an important small scale industry in Tamil Nadu and the sago is consumed all over the country. It is eaten generally by cooling with milk and sugar.

History: Cassava was introduced into India by the Portuguese when they landed in the Malabar region, presently part of Kerala state during the 17th century, from Brazil.

The popularization of the crop in the state of Kerala was attributed to the famous king of Travancore State, Sri Visakham Thirunal by introducing popular varieties from Malaya and other places. In 1943, Mr. Manickam Chettiyar of Salem, Tamil Nadu found cassava flour as a good substitute for American corn flour and marketed at Chennai. Sago production has commenced in Tamil Nadu by 1945 with the technical know-how from Malaysia. The Centro Internacional de Agricultura Tropical (CIAT) has been formally established in Cali, Colombia during 1967 and began its research in 1969 while the International Institute of Tropical Agriculture (IITA), in Nigeria, has been created around 1967 to do research on Cassava.

Origin: The centre of origin of Tapioca is southern border of the Amazon basin, South America (Brazil and Paraguay) and is domesticated about 5000 years ago.

Distribution: Tapioca is cultivated in Brazil, Nigeria, West Africa, Zaire, Mozambique, Indonesia, Thailand, Vietnam and India. In India, it is grown in Kerala, Tamil Nadu and Andhra Pradesh.

Botany: Tapioca is a perennial shrub, with latex in all its parts, which produces enlarged tuberous roots. The plant height ranges from about 1 to 3 m or even more. There are two distinct plant types, erect with or without branching at the top and spreading types. The erect types are generally tall. The spreading types show branching at a very low level (in some clones even at a height of 20 cm) and the angle of divergence of the branch is much wider. The stems are usually slender and glabrous. The stem is 1 to 4 m long and woody with a thick bark. The old part of the stem bears evident scars of fallen first leaves. It produces two to three branches. Stems vary in colour, being grey or silvery, green, greenish yellow, reddish brown or streaked with purple. Tapioca roots are concentrated in the top 30 cm, although source roots go down to 140 cm depth. There are usually 5 to 10 tubers per plant. The tubers are cylindrical or tapering with 15 to 100 cm long and 3 to 15 cm in diameter, occasionally longer. The cortex, generally known as rind is about 14% of the total weight. The rind may be cream or pink in colour. The mature cassava storage root has four distinct tissues viz., bark (periderm), peel (cortex), parenchyma and central vascular xylem bundle. The leaves, which are spirally arranged and have petioles 5 to 30 cm long, usually longer than the blades; the blades are deeply palmately divided with 5 to 7 lobes, each 4 to 20 cm long and 1 to 6 cm wide. Leaves vary in colour from green to reddish; the petiole and midrib may be deep red. Older leaves are shed leaving the prominent leaf scars on the stem. It is a monoecious crop. The flowers are borne as axillary recemes near the ends of branches. The female flowers (staminate) occur in the base while male flowers (pistillate) at the top of the inflorescence. Flowers are pale yellow in colour and 1 to 1.5 cm in diameter. Tapioca is a cross pollinated by insects under natural conditions but considerable selfing may also occur. Male flower opens 7 to 8 days after female flowers. Pollen varies in fertility from almost sterile to 0.95 fertile. Pollen viability is reduced to about half the day after opening, and is lost after 2 days. Female flowers open 11 am to noon. Receptivity of the stigma occurs 6 h before flower opening. Induction of flowering in tapioca depends on long photoperiods up to 16-hour day length associated with temperatures of around 24ºC. Cassava pollen loses viability rapidly after it is shed. 97% seed set with newly-collected pollen, 56% seed set with pollen stored for 24 hours at 25ºC, and 0.9% seed set (one seed from 102 pollinations) after 48 hours of storage.

Fruits are globose capsule, trilocular, schizocarp, 1.5 cm long. Each locule contains a single caruncle seed. Thus the fruit is three seeded. The endocarp of capsule split explosively on ripening around 3 to 5 months after pollination and ejects seeds. The seeds are ovoid-ellipsoidal, approximately 10 mm long and 4 to 6 mm thick. Seed colour is

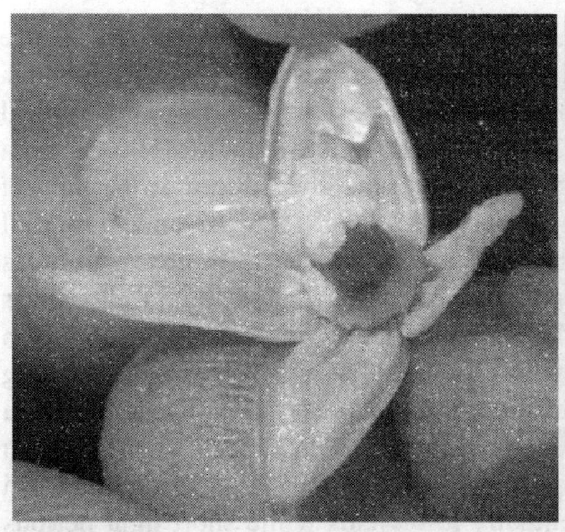

Female flower

Male flower

Flowering of cassava

Source: Mark E. Halsey *et al* 2008.

grey mottled with dark blotches. The seed weight varied from 95 to 136 mg per seed. A seed output up to 162 g per plant has been recorded. Newly harvested seeds are dormant, requiring 3 to 6 months of storage before they will germinate. Seeds can remain viable for up to 1 year, although germination percentages may decline substantially after 6 months. Seeds survive for up to 7 years with no loss of germination under cold storage conditions with 4°C and 70-80% relative humidity. Botanical seed is not typically used for commercial propagation. Tapioca is classified into sweet and bitter types. The sweet tapioca is known as *Manihot esculenta* Crantz (Syn: *M. utilissima* Pohl.) while the bitter tapioca is known as *Manihot palmata* Muell. Arg (Syn: *M. dulcis* Pax.). The sweet tapioca has low hydrocyanic acid (HCN) content which is confirmed to phylloderm of tubers. The high HCN in bitter cassavas is distributed in tubers and leaves. Cultivars containing cyanogenic glucoside with <100 mg to about 20 mg kg^{-1} fresh weight are called 'sweet' tapioca while cultivars with >100 to 500 mg kg^{-1} fresh weight are 'bitter' tapioca. Cyanogenic glycoside is distributed throughout the tuberous root; and the 'sweet' cultivars, in which the glycoside is confined mainly to the peel and is at a lower level. The flesh of sweet cultivars is therefore relatively free of glycoside, although it always contains some amount. In general, sweet cassava cultivars tend to have a short growing season, their tuberous roots mature in 6–9 months and deteriorate rapidly if not harvested soon after maturity. The bitter cassava cultivars require 12–18 months to mature and will not deteriorate greatly if not harvested for several months.

Climate: Tapioca is a typical tropical plant. It thrives best in warm humid tropical climate. Tapioca is grown between 30°N and 30°S with an altitude up to 1500 m from MSL in equatorial regions, though the growth is slow and yields are reduced at the highest altitudes. However, it is most productive between latitudes 15°N and 15°S. An annual rainfall of 1000 to 1500 mm is regarded as ideal, but the crop can be successfully grown in areas with rainfall ranging from 500 to 2500 mm. Occasionally it is irrigated if rainfall of less than 750 mm. It can withstand prolonged periods of drought and is therefore a useful crop in areas of low or uncertain rainfall. Tapioca can survive in areas with dry seasons as long as 8 months. Because of such hardiness, farmers in semiarid areas rely on cassava as a 'famine crop'. It is a sun loving crop. The tapioca grows well with a mean solar radiation level higher than 16 MJ/m^2. As it is a short-day plant, more than 12 hours of daylight can cause delays in tubering (starch storage) and eventually low yields while short light periods enhance flowering. Tuber yield is high at 12 hours photoperiod. Development of tapioca is affected by i) long days (greater than 12 hours) during the early growth period result in reduced yield, ii) low temperatures of < 19°C as well as high temperatures of >42°C delay growth of storage roots and iii) drought reduces leaf area index (LAI). Shading for short period of two months has little effect on shoot growth but reduces root growth. It grows poorly in cold climates and growth ceases at temperatures below 10°C. The cardinal minimum, optimum and maximum temperature for tapioca is 12°C, 25 to 30°C and 40°C respectively. Seed germinates with temperature of 24 to 30°C. Sprouting is inhibited at temperature of above 38°C, and decreasing markedly below 17°C. At optimal temperatures of about 30°C, after 5-8 days adventitious roots emerge from the base of the axillary buds and callus forms on the basal end of the original stem cutting. Simultaneously, the axillary buds expand and after ten days leaves begin to appear. Shoot growth is high at 28 to 32°C. The first forking is delayed at temperatures lower than 20°C and temperatures higher than 33°C. It cannot withstand cold or frost. The crop is typically grown in areas that are frost-free all year round. Frost causes necrosis in leaves and may kill the young green stem tissue. Maximum root production occurs within a temperature range of 25 to 32°C. When growth is restricted due to water or nutrient stress, fewer forks are produced and intervals between forks increase.

Soil: The crop grows well in well drained laterite, gravelly and sandy loam soils with good drainage. Heavy and rocky soils are less suitable because these restrict root development. Sandy loam soil with medium fertility is desirable for tapioca cultivation. The crop can also be grown on hill slopes and waste lands of low fertility. The crop cannot survive water logging and adequate measures have to be taken to provide drainage. It can be grown on soils ranging from clay loam soils with a pH of 8 to 9 and to sand or loose laterites with a pH of 5 to 5.5. The desirable soil pH for tapioca is 5.5 to 7.0. It is susceptible to salinity and alkalinity. Its growth is affected at EC 0.7 mmhos/cm and a sodium saturation of 2.5% is desirable. However, it tolerates up to an electrical conductivity of less than 4 dS/m and a sodium saturation of less than 15%. Swampy soils are unsuitable. When grown on clay soils, the plant produces stem and leaf growth at the expense of the roots and many cultivars give poor yields. Tapioca can tolerate soils of low fertility, especially if the feeder roots can penetrate to depths of 40 to 60 cm. Deep cultivation before planting is therefore recommended.

Season: Tapioca can be planted throughout the year under irrigation. Planting during the months of March/April, June-July and September/October are found to be the relatively popular planting seasons in Kerala. The crop is planted during December to February under irrigated conditions while it is planted in July- September as rainfed crop in Tamil Nadu and in June-August in Andhra Pradesh.

Varieties: The particulars of promising tapioca varieties are furnished in Table 1. Important tapioca varieties suited for hilly areas are Duggirula, Tekurpetta, Sugantham, Kodur, Armoor and Alleppey.

Table 1 Tapioca varieties in India

Variety	Duration (months)	Yield (t/ha)	Starch (%)	Special features	Source
CO 1	8.5 -9.0	32	32	Resistant to virus	Tamil Nadu
CO 2	8.5 -9.0	35	35	Suited for starch, oil extraction, resistant to virus	Tamil Nadu
CO 3	8.5	40-42	35.6	Resistant to virus	Tamil Nadu
CO (TP) 4	8.5	50	40	---	Tamil Nadu
Mullivadi (MVD1)	10	32	36	---	Tamil Nadu
M 4	10	20-25	25-27	Erect type, excellent cooking quality	Kerala
H 97	10	25-35	27-29	Semi-brancing type, tolerant to mosaic	Kerala
H 165	8-9	35	25	Non-branching type, susceptible to wilt	Tamil Nadu
H 226	9.5-10	30-35	29	Semi-brancing type, susceptible to mosaic	Kerala
Sri Vishakam (H 1687)	10	33-38	26	---	Madagascar
Sri Prakash (S 857)	7.5	35-40	30	Tolerant to drought, Cercospora leaf spot	Kerala
Sri Jaya (CI 649)	6.5	30-35	30	Direct consumption	Kerala
Sri Vijaya (CI 731)	6-7	40	30	Suited for chips	Kerala
Sri Sahaya (H 2304)	10	35-40	28	resistant to virus diseases	---
YTP 1	9-10	50	25-27	Resistance to cassava mosaic disease	Tamil Nadu
CTCRI - CO Tp) 5 (Sree Padmanabha)	9-10	38	28	Resistance to cassava mosaic disease	Kerala

Growth stages: Tapioca is generally propagated through stem cuttings of 20 cm in length. The axillary buds begin to sprout during the second week of planting. First leaves appear in about 10 days from planting of the stem cuttings. Leaves are usually folded at emergence and expand fully at 10 to 15 days after appearance of leaves. Fully expanded true leaves (seven lobes) could be noticed by about the 30th day after planting. Leaves size increases up to 4 months and then declines. Life of individual leaf is 60 to 120 days depending on environmental factors. The stem cuttings produce a rim of callus on the distal end within a week after planting. Roots are normally produced from the callus in about 10 days after planting and the nodes which are in contact with soil are also able to initiate root primordia. Thick roots (>1 cm diameter) are classified as tuberous roots in tapioca which can be noticed between 45th and 90th day after planting depending upon the cultivars. Root tubers develop by a process of secondary thickening as swellings on adventitious roots a short distance from the stem. Root thickening occurs from 21 to 42 days after planting. The roots may penetrate to a depth of 50 to 100 cm. Root enlargement is high from 5 to 7 months. Starch content increases with enlargement of tubers to a maximum from 7 to 10 months after planting and thereafter starch content decreases and fibre content increases. The total dry matter content of tuber is about 35%. Tuber dry matter yield of 21 t/ha has been obtained in crop duration of one year under optimum growing conditions. A potential yield of 90 t/ha (fresh tuber) in 12 months duration with LAI of 3.5 with 70% harvest index can be obtained. A mature tapioca plant has a distribution 6% leaves, 44% stems and 50% tubers in terms of weight. The flowering may occur as early as the sixth week after planting, although the exact time of flower initiation depends on the cultivar and on the growing environment. Branching cultivars starts flowering on 90 days after planting. Flowering occurs maximum between 11 a.m. and 2 p.m. The period between pollination and fertilization is about 8 hours. The female flowers open about 10 days before the male flowers. Though there is a time lag between male and female phase in the same inflorescence, it is common to find a female and male flowers open at the same time in the different branches of the same plant, particularly in the profusely spreading branching clones. Sometimes hermaphrodite flowers are produced. The fruit is a six winged capsule with septicidal dehiscence. It generally takes about 75 to 90 days for the capsules to mature. The seeds attain maturity in 48 days. There are about 620 capsules per plant. Store the seeds at 26°C for 16 hours and at 38°C for 8 hours for 35 days for breaking seed dormancy. Germination of the fresh seeds is up to 70% and takes about 16 days for germination. Most cultivars are ready for tuber harvest at 8 to 10 months after planting.

Growth stages of Tapioca:

S.No.	Description of stage	Days after planting (DAP)
0	Adventitious roots sprout from buds or cuts	5-7
1	Emergence. First sprout occurs. small leaves emerge. Adventitious roots and first leaves emerge	10-12
2	Seedling. Emergence of all sprouts. the growth of the plant depends on the food stored in the planting material 20 days after planting for its growth	15
3	True leaves begin to expand and photosynthesis begins from 25 to 40 DAP. Leaves can have 3-9 lobes. Within 2-3 weeks, adventitious roots develop at the base of the cuttings. Subsequently, roots develop into a fibrous root system. Fibrous roots may be up to 200 cm long. Fibrous roots start to grow. The root and shoot growth depends on food reserves of planted stem cutting /sett/stake	30
4	Some fibrous roots develop into 'storage roots' or 'tuberous roots' on 30-40 days after planting.	30-40

5	Few fibrous roots become storages roots. Top dressing of fertilizers. Stems, leaves, petioles and tubers accumulated more N,K, Ca, Mg, and S. Hence, cassava requires an application of fertiliser (compound) 60 DAP (setts provide nutrition for the plant in its early growth phase. At 75 DAP the storage roots represent 10–15% of total dry matter. the major sink from 60-90 DAP in the tuberous roots.	60-90
6	Canopy development with maximum growth of leaves and stems. Leaves intercept most of most of incident light on canopy. Most active vegetative growth. Hence any attacks by pests and diseases will reduce yield at this time. Rapid tuber bulking with accelerated translocation of assimilates to tubers. 4-8 of the fiberous roots thicken into storage roots.	90-150
7	Faster storage root bulking/enlargement. 4-8 of the fibrous roots thicken into storage roots. Photoassimilate partitioning from the leaves to the roots is accelerated. Tubers accumulated more K, than N, followed by P, Ca, Mg and S. hence, cassava requires a late application of potassium.	150-210
8	Highest dry matter/starch accumulation in enlarged storage roots. Leaf senescence increases and hastens leaf fall. Stem become lignified.	210-300
7	Maturity. Decreased rate of leaf production. Almost all the leaves fall and shoot vegetative growth is finished. Translocation of sugars and conversion into starch in roots continues. Growth cycle completed in 12 months.	300-360

Tapioca is generally planted as a cutting because seed germination is less than 50%.
Final plant height is about 200-350 cm tall.
LAI of 3-4 is optimal in the tropics
Leaf life ranges from 40 to 210 days, but is commonly 60 to 120 days.
Number of leaves retained per plant is 44 to 146.
The optimum LAI for storage root bulking rate is 3.0 to 3.5.
The peak of LAI is between 4-6 months after planting at which point leaves die back.
Flowering starts 6 weeks after planting with long days > 13.5 hour and temperature of 24°C. The female (staminate) flowers open 1-2 weeks before the male flowers in the same inflorescence. In heavier branching types, male and female flowers may open at the same time at different branching points. Fruit maturation occurs 75-90 days after pollination.
Total dry matter in leaves is more in leaves than stems and storage roots up to 60 to 75 days after planting.
Total dry matter in storage roots is 50 to 60% around 120 days after planting.
Maximum rate of dry matter accumulation occurs in 4 to 7 months.
Harvested long tapered root is 5-10 cm in diameter and 50-80 cm long. Yield is approximately 15-20 t/ha dry root yield and 40-50 t/ha fresh root yield. Harvest index 0.49 to 0.77 can be obtained after 10 to 12 months after planting.

Land preparation: Plough the field three to four times to get a fine tilth. The soil depth should be at least 30 cm. Ridges and furrows are formed with 75 cm × 75 cm and 90 × 90 cm under irrigated condition while 60 cm × 60 cm under rainfed conditions.

Planting: Tapioca is generally planted as a cutting because seed germination is less than 50%. Seed is used only for breeding work. Tapioca stem cuttings are called "stakes". The stems are normally cut when the mother plant is 8-12 months old. Stakes derived from the lower and middle part of the stem has higher germination rates than those derived from the upper part of the stem and 15-20 cm stakes had higher germination than shorter stakes of 5-10 cm length. Tapioca is normally propagated by means of stem cuttings, which are known as 'stakes'. Select healthy, mosaic free vigorous plants for taking planting materials. Storing stem branches vertically is preferable to horizontal storage in a semi-shaded place for 2–3 weeks before preparation of setts/stakes. This makes the stems sprout faster than when they are planted freshly cut from the field. Before storage, soil should be scuffed and dampened so that each stem makes good contact with the soil. Stored stems are moistened regularly, with the surroundings kept free from weeds. Handle the stems with care not to destroy the nodes that may result in losses. Do not make jagged cut surfaces or keep stems in the open (leading to drying). Twelve setts/stakes can be produced in tapioca plant which has 2 to 3 branches per plant @ 5 to 6 stakes per branch in 12 months of crop duration. Stakes taken from primary stems or basal or middle parts give rise to plants with higher yields than those developed from stakes taken from apical parts. Increases in yield in the older stakes may be due to a higher concentration of nutrients. Stem cuttings (sticks/stakes) about 10 to 15 cm or 15 to 20 cm length; 2.5 to 3.75 cm thick and with at least 5 to 7 viable buds, taken from the middle two thirds portion of the stem of 8 to 12 months old plants. As a general rule, the diameter of the stems should be not less than one half the diameter of the thickest part of the stem for the variety being used. Avoid mechanical damage while preparation and handling of setts. The cut end should be uniform. Tapioca multiplication rate is 10. Setts prepared from one ha nursery can be used for planting 10 to 12 ha main field. Planting materials for one ha (10,000 stakes) weigh about 0.7 t and occupy a volume of about 2 m^3.

Tapioca planting materials deteriorate during storage as stems dehydrate, reserves are lost through sprouting, and pests and other pathogens attack. The planting materials can be stored for as long as 6 months. In general, the sticks/stakes can be stored 2 to 8 weeks in cool, well ventilated conditions except when harvested under rainy conditions. Setts storage is normally reduced to 7 to 10 days. The sticks may be cut obliquely or at right angles to the axis.

Hand planting is done in one of three ways: vertical, flat below the soil surface or tilted/ inclined (slanted) at a 45 or 60° angle leaving 2-3 nodes above ground on the ridge or leveled field.

- Vertical method: The stake is pushed vertically and about ½ of its length into the ground. With this method the stake sprouts quicker than with the other two methods, but it produces deeper roots than the horizontal or inclined planting methods.
- Horizontal method: The entire stake is placed horizontally and buried at a depth of 7.5 to 10 cm in the ground. This method produces shallower roots than slanted and vertical planting.
- Inclined method: The stake is placed 2/3 of its length in the ground and at an angle ranging from about 45° to 60°.

Under low rainfall conditions, vertical planting may result in the desiccation of the cuttings, while in areas of higher rainfall; flat-planted cuttings may rot. In general, flat planting 7.5-10 cm below the soil surface is recommended in dry climates and when mechanical planting is used. Planting vertically or slanted generally produces higher yields than planting horizontally, especially during periods of drought, and the vertical method is suitable in sandy soils and under erratic rainfall. In sandy clay loam soils, planting vertically or inclined produced higher root yields than planting horizontally. If the soil condition is loose and friable, stakes can be planted vertically or slanted position by pushing the lower part of the stake about 5-10 cm into the soil. Planting stakes horizontally is common in heavy clay soils or with zero- or minimum-tillage methods of land preparation. When the soil is well prepared and friable, planting vertically or slanted is faster than planting horizontally, but care should be taken that the eyes or buds on the stakes face upward; with horizontal planting this is of no concern. Planting one stake per hill produces high tuber yield as compared to the traditional practice of planting two stakes per hill, slanted in opposite directions. Stem cuttings (sticks) are planted with half their length in the soil, or flat below the surface at a depth of about 10 cm. When sticks are planted vertically or inclined, tubers form only at the extreme end of the cut, forming a slanted cluster often of irregular size; when a stick is cut at right angles and planted vertically, the roots are evenly distributed around the circumference and are more uniform in size. Horizontally planted sticks produce tubers at each node, but lodging of the aerial parts is increased and yields reduced. With vertical or inclined planting, the roots penetrate more deeply and tubes may be formed at intervals along the planted portion, but in areas of low rainfall desiccation of the cutting may occur. Sprouting usually takes place in about a week. Dip the setts in Dimethoate @ 2 ml/litre of water + Carbendazim @ 2 g/litre of water for 15 minutes before planting to control scale insects and mealy bug. Plant the setts vertically with buds pointing upward on the sides of ridges and furrows. For rainfed conditions, treat the setts with 0.5% potassium chloride (5g/lit) for 20 minutes to induce drought tolerance. Dip the setts in super phosphate slurry before planting to ensure maximum root growth. A quantity of 75 kg super phosphate is sufficient to dip the setts required to plant one ha. Prepare super phosphate slurry 2 hours before planting. The super phosphate, soil and water should be mixed in the ratio of 1:1:2 (by weight). The basal portion of the setts should be dipped up to 5 cm (the portion which goes into the soil) and planted immediately after dipping without allowing the slurry to get dried. Dip the setts for 20 minutes in azosprillum @ 30 g/lit and phosphobacteria @ 30 g/lit of water to fix atmospheric nitrogen and to make the soil phosphorus available to plants. Normally erect and non-branching varieties are planted at 75 × 75 cm and branching or semi-branching varieties at 90 × 90 cm. In case, sets are dried after planting, 5% of stakes may be planted as reserve in field, separately at a closer spacing of 4 × 4 cm for gap filling after 20-25 days. The stake's growth into a new plant within 7 to 20 days after planting is exclusively at the expense of the reserves accumulated in the stem. Three weeks after planting, with the appearance of the first leaves and roots, photosynthesis begins to contribute to plant growth.

Cropping systems: Intercrops such as onion, coriander, pulses and vegetables can be grown.

After cultivation: Gap filling with new cuttings is substituted within 15 to 20 days of planting. One stake normally forms one to four shoots. Thin to two shoots per plant on 60th day. Nipping of excess sprouts after retaining two shoots per plant on opposite sides is advocated as it helps in the production of more number of uniform sized roots all-round the base of the plant. Nipping of excess sprouts after retaining two, 2-3 weeks after planting is advocated. High root yield can be obtained with two stems per plant compared to one or three stems per plant. Removal of 25% of lower leaves per plant enhance crop growth rate when top leaves cover the land area or when LAI is higher than 3.5. Earthing up is done on 120 to 150 days after planting to cover the base of the plant and to encourage tuber formation.

Weed management: Critical weed free period is from 60 to 90 days after planting to produce high tuber yield since weeds can reduce tuber yield as much as up to 50%.. The first weeding should be given 20 days after planting, followed with three to four weeding on 40, 60, 90 and 120 days after planting depending on field condition and thereafter the enlarged crop canopy limits weed growth. Weeding after 120 days after planting does not increase root production. Spray Alachlor @ 3.75 lit/ha or Pendimethalin @ 3.75 lit/ha or Flucholoralin @ 2.5 lit/ha mixed in 625 lit of water on third day after planting as pre-emergence using a fan or deflector type nozzle to control weeds. Otherwise Alachlor G @ 20 kg mixed in 50 kg sand can be applied just before planting and irrigation is given immediately after planting the setts (sticks).

Manures and fertilizers application: Tapioca crop producing a tuber yield of 30 tonnes per ha removes 147-32.4-175-55-22.5 kg of N, P_2O_5, K_2O, Ca and Mg. This shows the uptake pattern of nutrients and importance of nutrition to obtain maximum yield. Nitrogen deficiency shows stunted plant growth and yellowing of leaves starting from the bottom. Under severe deficiency conditions, necrosis of leaf tips and margins, and fine cracks in the stems are noticed. Phosphorus deficiency is very common in acid laterite and red soils containing high levels of Fe and Al. Phosphorus deficiency shows reduced growth with smaller leaves and premature leaf falling have been observed. High K applications resulted in Ca and Mg deficiency. Calcium (Ca) supports the supply and regulation of water in cassava plant, while Magnesium (Mg) is a basic component of chlorophyll for photosynthesis. Calcium and magnesium deficiencies are commonly encountered in acid infertile Oxisols and Ultisols. In cassava, Ca deficiency leads to tip burn and deformation of upper leaves, whereas Mg deficiency results in reduction of growth and interveinal yellowing of lower leaves. Sulphur deficiency symptoms first appear on upper young leaves, where fading of normal green colour and uniform yellowing are commonly observed. Zinc deficiency has been observed in acid and alkaline soils. Interveinal chlorosis and narrowing of young leaves are usual symptoms of Zn deficiency. Stunted growth and complete yellowing of the upper part of the plant have been reported to due to deficiency of copper. Iron deficiency has also been observed in tapioca. Fe deficiency can be induced with excess application of Cu, Mn and Zn.

For irrigated crop, apply FYM @ 12.5 to 25 t/ha and incorporate at the time of planting. Combined soil application of Azospirillum @ 4 kg/ha, Phosphobacteria @ 4 kg/ha, *Trichoderma viride* @ 2.5 kg/ha, *Aurbuscular mycorrhiza* (AM Fungi) @ 10 kg/ha and *Pseudomonas fluorescens* @ 2.5 kg/ha are mixed thoroughly in 500 kg of finely powdered FYM at the time of planting to increase the availability of nitrogen, phosphorus, exchangeable potassium, iron, manganese, zinc in soil to crops and also helps to control of pathogenic fungi such as *Fusarium, Phytopthara, Scelerotia*, etc. The same may be repeated during 3rd and 6th months after planting to get maximum benefits. Chemical fertilizers such as N, P_2O_5, K_2O are applied @ 120 : 60 : 175 kg ha^{-1} under irrigated conditions. The N: P: K ratio of 2:1:3, 2:1:3.5 or 2:1:4 are followed in a few places. Phosphorus can be applied as basal dressing or shortly after planting. Split applications of phosphorus @ 50 % as basal and the remaining 50 % on 30 or 60 days after planting is recommended. Nitrogen fertilizers are usually applied in four splits on 0, 60, 90 and 120 days after planting and as five splits @ 24 kg/ha on 0, 30, 60, 90 and 120 days after planting. Neem coated urea improves nitrogen use efficiency. Higher levels of nitrogen tend to increase HCN content of the tubers. K_2O fertilizers are applied as 5 splits @ 35 kg/ha on 30, 60, 90, 120 and 180 days after planting. K_2O fertilizers can also applied as 6 splits @ 30 kg/ha on 30, 60, 90, 120, 150 and 180 days after planting. Fertilizer application 120 days after planting enhances tuber bulking. Nitrogen application increases the tuber weight and yield of tapioca. Fertilizers can be placed 15 to 20 cm from the stem base at 10 to 15 cm depth. The fertilizer requirement through fertigation is 90: 90: 240 kg/ha of N, P_2O_5, K_2O which is carried out once in every three days throughout the cropping period. Gypsum ($CaSO_4$) is recommended @ 250 kg/ha as basal dressing. Application of $MgSO_4$ @ 20 kg/ha, $ZnSO_4$ @ 12.5 kg/ha, $FeSO_4$ @ 12.5 kg/ha, and $CuSO_4$ @ 4 kg/ha as basal dressing at the time of planting is recommended to enhance tuber yield and quality. Foliar spray of $FeSO_4$ @ 1%+ $ZnSO_4$ @ 0.5% at 60 and 90 days after planting 3 to 4 times at weekly intervals. Equal quantity of lime should be added for neutralizing the solution. Spray $MgSO_4$ @ 1% is recommended to correct magnesium deficiency. Application of borax @ 5 kg/ha as basal dressing at the time of planting or spray of borax @ 0.5% is recommended to correct boron deficiency. Spray $CuSO_4$ @ 0.2% is recommended to correct copper deficiency.

For rainfed crop, apply FYM @ 12.5 t/ha; 50-65-125 kg of N, P_2O_5 and K_2O per ha as basal and 2 kg of Azotobacter as soil application on 30 to 60 days after planting on receipt of showers (2.0 kg Azotobacter + 20 kg FYM + 20 kg soil per ha).

Water management: The crop's water requirement ranges from 400 to 750 mm for a 300-day production cycle. But higher yields have been obtained with much higher levels of water supply. The maximum root yields can be achieved with rainfall totaling about 1700 mm during the 4th to 11th month after planting. It responds well to irrigation. It is also susceptible to *excess* water if the soil becomes water-logged, sprouting and early growth is affected and yields fall. Tapioca is a drought hardy plant. It has the ability to withstand fairly prolonged periods of drought. However, the crop is very sensitive to soil water deficit during the first three months after planting. Stakes will only sprout and grow well when the temperature is above 15°C and the soil moisture content is at least 30% of field capacity. Water stress at any time in that early period reduces the growth of roots and shoots, which impairs subsequent development of the storage roots, even if the drought stress is alleviated later. It is susceptible to excess water. Standing water for 24 hours may kill the crop. Water logging greatly reduces growth. First irrigation is given at the time of planting. Life irrigation is given on the third day followed by once in 7 to 10 days from third to seventh month and once in 20 to 30 days from eighth month to till harvest. Total number of irrigation is 35 to 40 in a ten month crop. Irrigation at 0.7 to 1.0 IW/CPE during the dry period resulted in 62 to 100% increase in tuber yield. Water stress during the period between 30 and 150 days after planting is most critical and water stress after six months of planting does not significantly reduce the tuber yield. Irrigation approximately at 75% available soil moisture throughout the growing season increases the tuber yield.

Plant protection: Select the planting materials from healthy plants. Soaking spray with Monocrotophos or Dimethoate @ 0.05 % at monthly intervals or Dicofol 18 EC @ 2.5 ml/l during third and fifth month will control mite infestation. For the control of vectors, spray neem oil 3 % or fish oil rosin soap 25 g/l or Methyl demeton 25 EC @ 2 ml/l for four times in 30 days interval commencing from 75th day. Mancozeb is sprayed @ 2 kg/ ha to control leaf spot. While using neem oil, teepol or sandovit should be added at 1 ml/l for better contact with foliage. Whitefly can be controlled with spray Dichlorvos 76 WSC @ 1 ml/l or Triazophos 40 EC @ 2 ml/l. Add wetting agent. *Cercospora* leaf spot can be controlled with spray Mancozeb @ 2 g/l twice at 15 days interval. Avoid water stagnation. Give good drainage facilities. Spot drench with Copper oxychloride 2.5 g/l or apply through soil *Trichoderma viride* @ 2.5 kg/ha as basal and at 3rd and 6th month after planting to control Tuber rot. Papaya mealy bug infestation in tapioca can be controlled with release of mealy bug parasitioid (*Acerophagus papayae*) @ 250 Nos./ha. Tapioca mealy bug can also be controlled with 300 male and female pairs of *Anagyrus lopezi* wasp per hectare. *Anagyrus lopezi*, the natural enemy of pink mealy bug, is native of South America. Australian lady beetle/grub *Cryptolaemus montrouzeri* (also known as mealy bug destroyer) can be released @ 10,000 per ha. Both adults and larvae kill mealy bugs. Single grub can feed 900-1500 eggs or 300 nymphs or 30 adults in its lifetime. Other predators like Green lace wing, *Chrysoperla carnea* and Lepidoptera predator, *Spalgius epius* are found effective. Parasitoids like *Anagyrus dactylopii* are found parasitizing mealy bug up to 70%. The tapioca mealy bug can be controlled by soaking of plant material with Thiamethoxam 25% WG (4 gm/lit of water), Thiamethoxam 35% FS (3 ml/lit of awter), Imidacloprid 70% WG (4 gm/lit of water), Imidacloprid 60% FS (5 ml/lit of water), Clothianidin 16% SG (30 gm/litre of water),

Harvest: Harvest early varieties at 7-8 months, medium varieties at 8-9 months and long duration varieties at 9-10 months after planting. The plants are normally topped two weeks before harvesting, leaving only 8–10 inches of the stem protruding from the ground. This allows the sugars in the tubers to be converted to starches, resulting in a cassava that will not deteriorate or rot as easily. 5 to 10 tubers are produced at the base of the stem having a length of 30 to 50 cm and thickness of 9 cm to 10 cm. Cut the plant at about 30 to 50 cm above the ground and the roots are lifted out of the ground by pulling the stem gently. Do not drag the roots. Dragging can cause bruises and cuts which may lead to early deterioration. If the soil is compact, loosen it but take care not to damage the roots. Separate the roots from the stem using a sharp knife or cutlass. Cut each root near to the stem. Do not break the roots from the stump by hand. This will cause injuries which lead to root rot. After harvesting, do not leave the roots under the sun. Too much heat causes weight loss and tubers deteriorate rapidly. The manual method of harvesting usually requires 40-60 persons for one ha in one day depending on the season. There are two distinct types of deterioration which occur during storage, one physiological and the other due to microorganisms. Physiological deterioration begins to appear within three days. It is essentially a humidity sensitive wound response with increase in enzyme activity leading to the production of phenols including *catechins* and *leucoanthocyanidins*. This leads to discolouration and forms condensed tannins. Visible signs of discolouration are at first blue, becoming brown which initially appear in the peripheral vascular bundles and spread to adjacent parenchyma. Physiological deterioration appears to be connected with enzyme activity. It can be controlled through withholding the roots at low temperature, or storing in air with low oxygen levels or in carbon dioxide. Mechanical damage to the roots also permits the entry of microorganisms, causing internal rotting. Pre-pruning of aerial portions of the plant 2 to 3 weeks before harvesting minimizes the physiological deterioration in the tubers. Coating with a fungicidal wax extends the storage life up to 1 to 2 tubers at 0 to 20°C and 85 to 90% RH is reported as satisfactory for periods of about 4 weeks. When the root is used as food, the best time to harvest is at about 8 to 10 months after planting; a longer growing period generally produces a higher starch yield. Tapioca roots are highly perishable and become inedible with 24 to 72 hours after harvest due to rapid physiological deterioration.

Yield: Tuber yield is 40-50 and 20-25 t/ha under irrigated and rainfed conditions respectively. Tubers are dark coloured, fleshy and cylindrical, varying in size and shape, with a starch content of 20 to 40%. Each plant normally yields 5 to 10 tubers, usually 30 to 45 cm long, with a diameter of 5 to 15 cm and weighing 0.9 to 2.3 kg. Tapioca roots are more than 60 percent water. However, their dry matter is very rich in carbohydrates, amounting to about 250 to 300 kg for every tonne of fresh roots. The peel accounts for 10 to 20% of the tuber and consists of an outer corky rind and an inner part which separates the peel from the flesh of the roots. Harvest index varies from 40 to 70% and the optimum is considered to be 60%. Tubers can be treated with Thiobendazole @ 0.4% solution and packing than in polythene bags to prevent spoilage up to three weeks.

Starch production: The mature roots are first washed to remove dirt and loose soil. The roots are then peeled to remove the skin and cortex. The roots are next sliced and put through a rasping or grating machine to produce a slurry or pulp. The slurry is then sieved to separate the fibrous tissue from the starch milk. Clean water is used to ensure efficient separation of the starch granules from the slurry. The surface layer of the starch mass is usually a yellowish green colour and contains impurities. It is therefore scraped off, leaving a creamy white mass below. Then it is stirred vigorously with water and then left to settle. This washing and settling process is repeated once or twice more until the starch is judged to be sufficiently pure. The starch cake is dried, either by spreading it out in trays in the sun or in factories in hot air driers. Finally, the hard lumps of starch are crushed into a powder and sieved. Since tapioca roots are relatively tough, the grinding process must be efficient in order to liberate all the starch granules. The commercial starch extraction rate is approximately 20 to 25% of the tubers.

Utilization: Tapioca is the staple food of the poorer section of the population of many tropical countries. The fresh peeled tubers are eaten as a vegetable after boiling or roasting. For edible purposes, cultivars with a high starch and protein content and a low HCN content are preferred. For starch manufacture, cultivars with high starch content are favoured and HCN content is of less importance. Tapioca tuber is used for manufacture of starch which is used in the foodstuff, textile and paper industries. The tapioca starch (amylose) content is about 17% compared with 22% for potato starch and 27% for maize starch. Good quality starch should be absolutely free from specks and have a pure white colour, a pH of 4.7 to 5.3, a moisture content of 10.0 to 13.5% and an ash content of less than 0.2%. The tuber constitutes 59.4% moisture, 0.31 to 1.0% protein, 0.1 to 0.4% fat, 38.1% carbohydrate, 0.6% fibre and 0.8 to 1.0% ash. Dried roots are being used for livestock feeding. The fried chips are used as snack foods. Tapioca flour is used in the preparation of bread, biscuits and confectionery. The flour can be stored up to six months, if kept free from moisture. Glucose is produced from tapioca in Kerala. Tapioca meal accounts 10 to 20% which is left after the extraction of starch. The tapioca meal can be used as a livestock feed, or as a raw material for the production of adhesives. Mature leaves have a protein content of 5 to 7% and can be used for animal feeding. Tapioca stems are used for the manufacture of particle board.

Toxic principles: Tapioca roots contain the cyanogenic glycosides (CNG) viz., linamarin and lotaustralian. Both the glucosides on hydrolysis releases the toxic meatabolite hydrocyanic acid or prussic acid (HCN) by the enzyme linamarase. HCN occurrence vary between plant parts (leaf >rind >flesh). The varieties are classified into bitter and non-bitter tubers which are primarily due to the bitter principle, *hydrocyanogenic glucoside* (HCN). In the non bitter clones, the HCN may be as low as 5 ppm or even non-detectable and in the bitter clones it is reported up to 495 ppm. The concentration of HCN above 100 µg/g of fresh tuber has been reported to be highly toxic. HCN confined to rind in sweet tapioca while widely distributed in bitter tapioca. HCN is activated when tubers are dug and wilt. Stale tubers are therefore poisonous. HCN content varies from 10 to 490 mg/kg of tuber. HCN amount varies from 10 to 370 mg/kg of fresh tuber. HCN content less than 50 mg is considered vivocuous, 50 to 100 mg is moderately poisonous and over 100 mg is highly poisonous. The mean value of hydrocyanic acid is 185 mg/g in tuber skin and 15 mg/g of tuber flesh. Cooking of tuber in water for five minutes removes more than 5% of the cyanide. Cyanogenicglucosides concentrations <50ppm are safe. Normal cooking process reduces the cyanide still further. Sun drying of the tuber slice destroys around 75% of the total cyanide. The cyanide content in leaves can be reduced to the safer limits by cooking, steaming and sun drying. Peeling and cooking gives a partial detoxification, but soaking of roots for long periods, repeated boiling with change of water, roasting, fermentation or fermentation followed by heat treatment, are all used for detoxification of hydrocyanic acid.

REFERENCES

Alves, A.C. 2002. Cassava botany and physiology. In: R.J. Hillocks, J.M. Thresh and A.C. Bellotti (eds.). Cassava: Biology, Production and Utilization. CABI Publishing, New York, USA. pp.67-89.

Anon. 1990. Cassava in Tropical Africa: A Reference Manual. International Institute of Tropical Agriculture. Ibadan, Nigeria.

Anon. 2010. Good agricultural practices for cassava. National Bureau of Agricultural Commodity and Food Standards. Ministry of Agriculture and Cooperatives. Bangkok 10900.

Anon. 2015. Cassava Handbook. Department of International Cooperation. Ministry of Agriculture, Forestry and Fisheries, Phnom Penh, Cambodia. p31-33.

Adebayo, B. Abass, Elifatio Towo, Ivor Mukuka, Richardson Okechukwu, Roger Ranaivoson, Gbassey Tarawali and Edward Kanju. 2014. Growing cassava: A training manual from production to postharvest. IITA, Ibadan, Nigeria.

Allem, A.C. 1994. The origin of *Manihot esculenta* Crantz (Euphorbiaceae). *Genet. Res. Crop Evol.* 41:133-150.

Allem, A.C. 1999. The Closest Wild Relatives of Cassava (*Manihot esculenta* Crantz), Euphytica. 107: 123 – 133.

Anon. 2015. Cassava Handbook. Department of International Cooperation. Ministry of Agriculture,

Forestry, and Fisheries, Phnom Penh, Cambodia. P31-33.

Ashokan, P.K. and C. Sreedharan. 1977. Influence of levels and time of application of potash on growth, yield and quality of tapioca (*Manihot esculenta* Crantz). J. of Root Crops 3: 1-4.

CIAT. 2011. The Cassava Handbook. A Reference Manual based on the Asian Regional Cassava Training Course, held in Thailand. Centro Internacional de Agricultura Tropical, Cali, Colombia.

Cock, J,H. 1985. Cassava: new potential for a neglected crop. Westview Press, Boulder, CO, USA.

Connor, D.J. and J.H. Cock. 1981. Response of cassava to water shortage. II. Canopy dynamics. Field Crops Res. 4: 285-296.

CPG. 2013. Crop Production Techniques of Horticultural Crops. Horticultural College and Research Institute, Tamil Nadu Agricultural University, Coimbatore – 641 003

Edison, S., Anantharaman, M. and T. Srinivas. 2006. Status of cassava in India: an overall view. Central Tuber Crops Research Institute, ICAR, Sreekariyam, Thiruvananthapuram, Kerala, India.

El-Sharkawy, M.A. 1993. Drought tolerant cassava for Africa, Asia and Latin America. Bioscience, 43: 441-451

Girija Ganeshan and A. Manoj Kumar. 2005. *Pseudomonas fluorescens*, a potential bacterial antagonist to control plant diseases, Journal of Plant Interactions, 1:3,123-134

Howeler, R. 1981. Mineral nutrition and fertilization of cassava. Centro Internacional de Agricultura Tropical (CIAT). Cali, Cclombia, 52p.

Howeler, R.H., Ezumah, H.C. and Midmore, D.J. 1993. Tillage systems for root and tuber crops in the tropics. Soil Tillage Res., 27: 211-240.

Howeler, R.H. 1996. Mineral Nutrition of Cassava. In: Craswell, E.T., Asher, C.J. and O'Sullivan, J.N. (eds.) Mineral Nutrient Disorders of Root Crops in the Pacific. Workshop Proceedings, Nuku'alofa, Kingdom of Tonga, 17-20 April 1995. ACIAR, Canberra, Australia. p. 110-116.

Howeler, R.H. 2001. Cassava agronomy research in Asia: Has it benefited cassava farmers? In: R.H. Howeler and S.L. Tan (Eds.). Cassava's Potential in Asia in the 21st Century: Present Situation and Future Research and Development Needs. Proc. 6th Regional Workshop, held in Ho Chi Minh city, Vietnam. Feb 21-25, 2000. pp. 345-382.

Howeler, R.H. 2002. Cassava mineral nutrition and fertilization. In: Hillocks, R.J, Thresh, J.M. and Bellotti, A.C. (Ed.). *Cassava: Biology, Production and Utilization.* CABI, Wallingford, UK.

IITA. 2014. Cassava training manual on Rapid Multiplication of Cassava Stems. International Institute of Tropical Agriculture. Ibadan, Nigeria.

IITA. 2015. Farmers' Guide to Cassava Cultivation: Frequently Asked Questions (FAQ). International Institute of Tropical Agriculture. Ibadan, Nigeria

Indira J. Ekanayake, David S.O. Osiru and Marcio C.M. Porto. 1997. Agronomy of cassava. IITA Research Guide 60. IITA, Ibadan, Nigeria.

Irikura, V., J.H. Cock and K. Kawano. 1979. The physiological basis of genotype-temperature interactions in cassava. Field Crops Res. 2: 227-239

Javier López. 2012. Cassava Planting Materials. pp 91-112. In: Bernardo Ospina and Hernán Ceballos (eds). Cassava in the third millennium: modern production, processing, use, and marketing systems. Centro Internacional de Agricultura Tropical (CIAT); Latin American and Caribbean Consortium to Support Cassava Research and Development (CLAYUCA); Technical Centre for Agricultural and Rural Cooperation (CTA).

Kay, D.E. 1973. TPI Crop and Product Digest, 2. Root Crops. Trop. Prod. Inst., London.

Keating, B.A. and J.B. Evenson. 1979. Effect of soil temperature on sprouting and sprout elongation of stem cuttings of cassava. Field Crops Res. 2: 241-252.

Keating, B. 1982. Environmental effects on growth and development of cassava (*Manihot esculenta* Crantz) with special reference to photoperiod and temperature. *Cassava Newsl.* 10:10-12.

Kenneth, M. Olsen. and Barbara A. Schaal. 1999. Evidence on the origin of cassava: Phylogeography of Manihot esculenta. Proc. Natl. Acad. Sci. USA Vol. 96, pp. 5586-5591

Kumar, M.B., R.C. Mandal and M.L. Magoon. 1971. Influence of potash on cassava. Indian J. Agronomy 16: 82-84.

Mark E. Halsey, Kenneth M. Olsen, Nigel J. Taylor, and Paul Chavarriaga-Aguirre. 2008. Reproductive Biology of Cassava (*Manihot esculenta* Crantz) and Isolation of Experimental Field Trials. Crop

Science, 48: 49-58

Muthuswamy, P., and K.C. Rao. 1979. Influence of Nitrogen and Potash Fertilization on Tuber Yield and Starch Production in Cassava (Manihot esculenta Crantz) Varieties. Potash Review. Subject 27: Tropical and Subtropical Crops No.6/1979, 91st suite. International Potash Institute, Switzerland.

Mutsaers, H.J.W., Ezuma, H.C. and Osiru, D.S.O. 1993. Cassava-based intercropping: A review. Field Crop Res., 34: 431-457

Nassar, N. M. A. and R. Ortiz. 2007. Cassava improvement: challenges and impacts. Journal of Agricultural Science. 145: 163–171

Nayar, T.V.R., Mohankumar, B. and Pillai, N.G. 1985. Productivity of cassava under rainfed and irrigated conditions. J. Root Crops, 11(1-2): 37-44

Olsen, K.M. and Schaal, B.A. 2001. Microsatellite variation in cassava (*Manihot esculenta*, Euphorbiaceae) and its wild relatives: further evidence for a southern Amazonian origin of domestication. Am. J. Bot. 88 (1): 131–142

Onwueme, I.C. 1978. The tropical tuber crops: yams, cassava, sweet potato, and cocoyams. Wiley, New York

PhilRootCrops. 2011. Technoguide on Cassava Production. Department of Agriculture. Agriculture and Fisheries Information Service. Elliptical Road, Diliman, Quezon City 1100. Philippines.

Reinhardt Howeler, 2014. Sustainable soil and crop management of cassava in Asia. Centro Internacional de Agricultura Tropical (CIAT), CIAT Publication No. 389. Cali, Colombia. 280 p.

Susan John, K., G. Suja, M.N. Sheela, and C.S. Ravindran. 2010. Potassium: The Key Nutrient for Cassava Production, Tuber Quality and Soil Productivity - An Overview. Journal of Root Crops 36:132-144.

Susan John, K., Ravindran, C.S., James George, Manikantan Nair, M. and G. Suja. 2013. Potassium: A Key Nutrient for High Tuber Yield and Better Tuber Quality in Cassava. Better Crops. South Asia. 26-27.

Tarawali, G., Ilona, P., Ojiako, I.A., Iyangbe, C., Ogundijo, D.S., Asumugha, G. and U.E. Udensi. 2013. A comprehensive training module on competitive cassava production. IITA, Ibadan, Nigeria. 40 pp.

Tongglum, A., P. Suriyapan and R.H. Howeler. 2001. Cassava agronomy research and adoption of improved practices in Thailand – Major achievements during the past 35 years. In: R.H. Howeler and S.L. Tan (Eds.). Cassava's Potential in Asia in the 21st Century: Present Situation and Future Research and Development Needs. Proc. 6th Regional Workshop, held in Ho Chi Minh city, Vietnam. Feb 21-25, 2000. pp. 228-258.

https://en.wikipedia.org/wiki/Cassava

http://agritech.tnau.ac.in/horticulture/horti_vegetables_tapioca.html

https://hort.purdue.edu/newcrop/CropFactSheets/cassava.html

http://www.iita.org/

https://www.uoguelph.ca/plant/courses/pbio-3110/documents/Cassava_08.pdf

http://agridr.in/tnauEAgri/eagri50/HORT281/pdf/lec24.pdf

https://www.researchgate.net/publication/227339792

http://www.oisat.org/downloads/field_guide_cassava.pdf

http://biblio.iita.org/documents/U90ManIitaCassavaNothomNodev.pdf-52ca5ad61b1fba9127fc40 6b6a6982a2.pdf

http://www.fao.org/agriculture/crops/agp-home/en/

CHAPTER

31

Yam / Dioscorea–*Dioscorea spp.*

Family: Dioscoreaceae

This genus Dioscorea belonging to family Dioscoreaceae with over 60 species is widely distributed in tropical world. The edible yams/tubers are derived mainly from about ten species. The most economically important species are:

(i) White yam (*Dioscorea rotundata* Poir): It is native of Africa and is the most widely grown and preferred yam species. The tuber is roughly cylindrical in shape, the skin is smooth and brown and the flesh usually white and firm.

(ii) Asian greater yam / Water yam (*Dioscorea alata* L.): It is origin of Indo-Malayan center, South East Asia. The common names are water yam or winged yam. It is the species most widely spread throughout the world and in Africa. It is second to white yam in popularity. It is cultivated in India, Papua New Guinea, Indonesia, South Pacific islands, West Africa, Caribbean and South America. *D. alata* is a glabrous vine that climbs by twining to the right. The length of the vine varies from 2 to 30 m. The fleshy stems are characterized by wings. The stems may be quite thorny, in which cases the thorns occur in straight lines representing the wing. Wings are believed to help the stem grasp smooth objects in twining. The leaves are large, glabrous and opposite on the stem. Male and female flowers are borne on different plants. The capsules, which develop rapidly after pollination, reach 20 to 30 mm in length. The capsules mature, dry, and split along the sutures of the wings to release two seeds from each of the three locules. *D. alata* is a polyploidy with 2n= 30, 50, or 70. The tuber shape is generally cylindrical, but can be extremely variable. Tuber flesh is white and watery in texture. It normally grows for 8 to 10 months, then goes dormant for 3-4 months, with aerial stems dying back during dormancy. Stems to 10 m or more in length, freely branching above; internodes square in cross section, with corners compressed into 'wings', these often red-purple tinged. Aerial tubers (bulbils) formed in leaf axils, elongate, to 10 × 3 cm, with rough, bumpy surfaces. Leaves long petioled, opposite (often with only 1 leaf persistent); blades to 20 cm or more long, narrowly heart shaped, with basal lobes often angular. Flowers small, occasional, male and female arising from leaf axils on separate plants (i.e., a dioecious species), male flowers in panicles to 30 cm long, female flowers in smaller spikes. Fruit a 3-parted capsule; seeds winged.

(iii) The lesser yam (*Dioscorea esculenta* Lour. Burk) (2n=40): It is origin of Indochina and in adjacent islands, including Papua New Guinea and those of the Pacific Ocean. *D. esculenta* has domesticated in Myanmar or Indochina (Indian center of origin). It is cultivated in India, China, West Africa, Papua New Guinea and other Southeast Asian countries. *D. esculenta* (Lour.) Burk. has also been named *D. aculeata* L. and *Oncus esculenta* Lour. It is often called the lesser yam, the potato yam, or the Chinese yam. *D. esculenta* is a thorny, climbing vine, seldom more than 3 m high. The stems are thin (1 to 3 mm diameter) and thorny to glabrous. The leaves are alternate (or occasionally subopposite at the base of the stem. The petioles are thickened at their proximal end and are armed with sharp prickles. The roots of *D. esculenta* often bear thorns, which may impede depredations by burrowing animals and wild pigs. These sharp thorns arise only from the fibrous roots and not from tuber-bearing stolons. The tubers appear to be tuberized roots, but are swellings of stolons arising from the crown of the plant. The stolons vary in length from 5 to about 50 cm. Each stolon bears only one terminal tuber. About 4 to 20 tubers are borne by each plant. The tuber is usually shaped like a potato.

(iv) Yellow yam (*Dioscorea cayenensis* Lam): It derives its common name from its yellow flesh, which is caused by the presence of carotenoids. It is also native to West Africa and very similar to the white yam in appearance. The tuber skin is firm and less extensively grooved, the yellow yam has a longer period of vegetation and a shorter dormancy than white yam.

(v) Bitter yam (*Dioscorea dumetorum*): It is also called trifoliate yam because of its leaves. It is native of Africa where wild cultivars also exist. One marked characteristic of the bitter yam is the bitter flavour of its tubers. Another undesired characteristic is that the flesh hardens if not cooked soon after harvest. Some wild cultivars are highly poisonous.

Climate: It requires warm and humid conditions with a mean temperature of 30ºC and a well distributed annual rainfall of 1200 to 2000 mm.

Soils: Sandy loam soil with a pH of 6.0 to 6.5 is preferred with good drainage and cool weather.

Season: May-June is suitable for planting.

Varieties:

Peruvalli (*Dioscorea alata*) : Co 1, Sree Roopa, Sree Keerthi, Sree Shilpa

Siruvalli (*Dioscorea esculenta*) : Sree Latha, Sree Kala

Seed rate: Use mature tubers or pieces of 250 to 300 g tubers taken from the previous crop as seed material at the rate of 1875 to 2500 kg/ha. Mini setts of 25 g are recommended for planting directly in the field or raising a nursery and planting plants after 60 days. For planting lesser yam, medium sized tuber of 100 to 150 g is sufficient. Planting is done in beds or in ridges or in mounds or in rows 75 cm apart either way.

Land preparation: Plough the field to a fine tilth and form ridges and furrows at 75 cm spacing for raising lesser yam. Ridges and furrows at 75 cm spacing or pits of 45 cm × 45 cm × 45 cm for planting greater yam at 90 cm × 90 cm. Fill the pits with top soil and FYM.

Planting: To plant greater yam, 3000-3700 kg and for lesser yam 1800-2700 kg of seed material is required. Mini setts weighing 25 g are recommended for planting directly in the field or raising a nursery and transplanted after 60 days. Planting is done in beds or in ridges or in mounds or in rows 75 cm apart either way.

After cultivation: The vines should be trained on bamboo poles. Weeding can be done as and when necessary.

Trailing: It is done within 15 days after sprouting by coir rope attached to artificial supports in the open area or to the trees where it is raised as an intercrop. Trailing is necessary to expose the leaves to sunlight.

Manures and fertilizers application: Apply FYM @ 25 t/ha at the time of last ploughing. Apply fertilizer dose of 40:60:120 kg of N, P_2O_5, K_2O per ha as basal. Apply 4 kg/ha of *Azospirillum* (mixed with 40 kg of soil) 30 days after planting. Apply 50 kg N and 120 kg K_2O per ha on 90 days after planting.

Water management: Copious watering once in a week is necessary.

Cropping systems: It can be intercropped profitably in coconut, arecanut, rubber, banana and robusta coffee plantations at a spacing of 90 × 90 cm. In Robusta banana + Dioscorea system, banana should be manured at the full recommended dosage and for yams, manuring at the 2/3rd recommended level is sufficient.

Pests and diseases management: Yam scale is found to occur on the tubers both under field and storage conditions. As a prophylactic measure, dip the planting material in Monocrotophos @ 0.05% or Methyl demeton @ 0.25%. Use scale free seed tuber for planting.

Harvest: Greater yam and white yam become ready for harvest by 9-10 months after planting. Lesser yam takes 8-9 months for attaining maturity. Carefully dig out the tubers without causing injury.

Yield: It can yield about 20 to 25 t/ha in 240 days of tubers.

REFERENCES

Anon. 1974. Tropical Yams and their potential. Part 1. *Dioscorea esculenta*. Agriculture Handbook No. 457. Agricultural Research Service, United States Department of Agriculture, in cooperation with U.S. Agency for International Development, Washington, D.C.

Anon. 1976. Tropical Yams and their potential. Part 3. *Dioscorea alata.* Agriculture Handbook No. 495. Agricultural Research Service, United States, Department of Agriculture, in cooperation with U.S. Agency for International Development, Washington, D.C.

Coursey, D.G. and Haynes, P.H. 1970. Root crops and their potential as food in the tropics. World Crops. July/Aug, 1970.

CPG. 2013. Crop Production Techniques of Horticultural Crops. Horticultural College and Research Institute, Tamil Nadu Agricultural University, Coimbatore – 641 003

http://guides.library.manoa.hawaii.edu/paccrops/yam

http://agridr.in/tnauEAgri/eagri50/HORT282/pdf/lec21.pdf

CHAPTER

32

Aroids

Aroid is a generic term that includes the edible species of the family Araceae, notably the genera, Colocasia, Alocasia, Amorphophallus, Cyrtosperma, and Xanthosoma. Araceae is sometimes known as the Philodendron or Arum family. The aroid plants are grown for edible corms which are used as food, medicine, animal fodder, ornamental plants and cut flowers. Aroids are the world's oldest food crops, and are the most widely distributed starchy food plants. The main centres of origin and diversity of aroids are tropical Asia and tropical America. The five most important cultivated edible aroids, used as food are:

(i) Elephant ear (*Alocasia macrorrhiza*)
(ii) Elephant foot yam (*Amorphophallus campanulatus*)
(iii) Swamp taro (*Cyrtosperma chamissonis*)
(iv) Taro or cocoyam or Dasheen (*Colocasia esculenta*)
(v) Tannia or New Cocoyam (*Xanthosoma sagittifolium*)

Taro and tannia are the most widely distributed and consumed aroid. The aroids produce edible starchy storage corms or cormels. All plant parts of aroids are eaten. The mature corms and young shoots of edible aroids are mostly used as boiled vegetables, but the corms are also roasted, baked, or fried. Roasted or boiled corms can be eaten alone or with stew. Tannia is used in small quantity as soup thickener after boiling and pounding to obtain a consistent paste. The roots and tubers are rich in carbohydrates, vitamins and minerals. The leaves, stems and petioles are frequently eaten as a green vegetable and represent an important source of vitamins, especially folic acid. These are important food crops in India, Southeast Asia, Africa, Latin America, the Caribbean and the Pacific Islands.

1. Elephant ear–*Alocasia macrorrhiza* (L.) G. Don.

Family: Araceae.

Vernacular names: Giant taro, elephant ear taro (English); Birah, Alavu, Alooku, Alu, Manaka, Mankachu, Mankanda (India).

Importance: It is grown as a subsistence food crop or for animal feed or ornamental crop.

Origin: Giant taro probably originated in Sri Lanka or India.

Distribution: It is grown in Southeast Asia, Malaysia, Indonesia, Solomon Islands and parts of tropical America.

Botany: Elephant ear taro is a tall succulent herbaceous plant, reaching 4.5 m in height, with a thick cylindrical stem arising from a basal corm. It is large-leaved aroid, with erect, bluntly triangular leaves. The sap contains needle-like crystals of calcium oxalate, and is well-known for being irritating (itch-inducing) to the skin. Corms are stem-like, mostly growing above ground, with only the bottom six inches or so rooted in soil. The corm is about 1 m long, 15-20 cm diameter and weighs up to 18 kg. The leaves are borne on long petioles. The blades are large, heart- or arrow-shaped leaves. The blades point upwards forming a straight line with the petiole. The leaf blades have a conspicuous midrib, raised on the upper surface, and grow up to about 1 m in length. They are usually green, but there are variegated forms which are blotched or mottled with white. The spathe has a glaucous, yellowish green blade. Cormels are formed around the basal corm. The plant contains latex.

Climate: It is found in humid tropical and subtropical climates. It is grown up to an altitude of 600 m from MSL. It is cultivated in areas receiving mean annual rainfall of 1,500 mm to 3,000 mm. It is frequently found naturally along river banks, but cannot stand waterlogging. It can tolerate up to 4 months of drought. It grows in areas with mean annual temperature ranges from 23°C to 31°C. The temperature below 10°C is detrimental to growth.

Soils: It grows well in well drained loamy to clay loamy soils.

Plant protection: It is a very hardy plant that is resistant to most pests and diseases.

Harvest: The crop life is usually 12 to 18 months, but harvesting can be delayed for up to four years. The plant may be allowed to grow and produce corms weighing about 18 kg. The giant taro is dug by hand. The plant ready for harvest can remain in the ground as 'field-stored' for about 3 months after reaching maturity without any deterioration. The flesh, which may be purplish, yellow or white depending on the cultivar, is edible after being boiled.

Yield: It produces 7 to 11 t/ha per crop (1.8-2.7 kg/plant) when harvested at 11 months. Theoretically, yields for pure stands could be almost 200 t/ha.

Utilization: The leaves as well as the stem are eaten. The stem tuber is peeled, cut into pieces and eaten as a vegetable after cooking, usually in curries or stews. Older stems may require prolonged cooking with several changes of water to remove acridity. The *Alocasia* corms contain oxalic acid crystals, and must be boiled for a prolonged period before becoming edible.

2. Elephant foot yam–*Amorphophallus campanulatus* Blume; (2n=28)

Synonym: *A. paeoniifolius*

Family: **Araceae**

Vernacular names: Elephant yam, Elephant bread, Elephant foot yam, Suran, Sweet yam (English), *Suranakanda, Suran, Jamikand, Zamim-kand, Gimmikand* (Hindi), *Suranah, Balukand* (Sanskrit); *Chena* (Malayalam), *Suvarna gadde* (Kannada); *Karnai-kilangu* (Tamil); *Kanda gadda* (Telugu); *Suran* (Marathi); *Ole* (Bengali).

Importance: Elephant foot yam is basically an underground stem tuber. It is a popular tuber crop and is grown as a vegetable in many parts of India. It is sometimes fed to pigs.

Origin: It is originated in India.

Distribution: It is grown in Philippines, Malaysia, Indonesia, Sri Lanka, India, Bangladesh and the South-East Asian sub-continent. In India, it is cultivated in Andhra

Pradesh, West Bengal, Gujarat, Kerala, Tamil Nadu, Maharashtra, Uttar Pradesh and Jharkhand.

Botany: A robust herbaceous plant, with an erect solitary stem usually 1-2.5 m in height and bearing at the top one or two tripartite leaves. Inflorescence is produced towards the end of the plant's cycle, usually 4-6 years. The inflorescence consists of a short stalk and spathe and a spadix, which emits a malodorous smell, reminiscent of rotten meat. Flowering is erratic. There is little or marginal overlapping of male and female flowering and female accessions flower earlier than male accessions. Male flowers are produced in terminal and axillary panicles. Female flowers are borne on long axillary spikes, 1-2 spikes arising per axil. Pollinations are carried out as and when the corms produce spadics, which is always a single spadix per corm. The flowers are extremely protogynous, stigma receptivity lasting only 24 to 78 hours before spathe opening. Pollen is yellowish in colour and sticky with more than 90% fertility. It is a cross pollinated crop and dioecious in nature. Fruit or seed set is not observed under natural condition. The corms are large globose depressed tubers, usually dull-yellow or brownish-yellow in colour. The plant produces 5 to 10 cormels at the end of each growing season.

Climate: It is a tropical and sub-tropical crop. It thrives well under warm humid climate. It is grown in areas receiving a well distributed rainfall of 1000 to 1500 mm spread over a period of 6–8 months. The crop performs well under irrigated condition and hence could be grown even in areas where the rainfall is scanty but has assured irrigation. It requires an average temperature of 25-35°C, preferably fairly equable during its growing period.

Soils: It grows well on a variety of soils but a well drained deep sandy loam or sandy clay loam soil with a near neutral soil pH 5.5 to 7.0. The soil should be rich in organic matter with adequate amount of available plant nutrients. The crop cannot stand waterlogging and heavy clays are therefore unsuitable.

Varieties: Gajendra, Sree Padma.

Season: It undergoes a dormancy period of 45 to 60 days. Traditionally farmers take advantage of the dormancy period by planting during February-March so that the setts would sprout with the pre-monsoon showers. April – May is the planting season.

Land preparation: The land is brought to fine tilth and beds of convenient size are formed.

Planting: The cormels and mini sett transplants of 100 g weight can be used as planting material at a closer spacing of 45 cm × 30 cm. The tender buds have to be removed before planting as they do not give vigorous growth. Since the buds are located in a ring at the centre of the corm, setts are made out in such a way that each sett has a piece of the central bud. An ordinary sized yam gives about 6 to 8 bits for planting. The cut pieces are dipped in cow dung solution and dried under shade for one or two days before planting to prevent evaporation of moisture from cut surface. In some places, the small round daughter corms are also planted. Whole corms of 500 g size can also be used as a planting material and planted with spacing of 90 × 90 cm. The cut pieces/ setts are planted in beds at 60 cm × 45 cm spacing or pit of 60 × 60 × 45 cm size is dug and planted. The pit should be filled with top soil and farm yard manure (2 kg/pit) prior to planting. The pieces are planted in such a way that the sprouting region (the ring) is kept above the soil. The planting material is placed vertically in the pits and is then covered with soil and compacted lightly. It is planted shallow, as deep planting would interfere with harvest operations and most of its feeder roots are found on the surface. About 3500 kg of corms will be required to plant one hectare. Sprouting takes place in about a month.

After cultivation: Weeding and earthing up as and when necessary.

Mulching: Mulching immediately after planting not only conserves soil moisture and regulates soil temperature but also suppresses weed growth.

Weed management: Generally two weedings coinciding with the requirement of top dressing are recommended. The first weeding may be done about 30 days after planting. The second weeding may be done about 75 days after planting.

Manures and fertilizers application: Apply 25 tonnes of FYM/ha during last ploughing. Fertilizer of 150:60:150 kg/ha of N, P_2O_5 and K_2O is recommended. At the time of planting, the full dose of P and half N and K are to be applied in the pits. The remaining dose of N and K is applied on 35 and 75 days after planting at the time of weeding and earthing up.

Water management: It is mostly raised as a rainfed crop. However, irrigation is required when monsoon fails where it is grown on a large scale. Water stagnation is harmful to the crop. Wherever irrigation facility is available, irrigation can be given once a week.

Cropping systems: Vegetable cowpea var. CO 2 is recommended as suitable intercrop in elephant foot yam. It can be intercropped profitably in coconut, arecanut, rubber, banana and robusta coffee plantations at a spacing of 90 × 90 cm. Half the quantity of FYM @ 12.5 t/ha and one third of NPK @ 27:20:33 will be sufficient for the intercrop.

Crop protection: Collar rot disease is caused by a soil borne fungus *Schlerotium rolfsii.* Water logging, poor drainage and mechanical injury at collar region favour the disease incidence. Brownish lesions first occur on collar regions, which spreads to the entire pseudostem and cause complete yellowing of the plant. In severe case, the plant collapses leading to complete crop loss. Use disease free planting material, remove infected plant materials, improve drainage conditions, incorporate organic amendments like neem cake, drench the soil with carbendazim or apply biocontrol agents like *Trichoderma harzianum* @ 2.5 kg/ha mixed with 50 kg of FYM (1 kg/l of water).

Harvest: Maturity is indicated by yellowing and total senescence of the leaves and corms weigh from 3 to 9 kg. The growth cycle of the corms normally takes about 7 to 12 months but the tubers are small and unmarketable after only one season and 3-4 seasons are required for an economic crop, except when planted from four year old corms. The corms are dug by hand. Simple tools such as hoe, knife and shovels are used to remove the soil around the corm. Physical damage such as punctures, cuts or abrasion lead to high rates of moisture loss and provide avenues for microbial infection. These conditions lead to high incidence of shrinkage and post-harvest losses. The corms are cleaned and are stored in heaps preferably in well-ventilated sheds. Corms can lose as much as 25% of their initial weight in the first month of storage, but can be successfully stored at 10°C for several months. Corms dipped for one minute in a 4% fungicidal emulsion can be stored at room temperature for about 2 months with minimal loss of weight or sprouting.

Yield: Corm yield is 20 to 30 t/ha in 240 days. The average yield is 10 to 12 t/ha as an intercrop with coconuts. Corms often have a diameter of 30 cm or more and under good cultural conditions can weigh 7-9 kg by the fourth season. For seed purpose, the yams can be left in the field itself till planting the next crop or the lifted yams can be stored in sand or paddy straw.

Utilization: The tuberous roots of the plant are used traditionally for the treatment of piles, abdominal pain, tumors, enlargement of spleen, asthma and rheumatism. The tuberous roots of the plant possess blood purifier properties and have been used traditionally for the treatment of piles, abdominal disorders, tumours, enlargement of spleen, asthma and rheumatism. The corms are acrid, astringent, anti-inflammatory, anti-haemorrhoidal, haemostatic, expectorant, digestive, appetizer, stomachic, liver tonic, rejuvenating and tonic. It is used in treatment of anorectal abscess, hemorrhoids.

3. Swamp taro–*Cyrtosperma chamissonis* (Schott) Merr.

Synonym: *Cyrtosperma merkusii* Hassk.

Family: Araceae.

Vernacular names: Swamp Taro, Gallan, Giant swamp tarot, keladi pari, gele-gele (English).

Origin: The giant swamp taro probably originated in Indonesia.

Distribution: The swamp taro is grown in Indonesia, Philippines, Papua New Guinea, Sumatra, Java and the Pacific Islands.

Botany: The swamp taro is a giant herbaceous perennial with typically 6-8 huge leaves arising from a short subterranean stem. The leaf blades are arrow shaped, 1-2 m in length, and are borne on stout petioles, 1-2 m long and tapering from about 10 cm in diameter; in some cultivars the lower parts are covered with spines. A mature plant may reach 3-4 m in height. The inflorescence has a long, thick yellowish spathe and a purplish spadix, though the seeds are often not fertile. The stem thickens rapidly at the base becoming a large corm, varying in shape from cylindrical to conical or almost spherical. The size varies with cultivar and age. The corm weigh is 15 to 25 kg is common, but it can weigh up to 90 kg or more in a 10 year old plant. The giant swamp taro is believed to be the largest plant in the world which produces edible corms. Cormel produces leaves which develop into suckers as side shoots on the parent corm after about three years.

Climate: It is grown between 18ºN to 20ºS latitude. It is commonly grown in freshwater swamps or artificial pits excavated to reach the water table. It grows in the elevation range from sea level to 600 m. It needs a constant supply of fresh water and can be grown in slightly brackish water, thrives in freshwater swamps and can even be found in swiftly flowing rivers and streams. The cardinal minimum, optimum and maximum temperature is 15ºC, 23 to 31ºC and 38ºC respectively.

Soils: It is a water loving plant (hydrophyte) adapted to fresh to brackish water conditions in coastal marshes and swamps. Deep soils are preferable. The desirable pH is 4.5 to 8.5.

Pest management: It is relatively free of pest problems. Taro beetle causes severe losses in some areas.

Harvest: Harvesting may occur in less than one year for some cultivars or may be delayed up to 6 years and even longer. The giant taro usually requires 2 to 3 years to produce a reasonably-sized tuber, younger than this the tubers of some cultivars have an unpleasant taste and some early-maturing types are harvested after about one year. The crop should be harvested when the plants are 3 to 6 years old for good flavour, edibility and yield. The tubers are dug by hand. Storage is not usually practiced, but tubers are sometimes buried in a damp place where they may be kept for up to 6 months. The corms weigh 1-4 kg.

Yield: Corms yield is 7 to 10 t/ha for a crop between 18 months and 2 years of age.

Utilization: The corms may be acrid and contain high levels of calcium oxalate. The tubers are normally eaten as soon as harvested. The tubers are the staple carbohydrate foodstuff in many countries, where they are eaten boiled, steamed or roasted, sometimes with the addition of coconut milk, or they may be sliced and fried and eaten with sugar. The young leaves and young inflorescences are used as a vegetable.

4. Taro–*Colocasia esculenta* (L.) Schott. (2n=28, 42)

Family: Araceae.

Common names: Taro, Dasheen, Eddoe, old Cocoyam, colocasia, (English); *Arvi, Guinya* (Hindi), *Kusu, Kachu, Kuchu* (Bengali), *Kuchu, Kachu, Garo kachu* (Assamese), *Kattuchembu, Velichembu, Chembu, Thalu* (Malayalam); *Kadukesu, Shama gadda* (Kannada); *Seppaikilangu, Kattu- kilangu* (Tamil); *Saru, Kachu* (Oriya).

The genus Colocasia has been derived from the word 'Colcas', an Egyptian word for taro. The Greek word 'Colocasia' has its origin from the Arabian word 'quolqas' for taro. The Greek word 'quolcas' is derived from the root Malayan word 'thallas', for taro, which is also used in Java. There is relationship of the Malayan word 'Thalla' for taro and the Malayalam word 'Thal' for wild taro, 'thalu' for petiole of taro. Similarly the Oriya word 'Saru' is similar to Taro. In certain places in Kerala the black taro is still known as 'Karinthal' ('kari' means black and 'thal' means taro). The new world name is 'Dasheen' which means 'from China' (da Chin) in French. The African name is Old Cocoyam. *Colocasia antiquorum* is known as taro, eddoe, old cocoyam, keladi, arvi or arbi and dasheen of West Indies.

Importance: Taro is mainly cultivated for its mother tubers, (corms), lateral tubers (cormels), leaves and petiole. Both corms and cormels are edible in the case of *Colocasia esculenta* var. esculenta.

Origin: The plant appears to have originated in India and spread eastwards to Myanmar and China, and southwards to Indonesia. Subsequently, it has been taken to Japan, Melanesia, Polynesia and Hawaii; in historical times it spread to Egypt and the eastern Mediterranean, thence to Africa, the Guinea coast, and, along with the African slaves, to the Caribbean.

Distribution: Taro is either cultivated or occurs as wild in Australia, Bangladesh, Brazil, Cameroon, China, Japan, Hawaii, Columbia, Caribbean, Costa Rica, Cuba, Egypt, Fiji, Guatimala, India, Myanmar, Indonesia, Japan, Malaysia, Nepal, Nigeria, Egypt, Panama, Papua New Guinea, Peru, Philippines, Solomon Islands, Sri Lanka, Thailand, Togo, United States of America and Vietnam. In India, it is cultivated in Assam, West Bengal, Odisha, Andhra Pradesh, Tamil Nadu, Kerala, Karnataka and Maharashtra.

Botany: It is an herbaceous perennial. It grows to 0.5 to 1.5 m height. It has an underground starchy corm which produces at its apex a whorl of large leaves with long robust petioles. Taro leaves have two main parts, the petiole (i.e., leaf stem) and blade (large and heart shaped). The petiole is large, succulent, often purplish near top. The leaves are heart-shaped, 20 to 50 cm long, with rounded basal lobes; the leaf stalk joins the blade some distance inward from the notch between the lobes. Inflorescence is a fleshy stalk which is enveloped by a long yellow bract (spathe) about 20 cm long. Flowers tiny, densely crowded on upper part of fleshy stalk, with female flowers below and male flowers above. Fruit is a small berry, in clusters on the fleshy stalk. The seeds are produced extremely rare. It has edible underground stems called corms which are round/cylindrical, up to 35 cm long and 15 cm in diameter, which are wrapped in scale-like leaves and have bud-like smaller corms, called cormels, coming off them. The root system is superficial and fibrous. Above ground, there are other leaves which can also be eaten. There are two species of cultivated taro viz. the dasheen type of taro *Colocasia esculenta* var. *esculenta* with 2n=2x=28 and the eddoe type of taro *Colocasia esculenta* var. *antiquorum* with 2n=3x=42 chromosomes.

Growth stages: The growth is slow initially for two months. The most rapid leaf growth occurs 3 to 5 months after planting. Maximum leaf size and area occurs on fifth month while high numbers of leaves are produced from 3 to 7 months after planting. The corm growth occurs 5 to 9 months after planting while the corm growth continues up to 12 months.

Climate: Taro is a tropical root crop. It is grown as a staple or subsistence crop, throughout the tropics, subtropics, and in many warmer regions of the temperate zone. It is widely cultivated in cool-temperate latitudes in both hemispheres. It can be grown from sea level to an altitude of 1800 m. It is cultivated in a variety of agro-ecological situations from swamps to uplands under rainfed and irrigated conditions. It is well adapted to humid conditions. It is grown in areas receiving an annual rainfall up to 2500 mm. It requires a fairly distributed rainfall of about 1200 to 1500 mm for rainfed cultivation. It is adapted to various conditions such as swamps, tropical wet rainforest, dry uplands and to the foothills of the Himalayas. In general the eddoes are hardier than the dasheens and can be grown in drier conditions on poorer soils. The cardinal minimum, optimum and maximum temperature is 10°C, 21 to 27°C and 35°C respectively. It requires a 6-7 month frost-free period when grown in temperate areas or at high altitudes.

Soils: It comes up well in loamy soils with a pH range of 5.5-7.8. It grows well on deep, well-drained, sandy loams, loams, particularly alluvial loams, with a high water-table. Yield may be reduced at pH 4.2 with aluminum saturations.

Season: Planting is done during June-July under rainfed condition. It can be raised throughout the year, if grown as irrigated crop.

Varieties: Co 1, Panchamukhi and Satamukhi (Kovvur), Sree Pallavi, Sree Rashmi, Sree Pallavi, Sree Kiran, Muktakesi, Panisaru-1, Panisaru-2. The taro varieties such as Ahina Poonam Pat, Salem V, Bhadia Kachu, Naga Kachu, Pusa Sakin II and Simla are resistant to blight.

Land preparation of field: Plough the field to a fine tilth and form ridges and furrows at a spacing of 60 × 45 cm or 45 × 45 cm.

Propagation: Cormels weighing 15 to 20 g or 'hulis' which consists of the upper 1 cm section of the corm or cormel and first 20 to 25 cm of the petiole are used as planting material. Large propagules give high yields. The highest corm yield is produced with 56 g seed size while 28 g seed size is found to be the most economical. Under flooded culture of taro, planting is done in 2.5 cm of standing water at a depth of 15 to 25 cm. A spacing of 60 × 45 cm, 60 × 30 cm, 90 × 30 cm or 30 × 30 cm with a planting depth of 2.5 to 7.5 cm is practiced in uplands. Seed rate of 800 kg/ha is required. Planted seed tubers take 30 to 45 days for sprouting.

After cultivation: Under field condtions, 5-10 per cent of the seeds fail to sprout. To overcome this situation, about 2000-3000 corms / cormels per ha may be planted in a nursery at a close spacing so that sprouted tubers from the nursery can be used for gap filling.

Weed management: Weeding and earthing up should be done 45 to 60 days after planting. Deep cultivation should be avoided. Small suckers have to be removed from the mother plants along with the sécond weeding. It requires profuse irrigation and shade.

Manures and fertilizers applications: Apply 25 tonnes of FYM, 20-30-60 kg/ha of N, P_2O_5 and and K_2O per ha as basal dressing and same quantity of NPK on 45 days after planting.

Water management: Irrigation is given once in a week. It requires a fairly distributed rainfall of about 1200 to 1500 mm for rainfed cultivation. Dasheen grown under erratic moisture conditions show peculiar dumbell-like shapes, reflecting constrictions in growth during dry periods while eddoes produce few corms under water stress.

Cropping systems: Taro is intercropped with various crops such as coconut, grain amaranth, elephant-foot yam, turmeric, chillies, ginger, maize, yam, okra, cassava, new coccoyam and rubber. Taro is a shade tolerant crop.

Pests and diseases: Taro is affected by pests such as aphids, Japanese rose beetle, mealy bug, taro plant hopper, white fly, taro beetle, taro hornworm or hawkmoth, army worm or cluster caterpillar, and spidermites. Root-knot nematode can be a problem in taro. Taro leaf blight is the most serious. Dasheen Mosaic Virus (DMV) also occurs. Spray Quinalphos or Dimethoate @ 0.05% to control aphids and worms which attack the leaves. Mealy bugs and scale insects damage cormels and corms. If infested, the seed cormels should be dipped in Dimethoate or Monocrotophos @ 0.05% solution for 10 minutes. Leaf blight causes oval or irregular purplish or brownish necrotic lesions with watersoaked periphery appear on leaves. Spray with fungicides *viz.,* Mancozeb @ 0.2% or Ridomil MZ 72 @ 2 g/l of water and treating the seed tubers with biocontrol agents *viz., Trichoderma viride* to control leaf blight.

Harvest: Taros are ready for harvesting when the leaves begin to turn yellow and start to wither. The maturation period varies according to the cultivar, and ranges from 6 to 15 months. The shortest crop duration ranges from 3 months, 5-6 months (eddoes) and 8-10 months (dasheens). One month prior to harvest, all the suckers may be wrapped around the base of the mother plant and covered with soil by earthing up, for arresting further vegetative growth and sprouting of tubers. After this, irrigation has to be withheld to hasten maturity. Harvesting is done by carefully uprooting the plants and the mother corms and cormels are separated.

Yield: The corm yield ranges from 4-6 t/ha under rainfed and 15 to 20 t/ha under irrigated conditions.

Corm curing and storage: The storage life of taro corms is up to 4 weeks. However, Taro can be stored in shaded pits for about 4 months. Edible aroids can be stored at tropical ambient conditions (24-29°C with relative humidity of 86-98%) for at least 2 weeks without changes in nutritional values such as crude protein content and total amino acids. Corm storage life is improved under refrigerated storage conditions. Storage life is generally improved at conditions of lower temperature and high humidity. If storage environment can be maintained at 11-13°C and 85-90% relative humidity, the length of storage of taro can be extended to about 150 days. At low temperature of 15°C and high humidity of 85%, taro can be stored for 5-6 weeks. Storage of taro packed in soil in brick-built containers or in pits stored for up to 5 months at ambient temperature of 8-15°C or lower. The corms remain in good condition for up to 6 weeks, but once they are exposed to ambient conditions they deteriorate rapidly after 24 hours.

Utilization: The main use of taro is as a vegetable. The leaves or corms (underground stems) are eaten after baking, roasting, or boiling. Its leaves and petioles can be dried and preserved and eaten in times of food scarcity. Taro has a number of medicinal uses. Its corm is used as an abortifacient, to treat tuberculosis ulcers, pulmonary congestion, crippled extremities, fungal abscesses in animals, and as an anthelmintic. Its foliage is used as a styptic and poultice. The stem sap is used as a treatment for wasp stings. Its petiole is known as 'thalu' in Kerala and is consumed as a vegetable especially during the famine stricken monsoon period in Kerala.

5. Tannia or new cocoyam–*Xanthosoma sagittifolium* (L.) Schott (2n=26)

Family: Araceae

Vernacular names: *Tannia, Tania; Yautia. New cocoyam, Tanier* (English); Palchempu (Malayalam), *Talas, Keladi* (India).

Origin: It is native of Central America (Caribbean to North Brazil). It has spread to South-East Asia, the Pacific islands and Africa. In India, it is a popular vegetable crop in Andhra Pradesh, Odisha, Bihar, Uttar Pradesh, West Bengal and the North Eastern states.

Botany: It is an herbaceous perennial crop. It has a corm or main underground stem in the form of a rhizome from which swollen secondary shoots, or cormels, sprout. The leaf blades point downwards to form an acute or right angle with the leaf stalk. It seldom flowers and some cultivars do not at all. When flowering does occur, the inflorescence arises in leaf axils, and depending on species, one or several species may form flowers, depending on the species are unisexual or bisexual. All are sessile and form on a spike (spadix) enclosed in large bract (spathe). In the unisex inflorescence, pistillate flowers are at the base of the spadix, staminate flowers at the top, and in between are a grouping of abortive sterile flowers. The female flowers occur on the lower portion, male flowers on the upper portion and sterile flowers in the middle portion of the spadix. Insects pollinate the flowers, but fruit and seed set are rare. Fruits are berries, each containing 2-5 small, oval and hard seeds. Fruits produce few viable seeds. The growth cycle lasts from 9 to 11 months. The corms and leaves develop during the first six months while the foliage remains stable and when it begins to dry, the plants are ready for the cormels to be harvested in the last four months.

Climate: Tannia found in tropical rain forest. It is grown up to an altitude of 1,500 m from MSL. It is grown in areas receiving the mean annual rainfall of 1000 mm to 3000 mm. The cardinal minimum, optimum and maximum temperature is 13°C, 21 to 25°C and 35°C respectively.

Soils: They require well-drained soils and do not tolerate the permanent presence of water. The plant roots within 15 to 20 cm of the soil surface. The crop performs well in well-drained loamy soil with pH of 5.5 to 6.5. It is not considered salt tolerant. It cannot be grown in flooded fields (e.g. on the margins of rice paddies).

Land preparation: It can be done by ploughing in the summer before the onset of monsoon. Ridges and furrows are formed 90 cm apart. Planting of setts or cormels is done on the ridges at a spacing of 60 cm to 1 m.

Propagation: Corms and cormels of 100 to 150 g with three or four buds and 20 to 25 cm long are the usual planting material. The setts from the top portion of the main corm with a thickness of 5-10 cm containing the apical bud are also used for propagation.

Weeding: Two weedings and earthing up operations have been suggested at 30th and 60th days after planting.

Manures and fertilizer application: Apply FYM @ 12.5 t/ ha and 80-60-80 kg/ha of N, P_2O_5 and K_2O is recommended. Full dose of phosphorous and half dose of nitrogen and potash are applied as basal a week after sprouting and the remaining half dose after a month.

Water management: It grows as an upland rainfed crop on well drained soils where rainfall is well distributed throughout most of the year.

Harvest: If the crop is intended for taking corms, then it can be harvested when older leaves start turning yellow. Its corms do not deteriorate if left unharvest in the soil. Harvesting can be done from 6 months onward up to 10 months. It is hand-harvested by digging out mature corms. If it is to be grown as a ratoon crop, the soil around the clump should be dug to remove the mature cormels. The main corm is retained along with roots and covered with soil. This can be practised for 4 to 5 years. If the crop is grown for leaf purpose, then harvesting of the leaves should be done immediately after they attain maturity. A leaf size of 20-30 cm × 20-30 cm is ideal for making roll. Thus 40-50 leaves can be harvested from one clump during its entire growth period. Leaves are harvested along with the petiole. Petioles are also used as a vegetable.

Yield: It matures and is ready for harvest within 9 -10 months after planting. It produces small edible cormels 15 to 22 cm in length attached to large corms. These cormels should be harvested before they produce new shoots. It yields about 6-12 t/ha of corms, but yields of 20 to 30 t/ha can be achieved.

Corm curing and storage: Tannia can be stored for 5-6 weeks at low temperature of 15°C and high humidity of 85%. Storage at 7°C and 80% relative humidity is found to maintain corms in good condition and good eating quality for about 120-130 days.

REFERENCES

Anuradha Singh and Neeraj Wadhwa. 2014. A Review on Multiple Potential of Aroid: *Amorphophallus paeoniifolius*. Int. J. Pharm. Sci. Rev. Res., 24(1): 55-60

Bradbury, D.H., W.D. Holloway. 1988. Chemistry of tropical root crops. ACIAR, Canberra.

Caesar, K. 1980. Growth and development of *Xanthosoma* and *Colocasia* under different light and water supply conditions. Field Crops Research, 3, 235-244.

Coursey, D. G. 1968. The edible aroids. World Crops, 20 (3): 25-30.

CPG. 2013. Crop Production Techniques of Horticultural Crops. Horticultural College and Research Institute, Tamil Nadu Agricultural University, Coimbatore – 641 003

Danilo Mejía (ed). 2003. Edible Aroids: Post-Harvest Operations. FAO, Rome.

Goenaga, R. 1995. Accumulation and partitioning of dry-matter in Taro [*Colocasia-esculenta* (L) Schott]. Annals of Botany 76(4): 337-341.

Goenaga, R. 1996. Taro yield and dry matter distribution under upland conditions in Puerto Rico. African Crop Science Journal 4(3): 289-294.

Gollifer, D.E. and R.E. Booth. 1973. Storage losses of taro corms in the British Solomon Islands

Greenwell, A. B. H. 1947. Taro with special reference to its culture and uses in Hawaii. Economic Botany. 1(3): 276-289.

Kay, D.E. 1987. Crop and product digest No. 2 Root crops. 2nd Edition. London: Tropical Development and Research Institute. 380 pp.

Moorthy, S.N. 1994. Tuber crop starches. Central Crops Research Institute, Sreekariyam, Thiruvananthaapurram.

NAS. 1975. Taro. pp. 37-43 In: Underexploited Tropical Plants with Promising Economic Value. Report of an Ad Hoc Panel of the Advisory Committee on Technology Innovation. National Academy of Sciences, Washington, D.C.

O'Hair, S.K. and M.P. Asokan. 1986. Edible aroids: botany and horticulture. Horticultural Reviews 8: 43-99.

Onwueme, I.C. 1978. The Tropical Tuber Crops. John Wiley and Sons, New York.

Plucknett, D. L., PE, R. S. de la and Obrero, F. 1970. Taro (*Colocasia esculenta*). Field Crop Abstracts, 23, 413-426

Plucknett, D. L. 1976. Edible Aroids: Alocasia, Colocasia, Cyrtosperma, and Xanthosoma (Araceae). Pp. 10-12 in Evolution of Crop Plants. Edited by N.W. Simmonds. Logman Group Limited, London, England.

Plucknett, D. L. 1977. Giant swamp taro, a little-known Asian-Pacific food crop. Proceedings of the 4th Symposium of the International Society for Tropical Root Crops (Colombia, 1976), IDRC-080e (Cock, J., MacIntyre, R. and Graham, M., eds), pp. 36-40. Ottawa, Canada: International Development Research Centre, 277 pp.

Sakai, W. S. 1983. Aroid root crops: Alocasia, Cyrtosperma and Amorphophallus. Handbook of Tropical Crops (Chan, H. T. (Jr.), ed.), pp. 29-83. New York: Marcel Dekker Inc., 639 pp.

https://en.wikipedia.org/wiki/Colocasia_esculenta

https://en.wikipedia.org/wiki/Arum

http://www.britannica.com/plant/Arum

CHAPTER

33

Narcotics, Masticatories and Beverages

Narcotics are crop plants or their products that are used for stimulating, numbing, drowsing effect or relishing effects and mild poisons. Narcotics relieve pain and produce sleep. The narcotic plants include the tobacco, ganja (*Cannabis sativa Linn.*), anise, opium poppy (*Papaver somniferum*), Indian hemp (*Cannabis indica*), the coca plant (*Erythroxylon coca coca* Lam. and *E. novogranatense* (D. Morris) Hieron.)), jimsonweed (*Datura stramonium*), *Datura innoxia* and *Physochlaina physaloides*. Fumitories are those substances which are used for smoking while masticatories are used for chewing. When substances are smoked because of the stimulating effect as in the case of tobacco, they are 'fumitories'. Substances which are chewed such as betelvine and arecanut are known as 'masticatories'. Tobacco comes under narcotics, a fumitory and also a masticatory. As recreational narcotic drugs, they may be used orally, but are also commonly smoked, snorted or self-administered by the more direct routes of subcutaneous (skin popping) and intravenous injection, depending on the precise substance in question. Narcotic use is associated with a variety of side effects, including drowsiness, itching, sleeplessness, inability to concentrate, apathy, lessened physical activity, constriction of the pupils, dilation of the blood vessels causing flushing of the face and neck, constipation, nausea, vomiting and most significantly, respiratory depression.

Man's earliest 'beverage' is probably the juice squeezed from fruits. Coffee, tea and cocoa are the important beverages of the world. In general, beverages are drinks, which form an essential part of the human diet because of their liquid content. The above mentioned beverages have stimulating effects. For juices like lemonades, orangeades, apple, pineapple and mango juices constitute for soft drinks. Coffee and tea are commercial crops which have potentials in the export market. Cocoa is gaining importance in the country and has great potential both as beverage and as confectionery. Two important groups of beverages (exclusive of soft drinks) can be recognized: alcoholic beverages and non-alcoholic beverages.

Alcoholic beverages: Alcoholic beverages are depressants and lowering the activity of the brain. These are classified into two main groups viz., i) the fermented ones in which alcohol is formed by the fermentation of sugar and ii) the distilled ones obtained by successive distillation of fermented liquors. Wines and beers are among the oldest and most cherished of man's fermented beverages. Present day beers are made by fermenting cereal starch, especially barley. It contains from 3 to 8 % alcohol and its nutritive value is due to the presence of sugars, dextrin, protein and phosphates. Minor fermented alcoholic beverages such as cider, made from the juice of apple; perry, from pear juice; palm wine made from the juice of palm inflorescences and chicha from maize kernels. Distilled spirits have a higher caloric content than merely fermented drinks, but less nutritional value. The chief distilled alcoholic beverages are; whiskey (from fermented mash of cereals and potatoes), brandy (from fermented juice of various fruits, especially grapes) and rum (from sugarcane juice or molasses). Gin is prepared by the distillation of the fermented malt of barley and, rye, flavoured with the coriander and other spices including cassia, nutmeg, lemon peel and cardamom seeds.

The types of alcoholic beverages are as follows:

(i) Wines are alcoholic beverages produced from fermenting grapes, apples, peaches, and plums.
(ii) Beers are alcoholic beverages produced from fermenting rice, barley and wheat.
(iii) Spirits are produced from fermenting barley, rye, corn, sugarcane, potato, etc.
(iv) Sake is produced from fermenting rice which is popular in Japan
(v) Cider is produced from fermenting apple juice which is popular in England
(vi) Mead is a wine like drink produced from fermenting honey, occasionally together with other fruits which has originated from medieval Europe. The term 'honeymoon' is coined from the practice of giving a gift of mead to a newly married couple: if they drank mead (honey) each night until the next moon, they would be given the gift of a new child.

Non-Alcoholic beverages: The refreshing and stimulating properties of non-alcoholic beverages are due to the presence of caffeine or related alkaloids. The three major non-alcoholic beverages are tea, coffee and cocoa. These are all tropical crops. The first two have little nutritive value while cocoa is a good source of energy. The types of non-alcoholic beverages are from i) seed: coffee, cacao and ii) leaf: tea.

CHAPTER

34

Tobacco–*Nicotiana tabacum* L. (2n=48)

Family: Solanaceae

Vernacular name: Tobacco (English), *Tambaku* (Hindi), *Tamaku* (Gujarati) *Pugayilai* (Tamil) and *Pogaku* (Telugu). The English word 'tobacco' originates from the Spanish and Portuguese word '*tabaco*'.

Importance: Tobacco is considered as the 'Golden leaf of India'. It is grown for its leaves, which are used as cured product. Flue-cured Virginia (FCV), Bidi, Hookah, Chewing, Cigar-wrapper, Cheroot, Burley, Oriental, HDBRG, Lanka, Pikka, Natu, Motihari, Jati, etc., are the different types of tobacco grown in the country. In India, Tobacco crop is grown in an area of 0.45 M ha (0.27% of the net cultivated area) producing 750 M kg of tobacco leaf. India is the 3rd largest producer of tobacco in the world after China and Brazil. India ranks 5th largest exporter of tobacco in the world after Brazil, USA, Malawi and Turkey. Indian tobacco accounts for 10% of the world area and 9% of the world total production. FCV, Burley and Oriental tobacco are the major exportable types. It is also a very large consumer of tobacco products. Tobacco is one of the important cash crops in the country, and makes a significant contribution to the Indian economy in terms of employment, income and government revenue. In India, tobacco is an important commercial crop fetching more than Rs. 4,400 crores of foreign exchange and generates over Rs 14,000 crores excise revenue to the exchequer. In India, tobacco provides livelihood security to 36 million people including 6 million farmers and 20 million farm labour engaged in tobacco farming besides 10 million people working in processing, manufacturing and exports.

History: It is believed that tobacco began growing in the Americas about 6,000 B.C. Christopher Columbus found tobacco in 1492 during his voyage across Atlantic leading to the discovery of America. The rest of the world was ignorant of all species of Nicotiana until 1492 when Columbus encountered the New World. Columbus and other explorers brought tobacco back to Europe with them. In 1560, Jean Nicot de Villemain brought tobacco seeds and leaves as a 'wonder drug' to the French court. The botanical name of the plant Nicotiana and the word nicotine have been derived from his name. In 1571, a Spanish doctor named Nicolas Monardes wrote a book about the history of medicinal plants of the new world. In this he claimed that tobacco could cure 36 health problems. In 1586 the botanist Jaques Dalechamps gave the plant the name of *Herbanicotiana*, which was also adopted by Linné. It was considered a decorative plant at first, then a panacea, before it became a common snuff and tobacco plant. In 1588, a virginian named Thomas Harriet promoted smoking tobacco as a viable way to get one's daily dose of tobacco. Unfortunately, he died of nose cancer (because it was popular then to breathe the smoke out through the nose). Tobacco cultivation in India was introduced by Portuguese in 1605. John Rolfe is said to have brought to Virginia a more full-flavoured tobacco plant from the Caribbean island of Tobago (hence the name tobacco) as early as 1612. In 1814, seven species of *Nicotiana* imported from America were cultivated in botanical gardens of Calcutta. In 1826, the pure form of nicotine is finally discovered. In 1836, New Englander Samuel Green stated that tobacco is an insecticide, a poison, and can kill a man. In 1851, the Belgian chemist Jean Stas documented the use of tobacco extract as a murder poison. In 1875, a model farm was established at Pusa, Bihar for growing and curing tobacco. In 1875, R.J. Reynolds Tobacco Company (better known for its Reynolds Wrap Aluminum Foil) was established to produce chewing tobacco. Imperial Agricultural Research Institute, established in 1903, carried out botanical and genetic studies of tobacco. The first Director of Imperial Agricultural Research Institute (IARI), Dr. Howard isolated 52 lines of tobacco. Later Shaw and Kashiram added 18 more lines in the series. Among those lines NP-28, NP-58 and NP-63 were found most promising. IARI, New Delhi established a Cigarette Tobacco Research Station at Guntur in 1936. Tobacco Research Station was started at Naidabad in 1922 and at Nipani (1939) in Bombay province. In 1940, Dr. B.P.Pal, the Royal Economic Botanist of IARI identified a selection NP-70, which became very popular among the tobacco growing area of Bihar. The Indian Central Tobacco Committee (ICTC) is formed in 1945 which in turn established Central Tobacco Research Institute (CTRI) in Rajahmundry (Andhra Pradesh) in 1947. The administrative control (function) of CTRI was taken over by ICAR in 1965. All India Coordinated Research Project on Tobacco functions at Central Tobacco Research Institute, Rajahmundry for flue cured tobacco in 1970.

Origin: Tobacco is native to Central and South America. Tobacco originated in the Ecuadorean and Peruvian Andes mountain range, where it had been growing for at least five thousand years before the Incas began to use it.

Distribution: Tobacco is grown in Brazil, Indonesia, Thailand, Turkey and Russia. Oriental tobacco type is best grown in arid environments. There are three types of tobacco under cigarette tobacco viz., i) Virginia Flue cured Tobacco, ii) Natu tobacco and iii) Burely Tobacco. Virginia Flue cured Tobacco is used in cigarettes which is an export-oriented crop and accounts for about 21% of India's total production of tobacco. Virginia tobacco is grown mostly in Andhra Pradesh and Karnataka. Burely Tobacco is a cigarette type is grown in Telangana areas of Andhra Pradesh. Natu tobacco is used in the manufacture of cigarettes and tobacco mixtures for pipes. Bidi Tobacco is the most important Non-Virginia tobacco produced in the country. More than 85% of the worlds bidi production is accounted for by India. Bidi tobacco occupies 30-35% of the total area under tobacco in the country whereas in terms of production its share is around 35-40 per cent. Gujarat, Karnataka and Maharashtra are the main bidi tobacco producing States. Chewing tobacco is mostly grown is the States of Tamil Nadu, Uttar Pradesh, Bihar, West Bengal and Orissa. There are three distinct Cigar types of tobacco used in the manufacture of cigar *viz.* filler, binder and wrapper. In India, good quality filler tobacco is also used as a binder. Wrapper tobacco is mainly grown in West Bengal. Cigar filler tobacco is mainly grown in Tamil Nadu. There are two types of tobacco under Cheroot Tobacco *viz.,* i) Lanka tobacco is mainly grown in Andhra Pradesh and ii) Natu tobacco is mainly grown in Tamil Nadu. Hookah tobacco is used in smoking through the hookah or hubble-bubble which is grown in cold climate. Hookah tobacco is mostly *Nicotiana* rustica type and is cultivated in Uttar Pradesh, West Bengal and Orissa. Snuff tobacco used for snuff are grown in Andhra Pradesh, Kerala, West Bengal, Uttar Pradesh and Gujarat. Snuff Tobacco is mainly consumed as hookah, chewing and bidi. The important snuff manufacturing States in India are Tamil Nadu, Gujarat and Uttar Pradesh. Chennai in Tamil Nadu State is an important centre for manufacture of snuff.

Botany: Tobacco belongs to the family of Solanaceae. Among the different species of tobacco, *N. tabacum* and *N. rustica,* the two species of commercial importance. Both species have got a sturdy or more or less woody habit of growth with evenly spaced leaves. *N. rustica* is a much smaller plant, about 0.6 to 1.2 m in height and usually developing suckers (lateral shoots). *Nicotiana tabacum* is highly polymorphic. It is a stout, sticky annual herb attaining a height of about 1.2 to 2.7 m when not topped and produces 10 to 20 leaves from its central stalk. Plant height of 40 to 50 cm is maintained under field condition. The leaves of *N. rustica* are usually petiolate and of regular ovate or cordate shape with a dark green, shiny surface. The leaves are somewhat uneven (puckered) surface. Sessile, ovate or oblong-lanceolate shaped leaves are most common in *N. tabacum* while the leaves are thicker, dark green and have uneven surface in *N. rustica*. The *N. tabacum* types are taller than *N. rustica* types. The inflorescence is usually a compact `Thyrse' represented by a thickened central axis, shortened branches and monopodial-sympodial development of peripheral flowers. *N. tabacum* is semi-xerophytic to mesophytic whereas *N. rustica* is a mesophyte. In *N. rustica* the pedicels are short at first (3-4 mm long) but extend to more than 6 mm when the flower opens. Calyx is cylindrical, pubescent, membranous and narrow with pointed sepals with a longer odd sepal. Corolla is greenish yellow, 1.2-1.5 cm long, pubescent with tube proper 3 mm long and 2 mm wide. Throat is broadly obconic with contraction near the mouth. Stamens white, with four of them long and one short. Capsule is elliptic, ovoid or subglobose, 7-15 mm long. Seeds are elliptic, oval and dusky brown. In *N. tabacum* the pedicels are 5-10 mm long during early stages but elongate to 10-20 mm later. Floral calyx is cylindric or cylindric campanulate; 12-20 mm long. Sepals are triangular, acuminate, shorter than calyx tube, unequal in length. Corolla is pubescent on the outer surface and tube proper 10-15 mm long and 2-3 mm wide, throat 25-40 mm long, lower half cylindric, pale green or creamy, upper half similar in colour or pink to red, expanded into deep cup 7-12 mm wide. Limb 10-15 mm wide lobed or pentagonal, white, pink or red, stamens inserted on base of corolla throat, erect with 4 long and 1 short stamen. Capsule is narrow, elliptic, ovoid or orbicular, acute or blunt 15-18 mm long. Seeds are nearly spherical or elliptic and light brown.

N. tabacum is grown all over the country while *N. rustica* is grown in Punjab, Uttar Pradesh, Bihar, West Bengal and Assam only since it requires cooler climate. *N. tabacum* varieties are known as '*desi*' types which have tall plants with broad leaves and have pink flowers. *N. rustica* varieties known as 'vilayati' and '*calucttia*' *which* are characterized by short plants with round puckered leaf and yellow flowers. *N. rustica* has been replaced to some extent, as a commercial source of tobacco for smoking, snuffing and chewing by *N. tabacum*. With its high nicotine content, *N. rustica* is much stronger than *N. tabacum*. *N. tabacum* is used mainly in the production of tobacco products which are smoked, chewed or sniffed while *N. rustica* is used for smoke. *N. affinis* is grown for ornamental purposes. *N. tabacum* is suitable for all uses while *N. rustica* is suited for hookah, chewing and snuff and extraction of nicotinic acid and citric acid for pharmaceutical industry. The flowers are cut off as to encourage the leaves to grow further down the stem. It is smoked as pipe, cigar, cigarette or hookah, used as snuff or chewed as a quid in many forms and to prepare nicotine sulphate, an insecticide.

In general, leaf petiole is 5 to 6 cm long and lamina ovate. Leaf length ranges from 10 to 30 cm while leaf width ranges from 8 to 25 cm. Glandular hairs on the surface of the leaves secrete gums and oils, which make the surface sticky. The flowers are tabular borne in a terminal panicle and are usually pink or white. Tobacco is normally a self-fertilised crop, but cross fertilization ranges from 4 to 10% which is carried out by insects. The 1000 seed weight is only 80 mg. A single fruit may contain 8000 seeds. One ounce (28 g) will contain as many as 3 lakhs seeds.

Climate: Tobacco is tropical in origin, but it is grown successfully under tropical, sub-tropical and temperate climates. Ideal conditions required for successful production of high quality leaf are i) a liberal and well-distributed rainfall during active vegetative growth stage, ii) long day lengths, iii) a mean temperature of 26°C during growing season and iv) a high relative humidity of 70-80%. Normally it requires about 100 to 120 days, frost-free climate with an average temperature of 27°C, to mature. Tobacco grows between 40°N and 40°S. In India, tobacco is grown under a very wide range of conditions from the coast-line to an altitude of 3,000 feet. *Nicotiana rustica* adapted for growth in cool climate. Tobacco is a day neutral plant. The optimum light intensity required for tobacco is 32.3-86.1 klux. Tobacco seeds require about 21°C temperature for germination. Temperature below 13°C is undesirable for tobacco growth. Lower temperature increases the growth period. The optimum air temperature for the growing season is 20 to 32°C. Temperature above 35°C on bright days may result in leaf scalding. A well distributed rainfall of 1270 to 2500 mm throughout the crop growing season is required. Excess rainfall results in thin, lightweight leaves. Rainfall is undesirable at the time of maturity of the crop as gum and resins on the leaf get washed. Tobacco suffers severe injury from strong winds and hail. It requires at least 90 to 120 frost free days after transplanting. High quality tobacco has large leaves with a high proportion of lamina to vein and low nitrogen content at maturity. Continuous rain during the growing season leads to diseases and thin, light weight leaves. Tobacco will not tolerate waterlogging. The desirable relative humidity for growth and yield of tobacco ranges from 80 to 85% in the morning and 50 to 60% at midday. A dry period is required for ripening and harvest of the leaves. Tobacco requires a relatively dry climate to develop its full aroma.

Soils: Most of the tobacco growing soils are low in organic matter, available N and P, medium in K content and mostly alkaline in reaction (pH 8.2 to 8.4). However, light soils tend to produce a large and thin leaf, light in weight and colour, mild in strength and weak in aroma whereas leaf produced on heavy soils is usually thick and heavy, dark coloured, strong and aromatic. Tobacco is very sensitive to very wet soil conditions. A light soil is essential for flue-cured tobacco and light sandy loams are used ideally with about 70% fine sand and 6-8% of clay. Cigar and Wrapper also requires light soils. Fire cured, dark air cured and cigar-filler tobacco are grown on fertile heavier silt or clay loams. Light air-cured and cigar binder tobacco are grown on soils intermediate between the above two. Tobacco is adapted to moderately acidic soils, with a pH ranging from 5.5 to 6.5. Typical characteristics of soils best suited to the production of high quality Flue-cured tobacco are (i) a sandy surface soil up to 15 to 25 cm depth (ii) a yellowish or a reddish sandy clay subsoil extending up to a depth of 150 cm. (iii) an acidic soil with pH of 5.5 to 6.5 (iv) a low reserve of essential plant nutrients (v) a low organic matter content and (vi) very low chloride content (less than 100 ppm). Saline soils whose soil chloride content is greater than 100 ppm should be excluded from tobacco cultivation. Chewing tobacco is mostly grown in alkaline soils with waters having high salt content.

Growth and development stages: The description and coding of the phenological stages for tobacco is presented in Table 1.

Table 1 Description and coding of the phenological stages for tobacco

Stage code	Description	Length (days)
Initial I - emergence	Early post-transplant stage. Less than 5 unfolded leaves, stem reaches ≤ 0.15 m. Seedling emergence in 10 days. Leaf development in seedlings.	10
Initial II - transplant stage	Late post-transplant stage. 6 - 10 unfolded leaves, stem reaches ≤ 0.30 - 0.35 m. Leaf development and stem elongation in seedlings	20
Crop Development I - knee high stage	First growth stage (knee high). 11 - 20 unfolded leaves, stem reaches ≤ 0.55 - 0.6 m. Stem elongation in field plant. Crop cover within rows.	30-50
Crop Development II - elongation stage	Elongation and rapid growth stage. More than 21 unfolded leaves, stem reaches 1 m but there is no hint of reproductive organ formation. Crop cover between rows	
Pre-flowering I -bud formation stage	Bud Formation. Apical bud swelling but with inflorescence not yet visible or only visible between the apical leaves.	10-20
Pre-flowering II - bud emerging stage	Bud Emerging. Inflorescence emergence continuous till 1st corolla visible but still closed.	
Pre-flowering III - close flower stage	Close flower. First petals visible but not yet open.	10-20
Flowering I - open flower stage	Open flower. From beginning of flowering, first petals open, to 50% of flowers open	
Flowering II - flowering stage	Advanced flowering. Continuous stage until more than 90% of flowers open	
Development of fruit	% fully developed green capsules (i.e. seed capsules)	20
Ripening of seed	% darkened capsules	
Termination of crop (harvesting and curing of leaves)	0 Harvesting of leaves - % leaves harvested 1 Lamina colouring phase - % lamina coloured 2 Lamina drying phase - % lamina dried 3 Mid-vein drying phase - % mid-vein dried	10

Varieties: Tobacco varieties cultivated in India and Tamil Nadu are furnished in Table 2 and Table 3.

Table 2 Tobacco types and varieties for cultivation in India

S.No.	Tobacco type	Varieties released / identified	Regions
1	Flue-cured tobacco	Chatam, Delcrest, Kanakaprabha, Dhanadayi, CTRI Special, Jayasri, CTRI Spl. (MR), 16/103, FCV special, Godavari Special, Swarna, McNair 12, Jayasri (MR), Hema, Bhavya, Gauthami, CM 12 (KA), VT 1158, Kanchan, Thrupthi, Rathna, Kanthi, Hemadri, Siri, KST-28 (Sahyadri), CH-1, CH-3, N-98, FCH 222, Virginia gold.	Andhra Pradesh, Karnataka

2	Bidi tobacco	GT 4, NPN 190, Anand 119, Anand 2, Spoorthy (PL 5), GT 5, GT 7, GTH1, Bhavyasree, GT 9, NBD 43, MRGTH-1, ABT 10, Vedaganga 1, GABT-11, NBD-119	Gujarat, Karnataka
3	Chewing tobacco	Chama, Podali, DP 401, Gandak Bahar, Sona, Vairam, Thangam, Bhagyalakshmi, Maragadham, Prabha, PT 76, Meenakshi, Vaishali Special, Lichchavi, Manasi, Abirami, Kaviri, Meenakshi (CR), Sangami, Kamatchi, Abirami (CR), DJ-1, Monnai, Vazhaikappal, Vadamugam, Periya vadamugam	Uttar Pradesh, Bihar, Orissa, West Bengal, Assam, Gujarat, Tamil Nadu
4	Hookah and Chewing (*Rustica*) tobacco	DD 437, Sonar Motihari, GC 1, GT 6, GCT 2, GT 8, GCT 3, Dharla, Azad Kanchan,	Uttar Pradesh, Bihar, Tamil Nadu, Orissa, West Bengal, Gujarat
5	Oriental tobacco	Tungabhadra	
6	*Motihari* tobacco (West Bengal)	Manasi, Torsa	
7	Natu tobacco	Prabhat, Vishwanath, Natu Special, Gajapati, Bhairavi	Andhra Pradesh, Orissa
8	Cheroot tobacco	Bhavani Special, Lanka Special, Sendarapatty Special, Oosikappal	Tamil Nadu, West Bengal
9	Cigar filler	Krishna, Maryland, Havana	Andhra Pradesh
10	Cigar-wrapper tobacco	S5, Dixie Shade, Rangpur, Sumatra, Vellai vazhai (VV2) Karuvazhai (KV1), Krishna	Tamil Nadu, West Bengal
11	Burley tobacco	Burley 21, Banket A1, HDBRG (Dark burley)	

Source: Central Tobacco Research Institute, Rajahmundry

Table 3 Tobacco varietal features recommended for cultivation

S.No.	Variety	Duration (days)	Yield (kg/ha)
A	**Sun-cured chewing tobacco varieties**		
1	I-64 (Monnai)	110–120	3000
2	I-115(Meenampalayam)	100–110	2500
3	VTK-1	120–130	2800
4	VR-2	110–120	2500
5	Bhagyalakshmi	115–125	3500
6	Meenakshi	120–125	3750
7	Abirami	120–125	4000

B	**Smoke cured chewing tobacco varieties**		
1	PV-7	120–125	3200
2	Thangam	120–125	3600
3	Maragadham	100–105	3200
C	**Pit-cured chewing tobacco varieties**		
1	VD.1	110–120	2500
2	Vairam	110–120	3500
D	**Cigar tobacco varieties**		
1	VV-2	80–90	2100
2	(Vellai vazhai) KV-1	90–100	2400
3	(Karu vazhai)	100–110	2500
4	Krishna	85–90	1600
E	**Country cheroot tobacco varieties**		
1	I-737 (Narrow leaf Oosikkappal)	90–100	2000
2	Bhavani Special	90–100	2000
3	OK-1 (Broad leaf Oosikkappal)	90–100	2500
4	Sendarappatty Special	90–100	2500

Hybrid tobaccos: There are two Bidi hybrids, GTH-1 and MR GTH-1 (Mosaic resistance); one chewing hybrid, Kamatchi and two FCV tobacco hybrids (CH- 1 and CH-3) identified.

Seasons: In Gujarat, Bihar and West Bengal, it is grown between September and January. In Andhra Pradesh, Karnataka and Tamil Nadu, the crop is raised in winter from October to March when the temperatures are moderate, but in Punjab it is grown as an early summer crop.

Nursery and its management: Tobacco seeds are very small and egg-shaped with thick seed- coat. They are about 0.75 mm long, 0.53 mm broad and 0.47 mm thick. Depending on the variety and the conditions under which the seed is produced, the size and the weight of the seed vary considerably. In *N. tabacum* the average weight of the seed is 0.08 to 0.09 mg and there are 11,000 to 12,000 seeds per gram. In *N. rustica*, the seed is larger and about three times heavier. The emerging seedlings are tiny and delicate and therefore, the seeds are unsuitable for sowing directly in the field. Hence tobacco seeds are sown in small areas called nurseries or seed beds.

(i) **Selection of nursery site:** Selection of well drained and healthy soil is pivotal for tobacco nurseries. Nursery area should be very close to the water source. There should be ample drainage and protection from heavy wind. The area should be free from nematode infestation. Red sandy loam soils are preferred for tobacco nurseries. Sandy loams or loamy soil with high organic matter and good drainage are ideal for raising the nursery. To raise the nursery on heavy soils, they should be made porous by mixing about 200 cart loads (100 tones) of sand per hectare in the top layers. Nursery site should be changed every year to avoid accumulation soil borne pathogens and nematodes. The old sites can be sterilized by 'rabbing'. This will kill the soil borne diseases, neamtodes pests and also the weed seeds.

(ii) **Nursery bed preparation:** Three to four ploughings, harrowing will bring the soil to a fine tilth. Deep ploughing (12-15 inches) by tractor drawn disc plough is recommended in decomposition of plant residues prior to seed bed preparation. To raise the nursery on heavy soils, they should be made porous by mixing about 250 cart loads (100 tonnes) of sand per hectare. Raised beds of 10 m long, 1.2 m wide (to get 1.0 m wide bed after tapering) with 30 cm channels between the beds are formed. Seedlings should be raised in raised beds of 10-15 cm height to prevent from damping off and leaf blight. After preparation of the bed, 'rabbing' (spreading and burning of paddy husk at 15 kg per 2.5 m^2 bed or burrowing of farm waste such as tobacco stalks, sugarcane-trash, bajra or jowar straw) should be done to ensure destruction of pathogens, pupae and egg masses of insects and weed seeds to a depth of 5 to 10 cm. Application of farm yard manure or municipal compost (12.5 kg / 2.5 m^2 bed) and 80 g of super phosphate before sowing the seed is found to be highly beneficial in obtaining more number of seedlings. Application of Ridomil 25 WP @ 0.2% controls damping off. Dusting of Endosulphan @ 40 gm per 10 m^2 bed before sowing to control insects.

Seeds and sowing: Seed rate is 1.50 gm per 2.5 m^2 bed. 28 to 32 g seeds sown in 100 m^2 area is sufficient to plant one acre of tobacco main field. Higher seed rate results in over-crowding of seedlings which in turn lowers their quality and induces diseases like 'damping off'. Soaking seed with 50 ppm gibberelic acid for 48 hours and drying before commencement of sowing improves germination even under adverse weather conditions. After leveling the beds, form 0.25 cm deep rills (furrows) with 5 cm apart across the bed length. Seeds and carbofuran granules may be mixed with sand and then sown in rills. After sowing the top layer of the beds are raked by fingers for mixing the seed with the soil. The seed is sown in the rill which is covered by dragging small brooms. Then compact the bed lightly by rolling 22.5 cm diameter pipe. The beds are irrigated immediately after rolling the Iron pipe .

Care of beds after sowing: To prevent erosion and washing off seedlings during heavy rains, spread paddy straw as much for 25 days after the beds are sown and compacted. This mulch is thinned gradually when seedlings are 3 weeks old and exposed to the sunlight daily in the morning hours to avoid etiolation of the seedlings and to drive out excessive moisture. The cover is to be completely removed after another one week. Water the beds for 6 to 8 times per day to keep them moist but wet till germination is over. If the seedlings are over-crowded in sown places, they can be thinned out at about three weeks age and transplanted in places where the stand is sparse. Apply FYM @ 5 kg per m^2 mixed well with the top soil by light digging. Basal application of 50 gm NH_2SO_4, 50 gm of K_2SO_4, 300 gm of super phosphate and 100 gm of dolomite per 10 m^2 bed is recommended. The beds should be covered with coconut fibre/ dried grass or paddy straw @ 750 gm per 2.5 m^2 to protect the germinating seeds from beating rain and scorching sun and to conserve moisture. The covering should be removed in phased manner from 15th to 25th day of sowing. Watering the bed 5 to 8 times per day by rose cane ensures uniform germination of the seeds. It is imperative that the seed beds are neither allowed to dry nor retain excess moisture. At the end of nursery period, the left over seedlings are pulled and the area is immediately ploughed. Application of neem seed kernel suspension @ 5% or nuclear polyheterosis virus @ 200 ml per ha in four week old seedlings controls tobacco caterpillars. Spray monocrotophos 36 WSC 15 ml in one litre of water to control nursery pests. Drench the seed bed with 4% bordeaux mixture or copper oxychloride 0.2% to control damping off and blackening of stem and roots at ground level, 2 weeks after sowing. It is a good practice to drench immediately after each rain at 4 days interval. TMV mosaic virus is controlled by 1% Bougainvillea or Basella alba leaf extract 2 to 3 times at weekly intervals. Optimum seed rate of tobacco is 70 to 80 gm per ha. Seeds are sown in nurseries in 8 to 10 weeks advance of transplanting. Seeds should be broadcasted by mixing it with fine sand or ashes the proportion being one part of seed and 15 part of sand. Seeds are sown at a depth of 2.5 cm. After sowing, the seed should be mixed well with the soil by a rake or brush. About 50 m^2 seed bed area gives about 6000 transplantable seedlings. The seed is long lived and will remain viable for 20 years or more when stored under good conditions. Before sowing, seeds should be cleaned by a blast of air and disinfected for 15 minutes in 1:1000 solution of mercuric chloride or silver nitrate to control angular leaf spot and wild fire. Seed of good quality after cleaning gives 90 % or more germination, for which the optimum temperature is 22 to 27^0 C. Germination begins after 5 to 7 days, when primary root ruptures the testa, followed by the hypocotyls carrying with it the testa. The cotyledons then emerge and a few days later the first two leaves appear. Seedlings of pencil thickness and of 10 to 15 cm length are normally preferred. Short seedling may establish well under optimum conditions in heavy soils. In light soils longer seedlings are preferred for planting. Normally, the seedlings are ready for planting at the end of 7th week and in the first picking 30 to 40% of the total seedlings will be available. The seed bed is top-dressed after each pulling to make the remaining seedlings also grow to transplantable size. Seedlings are hardened off gradually by the removal of shade covers for a short period daily and ready for transplanting in 50 to 60 days after sowing with 15 to 20 cm plant height. Select healthy seedlings of uniform size free from pests, diseases and nematodes and plant them in cool hours of the evening for better establishment. When planting is delayed due to unfavourable field conditions, the overgrown seedlings in beds are clipped without damaging the growing point in order to retard the growth.

Cigar Tobacco

Main field preparation: Deep tillage to a depth of 25 cm is necessary. Four to five ploughings followed by two harrowings are necessary to get the required tilth. Summer ploughing reduces building up of the pest and diseases including nematode. Ridges and furrows are preferable than flat beds.

Spacing: The spacing for chewing tobacco, cigar tobacco, cheroot tobacco and FCV cigarette tobacco are 75 cm × 75 cm, 75 cm × 50 cm, 60 cm × 45 cm and 84 cm × 84 cm respectively. Broad leaf varieties belonging to *N. tabacum* are spaced wider apart than *N. rustica*, which has sharp narrow leaves.

After cultivation: Gap filling is done on 7 to 10 days after planting.

Cropping system: Onion and coriander are suitable intercrops without affecting tobacco yield.

Weed management: Two hand weeding are given on 21 and 45 DAT (days after transplanting). Subsequently ridges and furrows are formed on 52 DAT to facilitate heavy irrigation at grand growth stage of the crop. Fluchloralin @ 1.0 litres per ha or Oxyfluorfen @ 0.5 litres per ha in 750 litres of water sprayed as pre-emergence (one week prior to planting) control weeds up to 25 DAS and one hand weeding and earthing up on 45 DAT is sufficient for tobacco crop. Broomrape (*Orobanche aegyptiaca* L. Orobanchaceae) is an annual root parasites lacking chlorophyll, up to 1 m tall, usually parasitize Solanaceae and Fabaceae hosts reducing crop yield severely. High soil moisture due to irrigation or rain after planting, low soil temperature during winter months encourage heavy incidence of *Orobanche.* Seeds of *Orobanche* are irregular wedge shaped oblong, tiny dust like 0.2 to 0.5 mm long black to brown coloured. Seeds germinate in response to host root exudates and the seedlings must come in contact with host root immediately after germination. Shoots emerge in clusters and their basal portion is attached to tobacco roots through which it draws nourishment and depletes the host. Some species may produce flowers within a week of emergence from the soil. Affected plants become stunted, leaves turn pale and wilt. Initially the leaf tip droops and as the attack intensifies, all the leaves wilt with characteristic ribbing of the midribs. Orobanche is controlled by removing the plants as and when the shoot appear above the ground level before flowering and seed set. The removed shoots are to be buried or burnt. Trap crop of greengram, sesame or sorghum can be grown to reduce Orobanche infestation. *Orobanche* infestation reduces the yield and quality of tobacco about 24 to 52% respectively. *Orobanche* can be controlled/managed with deep ploughing twice or thrice in summer buries *Orobanche* seed to deeper depth and thereby helps in reducing the emergence of *Orobanche*. In *Orobanche* sick fields, skip off growing tobacco for one or two seasons. Avoid growing brinjal, tomato and bhendi crops in these sick fields. Periodical removal of *Orobanche* shoots before flowering and setting of seeds, reduces the menace. *Orobanche* shoots should be destroyed by burning. Grow trap crops such as jowar, gingelly, blackgram and greengram in *Kharif* which facilitate *Orobanche* germination but will not allow growing. This will reduce the *Orobanche* seed load in the soil. CTRI developed a spear with a 2 m stick at the end of which 18 cm length, 8 cm breadth and 0.5 cm thick sharp iron blade is arranged. Cutting the *Orobanche* shoots within 3-4 days after emergence out of the soil, before flowering of *Orobanche* either up to the soil level or 2-3 cm below the soil level with spear will effectively control *Orobanche*.

Topping and desuckering: 'Topping' is the removal of the tobacco flowers while 'suckering' is the pruning out of leaves that are otherwise unproductive. Topping consists of the removal of the terminal bud inclusive of inflorescence at the time of flowering to increase yield through leaf area and thickness. Topping is done leaving ten leaves on the plant. Soon after topping, suckers are produced due to the growth of auxiliary buds. 'Desuckering' is the removal of the 3 to 5 inches long buds found in the leaf axile. Desuckering is done weekly interval for 4 to 5 times. The time from transplanting to topping is 65 to 70 days. Chewing tobacco is topped at 10-12 leaves (including bottom sand leaves). The process of topping by removing apical dominance promotes growth of lateral buds. Second topping is to be completed within a week from first topping. This ensures uniform maturity of plants at the time of harvest. Desuckering is done to prevent nutrients being diverted to lateral shoots in turn or either removed by hand or suppressed by use of chemicals. Application of neem oil emulsion at 35 % immediately after topping completely suppresses the suckers (neem oil 1.75 litre + sandovit 0.5 lit + water 2.75 lit will suffice for 1000 plants). Application of Decanol 6% or Royalten or Power @10 ml/plant after topping controls suckers in the top 5-6 axils. If these chemicals are not available NOE 35 % is recommended. Unless desuckering is done regularly, the advantages of topping are lost. The objective of both topping and desuckering is to divert the nutrients from flower heads of the leaves which cause them to grow in size and weight. The actual practice of topping and desuckering depends on class of tobacco grown, types of soil and it's fertility and spacing. With fire cured tobacco 8 to 10 good leaves are usually left on the plant; 12 to 20 leaves in the case of flue-cured and air cured is intermediate. Low topping causes an increase in size and thickness of leaves, accumulation of nutrients and nicotin and gives a cured leaf which is heavy, oily, dark in colour, tough and somewhat leathery. If the leaf of flue-cured types is too heavy, topping may be deferred until the flowers open or even until the seeds have begin to set. Judicious topping and desuckering improved cured leaf production by 26%. Bud topping at 18 leaves followed by sucker control with 4% decanol results in additional cured leaf yield of 2 quintals per ha. The particulars of topping and desuckering operations practiced in different tobacco types are presented in Table 4.

Table 4 Topping and desuckering operations practiced in different tobacco types

Tobacco type	Time of topping	Topping interval	Time of desuckering	Desuckering interval / number of leaves left in plant
Cigarette	A week before or a day after flowering	Twice at weekly intervals	10 days after topping	NA
Bidi	40 to 56 days after transplanting	Twice at weekly intervals	10 days after topping	10 to 12 leaves
Cheroot	After emergence of flower head	Twice at weekly intervals	10 days after topping	6 days interval/ 14 leaves
Chewing	After emergence of flower head	Twice at weekly intervals	10 days after topping	8 to 15 leaves
Hookah	After emergence of flower head	Twice at weekly intervals	10 days after topping	NA

Note: NA means data not available
Source: Central Tobacco Research Institute, Rajahmundry

Manures and fertilizer application: A tobacco crop yielding about 1000 kg per ha removes approximately 90-22-120 kg NPK per ha per year. In an estimate 2.14, 2.5 and 1.85 kg of NPK are required respectively to produce 1 quintal of tobacco leaf. FYM or compost or night soil @ 25 tonnes per ha applied during last ploughing. Compost can well be substituted by basal application of 250 kg per ha of neem cake without any reduction in the yield. Sunnhemp as a green manure crop can be raised and ploughed *in-situ*. Tobacco responds well up to 200-100-100 kg N, P_2O_5 and K_2O per ha. However the application of 150-100-100 kg of N, P_2O_5 and K_2O per ha is optimum for economic returns. All fertilizer materials should be applied before rapid growth begins in the third to fourth week after transplanting. Applications after this time may be available in later stages of growth, delaying N depletion, delaying maturity, and adversely affecting quality. Normally the entire phosphorus is applied basally before transplanting. As the growth of transplanted tobacco in the initial stage is very slow, nitrogen and potassium are top dressed in the two equal splits in 45th and 60th day after transplanting. Nitrogen can be applied as urea or NH_4SO_4. Organic N is believed to prevent tobacco from ripening too rapidly by maintaining a more uniform N level. 100 kg N per ha is to be applied as NH_4SO_4. Application of NH_4SO_4 and groundnut cake, each to supply 50 kg N can be replaced safely with application of 100 kg N per ha as NH_4SO_4. Soil application of 50% N as urea and foliar application of the remaining 50% N as 2% urea three times at 10 days interval is found to be a better N management practice for maximizing the yield. Soil application and seedling dipping of azosprillium, along with the application of 75% of recommended nitrogen is the most efficient and economic nitrogen management practice in tobacco. Smoker's choice varies from wild tobaccos (low nitrogen products) like flue-cured tobacco to strong and pungent tobaccos like cheroot and hookah tobacco (high nitrogen products). Smoking tobacco giving mild and pleasant smoke require low nitrogen fertilization while strong tobaccos like cheroot and hookah require high nitrogen dose. Nitrogen affects yield and quality of tobacco. Low nitrogen reduces tobacco yield and the results in pale green leaf and slick with poor texture while excessive N may slightly increase yields, it also stimulates excessive suckering, delays maturity, and may result in dark colored, unripe cured leaf. When managed properly, soil N levels may decline rapidly as full flowering approaches. After topping, the plants may begin to break from a dark green colour to a lighter green colour as the N depletion nears completion. The recommended dose of nitrogen and potash for different types of tobacco is furnished in Table 5.

Table 5 Recommended dose of nitrogen and potash for different types of tobacco

S. No.	Types of tobacco	N (kg/ha)	K_2O (kg/ha)
1	Flue cured tobacco		
	a) traditional black soils	30	30 - 50
	b) northern light soils	60	80 - 120
	c) southern light soils	30	60
	d) Karnataka light soils	40	60 - 80
2	Lanka (deltaic soils)	50	40
3	Cherrot	50	100
4	Burley	80	40
5	HDBRG or brazilian tobacco (HD Burley)	100	50
6	Chewing	100	50
7	Hookah	112	112
8	Cigar wrapper	125	224
9	Cigar filler	150	100
10	Bidi	180	40
11.1	Natu irrigated (light soil)	200	50
11.2	Natu black soil	100	50

Source: Central Tobacco Research Institute, Rajahmundry

A maintenance dose of 50 to 100 kg of P_2O_5 per ha depending on type of tobacco and soil is applied to meet the phosphorous requirement. Magnesium is a constituent of chlorophyll. Mg deficient leaf produces chlorotic areas which shows relatively dark and of uneven colour symptoms after curing. Mg deficiency is called 'sand drown' and reported in very light soils. Application of 30 to 37.5 kg/ha of MgO as dolomite ha in two splits at 25 and 45 days after planting respectively to improve the burning and ash holding quality. MgO is used to correct Magnesium deficiency which mostly occurs under repeated irrigation or heavy rainfall. Small quantity of chloride (2% in fertilizer) improves the yield, colour, moisture content, burning and keeping quality of tobacco. Larger amounts of chloride produce muddy and uneven colour in cured leaf with excessive hygroscopicity. Burning quality of tobacco leaf is adversely affected if leaf chlorides content exceeded 2%. The tobacco leaf becomes in combustible of the chloride content exceeder 2.5%. For growing chewing tobacco the irrigation water should have potassium and moderate chloride contents. High chloride tobacco is called as saline tobacco which has the following characteristics *viz.*, (i) greater than 1.5% chloride content, (ii) masky or characteristic "linoleum smell", (iii) moisture absorption and deterioration in keeping quality during stiorage, (iv) poor combustability. The problem of high chloride content in tobacco can be overcome by (i) growing tobacco in soils with less than 0.01% chloride, (ii) irrigation water should contain less than 50 ppm chlorides, (iii) giving only 6 to 8 irrigations and (iv) avoiding application of fertilizers containing chlorides like NH_4Cl (ammonium chloride). Boron deficiency shows die back symptoms. Bud leaves exhibit a light green colour and distorted appearance with breakdown of the tissues. Application of 5 kg borax corrects boron deficiency. Borax application increases the yield by 200 kg per ha.

Water management: Tobacco has 80 to 85% water of the plant weight. The bulk of the roots of the tobacco plants is found in the top 30 cm of the soil and two third of the water loss is from this layer. In the black soils, maximum root activity is found in the top 30 cm layer of the soil under dry conditions. Most of the farmers grow tobacco under flat bed system which is uneconomical as far as the water usage is concerned. Wider ridges and with spacing of 60 cm × 30 cm shallow furrows and staggered planting at 50 cm apart on both side of ridges can be practical to increase water use and leaf yield in tobacco. Irrigating the crop with 12 mm of water per irrigation in the establishment stage (planting to 35 days), 36 mm per irrigation during growth phase (35 to 80 days) and 57 mm later on in wider ridge system is found to be useful in economizing the water to the tune of 28 to 32 per cent as compared to normal method. The crop requires minimum of 8 to 9 irrigations during the cropping period for the flue-cured tobacco. Generally 10 to 12 irrigation are given to hookah tobacco. For cigar and cheroot tobacco about 20 irrigations at four to five days interval are required in about 120 days. 25 to 30 irrigations are given to Chewing tobacco during its cropping period. Water requirement is 450 to 600 mm for a tobacco crop of 120 to 125 days. It requires 24 mm water per week. For irrigating tobacco crop, an IW/CPE ratio of 0.8 is found optimum for maximum yields. Highest tobacco yields are maintained when available moisture remains above 50 percent of field capacity. Tobacco roots begin to die within six to eight hours in saturated soils. This is the reason tobacco plants die after heavy rainfall occurrences. Water having excess of 50 ppm chloride causes leaf burn which leads to low pricing index.

Plant protection: Diseases such as damping-off, black shank and leaf blight, collar rot, leaf spot diseases such as anthracnose, frog-eye spot; root-knot nematode, tobacco mosaic occur in tobacco crop. The pests that infest the nursery are viz., i) caterpillars and cutworms, ii) stem borer, iii) grass hoppers and iv) whitefly. Before sowing seed beds, mix chlorpyrifos 1.5 dust @ 40 gm per 10 sq. m. bed in the top soil to prevent damage to seeds and tiny seedlings from soil pests like ants, mole crickets, earthworms etc. Rabbing the seed beds before sowing with paddy husk or any other slow burning farm waste material @ 6 kg/m^2 to destroy nematode larvae and egg masses gives initial protection to seedlings for about a month. Bordeaux mixture @ 0.4% or Fytolan or Blitox @ 0.2% or Ridomil MZ 72 WP @ 0.2% is to be applied to control diseases. Basamid @ 30-40 g/m^2 should be applied 10 days before sowing to control nematodes. Spray calyxin 0.05% or carbendazim @ 0.03% (3 g in 10 litres of water) or mancozeb @ 20 g in 10 litres of water at 15 days interval, commencing from 30 days after planting, for the effective control of the leaf spot diseases such as Frog eye spot, Anthracnose and Brown spot disease as well as Fusarium wilt. Prophylactic sprays with virus inhibitors of plant origin like *Basella alba* and *Bougainvilea spectabilis* and Neem leaf extracts @ 1% dilution on 30th, 40th and 50th day of planting tobacco are useful to control tobacco mosaic virus. Spray 1 % neem seed kernel suspension (NSKS) to control insect pests. If the tobacco seedlings damaged by pests such as stem borer, grasshoppers, whiteflies, aphids, leaf eating caterpillar, ground beetle, capsule borer, spray one of the insecticides *viz.*, Chlorpyrifos 20% E.C. @ 25 ml in 10 litres of water, Acephate 75% S.P. @ 10 g in 10 litres of water, Methomyl 24 L @ 18 ml or methomyl 12.5 L @ 30 ml in 10 litres of water from fourth week onwards. If 100 whiteflies stick to the trap, any one of the following insecticide such as Imidacloprid 200 SL @ 2.5 ml, Chlorpyrifos 20% E.C. @ 25 ml, Thiamethoxam 25 WG @ 2 g or Acephate 75% S.P.@ 10 g in 10 litres of water may be sprayed.

Harvest: The stage of maturity and method of harvesting differ with the type of tobacco. Bulging of the interveinal portions of the leaf and the appearance of the brown spot on the leaves and change of green leaf to yellowish green are the signs of maturity. The crop duration is 100 to 125 days. Normally 18 to 22 leaves are harvested with 2 to 3 leaves per week for 30 to 50 days. Good yields under commercial production with adequate water supply are in the range of 2 to 2.5 ton/ha of leaves. There are two methods of harvesting tobacco namely,

i) *priming* is a method of harvest practiced in flue cured tobacco or cigarette tobacco in which leaves are individually harvested from the bottom of the plant upwards as they mature. Harvesting normally commences 80 to 100 days from planting and 14 to 21 days after topping. Two to three leaves are harvested at weekly intervals. The leaves are strung back to back in alternate pairs on either side of sticks and are hung in the barn.

ii) *stalk-cut or whole plant-cut method*: For production of air and flue cured tobacco like hookah, bidi, cigar, cheroot and chewing tobacco, the whole plant is cut off at close to ground level when the most of leaves are ripe. This usually occurs from 45 to 55 days after topping. The leaves are cured on stems which are tied to poles and the leaves are wilted before transfer to the curing barn. Chewing tobacco matures in 120-130 days. Heavy puckering and spangle development on puckered leaf with shiny oily appearance are the indications of maturity. Puckering refers to the bulging of intervenal portion of leaf and spangling refers to the reddish brown spot on the puckered portion. Britling of midrib on touching the leaf is also a sign of maturity. A light irrigation is to be given a day prior to harvest and the plants are cut with a sickle close to the ground in the afternoon hours and to be left to wilt on the ground. Next day morning the wilted plants are arranged in small heaps in the field and covered with coconut leaves or palm mat. The plants are tied in pairs at the butt end with the help of jute string and strung on the bamboo scaffolds.

Characteristics of tobacco leaves based on stalk position

Leaf: One third of the plant (34.5%) is made up of the leaves on the middle to upper stalk. These leaves are firm, thick, and heavy bodied with pointed tips. They contain 3 to 3.5% nicotine and up to 15.5% sugars.

Cutters: The largest leaves on the plant, both in length and width, although only 8% of its weight. Thin to medium-bodied leaves from the middle of the stalk or below, cutters have rounded tips and a most desirable color when ripe. High in oil and resin content, cutters contain about 2.5% nicotine and 12% to 22% sugars.

Primings: The first leaves to ripen and to be harvested, primings make up 12% of the total plant weight and contain 1.5 to 2% nicotine and 5 to 10% sugars.

Tips: These leaves at the stalk top make up around 18% of the plant's total weight. Tip leaves are narrow and pointed, smaller than lower leaves, yet thicker and more full bodied. Tips of flue-cured tobaccos contain from 3 to 3.5% nicotine and 6 to 6.5% sugars.

Smoking leaf: The leaves just above the stalk middle are thinner than the 'bodied' leaves above them, and their tips are less pointed. About 7.5% of the plant, smoking leaf ripens to a rich orange color and contains about 3% nicotine and 12 to 20% sugars.

Lugs: These thin blunt-tipped leaves around the bottom of the stalk make up 13% of the plant's weight. Lugs contain about 2.5% nicotine and 12 to 20% sugars.

Source: http://ipm.ncsu.edu/

Yield: The yield of cured leaves is 2500 to 3500 kg per ha under irrigated conditions. Seed yield is 336 kg per ha. Tobacco seed contains about 35 per cent of semi drying oil which is nicotine-free. Tobacco seeds are crushed for oil in the same expellers as used in the groundnut oil mills. Yield of oil from the standard expeller is 25 to 30 per cent. Further 8 to 10 per cent oil can be extracted from the cake by using solvent extraction process. Tobacco seed oil is mostly light brownish yellow in colour. The colour of the refined tobacco seed oil is golden yellow. Tobacco seed oil is mostly used in the manufacture of paints and varnish. Tobacco seed oil industry is concentrated mostly in Prakasam district of Andhra Pradesh. The dehydrated tobacco seed oil-cake is used as cattle feed.

Curing: Tobacco leaves contains 85 to 95% moisture at maturity. Curing is the process by which the harvested tobacco leaf is made ready for the market. Curing is essentially a drying process whereby most of the moisture in the harvested leaf is removed. A bad leaf produced on field cannot be improved by curing; but a good leaf can be spoiled by bad and defective curing. The process of curing has an intimate bearing on the quality of the final produce. Depending on the type of the tobacco, four principal methods of curing can be distinguished, namely, i) flue-curing, ii) air-curing, iii) fire-curing and iv) sun-curing. During curing as the colour and aroma of the leaf are developed, the moisture content drops by 10 to 25%. If the leaf has been well grown at maturity the majority of the dry matter is carbohydrate. This fraction comprises mainly starch of 25% in cigar tobacco and 50% in flue-cured tobacco. Nicotine content in leaves of *N .tobaccum* is 1.5 to 4% and up to 10% in *N. rustica.* This alkaloid is both narcotic and addictive. The characteristic aroma of tobacco is derived from etherial oils and resins that are secreted by the glands of the leaf hairs. Tobacco leaf may also have a high ash content, which can affect both combustion and flavour qualities. Leaf with high chlorine content is not acceptable to smokers. Cigaratte tobaccos is flue-cured; cigar tobacco, burley and maryland tobacco are air cured; dark heavy chewing tobacco are fire cured; Turkish tobacco is sun cured. Cigar tobacco comprises three types of leaf viz., i) 'filler' which provides the bulk of the cigar, ii) 'binders' which ties the filler together and iii) 'wrapper' which envelopes the whole cigar.

Flue curing (Cigarette tobacco – Virginia tobacco) is done in brick barns in which heat from wood, oil, gas or other source is supplied in closed flues. Absolute control of temperature and humidity in the barn is essential. A barn of 12 × 12 × 16 ft is sufficient for 1 to 2 acres of the crop and takes 1500 to 2000 lbs of green leaf at each filling. 5000 lbs of green leaf produces about 800 lb of cured leaf and the number of curing per crop is 4 to 8. Flue-curing tobacco is cured in 'ovens/barn' at temperatures of around 38–54°C over 5-10 days, but it is depending on the leaf type. Heavier leaf takes higher time than the light leaf. There are three steps in flue curing *viz.*, i) Yellowing of leaf: the leaves turn golden yellow colour in 36 to 48 hours at a temperature of 46 to 52°C and the curing process takes 12 to 20 hrs to fix the yellow colour in the leaf and iii) Drying the leaf:web is completed at 57 to 60°C and mid rib at 71°C. This process takes 50 hours.

Sun curing: Sun curing is done for cigar tobacco, chewing tobacco, bidi tobacco and natu tobacco. Ferment the leaves in heaps for 24 to 36 hours, before exposing it to the sun. 25 to 30 leaves are hung on a 3 m long pole for 2 to 3 weeks to expose to the sun. On alternative days the stalks are turned to expose the plants uniformly to sun on all sides for 2 to 3 weeks depending on moisture content of leaves and weather conditions. For reducing the cost, stringing can be done on wires at 15 to 22 cm distance. After sun-curing, when the plants are dried completely, are unloaded in the morning time when the leaves are pliable. These plants are arranged in the shed layer by layer in a rectangular bulk of 5 feet length, 5 feet breadth and 4 feet height with butt ends pointing out side. The bulk is covered with date palm mat or Hessian cloth. Turnings are given once in 3 to 4 days depending upon the temperature of the bulk. After 4 or 5 turnings the bottom leaves of the plants will show signs of separation from the stalk and the colour of the leaves turn to dark brown. It is an indication for stripping the leaves. The leaves are to be stripped individually from each stalk and graded into Rasi long, Rasi short and Kruz. The graded leaves are made as a bundle having 40 leaves and tied at the butt ends. The leaf bundles are again arranged in bulks and turned once in 3 days up to 4 to 5 turnings and later once in a week. During this process, the leaves become dark brown. When the temperature of the bulk cools down the leaves exhibit good smell and signs of whitish incrustations called 'Pupal' which is an indication for

the proper handling (fermentation) of the leaves. The leaves become ready for disposal one month after stripping. The particulars grading in Chewing tobacco are furnished in Table 6.

Table 6 Chewing Tobacco- Grading

S.No.	Basic Criteria / Leaf Integrity	Grade	Sub-criteria length	Sub-Grade
1	Whole leaves without damage	Rasi	60-70 cm	Long I
			50-59 cm	Medium II
			40-49 cm	Short III
2	Leaves damage 4d / Broken (mostly bottom leaves) and bits	Kruz	> 40 cm	Long
			<40 cm	Short

Air curing: Air curing method is practiced in cheroot, cigar and chewing tobacco. The leaf should be yellow before it dries out. Then the rate of drying is gradually increased by increasing the ventilation. The best temperature for the early wilting is 21 to 24°C and it should never exceed 43°C even in the final stages. The time taken for curing is 6 to 8 weeks.

Smoke curing or Fire curing (for chewing tobacco): The harvested plants are allowed to wilt one day in the field. The leaves are cut individually from the wilted plants along with a portion of stalk (stem) and tied in a bundle of three leaves and hung on the nails provided in the barn used for smoke-curing. The smoke-curing barn is a brick structure having 3.0m × 2.7m × 1.8m dimensions with opening of 0.9m × 0.75m to facilitate the insertion of smoke-cured materials as well as the harvested leaf for smoke-curing process. The top of the barn is also open having a bamboo structure with nails for hanging the leaf. Generally coconut husks are used for smoke-curing. The coconut husks are put in fire after placing in the barn and immediately the fire is put off by sprinkling water so that only smoke is emitted from the coconut husks. Care is to be taken that smoke does not escape out from the barn. The top portion of the barn is also covered with waste materials to avoid escape of smoke from the barn. Each bamboo structure in the barn is able to hold 40 pairs of bundles. A total of 12 hours is needed for smoke-curing. After smoke-curing, the leaves are taken out from the barn and arranged in a rectangular bulk and turnings are attended periodically. Again the leaves are kept in the barn and smoke-cured for 12 hours. Hence smoke-curing and bulking process is to be repeated 4 times. Later, the leaves are hung outside for drying about 2 to 3 weeks. The cured leaves are then bundled according to colour weighing one kg each and the base of the leaf is tied with the help of plantain sheath. Then the leaves are dipped in salt water. The excess salt water is drained out and the bundles are opened up. The leaves are spread out and rolled around the mid rib of each leaf like a tape. Then the rolled leaves are tied in a bundle of five each and bulked for one month. The bulk is turned every fortnightly interval. Appearance of whitish incrustations on the leaf indicates proper curing and processing of leaf. Then the smoke cured leaves are tied in a bundle of five kg each. The tip of the leaf is tied with a help of a palmyra string and the bundle will look a 'down's cap' which is unique of smoke-curing tobacco.

Pit curing (for hookah, lanka tobacco and chewing tobacco): The harvested plants are allowed to wilt overnight and are kept in small heaps for a few days and then transferred to a pit. The pit size will be generally 3' × 2'× 3'. The heaped plants are gently pressed so that the plants are filled up uniformly. After complete heaping, the top portion of the pit is to be covered with hessian cloth or leaves of the plants in the top and plastered with mud. The pit is opened after 12 to 14 days. Then the leaves are stripped for hardening. After stripping, the stripped plants are hung on the bamboo scaffolds for 3 to 4 days for complete sun drying. Later the cured leaves are bulked in the godown. The bulk is turned periodically at 4 to 5 days interval. Whitish incrustations and fruity odour can be observed in the cured leaf after bulking which is a characteristic feature of pit-cured produce.

Bulking and Grading: After curing, the leaf is taken from the barn and is bulked in platforms up to 6 feet height. Weight are placed on bulks to allow some fermentation. The tobacco should be bulked for at least a month before grading and marketing. The leaves are then graded according to size, colour and texture. The grade will depend partly on the position of the leaf on the plant. The bottom leaves are called as *lugs*, the lower middle leaves are called as *cutters;* the upper middle and top leaves are called *tips* and *green tips* respectively. The lug is further classified as good lugs, sand lugs, trash lugs and scrap trash. After grading, the leaves are tied with hands with 12 to 15 leaves per hand for fire cured and 20 to 30 leaves for flue-cured. This is done by wrapping the leaf of the same grade folded into a band round the top of the petioles and then tucking the stem of the leaf inside the hand. Sometimes the leaf is stripped by removing the petiole and midrib before selling. Then the leaf is conditioned when it is dried thoroughly and steamed for a few seconds to acquire the maximum moisture content which should be about 12% for local sale and 10% for export market. The tobacco is then boiled using hessian cloth (jute cloth). The tobacco is left to mature in the bales for 1 to 2 years before being used for manufacture. The best flue cured tobacco should have leaves of medium texture with good elasticity, relatively free from gum, slightly oily, mildly aromatic and a bright yellow colour without blemish. Fire cured tobacco should have large, broad leaves with heavy body and texture, good elasticity and should be dark mahogany in colour without blemish. Light air-cured tobacco should be relatively light in body and texture with light or reddish brown colour. Cigar-filler should be heavy bodied with a considerable aroma and a satisfactory burn. Cigar-binder should have a fine texture and more elasticity. Cigar-wrapper, the elite of tobacco, should have thin, silky leaves with fine veins and texture, good elasticity, free from injury and blemish and with a satisfactory even colour, aroma and burn. The farm grades of Flue cured tobacco for light soils and black soil in Table 7 and Table 8 respectively.

Table 7 Farm grades of Flue cured tobacco for light soils:

Group (plant position)	Grade	Body	Colour	Spot/ blemish/ injury/waste in terms of percentage
Primigs (P)	P1	Very thin	Bright lemon or orange	< 20
	P2	-do-	lemon or orange	20 to 30
	P3	-do-	lemon or orange	30 to 55
	P4	-do-	lemon or orange	55 to 80
	P5	-do-	lemon or orange	> 80
Lugs and cutters	X1	Thin to medium	Bright lemon or orange	Up to 20
	X2	-do-	lemon or orange	20 to 30
	X3	-do-	lemon or orange	30 to 55
	X4	-do-	lemon or orange	55 to 80
	X5	-do-	lemon or orange	> 80
Leaf (L)	L1	Medium to heavy	Bright lemon or orange	< 20
	L2	-do-	lemon or orange	20 to 30
	L3	-do-	lemon or orange	30 to 55
	L4	-do-	lemon or orange	55 to 90
	L5	-do-	lemon or orange or mahagony	> 90
Tips (T)	T1	Medium to heavy	Bright lemon or orange	< 20
	T2	-do-	lemon or orange	20 to 30
	T3	-do-	lemon or orange	30 to 50
	T4	-do-	orange or mahagony	55 to 80
	T5	-do-	orange or mahagony	> 80

Source: Central Tobacco Research Institute, Rajahmundry

Table 8 Farm grades of tobacco for black soil

Grade	Colour	Body	Texture	% Blemish
F-1	Bright lemon or orange	Thin to medium	Soft	25
F-2	Light brown or brownish lemon	Medium	Good	25 (white to yellow blemish allowed)
F-3	Light brown	Good to medium	Medium	50
F-4	Brown	Heavy body	Medium to coarse	50 (brown blemish allowed)
F-5	Dark brown	Heavy body	Medium to coarse	50
F-6	Light greenish orange	Good	Soft to medium	10

F-7	Light medium green	Heavy	Medium to coarse	25
F-8	Medium green	Heavy	Medium to coarse	35
F-9	Dark green	Coarse	Coarse	---
F-10	Yellow, green and / or brown	Variable	---	---

Source: Central Tobacco Research Institute, Rajahmundry

Tobacco products: Products entirely or partly made of the leaf tobacco as raw material, and which are manufactured to be used for smoking, sucking, chewing or snuffing. Smoked forms of tobacco products are *Bidis, Cigarettes, Cigars, Cheroots, Chuttas, Dhumti, Pipe, Hooklis, Chillum and Hookah.*Smokeless forms of tobacco products are *Paan (betel quid)* with tobacco, zarda, *Paan masala* with tobacco; Tobacco, areca nut and slaked lime preparations, *Mainpuri tobacco, Mawa, Khaini, chewing tobacco, snus, gutkha, creamy snuff, Tobacco water, Nicotine chewing gum.* Beedis are the most popular smoking form of tobacco in India. Thirty-four per cent of the tobacco produced in India is used for making beedi (bidi). Beedi is made by rolling a dried, rectangular piece of tendu leaf with 0.15 to 0.25 g of sundried, flaked tobacco. The collection of tendu leaf that is used to wrap bidi forms an important link for the bidi industry. Tendu leaf is almost wholly grown on government-owned forestland, with around 62% of tendu leaf being grown in Madhya Pradesh. Tendu leaf accounts for 74% by weight of bidi. Dark and sun-dried tobacco varieties are used in bidi production. Almost 80% of bidi tobacco comes from Gujarat, and the rest comes from Karnataka. Bidi rolling is concentrated in the states of Madhya Pradesh, Andhra Pradesh, Tamil Nadu, Uttar Pradesh and West Bengal. Bidis account for over 50% of total tobacco use, compared with less than 20% by the cigarette segment. There are an estimated 290 000 growers of bidi tobacco in India. Bidi tobacco is considered the poor person's tobacco. Cigarette smoking is the second most popular smoking form of tobacco used in India after beedis. Currently, there are four major cigarette manufacturers in India: ITC Limited (formerly Imperial Tobacco Co.); VST Industries Limited (formerly Vazir Sultan Tobacco Co.); Godfrey Philips India Ltd; and GTC Industries Limited (formerly Golden Tobacco Co., Ltd.). Cigars are made of air-cured, fermented tobacco, usually in factories. A cheroot is a roll made from tobacco leaves. Chuttas are coarsely prepared cheroots. Chutta smoking is widespread in the coastal areas of Andhra Pradesh, Tamil Nadu and Orissa. Pipe smoking is one of the oldest forms of tobacco use.

Phytochemicals: All parts of the plant contain nicotine, which can be extracted and used as an insecticide. The dried leaves can also be used; they remain effective for 6 months after drying. The juice of the leaves can be rubbed on the body as an insect repellent. The leaves can be dried and chewed as an intoxicant. The dried leaves are also used as snuff or are smoked. Nicotine sulphate is a powerful insecticide/pesticide and is used in agricultural sprays. Tobacco leaf contains about 0.8 per cent solanesal, an isoprenoid alcohol used in cardiac drug. Solaneasal in a crude in form is exported to Japan. Tobacco Specific Nitrosamines (TSNA's) have been identified as being harmful and causing cancer. A drying oil is obtained from the seed. Smoke tar (nicotine free total particulate matter), carbon monoxide and tobacco specific nitrosamines (TSNA) are the chemical substances responsible for tobacco related health problems. The particulars of phytochemicals produced in tobacco, their products and uses are furnished in Table 9.

Table 9 Phytochemicals produced in tobacco, their products and uses

Phytochemical	Content (%)	Value added product (s)	Use (s)
Nicotine	1.5-3.0	40% Nicotine sulphate	Botanical pesticide
		Pure nicotine	Drugs/Tobacco cessation products
Solanesol	0.2-2.0	Coenzyme Q9 Coenzyme Q10	Cardiac drug
		Vitamin K	Anti-haemorrhagic vitamin
		Vitamin E	Anti-sterility vitamin
Organic acids		Crude organic acid fraction	Solubilisation of Rock phosphate
	4.0-4.5	Malic acid	Foods & Beverages
	0.5-2.0	Citric acid	Foods & Beverages
Proteins (Green leaf)	0.3-0.5	Crude protein	Feed supplement
Seed oil	32-36	Crude oil	Paint and soap industry
		Refined oil	Edible oil

Source: Sarala et al. 2013

Nicotine: Nicotine ($C_{10}H_{14}N_2$) is the principal alkaloid synthesized in roots and occurs throughout the plant, but reaches its highest concentration in the leaves. Nicotine at 40–60 mg can be a lethal dosage for an adult which is an extremely deadly poison. Nicotine occurs in plant in the form of 40% Nicotine sulphate. Nicotine in pure form is a colourless liquid. Nicotine can produce dizziness, nausea, and hallucinations in excess, and is physiologically addicting. Nicotine has insecticidal property. Nicotine is used to produce nicotinic acid and nicotinamide, the important vitamin B complex constituents. The nicotine content of types of tobacco and in different curing methods is presented in Table 10 and Table 11 respectively.

Table 10 Nicotine content of different types of tobacco

Tobacco type	Nicotine (%)	Ash (%)
Virginia tobacco	1.2 - 3.6	$\leq$ 18.0
Bidi tobacco (Anand)	9.71	19.0
HDBRG (Guntur)	3.89	
Natu tobacco (Black soils)	2.79	
Natu tobacco (Light soils)	3.5	
Burley tobacco	1.26	
Hookah	3.0-7.0	<25.0
Chewing tobacco (Tamil Nadu)	2.93	
Cigar tobacco (Tamil Nadu)	1.24	16.0-22.5
Chewing tobacco(Bihar)	3.7	
Cigar filler (West Bengal)	2.05	
Cigar wrapper (West Bengal)	1.44	

Jati-Chama (West Bengal)	3.69	
Jati-Podali (West Bengal)	4.02	
Motihari-Hemti (West Bengal)	4.83	
Motihari-Bitri (West Bengal)	6.64	

Source: Central Tobacco Research Institute, Rajahmundry

Table 11 Nicotine content of tobacco types under different curing methods

Tobacco type	Type of curing	Nicotine (%)
Burley	Air-curing	0.5-1.0
FCV	Flue-curing	2-3
Lanka/Natu	Sun-curing	3-4
Hookah/chewing	Air curing	4-6
Bidi	Sun-curing	6-10

Source: Sarala *et al.* 2013

Solanesol: Solanesol ($C_{45}H_{74}O$), a tri-terpene alcohol, is a ubiquitous compound present in plant kingdom, is a major component of tobacco and ranges from traces to 4.7%. Solanesol has gained importance because of its value as a source of isoprene units for the synthesis of metabolically active quinones, vitamin K analogs, vitamin E, coenzyme Q9 (CoQ9) and coenzyme Q10 (CoQ10). Chewing tobacco variety, Abirami and HDBRG tobacco are found with high solanesol content (2.0-3.5%).

Seed oil: Tobacco plant is a prolific producer and the tiny tobacco seed contains 35 per cent oil. Tobacco seed oil is found to be rich in linoleic acid and have high PUFA value and 1.5% W-3-fatty acid (an essential fatty acid). Chemical quality of the oil is comparable to safflower oil and is superior to groundnut oil. The peroxide value increased from 8.24 to 84.24 me peroxide/ kg within 90 days from the date of extraction in tobacco seed oil whereas it varied from 6.24 to 50.12 me peroxide/kg within 90 days in sunflower oil.

Seed production: The nucleus seed is produced by the breeder who develops a particular variety where the variety is developed. The breeder seed provides the basis for multiplication of pure seed. A single tobacco capsule may yield 2000 to 4000 small seeds. A single plant can produce several thousand seeds. Thus, a single plant is selected from the original source crop based on the morphological identity, uniformity and genetic purity. The panicle of each selected panicle is bagged. The seed of entire panicle is to be collected and trashed carefully. The seed must be dried to 4% or below 4% moisture and stored. The crop is raised from October to March when the day temperatures are moderate. The duration of the crop is 100-160 days. One single plot is earmarked for a particular variety for nucleus seed production. Being a self pollinated crop, an isolation distance of 4.0 m from the neighbouring plot is sufficient for nucleus seed production. In tobacco, 350-400 kg/ha of seed can be produced.

Utilization: The term .smokeless tobacco is used to describe tobacco that is consumed without heating or burning at the time of use. Smokeless tobacco can be used orally or nasally. For nasal use, a small quantity of very fine tobacco powder mixed with aromatic substances called dry snuff is inhaled. This form of smokeless tobacco use, although still practised, is not very common in India. Mature green leaf of bidi tobacco is found to be good source for extraction of nicotine and solanesol. Bidi tobacco varieties *viz.*, Kumkumardri,

GT 5 and selection 9-14 are rich in nicotine (above 9%), while varieties TI 163 and TI 1112 posses high levels of solenesol and intermediary cardiac drug. Tobacco seed contains 35% semidrying oil which is nicotine free and used in paint and soap industries. Seed cake contains 6% nitrogen which could be used as manure.

REFERENCES

CPG. 2012. Crop Production Guide. Department of Agriculture, Government of Tamil Nadu, Chennai and Tamil Nadu Agricultural University, Coimbatore.

Francesca Orlando, Marco Napoli, Anna Dalla Marta, Francesca Natali, Marco Mancini, Camillo Zanchi, Simone Orlandini. 2011. Growth and Development Responses of Tobacco (Nicotiana tabacum L.) to Changes in Physical and Hydrological Soil Properties Due to Minimum Tillage. American Journal of Plant Sciences. 2: 334-344.

Nayanatara S. Nayak. 2013. Tobacco Curing and Fuel Efficiency in Karnataka, India. SANDEE Working Paper No. 77–13. South Asian Network for Development and Environmental Economics (SANDEE). PO Box 8975, EPC 1056, Kathmandu, Nepal.

Sarala, K, Murthy, T.G.K., Prabhakara Rao, K, and Ravisankar, H. 2013. Tobacco Research in India: Trends and Developments. Agrotechnol 2: 113.

http://www.ctri.org.in/

http://www.ct.gov/dph/lib/dph/hems/tobacco/tobacco_products.pdf

https://en.wikipedia.org/wiki/Tobacco

http://www2.ca.uky.edu/agc/pubs/id/id160/id160.pdf

http://www.uky.edu/Ag/Tobacco/

http://www.tobaccoleaf.org/default.asp

http://www.coresta.org/Guides/Guide-No07-Growth-Stages_Feb09.pdf

http://www.fao.org/docrep/006/y4997e/y4997e0h.htm

http://academic.udayton.edu/health/syllabi/tobacco/history.htm#begin

http://www1.american.edu/ted/turkish-tobacco.htm

https://www.hort.purdue.edu/newcrop/ncnu07/pdfs/langham144-182.pdf

https://en.wikipedia.org/wiki/Nicotiana_tabacum

http://www.fao.org/nr/water/cropinfo_tobacco.html

http://www.uky.edu/Ag/TobaccoProd/publications.html

CHAPTER

35

Tea–*Camellia sinensis* (L) O. Kuntze (2n=30)

Family: Theaceae

Vernacular name: Tea (English), Cha, Chai, *Chaie* (Hindi), Teyila (Malayalam), *Cha* (Bengali), Tea (Kannada), *Theyilai* (Tamil), *Cha, Chah* (Assamese), Teyaku (Telugu, Kannada), Syamaparni (Sanskrit).

Importance: Tea is one of the most important beverage crops in the world. The word '*tea*' is derived from 'Te' in Amoy language and cha is derived from Cantanse language. India is the largest producer and exporter of tea. Darjeeling tea is called 'Champagne of Teas'. Tea earns 7% of total foreign exchange. Tea is a labour intensive crop providing employment opportunities. India is the largest producer of black tea as well as the largest consumer of tea in the world. Currently, India produces 23% of total world production and consumes about 21% of total world consumption of tea. Nearly 80% of the tea produced is consumed within India. Over the last 20 years, India's world ranking fourth as an exporter in the face of stiff competition from Sri Lanka, Kenya, and China.

History: The Emperor Shen Nung of China has discovered tea one day while drinking hot water in his garden in 2727 BC. Tea cultivation has gained popularity by about 650 A.D. during the TS'ang dynasty in most provinces in China. The first book on tea, the Ch'a Ching (The Classic of Tea), written by the Chinese Poet Lu Yu is published in 780. Tea is introduced to England by the Dutch East India Company during 1652. The knowledge of tea travelled slowly from East to West. Col.Kyd, a resident of Calcutta and a famous botanist, has seen tea plants growing in his garden in 1780. In 1788, Sir Joseph Bank has recorded the existence of indigenous tea growing wild in Coochbehar and Rangpur districts of Bengal. Robert Bruce, a British national discovered tea plants growing in the upper Brahmaputra valley in Assam and adjoining areas. In the early 1820s, the British East India Company began large-scale production of tea in Assam. The birth of Indian tea industry has been marked by the discovery of indigenous tea plants in Assam in 1823 by Robert Bush. The 'Tea Committee', appointed during 1834 by the Governor-General Lord William Cavendish Bentinck, reports that tea can be successfully grown in India. In 1836, C.A. Bruce has been appointed as the Superintendent of Tea Forests. In 1837, the first English tea garden was established at Chabua in Upper Assam. In 1838, Indian tea that was grown in Assam was sent to the UK for the first time for public sale. The Assam Tea Company began the commercial production of tea in the region in 1840. The Tea planters formed an association named Indian Tea Association (ITA) in 18 May 1881, with its headquarters at Calcutta. The biggest research centre of tea in the world, now situated in Jorhat, was started in Calcutta in 1900 by the India Tea Association. In 1904, a laboratory was started at Heelea-kah Tea Estate near Mariani. In 1912, the laboratory was shifted to Tocklai (Jorhat) and was renamed as Tocklai Experimental Station. The paper tea bag is developed by the Tetley tea Company in 1953. In 1964, the experimental station became Tea Research Association (TRA). Government of India has set up of Indian Tea Industry Development Association in 1974.

Origin: Tea is native of Southern China in the high valley of the Brahmaputra, the Irrawaddy, the Salween and the Mekong rivers of the borders separating India, China and Myanmar; southeast to the hilly areas of Myanmar to Thailand and Vietnam. The Assamica (*ah-SAH-mee-ka*) strain is native to the Assam region in India.

Distribution: Tea is grown in China, India, Sri Lanka, Kenya, Malawi, Rwanda, Tanzania, Uganda, Turkey, Iran, Argentina, Brazil, Indonesia, Taiwan, Vietnam, Japan and Russia. In India, it is cultivated in Assam, West Bengal, Tamil Nadu, Kerala, Tripura and Karnataka. Tea is grown in a limited area in Kangra (Himachal Pradesh) Tripura and Dehradun Valleys (Uttaranchal), Arunachal Pradesh, Karnataka, Sikkim, Bihar, Manipur, Odisha and Nagaland.

Table 2 Tea growing regions in India

S.No.	Region	Area under tea (ha)	Elevation (m)	Rainfall (mm)	Production (million kg)
1	Assam	312210	45-60	2500-3000	507.0
2	Darjeeling (WB)	17820	90-1750	2500-5000	9.8
3	Anamallais, Tamil Nadu	12625	900-1600	3000-3800	30.0
4	Wayanad (Kerala)	5470	850-1400	2000-2500	16.0
5	Karnataka	2140	750-1000	2000-3500	6.0
6	Dooars and Terai (WB)	97280	90-1750	3000-5000	216.0
7	Munnar - HiahRanaes	13000	950-2600	1300-3000	27.0
8	Kangra (HP)	2348	700-1000	2300-2500	0.8
9	Travancore, Kerala	14000	750-1350	2000-3000	20.0
10	Nilgiris, Tamil Nadu	66175	1000-2634	1000-1500	135.0

Source: http://www.indiatea.org/tea_growing_regions.php

Botany: Tea plant is broadly classified as China, Assam and Cambodia types. The scientific nomenclature is as follows:

1. China tea: *Camellia sinesis* var. *sinensis* (L),
   1.1. Chinese big leaf tea (*Camellia sinensis* var. *macrophylla*).
   1.2. Chinese small leaf tea (*Camellia sinensis* var. *bohea*)
   1.3. Shan tea (*Camellia sinensis* var. *shan*)
2. Assam tea, Indian tea: *Camellia sinensis* var. *assamica* (Maston)
3. Cambodia tea: *Camellia assamica* ssp. *lasiocalyx* (Planch. M.S.)

The tea used for processing consists of the tender leaves and the leaf buds. The tea plant is an evergreen or semi-evergreen woody shrub, attaining a height of 9.1 to 15.2 m. It is never allowed to grow beyond plucking height under cultivation. The bushes are often pruned back to encourage maximum leaf production. The leaves are alternate, generally elliptic to lanceolate with toothed margins. The older leaves are leathery, bright green in colour and 5 to 30 cm long. The undersurface of young tender leaves is densely covered with soft hairs that vanish as the maturity advances. The characteristic fragrance and aroma of the leaves is due to the presence of numerous oil glands. Yellow-centred white or pinkish fragrant flowers are borne in leaf axils either singly or in groups of two to four and produce at maturity three-celled woody capsules, each compartment of which contains a brown seed, about 1.25 cm in diameter, flowers 1-3, in axillary or sub-terminal cymes, deflexed, 2-5 cm broad, aromatic, white or pinkish, actinomorphic, sepals and petals 5-7, pedicels 5-15 mm long; stamens numerous; ovary 3-5-carpellate, each carpel 4-6-ovulate; capsules depressed-globose, brownish, lobate, to 2 cm broad, valvate, with 1-3 subglobose seeds in each lobe; approximately 500 seeds per kg. It requires 4-12 years to bear seed. The tea plant has an economic life span of 50-60 years. However, some of the existing tea plantations are over 80-100 years old. The characteristic features of Assam and China tea are furnished in Table 1.

Table 1 Characteristic features of Assam and China tea types

Assam (*C. assamica*)	**China (*C. sinesis*)**
It is a tree (15 to 18 m height)	It is a shrub (3 to 6 m height)
Few robust branches	Branches abundant and whippy
Large, glossy leaves, (15-20 cm long)	Small, leathery leaves, (3-6 cm long)
Light to medium green colour	Dark green colour
High yield and medium quality	Low yield and good quality
Susceptible to drought and frost	Hardier ecotype and resistant to drought
Sparse flowering	Profuse flowering
Withstands the hotter temperatures	------
Adapted to lower altitudes	------
Assam-type is believed to have originated under the shade of humid, tropical forests	China-type is thought to have originated under open conditions in the cool, humid tropics.

Flowers of *Camellia sinensis*

Tea Seeds

Climate: Tea is cultivated under equatorial, humid and temperate climatic types. Generally, tea thrives within latitude 43° north and 27° south. Low-grown teas are produced from sea level up to 600 m, mid-grown from 600 to 1200 m while the high-grown teas are cultivated between 1,200 and 2,500 m from MSL. The performance of tea is excellent at elevations ranging from 1000 to 2500 m. The tea growing *teelas* of Cachar (Assam) are low hillocks which rarely raise above 150 m whereas the elevation of a few tea estates at the foot of the Himalayas goes up to 500 m from MSL. It grows in areas receiving rainfall of 1250 to 1500 mm distributed in 8 to 9 months in a year. In cloudy conditions and heavy, continuous rainfall, the yield drops. The desirable relative humidity is 80% for tea growth and should be never less than 40%. In dry air, the bud becomes dormant and the plant stops growing. Sunshine hours of 5 hours per day, on the average, are required by tea. The growth of tea shoots is depressed when the photoperiod is less than about 11 hours especially when combined with warm (20°C) nights. Tea growing areas has mean minimum temperature range of 5 to 13°C while the mean maximum temperature range of 24.9 to 33.9°C. The optimum temperature for tea growth is 16 to 20°C. The mean minimum temperatures should not fall below 13°C, nor maximum go above 30°C. The tea plant dies at 5°C. It does not tolerate frost.

Soil: Tea can be cultivated in well drained soil with high amount of organic matter and pH of 4.5 to 7.3. However, tea has good growth a pH between 4 and 5.5.

Season: Tea is planted during May–June or August–September months.

Varieties: The three most distinct known varieties of tea in India are:

(i) Assam tea (grown in Assam and other parts of NE India)
(ii) Darjeeling tea (grown in Darjeeling and other parts of West Bengal)
(iii) Nilgiri tea (grown in the Nilgiri hills of Tamil Nadu)

Assam tea varieties: TV2, TV12, TV13, TV17, TV21, TV28, S3A1, S3A3, Tingamira, TA 17, T3E3, Kaliapani 25, Borahi 33, Koomsong 23, Koomsong 29, Bagmari 10, Nagrijuli 6/24, Seajuli 16, Seajuli 19, Bagmari 20, Gohpur 33, Bormajan 2, Longai 17, and Lalamookh 7 (where TV stands for Tocklai Vegetative and TA stands for Tin Ali).

China tea varieties: TV7, 14/13/3, 14/100/10, 14/100/16, 14/100/6, 317/1, 317/2, 317/3, 317/4, and P126 (P stands for Panitola).

Cambod tea varieties: TV9, TV18, TV22, TV23, TV25, TV26, TV30, Manipuri, Lushai and Betjan.

Tea varieties for Tripura: Meghlibundh 11, Meghlibundh 20, Meghlibundh 25, Huplongcherra 18 and Huplongcherra 26

Tea varieties for West Bengal: AV2 (Balai), Tukdah 246, Tukdah 78, Tukdah 383, Bunnockburn 668, Phoobsering 312, Bannockburn 157, Kopati 1/1, Mohargung Gulma 25, Sanyasithan 8, Sanyasithan 9, Sanyasithan 10, Sanyasithan 27, Sukna 7, Sukna 23, Sukna 25, Kamalpur 6, Kamalpur 17, Hantapara 12, Hantapara 30, Turturi 22, Leesh River 9/34, and Huldibari 19.

Tea varieties in Kerala: UPASI-3, UPASI-8, UPASI-17 UPASI-25, UPASI-27, UPASI-28, TRF 1, BSS 1, BSS 2, BSS 3, BSS 4 and BSS 5. (UPASI stands for United Planters Association for South India while BSS stands for Biclonal Seed Stock).

Tea varieties in Tamil Nadu: Pandian, Sundaram, Golconda, Jayaram, Evergreen, Athrey, Brookeland, BSS 1, BSS 2, BSS 3, BSS 4, BSS 5, Biclonal seed stocks and Grafts.

Table 2 Tea varieties in Tamil Nadu

S.No.	Clone	Important features
1	UPASI 1 (Ever green)	Hardy, Quality-Above average
2	UPASI 2 (Jayaram)	Hardy, Quality-Above average, tolerant to drought and wind
3	UPASI 3 (Sundaram)	Natural triploid quality clones and very high yielding
4	UPASI 6 (Brooklands)	Suited to mid and higher elevations
5	UPASI 8 (Golconda)	Suited to all elevations, high yielding
6	UPASI 9 (Arthrey)	Firly tolerant to drought and withstand slightly high pH, high yielding
7	UPASI 10 (Pandian)	Hardy, Quality-Above average, tolerant to drought and wind
8	UPASI 14 (Singara)	Suited to higher elevations , High yield
9	UPASI 15 (Spring field)	Flushes throughout the year
10	UPASI 17 (Swarna)	Flourishing well at mid and high elevations
11	UPASI 24	Hardy
12	UPASI 25	High yielding
13	UPASI 16	High yielding
14	UPASI 27	Drought tolerant
15	UPASI 28	Biclonal, Good strength and high quality

Table 3 Special characters of tea cultivars in Tamil Nadu

S.No.	Special characters	Clone	Originators
1	Wind Tolerance	UPASI-2 UPASI-10	UPASI-TRF India
2	Drought resistance	UPASI-9	UPASI-TRF India
3	Frost resistance	B-26	HPKV-TES India
4	Smallest leaf	CH-1	IHBT India
5	Biggest Leaf	Betjan	BETJAN T.E, India
6	Blister blight tolerance	TRI-2043,	TRI, Sri Lanka
7	High pubescence	TRI-2043	TRI, Sri Lanka
8	High anthocyanin pigmentation	TRI-2025	TRI Sri Lanka
9	High tolerance to pH	TN-14-3	TRF, Kenya
10	Poor fermenter	12/2	TRF, Kenya
11	Mite tolerance	7/9	TRF, Kenya
12	Scale insect tolerance	TN-14-3	TRF, Kenya
13	High polyphenol content (53.7%)	Luxi white tea	TRI, China
14	High amino acid content (6.5%)	Anji white tea	TRI, China
15	Low caffeine content (0.14%)	Guangdong tea	TRI, China
16	High caffeine content (6.96%)	Wild tea at Yunnan	TRI, China
17	Water logging tolerance	TV-9	TES, India
18	Good strength and high quality	UPASI 28	(UPASI 10 × TRI2025)

Seed propagation: Tea is propagated either from seeds or by vegetative means. One kg of seeds contains 250 to 400 seeds. Flowering in tea occurs during October to November. The duration from flowering to seed is one year. Seed viability is maximum six months. Seeds germinate in 3 to 4 weeks. The seedlings are ready for planting in 6 to 9 months.

Nursery management: Soak the tea seeds in water overnight prior to planting. Pot is filled up with pot mixer provided for good drainage. Sow the seeds in a small pot or seed starting containers, and cover with approximately 2.5 cm of soil. Watering is done frequently in the form of mist to keep the soil moist, but not soaked. A misting spray bottle works well. Place the plants in a shaded area until the seeds begin to sprout, then gradually move outdoors into sunlight. If using grow lights inside or in a greenhouse, make sure the plants receive approximately 12-18 hours of light. Water approximately two times per week in the first year as the roots become established. The nursery soil should be well drained and deep loam in texture with pH of 4.5 to 4.8. The soil and sand used in raising seedlings are tested for pH and nematode infestation. Soil pH can be reduced by treating with aluminium sulphate. For this purpose, beds are formed with one metre width and about 8 cm height and of a convenient length. Then the beds are drenched with 2% NH_4SO_4 @ 10 litres/2.5 m^2 of area. Over this, another layer of soil of 8 cm height is spread and again drenched with equal quantity of water twice. Then the soil is allowed to dry and the pH is checked before use in the nursery. Polythene sleeves of 150 or 200 gauge and 10 cm width and 30 to 45 cm length may be used. The lower ¾ of the sleeves should be filled with 3:1 sand and soil mixture and the top ¼ with 1:1 sand and soil mixture and staked in rows. Overhead shade is provided. Healthy and vigorously growing high yielding bushes should be selected. Apply 60 g of young tea mixture (NPK 4:3:3) up to 5 years to each selected bush. In addition 60 g of pruning mixture (NPK 1:1:2) should also be given after each cutting in addition to routine manuring. Cuttings are taken on April–May and August–September. Semi-hard wood cuttings are prepared with one full leaf and an internode with a slanting cut at the bottom. The sleeves are watered thoroughly and holes are made in the soil. The cuttings are inserted in the hole and the soil around is pressed firmly to avoid air space followed by watering. Small polythene tents may be provided which maintain high humidity and regulate the temperature inside. Cuttings may take 10 to 12 weeks for rooting. When all the cuttings have rooted the polythene tent may be removed gradually after 90 days over a period of 10 to 15 days. After the tent is removed, the cuttings are sorted and staked. 30 g of fertilizer mixture of the following composition dissolved in 10 litres of water may be applied over an area of 4 m^2. Composition of the soluble tea nursery fertilizer mixture includes 35 parts of ammonium phosphate (20:20), 15 parts of potassium sulphate, 15 parts of magnesium sulphate, 3 parts of zinc sulphate with a total of 68 parts. This should be done fortnightly. Hardening of 4 to 6 months old young plants should be done by removing shade gradually in stages over a period of 4 to 6 weeks starting from a few hours exposure to sun every day initially and extending the time of exposure gradually.

Land preparation and clonal planting: Deep cross ploughing followed by harrowing and leveling is done and small drains are dug as per specification (pit size 45 cm × 30 cm) at 105 cm spacing between drains. Keep the top and bottom soil separately. In clayey soil and drought-prone areas, deeper pits (60 cm) or trench planting will be advantageous. The rehabilitation crops are also sown with same spacing as that of tea after 18-24 months. The planting density varies from 14,000 to 17,000 bushes per hectare. Vegetative propagated mother bush is known as 'clones'.

Methods of planting

(i) Single hedge system: In this method, the plants are spaced at 1.2 m × 0.75 m accommodating 10,890 plants/ha.

(ii) Double hedge system: In this method, the spacing is 1.35 m × 0.75 m × 0.75 m accommodating 13,200 plants/ha. Sleeves should be opened lengthwise without injuring the roots and planted in the pit and the soil is gently pressed.

Weed management: Weeds are controlled with spray of Diuron @ 1 kg/ha as pre-emergence; Paraquat @ 1.5 lit/ha to control both dicots and monocots; 2,4-D Na salt or amine salt @ 1.5 lit/ha as post emergence to control dicot weeds and Dalapan @ 5.6 kg/ha with Kaolin @ 2.3 lit per ha to induce abrasive action to control grasses. Control perennial grass weeds can be achieved by spraying a combination of Ansar-529 @ 3.75 lit+ Dalapon 3.75 kg + Fernaoxone 700 g in 450 lit of water per ha.

Manuring: Manuring should be done 2 months after planting. Phosphorus should be applied at 80 to 100 kg/ha as rock phosphate once in a year by placement at 15 to 25 cm depth upto the first pruning and thereafter once in two years. N/K ratio in 2:3 should be adopted for the first 2 years and a ratio 1:1 thereafter. Application of fertilizers should be done before the onset of monsoon. Fertilizers should be broadcasted around the drip circle avoiding contact with the collar.

Rates of fertilizer application for young tea in soils with pH below 4.5 are given below:

Age	$N:K_2O:MgO$ kg/ha/annum	No. of splits
1st year	180:270:30	5
2nd year	240:360:40	6
3rd year	300:450:50	6
4th year & above	300:300:50	6

Apply phosphorus at 90 kg/ha every year in one application.

Rates of fertilizer application for young tea in soils with pH between 4.5 and 5.5.are given below

Age of the clearing	kg/ha/year		No. of applications	Quantity per plant (g)
	N	K_2O		
1st year	180	270	5	21
2nd year	240	360	6	23
3rd year	300	300	6	26

Rates of application for soils with pH above 5.5 are as follows (use water soluble P):

Age	$N:P_2O_5:K_2O$ kg/ha/annum	No. of splits
1st year	180:60:180	5
2nd year	240:80:240	6
3rd year onwards up to 1st pruning	300:100:300	6

Water management: Irrigation should be given during summer season. Irrigation is done either with sprinklers or drip system in the morning or night. The drains of 2 ft wide and 3 ft deep are provided over the area of 40 ft intervals along with the slope of the land.

Shade regulation in tea plantation: Tea is a shade loving plant, which is usually grown as a mono crop, under a canopy of shade trees throughout its life. Various physiological functions of tea are facilitated by shade, leading to sustain its vigour, yield and quality of the final produce. Conventionally, tea is grown under shade. These shade trees help in regulating the temperature and humidity at the bush level and minimising the loss of water through evaporation and transpiration. Temporary shade can be provided with for frost prone areas with *Acacia mearnsii,* for mid elevations areas with *Indigofera teysmanii,* for higher altitude: *Sesbania cinerescens, Crotalaria agathiflora* and *Acacia elata.* Temporary shade should be removed after establishment of *Grevilliea* after 3 years. Remove old trees after establishing new shade. The shade trees species recommended are viz., *Grevillea robusta, Albizzia moluccana* and *Albizzia chinensis* as high shade and *Acacia pruinosa, Acacia decurrens, Erythrina lithosperma, Calliandra calothrysus* and *Gliricidia sepium* as medium shade. One species each from the two categories is usually grown at each location. The best permanent shade tree for tea plantation in South India is silver oak *(Grevillea robusta).* Potential age of *Grevillea* is 40-60 years. The silver oak can be propagated through seeds. The seeds should be sown within six months after collection. Seed should be sown in raised beds of one m width and of convenient length using sandy loam soil with a pH around 6.0. Seed should be covered with thin layer of sand / ash. Germination takes place in 2-3 weeks. Use 6-9 month old seedlings for planting. Plant the seedlings at a spacing of 6 m × 6 m (275 plants/ha). Apply a mixture of 100 g rock phosphate and 400 g dolomite per pit and thoroughly mix with soil prior to planting. Apply NK mixture @ 100 g/tree twice in a year; rock phosphate at 250 g/tree and borated lime 1.1 kg (1 kg dolomite + 100 g boric acid) during alternate years.

Since tea crop requires only sparse shade, retain optimum stand of shade based on the growth of the shade tree, altitude of the garden and aspect of the field (south and west slopes require more shade). Thin out shade initially to 12 m × 6 m after 8-10 years of planting and if required further thinning may be done to 12 m × 12 m at later stages (12 years from planting). Always thin out shade prior to pruning. Cutting the main stem with the objective of developing lateral branches is 'pollarding'. Commence pollarding when the trees attain a girth of around 50 cm at elbow level. Pollarding depends on altitude (8 m height for higher altitude, 9 m for low elevation). Leave one branch in each direction and 3 to four tiers of branches, below the pollarding height. Cutting the erect growing branches on the laterals is lopping, which should be done before the onset of monsoon and lop only the erect branches and retain the laterals. Pollarding and periodic lopping of high and medium shade respectively are practiced to ascertain the optimal shade levels of 10-40%. Pollarding of shade trees should be done prior to heavy rains at a height of 8-10 m from the ground level.

Windbreaks: Windbreaks are essential, particularly in exposed regions at high altitudes. The species normally used include *Hakea saligna, Grevillea robusta, Accacia spp., Cupressus spp., Eucalyptus spp.* and even tea itself.

Training of the plants

(i) **Centering:** To facilitate better development of branches, centering should be done from 3 to 6 months after planing. The main stem should be cut leaving about 8 to 10 matured leaves.

(ii) **Cut across:** This is generally done in April-May at about 45 cm from the ground level to stimulate the production of more young shoots.

(iii) **Tipping:** Tipping is done from the cut across level of 75 cm to a height of 15 cm during the first year to minimize the height for easy plucking of young buds, tender leaves and shoots. The tipping height is 12.5 cm, 10 cm, 7.5 cm and 5 cm during the second, third, fourth and fifth year of planting from the cross level.

Pruning: Pruning is to renew the actively growing branches so that replacement of healthy wood and foliage keep space with ravages of death or damage. It is cutting the branches at a predetermined height at a specified interval. Pruning should be done in April-May or August-September. Factors governing pruning height are viz.,

(i) existing height of bushes,

(ii) last pruning height,

(iii) health and vigour of bushes,

(iv) incidence of pests and diseases and anticipated crop during first year.

Pruning height:

Rejuvenation pruning - <30 cm

Hard pruning - 30-45 cm

Medium pruning - 45-60 cm

Light pruning - 60-65 cm

Skiffing - >65 cm

Types of pruning

(i) Collar pruning: The whole bush should be cut near the ground level with a view to rejuvenate the bushes.

(ii) Medium pruning: To check the bush growing to an inconvenient height, this type of pruning is done in order to stimulate new wood and to maintain the foliage at lower levels.

(iii) Top pruning: Top pruning depends on the previous history of the bush raising the height of medium pruning by an inch or less to manageable heights for plucking.

(iv) Skiffing: This is the lightest of all pruning methods. A removal of only the top 5 to 8 cm new growth is done so as to obtain a uniform level of pruning.

(v) Shade regulation: Pollarding of shade trees should be done prior to heavy rains every year at a height of 8 m to 10 m from the ground level.

Post pruning care: It is done with application of fungicide to cut ends, burial of pruning and lime washing.

Application of fungicide to the cut ends: Copper oxychloride/sulphur +linseed oil (1:1) ratio has to be applied to large cut ends for rejuvenation and hard pruned bushes and to prevent the entry of fungal pathogen. Lime washing kills epiphytic growth (moss and lichen), helps in early and even bud break. It minimizes sun scorch to bush frame. 115 kg of shell lime (micronised) in 1150 litres of water per ha is recommended.

Plant protection: Nematodes in nursery: Root knot nematode is widespread in south Indian tea areas. The nematodes are eradicated by heat and chemical treatment through injection and drench methods.

(a) Heat treatment: Heat the soil on a metal sheet with fire below up to 60° C. The thickness of soil spread on the sheet should not exceed 8 cm height. Sprinkle water occasionally and stir the soil constantly.

(b) Chemical treatment:

(i) Injection method: The nursery soil should be made into beds of 7.50 m × 1.20 m × 0.3 m. Make holes to a depth of 20 to 22.5 cm at 30 cm intervals both lengthwise and breadthwise leaving 1.5 cm as margin on all sides. The required quantity of chemicals may be poured into the holes through a calibrated dropper. Then compact the beds by rolling a light roller. Sprinkle water all over the bed and mix it thoroughly. The soil can be used a month after treatment with DD (Shell/Nocil) @ 4 ml per 30 cm^2, Nemegon (Shell.Nocil) @ 1 ml per 30 cm^2, Tapem (Methem sodium) @ 2.5 ml per 30 cm^2 and Hexenema (B.P.M) @ 2.5 ml per 30 cm^2.

(ii) Drench method: Soil should be made into beds of 7.5 m × 1.20 m × 0.3 m size. Nemagon @ 100 ml or Vapam @ 250 ml or Hexanema @ 250 ml should be mixed with 225 lit of water. Drench the beds with this solution and allow it to soak. Rake and mix the soil thoroughly and remake the bed and compact it. After 2 weeks, turn up the soil mix well and leave it for 2 weeks before use.

Tea mosquito bug is controlled with a spray of Malathion 50EC @ 2 ml per litre of water or lambda cyhalothrin 0.005%, neem formulation at 2ml/lit. Thrips and aphids are controlled with a spray of Phosalone 35EC or Endosulfan 35 EC @ 1.5 ml per litre of water. Mites are controlled with spray of Dicofol 18EC or Phosalone 35EC @ 1.5 ml per litre or Ethion50 EC @ 1 ml per litre or Tetradifon 8EC @ 2 ml per litre to control mites.

Blister blight: Manure should not be applied after July. If blisters appear on new growth in April, remove them before they mature and burry them there itself. Pruning should not be done from 15th February to till end of August. Spray copper oxychloride @ 0.25% on 7 to 10 days interval from 15th March to 15th May. Spray 0.25% copper oxychloride and 0.25% nickel chloride per ha at 5 days interval from June–September and 11 days intervals in October and November.

Pink disease: Remove badly affected bushes. Spray 1% bordeaux mixture or 0.2% copper oxychloride immediately after pruning.

Root disease: Prepare isolation trenches 120 cm deep 45 cm wide along with one row of healthy bushes. Keep trenches clean to allow sunlight. Replant only after two years or soil fumigation with Durofume (Methyl bromide + Ethylene dibromide 1:1) 450/526 sq.ft by drilling holes 20 cm deep at 30 cm intervals and 20 to 25 cm away from collar of healthy plants and pour the chemical. Replant after 12 weeks.

Harvest: Plucking of young tea shoots with two to three leaves and bud commences when the tea bush is 3 years old. The plucking of extreme tip of the growing branch consists of an unopened bud together with two leaves is popularly known as 'two leaves and a bud'. The shoots arising out of the sticks (frame) of pruned bushes are called primaries. After plucking, shoot growth starts from the auxiliary buds just below the point of plucking. There small appendages known as scale leaves which fall off few days after opening. There is one or two small to medium size unfolded leaves above the scale leaves. The small leaf immediately above the small scale leaves is termed as janam leaf and the other one is called the fish leaf. Then the bud starts to produce normal leaves which are included in the harvest. Plucking continues throughout the year in South India at weekly intervals during March-May and at intervals of 10 to 14 days during the other months. Tea shoots are plucked in the months of mid February to mid December and plucking round (the number of days between two successive plucking) of 4 - 14 days is followed by small tea growers and 7 day plucking round is most common in Assam and West Bengal. Harvesting is most expensive and labour intensive. It influences bush health, yield and quality of tea.

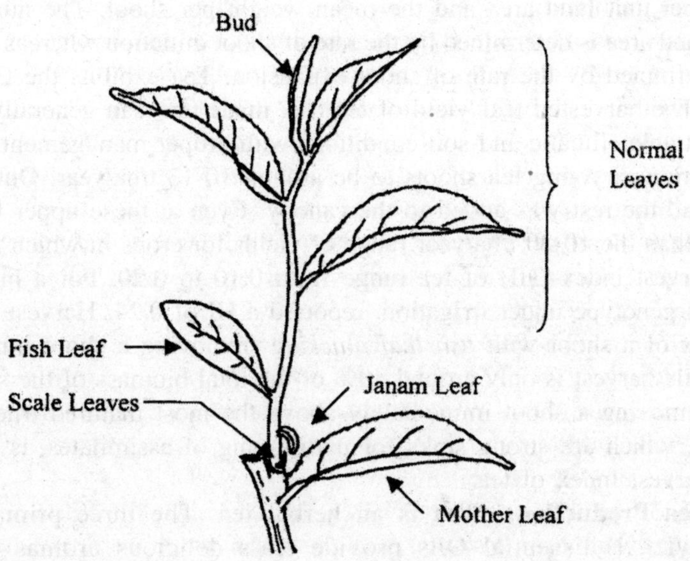

Plucking standards

Factory	Fine	2 leaves and a bud or single leaf banji
	Medium	3 leaves and a bud or 2 leaves and a banji
	Coarse	More than 3 leaves and a bud
Bush	Light	Mother leaf plucking
	Hard	Below mother leaf

Frequency of plucking: Frequency of plucking is 7 to 10 days during growing season and 12 to 15 days during lean season. Closer rounds enhance the productivity and quality. On an average the plucking frequency is 10 days interval and the number of rounds per annum 365/10=36 rounds. Normally 30 to 36 rounds of plucking is carried out depending on the age of the field from pruning.

Plucking rounds

Month	No. of rounds	Month	No. of rounds
January	3	July	2
February	2	August	2
March	2	September	2-3
April	2-3	October	3-4
May	3-4	November	3-4
June	3-4	December	3

Yield: The productivity of tea is quantified in terms of the weight of 'made tea' per unit land area per year. 'Made tea' refers to the form of tea obtained after the harvested (or 'plucked') shoot has gone through the manufacturing process (i.e. withering, fermenting and drying). Weight of made tea is directly related to the fresh weight of plucked shoot (2-3 leaves and a bud) by a factor of 0.2. Therefore, yield components of tea are the number of plucked shoots per unit land area and the mean weight per shoot. The number of plucked shoots per unit land area is determined by the rate of shoot initiation whereas the mean weight per shoot is determined by the rate of shoot expansion. Tea exhibits the C_3 mechanism of photosynthesis. The harvested leaf yield of tea (i.e. made tea) can generally reach 4-5 t/ha/year under favourable climatic and soil conditions with proper management. The annual dry matter accumulation in young tea shoots to be around 10-13 t/ha/year. Out of this 48-58% was harvested and the rest was added to the canopy. Even at these upper limits, tea yields are much lower than the 10-20 t/ha/year range of yields for crops in which a vegetative part is harvested. Harvest index (HI) of tea range from 0.10 to 0.20. For a high yielding (i.e. 5600 kg/ha/year) genotype under irrigation, reported a HI of 0.24. Harvest is about 80% of the total biomass of a shoot with *fish leaf plucking* (removing a shoot immediately above the fish leaf) while harvest is only around 40% of the total biomass of the shoot with *single leaf plucking* (removing a shoot immediately above the most matured true leaf). Plucking of young shoots, which are strong sinks for partitioning of assimilates, is the main reason for the lower Harvest Index of tea.

Types of Tea Production: Tea is an herbal tea. The three primary components of brewed tea viz., i) Essential Oils provide tea's delicious aromas and flavors, ii) Polyphenols provide the "briskness" or astringency in the mouth and are the components that also carry most of the health benefits of tea and iii) Caffeine is found naturally in coffee, chocolate, tea and Yerba Mate, caffeine provides tea's natural energy boost. Fresh leaves contain 4% caffeine. Tea is commercially classified into six types based on the degree or period of fermentation (oxidation) the leaves have undergone.

(i) **White Tea:** White tea is produced when two leaves and a bud are picked just before sunrise to preserve the moisture in the leaf which has undergone no oxidation. White tea are from two varieties of the Chinese tea bush *viz.*, the Shui Hsien and the Dai Bai. This tea is characterized by a delicate flavour with very little colour. However, it is highly priced because a day's picking produces only about 1-2 kgs. Although it is called white tea, the tea does have some light green colour characteristic of the newest buds originating on the bush. The buds are steamed to destroy the enzymes that would otherwise destroy the tea and dried either in the dryer or in the sun. It is popular in China.

(ii) **Black tea:** Freshly-picked green tea leaves are withered, spread out on racks to dry, and then crushed by rollers to release the juices from them (fermented or oxidised). The leaves turn brown and are then fired (or dried) by hot air and sorted into grades. The tea leaves are allowed to oxidise completely. If the tea is fermented long enough, the leaves turn black, hence the term 'black tea'. Black Tea is the most common form of tea in Kenya, Malawi, Tanzania, Uganda and Mozambique and South Asia (India, Sri Lanka, Bangladesh, etc). Black tea is further classified as either orthodox or CTC (Crushing, Tearing and Curling) tea. Black tea is mainly produced in India, Sri Lanka and China. India is the largest producer of black tea as well as the largest consumer of tea in the world. It has been thought for centuries that black and green teas came from different plants. In fact they come from the same species, but black tea is fermented. 98% of the international trade is in black tea.

(iii) **Green Tea:** Green tea is produced by steaming the leaves to destroy the enzymes that might otherwise ferment the leaves. The leaves are then rolled either by hand or by mechanical rollers, to bring out the juices in the leaves that are responsible for its flavour. The rolled leaves are then fired to dry them. The entire process of rolling and firing is repeated several times until the leaves are completely dry. Variation in the drying time can result in fermentation of the leaves which spoils its flavour. Naturally low in caffeine, the brew is very light in colour. Green teas range from a light, fragrant taste to a very bold vegetal flavour. Green tea is mainly produced in China, Japan, Taiwan and India.

(iv) Oolong tea: The process of producing Oolong tea begins with picking of the two leaves and a bud, generally early in the morning. When the leaves start turning red - at a stage, when 30% of the leaves are red, and the rest 70% are green, the leaves are rubbed repeatedly by hand or mechanically to generate flavour and aroma and finally dried over charcoal. Oolong is a compromise between black tea and green tea. It is more delicate than black tea and stronger than green tea. It is intermediate between green and black tea, having the flavour of the former but the colour of the latter. It is prepared in Taiwan from China tea 'Chesima'. Oolong tea comes almost from Taiwan and the bulk of it is exported to the USA.

(v) Scented tea/Flavoured tea: Green, semi-fermented or black tea that has been flavoured by the addition of flowers, fruit or essential oils. Earl Grey is one of the most famous of scented black tea. Madame Butterfly is a green tea scented by the addition of flowers and flavouring. It includes fruit teas and perfumed teas (e.g. containing anise or cinnamon flavour). These are predominately blended in the Europe Union and mainly concern black and green tea.

(vi) Brick tea: This is prepared from the waste left after the preparation of black and green teas. It may consist of leaf, stalks and even twigs, or mainly coarse tea dust. The bulk is softened with steam and then compressed into blocks or bricks. It is mostly consumed in Tibet.

Green tea is essentially unfermented, Oolong tea is partially fermented and Black tea is fully fermented. Black Tea or Green Tea is packed either in original form or in blended form in small consumer packs. These are known as 'Packet tea'. Iced tea was initially introduced in Belgium as a sports drink. It is a particularly popular beverage in Germany and Italy.

Tea Active Components: Tea contains about 180–360 mg/g of polyphenols in one gram leaves, among which 70–80% are flavanols. Flavonoids, the most prominent of which is catechins and their derivative polyphenols, are the most abundant and most biologically active molecules that are responsible for most of the health promoting properties of tea. Catechins, (particularly, epigallocatechin) interact with an enzyme in human intestines to suppress glucose uptake, by inhibiting sodium-dependent glucose transporter mechanism, thus preventing diabetes. Caffeine/Theine are technically same substance. Caffeine has been first discovered from coffee bean. The caffeine substance is also found in tea leaves and other plants. The term theine is used to refer tea caffeine. The black tea is orange to dark-red in colour which sometimes gives scented infusion. It contains low level of Catechins (4.0 g/100g) and high level of Theaflavins (0.94 g/100g). The green tea, on the other hand, gives insipid beverage with yellowish colour and it contains highest level of Catechins (14.2 g/100g) and zero Theaflavins. Tea contains 20–30 mg of caffeine per 100 ml and daily caffeine intake must not exceed 600 mg/day which is equivalent of 2 to 3 litres of tea/day. Caffeine in tea may adversely affect their sleep. Tea also contains a unique amino acid, L-theanine which has the ability to promote quality sleep.

REFERENCES

Anon. 2009. Tea Industry. Ministry of Agriculture. Government of India. New Delhi.

Arunachalam, K. 1995. A Hand Book of India Tea. The Tamil Nadu Tea Plantation Corporation Ltd. Coonoor.

Barman, T. S. 2011. Tea cultivars for north east India. Science and Culture. 77(9–10): 391 - 395

Basu Majumder A., Bera B. and Rajan, A. 2010. Tea Statistics: Global Scenario. *Inc. J. Tea Sci.* 8 (1): 121-124

Bhagat, R.M., Deb Baruah, R. and Safique, S. 2010. Climate and Tea [*Camellia sinensis* (L.) O. Kuntze]. Production with special reference to north eastern India: a review. Journal of Environmental Research and Development. 4(4): 1017-1028

CPG. 2013. Crop Production Techniques of Horticultural Crops. Horticultural College and Research Institute, Tamil Nadu Agricultural University, Coimbatore – 641 003

Janendra, W.A., De Costa, M., Janaki Mohotti, A. and Madawala A. Wijeratne. 2007. Ecophysiology of tea. Braz. J. Plant Physiol., 19(4):299-332, 2007

Karmakar, K.G. and G.D. Banerjee. 2005. The Tea Industry in India: A Survey. National Bank for Agriculture & Rural Development. Mumbai

Nizara Arya. 2013. Indian Tea Scenario. International Journal of Scientific and Research Publications. 3 (7): 1-10

Parag Shil. 2013. Export scenario and challenges in front of Indian Tea industry in the 21st century. Indian Journal of Applied Research. 3(6): 378-379

Gadapani Sarma, 2013. A Historical Background of Tea in Assam. Journal of Humanities and Social Science. 1(4): 123-131

Santanu Sabhapondit, Tanmoy Karak, Lakshi Prasad Bhuyan, Bhabesh Chandra Goswami, and Mridul Hazarika. 2012. Diversity of Catechin in Northeast Indian Tea Cultivars. The ScientificWorld Journal. p 1-8.

https://en.wikipedia.org/wiki/Indian_tea_culture

http://www.upasitearesearch.org/

http://www.agritech.tnau.ac.in/horticulture/horti_plantation%20crops_tea.html

http://www.tocklai.net/activities/tea-classification/

http://www.nitm.in/

http://www.flowersofindia.net/catalog/slides/Tea.html

http://www.kew.org/science-conservation/plants-fungi/camellia-sinensis-tea

https://en.wikipedia.org/wiki/Japanese_tea_ceremony.

http://www.vegetableipmasia.org/www.communityipm.org/docs/Tea_Eco-Guide/01_History.PDF

http://www.celkau.in/Crops/Plantation%20Crops/Tea/cultivation_practices.aspx

http://www.teaclass.com/lesson_0111.html

https://hort.purdue.edu/newcrop/duke_energy/Camellia_sinensis.html

http://www.agritech.tnau.ac.in/horticulture/horti_plantation%20crops_tea.html

http://agritrade.iift.ac.in/html/Training/ASEAN%20%E2%80%93%20India%20FTA%20%20 Emerging%20Issues%20for%20Trade%20in%20Agriculture/Tea.pdf

http://www.teaboard.gov.in/

http://www.indiatea.org/#/

http://www.plantauthority.gov.in/pdf/DTea.pdf

http://dwr.org.in/slider.aspx

http://www.indiatea.org/history_of_indian_tea.php

CHAPTER

36

Betel Vine–*Piper betle* L. (2n=26)

Family: Piperaceae

Vernacular name: Betel vine's common name is betel in English. It is known as *Nagavalli, Nagavallari, Nagini* (Sanskrit); *Paan* (Hindi, Bengal, Gujraji, Urdu); *Vilya, Veeleya, Villayadel* (Kannada); *Phodipaan* (Konkani); *Vettila, Vettilakkoti* (Malayalam); *Vidyache pan* (Marathi); *Vetrilai* (Tamil); *Tamalapaku* (Telugu); *Bulath* (Sinhalese); *Sirih*in Malaysia and Indonesia; *Tanbol* (Arabic); *Burg-e-Tanbol* (Persian).

Importance: Betel vine is grown for its leaves which are widely used as a mouth freshener after meal. Betel leaf is an aromatic-stimulo-carminative, astringent and aphrodisiac. Betel leaves are mainly used for mastication. It is usually chewed with slice of betel nut, slaked lime, cardamom and clove in many Asian countries. The fresh leaves of betel leaves have been wrapped together with the areca nut, mineral slaked lime, catechu, flavouring substances and spices are chewed since the ancient time. Marcopolo (1295 AD) took notice of the pan chewing habit of the people in South India. Over the centuries, pan chewing had become so prevalent that serving and chewing of pan had been raised to the level of a fine art at the Mughal Darbar, particularly during the Akbar's regime. In course of time, offering the 'bida' of betel vine has become a symbol of offering and acceptance of mutual love and friendship. The betel leaf produces an aromatic volatile oil, contains a phenol called 'chavicol', which has a powerful antiseptic properties. The alkaloid 'arakene' has properties resembling cocaine in some respect. Pharmacological effects of betel chewing include abundant flow of saliva, temporary dulled of taste perception, stimulation of muscular and mental efficiency. The harvested leaves are used both for domestic consumption and for export to UK, USA, Canada, Pakistan, Bangladesh, Myanmar, Malaysia, Singapore, Sri Lanka and other Arabian countries. Betel vine is a highly capital and labour intensive crop.

Origin and distribution: It is a native of Malaysia. This plant is extensively grown in India, Bangladesh, Sri Lanka, Malaysia, Myanmar, Pakistan, Thailand, Taiwan, Indonesia, Singapore, Cambodia, Vietnam and Philippines. In India, it is widely cultivated in Karnataka, Odisha, Tamil Nadu, West Bengal, Bihar, Madhya Pradesh, Maharashtra and Uttar Pradesh.

Botany: The plant is evergreen and perennial climber/creeper, with glossy heart-shaped leaves and white catkins, and grows to a height of about one metre. It is a woody climber with adventitious roots at swollen nodes. Leaf is simple, alternate, cordate, 8-12 cm wide, 12-16 cm long, with odour and spicy taste. Inflorescence is an axillary spike; flowers unisexual, white. Fruit is a globose berry.

Climate: It grows best under the shaded, tropical forest ecological conditions with a rainfall of about 2250 to 4750 mm, relative humidity of 60-80% and temperature ranging from 15 to 40°C, respectively. Temperature below 10°C and above 40°C causes wilting. However, in the areas with lower rainfall of 1500-1700 mm the crop is cultivated with small and frequent irrigations, i.e. every day in summer and every 3-4 days in winter, whereas adequate drainage is required during the rainy season. It is a shade loving plant. The plant betel vine grows fast and their vegetative growth is good under shade which provides low light intensity and mild temperature. Hot wind burns the tender leaves and causes wilting while cold wave causes yellowing of leaves.

Soil: A well-drained fertile sandy or sandy loam or sandy clay soil with pH range of 5.6 –7.5, rich in humus, is the best for betel vine cultivation. The clay soil is not good for the crop because it favours disease during the rainy season.Soil with good waterholding capacity and organic matter content is considered ideal. However, this crop is very sensitive to stagnant water, waterlogged, saline and alkaline soils.

Season: The cuttings of the betel creeper are planted in the months of May and June, rainy season (July-August), post-monsoon months (October-November-December). It is also planted during the last week of January to first week of March.

Land preparation: A well prepared land with thorough and deep cultivation is recommended for the crop. Deep ploughing is done during early summer (end of April month). After ploughing, upper soil is left exposed in sun for two months (April-May), because it reduces the microorganism population as well as soil borne pathogens. During the first week of June, two or three ploughing with harrowing is done. After that, about 350 quintals of compost on per hectare basis is mixed in the soil. Thereafter, field is again ploughed fifteen days before planting. In such ploughed fields, steps have to be taken for bringing insects and pests population below the economic threshold level. Thereafter bed is raised. The optimum size of bed is raised from place to place by 30 to 50 cm. The main aim for raising bed is to facilitate drainage from the field. The field is prepared to a fine tilth and beds of 2 m wide are formed to a convenient length. Provide drainage trenches of 0.5 m width by 0.5 m depth in between two adjoining beds. Plant the seeds of *Sesbania sp.* (*Sesbania grandiflora*) in long rows. About 750 banana suckers are planted at the edges of the beds, which are used, for tying the vines on the live support and for packing the betel leaf. When the *Sesbania sp.* plants reach 4 m height, they are topped off for maintaining the height. The crop is planted in two rows in beds of 180 cm width on *Sesbania sp.* plants with a spacing of 45 cm between plants in the row. In garden land condition, the soil is dug well and laid out into furrows of 10-15 m length, 75 cm width and 75 cm depth. A typical farm will be fenced with bamboo sticks and coconut leaves, with the husk of paddy laid over it.

Varieties: Betel vine varieties popular in West Bengal are Desawari, Bangla, Kapoori, Meetha, Sanchi, Banarasi and Maghi. In Bihar, Bangla, Deshi, Calcuttia, Maghi, Kapoori and Semehi varieties are cultivated. Anthiyurkodi, Bangla, Kanyurkodi, Thulasi, Venmani, Arikodi, Kalkodi, Kallarkodi, Karilanchi, Karpurakodi, Chelanthikarpuram, Koottakkodinandan, Perumkodi, Amaravila and Pramuttan, Kallarkodi, Revesi, Kapoori, SGM 1, SGM (BV) 2, Vellaikodi, and Pachaikodi varieties are grown in Tamil Nadu. *Kariyele, Mysore yele, Rani yele, Kalasavalli, Panchavalli, Gobbarpanchavalli, Lavangavalli, Sarabyele,* Godi Bangla and Swarna Kapoori are cultivated in Karnataka. Betel vine cultivars viz., Cheelanthikarpuram-Red, Thulasikodi, Venmony Vettila, Mulamkodi, Cheelanthivella, Chettankodi, Nadankodi are cultivated in Kerala. Betel leaf Hinjilicut cultivated in Odisha is of fine quality.

SGM 1: It is a clonal selection from a Palghat type. It is adaptable to all betel vine-growing areas of Tamil Nadu. It produces a higher leaf yield of 109 lakh leaves per hectare in a crop duration of 2 to 2½ years. The vines are dwarf stature with vigorous bushy growth having thick hardy stem with short internodes and multilateral. Leaves are attractive yellowish green colour with desirable pungency.

SGM (BV) 2: This is a pure line selection from Dindigul local, Tamil Nadu. It possesses multilateraled vines (17-20/vine) with long petioles and attractive deep green leaves. The leaves are moderately pungent with good chewing quality. It is a high yielder with good market appeal. The duration of the crop is 2-2½ years. The suitable season for cultivation was January-March and June-August for *Sesbania sp* and March-May and August- October for betel vine. The crop is moderately resistant to phytopthora wilt, blight and nematodes. It yields about 49 lakh leaves / ha / year which is 25.4% increase over SGM 1, 33.8% increase over Karpoori and 62.0% increase over Vellaikodi. It can be cultivated all over Tamil Nadu and suitable for open trench cultivation.

Seed materials and planting: Setts with vigorous apical buds and nodal adventitious roots / terminal stem cutting or setts about 30-45 cm long of the 3-5 years old healthy, disease freecrop are selected for propagation.The top portions of the vines is easy to root and establish. Terminal cuttings having 3-5 nodes can also be raised in polybags and later on shifted to main field. Sometimes, the vine is raised by vegetative propagation from the cuttings under partially shaded and humid environment inside the *Boroj,* which is a small hut like structure of approximately 2 m in height and 0.02 ha in area. It is constructed with the locally available materials like bamboo stems, jute sticks, paddy straw, petioles and leaves of banana, etc. wherein the vines are grown on elevated beds imitating the natural ecological conditions suitable to the crop. However, height of the beds goes on increasing due to frequent earthing up of the vines. If there is scarcity of planting material, single node cuttings with a mother leaf can also be used. One lakhs setts of are required for one hectare of planting. Cuttings are to be treated with copper fungicide.Two sprayings of copper fungicide at an interval of 15 days is essential as a precautionary measure. Further, these are planted in the furrows (8-10 cm deep) of sterilized soil with spacing of 50-60 cm × 15-20 cm during rainy or autumn season.

Cultivation systems

(i) **Open cultivation with support trees:** It is grown with a support, shade loving and fast growing plants such as *Sesbania grandiflora (Sesbania sp.)*, *Sesbania sesban*, *Erythrina variegate* (Dadap), *Moringa oleifera* (Drumstick), *Ceiba pentandra* (Silk cotton tree). Support trees are spaced at 1.2 × 1.2, 1.2 × 0.6, 1.5 × 0.75 with a plant population of 6719 to 27,000 plants per ha.

(ii) **Mixed crop in Areca nut**: Betel vine is cultivated as mixed crop with Arecanut which is grown at a spacing of 2.7 × 2.7m allowing 1370 vines per ha.

After cultivation: Since the betel plant is a creeper, it needs a compatible tree or a long pole for support. The side branches of *Sesbania sp.* trees up to a height of 2 m are removed for early creeping of the vines before the establishment of vines. The vines grow to height of 3 m in one year period under normal cultivation. When they reach this height, their vigour to produce normal size leaf is reduced and they need rejuvenation.

Training of the vines: The cuttings sprout and creep in about a month. At this time, they must be trailed on the standards. Training is done by fixing the vine at intervals of 15 to 20 cm along the live standards loosely with the help of banana fibre. Training is done at every 15 - 20 days interval depending upon the growth of vines. Instead of live standards sometimes bamboo standards are erected at intervals and linked by tying at heights of 30 cm and 150 cm using coir rope. In the initial stages training is done on coir tied for the purpose. Training is done further by tying the vines, at intervals of 15-20 cm along the standards loosely with the help of banana fibre. When vines come in contact with standards, they produce adventitious roots using which they cling to support. Training is done every 15-20 days depending on the growth of vines.

Lowering of vines: Betel vine grows to height of 3 metre in a year. Afterwards the vine's vigour to produce normal size leaf is reduced and needs rejuvenation by lowering during March-April months. After the vine is lowered, the tillers spring up from the nodes at the bends of the coiled vines at the ground level and produce many primary vines. Irrigation should be given after each lowering. However under arecaunt only lower portion of 1-2 m length of basal portion of the vine is buried inthe soil and rest is tie to arecanut. Leaf quality and yield will improve after lowering of vines.

Manures and fertilizer application: Oil cakes, farm yard manure and leaves are thoroughly incorporated with the top soil of the furrows along with wood ash. Apply 150 kg N per ha per year through Neem cake (75 kg N) and Urea (75 kg N) and 100 kg P_2O_5 through Super phosphate and 30 kg Muriate of potash in three split doses first at 15 days after lifting the vines and second and third dose at 40-45 days intervals. Apply on beds shade dried neem leaves or Calotropis leaves at 2 t per ha and cover it with mud (2 tons in 2 split doses). Basal dressing of 37.5-100-50 of N, P_2O_5 and K_2O kg/ha as basal dressing and top dressing 112.5 of N kg/ha into 3 equal split doses, first at 15 days after lifting the vines and second and third dose at 40-45 days intervals. Since betel vine is cultivated for its green leaf, nitrogen plays a direct role on growth, yield and keeping quality. Application of neem cake / poultry manure + urea in 1:1 or 2:1 ratio is the best integrated nutrient management way for obtaining high yield of 30.72 lakh leaves per ha with quality of betel vine which is considered as economically viable options. The excess use of nitrogenous fertilizers aggravates the foliage diseases and deteriorating the keeping quality of the leaves.

Water management: Irrigate the field immediately after planting and afterwards once in a week. Betel needs constantly moist soil, but there should not be excessive moisture. Betel vine requires frequent but light irrigation all round the year. Excess of irrigation causes decay of roots and dropping of leaves.The standing water does not remain for more than half an hour in the bed. If water logging due to heavy rains or excess irrigation occurs, drainage should be arranged immediately. The best time for irrigation is in the morning or evening. Dried leaves and wood ash are applied to the furrows at fortnightly intervals and cowdung slurry sprinkled.

Plant protection: White fly, black fly and scale insects are the major pests of betel vine. Botanicals are recommended for controlling the pests. Nematode problem is becoming serious. There are many leaf and shoot / stem diseases of betel vine. The incidence of diseases (mainly caused by *Phytophthora* sp. and *Colletotricum* sp.) is the highest with application of commercial chemical N- fertilizer and the lowest in the treatment receiving quality organic manure. Select scale-free seed vines. Spray Chlorpyriphos 20 EC @ ml/lit when one or two scales are noticed on the basal portion of the stem/leaves. Direct the spray solution to the basal portion of the vines. Spray NSKE @ 5% or Malathion 50 EC @ 1 ml/lit. Mites can be controlled by spraying Wettable sulphur 50 WP @ 1 g/lit or Dicofol 18.5 EC @ 0.5 ml/lit. To control aphids, spray Chlorpyriphos at 2 ml/lit on *Sesbania sp.* leaves and clip off excess *Sesbania sp.* leaves. Mealy bugs can be controlled by spraying Chlorpyriphos 20 EC @ 2 ml/lit or Dimethoate 30 EC @ 2ml/lit. Concentrate the spray towards the collar region. Application of Neem cake @ 1 t/ha or shade dried *Calotropis* leaves @ 2.5 t/ha can be applied to soil for controlling the nematode populations. Soil application of *Bacillus subtilis* (BbV 57) or *Pseudomonas fluorescens* @ 10 g / vine for the control of root knot nematode and quick wilt of betel vine

Most of the diseases infected betel vine crop are soil borne, thus soil treatment is an important management practice. Soil treatment may be done either by solarization, or chemical means using formalin solution / copper fungicide. Spray Streptocycline @ 400 ppm

+ Bordeaux mixture @ 0.25% at the time of first bacterial leaf spot, blight and bacterial stem rotdisease symptoms appear. Continue spraying at 20 days intervals. Always spray the chemical after plucking the leaves. Spray 0.2% Ziram or 0.5% Bordeaux mixture after plucking the leaves after the first appearance of the Anthracnose (Theechal) symptom. The variety Karpoori is susceptible to the disease. Powdery mildew can be controlled by spraying 0.2% Wettable sulphur after plucking the leaves.

The integrated method for the management of *Phytophthora* wilt is as follows:

- Select well matured (more than one year old) seed vines free from pest and diseases.
- Soak the seed vines for about 30 minutes in Streptocyline 500 ppm or Bordeaux mixture 0.5 %.
- Apply 150 kg N/ha/year through Neem cake (75 kg N) and Urea (75 kg N) and 100 kg P_2O_5 through Super phosphate and 30 kg Muriate of potash in 3 split doses first at 15 days after lifting the vines and second and third dose at 40 - 45 days intervals. Apply on beds, shade dried neem leaf or *Calotropis* leaves at 2 t/ha and cover it with mud (2 t in 2 split doses).
- Drench Bordeaux mixture 0.25% in basins formed around the vine at monthly intervals starting from October – January, three times soil drench and six times spray from June - July.
- During winter season avoid frequent irrigation.
- Remove the affected vines away from the garden and burn them.
- Application of *Trichodermaviride* @ 5 g/vine.

Harvesting: Leaves are harvested with a portion of petiole by hand or Iron nail specially made for harvesting betel leaves. Leaves are plucked from side shoots (laterals). In about 3-6 months time, vines grow to a height 150-180 cm. Harvesting is done at 15-30 days interval to about a month till the next lowering of vines. Harvest depends upon the growth of the vines and market condition. Once harvesting starts, it continues almost every day. The leaves of the plant become ready for plucking after one year of planting and the creeper remains productive for several years from the date of planting. Harvesting of leaves starts after 6-12 months from the time of planting. In the rainy season frequent harvesting is done. But the leaves are picked throughout the year when it matures. The most common method of picking is hand picking. During the picking of leaves special care is taken by the growers that leaves are not harvested within 10 to 15 days of spray of pesticides. Roughly 30 lakh leaves are harvested annually from one hectare of land. In about three to six months time, betel vine grows to a height of 150-180 cm. At this stage, branching is noticed. Leaves are plucked along with the petiole, with the right thumb. Once harvesting commences, it continues almost every day or week.

Yield: The annual yield of a good crop is about 60-70 leaves/plant. The yield of betel leaves varied in various months of the year. It is comparatively low during the winter months than in the summer months. About 75 to 100 lakh leaves/ha/year can be obtained. The cost benefit ratio is 1:1.85. The betel leaf is a very perishable commodity and therefore, always subject to wastage by quick spoilage due to dehydration, fungal infection, dechlorophyllation, etc. This may cause a post harvest loss ranging from 35 to 70% during transport and storage. Such wastage may also be minimized by extracting essential oil from the stranded or unsold leaves be it fresh or stale ordechlorophylled or even partially decayed, by anapparatus called 'Betel leaf oil extractor'. the Mitha, Bangla and Sanchi varieties of betel leaves contained about 2.0%,1.7% and 0.8% essential oil respectively, on dry weight basis. The oil contained in the leaves possesses antibacterial, antiprotozoan and antifungal properties.

Utilization: Betel leaves are folded for chewing with some calcium hydroxide is smeared inside. Slices of the dry Arecanut are on the upper left of the leaf and slices of the tender arecanut on the upper right. Betel leaf juice mixed with dilute milk and sweetened slightly helps in easing the passage of urine. The betel is a spice whose leaves have medicinal properties. The leaves contain vitamins and significant amount of all essential amino acid except lysine, histidine and arginine which occurs in traces. In Indian folkloric medicine, betel leaf is popular as an antiseptic and is commonly applied on wounds and lesions for its healing effects. The aroma of betel leaf is due to the presence of essential oils, consisting of phenols and terpenes. Betel leaves are beneficial to the throat and remove viscidity in human beings. Leaves help in digestion and tend to remove thebad smell of the mouth. The juice of betel leaves is used as an adjunct to pills administered in the Ayurvedic medicines. The fresh crushed leaves are used asantiseptic for cuts and wounds. It is also good for the respiratory system and is usedin treatment of bronchitis, cough and cold. The leaves of the pan plant have been traditionally used for chewing. Pan chewing is considered as a good and cheap source of dietary calcium. It increases digestive capacity when used with lime. Besides, it neutralizes the acidity and acts as blood purifier. Main constituents of betel leaves are vitamin B and C, carotene, and other elements.

Medicinal properties of betel leaves

(i) In case of indigestion, stomach ache and diarrhoea: Boil 30-60 g fresh leaves in 600 ml of water; decoct to 300 ml; take in two equally divided doses, at morning and evening.

(ii) In case of wounds: Crush fresh leaves and apply directly over the wound.

(iii) In case of eczema: Apply sap of leaves to alleviate eczema.

(iv) In case of lice infestation: Apply sap of leaves to the scalp to remove lice.

(v) In case of sore of throat: Crush fruit or berry, mix with honey, take in, it will relieve irritating cough.

(vi) In case of respiratory disorder: Soak leaves in mustard oil, warm and apply to chest to relieve cough and for easy breathing.

(vii) In case of scanty or obstructed urination: Mix sap of fresh leaves with dilute slightly sweetened milk and take it.

(viii) In case of flatulence: Apply a slightly heated fresh leaf over the abdomen.

REFERENCES

Anonymous, 2005. The wealth of India raw materials. Vol VIII, National Institute of Science communication and information resources (NISCAIR), NewDelhi, p 84-94.

Balasubrahmanyam, V. R. and A. K. S. Rawat. 1990. Betel vine (Piper betel, Piperaceae). Economic Botany. 44(4): 540-543.

Balsubrahmanyam,V.R. 1992. Irrigating betel vine plantation the right way. Indian Horticulture. Vol. 136, No. 4, 1992.

Balasubrahamanyam, V.R. 1994. Betel vine. National Botanical Research Institute, Lucknow. pp 6-7.

Chaurasia, R. S. and Johri, J.K. 1990. Production analysis of betel leaves. Agricultural Situation in India. 44 (1): 23-25.

CPG. 2013. Crop Production Techniques of Horticultural Crops. Horticultural College and Research Institute, Tamil Nadu Agricultural University, Coimbatore – 641 003

Debanath, P.K., Sengupta, K., Sengupta, C. and Chatterjee, B.N. 1985. Betel vine cultivation in West Bengal. Indian Fmg. 25 (7):22-23

Guha, P. 2006. Betel leaf: The neglected Green Gold of India. J. Hum. Ecol. 19(2): 87-93.

Mandal, B.K., Choudhuri, S.K., Sengupta, K. and Dasgupta, B. 1994. Response of betel vine varieties to organic manure and prilled urea. Indian J. Agric. Sci. 64:297-301

Meenakshi Sundaram, 1987. An Economic Analysis of Production of Betel vine in Tamil Nadu, Agriculture Situation in India, Vol. XLII, No. 1, December 1987.

Usha C. Thomas, S. Chandini, S. Anil Kumar, Allan Thomas, and T. Satyanarayana, 2010. Evaluation of different nutrient management options for leaf yield, quality, and economics of betel vine cultivation. Better Crops–South Asia.p16-17

http://agritech.tnau.ac.in/horticulture/horti_plantation%20crops_betelvine.html

http://www.spices.res.in

CHAPTER 37

Spices and Condiments

Spice plants are plants used for seasoning, spicing, flavouring and colouring foods, drinks and different products of the food processing industry. Spices are defined as the aromatic vegetable product used in food seasonings such as pepper, cinnamon, nutmeg, mace, allspice, ginger, cloves, etc., used in cooking to season food and to flavour sauces, pickles, etc and as a vegetable condiments collectively'. A spice is 'one which enriches or alters the quality of a thing, especially the taste of food which gives zest or pungency'. The American Spice Trade Association uses the broad definition for spice as 'products of origin which are used primarily for the purpose of seasoning food'. The spices and condiments is defined as the natural vegetable products, or mixtures thereof, without any extraneous matter as are used for flavouring, seasoning and imparting aroma to foods; the term applies to the product either in the whole form or in the ground form. The Geneva-based International Standards Organisation defines spices and condiments as: Vegetable products or mixtures thereof, free from extraneous matter, used for flavouring, seasoning and imparting aroma in foods. The Webster describes spices as: Any of various aromatic vegetable productions as pepper, cinnamon, nutmeg, mace, allspice, ginger, cloves, etc., used in cookery to season and to flavour sauces, pickles, etc.; a vegetable condiment or relish, usually in the form of a powder; also, as condiments collectively. The condiment is defined as: 'A condiment shall be a prepared food compound, containing one or more spices or spice extractives, which when added to a food after it has been served, enhances the flavour of food'. Condiments can be of two types viz., simple and compound. Simple condiments include celery salt, garlic salt, onion salt, pepper salt etc. Compound condiments include products such as chilli sauce, chutney, horseradish sauce, meat sauce, mint sauce, prepared mustard, soya sauce, sweet and sour sauce, tomato ketchup, etc. The secretions of spices can be broadly classified into essential oils, oleoresins, and oleo-gum-resins, where 'oleo' or oil, refers to the essential or volatile oil, in contrast to fixed oil (the glycerides of fatty acids). Oleoresin cells are distributed in turmeric rhizomes. The oleoresins are prepared from the turmeric through extraction with a suitable solvent or solvents. Turmeric oleoresin acts as food flavouring agent. 'Alleppey' turmeric variety is very rich in curcumin. It contains about 6.5% of curcumin as against 3 to

4% in other varieties. In turmeric, the primary consideration for extraction is the curcumin content, the volatile oil being not regarded as very important for food flavouring agent. There are different types of classification of spices and condiments.

Classification based on season of growth:

(i) Annual spices (life cycle in one growing season): coriander, cumin, fennel, fenugreek, ajowan and black cumin.

(ii) Biennial spices (life cycle in two growing season): onion and parsley.

(iii) Perennial spices (life cycle in more than two years): Black pepper, saffron, clove, nutmeg and cinnamon

Classification of Spices based on climate requirement

(i) Tropical spices: ginger, turmeric, black pepper, cinnamon, kokam, galangal, small cardamom and clove.

(ii) Subtropical spices: spices grown in winter are cumin, fennel, coriander, fenugreek, onion and garlic and grown during summer are turmeric and ginger.

(iii) Temperate spices: thymes, saffron, savoy, caraway seed and asafoetida.

Classification based upon the parts:

(i) Rhizomes and root spices: ginger, turmeric, galangal and garlic.

(ii) Seed spices: nutmeg, mustard, coriander, fennel, cumin, fenugreek, ajowan, poppy, dill, aniseed, celery and bishopweed.

(iii) Bark spices: Cinnamon and cassia

(iv) Fruit spices: Chilli, cardamom, black pepper, vanilla, pimento, cassia, tamarind, mace (aril), allspice and kokam, etc.

(v) Leaf spices: Mint, bay leaf, curry leaf, chive, rosemary and savory etc.

(vi) Flower spices: clove, saffron, asafoetida, savory, caper and marjoram.

(vii) Pod spices: Vanilla and tamarind

(viii) Kernel spices: Nutmeg

(ix) Bud spices: Clove and caper

(x) Latex spices: Asfotida

(xi) Aril spices: Mace and anardana

(xii) Berry spices: Black pepper, juniper and allspice

Classification based on origin and flavour

(i) Aromatic spices: Cardamon, aniseed, clery, cumin, coriander, fenugreek and cinnamon.

(ii) Pungent spices: Ginger, chilli, black pepper and mustard

(iii) Phenolic spices: Clove and allspice

(iv) Coloured spices: Turmeric, saffron and paprika

Classification of spices based on economic importance: There are two groups viz., major and minor spices in the spice trade industry of the world.

a) Major spices are cardamom, black pepper, chilli, turmeric and ginger.

b) Minor spices are further divided into five sub groups such as

4% in other varieties. In turmeric, the primary consideration for extraction is the curcumin content, the volatile oil being not regarded as very important for food flavouring agent. There are different types of classification of spices and condiments.

Classification of spices based on season of growth:

(i) Annual spices (life cycle in one growing season): coriander, cumin, fennel, fenugreek, ajowan and black cumin.
(ii) Biennial spices (life cycle in two growing season): onion and parsley.
(iii) Perennial spices (life cycle in more than two years): Black pepper, saffron, clove, nutmeg and cinnamon

Classification of Spices based on climate requirement

(i) Tropical spices: ginger, turmeric, black pepper, cinnamon, kokam, galangal, small cardamom and clove.
(ii) Subtropical spices: spices grown in winter are cumin, fennel, coriander, fenugreek, onion and garlic and grown during summer are turmeric and ginger.
(iii) Temperate spices: thymes, saffron, savoy, caraway seed and asafoetida.

Classification of spices based upon the parts:

(i) Rhizomes and root spices: ginger, turmeric, galangal and garlic.
(ii) Seed spices: nutmeg, mustard, coriander, fennel, cumin, fenugreek, ajowan, poppy, dill, aniseed, celery and bishopweed.
(iii) Bark spices: Cinnamon and cassia
(iv) Fruit spices: Chilli, cardamom, black pepper, vanilla, pimento, cassia, tamarind, mace (aril), allspice and kokam, etc.
(v) Leaf spices: Mint, bay leaf, curry leaf, chive, rosemary and savory etc.
(vi) Flower spices: clove, saffron, asafoetida, savory, caper and marjoram.
(vii) Pod spices: Vanilla and tamarind
(viii) Kernel spices: Nutmeg
(ix) Bud spices: Clove and caper
(x) Latex spices: Asfotida
(xi) Aril spices: Mace and anardana
(xii) Berry spices: Black pepper, juniper and allspice

Classification of spices based on origin and flavour

(i) Aromatic spices: Cardamon, aniseed, clery, cumin, coriander, fenugreek and cinnamon.
(ii) Pungent spices: Ginger, chilli, black pepper and mustard
(iii) Phenolic spices: Clove and allspice
(iv) Coloured spices: Turmeric, saffron and paprika

Classification of spices based on economic importance: There are two groups viz., major and minor spices in the spice trade industry of the world.

a) Major spices are cardamom, black pepper, chilli, turmeric and ginger.
b) Minor spices are further divided into five sub groups such as

CHAPTER

38

Turmeric–*Curcuma domestica* Val. (2n=63)

Synonymous: *Curcurma longa* L.

Family: Zingiberaceae

Vernacular name: Turmeric, Indian saffron (English); *Halda, Haldi* (Hindi); *Halud, Pitras* (Bengali); *Haldhar* (Gujarati), *Arishina* (Kannada), *Manjal* (Malayalam), *Haladi, Harita* (Sanskrit), *Manjal* (Tamil), *Pasapu* (Telugu), *Haladi* (Urudu), *Lidar* (Kashmiri), *Curcuma* (Spanish, French, Italian), *Kurkum* (Arabic).

Importance: Turmeric is an important spice and condiment. Turmeric generic name originated from the Arabic word '*kurkum*' meaning 'yellow' and most likely refers to the deep yellow rhizome colour of the true turmeric. It is known as 'Indian Saffron', 'earthy herb of the Sun', 'herb of the Sun' '*Oushadhi*, the healing herb' and 'Golden Spice'. Global production is around 8 to 9 lakh tonnes. In India, it is grown in an area of 104,500 ha producing annually 3, 28,800 tonnes. Indian turmeric industry contributes about 78% of world production and 60% of the exports of Turmeric. Annually 18 to 20 crores worth of turmeric are exported. Indian turmeric is considered the best in the world market because of its high curcumin content. Asian countries consume much of their own turmeric production nearly 90%. Turmeric is second largest foreign exchange earner among Indian spices. India consumes nearly 80% of turmeric.

Turmeric is used extensively by all classes of people in the preparation of tasty curried dishes. Turmeric is bitter in taste and its action is "pungent-like" after digestion and metabolism. Being hot, light, acrid, and irritant, it is able to reduce corpulence and stimulate all functions. The use of turmeric as a spice, a dye, or a cosmetic is well known the world over. The Sanskrit meaning for turmeric is yellow colour. The rhizome has 1.8 to 5.4% curcumin, the pigment and 2.5 to 7.2% of essential oil. It is used for painting the body in rites connected with birth, marriage, death and agriculture. Turmeric rhizome and its powder is an auspicious article in the Hindu religious functions. It is a colouring matter in confectionery, cosmetic food material and pharmacy industries. It is used as a dye in textile industries for dyeing cotton, silk and wool. It is used as a food adjunct in vegetable, meat and fish preparations. It is used to flavour and at the same time to colour butter, cheese, margarine, pickles, mustard and other food stuffs. It is also used to colour liquor, fruit drinks, cakes and table jellies. It acts as an appetizer and aids digestion. A pinch of turmeric is usually added to our savories to impart simultaneously an agreeable flavour and colour to improve keeping quality. It occupies an important place in the preparation of medicinal oils, ointments etc,. It is a curative, stomachic, carminative, tonic, blood purifier, vermicide and an antiseptic. It is indicated in case of diabetes and leprosy. The first evidences of the use of turmeric, known as *Haridra,* are found in *Atharvaveda* (a collection of Vedas and mantras) and it was considered a curative drug for skin disease, graying of hair, and for charming away jaundice. In Tibetan medicine also, the term "Haridra" is given for turmeric.

Origin: Turmeric is origin of southern and southeastern Asia (India and Indonesia).

Distribution: It is extensively cultivated in India, China, Myanmar, Nigeria, Bangladesh, Pakistan, Sri Lanka, Taiwan, Burma, Indonesia, Malaysia, Vietnam, Thailand and Central America. In India, Andhra Pradesh is the leading state followed by Tamil Nadu, Orissa, Karnataka, West Bengal, Gujarat, Kerala, Maharashtra and Bihar in turmeric cultivation. Andhra Pradesh is called as 'Turmeric bowl of India' as it has highest share of 38% and 15% in total India's turmeric area and production respectively

Botany: There are four important species in Turmeric. They are (a) *Curcuma longa*, the widely cultivated type (b) *C. aromatica*, the cochin turmeric or kasturi manjal (c) *C.*

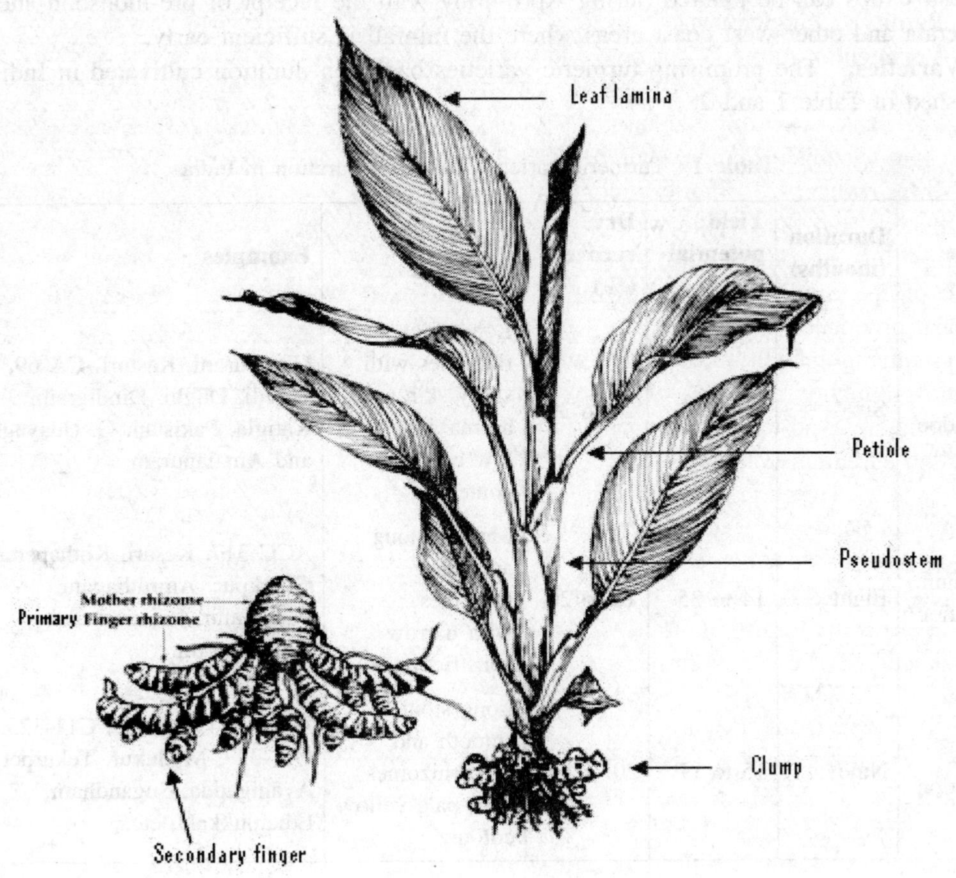

Source: http://www.plantauthority.gov.in/pdf/Turmeric.pdf

angustifolia, the East Indian Arrow root having plenty of starch in its rhizome and (d) *C. amada*, mango ginger, which had the taste and flavour of raw mango. Turmeric plant is a herbaceous perennial, 60-90 cm high, with a short stem and tufted leaf. There is a knobby rhizome at the base of the stem. There are 7 to 12 leaves, the leaf sheaths forms the pseudo-stem. The lamina is green above and pale green below and has a length of 30-40 cm and width 8-12 cm. The leaves are broadly lanceolate with long leaf stalks. Inflorescence is a central spike of 10-15 cm length. Flowers are concealed by the sheathing petiole; flowering bracts are pale green; bracts of coma tinged with pink. 1-4 flowers are born in axil of the bract opening one at a time. About 30 flowers are produced in a spike. Seeds are produced in capsules and there will be one to numerous sunken capsules in an inflorescence. It's underground rhizome give rise to primary and secondary rhizomes called 'fingers'. The rhizome is pungent, bitter, heating.

Climate: Turmeric can be grown from sea level to 1500 m above MSL in the tropics. It is grown in areas receiving mean annual rainfall of 1000 to 2250 mm. It grows well at a mean temperature range of 20 to 35°C.

Soil: It is grown in different types of soils such as red, black and alluvial soils in irrigated and rainfed conditions. It thrives best in well drained sandy or clay loam soils with a pH range of 4.5-7.5 and with high organic matter content status. The crop cannot stand waterlogging or alkalinity.

Season: In Andhra Pradesh and Tamil Nadu, sowing is done during May-June or July-August. Crops can be planted during April–May with the receipt of pre-monsoon showers in Kerala and other west coast areas where the rainfall is sufficient early.

Varieties: The promising turmeric varieties based on duration cultivated in India are furnished in Table 1 and 2.

Table 1 Turmeric varieties based on duration in India

Type	Duration (months)	Yield potential (t/ha)	Dry recovery (%)	Nature of rhizomes	Examples
Short duration	Six	8 to 20	26 to 30	Long thick rhizomes with shiny skin and aroma and low curcumin content	GL, Puram, Kasturi, CA 69, Jobedi, Dughi, Dindigram, Katigia, Pakistan, G. Udayagiri and Amalapuram
Medium duration	Eight	14 to 35	18 to 20	Medium long and thick rhizomes with narrow constrictions	ACC 317, Kesari, Kothapetta, Gorakpur, Amruthapani, Panamalur, Rajapuri and Amalapuram
Long duration	Nine	15 to 17	20	Long stout, smooth and hard rhizomes with pale yellow colour	Armoor, Duggirala, C11-325, C11-327, Mydukur, Tekurpet, Avanigadda, Sugandham, Ethamukkala, etc.

Table 2 Turmeric varieties particulars cultivated in India

Variety	Duration (days)	Curcumin (%)	Oil (%)	Rhizome colour	Rhizome stature	Rhizome yield (t/ha)	Cured rhizome yield (%)	Source
Co 1	285	3.2	3.2	Bright orange	Big size	30.0 (5.85)	19.5	Tamil Nadu
BSR 1	285	4.2	3.7	Bright yellow	---	30.7 (3.96)	18.5	Tamil Nadu
BSR 2	240	4.2	3.8	Bright yellow	---	40.5 (4.05)	20.0	Tamil Nadu
Suguna	190	4.9	6.0	Orange	Thick and plumpy	29.3 (6.03)	20.4	Kerala
Suvarna	210	4.0	7.0	Dark orange	Medium size	17.4 (4.60)	20.0	Kerala
Sudharshana	190	7.9	7.0	Orange	Thick and plumpy	28.8 (5.49)	20.6	Kerala
Krishna	255	2.8	2.0	---	Long and plumpy	9.2 (4.00)	16.4	Maharashtra
Sugundham	210	3.1	2.7	Reddish yellow	Stout and long	15.0 (4.00)	23.3	Gujarat
Roma	253	9.3	4.2	Bright Yellow	---	20.7 (4.00)	31.0	Odisha
Suroma	253	9.3	4.4	Reddish Yellow skin	Round and plumpy	20.0 (4.50)	26.0	Odisha
Rajendra sonia	225	8.4	5.0	Deep orange	Stout and plumpy	42.0 (4.50)	18.0	Bihar
Ranga	250	6.3	4.4	Orange yellow	Bold and spindle	29.0 (7.00)	24.8	Odisha
Rasmi	240	6.4	4.4	---	Round and plumpy	31.3 (7.80)	23.0	Odisha
Prabha	195	6.5	6.5	Reddish yellow	---	37.5 (3.75)	19.5	Kerala
Pradhiba	188	6.2	6.2	Reddish yellow	---	39.1 (3.91)	18.5	Kerala
Alleppey Supreme	210	6.0	4.0	---	---	35.4 (6.72)	19.3	IISR
Kedaram	210	5.5	3.0	---	---	34.5 (6.52)	18.9	IISR

Growth stages: Leaves appear above the ground in about 4 weeks. Crop duration is 6 to 10 months.

Land preparation: Land is prepared to a fine tilth with four ploughings. Hydrated lime @ 500 kg/ha has to be applied for laterite soils and thoroughly incorporated. Ridges and furrows are formed.

Seeds and planting: The crop is propagated through rhizomes or fingers of 4 cm length with one or 2 buds. Whole or split mother rhizomes weighing 35 to 44 g are used for planting. Mother rhizomes are better yielders than finger rhizomes. Setts requirement is 2000 to 2500 kg per ha. As an intercrop in a fruit-garden, seed rate may be as low as 300-500 kg per ha. Well developed, healthy and disease free rhizomes are to be selected. Rhizomes are treated with 0.3% Dithane M-45 or 0.25% of Agallol and 0.5% Malathion for 30 minutes before planting. Treat the setts with *Trichoderma viridi* @ 4 g per kg of setts. Afterwards the chemical insecticides or fungicides should not be used. The rhizomes are to be dibbled in the sides of ridges. Spacing is 45 cm × 30 cm or 45 cm × 15 cm. It is planted at 5.0 cm depth.

Weed management: Spray glyphosate @ 3.75 lit/ha to control perennial weeds such as *Cynodon, Cyperus*, etc., as pre-plant application before planting of turmeric. Spray Alachlor @ 3.75 lit/ha in 625 lit of water on third day after planting as pre-emergence using a fan or deflector type nozzle to control weeds. Otherwise, Alachlor G @ 20 kg mixed in 50 kg sand can be applied on the third day of planting and irrigation is given immediately. Weeding may be done thrice at 30, 60 and 120 days after planting depending upon weed intensity, if herbicides are not used.

Manures and fertilizer application: FYM @ 10 tonnes/ha, neem or groundnut cake @ 500 kg/ha, 25, 60 and 18 kg of N, P_2O_5 and K_2O per ha; $FeSO_4$ @30 kg/ha, $ZnSO_4$ @15 kg/ha and borax @ 10 kg/ha are applied as basal dressing. Top dressing is carried out with 25 kg N and 18 kg K_2O per hectare each time on 30, 60, 90 and 120 days after planting. For correcting deficiency of micronutrients especially iron, zinc and boron at rhizome development stage, apply 375 g ferrous sulphate, 375 g zinc sulphate, 375 g borax and 375 g of urea in 250 lit of water per ha. Spraying is carried out twice at 25 days intervals. Fertigation is done as per the recommended dose with 150:60:108 kg of NPK/ha and is applied throughout the cropping period once in three days. 75 % of the recommended dose of phosphorous is applied as basal dose. Water soluble fertilizers like 19:19:19, Mono ammonium phosphate (12:61:0), Multi K (13:0:45) and urea are used. Fertigation schedule for turmeric is presented in Table 3.

Table 3 Fertigation schedule for turmeric

Crop Stage	Duration (in days)	Nutrients requirement (%)			Quantity applied (kg/ha)	
Planting to establishment stage	15	10	20	10	19:19:19 Multi K Urea	15.78 17.33 21.20
Vegetative stage	60	40	30	20	19:19:19 Multi K Urea	9.83 96.00 100.57
Rhizome initiation stage	60	30	30	30	19:19:19 Multi K Urea	4.91 71.28 76.29
Rhizome maturation stage	135	20	20	40	19:19:19 Multi K Urea	15.78 40.42 47.06
Total duration	270	100	100	100		

After cultivation: Earthing up is carried out each time after top dressing.

Mulching: The crop can be mulched immediately after planting with green leaves or banana pseudostem or sugarcane trash @ 12 to 15 t/ha. It may be repeated for second time after 50 days with the same quality of green leaves after weeding and application of fertilizers.

Water management: Depending on the soil types, crop requires 15 to 20 irrigations in heavy soils and 35 to 40 irrigations in light soils. Moisture stress affects the growth and development of the plant especially during the rhizome bulking stage. It cannot withstand waterlogging.

Cropping system: Turmeric is often rotated with rice, sugarcane, banana, chilli, garlic, elephant foot yam, vegetables, pulses, wheat, finger millet, maize, etc. It can be grown as an intercrop in coconut and arecanut plantations. Onion, coriander and fenugreek can be planted as intercrop in the sides of the ridges 10 cm apart. Since turmeric is a shade loving crop, castor, redgram or *Sesbania grandiflora* (agathi) can also be planted at wider spacing or may be raised along the border lines in the field to provide shade to turmeric crop. It is cultivated as a subsidiary crop to ginger in some areas and in other areas with quick-growing vegetables.

Plant protection: The seed rhizomes are dipped in Emisan @ 0.1% and Phosalone 35EC @ 2 ml/lit or Monocrotophos 36WSC @ 1.5 ml/lit for controlling rhizome rot and scales. Spray Dimethoate 30EC or Methyl demeton 25EC @ 750 ml/ha in 250 litres of water to control thrips. If shoot borer incidence is noticed, such shoots may be cut open and larve picked out and destroyed. If necessary neem oil 0.5% may be sprayed at fortnightly intervals. The nematode problem can be overcome by avoiding turmeric planting after banana or other solanaceous vegetables. Plant the rhizomes only after taking suitable control measures. Apply Carbofuran 4 kg a.i/ha twice on the third and fifth month after planting the rhizomes. Leaf spot and leaf blotch can be controlled by restricted use of Bordeaux mixture 1%. Drench with Bordeaux mixture 1% or Copper oxychloride 0.25% to control rhizome rot. Application of Trichoderma at the time of planting can check the incidence of rhizome rot. Treat the seed rhizomes with 0.3% Copper oxychloride for 30 min before storage. Spray Carbendazim 500 g/ha or Mancozeb 1.0 kg/ha or Copper oxychloride 1.25 kg/ha to control leaf spot.

Harvest: The crop becomes ready for harvest in seven to nine months depending upon the variety. Early varieties mature in 7 to 8 months, medium varieties in 8 to 9 months and long duration varieties in 9 to 10 months after planting. Rhizome is harvested when the leaves turns yellow and dry. The plants will start to lodge as it reaches maturity. Cut the plants 10 to 15 cm above the ground level. Keep the field in the same condition for two weeks to allow the rhizomes to attain maturity. Irrigate the field for easy harvesting. Then the land is ploughed and the rhizome is gathered by hand picking or the clumps are carefully lifted with a spade. The rhizomes are also collected after digging deeply with spade or digging forks. Harvested rhizomes are cleaned of mud and other extraneous matter adhering to them. Fresh wet rhizomes yield is 25 to 30 t/ha while the cured rhizomes yield is 5 to 6 t/ha.

Seed rhizomes: Rhizomes for seed purpose are generally stored after heaping under the shade of a tree in well ventilated shed and covered with turmeric leaves. Sometimes the heap is plastered with earth mixed with cow dung. The seed can be covered with wooden planks with one or two holes for aeration. Seed rhizomes can be stored in open sand media with a partially closed pandal /trellies. The rhizomes are to be dipped in quinalphos (0.075%) solution for 15 minutes if scale infestations are observed and in mancozeb (0.3%) to avoid storage losses due to fungi.

Processing: It involves three steps viz., curing, polishing and colouring.

(i) Curing: Harvested turmeric rhizomes are cleaned of mud and other extraneous matters adhering to them. Fingers are separated from mother rhizomes and are usually kept as seed material. Only good fingers separated from the rhizomes are used for curing. The fresh turmeric is cured before marketing. Curing involves boiling of fresh rhizomes in water and drying in the sun. The boiling/ cooking of turmeric is to be done within two or three days after harvesting. The boiling should last for 45-60 minutes when the rhizomes turn soft. Boiling the turmeric rhizome in the curing process reduces the microbial load on the rhizomes. Boiling in alkaline water by adding 0.05% to 1% sodium carbonate, or lime instead of cow dung slurry, may improve the colour. The cooked fingers/mother rhizomes are spread on bamboo mats or cement floor under the sun for drying. The rhizomes are spread in 5-7 cm thick layers for desirable colour of the dried product. During night time the material should be heaped or covered. It may take 10-15 days for the rhizomes to become completely dry. Artificial drying using cross-flow hot air at a maximum temperature of 60°C is also found to give a satisfactory product. The mother rhizomes and the fingers are generally cured separately. In the traditional method, the cleaned rhizomes are boiled in copper or galvanized iron or earthern vessels with water just enough to soak them. In certain places, cow dung slurry is used as boiling medium. From hygienic point of view, such rhizomes fetch poor market value. Boiling is stopped when froth comes out and white fumes appear giving out a typical odour. The boiling lasts for 45 to 60 minutes when the rhizomes are soft. Over cooking spoils the colour of final product while under cooking renders the dried product brittle. In the improved scientific method of curing, the cleaned fingers (approximately 50 kg) are taken in a perforated trough of size 0.9 m × 0.55 m × 0.4 m, made of GI or MS sheet with extended parallel handle. The alkaline solution (0.1% sodium carbonate or sodium bicarbonate) (100 g of sodium bicarbonate or sodium carbonate in 100 liters of water) is poured into the trough so as to immerse the turmeric fingers. In improved method (CFTRI, Mysore) of curing, the rhizomes are boiled in lime water or sodium carbonate. A water solution containing 20 g of sodium bisulphite and 20 ml of HCL acid per 45 kg of tubers is recommended to give them desired yellow tint. The whole mass is boiled till the fingers become soft and yield when pressed between fingers. The cooked fingers are taken out of the pan by lifting the trough and draining the solution into the pan. The drained solution in the pan can also be used for boiling another lot of turmeric along with the fresh solution prepared for the purpose. Alkalinity of the boiling water helps in imparting orange yellow tings to the core of turmeric. The cooked fingers are dried in the sun by spreading 5 to 7 cm thick layers on bamboo mat of drying floor. A thinner layer is not desirable, as the colour of the dried product may be adversely affected. During night time, the materials should be heaped or covered. Drying may take 10 to 15 days till they become hard and brittle and produce metallic sound on breaking. The yield of the dry product varies from 20 to 30% of the freshly harvested green turmeric depending upon the variety and the location where the crop is grown. Mother rhizomes give a higher curing percentage than fingers.

(ii) Polishing: Dried turmeric has poor appearance and a rough dull outer surface with scales and root bits. The appearance is improved by smoothening and polishing outer surface by manual or mechanical rubbing. Manual polishing consists of rubbing the dried turmeric fingers on a hard surface or trampling them under feet, wrapped in gunny bags. The improved method is by using hand operated barrel or drum mounted on a central axis, the sides of which are made of expanded metal mesh. When the drum filled with turmeric is rotated at 30 rpm, polishing is effected by an abrasion of the surface against the mesh as well as by mutual rubbing against each other as they roll inside the drum. The turmeric is also polished in power operated drums. The yield of polished turmeric from the raw materials varies from 15 to 25%.

(iii) Colouring: It is done to give a good appearance and better finish to the product. The yellow colour of turmeric always attracts buyers. This is done to half polished rhizomes in two ways, known as dry and wet colouring. Turmeric powder is added to the polishing drum in the last 10 minutes in dry process. In wet process, turmeric powder is suspended in water and mixed by sprinkling inside the polishing basket. For giving a brighter colour, the boiled, dried and half polished fingers are taken in baskets which are shaken continuously when an emulsion is poured in. When the fingers are uniformly coated with the emulsion, they may be dried in the sun. The composition of the emulsion as recommended by the CFTRI, Mysore, required for coating 100 kg of half boiled turmeric in 40 g alum, 2 kg turmeric powder, 140 g castor seed oil, 30 g sodium bisulphate and 30 ml conc HCL acid.

Storage: The moisture content of the dried turmeric is kept at 8 to 10% for better storage. The curcuminoid pigments in turmeric powder and oleoresin are stable if stored away from light and in a very dry environment. All water or ethanol solvent should be removed from the oleoresin to assure pigment stability.

Marketing: The major trading centers of turmeric are Nizamabad, Dugirala in Andhra Pradesh, Sangli in Maharshtra and Salem, Erode, Dharmapuri and Coimbatore in Tamil Nadu.

Utilization: The different products of turmeric are hard fingers, turmeric powder, volatile oil, turmeric pigment and oleoresin. 'Kum-kum', a type of starch by-product of turmeric is popular with every house wife. It finds a place in offerings on religious and ceremonial occasions. It is ideal produce as a food colourant. Turmeric (whole) is a unique, colourful and versatile natural plant product combing properties of a spice / flavourant, a colourant; brilliant yellow hue, a cosmetic and a drug useful in many diseases. It is used as a food adjunct in vegetable, meat and fish preparations. It is used to flavour and at the same time to colour butter, cheese, margarine, pickles, mustard and other food stuffs. It is also used to colour liquor, fruit drinks, cakes and table jellies. It acts as an appetizer and aids digestion. A pinch of turmeric is usually added to impart simultaneously an agreeable flavour and colour to improve keeping quality. It is largely used for dyeing wool, silk, cotton to impart an yellow shade. It occupies an important place in the preparation of medicinal oils, ointments and poultice. It is a curative, stomachic, carminative, tonic, blood purifier, vermicide and an antiseptic. Smearing turmeric powder on face, limbs during a bath is found to clear the skin and beautify the face. It is used as medicine in case of diabetes and leprosy. It is applied as a paste with sesame oil and neem leaves to cure small pox. Burnt turmeric is also used as tooth powder which relieves dental troubles. Turmeric is an important constituent in the curry powder formulations and pickles and soups. The powder is mostly consumed domestically for culinary purposes. A moisture level of above 12.1% (dry weight basis-DWB) is critical with respect to the free flow characteristics of the turmeric powder. Turmeric powder contains 1.3 to 5.5% volatile aromatic orange red fluorescent oil called termoerones ($C_{15}H_{22}O$). The colouring matter is curcumin ($C_{12}H_{20}O_6$). Turmeric contains 5.8% moisture, 8.6% protein, 8.9% fat, 63.0% carbohydrates, 6.9% fibre and 6.8% mineral matter. It has calorific value or food energy 390 calories per 100 gram. The dried rhizomes yield 5-6%, while fresh ones give 0.24% essential oil. About 58% of the oil is composed of turmerones (sesquiterpene ketones) and 9% tertiary alcohols. Turmeric oleoresin is described as a 'deep red or orange red, somewhat viscid liquid, with characteristic odour. Turmeric oleoresin is valued for its curcumin content.

Turmeric principles: Curcumin, demethoxycurcumin, bis-demethoxycurcumin, and aromatic-turmerone are four major active components of turmeric. Turmeric contains protein (6.3%), fat (5.1%), minerals (3.5%), carbohydrates (69.4%) and moisture (13.1%). The essential oil (5.8%) obtained by steam distillation of rhizomes. Curcumin (diferuloylmethane) (3–4%) is responsible for the yellow colour, and comprises curcumin I (94%), curcumin II (6%) and curcumin III (0.3%). The yield of oleoresin from dried root is typically 10-12%. Oleoresin is a highly viscous orange brown product containing 30 to 35% curcumin, 15 to 20% volatile oil and has a characteristic turmeric aroma. Oleoresin cells are distributed in turmeric rhizomes. The oleoresins are prepared from the turmeric through extraction with a suitable solvent or solvents. If one extracts turmeric with petroleum ether, the product is a light fluid oleoresin which is highly aromatic and smells strongly of ground turmeric, but has little of its yellow colouring power. If, however, one extracts it with acetone, one gets a brilliant yellow extractive which is a hard solid with only a very small amount of the characteristic odour. Both of these products are strictly oleoresins, but the difference in their physical condition, aroma and colouring power, is determined entirely by the nature of the solvent which is used in their preparation. 'Alleppey' turmeric variety is very rich in curcumin. It contains about 6.5% of curcumin as against 3 to 4% in other varieties. In turmeric, the primary consideration for extraction is the curcumin content, the volatile oil being not regarded as very important for food flavouring agent. The volatile oil derived form steam distillation of crushed turmeric tubers is an orange yellow liquid, occasionally fluorescent with an odour reminiscent of the tuber. The rhizomes yield 5 to 6% of oil and fresh ones give 0.24% essential oil. The oil is composed of turmerones (sesquiterpene ketones) and teritiary alcohols. Their volatile oil content ranged from 2.5 to 7.2% and their curcumin content from 1.8 to 5.4%. Acetone, alcohol and ethylene dichloride are found suitable for extracting oleoresin from turmeric. Both soxhelt and the cold percolation extraction methods give good yields of oleoresin, with a good recovery of curcumin (85%) when finely ground spice (60 mesh) is used.

REFERENCES

Angles, S., A. Sundar and M. Chinnadurai. 2011. Impact of Globalization on Production and Export of Turmeric in India – An Economic Analysis. Agricultural Economics Research Review. Vol. 24 July-December 2011 pp 301-308

CPG. 2013. Crop Production Techniques of Horticultural Crops. Horticultural College and Research Institute, Tamil Nadu Agricultural University, Coimbatore – 641 003

Khanna, N.M. 1999. Turmeric: nature's precious gift. Current Sci., 76(10):1351-1356.

Remadevi, R. and Ravindran, P.N. 2005. Turmeric: Myths and Traditions. Spice India. 18(8):11-17.

Subramanian, K.S., N. Sivasamy, and T. Thangaraj. 2001. Integrated nutrient management for turmeric. Spice India, 14 (12):25-26.

Weiss, E.A. 2002. Spice Crops. CAB International publishing, Oxon, UK.

Velayudhan, K.C., Muralidharan, V.K., Amalraj, V.A., Gautam, P.L., Mandal, S. and Dinesh Kumar. (1999). *Curcuma* Genetic Resources. Scientific Monograph No. 4. National Bureau of Plant Genetic Resources. New Delhi. pp. 149.

http://www.plantauthority.gov.in/pdf/Turmeric.pdf

www.nhm.nic.in/Conf-PPT/25-IISR-Calicut.ppt

http://ijsrm.in/v1-i4/6%20ijsrm.pdf

http://www.indianspices.com/pdf/state_prd.pdf

http://agriexchange.apeda.gov.in/Market%20Profile/MOA/Product/Turmeric.pdf

http://kaubic.in/spicesdatabase/Turmeric/varities.htm
http://www.fda.gov/ForIndustry/ImportProgram/default.htm
http://agmarknet.nic.in/spices.pdf
http://agritech.tnau.ac.in/horticulture/horti_spice%20crops_turmeric.html
http://agritech.tnau.ac.in/banking/PDF/Tumeric.pdf

CHAPTER

39

Aromatic Crops

Aromatic plants are those plants which produce a certain type of aroma and the taste of the material (aromatic herbs). Their aroma is due to the presence of essential oil with chemical constituents. The essential oils are 'distilled essence' which represent the active constituent of plants. The term essential oil is concomitant to fragrance or perfumes because these fragrances are oily in nature and they represent the essence or the active constituents of the plants. The essential oil or volatile oil evaporates when exposed to the air at ordinary temperatures. These aromatic plants accumulate oils in certain specific organs or plant parts which are then used for the production of essential (ethereal) oils. Essential oils are highly concentrated, low volume, high value products. The essential oils or volatile oils are produced from 400 plant species and from 67 families. The most important families are Asteraceae, Lamiaceae, Apiaceae, Fabaceae, Rutaceae, Lauraceae, Cupressaceae, Pinaceae, Santalaceae, Zingiberaceae, Myrtaceae, Rubiaceae and Burseraceae. Aromatic plants, their extracts and essential oils contain volatile substances which occur as gum exudate, balsam and oleoresin in one or more parts, namely, root, stem, wood, twigs, bark, buds, foliage, flower and fruit. Aromatic plants are used as raw materials for extraction of essential oils (which, in turn, are used in the flavour and fragrance industries), as well as the sources of spices, herbs, and other natural products such as cosmetics, botanical pesticides, insect repellents, herbal teas/drinks, etc.

Importance of aromatic plants: There is demand for aromatic plants in both the domestic market and in overseas markets. Aromatic plants form a large group of economically important plants, which provide basic raw materials for perfumes, flavours and cosmetics. These plants and their products not only serve as valuable source of income for small land holders and entrepreneurs but also earn valuable foreign exchange by way of export. Essential oils and aroma chemicals constitute a major group of industrial products. These oils form indispensable ingredients of the necessities in many spheres of human activity. The essential oils are adjuncts of cosmetics, soaps, pharmaceuticals, perfumes, scents, confectionery, ice-creams, aerated waters, disinfectants, tobacco, agarbathis, beverages, toiletries, etc. Some of the categories of plant-derived products include natural health products, phytocosmetics and personal care products.

History: Plants are primary source of perfumes in the ancient civilizations of Egypt, India, Greece and Rome. The Egyptian, the Persian, and the Babylonian were known to grow and use aromatic plants in making perfumes and other scented waters from a distillation of rose petals and orange blossoms. Oriental people were also fond of aromatic plants. The aromatic plants were grown in the palace compounds and used as raw materials to make perfumes, scented water, and a dozen of other aromatic products. In India, plants such as cinnamon, ginger and sandalwood are mentioned in Vedic literature around 2000 BC. Dioscorides' treatise De materia medica deals with aromatic plants. Asia is known as the 'Land of Spices' and the 'Land of Herbs' (both of which are also aromatic plants), 'Land of Traditional Perfumes and Cosmetics' as well as the 'Land of Traditional Medicines'.

Classification of aromatic plants

a) Classification of aromatic plants based on growth habit: Aromatic plants are classified based on growth habit as i) Aromatic grasses: Lemongrass, palmarosa, citronella, vetiver; ii) Aromatic herbs and shrubs: Mints, ocimum, patchouli, rosemary, clary sage (*Salvia sclarea*), thyme, celery, coriander, cumin, fennel, ajowan, davana, chamomile, geranium, cardamom, ginger, kacholam; iii) Aromatic trees: Sandalwood, eucalyptus, clove, camphor, cinnamon, nutmeg, linaloe (*Bursera delpechianum*).

b) Classification of aromatic plants based on plant parts used as i) flowers (rose, jasmine, tuberose, marigold, champaca, hyacinth); ii) leaves (lavender, peppermint, eucalyptus); iii) barks (cinnamon, cassia); iv) stem/twigs (citronella, geranium, patchouli); v) Heart wood (sandalwood, cedar, bursera); vi) roots (vetiver); vii) rhizomes (ginger, calamus), anise, fennel; viii) seeds (nutmeg, cumin, celery, fennel) and ix) fruits (lemon, grape seeds, olive, melons and strawberry).

c) Classification of aromatic plants based on how they are utilized as i) raw materials for essential oil extraction, ii) spices in which their non-leafy parts are used as a flavouring or seasoning, iii) herb plants in which their leafy or soft flowering parts are used as a flavouring or seasoning and iv) miscellaneous group of aromatic plants used as cosmetics, dyes, air fresheners, disinfectants, botanical pesticides, herbal drinks/teas, insect repellents, etc.

Cropping systems: Aromatic plants are collected from the wild forests. Due to overexploitation, many species have become extinct or scarce so that they have to be cultivated. However, a few countries, viz. Nepal, Bhutan, and Lao PDR, still collect aromatic plants from the wild. Few countries in Asia such as China, India, Indonesia, Nepal, Sri Lanka and Thailand produced essential oils on an industrial basis. There is great scope of cultivation of aromatic plants on marginal soils/ wastelands without disturbing the food priorities. In order to obtain maximum benefits of existing space, season, soil moisture and nutrients, several cropping systems involving aromatic plants have been employed. These include intercropping and crop rotation. Several cash crops, e.g. vegetable, legumes, cereals, root crops, etc. can be grown together with aromatic plants. This practice is particularly recommended for aromatic plants with slow growth like vetiver and palmarosa.

CHAPTER

40

Citronella–*Cymbopogon winterianus* Jowitt (2n=20)

Family: Poaceae

Vernacular names: Citronella, Citronella grass, Java citronella grass (English).

Importance: Citronella oil is one of the essential oils obtained from the leaves and stems of the plants. The oil is rich in citronellal, geraniol and citronellol which are used for making high grade blended perfumes. It is classified in trade into two types *viz.*, Ceylon citronella oil, obtained from *Cymbopogon nardus* (inferior type), while Java type citronella oil obtained from *Cymbopogon winterianus* (superior type).

Origin: It is native of Nepal and Sri Lanka.

Distribution: Citronella is cultivated in China, Vietnam, Indonesia, Argentina, Taiwan, Sri Lanka, Brazil, Madagascar, India, Malaysia, Argentina, Ecuador, Mexico, Guatemala, Honduras and West Indies. The crop is grown in the states of Assam, Gujarat, Jammu & Kashmir, Madhya Pradesh, Karnataka, Maharashtra, Tamil Nadu, Andhra Pradesh, Odisha, West Bengal and Uttar Pradesh.

Botany: It is a tufted aromatic perennial herb with fibrous roots, erect over 2 m tall, with smooth leaves and bearing a large inflorescence.

Climate: Citronella thrives well under the tropical and subtropical conditions. Although 180-250 m altitude is optimum, the plants are reported to grow well at the altitudes between 1000-1500 m. Rainfall of about 2000-2500 mm well spread over the year and high atmospheric humidity of 70 to 90% favour the growth of the plant, yield and quality of the oil. In areas where rainfall is less it can be grown with supplemented irrigations.

Soils: The sandy loam soil with abundant organic matter is the most suitable. Heavy clayey soils, which tend to water log and light sandy soils are not suitable for this crop. It does not tolerate water stagnation. The plant has been found to grow well under a pH range of 5.0-8.5.

Season: The onset of monsoon is the best time. The grass is planted in the beginning of monsoon (June-July). Under irrigated condition, February-March is also suitable. Life span is 3 to 4 years

Varieties: Mandakini, Manjusha, Manjiri, Bio-13, Java-2, Jorhat-2, Jalpallavi, Java citronella, Java II, Ceylon citronella.

Land preparation: The land should be prepared to have fine tilth by discing and tilling. Ridges and furrows are made at 60 cm or 45 cm intervals to avoid waterlogging. It is also beneficial to add 20-25 t/ha of FYM or compost and mixed properly in the soil at the time of final tillage.

Propagation: Citronella is a perennial grass, however, it does not produce viable seeds, therefore, the species can be propagated only vegetatively by slips. Slips are obtained by splitting well-grown clumps of vigorously growing plant of 6-months to one year old into slips. The clump is gently dug out and separated into number of slips containing 2- 3 tillers/ slip. A year old clump yields on an average of about 50 slips. Each clump is separated into slips, each containing 2 to 3 tillers. Fibrous roots and leaves in slips should be trimmed off at 25-30 cm height before planting. Slips are planted at 60 cm spacing about half way down the slope of the ridge. Before planting, the old roots on the slips are clipped off and leaves are completely removed. The slips are planted at 45 × 45 cm or 60 × 45 cm or 60 × 60 cm at 5-8 cm depth on one side of the ridges half way up the slope. One slip is inserted in each pit, latter filled with loose soil and firmly pressed keeping the slip vertical. Planting may be done at closure spacing at 45 × 22.5 cm and after first harvest every alternate plant from the row is uprooted and used as planting material. Plants get established within 15-20 days and after 30-40 days complete green colour appears in the field. The field should be irrigated immediately after planting if there are no rain within next 24 hours.

Weed management: The plantation should be kept weed free for 60-75 days, which is critical period for weed competition. Generally two weeding are given, one at 20-25 and another at 40-45 days after planting. Simazine @ 1.25 kg or Diuron@ 1.25 kg in 1250-1500 litre water or 625 g Oxyflorofen as pre-emergence may be applied at least 15 days before planting.

After cultivation practices: Interculture after each harvest is necessary. Earthing up is done after about 4 months of planting and again after every harvest.

Manures and fertilizer application: Java citronella generally requires high dose of nitrogen for good growth. A fertilizer dose of 80 to120-40-40 kg of N, P_2O_5 and K_2O per ha is recommended in North-Eastern India and North India. It is beneficial to apply N in 4 equally split doses, the first about a month after planting and then after each harvest, at an interval of about three months. In poor soils, Farm Yard Manure @ 12.5 t/ha and fertilizer dose of 200-80-40 kg of N, P_2O_5 and K_2O per ha is recommended. Nitrogen is applied in 5 split doses starting from one month after planting and the remaining in 4 doses one after each harvest. P and K are applied fully as basal at one time. Iron deficiency may be rectified by spraying 0.25% Ferrous sulphate solution at 10-15 days intervals for 3-4 times.

Water management: In the areas with well distributed annual rainfall of about 2000-2500mm, supplemental irrigation is not necessary. In drier months, irrigation once in 10-15 days is required. For undulating areas sprinkler irrigation is recommended.

Cropping systems: The grass can be intercropped with arecanut and coconut in the initial 3-4 years of growth as these crops attain bearing stage after about 6-7 years.

Pest and diseases management: Mites may be controlled by spraying a mixture of Docofol (Kelthane) @ 2 ml/lit and Ethion (Tafethion 50 EC) @ 1 ml/lit along with a wetting agent like Triton AE or Enditron AE @ 0.5 ml/lit. Stem borer can be controlled by applying Furadan @ 20 kg/ha or by spraying 0.5 % Dimethoate or 400 to 500 kg Neem oilcake per ha. Drench planted slips with sumicidin at the rate of 30 ml/10 litre of water into the soil per hectare before planting or spray ethion 10 ml/10.1itre to control termites. Leaf blight or

Carbularia leaf blight can be controlled by Carbendazim (Bavistin at 1.0 g/litre) followed by Benomyl of Benzimidazole group (Benlate 50 WP) @ 0.2% at 10 days interval.

Harvest: The crop is ready for the first harvest after about 6 months of planting. Harvesting can be done 4 times a year. It can be taken at an interval of 3 months. Harvesting is done by sickle above the first node at 20-45 cm above the ground. Generally leaf blades are cut and sheath are left out. This is because the sheath contains only little and poor quality oil. The harvested grass sometimes contains dead leaves. These should be removed before packing into the vessel. After cutting, the herb is allowed to wilt for 12-24 hours to remove the excess moisture. This wilting allows better packing in the vessel and saving of steam and fuel. Wilting more than 24 hours results in loss of essential oil. Cutting the grass into shorter length also gives 10-15 % higher recovery. Harvested material is distilled by steam distillation unit within 24 hours of harvest. Flowering should be discouraged as it causes ageing in plants and reduces their life span. Generally the crop once planted yields profitable returns for about 3-4 years and needs replanting only afterwards.

Yield: The average herbage yield is 20 to 25 t/ha and oil yield is 120 to 150 kg/ha/ year. Under very favourable conditions, yield of 200-250 kg oil/ha can be obtained. On an average, the oil content is about 1% on the basis of fresh weight of leaves. The aromatic oil has odour like lemon due to presence of Citral in the oil.

Utilization: Soaps, soap flakes, detergents, household cleansers, insecticides, etc are often perfumed exclusively with this oil. It is also a valuable constituent in perfumery for soaps and detergents. Citronella may irritate sensitive skin.

REFERENCES

Aakanksha Wany, Shivesh Jha, Vinod Kumar Nigam and Dev Mani Pandey. 2013. Chemical analysis and therapeutic uses of citronella oil from *Cymbopogon winterianus*: a short review. International Journal of Advanced Research, 1 (6): 504-521

Priyadarshini, T. and Rakesh, M. 2014. Cultivation of citronella and its oil extraction. International Journal of Sciences. 1(3): 1-2.

Ranjana Katiyar, Somesh Gupta and K. R. Yadav. 2010. *Cymbopogon winterianus*: An Important Species for Essential Java Citronella Oil and Medicinal Value. National Conference on Forest Diversity: Earth's Living Treasure, 22, May 2011. pp115-118.

http://horticulture.kar.nic.in/APMAC_website_files/citronella.htm

CHAPTER

41

Geranium–*Pelargonium graveolens* (L) Hervitt. (2n=88)

Family: Geraniaceae

Vernacular names: Geranium (English, Hindi); *Pannir soppu, Pannir patre* (Kannada), Geranium (Tamil).

Importance: Geranium is yielding an essential oil which has strong rose-like odour. The plant is also known as rose geranium. The name Pelargonium is derived from the Greek word 'palargos', meaning 'stork', since the fruit is thought to resemble a stork's bill.

Origin: Geranium is a native of the Cape Province in South Africa.

Distribution: It is cultivated in France, Belgium, Spain, Italy, Morocco, Madagascar, South Africa, Egypt, Reunion Island, Congo, China, India and Russia. It is cultivated in the Nilgiris and Kodaikanal hills of Tamil Nadu; in and around Bangalore in Karnataka.

Botany: Geranium is a bushy, aromatic plant. The stem is cylindrical, woody at the base, pubescent, green when young and turning brown with age. The leaves are alternate, stipulate, simple, with 5 primary lobes and secondary lobes which are densely pubescent. The leaves are highly aromatic in nature. The inflorescence is umbellate and hairy. The flower is bisexual, hypogenous, with a pink corolla; the two posterior petals are larger with reddish to purple markings.

Climate: Geranium can be grown in temperate, subtropical and tropical climates at various altitudes from 1000 to 2200 m. The crop flourishes in areas having annual rainfall ranging from 700 to 1500 mm. However, heavy rainfall results in water-logging, causes root-rot and stunted growth. It grows well with a temperature ranging from 10 to 33ºC. It requires optimum daytime temperature of 20-25ºC. However, temperatures below 3ºC will kill the plant. It prefers frost free conditions. It tolerates higher temperatures up to 43ºC in the plains when grown under irrigated conditions.

Soils: Geranium is shallow-rooted crop which requires well drained loam soil with pH of 5.8 to 8.5. The soils either saline or alkaline with poor drainage are unsuitable for its cultivation. The crop is found to perform well in red lateritic soils, though a calcium rich porous soil is the best. The ideal soil should be rich in organic matter and have clay content of not more than 40%.

Season: The onset of monsoon is the best time. The grass is planted in the beginning of monsoon (June-July). Under irrigated condition February-March is also suitable. November to January months are best suited for raising the nursery.

Land preparation: The land is prepared with disc followed by cultivator to bring into fine tilth. Ridges and furrows are made with a spacing of 60 cm.

Varieties: KKL-1, Sel-8, Hemanti, Bipuli, Kunti, IIHR-8, Kodaikanal 1, Hemanti, Egyptian, Algerian or Tunisian, Reunion or Bourbon.

Propagation: As there is no seed setting in this crop, it is propagated vegetatively by cuttings. Terminal cuttings of 20 cm length consisting of about 8 nodes and 3-4 leaves from the top are taken for propagation. A slant cut is made with a sharp knife just below 6th to 7th node. The cuttings are dipped in 0.1%, Benlate solution for 10 to 20 seconds. They are then planted at 8-10 cm spacing in the nursery beds of 3 m long and 1 m wide. Beds are provided with proper shade and watered twice daily for about 3-4 days and once in a day, subsequently. The cuttings are treated with IBA or IAA at 200 ppm for 6 minutes to enhance rooting. The cuttings can also be rooted in polythene bags of 10 × 10 cm size, which help to avoid damage to the root-system while planting in the main field. The cuttings will be ready for transplanting in about 2 months from planting. The rooted cuttings are carefully dug out from the nursery. Pits of 30 × 30 cm size are dug and cattle manure at 250 g/pit is applied and rooted cuttings of 2 months age are planted at a spacing of 45 × 45 cm or 60 × 60 cm.

After cultivation: Pruning of the bushes is necessary when the bush shows signs of decline. The branches are cut back leaving 15 to 20 cm once in 4 to 5 years. A deep soil forking around the plants is given to improve the growth of more suckers from the second year onwards.

Weed management: Generally two weeding are given, one at 20-25 and another at 40-45 days after planting.

Manures and fertilizer application: Application of FYM @ 12.5 t/ha and a fertilizer dose of 200-35-35 kg of N, P_2O_5 and K_2O per ha is recommended. Nitrogen is applied in 6 equal split doses. The first dose is given as a basal dose and thereafter at bio-monthly intervals. The application of zinc sulphate @ 20 kg/ha and borax @ 10 kg/ha are recommended.

Water management: The crop is irrigated immediately after planting. Irrigation is continued on at intervals of 7 to 10 days. It tolerates short periods of drought. Water-logging of the crop must be completely avoided. It is usually grown as a rainfed crop.

Pest and diseases management: Termites can be easily controlled by mixing into the soil Heptachlor @ 25 kg/ha. Spray the crop with Benlate @ 0.2% solution two weeks before the harvest or drench with Carbendazim @ 1 g/lit at monthly intervals to control wilt. Application of Carbofuran @ 2 kg a.i./ha or Aldicarb @ 20 kg/ha to the soil reduces the incidence of root-knot nematodes.

Harvest: The maturity symptoms for harvest when the leaves begin to turn light green to yellow in colour and exhibit a change from lemon like odour to that of rose when the leaves pressed between fingers. The crop is ready for harvest after about 4 months of transplanting. The green leafy shoots are harvested with a sharp sickle. Harvesting is done 3-4 times per year. As the crop is perennial, it can give harvests for about 3 to 4 years. The terminal portion with 6 to 12 leaves contains more oil than the middle and basal portions.

The harvested material is withered in shade for two to three hours and distilled.

Yield: The yield of fresh herbage from all the 3 to 4 harvests may be about 15 t/ha which on distillation may yield oil of about 15-20 kg per ha per year. The recovery of the oil ranges from 0.08 to 0.15%, depending upon the season of harvest and type of material.

Utilization: The chief constituent of the oil are geraniol and citronellol. The pure geranium oil is almost a perfume by itself and blends well with all other perfumes.

CHAPTER

42

Lemon Grass–*Cymbopogon flexuosus* (Steud) Wats. (2n=20, 40) and *Cymbopogon citratus* (DC.) Stapf

Family: Poaceae

Vernacular names: Lemon grass, Cochin grass, Malabar grass, East Indian Lemongrass, Nimbu grass (English); *Lemon grass, Gandhatran* (Hindi); *Bhustrina, Karpoorathrina, Takratani* (Sanskrit); *Nimbe hullu* (Kannada); *Nimma Gaddi* (Telugu); *Elumichai Ennai Pullu, Wella poolu* (White grass), *Choomana pullu* (Red grass), *Elumichai-Ennai Pullu, Karpoorappullu* (Tamil); *Inchippullu, Theruvappullu,* Malabar grass (Malayalam).

Importance: It is a perennial and multicut aromatic grass. The prefix 'lemon' owes to its typical lemon like odour, which is mainly due to the presence of citral in the oil. The characteristic smell of oil makes its use in scenting of soaps, detergents, insect repellent preparations. However, the major use of oil is as a source of citral, which goes in perfumery, cosmetics, beverages and is a starting material for manufacture of ionones, which produces vitamin A. The citral rich oil has germicidal, medicinal and flavouring properties. Lemongrass oil is used for synthesis of β-ionone used for synthesis of a number of useful aromatic compounds and Vitamin- A. Lemongrass oil is used as a main substitute for 'Cod liver oil'. Synthetic citral is also available which competes with this oil and natural citral in the market. Cochin and Mumbai are the major trading centres for lemon grass oils. India is the largest producer of lemon grass and about 80% of the produce is being exported to Europe, USA and Japan.

Origin: The East Indian lemon grass, Malabar or Cochin grass (*Cymbopogon flexuosus*) is native to India, Sri Lanka, Myanmar and Thailand while the West Indian lemon grass (*Cymbopogon citratus*) is native of Malaysia. Jammu lemongrass (*Cymbopogon pendulus*) is white stemmed and dwarf in stature. The plant is frost resistant and suited to Sub-Himalayan areas of Northern India.

Distribution: *Cymbopogon flexuosus* is cultivated in Ethiopia, Guatemala, Indonesia, Japan, Madagascar, Seychelles, Sri Lanka and Thailand. *Cymbopogon citratus* is grown in Argentina, Brazil, Cameroon, Cuba, Gautemala, Haiti, Indonesia, Jamaica, Japan, Kenya,

Mexico, Philippines, Seychelles, Somalia, Surinam, Tanzania, Thailand, Uganda and Zaire. In India, the crop grows in an area of about 3000 ha, largely in the states of Kerala, Karnataka, Tamil Nadu, Andhra Pradesh, Maharashtra, Uttar Pradesh, Uttaranchal, Arunachal Pradesh, Sikkim and Assam.

Botany: Lemon grass is a perennial and multicut aromatic grass. The culm is stout, erect, up to 1.8 meter high. The bulblike stems consist of terete and glabrous linearly venated sheathed leaves with narrow base and acute apex. Leaves are long of about about 100 cm in length and 2 cm in width, glaucous, green, linear tapering upwards and along the margins; ligule very short; sheaths terete, those of the barren shoots widened and tightly clasping at the base, others narrow and separating. The inflorescence is a long spike about one metre in length. Flowers borne on decompound spatheate and panicles are 30 to over 60 cm long. The seeds are very light and are covered with husk. One gram of the seed material, named as 'fluff', contains about 1700 seeds.

There are two main types of lemongrass namely East Indian and West Indian. The East Indian lemongrass oil is obtained from *Cymbopogon flexuosus* Stapf. The species is considered to have originated in Kerala, India. According to the colour of the stem, this is again divided into two types based on colour of stem. The 'red grass' is identified as *Cymbopogon flexuosus* var. *flexuosus* which is known as 'choomanna poolu' in Tamil. Stem and leaf sheath are reddish or purple. The oil content is 0.3 to 0.5% and the citral content is 80-86%. The oil has good solubility in alcohol and hence is superior in quality. It is recognized as true lemongrass and is commercially cultivated. The 'white grass' known as 'wella poolu' has been identified as *Cymbopogon flexuosus* var. *albescens*. Stem and leaf sheath are white in colour. The oil content is 0.4 to 0.7% and the citral content is 55-70%. The oil has has very low aldehyde content, poor solubility in alcohol and hence is inferior in quality. The West Indian oil is extracted from Cymbopogon citratus (DC) Stapf that is mainly cultivated in Central and South America, in parts of Africa, South East Asia and the Indian Ocean Islands. A third species, *Cympobogon pendulus* (Nees ex Steud) is popularly known as Jammu Lemongrass.

(i) *Cymbopogon flexuosus*: It is 3 m tall grass arising from a woody rhizome. Leaves are linear and lanceolate. Leaves are aromatic and have 1-2 cm width and 75 cm length. Leaf sheath is glabrous, hairy at the junction blade. .Leaf blade is 1 m long, 1.5 cm wide, linear, acuminate, glaucous. Inflorescence is large, drooping, lax, greenish when tender and becomes pinkish on maturity.

(ii) *Cymbopogon citratus*: It is a perennial aromatic grass having dense fasicles of leaves from a short/oblique annulate, sparingly branched rhizome. Leaf blade is linear, long attenuated towards the base and tapering upwards, approx. 90 cm long, 5 cm wide, smooth or rough upwards and along the margins, glabrous, glaucus green, base narrow. Sheaths are terete, those of the barren shoots much widened at the base. Inflorescence spatheate panicle, lose, 30 to 60 cm long.

(iii) *Cymbopogon pendulus*: It is a perennial robust grass. Clump is erect, 120 to 150 cm high and glabrous. Leaf blades are up to 80 cm long, 11 mm wide, slightly rough on the low surface, glabrous, leaf-sheaths tomentose below. Spathate is panicle, 60 cm long, more or less interrupted, each tier composed of three rays in cluster spatheole 15 to 26 mm long.

Climate: Lemongrass is distributed in tropical and subtropical countries. It is grown up to an altitude of 300m. However in Kerala, it grows well at altitudes between 900 and 1200 m. It requires warm and humid climate with sufficient sunshine and rainfall, ranging from 1800-3000 mm, uniformly distributed throughout the year. It is a short day plant. It is cultivated with temperature range of 10-33°C. The optimum temperature for growth of the grass is from 23 to 30°C.

Soils: It is best suited to well drain sandy loam with soil pH of 4.5 to 7.5. It can also be cultivated in marginal soils, laterite soil, wastelands and alkaline soils having pH 9.6. It is suitable to 'jhum fallow', hill slopes and flood free degraded land. Calcareous and water logged soils are not suitable for cultivation.

Season: Seedlings are to be raised and transplanted during June-July or October-November.

Varieties: Sugandhi (OD 19), Pragati (LS48), Praman (Clone 29), Jama rosa, Cauvery, CKP-25, OD-408, RRL 16, RRL-39, NLG 84, Krishna, Jor Lab L-2, Kalam, Nima.

Land preparation: The land should be prepared with two to three ploughing to have fine tilth. Ridges and furrows are made at 60 cm or 45 cm intervals to avoid waterlogging.

Propagation: The crop is propagated through seed and slips. The transplanting of nursery raised seedlings is found to be superior to direct sowing of seeds.

Nursery management: Seed rate is 3 to 5 kg/ha. The seeds are sown on raised beds of 1m to 1.5m width and are covered with a thin layer of soil. The bed should be irrigated after sowing and care should be taken to maintain adequate moisture in the soil. Seed germinates in 5-6 days and the seedlings are ready for transplanting after a period of 60 days.

Preparation of root slips: It is propagated vegetatively with root slips. Tops of culms are cut off within 20-25 cm above ground. The culm (root portion) is divided into slips containing 2-3 tillers. The lower sheath is removed to expose young roots and the old roots are clipped off keeping the slip 25-30 cm long.

Planting: Seedlings or slips are planted with a spacing of 45 × 30 cm, 45 × 45 cm, 60 × 30 cm or 60 × 45 cm depending on soil productivity and soil slope. It is better to plant on ridges in areas receiving high rainfall. In case of rooted slips one or two slips are placed into each hole, about 5-8 cm deep. Seedlings/slips are established well within 25-30 days after planting.

Weed management: The field is kept free of weeds for the first 3 - 4 months after plating. Similarly, weeding cum hoeing is done up to 1 month, after every harvest. Generally, one hand weeding at 25-30 days followed by one hoeing at 40-60 days after planting is enough to control weeds. Diuron @ 1.5 kg ai/ha and Oxyfluorfen @ 0.5 kg ai/ha are used to control weeds.

Manures and fertilizer application: FYM @ 12.5 t/ha to be applied and mixed well at the time of final land preparation. Fertilizer dose of 150-60-60 of N, $P_2 0_5$ and $K_2 0$ kg/ ha/year is recommended. The nitrogen is applied in six equal split doses at two monthly intervals. Iron chlorosis can be controlled with foliar application of Ferrous sulphate @ 3%. In Zinc deficient soils, Zinc sulphate @ 25 kg per ha is applied.

Water management: Irrigation is given immediately after planting and thereafter irrigations are given at 10 days interval. It is cultivated on hill slopes as a rainfed crop. If rains are erratic, the field is irrigated at an interval of 3 days during the first month and 7 to 10 day intervals subsequently. After the establishment of plants, irrigation schedule is adjusted depending on water holding capacity of the soil and weather conditions. In northern India, 4 to 6 irrigations are required during February-June. As the plant cannot withstand any amount of water-logging, planting on ridges or open hill slopes is recommended.

Cropping systems: The plant does not tolerate shade and oil yields are drastically reduced when the crop is grown under diffused light. However, intercropping of lemon grass in regularly pruned cinnamon plantation and newly planted cashew fields during initial 4-5 years is being widely practiced.

Pest and diseases management: In general, pests and diseases may not affect the plant. If there is any sucking pest spray Methyl demeton 25 EC or Dimethoate 30 EC @

1 ml/lit. Spray Phosalone 35 EC or Monocrotophos 36 EC 2 ml/lit to control caterpillars. The scale insect produces yellow spot on the stem and sucks the sap of the leaves and stem. The scale insect can be controlled by spraying Dimethoate @ 0.5 %. Leaf blight can be controlled with spray benzimidazole like Benlate 50 WP at 0.2% or Dithane Z-78 (0.2%) or 0.3% Copper oxychloride at an interval of 10 to 15 days interval. Little Leaf or Grassy Shoot is controlled with spraying Dithane Z-78 @ 0.3% just before flowering stage at an interval of 10-12 days. Nematodes are controlled with application of Fenamiphos @ 12.5 kg/ha or Phorate @ 10 kg/ha or neem cake @ 10 kg/ha at the time of last ploughing is recommended.

Harvest: The crop should be harvested before flowering for both quality and oil yield. The optimum period of harvesting when grown on hilltop or upper slopes is 75 days while at foothill and plains it is 60 days. First harvest is taken at 4 months of transplanting seedlings and subsequent harvests are at 2 to 3 months intervals. Harvesting is done by cutting the grass 10 to 15 cm above the ground level. During first year 3 cuttings and subsequently 3-5 cuttings per year can be taken. Grass is harvested when individual tiller has 4-5 fully opened leaves. Harvesting is done with the help of sickles. The plants are cut about 10-15 cm above the ground level with sickles.

Yield: Average herbage yield from 4 to 5 cuttings is 15 to 20 t/ha/year. The oil yield ranges from 80 to 100 kg per hectare with 82-88% citral under rain-fed conditions. The average oil content is 0.2 to 0.4%. In general, the recovery of oil is lower in rainy seasons (June-August) than in summer season.

Oil extraction: The oil is extracted by hydro (water) distillation or steam distillation from the dried or fresh leaves and flowering tops of the plant. The grass is allowed to wilt for 24 hours before distillation as it reduces the moisture content by 30% to improve oil yield. The crop is chopped into small pieces before filling in the stills. The wilted leaves are steam distilled which takes about 3 to 4 hours. Dipping the chopped lemongrass in sodium chloride solution for 24 hours at 1-2 % concentration before distillation has been found to increase the citral content. Herbage yield from 8 ha area is required for effective running of 500 l capacity stem distillation unit. The oil recovery is 0.2 to 0.4 %. The oil has a strong lemon like odour. The oil is yellowish in colour.

Utilization: Lemongrass oil obtained from the leaves and shoot of the plant. The oil of lemongrass is primarily used for the isolation of citral for manufacturing Vitamin-A. It is used for confectionery, culinary purposes, drugs, flavouring, insect repellents, liquors and perfumery purposes. The spent lemongrass is suitable for making paper. It is also used as fuel for the distillation of the grass. It is an excellent source of manure. It may be used for mulching coffee. It is a good crop for checking soil erosion.

Seed Production: Lemongrass is cross pollinated. Hence to maintain genetic purity, isolation distance of 300-400 m should be given between two varieties. The crop flowers during November-December and seeds mature in next two months viz. February-March (dry season in Kerala). For collection of seeds, the plants are maintained in good health as the yield of seeds from plants subjected to regular harvest is low. On an average, a healthy plant gives about 100-200 g of seeds. At the time of seed collection, the whole inflorescence is cut and sun dried for 2-3 days. These are then threshed and seeds are again dried in the sun and the seed remain attached with fluffy mass which is removed by beating of seed bag at sowing. Seed mass which is known as 'fluff' is dried and stored in gunny bags lined with polythene. Seed yield is about 60 kg/ha. The seeds lose their viability if stored for a period more than one year.

REFERENCES

NMPB. 2009. High yielding varieties of some medicinal and aromatic plants with general guidelines for seed production and certification. National Medicinal Plants Board, Alakananda Advertising Pvt. Ltd., New Delhi. p. 19-21

Puraima Jayasinha. 1999. Lemon grass - a literature survey. Industrial Technology Institute, 363, Baudhaloka Mawatha, Colombo -7, Sri Lanka.

http://nhb.gov.in/Horticulture%20Crops%5CLemongrass%5CLemongrass1.htm

http://agritech.tnau.ac.in/horticulture/horti_aromatic%20crops_lemongrass.html

http://agridr.in/tnauEAgri/eagri50/AGRO301/lec19.html

http://www.jnkvv-aromedicinalplants.in/front/Plant_Details/index.php?pid=1

http://vikaspedia.in/agriculture/crop-production/lemon-grass

CHAPTER

43

Menthol Mint–*Mentha arvensis* L. (2n=86)

Family: Lamiaceae

Vernacular names: Menthol mint/Japanese mint/Corn mint (English); *Pudina* (Hindi, Kannada, Tamil), *Putiha* (Sanskrit). Mentha name is also known as mint. The word mint is derived from from Greek word míntha, Linear B mi-ta.

Importance: There are four cultivated species are viz., i) Japanese Mint/Menthol Mint (*Mentha arvensis*), ii) Peppermint (*M. piperita*), iii) Spearmint (*M. spicata*) and iv) Horsemint/Bergamot mint (*M. citrata*).

Origin: It is native to Mediterranean regions, the temperate regions of Europe, Western and Central Asia, Eastern Himalaya and Siberia.

Distribution: Menthol mint is cultivated in India, Brazil, Argentina, Paraguay, USA, Brazil, Paraguay, France, Russia, Bulgaria, Czechoslovakia, Hungary, Bulgaria, Italy, Switzerland, Argentina, China, Japan, Thailand, Vietnam, Angola, Morocco, Australia and India. In India, it is grown in Uttar Pradesh, Himachal Pradesh, Kashmir, Punjab, Haryana, Rajasthan and Bihar.

Botany: It is a perennial herb growing about 60-80 cm in height. The stem is rigid, pubescent and highly branched. Leaves are lanceolate-oblong in shape and with a coarsely serrated margin. The leaf surfaces, on both sides, appear hairy and have glandular trichomes. There are more trichomes on the ventral surface as compared to the dorsal. The inflorescence is verticillate known as cyme. Flowers are borne in axillary and terminal and purplish in colour. The flowers are small. The plants blossom profusely but rarely set seed. The roots are shallow and creeping. The plants have runners (rainy season) and stolons (winter), which develop new roots and shoots at the nodes and form plants. The propagation is mainly by its stolons and suckers. These suckers grow underground at a depth of 15 cm and spread in all directions.

Climate: It can be cultivated both in tropical, subtropical and temperate areas. It needs a well distributed rainfall of 1000-2500 mm. This herb grows very fast in cool, moist soils and prefers partial shade. It requires bright sunshine at the time of harvesting. It is grown with mean temperature from 20 to 41°C during its growing season. A temperature of 20-25°C promotes vegetative growth, but the essential oil and menthol increase at a high temperature of 30°C. It does not tolerate damp winters which cause root rot.

Soils: Well drained loam or sandy loam soils rich in organic matter having pH between 6 and 8.2 are desirable. It can also be cultivated on both red and black soil. In case of acidic soil having pH less than 5.5, liming is recommended. The soil should have a good water holding capacity but waterlogging should be avoided.

Season: The planting is done during June - July or February- March. The crop duration is 5-6 months.

Varieties:

Japanese mint varieties: Himalaya (MAS-1), MA-2, Kalka (Hybrid-77), Shivalik and EC-41911, Damroo, Saksham, Kosi.

Spear mint varieties: MSS-1, MSS-5 and Punjab Spearmint-1, Ganga, CIM-Indus

Bergamot mint varieties: Kiran

Pepper mint varieties: Kukrail, CIM-Madhuras

Land preparation: Land is ploughed, harrowed to have fine tilth. It is planted on flat land or ridges.

Propagation: Mint rarely produce viable seeds. Hence, propagation is done mainly by divisions of 8 to 10 cm long stolons and runners (suckers). The seed rate used is 80 kg for nursery and 400 to 500 kg of stolons/suckers for planting one ha. An area of 200 sq.m. is required to produce stolons for one hectare. These stolons can be used immediately or within a fortnight. The suckers should be washed and dipped into 0.1% Baviston 50 WP solution (1 g/litre of water) for 5-10 minutes before planting. Rooted stem cuttings are planted with 30 × 30 cm, 45 × 30 cm or 60 × 30 cm spacing. The suckers may be placed end to end, 5 to 7 cm deep in furrows in beds. The stolons are planted half-way down on the inner sides of the ridges while planting on ridges. The plot is irrigated immediately after planting. The stolons sprout in about 2 to 3 weeks.

Weed management: The crop should be kept free from weeds at all the stages of growth. In the early stages of growth, a wheel-hoe may be used. Weeding is done up to 75 days of planting. Oxyflurofen @ 0.5 kg a.i./ha, Pendimethalin @ 0.75 a.i./ha, Simazine @ 1 kg a.i./ha, Isoproturon 75 WP @ 1 kg/ha, Diuron @ 2 kg a.i/ha or Atrazine (1 kg a.i. / ha can be applied as pre-emergence spray is recommended. Application of pre-emergence herbicide followed by manual or mechanical weeding at 8 to 10 weeks is recommended. Dalapon @ 4 kg/ha, or Gramaxone @ 2.5 l/ha as post-emergence is recommended.

Manures and fertilizer application: Application of FYM @ 12.5 t/ha at the time of land preparation is recommended. Sunnhemp is an ideal green manure crop before the mint is planted. A dose of 120-50-40 kg of N, P_2O_5 and K_2O/ha is recommended. The entire quantity of P and K along with one-fifth of N is mixed with the soil at the time of planting, the remaining four-fifth of N is applied in 2 split doses at 30 and 60 days after planting and similar quantities for ratoon crop at 25 days and 45 days of the harvest

Water management: Ten irrigations are given during summer season at intervals of 10-12 days whereas 4-6 irrigations for July to October season. Waterlogging during rainy season should be avoided by providing adequate drainage. In case of heavy soils and the soils prone to waterlogging, it is preferable to cultivate mint on ridges.

Cropping systems: Farmers consider mentha as a bonus crop as it does not disturb or replace the cultivation of any major winter (*Rabi*) or rainy season (*Kharif*) crop. Mentha is grown in rotation such as mentha-potato, mentha–*toria*, mentha–oats (for fodder purpose), mentha-Basmati rice, mentha–wheat–maize–potato, mentha-maize-potato, mentha-maize, mentha-early paddy and potato, mentha-late paddy and sweet pea, mentha-maize, or menthe-paddy rotation.

Pest and diseases management: Termites and cutworms are controlled with application of Dursban/Radar 20 EC (Chlorpyriphos) @ 5 litres/ha is mixed in 25 kg soil/sand and broadcast uniformly in the field followed by light irrigation. Jassid and whitefly are controlled with spray of Rogor 30 EC (Dimethoate) or metasystox 25 EC (oxydemeton methyl) @ 250 ml per acre. The foliage-eating insects are controlled by spraying the crop with 2.5 kg of Sevin 50 WP (Carbaryl) or 800 ml of Ekalux 25 EC (Quinalphos) per ha. Dip the planting-stock into 0.1% solution of Bavistin 50 WP for 5-10 minutes to control root rot and stem rot.

Harvest: The crop should preferably be harvested at the flower initiation stage or when the lower leaves of the plants turn yellow and start shedding. Harvest the crop, leaving 6-8 cm long stumps to secure better sprouting. First cutting starts in about 3 to 4 months after planting and subsequently at 3 months interval. Under good management conditions, the crop will give economic yield for about four years. Further, harvesting should be done in bright sunny weather since harvesting on cloudy or rainy days decrease the menthol content in the oil.

Yield: The average fresh herbage yield is 20 to 25 t/ha in two harvests, which, in turn, yields around 100 to 150 kg/ha of oil in a year.

Post-harvest management: Mint herbage should be shade dried for about a day before it is distilled. There would be some reduction in oil yield if wilted herbage crop is stored for a longer period of 2-3 days. The fresh herbage contains 0.5 to 0.8% oil. The oil contains 78-80 % menthol and 12- 17 % isomenthone. Oil is obtained through steam distillation. The oil is of golden yellow colour.

Utilization: Mint leaves can be harvested any time and used fresh or dried. These aromatic leaves are used as ingredient of many kitchen recipes like chutneys, jellies, syrups, beverages and even candies and ice creams. Moreover, the mint oil is popularly used in breath freshners, mouth rinses, toothpastes, shampoos, chewing gums and also chocolates. The oil is used in aromatherapy. Mint oil also serves as an insecticide to control wasps, ants and cockroaches. The oil contains aromatic chemicals like menthol, menthyl acetate, menthone, isomenthone, menthofuran, carvone, linalool, linalyl acetate and piperitenone oxide.

CHAPTER

44

Palmarosa–*Cymbopogon martinii* [Roxb.] Wats. (2n=40)

Family: Poaceae

Vernacular names: Palmarosa, Rosha Ghas (English); *Rosha grass, Rusha ghas* (Hindi); *Rauns, Rosdo* (Gujarati); *Rohisha, Rohisa* (Sanskrit).

Importance: The palmarosa, is an essential oil-bearing crop. The oil is obtained from the floral shoots and above ground parts. The oil has natural geraniol with 75 to 90% content which is used in perfumery, cosmetics and flavour industries. It is a hardy crop which can be grown under varying agro-climatic conditions and different types of wastelands.

Origin: This plant is a native of India.

Distribution: It is grown in Madhya Pradesh, Maharashtra, Andhra Pradesh, Karnataka, Uttar Pradesh, Odisha, Rajasthan, Madhya Pradesh, Gujarat and Tamil Nadu.

Botany: It is a tall perennial grass, which contain an oil of rose like odour in its flowering tops and foliage. It grows to a height of 1.5 to 2.5 m. The roots are shallow and fibrous. The culms are erect and nodes swollen. It has dark green leaves, which are leathery, prominently mid ribbed, roundish at the base and form an obtuse to right angle with the stem. Inflorescence is large compound panicle. The mature seeds are brown, fine, hairy and easily disposed by air.

Climate: It comes up well in warm humid areas under tropical conditions with an annual rainfall of about 900 to 1500 mm. It is grown up to an elevation of 300 m from MSL. It grows well with temperature of 15 to 35ºC. It is susceptible to frost. It tolerates drought.

Soils: A well-drained loamy soil with pH 6 to 7 and rich in organic matter is suitable. It tolerates up to a pH of 9.5. It can be grown on marginal waste lands including saline/ sodic/alkaline soils.

Season: The crop is sown/planted in June to August. The life span is 3 to 4 years.

Varieties: Motia, Sofia, Tripta, Trishna, Vaishnavi, Jamrosa, PRC-1, IW-31243, IW 31245, IW 3629, IW 3244, OPD-1, OPD-2, RRL (B)-77, Cim-Harsha are the popular varieties under cultivation.

Land preparation: The land is ploughed 2-3 times to produce a fine tilth before the seedlings or slips are transplanted.

Propagation: The crop can be propagated by both seeds and slips. It is advisable to plant two seedlings/slips per hill to avoid seedling mortality. The optimum seed rate is 2.5 to 3.0 kg of seeds/ha. The seeds are sown in 100 sq.m nursery area. The seeds are mixed with fine sand or soil in a ratio of 1:10 for even distribution and ease in sowing. The seeds are sown in raised nursery beds in lines at 15 to 20 cm apart. The seeds start germinating within 3-4 days. Spray urea 0.2-0.5% solution for good vegetative growth. 3 to 4 weeks old seedlings are used for transplanting in the main field in ridges at a spacing of 60 × 60 cm. Clumps are trimmed from 20-25 cm above ground and dug out without injuring the roots. The rooted slips are planted at 60 × 60 cm spacing. 28,000 slips will be required to plant one hectare. Gap filling should be done within 8-10 days of planting.

Weed management: One or two weeding may be taken. Diuron @ 1.5 kg ai/ha or Oxyfluorfen @ 0.5 kg ai/ha are used as pre-emergence to control weeds.

Manures and fertilizer application: FYM 12.5 t/ha and fertilizer dose of 100-50-50 kg/ha of N, P_2O_5 and K_2O is recommended. One fourth of N and entire P and K are applied as basal dose. Top dressing is done in 3 splits at 3, 6 and 9 months after planting. Application of $ZnSO_4$ @ 25 kg/ha is recommended.

Water management: The crop is highly sensitive to waterlogging. In general, the field is to be irrigated at 7 to 10 days interval during summer season. Irrigation should be discontinued 10 days before harvesting.

Pests and diseases management: Sucking pests such as aphids and thrips can be controlled with spray of Azadirachtin 1% @ 5 ml/l. Flooding with irrigation water kills the white grubs and termites. The blight disease can be managed by foliar spraying of Bordeaux mixture @ 1% at 15 days interval. Curvularia blotch disease may be controlled with spray of Bordeaux mixture @ 1% or Mancozeb @ 0.3% at 15 days interval.

Harvest: The crop should be harvested at full flowering to seed production stage in order to obtain maximum and good quality oil. Harvesting is usually done with a sickle at 15-20 cm above the ground surface. The first harvest commences at 3 to 4 months of planting. Subsequent harvests are done at 3 to 4 months interval. It can be harvested at 1-2 times during the first year and 3-4 times from second year onwards. This crop can be maintained for five years. Cut the bush by leaving 10-15 cm from the ground level. Dry the leaves for 2 to 3 days under shade and then used for steam distillation for oil extraction.

Yield: The average herbage yield is 20 to 30 t/ha/year while the oil yield is 100 to 125 kg/ha/year. The herbage contains 0.3-0.4% of oil which is rich in geraniol content (75-90%) and geranyl acetate (17.89%).

Post-harvest technology: Floral shoots and above ground parts of the plant are used for distillation of essential oil.

Seed Technology: Seeds attain physiological maturity at 40th day after 50% flowering when the fluff (seed) moisture is around 20%. Leaching of fluffs in running water for 8 hours followed by soaking in KNO_3 at 0.5% for 6 hours recorded maximum germination.

Utilization: It is used in perfumery, food flavouring and medicinal pharmaceutical industry..

CHAPTER

45

Patchouli–*Pogostemon patchouli* Hook

Synonyms: *Pogostemon cablin* (Blanco) Benth. *Mentha cablin* Blanco

Family: Lamiaceae

Vernacular names: *Patchouli* (English, Hindi); *Patchpan* (Sanskrit); *Xukloti* (Assamese). The name Patchouli is derived from the Tamil word *patchai* (green), *ellai* (leaf).

Importance: Patchouli is cultivated for its highly fragrant leaves which contain a very sweet smelling oil of lasting sticky camphoraceous fragrance.

Origin: Patchouli is native to India, Philippines and Malaysia.

Distribution: It is grown India, Indonesia, Malaysia, Vietnam, China, Brazil and Caribbean countries. In India, it is cultivated in coastal regions of Tamil Nadu, Kerala, Karnataka, Assam and West Bengal.

Botany: Patchouli is an erect branched, bushy herb to under-shrub growing 1 to 1.2 m tall. It has quardiangular stems. The leaves are simple, ovate to oblong ovate, leathery, dentate margins pale to purplish green in colour. The leaves are covered with trichomes all over the epidermis, which contains the essential oil. The flowers are small, borne in spikes at the end of branches. Its flowers are violet in colour. Seeds are not produced. The propagation of the crop is through stem cutting.

Climate: It is a tropical crop which can also be grown under sub- tropical conditions. It prefers a warm and humid climate. It is grown up to an altitude of 800-1000m above the MSL. It is a shade loving plant. The plant grows well as an intercrop in partial shade, but complete shade should be avoided. It is a short day plant. The crop can be grown under an evenly distributed annual rainfall of 800 to 3000 mm. The temperature between 25 to 35°C is found to be ideal.

Soils: It is a hardy plant and adapts itself to a wide range of soil conditions. It requires deep, well-drained, fertile, slightly acidic, deep loamy soil, rich in humus and nutrients. The pH of the soil should range from 5.5 to 7.5 for good growth. Waterlogged conditions is found to be detrimental for the crop and must be avoided because they are susceptible to nematode attack.

Season: Planting time is June- September under rainfed and irrigated condition while February to March under irrigated condition. The plants give good yield of leaves for 3 years.

Varieties: The promising varieties are Java, Singapore, Johore, Singapore, Indonesia and Malaysia.

Land preparation: The land is prepared to have good tilth by ploughing twice followed by harrowing. The ridges and furrows are formed with spacing of 60 cm.

Propagation: It is vegetatively propagated with rooted stem cutting of 15-20 cm. The raised beds of 75 × 45 cm are prepared. The stem cuttings are planted in nursery at 10 × 10 cm spacing. The nursery should be located under partial shade. Under favourable conditions, about 85-90 % cuttings put forth roots in a fortnight and they are ready for planting in the field in next six to eight weeks. . The rooted cuttings are planted at a spacing of 30 × 30 cm, 45 × 45 cm or 60 × 30 cm.

Weed management: The field should be kept weed free during the first 2 to 3 months of crop growth either by wheel hoeing or by hand weeding two to three times. Weeding is also necessary after about a month of each harvest.

Manures and fertilizer application: FYM 12.5 t/ha and fertilizer dose of 50-50-50 kg/ha of N, P_2O_5 and K_2O is recommended. The entire quantity of P and K along with 25 kg of N is applied at the time of planting. Top dressing of 25 kg N per ha is done after two months after planting. After each harvest, 50 kg N is applied in two split doses, the first dose just after the harvest and the other about two months later. In zinc deficient soil, zinc sulphate @ 25 kg/ha is applied.

Water management: It is grown as a rainfed crop in hills. In plains, irrigation is to be given for 3-4 days interval for the first 15-20 days after planting in the field and later done at 7 to 10 days interval. Waterlogging should be avoided. Under drip system of irrigation, irrigation is provided withdischarge at a rate of 2 litres of water/hour. Irrigation for 30 minutes per day is adequate.

Cropping systems: It is a shade loving plant. It has been successfully grown as an understory crop in arecanut and coconut orchards under irrigated conditions. Gliricidia or Erythrina could be planted well in advance at 5 × 5 m spacing in patchouli field in order to provide the necessary shade. It thrives best in hot and humid conditions, under shade of rubber, coconut, arecanut, coffee, etc.

Pests and diseases management: Nursery should be raised from healthy mother stock under nematode-free conditions. Application of Furadan @ 20 kg/ha (2% a.i.) or Dasanit 150 kg /ha (5% a.i.) checks the nematode infection. Pre-planting treatment with nematocide as first dose and the second dose after one year of transplanting is recommended. Soil application of neem cake @ 1 t/ha is recommended to control of nematodes and Fusarium wilt. Application of Dithane Z-78 0.5%, at one-month interval is the recommended control leaf blight.

Harvest: The right time for harvesting is when the plants are about 1m high, leaves turn pale green or slightly brown in colour. The first crop is ready for harvesting 4 to 6 months after transplanting. Subsequent harvests can be taken after every 3-4 months. Harvesting is preferably done during drier months. The crop can be maintained for 3 years. The harvested leaves are dried in thin layers in shade for 3-4 days when these develop their characteristic odour. Mature thick stalks are removed as these contain no oil. Ageing on storage improves odour of the leaves and therefore, the crop is stored for six months before distillation.

Yield: A good crop stand yields about 2 to 4 t/ha of dry leaves per annum. The oil is found in the leaf and small quantity is present in the tender parts of the stem. The yield of fresh leaves is about 20 t/ha/year which on shade drying reduces to 4 t/ha/year. The oil yields about 40 to 60 kg/ha/year. The oil content varies from 2.5 to 3.5% in shade dried leaves. An yield of 2.5% may be considered satisfactory in commercial distillation. The herbage moisture of 10% is good for high recovery in oil extraction.

Post-harvest management: The harvested material is spread out under shade in thin layers and is turned periodically to ensure proper drying. For higher recovery and good quality of oil, moisture content of herbage should be between 8-10 %. Drying normally requires 3-6 days. The dried leaves has patchouli aroma, which is less noticeable in fresh leaves. The oil is distilled from the air-dried herb by using steam or hydro distillation process. The oil yield is from 2.5 to 3.5% on dry weight basis.

Utilization: It is an essential oil bearing aromatic plant. It is used in perfume industry. The alkaloids found in Patchouli are Patcholol, Sesquiterpenes, patchoulene and γ-guaislene etc.

CHAPTER

46

Medicinal Crops

Herbal medicines are known as botanical medicines or herbal medicinal product. It makes use of herbs, shrubs and trees for therapeutic or medicinal purposes. Herbal medicines are safe, acceptable culturally and have lesser side effects. Herbal medicines are perceived to be healthier than manufactured medicine. According to World Health Organization (WHO) about 80% of population relies on traditional or herbal medicines for their primary health-care needs and 20% of the drugs in pharmaceutical firms are of plant origin, either extracted from the plants or synthetic derivatives of these plant species. The goal of 'Health for All' cannot be achieved without herbal medicines. The world is turning back to natural herbal medicines with hope of safety and security, realising the drawbacks of allopathic medicine. The interest in medicinal plants has increased considerably because the plant drugs are used for the treatment of mental diseases, cancer tumors, hypertension in pregnancy. Ancient literature also mentions herbal medicines for age-related diseases namely memory loss, osteoporosis, diabetic wounds, immune and liver disorders, etc. for which no modern medicine or only palliative therapy is available.

History: Medicinal plants are component of all indigenous or alternative systems of medicine. The term nutraceutical refers to the substance which may be a food or part of a food, that provide medical and/or health benefits, including the prevention and treatment of diseases. India and Tibet has a rich heritage of plant based healthcare systems like Ayurveda, Unani, Siddha and Amchi system of medicine. Ayurveda medicines are derived from herbs. Unani medicines have herbal, animal and mineral origins. Siddha medicines are based on herbs. Amchi System of medicine based on plants which is practiced in Ladakh, India and Tibet. Homeopathic medicines based on plants, minerals, water and alcohol. Oriental medicine is based on herbal medicine, acupressure, diet, exercise, breathing and massage. Unani, Amchi and Siddha systems utilize 700, 600 and 600 medicinal plants respectively. Naturopathy therapies include special diets, mud packs, acupuncture, acupressure and magnet therapy. The use of plants for curing various human ailments figured in ancient manuscripts such as The Bible, The Rig-Vedas. The Charaka Samhita, Susruta Samhita, Bela Samhita, Kashyap Samhita, Agnivesh Tantra, Vagbhata's Ashtang hridaya and Dioscorides's De

Materia Medica'.

Medicinal plants: Medicinal plants or plant parts, which contain a substance or substances of medicinal properties, which have been proven to be useful as drugs or contain drug constituents, used for therapeutic purposes or those that synthesize metabolites to produce useful drugs. Out of 17,000 plants known to have medicinal properties in world, about 8000 plants are being used in Indian System of Medicine. Charak Samhita, Sushruta Samhita and Vagbhata describe 700 herbal drugs. The Siddha system of medicine uses around 600, Ayurveda 700, Unani 700 and modern medicine about 30 plants species. The plant parts used from a medicinal plant are viz., stem, wood, root, rhizome; tubers, bulb, leaf, buds, flower, fruit/berries, seed, bark on the trunk or on the root and in some cases whole plants or their extracts. The medicinal property of drugs is due to the presence of substances such as alkaloids, glycosides, resins, volatile oils, gums, tannins, etc.

Importance of medicinal plants Medicinal plants play an important role in many ancient traditional systems of medicine. They generate income to the people who earn their living from selling of the materials collected from wild or through cultivation. They are cultivated at the farm level to meet their ever increasing demand. Medicinal plants and their products serve as valuable source of income for small land holders and entrepreneurs. Collection, processing and marketing of medicinal plants and their products provide employment to the tribal communities in forest areas. Medicinal plants earn valuable foreign exchange by way of export. About 25 to 30% of all modern medicines are directly or indirectly derived from medicinal plants. The herbal medicine industry is one of the fastest growing industries in the world. The industrial demand on medicinal plant resources increases due to the production of herbal health care formulations; herbal based cosmetic products and herbal nutritional supplements. Medicinal plants play a key role in religious ceremonies. For example, Datura is associated with the worship of Lord Shiva and Tulasi with Lord Vishnu.

Classification of medicinal plants

(a) Classification of medicinal plants based on the plant and plant parts usage: Botanical drugs are classified based on the plant and plant parts from which they are derived viz., i) drug obtained from underground parts: liquorice, belladonna, sarpagandha, ashwagandha, *dioscorea* sp.; ii) drug obtained from bark: cinchona, ashoka tree; iii) drug obtained from stem: Sandal wood, *Acacia catechu;* iv) drug obtained from leaves: belladonna, senna, aloe, adhathoda, brahmi, tulsi; v) drug obtained from flowers: saffron, *Mesua ferra*; vi) drug obtained from fruits: *Aegle marmelos*, opium, long pepper, chebulic myrobalan and vii) drug obtained from seeds: isabgol, *Croton tiglium*.

(b) Classification of medicinal plants based on their usage: The herbs are as (i) medicinal herbs have curative powers and are used in making medicines because of their healing properties and (ii) Culinary herbs are used as cooking herbs because of their strong flavours like parsley, basil.

(c) Classification of medicinal plants based on the active constituents: The herbs are classified based on the active constituents as Aromatic herbs (volatile oils), Astringents herbs (tannins), Bitter herbs (phenol compounds, saponins, and alkaloids), Mucilaginous herbs (polysaccharides), and Nutritive herbs (food stuffs). The aromatic herbs are divided into two subcategories: stimulants and nervines. Stimulant herbs increase energy and activities of the body, or its parts or organs, and most often affect the respiratory, digestive, and circulatory systems. The examples are fennel, ginger, garlic, lemon grass. Nerving herbs are often used to heal and soothe the nervous system, and often affect the respiratory, digestive, and circulatory systems as well. The examples are ginger, catnip. The astringent herbs have tannins. They are analgesic, antiseptic and astringent. The examples are peppermint, red raspberry. Bitter herbs are divided into four subcategories: laxative herbs, diuretic herbs, saponin containing herbs, and alkaloid-containing herbs. Laxative bitter herbs are used as purgative, hypotonic, vermifuge, and blood purifier. The examples are aloe, liquorice, pumpkin and senna. Diuretic herbs are used to clean the vascular system, kidneys, and liver. These are antibiotic, antiseptic, and blood purifier in nature. The examples are asparagus, corn silk and parsley. Saponin-containing herbs are cardiac stimulant and increased longevity in nature. The examples are yam root, liquorice, and ginseng. Mucilaginous herbs eliminate the toxins from the intestinal system, help in regulating it and reduce the bowel transit time. The examples are aloe, fenugreek and psyllium. Nutritive herbs are true foods and provide some medicinal effects as fibre, mucilage and diuretic action. The examples are asparagus, onion and stevia.

Marketing of medicinal plants: Medicinal plants procured from forest areas and cultivated in private fields which account 90% and 10% of the total medicinal plants in trade. As world demand for medicinal plants is continuously increasing, there is an ample opportunity for these countries to expand their global export. Medicinal plants and their by-products constitute an important part of the foreign trade showing continuous growth in export over import. Most of the medicinal plant material in the world market originates from developing countries. India and China are the largest users of herbal medicines as well as the leading exporting countries of medicinal plant material to the world market. The medicinal plants are produced in Germany, Singapore, Egypt, Chile, the USA, Morocco, Mexico, Pakistan, France, Thailand, Albania, Brazil, Spain, Bulgaria, Hungary, Argentina, Greece, the Netherlands, Poland, Yugoslavia, Zaire Korea, Turkey, Kenya, Mauritius and Indonesia. Hamburg is the world's leading trade center in medicinal plant materials. The Europe Union, the USA, Russia and Japan are the biggest consumer markets of medicinal plant materials. Germany dominates the European trade in medicinal plants as it dominates the European market for phyto-pharmaceuticals. The major importers of medicinal plants in Asia are Hong Kong, Japan, Singapore and Malaysia. China and Japan have high numbers of patents on herb or herbal based or related medicine. India's share in world market of medicinal plant and products is 2.5%. The export market is growing faster than the domestic market. However, the export is being carried out in form of plants, or their parts and not as value added products. The Indian exporters face major constraints while exporting medicinal plants. The cyclical nature of medicinal plant market makes it difficult for them to determine supply and demand in the markets.

Constraints in trade of medicinal plants:

(i) No inventories of medicinal plants at All India basis
(ii) No reliable system of matching trade names to botanical names
(iii) Medicinal plants are harvested and traded in their raw form. Only the experienced people in the trade are able to recognize the species by their parts used.

REFERENCES

Alok Sharma, C. Shanker, Lalit Kumar Tyagi, Mahendra Singh and Ch.V.Rao. 2008. Herbal Medicine for Market Potential in India: An Overview. Academic Journal of Plant Sciences 1 (2): 26-36

Ameenah Gurib-Fakim. 2011. Traditional Roles and Future Prospects for Medicinal Plants in Health Care. Asian Biotechnology and Development Review. 13 (3): 77-83

Bannerman, R.H.1982, Traditional Medicine in Modern Health-care, World Health Forum. 3 (1): 8-13

Dubey NK and Rajeshkumar, Tripathi P. Global promotion of herbal medicine: India's opportunity Curr Sci ,2004; 86: 37–41

Gauniyal A.K., Singh A.K. and O.P. Virmani. 1991. Major Medicinal Plants as Foreign Exchange Earner. Yojana, July: 14-18

Sangita Kumari, Govind Shukla and A. Sambasiva Rao. 2011. The Present Status of Medicinal Plants-Aspects and Prospects. International Journal of Research in Pharmaceutical and Biomedical Sciences. 2 (1): 19-22

Sanjoy Kumar Pal and Yogeshwer Shukla. 2003. Herbal Medicine: Current Status and the Future. Asian Pacific J Cancer Prev, 4, 281-288

http://www.dmapr.org.in/Publications/E-book%20on%20Bulletins.html

http://www.cals.ncsu.edu/plantbiology/Faculty/dxie/Chapter1-1.pdf

http://www.crida.in/agrl_martng/ISAM/PDF%20FILES/T-I/Komaraiah.pdf

CHAPTER

47

Aloe–*Aloe vera* (L.) Burm. (2n=14)

Synonyms: *Aloe barbadensis* Miller, *Aloe indica, Aloe Barbados, A. perfoliata var. vera L., A. vulgaris Lam.*

Family: Xanthorrhoeaceae

Vernacular names: Chinese Aloe, Indian Aloe, True Aloe, Barbados Aloe, Burn Aloe, First Aid Plant, Aloe, *Aloe vera*, Barbados aloe, Curacao aloe, Jaffarabad aloe, Sabila (English); *Ghee kunwar*, Ghrit kumari (Sanskrit), Kunvar pathu, Ghikumar, Gheekanvar (Hindi); *Kathalai* (Tamil). The name, aloe, is derived from the Arabic 'alloeh' or Hebrew 'halal' meaning shining bitter substance.

Importance: Aloes are planted in gardens due to their architectural forms and coloured flowers. Aloe vera is used externally to treat various skin conditions such as cuts, burns and eczema. Its sap eases pain and reduces inflammation. Its gel is useful for dry skin conditions, especially eczema around the eyes and sensitive facial skin. The major markets for Aloe vera are Australia, USA and the entire Europe.

Origin: It is indigenous to North Africa, the Mediterranean region of southern Europe, and to the Canary Islands.

Distribution: It is grown in Australia, Bangladesh, Cuba, the Dominican Republic, China, Mexico, India, Jamaica, Kenya, Tanzania, South Africa, Madagascar, USA, Saudi Arabia, Malagasy, Arabia, Greece, Iran, Cyprus, Malta, Sicily, Canary cape, Cape Verde, Venezuela, the Pacific Rim countries, South America, Central America and the Caribbean. It is under cultivation in fairly large areas in many parts of India viz., Tamil Nadu, Gujarat, Rajasthan, Maharashtra, Madhya Pradesh, etc.

Botany: It is a perennial, xerophytic plant, growing to the height of 30 to 60 cm. It is short-stemmed plant spreading by offsets and root sprouts. The leaves grow from the base in the rosette pattern. Each plant usually has 12-16 leaves which may weigh up to 1.36 kg at maturity. The length of the leaves ranges from 25-30 cm while the breadth ranges from 3-5 cm. The leaves are lanceolate, thick and fleshy, green to grey-green, with a serrated margin. The two sides of the leaves have thorny structure with a thorny tip. A waxy coating in bluish colour on the surface of the leaves of aloes minimizes evaporation of water from cells on the surface of the leaves and limiting water loss to the stomata. The inner substance of the leaves is jelly like, with bad odour and bitter in taste. The inflorescence is produced on a spike up to 90 cm tall which has a number of small pink or yellow or orange flowers all around. The fruit is a capsule. It is normally not propagated through seeds.

Climate: It is cultivated in tropical and subtropical climates. However, it adapted to desert, grassland, and coastal or even alpine environments. The plant grows well in areas receiving annual rainfall of 500 to 2000 mm. It is highly sensitive to water stagnation. It is a drought-resisting plant. It is frost sensitive crop.

Soils: It is grown in well drained sandy loam, loam and clay loamy soil with high organic matter. The plants tolerate high pH up to 10.

Season: Suckers should be planted in July-August during monsoon season. However, planting can be done throughout the year except in winter months (November-January) under irrigated condition. Aloe is a 6 to 8 year crop.

Land preparation: The soil should not be ploughed deep as the root system does not penetrate below 20-30 cm. One or two ploughing is given followed by leveling may be done. Beds and channels are formed.

Propagation: Aloe does not produce seeds. The plants are generally propagated by root-suckers or offshoots or pups rhizome cuttings. The underground rhizome is also dug out and made in 5-6 cm long cuttings with at least 2-3 nodes on them. The suckers/ cuttings are planted at a spacing of 60 × 30 cm or 60 × 45 cm or 60 × 60 cm. After planting of suckers, the soil around the root zone must be firmly pressed and drainage must be made proper to avoid water stagnation.

Weed management: Weeding is given after 40 days. Earthing up is done after weeding and fertilizer application.

Manures and fertilizers application: During the first year of plantation, FYM @ 20 t/ha is applied at the time of land preparation and the same is continued in subsequent years. Besides vermicompost @ 2.5 t/ha can also be applied. A fertilizer dose of 35-25-25 kg of N, P_2O_5 and K_2O per ha is applied as basal dressing.

Water management: Aloe can be cultivated both under irrigated and rainfed conditions. Irrigation is given once in 7 to 10 days interval depending on soil and climatic condition. It is slightly tolerant to drought, but very sensitive to water stagnation. Therefore, proper drainage is more important than irrigation.

Pests and diseases management: Pests and diseases such as spider mites, mealy bug, scale insects, aphid, anthracnose and leaf spots affect aloe. Neem cake @ 350-400 kg / ha may be applied to control termites.

Harvest: The older outer leaves are generally harvested leaving the fresh and young leaves from the top. Offshoots are grown spontaneously next to the mother plant. Care also has to be taken to reduce the loss of juice from the cut portion. Leaves harvest starts after 7-8 months of planting suckers/cuttings. The plants can be harvested every 6 to 8 weeks by removing the outermost 3 to 4 leaves per plant at the white base of the plant. Harvest can be made 3 to 4 times per year. Sharp knife is used for harvesting. An aloe plantation gives a commercial yield from the second year up to the age of five years, after which it needs replanting. Harvested leaves are carefully stacked and then transferred to a refrigeration or processing facility.

Yield: After 1½ years, 10 to 12 kg of leaves can be harvested per plant per year. Afterwards, 22 to 24 leaves are harvested per plant per year. The leaves yield of the crop on a fresh-weight basis will be around 10 to 12 t/ha.

Post-harvest technology: Aloe plant has three distinct raw material components namely leaf juice, inner leaf juice and aloe latex. The entire leaf is ground or macerated to obtain leaf juice. The inner leaf juice is obtained from the gel after removing the outer rind of the plant which can be pulp-free or contain pulp. Aloe latex is a sap-like material that bleeds when the rind is cut. Aloe latex is yellow-brownish or reddish in colour.

Utilization: Aloe leaves are used for medicinal and cosmetic uses. Aloe leaf possesses many medicinal activities, including antimicrobial, anticancer, antioxidant, antidiabetic, antiulcer, hepatoprotective, immunomodulatory. Aloe latex is taken orally for conditions ranging from glaucoma to multiple sclerosis. The Aloe juice repels attacking insects, rodents, snakes by means of the bitter latex substance called Aloin (the yellow coloured part of the sap), aloe-emodin, and barbaloin just beneath the rind. Aloe contains two classes of Aloins: (i) nataloins, which yield picric and oxalic acids with nitric acid, and do not give a red colouration with nitric acid; and (ii) barbaloins, which yield aloetic acid, chrysammic acid, picric and oxalic acids with nitric acid, being reddened by the acid. Lupeol and salicylic acid present in the juice are two very effective pain-killer. The gel from inside the leaf may help to heal burns and abrasions.

REFERENCE

Jat, R.S., Nagaraja Reddy, R., Ruchi Bansal and P. Manivel. 2015. Good Agricultural Practices for Aloe. Directorate of Medicinal and Aromatic Plants Research, Boriavi, Anand - 387 310, Gujarat.

Karkala, M. and B. Bidya. 2014. *Aloe vera*: a wonder plant its history, cultivation and medicinal uses. Journal of Pharmacognosy and Phytochemistry. 2 (5): 85-88

Sampath Kumar, K. P., Debjit Bhowmik, Chiranjib, Biswajit. 2010. *Aloe vera*: A Potential Herb and its Medicinal Importance. J. Chem. Pharm. Res. 2(1): 21-29

http://www.iasc.org/pdfs/IASC_Aloe_vera_A_Scientific_Primer.pdf

http://www.iasc.org/faq.html

CHAPTER

48

Ashwgandha–*Withania somnifera* (Linn.) Dunal

Synonyms: *Withania ashwagandha* Kaul, *Physalis flexuosa* Linn, *Physalis somnifera* Linn.

Family: Solanaceae

Vernacular names: Ashwagandha, Indian ginseng, Indian Winter Cherry, Winter Cherry (English); *Asgand, Asgandh, Askandhatilli, Punir* (Hindi); *Ashwagandha, Varahakarni, Hayagandha, Vajigandha* (Sanskrit); *Amukkira, Amukkirankizhangu* (Tamil); *Amukkiram* (Malayalam); *Asundha* (Gujarat); *Asagandha, Askagandha* (Marathi); *Amangura, Angarberu,* (Kannada); *Ashvagandha, Pennerugadda* (Telugu).

Importance: Ashwagandha roots, leaves and seeds have medicinal properties and are used in preparation of various drugs. It is a stress reliever and is used in treating senile dysfunctions. The species name *somnifera* means 'sleep-bearing' in Latin, indicating it was considered a sedative, but it has been also used for sexual vitality and as an adaptogen.

Distribution: It is found in Congo, South Africa, Egypt, Morocco, Jordan, Pakistan, Sri Lanka, Baluchistan, Afghanistan, Sind and Mediterranean regions. It is cultivated in Rajasthan, Punjab, Haryana, Uttar Pradesh, Gujarat, Maharashtra and Madhya Pradesh.

Botany: The plant is an evergreen, erect, branching, low lying shrub reaching a height of about 0.75 to 1.50 m. It has a central stem from which branches extend radially in a star pattern (stellate) and covered with a dense matte of wooly hairs (tomentose). Leaves are simple, alternate or sub-opposite, ovate, glabrous, 10 cm long, dense beneath and sparse above. The flowers are small, green or greenish-yellow. The mature fruits are orange-red berries. Seeds are yellow, reniform and 2.5 mm in diameter. The roots are stout, long tuberous, fleshy, whitish brown, gradually tapering down, 10 to 17.5 cm long and 6 to 12 mm in diameter. The main roots bear fiber-like secondary roots.

Climate: The plant grows in dry parts in sub-tropical regions. It can be cultivated up to an altitude of 1200 m from MSL. It is grown in areas receiving annual rainfall of 500 to 750 mm. It is a xerophytic plant. The cardinal minimum, optimum and maximum temperature are 10°C, 20 to 35°C and 40°C respectively.

Soils: It grows well in sandy loam or light red loamy soil having pH 7.5-8.0 with good drainage.

Season: The seeds are sown in June-July and in August. The crop is ready for harvest in January-March at 150 to 180 days after sowing.

Varieties: Ashwagandha commercial varieties are Jawahar Asgandh-20, Jawahar Asgand-134, Poshita and Raj Vijay Ashwagandha-100. Two varieties referred in Unani literatue are (i) *Asgand Nagori* and (ii) *Asgand Dakani*.

Land preparation: The land is ploughed once with mould board plough and harrowed twice to bring the soil to fine tilth. The field is then leveled by planking.

Seeds and sowing

Direct sowing: Seed rate is 10 to 12 kg/ha for direct sowing. Seeds can be treated with Thiram or Indofil or Dithane M 45 (@ 3 g/kg of seeds. The seeds are sown by broadcasting or in the line in furrows with spacing at 30 × 15/10 cm. The seeds are sown at about 1-3 cm depth. Seeds germinate in 6 to 10 days after sowing. It should be thinned out by hand at 25-30 days after sowing. Line sowing facilitates better intercultural practices.

Nursery management: Seed rate is 500 to 750 g/ha for raising seedlings for one hectare crop. Seeds can be treated with Dithane M-45 @ 3 g/kg or Thiram-45 @ 3gm/kg of seeds before sowing. Seeds are sown in the nursery about 1-3 cm deep and are covered with light soil. Seeds germinate in 6-7 days after sowing.

Transplanting: The seedling after 25-35 days after sowing can be transplanted in the field with spacing of 30 × 15 cm or 60 × 60 cm.

After cultivation practices: The directly sown crop is thinned at 25 to 30 days.

Weed management: Two hand weeding are given on 25-30 days and 60-70 days after sowing. Isoproturon at 0.5 kg/ha, Glyphosate at 1.50 kg a.i./ha and Trifluralil (48% EC) @ 4.1 a.i/ha is used to control weeds.

Manures and fertilizers application: Mostly organic manure should be applied. Apply 12.5 t/ha of FYM and 30-25-20 kg/ha of N, P_2O_5 and K_2O as basal dressing. All the quantity of P, K and ½N should be applied at the time of sowing or planting and the remaining ½N on 30 days after planting.

Water management: It is usually grown as rainfed crop. Excessive rainfall/water is harmful to the crop. Under irrigated conditions, the crop can be irrigated once in 7 to 10 days depending on soil and climatic conditions.

Cropping systems: It is planted as intercrop with newly planted coconut, mango, teak, simaruba, jatropha, pine and populus.

Insect pests and diseases management: The major insect pests are shoot borer and mite. Neem seed kernel extract will have more effect on mites. Neem cake can also be applied in soil to control the diseases. Shoot borer can be controlled with Sumicidin @ 10 ml per liter. The fungus disease like damping of fungus, seedling blight, seed rotting, die-back, etc, occur in seedling stage. Treat the seeds before sowing with Captan at the rate of 5 g/kg of seed to reduce seedling blight and leaf blight and drench Calphomin at 3 ml/litre in nursery before sowing to control soil bone pathogens. The crop should be sprayed with Dithane M-45 at the rate of 3 g/litre of water, when 30 days old, and the spray should be repeated at 15 days interval to control diseases.

Harvest: The maturity of crop is judged by drying out of leaves and berries become yellow red. The entire plant is uprooted for roots which are separated from aerial parts by cutting the stem 1 to 2 cm above the crown roots. The roots are then either cut transversely into small pieces of 7 to 10 cm. The roots are sun dried for 3 to 4 days. The berries are plucked from the dried plants and are threshed to obtain the seeds. Root size, root and shoot biomass and alkaloid content are found maximum in 180 DAS which should be considered as harvesting time. The whole plant is uprooted and roots are separated.

Yield: The average yield of dried roots is 650 to 800 kg/ha which comes to 350 to 450 kg on drying. Seed yield is 40 to 50 kg/ha. It yields 3 to 5 kg/ha of Withanoloide.

Utilization: The roots, leaves and fruits (berry) possess tremendous medicinal value. The medicinal activity of the root is attributed to the alkaloids, withanoloids and withanins (steroidal lactones). Alkaloid percentage in roots ranges from 0.13 to 0.31%. It is rejuvenative and helps in nourishment of the tissues, particularly muscle and bones, while supporting the proper function of the adrenals and reproductive system. It improves blood circulation, increases haemoglobin (red blood count) and hair melanin. It stabilizes blood sugar and lowers cholesterol. Ashwagandha based product increases semen quantity, sperm count and mobility. It is effectively used in erectile dysfunction, low libido and premature ejaculation. It is used for centuries for vivid health disorders.

CHAPTER

49

Ashoka–*Saraca asoca* (Roxb.) de Willde

Synonym: *Saraca indica* Roxb.

Family: Fabaceae/Caesalpiniaceae

Vernacular names: Ashoka, Sorrowless tree (English), *Sita Asoka* (Hindi), *Sita-Ashoka, Kankeli* (Sanskrit), *Devdaru, Ashoka* (Bengali), *Ashoka* (Assamese), *Ashok* (Bengali), *Ashoka* (Oriya), *Asokam* (Tamil), *Asokam, Gapis, Tengalan* (Malayalam), *Asokamu, Ashokapatta, Vanjulamu* (Telugu), *Ashokadamara, Achenge, Akshth, Ashanke, Kankalimara* (Kannada), *Jasundi* (Marathi), *Ashopalava, Ashoka* (Gujarati).

Importance: It is grown as an avenue tree due to its foliage and fragrant flowers. Asoka, the sacred tree of Hindus and Buddhists, possesses varied medicinal uses. Ashoka is a Sanskrit word meaning 'without grief (sorrow)' or that which gives no grief. *Polyalthia longifolia* is also known as Ashoka tree which is the false Ashoka tree.

Origin: It is native of India.

Distribution: It is distributed throughout India, Myanmar, Malaysia and Sri Lanka. It is grown in Kerala, West Bengal, Orissa, Andaman Islands and Assam.

Botany: It is a medium-sized, evergreen tree 7-10 m high with numerous spreading, drooping glabrous branches. The bark is dark brown to grey or black with a warty surface. The thickness varies from 5 mm to 10 mm. The cut surface turns reddish on exposure to air. Leaves are parpinnate 15 to 20 cm long and the leaflets 6 to 12, oblong. Leaves are narrowly lanceolate, cork like at the base. Leaves are 30 to 60 cm long, copper red in colour when young and green when mature. The bark is dark green, dark brown or grey or almost black with warty surface. Stem bark are rough and uneven due to the presence of rounded or projecting lenticles. The tree starts flowering from 6-7 years of growth. Flowers are orange or yellowish orange turning to scarlet in colour with beautiful fragrant flowers. Fruit is a four to eight seeded, flat, black coloured and leathery pod. Pods are flat, black and leathery, compressed and tapering at both ends. Seeds are 4 to 8, ellipsoid-oblong and compressed. 100 seeds weigh one kg.

Climate: It grows well in moist tropical climate with well-distributed rainfall. It occurs in evergreen forests of central and eastern Himalayas up to 750 m altitude. It also thrives well in partially shaded locations.

Soils: The plant requires slightly acidic to neutral soils

Season: Planting may be done in May-June or July-August. It can also be planted in summer, if irrigation facilities are available.

Propagation: Seeds are used for propagation. Mature seeds are collected from 5 to 6 year old trees in December–January.

Nursery management: Seeds are collected from trees above 6-7 years. Since the viability of the seeds is short, it should be used immediately for seedling production. Two kg seeds are required for raising seedlings for planting in one hectare of land. The seeds may be soaked in water for 12 hours before sowing to improve the germination. The seeds are sown in mother beds or polybags of 25 × 20 cm size. The potting mixture consists of equal quantities of soil, sand, and FYM. The seeds germinate in about 15 days. One year old seedlings are used for field planting. Self-sown seedlings are abundantly seen under the trees through germination of fallen seeds. These can be uprooted and field planted directly during rainy season. Air layering can be done during May-June on greenish brown shoots of current season growth. The layers can be separated from the tree after 2-3 months and established in polybags before field planting.

Land preparation: Pits of size 45 cm × 45 cm × 45 cm are prepared at a spacing of 3 m × 3 m. The pits are weathered and refilled with topsoil after mixing with 10 kg FYM per pit.

Planting: Two month old seedlings are transplanted in the pits. Thinning of the plants may be done after 15-30 days.

Weed management: Three weeding are taken up at one month interval. The interspaces are kept weed-free either by hand weeding or by spraying of paraquat @ 0.8% or glyphosate @ 0.4%.

Manures and fertilizer application: FYM at the rate of 10 kg/tree/year is applied twice: first in May–June while filling the pits and again in October–November at the time of second weeding. During initial years, apply 2 kg FYM, ½ kg bonemeal and ½ kg neemcake each during May June and October November.

Cropping systems: It is grown as a mixed crop with perennial trees like coconut, which provide partial shade to the crop.

Water management: The crop is raised as rainfed crop in high rainfall tracts. During the months without rains, the seedlings require frequent irrigation. Watering of grown-up trees is done by forming a ring channel around trees' base to hasten growth.

Pests and diseases management: Pod borers and caterpillars can be controlled by spraying 3% Neem kernel extract. Spraying tobacco decoction controls aphids.

Harvest: The tree flowers profusely at six to eight years of age. The tree survives for about 50 years. It is often felled after it reaches 20 years of age for collecting bark. It is cut at a height of 15 cm from the soil level. Sometimes, the bark can be collected without cutting down the tree. The bark can be peeled off from main stem from tenth year onwards. Bark should be stripped off carefully without damaging the cambial layer. The bark is peeled off in vertical strips with 6 cm interspaces between each strip. The peeled off area is renewed with fresh bark in one to two years. Then, the bark on the other areas can be peeled off without cutting the tree. This nondestructive method should be preferred for harvesting. Bark is removed from about ten years or older tree and then it has to be sun dried. The bark is dried in the shade, packed and stored in containers. If sufficient irrigation and fertilizers are provided, the stumps will regenerate new coppice shoots, which can be harvested again after 10 years.

Yield: One tonne of dry stem bark per ha is produced from a sole crop. When grown as mixed crop with coconut, yield is 0.6 t/ha. The economic parts used are stem, bark, root, flowers, fruits and seeds.

Utilization: Stem bark of Asoka tree is strongly astringent and a uterine sedative, uterine tonic, and styptic. It has a stimulating effect on endometrial and ovarian tissue. The bark is also useful in dyspepsia, fever and burning sensation. It is also used to treat menorrhagia, leucorrhoea, internal bleeding, hemorrhoids and hemorrhagic dysentery. The bark decoction taken internally prevents uterine disorders. The crushed flowers and leaves are rubbed on the skin to get relief from skin diseases. *Saraca asoca* is considered as best female tonic.

CHAPTER

50

Bael–*Aegle marmelos* (L.) Correa

Synonyms: *Belou marmelos* (L.) A. Lyons, *Crateva marmelos* L.

Family: Rutaceae

English names: Bengal quince, Indian quince, Wood apple, Golden apple, Stone apple, Elephant apple, Bael fruit, Bek fruit, Indian bael, Holy fruit (English); *Bilva, Sriphala* (Sanskrit); *Bael Tree, Baelputri, Bela, Sirphal, Sriphal, Kooralam* (Hindi); *Billi, Bilum, Bilivaphal* (Gujarati); Kaveeth (Marathi); *Bilpatra, Kumbala, Malura* (Kannadam) *Vilwam, Kuvalam, Koovalam* (Malayalam), *Vilvam*, *Kuvalum* (Tamil); *Maredu* (Telugu), *Bael, Shreefal* (Bengali), *Bael, Vael* (Assamese); *Belo, Bel*a (Oriya). The English name for Bilva is Bael, also called 'stone apple' as its rather large fruit is like pale yellow suns when ripe. In Sanskrit it is also called Bilva or Sriphal. The word Bilva is derived from 'bil' to split. Yaska derived this word from bhr 'to support' or 'to nourish', or from bhid 'to split'.

Importance: Bilva is one of the sacred trees of India. Felling of Bilva tree is strictly prohibited. Plantation of the Bilva tree is highly recommended in the Vrksayurveda. It is grown in most temples especially of Lord Siva and house gardens. Shiva is always worshipped with its leaves. Even a sight of this tree helps in absolving all sorts of sin. It is a healing tree, which cures all diseases caused by vata (wind) and gives strength to the body. The roots, trunk, bark, leaves, fruits and seed of Bilva tree are used for medicinal purposes.

Origin: It has originated from Eastern Ghats and Central India.

Distribution: It is also cultivated in India, Pakistan, Nepal, Myanmar, Tibet, Vietnam, Laos, Cambodia, Egypt, Sri Lanka, Bangladesh, Thailand, Java Islands, Indonesia, Malaysia, the drier areas of Java, Fiji and Philippine. It is found in the states of Himachal Pradesh, Jammu and Kashmir, Punjab, Rajasthan, Uttar Pradesh, Bihar, West Bengal, Tripura, Madhya Pradesh, Maharashtra, Andhra Pradesh, Karnataka, Kerala, and Tamil Nadu.

Botany: It is tree growing to a height of 6.0 to 9 m. The stems are lenticelled and show conspicuous woody thorns. Its branches have many knots all along and bear straight and strong thorns. The trunk bark is bluish grey. A clear, gummy sap, resembling gum arabic, exudes from wounded branches and hangs down in long strands, becoming gradually solid. Taste of this gum is sweet at first but later irritating to the throat. The leaves are alternate and trifoliate aromatic leaves. Lamina or the leaf blade has numerous translucent pellucid glands all over are filled with fragrant, volatile oil. Flowers are greenish white and sweet scented. Fruit is hard, greenish and egg like. They are 5 to 10 cm in diameter. Pulp is sweet, thick and orange coloured. The golden coloured bael fruit resembles a golden apple and hence the generic name Aegle. The specific name marmelos is derived from marmelosin contained in the fruit. Root, bark, leaves and fruits are hypoglycaemic, astringent and febrifuge. Fruits is seen on the tree round the year. Fruits are large, globose, aromatic and carminative. Fruit has hard rind. Fruit weighs 0.6 to 0.8 kg. Seeds, numerous, embedded in the mucilaginous pulp, oblong, compressed, white, having cotton-like hairs on their outer surface.

Climate: It is a subtropical species. The tree grows wild in in mixed deciduous and dry dipterocarp forests, dry forests on hills and plains of central and southern India. It grows up to an altitude of 1,200 m. It is grown in areas receiving mean annual rainfall of 570-2000 mm. It is grown areas with temperature of 48ºC in summer and descends to -6ºC in the winter season.

Soils: It can be grown in well drained soil. It also grows well in swampy, alkaline or stony soils having pH range from 5 to 8.

Season: Sowing is done in June or July.

Varieties: The most promising varieties are Mirzapuri, Darogaji, Ojha, Rampuri, Azamati, Khamaria, Narendra Bael-4 'Narendra Bael-5, Narendra Bael-6, Narendra Bael-7, Basti No.1, Gonda No.1, Gonda No 2, Gonda No 3, Kagzi Etawah, Sewan Large, Deoria Large, Chakaiya, Lamb, Baghel, Pant Shivani and Pant Aparna.

Nursery management: Bael is generally propagated by seeds. Seeds can be hydro primed (soaked in water and dried back to original moisture content) for 6 hours, adopting the seed to solution ratio of 1:1 to obtain uniform emergence, high germination percent and seedling growth. The development of seedlings is very slow. One year old seedlings should be transplanted in rainy season. It is also propagated by root cuttings and stem cuttings treating with IBA @ 4000 ppm using quick dip method.

Planting: Seedlings or budded plants are transplanted in the field at a spacing of 6 to 10m. Budded plants start bearing fruits at the age of 4-5 years, whereas seedling trees require 7-8 years.

Water management: The field after plantation should be irrigated weekly or fortnightly as and when required.

Harvest: The fruits are deep green initially and become yellow gradually at ripening. The fruits are harvested along with it's a portion of fruiting stalk as it serves as a signal of ripening. The stalk is easily detached in the ripe fruits. The fruits require about a year for ripening.

Yield: The average fruits yield is 62.5 kg/tree. The average number of fruits per tree is 300 to 400.

Utilization: The plant is regarded as sacred and the leaves are used for the worship of Lord Shiva. Fruits are favourite to monkeys. Root, stem and bark, flower, fruits and seeds are used for medicinal purposes. Fruits are useful in bowel ailments and veterinary practice for abdominal complaints. The fruits offer a remedy for dysentery. The ripe fruits are used to promote digestion and to soothe inflammation of the rectum. The young leaves are used externally to soothe inflammation, sores, and to deflate swellings. Leaves are also eaten to induce abortion. The roots are infused in hot water to make a drink that is said to calm palpitation of the heart. The fruits are used to treat tuberculosis and liver dysfunction. The active principles present are ethyl cinnamide, aegeline and marmeline.

REFERENCES

Dhankhar S., Ruhil S., Balhara M., Dhankhar S. and Chhillar A. K. 2011. *Aegle marmelos* (Linn.) Correa: A Potential Source of Phytomedicine, Journal of Medicinal Plants Research, 5 (9): 1497-1507.

Jauhari O.S. and Singh R.D. 1971. Bael-The Valuable Fruit., Indian Horticulture, 16 (1): 9-10

Kirtikar K. R. and Basu B. D.1935. A Review on *Aegle marmelos* (L.) Correa, Indian Medicinal Plants, 1- 4, 499

Sekar D.K., G.K., Karthik L. and Bhaskara Rao K. V. 2011. A Review on Pharmacological and Phytochemical Properties of *Aegle marmelos* (L.) Corr. Serr. (Rutaceae), Asian Journal of Plant Science and Research, 1 (2): 8-17

Srivastava K.K. and Singh H.K. 2004. Physico-Chemical Quality of Bael (*Aegle marmelos* Correa) Cultivars, Agric. Sci. Digest, 24 (1): 65 – 66

Sinha, B. C. 1979. Tree Worship in Ancient India, Books Today, New Delhi.

Dhananjay, V. D. 2012. Bilva in Indian tradition. Indian Journal of History of Science, 47(1): 37-61.

Venudevan, B. and Srimathi, P. 2013. Conservation of endangered medicinal tree bael (*Aegle marmelos*) through seed priming. Journal of Medicinal Plants Research. 7(24): 1780-1783.

Nigam, V. and Nambiar, V. S. 2015. Therapeutic potential of *Aegle marmelos* (L.) Correa leaves as an antioxidant and anti-diabetic agent: a review. International Journal of Pharma Sciences and Research. 6(3): 611-621

https://www.hort.purdue.edu/newcrop/parmar/01.html

CHAPTER

51

Belladonna–*Atropa belladonna* L.

Family: Solanaceae

Vernacular names: Deadly Nightshade, Devil's cherries, Death cherries, Great Morel, Divale, Apples of Sodom, Black Cherry, Belladona, Naughty Man's cherries, Sorceror's Berry, Witch's Berry (English); *Suchi* (Sanskrit); *Angur Shefa, Luckmuna* Luckmunee (Hindi); *Bellatona, Pelletonacceti* (Tamil); *Sagangur* (Kashmiri); *Yebruj* (Bengali). The genus name 'Atropa' is from the Greek God Atropos which refers one of the fates that cut the thread of life. The Venetians called the plant as 'Herba Belladonna' because their ladies used a distilled water of the plant as a cosmetic. Hence the name, 'herba bella-donna' which means 'herb of the beautiful lady' in Italian language. This herb has been used in eye-drops by women to dilate the pupils of the eyes to make them appear seductive.

Importance: The commercial drug is obtained from the leaves, flowering tops and roots. Leaves and roots of belladonna contain tropane alkaloids whose concentration varies from 0.13 to 0.70% (average 0.45%) which include hyoscine, hyoscyamine and atopine. These are used in medicine because of their antipholinergic and roots of are used in the pharmaceutical industry.

Origin: The plant species is native to Europe (Germany, Austria, Ukraine, and Albania), Northern Africa (Algeria, Morocco) and Western Asia (Iran, Turkey).

Distribution: It is cultivated in Europe, Russia, Northern and Southern America and in few parts of India and Pakistan.

Botany: It is a perennial herb with purplish coloured stem, thick roots, darkish colour leaves with short petiole. It grows to a height of 60 to 90 cm. The flowers are purple-brown, bell-shaped, and drooping about 2.5cm long. The leaves are smooth and long with prominent veins. The fruit is shiny, smooth, black colour fruit berry with disperse seeds.

Climate: It is a temperate cool season crop. It behaves as a perennial in temperate climates and gives maximum herbage and alkaloid yield. In sub-tropical areas, it can be grown as a winter crop. However, the plant behaves as an annual as it dies during the summer months. The seeds germination is 82.5% in 6 h at 30°C and 18 h at 15°C.

Soils: The plant grows well in moist environment in calcareous and saline in soil.

Heavy clay soils which are water-logged may be avoided.

Season: The nursery may be raised from the second week of May to September under sufficient shade. The ideal time for planting in the field is February-March or October-November.

Propagation: The crop is propagated through seeds extracted from berries. It can be cultivated by direct sowing, but raising seedlings in nursery is also practiced.

Nursery management: The nursery may be raised under sufficient shade. The land should be ploughed well to provide a fine tilth. Raised beds of size 3m × 1m surrounded by drainage and irrigation channels to be made. FYM may be applied to the soil in nursery. Seeds are sown in raised beds at the rate of 1 kg per 100 m^2. Seeds may be treated with 80% sulphuric acid at the time of sowing for 2 minutes to improve the germination. Germination of stored seed is slow and erratic, usually takes 1 to 6 months at 5.5°C. A maximum germination of 90% can be obtained with treatment of gibberellic acid @ 1mg/ litre of water. Seeds are treated with Dithane M-45 or Agallol @10 g per kg of seeds may be mixed with fine soil in 1:4 ratios and the seeds are broadcasted in the nursery beds. Cover the seed beds with a layer of FYM and then with straw. Watering of beds should be done immediately after sowing with a rose can. The seeds germinate in 3 weeks' time. Seedlings are ready for planting in the field when they attain a height of 15-20 cm after 8-12 weeks.

Planting: Seedlings are planted at a spacing of 60 × 60 cm or 60 × 45 cm. It is always safer to plant the seedlings on raised beds with 1 m wide strips or ridges as it avoids water logging and facilitates irrigation. The field may be irrigated immediately after transplanting.

Manures and fertilizer application: A basal dose of FYM @ 12.5 t/ha and 60-25-25 kg N, P_2O_5 and K_2O as basal dressing is recommended. Again 60 kg N/ha is given in 4-5 splits as top dressing after every harvest.

Harvest: The first harvest of the leaves is available three month after planting. Harvesting should be done as soon as the plants start flowering. The alkaloid content is high at flowering stage. A total of 5-6 cuttings per year can be obtained. The leaves are cut about 30 cm above ground level using pruning scissors. Leaves are dried immediately after the harvest under shade or sun or artificial heat with or without fans for air circulation. Leaves should be turned over frequently while drying. The woody stems are discarded before drying. The roots are also harvested after 3 years. After the harvest, they are washed, cut into 10 cm length, split length wise if thick and shade or sun dried. Berries are collected and crushed gently to separate the seeds and dry them.

Yield: An average dry herb yield of 1000 kg/ha is obtained during the first year. The yield increases to 1500 kg per hectare during 2nd and 3rd year. The yield of dry roots will vary from 170 to 335 kg/ha.

Utilization: The berries, leaves and roots are toxic. All parts of the plant contain tropane alkaloids: atropine and scopolamine. The plant is used in herbal medicine as a pain reliever, muscle relaxer, anti-inflammatory and to treat menstrual problems, peptic ulcer disease, histaminic reaction, and motion sickness. Leaves are used for the manufacture of tinctures and plasters. The extract of belladonna helps to mitigate of mental illnesses.

CHAPTER

52

Brahmi–*Bacopa monnieri* (L.) Wettst.

Synonyms: *Bramia indica, Bacopa monnieria, Herpestis monnieri* L.

Family: Scrophulariaceae (Dog flower family)

Vernacular names: Brahmi, Water hyssop, Indian pennywort (English); *Brahmi* (Hindi, Assamese, Marathi); *Nira-Brahmi, Gundala, Indravalli, Jalasaya* (Sanskrit); Brahmi-sak (Bengali, Manipuri); *Jalanevari* (Gujarati); *Brahmi, Jalabrahmi, Neer brahmi* (Kannada); *Nirbrahmi* (Tamil); *Sambrani chettu, Neeri sambraani mokka* (Telugu).

Importance: The name Brahmi is derived from the word 'Brama' the mythical 'creator' in the Hindu pantheon. The plant is prescribed for nervous disorders such as insanity, epilepsy, nervous breakdown, *etc*. It is considered a bog plant that likes wet spots, but will grow in pots, or in the garden in shade, or even full sun, provided ample water is provided in dry spells as it is a shallow rooting plant. It is grown in hanging pots and as the stems cascade over the sides, they are easily nipped off for use.

Distribution: It is grown in Uttar Pradesh, Punjab, Haryana, Bihar, Bengal, Tamil Nadu, Kerala, Karnataka, Foot hills of Himachal Pradesh and Uttaranchal.

Botany: Brahmi is a perennial, creeping herb with rooting at the nodes. The leaves of this plant are succulent and relatively thick. Leaves are succulent, sessile, opposite decussate, obovate-oblanceolate on the stem. The edible leaves and stems have a strong bitter taste. Small flowers are borne in leaf axils. Flower stalk is 0.5-3.5 cm long. Flowers are blue, purple, or white. Seeds are yellow-brown, ellipsoid, truncate at one end, longitudinally channeled.

Climate: It grows well in sub-tropical conditions. The herb is found at elevations from sea level to altitudes of 1400 m. The plants grow at high temperatures of 30-40°C and humidity of 65-80%. It is frost sensitive.

Soils: It is found in wetlands, sandy areas, muddy shores, damp or marshy areas near streams or on the border of ponds, throughout India.

Propagation: It is propagated by seed, cuttings and by root division.

Season: It is cultivated both as summer and rainy season crop.

Land preparation: The field should be ploughed thoroughly to have fine tilth.

Varieties: Subodhak and Pragyashakthi.

Planting: Plant cuttings of about 5 to 8 cm long, each containing a few leaves, nodes and roots are used as planting materials. The cutting are transplanted at spacing of 40 cm × 40 cm.

Weed management: Hand weeding is taken 15-20 days interval.

Manures and fertilizer application: It is fertilized with seaweed or other organic manures.

Water management: Irrigation immediately after transplanting is essential for the survival of the plants. Subsequently, the fields are irrigated once in 7-10 days. There is no need for irrigation during the monsoon. It is a succulent and water loving plant.

Harvest: The plants are harvested so that the upper portions of the stem 4-5 cm from the base are removed and the rest left for subsequent regeneration. The plants can be dried on the ground under shade at room temperature. The material is to be cleaned free of any external matter. The dry material should be stored in a cool dry room packed in bags/boxes.

Yield: The fresh and dry herb yields of Brahmi go up to 20 t/ha and 4 t/ha, respectively in a year while bacoside-A yield is 85 kg/ha.

Utilization: The herb is primarily used for its ability to enhance memory capacity, improve intellectual and cognitive functions, reduce stress-induced anxiety and increase concentration. The active constituents of the plant are brahmine, herpestine, alkaloids and saponins. The memory-enhancing effects have been attributed to the saponins, bacosides A and B.

REFERENCES

Ali Esmail Al-Snafi. 2013. The pharmacology of *Bacopa monniera*. A review. International Journal of Pharma Sciences and Research. 4 (12): 154-159

Anon, 2004. *Bacopa monniera*. Monograph. Altern Med Rev. 9(1):79-85.

Russo A, Borrelli F. 2005. *Bacopa monniera*, a reputed nootropic plant: an overview. Phytomedicine. 12(4): 305-317.

Singh, H. K., Rastogi, R. P., Sriman, R. C. and Dhawan, B. N., 1988. Effect of bacoside A and B on avoidance response in rats. Phytother. Res. 2, 70–75.

Warrier PK, Nambiar VPK, Ramankutty C, Ramankutty VPK, Vasudevan Nair R. 1996. Indian Medicinal Plants: A Compendium of 500 Species. Orient Blackswan.

Nadkarni KM. 1988. The Indian Materia Medica. Columbia, MO: South Asia Books: 624-625.

CHAPTER

53

Chethikoduveli–*Plumbago indica* Linn.

Synonym: *Plumbago rosea* Linn.

Family: Plumbaginaceae

Vernacular names: Red flowered leadwort, Indian leadwort, Chitrakmool, Fire plant, Scarlett leadwort, Radix Plumbago (English); *Koduveli, Chuvana-koduveli, Chivappu-koduveli, Chethikoduveli, Thiruvathirapoovu* (Malayalam); *Lalcitra, Raktacitra* (Hindi); *Citraka* (Sanskrit); *Citramulam, Cenkodiveli* (Tamil); *Eracitramulam* (Telugu).

Importance: The synonyms of fire like agnih, vahnih, etc. attributed to this drug to indicate the burning action of the root, causing blisters on the skin.

Origin: It is native of South Asia.

Distribution: It is grown in India and Sri Lanka. It is largely cultivated in Kerala, Andhra Pradesh, Maharashtra, Madhya Pradesh, West Bengal and Chattisgarh.

Botany: A perennial spreading evergreen shrub of 0.5 to 1.5 m height, with semi-woody stems. The tuberous roots are irregularly bent and yellowish brown. A healthy plant may produce 5-6 stout roots of 50 cm long and 5-6 mm diameter. The branches are sometimes rooting, cylindrical, striate, without fascicled leaves in the axils, young stems flaccid. Leaves are alternate, oblong-elliptic, broadest towards the middle, acute at apex, attenuate at base, glabrous, finely scurfy underneath with reddish petiole. Flowers are red, purplish red to scarlet racemes, rachis and bracts without glands. Sepals are red, glandular all over. Corolla tube is up to 3.5 cm long.

Climate: It grows in warm humid tropical and subtropical climates or a greenhouse in cool climates with a minimum temperature of 7°C. The plant prefers partially shaded locations. The open and sunny conditions are not favourable for its growth.

Soils: It can be grown in a variety of soils, ranging from red laterite soil to black soil. It prefers well drained deep sandy loam to clayey loam soil with high organic content. In natural habitats, the plant prefers moist soil with high organic content. The crop is very sensitive to waterlogging.

Season: The crop is planted in main field at the onset of monsoon (May to June).

Varieties: Agni and Mridhula

Land preparation: The field should be prepared by operating mould board plough once, followed by disc ploughing twice and levelling to obtain a uniform fine tilth.

Propagation: It can easily be propagated through stem cuttings or seeds. Viable seeds are not produced in the plant. Hence propagated by stem cuttings. Since it is grown as a two year crop, shoots are collected from the existing crop at the end of first year and used for new planting. Stem cuttings of 10–15 cm length, having with two to three nodes, can be obtained from mother plants to raise the seedlings. The maximum success rate is obtained from the basal cuttings and it reduces gradually towards stem apex. Tender and over mature stem portions are discarded.

Nursery management: The cuttings can be treated with NAA 500 ppm or IBA 0.05% to enhance rooting. The cuttings are planted within 24 hours in raised nursery beds (15 cm) during rainy season and flat nursery beds during winter and summer. The beds of size 10 m × 1 m should be made under partial tree shades. In mist chamber, the cuttings should be planted in trays filled with sand throughout the year. Out of three nodes, one node must be buried in the soil/sand, as the roots would sprout from this node. The cuttings should be planted in 10 × 5 cm spacing. Rooting takes 2-2½ months' time. These beds should be irrigated regularly. The cuttings start taking root within one month of planting in nursery. The sprouting percentage and growth are better in mist chamber than in the open conditions. Seeds show poor germination percentage and should be scarified. The seeds are sown in polybags filled with equal amounts of sand, soil, and FYM. The seeds show about 70% germination in 10–12 days.

Planting: Transplanting can be done with 60 to 75 day's old growth of cuttings. The spacing is 50 cm × 25 cm.

Intercultural practices: Two to three weeding can be done out till the crop establish and cover the ground. Pruning can be done before harvest of the crop.

Manures and fertilizer application: FYM @ 12.5 t/ha is applied 30 days before planting at the time of land preparation. A fertilizer dose of 40-25-25 kg of N, P_2O_5 and K_2O per ha is applied as basal dressing. Top dressing of 20 and 10 kg/ha of N and K_2O is done on 60 and 120 days of planting. Application of 200 kg neem cake, and 1 kg each of biocontrol agents like *Trichoderma*, *Arbuscular mycorrhiza*, *Pseudomonas* along with farmyard manure helps to minimise pest and disease problems.

Water management: It is cultivated under rainfed condition. Irrigation is given once in 7 to 10 days interval depending on soil and climatic condition.

Cropping system: It is grown in open condition as well as in aged coconut plantations as intercrop. It is also grown as an intercrop with many fruit trees (guava, mango or citrus orchards) and within *Gmelina arborea, Oroxylum indicum* or other medicinal tree species as ground crop.

Pests and disease management: The plants get infested with mealy bug, thrips, nematodes and hairy caterpillar. Sucking pests can be controlled by spraying malathion at the rate of 2 ml/litre of water twice at an interval of 15 days when they appear on the crop. Mealy bug and thrips are controlled with spray of 2% neem oil suspension or 0.05% Quinalphos. Nematodes are controlled with soil application of neem cake @ 2 kg/cent at planting and after six months. Dip cuttings in *Pseudomonas fluorescens* @ 10 to 20 g/litre water at planting. Blight disease is controlled with spraying of 1% Bordeaux mixture.

Harvest: Plant attains maturity in 10–12 months after transplanting. Care should be taken to wear gloves while handling roots to avoid skin burn. The tubers are cleaned of adhering soil, tied into bundles and commonly marketed as fresh.

Yield: The fresh tuber yield is 10 t/ha while the dry root yield varies from 2.5 to 3.0 t/ha. The root bark contains an orange yellow crystalline pigment named plumbagin, a hydroxy naphthaquinone. Plumbagin content is 0.22 to 0.80% which is responsible for the burning action of roots on skin.

CHAPTER

54

Cinchona–*Cinchona pubescens* (Vahl) Kuntze (2n=34)

Synonyms: *Cinchona asperifolia* Wedd., *Cinchona caloptera* Miq., *Cinchona chomeliana* Wedd., *Cinchona hirsuta* Ruiz & Pavon, *Cinchona lechleriana* Schlecht., *Cinchona succirubra* Pav. ex Klotzsch, *Cinchona ledgeriana, Cinchona officinalis.*

Family: Rubiaceae (Coffee family)

Vernacular names: Quinine tree, Quinoa, Jesuits' Peruvian-bark, red Cinchona, Countess's bark, Brown peru bark, red Peruvian-bark; sacred Peruvian-bark, Crown bark, Fever tree, sacred bark and Jesuits bark (English).

Importance: Cinchona is cultivated as the source of quinine, an anti-malarial drug. The natives of Peru and Bolivia had been using the bark of cinchona as early as 1600 against such fever. The plant is named in honor of Ana de Osorio, Countess of Cinchon, wife of the viceroy of Peru, who has been cured of a fever in 1638 by a remedy which has derived from Cinchona plants. Two scientists, Pelletier and Caventou, isolated a quinoline alkaloid in the bark which provided the highest anti-malarial effect and name it 'Quinine' in 1820. Clement Markham, an English civil servant in India Office visited India in October 1860 with his small collection of seeds and plants of cinchona and entrusted them to W.G. Mclvor, Superintendent of the Botanical garden at Ootacamund in Nilgris hills.

Origin: It is native of Peru, Costa Rica, Venezuela and Bolivia. The native area spread in Amazonian area of the Andes (Peru, Ecuador, Colombia, Costa Rica, Bolivia).

Distribution: It is grown in India, Myanmar, Sri Lanka, Indonesia, Ecuador, Peru, Bolivia, Colombia, Costa Rica, Venezuela, Guatemala, Panama, Australia, Cameroon, Papua New Guinea, Philippines, USA, Hawaii, Congo, Kenya, Nigeria, Tanzania and Uganda. It is cultivated on a large scale Nilgiris, Darjeeling, Assam, Belgaum, Coorg, Palani hills, Travancore, Marcara in India. It is cultivated in moist montane habitats in the Indian states of Tamil Nadu (Nilgiri Hills) and Sikkim (Himalayan foothills) at 700-1800 m altitude, but is now cultivated through a broad range in the tropics.

Botany: Cinchona is a large, erect, rapidly growing evergreen tree. It attains a height of 4-10 m); sparsely branched as lower branches shed, with young branches glabrous or pubescent. This broad leaf tree with a dense canopy grows rapidly, about 1-2 m per year, out-competes, shades out and replaces desirable native vegetation. Bark is brown, smooth; inner bark reddening when cut. The bark is spongy with a slight odour and strongly bitter. Leaves deciduous, opposite, large, broadly elliptic-ovate to broadly oblong; upper surface light green, puberulent or glabrate, thin, membranous to papery; lower surface with tufts of hairs in axils of lateral veins; old leaves reddish-orange. Flower numerous in panicles up to 20 cm long, rose-pink, fragrant; pale pink corolla tube comprising 5 spreading lobes. Fruits/capsules lanceoloid to oblong, contains 40-50 seeds, 1-4 cm long, dehiscent from base to apex. Seeds are 2-4 mm long, with a broad ciliate wing. It starts flowering in 3-4 years. Trees reach seed production within four years. The sweetly scented, tube shaped flowers are pollinated mainly by bees and butterflies. Fruits mature about 7-8 months after flowering. Seeds are surrounded by a papery wing, facilitating wind pollination.

Climate: It is mostly found in tropical and subtropical climates. Cinchona thrives best on steep mountain slopes of tropical forests. It is found in between 28°N and 21°S latitude and up to an altitude of 3500 m. It is grown in areas receiving mean annual rainfall of 600 to 1500 mm and average monthly temperature of coldest month is 18°C. The cardinal minimum, optimum and maximum temperature is 12°C, 18 to 25°C and 30°C respectively. In its natural habitat, predominately montane rainforest, it is grown in well-drained slopes with a well-distributed annual rainfall of over 2000 mm with an average minimum temperature of 14°C.

Soils: It grows best on light loamy to loamy soil, virgin forest soil with soil pH 4.5 to 6.5, rich in organic matter and preferably of volcanic origin. It cannot tolerate waterlogging and grows poorly on soils exposed to fire. It appears to favour mesic wet forest and moist tropical montane scrub always based on volcanic soil.

Propagation: It can be propagated from seeds or cuttings of ripe wood. Seeds take 2-3 weeks for germination. It is a hermaphrodite and insect-pollinated (bees, butterflies). Seeds are surrounded by a paper-like wing which aids wind dispersal. The seeds do not travel long distances since most seedlings are found within 100 m of the parent tree. Seeds are short-lived and quickly lose viability. It can re-grow from root fragments. It also produces root suckers and readily re-sprouts from cut or damaged stems.

Intercultural practices: Weeding is done mostly by hand for the first three years. Thinning is done until the tenth year when a stocking rate of 3000 stems/ha is achieved. Pruning is only carried out to shape the trees. All the trees are coppiced after 7-8 years to a height of 15-20 cm to minimize competition for light. Two to three shoots per stool are maintained to take over the new cycle.

Harvest: Cinchona is a rapidly growing broad-leaved tree (about 1-2 m height increment/year) with a dense canopy, achieving harvestable bark within 6 years. The tree trunks are beaten and the peeling stem bark is removed from the tree (at least 40 cm up from the soil level) in strips and dried. The bark partially regenerates on the tree and, after a few years and several cycles of bark removal, the trees are uprooted and new ones are planted.

Yield: The bark of wild species may yield a quinine content of as high as 7%, whereas cultivated crops yield contents up to 15%.

Utilization: It is suitable under shade of rainforest trees for preventing soil erosion. The alkaloid 'quinine' is present in the bark of the tree. Quinine bark has been the most used treatment for malaria. Quinine is also used as an additive in some carbonated beverages.

REFERENCES

Fosberg, F. R. 1947. Cinchona plantation in the New World. Econ. Bot. 1:330–333

George King. 1880. A manual of cinchona cultivation in India. Office of the Superintendent of Government Printing. Calcutta.

Heinke Jäger. 2014. Biology and impact of Pacific Island invasive species. #11. *Cinchona pubescens* (red quinine tree) (Rubiaceae). Pacific Science, 69 (2): 1-40

Hill, A. F. 1952. Economic Botany: A Textbook of Useful Plants and Plant Products. McGraw-Hill, New York.

Hodge, W. H. 1948. Wartime Cinchona procurement in Latin America. Econ. Bot. 2:229–257.

Martin, W. E., and J. A. Gandara. 1945. Alkaloid content of Ecuadorian and other American Cinchona barks. Bot. Gazette 2:184–199

Prendergast, R. and D. Dolley. 2001. Jesuit's bark (*Cinchona* [Rubiaceae]) and other medicines. Economic Botany, 55:3-6.

Purseglove JW, 1968. Tropical crops. Dicotyledons. 1. London: Longmans, Green & Co.Ltd., 225-236.

http://www.cabi.org/isc/datasheet/13484

http://www.worldagroforestry.org/resources/databases/agroforestree

CHAPTER 55

Coleus–*Coleus forskohlii* Briq. (2n=30)

Synonym: *Coleus barbatus* (Andr.) Benth.
Family: Lamiaceae
Vernacular names: Coleus, Hausa potato, Chinese potato, Country potato or Coleus potato (English); *Pashan Bhedi, Mayani, Makandi* (Sanskrit); *Garmai, Makandi, Patharchur, Pashanbhedi, Koorka, Gandhmoolika* (Hindi); *Makandiberu, Mangana beru* (Kannada); *Garmalu* (Gujarati); *Maimnul* (Marathi); *Koorkan kilangu, Marundu Koorkan, Gandhmoolika* (Tamil); *Makandiberu* (Kannada). The genus *Coleus* name was derived from the Greek word 'Coleos' meaning sheath.

Importance: Coleus is grown for its edible tubers, which have flavour and sweet taste and used as vegetable. The tuberous roots are found to be rich source of forskolin (syn.Coleonol) which is being developed as a drug for hypertension, glaucoma, asthma, congestive heart failures and certain types of cancers. The tubers are used as condiments in the preparation of pickles.

Origin: Coleus is a native of Africa and India.

Distribution: The crop has been distributed all over the tropical and subtropical regions of India, Pakistan, Sri Lanka, Nepal, Myanmar, Thailand, Brazil, Egypt, Arabia, and Ethiopia. In India, it is found to be in the subtropical Himalayan regions from Kumaon to Nepal, Bihar and the Deccan Plateau of Southern India. It is cultivated in parts of Gujarat, Rajasthan, Madhya Pradesh, Maharashtra, Bihar, Kerala, Karnataka and Tamil Nadu.

Botany: Coleus is an aromatic, perennial plant, 0.5 m tall with thick tubers. The entire plant is aromatic (whether fresh or dried). It has four angled square stems, branched, the nodes are often hairy. The leaves are usually pubescent, narrowed into petioles. Inflorescence is raceme, 15 to 30 cm in length. The flowers borne on racemes are perfect; the calyx is fine toothed and deflexed in the front. The corolla is pale-blue, lilac to pale lavender in colour, the lower lobes are elongated and concave. The ovary is four parted and stigma is two lobed and the flower is cross-pollinated by wind or insects. The roots are tuberous, golden brown in colour, fasiculated, up to 20 cm long and 0.5-2.5 cm thick, they are conical, fusiform, straight, orangish and strongly aromatic. The tuberous root resembles a carrot in shape and brown in colour. The forskolin is concentrated in roots.

Climate: Coleus is a crop of the tropics. It is grown in latitudes between 8 and 31°N. It is found growing well on barren hills at an attitude of about 2400 m. It prefers humid climate with relative humidity of 60 to 95% and a temperature of 10 to 25°C is suitable. It is grown in areas receiving mean annual rainfall of 1000-1600 mm. It requires a reasonably good evenly distributed rainfall and cannot withstand drought conditions.

Soil: It can be grown in well-drained soils. Sandy loam to alluvial soil, rich in organic matter is ideal while heavy clay soils are not suitable. It can be grown on soils with marginal fertility. The red sandy loam soils having a pH of 5.5-7 are ideal for the cultivation. The crop cannot withstand waterlogging or flooded soil conditions as excess soil moisture reduces tuber yield considerably.

Season: The best period for planting is during June-July and September-October.

Varieties: Manganiperu, Maimul, Garmai, Selection K-8, Sree Dhara.

Land preparation: The soil is ploughed to a depth of 20–25 cm two to three times. Ridges and furrows can be formed.

Propagation: The crop is propagated through seeds, suckers, stem cuttings or tubers. Coleus can be grown by seeds and stem cuttings of 10-15 cm long cuttings comprising 3-4 pairs of leaves. Normally 0.4-1.2 tonnes of tubers are planted in 0.2 ha of land to produce planting material (sucker) for a hectare crop. Nursery beds 15 cm high and 1.0 m wide are prepared and tubers are planted at 5 × 15 cm spacing, 4 cm below the soil surface. Sprouting starts in 15 days and it grows 15-20 cm long after 3 weeks. The cuttings initiate sprouting and ready for transplanting within a month. These rooted plants are suitable for planting in the main field at the distance of 60 × 45 cm, 60 × 30 cm or 20 × 20 cm. The rooted propagules are planted in the soil up to a depth of 15 cm using the crowbar with atleast three nodes inside the soil and remaining one to two nodes above the soil.

Weed management: Weeding and hoeing is to be done at regular interval. Weeding and earthing up on 30 and 60 days after planting. The second earthing up may be given one month after the first earthing up.

Manures and fertilizer application: Compost/ FYM @ 12.5 t/ha or vermicompost at 4 t/ha is recommended. A fertilizer dose of 50-75-50 kg of N, P_2O_5 and K_2O per ha is applied as basal dressing. Half the dose of N, the whole P and whole K may be applied as the basal dose followed by the remaining half N at 30 days after planting as top-dressing. Apply $ZnSO_4$ @ 10 kg/ha to avoid deficiency.

Water management: Irrigation is given immediately after transplanting, if there are no rains. During the first 2 weeks after planting, the crop is irrigated once in 3 days. Afterwards the crop is irrigated as and when required weekly or fortnightly interval.

Pest and diseases management: The major insects infesting coleus are leaf eating caterpillar, mealy bug and root knot nematodes and the major diseases are bacterial wilt, leaf spots, leaf blight and root rot. The leaf eating caterpillars, mealy bugs can be controlled by spraying the plants with 0.1% methyl parathion. *Trichoderma viride* @ 5 kg/ha is mixed with FYM and applied twice at 20 days interval to control root rot. Dipping the terminal cuttings in Carbendazim solution (1 gram per litre) before planting or Mix 5 kg of *Trichoderma viride* in 250 kg of compost and apply around the roots in every 20 days interval to control Fusarial wilt. Drenching Streptocyclin 300 ppm solution around the roots or application of *Pseudomonas fluorescens* @ 5 kg/ha mixed with 300 kg of FYM/Compost to control bacterial wilt. Apply 500 kg/ha of neem cake before planting or apply 15 to 20 kg/ha of Carbofuran to control the nematode infestation.

Harvest: Coleus is a short duration crop of 4-6 months. Flowers if any may be nipped off during the growing period to obtain more biomass of roots. The tubers can be harvested when all leaves and stem begin to wither. The crop is harvested manually by uprooting the individual plants. The harvesting should not be delayed as the mature tubers decay rapidly if left in the field. Harvested tubers are cut into small pieces and dried under shade until it reaches 8 to 12 % moisture level.

Yield: The tubers are separated, cleaned, chopped into pieces and shade dried yielding about 12% of the dry matter. The tubers contain 0.3 to 0.5% forskolin. The fresh tuber yield is 15 to 20 t/ha over a period of 6 months while the dry tuber roots yield is 1.5 to 2.2 t/ha. The roots contain 0.5% to 1.4% oil, on air dry weight basis. The seed yield is 4.47 g per plant.

Utilization: The tuber can also be baked or made into chips. Coleus is used to treat dysentery and digestive disorders. Its tuberous roots are found to be rich source of an alkaloid called Forskolin. Forskolin is reported to be useful in the treatment of congestive heart failure, glaucoma, asthma and certain type of cancers. Forskolin is used in preventing hair greying and restoring grey hair to its normal colour.

REFERENCES

Ammon, H.P.T. and Muller, A.B. 1985. Forskolin: from an ayurvedic remedy to a modern agent. Planta Med., 46: 473-477.

Kavitha, C., Rajamani, K. and E. Vadivel. 2010. *Coleus forskohlii*: A comprehensive review on morphology, phytochemistry and pharmacological aspects. Journal of Medicinal Plants Research, Vol. 4(4): 278-285

Rajamani, K. and Vadivel, E. 2009. Marunthu Kurkan –Medicinal Coleus. In: Naveena Mulikai Sagupaddi Thozhil Nuttpangal, Tamil Nadu Agricultural University, Coimbatore, pp. 17-22

Rakshapal Singh, Surendera P. Gangwar, Deepmala Singh, Rachana Singh, Rakesh Pandey and Alok Kalra. 2011. Medicinal plant *Coleus forskohlii* Briq.: Disease and Management. Medicinal Plants. 3(1): 1-7.

Reddy, N.S. 1952. Chromosome numbers in Coleus. J. Hered. 43:233-237.

Selima Khatun, Narayan Chandra Chatterjee and Ugur Cakilcioglu. 2011. Antioxidant activity of the medicinal plant *Coleus forskohlii* Briq. African Journal of Biotechnology. 10(13): 2530-2535.

Valdes, L.J., Mislankars, S.G. and Paul, A.G. 1987. *Coleus barabatus* (Lamiaceae) and the potential new drug forskolin (Colenol). Eco. Bot., 41: 474-483

Veeraragavathatham, D., Venkatachalm, R. and Sundararajan, S. 1985. Performance of two varieties of *Coleus forskohlii* under different spacing levels. South Indian Hortic. 33: 389-392.

CHAPTER

56

Dill–*Anethum graveolens* L.

Synonyms: *Peucedanum graveolens* L.

Family: Umbelliferae/Apiaceae

Vernacular names: Dill is known as Sowa. The genus name 'Anethum' is derived from Greek word 'aneeson' or 'aneeton' which means strong smelling.

Importance: Dill is used as pot herb. The seeds are used as a spice and its fresh and dried leaves are used as condiment. The aromatic herb is commonly used for flavouring and seasoning of various foods such as pickles, salads, sauces and soups. It has been used in medicinal purposes. The leaves, stems and roots were rich in tannins, terpenoids, cardiac glycosides and flavonoids. Two flavonoids quercetin and isoharmentin, have been isolated from *A. graveolens* L. seed which have antioxidant activity and may help to prevent peptic ulcer.

Origin: It is indigenous to South-east Europe, Russia and Mediterranean.

Distribution: It is cultivated in India, Malaysia, Turkey, Russia, USA, Japan and Europe. It is cultivated commercially throughout the country.

Botany: It is an annual herb, grows up to 40-60 cm tall, with slender hollow stems; leaves are alternate, finally divided three or four times into pinnate sections. The flowers are white to yellow. The flower develops into umbels. Insects, bees and wasps are attracted to the yellow flowers of anethum for plant resources like nectar and pollens. The seeds are not true seeds. The dry fruits called schizocarps. Dill fruits are oval, compressed, winged about one-tenth inch wide. The seeds are smaller, 4–5 mm long and 1 mm thick in size, flatter and lighter than caraway and have a pleasant aromatic odour.

Climate: It is a cold weather crop. It requires warm to hot summers with huge sunshine levels at maturity. Even partial shade will reduce the yield substantially. The plant quickly runs into seeds in dry weather.

Soils: Dill prefers rich well-drained, loose soil and full sun. It tolerates a pH in the range 5.3 to 7.8.

Season: It can be sown in September-October.

Land preparation: The land was ploughed once with mould board plough and harrowed twice to bring the soil to fine tilth.

Propagation: It often self-sown when growing in a suitable position. Propagation is through seeds. Seed rate is 5 kg/ha. Seeds are viable for 3–10 years. Spacing is 45 × 15 cm.

Weed management: Two hand weeding are given on 20-25-days and 40-45 days after sowing.

Manures and fertilizer application: FYM @ 20 t/ha is applied at the time of land preparation. A fertilizer dose of 50-20-20 kg of N, P_2O_5 and K_2O per ha is applied as basal dressing.

Water management: The crop can be irrigated once in 7 to 10 days depending on soil and climatic conditions.

Pests and diseases management: Root rot, powdery mildew are controlled with dusting sulphur @ 25 kg/ha.

Harvest: The seed is harvested by cutting the flower heads off the stalks when the seed is beginning to ripe or milky wax stage.

Yield: Seed yield is 0.8 to 1.0 t/ha. The dill oil which is pale yellow in colour, darkens on keeping, with the odour of the fruit. The oil content is 2.5 to 3.5% in fruits and 0.55-0.60% in whole plant; its specific gravity varies between 0.895 and 0.915. Carvone and limonene are monoterpenes are present in dill oil obtained from fruits. The odour of dill herb is caused α-phellandrene, dill ether and myristicin are the compounds. Dill oil is extracted from seeds, leaves and stems, which contains an essential oil used as flavouring in food industry.

Utilization: It is used as medicine is in abdominal discomfort, colic and for promoting digestion. It cures 'vata', 'kapha', ulcers, abdominal pains, eye diseases and uterine pains. It is used in perfumery to aromatize detergents and soaps. It is thought to protect against witchcraft. Greeks covered their heads with dill leaves to induce sleep. The oil is used as repellent and toxic to growing larvae and adults insect pest. It stimulates milk flow in lactating mothers, and is often fed to cattle for this reason.

REFERENCES

Blank I, Grosch W. 1991. Evaluation of potent odorants in dill seed and dill herb (*Anethum graveolens* L.) by aroma extract dilution analysis. J. Food Sci.56: 63–67.

Carrubba A, Torre R, Saiano F, Aiello P. 2007. Sustainable production of fennel and dill by intercropping. Agro Sust. Develop. 28:247–256.

Huopalathi R, Linko RR. 1983. Composition and content of aroma compounds in dill, *Anethum graveolens* L., at three different growth stages. J. Agri. Food Chem. 31:331–333.

Jana, S. and Shekhawat, G. S.. 2010. *Anethum graveolens*: An Indian traditional medicinal herb and spice. Pharmacogn Rev. 4(8): 179–184.

Raghvan B, Abrahman KO, Koller WD. 1994. Shankarnarayanan ML. Studies on flavor changes during drying of Dill (*Anethum sowa*. Roxb) leaves. J. Food Qual. 17:457–66.

CHAPTER

57

Glory Lily–*Gloriosa superba* Linn

Family: Liliaceae

Vernacular names: Malabar glory lily, Superb lily, Climbing-lily, Creeping-lily, Flame lily, Glory-lily, Gloriosa lily, Tiger claw, Isimiselo, Vlamlelie (English); *Kalihari, Languli* (Hindi); *Agnimukhi, Agnisikha, Langali, Visalya* (Sanskrit); *Kazhappai kizhangu, Nabhikokodi* (Tamil); *Adavi-nabhi, Kalappagadda, Ganjeri* (Telugu); *Agnishike, Nangulika, Agnishike, Gowrihoovu, Akkatangiballi* (Kannada); *Manthori khizangu, Mettoni, Kithonni (Malayalam); Kallavi, Indai, Karianag, Khadyanag* (Marathi); *Bishalanguli, Ulatchandal* (Bengali); *Dudhio, Vacchonag* (Gujarati); *Kariari, Mulim* (Punjabi).

Importance: It is propagated as an ornamental in temperate countries. It is best suited to greenhouses. All parts of the plant especially the tubers and seeds contain alkaloids such as colchicines and gloriosine.

Origin: It is native of tropical Africa.

Distribution: It is grown in India, Bangladesh, Malaysia, Myanmar, China, Madagascar, Sir Lanka and USA. In India, it is grown in Tamil Nadu, Andhra Pradesh, Karnataka, Kerala, Maharashtra and West Bengal.

Botany: It is an herbaceous, tendril climber, perennial, growing between 3.5 to 6 m in length, usually trained at 1.5 m above the ground level. The vines are semi hard stemmed with white tuberous roots/rhizome. Leaves are sessile and alternate with tendrils formed at the tip of the leaves. Flowers are bright, solitary, at first greenish later becoming yellow and finally scarlet. The fruit is oblong containing about 20 globose red colored seeds in each valve. A single plant produces 75 -100 flowers and a single fruit contains 70 -100 seeds.

Climate: It is a tropical plant and comes up well in warm, humid regions. An annual rainfall of 2000 to 3500mm, well distributed throughout the year, is suited to the crop. It occurs and cultivated up to 2400 m from MSL.

Soils: The plant grows well in sandy loam soil, having pH 5.5 to 7.0. Good drainage is essential. Waterlogged soils should be avoided. It is grown in red or black loamy soils with medium water holding capacity and good drainage.

Season: Planting is can be done during July-August.

Varieties: Singaleri selection possesses 0.25 % Colchicine.

Land preparation: The field should be ploughed and harrowed three to four times to bring the soil to a fine tilth. The field must be leveled properly for making adequate drainage arrangements. Ridges and furrows are formed with 60cm spacing.

Propagation: It can be grown by seeds and tubers but plants are mainly raised from tubers. The vigour of the vine and flowering and fruiting ability depends on the size of the rhizomes. The optimum weight of the rhizomes should be about 50-60 g. The plants from small rhizomes do not flower during the first year. The rhizomes can be treated with Carbendazim @ 0.1% to control tuber rot. Treated rhizomes are planted in furrows with 60 cm × 45 cm or 60 cm × 30 cm at a depth of 6-8 cm. The tubers/rhizomes of 1.5 to 2.0 t/ha are required to plant one ha.

After cultivation practices: The plant requires support, since the stem is tender. When the plants are about 30-40 cm tall, they should be staked or tied to wires or allowed to climb on some sort of frame.

Hand pollination: Hand pollination is required due to peculiar position of stigma and anthers. Muslin cloth or cotton tied sticks can be used for pollination. Morning time between 7-10 am is preferred for hand pollination.

Weed management: It requires 4-5 weeding in initial stages. Utmost care should be taken to avoid any damage to the growing tips as once damaged it does not sprout again during the season.

Manures and fertilizer application: Apply 12.5 t/ha of FYM and 60-25-40 kg/ha of N, P_2O_5 and K_2O is recommended. Of the nutrients, the whole P_2O_5 and K_2O and one-third of nitrogen are applied as a basal dose and the remaining two-third of nitrogen is given in the first six to eight weeks after planting.

Water management: Irrigation is given immediately after planting. An irrigation interval of 7 days during initial period and later on at an interval of 15 days is recommended. It does not withstand continuous moisture stress. Excess watering is harmful to the plant. On an average basis, a plant requires 5 litres of water per day. No irrigation is required after flowering.

Pests and diseases management: Lily caterpillar and Green caterpillar can be controlled with spray of Metacid @ 0.2% or Dichlorovas @ 0.2 % at fortnightly intervals. Proper drainage can be given during rainy season and drench the soil with Bavistin at 0.2% in order to control rhizome rot/tuber rot. Leaf blight can be controlled by spraying 0.3 per cent Dithane M-45 (or contaf 10 ml/10 litres of water) at fortnightly intervals.

Harvest: Harvest the pods when its colour starts turning to light green from dark green and the skin of the fruit shrinks. At this stage, when pressed the pod gives a cracking sound. Pods are harvested at 160 to 180 days after sprouting of tubers. Dry the pods in shade for 10-15 days after picking. Then, the fruits turn yellow and open out showing deep orange yellow coloured seeds. Collect the seeds, dry them in shade again for a period of one week before packing them for storage. The tubers are harvested after 5-6 years of plantation, cut into small pieces and dried in shade.

Yield: The yield of seed is 250-300 kg/ha/year. About one tonne of tuberous roots is harvested after five years of the plantation (i.e. at the end of economic life of the plantation).

Utilization: The economic part is seed and rhizome. The major plant constituents are colchicine (0.5-0.7%) and colchicoside. The colchicine content varies from 0.15 to 0.3% in the rhizomes and 0.7 % to 0.9% in the seeds. It's rhizomes are used as a tonic, anti-periodic, anti- helminthic and also against snake bites and scorpion stings. The leaves when applied in the form of a paste to the forehead and neck to cure asthma in children. The leaf juice is used against head lice.

REFERENCES

Alok Jain and Satish Suryavanshi, 2010. *Gloriosa superba* Linn: A pharmacological review, IJPRD, 2(8): 24-28.

BhargavBhide and Rabinarayan Acharya. 2012. Uses of Langali (*Gloriosa superba* Linn.): An Ethnomedicinal Perspective, Ayur. Pharm. Int. J. Ayur. Alli. Sci., 1(3), 65–72.

Bhushan Pawar, Vishal Wavhal, Nayana Pawar, Mohan Agarwal and Prashant Shinde. 2010. Anthelmintic Activity of *Gloriosa superba* Linn (Liliaceae), Int. J. Pharm. Tech. Research., 2(2), 1483-1487.

Kavithamani, D, Umadevi, M. and Geetha, S. 2013. A review on *Gloriosa superba* as a medicinal plant. Indian Journal of Research in Pharmacy and Biotechnology. 1(4): 554-557.

Rajendran K, Balaji P, JothiBasu M. 2008. Medicinal plants and their utilization by villagers in southern districts of Tamil Nadu, Indian J. Traditional Knowledge., 7(3), 417-420.

Ashok Kumar Khandel, Sujata Ganguly and Amarjeet Bajaj. 2012. *Gloriosa superba* L. (Glory lily) spotted for the first time in vegetation of Pachmarhi Biosphere Reserve (Hoshangabad district), Central India. Int. J. of Pharm. & Life Sci. 3(6): 1725-1732.

CHAPTER

58

Guggal–*Commiphora wightii* (Arn.) Bhandari

Synonym: *Commiphora mukul* Hook.

Family: Burseraceae

Vernacular names: Indian Bedellium tree, Corkwood, Commiphora (English); *Guggul, Guggal, Muqil* (Hindi); *Guggul* (Bengali); *Guggaru, Guggal* (Gujarati); *Gulgulu, Guggalu* (Malayalam); *Guggul, Mahishaksh* (Marathi); *Guggal, Kanthagana, Mahishaksha guggulu, Guggulugida,* (Kannada); *Maishakshi, Gukkal* (Tamil); *Makishakshi guggulu, Guggipannu (*Telugu). The generic name is derived from Greek 'kommis' and 'phora' meaning gum bearer'. The Sanskrit definition for the term 'Guggul' is 'one that protects against diseases'.

Importance: Guggul is known for oleogum-resin, which has immense medicinal value. The wood and bark are used as toothbrush which helps to cure of pyorrhoea and other tooth and gum troubles. Guggul resin is used for the treatment of asthma, boils, ulcers and headache. Fumigation and inhalation of resin is used to cure nasal catarrh, bronchitis and typhoid. Resin is used as mosquito repellent and its fumes as a cure of typhoid.

Origin: It is indigenous to India.

Distribution: It is distributed in India, Pakistan, Sri Lanka, Madagascar and South Africa. It is grown in semi-arid states of Karnataka, Gujarat, Rajasthan and Madhya Pradesh.

Botany: It is a deciduous tree. It is a slow growing shrub or small tree takes 8 to 10 years to come to a height of reaching up to 3 to 4 m height. It is much-branched, crooked and spirally ascending branches ending in sharp spines. Leaves are sessile, alternate, 1-3 foliate, highly aromatic, leathery, shinning green on top and greyish below with irregularly toothed edges. Flowers are small in fascicles, brownish red in colour. Fruit an ovoid green berry like drupe, reddish, 6-8 mm in diameter. Seed generally contain an under developed embryo.

Climate: It is distributed in tropical and sub-tropical regions. The tree is found in rocky and open hilly areas or rough terrain and sandy tracts in warm and semiarid to arid areas. It is a xerophyte plant. It is grown up to an altitude of 250-1800 m with mean annual rainfall of 250-500 mm. It grows well in areas with temperature range of 20 to 35°C.

Soils: It can be cultivated in sandy to silt-loam soils.

Season: Planting can be done during July-August.

Varieties: Marusudha,

Propagation: The plants are raised from stem cuttings or through air layering of 5 to 8 year old mother plants. The cuttings initiate sprouting in l0-15 days. These rooted plants are suitable for planting in the field during next rainy season. The cuttings give 80-95% sprouting. One to four seedlings emerge from a single seed due to polyembryony. Only the black-coloured seeds are viable. Seed germination is only 40%. However, due to poor germination, two seeds are sown in each polybags. Seeds germinate within 7 to 10 days after sowing in the rainy season. Approximately, 100 g of black seeds are required for raising plantation in 1 hectare of land at a spacing of 2 m × 2 m or 4 m × 4 m.

Weed management: Weeding is necessary for 2-3 years after planting.

After cultivation practices: Pruning or removal of branches in early stages helps to achieve good growth, increase in girth of growing branch and thereby high gum yield.

Manure and fertilizer application: Application of FYM @ 5 kg/plant/year and a fertilizer dose of @ 75:130:30 g/plant of N, P_2O_5 and K_2O is to be applied after every three months up to a period of one year.

Water management: It requires moderate irrigation. Even limited irrigation during summer season enhances the rate of growth. Irrigation with 8 litres of water per plant at an interval of 15 days is practiced.

Insect pests and diseases management: The plant is infected at collar rot when the water remains stagnant near the base at early growing stage. The condition may be controlled by avoiding stagnation of water around the base of the plant and spraying Diathane M-45 @ 2 g/litre of water on the crop. Termite can be controlled successfully by drenching the soil with chloropyrophos @ 4 ml/litre of water.

Harvest: Harvesting of Guggul is done twice in a year as its gets gum secreted in summer and winter seasons. As a traditional practice of gum tapping is very crude and destructive. Gum tapping leads to killing of trees. The oleo-gum resin is tapped by making a 7 to 10 cm long incision in the main stem near the base with the help of a sharp half circular knife. The exuded gum secreted is collected every week up to one month's time. If the sapling is irrigated and nurtured well, the secreted resin from the tree can be expected to be available from the 7th year onwards. However in normal course the gum tapping is expected after 15 year. The oozing of gum starts and continues till exhaust. The dried resin extract (gum) is then collected.

Yield: The bark incised to extract an oleo resin gum called 'guggul'. The dry gum yield is 500-800 g/plant/year. A yield of 120 to 130 kg oleo-gum resin is obtained per hectare after about eight years.

Utilization: The oleo-gum resin of guggal contains guggulosterol, guggulosterone, myrcene and dimyrcene which helps in the treatment of obesity and arthritis.

REFERENCES

Satyavati GV. 1988. Gum guggul (*Commiphora mukul*): The success of an ancient insight leading to a modern discovery. Indian J. Med. 87:327-35.

http://www.worldagroforestry.org/treedb/AFTPDFS/Commiphora_wightii.PDF

http://www.plantzafrica.com/plantcd/commiphora.htm

CHAPTER

59

Insulin Plant–*Costus igneus* Nak

Synonyms: *Costus pictus* D. Don, *Costus mexicanus* Liebm, *Costus congenitus* Rowle

Family: Costaceae

Vernacular names: Insulin plant, Spiral-Flag, Fiery costus, Step ladder, Crepe ginger (English); *Keukand, Keu* (Hindi and Bengali); *Kostum* (Tamil); *Kemuka, Kushta, Kashmira, Shura, Katar katar* (Sanskrit); *Pushpamoola* (Kannada); *Kashmeeramu* (Telugu); *Penava, Pushkarmula* (Marathi); *Pushkarmula* (Malayalam); *Pakarmula* (Gujarati); *Jom lakhuti* (Assamese).

Importance: It is cultivated for its medicinal uses and elsewhere as an ornamental. It is used as an herbal cure for diabetes and hence commonly called as 'insulin plant'. The diabetic people eat one leaf daily to keep their blood glucose low.

Origin: Insulin plant is native of South America and Central America.

Distribution: It is grown in South America, Central America, India, Indonesia, Malaysia and Sri Lanka. Within India, it is grown in Kashmir, Himachal Pradesh, Assam, Gujarat, Maharashtra, Karnataka, Andhra Pradesh, Tamil Nadu and Kerala.

Botany: It is a perennial, upright, grows up to 60 cm height, with the tallest stems falling over and lying on the ground. The underside of the dark green leaves have light purple shade. The leaves are spirally arranged around the stem, arching clumps arising from underground rootstocks. Leaves are simple, alternate, oblong, 10 to 20 cm in length with parallel venation. The flowers are 3 to 4 cm diameter, orange in colour and appear to be cone-like heads at the tips of branches. Fruits are green in colour.

Climate: It grows well in the climate with high humidity and low temperature. It needs sunshine. It grows well in shady area under mixed deciduous forests. It is moderately drought tolerant.

Soils: It needs fertile and moist loam to clayey loam soils. It is often planted near water. It is can tolerate acidic, alkaline, clay, sand or loamy soils.

Season: It is mainly cultivated in rainy seasons.

Propagation: Propagation is with rhizome or stem cuttings or offsets or plantlets that form below the flower heads. The spacing is 45 cm × 30 cm or 60 cm × 30 cm. The birds disperse seeds.

Weed management: Two hand weeding are given on 20-25-days and 40-45 days after sowing.

Manures and fertilizer application: FYM @ 20 t/ha is applied at the time of land preparation. A fertilizer dose of 50-20-20 kg of N, P_2O_5 and K_2O per ha is applied as basal dressing.

Water management: The crop can be irrigated once in 7 to 10 days depending on soil and climatic conditions.

Insect pests and diseases management: Mites, caterpillars and nematodes can be a problem, especially on light, sandy soil. The plant has no diseases are of major concern.

Utilization: The leaves are used to treat diabetes. The plant used to treat fever, rash, asthma, bronchitis and intestinal worms. It is used as an ingredient in a cosmetic to be used on the eyelashes to increase sexual attractiveness. The plant prevents the body from disease, protects mind and prolongs the longevity of life. The root is used as in the form of powder, decoction and oil. The plant contains resinoids, essential oil and alkaloid named saussurine, inulin and resin.

REFERENCES

Eevera, T., Pazhanichamy, K., Pavithra, S., Rubini, S., Lavanya, B. and Ramya, I. 2010. Morphological, anatomical and proximate analysis of leaf, root, rhizome of *Costus igneus*. Journal Int. Pharm. Res. 2010;3:747–52.

CHAPTER

60

Isabgol–*Plantago ovata* Forsk.

Family: Plantaginaceae

Vernacular names: Psyllium, Blonde Ispaghula, Indian Plantago, Isabgol, Ishagola, Isabghul, Spogel seed, (English); *Isabgol, Isabgul, Ispaghal* (Hindi); *Snigdhabijah, Snigdhajirakah* (Sanskrit); *Iskol, Isphogol* (Tamil).

Importance: The economic part is seed and husk and are used for medicine. The husk yields a colloidal mucilage consisting mainly xylose, arabinose and galacturonic acid. The name *Plantago ovata* comes from the Persian words 'isap' and 'ghol' that mean horse ear, which is descriptive of the shape of the seed. Isabgol is known as Psyllium, originated from a Greek word for a flea, referring to the size, shape, and whitish colour of the seed. The husk and peel are exported from India to USA, West Germany, the United Kingdom and France.

Origin: It is indigenous to the Persia, Mediterranean region and West Asia.

Distribution: It is grown in India, Pakistan, Baluchistan, Mediterranean countries, European countries, Russia, USA and Mexico. The main producing states are Gujarat, Rajasthan, Madhya Pradesh, Haryana, Punjab, Uttar Pradesh, Bihar and Karnataka in India.

Botany: It is a stem less annual herb that grows up to a height 30 to 40 cm. The root system has a well-developed tap root with few fibrous secondary roots. Leaves are alternate. The plant bears erect ovoid or cylindrical spike with white flowers. Fruit is a capsule. The seeds are enclosed in capsules which open at maturity. The seed husk is thin, boat shaped, white, translucent, odourless with mucilaginous taste.

Climate: The crop requires cool and dry climate during the growing season. Heavy dew or even a light shower will proportionately decrease the yield, at times leading to even total loss of the crop. It is grown in areas receiving annual rainfall of 500 to 1250 mm and average temperatures of 17 to 40°C. The optimum temperature for germination and growth is 20 to 30°C.

Soils: It grows well on light soils. Soil with poor drainage is not desirable. A well-drained sandy loam, loam and silty-loam soil having a soil pH from 4.7 to 8.0 is ideal for growth of plants and high yield of seeds.

Season: It is a *rabi* season crop. Sowing of seeds from last week of October to second week of December is recommended. Yield reduction is observed when sowing is delayed beyond first fortnight of December. The crop is grown from November-December to February-March.

Varieties: RI-87, RI-89, AMB-2, GI-1, GI-2, MI-4, MIB-121, HI-34, HI-2, HI-1, HI-5, JI-4, Niharika. Gujarat Isabgol-1, Gujarat Isabgol 2, Gujarat Isabgol 3, TS-1-10, EC-124345, Niharika, Haryana Isabgol -5 and Jawahar Isabgol-4.

Land preparation: The field is ploughed thrice followed by harrowing. Plot size of 8.0 m × 3.0 m is desirable.

Seeds and sowing: Seed rate is 4 to 6 kg/ha. Seed is sown after pretreatment with thiram @ 3 g/kg of seed to protect the seedlings from damping off. The seeds are broadcasted. After sowing, seeds are covered thinly by raking the soil. The sowing should immediately be followed by irrigation. Germination begins in four days after sowing. Line sowing at 30 × 5 cm spacing is recommended over broadcasting.

Weed management: Two hand weeding are taken on 20-25 days and on 60-70 days after sowing. Pre-emergence application of Isoproturone @ 0.5-0.7 kg ai per ha recommended to control the weeds.

Manures and fertilizer application: FYM @ 12.5 t/ha and fertilizer dose of 25-25-12.5 kg N, P_2O_5 and K_2O as basal dressing is recommended. Again 25 kg N/ha is given as top dressing at 30-40 DAS.

Water management: It is an irrigated crop and requires 7 to 10 irrigations. First irrigation should be given immediately after sowing. The seeds normally germinate in 6-7 days. Later on irrigations are given as and when necessary. Last irrigation should be given at the time when maximum number of spikes has reached the milk stage.

Cropping systems: Maize-Isabgul, Sorghum-Isabgul, Pearl millet-Isabgol, Groundnut-Isabgol, Soybean-Isabgol and Pigeon pea-Isabgol rotations are practiced.

Pests and diseases management: Spray of 0.025% oxydemeton methyl to control aphids. Application of 5% Aldrin or Lindane at 25 kg/ha at the time of last ploughing to control white grubs. Seed treatment with Captan @ 5 g/kg of seed followed by drenching the soil and spraying the plants with 0.2 per cent Captan solution and repeating the same a week after first application controls the spread of rhizoctonia wilt. Spray Bavistin at 0.1% or 0.2% wettable sulphur at 15 days interval two or three times to control powdery mildew.

Harvest: The crop will be ready in about 110-130 days after sowing. The plant bears the flowering spikes in about 60 days after sowing and matures in the next 2 months. When mature, the crop turns yellowish and the spike turns brownish. The yellowing of the lower leaves is an indication of maturity. The seeds shed when the spikes are pressed even slightly. At the time of harvest the atmosphere must be dry and there should not be any moisture on the plant. The crop can be harvested after 10 am. Harvested plants are sun dried for 2 days and then threshed with tractor or trampled by bullocks, winnowed and separated seed is collected. The dry seeds can be stored for 8-10 years.

Yield: Seed yield is 800-1000 kg/ha. The husk yield is 250-350 kg/ha. The husk:seed ratio is 25:75 by weight. Psyllium husk is obtained by milling the seed to remove the hulls. Mucilage yield amounts to approximately 25% or more (by weight) of the total seed yield. Plantago seed mucilage is often referred to as husk or psyllium husk. The milled seed mucilage is a white fibrous material that is hydrophilic (water-loving). Mucilage is colourless. The seeds are dry and can be stored for 8-10 years. The straw yield is twice of the seed yield which is about 1200-1600 kg/ha.

Utilization: The seed and seed husk (epicarp of seed) are used for medicinal uses. The seed and husk are used for cure of constipation, dysentery and genitourinary tracts and against irritation.

REFERENCES

Stephen C. M. and Aaron, L. 2008. The Biogeography of *Plantago ovata* Forssk. Int. J. Plant Sci. 169 (7): 954–962.

CHAPTER

61

Kacholam–*Kaempferia galanga* Linn (2n=54)

Family: Zingiberaceae

Vernacular names: Galanga, Cekor, Kencur, Aromatic Ginger, Sand Ginger, Resurrection lily, Cutcherry (English); *Chandramula, Abhuyicampa* (Hindi); *Chandramulika, Sugandhavacha, Bhucampaka* (Sanskrit); *Chandumula, Ekangi* (Bengali); *Chengazhinirkizhangu, Kacholam, Kacchuri, Kacchoram* (Malayalam); *Kacholam* (Tamil); *Chandramula* (Telugu).

Importance: It can be cultivated as an under crop in rubber or coconut plantations. The rhizomes are used in perfumes and cosmetics.

Origin: It is native only to India and Myanmar.

Distribution: It is cultivated in India, China, Sri Lanka, Indonesia, Java, Malaysia, Philippines, Nigeria and Mexico. In India it is cultivated mainly in Kerala, Karnataka, Tamil Nadu and West Bengal.

Botany: The plant is a perennial stem less herb. Leaves are spreading flat on the ground, round-ovate, deep green, with short petioles. Flowers are white with purplish spots in axillary fascicles and corolla tube 2.5 cm long. Fruits are oblong, 3-celled and 3-valved capsules. Seeds are arillate. The underground rhizome is globose with prominent secondary tubers and bulbous roots. The rhizomes has camphoraceous odour.

Climate: It is widespread in tropics and subtropics. It requires a warm humid climate. It thrives well up to an elevation of 1500 m from MSL. It is grown in areas with well distributed annual rainfall of 1500 to 2500 mm during growing period. It is a shade loving plant. Hence, partial shade condition should be provided for better vegetative growth.

Soils: The loamy soil with good drainage is preferable. Laterite soil with heavy application of organic matter is also suited. It is susceptible to waterlogging.

Season: Planting of the rhizome pieces during May-July is recommended.

Varieties: Kasthhuri and Rajani varieties available for cultivation.

Land preparation: Land is ploughed and beds of 1-2 m width, 25 cm height and convenient length can be taken.

Seeds and sowing: Seed rate is 500 to 750 kg/ha of tuberous rhizomes. Seed rhizomes are stored in cool dry place or in pits dug under shade. Smoking of rhizomes prior to planting is done for good germination and establishment of sprouts. Tubers are planted at 30 × 15 cm spacing.

Weed management: Weeding is done 25, 45 and 90 days after planting which is followed by fertilizer application and earthing up.

Manures and fertilizers application: Application of 25 t/ha of FYM or compost and a fertilizer dose of 50-50-50 kg of N and K_2O per ha is recommended.

Water management: The crop is irrigated once in 7 to 10 days interval depending on soil and climatic conditions.

Pest and diseases management: Leaf spot and rhizome rot diseases occur particularly during the rainy months which could be controlled by drenching and spraying with 1% Bordeaux mixture.

Harvest: Crop is harvested 6 to 7 months after planting when leaves start drying up. Rhizomes are dug out, cleaned and washed to remove adhering soil particles.

Yield: The rhizomes are dried for 3 to 5 days. The fresh rhizomes yield is 5 to 7 t/ha which on drying yields 1.7 to 2.2 t/ha. The sliced and dried rhizomes on steam distillation for 3 to 5 hours yield 2 to 3% of essential oil.

Utilization: The rhizome is stomachic and anti-inflammatory. The rhizome is used as a decoction or powder for treating indigestion, colds, swellings, pectoral and abdominal pains, headache and toothache. Leaves are used in lotions and poultices for sore eyes, sore throat, swellings, rheumatism and fevers. The oil is utilized in curry flavouring. It is also employed in cosmetics, mouth- washes and hair tonics. The major chemical constituents found in volatile oil of dried rhizome are as ethyl-p-methoxycinnamate (31.77%), methylcinnamate (23.23%), carvone (11.13%), eucalyptol (9.59%) and pentadecane (6.41%), respectively.

REFERENCES

Tewtrakul, S., Yuenyongsawad, S., Kummee, S., and Atsawajaruwan, L. 2005. Chemical components and biological activities of volatile oil of *Kaempferia galanga* Linn. Songklanakarin J. Sci. Technol. 27(Suppl. 2): 503-507

Sudipa Nag and Subrata Mandal. 2015. Importance of Ekangi (*Kaempferia galanga* L.) as medicinal plants- a review. International Journal of Innovative Research and Review. 3 (1): 99-106.

http://www.prota4u.org/protav8.asp?en=1&p=Kaempferia+galanga

CHAPTER

62

Kadukkai–*Terminalia chebula* Retz. (2n=24, 48, 72)

Family: Combretaeae

Vernacular names: Chebulic myrobalan, Black myrobalan, Yellow Myrobalan, Chebulic Myrobalan, Tropical almond, Gallnut, (English); *Harra, Harad, Halela zard* (Hindi); *Shilikha* (Bengali); *Haritaki* (Assamese); *Haritaki* (Sanskrit); *Harade* (Gujrati); *Alale* (Kannada); *Putanam, Katukka* (Malayalam); *Hirda* (Marathi); *Karadha* (Oriya); *Kadukkai* (Tamil); *Karaka, Karakkaya, Karkchettu* (Telugu);

Importance: It is popularly known as 'King of medicine' in Tibet and is always listed at the top of the list in Ayurvedic Materia Medica due to its extraordinary power of healing. Its fruit extract is exported which is used to improve leather quality. The generic name 'Terminalia' comes from Latin word 'terminus' or 'terminalis' (ending), and refers to the habit of the leaves being rowded or borne on the tips of the shoots.

Origin: It is native of India, Nepal, China, Malaysia, Vietnam and Sri Lanka.

Distribution: It is distributed throughout tropical and subtropical Asia, including India, Sri Lanka, Burma, Thailand, Malaysia, Java, China and Tibet. This tree is found in the forests of Uttar Pradesh, West Bengal, Assam, Madhya Pradesh, Tamil Nadu, Karnataka and Maharashtra.

Botany: A medium-sized, deciduous tree with umbrella-shaped crown, grows up to 25 m height. The bark is dark brown with some longitudinal cracks. Leaves are ovate and elliptical, with two large glands at the top of the petiole. The flowers are monoecious, white to yellow, with a strong unpleasant odour, borne in terminal spikes or short panicles. The fruit (drupe) is about 1-2 inches in size. Fruit is green when unripe and yellowish grey when ripe. Seed is globose, generally 2–6 cm long, and pale yellow in colour.

Climate: It is distributed throughout India. It grows up to an altitude of 1500 m from MSL. It is a light demander, but withstands some shade in early growth and may then benefit from bright sun light. It is tolerant of frost and drought, and withstands fire, recovering well from burning and also from coppicing. The average temperature ranging from 10 to 48ºC is suitable for its growth. It is grown in areas receiving annual rainfall is over 650 to 2500 mm, where maximum shade temperature varies from 32 to 48°C and minimum from -1 to 15°C. It is frost-hardy and drought-resistant crop. The plant tolerates flooding for 10-15 days after 4-5 years growth.

Soils: It is grown on different soils ranging from poor rocky ground to sandy, sandy loam, clayey loam, deep or shallow loam, lateritic loam and gravelly fertile alluvium soils.

Season: The planting is done during May-June with the onset of monsoon.

Propagation: It can be propagated through seeds. Seeds are can directly be sown or can be transplanted after raising seedlings in nursery. Direct sowing is risky due to predation Seeds possess 60% viable up to 12 months. Seeds can be soaked for 36 hours; also alternate wetting and drying helps to improve germination. Fermentation of the stones/ fruits improves germination. Treated seeds germinate on an average within 15-20 days of sowing. Seedling age of 75-90 days is used for planting in field. Transplanting of one-year-old seedling is more successful. Shading is desirable in early stages in the nursery and after transplanting.

Nursery management: The de-pulped seeds are sown in nursery beds at about 15 × 15 cm spacing. The nursery should be shaded against the sun. The fruit stones after removing the outer pulpy portion are dried and sown in nursery beds that are covered with soil and regularly watered.

Vegetative propagation: Propagation by cuttings is possible, but less successful than transplanting nursery raised seedlings. Grafting is helps to reduce the juvenile period and enhance early bearing. Bud grafting and cleft grafting are desirable. Grafting of young shoots on seedlings of the same species resulted success in root induction.

Spacing: Spacing is 8 m × 8 m in the main field.

Weed management: The weeds may be kept in check with the help of scythes or tractor-operated cutters.

Manures and fertilizer application: The soil of each pit is mixed with 15 kg FYM and a mixture of N, P_2O_5 and K_2O @ 75:30:30 g and refilled before transplanting of seedlings.

Water management: Irrigation in pit areas is required in the initial three to four years, depending on the soil moisture and season. The plants may be irrigated at least once a week in summers.

Cropping systems: Shade-loving crops like Curcuma, Zingiber, and Alpinia galangal may also be intercropped.

Pests and disease management: Fruits are much eaten by squirrels, rats, porcupines, hares and peacocks. Protection may be given against damages caused by rats, squirrels and rodents.

Harvest: Fruiting starts from 7-8 years after planting and the commercial harvest starts from 8 to 10th year. The trees live for more than 50 years and continue to yield fruits every year. Fruits are collected from the time they begin to turn yellow and ripe. Shaking the tree and picking up from the ground generally collects the fruits. The dropped fruits remain undamaged even after 5-7 days of fruit fall. The harvested fruits are dried in thin layers, preferably in shade with 10% moisture content and graded for marketing.

Yield: The tree yields up to 10 kg fruits/tree/year. The seed viability is retained for about one year. The kernels of the fruits contain 49 % fatty oil.

Utilization: Chewing the fruit causes increase in digestion power. If it is taken after food, it helps to eliminate all the toxic effects due to food poisoning. If it is taken along with salt, it balances Kapha; if taken with sugar, it balances Pitta and if taken with ghee, it balances Vata disorders. The fruits, the bark, the leaves are used to cure several ailments such as fever, cough, diarrhea, gastroenteritis, skin diseases, candidiasis, urinary tract infection and wound infections. The phytochemicals like polyphenols, terpenes, anthocyanins, flavonoids, alkaloids and glycosides are found in *Terminalia cbebula*.

REFERENCES

Jayaramkumar, K. 2006. Effect of geographical variation on content of tannic acid, gallic acid, chebulinic acid, and ethyl gallate in *Terminalia chebula* fruits. Nat Prod. 2 (3–4): 170–175.

Muhammad, S., Khan, B.A., Akhtar, N., Mahmood, T., Rasul, A. and Hussain, I. 2012. The morphology, extractions, chemical constituents and uses of *Terminalia chebula*: A review. J. Med Plants Res. 6 (33): 4772–4775.

Rathinamoorthy, R. and G. Thilagavathi. 2014. *Terminalia chebula* - Review on Pharmacological and Biochemical Studies. Int. J. Pharm Tech Res. 6 (1): 97-116.

http://database.prota.org/PROTAhtml/Terminalia%20chebula_En.htm

CHAPTER

63

Kuchla–*Strychnos nux-vomica* Linn.

Family: Loganiaceae

Vernacular names: Strychnine tree, Nux-vomica tree, Poison nut tree, Snake-wood, Strychnine Tree, Quaker Buttons (English), *Kuchla, Kupilu* (Sanskrit); *Bailewa, Chibbenge, Kajra, Azaraqi, Kuchila, Visamusti* (Hindi); *Eddikunchera, Kanchurai, Yetti* (Tamil); *Kushti, Musadi, Mushadi* (Telugu); *Kanhiram, Kanjera, Kanjiram* (Malayalam); *Hemusthi, Ittangi, Itti, Kajavara, Khasea, Yetti, Yettica* (Kannada); *Jharkatchura, Kajrakar, Kara, Karo (Marathi); Kuchla* (Gujarathi); *Kachila, Kuchila, Thalkesur* (Bengali). The Latin 'nux' is suggestive of a nut, but the fruit is a berry, botanically. the Latin 'vomica' for vomit means that they cause disgorging.

Importance: The dried seed is used for medicine. The use of strychnine is mostly used in baits to kill feral mammals, including wild dogs, foxes and rodents.

Origin: Kuchla is a deciduous tree native to Southeast Asia and Australia, especially India and Myanmar.

Distribution: It is grown in India, Sri Lanka, China, Vietnam, Taiwan, Cambodia, Malaysia Philippines, Southeast Asia countries and Australia.

Botany: It is a medium-sized tree with a short thick trunk. It grows to a height of 15 m. The wood is dense, hard white and close-grained. The young shoots are deep green in colour with a shiny coat. The matured branches are covered with a smooth ashen bark. Leaves are simple, opposite, orbicular to ovate, 6 to 10 cm long and 5 to 6 cm broad, coriaceous, glabrous, five-nerved, short stalked, oval shaped, shiny coated and smooth on both sides. The flowers are white or greenish white and fragrant. The flowers have a foul smell. Fruit are large with hard shell, glossy orange and 4-5 cm wide. The fruits are large, spherical, hard-rinded berries. The flesh of the fruit is soft and white with a jelly-like pulp. Fruit contains three to eight round, flattened, grayish seeds. These seeds are covered with silky hairs, are known as strychnine nuts and are hard and extremely bitter in taste. Seeds are flat, circular discs or slightly convex on one side, concave on other side. Seeds are 2.5 cm in diameter and 6 mm in thickness. Seeds are very hard, tough, difficult to pulverize.

Climate: Kuchla is a deciduous tree that grows in tropical areas and is distributed throughout in moist deciduous forests India and Southeast Asia. It is grown up to an altitude of 1400 m from MSL.

Soils: It grows over laterite, sandy and alluvial soil

Land preparation: The land is ploughed with disc harrow and tillers to achieve a fine tilth. Pits of size 45 cm × 45 cm × 45 cm are dug at a spacing of 5 m × 5 m which gives a stand of 400 trees per hectare. The pits are refilled with mixture of soil, sand and manure in 1:1:1 ratio.

Propagation: Seeds are the best material for propagation of kuchila plant. The collected seeds are dried in the sun after removing the pulp. Fresh seeds should be used for propagation. About 1 kg seeds are required to raise 1 hectare of plantation. Germination can be increased by treating the seeds with hot water of 50°C for 6 to 12 hours prior to sowing. Seeds are sown in polybags of size 25 cm × 20 cm, filled with soil, sand and FYM mixture. Seeds are directly sown in the polybags after appropriate pretreatment. The polybags are watered regularly so as to keep them moist. The seeds germinate in about 20 to 30 days. Sometimes the germination may continue up to 45 days. The seedling growth is very slow but roots grow very fast. The plant can also be propagated through cuttings. For vegetative propagation, stem cuttings can be prepared and planted under moist conditions after treating with rooting hormones.

Weed management: The area around the basin of the plant should be kept weed-free by hand weeding or spraying of Paraquat @ 0.8% or Glyphosate @ 0.4%.

Manures and fertilizer application: About 10 kg of FYM is mixed in the soil during refilling of pits before planting. Additional 10 kg manure may again be applied to the soil around the plants at the time of weeding. A total of 20 kg FYM/plant/year in subsequent years is recommended. This is to be applied in two split doses in June–July and September–October.

Water management: The crop needs no irrigation during the rainy season and saplings may be irrigated in dry weather, especially in the early years of growth. For matured trees, irrigation by ring method around tree base at a distance of 30 cm during summer months is desirable.

Pest and diseases management: No significant pest or disease is observed on the crop.

Harvest: The tree has a life span of 50 to 60 years. It takes about 15 to 20 years for the tree to initiate flowering. Mature fruits are manually collected and seeds from them are extracted, washed, dried in shade, and stored for trade. The seeds have the shape of a flattened disk completely covered with hairs radiating from the center of the sides.

Yield: The crop produces 50 to 75 kg of dry seeds per tree per year, a yield of 10 to 12 t/ha/year may be obtained from a 20 year old onwards.

Utilization: Kuchla is an herbal medicine, recommended for liver cancer, vomiting, abdominal pain, constipation, intestinal irritation, heart burn, eye diseases, depression, migraine headaches, problems related to menopause and respiratory diseases. It is used as a respiratory stimulant, rodenticide and for killing stray dogs, even wild animals. Strychnine and brucine are two poisonous/toxic alkaloids present in the dried seeds of Nux vomica. The strychnine and brucine content are 0.4% and 0.6% in the seeds; 1.7% and 2.8% in root bark, 0.3% and 0.4% in root-wood, 0.9% and 2.1% in stem bark, 0.5% and 0.01% in stem wood and 0.2% and 0.5% in leaves, respectively. Strychnine is a deadly poison with a lethal dose to humans of about 30 to 120 mg. Strychnine occurs as colourless, odourless, rhombic prisms, having an intensely bitter taste. Strychnine is a highly dangerous poison. Strychnine must only be used for the land baiting of wild dogs and foxes. The ancient texts of Ayurveda quoted that the *Visha* (poison) act as an *Amrita* (nectar) if utilized legitimately.

CHAPTER

64

Liquorice (Mulethi)–*Glycyrrhiza glabra* Linn.

Synonym: *Glycyrrhiza glandulifera*

Family: Fabaceae

Vernacular names: Liquorice, Sweet wood, Sweet licorice (English); *Mithilakdi, Jothi-madh, Mulathee, Mulethi* (Hindi); *Yashtimadhu, Madhukah* (Sanskrit); Yashtomadhu, Jashtimadhu, Jaishbomodhu (Bengali); *Jeshthamadh* (Marathi); *Atimadhuramu, Yashtimadhukam* (Telugu); *Jestamaddu* (Kannada); *Jethimadhu* (Gujarati); *Atimadhuram* (root) (Tamil); *Itarttimadhuram, Erattimadhuram* (Malayalam); *Jatimadhu* (Oriya);

Importance: It is also used as a flavouring shrub. The economic parts used are peeled roots, leaves and flowers. It is used as a diuretic, choleretic and used as insecticide. It is a traditional medicine for coughs, colds and painful swellings. Glycyrrhiza is derived from the ancient Greek term 'glykos', meaning sweet, and 'rhiza', meaning root.

Origin: It is native to the Mediterranean region, the Near East, southern Europe, Central Asia and Western Siberia.

Distribution: It is distributed in southern Europe, UK, USA, France, Germany, Spain, Italy, Syria, Iran, Afghanistan, Russia, China, Iraq, Uzbekistan, Turkey, Turkmenistan, Sicily, Afghanistan, Pakistan and India. It is also cultivated in Punjab and Sub-Himalayan tracts in India.

Botany: It is a perennial under shrub, which reaches up to 1.2 m height. The diameter of the root varies from 0.75 to 2.5 cm, gray-brown exterior and yellow interior and externally it is longitudinally wrinkled with patches of cork with a characteristic pleasant sweet taste. The long woody stems bear compound pinnate leaves. Flower is an axillary spikes, lavender to violet in colour. Flowers are borne at the age of 2/3 years and onwards. Fruit is 2 to 2.5 cm long pods containing 2 to 5 seeds. There are three types of species viz., (i) *G. glabra* var. *typica*: (Spanish liquorice): This plant has purplish blue coloured papilionaceous flowers. It gives out large number of stolons. (ii) *G. glabra* var. *glandulifera* (Russian liquorice): It has a big root stock along with a number of elongated roots, but does not bear stolons and (iii) *G. glabra* var. *violacea* (Persian liquorice): This plant shows violet flowers.

Climate: It is widely distributed from 5°W to 100°E longitude and 20 to 50°N latitude. It is grown in areas receiving mean annual rainfall of 500 to 1000 mm. It inhabits dry cold temperature to Mediterranean climates where annual temperature varies from 25°C (summer) and 5°C in winter season.

Soils: It is a hardy plant. It is a halophytic shrub. It thrives best in fertile, sandy, sandy loam or clay loam soil with pH 5.5 to 8.2. It is found near a river or stream or under cultivation where it can be irrigated. It tends to grow on the riverbanks and in the flooded fields.

Season: Planting can be taken up during February-March, July-August or October seasons.

Varieties: Haryana Mulhati-1

Land preparation: This is a long duration crop and the preparation of field should be of good tilth and the field is to be leveled well to avoid stagnation of water. Ridges and furrows are formed at a distance of 60 or 90 cm to facilitate irrigation.

Propagation: The cuttings of the underground stem/root of 15-25 cm possessing 2-3 eye buds are planted directly in the field 6-8 cm deep in the soil with a spacing of 60 × 45 cm or 90 × 45 cm. Stem cutting of 250 to 300 kg of wet weight is required for plantation of one hectare land. The cuttings begin sprouting in 15 to 20 days after planting.

Weed management: Three to four hoeing cum weeding are required in the first year of planting and in subsequent years two hand weeding-cum-hoeings are considered to keep the fields weed free condition.

Manures and fertilizer application: FYM @ 20 t/ha is applied at the time of land. A fertilizer dose of 40-25-20 kg of N, P_2O_5 and K_2O per ha is applied as basal dressing. Top dressing of nitrogen @ 20 kg/ha for every year is recommended.

Water management: The crop requires irrigation at an interval of 10 to 15 days until the cutting sprout and establishes themselves in the field depending on soil and climatic conditions. It is important to avoid water-logging in field as stagnation of water in the field will cause-root rotting due to infection of soil borne diseases.

Harvest: Manual digging or one disc harrow followed by digging in 2½-3 year old crop can be practiced to harvest the roots. It overturns the soil, which is left in field for sun-drying; later the roots are sorted out and cleaned. The crop is harvested in winter season i.e. November or December months to obtain roots of high glycyrrhizic acid. At harvest, the roots contain 50-60% moisture and should be dried in the sun for 2-3 days and then in shade for next 10-12 days. The dry roots should possess not more than 10% moisture when these are ready to be stored in polythene lined bags. The roots are cut into pieces of convenient size and shorted into grades, based on thickness.

Yield: The yield of dry root at Hissar (Haryana) is recorded between 7.0 to 8.0 t/ha. At Anand (Gujarat) 10 to 20 months crop has given an average yield of 2.0 to 2.5 t/ha.

Utilization: Liquorice is used in confectionery, cough mixtures, lozenges, medical remedies, plug tobacco and in brewing stout. The rhizomes and roots comprise main active component glycyrrhizin, utilized commercially as a non-nutritional sweetener and flavouring agent in some candies and pharmaceuticals including anti-allergy, anticarcinogenesis, anti-diabetic and anti-inflammatory properties. The glycyrrhizin is responsible for sweet taste. The roots are sweet, refrigerant, tonic and mild laxative.

REFERENCES

Monica Damle. 2014. *Glycyrrhiza glabra* (Liquorice) - a potent medicinal herb. International Journal of Herbal Medicine. 2(2): 132-136

Sanjai Saxena. 2005. *Glycyrrhiza glabra*: Medicine for the Millennium. Natural Product Radiance. 4(5): 359-367

Sheetal, V. and Ashlesha, K. 2011. *Glycyrrhiza glabra* Linn. 'Klitaka': A review. International Journal of Pharma and Bio Sciences. 2 (3): 42-51

Varsha Sharma and Agrawal, R. C. 2013. *Glycyrrhiza glabra.* A Plant for the Future. Mintage Journal of Pharmaceutical & Medical Sciences. 2 (3): 15-20.

CHAPTER

65

Long Pepper–*Piper longum* Linn.

Family: Piperaceae

Vernacular names: Long pepper, Pipal (English); *Pippali, Magadhi, Upakalya,* (Sanskrit); *Pippali, Piplamul* (Hindi); *Hippali* (Kannada); *Pippili, Tippili* (Tamil), *Thippali* (Malayalam).

Importance: Long pepper is known as the 'queen' among the medicinal plants. The economic part of Pipal is the unripe spike. The spikes contain piperine and piplartine alkaloids. The roots and fruits are used in palsy, gout and lumbago. The fruit and root both have a bitter taste. It is the source of pippali (dried fruits) and pippalimulam (roots) as Ayurvedic medicine.

Origin: It is a native of Indo-Malayan region.

Distribution: It grows wild in the tropical rain forests of India, Nepal, Indonesia, Sri Lanka, Timor and Philippines. It is also grown in small area in the Khasi hills, the lower hills of West Bengal, Assam, Uttar Pradesh, Madhya Pradesh, Maharashtra, Kerala, Karnataka and Tamil Nadu.

Botany: It is a perennial under-shrub with creeping stem. Leaves are simple, alternate, stipulate and petiolate or nearly sessile. It flowers throughout the year. Inflorescence is spike with unisexual flowers. The male spikes larger, slender and are 2.5 to 7.5 cm long while the female spikes are 1.25 to about 2 cm long. Female spikes are greenish yellow in colour when young, turns greenish black or black when ripe. Only female spikes are used as drug. Fruit are greyish green or darker grey or yellowish orange berries. The berries are ovoid, sunk in the thick rachis about 0.25 cm in diameter.

Climate: The plant requires a hot moist climate and an elevation between 100 and 1000 m for its cultivation. It can be grown successfully even in areas which receive heavy rainfall with high relative humidity. It needs partial shade to the tune of 20-30% for good growth. Hence it is especially suited as a under crop in coconut and arecanut gardens with 20-25% shade intensity.

Soils: It can be cultivated in well drained loamy soil, forest soils, limestone soil and laterite soils rich in organic matter content.

Season: The nursery can be raised during March and April. The cuttings planted in March-April will be ready for planting in the main field by the end of May.

Varieties: Viswam, Gol Thippali, Pipal Nonsori thippali, Asali thippali, Suvali thippali, Cheema thippalli, Sri Lankan type, Kattu thippali.

Land preparation: The area should be ploughed two to three times and leveled properly. Raised beds of size 3 m × 2.5 m are prepared and pits are dug at a distance of 60 cm × 60 cm. FYM @ 100 g per pit is applied and mixed with soil. Two rooted cuttings or suckers with roots are planted in each pit. To avoid any water stagnation in beds, channels are laid out to drain excess rain water.

Propagation: It can be propagated through seeds, suckers or rooted vine cuttings or by layering of mature branches at the beginning of rainy season. However, it can be easily propagated through the 15-20 cm long terminal stem cuttings with 3-5 internodes obtained from one year old growth. Vine cuttings can be rooted in polythene bags filled with the common pot mixture. The best time for nursery planting is March-April. Vine cuttings and suckers are transplanted soon after the setting in monsoon rains. The rooted cuttings will be ready for transplanting in 2 months time. The pits are gap filled one month after planting. Earthing up after two months of planting is found to enhance crop growth and spike production.

Weed management: In first year, regular weeding should be done and as when the weed growth is noticed.

Manures and fertilizer application: Application of FYM @ 20 t/ha and fertilizer of 50-20-20 N, P_2O_5 and K_2O kg/ha is recommended.

Water management: The crop should be irrigated once in a week if it is grown as a pure crop. In case the crop is grown as an intercrop with other crops, the irrigation provided to the main crop is sufficient.

Cropping systems: It is an ideal intercrop in coconut and arecanut plantations.

Pest and diseases management: Mealy bugs, leaf and vine rotting fungi affect the crop. Apply Rogor @ 0.2% to control mealy bugs. Application of neem kernel extract at 2.5% will control Helopeltis. Spray 1% Bordeaux mixture during the rainy season to control rotting of leaf and vine.

Harvest: The vines start bearing spikes from six to nine months after planting. The spikes will be ready for harvest after two months since formation of spikes. The spikes are harvested at full maturity. Generally, spikes are picked when they are blackish green and most pungent. If left without picking, the spikes ripe and their pungency will be lost to a great extent. Harvested spikes are dried in the sun for 7 to 10 days. The green spike to dry spike ratio is around 10:1.5. The dried spikes have to be stored in moisture proof containers. Thicker parts of lower stems/roots are cut and dried for producing Piplamool. Besides the spike, the thick parts of stems and roots which have medicinal value may also be harvested from 18 months after planting. The stem is cut close to the ground and roots are dug up. It is cut into pieces of 2.5-5 cm length, dried in shade and marketed as piplamool.

Yield: The dry spike yield is around 400 kg per hectare during the first year. The yield increases thereafter up to 3 years and it will be about 1000 kg per hectare during the third year. After three years, the productivity of the vines decreases and should be replanted. The average yield of dried roots is 500 kg per hectare.

Utilization: The roots and fruits have a bitter and hot sharp taste and are used in palsy, gout and lumbago. The fruits (unripe spikes) and roots contain piperine and piplartine alkaloids. The alkaloid content is 2.83%. It is a good remedy for treating gonorrhea, menstrual pain, tuberculosis, sleeping problems, respiratory tract infections, chronic gut-related pain and arthritic conditions.

REFERENCES

Manoj, P., Soniya, E.V., Banerjee, N.S. and Ravichandran, P. 2004. Recent studies on well-known spice, *Piper longum* Linn. Natural Product Radiance. 3 (4): 222-227

CHAPTER

66

Medicinal Yam–*Dioscorea sp.*

Family: Dioscoreaceae

Importance: There are medicinal plants in genus Dioscorea. Among this, *Dioscorea floribunda*, *Dioscorea villosa*, *Dioscorea composita,* etc. are widely grown for diosgenin production. The diosgenin, obtained from Dioscorea tubers, is the major base chemical for several steroid hormones including sex hormones, cortisone, and other corticosteroids and is the active ingredient in the oral contraceptive pill. The major Dioscorea producing countries are Mexico, Guatemala, Costa Rica, India and China.

1. *Dioscorea floribunda* Mart and Gal

Origin: *Dioscorea floribunda* is native of Mexico.

Family: Dioscoreaceae

Distribution: It is cultivated in Karnataka, Assam, Meghalaya, Tamil Nadu, Goa, Maharashtra and Andaman.

Botany: The plants are very vigorous with a stout, robust vine, bearing broad dark green leaves: The tuber branches are thick, broad and deep. It produces compact tubers at a shallow depth. The diosgenin content varies from 2 to 7% depending on the age of the tubers.

Climate: It is a tropical crop. It can be grown in tropical and sub-tropical climatic conditions up to 1500 m above mean sea level. It is adapted to moderate to heavy rainfall areas.

Soils: Dioscorea plants can be grown in a wide variety of soils, but loamy or clay loamy soil is preferred since harvesting of tubers is easier in such soils. Extremely heavy clay soils are, in general not recommended, as they restrict tuber growth and make harvesting difficult. The ideal soil pH is 5.5 to 6.5.

Season: The transplanting of the seedlings in the fields can be done in June-July.

Land preparation: Land should be prepared thoroughly till a fine tilth is obtained. Deep furrows should be made at 60 cm or 45 cm distance with plough.

Varieties: FB(C)-l, FB(C) – 2, Arka Upkar.

Propagation: It can be propagated by tuber pieces, single node stem cuttings or seed. Commercial planting is normally established by tuber pieces only. Propagation through seed may take longer time to obtain tuber yields. Propagation can be done with from tuber pieces weighing about 50 to 70 g each with one or two buds. There are three types of tuber pieces for propagation purpose *viz.*, i) crown, ii) median and iii) tip. Crowns produce new shoots within 30 days after planting, since they have pre-formed buds. Medians and tips may take up to 100 days to sprout. Crowns are therefore preferred for commercial planting. However, if there is a shortage of material, median and tip portions can be used for planting. Dipping of tuber pieces for 5 minutes in Benlate @ 0.3% solution of Benlate or Carbendazim @ 2 g/lit for 10 minutes followed by dusting the cut ends with Benlate @ 0.3% in talcum powder before planting or storage in moist sand beds checks the tuber rot.

(i) Propagation from single leaf node cutting: Single leaf node cutting consists of a single leaf with petiole and about 0.8 cm of the stem can be used for rapid multiplication of the elite materials in the initial stages. Such cuttings are prepared from non-flowering plants. The cuttings are quick dipped in 500 ppm of IBA solution and are planted in the mist chamber in sand beds for rooting. After about 8 to 10 weeks, these plant cuttings are transferred to polythene bags containing equal mixtures of sand, soil and farm yard manure. They will be ready for transplanting in the main field in five to six months' time at a spacing of 45 × 30 cm during June-July or September-October.

(ii) Propagation from seed: Fresh seed should be sown in 8 cm × 12 cm plastic bags in the month of February. Polythene bags may be filled up with a pot mixture containing equal parts of sand, soil and farm yard manure. Vermiculite should be used on the top. At least two well filled seeds may be sown in each plastic bag at a depth of 0.5 to 1.0 cm. The nursery may be protected from drying by light shade. The bags may be watered with care, lightly and frequently. The germination process completes within four weeks. The seedlings being a vine, it should be supported promptly with thin twigs. Vigorous seedlings may be alone transplanted and others may be discarded. The bottom and sides of bag may be cut before transplanting so as to transplant the seedlings without disturbing the root system. As the progenies raised from seeds are highly variable and their growth is slow, this method is not recommended for commercial plantations.

Planting: The stored tuber pieces, seedlings or single node stem cuttings should be planted in furrows with 30 cm between the plants for one year crop and 45 cm for two year crop. The tuber pieces are planted at about 0.5 cm below the soil level.

After cultivation: The vines need support for optimum growth. The new sprouts should be staked or trained over pandal system or trellis. Thin sticks or gunny twines may be used to lead the vines to the pandal or trellis. After sprouting is complete, the plants could be earthed up. Soil from the ridges may be used for earthing up so that the original furrows will become ridges and vice versa.

Weeding: One or two weeding is essential in the first year.

Manures and Fertilizers: It requires high organic matter for good tuber formation. Apply 20 t/ha of FYM or compost, 300-150-150 kg of N, P_2O_5 and K_2O per hectare. The entire dose of P is applied as basal dose. N and K can be applied in 4 equal splits at 2, 4, 6 and 8 months after planting.

Water management: Irrigation may be given at weekly intervals in the initial stage and afterwards at about 10 day's interval. However during rainy season no irrigation is needed.

Intercropping: Intercropping with legumes like cowpea, horse gram, cluster bean and French bean has been found to smother weeds up to 4 months of planting and also provide an extra income without adversely affecting the tuber yield and diosgenin content.

Plant protection: The major pests of Dioscorea are the aphids and red spider mites. Aphids occur more commonly on young seedlings and vines. They feed on the young leaves and stem. Red spider mites attack the underside of the leaves at the base near the petiole. Spray the crop with Dicofol @ 25 ml in 10 lit of water to control aphids and red spider mites.

Harvest: The tubers grow to about 25 to 30 cm depth. Harvest at 2 years of age using pick axes or deep ploughing with mould plough or manual labour. The best season for harvesting is February-March, coinciding with the dry period. On an average 50 to 60 t/ha of fresh tubers can be obtained in two years duration.

Yield: The fresh tubers yield is 50 to 60 t/ha. The diosgenin content tends to increase with age which is 2.5 to 3.0% during the first year and 3.0 to 3.5% in the second year.

2. *Dioscorea villosa*

Family: Dioscoreaceae

Common names: Wild yam, colic root, rheumatism root.

Importance: *Dioscorea villosa* is an herbal remedy for pains associated with rheumatism and arthritis, colic and intestinal cramps, proving itself a reliable antispasmodic and anti-inflammatory. The rhizomes and roots have been popularly used as a non-conventional treatment of the symptoms of menopause, rheumatoid arthritis and hypoprogesteronaemia.

Origin: It is native to eastern North America.

Distribution: It is grown USA. It prefers moist open woods, thickets and roadsides.

Botany: *D. villosa* is a deciduous perennial herbaceous twiner that grows counterclockwise over small and medium-size shrubs. The upper leaves are alternate, heart-shaped and shiny with long petioles, entire margins, prominent veins and acuminated apices. The lower most leaves are usually arranged in whorls. The plants are dioecious. Small staminate (male) flowers are white and perfumed, and arranged in panicles, while carpelate (female) plants produce small solitary flowers at the leaf nodes. The fruit is a membraneous three valved capsule with one or two chocolate-colored winged seeds in each locule. The long, cylindrical seldom branched rhizomes grow to 5-10 mm in diameter and often two feet in length, with numerous tough, slender roots attached underneath. It is oval, being flattened above and below as it creeps in a horizontal position beneath the surface of the ground. It is difficult to powder or crush rhizome. It has no odour. It has taste beyond a slight acridity after prolonged chewing.

Climate: Its light requirement ranges from bright sunlight to light shade. If open fields are used, structures to provide light shade and support for the climbing vines should be provided. The shade providing structure shall be 2 m tall and open to the prevailing winds.

Soils: It requires sandy loam to clay loam soil with good drainage.

Season: Seed are sown in March-April and June-July seasons.

Propagation: Wild yam has been propagated easily by both seed and root division. Spacing is 30 to 45 cm apart.

Seed propagation: Both male and female plants must be grown if seed is desired. Seeds are collected and separated from the capsules. Seed can be scattered over the bare ground and sprinkled with garden soil. It should be protected with screening or chicken wire to protect from squirrels or seed eating animals. If seeds are to be stored, do not let them dry out. Germination takes place in 7 to 15 days.

Vegetative propagation: The wild yam can be propagated by basal cuttings or by root division once the plants have gone dormant. Root cuttings may produce more than one shoot, which can be cut about 5 to 7.5 cm below the shoot keeping fibrous roots attached and planted separately. The remaining tuber may be used in medicine.

Harvest: Propagation from rhizome can produce harvestable roots in 2 to 3 years. Seed propagated plants take at least four years old to produce the harvestable roots. Roots are harvested after the aerial parts die back for optimum concentration of the medically active ingredients. Harvested roots should be cleaned and washed using a mesh and hose; moldy or discolored areas should be removed and roots can be cut into smaller pieces for drying. Dry these rhizomes at 21°C for one day, then at 43°C for two or more days. The dried roots can be stored in moisture and light proof bags for up to one year after which they begin to lose their medicinal value.

Utilization: The economic part used is the dried rhizome and roots. It is used as medicine to attenuate menopause symptoms as well as for the treatment of joint pain and rheumatoid arthritis.

3. *Dioscorea composita* is a species of yam in the genus *Dioscorea.* It is native to Mexico. It is notable for its role in the production of diosgenin, which is a precursor for the synthesis of hormones such as progesterone. It is cultivated in Tamil Nadu, West Bengal, and Jammu. It requires cool climate. Seed propagation is taken up in March-April. It is propagated by seeds, tubers, or leaf node cuttings. Seeds are pre-treated in cold water for 3 to 4 hours and then sown in nursery. Germination takes place in 14 to 21 days. Seedlings of 2 to 3 leaf stage are transplanted in polythene bags. Grown up seedlings of 3 to 4 months can be transplanted in the main field with a spacing of 90 × 60 cm. Vines are allowed to grow for three years. Tuber contains diosgenin of about 3%. Tuber yield is 3 to 4 kg per plant.

REFERENCES

Komesaroff, P., Black, C.V., Cable, V. and Sudhir, K. 2001. Effects of wild yam extract on menopausal symptoms, lipids and sex hormones in healthy menopausal women. Climacteric 2001, 4 (Suppl 2):144–150.

Sautour, M., Miyamoto, T. and Lacaille-Doubois, M. 2006. Steroidal saponins and flavan-3-ol glycosides form *Dioscorea villosa.* Biochem System Ecol 2006, 74:60–63.

Wojcikowski, K., Wholmuth, H., Johnson, D.W. and Gobe, G. 2008. *Dioscorea villosa* (wild yam) induces chronic kidney injury via pro-fibrotic pathways. Food Chem Toxicol. 46: 3122–3131.

Hooker, E. 2004. Final report of the amended safety assessment of *Dioscorea villosa* (wild yam) root extract. Int J Toxicol 2004, 23:49–54.

Bhandari, M.R. and Kawabata, J. 2004. Assessment of antinutritional factors and bioavailability of calcium and zinc in wild yam (*Dioscorea* spp.) tubers of Nepal. Food Chem. 2004, 85:281–287

CHAPTER

67

Opium Poppy–*Papaver somniferum* L. (2n=22)

Synonyms: *Papaver album* Mill., *Papaver setigerum* DC.

Family: Papaveraceae

Vernacular names: Opium poppy, Poppy, Poppy seed, Poppy capsule, poppy heads, Peon poppy, (English); *Aphim, Khashkhash, Khasabija, Posta dana* (Hindi); *Khasatilah, Khakhastilah, Khakhasah, Aaphuka, Ahifen, Ahiphena* (Sanskrit); *Aapheen, Aphini, Biligasgase, Gasagase, Kasakase (Kannada); Avin, Aalan, Karappu, Kasakasa* (Malayalam); *Abini, Kasakash, Kasakasa, Postakai (Tamil); Abhini, Gasagasalu, Gasagasala chettu, Nallamandu* (Telugu); *Aaphim, Postadaanaa, Postabeej* (Bengali:); *Aapu* (Oriya); *Afu, Aphu, Khaskhas, Khuskhus, Posta* (Marathi); *Aphina, Apphou, Khuskhus, Khaskhas na doda,* (Gujarati); *Aaphin, Aphen* (Kashmiri); *Afim, Khashkhash, Kaishkhas* (Punjabi).

Importance: Opium poppies are grown as an ornamental plant and for seeds in the United States. The genus, Papaver, is the Greek word for 'poppy'. The species, 'somnijerum', is Latin for 'sleep inducing'. The opium poppy is used for anesthesia and for its sedative effects. The Sumerians refer to it as the 'joy plant'. The word 'opium' is derived from Greek word 'Opion' meaning 'liquid' or 'poppy-juice'. In 1803, the German Pharmacist Serturner isolated and described the principal alkaloid in opium, which he named morphium after Morpheus, the Greek God of Dreams. Khasaphalasheera is the exudates obtained from the fruit which is bearing the khasakhas seeds. Aphukam is the Arabic name for the drug. It was known to Indian physicians in this name because of its magical properties. Opium is the source of many drugs, including morphine (and its derivative heroin), thebaine, codeine, papaverine and noscapine. The opium poppy is not only grown in fields but also as an ornamental flower in gardens. The flowers of the opium poppy can be found in many different colours, but are most frequently seen as red, white or pink flowers.

Origin: It is native to northern Africa (i.e. Algeria, Libya, Morocco and Tunisia), the Azores, the Madeira Islands, the Canary Islands, southern Europe (i.e. Greece, Italy, France, Portugal and Spain) and Mediterranean region.

Distribution: Opium poppy is grown in Australia, Mediterranean region, Austria, France, Hungary, the Netherlands, Poland, Czech Republic, Slovenia, Spain Iran, Turkey, Afghanistan, Mexico Yugoslavia, Macedonia, Bulgaria, China, Manchuria, Japan, India, Pakistan, Myanmar, Thailand, Vietnam, Laos, Lebanon, Guatemala and Colombia. It is cultivated at higher elevations in Tamil Nadu, Kerala, Bihar, Uttar Pradesh, Madhya Pradesh, West Bengal, Assam, Rajasthan and the Himalayan regions.

Botany: The opium poppy is an annual plant. It is an erect, rarely branched, glaucous annual, growing to a height of 60 to 120 cm. The leaves are ovate, oblong or linear oblong. The stem and leaves are sparsely covered with coarse hairs. The leaves are lobed and clasp the stem at the base. Opium poppies flower in about 90 days and continue to flower for two to three weeks. The entire growth cycle takes about 120 days. Flowers are large usually bluish with a purplish base or white, purple or variegated. Plants produce three to five mature pods per plant. These fruits or pods (also called capsules, bulbs, or poppy heads) are oblate, elongated, or globular, hairless and mature to about the size of a chicken egg. It produces capsular type of fruits from which the latex known as opium is obtained on lancing. The fruits are about 2.5 cm in diameter, globose in shape. Capsule are brown when mature, spherical or oblong-elliptic, 4-9 × 4-5 cm. The seeds are reniform with white, yellow, coffee-colour, gray, black, or blue in colour. There are 200 seeds per pod. All parts of the poppy plant contain a white milky juice or latex while the unripe capsules contains the white milky latex or juice in abundance are used for extraction of morphine and other alkaloids.

Climate: It is cultivated in tropical, subtropical and warm temperate regions with low humidity. It is grown up to an elevation of 1000 m above sea level. It is a long day photo-responsive plant. Poppy is grown in areas receiving annual precipitation of 500 to 2000 mm and annual mean temperature of 5.6 to 23.5°C. Seeds germinate well at 15°C. The crop needs long cold season with temperature of 20°C in the early season for a healthy vegetative growth while warm, dry weather with a temperature of 30-35°C is required during the reproductive period. Cloudy weather, frost, hailstorms and high gusty winds, particularly during lancing, causes immense damage to the growing crop. Frosty or desiccating temperature, cloudy or rainy weather tends to reduce not only the yield but also the quality of opium. If the field becomes waterlogged, the plants wither and die. Dew during the lancing period increases the apparent yield. Rain at the time of harvest washes away the latex and reduces the yield considerably. Dry, warm weather conditions favour a good flow and high yield of latex.

Soils: It thrives in deep, warm, moderately moist, loamy or clayey loam soil rich in organic matter with pH of 6 to 8.2.

Season: The best time for sowing is October or November. The crop flower in April and May and the capsules are ripe in June to July.

Varieties: The white flowered variety of opium poppy is cultivated in Rajasthan and Central India while red flowered variety with dark seeds is cultivated in the Himalayas. The promising varieties are with crop duration from 140 to 160 days. The important varieties are Talia, Ranghatak, Dhola Chota Gothia, MOP-3, MOP-16, Shama, Shweta, BROP 1, NBRI-3, Kirtiman (NOP-4), Chetak (U.O.285), Jawahar Aphim 16 (JA-16), Sujatha

Land preparation: The field should be ploughed 3 or 4 times to produce fine tilth. The field is then prepared into beds and channels of convenient size.

Seeds and sowing: Seed rate is 7-8 kg/ha for broadcast method and 4-5 kg/ha for line sowing. The seeds may be treated with Dithane M.45 @ 4 g per kg of seeds. Seed is usually mixed with fine sand before broadcasting to ensure uniform spread in the bed. Line sowing is preferred to broadcasting. A spacing of 30 × 30 cm or 45 × 30 cm is normally adopted.

After cultivation: Germination takes five to ten days depending upon the moisture content of the soil. Thinning is done within a period of 3-4 weeks after sowing at 5-6 cm height, having 3-4 leaves.

Weed management: Two weeding are given on 20 and 35-40 DAS.

Manures and fertilizer application: Farm yard manure @ 20-30 t/ha is applied by broadcasting while the field is prepared for sowing. A fertilizer dose of 80-40-0 kg of N, P_2O_5 and K_2O per ha is recommended. Half of N and entire P are applied at sowing time and another 50% dose on 35 DAS.

Water management: A light irrigation is given immediately after sowing followed by irrigation at an interval of 12-15 days are given up to pre-flowering stage and then irrigation at an interval of 8-10 days during flowering and capsule formation stage. Normally, 12-15 irrigations are given during the entire crop period.

Pest and diseases management: The opium poppy is infested with insect pests are root weevil, aphids, thrips, sawfly, Head gall fly, capsule weevil and capsule borer and the diseases such as downy mildew, powdery mildew, leaf blight, leaf spots, capsule rot, wilt and root rot.

Lancing and latex collection: Opium starts flowering in 95-115 days after sowing. The petals start shedding after 3-4 days of flowering. The capsules mature after 15-20 days of flowering. Lancing of the capsules exudes maximum latex at this stage. This stage can be visually judged by the compactness and a change in the colour from greenish to light green coloured ring in the capsule. The stage is called as industrial maturity. Lancing may be done with a knife having three or four equispaced pointed ends which does not penetrate more than 1-2 mm in the capsule. Too deep or too shallow incision is not advisable. Lancing may be done early in the morning before 8.00 a.m. at two days interval in each capsule. The length of the incision should be 1/3 or less than the full length of capsule.

Harvest: The crop is left for drying for about 20-25 days when the last lancing on the capsules stops exudation of latex. The capsules are then picked up and the plant is cut with sickles. Harvested capsules are dried in open yard and seeds are collected by beating with a wooden rod. The opium is tapped from each pod while it remains on the plant. High-quality raw opium will be brown (rather than black) in colour and will retain its sticky texture. After the opium is scraped, the pods are cut from the stem and allowed to dry. Once dry, the pods are cut open and the seeds are removed and dried in the sun. Poppy seed oil is straw-yellow in colour, odourless, and has a pleasant, almond-like taste.

Yield: The opium yield from a single pod varies greatly, ranging from 10 to 100 mg of opium per pod. The average yield is about 80 mg. The milky latex yield is 40 to 60 kg/ha. The dried opium yields ranges from 8 to 15 kg/ha with 0.4 to 0.9% dry morphine. The seed yield is 0.6 to 1.0 t/ha. The seed contains 45–50% oil. The straw yield is 6 to 8 t/ha. The opium latex contains morphine (12-13%), codeine (2.05-2.76%), thebaine (1.84-2.16%), papaverine (0.82-1.0%) and Narcotine (5.89-6.32%). Morphine content is highest during period 10 to 30 days after flowering. Diacetylmorphine (heroin) is synthesized from morphine. For each kg of morphine, about 680 g of crude heroin base is produced. Heroin has the appearance of light brown, granulated sugar. However, the white, fluffy powder or crystals form of heroin is produced from Southeast Asia or China heroin. Heroin is injected, snorted or smoked.

Utilization: The poppy seeds and poppy seed oil provide a unique flavour and are used as food ingredients. The seeds can be dry roasted and ground to be used in wet curry (curry paste) or dry curry. The opium serves as an analgesic. Opium can be smoked, intravenously injected, or taken in pill form. The seeds pounded with milk and pasted destroy dandruff. The alkaloids from the opium poppy are papaverine, cryptopine, codiene, amorphine and morphine which are used to treat the diseases. One of the most addictive drugs, heroin, comes from the opium poppy. Heroin, or diacetylmorphine, is a semi synthetic drug created from morphine. Heroin is also used to treat severe acute pain. Morphine and codeine can help alleviate pain, spasms, coughs and many other conditions.

REFERENCES

Anon. 1992. Opium poppy cultivation and heroin processing in Southeast Asia. U.S. Department of Justice. Washington, DC 20537.

Duke, J.A. 1973. Utilization of Papaver. Econ. Bot. 27(4): 390-391

Loof, B. 1966. Poppy cultivation (Review article). Field Crop Abstracts 19(1): 1-5.

https://www.hort.purdue.edu/newcrop/duke_energy/Papaver_somniferum.html

CHAPTER

68

Periwinkle–*Catharanthus roseus* (Linn.) G. Don.

Synonyms: *Vinca rosea, Vinca multiflora* and *Lochnera rosea.*

Family: Apocynaceae

Vernacular names: Periwinkle, Madagascar Periwinkle, vinca, pervinca, vincapervinca (English); *Sadabahar* (Hindi); *Nityakalyani* (Tamil); *Billaganneru* (Telugu); *Shavam Naari* (Malayalam); *Sadaphuli* (Marathi); *Nayantara* (Bengali). The Greek word 'Catharanthus' means 'clean or pure flower' while the Latin word 'roseus' means 'rose-coloured'. It is known as 'Sadabahar'in Hindi meaning 'always in bloom'.

Importance: Periwinkle is exported to developed countries. Most of the exports take place from Tuticorin port in Tamil Nadu. The flower colour is lavender blue, pale tint of blue or a 'pastel blue'. It is a popular ornamental plant found in gardens and homes. USA, Hungary, Italy, Netherlands and Germany are the major consumer of its leaves. It is also cultivated as an ornamental plant.

Origin: This periwinkle is native to Madagascar and southern Europe.

Distribution: It is grown in Australia, Africa, India, Malaysia, Indonesia, China, Philippines, South Africa, Mozambique, Israel, USA and southern Europe. In India, it is being grown in Tamil Nadu, Karnataka, Andhra Pradesh, Madhya Pradesh, Gujarat and Assam.

Botany: It is an evergreen, herbaceous plant growing to a height of 1 m. The leaves are oval to oblong, glossy green hairless with a pale midrib and a short petiole. Flowers have five petals and are soft pink, tinged with red. The fruit is a pair of follicles about 2-4 cm long and 3 mm broad.

Climate: It is cultivated in tropical and subtropical regions. It is naturalized in arid coastal locations. It can be grown up to an elevation of 1300 m above sea level. A well distributed rainfall of 1000 mm or more is ideal for raising this crop under rainfed conditions. It is tolerant to heat and drought. The seed germinates in one week at a temperature of 21 to 24°C.

Soils: The crop grows well on a wide variety of soils. It prefers well-drained sandy loam to loam soils. It tolerates saline and alkaline soils. It does not tolerate water-logged soil.

Varieties: There are three local types based on the colour of the flower viz., white flowers, pink rose coloured flowers and white flowers having rose-purple spot in the centre.

Season: Sowing is done during February-March and July-August seasons.

Propagation: Propagation is through seeds. Seed rate is 2 to 3 kg/ha. Seed germinates in 10 days after sowing. Seeds are sown with spacing of 45 cm × 30 cm or 45 cm × 45 cm. About 500 gram of seed is sown in 200 m^2 nursery bed to produce seedlings required for one hectare.

Cultural practices: Thinning is done 15 days after sowing. Detopping of plants by 2 cm at 50% flowering improves the root yield and alkaloid content.

Weed management: This crop requires two weeding on 60 and 120 days of sowing. Pre-emergence application of Fluchoraline @ 0.75 kg ai/ha is recommended to control weeds. Application of a mixture of 2-4-D and Grammaxone @ 25 kg/ha to the soil before sowing control the weeds.

Manures and fertilizers application: Compost/ FYM @ 12.5 t/ha and a fertilizer dose of 80-50-50 kg of N, P_2O_5 and K_2O per ha is applied as basal dressing. FYM @ 12.5 t/ha and 40 kg N per ha are recommended under rainfed conditions.

Water management: The plants do not require any irrigation in places where rainfall is evenly distributed throughout the year. However, irrigation is applied at interval of 7 to 10 days under irrigated condition.

Insect pests and diseases management: The plant is sufficiently hardy and practically free from the attack of insect pests and diseases. However, the little leaf disease, 'die back' or twig blight or top rot have been found to affect the crop during the monsoon. It can be controlled by spraying Dithane Z-78 at an interval of 10-15 days.

Harvest: The crops are harvested after 12 months of sowing. The plants are cut about 7.5 cm above the ground level and dried for the stems, leaves and seeds. The field is then copiously irrigated, the roots are collected through digging or ploughing. The roots are washed and dried in the shade. If there is a demand for leaves, two leaf stripping, the first after 6 months and the second after 9 months of sowing can be taken. A third leaf stripping is also obtained when whole plant is harvested. After the plant is harvested it is dried in the shade.

Yield: Under irrigated conditions, about 3-4 t/ha of leaves, 1.5 t/ha of stem, and 1.5 t/ha of roots on an air-dried basis can be obtained. Whereas, under rainfed conditions, about 2 t/ha of leaves and 0.75 t/ha each of stem and roots on an air-dried basis may be obtained. One tonne of leaves yields 50 g of vincristine sulphate in crude form. On further purification, 40 grams of vincristine is obtained.

Utilization: The leaves and stems are the sources of dimeric alkaloids, vincristine and vinblastine, while roots have antihypertensive, ajmalicine, reserpine and serpentine. The leaves are used to control diabetes. Vincristine sulphate is used against acute leukemia and vinblastine sulphate as 'Velbe' to cure Hodgkin's disease. The root bark contains the alkaloid, Alstonine which is used to reduce blood pressure. The alkaloids like vinblastine and vincristine are used to produce, anticancer drug. The total alkaloid content in the leaf varies from 0.15 to 1.34%, of which the average content of vinblastine is 0.002%, while that of vincristine is 0.005%.

CHAPTER

69

Sacred Basil / Tulsi–*Ocimum sanctum* Linn (2n=32)

Family: Lamiaceae

Vernacular names: Sacred basil / Holy basil (English), *Tulasi* (Hindi, Gujarati, Bengali, Assamese, Sanskrit, Tamil); *Surasa, Bana Tulasi* (Sanskrit); *Tulsi-chettu* (Telugu); *Tulasi, Shree Tulasi, Vishnu Tulasi, Amli, Huli* (Kannada); *Pachcha tulasi, Sivatulasi* (Malayalam); *Sabje* (Gujarati); *Babui tulsi* (Bengali).

Importance: There are two types of Ocimum sanctum under cultivation viz., the green type, Sri tulsi (Ram tulsi) is the most common; the second type, Krishna tulsi bears purple leaves. The sacred basil oil has about 71% eugenol. Eugenol is widely used in perfumery, cosmetics, pharmaceuticals and confectionary industries. It is one among the few plants which purifies the atmosphere. Basil is derived from Greek word 'Basilica' which means royal plant.

Origin: It is indigenous to the lower hills of Punjab and Himachal Pradesh of India.

Distribution: It is also cultivated in USA, Egypt, France, Belgium, Hungary and other Mediterranean countries. It is distributed in the Himalayas and in Andaman and Nicobar Islands of India.

Botany: It is a biennial shrub and grows to a height of 30 to 90 cm. Stems and twigs are usually quadrangular. Young twigs are greenish, purplish or brownish in colour. The leaves are simple, petiolate and ovate. The leaves have numerous oil glands with aromatic volatile oil. It has racemose type of inflorescence. The herb bears small white flowers in racemes. The freshly picked bright green leaves turn brownish green when dried and become brittle and curled. Seeds are globose, shining and brownish in colour.

Climate: It has a wide adaptability and can be grown in tropical and sub-tropical climates. It is grown up to an altitude of 1800 m from MSL. It flourishes well under fairly high rainfall and humid conditions. Long days and high temperatures have been found favourable for plant growth and oil production. The plant is moderately tolerant to drought and frost. The plant can be grown under partially shaded conditions but with low oil contents.

Soils: It can be grown on any type of soil including acidic, saline and alkaline soils. However, sandy loam soil with good organic matter is suitable. It is susceptible to water logged conditions, since it cause root-rot and results in stunted growth.

Season: Planting can be taken up in February-March and August-September. The crop matures at about 75-90 days after transplantation while the short duration variety comes to maturity within 60-80 days.

Varieties: CIM-Saumya, CIM-Ayu, CIM-Kanchan, Vikarsudha, RRL-011 and Kusumohak.

Land preparation: Two to three ploughing followed by harrowing are carried out to have a fine tilth. Incorporate FYM/compost during last ploughing.

Propagation: Tulsi is propagated through seeds. Seeds will get deteriorated over generations, due to its high cross-pollination. Hence, the growers have to take fresh seeds for fresh plantings. Raised seed beds of convenient size are prepared with the addition of farm yard manure @ 2 kg/sqm to improve organic matter. Seed rate is 200 to 300 g/ha. Since the seeds are very small, it should be mixed with sand in the ratio of 1:4 and sown in lines of 6 cm apart at a depth of 2 cm. After sowing, the seeds in the nursery, a mixture of farm yard manure and soil should be spread in a thin layer over the seeds and irrigate with a sprinkler hose. The seeds germinate in 8-12 days and the seedlings are ready for transplanting in about 6 weeks' time at 4-5 leaf stage. A spray of 2% urea solution on seedlings at 15 to 20 days before transplanting is done to get healthy seedlings. Seedlings are transplanted at a spacing of 45 × 30 cm, 60 × 30 cm or 60 × 60 cm to get high herbage and oil yield. It can also be propagated by terminal cuttings with first 2-3 pairs of leaves, 8-10 nodes and 10-15 cm length during October-December months. The rooting is complete in about 4-6 weeks' time. The plants are transplanted at a spacing of 45 × 30 cm.

Weed management: Two weeding are carried out on 30 and 60 days after planting.

Manures and fertilizer application: Farm yard manure / compost are to be applied at 12.5 t/ha before planting. A fertilizer dose of 120-60-60 kg of N, P_2O_5 and K_2O per ha is recommended. Half the dose of N and the entire dose of P_2O_5 and K_2O should be given as a basal dose, whereas, the remaining N is applied in two split doses after first and second cuttings. Application of cobalt and manganese at 50 and 100 ppm can be done to increase the oil yield.

Water management: Irrigation is provided immediately after transplanting. Later, it is given at weekly interval depending upon the rainfall and soil moisture status.

Pest and diseases management: Leaf rollers, tulsi lace wing, powdery mildew, seedling blight, root-rot are infesting Tulsi plant. Spray with 0.2% malathion or 0.1% methyl parathion to control leaf rollers. Spray Azadirachtin 10,000 ppm @ 5 ml/l to control tulsi lace wing. Spray with 0.3% wettable sulphur and drench the nursery beds with 0.1% solution of mercurial fungicide to control powdery mildew. The seedling blight and root-rot is controlled with drenching the nursery beds with Bavistin @ 1%.

Harvest: The crop is harvested at full bloom stage by cutting the plants at 15 cm from ground level to ensure regeneration of the crop. The first harvest is done after 90 days of planting a at full bloom stage and the lower leaves start turning yellowish. Subsequently, it may be harvested at every 65-75 days interval. Three to four floral harvest can be obtained. Harvest the crop on bright sunny days to get good yield and oil quality. The harvested produce may be allowed to wilt in the field itself for 4-5 hours so as to reduce the moisture and also the bulkiness. The oil quality and its yield do not diminish up to 6-8 hours after harvest.

Yield: The fresh herbage and flower yield is 10 to 15 and 3-4 t/ha/year respectively. The whole herb at full bloom stage is used for extraction of aromatic oil. The inflorescence contains 0.4% oil while the whole herb contains 0.10 to 0.25% oil.

Utilization: The dried leaves and tender four sided stems are used as spice for flavouring and for extraction of essential oil. The oil is widely used in perfumery compounds. It is used as medicine, insecticide and bactericide. The oil contains 40% linalool and 35% methyl chavicol. The juice of the leaves possesses antiseptic, diaphoretic, antiperiodic, stimulating, expectorant, anti-pyretic and memory improving properties.

REFERENCES

http://horticulture.kar.nic.in/APMAC_website_files/Tulsi.htm

CHAPTER

70

Safed Musli–*Chlorophytum borivilanium* Ker.

Family: Liliaceae

Vernacular names: India spider plant (English); *Safed Moosli, Dholi Musli, Khiruva* (Hindi); *Swetha musli* (Sanskrit, Telugu); *Taniravi Thang* (Tamil); *Dholi Musli* (Guajarati).

Importance: It is a medicinal plant, used to treat male impotency. It is considered as an alternative 'Viagra'. The global markets for Safed musli are USA, Britain, European countries, Australia, Japan and China.

Origin: It is native of India.

Distribution: It is grown in India, America and Australia. In India, it is naturally distributed in the forest areas of Maharashtra, Gujarat, Rajasthan and Madhya Pradesh. It is also cultivated in Himachal Pradesh, Uttar Pradesh, Bihar, West Bengal, Andhra Pradesh, Kerala and Tamil Nadu.

Botany: It can grow up to a maximum height of 45 cm. Roots are fasciculate, sessile, cylindrical or ellipsoidal, number and size increases with the age of the plant. Tubers can grow up to a depth of 25 cm. Leaves are sessile or short petiolate, with sheathing bases, 15 to 45 cm long and 1.5 to 2.5 cm wide. Leaves are linear lanceolate, membranous, glabrous or sparsely softly hairy. The leaf tip in contact with soil produces adventitious roots. Flowers are white born in clusters on sparse panicles up to 120 cm long. Fruit is a capsule, 4- seeded with a slender beak and spongy septa, seeds oblong, black and shiny testa. The seeds are black in colour and with angular edges.

Climate: It is distributed in tropical and sub-tropical climate of the world. It prefers warm and humid climate during its crop growth period. Temperatures more than 35°C and less than 15°C are not suitable for its growth and yield. Areas receiving 500 to 1500 mm rainfall from June to October are considered to be suitable for its cultivation.

Soils: It can be cultivated on various types of soils such as acidic soils, black cotton soils, red lateritic soils and calcareous soils. It is grown in soils from sandy loam, loam and clayey loam in texture. However, crop prefers well drained loamy soils.

Season: Planting is done during the February-March and June-July under irrigation. The optimum time for its planting is during the onset of monsoon under rainfed condition.

Varieties: Jawahar Safed Musli 405 and Rajvijay Safed Musli 414.

Land preparation: Land is prepared with ploughing and harrowing. Ridges and furrows or beds and channels are formed. Apply and incorporate FYM at the time of last ploughing.

Propagation: The high cost of the vegetative planting material has forced to raise the crop through true seeds. However, due to low seed germination percentage, farmers use root cuttings weighing 3-4 g/root as propagating material. Prepare nursery beds of 1.2 m width and desirable length depending on soil and irrigation facility. About 8.5 kg seed is required for raising seedlings to transplant one hectare of land. Seeds are sown by broadcasting and covered with light soil layer. Seeds germinate in the nursery within 5-6 days. Seedlings of about one month old are ready for transplanting in the main field at 30 × 15 cm spacing. Vegetative seed root stock requirement is 0.3 to 0.5 t/ha. Fleshy roots are planted on top of the ridges of 15-20 cm height at 30 × 15 cm spacing.

Weeding: Two to three weeding-cum-hoeings are required during the crop growth to keep the soil porous and free of weeds.

After cultivation practices: In the rainy season, earthing-up is required to avoid exposure of fleshy roots which often occur when crop is grown on raised beds or ridges. Inflorescence should be removed when it appears to improve fleshy root yield.

Manures and fertilizer application: FYM @ 15 to 30 t/ha or vermicompost @ 5 t/ha is applied at the time of last ploughing. A fertilizer dose of 50-40-40 kg of N, P_2O_5 and K_2O per ha is applied as basal dressing.

Water management: The crop may be sown/planted after receipt or rains. If there is no rain after sowing of seeds or planting of fleshy root propagates, then on irrigation is to be provided immediately. Later irrigation may be done after 10 to 15 days interval depending upon the soil moisture retention capacity and weather conditions. One light irrigation is given before harvesting.

Insect pests and disease management: The diseases like leaf spot and leaf blight infects the crop. The incidence of orange banded blister beetle and root-knot nematode also infest fleshy roots.

Harvest: The crop matures in about 90 days under cultivation. At maturity, the leaves start yellowing and ultimately dry up from the collar part and fall down on soil surface. During digging of plants, fleshy root bunches should be lifted from the soil. The harvested fleshy roots are cleaned and skin is removed. The white tubers are spread and dried in the shade for about 5 to 7 days. Plants can be harvested at this stage if the planting material is not required for next season. The crop goes in dormancy on the onset of winter in the month of October-November. However, for seed crop, fleshy root bunches are allowed to remain in soil up to January-February and harvested in March-April.

Yield: The fleshy root yield is one tonne per hectare which is reduced to 200 kg/ha on drying. However, a yield of 2 to 3 t/ha of fresh root yield can be obtained under favourable growing conditions which gives about 0.4 to 0.6 t/ha of dry root yield.

REFERENCES

Elizabeth, K. G. 2001. Safed musli a promising medicinal plant. Indian Journal of Areca nut, Species and Medicinal Plants. 5 (2): 65-69.

Kothari, S. K. and Sigh, K. 2003. Production technique for the cultivation of safed musli (*Chlorophytum borivilianum*). J. Horticulture Sci. Biotech. 8 (2): 261-264.

Singh, A. and Chauhan, H.S. 2003. Safed musli (*Chlorophytum borivilianum*) distribution, biodiversity and cultivation. Journal of Medicinal and Aromatic Plant Sciences. 25: 712-719

CHAPTER

71

Sarpagandha–*Rauvolfia serpentina Beth.* ex Kurz (2n=22)

Family: Apocynaceae

Synonyms: *Ophioxylon salutiferum* Salisb., *Ophioxylon obversum* Miq., *Ophioxylon serpentinum* L., *Rauvolfia obversa* (Miq.) Baill., *Rauvolfia trifoliata* (Gaertn.) Baill.

Vernacular names: Devil-pepper, Snake-root, Sarpagandha, Rauvolfia-root, Indian Snake-root, Serpentina root (English); *Chandra, Chandrabagha, Chota-chand* (Hindi); *Chandrika, Chandramarah, Naakuli* (Sanskrit); *Chandra* (Bengali); *Sarpaganthi, Chivan amelpodi, Chovannamilpori* (Tamil); *Patalagaruda, Patalagandhi, Sarpagandhi, Sutranabhi* (Kannada); *Chuvannavilpuri, Chuvannavilpori, Amalpori* (Malayalam); *Paataalagani, Patalaagandhi, Sarpagandhi,* (Telugu); *Harkaya, Harki, Adkai* (Marathi); Amelpodee (Gujrati); *Dhanbarua, Patalgarur, Sanochada* (Oriya).

Importance: Sarpagandha contains alkaloids like reserpine, desrpidine and reseinamine. The roots, leaves and juice are of medicinal importance.

Origin: It is native to India and Bangladesh.

Distribution: It is distributed in India, Pakistan, Bangladesh, Nepal, Bhutan, China, Myanmar, Malaysia, Thailand, Laos, Vietnam, Indonesia and Sri Lanka. The plant is cultivated in Himachal Pradesh, Jammu & Kashmir, Punjab, Uttaranchal, Uttar Pradesh, Uttarakhand, Bihar, Odisha, West Bengal, Assam, Arunachal Pradesh, Meghalaya, Tamil Nadu, Kerala, Karnataka, Maharashtra, Gujarat, Lakshadweep and Andaman Islands.

Botany: Sarpagandha is an evergreen, woody, glabrous and perennial shrub with maximum height up to 60 cm. The tuberous root has pale brown cork. The root bark constitutes 40-60% of the whole root and possesses high alkaloid concentration. The fresh roots emit a characteristic acrid aroma and are bitter in taste. Leaves are lanceolate to oblanceolate or obovate in whorls of three. Flowers are 1.5 cm long, white or pinkish in cymes; peduncles deep red. The number of inflorescences per plant ranged from 50 to 60 while the flowers in the inflorescences ranged from 35 to 50. Anthesis takes place in the morning when atmospheric temperature ranges from 25-29°C and anther dehiscence from

28-31°C. Flowers are protogynous preventing selfing. Each flower produces four seeds on an average. A plant has absolute/maximum reproductive potential to produce 7350 seeds. However, the ecological/realized reproductive potential is 43.2% (3160 seeds per plant) of its potential capacity. Drupes are purplish-black, obliquely ovoid or rounded, connate.

Climate: It grows well in frost-free tropical to subtropical situations. It is grown up to an altitude of 1400 m from MSL. It grows well with temperature of 10 to 38°C. It prefers partial shade. It is grown areas receiving mean annual rainfall of 600 to 2500 mm.

Soils: The plant requires well drained silt-loam to clay-loam soil rich in organic matter content. It prefers sandy alluvial loam to red lateritic loam or stiff dark loam. The ideal soil pH is from 4.6 to 6.5.

Season: April-May is found to be suitable for sowing seeds in nursery and transplanting is done during June-July. Hard wooded stem cutting are closely planted during June in the nursery beds. Root cuttings are planted in March-June. The root stumps are transplanted in May-July.

Propagation: It is a long duration crop of 18 months duration and slow growing crop particularly in the initial stage. The crop can be propagated by seed, stem cutting, root cuttings and root stumps. Seed propagation is the best method for raising commercial plantation.

(i) **Seed propagation:** Seeds of 6 kg sown in a 500 m^2 bed will be sufficient for one-hectare area. Seeds stored more than 7 to 8 months do not germinate. The germination of seed is 10% due to the presence of cinnamic acid derivatives. Seeds are soaked in water for 24 hours improves germination. The nursery is prepared with raised beds preferably under partial shade. Seeds are sown in lines 2-3 cm apart in shallow furrows. The furrows are then covered with a fine mixture of soil and FYM. The bed should be kept moist by light watering. Germination starts after 15-20 days and continues up to 40-50 days which have 4-6 leaves and about 10-12 cm height. Seedlings may be raised in poly bags also. The seedlings are transplanted at spacing of 45 × 30 cm.

(ii) **By stem cutting:** Hard wooded stem cuttings of 15 to 20 cm length are treated with β-indole acetic acid @ 30 ppm for 12 hours to induce rooting. The treated cuttings root within 15 days.

(iii) **By root cutting:** Root cuttings of 100 kg are required to plant one ha. Roots are cut at 2.5 to 5 cm length and planted in nursery beds containing well matured FYM, sand and saw-dust. The beds are kept moist through watering. The root cuttings sprout within 3 weeks. The seedlings are transplanted at spacing of 45 × 30 cm.

(iv) **By root stumps:** The root stumps of about 5 cm root with a portion of the stem above the collar are prepared. The limitation is only one plant can be raised from a single stump. The root stumps are directly transplanted in the field having irrigation facilities. This method may be used to establish initial mother stock.

Weed management: Five to six weeding are taken to keep the crop is weed free condition.

Manures and fertilizers application: FYM @ 20-25 t/ha should be applied during land preparation. Fertilizer dose of 30:60:30 kg/ha of N, P_2O_5 and K_2O is recommended. A dose of 10:60:30 kg/ha of N, P_2O_5 and K_2O is applied as basal dose. Later two equal doses of N each of 10 kg /ha in moist soil may be applied at 50 days and 170 days after planting.

Water management: It is grown as rainfed crop in areas which receive rainfall of 1500 mm or above well distributed throughout the growing season especially in Assam and

Kerala. It requires 15 to 16 irrigations at 20 days interval in summer and at 30 days interval in winter season under irrigated condition.

After cultivation practices: Flowering and fruiting starts from 6 months onward after transplanting. If the plant is allowed to flower and bear fruits, it depresses both shoot and root growth. Hence defloration is recommended for root growth and yield. For collection of seeds, plants in small portion of plantation are allowed to flower and bear fruits. Mature seeds bear a purplish black fleshy covering. Since all the seeds do not mature at a time, these are picked periodically, otherwise ripe seeds drop off. After the collection, fleshy covering should be removed by washing the seeds in water. After washing, the seeds are dried in shade and are stored.

Cropping systems: It is intercropped with patchouli in the first year under irrigated condition. Since it is shade loving crop, it can be planted as intercrop in orchards or some other plantation crops.

Pests and diseases management: Caterpillar, grubs, root knot nematode, *Cercospora* leaf spot and *Alternaria* leaf spot are infesting the crop. Application of Carbofuran 3 G @ 25 kg/ha or Phorate granules 10 G @ 20 kg/ha will control root knot nematode and grubs. Caterpillar can be controlled by spraying 0.2 % Rogor. Leaf spot is controlled with Dithane M-45 @ 0.2% while Alternaria leaf spot is controlled with spray of Blitox @ 30 g in 10 litre water.

Harvest: The roots are collected 18 to 30 months after planting. At this stage root contains maximum concentration of total alkaloids. During harvest the roots are found to go up to 40 cm deep in the soil. During root harvesting the thin roots are also collected. The bark constitutes 40-56% of the whole root. A light irrigation is necessary to make digging easier. After digging, the roots are cleaned, washed and cut into 12-15 cm pieces for drying and storage. The dry root possesses up to 10-12% moisture.

Yield: The root yield vary from 1.5 to 2.5 t/ha of dry weight under irrigation depending upon soil fertility, crop stand and management.

Utilization: The plant parts, root and rhizome have been used in Ayurvedic medicines for curing blood pressure, mental agitation, epilepsy, traumas, anxiety, excitement, schizophrenia, sedative insomnia and insanity. The root contains alkaloid such as 'reserpine'.

REFERENCES

Sihag, R.C. and N. Wadhwa. 2011. Floral and reproductive biology of Sarpagandha, *Rauvolfia serpentina* (Gentianales: Apocynaceae) in semi-arid environment of India. Journal of Threatened Taxa 3(1): 1432-1436.

Reeta Kumari, Brijesh Rathi, Anita Rani and Sonal Bhatnagar. 2013. *Rauvolfia serpentina* L. Benth. ex Kurz.: Phytochemical, Pharmacological and Therapeutic Aspects. Int. J. Pharm. Sci. Rev. Res., 23(2): 348-355

Dey A, De JN. 2010. *Rauvolfia serpentina* (L). Benth. Ex Kurz. - A Review, Asian Journal of Plant Sciences, 9(6), 2010, 285-298.

CHAPTER

72

Senna–*Cassia angustifolia* Vahl. (2n=28)

Synonyms: *Cassia senna* L., *Cassia acutifolia* Delile and *Senna alexandrina* Mill.
Family: Fabaceae
Vernacular names: Senna, Aden senna, Alexandrian senna, Indian senna, Nubian senna, Khartoum senna, True senna (English); *Senna, Sanaka patt, Sonamukhi* (Hindi); *Swarn patri, Sanai* (Sanskrit); *Tinnevelly senna, Nilavarai, Nelavakari* (Tamil).

Importance: Senna can be grown in wasteland development, desertification control and sand dune stabilization areas. India is the largest producer and exporter of senna leaves, pods and total sennosides in the world market. Germany, Hungary, Japan, Netherlands and USA are the major markets of senna.

Origin: Senna is native of India, Pakistan, Sri Lanka, Somalia, Arabia, Sudan and Sinai.

Distribution: It is grown in India, Sudan, Egypt, Sudan, Somalia, Pakistan, China and Korea, the Caribbean and Mexico. In India, it is cultivated in Gujarat, Rajasthan, Karnataka and Tamil Nadu.

Botany: Senna is a small perennial under shrub, grows to a height of 1 to 2 m. The stem is erect, smooth, and pale green to light brown with long spreading ascending branches. It is a deep rooted hardy plant. The leaves are large, compound and pinnate with four to eight pairs of leaflets. The full grown leaflets are bluish-green to pale-green in colour and emit a smell when crushed. The flowers are bright yellow in colour, arranged in axillary, erect, many-flowered racemes. It is predominantly self-pollinated crop but outcrossing can be up to 20% through beetles. The pods appear immediately after flowering. The pods are 2.0 to 3.5 in long, about 0.8 in wide. The pods are green in the beginning changing to greenish-brown to dark brown on maturity and drying. Each pod has 5 to 7 ovate, compressed, smooth, dark-brown seeds. The weight of 1000 seeds is about 30 g.

Climate: It is grown in semi-arid tropical and subtropical climatic conditions. It occurs from sea-level up to 1300 m altitude. Senna is a sun-loving crop. Heavy rains and cloudy weather during growth are harmful to the crop. An average rainfall of 250 to 400 mm, distributed during its growing season is found to sufficient to have a good crop. It is highly sensitive to heavy rainfall and water logging conditions. The crop cannot survive in water logging condition even for a day. It sheds leaves with the onset of cold weather in north-western India.

Soils: The crop can be grown on a variety of soils including on sandy loam lateritic soils, red loams, on alluvial loams with pH of 7.0 to 8.5. The texture of the soil varies from sandy loam to loam, while the black cotton soils are heavier and more fertile. The crop thrives over well-drained since it is sensitive to waterlogging. It is a salt-tolerant crop.

Season: The main crop is largely rainfed, sown with monsoon rains (July in north-western India and September-October in south India), whereas February sown senna is restricted to Tamil Nadu where it is sown with spring rains as an irrigated crop. It is mostly grown as an annual crop, but can be left standing to produce for another 2 to 3 years.

Varieties: ALFT-2 (Gujarat), Tinnevelly Senna (TNAU, Coimbatore), KKM-Se 1 (TNAU, Killikulam), Sona (Lucknow, Uttar Pradesh). As an annual crop, it remains in field for 110–140 days.

Land preparation: The land is prepared with ploughing twice and harrowing once. Senna does not require fine tilth. However, weeds and pebble free land is desirable. FYM is incorporated into the soil at the time of the final cross-ploughing. Then the land is laid out into plots of convenient sizes with irrigation channels.

Seeds and sowing: Seeds are sown either broadcasting or line sowing. Seed rate is 5-6 kg/ha and 7.5 kg/ha under irrigated and rainfed condition when sown through broadcasting. Since the seeds have a hard and tough seed-coat. Soaking of seeds in water for 10-12 hours before sowing provides 90 to 100% germination and even crop stand. Then seed treatment with Captan or Thiram at 3g per kg seed is recommended to avoid seedling diseases. Line sowing of seeds in wet or air-dried condition at 30 × 30 cm or 45 × 30 cm row to row and plant to plant spacing is recommended. Seeds should be placed at a depth of 1-2 cm in soil for uniform germination.

After cultivation practices: The first flower stalks that appear are cut, which induces branching and increases the sennoside content of the leaves.

Weed management: The growth of senna is very slow at initial stages and facilitates the growth of weeds. Once the plants attain 20-25 cm height, they are capable to suppress weed growth. The first weeding cum hoeing is done at 25-30 days of sowing, a second at 75-80 days and a third at 110 days after sowing to keep the crop free from weeds. Pre-emergent spray of Teeflan @ 4 kg/ha is recommended.

Manures and fertilizer application: FYM @ 12.5 t/ha and fertilizer dose of 50-40-20 kg N, P_2O_5 and K_2O is recommended. 50 % N and entire P_2O_5 and K_2O is applied as basal dressing and the remaining 50% N is given as top dressing at 60 to 75 days after sowing.

Water management: It is grown as an annual crop of 5 to 7 months duration both under rainfed and irrigated conditions. One light irrigation should be given immediately after sowing which improves germination to maintain proper plant stand. About 15 to 18 light irrigations are enough to raise under irrigated condition.

Cropping systems: Senna fits well in crop rotation as *kharif* crop in commercially cultivated areas. In Southern India, it is grown after paddy, cotton and sorghum and in North as well as in Western India, senna is followed by mustard, chickpea or coriander. Crop rotation such as Senna-mustard and Senna-coriander are practiced.

Insect pests and disease management: The larvae of leaf eating caterpillars, pod borer, cut-worms and termites are infesting the crop. The release of *Trichogramma chilonis* at 150 thousand populations per hectare per week at the adult stage coinciding with the egg laying stage of the pest controls caterpillars. Spraying of Carbaryl @ 4 g/litre controls the caterpillars infestation. Pod borer can be controlled by spraying Endosulphan @ 0.05% or Carbaryl @ 0.25% at an interval of 10-15 days. Application of 25 kg of Aldrin @ 5% or BHC @ 5% is recommended to manage cut-worms and termites. Seed treatment can be done with Thiram or Captan @ 3 g per kg seeds controls damping off, leaf spot and leaf blight. Two to three sprayings of Dithane M-45 at 0.15% at fortnightly interval also controls these diseases. The fields can be drenched with 0.2% Brassicol or 0.5 to 01. %. to control damping off. The crop should be harvested 25-30 days after spray of pesticides.

Harvest: Senna flowers are borne in 35 to 70 days after sowing and the first flush of flowering stalks should be removed to encourage a higher degree of vegetative growth. Harvesting is done by plucking the leaves when leaves are fully grown, thick and bluish in colour. The first harvest is usually done 90 days after sowing and subsequently second and third harvests at 150 and 210 days after sowing. Under rainfed condition, cutting or uprooting of whole plants after 4-5 months is practiced for harvesting. The colour of dry leaves and pods should be ensured to remain light green to yellow in colour. Harvesting is recommended in dry season to avoid spoilage of leaves due to fungal infection during storage. One to three days sun drying of leaves and pods is advisable. It takes 10-12 days to dry completely in well-ventilated drying sheds. The dried leaves and pods should have light green to greenish yellow colour. Senna plants produce foliage containing higher sennosides between 50-90 days of sowing. The young senna leaves and pods contain a high sennoside content but since the produce is sold on the basis of weight, a balance between weight and content has to be made, to choose the right stage for harvest. The harvested produce should have not more than 8.0% moisture at storage.

Yield: The crop duration ranges from 115 to 140 days A good crop of senna can give 1.5 to 2.0 t/ha of dry leaves and 0.7 to 0.8 t/ha of pods, under irrigated with good management conditions. The yield is about 0.7 to 1.0 t/ha of leaves and 0.2 to 0.4 t/ha of pods under rainfed conditions.

Utilization: Senna leaves and pods both are used for medicinal uses. The leaves are used as natural laxative to treat constipation. It is also useful in the treatment of gonorrhoea, skin diseases, dyspepsia, fevers and haemorrhoids. The leaves, stems, pods, buds and flowers contain sennosides. But the sennosides is not found in the seeds. In general, mature leaves containing 2.0 to 2.5% and pods containing 2.5 to 3.0% of sennosides. The sennoside content is higher when plants are under stress, moderate irrigation and fertilization.

REFERENCES

Balasankar, D., Vanilarasu, K., Selva Preetha, P., Rajeswari, S. Umadevi, M. and Debjit Bhowmik. 2013. Senna: A medical miracle plant. Journal of Medicinal Plants Studies. 1 (3): 41-47.

Pareek, S.K. 1983. Investigation in agronomic parameters of Senna (*Cassia angustifolia* Vahl) as grown in north western India. Int. J. Trop. Agric., 1: 139–144.

Ratnayaka, H.H., Meurer-Grimes. B. and D. Kincaid. 2002. Sennoside yields in Tinnevelly Senna affected by deflowering and leaf maturity. Hort. Sci. 37(5): 768-772.

Singh, S.P., Sharma, J.R., Misra, H.O., Lal, R.K., Gupta, M.M. and S. Tajuddin. 1997. Development of new variety Sona of Senna (*Cassia angustifolia*). J. Med. Aromatic Plant Sci., 19(2).

Tripathi, Y.C. 1999. *Cassia angustifolia*: A versatile medicinal crop. Int. Tree Crops J., 10(2): 121-129.

CHAPTER

73

Stevia–*Stevia rebaudiana* (Bert.) Bertoni

Family: Asteraceae

Vernacular names: Stevia, Sweet leaf, Sweet-herb, Honey leaf, Candy leaf (English); *Madhu patra* (Sanskrit); *Cheeni Tulsi, Mou Tulsi* (Tamil).

Importance: Stevia is commonly called 'sweetleaf' or 'sugarleaf'. The sweetness of stevia comes from its leaves which produce the steviol glycosides. Steviol glycosides are white or slightly yellowish white, crystalline, odourless or having a slight characteristic odour, water soluble powders. There are four main glycosides: stevioside, rebaudioside A, rebaudioside C and dulcoside A. Ground-up stevia leaves, as well as crude leaf extracts, have been used as a natural sweetener all over the world. The crude Stevia leaves and herbal powder (green) are reported to be 10 to 15 times sweeter than table sugar. The refined extracts of Stevia called steviosides are 200 to 300 times sweeter than sucrose/table sugar. The refined Stevia extracts are considered to be non-caloric. It is recommended for diabetes. It is an alternative to sugar and artificial sweeteners like saccharin, aspartame, asulfam, etc. Stevia can be utilized as a source of chlorophyll, phytosterols (non-food: medicine).

Origin: It is native of Paraguay and Brazil.

Distribution: It is grown in Paraguay, Brazil, Colombia, Venezuela, Israel, Canada, USA, China, Korea, Taiwan, Japan, and United Kingdom. Indonesia, Tanzania,

Botany: Stevia is a tender perennial crop, grows up to a height of 65 cm. Leaves are sessile, 3-4 cm long, elongate-lanceolate or spatulate shape with blunt tip lamina, serrate margin from the middle to the tip and entire below. The flowers are white in colour with a pale purple throat. Fruit is a five-ribbed, spindle shaped achene.

Climate: It grows well in humid subtropical regions. It is grown in latitudes from 45°N to 40°S. Plants grown at higher latitudes have a higher percentage of sweet glycosides. Sweetness is intensified by cooler temperatures and short days; however, sugar levels decline after flowering. It is cultivated in areas receiving mean annual rainfall of 900 to 1700 mm. It is a short day plant. Long days favour leaf growth. It shows higher leaf production under high light intensity and warm temperature. Day length is more critical than light intensity. Stevia prefers partial shade during summer season. The cardinal minimum, optimum and maximum temperature is 15, 20-25 and 43°C respectively. The sweet stevia is a strong water loving plant.

Soils: Stevia prefers a well-drained fertile sandy loam or loam soil, high in organic matter with ample supply of water. It prefers acidic to neutral with pH of 6 to 7.5. It requires a consistent supply of moisture. It is sensitive to waterlogged condition, flooding or soft, pudding like consistency or in saline soils.

Varieties: S.R.B-123, S.R.B-512 and S.R.B-128.

Season: The cuttings are planted during the month of February-March.

Land preparation: Land is ploughed and /or harrowed to break down the clods and to prepare fine tilth. With proper drainage and irrigation channels, the field is divided into plots of convenient size. Forming raised beds is the most economical way to grow Stevia. The raised bed should be of 15 cm in height and 60 cm in width.

Propagation: Stevia is usually propagated by stem cuttings of 5 to 10 cm long with two to three leaf buds. Cuttings are treated with Paclobutrazol @ 100 ppm to induce the root initiation. This species can also be propagated by seed. The time from seed-to-transplant is approximately seven to eight weeks. The distance between two rows should be 40 cm and that between each plant 25 cm in a raised bed system.

After cultivation practices: Since the crop is grown in raised beds, inter-cultural operations are easier by manual labour. Flowering of the plant should be avoided. Pinching of the apical bud would enhance busy growth of the plant with side branches.

Manures and fertilizer application: FYM @ 25 t/ha has to be applied as a basal dressing during the last ploughing to incorporate the manure with the soil and fertilizer dose of 60-45-45 kg of N, P_2O_5 and K_2O per ha is recommended. Entire phosphorus is applied as basal dressing. The nitrogen and potash fertilizer can be split and applied as ten doses in every month. Boron deficiency can be rectified by spraying Borax @ 6%.

Water management: Stevia cannot withstand drought. The plants do no wilt for want of water. Micro-sprinklers / drip method of irrigation may be adopted to avoid damage due to excess moisture.

Insect and diseases management: Neem oil diluted in water may be sprayed against aphids if appear. Deer and rabbits are fond of Stevia due to sweet taste. Fencing is necessary.

Harvest: Harvesting is done manually at 8 to 10 cm stem height from the ground level. It is harvested when plants are mature and blooms have just begun to form. Sweetness is intensified by cooler temperatures and short days; however, sugar levels decline after flowering. A rotary mower with a bag attachment can be used for larger areas instead of manual harvesting. First harvest can be made after 4-5 months while the subsequent harvests can be made every 3 months for 3 years. The plants are cut just before flowering as the sweetener in the leaf is the maximum at that time. Leaves are harvested by plucking in a small quantity, or the entire plant with the side branches is cut leaving 10 to 15 cm from the base. Harvested plants are dried. It generally takes 24 to 48 hours to dry stevia at 40 to 50°C depending on weather conditions and density of loading. Dried plants are then threshed to separate the sweet leaves from the bitter stems. Once dried, leaves can be stored.

Yield: The biomass production is 4 to 7 t/ha/year. Of this, the roots, stems and leaves constitute 26 %, 35 % and 39 % respectively. The green to dry biomass ratio is 0.3%.

Utilization: Stevia leaves are 20-30 times sweeter than sugar. Stevia leaves can be dried and stored. Stevia is short duration crop. It is harvested 3 to 4 times a year. The yearly yields can be in the range of 3-4 tonnes. The glycoside content is 9-12% on dry weight basis. Stevia consists of eight glycosides named as stevioside, steviolbioside, rebausiosides A-E, and dulcoside A. Of these eight glycoside, one called stevioside which is 300 times sweeter than sugar.

REFERENCES

Brandle, J.E. and N. Rosa. 1992. Heritability of yield, leaf-steam ratio and stevioside content estimated from a ladrace cultivar of *Stevia rebaudiana*. Can. J. Plant Sci. 72: 1263-1266.

Goyal, S. Samsher, K. and R.K. Goyal. 2010. Stevia (*Stevia rebaudiana*) a bio-sweetener- a review. Int. J. Food Sci. Nutr. 61(1):1-10.

Reshu Gupta, Vidushi Yadav and Manvi Rastogi. 2014. A review on importance of natural sweetener, a zero calorie plant-stevia having medicinal and commercial importance. International Journal of Food and Nutritional Sciences. 3 (3): 90-94

Singh, S.D. and G.P. Rao. 2005. Stevia: The herbal sugar of 21st century. Sugar Tech. 7:17-24.

http://www.uky.edu/Ag/CCD/introsheets/stevia.pdf

CHAPTER

74

Sweet Flag–*Acorus calamus* L. (2n=24, 36)

Family: Acoraceae

Vernacular names: Sweet flag, Sweet calomel, sweet myrtle, myrtle flag, calamus root, flag root, rat root (English); *Bach, Gora-bach, Vasa-bach, Safed-bach* (Hindi); *Haimavati, Bhutanashini, Jatila, Ugragandha* (Sanskrit); *Themepru* (Assamese), *Ganghilovaj, Ghodvach* (Gujarati); *Varch, Ghodavaca* (Punjabi); *Bajegida* (Kannada); *Vashampe, Vayambu* (Malayalam); *Vaca, Vekhandas*, (Marathi); *Bhuta-nashini* (Sanskrit) *Vasambu* (Tamil); *Baje, Narru Berua* (Kannada); *Vacha bach* (Unani) and *Vasa* (Telugu). The generic name is the Latin word acorus is derived from the Greek word 'achorou' and 'kori' which means pupil (of an eye), due to the juice from the root of the plant being used as a remedy in diseases of the eye (darkening of the pupil). The specific name calamus (meaning 'cane') is derived from Greek kálamos, meaning 'reed', which is cognate to Latin culmus (meaning 'stalk'). The term 'sweet' refers to the pleasantly aromatic odour of the plant while the term 'flag' refers to the drooping and prostrate leaves. The name sweet flag refers to its sweet scent.

Importance: It is known for its medicinal value. The leaves have a lemony scent as well as the roots have a sweet fragrance. This plant is used as an ornamental pond plant in horticulture. The fragrant oil of sweet flag has been used for many centuries in perfumes. The economic part is the rhizome (root).

Origin: Sweet flag is native to North America and northern and eastern Asia, and naturalised in southern Asia and Europe.

Distribution: Sweet flag is cultivated in China, India, England, France, Belgium, Sweden, Germany, Russia, Poland, Argentina, Brazil, Pakistan, Sri Lanka, Taiwan, Indonesia, Java, Japan, Philippines, South Africa, Thailand, Turkey and USA. In India, it is grown in Karnataka and the Himalayan region.

Botany: Sweet flag is a perennial herb, semi-aquatic and smelly plant. It grows up to 1.5 to 2.0 m in height. It is aromatic, sword-shaped leaves emerge from a tortuous, branched, underground rhizome with V-shaped leaf scales. The rhizome is whitish-pink internally, cylindrical, 1-2 cm thick and up to a metre long. It has small, yellow/green flowers on a spike-like spadix, which is subtended by a leaf-like spathe. The spadix will turn brown as the seed ripens.

Climate: It is grown in tropical, sub-tropical climate and temperate zones. Plants are hardy to about -25°C. Sweet flag is semi-aquatic, occurring in wetlands, swamps and the edges of streams, marshes, ponds and lakes. Like many other marsh plants, they depend upon aerenchyma to transport oxygen to the rooting zone. These frequently occur on shorelines and floodplains where water levels fluctuate seasonally.

Soils: It prefers, growing in shallow water or in a very moist loamy soil with a pH range of 5.5 to 7.5.

Season: The planting season is February-March and June-July. The crop duration is 10 to 11 months.

Land preparation: The field is prepared similar to irrigated rice. The field is puddled with incorporation of FYM or green leaf manure.

Propagation: This plant is propagated through roots/rhizomes and seeds. Seed should be planted in a greenhouse. Scatter seed sparsely on the surface and press firmly into the soil. Do not bury further than 3-5 mm deep. Keep the soil from moist to saturate. Seed does not require stratification and germinates in 2 weeks. When plants reach 7.5 to 10 cm height, transplant them into individual 10 cm pots. Pots can be placed in shallow water or irrigated frequently to maintain very moist to saturated conditions. Keep soil very moist to saturated, sweet flag does not tolerate droughty conditions. It should be planted with spacing of 30 × 30 cm where it will be in full sun to partial shade.

Weed management: Weeding is done once in a month for the first 4 to 5 months.

Manures and fertilizer application: Starter fertilizers may be used indoors to improve early growth. Apply 12.5 t/ha of FYM and 45-12.5-12.5 kg/ha of N, P_2O_5 and K_2O as top dressing in 2 to 3 splits after 60 days after planting is recommended.

Water management: The field is flooded with 5 cm. It grows well under seasonal shallow inundation. However, avoid flooding of newly established plants or seeded areas.

Harvest: The leaf tips begin to turn yellow indicates maturity. Rhizomes grows 30 to 60 cm long. Collect the rhizomes when large and firm, generally after 2 to 3 years of growth, before becoming hollow. Rhizomes are cut into 5 to 7.5 cm long pieces. It is dried in sun and later beaten and rubbed to 2 to 3 times to remove the leaf scales and fibrous roots.

Yield: The dry rhizome yield is 2.5 to 4 t/ha. The dried rhizomes and matured leaves at the time of harvest contain 1.85 and 0.1% calamus oil respectively.

Utilization: Sweet flag rhizome is used in perfumes. Sweet flag is used as an insecticide, an antifungal agent, an antibacterial agent, and a fish toxin. The plant has also been used as an insecticide to repel fleas and protect clothes and grain from moths and the powder is sprinkled around the bases of trees to protect them from infestation of termites. The rhizome, mixed with garlic, cumin seeds, salt, sugar, and butter, is used in veterinary medicine to strengthen horses. It is one of the ingredients used in the treatment of foot and mouth disease. Its reputed ability to arouse sexual desire resulted in sweet flag being tagged with the name 'Venus plant'. Sweet flag is one of many ingredients used in a love potion in medieval times. Women are given the rhizome for painful menstruation. The rhizome is used to treat bronchitis, coughs, colds and to clear the voice. Sweet flag contains toxic alkaloid namely, carcinogenic beta-asarone. The leaves also possess the aromatic properties of the rhizome. Leaves are not employed as a medicine.

REFERENCES

Agarwal, S. L., P. C. Dandiya, K. P. Singh, and R. B. Arora. 1956. A note on the preliminary studies of certain pharmacological actions of Acorus calamus. Journal of the American Pharmaceutical Association 45:655-656.

Alankararao, G. S. J. G., and Y. Rajendra Prasad. 1981. Antimicrobial property of *Acorus calamus* L.: invitro studies. Indian Perfumer. 25:4-6.

Bhaskar, F., P. Saxena, 0. Koul, and K. Tikku. 1977. Inhibition of embryonic development by oil of *Acorus calamus* L. treated males in *Dysdercus koenigii*. Indian Perfumer. 21:73-78.

Bhattacharya, I. C. 1968. Effect of vacha (*Acorus*) oil on the amphetamine induced agitation, hexobarbital sleeping time, and instrumental avoidance. Journal of Research in Indian Medicine 2:195.

Bose, B. C., R. Vijayvargiya, A. Q. Saifi, and S. K. Sharma. 1960. Some aspects of chemical and pharmacological studies of *Acorus calamus*. Journal of the American Pharmaceutical Association 49:32.

Motley, T.J. 1994. The ethnobotany of sweet flag, *Acorus calamus* (Araceae). Econ. Bot. 48: 397-412.

Panchal, G.M., Venkatakrishna-Bhatt, H., Doctor, R.B., and S. Vajpayee. 1989. Pharmacology of *Acorus calamus* L. Indian J. Exp. Biol. 27: 561-567.

CHAPTER

75

Vasa–*Adhatoda vasica* Nees

Synonym: *Adhatoda zeylanica* Medic

Family: Acanthaceae

Vernacular names: Adhatoda, Malabar nut tree (English); *Adulsa, Adalsa, Vasaka* (Hindi); *Amalaka, Bashika, Vasaka* (Sanskrit); *Basaka* (Bengali); Titabahak, Bahak, Vachaka (Assamese); *Adhatoda* (Tamil); *Vasuka* (Marathi); *Adasaramu* (Telugu); *Adsale, Adusoge, Atarusha, Adsole, Adasale* (Kannada); *Atalotakam* (Malayalam).

Importance: It is an important and widely used medicinal plant. Cattle do not eat this plant as the leaves emit an unpleasant smell. It can be used as green leaf manure crop.

Origin: The plant is distributed all over the plains of India and in lower Himalayan ranges, ascending to a height of 1,500m. Vasaka is indigenous to India.

Distribution: It is grown in India, Pakistan and Nepal. It grows all over the India and especially in the lower Himalayan ranges.

Botany: It is an evergreen shrub of 2 to 3 m in height. Stem herbaceous above and woody below. Leaves are simple, opposite, ovate-lanceolate, acute and shiny. Inflorescences in axillary spicate cymes, densely flowered; peduncles short; bracts broadly ovate, foliaceous. The flowers are either white or purple in colour. It has capsular four seeded fruits.

Climate: It is distributed throughout India up to an altitude of 1300 m from MSL. The plant tolerate shade but is susceptible to water logging condition.

Soils: It can be grown on light loamy to loamy soil. It is grown well in alluvial soils.

Season: Nursery can be raised in March. It can be planted is during July-August.

Propagation: The crop is propagated through stem cuttings of 20 to 30 cm long with 3 to 4 nodes. Planting is done with 60 × 30 cm preferably in ridges and furrows.

Weed management: Weeding and hoeing is to be done at regular interval as and when necessary.

Manures and fertilizer application: Compost/ FMY @ 12.5 t/ha is recommended. A fertilizer dose of 50-50-50 kg of N, P_2O_5 and K_2O per ha is recommended. Half the dose of N, the whole P and whole K may be applied as the basal dose followed by the remaining half N at 30 days after planting as top-dressing.

Harvest: The first harvest of the biomass (leaf, stem, roots) is harvested two years of planting and subsequently once in a year.

Yield: The biomass yield is 10 to 12 t/ha/year.

Utilization: The leaves, flowers, fruit and roots are extensively used for treating cold, cough, whooping cough, chronic bronchitis and asthma, as sedative, expectorant and antispasmodic. The powder of herb boiled with sesame oil is used to heal ear infections. Boiled leaves are used to relieve the pain of urinary tract infections and to stimulate uterine contractions, thus speeding child birth. A decoction of the leaves may be used to help with cough and colds. The leaves contain alkaloids such as Quinazoline, vasicine, vasicinone, vasicol, adhatodinine and vasicinol.

Annexure I
Scientific Measurements in Crop Production

Weights and measures are the necessaries of life to every individual of human beings. Measurement is among one of the first intellectual achievements of early humans. The English System of measures had their origins in a variety of cultures viz., Babylonian, Egyptian, Roman, Anglo-Saxon, and Norman French. English units are the historical units of measurement used in England up to 1824. English system of measurement units became established in many parts of the world, including the American colonies. The term 'English units' is ambiguous, as it could refer either to the imperial units used in the UK, or to United States customary units, which retains some unit names but has some different definitions. (The terms imperial units or imperial measurements are used in the UK to refer to the non-metric system since they were used as a standard throughout the British Empire and the Commonwealth). Our words 'inch' and 'ounce' are both derived from that Latin word. The English system of measurement grew out of the creative way that people measured for themselves. Familiar objects and parts of the body are used as measuring devices. For example, people measured shorter distances on the ground with their feet. In England distances have been defined with reference to body features of the king. The 'yard' as a measure of length can be traced back to early Saxon kings. The word 'yard' comes from the Saxon word 'gird' meaning the circumference of a person's waist. A 'yard' is the circumference of his waist, an 'inch' is the width of his thumb, and a 'foot' the length of his foot. English farmers, however, estimated lengths in something they could more easily relate to: "furlongs", the length of an average plowed furrow. There are, for example, many units in which distance can be measured in the Customary system, but they bear no logical relationship to each other: 1 statute mile = 0.8688 nautical miles = 1,760 yards = 320 rods = 8 furlongs =5280 feet = 63360 inches = 880 fathoms = 15840 hands. Some of the most well-known of the early units of measurement are:

- inch - the width of the thumb.
- digit - the width of the middle finger (about 3/4 inch)
- palm - the width of four fingers (about 3 inches)
- span - the distance covered by the spread hand (about 9 inches)
- foot - the length of the foot. Later expressed as the length of 36 -barleycorns taken from the
- middle of the ear (about 12 inches).
- cubit - distance from the elbow to the tip of the middle finger (about 18 inches).
- yard - distance from the center of the body to the fingertips of the outstretched arm (about 36 inches).
- fathom - distance spanned by the outstretched arms (about 72 inches).

The particulars of conversion of English to Metric units are furnished in Table 1.

Table 1 Conversion of English to Metric Units

Length:	Weight:	Capacity:
1 in = 2.54 cm 1 ft = 30.48 cm = 0.305 m 1 yd = 0.914 m 1 mi = 1.609 km 12 inches (in) = 1 foot (ft) 3 feet = 1 yard (yd) 5280 feet = 1 mile (mi) 660 Feet = 1 Furlong 5,280 Feet = 1 Mile 1,760 Yards = 1 Mile 1 Mile = 8 Furlongs 1 mile = 1.6 kilometers 22 Yards = 1 Chain	1 oz = 28.350 g 1 lb = 0.453 kg 1 lb = 453.6 g 16 ounces (oz) = 1 pound (lb) 2000 lb = 1 ton Area: 1 Acre = (1 Chain) × (1 Furlong)	80 drops = 1 teaspoon or about 1/6 fluid ounce = 5 milliliters or cubic centimeters (cc); 1 tablespoon = 3 teaspoons = 15 milliliters (ml) or cubic centimeters (cc) = 1/2 fluid ounce; 1 tsp = 5 ml 1 c = 236 ml 1 qt = 0.946 l 1 gal = 3.785 l 3 teaspoons (tsp) = 1 tablespoon (tbsp) 1 cup (c) = 16 tablespoons 1 cup =8 fluid ounces 1 cup =236.6 milliliters (ml) or cubic centimeters (cc) 1 cup =1/2 pint 1 pint = 16 fluid ounces 1 fluid ounce = 30 milliliters 8 ounces (oz) = 1 c 2 c = 1 pint (pt) 2 pt = 1 quart (qt) 4 qt = 1 gallon (gal) 1 pint = 0.47 liter 1 quart = 0.95 liter 1 milliliter = 1 cubic centimeter (cc) = 0.2 teaspoon; 1 liter = 1,000 milliliters (ml) or cubic centimeters (cc)

http://www.regentsprep.org/regents/math/algebra/am2/MetEng.htm

Measurement Equivalency Chart

A teaspoon or tablespoon throughout this table refers to a level, standard measuring teaspoon or tablespoon.

1 cup = 16 tablespoons

1 cup = 8 fluid ounces

1 cup = 236.6 milliliters (ml) or cubic centimeters (cc)
1 cup = 1/2 pint
1 pint = 16 fluid ounces
(NOTE: 1 pint or quart dry measure is about 16 percent larger than 1 pint or 1 quart liquid measure.)
473.2 milliliters (ml) or cubic centimeters (cc)
1 fluid ounce = 2 tablespoons or 29.6 milliliters (ml)
or cubic centimeters (cc)
1 U.S. gallon = 4 quarts
= 8 pints
= 3,785 milliliters (ml) or cubic centimeters (cc)
= 8.3 pounds (lb) water
1 pound = 16 ounces = 453.59 grams
1 kilogram = 1,000 grams, approximately 2 pounds 3 ounces =1 ounce = 28.4 grams
1 bushel of soil = 1.25 cubic feet
1 mile = 5,280 feet
= 320 rods
= 1,609.4 meters
1 acre = 43,560 square feet
= 160 square rods
= 0.4047 hectare
10 millimeters (mm) = 1 centimeter (cm)
= 0.3937 inch
= 100 centimeters = 1 meter (m)
= 39.37 inches

Metric system: The modern Metric System is developed by the French in the late 1700's based on powers of ten. The abbreviated International System of Units (SI), which is French for 'Systeme International'. The term 'SI' is an abbreviation for Le Systeme International d'Unites (from French) or The International System of Units. The SI has been established in 1960 by the 11th General Conference on Weights and Measures (CGPM, Conférence Générale des Poids et Mesures). The CGPM is the international authority that ensures wide dissemination of the SI and modifies the SI as necessary to reflect the latest advances in science and technology. American Society of Agronomy committee has recommended adoption of SI units as a uniform method of reporting quantitative measurements (Thein and Oster, 1981).

Why Metric system?

The metric system is much easier than English (customary) system. Nearly the entire world (>95%), except the United States, now uses the metric system. United States use the antiquated English system of pounds, inches, feet, and so on. Metric is used exclusively in science. All metric units are related by factors of 10. Because the metric system uses units related by factors of ten and the types of units (distance, area, volume, mass) are simply-related, performing calculations with the metric system is much easier. The International System of Units of measurement are based upon seven base units: meter (distance), kilogram (mass), second (time), ampere (electrical current), Kelvin (temperature), mole (quantity), and candela (luminous intensity). The particulars of conversion of Metric to English units are furnished in Table 2.

Table 2 Common Prefixes Used in the Metric System

The Metric System

Prefix	Scientific Notation	Decimal equivalents	Example Units
mega (M)	$= 10^6$	= 1,000,000	
kilo- (k)	$= 10^3$	= 1000	kilogram (kg); kilometer (km)
hecta (H)	$= 10^2$	= 100	
deca (D)	$= 10^1$	= 10	
deci (d)	$= 10^{-1}$	= 0.1	
centi- (c)	$= 10^{-2}$	= 0.01	centimeter (cm)
milli- (m)	$= 10^{-3}$	= 0.001	milligram (mg); millimeter (mm)
micro- (μ)	$= 10^{-6}$	= 0.000001	microgram (μg) microliter (μL)
nano- (n)	$= 10^{-9}$	= 0.000000001	nanogram (ng) nanoamperes (nA)
pico- (p)	$= 10^{-12}$	= 0.000000000001	picogram (pg) picoamperes (pA)
1 meter = 100 centimeters= 1000 millimeters			

Scientific notation and significant figures: When dealing with very large or small numbers, it is best to use scientific notation. In scientific notation, numbers are expressed as the product of a number between 1 and 10 and a whole-number power (exponent) of 10. The exponent indicates how many times a number must be multiplied by itself. Some examples follow: $10^1 = 10$, $10^2 = 10 \times 10 = 100$, $10^3 = 10 \times 10 \times 10 = 1000$.

Exponents may also be negative. For example, $10^{-1} = 1/10 = 0.1$. Also, $10^{-2} = 1/100 = 0.01$ and $10^{-3} = 1/1000 = 0.001$. Following are some examples of numbers written in scientific notation:

$30 = 3 \times 10^1$ $\quad 150 = 1.5 \times 10^2$ $\quad 60,367 = 6.0367 \times 10^4$

$0.3 = 3 \times 10^{-1}$ $\quad 0.046 = 4.6 \times 10^{-2}$ $\quad 0.000\,002 = 2 \times 10^{-6}$

Length: Length is the distance between two points. The SI base unit for length is the meter.

Mass: Mass is the amount of matter that makes up an object. The SI unit for mass is the gram.

Weight: Weight is a measure of the force of gravity on an object. Gravity is the force of attraction between any two objects with mass. The force depends on two things:

more distance = less gravity = less weight

less distance = more gravity = more weight

more mass = more gravity = more weight

less mass = less gravity = less weight

The SI unit for weight is the Newton (N). The English unit for weight is the pound.

Volume

Volume is the amount of space contained in an object.

Volume = length × width × height

The base unit for volume is the Liter.

Water Mass and Volume: 1cm^3 water =1 ml of water = 1 gram

So what would be the mass of 50 mL of water be?

50 grams

So what would be the mass of 1 liter of water be?

1 L = 1000 mL so its mass would be 1000 grams or a kilogram.

Density: Density is the amount of matter (mass) compared to the amount of space (volume) the object occupies. We will measure mass in grams and volume in ml or cm^3.

Density is mass divided by volume.

Density = mass/volume

Remember, all fractions are division problems. Since the unit for mass is grams, and the unit for volume is ml or cm^3, then the unit for density is g/ml, or g/ cm^3

Water and Density: Since 1 gram of water has a volume of 1 mL, then the density of water will always be 1 gram/ml. 50 mL of water will have a mass of 50 grams, so again the density of pure water will be 1 g/ml. A kg of water will have a volume of 1000 mL, so it's density will be 1 gram/ml.

Floating and Sinking: Less dense materials will float on top of more dense materials. Objects with a density of less than 1 g/mL will float on top of water. Objects with a density greater than 1 g/mL will sink in water.

Neutral Buoyancy: Objects with a density equal to the density of water will float in mid water, at whatever level you place the object. Fish and submarines control their depth by changing their density.

Objects that Sink: Objects with a density greater than 1 g/mL will sink in water.

Table 3 SI Base Units

Quantity	Unit Name	Symbol
length	meter	m
mass	kilogram	kg
time	second	s
electric current	ampere	A
thermodynamic temperature	kelvin	K
amount of substance	mole	mol
luminous intensity	candela	cd

Table 4 SI Base Units equals

Quantity	Unit	1 such unit equals:
Area	square metre	1 metre × 1 metre
Volume	cubic metre	1 metre × 1 metre × 1 metre
Speed	metre per second	1 metre ÷ 1 second
Acceleration	metre per second per second	1 metre ÷ 1 second ÷ 1 second
Force	newton	1 kilogram × 1 metre ÷ 1 second ÷ 1 second
Energy	joule	1 kilogram × 1 metre × 1 metre ÷ 1 second ÷ 1 second

Power	watt	1 kilogram × 1 metre × 1 metre ÷ 1 second ÷ 1 second ÷ 1 second
one kilobit	1 kbit	10^3 bit = 1000 bit
one megabyte	1 MB	10^6 B = 1 000 000 B
one gigabyte	1 GB	10^9 B = 1 000 000 000 B

For ease of understanding and convenience, SI derived units have been given special names and symbols, as shown in Table 5 and Table 6.

Table 5 Examples of SI derived units

Derived quantity	**Name**	**Symbol**
area	square meter	m^2
volume	cubic meter	m^3
speed, velocity	meter per second	m/s
acceleration	meter per second squared	m/s^2
wave number	reciprocal meter	m^{-1}
mass density	kilogram per cubic meter	kg/m^3
specific volume	cubic meter per kilogram	m^3/kg
current density	ampere per square meter	A/m^2
magnetic field strength	ampere per meter	A/m
amount-of-substance concentration	mole per cubic meter	mol/m^3
luminance	candela per square meter	cd/m^2
mass fraction	kilogram per kilogram, which may be represented by the number 1	kg/kg = 1

Table 6 SI prefixes and symbols

Factor	**Decimal Representation**	**Prefix (name)**	**Symbol**	**Meaning in US**	**Meaning in other countries**
10^{24}	1 000 000 000 000 000 000 000 000	Yotta	Y		
10^{21}	1 000 000 000 000 000 000 000	Zetta	Z		
10^{18}	1,000,000,000,000,000,000	exa	E	one quintillion times	trillion
10^{15}	1,000,000,000,000,000	peta	P	one quadrillion times	thousand billion
10^{12}	1,000,000,000,000	tera	T	one trillion times	billion
10^9	1,000,000,000	Giga (jiga)	G	one billion times	milliard
10^6	1,000,000	mega	M	one million times	

10^3	1,000	kilo	k	one thousand times	
10^2	100	hecto (heck toe)	h	one hundred times	
10^1	10	deka	da	ten times	
10^0	1				
10^{-1}	0.1	deci	d	One tenth of	
10^{-2}	0.01	centi	c	one hundredth of	
10^{-3}	0.001	milli	m	one thousandth of	
10^{-6}	0.000 001	micro	m	one millionth of	
10^{-9}	0.000 000 001	nano	n	one billionth of	milliardth
10^{-12}	0.000 000 000 001	pico	p	one trillionth of	billionth
10^{-15}	0.000 000 000 000 001	femto	f	one quadrillionth of	thousand billionth
10^{-18}	0.000 000 000 000 000 001	atto	a	one quintillionth of	trillionth
10^{-21}	0.000 000 000 000 000 000 001	zepto	z		
10^{-24}	0.000 000 000 000 000 000 000 001	yocto	y		

Source: http://www.csun.edu/~vceed002/herr/hands_on_science.htm

Table 7 Units outside the SI that are accepted for use with the SI

Name	**Symbol**	**Value in SI units**
minute (time)	min	1 min = 60 s
hour	h	1 h = 60 min = 3600 s
day	d	1 d = 24 h = 86 400 s
degree (angle)	°	$1° = (\pi/180)$ rad
minute (angle)	′	$1' = (1/60)° = (\pi/10\,800)$ rad
second (angle)	″	$1'' = (1/60)' = (\pi/648\,000)$ rad
liter	L	$1\ L = 1\ dm^3 = 10^{-3}\ m^3$
metric ton	t	$1\ t = 10^3$ kg

Table 8. Other units outside the SI that are currently accepted for use with the SI

Name	**Symbol**	**Value in SI units**
nautical mile		1 nautical mile = 1852 m
knot		1 nautical mile per hour = (1852/3600) m/s

are	a	$1\ a = 1\ dam^2 = 10^2\ m^2$
hectare	ha	$1\ ha = 1\ hm^2 = 10^4\ m^2$
bar	bar	$1\ bar = 0.1\ MPa = 100\ kPa = 1000\ hPa = 10^5\ Pa$
ångström	Å	$1\ Å = 0.1\ nm = 10^{-10}\ m$
barn	b	$1\ b = 100\ fm^2 = 10^{-28}\ m^2$
Fraction values		
per cent	%	10^{-2}
parts per million ppm 10^{-6}		
parts per million	ppm	$\mu L\ L^{-1}$, $mg\ L^{-1}$, or $mg\ kg^{-1}$

Table 9 Units officially accepted for use with the SI

Name	Symbol	Quantity	Equivalent SI unit
minute	min	time (SI unit multiple)	1 min = 60 s
hour	h	time (SI unit multiple)	1 h = 60 min = 3600 s
day	d	time (SI unit multiple)	1 d = 24 h = 1440 min = 86400 s
degree	°	plane angle (dimensionless unit)	$1° = (\pi / 180)$ rad
minute	′	plane angle (dimensionless unit)	$1' = (1 / 60)° = (\pi / 10800)$ rad
second	″	plane angle (dimensionless unit)	$1'' = (1 / 60)' = (1 / 3600)° = (\pi / 648000)$ rad
hectare	ha	area (decimal unit multiple)	$1\ ha = 100\ a = 10000\ m^2 = 1\ hm^2$
litre	l or L	volume (decimal unit multiple)	$1\ L = 1\ dm^3 = 0.001\ m^3 = 1000\ cc = 1000\ cm^3$
tonne	t	mass (decimal unit multiple)	$1\ t = 10^3\ kg = 1\ Mg$
bar	bar	pressure	$1\ bar = 10^5\ Pa$
millibar	mbar	pressure	1 mbar = 1 hPa = 100 Pa
atmosphere	atm	pressure	1 atm = 1013.25 mbar = 1013.25 hPa = 101325 Pa exactly

Table 10 Customary units of measurement deemed to be derived from the International System of Weights and Measures

Unit	Abbreviation or Symbol	Base SI Unit from which derived	Value
Area			
square mile	------	metre	2.5899 sq kilometres
acre	------	metre	4046.86 sq metres
square rod	Sq rd	metre	24.687 sq metres
square yard	sq yd	metre	0.8361 sq metre
square foot	sq ft	metre	0.0929 sq metre
square inch	sq in	metre	645.16 sq millimetres

Capacity measurement / Volume			
barrel	bbl	metre	0. I5987294 cubic metre
gallon	gal	metre	0.00454609 cubic metre
Minim (1 /76800 gal)	-----	metre	value of the units is proportionate to the value of gallon
fluid drachm (1/1280 gal)	-----	metre	
pint (U8 gal)	pt	metre	
quart (114 gal)	qt	metre	
Length			
Yard	yd	metre	0.9 I44 metre
mile (1,760 yds)	m	metre	value of the units is proportionate to the value of yard
chain (22 yds)	ch	metre	
foot (▯ yd)	ft	metre	
rod, pole or perch (5½ yds)	-----	metre	
inch (1136 yd)	in	metre	
Mass			
pound	lb	kilogram	0.4535 I237 kg
grain (l/7OOO Ib)	gr	kilogram	value of the units is proportionate to the value of pound
ounce (]/I6 Ib)	oz	kilogram	
stone (14 Ibs)	st	kilogram	
quarter (28 Ibs)	qr	kilogram	
hundredweight (112 Ibs)	cwt	kilogram	
ton (2,240 Ibs)	t	kilogram	
Volume			
cubic yard	cu yd	metre	0.764554857 cubic metre
cubic foot (1/27 cu yd)	cu ft	metre	value of the units is proportionate to the value of cubic yard
cubic inch (1/46,656 cu yd)	cu in	metre	
Troy Series			
ounce troy (12/175 Ib)	oz tr	kilogram	------

Table 11 Non-SI units accepted for use with the International System of units

Quantity	**Name of unit**	**Symbol for unit**	**Value in SI units**
time	minute	min	1 min = 60 s
	hour	h	1 h = 60 min = 3600 s
	day	**d**	1 d = 24 h = 86 400 s
plane angle	degree	o	1o = (π/180) rad
	minute	′	1′ = (1/60)o = (π/ 10 800) rad
	second	″	1″ = (1/60)′ = (π/ 648 000) rad
area	hectare	ha	1 ha = 1 hm2 = 104 m^2
volume	liter	L	1 L = 1 dm3 = 103 cm^3 = 10^{-3} m^3

mass	metric ton	t	1 t = 103 kg
pressure	bar	bar	1 bar = 0.1 MPa = 100 kPa = 105 Pa
	millimeter of mercury	mmHg	1 mmHg ≈ 133.322 Pa
length	ångström	Å	1 Å = 0.1 nm = 100 pm = 10–10 m
distance	nautical mile	M	1 M = 1852 m
area	barn	b	1 b = 100 fm^2 = (10–12 $cm)^2$ = 10–28 m^2
speed	knot	kn	1 kn = (1852/3600) m/s

Table 12 Conversion of Metric to English units

Length: 1 mm = 0.04 in 1 cm = 0.39 in 1 m = 39.37 in = 3.28 ft 1 m = 1.09 yd 1 km = 0.62 mi 1 kilometer (km) = 1000 meters (m) 1 centimeter (cm) = .01 meter (m) 1 millimeter (mm) = .001 meter (m)	**Volume (Capacity):** 1 milliliter = .001 liter (l) 1 ml = 0.2 tsp 1 1 = 1.057 qt (1L = 1.057 qt) 1 qt = 946 mL
Weight: 1 g = 0.035 oz 1 L = 1.057 qt 1 kilogram (kg) = 1000 grams (g) = 10^3 g 1 milligram (mg) = .001 gram (g) 1 mg = 0.001 g = 10^{-3}g	Kilo means thousand (1000) Hecto means hundred (100) Deca means ten (10) Deci means one-tenth (1/10) Centi means one-hundredth (1/100) Milli means one-thousandth (1/1000)

Note: 10^{-6} kg = 1 mg (one milligram), **but not** 10^{-6} kg = 1 μkg (one microkilogram)

Table 13 Conversion operations in the English (customary) and Metric systems

English System	**Metric System**
1. Units of Distance	
12 in = 1 ft	
3 ft = 1 yd	10 mm = 1 cm
1760 yds = 1 mi	100 cm = 1 m
5280 ft = 1 mi	1000m = 1 km
1 inch = 2.54 cm; 1 mile = 1.61 km	
1 foot = 0.3048 meter (m)	
2. Units of Area	
144 in^2 = 1 ft^2	10,000 cm^2 = 1 m^2
43,560 ft^2 = 1 acre	10,000 m^2 = 1 hectare
640 acres = 1 mi^2	100 hectare = 1 km^2
1 in^2 = 6.45 cm^2; 1 mi^2 = 2.59 km^2	

3. Units of Volume	
57.75 in^3 = 1 qt	1 cm^3 = 1 ml
4 qt = 1 gal	1000 ml = 1 liter
42 gal (petroleum) = 1 barrel	1000 liter = 1 m^3
32 qt = 1 bushel	
16.39 cm3 = 1 in^3	3.79 liters = 1 gal
4. Units of Mass	
437.5 grains = 1 oz	1000 mg = 1 g
16 oz = 1 lb	1000 g = 1 kg
2000 lb = 1 short ton	1000 kg = 1 metric ton
453 g = 1 lb; 2.2 lb = 1 kg	

Note: (Mass-volume conversions for water, or material of equal density, are also easy because 1 kg of water = 1 liter = 1/1000 m^3 and 1 g of water = 1 cm^3).

Table 14 Conversion operations in the English (customary) and Metric systems

To Convert From	To	Multiply By
Quantities of Space and Time		
Plane angle		
Radian	degree arc	57.29578
Solid angle		
Length		
angstrom	nanometer (nm)	0.1
Fathom	meter (m)	1.828 804
foot (ft)	meter (m)	0.304 8
foot [U.S. survey] [12]	meter (m)	0.304 800 6
inch (in)	centimeter (cm)	2.54
inch (in)	millimeter (mm)	25.4
microinch (μin)	micrometer (μm)	0.025 4
mil (0.001 inch)	millimeter (mm)	0.025 4
mil (0.001 inch)	micrometer (μm)	25.4
yard (yd)	meter (m)	0.914 4
mile, international (5280 ft) (mi)	kilometer (km)	1.609 344
nautical mile[13]	kilometer (km)	1.852
point (printer's)	millimeter (mm)	0.351 46
pica	millimeter (mm)	4.217 5

Table 15 Conversion operations in the English (customary) and Metric systems

Section	To Convert From	To	Multiply By
Area			
acre[14]		square meter (m^2)	4 046.873
acre		hectare[15](ha)	0.404 687 3
circular mil		square millimeter (mm^2)	0.000 506 708
square inch		square centimeter (cm^2)	6.451 6
square inch		square millimeter (mm^2)	645.16
square foot		square meter (m^2)	0.092 903 04
square yard		square meter (m^2)	0.836 127 36
square mile		square kilometer (km^2)	2.589 988
Volume			
acre-foot		cubic meter (m^3)	1 233.489
barrel, oil[16] (42 U.S. gallons)		cubic meter (m^3)	0.158 987 3
barrel, oil (42 U.S. gallons)		liter (L)	158.987 3
cubic yard		cubic meter (m^3)	0.764 555
cubic foot		cubic meter (m^3)	0.028 316 85
cubic foot		liter (L)	28.316 85
board foot		cubic meter (m^3)	0.002 359 737
register ton[17]		cubic meter (m^3)	2.831 685
bushel[18]		cubic meter (m^3)	0.035 239 07
gallon		liter (L)	3.785 412
quart (liquid)		liter (L)	0.946 352 9
pint (liquid)		liter (L)	0.473 176 5

Table 16 Conversion operations in the English (customary) and Metric systems

To Convert From	To	Multiply By
fluid ounce[19]	milliliter (mL)	29.573 53
cubic inch	cubic centimeter (cm^3)	16.387 064
Time		
Velocity		
foot per second	meter per second (m /s)	0.304 8
mile per hour	kilometer per hour (km/h)	1.609 344
knot[21] (nautical mile per hour)	kilometer per hour (km/h)	1.852
Acceleration		
inch per second squared	meter per second squared ($m \cdot s^{-2}$)	0.025 4
foot per second squared	meter per second squared ($m \cdot s^{-2}$)	0.304 8
standard acceleration of gravity (*g*)	meter per second squared ($m \cdot s^{-2}$)	9.806 65
Flow rate		
cubic foot per second	cubic meter per second (m^3 /s)	0.028 316 85

cubic foot per minute	cubic meter per second (m^3 /s)	0.000 471 9474
cubic foot per minute	liter per second (L/s)	0.471 947 4
cubic yard per minute	liter per second (L/s)	12.742 58
gallon per minute	liter per second (L/s)	0.063 090 2
gallon per day	liter per day (L/d)	3.785 412
Fuel efficiency		
mile per gallon	kilometer per liter (km/L)	0.425 143 7
Quantities of Mechanics		
Mass (weight)		
ton (long)[24] (2240 lb)	kilogram (kg)	1 016.047
ton (long)	metric ton (t)	1.016 047

Note: In the United States, the cup, tablespoon, and teaspoon are defined as 8, ½, and 1/6 fluid ounces, respectively. For practical usage the metric equivalents are 250 mL, 15 mL, and 5 mL.

Fuel consumption (e.g., litre/kilometer) is the reciprocal of fuel efficiency. Thus, 20 mile/gallon fuel efficiency is equal to 20 (0.42514) = 8.503 km/L, which is equivalent to a fuel consumption of 1/8.503=0.1176 L/km, or more conveniently 11.76 L/ 100 km.

The hectare, equal to 10 000 m^2

The liter, equal to 0.001 m^3

The day (d = 86 400 s), hour (h = 3600 s), and minute (m=60 s). (The year is not an admitted unit, though for rough calculations it's usually adequate to assume 1 year ‘ 365.25 d.)

Table 17 Conversion operations in the English (customary) and Metric systems

To Convert From	To	Multiply By
ton (short) (2000 lb)	kilogram (kg)	907.184 74
ton (short)	metric ton (t)	0.907 184 74
slug	kilogram (kg)	14.593 9
pound (avoirdupois)	kilogram (kg)	0.453 592 37
ounce (troy)	gram (g)	31.103 48
ounce (avoirdupois)	gram (g)	28.349 52
grain	Milligram (mg)	64.798 91
Moment of mass		
pound foot	kilogram meter (kg · m)	0.138 255
Density		
ton (2 000 lb ([short]) per cubic yard	kilogram per cubic meter (kg /m^3)	1 186. 553
	metric ton per cubic meter (t /m^3)	1.186 553
pound per cubic foot	kilogram per cubic meter (kg /m^3)	16.018 46
Concentration (mass)		
pound per gallon	gram per liter (g /L)	119.826 4
ounce (avoirdupois) per gallon	gram per liter (g /L)	7.489 152

Momentum		
pound foot per second	kilogram meter per second (kg · m /s)	0.138 255 0
Moment of inertia		
pound square foot	kilogram square meter (kg · m^2)	0.042 140 11
Force		
pound-force	newton (N)	4.448 222
poundal	newton (N)	0.138 255 0
Moment of force, torque		
pound-force foot	newton meter (N · m)	1.355 818
pound-force inch	newton meter (N · m)	0.112 984 8
Pressure, stress		
standard atmosphere[25]	kilopascal (kPa)	101.325

The SI unit for pressure and stress is the pascal, which is equal to the newton per square meter.

Crop	Weight per bushel (kg)
Barley	21.8
corn, shelled	25.4
oats	14.5
potatoes, soybeans, wheat	27.2

Source: U.S. Department of Agriculture

Table 18 Conversion operations in the English (customary) and Metric systems

To Convert From	To	Multiply By
bar	kilopascal (kPa)	100
millibar	kilopascal (kPa)	0.1
pound-force per square inch (psi)	kilopascal (kPa)	6.894 757
kilopound-force per square inch	megapascal (MPa)	6.894 757
pound-force per square foot	kilopascal (kPa)	0.047 880 26
inch of mercury (32 °F)	kilopascal (kPa)	3.386 38
foot of water (39.2 °F)	kilopascal (kPa)	2.988 98
inch of water (39.2 °F)	kilopascal (kPa)	0.249 082
millimeter of mercury (32 °F)	kilopascal kPa)	0.133 322 4
torr (Torr)	pascal (Pa)	133.322 4
Viscosity (dynamic)		
centipoise	millipascal second (mPa · s)	1
Viscosity (kinematic)		
centistokes	square millimeter per second (mm^2 /s)	1
Energy, work, heat		
Kilowatt/hour	megajoule (MJ)	3.6

calorie (as used in physics)	joule (J)	4.184
calorie (as used in nutrition)	kilojoule (kJ)	4.184
Btu	kilojoule (kJ)	1.055 056
therm (U.S.)	megajoule (MJ)	105.480 4
horsepower hour	megajoule (MJ)	2.684 520
foot pound-force	joule (J)	1.355 818

Table 19 Conversion Factors for International System of units (SI) and non-SI units

To convert Col. 1 to Col. 2, multiply by:	Column 1 (SI Units)	Column 2 (non-SI Units)	To convert Col. 2 to Col.1, multiply by:
Length			
0.621	kilometer, km (10^3 m)	mile, mi	1.609
1.094	meter, m	yard, yd	0.914
3.28	meter, m	foot, ft	0.304
1.0	micrometer, μm (10^{-6} m)	micron, μ	1.0
3.94×10^{-2}	millimeter, mm (10^{-3} m)	inch, in	25.4
10	nanometer, nm (10^{-9} m)	Angstrom, Å	0.1
Area			
2.47	hectare, ha	acre	0.405
247	square kilometer, km^2 $(10^3\ m)^2$	acre	4.05×10^{-3}
0.386	square kilometer, km^2 $(10^3\ m)^2$	square mile, mi^2	2.590
2.47×10^{-4}	square meter, m^2	acre	4.05×10^3
10.76	square meter, m^2	square foot, ft^2	9.29×10^{-2}
1.55×10^{-3}	square millimeter, mm^2 $(10^{-3}\ m)^2$	square inch, in^2	645
Volume			
9.73×10^{-3}	cubic meter, m^3	acre-inch	102.8
35.3	cubic meter, m^3	cubic foot, ft^3	2.83×10^{-2}
6.10×10^4	cubic meter, m^3	cubic inch, in^3	1.64×10^{-5}
2.84×10^{-2}	liter, L (10^{-3} m^3)	bushel, bu	35.24
1.057	liter, L (10^{-3} m^3)	quart (liquid), qt	0.946
3.53×10^{-2}	liter, L (10^{-3} m^3)	cubic foot, ft^3	28.3
0.265	liter, L (10^{-3} m^3)	gallon	3.78
33.78	liter, L (10^{-3} m^3)	ounce (fluid), oz	2.96×10^{-2}
2.11	liter, L (10^{-3} m^3)	pint (fluid), pt	0.473
Mass			
2.20×10^{-3}	gram, g (10^{-3} kg)	pound, lb	454
3.52×10^{-2}	gram, g (10^{-3} kg)	ounce (avdp), oz	28.4
2.205	kilogram, kg	pound, lb	0.454
0.01	kilogram, kg	quintal (metric), q	100

1.10×10^{-3}	kilogram, kg	ton (2000 lb), ton	907
1.102	megagram, Mg (tonne)	ton (U.S.), ton	0.907
1.102	tonne, t	ton (U.S.), ton	0.907
Yield and Rate			
0.893	kilogram per hectare, kg ha^{-1}	pound per acre, lb $acre^{-1}$	1.12
7.77×10^{-2}	kilogram per cubic meter, kg m^{-3}	pound per bushel, bu^{-1}	12.87
1.49×10^{-2}	kilogram per hectare, kg ha^{-1}	bushel per acre, 60 lb	67.19
1.59×10^{-2}	kilogram per hectare, kg ha^{-1}	bushel per acre, 56 lb	62.71
1.86×10^{-2}	kilogram per hectare, kg ha^{-1}	bushel per acre, 48 lb	53.75
0.107	liter per hectare, L ha^{-1}	gallon per acre	9.35
893	tonnes per hectare, t ha^{-1}	pound per acre, lb $acre^{-1}$	1.12×10^{-3}
893	megagram per hectare, Mg ha^{-1}	pound per acre, lb $acre^{-1}$	1.12×10^{-3}
0.446	megagram per hectare, Mg ha^{-1}	ton (2000 lb) per acre, ton $acre^{-1}$	2.24
2.24	meter per second, m s^{-1}	mile per hour	0.447
Specific Surface			
10	square meter per kilogram, m^2 kg^{-1}	square centimeter per gram, cm^2 g^{-1}	0.1
1000	square meter per kilogram, m^2 kg^{-1}	square millimeter per gram, mm^2 g^{-1}	0.001
Density			
1.00	megagram per cubic meter, Mg m^{-3}	gram per cubic centimeter, g cm^{-3}	1.00
Pressure			
9.90	megapascal, MPa (10^6 Pa)	atmosphere	0.101
10	megapascal, MPa (10^6 Pa)	bar	0.1
2.09×10^{-2}	pascal, Pa	pound per square foot, lb ft^{-2}	47.9
1.45×10^{-4}	pascal, Pa	pound per square inch, lb in^{-2}	6.90×10^3
Temperature			
1.00 (K - 273)	Kelvin, K	Celsius, °C	1.00 (°C + 273)
(9/5 °C) + 32	Celsius, °C	Fahrenheit, °F	5/9 (°F - 32)
Energy, Work, Quantity of Heat			
9.52×10^{-4}	joule, J	British thermal unit, Btu	1.05×10^3
0.239	joule, J	calorie, cal	4.19
10^7	joule, J	erg	10^{-7}
0.735	joule, J	foot-pound	1.36

2.387×10^{-5}	joule per square meter, J m^{-2}	calorie per square centimeter (langley)	4.19×10^{4}
10^{5}	newton, N	dyne	10^{-5}
1.43×10^{-3}	watt per square meter, W m^{-2}	calorie per square centimeter minute (irradiance), cal cm^{-2} min^{-1}	698
Transpiration and Photosynthesis			
3.60×10^{-2}	milligram per square meter second, mg m^{-2} s^{-1}	gram per square decimeter hour, g dm^{-2} h^{-1}	27.8
5.56×10^{-3}	milligram (H_2O) per square meter second, mg m^{-2} s^{-1}	micromole (H_2O) per square centimeter second, μmol cm^{-2} s^{-1}	180
10^{-4}	milligram per square meter second, mg m^{-2} s^{-1}	milligram per square centimeter second, mg cm^{-2} s^{-1}	10^{4}
35.97	milligram per square meter second, mg m^{-2} s^{-1}	milligram per square decimeter hour, mg dm^{-2} h^{-1}	2.78×10^{-2}
Plane Angle			
57.3	radian, rad	degrees (angle), °	1.75×10^{-2}
Electrical Conductivity, Electricity, and Magnetism			
10	siemen per meter, S m^{-1}	millimho per centimeter, mmho cm^{-1}	0.1
10^{4}	tesla, T	gauss, G	10^{-4}
Water Measurement			
9.73×10^{-3}	cubic meter, m^3	acre-inches, acre-in	102.8
9.81×10^{-3}	cubic meter per hour, m^3 h^{-1}	cubic feet per second, ft^3 s^{-1}	101.9
4.40	cubic meter per hour, m^3 h^{-1}	U.S. gallons per minute, gal min^{-1}	0.227
8.11	hectare-meters, ha-m	acre-feet, acre-ft	0.123
97.28	hectare-meters, ha-m	acre-inches, acre-in	1.03×10^{-2}
8.1×10^{-2}	hectare-centimeters, ha-cm	acre-feet, acre-ft	12.33
Concentrations			
1	centimole per kilogram, cmol kg^{-1}	milliequivalents per 100 grams, meq 100 g^{-1}	1
0.1	gram per kilogram, g kg^{-1}	percent, %	10
1	milligram per kilogram, mg kg^{-1}	parts per million, ppm	1
Radioactivity			
2.7×10^{-11}	becquerel, Bq	curie, Ci	3.7×10^{10}

2.7×10^{-2}	becquerel per kilogram, Bq kg^{-1}	picocurie per gram, pCi g^{-1}	37
100	gray, Gy (absorbed dose)	rad, rd	0.01
100	sievert, Sv (equivalent dose)	rem (roentgen equivalent man)	0.01
Plant Nutrient Conversion			
	Elemental	*Oxide*	
2.29	P	P_2O_5	0.437
1.20	K	K_2O	0.830
1.39	Ca	CaO	0.715
1.66	Mg	MgO	0.602

Source: http://agron.scijournals.org/

Table 20 Conversion Factors for Metric Units and English Units

To convert Col. 1 to Col. 2, multiply by:	Column 1 (Metric Units)	Column 2 (English Units)	To convert Col. 2 to Col.1, multiply by:
Length			
0.0394	millimeter, mm (0.001 m)	inch, in	25.4
0.394	centimeter, cm	inch, in	2.54
39.37	meter, m	inch, in	0.0254
0.0328	centimeter, cm	foot, ft	30.48
3.28	meter, m	foot, ft	0.3048
1.094	meter, m	yard, yd	0.914
0.621	kilometer, km (1000 m)	mile, mi	1.609
Area			
0.00155	sq. millimeter, mm2 (0.001 m)2	square inch, in2	645
0.155	sq. centimeter, cm2 (0.01 m)2	square inch, in2	6.45
0.000247	square meter, m2	acre	4050
10.76	square meter, m2	square foot, ft2	0.929
2.47	hectare, ha	acre	0.405
247	sq. kilometer, km2 (1,000 m)2	acre	0.00405
0.386	sq. kilometer, km2 (1,000 m)2	square mile, mi2	2.59
Volume			
0.203	milliliter, ml	teaspoon, tsp	4.93
0.0676	milliliter, ml	tablespoon, tbs	14.79
0.0338	milliliter, ml	ounce (fluid), oz	29.58
0.00973	cubic meter, m3	acre inch, acre-inch	102.8
35.3	cubic meter, m3	cubic foot, ft3	0.0283

61,0000	cubic meter, m3	cubic inch, in3	0.0000164
0.0284	liter, L (0.001 m3)	bushel, bu	35.24
1.057	liter, L (0.001 m3)	quart, (liquid), qt	0.946
0.0353	liter, L (0.001 m3)	cubic foot, ft3	28.3
0.265	liter, L (0.001 m3)	gallon	3.78
33.78	liter, L (0.001 m3)	gallon	0.0296
2.11	liter, L (0.001 m3)	pint (fluid), pt	0.473
Mass			
0.00220	gram, g (0.001 kg)	pound, lb	454
0.0352	gram, g (0.001 kg)	ounce, (avdp), oz	28.4
2.205	kilogram, kg	pound, lb	0.454
0.01	kilogram, kg	quintal (metric), q	100
0.0011	kilogram, kg	U.S. ton (2,000 lb), ton	907
1.102	megagram, Mg (tonne)	U.S. ton (2,000 lb), ton	0.907
1.102	tonne, t	U.S. ton (2,000 lb), ton	0.907
Yield and Rate			
0.893	kilogram per hectare, kg/ha	pound per acre, lb/ acre	1.12
0.0777	kilogram per cubic meter, kg/ m3	pound per bushel, lb/ bu	12.87
0.0149	kilogram per hectare, kg/ha	bushel per acre, 60 lb	67.19
0.0159	kilogram per hectare, kg/ha	bushel per acre, 56 lb	62.71
0.0186	kilogram per hectare, kg/ha	bushel per acre, 48 lb	53.75
0.107	liter per hectare, L/ha	gallon/acre	9.35
893	tonne per hectare, t/ha	pound per acre, lb/ acre	0.00112
893	megagram per hectare, Mg/ha	pound per acre, lb/ acre	0.00112
0.446	megagram per hectare, Mg/ha	ton (lb) per acre, ton/ acre	2.24
2.24	meter per second, m s-1	ton (lb) per acre, ton/ acre	0.447
Water Measurement			
0.00973	cubic meter, cm3	acre-inch, acre-in	102.8
0.00981	cubic meter per hour, cm3/ha	cubic foot per second, ft3/s	101.9
4.40	cubic meter per hour, cm3/ha	U.S. gal. per minute, gal/min	0.227
8.11	hectare meter, ha/m	acre-foot, acre-ft	0.123
97.28	hectare meter, ha/m	acre-inch, acre-in	0.0103

0.081	hectare centimeter, ha/cm	acre-foot, acre-ft	12.33
Temperature			
1.00 (K-273)	Kelvin, K	Celsius, C	1.00 (C + 273)
(1.8 C) + 32	Celsius, C	Fahrenheit, F	0.56 (F – 32)

Common Conversion Factors

Powers

Mega (M) = million
Kilo (k) = thousand
Hecto (h) = hundred
Deka (da) = ten

Fractions

Deci (d) = one-tenth
Centi (c) = one-hundredth
Milli (m) = one-thousandth
Micro = one-millionth

Units

1 crore = 100 lakh
1 crore = 10 million
1 million = 10 lakh
1 lakh = 100000
1 billion = 1000 million

Table 21 Length notations

Measure	Symbol	Relative Length	Exponential Notation
Meter	M	1	10^{0}
Decimeter	dm	.1	10^{-1}
Centimeter	cm	.01	10^{-2}
Millimeter	mm	.001	10^{-3}
Micrometer or micron	μ	.000001	10^{-6}
Nanometer	nm	.000000001	10^{-9}
Angstrom	Å	.0000000001	10^{-10}

Table 22 Length conversion factors

1 nanometre (nm)	0.000000001 meters
	10 Å
1 millimeter (mm)	0.001 meters (1/1,000 of a meter)
1 mm = micrometer	0.000001 meters = 0.04 inches
	1000 nm
1 centimetre (cm)	0.3937 (0.4) inches
	0.01 meters (one hundredth of a meter)
	10 mm= 10^{-2} m

1 inch	2.54 cm = 2.54 × 10^1 mm = 2.54 × 10^4 μ = 2.54 × 10^7 nm
	25.4 mm = 25,400 μ = 25,400,000 nm
	0..254 metre
1 foot	0.3048 metre = 30.48 (30) centimeters
1 yard	0.9144 (0.9) metre
1 metre (m)	1.0936 yards =3.2808 feet (3.3) feet = 39.370 inches
	100 cm = 1000 mm = 0.001 km = 1 × 10^{-3} km
1 kilometre (km)	0.6214 (0.6) miles
	1,000 meters = 10^3 m
1 mile	1.6093 (1.6) kilometres = 1760 yards

Area: The basic metric unit of land area measurement is a square with each side 100 meters long, covering an area of 10,000 square meters. This unit of land is called the hectare (ha) and is equal to approximately 2.5 acres.

Table 23 Area conversion factors

1 square meter (sq m/ m^2)	1.196 (1.2) square yards
	10.7639 square feet
1 square yard	0.8361 (0.8) square meters = 9 square feet
1 square foot	0.0929 m^2
1 hectare (ha)	10,000 square meters
	2.471 (2.5) acres
	0.1 square kilometres
	11 960 square yards
1 acre (ac)	4,046.86 square meters
	0.4047 (0.4) hectares
	4840 square yard
	43560 sq. feet
1 square kilometer (sq km)	0.3861 (0.4) square miles
	100 hectares
	247.1 (250) acres
1 square mile	2.5898 (2.6) square kilometers
	259 (260) hectares
	640 acres
1 cm^2	100 mm^2 (i.e., 10 mm × 10 mm = 100 mm^2)

Volume (capacity)

The basic unit of volume in the metric system is a cube, 10 centimeters on each side. Contained in this cube are 1,000 cubic centimeters or one liter. A liter contains slightly more liquid than a quart. Very large volumes may be measured in cubic meters (1 cubic meter = about 264 gallons).

Table 25 Volume (capacity) notations

Measure	Symbol	Relative Volume	Exponential Notation
Liter	L	1.0	10^7
Deciliter	dl	0.1	10^{-1}
Millimeter	ml	.001	10^{-3}
Microliter	μ l	.000001	10^{-6}

Table 26 Volume (capacity) conversion factors

Microliter (mL)	**0.000001 liters**
1 ml	1 cm^3
	1,000 μ l
1 cm^3	0.000001 m^3
1 litre	1000 cm^3
	1 000 millilitres
	61.026 cubic inches
	0.21998 imperial gallons
	0.26418 U.S. gallons
	1.057 quarts
Kiloliter (kL)	1,000 liters
	265 gallons
1 imperial gallon	4.5460 litres
	1.20096 U.S. gallons
1 U.S. gallon	0.83267 imperial gallons
	3.78528 litres
1 U.S. barrel	42 U.S. gallons
	34.972 imperial gallons
	0.15899 cubic metres
1 cubic metre (m^3)	1 000 litres
	35.3148 cubic feet
	1.30795 cubic yards
	219.97 imperial gallons
	264.18 U.S. gallons
	1.308 (1.3) cubic yards
	28.38 (30) bushels
1 m^3 solid	750 kg. fuelwood with 40% moisture
1 dekaliter (dal)	2.642 (2.5) gallons
1 quart	0.9464 (1) liter
1 cubic yard	0.7646 (0.76) cubic meters
1 bushel	1.244 (1.25) cubic feet
1 bushel	0.0352 (0.035) cubic meters
1 liquid ounce	29.6 ml

Weight: Weight (or more technically accurate, mass) in the metric system is measured in kilograms (kg). One kilogram equals about 2.2 pounds. A gram (1/1,000 kilogram is approximately the weight of a paper clip), and is used for making very small measurements. One liter of water weighs one kilogram at sea level.

Table 27 Weight notations

Measure	Symbol	Relative Weight	Exponential Notation
Kilogram	Kg	1000	10^3
Gram	g	1	10^0
Milligram	mg	.0001	10^{-3}
Microgram	μ g	.0000001	10^{-6}

Table 28 Weight conversion factors

1 gram (g)	0.0022 pounds
1 megagram (Mg)	1,000 kilograms = 1 metric tonne = 1.10 U.S. tons
	1,000,000 grams = 10^6 gram = 0,001 gigagram
1 gigagram	1000 megagram
1 milligram (mg)	0.001 grams
1 kilogram (kg)	2.2046 (2.2) pounds (lb)
	1 000 grams
	35.274 ounce (oz)
	0.001 megagram
1 ounce (oz)	28.3495 gram
	0.028 349 5 kilogram
1 pound	453.592 grams
	0.4536 (0.45) kilograms
1 tonne	1 000 kilograms
	0.98421 tons (English)
	1.10231 U.S. tons
	2 204.62 (2200) pounds
	10 quintals
1 U.S. ton	2 000 pounds
	17.8572 hundred weight (cwt)
	907.184 kilograms
	0.907184 tonnes
	0.89286 tons (English)
1 ton, English	2 240 pounds
	1 016.05 kilograms
	1.01605 tonnes (.netric tons)
	1.12 U.S. tons
	20 hundred weight (cwt)
1 quintal (q)	220.5 (220) pounds

	100 kilograms
1 bushel corn (56#)	25.40 (25) kilograms
1 bushel wheat/soybeans (60#)	27.22 (27) kilograms
1 quintal	3.937 (4) bushels corn (56# bu)
1 quintal	3.674 (3.7) bushels wheat/soybeans (60# bu)
1 metric ton	39.37 (40) bushels corn (56# bu)
1 metric ton	36.74 (37) bushels wheat/soybeans (60# bu)
1 grain	0.000 064 799 kilogram
1 cwt	50.802 kilogram
1 ton	1016.05 kilogram
1 tola	0.011 663 8 kilogram
1 seer	0.933 10 kilogram
1 maund	137.324 2 kilogram
1 pound-force	4.448 Newtons
1 kilogram-force	9.807 Newtons
1 kilogram-force	2.2046 pounds-force

Concentration

$g\ L^{-1}$ = grams per liter = parts per thousand
$mg\ L^{-1}$ = milligrams per liter = parts per million
$mL\ L^{-1}$ = microliters per liter = parts per million
$kg\ m^{-3}$ = kilograms per cubic meter = 0.062 pounds per cubic foot
$g\ kg^{-1}$ = grams per kilogram = percent divided by ten
$mg\ kg^{-1}$ = milligrams per kilogram = parts per million

Percent solutions

There are three means of expressing concentration in the form of a percent figure:
Percent by weight (w/w); gm solute / 100 gm solvent
Percent weight by volume (w/v); gm solute /100 ml solvent
Percent by volume (v/v); ml solute / 100 ml solution

For dilute solutions, these differences are not significant, but at higher concentrations, they are. Chemists (when they use Percent designations) usually use w/w. Biochemists and physiologists more often use w/v. Both use v/v if the solute is a liquid.

Radioactivity

$pCi\ g^{-1}$ = picocurie per gram = 2.22 radioactive disintegrations per minute per gram of material

Table 29 Energy conversion factors

1 kilowatt	**1.3405 horse power**
1 horsepower	0.746 kilowatts
1 kilojoule	0.2389 kilogram calories
	0.948 British thermal units (BTU)
	0.001 megajoules
	0.00027778 kilowatt hours

1 kilowatt hour	3 412 British thermal units
	1.34 horse power hours
	3 600 kilojoules
	3.6 megajoules

Table 30 Density conversion factors

Bulk density of commercial charcoal	**= 250 to 300 kilograms per cubic metre**
Approx. weight of a stacked metre (stere) of:	
- Plantation grown radiata pine (partly seasoned)	= 550 to 650 kilograms
- Plantation grown Eucalypt wood (partly seasoned)	= 600 to 700 kilograms
- Medium density tropical hardwood (partly seasoned)	= 700 to 800 kilograms
- Dense tropical hardwood	= 900 kilograms

Table 31 Heating value conversion factors

Fuel	**High Heating Value KJ/Kg**
Green wood*	15000
Dry wood*	19000
Charcoal	31000
Coke	30000
Bituminous coal	27000
Fuel oil	44000
Kerosene	46000
wood tar	20000
Natural gas	45000
Producer gas	5000
Wood retort gas	6000

Note: * Influence of moisture on heating value of wood: Net heating value (MJ/kg) = 19000 – 220 M where M is moisture content in percentage of total weight.

Plant growth conversion factors

Table 32 Typical ranges of values for plant growth parameters.

Parameter	**Unit**	**Typical value**	**Trend over Time**
relative growth rate or specific growth rate *(RGR)*	$g.kg^{-1}.d^{-1}$	10 to 400	decrease
net assimilation rate or unit leaf rate	$g.m^{-2}.d^{-1}$	1 to 30	decrease
leaf area ratio	$m^2.kg^{-1}$	10 to 60	decrease
specific leaf area	$m^2.kg^{-1}$	5 to 30	decrease
leaf mass ratio (leaf weight ratio)	$kg.kg^{-1}$	0.4 to 0.8	decrease

crop growth rate	$g.m^{-2}.d^{-1}$	1 to 40	rapid increase, then gradual decrease
single leaf photosynthetic rate (CO_2 exchange)	$\mu mol.m^{-2}.s^{-1}$	1 to 40	decrease
leaf area index	m^2 leaf. m^{-2}ground	0.01 to 10	large increase (sigmoid curve)

Source: Frank B. Salisbury, 1996

Table 33 Meteorology units

Distances	**Nautical miles and tenths**
Altitudes, elevations and dimensions on aerodromes	Metres
Horizontal speed	Knots
Vertical speed	Metres per second
Wind velocity	Degrees and knots
Cloud height	Metres
Visibility	Metres
Altimeter setting	Millibar
Temperature	Degrees Centigrade
Time	24 hours, the day beginning at midnight Greenwich time

Table 34 Climate PET

Climate	**Millimeters Daily**
Cool Humid	3 to 4 mm
Cool Dry	4 to 5 mm
Warm Humid	4 to 5 mm
Warm Dry	5 to 6 mm
Hot Humid	5 to 8 mm
Hot Dry	8 to 11 mm 'worst case'

Cool = under 21˚C as an average midsummer high
Warm = between 21˚ and 32˚C as midsummer highs
Hot = over 32˚C
Humid = over 50% as average midsummer relative humidity [dry = under 50%]

Temperature

Temperature is measured in the Celsius (C) scale. The temperature at which water freezes is 0° and is equivalent to 32° F (Fahrenheit). A reading of 37° on the Celsius scale is the approximate equivalent to human body temperature (98.6° F) and water boils at 100°C (212°F).

$C = (F-32) \times 5/9$
$F = 9/5\ C + 32$
F = C
425 = 218.33
350 = 176.67

100 = 37.78
70 = 21.11
32 = 0
0 = -17.78

Fahrenheit-Celsius-Kelvin conversions:

T(ºC) = [T(ºF) – 32] × 5/9
T(ºF) = T(ºC) × 9/5 + 32
T(K) = T(ºC) + 273

Table 35 Convenient equivalences to help in remembering celsius scale

T(°F)	T(°C)	Conditions
-459	-273	Zero Kelvin (absolute zero)
-40	-40	extremely cold
-4	-20	temperature in a freezer
0	-17.8	
32	0	water freezes
68	20	room temperature
82	28	warm day
87.08	30.6	butter melts
98.6	37	Human body temperature
104	40	hot day
167	75	hot coffee
212	100	water boils

Application Rate or Crop Yield

1 U.S. ton per acre = 2.24 metric tons per hectare
1 U.S. ton per acre = 2.47 U.S. tons per hectare
1 metric ton per hectare = .446 U.S. tons per acre
1 metric ton per hectare = .405 metric tons per acre
1 metric ton per hectare = 892 pounds per acre
1 metric ton per hectare = 100 grams per square meter

Application rates are often given in weight of material per unit of area covered (pounds per acre) or volume of material per unit of area covered (quarts per acre).

1 kilogram/hectare (kg/ha) = .8922 (.9) pounds/acre
1 pound/acre = 1.121 (1.1) kilograms/hectare
1 liter/hectare (L/ha) = .4276 (.4) quarts/acre
1 quart/acre = 2.338 (2.3) liters/hectare
kg ha^{-1} = kilograms per hectare = 0.893 pounds per acre
Mg ha^{-1} = megagrams per hectare = 893 pounds per acre = 0.446 U.S. tons per acre
bushels ha^{-1} = bushels per hectare = 0.405 bushels per acre
m^3 ha^{-1} = cubic meters per hectare = 14.3 cubic feet per acre

Plant nutrients

Phosphorus (P) × 2.29 = P_2O_5
P_2O_5 × 0.437 = P

Potassium (K) × 1.2 = K_2O
K_2O × 0.830 = K
parts per million (ppm) × 2 = pounds per acre (lb/A)

Water Measurement Units and Conversion Factors

Water Measurement Units: There are two conditions under which water is measured *viz.*, water at rest and water in motion. Water at rest is measured in units of volume. Water in motion is measured in units of flow which is a unit of volume for a convenient time unit. It is important that the difference between a unit of volume and a unit of flow be kept in mind.

Volume Units: Water at rest; *i.e.*, ponds, lakes, reservoirs, and in the soil, is measured in units of volume *viz.*, gallon, cubic foot, acre-inch, and acre-foot.

Cubic Foot: The volume of water that would be held in a container one foot wide by one foot long by one foot deep.

Acre-inch: The volume of water that would cover one acre (43,560 square feet) one inch deep.

Acre-Foot: The volume of water that would cover one acre one foot deep.

Flow Units: Water in motion; i.e., flowing in streams, canals, pipe-lines, and ditches, is measured in units of volume per unit of time viz., gallons per minute (gpm), cubic feet per second (cfs), acre-inches per hour and acre feet per day. Cubic feet per second, sometimes written second-feet (sec. ft. or cusec) is most commonly used for measuring ow of irrigation water moving by gravity from streams and reservoirs. Gallons per minute are most commonly used for measuring flow from pumps.

Cubic foot per second: The quantity of water equivalent to a stream one foot wide by one foot deep owing with a velocity of one foot per second.

Gallon per minute: The quantity of water equivalent to a stream which will be a gallon measure once each minute of time.

A flow of one cubic foot per second is approximately equal to either 450 gallons per minute, one acre-inch per hour, or two acre-feet per day (24 hours).

List of Equivalents: The following equivalents are useful for converting from one unit to another and for calculating volumes from flow units.

Units of Water Measurement

FLOW: Gallons per Minute (GPM)

452.5 GPM = one acre-foot in 12 hours
452.5 wGPM = one acre-inch per hour
448.8 GPM = one cubic foot (7.48 gal) per second
694.4 GPM = 1,000,000 gallons per day

HEAD:

1 Pound per square inch (psi) = 2.31 ft. head of water
1 Foot of water = 0.43 psi
1 Atmosphere (at sea level) = 14.7 psi

WEIGHT:

1 U.S. Gallon water = 8.34 pounds
1 Cubic Foot of water = 62.4 pounds
1 Acre-Foot of water = 2,719,226 pounds

Water Measurement

1 cubic foot = 7.48 gallons = 62.4 pounds of water

1 acre-foot = 43,560 cubic feet = 325,851 gallons = 12 acre-inches 1 acre-foot covers 1 acre of land 1 foot deep;

1 acre-inch = 27,154 1 cubic meter = 1,000 liters = 264.18 gallons

1 acre-inch hour = 450 gallons per minute (GPM) or 1 cubic foot per second (cfs)

1 gallon = 128 ounces = 3,785 milliliters 1 pound = 454 grams

Volume Units

One gallon = 231 cubic inches = 0.13368 cubic foot weighs approximately 8.33 pounds

One cubic foot = 1,728 cubic inches = 7.481 gallons (7.5 for ordinary calculations) weights 62.4 pounds (62.5 for ordinary calculations)

One acre-inch

= ,630 cubic feet

= 7,154 gallons (27,200 for ordinary calculations) = 1/12 acre-foot weighs approximately 113.1 tons

One acre-foot = 43,560 cubic feet = 325,851 gallons = 12 acre-inches weighs approximately 1,357 tons = 325,851 gallons

Rate of Flow Units

One gallon per minute = 0.00223 (approximately 1/450) cubic foot per second = 0.00221 acre-inch per hour = 0.00442 acre-foot per (24 hour) day = 1 acre-inch in 452.6 hours (450 for ordinary calculations) = 1 acre-foot in 226.3 days

One cubic foot per second = 448.83 gallons per minute (450 for ordinary calculations) = 1 acre-inch in 1 hour and 30 seconds (1 hour for ordinary calculations) = 1 acre-foot in 12 hours and 6 minutes (12 hours for ordinary calculations) = 1.984 acre-feet per (24 hours) day (2 acre-feet for ordinary calculations)

Conversion formulas: The following formulas are useful for computing the approximate depth of water applied to a field.

Cubic feet per second × hours

acres = acre-inches per acre, or average depth in inches.

Gallons per minute × hours

450 × acres = acre-inches per acre, or average depth in inches.

The volume of water that cover an acre of land one foot or one inch deep

One acre = an area of land that is 43560 ft^2

One cubic foot (ft3) = 7.48 gallons

One acre-foot = 43560 ft^2 × 1 foot water depth = a volume that is 43560 ft^3

One acre-foot = 43560 ft^3 × 7.48 gallons/ft3 = 325,851 gallons

One acre-inch = 43560 ft^2 × .0833 ft (1 inch) a volume that is 3630 ft^3

One acre-inch = 3630 ft^3 × 7.48 gallons/ft^3 = 27,154 gallons

The amount of water applied to a field is usually reported in acre-inches of water, and the rate of ET (evapotranspiration) for crops is usually given in acre-inches per day. However, pump discharge is usually given in gallons per minute, and here in lies some confusion when calculating pumping time for an irrigation.

A four inch irrigation is about 110,000 gallons per acre (4 inches × 27,154 gallons/inch = 108,616 gallons); a six inch irrigation is about 160,000 gallons per acre.

Example: Assume you are going to irrigate a 50 acre field and apply 110,000 gallons per acre (4 inch irrigation), and your pump discharges 750 gallons per minute. How long will you have to run the pump. The answer is 5.1 days.

Calculated as follows:

(1) 50 acres × 110,000 gallons per acre = 5,500,000 gallons

(2) 5,500,000 gallons/750 gallons per minute pumping discharge = 7333 minutes pumping time

(3) 7333 minutes/60 minutes per hour = 122 hours; 122 hours/24 hours per day = 5.1 days.

Pressure (air pressure, vapour pressure)

1 pound per square inch (psi) = 2.31 feet of water

A column of water 2.31 feet deep exerts a pressure of 1 psi feet of head = psi × 2.31

Total Dynamic Head (TDH) includes: Pumping Lift, Elevation Change, Friction Loss, and Irrigation System Operating Pressure

TDH = Lift + Elevation + Friction + System Pressure

millibar (mbar)	1 mbar = 0.1 kPa (kilopascal)
bar	1bar = 100 kPa
centimetre of water (cm)	1 cm of water = 0.09807 kPa
millimetre of mercury (mmHg)	1 mmHg = 0.1333 kPa
atmospheres (atm)	1 atm = 101.325 kPa
pound per square inch (psi)	1 psi = 6.896 kPa

WIND SPEED

Standard unit: metre per second (m s^{-1})

kilometre per day (km day^{-1})	1 km day^{-1} = 0.01157 m s^{-1}
nautical mile/hour (knot)	1 knot = 0.5144 m s^{-1}
foot per second (ft s^{-1})	1 ft/s = 0.3048 m s^{-1}

RADIATION

Standard unit: megajoule per square metre and per day (MJ m^{-2} day^{-1}) or as equivalent evaporation in mm per day (mm day^{-1})

equivalent evaporation (mm/day)	1 mm day^{-1} = 2.45 MJ m^{-2} day^{-1}
joule per cm^2 per day (J cm^{-2} day^{-1})	1 J cm^{-2} day^{-1} = 0.01 MJ m^{-2} day^{-1}
calorie per cm^2 per day (cal cm^{-2} day^{-1})	1 cal = 4.1868 J = 4.1868 10^{-6} MJ
	1 cal cm^{-2} day^{-1} = 4.1868 10^{-2} MJ m^{-2} day^{-1}
watt per m^2 (W m^{-2})	1 W = 1 J s^{-1}
	1 W m^{-2} = 0.0864 MJ m^{-2} day^{-1}

EVAPOTRANSPIRATION

Standard unit: millimetre per day (mm day^{-1})

m^3 per hectare per day (m^3 ha^{-1} day^{-1})	1 m^3 ha^{-1} day^{-1} = 0.1 mm day^{-1}
litre per second per hectare (1 s^{-1} ha^{-1})	1 1 s^{-1} ha^{-1} = 8.640 mm day^{-1}
equivalent radiation in megajoules per square metre per day (MJ m^{-2} day^{-1})	1 MJ m^{-2} day^{-1} = 0.408 mm day^{-1}

Area/Length

1 acre = 0.405 hectare (ha) = 43,560 feet2 1 inch = 2.54 centimeters

Horse power

Water Horse power (WHP) — power required to lift a given quantity of water against a given total dynamic head.

WHP = (Q × H)/E where: Q = flow rate, GPM 3,960 H = total dynamic head, feet

Brake horse power (BHP) — required power input at the pump.

BHP = WHP/E where: E = pump efficiency

Units of Distance and Length

Metric Equivalents		
1 km = 1000 m; 1 m = 100 cm = 1000 mm = 10^6 um = 10^9 nm		
Metric Unit	**Multiplied by**	**= English Unit**
millimeters	**0.0394**	inches (in)
centimeters	**0.394**	inches (in)
centimeters	**0.0328**	feet (ft)
meters	**39.4**	inches (in)
meters	**3.28**	feet (ft)
meters	**1.1**	yards (yd)
kilometers	**3281**	feet (ft)
kilometers	**0.621**	miles (mi)
English Equivalents		
1 mi = 1,760 yd = 5,281 ft; 1 yd = 3 ft = 36 in		
English Unit	**Multiplied by**	**= Metric Unit**
inches (in)	5.4	millimeters
inches (in)	2.54	centimeters
inches (in)	0.254	meters
feet (ft)	30.48	centimeters
yards (yd)	91.44	centimeters
yards (yd)	0.9144	meters
miles (mi)	1.609	kilometers

Units of Mass

Metric Equivalents		
1 mt (metric ton) = 1000 kg = 2,205 lb 1 kg = 1000 g = 2.205 lb = 35.2802 oz 1 g = 1000 mg = 10^6 ng = 10^9 pg		
Metric Unit	**Multiplied by**	**= English Unit**
gram (g)	0.035	ounces (oz)
kilogram (kg)	2.2	pounds (lb)
metric ton (mt)	1.102	ton (t)
English Equivalents		
1 ton = 2000 lb = 907.2 kg or 0.9072 mt 1 lb = 16 oz = 0.4536 kg = 453.6 g		

English Unit	Multiplied by	= Metric Unit
ounce (oz)	28	grams (g)
pound (lb)	0.4536	kilograms (kg)
tons (t)	0.9072	metric tons (mt)

Units of Volume

Metric Equivalents
1 cm^3 (or cc) = 1000 mm^3 = 0.061 in^3
1 m^3 = 10^6 cm^3 = 61,024 in^3 = 35.31 ft^3 = 1.308 yd^3

Metric Unit	Multiplied by	= English Unit
cubic centimeters (cm^3)	0.061	cubic inches (in^3)
cubic meters (m^3)	35.31	cubic feet (ft^3)
cubic meters	1.308	cubic yards (yd^3)

English Equivalents
1 ft^3 = 1,728 in^3 = 28,317 cm^3 = 0.02832 m^3
1 yd^3 = 27 ft^3 = 0.7646 m^3

English Unit	Multiplied by	= Metric Unit
cubic inches	16.393	cubic centimeters
cubic feet	0.03	cubic meters
cubic yards	0.76	cubic meters

Units of Liquid Volumes

Metric Equivalents

1 L = 1000 ml = 2.113 pt (pints) = 1.06 qt (quarts) = 0.264 US gal
1 ml (or cm^3) = 1000 ul = 0.03 fl oz (fluid ounces)

Metric Unit	Multiplied by	= English Unit
milliliters (ml)	0.02957	fluid ounces (fl oz)
liters (L)	2.13	pints (pt)
liters (L)	1.0567	quarts (qt)

English Equivalents

1 US gal = 4 qt = 8 pt = 128 fl oz = 3.785 L
1 qt = 2 pt = 32 fl oz = 946.4 ml or 0.9464 L
1 pt = 16 fl oz = 473.2 ml or 0.213 L
1 fl oz = 29.57 ml

English Unit	Multiplied by	= Metric Unit
teaspoons (tsp)	5	milliliters (ml)
tablespoons (tbsp)	15	milliliters
fluid ounces (fl oz)	29.57	milliliters
cups (c)	0.24	liters (L)
pints (pt)	0.4732	liters
quarts (qt)	0.9464	liters
US gallons (US gal)	3.785	liters

Units of Area

Metric Equivalents		
1 km^2 = 100 ha = 1,000,000 m^2 = 270 A (acres) = 0.3861 mi^2 1 ha (hectare) = 10,000 m^2 = 107,600 ft^2 = 2.471 A 1 m^2 = 10,000 cm^2 = 1,000,000 mm^2 = 1,550 in^2 = 1.196 yd^2 1 cm^2 = 100 mm^2 = 0.155 in^2 1 mm^2 = 1,000,000 um^2		
Metric Unit	**Multiplied by**	**= English Unit**
centimeters squared	0.155	inches squared
meters squared	1.196	yards squared
kilometers squared	0.3861	miles squared
English Equivalents		
1 mi^2 = 640 A = 27,878,400 ft^2 = 259 ha = 2.59 km^2 1 A = 4,840 yd^2 = 43,560 ft^2 = 4,407 m^2 = 0.405 ha 1 yd^2 = 9 ft^2 = 1296 in^2 = 8,361 cm^2 = 0.836 m^2 1 ft^2 = 144 in^2 = 929 cm^2 1 in^2 = 6.452 cm^2		
English Unit	**Multiplied by**	**= Metric Unit**
inches squared	6.452	centimeters squared
feet squared	0.0929	meters squared
yards squared	0.836	meters squared
acres (A)	0.405	hectares (ha)
miles squared	2.59	kilometers squared

Units of Time: The most frequent units time used in the biological sciences are:

Years	days	Hours	Minutes	Seconds	Milliseconds
yr	d	h	min	Sec or s	msec

Source: http://abacus.bates.edu/~ganderso/biology/resources/writing/HTWabbr.html#topofpage

REFERENCES

Barry N. Taylor and Ambler Thompson. 2008. The International System of Units (SI). NIST Special Publication 330. U.S. Department of Commerce and National Institute of Standards and Technology, Gaithersburg, Maryland 20899-2600. USA.

BIPM. 2006. The International System of Units (SI), Eighth Edition, (Paris: Organisation Intergouvernementale de la Convention du Mètre)

available at: http://www.bipm.org/utils/common/pdf/si_brochure_8_en.pdf

BIPM. 2008. International vocabulary of metrology — Basic and general concepts and associated terms available at:

http://www.bipm.org/utils/common/documents/jcgm/JCGM_200_2008.pdf

Causton, David and Jill Venus. 1981. The Biometry of Plant Growth. Edward Arnold, London.

Hunt, R. 1978. Plant Growth Analysis. Edward Arnold, London.

Dietrich, C.F. 1991. Uncertainty, calibration and probability, Adam Hilger. Bristol.

EURAMET. 2008. Metrology – in short available at:

http://resource.npl.co.uk/international_office/metrologyinshort.pdf

Frank B. Salisbury. 1996. Units, symbols, and terminology for plant physiology. Oxford University Press. New York. 249 p

Hunt, R. 1982. Plant Growth Curves: The Functional Approach to Plant Growth Analysis. Univ. Park Press. Baltimore.

Hunt, R. 1990. Basic Growth Analysis. Unwin Hyman Ltd., London, UK; and Winchester, MA.

Ian Mills, Tomislav Cvitas, Klaus Homann, Nikola Kallay and Kozo Kuchitsu. 1993. Quantities, Units and Symbols in Physical Chemistry. Blackwell Science.

Jerrard, H.G., and D.B. McNeill. 1992. A Dictionary of Scientific Units: including Dimensionless Numbers and Scales, 6th ed., Chapman and Hall, New York, 255 p.

Kenneth S. Butcher, Linda D. Crown, Elizabeth J. Gentry and Carol Hockert. 2006. The International System of Units (SI) -Conversion Factors for General Use. U.S. Department of Commerce and National Institute of Standards and Technology, Gaithersburg, Maryland 20899-2600. USA. Spec. Pub. 1038, 24 pages

Pennycuick, C.J., 1988. Conversion Factors, Univ. of Chicago Press, USA.47 p.

Reifsneider, W.E., McNaughton, K.G. and Milford, J.R. 1991. Symbols, units, notation. A statement of journal policy. Agric. and For. Meteorol. 54:389-397.

Robinson, A. 2007. The Story of Measurement. Thames and Hudson. London

Salisbury, F.B. 1996. Units, Symbols and Terminology for Plant Physiology. Oxford University Press, ISBN 0-19-509445-X, 234 pp.

Taylor, B.N.1995. Guide for the use of the International System of Units (SI). NIST Spec. Pub. 811. Physics Lab., National Institute of Standards and Technology, Gaithersburg (MD), USA.

Thein, S.J. and J.D. Oster 1981. The International System of Units and its particular application to soil chemistry. J.Agron. Educ. 10: 60-72

U.S. Department of Agriculture. 2000. Agriculture Fact Book. Office of Communications. Online and available at http://www.usda.gov/news/pubs/fbook00/contents.htm, Washington DC.

Weast, R.C. (ed). 1974. Handbook of Chemistry and Physics, CRC Press, Cleveland.

Fenna, D. 2009. Oxford Dictionary of weights, measures and units. Oxford University Press. Oxford.

Stout, K. J. 1998. From Cubit to Nanometre, a History of Precision Measurement. Penton Press. London.

http://en.wikipedia.org/wiki/Metric_system

http://lamar.colostate.edu/~hillger/ www.metric.org

http://www.metric-conversions.org/conversion-calculators.htm

http://www.sciencemadesimple.com/conversions.html

http://www.nist.gov/metric

http://physics.nist.gov/cuu/Units/index.html

www.extension.iastate.edu/agdm

www.pnas.org

http://homepages.gac.edu/~cellab/appds/appd-a.html

https://www.nde-ed.org/GeneralResources/Units/USCustomarySystem.htm

http://homepages.gac.edu/~cellab/appds/appd-a.html

http://www.nist.gov/pml/wmd/metric/index.cfm

http://www.nist.gov/pml/

http://physics.nist.gov/cuu/index.html

https://en.wikipedia.org/wiki/Conversion_of_units

http://abacus.bates.edu/~ganderso/biology/resources/writing/HTWabbr.html

https://www.sdstate.edu/ps/extension/crop-mgmt/upload/Conversion-factors-for-selected-weights-and-measures.pdf

Annxure II
List of Important National and International Organizations

A. NATIONAL ORGANIZATIONS

Indian Council of Agricultural Research Directorate of Information and Publications of Agriculture (DIPA) Krishi Bhavan, New Delhi 110 001(INDIA) http://www.icar.org.in/

ICAR Central Research Institutes

Central Agricultural Research Institute (CARI), PB 181, Andaman Nicobar & Lakshwadeep Group of Islands, Port Blair, http://icar-ciari.res.in/

Central Arid Zone Research Institute (CAZRI), Jodhpur, Rajasthan. Web: http://www.icar.org.in/cari/

Central Avian Research Institute (CARI) Izatnagar, Uttar Pradesh. http://www.icar.org.in/cari/
A premier institute in the field of poultry research, education, extension and training in India.

Central Inland Capture Fisheries Research Institute (CICFRI), Barrackpore, West Bengal, http://www.ifsi.in/cifri/cifri.htm

Central Institute for Research on Goats (CIRG), PO Farah, Makhdoom, Mathura, Uttar Pradesh, http://www.cirg.res.in/

Central Institute for Sub-Tropical Horticulture (CISTH), Rae Bareli Road, PO Dilkusha, Lucknow, Uttar Pradesh. http://cish.res.in/

Central Institute for Research on Buffaloes (CIRB), Sirsa Road, Hissar, Haryana www.cirb.res.in

Central Institute for Research on Cotton Technology (CIRCOT), Adenwala Road, Matunga, Mumbai, Maharashtra. www.circot.res.in

Central Institute for Cotton Research (CICR), PO No. 225, GPO Panjari, Wardha Road, Nagpur, Maharashtra. www.cicr.org.in

Central Institute for Freshwater Aquaculture (CIFA), PO Kausalyaganga, Bhubaneshwar, Orissa. www.cifa.in

Central Institute of Post-Harvest Engineering & Technology (CIPHET), PAU Campus, Ludhiana, Punjab. http://ciphet.in/default.asp

Central Institute of Agricultural Engineering (CIAE), Nabibagh, Berasia Road, Bhopal, Madhya Pradesh. http://www.ciae.nic.in/Content/index.aspx

Central Institute of Brackishwater Aquaculture (CIBA), 141, Marshal's Road, Egmore, Chennai, Tamil Nadu. http://www.ciba.res.in/

Central Institute of Fisheries Education (CIFE), Jaiprakash Road, Seven Bungalows, Versova, Mumbai, Maharashtra. http://www.cife.edu.in/cifemod2/index.php

Central Institute of Fisheries Technology (CIFT), Willingdon Island, PO, Matsyapuri, Cochin, Kerala. http://www.cift.res.in/

Central Institute of Temperate Horticulture (CITH), PO Sanatnagar, Srinagar, Jammu & Kashmir. http://www.cith.org.in/

Central Marine Fisheries Research Institute (CMFRI), PO 1603, Ernakulam, Cochin, Kerala, http://www.cmfri.org.in/

Central Plantation Crops Research Institute (CPCRI), Kasaragod, Kerala http://www.cpcri.gov.in/

Central Potato Research Institute (CPRI), Shimla, http://cpri.ernet.in/

Central Research Institute for Jute & Allied Fibres (CRIJAF), 24 Paraganas, Barrackpore, West Bengal. http://www.crijaf.org.in/

Central Research Institute for Dryland Agriculture (CRIDA), Santosh Nagar, PO Saidabad, Hyderabad, Andhra Pradesh. http://www.crida.in/

Central Rice Research Institute (CRRI), Cuttack, Orissa, http://www.crri.nic.in/

Central Soil & Water Conservation Research & Training Institute (CSWCRTI), 218 Kaulagarh Road, Dehradun, Uttar Pradesh. http://www.cswcrtiweb.org/

Central Soil Salinity Research Institute (CSSRI), Zarifa Farm, Kachwa Road, Karnal, Haryana, http://www.cssri.org/

Central Tobacco Research Institute (CTRI), Rajahmundri, Andhra Pradesh, http://www.ctri.org.in/

Central Tuber Crops Research Institute (CTCRI), Sreekartiyam, Thiruvananthapuram, Trivandrum, Kerala. http://www.ctcri.org/

Central Sheep & Wool Research Institute (CSWRI), Avikanagar, Tehsil Malpura, Tonk, Rajasthan. http://www.cswri.res.in/

ICAR Research Complex for Goa (ICARRCG), Ela Old, Goa, http://icargoa.res.in/

ICAR Research Complex for NEH Region (ICARRCNEHR), Umroi Road, Barapani, Meghalaya. http://www.icarneh.ernet.in/

Central Institute for Women in Agriculture, Bhubaneshwar, http://icar-ciwa.org.in/

Indian Institute of Rice Research, Rajendranagar, Hyderabad-500 030, http://www.drricar.org/

Indian Institute of Natural Resins and Gums (IINRG) (formerly Indian Lac Research Institute (ILRI), Namkum, Ranchi, Bihar, http://ilri.ernet.in/~iinrg/

Indian Veterinary Research Institute (IVRI), Izatnagar, Uttar Pradesh, http://www.ivri.nic.in/

Indian Agricultural Research Institute (IARI), Pusa, Dr K. S. Krishnan Marg, New Delhi, http://www.iari.res.in/

Indian Agricultural Statistics Research Institute (IASRI), Library Avenue, Pusa, New Delhi, http://www.iasri.res.in/

Indian Grassland & Fodder Research Institute (IGFRI), Pahuj Dam, Jhansi-Gwalior Road, Jhansi, Uttar Pradesh. http://www.igfri.res.in/

Indian Institute of Pulses Research (IIPR), Kalyanpur, Kanpur, Uttar Pradesh, http://www.iipr.res.in/

Indian Institute of Oilseeds Research, Rajendranagar, Hyderabad–500 030 http://icar-iior.org.in/

Indian Institute of Horticulture Research (IIHR), Hassaraghatta, Lake Post, Bangalore, Karnataka. http://www.iihr.ernet.in/

Indian Institute of Soil Sciences (IISS), Nabi Bagh, Berasia Road, Bhopal, Madhya Pradesh, http://www.iiss.nic.in/index.html

Indian Institute of Spices Research (IISR), PO 1701, Marikannu PO, Calicut, Kerala http://www.iiss.nic.in/index.html

Indian Institute of Sugarcane Research (IISR), Rae Bareli Road, PO Dilkusha, Lucknow, Uttar Pradesh. http://www.iisr.nic.in/

Indian Institute of Vegetable Research (PDV), Gandhinagar, Naira PO, Varanasi, Uttar Pradesh. http://www.iivr.org.in/

Indian Institute of Agricultural Biotechnology, Namkum, Ranchi - 834010 http://ilri.ernet.in/~iiab/

Indian Institute of Maize Research, Pusa Campus, New Delhi-110 012 http://www.iimr.res.in/

Indian Institute of Wheat and Barley Research, (Erstwhile Directorate of Wheat Research), Karnal, http://www.dwr.in/

Indian Institute of Farming Systems Research, Modipuram, http://www.pdfsr.ernet.in/

Indian Institute of Millets Research (IIMR) upgraded from Directorate of Sorghum Research (DSR), Rajendranagar, Hyderabad-500 030 Telangana. India. http://www.millets.res.in//

Indian Institute of Oilseeds Research (IIOR) formerly operating as All India Coordinated Research Project on Oilseeds(AICORPO), (DSR), Rajendranagar, Hyderabad-500 030 Telangana. India. http://icar-iior.org.in/

Indian Institute of Oil Palm Research, Pedavegi ,West Godavari District, Andhra Pradesh, Pincode- 534 450, http://dopr.gov.in/

Indian Institute of Water Management, Opp. Rail Vihar, Chandrasekharpur Bhubaneswar, Odisha 751023, http://www.iiwm.res.in/

National Academy of Agricultural Research Management (NAARM), Rajendranagar, Hyderabad, Andhra Pradesh. http://www.naarm.ernet.in/

National Institute of Biotic Stress Management, Baronda, Raipur 493 225, Chhattishgarh. http://www.nibsm.org.in/

National Institute of Abiotic Stress Management, Malegaon, Baramati 413 115, Pune Maharashtra, India, http://www.niam.res.in/

National Institute of Agricultural Economics and Policy Research (NIAP), New Delhi http://www.ncap.res.in/

National Institute for Research on Jute & Allied Fibres Technology (NIRJAFT), 12 Regent Park, Calcutta, West Bengal. http://www.nirjaft.res.in/

National Institute of Animal Nutrition & Physiology (NIANP), NDRI Campus, Adugodi, Bangalore, Karnataka. E-mail: director@nainpbng.kar.nic.in http://www.nianp.res.in/

National Dairy Research Institute (NDRI), Karnal, Haryana http://www.ndri.res.in/ndri/Design/Index.html

Central Institute for Research on Cattle - Meerut http://www.circ.org.in/

Sugarcane Breeding Institute (SBI), Coimbatore, Tamil Nadu http://www.sugarcane.res.in/index.php/en/

Vivekananda Parvatiya Krishi Anusandhan Sansthan (VPKAS), Almora, Uttarakhand http://www.vpkas.nic.in/

ICAR National Bureaux

National Bureau of Fish Genetic Resources (NBFGR), 351/28, Radha Swami Bhawan, Duriyapur PO, Rajendranagar, Lucknow, Uttar Pradesh. http://www.nbfgr.res.in/

National Bureau of Animal Genetic Resources (NBAGR), PB 129, Karnal, Haryana http://www.nbagr.res.in/

National Bureau of Plant & Genetic Resources (NBPGR), Pusa Complex, Indra Puri, New Delhi. http://www.nbpgr.ernet.in/

National Bureau of Soil Survey & Land Use Planning (NBSSLUP), Shankar Nagar, Amravati Road, Nagpur, Maharashtra. http://nbsslup.in/

National Bureau of Agriculturally Important Microorganisms (NBAIM) Post Box. No. 6, Kushmaur, Mau Nath Bhanjan – 275103, Uttar Pradesh, http://nbaim.org.in/

National Bureau of Agricultural Insect Resources, P.Bag No:2491, H.A. Farm Post Bellary Road, Bengaluru - 560 024. Karnataka, http://www.nbair.res.in/

ICAR National Research Centres:

National Research Centre for Agroforestry (NRCAF), Jhansi, Uttar Pradesh http://www.nrcaf.res.in/

National Research Centre for Citrus. PB No.464, Shankar Nagar P.O., Nagpur – 440010, http://nrccitrus.nic.in/

National Centre for Agricultural Economics & Policy Research (NCAP), IASRI Campus, Library Avenue, Pusa, New Delhi. http://www.ncap.res.in/

National Centre for Integrated Pest Management (NCIPM), Lal Bahadur Shastri Bhawan, Wing L-1&M-1, Block F, IARI Campus, New Delhi http://www.ncipm.org.in/

National Research Centre for Agroforestry (NRCAF), IGFRI Campus, Pahuj Dam, Jhansi-Gwalior Road, Jhansi, Madhya Pradesh, E-mail: nrcaf@x400.nicgw.nic.in

National Research Centre for Arid Horticulture (NRCAH), 10th Milestone, Ganganagar Road, Beechwal, Bikaner, Rajasthan. E-mail: nrcah@x400.nicgw.nic.in

National Research Centre for Banana (NRCB), Thogamalai Road, Thayanur Post, Tiruchirapalli- 620102, Tamil Nadu. http://www.nrcb.res.in/

National Research Centre for Cashew (NRCC), Kamminje, Puttur, Karnataka E-mail: root@nrcashew.kar.nic.in

National Research Centre for Camel (NRCC), Jorbeer, PO Box 7, Bikaner, Rajasthan http://www.nrccamel.res.in/

National Research Centre for DNA Fingerprinting (NRCDF), NBPGR, IARI Complex, Pusa, New Delhi. E-mail: pdnre@nbpgr.delhi.nic.in

National Research Centre for Equines (NRCE), Sirsa Road, Hissar, Haryana http://nrce.nic.in/

National Research Centre on Meat, Chengicherla, P.B.No. 19, Boduppal post, Hyderabad – 500 092. http://nrcmeat.org.in/

National Research Centre for Grapes (NRCG), P.B. No. 3, P.O. Manjri Farm, Solapur Road, Pune – 412307, Maharashtra. http://nrcgrapes.nic.in/

National Research Centre for Groundnut (NRCG), Ivanagar Road, PB 5, Junagarh, Gujarat. E-mail: director@nrcg.guj.nic.in

National Research Centre for Meat & Meat Products (NRCMMP), c/o PD Poultry Science, APAU Campus, Rajendranagar, Hyderabad, Andhra Pradesh E-mail: nrcmeat@x400.nicgw.nic.in

National Research Centre for Medicinal & Aromatic Plants (NRCMAP), Borivai, Anand, Gujarat. E-mail: nrcmap@wilnetonline.com

National Research Centre for Mithun (NRCM), ICAR Research Complex, Jharnapani, Medziphema, Nagaland. http://www.nrcmithun.res.in/

National Research Centre for Mushroom (NRCM), Chambaghat, Solan, Himachal Pradesh, E-mail: root@nrcmrt.chd.nic.in

National Research Centre for Oilpalm (NRCO), Pedavegi, West Godavari District, Andhra Pradesh. E-mail: oilpalm@ap.nic.in

National Research Centre for Onion & Garlic (NRCOG), Rajgurunagar, Distt Pune, Maharashtra. E-mail: nrcogrpn@mah.nic.in

National Research Centre for Orchids (NRCO), Pakyong, Sikkim http://nrcorchids.nic.in/

National Research Centre for Rapeseed Mustard (NRCRM), Sewar, Bharatpur, Rajasthan, E-mail: root@nrcseed.raj.nic.in

National Research Centre for Weed Science (NRCWS), PB 17, Maharajpur, Adhartal, Jabalpur, Madhya Pradesh. E-mail: root@nrcweed.mp.nic.in

National Research Centre for Women in Agriculture (NRCWA), 1199 Jagamana, PO Khandagiri, Bhubaneswar, Orissa. E-mail: nrcwa@nrcwa.ori.nic.in

National Research Centre for Yak (NRCY), Dirang790101, West Kameng district, Arunachal Pradesh, http://www.nrcy.org.in/

National Research Center on Pig, Indian Council of Agricultural Research. Rani (Near Airport), Guwahati- 781 131, http://www.nrcp.in/

National Research Centre on Coldwater Fisheries (NRCCWF), Saurabh Cottage, Thandi, Sarak, Bhimtal, Nainital, Uttar Pradesh. E-mail: nrccwf@x400.nicgw.nic.in

National Research Centre on Plant Biotechnology (NRCPB), IARI Campus, Pusa, New Delhi, http://www.nrcpb.org/

National Research Centre on Sorghum (NRCS), Rajendranagar, Hyderabad, Andhra Pradesh, E-mail: nrcshyd@ap.nic.in

National Research Centre on Soyabean (NRCS), Khandwa Road, Indore, Madhya Pradesh, E-mail: nrcsoya@hub1.nic.in

National Research Centre for Litchi (ICAR), Mushahari, P.O. Ramna, Muzaffarpur-842 002, Bihar, http://www.nrclitchi.org/

National Research Centre on Pomegranate, NH-9, Solapur-Pune Highway, Kegaon (PO), Solapur District, Maharashtra State, http://www.nrcpomegranate.org/

National Research Centre on Seed Spice, Tabiji, Ajmer – 305206, Rajasthan. http://www.nrcss.org.in/

ICAR-Directorates/Project Directorates

Directorate of Rice Research, Rajendranagar, Hyderabad-500 030, Telangana http://www.drricar.org/

Directorate of Seed Research, P.O.: NBAIM, Pin: 275 103, Maunath Bhanjan, Uttar Pradesh, http://seedres.in/

Directorate of Groundnut Research, Post Box-5,Ivnagar Road,Junagadh 362 001, Gujarat, http://www.nrcg.res.in/

Directorate of Soybean Research, Indore Indore- 452 001, Madhya Pradesh, http://www.nrcsoya.nic.in/

Directorate of Rapeseed & Mustard Research, Sewar- 321303, Bharatpu, Rajasthan http://www.drmr.res.in/

Directorate of Mushroom Research, Solan Chambaghat- 173213, Solan, Himachal Pradesh, http://www.nrcmushroom.org/

Directorate on Onion and Garlic Research, Pune, Maharashtra. http://www.dogr.res.in/

Directorate of Cashew Research, Post Darbe, Puttur-574202, Karnataka. http://www.cashew.res.in/

Directorate of Medicinal and Aromatic Plants Research, Boriavi - 387 310, Anand District, Gujarat, http://www.dmapr.org.in/

Directorate of Floricultural Research, Pune, Maharashtra. http://dfr.res.in

Directorate of Weed Research, Maharajpur, Jabalpur - 482004, Madhya Pradesh, http://dwr.org.in/slider.aspx

Project Directorate on Foot & Mouth Disease, Mukteshwar, Project Directorate on Foot and Mouth Disease, IVRI Campus, Mukteswar, District Nainital Uttarakhand http://www.pdfmd.ernet.in/

Directorate of Poultry Research, Rajendranagar, Hyderabad 500 030, Telangana http://www.pdonpoultry.org/

Directorate of Knowledge Management in Agriculture (DKMA), New Delhi Project Director (DKMA), 5th Floor, Krishi Anusandhan Bhawan-I, Pusa, New Delhi 110 012, E-mail: pddkma@icar.org.in

Directorate of Cold Water Fisheries Research, Anusandhan Bhawan, Industrial Area, Bhimtal-263136, Distt: Nainital, Uttarakhand. http://www.dcfr.res.in/

B. RICE RESEARCH INSTITUTES

International Rice Research Institute (IRRI), Los Baños, Philippines

Bangladesh Rice Research Institute (BRRI), Joyebpur, Gazipur-1701, Bangladesh Website: www.brri.gov.bd, www.knowledgebank-brri.org

China National Rice Research Institute (CNRRI), Mail: No. 359, Tiyuchang Rd., Hangzhou, Zhejiang, 310006, P. R. China, http://www.chinariceinfo.com/english/

Philippine Rice Research Institute, Philippines, http://www.philrice.gov.ph/

Africa Rice Center (AfricaRice), West Africa Rice Development Association (WARDA) 01 BP 4029, Abidjan 01, Côte d'Ivoire, http://www.africarice.org http://www.warda.cgiar.org

C. COTTON INSTITUTIONS

International Cotton Advisory Committee (ICAC), 1629 – K Street, N.W., Suite 702, Washington DC 20006-1636 (USA), https://www.icac.org/

National Cotton Council of America, PO Box 2995, Cordova, TN 38088-2995, USA. www.cotton.org

National Cottonseed Products Association, Inc., 866 Willow Tree Circle, Cordova, TN 38018, USA. http://www.cottonseed.com/

Cotton Australia, Suite 4.01, 247 Coward Street, Mascot NSW Australia 2020 http://cottonaustralia.com.au/

Cotlook Limited, Outlook House, 458 New Chester Road, Rock Ferry, Merseyside CH42 2AE, United Kingdom, https://www.cotlook.com/

Cotton Research Institute (CRI), 9 El-Gamaa St. Giza, Egypt. http://www.arc.sci.eg/InstsLabs/Default.aspx?OrgID=2

American cotton Shippers Association, http://www.acsa-cotton.org/

D. JUTE & ALLIED FIBRES

Central Research Institute for Jute & Allied Fibres, (CRIJAF), Barrackpore, Kolkata 700120, West Bengal (India), www.crijaf.org.in

National Institute of Research on Jute & Allied Fibres Technology (NIRJAFT) www.nirjaft.res.in

Office of the Jute Commissioner, www.jutecomm.gov.in

National Jute Board, www.jute.com

The Jute Corporation of India (JCI), www.jci.gov.in

Indian Jute Industries' Research Association (IJIRA), www.ijira.org

Jute Manufactures Development Council (JMDC), www.jmdcindia.com

Indian Jute Mills Association (IJMA), www.ijma.org

Institute of Jute Technology, www.ijtindia.org

Bureau of Indian Standards (BIS), http://www.bis.org.in/index.asp

Bangladesh Jute Research Institute, www.bjri.gov.bd

International Jute Study Group (IJSG), www.jute.org

E. INTERNATIONAL ORGANIZATIONS

Food and Agriculture Organization, 00100 Rome, Italy, http://www.fao.org/

Global Forum on Agricultural Research, GFAR Secretariat, c/o FAO/SDR, viale delle Terme di Caracalla 00100 Roma, ITALY, http://www.egfar.org/

Australian Centre for International Agricultural Research (ACIAR), GPO Box 1571, Canberra, http://www.aciar.gov.au/

Caribbean Agricultural Research and Development Institute (CARDI), University of the West Indies, University Campus, St. Augustine, Trinidad and Tobago. http://www.cardi.org/

International Centre for Advanced Mediterranean Agronomic Studies (CIHEAM) 11, rue Newton - 75116, Paris, France, Tel: 33(0)1 53 23 91 00 secretariat@ciheam.org, www.ciheam.org, http://www.iamm.fr/

International Center for Tropical Agriculture (CIAT), Cali, Colombia http://www.ciat.cgiar.org/

CGIAR (Consultative Group on International Agricultural Research), Washington, DC 20433, USA, http://www.cgiar.org/

Centro Internacional de Agricultura Tropical (CIAT), 6713 Cali, Colombia. https://ciat.cgiar.org/

Center for International Forestry Research (CIFOR), P.O. Box 0113 BOCBD Bogor 16000, Indonesia, http://www.cifor.org/

International Maize and Wheat Improvement Center (CIMMYT), Apdo. Postal 6-641 06600 Mexico, D.F., MEXICO, http://www.cimmyt.org/en/

International Potato Center (CIP), Avenida La Molina 1895, La Molina, Apartado Postal 1558, Lima, Peru. http://cipotato.org/

International Center for Agricultural Research in the Dry Areas (ICARDA), Dalia Building 2nd Floor, Bashir El Kassar Street, Verdun, Beirut, Lebanon 1108-2010 http://www.icarda.org/

World Fish Center, [International Center for Living Aquatic Resources Management (ICLARM)], http://www.worldfishcenter.org/

International Crops Research Institute for the Semi-Arid Tropics (ICRISAT), 401/5, Patancheru 502324, Andhra Pradesh, India, www.icrisat.org

International Food Policy Research Institute (IFPRI), 2033 K St, NW, Washington, DC 20006-1002 USA, www.ifpri.org

International Institute of Tropical Agriculture (IITA), PMB 5320, Ibadan, Oyo State, Nigeria, http://www.cgiar.org/iita/research/ephta1.htm

International Livestock Research Institute (ILRI), P.O. Box 30709, Nairobi 00100, Kenya, https://www.ilri.org/

Bioversity International, [International Plant Genetic Resources Institute (IPGRI)], Via dei Tre Denari, 472/a, 00057 Maccarese (Fiumicino), Italy http://www.bioversityinternational.org/

International Water Management Institute (IWMI), P. O. Box 2075, Colombo, Sri Lanka., 127, Sunil Mawatha, Pelawatte, Battaramulla, Sri Lanka http://www.iwmi.cgiar.org/

World Agroforestry Centre, [International centre for research in agroforestry (ICRAF)] United Nations Avenue, Gigiri, PO Box 30677, Nairobi, 00100, Kenya www.worldagroforestry.org

CIGR-FAO Global Network on Agricultural Engineering, http://www.cigr.org/

CIRAD (Scientific and Technical Information Service), France, France http://www.cirad.fr/

Technical Centre for Agricultural and Rural Cooperation, 6700 AJ Wageningen, Netherlands, http://www.agricta.org/

EMBRAPA (Brazilian Agricultural Research Corporation), Brazil, http://www.embrapa.br/

European Consortium for Agricultural Research in the Tropics, NRI, Central Avenue, Chatham Maritime, Kent ME4 4TB, United Kingdom, http://ecart.iao.florence.it/

Instituto de Investigação Científica Tropical (IICT), Portugal

Natural Resources Institute (NRI), University of Greenwich, Central Avenue, Chatham Maritime, Kent, ME4 4TB,United Kingdom, http://www.nri.org/

Deutsche Gesellschaft Für Internationale Zusammenarbeit (GIZ) GMBH, Friedrich-Ebert-Allee 36 + 40, 53113 Bonn, Germany https://www.giz.de/en/worldwide/germany.html

IAO – Istituto Agronomico per l'Oltremare / Agronomic Institute for Overseas, via A. Cocchi, 4, [entrance: Training Centre, via A. Baldesi, 14], 50131 – Florence, Italy, http://www.iao.florence.it/?lang=en

Inter-American Institute for Cooperation on Agriculture (IICA), PO Box 55, 2200 Coronado, Costa Rica. http://www.iicanet.org/

Inter-American Institute for Cooperation on Agriculture, Jamaica Office (IICA-Jamaica), PO Box 349, Kingston 6, Jamaica. http://www.agroinfo.org/caribbean/iicacarc/jamaica/

International Centre for Agricultural Research in the Dry Areas (ICARDA), PO Box 5466, Aleppo, Syria. http://www.icarda.cgiar.org/

International Commission of Agricultural Engineering (CIGR), CIGR General Secretariat, Institut Für Landtechnik, Universität Bonn, Nussallee, 5, D-53 115 Bonn, Germany, http://www.cigr.org/

International Farming Systems Association (IFSA), Institute of Community Education, University of the Philippines, Los Banos College, Laguna 4031, Philippines http://www.laguna.net/ifsa/

International Institute of Tropical Agriculture (IITA), Oyo Road, PMB 5320, Ibadan, Nigeria, http://www.iita.org

International Service for National Agricultural Research (ISNAR), PO Box 5689, Addis Ababa, Ethiopia., http://www.isnar.cgiar.org

Koninklijk Instituut voor de Tropen (KIT-Agriculture), PO Box 95001, 1090 HA Amsterdam, Netherlands., http://www.kit.nl

Malaysian Agricultural Research and Development Institute (MARDI), Petit Surat 12301, 50774 Kuala Lumpur, Malaysia, Web: http://www.mardi.my/

National Agricultural Research Institute (NARI), PO Box 4415, Lae 411 Morobe, Papua New Guinea., http://www.nari.org.pg/

Centre for International Environment and Development Studies (Noragric), PO Box 5001, N-1432, Norway, http://www.nlh.no/noragric/

Caribbean Agricultural Information Service (CAIS) to ensure information flow through the Networks (http://www.caisnet.org/).

Resource Centres on Urban Agriculture and Forestry (RUAF), ETC Foundation, PO Box 64, 3830 AB Leusden, Netherlands, http://www.ruaf.org/

The Rodale Institute, 611 Siegfriedale Road, Kutztown, PA 19530, USA., http://www.rodaleinstitute.org/

Tropical Agriculture Association (TPA), c/o Membership Secretary, Rothes, Frankscroft, Peebles, EH45 9DX, United Kingdom, Web: http://www.taa.org.uk/

West and Central African Council for Agricultural Research and Development (CORAF/WECARD), BP 48 Dakar RP, 7 Av. Bourguiba, Dakar, Senegal., Web: http://www.coraf.org/

Bees for Development, Troy, Monmouth, NP25 4AB, United Kingdom
Web: http://www.beesfordevelopment.org

Camel Applied Research and Development Network, PO Box 2440, Damascus, Syria.
Web: http://www.acsad.org/irp/camel/page.htm

Institute for Animal Health (IAH), Compton Laboratory, Compton, Newbury, Berks, RG20 7NN, United Kingdom, Web: http://www.iah.bbsrc.ac.uk/

International Livestock Research Institute (ILRI), PO Box 30709, Nairobi, Kenya
Web: http://www.ilri.cgiar.org/

Livestock, Environment and Development Initiative (LEAD), Animal Production and Health Division, FAO , Viale delle Terme di Caracalla, Rome, ITALY
Web: http://lead.virtualcentre.org

Network for Smallholder Poultry Development (NSPD), The Royal Veterinary and Agricultural University, Dyrlaegevej 2, DK-1870 Frederiksberg C, Denmark
Web: http://www.poultry.kvl.dk/

Centre for Tropical Veterinary Medicine (VETAID), Pentlands Science Park, Bush, Penicuik, Midlothian EH26 OPZ, Scotland, United Kingdom
Web: http://www.vetaid.org/

Asian and Pacific Coconut Community (APCC), 3rd Floor Wisma Bakri, Jl. H. R. Rasuna Said Kuningan, Jakarta 12920, PO Box 1343, Jakarta 10013, Indonesia
Web: http://www.apcc.org.sg/

AVRDC - the World Vegetable Center, PO Box 42, Shanhua 741, Taiwan 74199, Taiwan
Web: http://www.avrdc.org/

Bureau for the Development of Research on Tropical Perennial Oil Crops (BUROTROP), 34394 Montpellier Cedex 5, Avenue Agropolis, France
Web: http://www.burotrop.org/database.htm

CIDICCO (Cover Crop International Clearing House), Apartado Postal 4443, Tegucigalpa MDC, Honduras, http://cidicco.hn.nerdydata.com/

International Maize and Wheat Improvement Centre (CIMMYT), Apartado Postal 6-641, 06600 Mexico, DF, Mexico, Web: http://www.cimmyt.org/

CIP (Centro Internacional de la Papa) [International Potato Center], Centro Internacional de la Papa, PO Box 1558, Lima 12, Peru, Web: http://www.cipotato.org/

GRAIN (Genetic Resources Action International), Girona 25, pral., E-08010 Barcelona, Spain, Web: http://www.grain.org/

International Centre of Insect Physiology and Ecology (ICIPE), PO Box 30772-00100, Nairobi, Kenya, Web: http://www.icipe.org

International Coconut Genetic Resources Network (COGENT), IPGRI Regional Office for Asia, the Pacific and Oceania, PO Box 236, UPM Post Office, 43400 Serdang, Selangor, Malaysia, Web: http://www.ipgri.cgiar.org/networks/cogent/

International Crops Research Institute for the Semi-Arid Tropics (ICRISAT), PO Patancheru -502 324, Andhra Pradesh, India, Web: http://www.icrisat.org/

International Network for Bamboo and Rattan (INBAR), Beijing 100102-86, Beijing 100102, China, Web: http://www.inbar.int/

International Network for the Improvement of Banana and Plantain (INIBAP), Parc scientifique Agropolis II, 34397 Montpellier cedex 5, France
Web: http://www.inibap.org/

International Plant Genetic Resources Institute (IPGRI), Via dei Tre Denari 472/a, 00057 Maccarese (Fiumicino) Rome, Italy, Web: http://www.ipgri.cgiar.org/

International Rice Research Institute (IRRI), DAPO Box 7777, Metro Manila, Philippines, Web: http://www.irri.org

IPMEurope, Rural Development Division, Deutsche Gesellschaft für Technische Zusammenarbeit (GTZ), Dag-Hammarskjöld-Weg 1-5, Postfach 5180, 65726 Eschborn, Germany, Web: http://www.ipmeurope.org/

Southern African Root Crops Research Network (SARRNET), Chitedze Research Station, PO Box 30258, Lilongwe 3, Malawi, E-mail: SARRNET@malawi.net
Web: http://www.iita.org/sarrnet/sarrnet/asanet.htm or http://www.iita.org/sarrnet/

Survey of Economic Plants for Arid and Semi-Arid Lands (SEPASAL), Royal Botanic Gardens, Kew, Richmond, Surrey TW9 3AB, United Kingdom
Web: http://www.kew.org.uk/ceb/sepasal

Food & Fertilizer Technology Centre (FFTC), 5F.14 Wenchow St, Taipei 10616, Taiwan, Web: http://www.fftc.agnet.org/

Centre for International Forestry Research (CIFOR), PO Box 6596, JKPWB, Jakarta, Indonesia, Web: http://www.cifor.cgiar.org/

Alternative Farming Systems Information Center (AFSIC), National Agricultural Library, Rm 132, 10301 Baltimore Ave, Beltsville, MD 20705-2351, USA
Web: http://www.nal.usda.gov/afsic/afsabout.htm

Center for Indigenous Knowledge for Agriculture and Rural Development (CIKARD), 318 Curtiss Hall, Iowa State University, Ames, Iowa 50011, USA Web: http://www.ciesin.org/IC/cikard/CIKARD.html

Centre for Information on Low-External-Input and Sustainable Agriculture (ILEIA), PO Box 2067, 3800 CB Amersfoort, Netherlands, Web: http://www.ileia.org/

Entomology Library and Information Network (ELIN), Natural History Museum (UK) Library, Cromwell Road, London SW7 5BD, United Kingdom Web: http://www.nhm.ac.uk/hosted_sites/elin/

InfoAgrar, c/o Swiss College of Agriculture, Langgasse 85, CH-3052 Zollikofen, Switzerland, Web: http://www.infoagrar.ch/

International Veterinary Information Service (IVIS), PO Box 4371, Ithaca, NY 14852, USA, Web: http://www.ivis.org/

International Federation of Organic Agricultural Movements (IFOAM), Charles-de-Gaulle-Str. 5, 53113 Bonn, Germany, Web: http://www.ifoam.org/

Natural Resources Institute (NRI), Central Avenue, Chatham Maritime, Kent ME4 4TB, United Kingdom, Web: http://www.nri.org

Permaculture Association (Britain), London WC1N 3XX, United Kingdom Web: http://www.permaculture.org.uk/

International Centre for Integrated Mountain Development (ICIMOD), 4/80 Jawalakhel, Kathmandu, Nepal, Web: http://www.icimod.org.sg/

Asia Soil Conservation Network for the Humid Tropics (ASOCON), Manggala Wanabakti Building, Block I Floor 13, Jl. Gatot Subroto, Senayan, Jakarta 10270,, Indonesia, Web: http://www.asocon.org/main.htm

International Fertilizer Development Center (IFDC), PO Box 2040' Muscle Shoals, AL 35662, USA, Web: http://www.ifdc.org/

Institute of Grassland & Environmental Research (IGER), Plas Gogerddan, Aberystwyth, SY23 3EB, United Kingdom, Web: http://www.iah.bbsrc.ac.uk/

Inter-American Water Resources Network (IWRN), c/o Organization of American States, Unit for Sustainable Development and Environment, 1889 F Street, NW Washington, DC 20006, USA, Web: http://www.iwrn.net/mainenglish.html

International Commission on Irrigation and Drainage (ICID), 48 Nyaya Marg, Chanakyapuri, New Delhi 110021, India Web: http://www.icid.org/ or http://www.ciid-ciid.org/

International Institute for Land Reclamation and Improvement (Alterra-ILRI), PO Box 47, 6700 AA Wageningen, Netherlands Web: http://www.ilri.nl / http://www.alterra.wur.nl/UK/ilri/

International Programme for Technology and Research in Irrigation and Drainage (IPTRID), Land and Water Development Division, Food and Agriculture Organisation of the United Nations, Viale delle Terme di Caracalla, 00100 Rome, Italy Web: http://www.fao.org/iptrid

World Resources Institute (WRI), 10 G Street, NE (Suite 800), Washington, DC 20002, USA, Web: http://www.wri.org/wri.html

AQUASTAT, Land and Water Development Division, Food and Agriculture Organisation, Viale delle Terme di Caracalla, 00100 Rome, Italy
Web: http://www.fao.org/ag/agl/aglw/aquastat/aquastat.htm

International Soil Reference and Information Centre (ISRIC), PO Box 353, 6700 AJ Wageningen, Netherlands, Web: http://www.isric.nl/

International Development Research Centre (IDRC), 250 Albert Street, PO Box 8500, Ottawa, Ontario K1G 3H9, Canada, Web: http://www.idrc.ca/

Farmers' World network (FWn), Arthur Rank Centre, National Agricultural Centre, Stoneleigh, Warwickshire CV8 2LZ, United Kingdom, Web: http://www.fwn.org.uk/

German Appropriate Technology Exchange (GATE), Postfach 5180, D-65 726 Eschborn, Germany, Web: http://www.gtz.de/gate

Service for Information Technology in International Agriculture (SITIA), 1024 Hamilton Court, Menlo Park, California 94025, USA, Web: http://www.sitia.org/

International Sunflower Association (ISA),11 rue de Monceau, CS 60003, 75378-Paris, France, http://isasunflower.org/home.html

National Institute of Agricultural Botany, Huntingdon Road, Cambridge CB3 0LE, United Kingdom, http://www.niab.com/

National Institute of Agricultural Technology (INTA), Casilla de Correo 21, 2580 Marcos Juárez (Cordoba), Argentina. Website: http://www.inta.gov.ar/

All Russian Research Institute of Oil Crops (VNIIMK), Russia, http://vniimk.ru/-en

LegumePlus: http://sainfoin.eu/

French National Institute for Agricultural Research (INRA), http://www.inra.fr/en/

Institute of Nature Fibres, Poland, http://www.pi.gov.pl/eng/chapter_95285.asp

Institute of Natural Fibres and Medicinal Plants, 71 B Wojska Polskiego, 60-630 Poznań, Poland, Tel.: +48 61 848 00 61; Fax: +48 61 841 78 30, www.iwnirz.pl

Commonwealth Scientific and Industrial Research Organization, Canberra, Australia. http://www.csiro.au/

All- Russian Flax Institute, Torzhok, Russia, http://www.carbo-extreme.eu/index.php/Consortium/VNIIL

N.I. Vavilov Research Institute for Plant Industry, St. Petersburg, Russia
http://www.vir.nw.ru/

National Agricultural Research and Development Institute, 915200 Fundulea, 1 Nicolae Titulescu street, Călăraşi, ROMANIA
http://www.incda-fundulea.ro/informatii_en.htm

International Institute for Sustainable Development, 161 Portage Avenue East, 6th Floor, Winnipeg, Manitoba, Canada R3B 0Y4, Web site: http://www.iisd.org/

International Organization for Standardization (ISO), http://www.iso.org/iso/home.html

F. DIRECTORIES, GATEWAY SITES AND PORTALS

Agroweb Network: www.agrowebcee.net

AgNIC: http://www.agnic.org/

AGRIS: http://agris.fao.org/agris-search/index.do

Agriculture Knowledge and Information Systems (AKIS): http://www.agriculturesnetwork.org/

Global Online Research in Agriculture (AGORA)
http://agora.aginternetwork.org/content/en/journals.php
http://www.fao.org/agora/en/

AgriFor: http://www.agrifor.com.au/

AGRICOLA: http://agricola.nal.usda.gov/

WAICENT - World Agricultural Information Centre Portal
http://www.fao.org/waicent/st/level_1.asp?main_id=8

TEEAL: The Essential Electronic Agricultural Library: http://teeal.org/.

AROW: Agricultural Research Organizations on the Web
http://www.eldis.org/go/home&id=25946&type=Document#.VmrvJ0p97IU

Weed Information Management System (WIMS)
http://tncinvasives.ucdavis.edu/wims.html

British Society of Soil Science
http://soils.org.uk/

DRAiNET: http://www.drainnet.ca/

Conservation Agriculture Knowledge Resources
http://mulch.mannlib.cornell.edu/

Worldwide Portal to Information on Soil Health
http://www.efita.org/Agriculture/Soils/Worldwide-Portal-to-Information-on-Soil-Health-details-2505.html

FAO Soils Portal: http://www.fao.org/soils-portal/en/

G. NATIONAL AND INTERNATIONAL IMPORTANT WEBSITES

Institute	Website
ARS (USDA)	http://www.ars.usda.gov/main/main.htm
National Agricultural Library, USA	http://www.nal.usda.gov/
Information system for genetic resources	http://biodiversity.europa.eu/topics/genetic-resources http://www.planttreaty.org/content/gis http://www.fao.org/genetic-resources/en/
GENRES - Information System Genetic Resources	http://www.genres.de/en/

Federal Ministry of Food and Agriculture (BMEL), Deichmanns Aue 29 53179 Bonn, Germany	http://www.ble.de/EN/00_Home/homepage_node.html
APEDA	http://apeda.gov.in/apedawebsite/index.html
Global Information Services for Seed Professionals	http://www.seedquest.com/
Solvent Extractors' Association of India (SEA), 142, Jolly Maker Chambers No 2, 14th Floor, 225, Nariman Point, Mumbai – 400021, India	http://www.seaofindia.com
http://www.iopepc.org/	http://www.iopepc.org/
Soybean Processors Association of India (SOPA), India	http://www.sopa.org
ICRISAT	http://www.icrisat.org
DAC, GOI, New Delhi	http://agricoop.nic.in
Sunflower	http://www.helianthus.com
Agricultural Statistics	http://www.Indiaagristat.com
Dicle University, Faculty of Agriculture, Department of Field Crops, Diyarbakır, Turkey	http://www.dicle.edu.tr/
Çukurova University, Faculty of Agriculture, Department of Field Crops, Adana, Turkey	http://www.cu.edu.tr/eng/
National Institute for Agriculture and Food Research and Technology, Spain	http://www.eco-itn.eu/index.html
Agriculture and Agri-Food Canada	www.agr.gc.ca
Institute of Sustainable Agriculture (IAS), CSIC, Apartado 4084, Córdoba, Spain.	http://www.ias.csic.es/
Andalusian Institute of Agricultural Research and Training, Spain	http://www.claimproject.eu/index.aspx
Institute of Economic Crops, Xinjiang Academy of Agricultural Sciences, China.	http://www.xaas.ac.cn
Chinese Academy of Sciences (Academia Sinica) China.	http://www.caas.cn/en/
CSIRO, Australia	www.csiro.au
Linseed	http://flaxcouncil.ca, www.saskflax.com, www.goldenflax.com, www.healthyflax.com
Library of Crop Technology Lesson Modules	http://croptechnology.unl.edu/pages/
Environmental Challenges in Farm Management	http://www.ecifm.rdg.ac.uk/
Soil Fertility and Plant Nutrition Portal	http://www.fao.org/nr/aboutnr/nrl/en/
Weed Identification	http://weeds.cropsci.illinois.edu/weedid.htm
PlantStress	http://www.plantstress.com/
Food Standards Agency (FSA)	http://www.food.gov.uk/

H. METEOROLOGICAL WEBSITES

World Meteorological Organization: http://www.worldweather.org

India Meteorological Department: http://www.imd.ernet.in

NCMRWF, DST, India: http://www.ncmrwf.gov.in/

Indian Institute of Tropical Meteorology: http://www.tropmet.res.in

UK Meteorological Department: http://www.met.rdg.ac.uk

Australia Meteorological Department: http://www.bom.gov.au

Canada Meteorological Organization: http://www.msc-smc.ec.gc.ca

USA Meteorological Department: http://www.nws.noaa.gov

US National Center for Atmospheric Research data: http://dss.ucar.edu/datasets/

Annexure III
References for Further Reading

Anonymous, 1990. Status of Indian Sugar Industry. XXIII ISSCT Congress. International Society of Sugarcane Technologists, New Delhi - 110 065.

Alexander, A.G. 1973. Sugarcane Physiology, Elsevier, Amsterdam, 752p.

Babu, C.N. 1990. Sugarcane. Allied Publishers Ltd., Madras.

Batta, S.K. and Singh, R. 1991. Post-harvest deterioration in quality of sugarcane. Bharatiya Sugar. 16(4): 49-50.

Bakker, 1999. Sugarcane cultivation and management. Kleuver Academic / Plantation Publishers, New York.

Berger, J. 1969. The World's Major Fibre Crops: their cultivation and manuring. Centre D'etude de L'azote, Zurich, Switzerland.

Blackburn, F. 1984 . Sugarcane. Longman, London.

Chavan, V.M. 1961. Niger and Safflower. Indian Central Oilseeds Committee, Hyderabad.

Clements, H.F. 1980. Sugarcane Crop logging and crop control: principles and practices. University Press of Hawaii, Honolulu. 520p

Cobley, L.S. 1957. The Botany of Tropical Crops. Longmans Green and Co., London.

Crane, J.C. 1947. Kenaf: Fibre plant rival of Jute. Economic Botany. 1: 334-350

CPG. 2012. Crop Production Guide. Department of Agriculture, Government of Tamil Nadu, Chennai and Tamil Nadu Agricultural University, Coimbatore.

CPG. 2013. Crop Production Techniques of Horticultural Crops. Horticultural College and Research Institute, Tamil Nadu Agricultural University, Coimbatore – 641 003

Daisy E. Kay. 1979. Food legumes. Food Tropical Products Institute, London. 435p

Fageric, N.K. 1992. Maximizing Crop Yields. Marcel Dekker Inc., New York.

Husz, George Stefab. 1972. Sugarcane: Cultivation and fertilization, Ruhr Stickstoff A.G., Bochum, West Germany.

Ghosh, S.P., Ramajunjam, T., Jos, J.S., Moorthy, S.N., and Nair, R.G. 1988. Tuber Crops. Oxford and IBH Publishing Co. Pvt. Ltd., New Delhi. pp 3-148.

Hartley, C.W.S 1988. The Oilpalm. Longman, London. pp 761.

Humbert, R.P. 1986. The Growing of Sugarcane. Elsevier Publishing Company, New York.

John M. Munro. 1987. Cotton, Longman Scientific and Technical, NewYork.

Kakde, J.R. 1985. Sugarcane Production. Metropolitan Book Co. Pvt. Ltd., New Delhi.

Keeting, B.A., and Wilson, J.R. (eds). 1997. Intensive sugarcane production: Meeting the challenges beyond 2000. CAB International, Wallingford, United Kingdom.

Kirby, R.H. 1963. Vegetable Fibres. Leonard Hill (Books) Limited, London.

Mathur, P.S., and Behari, C. 1980. Growing sugarcane in different states of India. Directorate of Sugarcane Development, Ministry of Agriculture. Government of India, New Delhi.

Menon, K.P.V. and Pandali, K.M. 1960. The Coconut Palm: A Monograph. Indian Central Coconut Committee, Ernakulam. 384 p.

Mohan Naidu, K. and Arulraj, S. 1987. Sugarcane Technologies. Sugarcane Breeding Institute, Coimbatore

Muller, G. 1968. Cotton: Cultivation and Fertilisation. Ruhr-Stickstoff, Bochum, West Germany.

Norman, M.J.T., Pearson, C.J., and Searle, P.G.E. 1984. The Ecology of Tropical Food Crops. Cambridge University Press, Cambridge. 369p

NRCOP. 2000. Oil Palm Production. UNDP/ TMO & P Trainers Training Programme. National Research Centre for Oil Palm. Pedavegi, India.

Padwick, G.W. 1979. Growth phases in plants and their bearing on Agronomy. Experimental Agriculture. 15: 15-26

Parthasarathy, S.V. 1972. Sugarcane in India. K.C.P. Ltd, Madras.

Pruthi, J.S. 1998. Major Spices of India. ICAR, New Delhi. pp 289-310.

Purseglove, J.W. 1974. Tropical Crops: Dicotyledons. English Language Book Society, Longman. 719p

Ramasamy, C. 1983. Essential elements in coconut nutrition and their deficiency symptoms. Indian Coconut Journal. 14 (9): 8–11.

Rao, A.S. 1989, Water requirement of young coconut palms in humid tropical climate. Irrigation Science. 10: 245–249

Rethinam, P. 1987. Management of drought situations in coconut plantations. Indian Coconut Journal. 18 (6): 3-5.

Rethinam, P. and Nagendra Rao, T. (eds). 1997. Nutrient Management of Oil Palm Plantations in South India. National Research Centre for Oil Palm. Pedavegi, India.

Rethinam, P. and Suresh, K. (eds) 1998. Oilpalm Research and Development. National Research Centre for Oil Palm. Pedavegi. India.

Shahi, H.N. and Sinha, O.K. (eds) 2002. Food, nutrition and economic security through diversification in sugarcane production and processing systems. The Association of Sugarcane Technologists of India. Indian Institute of Sugarcane Research, Lucknow - 226 002.

Singh, R., Biswas, B.C., Maheswari, S.C, and Srivatsava, S.C. 1981. Sugarcane. Fertilizer Association of India, New Delhi.

Singh, J. 1998. Jaggery and Khandsari Research Digest, IISR (JK-Cell) / Technical Bulletin. 98-39, IISR, Lucknow.

Solomon, S. 2000. Post-harvest Cane deterioration and its milling consequences selected for ancillary uses. Sugar Tech. 2 (1& 2): 1-18.

Srivatsava, S.C., Joshi, D.P. and Gill, P.C. 1988. Manual of Sugarcane Production in India. ICAR, New Delhi.

Sundara, B. 1998. Sugarcane cultivation. Vikas Publishing House Pvt. Ltd., New Delhi.

Thuljanam Rao, J. and Srinivasan, T.R. 1980. Training Manual for Sugarcane Production, Directorate of Extension, Ministry of Agriculture, Government of India, New Delhi.

Verma, R.S. 2000. Weed Management in sugarcane. India Institute of Sugarcane Research, Lucknow - 226 002.

http://www.ipmcenters.org/cropprofiles/docs/carice.html